Speleogenesis
Evolution of Karst Aquifers

January 2000 Edition

Speleogenesis
Evolution of Karst Aquifers

January 2000 Edition

Alexander B. Klimchouk
Derek C. Ford
Arthur N. Palmer
Wolfgang Dreybrodt
Editors

With Members of the Commission on Karst Hydrogeology and Speleogenesis
International Speleological Union

Contributing Authors

• Thomas Aley	Protem, Missouri, U.S.A.
• Philippe Audra	Aix-en-Provence, France
• Patricia Beddows	Bristol, England, UK
• Thomas Bitterli	Peseux, Switzerland
• Simon Bottrell	Leeds, England, UK
• Jose Maria Calaforra	Almería, Spain
• James Carew	Charleston, South Carolina, U.S.A.
• Atlas Corrêa Neto	Rio de Janeiro, Brazil
• Wolfgang Dreybrodt	Bremen, Germany
• Victor Dublyansky	Perm, Russia
• Yuri Dublyansky	Novosibirsk, Russia
• Laurent Eisenlohr	Bremen, Germany
• Ralph Ewers	Richmond, Kentucky, U.S.A.
• E. Fernández-Gibert	Madrid, Spain
• Andrej Filippov	Irkutsk, Russia
• Derek C. Ford	Hamilton, Ontario, Canada
• Amos Frumkin	Jerusalem, Israel
• Franci Gabrovšek	Postojna, Slovenia
• John Gunn	Huddersfield, England, UK
• Philippe Häuselmann	Fribourg, Switzerland
• Carol Hill	Albuquerque, New Mexico, U.S.A.
• Louise Hose	Fulton, Missouri, U.S.A
• Peter Huntoon	Las Vegas, Nevada, U.S.A
• Pierre-Yves Jeannin	Neuchatel, Switzerland
• Alexander Klimchouk	Kiev, Ukraine
• Stein-Erik Lauritzen	Bergen, Norway
• Rudolf Liedl	Tübingen, Germany
• David Lowe	Huddersfield, England, UK
• Joyce Lundberg	Ottawa, Ontario, Canada
• J.E.J. Martini	Pretoria, South Africa
• John Mylroie	Mississippi State, Mississippi, U.S.A.
• R.A.L. Osborne	Sydney, Australia
• Arthur N. Palmer	Oneonta, New York, U.S.A.
• Margaret Palmer	Oneonta, New York, U.S.A.
• A. Pulido-Bosch	Granada, Spain
• C. Rossi	Madrid, Spain
• Martin Sauter	Jena, Germany
• Trevor Shaw	Bath, England, UK
• Jorg Siemers	Bremen, Germany
• France Šušteršič	Ljubljana, Slovenia
• Elizabeth White	University Park, Pennsylvania, U.S.A.
• Susan White	Bundoora, Victoria, Australia
• William White	University Park, Pennsylvania, U.S.A.
• Stephen Worthington	Hamilton, Ontario, Canada

NSS Special Publications Committee

• Design and Production	David McClurg, Chairman
• Text Style Editors	Margaret Palmer and George W. Moore
• Bibliography Editor	Gail McCoy
• Index Editor	Elizabeth White

National Speleological Society, Inc. **Huntsville, Alabama, U.S.A.**

Speleogenesis
Evolution of Karst Aquifers
January 2000 Edition

Published by the National Speleological Society, Inc.
2813 Cave Avenue
Huntsville, Alabama 35810, U.S.A.
256 852-1300
Email: manager@caves.org   Web page: http://www.caves.org

Library of Congress Catalog Card Number: 99-74622

ISBN: 1-879961-09-1

Printed in the U.S.A.

Conservation Alert: Speleothems (cave formations) and cave minerals are fragile, extremely fragile. Delicate soda straws and eccentric helictites, for example, that took untold thousands of years to create can be destroyed in an instant by a careless or thoughtless act on our part. The conservation ethic of the National Speleological Society—and other speleological associations worldwide—is to enjoy the natural beauty of speleothems in place without disturbing them in any way. Remember that cave conservation is more a state of mind than a set of rules. Use your common sense and always remember to take the action that will cause the least damage to the cave and its rare and beautiful speleothems.

Front Cover: Safari Straight, a river passage in Sof Omar Cave, Ethiopia. Photo by A. Klimchouk and V. Kisseljov.

Back Cover: Kizil Coba Cave, Crimea, Ukraine. Photo by V. Kisseljov.

Dedication

This book is dedicated to the memory of these three colleagues and to their outstanding lifetime achievements:

Alfred Bögli, Switzerland —Cave and karst studies
Vladimir Kisseljov, Russia—Exploration
Jim Quinlan, United States—Practical applications

Together they symbolize the unity of the three essential parts of cave science.

In addition, we express our respect for the other enthusiasts now departed who also devoted their lives to the exploration and study of the underground world.

This book is also dedicated to the memory of Fred L, Wefer, who initially as a vice president and later as the president of the National Speleological Society, strongly supported the publication of this book by the Society.

Contents

Preface

This book is a pioneer attempt by an international group of cave scientists to summarize modern knowledge about the cave origins in various settings, to examine the variety of approaches that have been adopted, and to outline the role of speleogenesis in the evolution of karst aquifers. Selected contributions from 44 authors in 15 nations are combined in an integrated volume. These were prepared between 1994 and 1998 as an initiative of the Commission of Karst Hydrogeology and Speleogenesis, International Speleological Union.

Despite a desire to produce an integrated book, rather than a mere collection of papers, the editors' policy has not been directed toward unifying all views. Along with some well-established theories and approaches, the book contains new concepts and ideas emerging in recent years. We hope that this approach will stimulate further development and exchange of ideas in cave studies and karst hydrogeology.

Préface

Cet ouvrage est l'aboutissement du travail d'un groupe international de spéléologues scientifiques faisant le point des connaissances récentes sur l'origine des cavités. Elles sont étudiées dans divers contextes, en prenant en compte les approches variées qui ont été développées et en esquissant le rôle de la spéléogénèse dans l'évolution des aquifères karstiques. Elaboré entre 1994 et 1998, à l'initiative de la Commission d'Hydrogéologie du Karst et de Spéléogénèse (KHS), rattachée à l'Union Internationale de Spéléologie (UIS), cet unique volume rassemble les contributions de 44 auteurs émanant de 15 nations différentes.

Tout en souhaitant produire un ouvrage global, plutôt qu'un assemblage d'articles divers, la politique éditoriale n'a toutefois pas pour autant uniformisé les divers points de vue. Au côté de quelques théories désormais bien établies, cet ouvrage recèle les nouveaux concepts et idées ayant récemment émergé. Nous souhaitons que cette approche stimulera les futurs développements et l'échange des idées sur l'étude des cavités et sur l'hydrogéologie karstique.

Translation – Phillipe Audra

Vorwort

Dieses Buch ist ein bahnbrechender Versuch einer internationalen Gruppe von Karstforschern und Speläologen, das gegenwärtige Wissen über die Entstehung von Höhlen unter verschiedenen geologischen Bedingungen zusammenzufassen. Es behandelt eine Vielfalt von neuen Ideen und gibt einen Abriss der Bedeutung der Speläologenese für die Entwicklung von Karstaquiferen. Ausgewählte Beiträge von 44 Autoren aus 15 Nationen sind in einem

Band aufeinander abgestimmt. Sie entstanden zwischen 1994 und 1998 durch eine Initiative der Kommission für Karst-Hydrogeologie und Speläogenese, Internationale Union für Speläologie.

Trotz des Ziels eher einen in sich abgeschlossenen Band vorzulegen, als eine bloße Sammlung von Beiträgen, ist es nicht Absicht der Herausgeber gewesen, alle verschiedenen Ansätze zu vereinen. Neben gut abgesicherten Theorien enthält das Buch neue, in der letzten Zeit entstandene Ideen und Konzepte. Wir hoffen, daß es zum Austausch von Ideen und zur weiteren Entwicklung in der Speläologie und Karst-Hydrogeologie anregen wird.

Translation—Wolfgang Dreybrodt

ПРЕДИСЛОВИЕ

Эта книга является новаторской попыткой международного коллектива ученых-спелеологов обобщить современные знания о происхождении пещер в различных обстановках, исследовать различные используемые подходы и выявить роль спелеогенеза в эволюции карстовых водоносных горизонтов. Работы 44 авторов из 15 стран объединены в единый том, подготовленный в период 1994-1998 годов по инициативе Комиссии по гидрогеологии карста и спелеогенезу Международного Спелеологического Союза.

Несмотря на желание произвести цельную книгу, а не просто сборник статей, политика редакторов не была направлена на унификацию всех взглядов. Наряду с некоторыми широко признанными теориями и подходами, книга содержит новые концепции и идеи, появившиеся в последние годы. Мы надеемся, что такой подход будет стимулировать дальнейшее развитие и обмен идеями в области пещерных исследований и гидрогеологии карста.

Александр Климчук

Prefacio

El presente libro es la tentativa pionera de un grupo internacional de investigadores relacionados con el estudio de las cavidades de resumir el conocimiento actual sobre el origen de cavidades en distintas situaciones, examinar las diversas aproximaciones teóricas que pueden realizarse y describir papel de la Espeleogénesis en la evolución de los acuíferos kársticos. Las contribuciones seleccionadas, realizadas por 44 autores provenientes de 15 nacionalidades distintas, se han combinado de en un volumen preparado entre 1994 y 1998 gracias a la iniciativa de la Comisión sobre Hidrogeología del Karst y Espeleogénesis perteneciente a la Unión Internacional de Espeleología.

Se ha puesto especial énfasis en producir un libro que constituya un conjunto homogéneo, mas que una simple recopilación de artículos; pero sin embargo, no ha estado en el ánimo de los editores una dirección que unificara los distintos puntos de vista expuestos. Además de algunas teorías y aproximaciones ya bien fundadas, el libro abarca también nuevos conceptos e ideas que han surgido en los últimos años. Esperamos que este intento estimule el futuro desarrollo e intercambio de ideas en la investigación sobre cavidades e hidrogeología kárstica.

Translation– Jose Maria Calaforra

Acknowledgements

We sincerely thank all contributors for their willing cooperation in the long and difficult process of preparing this book, for their participation in developing its logic and methodology, and for their cheerful response to numerous requests. We thank all colleagues who discussed the work with us and encouraged it in many ways, even though not contributing to its content as authors. We are particularly grateful to Margaret V. Palmer for invaluable help in editing the English in many contributions, to Nataly Yablokova for her help in performing many technical tasks, and to Elizabeth White, who prepared the comprehensive index. Our thanks are due to David Drew, Philip LaMoreaux, and George W. Moore for reviewing the manuscript and producing constructive notes and comments on improvement of the final product. We are also especially indebted to George W. Moore for his able assistance in style editing the final text.

The organizational costs and correspondence related to the preparation of the book were partially sponsored by the National Speleological Society, the publisher. We thank David McClurg, the Chair of the NSS Special Publication Committee, for his extensive technical and organizational support in the preparation and publishing process.

Alexander Klimchouk
Derek Ford
Arthur Palmer
Wolfgang Dreybrodt
Editors

1.0 Plate 1. *Delicate Arch and its surrounding landscape, in the Jurassic Entrada Sandstone, Arches National Park, Utah, U.S.A. Note person sitting at bottom of arch, at lower left. (Photo by A. Palmer)*

Part 1 Introduction

1.0 Plate 2. *Sinking stream (ponor) at the entrance of Onesquethaw Cave, New York, U.S.A. (Photo by A. Palmer)*

1.0 Plate 3. *Chamber of Columns, Sof Omar Cave, Ethiopia. (Photo by A. Klimchouk)*

About This Book

The aim of this book is to present the advances made in recent decades in our understanding of the formation of dissolutional caves, and to illustrate the role of cave genetic (*speleogenetic*) processes in the development of karst aquifers.

From the perspective of hydrogeology, karst groundwater flow is a distinct kind of fluid circulation system—one that is capable of self-organization and self-development due to its capacity to dissolve significant amounts of the host rock and transport them out of the system. Fluid circulation in soluble rocks becomes more efficiently organized by creating, enlarging, and modifying patterns of cave conduits, the process of speleogenesis. We can assert that *karst groundwater flow is a function of speleogenesis* and *vice versa.*

The advances in cave science are poorly appreciated in what may be termed "mainstream hydrogeology," which retains a childlike faith in flow models developed in the sand box. Many karst students also will not be aware of all emerging concepts of cave origin. This is because discussions of them are scattered through journals and books in different disciplines and languages, including publications with small circulation.

An understanding of the principles of speleogenesis and its most important controls is indispensable for proper comprehension of the evolution of karst systems in general and of karst aquifers in particular. We hope this book will be useful for both karst and cave scientists, and for general hydrogeologists dealing with karst terranes.

Organization of the Book

This book is organized in eight parts, each with chapters and subchapters.

Part 1

Part 1 is this Introduction: About This Book.

Part 2

Part 2, The Historical Perspective, gives a history of speleogenetic studies.

Shaw (Chapter 2.1) traces the development of the most important ideas from previous centuries through the early modern period in the first half of this century.

Lowe (Chapter 2.2) traces the development up to the threshold of modern times.

W. White (Chapter 2.3) brings the story right up to the present time.

The current state of the art is best illustrated by the entire contents of this book.

Part 3

Part 3 reviews the principal geologic and hydrogeologic variables that either control or significantly influence the differing styles of cave development that are found.

In Chapter 3.1, Klimchouk and Ford introduce an evolutionary approach to the typology of karst settings, which is taken as a base line for the book. Extrinsic factors and intrinsic mechanisms of cave development change regularly and substantially during the general cycle of geologic evolution of a soluble rock, and more specifically, within the hydrogeologic cycle.

The evolutionary typology of karst presented in this chapter considers the life cycle of a soluble formation, from deposition (syngenetic karst), through deep burial, to exposure and denudation. It helps to differentiate between karst types that may concurrently represent different stages of karst development, and is also a means of adequately classifying speleogenetic settings.

The different types of karst are marked by characteristic associations of the structural prerequisites for groundwater flow and speleogenesis, flow regime, recharge mode and recharge-discharge configurations, groundwater chemistry, and degree of inheritance from earlier conditions. Consequently, these associations make a convenient basis to view both the factors that control cave genesis and the particular types of caves.

In Chapter 3.2, lithological and structural controls of speleogenesis are reviewed in general terms (Klimchouk and Ford). Lowe in Chapter 3.3 discusses the role of stratigraphic elements and the speleoinception concept. Palmer in Chapter 3.4 reviews the hydrogeologic controls of cave patterns and demonstrates that hydrogeologic factors, the recharge mode, and the type of flow in particular, impose the most powerful controls on the formation of the gross geometry of cave systems.

Hence, analysis of cave patterns is especially useful in the reconstruction of environments from paleokarst and in the prediction and interpretation of groundwater flow patterns and contaminant migration. Any opportunity to relate cave patterns to the nature of their host aquifers will assist in these applied studies as well. Osborne in Chapter 3.7 examines the significance of paleokarst in speleogenesis.

More specific issues are treated such as the nature of epikarst and its role in vadose speleogenesis (Klimchouk, Chapter 3.5), and the role of condensation processes (Chapter 3.6, V. Dublyansky and Y. Dublyansky).

Part 4

Part 4 outlines the fundamental physics and chemistry of the speleogenetic processes in Chapter 4.1 and presents a variety of different approaches to modeling cave-conduit development in Chapter 4.2.

In Chapter 4.1, the chemical reactions during the dissolution of the common soluble minerals, calcite, gypsum, salt, and quartz, are discussed with the basic physical and chemical mechanisms that determine their dissolution rates. As limestone is the most common karst rock and its dissolution is the most complex in many respects, it receives the greatest attention.

Dreybrodt (Chapter 4.1.1) and Dreybrodt and Eisenlohr (Chapter 4.1.2) provide advanced discussion and report the most recent experimental data. These are used to obtain realistic dissolution rates for a variety of hydrogeologic conditions and as input for modeling the evolution of conduits.

Although direct comparison between theoretical or analytical dissolution rates and those derived from field measurements is difficult, a very useful comparison is provided by W. White (Chapter 4.1.3). The bulk removal of carbonate rock from karst drainage basins can be evaluated either by direct measurement of rock-surface retreat or by mass balance within known drainage basins. All of these approaches make sense and give roughly accurate results that are consistent with theoretical expectations.

It is well recognized today that the earliest incipient phases of speleogenesis are crucial in building up the pattern of conduits that evolve into explorable cave systems. To establish the major controls on these initial stages by purely analytical or intuitive methods is difficult, so modeling becomes particularly important.

Various approaches are presented in Chapter 4.2. Ford, Ewers, and Lauritzen present the results of a systematic study of the propagation of conduits between input and output points in an anisotropic fissure.

They use a variety of hardware and software models in series (Chapter 4.2.1) discussing the cases of:

- single input
- multiple inputs in one rank
- multiple inputs in multiple ranks.

The results indicate important details of the competitive development of protoconduits and help to explain branching cave patterns. In the competition between inputs, some principal tubes in near ranks first link ("breakthrough") to an output boundary. This reorients the flowfields of failed nearby competitors, which then extend to join the principal via their closest secondaries. The process extends outward and to the rear, linking up all inputs in a "cascading system."

The exploding growth of computer capability during the past two decades has greatly enhanced possibilities for digital modeling of early conduit development. Investigating the growth of a single conduit is a logical first step in understanding the evolution of caves, realized here by Dreybrodt and Gabrovšek in the form of a simple mathematical model (Chapter 4.2.2) and by Palmer by numerical finite-difference modeling (Chapter 4.2.3).

The models show that positive feedback loops operate. Widening a fracture increases flow through it. Therefore dissolution rates increase along it, and so on, until finally a dramatic increase of flow rate permits a dramatic enhancement of the widening. This breakthrough event terminates the initial stage of conduit evolution. From then on the water is able to pass through the entire conduit while maintaining sufficient undersaturation to preserve low-order kinetics.

So the growth rate is very rapid, at least from a geologic standpoint—usually about 0.001–0.1 cm/yr. The initiation ("break-through") time depends critically on the length and the initial width of the fracture, and for the majority of realistic cases, it covers a time range from a few thousand years to ten million years in limestones.

The modeling results give a clear explanation of the operation of selectivity in cave genesis. In a typical unconfined karst aquifer there is a great range of enlargement rates along the competing flow routes, and only a few conduits will grow to enterable size. The modeling also provides one starting point (others are discussed in Chapter 5.2) to explain uniform maze patterns. These are favored by enlargement of all openings at comparable rates where the discharge to length ratio is great enough.

Single-conduit modeling has the virtue of revealing how the cave-forming variables relate to each other in the simplest possible way. Although it is more difficult to extend this approach to two dimensions, many investigators have done so. Examples are Groves and Howard, 1994; and Howard and Groves, 1995. And in this volume—Ford, Ewers, and Lauritzen, Chapter 4.2.1; Dreybrodt and Siemers, Chapter 4.2.4; and Sauter and Liedl, Chapter 4.2.5.

The modeling performed by Dreybrodt and Siemers shows that the main principles of breakthrough derived from one-dimensional models remain valid. The evolution of karst aquifers has been modeled for a variety of different geologic settings, including also variation in lithology with respect to the dissolution kinetics. Sauter and Liedl simulate the development of conduits at a catchment scale for fissured carbonate rocks with rather large initial openings (about 1 mm). The approach is based upon hydraulic coupling of a pipe network to matrix continuum in order to represent the well-known duality of karst aquifer flow systems.

It is also shown how understanding of the genesis of karst aquifers and modeling of their development can assist in characterization of the conduit system, which dominates flow and transport in karst aquifers.

Part 5

An important point that has emerged from cave studies of the three decades is that no single speleogenetic model applies to all geologic and hydrologic settings. Given that settings may also change systematically during the evolutionary geologic cycles outlined above (Chapter 3.1), an evolutionary approach is called for.

This is attempted in Part 5, which is organized to give extended accounts of speleogenesis in the three most important settings that we recognize:

- Coastal and oceanic (Chapter 5.1).
- Deep-seated and confined (Chapter 5.2).
- Unconfined (Chapter 5.3).

Each chapter begins with a review of modern ideas on cave development in the setting, followed by representative case studies. The latter include new accounts of some "classic" caves as well as descriptions of other little-known cave systems and areas. Readers may determine for themselves how well the real field examples fit the general models presented in the introductory sections.

Mylroie and Carew in Chapter 5.1 summarize specific features of cave and karst development in young rocks in coastal and island settings that result from the chemical interactions between fresh and salt waters, and the effects of fluctuating sea level during the Quaternary. The case studies include a review of syngenetic karst in coastal dune limestones, Australia (S. White, 5.1.1) and an example of speleogenesis on tectonically active carbonate islands (Gunn and Lowe, 5.1.2).

Klimchouk in Chapter 5.2 reviews conditions and mechanisms of speleogenesis in deep-seated and confined settings, one of the most controversial but exciting topics in modern cave research. Conventional karst and speleogenetic theories are concerned chiefly with shallow unconfined geologic settings. These suppose that the karstification found there is intimately related to surface conditions of input and output, with the dissolution being driven by downward meteoric water recharge. The possibility of hypogenic karstification in deeper environments has been neglected for a long time, and the quite numerous instances of karst features found at significant depths have usually been interpreted as buried paleokarst.

However, the past decade has seen a growing recognition of the variety and importance of hypogene dissolution process, and of speleogenesis under confined settings that often precedes unconfined development. These include Hill, 1987, 1995; Klimchouk, 1994, 1996, 1997; Lowe, 1992; Lowe and Gunn, 1995; Mazzullo and Harris, 1991, 1992; Palmer, 1991, 1995; Smart and Whitaker, 1991; Worthington, 1991, 1994; and Worthington and Ford, 1995.

Confined (artesian) settings were commonly ignored as sites for cave origin. This is because the classic concept of artesian flow implies long lateral travel distances for groundwater within a soluble unit, resulting in a low capacity to generate caves in the confined area.

However, the recognition of nonclassical features in artesian flow—namely the occurrence of cross-formation hydraulic communication within artesian basins, the concepts of transverse speleogenesis, and of the inversion of hydrogeologic function of beds in a sequence—calls for a revision of the theory of artesian speleogenesis and of views on the origin of many caves.

It is proposed that artesian speleogenesis is immensely important to speleoinception and also accounts for the development of some of the largest known caves in the world. Typical conditions of recharge, the flow pattern through the soluble rocks, and groundwater aggressiveness favor uniform, rather than competing, development of conduits, resulting in maze caves where the structural prerequisites exist. Cross-formational flow favors a variety of dissolution mechanisms that commonly involve mixing.

Hydrogeochemical mechanisms of speleogenesis are particularly diverse and potent where carbonate and sulfate beds alternate and within or adjacent to hydrocarbon-bearing sedimentary basins.

Hypogene speleogenesis occurs in rocks of varied lithology and can involve a variety of dissolution mechanisms that operate under different physical constraints but create similar cave features.

Case studies include:

- The great gypsum mazes of the western Ukraine (Klimchouk, Chapter 5.2.1).
- Great maze caves in limestones in the Black Hills, South Dakota (Palmer, Chapter 5.2.2), and Siberia (Filippov, Chapter 5.2.3).
- Karstification in the Redwall Aquifer, Arizona (Huntoon, Chapter 5.2.4).
- A summary of the peculiar features of hydrothermal speleogenesis (Y. Dublyansky, Chapter 5.2.5).
- Hydrothermal caves in Hungary (Y. Dublyansky, Chapter 5.2.6).
- Sulfuric acid speleogenesis (Lowe, Bottrell, and Gunn, Chapter 5.2.7, and Hill, Chapter 5.2.8).
- An outstanding example of a hydrothermal cavity—in fact the largest ever recorded by volume—in the Rhodope Mountains (V. Dublyansky, Chapter 5.2.9).

Recognition of the scale and importance of deep-seated speleogenesis and of the hydraulic continuity and cross-formational communications between aquifers in artesian basins is indispensable for the correct interpretation of the evolution of karst aquifers, speleogenetic processes and associated phenomena, regional karst water-resource evaluation, and the genesis of certain karst-related mineral deposits. These and other theoretical and practical implications still have to be developed and evaluated, which offers a wide field for further research.

Ford in Chapter 5.3 reviews the theory of speleogenesis that occurs where normal meteoric water sinks underground through the epikarst or dolines and streamsinks, and circulates in the limestone or other soluble rocks without any major artesian confinement. These are termed *common caves* (Ford and Williams, 1989) because they probably account for 90% or more of the explored and mapped dissolutional caves that are longer than a few hundred meters. This estimate reflects the bias in exploration. Caves formed in unconfined settings and genetically related to surface recharge are the most readily accessible and hence form the bulk of documented caves.

Common caves chiefly display branchwork forms where the dissolutional conduits occupy only a tiny proportion of the total length or area of penetrable fissures that is available to the groundwater. The rules that govern the selection of the successful linkages that will be enlarged into the branchwork pattern are supported in the models presented in Chapter 4.2.

In longitudinal section, caves may be divided into deep phreatic, multiloop, mixed loop and water table, and ideal water table types, with drawdown vadose caves or invasion vadose caves above them. Many large systems display a mixture of the types.

The concepts of plan-pattern construction, phreatic, water table, or vadose state, and multiphase development of common caves are illustrated in the case studies that follow the introduction.

They are organized broadly to begin with examples of:

- Comparatively simple deep phreatic and multiloop systems—El Abra, Mexico (Ford, Chapter 5.3.1), and Castleguard Cave, Canada (Ford, Lauritzen, and Worthington, Chapter 5.3.2).
- Large and complex multiphase systems such as the North of Thun System, Switzerland (Jeannin, Bitterly, and Häuselmann, Chapter 5.3.3), and Mammoth Cave, Kentucky (Palmer, Chapter 5.3.8).
- Representatives of mixed vadose and phreatic development in mountainous regions—the Alps (Audra, Chapter 5.3.4), the Pyrenees (Fernandez, Calaforra, and Rossi, Chapter 5.3.5), Mexico (Hose, Chapter 5.3.6).
- Finally, cases where there is strong lithologic or structural control—folded Appalachians (W. White, Chapter 5.3.7), and gypsum caves in southern Spain (Calaforra and Pulido-Bosch, Chapter 5.3.10).

Two special topics are also considered:

- Speleogenesis of vertical shafts in the eastern United States (W. White in Chapter 5.3.9).
- Maze origin by diffuse recharge through an overlying formation (Palmer, Chapter 5.3.11).

The set concludes with:

- Two instances of nearly ideal water table cave development in Belize and Hungary (Ford, Chapter 5.3.12).
- A review of the latest models of speleogenesis from the region where modern karst studies began, the Classical Karst of Slovenia and Trieste (Šušteršič, Chapter 5.3.13).

Part 6

In Parts 2 to 5 attention is directed primarily on how the gross geometry of a cave system is established. Part 6 switches focus to the forms at meso- and microscales that can be created during enlargement of the cave.

Lauritzen and Lundberg in Chapter 6.1 summarize the great variety of erosional forms (*speleogenetic facies*) that can be created by a wide range of *speleogenetic agents* operating in the phreatic or vadose zones. Some forms of cave passages have been subject to intensive research and may be interpreted by means of simple physical and chemical principles, but many others are polygenetic and hence difficult to decipher with certainty. However, in addition to the analysis of cave patterns (see Chapter 3.4), each morphological element is a potential tool that can aid our inferences on the origin of caves and on major characteristics of respective past hydrogeologic settings.

In Chapter 6.2, E. White and W. White review breakdown morphology in caves, generalizing that the processes are most active during the enlargement and decay phases of cave development. Early in the process, breakdown occurs when the flow regime shifts from pipe-full conditions to open channel conditions (that is, when the roof first loses buoyant support). Later in the process breakdown becomes part of the overall degradation of the karst system. The chapter addresses the mechanism of breakdown formation, the geologic triggers that initiate breakdown, and the role that breakdown plays in cave development.

Part 7

As the great majority of both theoretical considerations and case studies in this book deal with speleogenesis in carbonate rocks, it is useful to provide a special forum to examine dissolution cave genesis in other rocks. This is the goal of Part 7.

Klimchouk (7.1) provides a review of speleogenesis in gypsum. This is a useful arena for testing the validity and limitations of certain general speleogenetic concepts. A difference in solution kinetics between gypsum and calcite impose limitations and peculiar features on the early evolution of conduits in gypsum. These peculiarities seem to be an extreme and more obvious illustration of some rules of speleogenetic development devised from conceptual and digital modeling of early conduit growth in limestone.

For instance, it was shown (Palmer, 1984, 1991; Dreybrodt, 1996; and also see Chapters 3.4 and 4.2.2) that initiation of early, narrow, and long pathways does not seem feasible under linear dissolution rate laws (n=1) due to exponential decrease of the dissolution rate. Although the dissolution kinetics of gypsum are not well known close to equilibrium, it is generally assumed that they are controlled entirely by diffusion and therefore are linear. If dissolution of gypsum is solely diffusion-controlled, with no change in the kinetic order, conduit initiation could not occur in phreatic settings or by lateral flow through gypsum from distant recharge areas in artesian settings. Hence, the fact that maze caves are common in gypsum under artesian conditions (see Chapter 5.2.1) gives strong support to a general model of "transverse" artesian speleogenesis where gypsum beds are underlain by or sandwiched between insoluble or low-solubility aquifers (Chapter 5.2). And it suggests that similar processes may be applicable to cave development in carbonates.

In unconfined settings, speleogenesis in gypsum occurs along fissures wide enough to support undersaturated flow throughout their length. Linear or crudely branching caves overwhelmingly predominate. These rapidly adjust to the contemporary geomorphic setting and to the maximum available recharge. Also, if considerable conduit porosity has been created in deep-seated settings, it provides ready paths for more intense groundwater circulation and further cave development when uplift brings the gypsum into the shallow subsurface.

Speleogenesis in salt, reviewed in general and exemplified by the Monte Sedom case in Israel (Frumkin, Chapter 7.2), has been documented only in open, unconfined settings, where it provides a model for simple vadose cave development. Chapter 7.3 deals with speleogenesis in quartzite, illustrated by case studies from southeastern Minas Gerais, Brazil (Corrêa Neto, 7.3.1), and South Africa (Martini, 7.3.2). The process involves initial chemical weathering of the quartzite to create zones of friable rocks (sanding or arenisation), which then are removed by piping, with further conduit enlargement due to mechanical erosion by flowing water.

Part 8

This part combines the theoretical with some applied aspects of speleogenetic studies. Worthington, Ford, and Beddows (Chapter 8.1) show the important implications of what might be termed "speleogenetic wisdom" when studying groundwater behavior in karst.
They examine some standard hydrogeologic concepts in the light of knowledge of caves and their patterns, considering a range of case studies to identify the characteristic enhancement of porosity and permeability due to speleogenesis that occurs in carbonate rocks. The chapter focuses on unconfined carbonate aquifers, as these are the most studied from a speleologic perspective and are most important for water supplies.

Four aquifers, differing in rock type, recharge type (allogenic and autogenic), and age (Paleozoic, Mesozoic, and Cenozoic), are described in detail to demonstrate the extent of dissolutional enhancement of porosity and permeability. All four cases are similar in hydraulic function, despite the fact that some of them were previously characterized as different end members of a "karst to non-karst" spectrum. Enhancement of *porosity* by dissolution is relatively minor. Enhancement of *permeability* is considerable, because dissolution has created dendritic networks of channels able to convey 94% or more of all flow in the aquifer, with fractures providing a small proportion and the matrix a negligible amount.

These conclusions may be viewed as a warning to hydrogeologists working in carbonate terranes: probably the majority of unconfined aquifers function in a similar manner. Sampling is a major problem in their analysis. This is because boreholes (the conventional exploration tool in hydrogeology) are unlikely to intersect the major channels that are conveying most of the flow and any contaminants in it. It is estimated—using examples of comprehensively mapped caves—that the probability of a borehole intersecting a conduit ranges from 1 in 50 to 1 in 1000 or more. Boreholes simply cannot be relied upon to detect the presence of caves or to "characterize" the hydrologic functioning of cavernous aquifers.

Wherever comprehensive evidence has been collected in unconfined carbonate aquifers (cave mapping, plus boreholes, plus lab analysis of core samples), dissolution inexorably results in a similar structure. Specifically, channel networks providing most of the permeability of the aquifer, yet occupying a very minor fraction of its volume (Worthington, Ford, and Beddows, Chapter 8.1).

Lowe (Chapter 8.2) focuses on developments in understanding the vital role played by karstic porosity (broadly viewed as being a product of speleogenesis), in the migration of mineralizing fluids (or hydrocarbons) and in their deposition (or storage).

He also comments on the potential role of new speleogenetic concepts in developing greater understanding in the future. Although some early workers were clearly aware of evidence for a relationship between karstic porosity and economic deposits—and others noted its theoretical likelihood—it has been ignored by many until relatively recent times.

This shortfall has gradually been redressed. New understanding of the extent and variety of karst processes is ensuring that new relationships are being recognized and new interpretations and models are being derived. The chapter clearly demonstrates the wide applicability of speleogenetic knowledge to issues in economic geology.

In Chapter 8.3 Aley provides an overview of the water and land-use problems that occur in areas with conduit aquifers. He stresses that sound land management must be premised on an understanding that karst is a three-dimensional landscape where the surface and subsurface are intimately and integrally connected. Failure to recognize that activity at the surface affects the subsurface, and the converse, has long been the root cause of many of the problems of water and land use in karst regions.

Karst areas have unique natural-resource problems, whose management can have major economic consequences. Although there is an extensive literature on the nature of particular problems, resource protection, and hazard minimization strategies in karst, it rarely displays an advanced understanding of the processes of conduit formation and their characteristic. Yet, these are always involved.

• • •

This book does not pretend to be a definitive text on speleogenesis. However, we hope that readers will find it to be a valuable reference source, that it will stimulate new ideas and approaches to develop and resolve some of the remaining problems. We also hope that it will promote an appreciation of the importance of speleogenetic studies in karst hydrogeology and applied environmental sciences.

Alexander Klimchouk,
Derek Ford ,
Arthur Palmer,
Wolfgang Dreybrodt, Editors

References

Dreybrodt, W., 1996, Principles of early development of karst conduits under natural and man-made conditions revealed by mathematical analysis of numerical models: *Water Resources Research* , v. 32, p. 2923-2935.

Ford, D.C., and Williams, P.W., 1989, *Karst geomorphology and hydrology:* London, Unwin Hyman, 600 p.

Groves, C.G., and Howard, A.N., 1994, Early development of karst systems, 1.: Preferential flow path enlargement under laminar flow: *Water Resources Research,* v. 30, p. 2837-2846.

Hill, C.A., 1987, Geology of Carlsbad Cavern and other caves in the Guadalupe Mountains, New Mexico and Texas: *New Mexico Bureau of Mines and Mineral Resources Bulletin,* no. 117, 150 p.

Hill, C.A., 1995, Sulfur redox reactions, native sulfur, Mississippi Valley-type deposits and sulfuric acid karst, Delaware Basin, New Mexico and Texas: *Environmental Geology,* v. 25, p. 16-23.

Howard, A.D., and Groves, C.G., 1995, Early development of karst systems, 2. Turbulent flow: *Water Resources Research,* v. 31, p. 19-26.

Klimchouk, A.B., 1994, Speleogenesis under confined conditions, with recharge from adjacent formations: *Publ. Serv. Geol. Luxembourg*, Comptes Rendus du Colloque International de Karstology a Luxembourg, v. 27, p. 85-95.

Klimchouk, A., 1995

Klimchouk, A., 1997

Lowe, D.J., 1992, A historical review of concepts of speleogenesis: *Cave Science* , v. 19, p. 63-90.

Lowe, D.J., and Gunn, J., 1995, The role of strong acid in speleo-inception and subsequent cavern development, *in* Barani-Kevei, I., ed., Environmental Effects on Karst Terrains: *Acta Geographica* (University of Szeged), v. 34, p. 33-60.

Mazzullo, S.J., and Harris, P.M., 1991, Overview of dissolution porosity development in the deep-burial environment, with examples from carbonate reservoirs in the Permian Basin, *in West Texas Geological Society Symposium Publication,* no. 91-89, p. 125-138.

Mazzullo, S.J., and Harris, P.M., 1992, Mesogenetic dissolution: Its role in porosity development in carbonate reservoirs: *American Association Petroleum Geologists Bulletin,* v. 76, p. 607-620.

Palmer, A.N., 1984, Recent trends in karst geomorphology: *Journal of Geological Education,* v. 32, p. 247-253.

Palmer, A.N., 1991, Origin and morphology of limestone caves: *Geological Society of America Bulletin,* v. 103, p. 1-21.

Palmer, A.N., 1995, Geochemical models for the origin of macroscopic solution porosity in carbonate rocks, *in* Budd, A.D., Saller, A.H., and Harris, P.M., eds., Unconformities and porosity in carbonate strata: *American Association Petroleum Geologists Memoir,* no. 63, p. 77-101.

Smart, P.L., and Whitaker, F.F., 1991, Karst processes, hydrology and porosity evolution, *in* Wright, V.P., Esteban, M., and Smart, P.L., eds., Paleokarsts and paleokarstic reservoirs: *P.R.I.S. Occasional Publication Ser.,* no. 2, University of Readings, p. 1-55.

Worthington, S.R.H., 1991, *Karst hydrogeology of the Canadian Rocky Mountains:* McMaster University, Ph.D. thesis, 380 p.

Worthington, S.R.H., 1994, The possible importance of sulfur minerals in initiating epigenic caves, *in* Sasowsky, I.D., and Palmer, M.V., eds., Breakthroughs in karst geomicrobiology and redox geochemistry: abs. and field-trip guide, *Karst Water Institute Special Publication,* no. 1, p. 80-82.

Worthington, S.R.H., and Ford, D.C., 1995, High sulfate concentrations in limestone springs: An important factor in conduit initiation?: *Environmental Geology,* v. 25, p. 9-15.

Part 2 Historical Perspective

2.0 Overview

The Historical Perspective, Part 2, gives a history of speleogenetic studies.

Shaw (Chapter 2.1) traces the development of the most important ideas from previous centuries through the early modern period in the first half of this century.

Lowe (Chapter 2.2) traces the development up to the threshold of modern times.

W. White (Chapter 2.3) brings the story right up to the present time.

The current state of the art is best illustrated by the entire contents of this book.

2.0 Figure 1. A painting made by Beduzzi in the Sloup Cave, Moravia (Czech Republic), during Nagle's visit in 1748 (courtesy Österreichischer Nationalbibliotek, Vienna)

2.1. Views on Cave Formation Before 1900

Trevor R. Shaw

Abstract

Nearly all present-day explanations for the formation of caves in limestone had been suggested before 1900. Some had been developed in considerable detail. Speleogenesis by erosion and the role of carbon dioxide in the dissolution of cave passages were understood in the first half of the 19th Century. By the end of the century there already existed schools of thought which supported dissolution by vadose water and by phreatic water.

The most important names in all this are those of:

- Alexander Catcott who in 1756 recognized that caves resulted from water action (though he did attribute the water to the Biblical Flood)
- Charles Lyell and Charles Edouard Thirria who independently in 1830 appreciated the importance of carbon dioxide in enabling water to dissolve caves in limestone
- Franklen Evans who understood in 1870 that some caves were formed by solution in groundwater
- Edouard Martel who from 1890 emphasized the solutional activity of vadose water and made it well known among speleologists
- Ferencz Pošpný who described the movement of groundwater and its speleogenetic effect in 1893
- Edouard Dupont who developed that idea.

There were also some distinctly primitive explanations, mostly in earlier times, including the creation of cavities when beds were folded and mountains formed, erosion by water of rock before it had become hard, expansion of gases in the soft rock from decomposing animal bodies drowned in the Flood, and the solution of salt inclusions having the same shape as the resulting caves. These catastrophic theories were often inspired by the need to explain the formation of caves, as of all landforms, in the relatively short period since the creation of the world, which was commonly believed to have happened about 4000 years BC.

Introduction

Nearly all present-day explanations for the formation of caves in limestone were born before 1900. Some were already developed in considerable detail. Speleogenesis by erosion and the role of carbon dioxide in the solution of cave passages were understood in the first half of the 19th Century. By the end of the century, there already existed schools of thought which supported solution by vadose water and phreatic water.

The ideas commonly associated with the 1930's had their roots, and sometimes much of their substance, 40 to 50 years before. In earlier days there had been several primitive explanations for the way in which caves were formed, but many of these were not unreasonable in the state of knowledge at that time. When chemists required phlogiston to explain combustion and did not distinguish between solution and suspension, and when geologists were debating whether rocks originated from volcanoes or by deposition from the "soup" of a primaeval ocean, it is not surprising to find that caves were held to result from processes that now seem distinctly odd.

The various cave formation processes once, or still, current were either mechanical (tectonic or erosional) or chemical (solutional). They may be subdivided as follows:

- made by humans
- contemporary with the surrounding rock
- tectonic action
- once-only events such as the Biblical Flood
- other short-duration processes such as

erosion or dissolution of soft materials and dissolution in strong acids
- slow removal of limestone by erosion under present-day conditions
- dissolution of limestone under present-day conditions.

Speleogenesis is similar in this way to other aspects of geology and geomorphology, where conflicting explanations existed over the process of denudation, the deposition of sediments, and the uplift of mountains.

The interrelation of the main explanations is shown diagrammatically in 2.1 Figure 1 (see pages 24 and 25). They are linked there in a logical grouping but are not necessarily developed from their predecessors; the initiator of a new theory was not always aware of the earlier ones. Thus the diagram shows interrelationships but not interdependence. Explanations that have led to the current understanding are indicated by heavy black outlines. Not all the people named in the chart are discussed in this short review. A much more detailed study of the whole subject can be found in Shaw (1992).

Perhaps the most important names in all this are those of:

- Catcott who in 1756 recognized that caves resulted from water action (even though he did attribute the water to the Flood)
- Lyell and Thirria who independently in 1830 appreciated the importance of carbon dioxide in enabling water to dissolve caves in limestone
- Evans who understood in 1870 that some caves were formed by solution in groundwater
- Martel who from 1890 emphasized the solutional activity of vadose water and made it well known among speleologists
- Pošpný who described the movement of groundwater and its speleogenetic effect in 1893, and Dupont who developed that idea in 1893 and subsequent years.

Cvijić and Grund are of equal or greater importance. But most of Cvijić's and all of Grund's works were published after 1900, so they are described in the next chapter of this book.

Early Explanations

Aristotle (1952), writing in the 4th Century BC, had realized that the existence of underground cavities required explaining, but he had no clear understanding of how to do so:

> ... the rivers that are swallowed up ... prove that there are chasms and cavities in the Earth ... The reason is that ... there are no outlets from the valleys ... ; so ... the water flowing in from above forces its way out and finds a way through ...

Others avoided the need for a natural explanation because they thought caves were man-made. Thus Trissino (1550) and others regarded them as underground quarries or mines, and even in the 18th Century Rudder (1779) had the same idea.

Most of the early explanations of landforms, whether they be mountains, valleys, or caves, required short-duration processes of great violence. Those who proposed them became known as catastrophists. The main argument for catastrophism was that it was only by such large-scale occurrences that the magnitude of the changes could be accounted for in the time thought to have been available since the creation of the Earth, then held to have occurred about 4000 BC.

The catastrophists sought their explanations either in tectonic Earth movements or in upheavals associated with the Biblical Flood. The tectonic theories depended in some cases on rock deformation caused during folding of the strata (Virlet, 1834) or by centrifugal uplift of the Earth's crust near the equator (Buffon, 1778); but others associated cave development with the creation of the world (e.g. Defoe, 1727) or with the all-but-instantaneous "convulsions" which they supposed had formed mountain ranges (Whitehurst 1778).

Some of the Flood hypotheses (e.g. Burnet, 1684) amounted to little more than the same idea. It merely explained the break-up of the Earth's crust as a part of the Flood process. Others, especially Catcott (see below), recognized the probable role of water in cave formation and welcomed the generally accepted Flood as a means of providing sufficiently vast quantities of water to scour out even the biggest cavities.

Primitive as these catastrophic theories now seem, some of them embodied ideas which grew later into essential parts of more acceptable explanations. The belief that Earth movements can form caves directly (as well as being true in a very few exceptional cases) has led to the recognition that joints and bedding planes are normally a prerequisite for the formation of caves by dissolution. The underground water action once associated solely with the Flood, and the evidence put forward to demonstrate it, are directly applicable to the subsequent theories embodying erosion.

Another fairly short-lived catastrophic theory associated with the biblical Deluge did not lead to any similar advance. That was Esper's (1774) suggestion that caves were formed by expansion of the gases given off by the decomposition of the bodies of animals drowned by the Flood and deposited in the mass of soft mud which subsequently hardened around the resulting large gas bubble until it became rock.

The most important of these catastrophic theories, in that it incorporated several ideas that are now an integral part of modern explanations, is that of Alexander Catcott (b.1725; d.1779). He argued convincingly, from actual detailed observation, that caves must have been formed by the action of water. He personally explored many caves, mainly in the 1750s and 1760s, and was familiar with the karst landscape of the Mendip Hills in England. What he saw there in the way of large valleys and sinkholes was largely consistent with his contention that the retreating water of the biblical Flood had scoured out the valleys as well as the caves through which it drained away.

Catcott's first clear statement that cave passages were formed by water occurs in his diary for 17 August 1756, when he visited a limestone quarry at Pengillgoet in Wales:

> In several places there were round holes like pipes (leading into larger Swallows) that had undeniably been formed or bored by the water; probably there was some little crack which the water by degrees made larger & larger. ... The smooth worne sides of the larger Swallows ... afforded occular demonstration that they were worne & formed by the action of the water. (Catcott, 1748-74)

An ink sketch in the diary (2.1 Figure 2, p. 25) shows a cross section through two such passages. It is noteworthy that the shape of the passages, elongated along the bedding plane at midheight, would now be recognized as indicating their formation while completely filled with water. It is even more remarkable that Catcott was interested enough in their shape to draw the sections at all.

Catcott's views on cave formation were expressed more clearly and fully when he published his *Treatise on the Deluge* in 1761. He argued there that the water in question could only have come from the Flood, when the flow would have been much more violent than that of modern streams.

Mechanical Erosion and Limestone

In contrast to the catastrophist theories just considered, there were uniformitarian ones involving long-continuing processes occurring at present-day rates. Of the alternative processes of erosion (corrasion) and dissolution, erosion was the easier to comprehend and hence it was introduced several centuries earlier than the other. The actual process of this erosion was not different from the erosion involved in some of the Flood hypotheses; it was the recognition that it could be carried out by normal streams that constituted the advance. Realization that hard rock could be eroded in this way, at a rate sufficient to account for existing caves, did not occur until the 18th Century, but a long time before that there were intermediate ideas such as the removal of soft inclusions in the rock by erosion. Another such intermediate idea was John Hutton's (1780), which suggested that streams of present-day size had formed cave passages by eroding the rock while it was still new and hence relatively soft.

This section is concerned with cave formation by mechanical erosion (corrasion) alone. Its effect in association with dissolution in stream water is considered later.

An enlightened statement on the importance of water action appeared in a little book on mountain building by Valerius Faventies (1561):

> There are in the bowels of the Earth tremendous caves extending back to an immense distance... . It is quite certain that all this is mainly the work of water.

Thus although he did not deny a part to Earth movements and fissuring, by implication he relegated them to second place. This recognition of the importance of water was a marked step forward in thinking on speleogenesis, and Faventies was well ahead of his time. Probably because only a few copies of his book were printed, it was little known and had no detectable influence. In the 18th Century Buffon (1749) came down very firmly in favor of water action in at least some caves:

> It can clearly be seen that Devil's-Hole (Peak Cavern) and other caves from which issue strong springs or rivers, have been formed and shaped by water carrying sand and silt obtained from between the stones and rocks.

Flamichon, too, who explored Isturitz and other Pyrenean caves in the 1770s, considered that they were all formed by the action of streams, either those that still ran through them or which had subsequently "taken to a lower course" (Latapie, 1816).

The gradual acceptance in the 18th Century, by some people only, of the mainly erosional origin of caves has a close parallel in surface geomorphology. Weathering and general denudation of hillsides had long been accepted by a few writers, but fluvial erosion of surface river valleys was hardly considered at all before the 1780s. It was only with James Hutton (1788) that the concept of fluvialism, or the erosive power of surface rivers, was first put forward.

At the end of the 19th Century, by which time dissolution was accepted by many as at least a partial mechanism for speleogenesis, an all-erosion theory was strongly defended by Armand Flamache (1896). He objected vigorously to the dissolution theory of Dupont, to be described later, mainly because he maintained that underground water would be saturated with calcium bicarbonate everywhere except close to the surface. Thus he was forced to favor mechanical erosion:

> Underground valleys are excavated by the same causes and according to the same laws as surface valleys.

An underground flow of water will take place provided there is a head difference between the ends of the passages. This can occur in a number of ways, in the rock between two streams at different heights for example, or between the rainfall on a plateau and the stream valley below.

Special Dissolution Theories

To those who noticed the uneven size and shape of cave chambers and passages, their occasional lack of any apparent connection with the surface and, in many cases, the absence of any running stream, it was perhaps natural enough to suppose that they were the cavities remaining after a body of some readily soluble material had been dissolved away from inside a mass of limestone.

There were indeed two men quite early in the 19th Century—Brongniart (1817) and Aubuisson de Voisins (1819)—who both attributed the formation of caves to the removal by water of deposits or inclusions of common salt from within the surrounding rock. The deposits, they postulated, must have had the exact shape of the caves that resulted. This idea was somewhat similar to Martyn's (1729) belief that soft inclusions in the rock were removed by erosion.

The most convincing refutation of this salt theory came from Thirria (1830), who said:

> ...it is known that where such deposits do occur they are wrapped around by the strata which are distorted by their presence like the moulding sand around a casting; such a deformation is never seen in the limestone beds surrounding caves ...

Thirria will be noticed again as one of the first to appreciate the importance of dissolved carbon dioxide in the dissolution of limestone caves.

The dissolution of rock salt (halite) did in fact have some impact on the understanding of speleogenesis, but not by way of the hypothesis just considered. Massive beds of salt are just as liable to be dissolved as the postulated inclusions. Such dissolution is influenced by the water available, by its ability to drain away, by the fissures by which it gains access to the interior of the salt mass, and indeed by most of the parameters that also affect the dissolution of limestone. The morphology of its dissolved cavities is also remarkably similar to that of calcareous ones, but they are formed at a very much greater rate. Such caves in salt are usually short-lived so that it is only when the water table is altered by mining that their formation can be observed at all.

Ferencz Pošpný of Bohemia actually observed cave formation taking place by groundwater dissolution in rock salt in 1864. He used it as an analogy in the paper in which he advocated the same process as the cause of limestone caves (Pošpný 1894: 215).

The idea of limestone itself being gradually dissolved away by dissolution rather than mechanical erosion in water was mainly a 19th Century concept arising from the more detailed and more precise examination of caves at that time. In a few cases the water was supposed to be hot or fiercely acidic; in others, before the presence of carbon dioxide in rainwater was realized, it was thought to be ordinary pure water.

The original all-pervading water of the Neptunists, containing dissolved chemicals along with all its fragmental material, must have seemed an ideal agent in the earlier years of the 19th Century. It was hot and could easily be imagined to be suitably aggressive. Parandier (1833) and De Serres (1835) did make use of the water of such a Neptunist sea in their hypotheses, but the idea of dissolution had been rather tentatively put forward a little earlier by Aubuisson de Voisins (1819) and Omalius d'Halloy (1831), using simply water containing "original acid."

The concept of dissolution in specially aggressive water persisted in various forms throughout the 19th Century, but it had no direct effect on the subsequent understanding of normal speleogenesis. What it did do was to emphasize to later workers the fact that dissolution was often the best or the only way of explaining certain characteristics of caves.

The Action of Vadose Water

It was in 1830 that the importance of dissolved carbon dioxide in rendering the cave water acidic was appreciated by Lyell and by Thirria. Soon afterward the function of joints as the initiators of the dissolution process was recognized. At first dissolution was thought of as occurring only in the vadose zone, i.e. where trickling and flowing water could be actually seen. It was not until 1870 that phreatic dissolution was suggested and only in the 1890's and afterward was it advocated as a major factor in speleogenesis.

Before 1830 there were nevertheless some

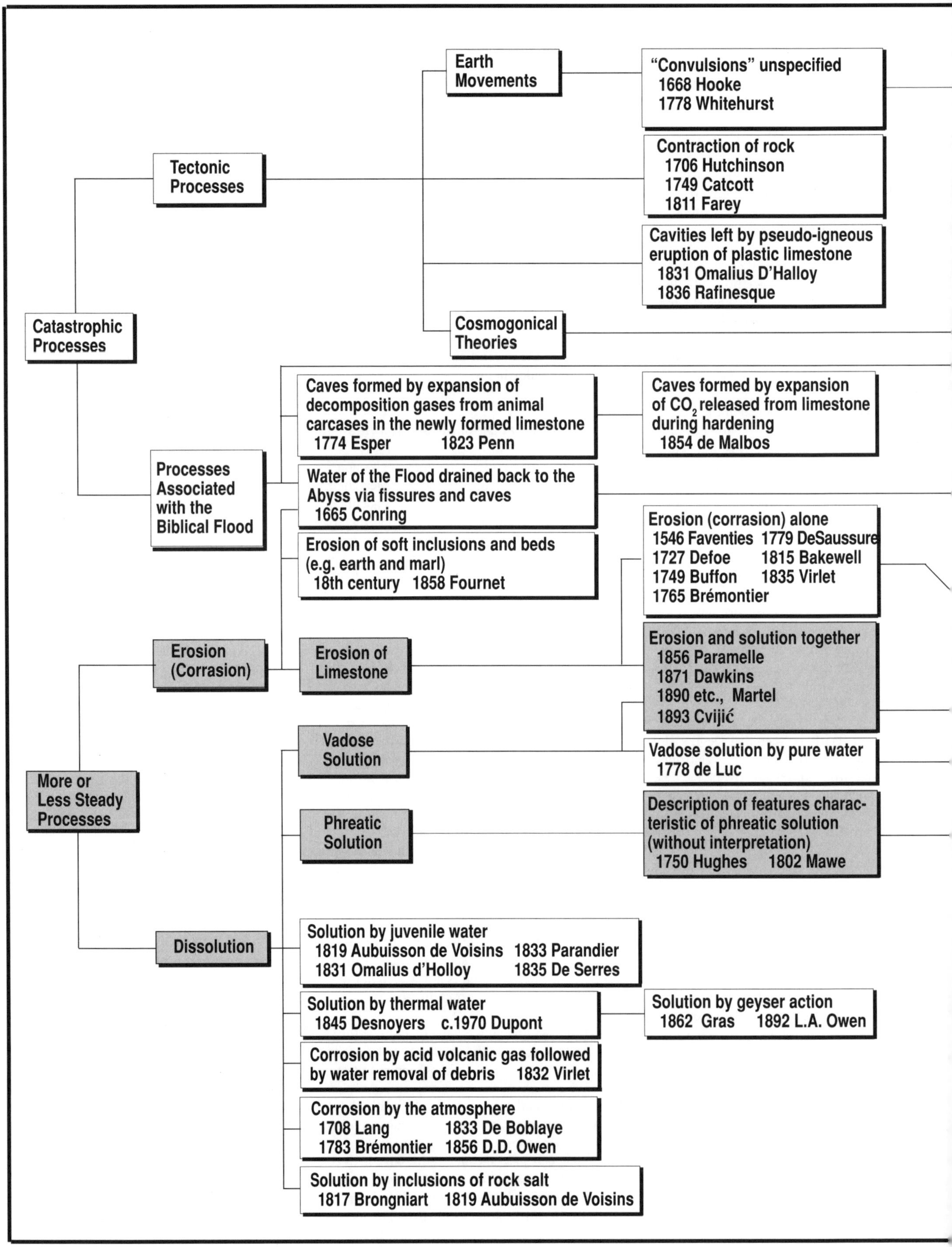
Catastrophic Processes
Tectonic Processes
Earth Movements
"Convulsions" unspecified 1668 Hooke 1778 Whitehurst
Contraction of rock 1706 Hutchinson 1749 Catcott 1811 Farey
Cavities left by pseudo-igneous eruption of plastic limestone 1831 Omalius D'Halloy 1836 Rafinesque
Cosmogonical Theories
Processes Associated with the Biblical Flood
Caves formed by expansion of decomposition gases from animal carcases in the newly formed limestone 1774 Esper 1823 Penn
Caves formed by expansion of CO_2 released from limestone during hardening 1854 de Malbos
Water of the Flood drained back to the Abyss via fissures and caves 1665 Conring
Erosion of soft inclusions and beds (e.g. earth and marl) 18th century 1858 Fournet
Erosion (corrasion) alone 1546 Faventies 1779 DeSaussure 1727 Defoe 1815 Bakewell 1749 Buffon 1835 Virlet 1765 Brémontier
More or Less Steady Processes
Erosion (Corrasion)
Erosion of Limestone
Erosion and solution together 1856 Paramelle 1871 Dawkins 1890 etc., Martel 1893 Cvijić
Vadose Solution
Vadose solution by pure water 1778 de Luc
Phreatic Solution
Description of features characteristic of phreatic solution (without interpretation) 1750 Hughes 1802 Mawe
Dissolution
Solution by juvenile water 1819 Aubuisson de Voisins 1833 Parandier 1831 Omalius d'Holloy 1835 De Serres
Solution by thermal water 1845 Desnoyers c.1970 Dupont
Solution by geyser action 1862 Gras 1892 L.A. Owen
Corrosion by acid volcanic gas followed by water removal of debris 1832 Virlet
Corrosion by the atmosphere 1708 Lang 1833 De Boblaye 1783 Brémontier 1856 D.D. Owen
Solution by inclusions of rock salt 1817 Brongniart 1819 Aubuisson de Voisins

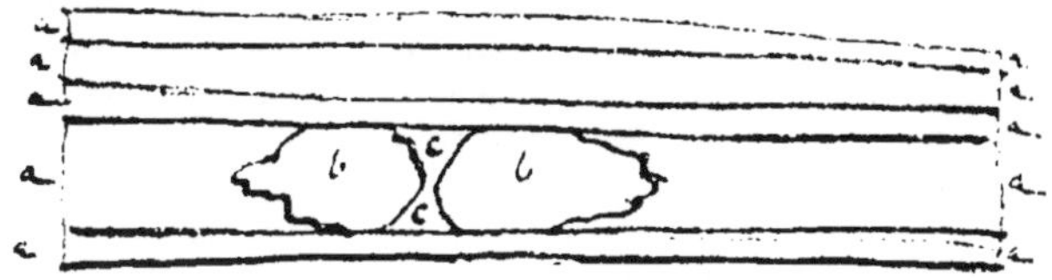

2.1 Figure 2. Passage cross sections sketched by Catcott on 17 August 1756 in his diary (1748-74). (Reproduced by permission of Bristol Reference Library.)

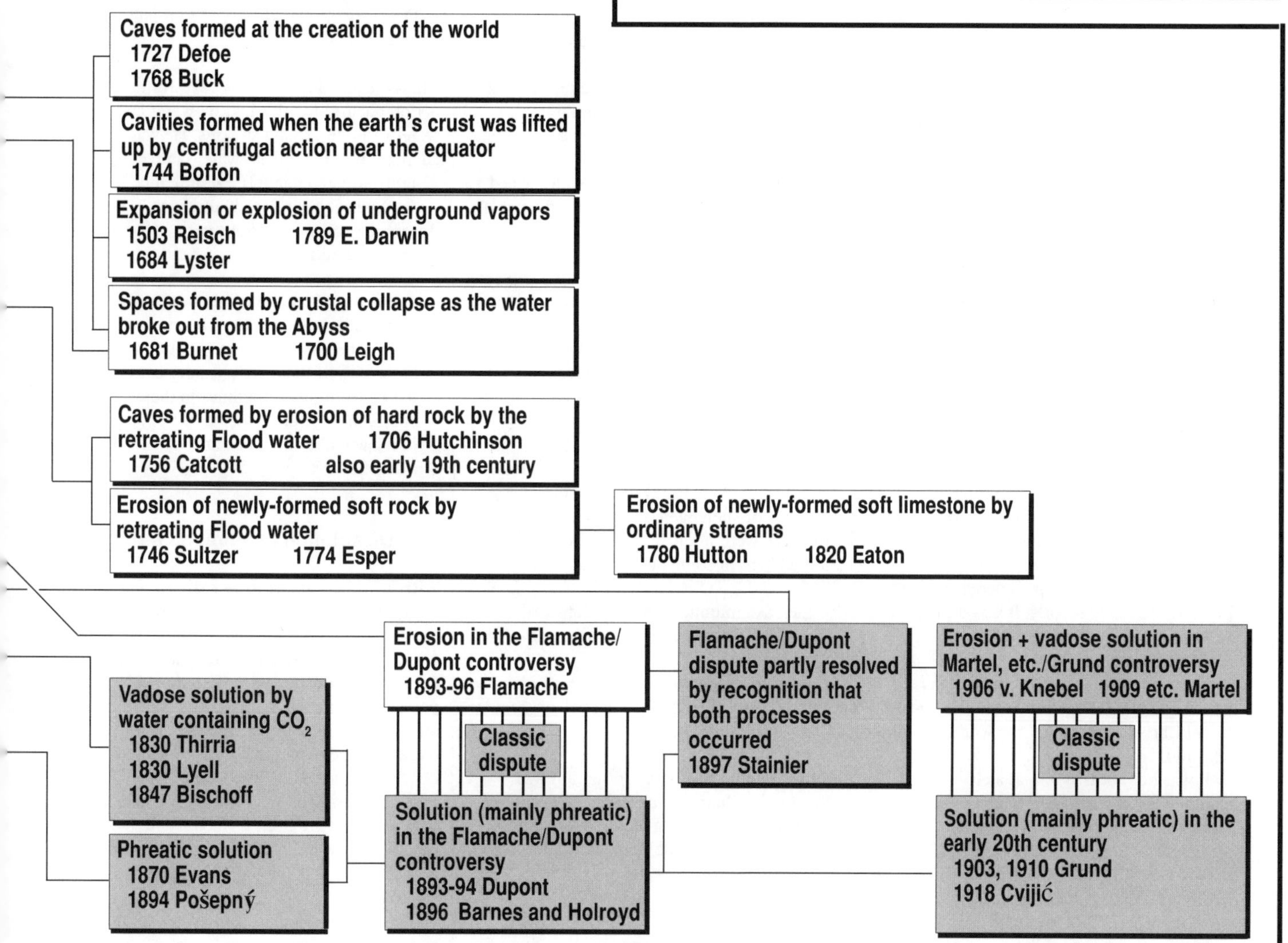

2.1 Figure 1. Chart summarizing the theories of speleogenesis.

who considered that water could somehow dissolve limestone. Lang (1708) described the manner in which water sinks from the surface by means of crevices and fissures. He then went on to suggest that the water flowing in these fissures and pores gradually corrodes them, preparing for itself a way of entry to the cavities of the interior. Then, exhalations and vapors of this corrosive water steadily dissolve the surface of the rock and convert it into lac lunae.

A real understanding of the action was not possible until the existence of phlogiston was disproved and oxygen identified in the 1780s. Observation did suggest dissolution before that time, and indeed the earlier concepts of unspecified "aerial acid"and "aerial nitre" did allow some approach toward understanding the chemistry involved. Quite early in the 18th Century, Lesser (1735: 16) had referred to a dissolution of the "aerial spirit" dissolving from certain rocks the material which was afterwards deposited in caves.

There were some, including Williams (1834) and De Malbos (1854), who accepted the fact that limestone was dissolved in carbonic acid to the extent necessary to account for stalactites, but could not bring themselves to believe that this effect could wholly explain the formation of large cavities and caves.

In these cases the process itself was understood and it is clear that there must have been some other reason preventing its acceptance for cave forming. It was possibly due to the psychological difficulty of scale. Whereas the relatively small amount of dissolution necessary for speleothems could be accepted fairly readily, it was difficult to visualize the same process occurring on the scale required to produce caves. The length of time involved was hard to grasp. Many people were still constrained by the conventional Christian view that the world was not very old.

It was not until 1830 that a clear explicit statement was made indicating the importance of carbon dioxide in water as a factor in speleogenesis. Both Charles Lyell (b.1797, d.1875) and Charles Edouard Thirria (b.1796, d.1868) published such statements in that year, apparently independently.

In the first edition of his *Principles of Geology,* Lyell (1830) wrote:

> To the solvent power of water, surcharged with carbonic acid, and percolating various winding rents and fissures, we may ascribe those innumerable subterranean cavities and winding passages which traverse the limestone in our own and many other countries.

In the same year Thirria (1830) gave a much fuller treatment of the same idea in a rather obscure provincial French journal. He was a mining engineer who explored and surveyed several caves in the department of Haute-Saône, including the Grotte d'Osselle.

One of the reasons why dissolution was first recognized as playing an important part in cave formation was the fact that many aspects of passage shape just could not be satisfactorily explained in any other way. The characteristic dissolutional rockform known as scalloping was described by De Serres (1835, p. 34):

> Moreover the walls of vertical rifts and of caves are as it were dented or hollowed with shallow depressions, rounded at the bottom and at the corners and edges. These depressions seem to have been made, not from rubbing by a solid material, but by the dissolving force of erosion by a liquid.

Again, Charles W. Wright (1858) described features which are in themselves evidence of the dissolutional origin of caves.

> There can be no doubt but that the solvent action of water holding carbonic acid in dissolution was the primary agency concerned in the formation of (Mammoth) Cave ... The little niches and recesses which are observed in various parts of the Cave ... were formed in this manner.

The distinguished French speleologist E.A. Martel (b.1859, d.1938) was a strong advocate of speleogenesis by stream action. His ideas were not original but they are studied here in some detail because of:

- the vast amount of observations and evidence amassed in the course of his explorations
- the immense influence of his many books and other publications and his world-wide personal contacts.

Martel's opinion on cave formation was stated quite clearly, even in his early publications. Thus in a lecture read to the Société Géologique de France on 1 December 1890 he said:

> Caves and potholes always show in the clearest manner evidence of the effect of water: mechanical effect and chemical effect (Launay and Martel, 1891).

Then in his English visit of 1895, the Great Chamber of Peak Cavern:

> ...is evidently the work of erosion and corrosion by subterranean water (Martel, 1897).

He emphasized:

> ...that vadose water containing carbonic acid from the air is perfectly sufficient to account for the progressive dissolution of limestone (Launay and Martel, 1891).

Three features are listed (Martel, 1896, p. 31-34) as characteristic of corrosion:

- "Cups (scallops) and grooves." He pointed out that somewhat similar dissolution features exist on limestone rocks on the surface.
- Red soil in caves, often the residue from the dissolution of impure limestone.
- Deposits of tufa where underground rivers emerge at the surface.

Martel insisted that all this cave-forming action occurred with trickling water and the water of underground streams. He was adamant that there is no water table but only discrete stream channels. Despite this, his wide experience and study of caves in many different terrains in various parts of the world led him to realize that:

> No theory about the origin of caves is universal: those that have been put forward have generally claimed to be too inclusive; almost all of them are partially correct; the whole truth lies sometimes in their combination, sometimes in the application of a particular one in a particular case. (Martel, 1896, p. 53-54).'

Dissolution by Phreatic Water

After a while it came to be realized that many caves cannot be explained by the erosive and dissolving action of vadose water alone. In particular there are those which contain no evidence of former stream passages, some where the floors are continually sloping up or down, and others where large chambers alternate with very small connecting passages. The shape of the passage cross sections is often inconsistent with their having been formed by streams or percolating water, and there are also some roof features that cannot be explained satisfactorily.

The realization that many caves are not stream-formed had inspired some of the earlier theories such as gas inflation, and the dissolution of salt and other readily soluble inclusions. Gradually it became appreciated that direct dissolution of the rock itself while it was totally submerged must have been involved.

It is to Franklen George Evans (b.1826, d.1904) (see 2.1 Plate 1) that the honor falls of being the first to indicate, in 1870, that cave formation takes place below the general groundwater level of a district. His statement appeared in the transactions of a provincial scientific society (Evans, 1872) and attracted

2.1 Plate 1. Dr. Franklen Evans who in 1870 stated that some caves are formed in the phreatic zone (a portrait from Vachell and Evans, 1905).

little or no attention either at the time or since. It was contained with no great prominence in a general paper on Carboniferous limestone read on 7 July 1870 to the Cardiff Naturalists Society, of which he was then vice president, and later president. Having explained that dissolution by rainwater occurred in limestone fissures, hollowing them out into a series of chambers, he went on:

> ...the cavities of limestone rock are met with in three different conditions, which in some districts may occur as successive periods in their geological history. They are seen as caverns full of water, like those of Pentyrch [Mine]; then, when the barriers are destroyed they form broken subterraneous channels for the passage of an engulphed stream, as at Ystradvellte [in Breconshire]; and finally, when rivers are diverted from any cause, they become dry caves ... [The] process [of dissolution] would be intensified at some depth underground where the superincumbent pressure of the rocks ... enabled the water to condense within itself a larger proportion of carbonic acid, and increase its solvent power.

Despite the importance of Evans' conclusions they have remained completely unknown even to historians of science, and exerted no influence whatsoever on future speleogenetic opinions.

Whereas Evans' statement had been hidden in a general paper in a local journal, subsequent publications on the subject were much more accessible. Two clear arguments in favor of phreatic dissolution were made in 1893 (by Pošpný and by Dupont). Both were

2.1 Plate 2. Dr. Edouard Dupont, the first influential exponent of speleogenesis in phreatic water, in 1885.

published in the following year. Pošpný has priority by a few days.

Ferencz Pošpný (b.1836, d.1895) realized that groundwater was not stagnant but moved both laterally near its surface and almost vertically at great depths (1894, p. 213-214).

> The water flowing at the bottom of a cave in limestone is unquestionably groundwater, and it follows that the whole complex group of cavities has been eaten out by it. If in another limestone cave we see no flowing water, the current must have found some lower outlet; and the cave represents for us an ancient groundwater channel.

Pošpný's conclusions, published in America, seem to have altogether escaped the attention of Cvijić, Grund, and Davis, who later adopted similar views. Nor was his work cited in any of the subsequent emerging discussions in Belgium.

The idea of phreatic dissolution was first taken seriously by members of the Société Belge de Géologie, de Paléontolgie et d'Hydrologie in the middle 1890's. It was argued with a wealth of detailed evidence by Dupont and stimulated a lively dispute especially with Flamache and his erosion-only theory.

Edouard François Dupont (b.1841, d.1911) (2.1 Plate 2) became the Director of the Musée Royale d'Histoire Naturelle de Belgique. His opinions on the phreatic origin of caves appeared in two places.

First they were presented verbally (and reported in Willems, 1894) to members of the Société de Géologie, de Paléontologie et d'Hydrologie, in the course of their summer meeting in August 1893. His principal paper on the subject (Dupont 1894) was presented at the meeting of the same society on 29 May 1894.

Having asserted that caves result from dissolution by water containing carbon dioxide, he pointed out that this could occur in the groundwater zone as well as above it. Thus in the Quaternary Period, caves already formed beneath the water table were intersected by surface valleys as these became deeper (Dupont, 1894, p. 238).

In arguing that many cave passages were formed entirely by dissolution, Dupont gave diagrams of passage cross sections (2.1 Figure 3) which showed clearly that they were completely filled with water at the time of their formation. The fissures in which the water circulates, he said, are most commonly the joints and sometimes the bedding planes.

Although cave passages are initiated and largely formed by dissolution, Dupont pointed out (in Willems, 1894, p. 308-309) that in some cases, when once a vadose channel is big enough to allow a faster flow, an additional erosive effect comes into play. This may be further aided by stones and gravel carried in by floods (see 2.1 Figure 4):

> In some of them, one can see toward the bottom of the side walls, evidence of mechanical erosion which has resulted in a partial enlargement of the channel previously formed by corrosion.

Despite this recognition of the fact that stream erosion can modify the form of passages already existing, Dupont (1894, p. 207-209) was adamant that stream action by itself cannot create a passage. Reasons he gave for this were:

- caves do not occur in schist, although it is softer than limestone and just as much fissured;
- many caves are not stream caves and never have been—all those, for example with only one entrance;
- erosion never acts sideways to any great extent as it would have to do if it were to form horizontal caves by direct action;
- the walls of stream-carrying caves do not show horizontal grooving, as they would if they had been made by stream erosion.

> From all points of view, mechanical action must be rejected as the cause of cave formation.

Around the turn of the century there flourished in central and southern Europe two most important geomorphologists who advanced the study of speleogenesis very significantly. The work of Cvijić and Grund forms the basis of much 20th Century thinking on the subject. They provided a transition from the 19th to the 20th Century and are dealt with in the next section of this book.

By 1900, therefore, three main theories of speleogenesis existed, each of them correct in certain circumstances. Flamache maintained that erosion alone was sufficient to form caves. Martel and others believed that a combination of both vadose dissolution and erosion provided the complete explanation. Dupont and Grund considered that phreatic dissolution was the principal factor.

Each of these schools maintained its own point of view as if it were applicable in all cases. Hence two major controversies arose, one between the advocates of erosion (corrasion) and dissolution and the other between the vadose and phreatic dissolutionists.

Gradually, very gradually in some cases, both disagreements were resolved by realization that the three processes were all valid—and they operated in combination, either together or successively, in varying proportions according to local circumstances.

References

Aristotle, 1952. Meteorologica: Trans. H. D. P. Lee. London, Heinemann. 432 p. (Loeb Classical Library) Bk. 1, ch. 13.

Aubuisson de Voisins, J. F. d'., 1819, Traité de géognosie...: Strasbourg, Levrault. v. 2, p. 382-383.

B[rongniart], [A.], 1817. Caverne. In*: Dictionnaire des sciences naturelles*. v. 7. Strasbourg. p. 298-309. (p. 302).

Buffon, G. L. Leclerc de, 1749, Preuves de la théorie de la terre. In: *Histoire naturelle, générale et particulière*.... v. 1: Paris: Imprimerie Royale, p. 125-612. (p.546).

Buffon, G.L. Leclerc de, 1778, *Histoire naturelle, générale et particulière ...* . Supplement 5: Paris, Imprimerie Royale, p. 463.

[Burnet, T.], 1684. The theory of the Earth ... : London: Kettilby, 2 v. (v. 1, p. 67).

Catcott, A., [1748-1774], Diaries of Tours Made in England & Wales: Bristol Reference Library MS, B. 6495. pt. 3, folio p. 16 (back)–17.

Catcott, A., 1761, A treatise on the deluge; containing...natural proofs of the deluge ... and...the cause of caverns or natural grottos... : London, Withers, 296 p.

[Defoe, D.], 1727, A tour thro' the whole island of Great Britain... v. 3: London, Strahan, (p. 59).

De Malbos, J., 1854, Mémoire sur les grottes du Vivarais, Mémoires de l'Académie des Sciences, Inscriptions et Belles-Lettres de Toulouse, sér. 4, v. 4, p. 96-114 (p. 100).

De Serres, [P.] M. [T.], 1835, Essai sur les cavernes à ossemens et sur les causes qui les y ont accumulés: Natuurkundige Verhandelingen van de Hollandsche Maatschappij der Wetenshappen te Haarlem, v. 22, p. [1]–222.

Dupont, E., 1894, Les phénomènes généraux des cavernes en terrains calcareux et la circulation souterraine des eaux dans la région Han-Rochefort: Bulletin de la Société Belge de Géologie, de Paléontologie et d'Hydrologie. Mém. v. 7 for 1893, p. 190-297.

Esper, J.F. ,1774, Ausfürliche Nachricht von neuentdeckten Zoolithen... : Nürnberg, Knorr. 148 p. (p. 100-107).

Evans, F. G., 1872, The Carboniferous limestone. Report and Transactions. *Cardiff Naturalists' Society, v.* 3, no. 1, for 1870-71, p. 39-47, p. 46-47).

Faventies, J., 1561, De montium origine: Venice, Academia Veneta. 16 folio p.

Flamache, A., 1896, Sur la formation des grottes et des vallées souterraines: *Bulletin de la Société Belge de Géologie, Paléontologie et d'Hydrologie Mém.* 9 for 1895, p. 355-367.

Hutton, James., 1788, Theory of the Earth... : *Transactions of the Royal Society of Edinburgh*, v. 1, p. 209-304.

[Hutton, John], 1780, A tour to the caves, in the environs of Ingleborough and Settle, in the West-Riding of Yorkshire... : London, Richardson & Urquhart, 49 p. (p.38-39).

Lang, K. N.. 1708. Historia lapidum figuratorum Helvetiae: Venice, for the author, 165 p. (p. 7).

Latapie, J., 1816, Théorie de la terre déduite de l'organisation des Pyrénées et pays adjacens redigée sur les manuscripts de M. Flamichon: Tonne, Pau. 311p. (p. 114-115).

Launay, L. de & Martel, E. A. 1891. Note sur quelque questions relatives à la géologie des grottes et des eaux souterraines .... *Bulletin de la Société Géologique de France.*. Sér.3, v. 19, Paris. 142-165 (pp. 144-145).

Lesser, F. C. & J. G., 1735, Anmerckungen über die Baumañs-Höhle wie er sie selbst anno 1734 den 21 May befunden... : Hamburg, Stromer, 28 p. (p. 16).

Lyell, C., 1830, Principles of geology..., v. 1: London, Murray, (p. 211).

Martel, E. A., 1896, Applications géologiques de la spéléologie... : Annales

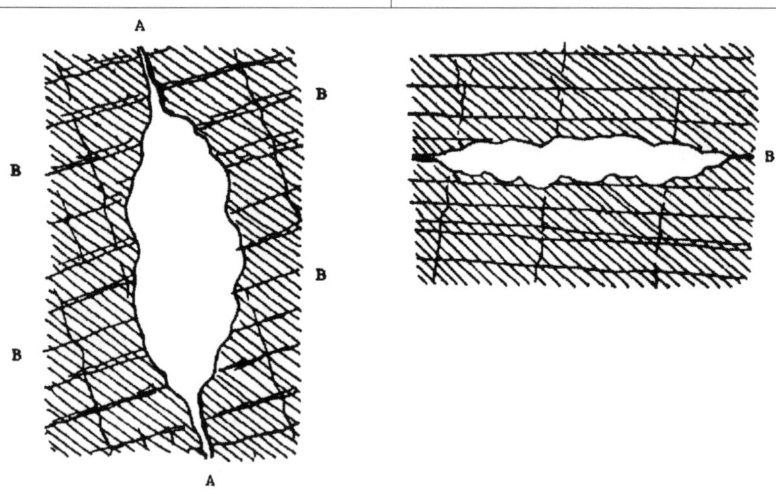

2.1 Figure 3. Diagrams by Dupont of typical cross sections of joint- and bedding-determined passages formed by solution when completely full of water. ***AA*** *is a joint,* ***BB*** *are bedding planes.*

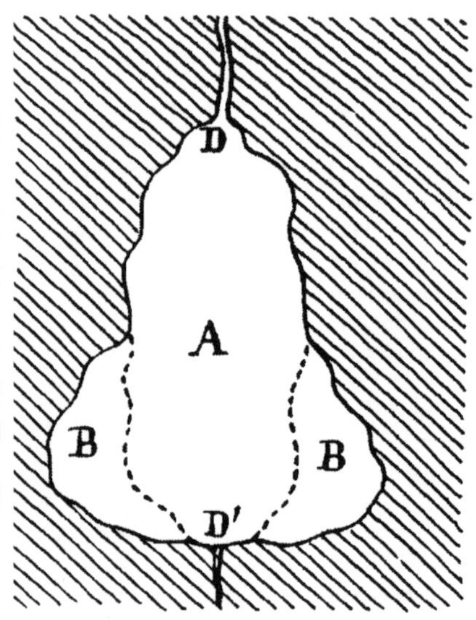

2.1 Figure 4. Diagram by Dupont showing cross section of solution passage ***A****, formed on the joint* ***DD'****, with lateral widening* ***BB*** *from subsequent stream erosion (from Willems, 1894, p. 309).*

des Mines… , Paris, sér. 9, v. 10, Paris. [5]–100.

Martel, E. A., 1897, Irlande et cavernes anglaises: Paris, Delagrave. 403 p. (p. 289).

Martyn, J., 1729, An account of some observations relating to natural history, made in a journey to the Peak in Derbyshire: *Philosophical Transactions of the Royal Society of London*, v. [36] (407). p. 22-32 (p. 27-28).

Omalius d'Halloy, J. B. J. d', 1831, Eléments de géologie: Paris, Levrault.

Parandier, A. N., 1833, Notice sur les causes de l'existence des cavernes… : Séance Publique de l'Academie des Sciences Belle-Lettres et Arts de Besançon, 28 January, p. 59-85.

Pošpný, F., 1894, The genesis of ore-deposits: *Transactions of the American Institute of Mining Engineers, v.* 23, New York. p. 197-[369].

[Rudder, S.], 1779, A new history of Gloucestershire…: Cirencester: Rudder. 855 p. (p. 798).

Shaw, T. R. 1992, History of cave science the exploration and study of limestone caves, to 1900. 2nd ed. : Sydney Speleological Society, v. 14, 338 p.

Thirria, [C] E., 1830, Notice sur le terrain Jurassique de département de la Haute-Sâone, et sur quelques-unes des grottes qu'il renferme. *Mémoires de la Société d'Histoire Naturelle de Strasbourg, v.* 1, p. [1]-62 (p. 41-42).

Trissino, G. G., 1550, [Letter dated 5 March 1537], folio p. 420-421. In: *Alberti, L. Descrittione di Tutte Italia…* : Bologna, Giaccarelli, 469 folio p., (folio p. 420, back).

Vachell, C. T. and Evans, P., 1905, Franklen George Evans, M.R.C.S., F.R. Met. Soc., F.R.A.S.: Report and Transactions. *Cardiff Naturalists' Society, v.* 37 for 1904.p. 19-25.

Virlet [d'Aoust], [P. T.], 1834, [Sur les formations des cavernes]: *Bulletin de la Société Géologique de France,* sér.1, v. 4. Paris. 345-347.

Whitehurst, J., 1778, An Inquiry into the Original State of Formation of the Earth… : London, Author and Bent, 195 p. (p. 49, 51).

Willems, J., 1894, Compte rendu de la session extraordinaire de 1893 du 4 au 9 août, étude de la circulation de l'eau dans les calcaires et visite des grottes de la Lesse: *Bulletin de la Société Belge de Géologie, de Paléontologie et d'Hydrologie Mém.* v. 7 for 1893, p. 298-377.

Williams, D., 1834, [Notebooks v. 5]. Somerset County Museum MS. TTNCM: 121/1993, Taunton, folio p. 35 (back)–36).

Wright, C. W., 1858, The Mammoth Cave, Kentucky: Vincennes, Harvey, 68 p. (p. 13-14, 17).

2.2. Development of Speleogenetic Ideas in the 20th Century: The Early Modern Approach

David J. Lowe

Abstract

During the first half of the 20th century, geomorphologists and geologists from eastern Europe and North America applied the modern scientific method to help refine earlier ideas of cave development. Preexisting concepts involving the developing role of vadose and phreatic waters were accepted and extended. These culminated in the establishment of partly conflicting and partly complementary models. Within these models it was hypothesized that cave development is focused above, below, or at the level of a supposed water table.

During this investigative and theoretical phase, relatively little attention was paid to the underlying question of cave origin, which could not be observed directly. However, theoretical ideas that appear to have gained only limited support at the time can now be shown to include concepts similar to explanations of cave origin that are being postulated and tested today.

Introduction

As pointed out in Chapter 2.1 and described in detail by Shaw (1992), ideas about speleogenesis in the 19th century and earlier included some that now appear ludicrous and bizarre. Far more importantly, however, it can be recognized that the three main themes that have formed the focus of speleogenetic study and debate in this century were already evolving before 1900.

In the light of the way that speleogenetic study has developed, as described in this and ensuing chapters, it is worth repeating a statement from 100 years ago by E.A. Martel (1896, p. 53-54):

> No theory about the origin of caves is universal: those that have been put forward have generally claimed to be too inclusive; almost all of them are partially correct; the whole truth lies sometimes in their combination, sometimes in the application of a particular one in a particular case.

This early view from Martel, who was arguably the preeminent cave explorer of his day, remains valid even today, as was effectively restated by White and Longyear (1962), Ford (1965), and Ford and Ewers (1978). It can only be contradicted on the semantic grounds that cave *origin* cannot necessarily be equated directly with cave *development* What Martel described in 1896 was the potentially multifaceted and concurrent nature of ongoing cave-development processes. Research might eventually confirm that the origin of caves is also multifaceted, but the facets need not be the same as those of later cave development. The nature and relative importance of a variety of potential contributory factors is, as yet, incompletely understood and is the subject of ongoing and vigorous debate.

Development of Speleogenetic Ideas in the 20th Century

A distinct phase of scientific thought is represented by theories of speleogenesis that matured in the late 19th and early 20th Centuries, growing from the more reasonable of the earlier ideas described above. Around the turn of the century, workers eminent in the then expanding field of geomorphology began to show greater interest in surface karst and, inevitably, there was an increase in speculation about the nature of underground karst and the movement of underground water.

In 1893 Jovan Cvijić published a monograph in which he discussed a sequence of surface landform development applicable to the Dinaric karst region. Some years later (Cvijić, 1918) he produced another contribution in which he related the development of karst landforms to subsurface hydrology.

Hydrographic zones that, by this or other names, have continued to play a major role in karst controversy throughout the 20th century, were discussed and their significance underlined in this paper. It was later translated into an English digest form by

Sanders (1921).

Cvijić's zones were termed, in descending order:

- The dry zone
- Transitional zone
- Saturated zone

This broad zonation was later to be promoted by Swinnerton (1932), and in various slightly more elaborate forms is still accepted by many workers today. According to Cvijić, most groundwater movement took place at or immediately below the water table. Hence, the transitional zone would be the major focus of cave development.

Between the appearance of Cvijić's 1893 and 1918 publications a meeting took place, in the field, between Cvijić's professor, Albrecht Penck, and the American William Morris Davis, both eminent geomorphologists of the day. The meeting had great significance for studies of karst and speleogenesis. Together with a group of Penck's students, Penck and Davis visited and studied areas within the Classic and Dinaric Karsts.

Such was their almost patriarchal influence in the wide field of geomorphology that the conclusions they reached, particularly their broad agreement on the importance and almost ubiquitous applicability of various geographical cycles, had repercussions that are still being felt today. In particular, at least some seeds of cyclic karst erosional and cave formational models that were described later were probably sown at this time, and much effort has since been wasted in attempts to make observed reality fit into the attractive but unreal erosional cycle concepts that developed.

From a wide range of contributions, two conflicting models began to evolve. Alfred Grund (1903) was of the view that karst aquifers could be divided broadly into two zones above and below a water table.

The lower zone was saturated by essentially stagnant water, whereas in the higher zone there was water circulation within open caves. The water table that separated the two zones was considered to be of regional extent. The misguided belief that water below a water table must be stagnant, or static, persisted among later workers who adopted Grund's views, perhaps reflecting a failure to grasp the minuscule nature of the changes in the flooded zone that might induce early speleogenesis.

Katzer (1909) considered that there was no zone of saturation and hence no water table, but that all the water within the karst aquifer flowed along open cave passages from surface sinks to topographically lower risings. Further, he considered that the passages were independent of each other, as the intervening rock mass was essentially impermeable. Edouard Martel (1921) was perhaps the most influential supporter of the Katzer viewpoint, on the basis of his own underground experience, which was second to none at that time. His idea of surface water sinking at a high level and carving routes through the easiest combination of fissures and bedding planes to valley bottom risings is not without its attractions. However, the stated importance of such a process to ongoing cave growth does not necessarily imply or prove a relevance to cave origin and the earliest stages of cave development, and the type of drainage that the model implies severely restricts the potential time scales of speleogenesis.

Following the translation of Cvijić's (1918) paper by Sanders (1921) and the appearance of Martel's *Nouveau traité des eaux souterraines* (also in 1921), little of general significance was published about speleogenesis until 1930. In the interim the work of exploring and observing caves continued and a number of short papers of local interest appeared, including early work by Malott and Swinnerton. Some of those originating in North America were consulted and quoted by the authors of several major publications during the next phase of cave research.

If similar studies were under way in eastern Europe they were largely overlooked or ignored by western workers. Thus it is not unreasonable to suppose that much of the ensuing controversy regarding cave development was fueled by regionally biased and narrow views. Theories based on observations in the relatively undeformed carbonate sequences of the middle U.S.A. would, inevitably, conflict with theories deduced from observation of relatively highly tectonized sequences in Alpine Europe.

Yet North American and British authors made little reference to results from these latter terrains. The only widely recognized major work to appear in Europe during the next phase, that of Otto Lehmann, expressed ideas quite unlike those in contemporary American publications.

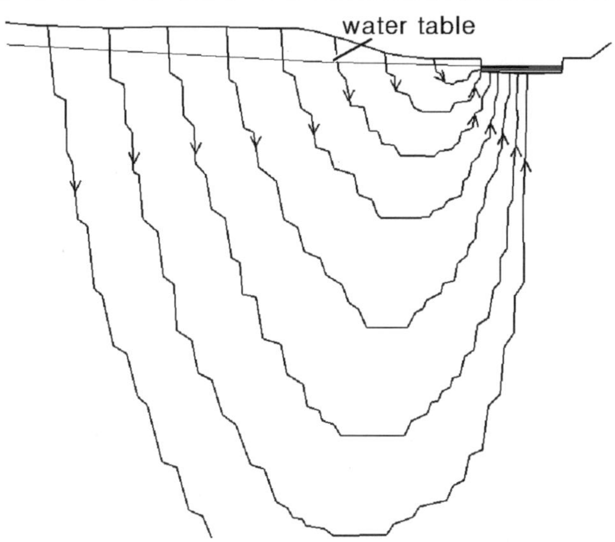

2.2 Figure 1. The pattern of groundwater flow deduced by W. M. Davis indicated a deeper movement than that proposed by most of his contemporaries. It also made some allowance for modification of routes along fractures and bedding planes (redrawn after Davis, 1930).

The Classic Period or Golden Age

Many aspects of ideas about speleogenesis that appeared between 1930 and 1942 have subsequently been forgotten, disputed or actually disproved. Even so this period may still be regarded as the Classic Period (Watson and White, 1985) or Golden Age (Lowe, 1992b) of cave-development theories.

Several major papers appeared and these, together with a number of shorter, but no less thought-provoking contributions, formed the foundation upon which modern cave study has been built. Despite the obvious weight of these publications, it is interesting to note that the central themes that they developed were those that already existed in somewhat less polished form—but were less widely circulated and read—at the end of the 19th century.

William. M. Davis

In 1930, William Morris Davis said:

> What is now most needed in the investigation of cavern origins is not so much a more detailed deduction of expectable features as a closer and more detailed observation of actual features (Davis, 1930, p. 566).

William Davis was 80 years old when his paper *The origin of limestone caverns* was published in 1930. It was toward the end of a long and distinguished scientific career (only three years before his death) that this classic essay on caves—with its broad applications to karst geomorphology and hydrology in general—appeared. By his own admission it lacked the detailed personal observations that would be achieved by some of his younger, more active successors.

However, it is interesting to note that, at the close of this phase of speleological thinking, Bretz (1942, p.676) stated, of the more modern studies:

> .... none [was] based on a really extensive acquaintance with caves.

Davis was probably the first worker to emphasize the potentially vast time scales that can be involved in cavern formation. His ideas followed Grund's broad theme. He believed that cave development would begin below the water table as soon as fresh water was able to replace primitive salt water, trapped within a limestone sequence during deposition and retained during diagenesis.

He visualized slow dissolutional development of groundwater routes, possibly at very great depths below the water table, and distinguished between this process and that of cave formation at shallower depth. The deep dissolutional processes were seen as possibly extending across tens, or even hundreds, of millions of years, with the shallower, but still sub-water table, stages of cave formation occurring in somewhat shorter time spans. The deep phreatic and subsequent shallow phreatic dissolution took place during the "first cycle of cavern development."

The major weakness of Davis' view was that it lacked detailed and convincing models for the guidance of the postulated deep drainage routes, the dissolution chemistry, and the mechanisms that drove the deep water movement. Davis, in common with many of his peers, was largely ignorant of the detailed chemistry of calcite dissolution. And, arguably, was misguided in assuming that carbonic acid dissolution of calcite alone was capable of achieving the development he proposed. He was unaware of the possible importance of salt water *(sensu lato)* or other potential corrosional agents in dissolutional processes, and was probably equally unaware of the nature and extent of buried paleokarsts.

Davis developed the view that cave passages of explorable size would enlarge close below the water table by modification of once deeper drainage routes when these were raised relative to the land surface. The relative rise could occur in response to regional uplift, as a result of the removal of overlying strata or when the drainage regime was altered by deep incision of surface valleys.

The flooded passages that were conceived at great depth and developed below the water table were in turn raised above the water table and drained. At this stage a new first erosional cycle would begin at depth, associated with a lower drainage route. This view enabled Davis to explain the presence of caves that he considered must have begun to develop below the water table but which were now observed to stand at levels well above the contemporary water table. Once the caves were above the water table the second cave-development cycle began, during which the caves could carry water from surface sinks along open passages. These passages might eventually become partially or totally blocked by carbonate deposits, rockfall, or clastic fill, and would ultimately be removed by erosion.

Outside the mainstream of his thesis there exists a plethora of linked or partially linked concepts. It is beyond the scope of this review to examine these in detail, even to the degree of discussing their potential relevance to modern theories of cave development. In passing, however, Davis' recognition of possible links between mineralization and preexisting caves (Davis, 1930, p. 619 *et seq)* was far sighted. Much of later written speculation on the subject of mineralization might have been unnecessary had this aspect of his work been more widely appreciated.

Of all the cave origin models to emerge during this phase, that of Davis has suffered the most adverse criticism in later years. Relatively recently, ideas have begun to emerge that suggest that there is significant truth in parts of the Davisian model. Davis knowingly left himself and his model open to attack by later workers, particularly those with views based upon more extensive underground experience. Many of these views, and the attacks on Davis' work they provoked, have been valid. But considering that Davis never claimed a formational model that was applicable to all caves, a creditable part of his original synthesis remains potentially acceptable.

Allyn C. Swinnerton

In 1929, Allyn Swinnerton stated:

> The whole problem of cave origin awaits systematic study and presentation." (Swinnerton, 1929, p. 82).

The quotation above appeared before the publication of Davis cave-origin model. Subsequently Allyn Swinnerton (1932) published another major and no less impressive consideration of cave development.

Throughout his paper Swinnerton's respect for Davis's ideas is apparent, though the main trend of his arguments was away from the concept of deep phreatic dissolution favored by Davis and toward cave origin close to and above the water table.

In many ways Swinnerton's work was a logical continuation of ideas developed by Cvijić. The latter's threefold division into hydrographic zones was reemphasized. Though admitting the possibility that dissolution can act in the deep phreatic zone, Swinnerton believed that its importance was minor, while the major locus of dissolutional cave development was in a fluctuating shallow-phreatic or water-table zone. Beyond this he recognized the importance of ongoing cave growth in the vadose zone, following drainage of previously flooded passages, by mechanisms similar to those outlined by Katzer, Martel, and other earlier workers.

Swinnerton's paper included several other significant elements in addition to his major hypothesis of shallow phreatic cave origin. He was particularly intrigued by his observations that many cave passages are subhorizontal and that, in a given area, they commonly occur at several levels, one above another, which he referred to as tiers.

A possible mechanism of tiered-cave development by sequential piracy of drainage to progressively lower levels in response to surface erosion cycles was suggested. Such an overview is superficially acceptable and is broadly in line with most observed relationships. But the mechanism by which the piracy is achieved was barely questioned or examined.

Tiered caves were reported to be present in folded as well as in unfolded, gently dipping sequences. That is, the subhorizontal trend of the passages appeared to truncate even extreme geologic structures. To explain this, Swinnerton postulated that the passages formed within a relatively narrow vertical range limited by the high and low positions of a fluctuating water table between wet and dry seasons.

Unexplained is a means whereby the water is able to fluctuate within an essentially

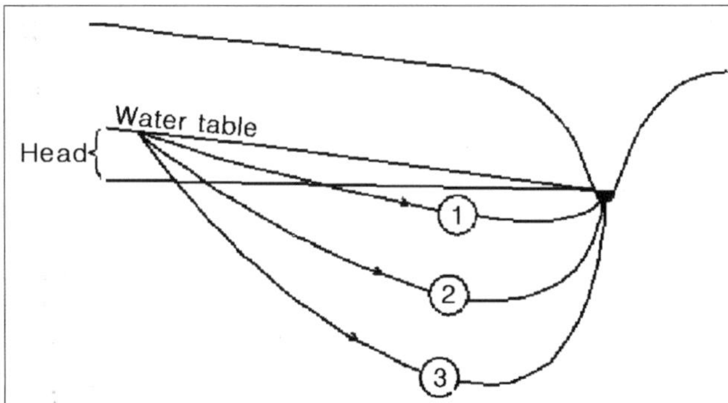

*2.2 Figure 2. Theoretical paths of groundwater flow according to Swinnerton. Below the water table, water follows many routes. But at a given time the shortest route **1** carries most water and hence is enlarged most rapidly. By implication, after uplift, route **1** may be above the water table, and rapid enlargement of route **2** (and later **3**) will begin to take place (redrawn after Warwick, 1962).*

impermeable rock mass and yet concentrate its dissolutional potential within a restricted volume that will ultimately become a cave passage. No illustration nor detailed example of a shallow phreatic conduit that ignores the constraints of geological structure was provided, and it is unclear what Swinnerton considered to be an extreme geologic structure. So it is impossible to comment fairly upon the viability of the suggested mechanism.

Apparently Swinnerton was a notable exception to the general rule that most early workers on speleogenesis were effectively ignorant of the finer points of hydrodynamics and dissolution chemistry (Watson and White, 1985). Nonetheless, modern views on these subjects are far removed from those that Swinnerton understood. Some 60 years later they are still evolving. Swinnerton's ideas were based on an understanding that was greater than that of his contemporaries. But it was still incomplete and certainly untested in the context of speleogenesis.

Thus, it is questionable whether his more knowledgeable consideration of these two fields represented a step forward that was any more significant than the advances of earlier and contemporary workers who had grasped and applied new, but only partially developed, concepts from other disciplines.

Otto Lehmann

Sweeting (1972, p. 129) described Otto Lehmann's (1932) *Die Hydrographie des Karstes* as a major work. It was based upon Lehmann's knowledge of the Classic and Dinaric Karst areas, the Alps, and the French Causses. It was the only European contribution to the speleogenesis debate to gain serious recognition at that time. In contrast to Davis' work, which was essentially deductive, that of Lehmann was inductive and reflected his wide-ranging field experience.

Even so, if citations are any indication, it seems that Lehmann's work was largely ignored by North American and British researchers. The reasons for this may be twofold. Firstly Lehmann's textbook was written in German, and it appears that neither an English translation, even in digest form, nor any English-language paper covering his main ideas appeared. Access to, and under–standing of, his ideas would have required significant effort that was perhaps considered unmerited in the light of the apparently definitive and readily accessible views on cave development that were emerging in the West.

Secondly, Lehmann had joined the still ongoing eastern European controversy regarding hydrographic zones. His work was strongly supportive of the views expressed by Katzer (1909) and Martel (1921). He believed that cave rivers are the major active agent of cavern formation and that a water table and associated hydrographic zones either do not exist or are unimportant in the context of speleogenesis.

It is possible that Lehmann's vision of the water table was unlike that of the workers he contradicted and, if so, some of his ideas can be viewed more sympathetically in the light of later understanding. He repudiated Cvijić's (1918) views, which had upheld Grund's (1903) groundwater theory and were being mirrored, at least in part, by the theories of Davis (1930) and Swinnerton (1932).

In the West, the Davis and Swinnerton models formed opposing faces of speleogenetic controversy, and the European dispute appeared not to impinge upon the problem.

In fact, so far as many British and North American workers were concerned, Lehmann's ideas must have been essentially out of favor as soon as they were published. The controversy that the ideas perpetuated seemed but a small part of the broader view of speleogenesis that was emerging.

In eastern Europe, however, Lehmann seems to have exercised a patriarchal, almost mesmeric, influence upon karst geomorphologists and hydrologists, not unlike that of W.M. Davis in the Western world in earlier years. According to Bögli (1980, p. vi):

> … the hypothesis of O. Lehmann was untenable. Thus the spell was broken which had blocked progress for a long time, but which, however, had also stirred up contradiction and thus encouraged new research.

The breaking of this supposed spell occurred in 1974, some 42 years after its publication. But earlier workers had noted supposed weaknesses within Lehmann's hypothesis (Zötl, 1961).

Lehmann considered that cave development could not commence within a rock mass unless initial cavities (Urhohlräume) with a minimum width of 2 mm were present. He also stated that for water motion to commence a minimal primary permeability (hydrographische Wegsamkeit) was necessary, as both ingress and egress of water to the rock mass are required. However, his main hypothesis, considered valid by most European workers until finally discounted by Zötl (1974) was that:

> … karst cavities are independent and have no relation to one another. (Lehmann, 1932, p. 15)

On the evidence of water tracing experiments in the Dachstein of Austria, Zötl (1961, p. 129) postulated "… a large connected body of karst water" in the test area. Zötl also pointed out that the number and width of joints decreases downward, and that:

> This observation reveals the basic fact that within massive karst there is in general a dense network in the upper regions with smaller open joints reaching much deeper, even below the base level." (Zötl, 1961, p. 130).

These and similar observations that were reiterated some years later (Zötl, 1974) were those that finally broke the hold of Lehmann's dogma on eastern European karst studies. However, Bögli (1980, p. iii) points out that he (Bögli) had been aware for some time prior to Zötl's work, that the physical principles upon which Lehmann's hypothesis was founded, and the actual hypothesis, were not borne out by underground observations.

Lehmann's ideas might have been misinterpreted by some later workers, and much of his basic philosophy is at least partially valid in terms of the needs of cave origin, if not of cave development in general. For example an obvious thread links Lehmann's concept of initial cavities (Urhohlräume) with the idea that at some stage a transition must take place between rock without conduits and rock with conduits (Lowe, 1992a; Chapter 3.3 in this volume).

Despite the fall from grace of some of Lehmann's more dogmatic statements regarding the general processes of underground karst formation, his work remains a major milestone of speleogenetic thinking. Among many observations he noted the so-called karst hydrological contrast and commented upon the mechanism whereby swallow holes could become springs, the dual function features being termed estavelles.

The karst hydrological contrast was a term used to describe this apparent discrepancy—the great number of potential input sinks and seepage points on the karst surface and the relatively small number of output points or springs at a lower level. Although such a discrepancy is commonly apparent, it is perhaps less extreme than superficial examination suggests—insofar as in most karst terrains there exist, in addition to the obvious major active risings, a very much larger number of normally inactive ones, many of which emit water in abnormally wet conditions.

Whatever the terrain, the number of potential inputs and outputs depends upon the interaction between geologic features (specifically bedding planes and various lithogenetic or tectonic fissures) and the landscape morphology in the input and output areas.

James H. Gardner

Of the many researchers who contributed to the rapid proliferation of cave development ideas in the 1930's, James H. Gardner (1935) came closest to describing concepts similar to those incorporated in some recently proposed speleogenetic models (Worthington, 1991, Lowe, 1992a). He is generally remembered for ideas about cave formation above a water table. But several seemingly insignificant "asides" are now seen to be prophetic and of importance to any overview of speleogenesis. His main concern was, however, in seeking an explanation of the origin of large caverns, and to this end he became somewhat preoccupied. The major significance of his suggested formational mechanisms to speleogenesis in general is smothered by a great weight of arguments to prove the applicability of his hypothesis to the genesis of Mammoth Cave and Carlsbad Cavern, systems that have little in common beyond large size.

Gardner's view was one of essentially down-dip passage formation commencing in carrier beds or aquifers, possibly of no great thickness, which form a relatively small part of a thick limestone sequence. These beds were considered to be especially susceptible to dissolutional cave development but, as pointed out by Warwick (1962, p. 62), he did not state the nature of the properties that determined this susceptibility. However, two sentences in particular (Gardner, 1935, p. 1260) are important in the context of cave inception. The second sentence is particularly thought provoking:

> In most cases the aquifer is porous limestone or dolomite. It may, however, consist of open spaces along bedding planes of limestone; or a sandstone bed intercalated in limestone might serve as the original circulating medium until solution acts on the adjacent calcareous beds.

Besides talking about these carrier beds or aquifers, Gardner described observations of stratigraphic guidance of cavern development and quoted the example of The Dry Branch, a sinking stream in Kentucky, where:

> .... the stream and its cavern follow down the bedding planes and remain confined to a zone of limestone strata which probably does not exceed fifty feet in thickness.

He considered that this cave was typical of the initial stage of cavern growth.

The triggering mechanism of cave development according to Gardner's hypothesis required downcutting by a surface valley to cut through the dipping limestone sequence (2.2 Figure 3). Steep dips were considered particularly suitable, but he pointed out that lower gradients were potentially susceptible and that the dip need only be: ".... enough to start the circulation of water by gravity."

Before valley downcutting the up- and downdip sides of the carrier bed (relative to the valley) were seen as being saturated by

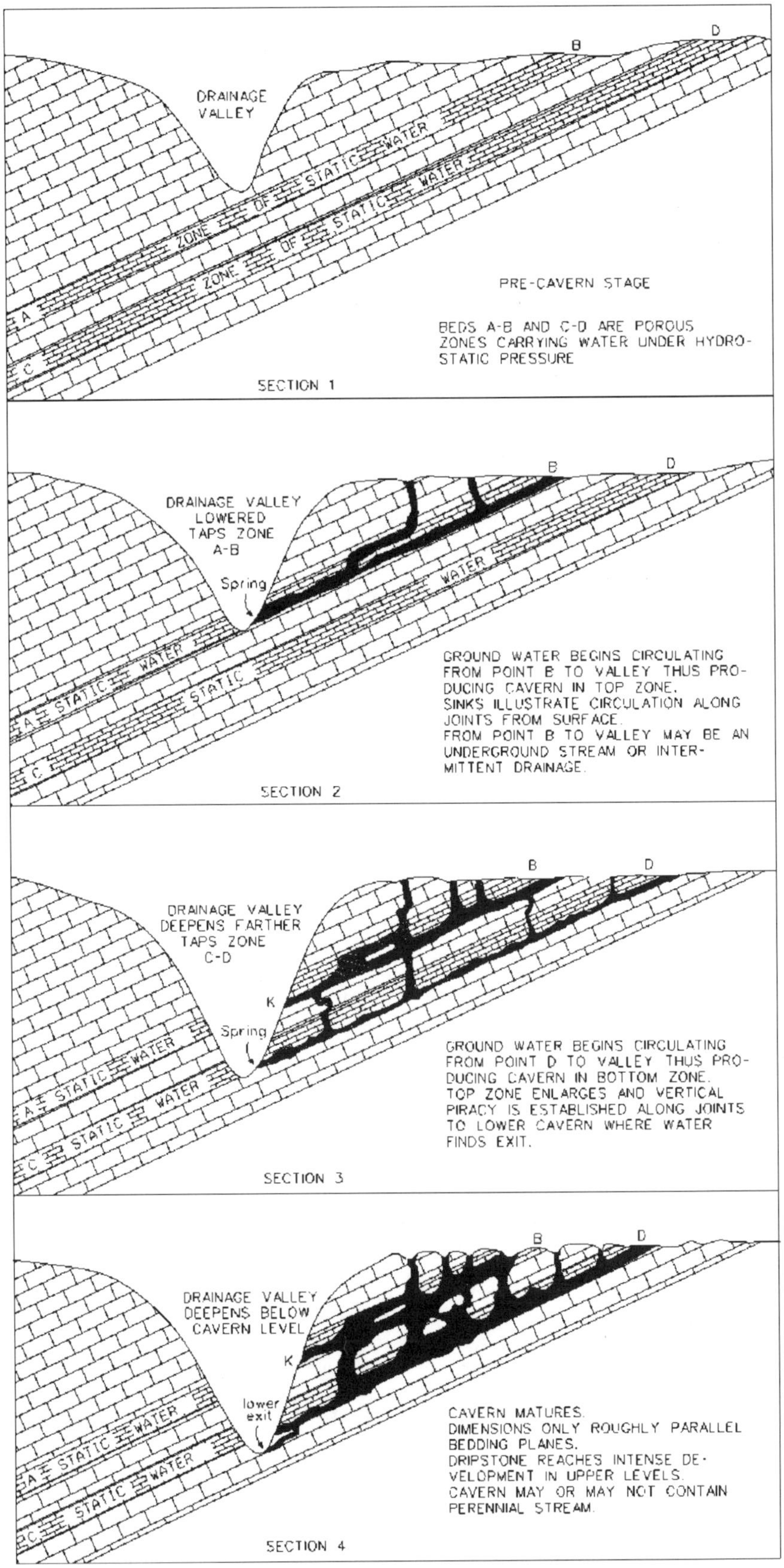

2.2 Figure 3. Sketch sections illustrating the origin and development of a large limestone cavern (redrawn after Gardner, 1935).

static water. Like Grund's (1903) view of the saturated zone, it must now be questioned whether the author meant truly static, or merely infinitesimally slowly moving. Breaching by the valley led to drainage of the updip saturated zone and subsequent vadose invasion and cave development within the carrier bed.

Continued downcutting might eventually lead to the intersection of a lower carrier bed, at which stage, according to the model, the upper level would be abandoned as its drainage was pirated by caves formed in the lower bed (2.2 Figure 3). The potential importance and regional significance of the downdip continuations of the intersected carrier beds were not pursued by Gardner.

In modern terminology Gardner described vadose cave modification within stratigraphically determined zones, wherein inception and initial enlargement must have been due to the dissolutional effect of essentially static water. Gardner admitted to the limited dissolutional effect of groundwater below the water table, but not to the extent required by Davis's (1930) two-cycle theory. Having accepted the limited dissolutional effects, Gardner went on to support Swinnerton's (1932) one-cycle theory, seeing most of the cave formational process as being by active streams above the water table.

Gardner's description of caverns occurring at several levels—five major levels in Mammoth Cave—was explained in terms of "cross-formational piracy" initiated by surface downcutting. (See discussion of "levels" in Mammoth Cave by Palmer, Chapter 5.3.8).

This is the major missing link in the Gardner model, insofar as if the strata between the carrier beds were impermeable, as required (Gardner, 1935, p.1258), there is no reason why drainage in a high-level carrier bed should be pirated by newly activated drainage in a lower horizon. (Or the corollary that if the intervening beds were permeable there is no reason that the carrier beds should exercise such strong guidance).

Gardner was close to providing the missing link, but did not follow the argument to its obvious limit. During the development of a multilevel system, Gardner (1935, p. 1263) believed that:

> Open joints are first drained of their water by circulation along the bedding below. In time, the water of the overlying cavern, whether a perennial stream or intermittent, is captured by a lower cavern stream and so on, resulting in complex development.

This statement begs the entire question of cross-formational piracy. Recognition of sub-water table dissolution along joints or faults between the separate carrier beds, before drainage, would have simply provided the necessary link.

Overall, Gardner's work could have been a far more significant milestone in the understanding of speleogenesis. Concepts related to initiation (sic), the precavern stage, multilevel caves, stratigraphic guidance, ore deposition, sulfur water and salt water all deserved recognition and follow-up work. In contrast other ideas were less inspired.

The laborious argument of similarity between Carlsbad Caverns and Mammoth Cave, pronouncements upon time scales of speleogenesis, and a consideration of the effects of earthquakes all diverted attention from the central theme of the paper.

Perhaps more obviously in terms of later ideas, Gardner (1935, p. 1273) showed unreserved support for Swinnerton's vadose formational model:

> Large caverns...originate and develop in thick terranes of limestone above the water -table.

This might, over the years, have obscured the major importance of the work as a whole.

C. A. Malott

The theory outlined in abstract form by Clyde Malott (1937) did not claim to discuss cave origins. Yet it came far closer to doing so than did the earlier work of Davis, Swinnerton, or even Gardner. Sadly the abstract was not followed up by a fully reasoned paper.

So only the briefest indication of Malott's ideas is available. The essence of Malott's "invasion theory of cavern development" was reproduced by Warwick (1962, p. 60):

> ... primitive, illy ("poorly" in Warwick's review) integrated, and somewhat selective three-dimensional passages are developed in limestones *below* the water-table.

There is no doubt that Warwick appreciated the significance of this rather stark statement, and he emphasized the importance of the word below by the use of italics.

After relative uplift and water-table lowering, percolating surface-derived water would tend to concentrate in the primitive openings and be channeled down toward the water table, within "... selected parts of the earlier prepared primitive systems," selectively enlarging its preferred routes. Later, surface streams would be captured and enter the existing underground routes, and these underground waters would ultimately "....carve out and integrate the large caverns at or near the water-table."

On the strength of his studies in southern Indiana, Malott believed that this capture of surface drainage was the major factor in cave development.

The significance is twofold. First Malott appeared to stress that although preuplift development is poorly integrated, it is three-dimensional. Thus, its guidance in most sedimentary successions is presumably related to both stratigraphic and structural factors. Also presumably, early dissolution can take place simultaneously at more than one level in the rock mass. Second, Malott considered that the major growth of most cave systems is achieved above but close to the water table and is due to erosion by invading surface streams. Despite this possible misconception, his ideas clearly indicate an awareness of the problems of explaining cavern inception. His description confirms, if only tacitly, that caves must originate in the saturated zone.

Ideally a follow-up paper would have described the chemical processes and drive mechanisms that he deduced to be involved in the saturated zone during the earliest development phase. Perhaps such a paper never appeared because Malott was misled by contemporary dogma and peer pressure into believing that accepted processes and mechanisms could not provide the framework that his theory predicted.

F. F. Laptev

As described above, Otto Lehmann was the only researcher outside North America whose ideas about karstification and cave development made a significant impact in the Western world during the explosion of theories that emerged in the 1930s. This lack of recorded recognition gives a misleading impression. It suggests that after the leap forward in the understanding of karst in Europe in the early part of the 1900's, karst science was largely moribund outside North America during the 1930's. This was not the case.

Arguably, just as much research took place outside as inside North America. It is inevitable that significant advances were made in the classical karst regions of Europe and, eastward, in the Soviet Union. However, due to the obvious disadvantage of language barriers—as discussed for Lehmann, and more particularly in the case of the Soviet Union—the multiple barriers of language, alphabet, and eventual isolationism, little of the Eastern output was either noticed or acknowledged in the West.

Only with the relatively recent removal of the previously existing artificial political barriers, even within parts of the scientific community, has it become increasingly obvious that much research was duplicated. This was the case not only between, but also within, the broadly western and eastern areas. But was perhaps most conspicuous between the USSR (and later the Soviet satellites) and the rest of the world.

In no case recognized to date is this inadvertent duplication better displayed than that of the work of F.F. Laptev. In 1939 Laptev published a paper, in Russian, describing the aggressive action of water on

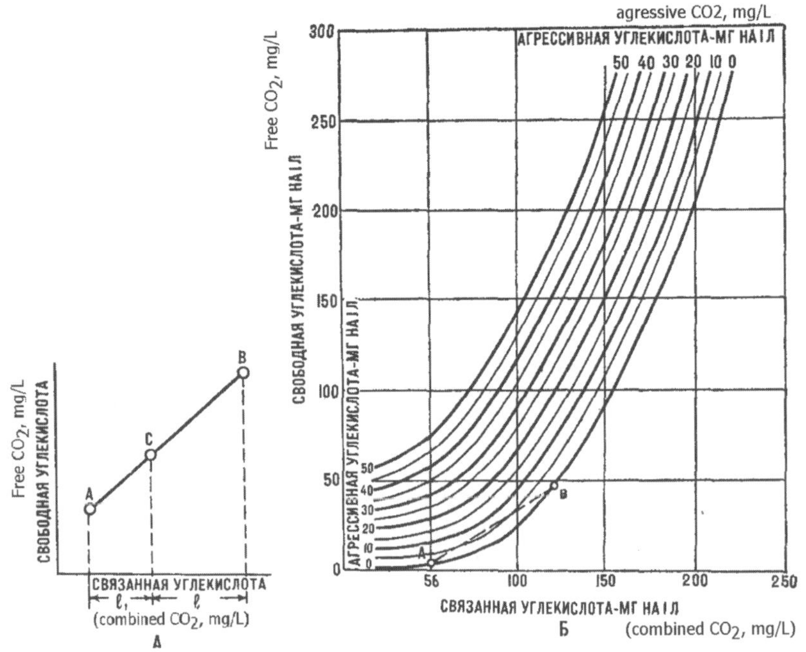

2.2 Figure 4. The diagram by F.F. Laptev illustrating the nature of the mixture corrosion effect (redrawn after Gvozdetsky, 1972).

carbonate rocks, gypsum, and concrete. The paper included the results of a detailed study of the equilibrium chemistry of calcite, and a description of the effect of mixture corrosion.

In this, Laptev described and illustrated in 2.2 Figure 4 the effect whereby the mixing of two waters saturated with respect to calcite under different carbon dioxide partial pressures produces a mixed solution that is less saturated and hence capable of further calcite dissolution.

It is surprising to realize that a process first described in Western literature in 1964 was described in detail by Laptev in 1939. It is even more interesting that the same process was mentioned, also in a Soviet publication, by A.N. Buneev, in 1932. However, Laptev's consideration was more detailed and more directly related to the potential significance of mixture corrosion to cave development.

In particular he stressed that the process could affect the development of subsurface karst forms in deep-seated, buried conditions, especially in situations where waters of different compositions mix at the points of intersection of tectonic fissures. He also pointed out that the mixing effect could probably account for the development of the large cavities that were known to occur at great depths, with no obvious relationship to shallow underground flow systems.

Subsequently, during the next phase of karst research development (described in Chapter 2.3), in a Russian-language text—*The principal conditions of karst development*—Sokolov (1962) referred to the 1932 work of Buneev and, more extensively, to Laptev's (1939) publication. In doing so he stressed the importance of the mixing effect, while discussing the solubility of common karst rocks in natural waters. Still later, Gvozdetsky (1972), also in Russian, referred extensively to Laptev's work in his *Problems of karst study and practice.*

Gvozdetsky raised the question of the historical precedence of Laptev's results over those published in the west by Alfred Bögli in 1964. Bögli's ideas were eagerly grasped by western karst scientists as providing potential answers to some of the then most puzzling questions of karst development (see Chapter 2.3 in this volume). However, this had not been appreciated in the international literature.

In raising the question of Laptev's precedence, Gvozdetsky did not attempt to deprecate the significance or merits of Bögli's work. It seems certain that Bögli's results and his subsequent description of mixture corrosion were arrived at totally independently of the much earlier work in the Soviet Union.

As former political barriers are removed and the barriers of language differences are gradually eroded by technology and greater human integration, it seems inevitable that other examples of "East-West" duplication will be unearthed.

Hopefully, however, a new era of international scientific openness and co-operation will see the end of such confusion and wasted effort in the future.

Berlen C. Moneymaker

The contribution made to cave development studies by Berlen C. Moneymaker was dominantly one of observation rather than of revolutionary theory. During the course of his work for the Tennessee Valley Authority during the 1930's he had access to the records and cores of many small- and large-diameter boreholes that were drilled for purposes of exploration and grouting across a variety of potential dam sites. He was also able to visit and examine many opencut excavations that were made for the same purposes.

Moneymaker noted evidence of many dissolutional cavities, some of large size, both above and below the supposed water table, which he considered to approximate the local river level.

He recognized that the details apparently revealed by small-diameter boreholes were of limited value, but presented some basic statistics dealing with the dimensions of the cavities encountered, their abundance at various depths, and their abundance in specific rock formations. The values presented are only relevant in the context of individual rock beds or at individual sites, but the broad limits were of cavities to 20 m (64 feet) high, at depths to 68 m (222 feet) below river level.

Among Moneymaker's many observations are descriptions of several situations where the "downward limit" of cavern formation was marked by beds of bentonite and/or shale. Other non-carbonate or impure carbonate lithologies were also reported within the formations that he listed. For example (Moneymaker, 1941, p. 80):

> The lower limit of cavernous rock was, with a few exceptions, controlled by the position of a thin bed of bentonite and shale near the base of the Cannon limestone.

Moneymaker's view of the function of these beds, the shales and bentonites in particular, was that they acted as aquicludes and had "protected" the underlying rock from dissolution. Such a function, plausible as it seems at face value, is not the only possible interpretation of the role of these noncarbonate and impure lithologies.

Moneymaker also noted that cavities appeared to be more numerous and of greater average size above than below the water table. He argued that this indicated initial dissolution below the water table followed by enlargement above the saturated zone after the drainage that followed downcutting by surface streams. He pointed out that all dissolutional cavities above the existing water table were related to earlier water tables, having occupied positions below, coincident with, and at various levels above the level of saturation. However, he was unsure about how much enlargement could be attributed to activity at these various levels.

Though not formally stated, the view of speleogenesis summed up by Moneymaker was not far removed from the elegant rationalization of earlier controversy that was supplied by Ford and Ewers (1978).

Considered in tandem with the implications of his observation and recording of simultaneous dissolutional development at more than one level in rock formations within the saturated zone, the overall, unformalized view was remarkably close to ideas that have reemerged during much more recent times (e.g. Worthington, 1991; Lowe, 1992a).

R. Rhoades and M. N. Sinacori

In 1941 Rhoades and Sinacori published a paper that may be regarded, if not as *the* first, at least as one of the first truly modern scientific considerations of the problems that underlie speleogenesis. Their paper was essentially theoretical and provides an early example of a mathematical consideration of underground water flow. By making use of a differential Laplace equation derived from Darcy's law of flow, together with the equation of continuity, they were able to define two sets of curves. In certain specified conditions these two sets of curves, which intersect at right angles and are known as flow lines and equipotential lines, define a flow net, a valuable hydrological tool that had been described in detail by Hubbert (1940) on the basis of earlier work. Using the flow nets that they constructed, Rhoades and Sinacori were able to model graphically the solutions of a wide variety of theoretical *two-dimensional* seepage problems.

Broadly they demonstrated not only that the deep and shallow flow paths and dissolutional developments postulated by Davis (1930) and Swinnerton (1932) were likely to occur, but also that they could co-exist at the same locality. Their model suggested that there could be significant dissolutional development at depth in the early stages of the drainage of a limestone aquifer towards a specific discharge point.

With time the drains closest to the discharge point would be enlarged preferentially and a master conduit would cut back, headwards, from the rising along a line that would follow the local water table. In turn this backcutting would modify the flow lines so that deep-seated water movement and dissolution would become proportionally less important and shallow alternatives would become relatively more important.

Rhoades and Sinacori pointed out that assumptions had to be made in order to fit the behavior of water in a limestone terrain into the idealized pattern suggested for flow in a more isotropic aquifer. They summed up (p. 791) on the basis that, if suitable openings did exist within a limestone mass:

> ...the flow will pass from joint to joint in the closest approximation to the ideal arcuate path which the configuration of the joint system will permit.

This observation is particularly significant, but even more so in that the authors went on to caveat their model by commenting:

> Because the spacing, open character, size, distribution and intersection of the joints depart from uniformity throughout the rock, the analogy between natural flow through rocks and flow through an ideal homogeneous medium departs from perfection to a variable but significant degree.

This seemingly open-ended caveat in no way belittles the overall importance of their statement, but adds a degree of flexibility and practicality to the overall strength of the concepts that they presented.

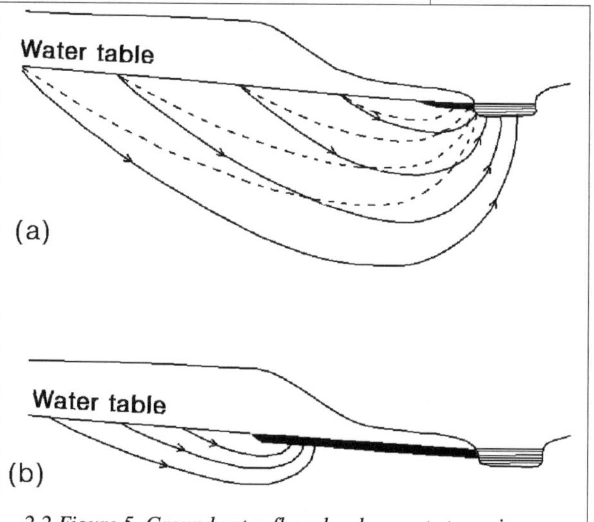

*2.2 Figure 5. Groundwater flow development stages in limestone according to Rhoades and Sinacori. Broken lines in **a** indicate the changes produced by the backward growth of a water-table cave from the discharge point. Diagram **b** shows the theoretical effect after significant conduit growth (redrawn after Rhoades and Sinacori, 1941).*

Conspicuous by its absence was a consideration of the actual chemistry of the dis-solutional development implied by the authors. It must be assumed that they envisaged the process as being due to the action of carbonic acid. It was to be some years before Alfred Bögli (1964) put into words the fact (that must have been noted by many, but stoically sidestepped) that most cave development could not be explained by that one simple mechanism.

However, the inadequacy of the dissolution chemistry tacitly assumed by Rhoades and Sinacori does not diminish their theoretical model.

The alternative chemistry suggested by Bögli (1964)—and similar reactions deduced by later authors—might go all or part of the way toward explaining dissolution in a zone where only nonaggressive water would be expected to flow.

J H. Bretz

Working slightly later than most of the authors discussed above, Bretz (1942) based his contribution to the classic speleogenesis debate upon a much wider observational base, having studied many caves within rocks of Cambrian to Permian age in a variety of structural situations.

In general Bretz supported the W.M. Davis cave development model, but his ideas added significantly to the earlier work. Bretz was the first to describe caves in terms of their development below or above the water table—and borrowing terms used by earlier hydrologists—he referred to these situations as the phreatic and vadose zones. He listed and described morphologic features that he considered diagnostic of cave development in each zone, and many of his conclusions remain at least partially valid today.

Though agreeing with and supporting the broad Davisian model, Bretz added to it by inserting a third phase between Davis's pre- and postuplift cycles. He believed that during this supposed intermediate phase the early passages remain below the water table, but contain stagnant water. Why initially mobile water should become stagnant was not made clear. As such stagnation seems unlikely to persist through the time spans required for significant cave development, this particular criterion must be treated with suspicion.

Deposits of "unctuous" red clay and silt, partly representing insoluble residues of limestone dissolution, are supposedly laid down by slow settlement through the stagnant water, ultimately filling or almost filling the flooded voids.

After relative uplift (initiating Davis's second cycle) the passages are drained and the fill either consolidated or desiccated, leaving space above for invasion by surface streams. Subsequently vadose stream activity removes all or part of the clay, leaving remnant fill forms, which Bretz had observed and regarded as visual proof of his concept.

There is ample evidence that fills such as those described by Bretz are emplaced under phreatic conditions, though the process is probably not ubiquitous. Similar fills are formed close to local saturation level, under so-called epiphreatic conditions. These epiphreatic fills, like those described by Bretz, are periodically cut into and redistributed by vadose stream activity during times of low flow. It seems therefore that there is no requirement for water to be continuously stagnant during the type of essentially depositional phase described by Bretz.

Conclusions

By the start of the 20th century the basic concepts of speleogenesis by vadose water and by deep or shallow phreatic water had already been discussed. These ideas were adopted and developed by eastern European and North American geologists and geomorphologists during the early part of the 20th century, and several imposing models for both karst and cave development were proposed. Each of the major theories was described as providing an explanation of cave origin, whereas in reality the question of this earliest phase of speleogenesis was barely considered.

The controversy surrounding the question of whether cave development took place above, below, or at a supposed water table had begun in the 19th century and was set to continue for some time to come, as described in the following chapters. Putting aside this controversy, the major papers that appeared between the late 19th century and 1942 included most of the concepts that have now been assimilated into modern cave development models—as well as many other interesting lateral thoughts.

References

Bögli, A., 1964, Mischungskorrosion—ein Beiträg zur Verkarstungsproblem: *Erdkunde* v. 18. p. 83-92.

Bögli, A., 1980, *Karst Hydrology and Physical Speleology*.: Berlin, Heidelberg, New York: Springer-Verlag, 284 p.

Bretz, J H., 1942, Vadose and phreatic features of limestone caves: *Journal of Geology* v. 50. p. 675-811.

Buneev, A.N., 1932, Aggressive carbon dioxide in mineral waters: Trudy Tsentral'nogo Instituta Kurortologii, v. 4.

Cvijić, J., 1893, Das Karstphänomen: *Geographische Abhandlungen herausgegeben von A. Penck* , v. 5, pt. 3, p. 218-329.

Cvijić, J., 1918, Hydrographie souterraine et évolution morphologique du Karst: *Recueil des Travaux de l'Institute de Géographie Alpine* , v. 4, p. 375-426.

Davis, W. M., 1930, Origin of limestone caverns: *Geological Society of America Bulletin* , v. 41, p. 475-628.

Egemeier, S. J., 1981, Cavern development by thermal waters. *National Speleological Society Bulletin* , v. 43, p. 31-51.

Ford, D C., 1965, The origin of limestone caverns: a model from the central Mendip Hills, England. *National Speleological Society Bulletin,* v. 27, p. 109-132.

Ford, D. C. and Ewers, R. O., 1978, The development of limestone cave systems in the dimensions of length and depth: *Canadian Journal of Earth Sciences* , v. 15, p. 1783-1798.

Gardner, J. H. ,1935, Origin and development of limestone caverns:*Geological Society of America Bulletin* , v. 46. p. 1255-1274.

Grund, A., 1903, Die Karsthydrographie. Studien aus Westbosnien: *Geographische Abhandlungen herausgegeben von A. Penck* 7. p. 103-200.

Grund, A., 1910, Beiträge zur Morphologie des dinarischen Gebirges:*Geographische Abhandlungen herausgegeben von A. Penck* 9.

Grund, A.,1914, Der geographische Zyclus im Karst: *Zeitschr. Ges. Erdkunde* , v. 52, p. 621-640.

Gvozdetsky, N.A. 1972. Problems of karst study and practice: Moscow:Mysl'. 390 p.

Hubbert, M. K., 1940, The theory of groundwater motion: *.Journal of Geology* v. 48, p. 785-944.

Katzer, F., 1909. Karst und Karsthydrographie. *Zur Kunde der Balkanhalbinsel, Reisen und Beobachtenungen.* v. 8, *no. 3, 94 p.* Sarajevo: Kajon.

Laptev, F.F., 1939, Aggressive action of water on carbonate rocks, gypsum and concrete: Trudy Spetsgeo, v.1. Moscow-Leningrad: GONTI.

Lehmann, O, 1932, Die Hydrographie des Karstes: in *Enzyklopädie der Erdkunde*. Leipzig and Vienna, Frantz Deuticke, 212 p.

Lowe, D. J., 1992a, The origin of limestone caverns: An inception horizon hypothesis. Unpublished PhD thesis: Manchester Metropolitan University, 512 p.

Lowe, D. J., 1992b, A historical review of concepts of speleogenesis:*Cave Science* , v. 19, no. 3, p. 63-90.

Malott, C. A., 1937, Invasion theory of cavern development (abstract). *Proceedings of the Geological Society of America for 1937.* 323 p.

Martel, E. A., 1896, Applications géologiques de la spéléologie.... : *Annales des Mines ...* Paris, v. 10, p. 5-100.

Martel, E A., 1921, Nouveau traité des eaux souterraines: Paris, Delagrave.

Moneymaker, B. C., 1941, Subriver solution cavities in the Tennessee Valley. *Journal of Geology* , v. 49. p. 74-86.

Rhoades, R. and Sinacori, N. M., 1941, Patterns of groundwater flow and solution. *Journal of Geology* , v. 49, p. 785-794.

Sanders, E. M. 1921. The cycle of erosion in a karst region (after Cvijić). *Geographical Review* . v. 11, p. 593-604. [see Cvijić, 1918.]

Shaw, T. R., 1979, History of cave science: the exploration and study of limestone caves, to 1900, 2nd Edition: Sidney Speleological Society, 338 p.

Sokolov, D.S., 1962, The principal conditions for karst development: Moscow, Gosgeoltekhizdat, 320 p.

Sweeting, M. M. ,1972, Karst Landforms. London: Macmillan.

Swinnerton, A. C., 1929, The caves of Bermuda: *Geological Magazine* , v. 66. p. 79-84.

Swinnerton, A. C., 1932, Origin of limestone caverns:*Geological Society of America Bulletin* , v. 43, p. 662-693.

Warwick, G T., 1962, The origin of limestone caves: In Cullingford, C. H. D., ed. British Caving. 2nd edition: London, Routledge and Kegan Paul. p. 55-82.

Watson, R. A and White, W. B., 1985, The history of American theories of cave origin: *Geological Society of America, Centennial Special Volume 1.* p. 109-123.

White, W. B and Longyear, J., 1962, Some limitations on speleo-genetic speculation imposed by the hydraulics of water flow in limestone: *National Speleological Society, Nittany Grotto Newsletter* , v. 10. p. 155-167.

Worthington, S. R. H., 1991, Karst hydrogeology of the Canadian Rocky Mountains: Unpublished PhD thesis, McMaster University, 217 p.

Zötl, J. G., 1961, Die Hydrographie des nordostalpinen Karstes: *Steierische Beiträg zur Hydrogeologie* (Graz) v. 2, p. 54-183.

Zötl, J. G., 1974, Karsthydrogeologie. Wien: Springer, 291 p.

2.3 Development of Speleogenetic Ideas in the 20th Century: The Modern Period, 1957 to the Present

William B. White

Abstract

The modern period begins—somewhat arbitrarily—with a shift in emphasis from water table controls to investigations of underlying processes and mechanisms of cave development. Important conceptual shifts were:

- renewed emphasis on controls by details of the geologic setting.
- the recognition of caves as an important part of carbonate aquifer hydrogeology.
- the greatly improved understanding of the chemical equilibria and chemical kinetics of carbonate rock dissolution.

A further development of the modern period was the recognition of halocline caves, hydrothermal caves, and sulfuric acid caves, which resulted from new mechanisms different from shallow groundwater, carbonic acid cave development.

Introduction

The study of the origin of caves has a long history in Europe, particularly in eastern Europe. Extensive commentary on caves and speleogenesis appear in the work of Cvijić, and others in the late 19th century (see Cvijić, 1960). The concepts appear in a conceptually complete form in Kyrle's (1923) textbook. In the United States, the study of the origin of caves has a more complicated history. In the early years of this century there was little interest in caves. The origin of caves was considered an obvious matter of sinking streams and underground rivers much in the tone that E.A. Martel set for French thinking on speleogenesis (Trombe, 1952; Gèze, 1965).

Cave origins attracted attention in the 1930's in the United States with the work of William Morris Davis (1930) and the counter-commentary by Gardner (1935), Swinnerton (1932), Malott (1937), and others in what might be called the "classic period" of cave origin studies.

This period of investigation in the United States came to a close in 1942 with the publication of J Harlan Bretz's classic monograph (Bretz, 1942). There followed a hiatus from about 1942 to about 1957 when new thinking concerning speleogenesis was relatively sparse. Partly this was due to the Second World War, but it had a deeper scientific meaning in the English speaking countries. The study of cave origins had become a sterile debate concerning the position of cave development with respect to the water table and whether some particular cave was of a vadose origin or a phreatic origin. The scientific community appeared to have simply become bored and lost interest (Watson and White, 1985).

Although there were many important karst papers in the English speaking countries prior to 1957 and many "modern" concepts have their roots in these papers, 1957 marks the beginning of a new approach to speleogenesis in which the mechanisms of the cave-forming processes become the focal point.

A specific marker is the 1957 paper of Clifford Kaye. Kaye was concerned with relative rates of dissolution as a function of fluid flow. He was looking at the dissolution-cut notches in carbonate islands and he simulated these with jets of acid against limestone blocks. Kaye then extended these concepts to caves. He may have been the first to suggest the runaway process that leads to single conduit caves.

The present account attempts to summarize some of the highlights of speleogenetic discoveries that took place during the 40-year interval, 1957-1997, that we may call the *modern* or *scientific* period of speleogenetic studies. Of necessity, the account is abbreviated and does not purport to be a comprehensive review of a very large literature. A detailed and more critical analysis of speleogenetic history is provided by Lowe (1992).

The subject of speleogenesis has been approached from three different directions:

- From the perspective of cave geology: an interpretation of cave development deduced from cave morphology, sedimentation, and geologic setting.
- From the perspective of karst hydrology in which the development of caves is only one component of the evolution of karst aquifers and karst drainage basins.
- From the perspective of the chemistry and fluid mechanics of the cave-forming process.

Clearly, all three perspectives are needed for a complete understanding of speleogenesis.

The Geologic Interpretation of Cave Development

The most obvious flaw in the classic period interpretations of cave development is the absence of much detail concerning the geologic environment in which the caves developed. A new outlook was that the development of any particular cave must be interpreted in terms of local stratigraphic, lithologic, structural, and physiographic constraints. Geologic interpretation has been greatly aided by a growing data base of good cave descriptions and detailed maps which were not available in the classic period.

In terrains of moderate relief, caves and passages within large caves are observed to frequently occur in tiers (or levels). These were shown to relate to stream terraces in nearby surface valleys and thus tie cave development to the evolution of nearby surface drainage (Sweeting, 1950; Droppa, 1957; Davies, 1960). Such interpretations have to be made with some care, taking account of structural and stratigraphic controls as well as base level controls (Palmer, 1987).

The argument concerning whether caves formed above, at, or below the water table that so preoccupied researchers of the classic period, was definitely put to rest by Ford and Ewers (1978). The answer was, "yes."

The Ford-Ewers model, based on fracture frequency within the bedrock, provides an explanation within a common conceptual framework of:

- Caves with deep flow loops (e.g. Hollöch in Switzerland).
- nearly horizontal base level conduits with alternating reaches of free surface streams and water-filled conduits (many caves in moderate relief terrain, e.g. the Mammoth Cave System, Kentucky, U.S.A.).
- Deep shaft systems characteristic of high relief terrains such as the high plateaus of Mexico (e.g. the Huautla System) and many alpine caves in Europe and elsewhere.

A most important new development of the modern period is the discovery of techniques for obtaining absolute ages for cave deposits. Age dating by uranium/thorium ratios in speleothems was developed mainly by Henry Schwarcz, Derek Ford, and their students and collaborators at McMaster University in Canada (Ivanovich and Harmon, 1982).

U/Th dating can be used for ages of up to 350,000 years with alpha-counting methods and perhaps to 500,000 or more years with more accurate isotope ratios determined by mass spectrometry. Speleothem ages give minimum ages for the cave passages in which the speleothems are formed.

Measurement of the occasional reversals of the Earth's magnetic field as recorded in the minor amounts of magnetic minerals incorporated in clastic sediments provides fixed time markers rather than continuous ages. Clastic sediments are deposited when the cave passages are actively forming. Suspended grains of magnetic minerals rotate to orient their magnetic poles with the Earth's field as it exists at that time. The grains become locked in place as the sediment pile consolidates and record the Earth's polarity at the time of deposition. The chronology of reversals is well established from other geophysical investigations. Because cave sediments are very weakly magnetic, special high sensitivity magnetometers are required for the measurements. Schmidt (1982), one of the pioneers of cave paleomagnetic studies, was able to establish a chronology for Mammoth Cave that showed the highest passages in the system to be at least of late Pliocene to early Pleistocene age.

Other new methods for dating caves and cave deposits are under development. Polyak et al. (1998) were able to apply $^{40}Ar/^{39}Ar$ dating to the mineral alunite:

$$KAl_3(SO_4)_2(OH)_6,$$

found in sediment in the Big Room of Carlsbad Cavern. The Big Room seems to have formed 4.0 to 3.9 million years ago, making it older than most of the shallow groundwater caves so far investigated.

Even greater ages were found for other caves in the Guadalupe Mountains. Dating of cave sediment by cosmogenic isotopes in quartz pebbles is also under development and shows great promise.

Speleogenesis as Karst Hydrology

Each country seems to have had its own history and independent approach to karst hydrology prior to the modern period. The modern period was well underway, in the 1960's, when these disparate points of view began to come together.

In France, the pioneering cave exploration of E.A. Martel in the high gradient river systems of the Pyrenees impacted French thinking for many years. Karst waters were widely regarded as rivers with a roof flowing independently through the karst. The interconnections with the fracture system were rarely discussed (Gèze, 1958).

In Germany and eastern Europe there was an ongoing debate between proponents of caves as independent underground rivers and proponents of caves draining to an underlying groundwater. In Germany, Otto Lehmann (1932), as an associate of W. Prandl and other pioneers of fluid mechanics, treated caves as examples of pipe flow and subject to the laws of conduit and open channel flow. Lehmann's work has been underappreciated in part because of the abstruseness of his concepts and in part because Lehmann did not extend his concepts to an all encompassing model for karst hydrology.

In England there was much discussion of cave origins but little in the way of hydrologic interpretation. Karst studies in England were firmly in the hands of the geomorphologists. They had become enamored of Davisian concepts of speleogenesis, which by their nature exclude hydrologic interpretation. Indeed the Davisian approach was so strong that a criticism of Davis' speleogenetic ideas (Watson and White, 1985) brought a strong reaction (Lowe, 1992).

In the United States, a promising start was squelched by an unfortunate diagram. Allyn C. Swinnerton was a hydrologist who worked tirelessly through the newly-formed American Geophysical Union to establish karst hydrology as a legitimate part of the rapidly developing science of hydrology.

Figure 1 in Swinnerton's 1932 paper shows his ideas of alternate flow paths in a karst aquifer. It was an unfortunate drawing, however, in that it seems to show water emanating from a single point—water being created out of nothing (see 2.2 Figure 2). M. King Hubbert (1940), in his classic paper that established the scientific basis for fluid flow in porous media, seized on Swinnerton's paper with searing criticism. Swinnerton's ideas were discredited. For the next 20 years it was held that karst aquifers were not intrinsically different from any other porous aquifer, and that caves had little if anything to do with the movement of groundwater.

Karst hydrology began its recovery in the United States in the late 1960's with the gradual realization that many aquifers contain an integrated system of conduits that carry water from recharge areas to discharge areas. The accessible population of caves became understood to represent only a sampling of a much larger conduit system. The International Hydrologic Decade (1964-1974) served to focus new attention on karst hydrology and began the task of merging the many disparate points of view (Stringfield and Legrand, 1969).

Use of tracers, most commonly fluorescein dye, to establish underground drainage connections and flow paths dates from the early work of Martel (1913). Major advances were:

- The invention of the charcoal dye detector by J.R. Dunn in 1957 (see Haas, 1959).
- The introduction of optical brightners as tracers (Glover, 1972)
- The use of dyed lycopodium spore tracers (Zötl, 1974).

Quantitative tracing with information gleaned from the analysis of dye-breakthrough curves became possible with the use of automatic water samplers combined with the measurement of dye concentration by spectrofluorimetry. On-site spectrofluorimetry is now coming into use. Smart and Laidlaw (1977) published comprehensive data on the tracer properties, advantages and disadvantages, and harmfulness or safety of a great variety of tracer dyes.

Comprehensive tracer studies revealed the overall patterns of conduit systems. The development of caves could then be analyzed with regard to their role in an integrated drainage system and not as isolated fragments. From the tracer studies came also the concept of a subsurface drainage basin with a distinct drainage divide.

One of the earliest comprehensive karst drainage basin evaluations was the classic piracy from the Danube to Aach Spring and thus to the Rhine in southern Germany (Batsche et al., 1970). Many basins were delineated in Yugoslavia (Gospodarič and Habič, 1976). In the United States, early examples of subdividing karst areas into their subsurface drainage basins were the Greenbrier Karst in West Virginia (Jones, 1973) and the Mammoth Cave area of Kentucky (Quinlan and Ray, 1981; Quinlan and Ewers, 1989).

Thrailkill (1968) showed that different types of cave waters, which he called vadose drips and vadose flows, have a distinctly different chemistry and that many cave waters are undersaturated with respect to calcite. Geochemical criteria were introduced to distinguish between aquifers dominated by conduit flow and aquifers dominated by fracture (diffuse) flow (Shuster and White, 1971). However, this binary classification was soon found to be over simplistic and more comprehensive classifications have been introduced (Smart and Hobbs, 1986).

If the objective is to understand the origin of caves, one must have a clear description of what a cave is. In the classic period, the features that characterized caves were:

- The patterns and shapes of cave passages, usually displayed in the form of maps and cross sections.
- Detailed features sculptured on the cave walls, displayed as sketches and photographs.
- The breakdown, clastic sediments, and speleothems within the cave also displayed as sketches and photographs.

This was the body of factual information that a speleogenetic theory was supposed to interpret. Karst hydrology, in contrast, was concerned with entire aquifers and their associated subsurface drainage basins. New methods for the characterization of the conduit systems were introduced. These included, in addition to tracer studies, measurement and analysis of the hydrographs of karst springs, and measurement and analysis of chemical variation of karst springs.

Major contributors to the analysis of spring hydrographs were D.J. Burdon (1963) working in the Mediterranean region, Alain Mangin (1970, 1971) working in France, and workers in the former Yugoslavia (Milanovič 1981). Storm-pulse travel times, ratio of storm peaks to base flow, and the mathematical fitting of the recession limb of the hydrograph give information on the response time of a conduit system, on the storage of groundwater within the aquifer, and on the drainage of tributaries into master conduits.

The chemistry of many karst springs varies with discharge. If the chemistry is monitored continuously, either by recording specific conductance (Bakalowicz, 1975), or by frequent sampling and chemical analysis (Dreiss, 1989), the resulting chemical composition to time plot, is termed a "chemograph." Chemographs, particularly in parallel with hydrographs, give information on the movement of storm pulses through conduits, the discharge of water from small tributaries and the fracture system into the main conduit, and the return of the system to prestorm conditions.

Cave development is effectively a runaway process, requiring that an evolving system of enlarging fractures exceed thresholds in kinetics, in the onset of turbulence, and in the onset of clastic sediment transport. Although independent, all three thresholds are reached by coincidence, when the aperture of the enlarging fractures reaches about 1 centimeter. The 1 centimeter critical aperture provides a convenient boundary between fracture aquifers and karstic aquifers (White, 1993).

Cave Forming Processes

Concern with the actual mechanisms by which fractures and bedding plane partings become enlarged into caves marked the beginning of the modern or scientific era of speleogenetic theory. However, like geologic interpretation and hydrologic interpretation of cave development, speleogenetic processes also have roots that extend back into the classic period and before. It was generally agreed that limestone was removed by the solvent action of carbonic acid, and that much of the cave forming process could be encapsulated in the single reaction

$$CaCO_3 + CO_2 + H_2O \leftrightarrow Ca^{2+} + 2HCO_3^-$$

Although the theorists of the classic period made little use of it, the equilibrium chemistry of limestone dissolution was well known as exemplified by Julius Pia's extensive monograph (1933). From the detailed chemistry, the solubility of calcite, aragonite, and dolomite could be calculated as a function of carbon dioxide partial pressure. These curves allowed an understanding of the concentrations of dissolved carbonate in karst waters, the deposition of speleothems, and the rate of karst denudation. The solubility curve for calcite was used widely in the 1950's and 1960's as a frame on which to display the experimentally determined compositions of karst waters.

An innovation of the modern period was the use of aqueous chemistry as an interpretive tool. Indeed, the application of aqueous chemistry to geology at large dates only from the late 1950's and early 1960's. The marker is often taken as the publication of Garrels' *Mineral Equilibria* in 1960 and its sequel in 1965 (Garrels and Christ, 1965). Many karst waters are out of equilibrium, undersaturated in springs, and cave streams and supersaturated in speleothem drip waters. The calculated saturation index introduced in the early 1970's (Langmuir, 1971) proved to be a useful interpretative tool, as was the calculation of carbon dioxide pressure from karst water chemistry.

It had been realized for a long time that the initiation and development of conduits was a matter of kinetics rather than equilibrium. Some mechanism had to be found that would allow undersaturated water to penetrate deep within the carbonate aquifer. Otherwise reaction at the bedrock surface would bring the infiltrating water to equilibrium. There would be surface karst but there would be no caves.

Many experimental investigations studied the dissolution kinetics of calcite under different experimental conditions and used different rate equations. There was not even agreement on the rate-controlling mechanism. Both mass-transport control by diffusion through a laminar boundary layer and surface reaction control were advocated. The most commonly used rate equations were developed by Plummer and Wigley (1976) and by Plummer, Wigley and Parkhurst (1978). An important discovery was that the rate of limestone dissolution was a highly nonlinear function of the degree of undersaturation. Dissolution proceeds rapidly until the karst water reaches a saturation index of about -0.3. As the reaction between the water and the limestone approachs equilibrium, the rate slows down by several orders of magnitude. This is the feature that allows dissolution to take place deep within an aquifer (see Chapter 4.1.2 in this volume for detailed discussion).

By the mid-1970's it was becoming understood that cave development arises from the interaction of two processes—the rate at which water can move through the fractures and developing cave passages, and the rate at which limestone dissolves. A successful model for cave development has to combine flow dynamics with dissolution kinetics. Of many calculations, models constructed by Palmer (1991) and Dreybrodt (1996) have the most contemporary relevance (see Chapters 4.2.2 and 4.2.3 in this volume).

Alternative Mechanisms for Cave Development

Throughout the classic period, and well into the modern period, the one paradigm that was agreed upon was that solution caves are formed by the dissolution of carbonate rocks by shallow circulating groundwater carrying dissolved carbon dioxide. One of the most significant discoveries of the latter part of the modern period is that at least three alternative mechanisms lead to the formation of caves, and the caves formed by these processes are by no means rare.

Mixing-Zone and Halocline Caves

The idea that additional dissolution can be achieved by mixing waters saturated at different carbon dioxide partial pressures was advanced by Laptev (1939) and independently discovered by Bögli (1964), who extensively developed the idea. The concept of *Mischungskorrosion* (mixing corrosion) provided another mechanism for bedrock dissolution deep within a limestone mass. It was only later recognized that mixing of carbonate saturated fresh water with carbonate saturated salt water also produces an undersaturated solution capable of further dissolution. This latter concept was applied by Back et al. (1979) to interpret lagoons and crescent beaches on the east coast of the Yucatan Peninsula.

Cave development also takes place in the active mixing zone along the halocline in coastal karst. Mixing-zone (or halocline, or flank-margin) caves were recognized as a distinct new speleogenetic type with a characteristic morphology by Mylroie and Carew (1990; Chapter 5.1 in this volume).

Hydrothermal Caves

Caves formed by the upwelling of deep-source thermal waters were recognized as originating by a mechanism distinctly different from caves formed from cold groundwater (V. Dublyansky, 1980; Y. Dublyansky, 1990). Such caves are often associated with ore deposits. Hydrothermal caves have been recognized from many localities including caves near Budapest, Hungary, and large maze caves in the Black Hills, South Dakota, U.S.A. (Bakalowicz et al., 1987).

Sulfuric Acid Caves

Limestone can also be dissolved by sulfuric acid. One source of sulfuric acid is the oxidation of sulfide minerals, mainly pyrite, that occur dispersed in many carbonate rocks. Advocates of sulfide oxidation as a source of acid for speleogenesis date back at least to the work of Durov (1956). Sulfide oxidation was never considered very important as a primary mechanism. This was because of the mass of sulfide minerals that would be required to dissolve the mass of limestone that must be removed to form a large cave.

Of more importance was the discovery of a more important sulfuric acid source, the oxidation of H_2S in deep waters seeping upward into carbonate rock masses from nearby petroleum deposits. It is now clear that Carlsbad Cavern, Lechuguilla Cave, and other caves of the Guadalupe Mountains, New Mexico, U.S.A., formed by this mechanism (Hill, 1987; Chapter 5.2.8, in this volume). Other examples are now being identified.

Integration

Quite possibly the most important accomplishment of the modern scientific period has been the integration of speleogenetic data and concepts. Earlier periods were marked by national isolation brought on by both geographic and linguistic barriers. Thus there evolved almost independently American-British concepts, French concepts, German concepts, Eastern European concepts, Soviet concepts, Chinese concepts, and probably others.

Much credit is due to the International Congresses of Speleology. Beginning in 1953, it overcame the barriers by bringing together researchers, first mainly from Europe, and more recently from the entire world. Not all of the problems of speleogenesis have been solved. But the solutions now are at least being pursued from a common conceptual framework.

References

Back, W., Hanshaw, B.B., Pyle, T.E., Plummer, L.N., and Weidie, A.E., 1979, Geochemical significance of groundwater discharge and carbonate solution to the formation of Caleta Xel Ha, Quintana Roo, Mexico: *Water Resources Research*, v. 15, p. 1521-1535.

Bakalowicz, M., 1975, Géochimie des eaux karstiques et karstification: *Annales de Spéléologie*, v. 30, p. 581-589.

Bakalowicz, M.J., Ford, D.C., Miller, T.E., Palmer, A.N., and Palmer, M.V., 1987, Thermal genesis of dissolution caves in the Black Hills, South Dakota: *Geological Society of America Bulletin*, v. 99, p. 729-738.

Batsche, H., Bauer, F., Behrens, H., Buchtela, K., Dombrowski, H.J., Geisler, R., Geyh, M.A., Hötzl, H., Hribar, F., Käss, W., Mairhofer, J., Maurin, V., Moser, H., Neumaier, F., Schmitz, J., Schnitzer, W.A., Schreiner, A., Vogg, H. and Zötl, J., 1970, Kombinierte Karstwasseruntersuchungen in Gebiet der Donauversickerung (Baden-Württemberg) in den Jahren 1967-1969: *Steirische Beiträge zur Hydrogeologie*, for 1970, p. 5-165.

Bretz, J.H., 1942, Vadose and phreatic features of limestone caverns: *Journal of Geology*, v. 50, p. 675-811.

Bögli, A., 1964, Mischungskorrosion-Ein Beitrag zum Verkarstungsproblem: *Erdkunde*, v. 18, p. 83-92.

Burdon, D.J., and Papakis, N., 1963, *Handbook of karst hydrogeology:* Institute for Geology and Subsurface Research, Athens, 276 p.

Cvijić, J., 1960, La géographie des terrains calcaires: *Academie Serbe des Sciences et des Arts Monograph,* no. 26, 212 p.

Davies, W.E., 1960, Origin of caves in folded limestone: *National Speleological Society Bulletin*, v. 22, p. 5-29.

Davis, W.M., 1930, Origin of limestone caverns: *Geological Society of America Bulletin*, v. 41, p. 475-628.

Dreiss, S.J., 1989, Regional scale transport in a karst aquifer. 1. Component separation of spring flow hydrographs: *Water Resources Research*, v. 25, p. 117-125.

Dreybrodt, W., 1996, Principles of early development of karst conduits under natural and man-made conditions revealed by mathematical analysis of numerical models: *Water Resources Research*, v. 32, p. 2923-2935.

Droppa, A., 1957, *Demänovské Jaskyne:* Bratislava, Slovak Academy of Sciences,

287 p.

Durov, S.A., 1956, On the question about the origin of the salt composition of karst waters: *Ukrainian Chemical Journal*, v. 22. (English translation in *Cave Geology*, v. 1, p. 186-190, 1979).

Dublyansky, V.N., 1980, Hydrothermal karst in the Alpine folded belt of the southern part of USSR: *Kras i Speleologia* (Poland), v. 3, no. 12, p. 18-36.

Dublyansky, Y.V., 1990, *Mechanism of formation and modeling of hydrothermal karst:* Novosibirsk, USSR Academy of Sciences, 150 p.

Ford, D.C., and Ewers, R.O., 1978, The development of limestone cave systems in the dimensions of length and depth: *Canadian Journal of Earth Sciences*, v. 15, p. 1783-1798.

Gardner, J.H., 1935, Origin and development of limestone caverns: *Geological Society of America Bulletin,* v. 46, p. 1255-1274.

Garrels, R.M., and Christ, C.L., 1965, *Solutions, minerals, and equilibria:* New York, Harper and Row, 450 p.

Gèze, B., 1958, Sur quelques caractères fondamentaux des circulations karstiques: *Annales de Spéléologie*, ser. 3, v. 13, p. 1-22.

Gèze, B., 1965, Les conditions hydrogéologiques des roches calcaires: Bureau de Recherches Geologiques et Minieres, *Chronique d'Hydrogeologie*, v. 7, p. 9-39.

Gospodarič, R., and Habič, P., 1976, *Underground water tracing:* Ljubljana, Institute of Karst Research, 309 p.

Glover, R.R., 1972, Optical brightners-A new water tracing reagent: *Cave Research Group of Great Britain Transactions*, v. 14, p. 84-88.

Haas, J.L., 1959, Evaluation of ground water tracing methods used in speleology: *National Speleological Society Bulletin*, v. 21, p. 67-76.

Hill, C.A., 1987, Geology of Carlsbad Cavern and other caves in the Guadalupe Mountains, New Mexico and Texas: *New Mexico Bureau of Mines and Mineral Resources Bulletin,* no. 117, 150 p.

Hubbert, M.K., 1940, The theory of ground-water motion: *Journal of Geology*, v. 48, p. 785-944.

Ivanovich, M., and Harmon, R.S., 1982, *Uranium series disequilibrium: Applications to environmental problems:* Oxford University Press, 571p.

Jones, W.K., 1973, Hydrology of limestone karst: *West Virginia Geological and Economic Survey Bulletin,* no. 36, 49 p.

Kaye, C.A., 1957, The effect of solvent motion on limestone solution: *Journal of Geology*, v. 65, p. 35-46.

Kyrle, G., 1923, *Grundriss der Theoretischen Speläologie:* Wein, Östereischischen Staatsdruckerei, 353 p.

Langmuir, D., 1971, The geochemistry of some carbonate ground waters in central Pennsylvania: *Geochimica et Cosmochimica Acta*, v. 35, p. 1023-1045.

Lehmann, O., 1932, *Die Hydrographie des Karstes:* Leipzig, Franz Deuticke, 212 p.

Lowe, D.J., 1992, A historical review of concepts of speleogenesis: *Cave Science*, v. 19, no. 3, p. 63-90.

Malott, C.A., 1937, The invasion theory of cavern development: *Geological Society of America Proceedings for 1937,* p. 323.

Mangin, A., 1970, Contribution a l'étude des aquifères karstiques a partir de l'analyse des courbes de décrue et tarissement: *Annales de Spéléologie*, v. 25, p. 581-610.

Mangin, A., 1971, Étude des débits classés d'exutoires karstiques portant sur un cycle hydrologique: *Annales de Spéléologie*, v. 26, p. 283-329.

Martel, E.A., 1913, On fluorescein tests at great distances: *Comptes Rendus de l'Academie des Sciences*, v. 157, p. 225-227. (English translation in *Cave Geology*, v. 1, p. 279-281, 1983).

Milanovič, P.T., 1981, *Karst Hydrogeology:* Fort Littleton, Colorado, Water Resources Publication, 281 p.

Mylroie, J.E., and Carew, J.L., 1990, The flank margin model for dissolution cave development in carbonate platforms: *Earth Surface Processes and Landforms*, v. 15, p. 413-424.

Palmer, A.N., 1987, Cave levels and their interpretation: *National Speleological Society Bulletin*, v. 49, p. 50-66.

Palmer, A.N., 1991, Origin and morphology of limestone caves: *Geological Society of America Bulletin*, v. 103, p. 1-21.

Pia, J., 1933, Die Theorien über die Löslichkeit des kohlensauren Kalkes: *Mitteilungen der Geologischen Gesellschaft in Wien*, v. 25, p. 1-93.

Plummer, L.N., Wigley, T.M.L., and Parkhurst, D.L., 1978, The kinetics of calcite dissolution in CO2-water systems at 5° to 60°C and 0.0 to 1.0 atm CO_2: *American Journal of Science*, v. 278, p. 179-216.

Polyak, V.J., McIntosh, W.C., Guven, N., and Provencio, P., 1998, Age and origin of Carlsbad Cavern and related caves from $^{40}Ar/^{39}Ar$ of alunite: *Science*, v. 279, p. 1919-1922.

Quinlan, J.F., and Ray, J.A., 1981, Groundwater basins in the Mammoth Cave Region, Kentucky: *Friends of Karst Occasional Publication*, no. 2, 1 sheet.

Quinlan, J.F., and Ewers, R.O., 1989, Subsurface drainage in the Mammoth Cave Area, *in* White, W.B., and White, E.L., eds., *Karst Hydrology: Concepts from the Mammoth Cave Area*, New York, Van Nostrand Reinhold, p. 65-103.

Schmidt, V.A., 1982, Magnetostratigraphy of sediments in Mammoth Cave, Kentucky: *Science*, v. 217, p. 827-829.

Shuster, E.T., and White, W.B., 1971, Seasonal fluctuations in the chemistry of limestone springs: A possible means for characterizing carbonate aquifers: *Journal of Hydrology*, v. 14, p. 93-128.

Smart, P.L., and Laidlaw, I.M.S., 1977, An evaluation of some fluorescent dyes for water tracing: *Water Resources Research*, v. 13, p. 15-33.

Smart, P.L., and Hobbs, S.L., 1986, Characterization of carbonate aquifers: A conceptual base: *Environmental Problems in Karst Terranes and Their Solutions Conference*, Bowling Green, Kentucky, Proceedings, p. 1-14.

Stringfield, V.T., and LeGrand, H.E., 1969, Hydrology of carbonate rock terranes-A review: *Journal of Hydrology*, v. 8, p. 349-417.

Sweeting, M.M., 1950, Erosion cycles and limestone caverns in the Ingleborough District: *Geographical Journal*, v. 115, p. 63-78.

Swinnerton, A.C., 1932, Origin of limestone caverns: *Geological Society of America Bulletin*, v. 43, p. 663-694.

Thrailkill, J., 1968, Chemical and hydrologic factors in the excavation of limestone caves: *Geological Society of America Bulletin*, v. 79, p. 19-46.

Trombe, F., 1952, *Traité de Spéléologie:* Paris, Payot, 376 p.

Watson, R.A., and White, W.B., 1985, The history of American theories of cave origin: *Geological Society of America Centennial Special Volume,* no. 1, p. 109-123.

White, W.B., 1993, Analysis of karst aquifers, *in* Alley, W.M., ed., *Regional Ground-Water Quality*, New York, Van Nostrand Reinhold, p. 471-489.

Zötl, J., 1974, *Karsthydrogeologie:* Wein, Springer-Verlag, 291 p.

Part 3 Geologic and Hydrogeologic Controls of Speleogenic Processes

3.0 Overview

Part Three overviews the principal geologic and hydrogeologic variables that either control or significantly influence the differing styles of cave development that are found.

In Chapter 3.1, Klimchouk and Ford introduce an evolutionary approach to the typology of karst settings, which is taken as a base line for the book. Extrinsic factors and intrinsic mechanisms of cave development change regularly and substantially during the general cycle of geologic evolution of a soluble rock, and more specifically, within the hydrogeologic cycle.

The evolutionary typology of karst presented in this chapter considers the life cycle of a soluble formation, from deposition (syngenetic karst), through deep burial, to exposure and denudation. It helps to differentiate between karst types that may concurrently represent different stages of karst development, and is also a means of adequately classifying speleogenetic settings.

The different types of karst are marked by characteristic associations of the structural prerequisites for groundwater flow and speleogenesis, flow regime, recharge mode and recharge-discharge configurations, groundwater chemistry, and degree of inheritance from earlier conditions. Consequently, these associations make a convenient basis to view both the factors that control cave genesis and the particular types of caves.

In Chapter 3.2, lithological and structural controls of speleogenesis are reviewed in general terms (Klimchouk and Ford). Lowe in Chapter 3.3 discusses the role of stratigraphic elements and the speleoinception concept. Palmer in Chapter 3.4 overviews the hydrogeologic controls of cave patterns and demonstrates that hydrogeologic factors, the recharge mode, and the type of flow in particular, impose the most powerful controls on the formation of the gross geometry of cave systems.

Hence, analysis of cave patterns is especially useful in the reconstruction of environments from paleokarst and in the prediction and interpretation of groundwater flow patterns and contaminant migration. Any opportunity to relate cave patterns to the nature of their host aquifers will assist in these applied studies as well. Osborne in Chapter 3.7 examines the significance of paleokarst in speleogenesis.

More specific issues are treated such as the nature of epikarst and its role in vadose speleogenesis (Klimchouk, Chapter 3.5), and the role of condensation processes (Chapter 3.6, V. Dublyansky and Y. Dublyansky).

3.0 Plate 1. A bedding-plane passage in Archeri Cave, Armenia. (Photo by A. Klimchouk)

3.0 Plate 2. Sof Omar Cave, Ethiopia. (Photo by A. Klimchouk).

3.0 Plate 3. Windy City Lake, in the lowest level of Wind Cave, Wind Cave National Park, South Dakota. (Photo by A. Palmer).

3.1. Types of Karst and Evolution of Hydrogeologic Setting

Alexander Klimchouk and Derek Ford

Abstract

Karst is treated as a specific kind of fluid circulation system capable of self-development and self-organization. Active karst may evolve at a wide range of geologic environments, from deep-seated (without any apparent relation to the surface) to subsurface, and be represented by confined and unconfined circulation systems. Extrinsic factors and intrinsic mechanisms of karst development change regularly and considerably within the general cycle of geologic evolution of soluble rocks, or more specifically, within a hydrogeologic cycle. The latter encompasses a period of exposure between major marine transgressions and is characterized by progressively expanding meteoric groundwater circulation.

A broad evolutionary approach is therefore needed to differentiate between karst types, which concurrently represent distinct stages of karst development. This is also a means to adequately classify speleogenetic settings. Evolutionary typology of karst considers the whole cycle of a geologic formation's life, from deposition (syngenetic karst), through deep burial, to exposure and denudation. The group of intrastratal karst types includes deep-seated, subjacent, entrenched, and denuded karst. The latter also overlaps into the group of exposed karst types. Exposed karst also includes open karst, which represents the unmixed exposed development, that is, karst evolved solely when soluble rock is exposed to the surface. Exposed karst development can be interrupted by subsequent burial (buried karst), with paleokarst formed as a result, and also can be rejuvenated by exhumation.

The types of karst are marked by characteristic associations of:

- Structural prerequisites for groundwater flow and speleogenesis.
- Flow regime.
- Recharge mode and recharge/discharge configurations.
- Groundwater chemistry.
- A degree of inheritance.

Consequently, these associations generate particular types of caves.

Introduction: Definition and Classification of Karst

The purpose of Chapter 3.1 is to place the subject of dissolutional cave development within the broader context of the karst system processes and forms, and to introduce the particular approach that will be emphasized in the following chapters. A schematic picture of the system as it has been described in general karst textbooks is shown in 3.1 Figure 1 (see page 46).

Karst extends from the newest limestone, dolomite, gypsum, and salt accumulating around the margins of seas and lakes to very ancient rocks deep in continental interiors. The waters may be meteoric, marine, expelled from deep sedimentary basins or from magmatic rocks, or mixtures of two or more of these in differing proportions. It is true that some types of karst landforms are entirely superficial erosional features—for example, rillenkarren on bare rocks. But the essence of the system is that water circulates underground, creating caves or breccias there, because the rocks are soluble. Surface features such as sinkholes and poljes (if they occur), develop to supply water to the underground or to drain it; they are not essential to the fundamental circulation.

The term *karst* itself means "stony ground." It came to be adopted as a place name for a limestone region encompassing western Slovenia and the hinterland of

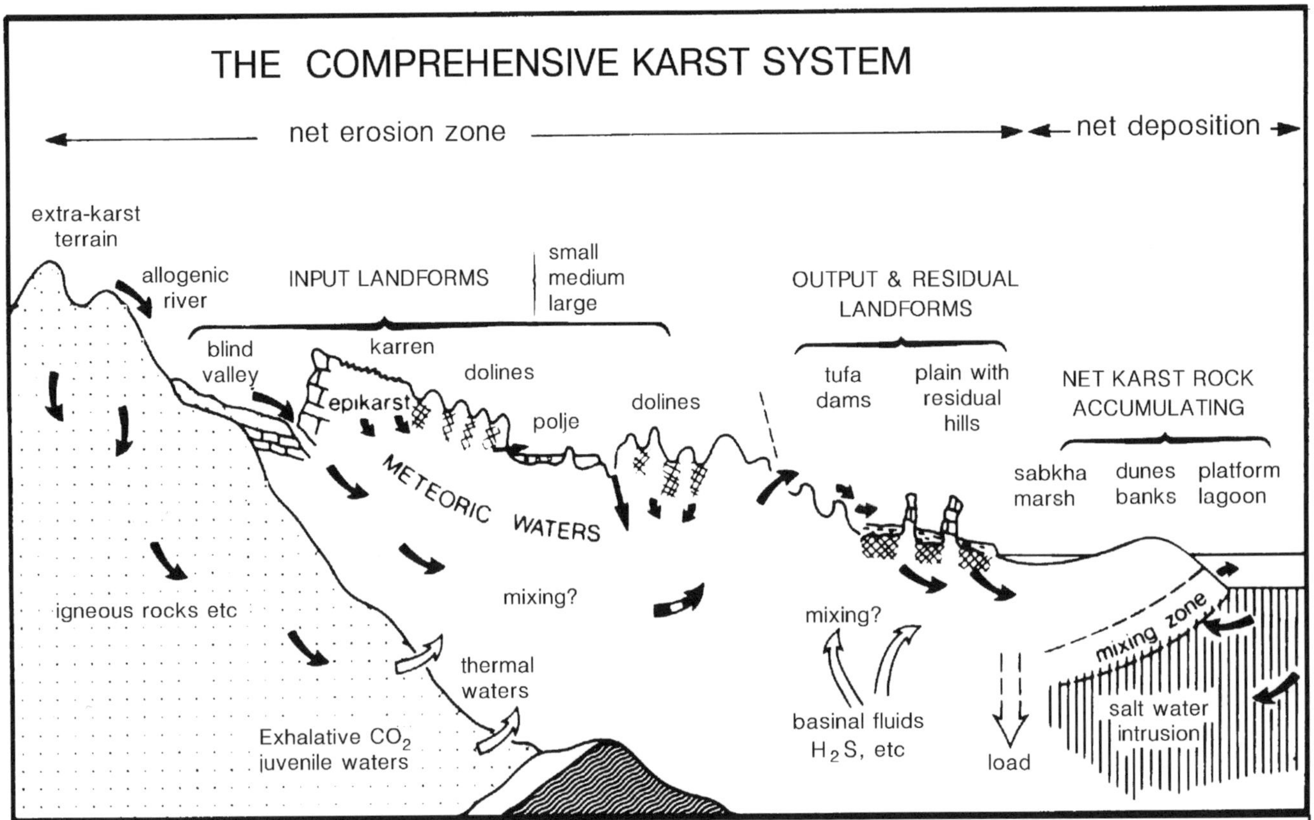

3.1 Figure 1. The comprehensive karst system. A composite diagram illustrating the major phenomena encountered in active karst terrains (From Ford and Williams, 1989).

Trieste. Here, historically, the land was stony and barren because deforestation and overgrazing had led to the loss of most soil into sinkholes and caves (Gams, 1993). Poljes with deep alluvium on their floors were the best agricultural lands remaining.

This geographic association—plus until the middle of this century, the lack of cave maps, extensive groundwater tracing, or information from deep drilling for oil or mineral exploration—led to a principal focus on karst as a specific surface landscape with distinctive hydrology and landforms (Sweeting, 1973; Jennings, 1985; White, 1988; Ford and Williams, 1989; Lowe, 1992; EUR 16526, 1995).

Today we have become increasingly aware that dissolution features originate at deep-seated horizons without any apparent relationship to the land surface. This awareness, plus a desire to extend the historical analysis to the earliest possible phases of karstification in a given rock formation, make it desirable that the speleological definition of karst be broadened and separated from a strict association with surface landforms. It should focus on the most universal and essential characteristics of the phenomena and be applicable to any environment where these may develop.

From the perspective of hydrogeology, karst flow is a specific kind of fluid circulation system capable of self-development and self-organization due to its capacity to mobilize and transport matter by dissolution of host rocks.

Adapting a proposal of Huntoon (1995), karst may be defined as follows:

> The karst system is an integrated mass-transfer system in soluble rocks with a permeability structure dominated by conduits dissolved from the rock and organized to facilitate the circulation of fluid.

This definition does not require that the rocks have a specific lithology, nor is one specific dissolutional process called for. The circulating fluid is not limited to water, although that is overwhelmingly predominant in known karsts. The definition is sufficiently broad to encompass circulation systems in unconfined and confined geologic settings, in the shallow subsurface and deep underground.

Whether karst is expressed at the surface is then not relevant. A karst system can operate in the subsurface without any relation to the surface, being represented exclusively by underground forms that draw their input water and discharge their output water from other nonkarstic rocks.

Many criteria can be used to distinguish the different types of karst that exist. In this volume, we attempt to approach the problem of karst types from the perspective of the geologic evolution of a bedrock formation—more specifically, from its hydrogeologic evolution. Such an approach seems promising for a better understanding of:

- genetic and functional relationships between numerous types and styles of karst forms,
- factors and mechanisms for their origin and development,
- and eventually for describing the average conditions of the evolution and functioning of karst circulation systems.

With this approach, certain types of karst can be viewed as successive stages in development. In each stage, major boundary con–ditions and the overall structure of the system may differ considerably from those of the preceding and succeeding stages. Despite the many local variations that may occur, on a global scale the major boundary conditions change between stages in a quite regular way over the course of geologic evolution. Therefore, such an evolutionary typology can also be used to classify karst and speleogenetic settings.

Speleogenesis can be viewed as the creation and evolution of organized permeability structures in a rock that have developed as the result of dissolutional enlargement of an earlier porosity. It is, therefore, an essential part of the concept of karst defined above. One can assert that:

karst is a function of speleogenesis,

a statement for which the validity is particularly evident in cases where there is no surface landscape component, the karst consisting solely of underground forms.

Hydrogeologic Structures

A clear scale hierarchy of water-bearing structures can be recognized in karst rocks. The smallest structures are the four types of elementary water-containing spaces:

- pores in the rock matrix—tiny intergranular or intercrystalline voids.
- fissures—planar discontinuities such as bedding planes, joints, and faults in which the opening (the aperture or width) is negligible in scale when compared to the length.
- conduits—elongated planar or tubular openings where width is a significant proportion of the length.
- vugs and caverns, that are seemingly isolated voids of irregular shape and with diameters several orders of magnitude greater than those of average matrix pores.

At a second scale, these several elementary types of voids combine in various proportions to form groundwater bodies termed *aquifers.* These are categorized by the predominant types of voids contributing to the overall porosity, such as porous aquifers, fissure aquifers, matrix-fissure aquifers, and fissure-conduit aquifers.

The third scale is the important hydrogeologic consideration as to whether these elementary aquifers are arranged in accordance with the sedimentary layering. That is, are they stratiform, intrastratal, or interstratal. Are they distributed across sequences of rocks with no respect to bedding and other layering. Are they within nonstratified rocks. In the stratified case, individual aquifers and their flow systems may correspond to individual stratigraphic units, such as geologic formations or members.

In the crosscutting case, the aquifers are commonly confined to individual tectonic blocks, or between lithologically determined boundaries such as between a lava flow and a coral reef. Various structural controls of cave development at this scale are considered in more detail in the following sections.

At a fourth and largest scale there are regional hydrogeologic structures. These are principally *artesian basins*. Here rocks are prominently layered and contain stratiform groundwater bodies (layered aquifers), and *hydrogeologic massifs*, where groundwater bodies are of the tectonic-block type, with an overwhelming prevalence of crosscutting fissure-conduit permeability (Zajtzev and Tolstikhin, 1971; Pinneker, 1977).

Transitional types include *disrupted basins* (layered structure prevails but is complicated by block and crosscutting structures) and *layered massifs*, with layered structures of lesser importance. The succession *artesian basins—disrupted basins and layered massifs—hydrogeologic massifs* can correspond to different stages in the evolution of sedimentary basins. Artesian basins are modern or ancient basins displaying only minor to moderate tectonic disturbance of the layered structure. Disrupted artesian basins contain considerably deformed strata in tectonically active and uplifted parts of cratons, or in folded regions. Hydrogeologic massifs are common in highly deformed and uplifted highland belts. These categories may display significant differences in the boundary conditions for groundwater circulation systems and speleogenesis.

Karst Evolution within the Hydrogeologic Cycle

The development of groundwater circulation systems is broadly cyclic. The hydrogeologic cycle begins with marine sedimentation that is succeeded by tectonic subsidence and the formation of connate waters (trapped seawater with further minerals dissolved in it). It then encompasses uplift, with denudation and the progressive invasion of meteoric waters into the reservoir. It may include the intrusion of magma with release of juvenile waters. It closes with a new marine transgression and sedimentation (Pinneker, 1982).

Groundwater history in sedimentary rocks can be considered as a part of the diagenetic history of the deposits. Diagenesis is the progressive alteration of a sediment by processes of compaction, expulsion of water, and cementation, with or without dissolution. In karst rocks it is closely linked with, and largely operates through, groundwater processes. Creation and loss of solutional porosity is treated by sedimentologists as a diagenetic process (Choquette and James, 1988).

Choquette and Pray (1970) divided carbonate diagenesis into three stages:

- *eogenesis* when the shallow marine sediment is first laid down and may be altered by the effects of periodic drying or exposure to rainwater.
- *mesogenesis* when this sediment is buried, compacted, and altered under later rocks.
- *telogenesis* when later uplift and erosion thin or remove the cover rocks and permit meteoric water to invade and overwhelm any remaining connate or juvenile waters.

With increasing burial, connate waters are expelled from clayey sediments into coarser grained (sandy) reservoirs, the fluid potentials (pressure fields) being guided mainly by compaction. With burial to depths greater than 1 to 2 km, the increasing temperature may cause dehydration of minerals containing combined water and the consequent release of liquid resurgent waters that partially replace connate waters in the sediments. Groundwater movement is directed from areas of greatest subsidence to the margins of basins where the rocks are at shallower depths. This type of circulation is termed *expulsion* circulation (or *elision* circulation in the Eastern European literature; Kartzev et al., 1969).

When meteoric water invades the system following uplift, it is driven by the hydrostatic head. This is *meteoric* (or *infiltration) circulation.* During the continental regime, meteoric groundwater increasingly flushes out connate and resurgent waters from a basin, although expulsion circulation may still predominate in deep environments. The full hydrogeologic cycle may thus include replacement and expulsion, meteoric and magmatic (endogenous) circulation (3.1 Figure 2A, see p. 48), although the latter often does not occur.

If there is a second hydrogeologic cycle the new marine waters may enter any surviving older lithified rocks, replacing meteoric or hypogene waters in them that were inherited from the interrupted previous continental regime. At the same time, in the deeper parts of the basin expulsion of connate and resurgent waters from the first burial stage may be continuing (3.1 Figure 2B).

Consequently, the composition and circulation of groundwater in a large basin may be quite complicated where several hydrogeologic cycles are superimposed. The situation is further complicated where an originally unified groundwater basin is deformed and subdivided by differential tectonic movements, especially where time lags in the adjustment of the flow systems to the changing conditions mingle groundwaters of differing ages (Töth, 1995).

In contrast to groundwater reservoirs of the stratiform class (artesian and disrupted artesian basins that predominate on continental platforms and along passive margins), massifs in orogenic regions are characterized by the overwhelming predominance of contemporary meteoric groundwater. This groundwater has almost entirely replaced waters of previous stages. Episodes of

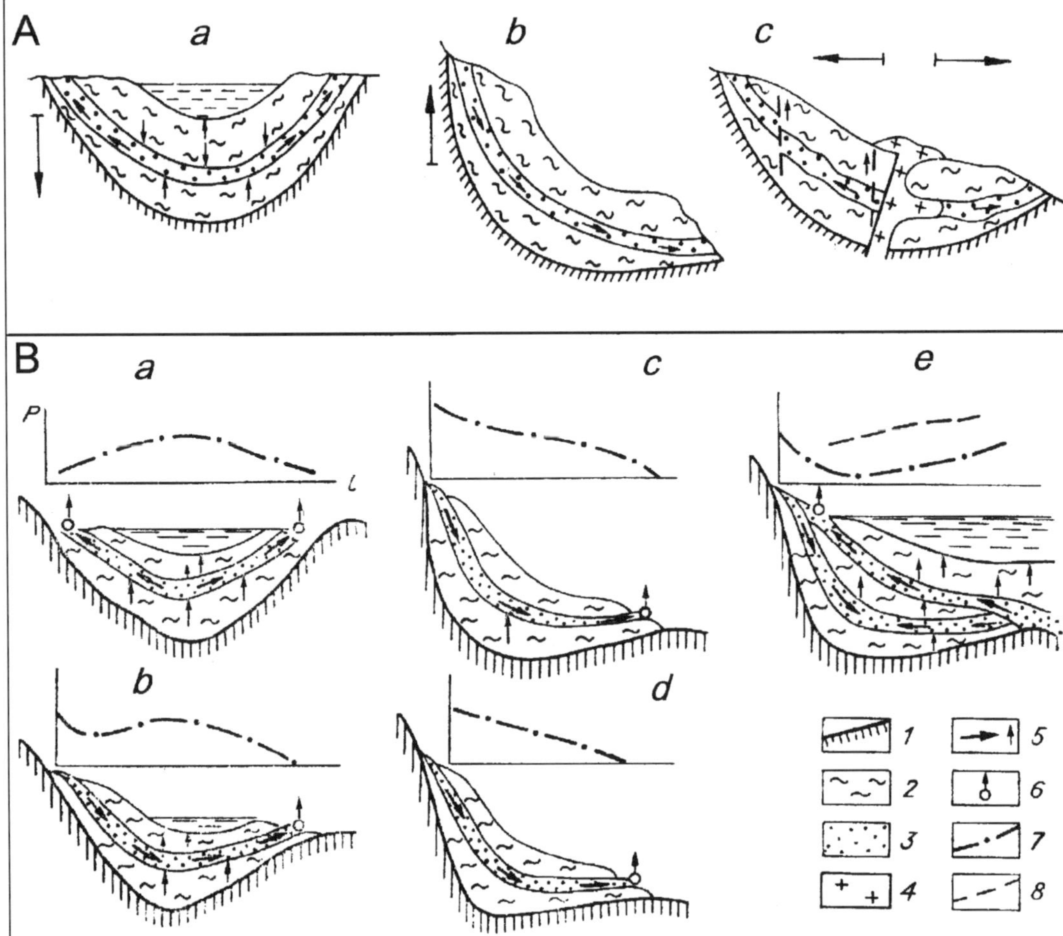

Upper diagrams:
Stages:
***a.** expulsion circulation (sedimentation).*
***b.** meteoric circulation.*
***c.** magmatic (endogenous) circulation. Large arrows indicate directions of tectonic movements.*

Lower diagrams:
***a** through **d**—Stages of the 1st hydrogeologic cycle:*
***a.** expulsion circulation*
***b.** meteoric circulation with inherited features of expulsion circulation.*
***c.** meteoric nonsteady regime circulation.*
***d.** meteoric steady regime circulation.*
***e.** expulsion circulation, with inherited features of meteoric circulation in the lower aquifer and expulsion circulation in the upper aquifer of the 2nd hydrogeologic cycle.*

Key: 1–basement (the lower aquifuge). 2–clay. 3–sandy sediment (aquifers). 4–magmatic rocks. 5–groundwater flow direction. 6–discharge points. 7 and 8–potentiometric surfaces of the aquifers: 7 of the lower aquifer, 8 of the upper aquifer

3.1 Figure 2. Diagram A illustrates the full hydrogeologic cycle (after Pinneker, 1982). Diagram B, the hydrogynamic evolution of groundwater systems (Kudrjakov, 1970).

magmatic groundwater invasion and circulation are often important during the deformation that creates such regions. It may still be occurring in tectonically active areas.

Baskov et al. (1996) have emphasized that on the continents, (especially in the little disturbed stratiform regions), the hydrogeodynamic circulation discussed above often can be described as displaying two distinct stories. The upper story has active meteoric water circulation signified by the presence of significant amounts of dissolved oxygen and nitrogen of meteoric origin. The lower story is dominated by connate and resurgent water lacking such gases. During uplift and denudation, sedimentary formations move progressively into the upper story.

Ezhov and Vdovin, 1970; Ezhov and Lysenin, 1986; and Ezhov et al., 1991, 1992, have enlarged our thinking about water in the crust of the Earth by introducing the further concepts of *exokarst* and *endokarst.* The hydrogeodynamic circulation (both stories, meteoric and expulsion groundwater systems) constitutes an upper exokarst zone of the continental hydrosphere. This is one in which fluid pressures are close to the hydrostatic pressures, not exceeding them by more than 10-15%.

Below it is the deepest zone where fluid pressures are created chiefly by the lithostatic load (the weight of overlying rocks) plus any tectonic forces. In between is a transition region, the mesozone, where pressures can be highly irregular in both space and time. This is marked by maximum compaction of rocks and serves as an isolator between the upper and lower zones. Local releases of pressure in it allow fluids from below to break through into the exokarst from time to time. In the lithostatic zone and deep mesozone there may be a high degree of dissolution and fluid rupturing of rocks—the endokarst. Besides dissolving carbonates, the strong thermo-baric conditions there may make fluids highly aggressive with respect to silicate and aluminosilicate minerals that are of low solubility under exokarst conditions. For example, the solubility of quartz at temperatures of 300°-350°C and pressures of 200-250 MPa is comparable to that of gypsum at the surface of the Earth. The endokarst is discussed further in Chapter 5.2.5.

The Hydrogeologic Cycle and Perspectives on Speleogenesis

Most of the karst development that has been well studied is known to be associated with continental subaerial exposure and meteoric groundwater circulation. Conventional karst and speleogenetic theories are concerned mainly with shallow unconfined geologic settings, supposing that the karstification is ultimately related to the surface and that dissolution is driven by downward meteoric water recharge.

The possibility of karstification in deeper environments (although still within the exokarst) has been neglected for a long time. So the quite numerous instances of karst features found at significant depth have usually been interpreted as buried paleokarst (as defined below).

However, some recent studies reflect increasing recognition of the importance to speleogenesis of processes operating earlier in the hydrogeologic cycle or deeper underground:

- Some of the features formed during sedimentation and eogenesis (changes during shallow burial) can influence subsequent speleogenesis. Caves of enterable size can be formed penecontemporaneously with carbonate sedimentation or dune formation in the littoral environment and freshwater/seawater mixing zone (Jennings, 1968; Ollier, 1975; Esteban and Klappa, 1983; Lowe, 1992; Mylroie and Carew, 1995, and many others. In this volume see Chapters 3.2, 3.3, and 5.1).

- Deep dissolution during mesogenesis or telogenesis can play an important role in the earliest phase of speleogenesis, preparing potential flow paths for later selective exploitation. A shortcoming of many studies of modern caves has been the lack of recognition of possible inheritance from earlier stages. Moreover, there can be full scale cave development (macroscopic conduit porosity and cave systems) in the deep environment due to hypogene processes (Klimchouk, 1994, 1996b, 1997; Lowe, 1992; Lowe and Gunn, 1995; Mazzullo and Harris, 1991, 1992; Palmer, 1991, 1995; Smart and Whitaker, 1991; Worthington, 1991, 1994; and Worthington and Ford, 1995; in this volume see Chapters 3.3, 3.4, and 5.2).

Further important points to emerge from cave studies of the past three decades are these:

- That no one speleogenetic model can be applied to all geologic and hydrologic settings.
- That a number of different dissolution mechanisms may operate in the different settings to produce organized karst permeability.

This makes it necessary to categorize speleogenetic settings (environments) in order to evaluate genetic theories and mechanisms. As settings change regularly during the hydrogeologic evolutionary cycles outlined above, an evolutionary approach again is a helpful means to order the theories and models and to clarify their relations.

A Hydrogeologic Evolutionary Typology for Karst

The sequence of karst settings that a given formation could experience during its history is outlined in 3.1 Figure 3. These are:

- Syngenetic to penecontemporaneous karst.
- Intrastratal (interstratal) karst.
  - Deep-seated karst.
  - Subjacent karst.
  - Entrenched karst.
- Exposed karst.
  - Open karst.
  - Denuded karst.

  (The subtypes open karst and denuded karst are not sequential categories. They are characterized by similarly exposed geomorphic settings but differ in their previous karstification history. The true denuded karst is a former intrastratal karst)
- Mantled karst.
- Buried karst.
- Exhumed karst.

In actuality, no individual karst is known that displays all of its possible sequence, but many have experienced several of the components. Karstification need not begin syngenetically. Many limestones, gypsums, and salt deposits are buried without suffering significant earlier dissolution. Where buried, karstification may be initiated at any of the stages of intrastratal development that are shown in the figure. Or they may be delayed until the rock is exposed by stripping of the cover (open type). The karst may be completely destroyed, along with its host formation, within the same stage that its development commenced. This is more common for karst in sulfates than in carbonates and is the fate of most salt. At the other extreme, carbonate karst can survive through several cycles of burial-exposure (and the accompanying hydrogeologic cycles), being repeatedly fossilized and rejuvenated.

Syngenetic Karst

Most limestones, dolostones, and gypsum deposits are laid down on shallow oceanic or coastal platforms experiencing slow subsidence or uplift. Eustatic and glacioeustatic changes of sea level may result in alternating deposition and intermittent subaerial exposure on them. Freshwater vadose and phreatic, and freshwater-marine mixing environments can cause substantial and rapid cave development in the young rocks. Jennings (1968) categorized this as *syngenetic karst*. Its occurrence is not very widespread at present, because global sea levels are exceptionally high. But its importance in the past—and hence its role in generating features that will fall into the paleokarst category after subsequent burial—is underlined by the fact that a majority of limestones were deposited in less than 10 m of water. Thus they were vulnerable to even minor changes in sea-level and the related meteoric diagenesis (Wright, 1991). Speleogenesis in oceanic and coastal settings is treated in detail in Chapter 5.1.

Where carbonate rocks are frequently exposed and acquire considerable diagenetic maturity without being buried, this type of karst may pass directly into the open type. It is more common, however, that any syngenetic karst is buried to various depths, becoming paleokarst. Karstification then may be restarted at any stage of the intrastratal or later sequence (see 3.1 Figure 3, p.50). But any inheritance is commonly of the posthumous type.

Intrastratal karst

The concept that an evolutionary karst sequence can be linked to the relationships with insoluble cover beds has been considered by many geomorphologists. For example, Ivanov (1956) distinguished covered, semi-open, and open *stages* of karst development as an evolutionary succession for the neotectonic epoch in many regions. More detailed classifications were proposed by Quinlan (1978). There is also further discussion in Bozák, Ford and Glazek (1989), Palmer and Palmer (1989) and others for limestone, and Klimchouk (1996a) for gypsum.

Quinlan (1978) originally proposed *interstratal karst* as the term to describe dissolution features that develop beneath cover rocks deposited before any karstification. Palmer and Palmer (1989) suggested that *intrastratal karst* is more appropriate, because most dissolutional processes at depth are not limited to the boundaries between different strata. Perhaps *interstratal* is better where a soluble stratum is being entirely removed (the fate of many salt beds) but *intrastratal* is used here as the more broadly applicable term.

On all continents the area of common karstifiable rocks that lie at some depth beneath insoluble sediments substantially exceeds the area of exposed karst. There is much evidence from exploration drilling and mining that karst features occur widely beneath cover rocks. But, as noted, these have generally been treated as inert paleokarst (*buried karst* in the classification adopted here) rather than actively evolving forms. However, the likelihood and mechanisms for intrastratal karstification to considerable depths are increasingly recognized, and occurrence of active deep-seated intrastratal karst is probably much greater than traditionally supposed.

The fluid potentials that drive groundwater flow in deep normally confined settings originate from different sources, including gravity, compaction, compression, dilatation, and heating. The deep chemical processes that can generate acids for dissolution and speleogenesis are also various and include maturation of hydrocarbons, sulfate reduction, and metamorphism. Nonacidic aggressiveness with respect to carbonates can be caused by the cooling of ascending water (increasing the solubility of CO_2), and, for sulfates, by the reduction process. Dissolution by mixing of waters with contrasting chemistries is of particular importance. Dedolomitization (dissolution of $MgCO_3$ with concomitant precipitation of $CaCO_3$) can contribute substantially to the creation of porosity in both sulfates and dolomites. Thus it can be important in interbedded sulfate-carbonate formations (which are very common). Settings and mechanisms of deep-seated speleogenesis are reviewed in more detail in Chapters 3.4 and 5.2.

Hydraulic continuity and vertical cross-

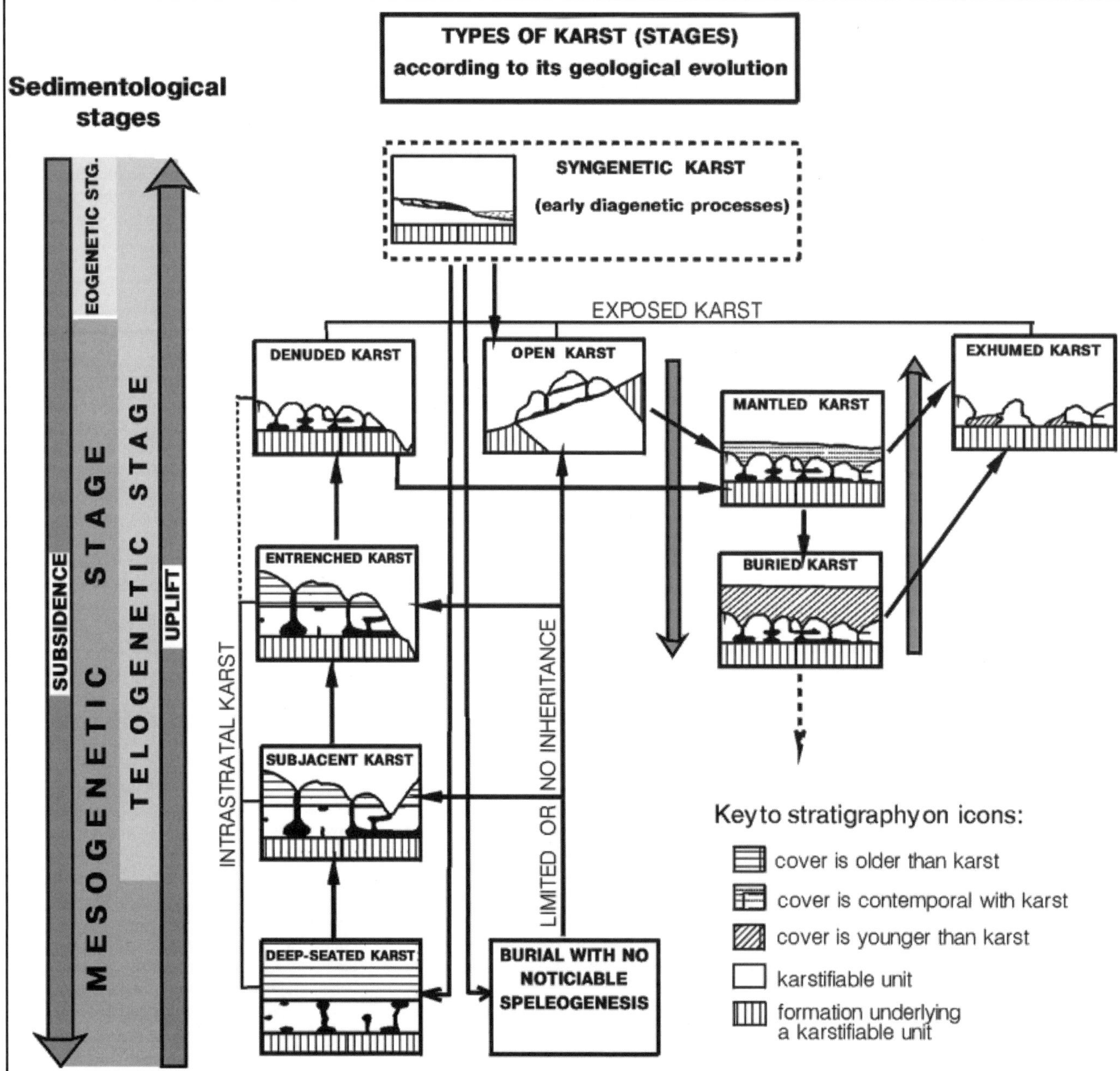

3.1 Figure 3. Evolutionary types of karst (modified from Klimchouk, 1996a).

formational hydraulic communication is of paramount importance for intrastratal karstification (Klimchouk, 1994, 1997b; Töth, 1995). This is because it helps to activate different chemical mechanisms and reinforce mixing corrosion effects. In most deeply covered karst strata, the only source of recharge is from an adjacent formation. Whether the recharge is focused or dispersed depends largely on the structural situation. Particularly favorable conditions for speleogenesis are created where expulsion waters and meteoric waters can mix and the meteoric water begins to flush out basinal brines from the aquifers.

Deep-seated intrastratal karstification tends to be more diffuse and less selective than that of shallow settings, due to both hydraulic and structural controls. As a consequence of standard denudation and uplift on the continents, the deep-seated rocks are shifted with time into progressively shallower positions. They pass into the upper hydrogeodynamic level where meteoric circulation is driven predominantly by topographic differences of head. At some stage *en route* to the surface, erosional incision into the cover rocks locally breaches the hydrogeologic confinement, and the aquifer is brought into direct hydraulic connection with the surface.

Further incision causes inversion of the circulation system, drastic changes in recharge-discharge configuration, and establishment of a vadose zone and water table conditions within the karstic strata. At this point the unit is still commonly capped by some insoluble beds over most of its area. Progressive denudation may eventually expose the rock entirely.

Various stages of this evolution are represented by the sequence of intrastratal karst types. The boundaries between them are in reality transitional, but can be drawn in the following way:

- *Deep-seated karst* is not evident at the surface and the soluble rock is not exposed.
- *Subjacent karst* occurs where the soluble rock is locally breached by erosion over a minor part of its thickness. Karst features may already be expressed at the surface at springs or collapse features.
- *Entrenched karst* is where the entire thickness of the soluble rock is entrenched along valleys, but the insoluble cap remains over most of the interfluves.
- *Denuded karst* is where the cap rocks are removed.

Where continuous karst develops from the deep-seated stage to the denuded stage, the role of inheritance can be quite important.

In some cases deep-seated karstifiable rocks, particularly gypsum and salt, may be entirely removed by dissolution before any breaching and exposure. Such dissolutional removal leaves characteristic karst breccias composed of the insoluble or less soluble remnants of intercalated beds (commonly carbonates) and broken fragments of the overlying rocks.

The different intrastratal karst stages are marked by characteristic changes:

- In the geologic controls of speleogenesis (see Chapter 3.2).
- In the dynamics of the flow system, recharge mode, and recharge-discharge configurations (see Chapter 3.4).
- In groundwater chemistry (see Chapters 3.4 and 5.2).

For instance, the recharge mode (which largely determines the style of speleogenesis) tends to switch from a predominantly diffuse, steady flow from adjacent formations in deep-seated karst, to a highly focused and variable flow from the surface in subjacent and entrenched karsts where caprocks are poorly permeable. However, the occurrence of diffusely permeable caprocks may still maintain diffused recharge in these settings. Recharge becomes less focused and variable in denuded karst, but underground flow patterns are largely inherited from the earlier stages. Confined circulation systems give way to unconfined phreatic flow when passing from deep-seated to entrenched and exposed stages, with consequent development of water-table and vadose zones. Therefore, the cave patterns vary between these settings (see Chapter 3.4).

The concept of a succession of intrastratal karst stages applies to the evolution of artesian basins and disrupted basins. In highly deformed and folded highland regions (many of which contain thick karstifiable formations), stratiform groundwater circulation characteristic of basins is overwhelmed by fault block circulation. Individual stages may be recognized where there are cover beds but not the regular stratiform sequence of karst development suggested by the term "intrastratal."

Exposed Karst

When the surface extent of soluble rocks exceeds that of any remaining caprock, the karst can be considered exposed. There are three paths to this condition (3.1 Figure 3. The first two correspond to the open karst, while the third one generates denuded karst:

- The karst rock has never been buried by other rocks.
- The rock has been buried but then stripped of cover before there was any significant development of karst circulation or features. these are "open" karsts.
- Earlier intrastratal karst is exposed and becomes denuded.

There may be considerable differences in mechanisms of development and styles of karst landforms and caves between the open and denuded categories.

Denuded karst was proposed by Quinlan (1978) to identify former subjacent or entrenched karst exposed by removal of its cover. It is characterized by the coexistence of features formed during the exposed stage and those inherited from previous stages. The latter may be largely relict or adjusted in varying degrees to the new hydrologic and hydrogeologic setting. Inherited cave patterns greatly influence the development of the karst landscape and hydrology.

Open karst is that developed largely or entirely during the exposure stage. As such it is unmixed and controlled by surface and shallow subsurface factors without any inheritance from previous karst stages. It is dominant in the great cockpit and tower karst highlands of the tropics and subtropics. It is common where evaporites or any soluble lithology are exposed in areas of very low relief, with aquicludes such as shales prohibiting vertical interformational groundwater flow. It also predominates on tropical islands and coasts where recent carbonates have been continuously exposed. The Yucatan Peninsula of Mexico is the great example.

Exhumed karst is earlier karst exposed by the erosion of sediment that once buried it (Quinlan, 1978; Bosak, Ford and Glazek, 1989). It represents former exposed karst, once (or repeatedly) interrupted by burial but now returned to an active state by the exhumation. Exhumed karst exhibits complicated polyphase inheritance and is characterized by the presence of many paleokarst features.

Exposed karst settings characteristically generate branchwork cave patterns in carbonates, and branched to linear patterns in sulfates. As fissure patterns and the relief and thickness of the vadose zone may range widely, the profiles of invasion vadose caves can be quite varied. Denuded karst can exhibit a wide range of cave patterns inherited from the preceding stages, utilized to varying extents by the active systems.

Mantled Karst and Buried Karst

Mantled karst is that covered by significant thicknesses of unconsolidated sediment that accumulates as the karst develops. Most common are soils formed from the insoluble residuum of impure limestone and dolostone (locally derived or "autochthonous" deposits). An "allochthonous" mantle is one carried in from outside the karst terrain, such as the volcanic ash layers that mantle the karst at Waitomo, New Zealand.

Mantled karst should be distinguished from *buried karst*. Buried karst is a complete infilling and burial by later rocks such as transgressive marine sediments, surpressing or (usually) terminating the karstification. Buried karst should not be confused with intrastratal karst, where the karstic rocks were buried before any karstification occurred. Buried karst has only the direct meaning—a karst that was exposed and then buried. We suggest that minor syngenetic karst in diagenetically immature rocks should not be classified as buried karst where burial followed almost immediately after deposition. Only karst that has undergone the pronounced exposed development before burial should be treated as truly buried karst.

When karst is buried, it is generally the most unambiguous case of true paleokarst, as described below.

Paleokarst

Paleokarst does not imply any particular type of karst but rather refers to a condition, namely fossilized. From the definition of karst, a paleosystem is karst that lost its mass transport function. It is hydrologically decoupled from contemporary systems, in contrast to *relict* systems that exist within contemporary systems but are displaced from the environment in which they formed (Ford and Williams, 1989).

At the extreme, a karst rock can be entirely removed (dissolved). It is replaced in the geologic cross section or at the surface by residual sediments or by sediment that was emplaced within karst forms, such as sinkholes that have since been destroyed. This means that paleokarst can be recognized even when the host rock no longer exists. This is a point of particular importance for the interpretation of interstratal gypsum karst.

Karst systems, or their components, can be fossilized at any of the stages of development described above. Fossilization is almost inevitable when syngenetic karst is buried. In deep sulfate beds a karst system can develop fully and then disintegrate, possibly to the extent of total removal of gypsum beds from a cross-section, without having been elevated to the shallow subsurface. As noted above, karst-breccia horizons of minor insoluble intercalations and fragmented overlying rocks are common in sulfate and sulfate-carbonate sequences, where they represent intrastratal paleokarst. However, it is worth reemphasizing that the occurrence of deep-seated karst does not, in isolation, indicate a paleokarst condition. Active karstification also occurs commonly in this environment, where it may develop intermittently over long timespans, being repeatedly activated and deactivated. Distinguishing between intrastratal karst and paleokarst is often difficult, as pointed out by Palmer and Palmer (1989, p. 337).

Intrastratal features are considered paleokarst only if they are out of adjustment with the present geologic setting. This criterion is often difficult to apply, for intrastratal processes tend to operate over a long time span, and many features that qualify as intrastratal paleokarst were formed by processes still operating at a diminished rate today.

Even those intrastratal features that are filled with collapse breccia are not necessarily paleokarst. Breccia pipes or "geologic organ pipes"—Klimchouk and Andrejchouk (1996) have suggested the generic term vertical through structures to denote such forms—are common features of upward stoping through overlying formations, induced by dissolution of underlying salt or gypsum beds. Breccia pipe development is not merely a breakdown process, but is maintained by active groundwater circulation through the pipes, with partial mass-removal by dissolution. They become paleokarst features only when they are fossilized to the extent that groundwater circulation is interrupted.

When karst passes into the entrenched stage, most earlier conduits (artesian, phreatic, water-table) become relict. This is because circulation in the now dominant vadose zone is highly localized. Further evolution into the exposed karst stage may reactivate many of these relict features as they become integrated into a karst landscape where the dissolution is more dispersed. Should there be subsequent burial, different parts of one contemporary system can become separated hydrologically, some becoming completely fossilized, others not. In some situations such paleokarst can be exhumed and reintegrated into active systems. Further discussion of the problem of paleokarst, as it relates to speleogenesis, is provided by Osborne in Chapter 3.7.

References

Bakalowicz, M., and Mangin, A., 1980, L'aquifere karstique, sa definition, ses characteristiques et son identification: *Memoires hors Serie Societe Geologique de France*, no. 11, 71 p.

Baskov, E.A., Petrov, V.V., and Ivanov, T.K., 1996, Hydrogeology of sedimentary basins, *in* Sedimentary basins of Russia, v. 2: St. Petersburg, VSEGEI. 50 p. (in Russian).

Bosák, P., Ford, D.C., and Glazek, J., 1989, Terminology, *in* Bosák, P., Ford, D., Glazek, J., and Horácek, I., eds., *Paleokarst-A systematic and regional review:* Praha, Academia, p. 25-32.

Choquette, P.W., and Pray, L.C., 1970, Geologic nomenclature and classification of porosity in sedimentary carbonates: *American Association of Petroleum Geologists Bulletin*, v. 54, p. 207-250.

Choquette, P.W., and James, N.P., 1988, Introduction, *in* James, N.P., and Choquette, P.W., eds., *Paleokarst:* New York, Springer, p. 3-21.

Esteban, M., and Klappa, C.F., 1983, Subaerial exposure, *in* Scholle, P.A., Bebout, D.G., and Moore, C.H., eds., Carbonate depositional environments: *American Association of Petroleum Geologists Memoir,* no. 33, p. 1-54.

EUR 16526, 1995, *Hydrogeologic aspects of groundwater protection in karstic areas-Guidelines:* Luxembourg, Office for Official Publications of the EC, 15 p.

Ezhov, Y.A., and Lysenin, G.P., 1986, Vertical hydrodynamic zoning of the Earth's crust: *Soviet Geology*, v. 8, p. 111-120 (in Russian).

Ezhov, Ju.A., et al., 1991, Deep-seated karst. Endokarst. Hydrothermal karst (abs.), *in* Karstogenetic settings: Deep-seated karst, endokarst, hydrothermal karst: Regional Symposium, Kungur, p. 4-13 (in Russian).

Ezhov, Ju.A., Lysenin, G.P., Andreychuk, V.N., and Dublyansky, Y.V., 1992, *Karst in the Earth's crust: Distribution and the main types:* Novosebirsk, United Institute of Geology, Geophysics, and Mineralogy, 76 p. (in Russian).

Ezhov, Ju.A., and Vdovin, Ju.P., 1970, On the question of vertical hydrodynamic zonality of the Earth's crust: *Soviet Geology*, v. 8, p. 66-76 (in Russian).

Ford, D.C., and Williams, P.W., 1989, *Karst geomorphology and hydrology:* London, Unwin Hyman, 600 p.

Huntoon, P.W., 1995, Is it appropriate to apply porous media groundwater circulation models to karstic aquifers? *in* El-Kadi, A.I., ed., *Groundwater models for resources analysis and management:* Boca Raton, Lewis Publishers, p. 339-358.

Ivanov, B.N., 1956, On the typology of karst landscapes of planes, on the example of the Podol'sko-Bulovinsky karst region, *in* Voprosy izuchenija karsta na juge Evropejskoj chasti SSSR: Yalta, p. 131-156 (in Russian).

Jennings, J.N., 1968, Syngenetic karst in Australia, *in* Contributions to the study of karst: Australian Natl. University, Dept. of Geography publications, no. G/5, p. 41-110.

Jennings, J.N., 1985, *Karst geomorphology:* New York, Basil Blackwell, 293 p.

Kartzev, A.A., Vagin, S.B., and Baskov, E.A., 1969, *Paleohydrogeology:* Moscow, Nedra, 151 p.

Klimchouk, A.B., 1994, Speleogenesis under confined conditions, with recharge from adjacent formations: *Publ. Serv. Geol. Luxembourg*, Comptes Rendus du Colloque International de Karstology a Luxembourg, v. 27, p. 85-95.

Klimchouk, A.B., 1996a, The typology of gypsum karst according to its geologic and geomorphological evolution, *in* Klimchouk, A., Lowe, D., Cooper, A., and Sauro, U., eds., Gypsum karst of the world: *International Journal of Speleology*, v. 25, no. 3-4, p. 49-60.

Klimchouk, A.B., 1996b, Hydrogeology of gypsum formations, *in* Klimchouk, A., Lowe, D., Cooper, A., and Sauro, U., eds., Gypsum karst of the world: *International Journal of Speleology*, v. 25, no. 3-4, p. 83-90.

Klimchouk, A.B., and Andrejchouk, V.N., 1996, Breakdown development in cover beds, and landscape features induced by intrastratal gypsum karst, *in* Klimchouk, A., Lowe, D., Cooper, A., and Sauro, U., eds., Gypsum karst of the world: *International Journal of Speleology*, v. 25, no. 3-4, p. 127-144.

Klimchouk, A.B., 1997a, The role of karst in the genesis of sulfur deposits, Pre-Carpathian region, Ukraine: *Environmental Geology*, v. 31, p. 1-20.

Klimchouk, A.B., 1997b, Artesian speleogenetic setting*: 12th International Congress of Speleology,* La Chaux-de-Fonds, Switzerland, Proceedings, v. 1, p. 157-160.

Kudrjakov, V.A., 1970, General features of the development of stratally confined groundwater systems, *in* Metodika paleogydrogeologicheskih issledovanij: Ashghabad, p. 88-93 (in Russian).

Lowe, D.J., 1992*, The origin of limestone caverns: An inception horizon hypothesis:* Manchester Polytechnic University, Ph.D. thesis, 512 p.

Lowe, D.J., and Gunn, J., 1995, The role of strong acid in speleo-inception and subsequent cavern development, *in* Barani-Kevei, I., ed., Environmental Effects on Karst Terrains: *Acta Geographica* (University of Szeged), v. 34, p. 33-60.

Mazzullo, S.J., and Harris, P.M., 1991, Overview of dissolution porosity development in the deep-burial environment, with examples from carbonate reservoirs in the Permian Basin, *in West Texas Geological Society Symposium Publication*, no. 91-89, p. 125-138.

Mazzullo, S.J., and Harris, P.M., 1992, Mesogenetic dissolution: Its role in porosity development in carbonate reservoirs: *American Association Petroleum Geologists Bulletin*, v. 76, p. 607-620.

Mylroie, J.E., and Carew, J.L., 1995, Karst development on carbonate islands, *in* Budd, A.D., Saller, A.H., and Harris, P.M., eds., Unconformities and porosity in carbonate strata.: *American Association Petroleum Geologists Memoir,* no. 63, p. 55-102.

Ollier, C.D., 1975, Coral island geomorphology—the Trobriand Islands: *Zeitschrift für Geomorphologie*, v. 19, p. 163-190.

Quinlan, J.F., 1978, *Types of karst, with emphasis on cover beds in their classi sfication and development:* Univ. of Texas at Austin, Ph.D. thesis, 325 p.

Palmer, A.N., 1991, Origin and morphology of limestone caves: *Geological Society of America Bulletin*, v. 103, p. 1-21.

Palmer, A.N., 1995, Geochemical models for the origin of macroscopic solution porosity in carbonate rocks, *in* Budd, A.D., Saller, A.H., and Harris, P.M., eds., Unconformities and porosity in carbonate strata: *American Association Petroleum Geologists Memoir,* no. 63, p. 77-101.

Palmer, M.V., and Palmer, A.N., 1989, Paleokarst of the United States, *in* Bosák, P., Ford, D.C., Glazek, J., and Horácek, I., eds., *Paleokarst: A systematic and regional review* Prague and Amsterdam, Academia and Elsevier, p. 337-363.

Pinneker, E.V., 1977, *Problems of regional hydrogeology (Regularities of distribution and formation of groundwaters):* Moscow, Nauka, 196 p. (in Russian).

Pinneker, E.V., 1982, Paleohydrogeologic reconstructions, *in* Principles of hydrogeology-geologic activity and history of water in the Earth's subsurface: Novosibirsk, Nauka, p. 153-188 (in Russian).

Smart, P.L., and Whitaker, F.F., 1991, Karst processes, hydrology and porosity evolution, *in* Wright, V.P., Esteban, M., and Smart, P.L., eds., Paleokarsts and paleokarstic reservoirs: *P.R.I.S. Occasional Publication Ser.*, no. 2, University of Reading, p. 1-55.

Töth, J., 1995, Hydraulic continuity in large sedimentary basins: *Hydrogeology Journal*, v. 3, no. 4, p. 4-16.

Tsykin, R.S., 1989, Paleokarst of the Union of Soviet Socialist Republics, *in* Bosák, P., Ford, D., Glazek, J. and Horácek, I., eds.*, Paleokarst-A systematic and regional review:* Praha, Academia, p. 337-365.

White, W.B., 1988, *Geomorphology and hydrology of karst terrains:* New York, Oxford University Press, 464 p.

Wright, P.L., 1991, Paleokarst: Types, recognition, controls and associations, *in* Wright, V.P., Esteban, M., and Smart, P.L., eds., Paleokarsts and paleokarstic reservoirs: *P.R.I.S. Occasional Publication Ser.* 2, University of Reading, p. 56-88.

Worthington, S.R.H., 1991*, Karst hydrogeology of the Canadian Rocky Mountains:* McMaster University, Ph.D. thesis, 380 p.

Worthington, S.R.H., 1994, The possible importance of sulfur minerals in initiating epigenic caves, *in* Sasowsky, I.D., and Palmer, M.V., eds., Breakthroughs in karst geomicrobiology and redox geochemistry: abs. and field-trip guide, *Karst Water Institute Special Publication,* no. 1, p. 80-82.

Worthington, S.R.H., and Ford, D.C., 1995, High sulfate concentrations in limestone springs: An important factor in conduit initiation?: *Environmental Geology*, v. 25, p. 9-15.

Zajtzev, I.K., and Tolstikhin, N.I., 1971, Classification of groundwaters and rocks-Basis for hydrogeologic mapping and regionalization, *in* Problemy gidrogeologicheskogo kartirovanija I rajonirovanija, Leningrad, p. 5-14. (in Russian).

3.2. Lithologic and Structural Controls of Dissolutional Cave Development

Alexander Klimchouk and Derek Ford

Abstract

This chapter summarizes the important general controls that lithology and geologic structure impose on most cave genesis: rock purity, the presence of interbedded clastic rocks and adjacent or interbedded sequences of sulfates and carbonates, and various kinds of initial porosity, fissures in particular.

Lithologic and structural conditions for speleogenesis evolve throughout sedimentation, eogenesis, mesogenesis, and telogenesis, and change drastically between these stages. Inheritance in the evolution of different kinds of prespeleogenetic porosity causes increasing heterogeneity in their distribution and parameters, which reaches the highest degree after burial at the stage of rock emergence to the shallow subsurface and the surface. The importance of fabric-selective porosity and stratigraphic elements diminishes with time in favor of fissure-network porosity.

Fissures evolve at different stages of the rock evolution. Networks are composed of complex planar and curvilinear surfaces interconnecting in three dimensions, constructed from fissures of various origins, generations, and ages. The initial structural conditions for speleogenesis thus can be highly varied depending at which particular stage speleogenesis commences. Conditions in deep-seated settings favor uniform speleogenetic development, whereas in shallow settings increased heterogeneity in fissure parameters can favor selective development. Modeling of conduit initiation and early development needs to take into account a great variability of initial permeability structures between common geologic environments and evolutionary stages, especially rather dynamic nondissolutional changes of these structures in shallow settings.

Introduction

The dissolutional cave systems that we explore will develop only in certain types of rock under particular hydraulic conditions.

The purpose of this chapter is to summarize the important general controls that lithology and geologic structure impose on most cave genesis. In Chapter 3.3 another author presents perspectives on the significance of bedding or other parting planes and contacts in cave development. Chapter 3.4 then switches the focus to the hydraulic setting. Chapter 5 includes many individual case studies, grouped according to major speleogenetic environments, in which lithology and structure are discussed. Structural controls of the shape of individual conduits and dissolutional microforms are also discussed in Chapter 6.1.

Looking at a great limestone cliff on the seacoast or in a river gorge, we often notice that caves appear to be developing preferentially in particular beds or other patches of the rock. This may be due to favorable variations in the composition of the rock itself. They are considered first in this chapter, under the heading "Factors of Lithology."

More often, when in a large passage inside a cave, we note that it straddles several beds whose slightly different resistances to dissolution are marked by some projection or recession. But much more apparent is the fact that the passage originated in some linear structural feature such as a bedding plane, joint, or fault which guides it.

The location and orientation of most segments of most caves are controlled by such fissures or the intersection of two or more of them. This is discussed at greater length in the second part of the chapter, "Fissures." Evolution of effective porosity in the rock and their fissures during diagenesis is considered in the final section of the chapter.

Factors of Lithology

Carbonate Rocks—Limestone and Dolostone

A popular classification of types of limestone is shown in 3.2 Table 1. Allochthonous limestones are made of calcium carbonate ($CaCO_3$) debris such as shells and skeletons (whole or fragmented) and fecal pellets, cemented by calcium carbonate mud precipitated from warm sea water ("micrite") or larger crystals that result from early dissolution and reprecipitation ("sparite").

They accumulate as detritus in environments ranging from the deep sea (many mudstones) to intertidal flats or even freshwater lake bottoms. Shallow lagoonal types tend to predominate in many well developed karsts, as noted. Here they usually display distinctive layering termed "beds" or "bedding." Autochthonous limestones are buildups of coral, alga, and other immobile marine organisms that secrete and trap calcium carbonate. Dunes and sandbanks form similar mounds, but of inorganic origin.

Some dolostones form as primary precipitates of the mineral *dolomite* ($CaMg(CO_3)_2$), usually in intertidal settings where they may be interbedded with gypsum. More are secondary, formed when limestone is first dissolved and then reprecipitated from water that is enriched in Mg ions, as seawater usually is. The extent of the conversion can range from very little to complete. Most is believed to occur near the sedimentation surface during eogenetic stages. But deeper seated mesogenetic dolomite and even incipitently-ended telogenetic dolomite forms in places where thermal water passes through limestone. Both kinds of rock may also contain nodules, stringers, or thin sheets of chert (precipitated silica), which are much less soluble.

The composition, depositional settings and diagenesis of the carbonate rocks are very varied. As a consequence it is claimed that there are more distinct types of limestone and dolostone than there are of all other sedimentary rocks combined (James and Choquette, 1988). For more details, interested readers should consult some of the many books that are dedicated to them, such as Bathurst (1975), Scholle et al. (1983), Moore (1989), Budd et al. (1995). The karst textbooks by White (1988) and Ford and Williams (1989) also give more details, with emphasis on their significance in karstification.

Evaporite Rocks

Gypsum, anhydrite, and salt are much simpler deposits, accumulating on lake and sea floors, in intertidal zones, and coastal marshlands, as a consequence of evaporative concentration of ions. Most evaporites display a strong layering (bedding) structure as a consequence.

The diagenetic alteration may be profound (for example, gypsum to anhydrite, or recrystallization) and may seal or destroy layering structure in some cases, overprinting other kinds of textural and structural heterogeneity at varying scales. Interested readers may consult some recent reviews for details (Klimchouk, 1996; Klimchouk and Andrejchouk, 1996a).

Rock Purity

This is a very important factor in cave and karst development. A good working picture of the carbonate and calcareous (carbonate-rich) rocks is given in 3.2 Figure 1, p. 56. Calcite is much more soluble than dolomite (see Chapter 4.1). In many karst areas, beds or zones of dolostone are seen to reduce or prohibit cave development. However, in some extensive caves in dolostone the fissuring and hydraulic setting are suitable; generally, this implies comparatively large fissure apertures, high hydraulic gradients, or long periods of dissolution.

Clay minerals and silica are the most common insoluble impurities in the carbonate rocks. Limestone with more than 20 to 30% clay (argillaceous limestones, 3.2 Figure 1) forms little karst, probably because the abundant clay particles clog any tiny precave or protocave voids. Sand grains, being larger, create pore openings between them and so may not obstruct water flow as effectively. Some well-formed caves are known to develop in calcareous sandstone.

However, most limestone and dolostone caves are associated with bulk purities of greater than 90% $CaCO_3$ or $CaMg(CO_3)_2$. Most individual beds or continuous formations of the sulfate and salt rocks do not have enough sand or clay to prohibit caves.

Interbedded Clastic Rocks. The beds of carbonate, sulfate, or salt in many formations are without any clastic interbeds (layers of clay, shale, sandstone, coal) of similar thickness for sequences of 10s to 100s of meters, even 1000's of meters of beds in

Allochthonous limestones original components not organically bound during deposition						Autochthonous limestones original components organically bound during deposition		
Less than 10% >2 mm components				Greater than 10%		By organisms which act as baffles	By organisms which encrust	By organisms which build a rigid framework
Contains lime mud (<.03 mm)			No lime mud	Matrix supported	>2 mm component			
Mud supported		Grain supported						
Less than 10% grains (>.03 mm <2 mm)	Greater than 10% grains							
Mudstone	Wackestone	Packstone	Grainstone	Floatstone	Rudstone	Bafflestone	Bindstone	Framestone

3.2 Table 1. The R.J. Dunham (1962) classification of carbonate rocks, as modified by Embry and Klovan (1971).

3.2 Figure 1. A bulk compositional classification of carbonate rocks by Leighton and Pendexter (1962) (by permission of American Association of Petroleum Geologists).

some carbonate groups. These strata generally yield the greatest cave development. However, many geologic formations are mixtures of beds or packages of beds of the clastic and soluble rocks. These grade to shale and sandstone sequences with limestone, dolostone, or gypsum interbeds, and the hydrogeologic systems grade from wholly karstic to nonkarstic.

Valid generalizations concerning cave development are difficult for intermediate conditions. As the frequency of shale beds increases in a formation, so does the likelihood that the intervening limestone will be argillaceous and thus without caves, but this is not always true.

Groundwater penetration to initiate caves is often easier at a contact between limestone and shale—because of diagenetic effects or differential slippage there—than it is at bedding planes within the limestone. Sandstone interbeds often provide even more effective initial flow paths within an otherwise carbonate sequence.

Dissolutional conduit systems may develop in adjoining sandwiched limestones or sulfates. Just a few meters of pure karst rock beds in a clastic formation may permit caves to form and attain substantial length or depth if the hydraulic setting is favorable. This "stripe karst" is the dominant style of cave development in Norway, for example. Ford (1967) described caves passing through mountains of schist and quartzite in Canada, contained within a few meters of marble. Another outstanding example is Botovskaya Cave in Siberia, described in Chapter 5.2.3.

Adjacent or Interbedded Sequences of Sulfates and Carbonates. These may influence fluid flow and speleogenesis in several ways:

- Because carbonates have higher initial porosity, they commonly provide an access for fluids to the more soluble but compact sulfates. Bedding planes (especially unconformities) between carbonates and sulfates are particularly important.
- Sulfates are much more readily dissolved in deep settings than carbonates, so dissolution and speleogenesis in sulfate beds becomes a factor in conveying fluids to adjacent carbonates. Collapse of discrete cavities causes local deformation and fracturing in overlying carbonates, and under certain conditions, will trigger the upward stoping of vertical structures termed *breccia pipes* or *geologic organs* that can become important elements in improving cross-formational fluid flow.
- Thin sulfate beds within carbonates or intercalated sequences can be entirely dissolved over large areas, producing prominent flow paths that may influence later speleogenesis. Dissolution of thicker gypsum or salt can remove the structural support from overlying carbonates, causing disturbance that ranges from enhanced fissuring to complete brecciation, depending on the rate and pattern of leaching and the thickness of the leached rock. It may greatly affect subsequent speleogenesis.

Such deformation and brecciation are widely reported (see for example the discussions in Klimchouk and Andrejchouk, 1996b; Ford, 1997; Friedman, 1997; James and Choquette, 1988; Palmer, 1995), although it is sometimes misinterpreted as results of cavern collapse in carbonate rocks alone.

Porosity in the Karst Rocks

The *porosity* of a given rock is the aggregate amount of void space in it, normally expressed as a percentage of the total volume. *Effective porosity* is that which is sufficiently interconnected that it can be drained under given physical conditions such as by gravity.

Matrix Porosity. It is common to view the fabric selective bulk porosity as being of negligible importance for speleogenesis. This is supported by observations that very porous but little fissured young limestones in exposed settings generally do not favor conduit development. An obvious exception is fresh water/sea water mixing speleogenesis, which relies on location at the contact between contrasting hydrogeochemical zones within a porous aquifer rather than on any specific structural prerequisites.

Initial differences in matrix porosity and its subsequent diagenetic redistribution and modification can play an important role in the guidance and focusing of fluid flow along certain horizons or zones. This further increases heterogeneity because of a kind of feedback mechanism between diagenesis and the flow pattern. Heterogeneity of porosity and permeability due to diagenetic processes particularly increases during the periods of transition from basinal circulation of connate and resurgent fluids to meteoric groundwater circulation, with corresponding changes in hydrogeochemistry.

Under favorable geochemical and hydrodynamic conditions—particularly where cross-formational hydraulic communication permits waters with contrasting chemistry to mix—this may lead to conduit inception and development, either along the mixing zones themselves or in adjacent more compact and fissured zones. Thus, enhanced heterogeneity in matrix porosity, either primary or due to diagenesis, can form types of structural guidance for subsequent conduit inception and development.

The development of macroscopic porosity of any kind can influence diagenetic processes in the rock matrix. For instance, Cander (1995) demonstrated that in the quite young carbonates of the Floridan aquifer, the development of conduit flow reduces diagenetic alterations in the matrix because water-rock interaction becomes focused in the conduits and decreased elsewhere. This changing relationship between matrix porosity and focused diagenesis around conduits (the latter falling into the realm of speleogenesis) is an example of a widespread characteristic of the hydrogeologic evolution of karst aquifers over time—that is, a decrease of the proportion of diffused flow in favor of conduit flow.

Porosity in diagenetically immature carbonates ranges between 25 and 80%. In well indurated carbonates it is greatly reduced and tends to be positively correlated with grain size and textural heterogeneity, although there is much variation. Porosity of micrite is generally less than 2%, that of sparite between 5 and 10%. Dolomitization increases porosity to 5 to 15% in most instances. The porosity of metamorphosed carbonates is low, being less than 1% in marbles.

Evaporites anneal readily and so tend to have negligible bulk porosity. The thermodynamic stability and solubility of gypsum and anhydrite are greatly affected by changes in the physical and chemical parameters that occur within common geologic environments. This is why conversions of gypsum to anhydrite and back to gypsum are common processes during diagenesis, as well as their recrystallization. Gypsum may acquire a considerable intercrystalline porosity as a result of these processes. The more important consequence is that substantial heterogeneity in rock texture, structure, and hence mechanical strength may develop across an individual bed or unit, which can cause networks of fissures to form within it that are largely independent of those in the adjoining beds (Klimchouk et al., 1995).

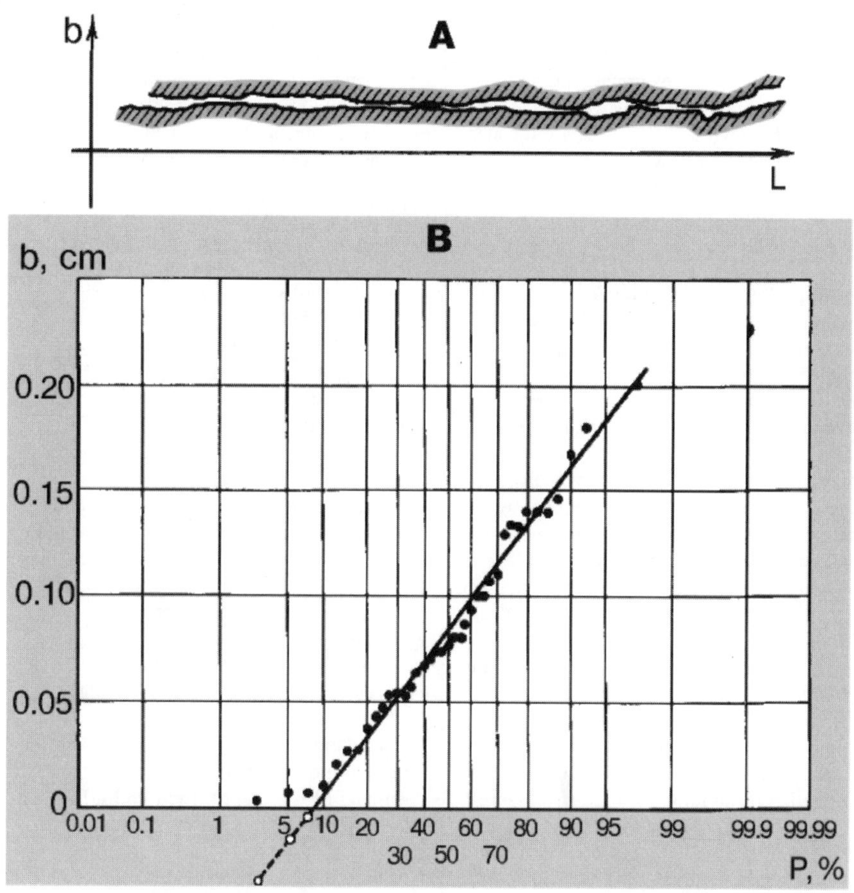

3.2 Figure 2. Distribution of fissure width: A is a sketch of a fissure profile. B is a probability plot (from Chernyshev, 1983).

Fissures

Bedding planes, joints, and faults are planar breaks that serve as the principal structural guides for groundwater flow in almost all karstified rocks. In hydrogeology it is customary to categorize all of them together as *fractures* forming *fracture aquifers*. This is somewhat misleading, however, because most bedding planes originate as primary depositional features in the rock rather than as later fractures through it. Here we use fissures as a general term. In common use it refers to breaks that are visibly open, or gaping. Here we use it:

> to encompass all planar breaks that can be significantly penetrated and modified (by dissolution or precipitation) by circulating groundwaters, past or present.

Structural engineers concerned with slope stability often apply the term *penetrative discontinuities* to such features (e.g. Cruden 1976).

Except where they are truly gaping, the physical nature of these breaks is complex and difficult to measure. In quantitative modeling it is standard practice to treat the hypothetical fissure as a fracture with strictly parallel walls spaced a fixed distance apart. In reality this will never apply. Separation in the fissure is spatially most variable, with some areas of it usually being completely closed and so prohibiting groundwater flow. For speleogenetic purposes, it is best to think in terms of the water in a fissure passing from more widely open areas to others down the hydraulic gradient via constrictions or *throats* highly variable in form and dimension.

Minimum throat apertures for cave development will usually exceed 10 micrometers, and are probably greater than 100 micrometers in most instances. In 3.2 Figure 2 is an example of apertures measured in sample joints given in Chernyshev (1983). The area of a fissure surface that is completely closed (i.e., there is bedrock-to-bedrock contact with no aperture) is estimated to be no more than 1%, at least to a depth of several hundred meters.

The genetic importance of the different types of fissures summarized below can also vary a great deal between (and even within) individual caves. In some, bedding planes or contacts are quite predominant, and joints or faults play a very subordinate role.

Mammoth Cave, Kentucky, (Chapter 5.3.8) is a prime example. Joint control is paramount in rectilinear maze caves such as those of the Ukraine (Chapter 5.2.1). Caves that are guided largely or entirely by faults seem to be comparatively rare (Grotte di Castellana, Italy, is an example) but fault planes are important in many. Perhaps a majority of the world's larger meteoric-water caves (Chapter 5) follow combinations of these different types in differing proportions, plus their intersections (e.g. joints with bedding planes).

Bedding Planes and Contacts

Bedding or parting planes in sedimentary rocks are produced by some change in sedimentation or an interruption of it. The change may be minor, e.g. from one size of carbonate grain to another a little bigger. Major changes are represented by big differences in grain size or homogeneity and, more often, by the introduction of clay by a storm or flood that leaves a paper-thin or thicker parting between the successive layers of karstic rock. Interruption is usually a brief marine emergence with some erosion and the start of meteoric diagenesis or beachrock formation, before there is submergence again.

Many of the smaller subaerial unconformities that occur in young rocks tend to be diminished during diagenesis because they are preferred sites for mineral infilling or for obliteration by stylolite formation (pressure solution). The more prominent interruptions persist as minor geologic disconformities and are widely used to divide layered sequences into individual *formations*, or *members* within formations. These may be referred to as *contacts*. The junctions between layered sediments and mound structures such as reefs in carbonate deposits are another type of contact that is often preferentially penetrated by groundwaters.

Only a minority of sedimentologic bedding planes are penetrable by groundwater under reasonable pressure gradients so as to function as fissures to guide cave initiation and development. The rock between any pair of such planes thus constitutes an individual bed for speleogenetic purposes (although some impenetrable planes within it may later rupture under mechanical stress in a breakdown). A standard sedimentologic classification of bed thickness (i.e. of bedding plane separation) and of joint spacing in beds or sequences of beds is given in 3.2 Table 2.

The areal extent of penetrable bedding planes or contacts varies considerably. Where bedding is thin to very thin, they may cover only a few square meters. Where it is medium to thick, the extent is normally 10^3 to 10^6 m^2 or much more. Truly major planes can be followed throughout a formation, sometimes for hundreds of kilometers. As a consequence, major bedding planes can be considered to be *continuous* entities when solution caves are propagating through them, whereas joints and most faults are *discrete* (they terminate in comparatively short distances). This enhances the significance of bedding planes in cave genesis (Ford, 1971).

Bedding planes and contacts most readily exploited by groundwater include those with substantial depositional disconformities, plus planes with shale laminae or thicker partings (often with disseminated pyrite) and planes with nodules or sheets of chert. Perhaps the most important are those that have served as surfaces of differential slippage during tectonic events. Even if the slippage is just a few centimeters there is some slickenside striation and brecciation that enhances openings. Most steeply tilted and all folded strata will display some measure of differential slip.

Where an overburden load is removed abruptly, previously impenetrable bedding planes may open as a consequence of pressure release. The retreat of a thick glacier is the principal unloading mechanism. The opened planes have been termed *bedding joints* or *sheeting fractures* by hydrogeologists. However, the latter term should more properly be limited to the less equivocal sheeting in plutonic rocks or in limestone around quarries. Pressure release opening appears to be limited to the top one meter of strata or less in the carbonate rocks of Ontario, which were buried by 1500-2000 m of ice during the last glaciation.

For caves of enterable size to develop, the bedding must normally be medium to massive. In thinly bedded rocks the dissolutional attack is too dispersed amongst the closely spaced planes and densely packed joints that interlink them. They also lack the mechanical strength to support large void spaces without breakdown. However, many cyclic depositional sequences in which bedding thickness varies in a systematic manner yield narrow packages of thin beds between greater units of massive ones. Penetration and cave initiation then often is in the thin package (3.2 Plate 1, p. 61).

Bed thickness (cm)		Joint Spacing (cm)	
100-1000	very thick bedded or massive	> 300	very wide
30-100	thick	100-300	wide
10-30	medium	30-100	medium
3-10	thin	5-30	close
1-3	very thin	< 5	very close
< 1.0	laminated		

3.2 Table 2. Terminology for bed thickness and joint spacing (from Ford and Williams, 1989).

Joints and Shear Fractures

Joints are simple pull-apart breaks in previously consolidated (or partly consolidated) rocks. In shear fractures there is some additional lateral or vertical displacement but it is too small to be recognized in the field or in hand specimens (Twiss and Moores, 1992). For all practical purposes they can be treated as joints. A general scale classification by Chernyshev, (1983) that stresses the continuum from microjoints through joints and shear faults to faults is presented in 3.2 Table 2. Veins are fractures filled with secondary minerals, normally calcite or quartz in the carbonate rocks; smaller ones are usually impenetrable for modern groundwater.

A joint is initiated by a rupture within a bed that rapidly extends to the full depth of the bed (or package of similar beds) and outwards in it. The two joint surfaces display hackling and cusps around the initial breakage and plumose (fan-like) striations away from it (Barton and Stephansson, 1990). But in the karst rocks this fine structure is usually destroyed by dissolution in any penetrable joints. Rupture in sedimentary rocks can be caused by a wide variety of effects, including weathering or shrinkage during early subaerial exposure, loading, compression, and heating during burial or magmatic intrusion, tectonic forces, and unloading, expansion, and cooling during uplift and erosion. Any of these effects can also extend preexisting joints and shear fractures. The consequence is that their extent, geometric patterns, and density can be very varied at all scales, local, regional, and continental. These characteristics vary also with time in the course of evolution of joint networks.

Historically, joints and shear fractures were classified first by their orientation, then by age based on their crosscutting relationships, and more recently by their relationships to diagenesis and microtectonics. Most karst areas are dominated by patterns of straight joints with identical or closely similar orientations. These are *systematic* joints, and those of the same or similar orientation constitute a *joint set*. Where two or more different sets intersect at constant (or nearly constant) angles they form a *joint system*. In most platformal sedimentary rocks, including limestones and dolostones, a rectangular system of tension jointing is the most common (e.g. 3.2 Plate 2, p. 62). Some authorities go so far as to describe it as "universal" in these rocks (Dennis, 1987). A second system with 60/120 degree intercepts is also very common, associated with compression and shear fracturing. Combination of these two systems is also frequent and creates more complex patterns (e.g. 3.2 Plate 3, p. 62). Curving joints are *nonsystematic* and are normally secondary in origin, terminating at the nearest systematic joint. In folded rocks strike-oriented joints are usually dominant along the fold crests and troughs, with mixtures of dip, strike, and 60 degree shear systems occurring on the limbs. More than 90% of measured joints make perpendicular intersections or terminations with the bedding planes, but oblique intersections are often found at fold crests and troughs.

The length and depth of joints and shear fractures range over several orders of magnitude, as indicated in 3.2 Tables 2 and 3. In all bedded sedimentary rocks, the size and spacing of individuals usually is closely and positively—but with some retardation—correlated to the thickness of the bed. This relation was noted by Bogdanov (1947), confirmed by later workers, and generalized by Chernyshev (1983) in a form $a = rMk$, where a is a distance between joints, M is a bed thickness, and r and k are empirical indexes. Thin beds thus have a very high density of joints. Where these joints are penetrable, the rock is rapidly removed by dissolution underground or reduced to rubble if subject to subaerial weathering.

Under increasing stress (e.g. from increasing load) the larger fractures tend to extend and deepen. In medium- to massively-bedded karst rocks particular sets can lengthen to hundreds of meters and breach many successive beds. They are often termed *master joints*. Where they occur, they will usually come to exercise important controls on karst development. (3.2 Plate 4, p. 63).

Chernyshev (1983) has developed a classification that links joint patterns to the stress fields that exist or have existed in the rock mass (3.2 Figure 3, p. 60). As suggested above, systematic or complex network patterns (Chernyshev's Type III and IV) predominate in most karstlands and karst aquifers. Spheroidal (Type I) patterns are associated chiefly with igneous intrusions, but are found occasionally in limestone patch and pinnacle reefs.

Polygonal patterns (Type II) are also most important in igneous or metamorphic rocks that have experienced considerable cooling but are quite common in thin- to medium-bedded fine-grained limestones, where they can resemble crazy paving. Polygonal networks also occur frequently in gypsum rocks. In all types, joint patterns are said to be *continuous* where most or all individuals in the pattern are extended until they terminate at others of the same or greater age (i.e. the network is well interconnected), and *discontinuous* to *fragmented* where the interconnection is incomplete.

Classes by length		Parameter values, cm	Classes by width	
Large faults		10^8		
		10^7		
		10^6		
Faults		10^5		
		10^4	Zones of crush breccia	
	Long	10^3		
Joints and shear faults	Medium	10^2		
	Short	10	Crevices	
		1		Wide
		10^{-1}	Joints	Medium
Microjoints		10^{-2}		Narrow
		10^{-3}		
		10^{-4}	Capillary	
		10^{-5}		
		10^{-6}		
			Subcapillary	

3.2 Table 3. Classification of fissures by length and width (from Chernyshev, 1983).

Faults

Faults are fractures with some visible displacement of the rock walls up, down, or laterally. At the extremes, displacement can exceed 1 km vertically and 100's of km laterally. *Normal faults* are pull-apart tensional features like joints. Wide openings are possible and can exist even at great depths, because the displacement prohibits complete closure of the ruptured surfaces. However, the fault planes are commonly filled with breccia or secondary calcite.

Reverse and *lateral* or *transcurrent faults* are compressional features and can be impenetrably tight in small examples. But brecciation, slickenside grooving, and displacements can permit flow through others.

Thrust faults are low-angle reverse faults. They are often particularly important because they are really extensive (the steep-to-verticalc normal and reversed faults are not) and so can emulate very penetrable bedding planes in their capacity to host interconnected solution conduits. In regions of mild tectonic

Type of joint network	Graphic appearance	Stress field configuration that generates joints	Anisotropy of a massif generated by a fissure network
Spheroidal **I**		$\sigma_1 = \sigma_2 = \sigma_3$	Isotropic
Polygonal, axis-symmetric **II**		$\sigma_1 = \sigma_2 > \sigma_3$	Transverse-isotropic
Systematic, equiangular **III**	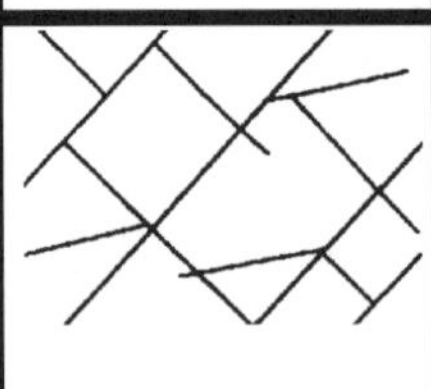	$\sigma_1 > \sigma_2 > \sigma_3$	Anisotropic
Complex, asymmetric **IV**		Changing stress field configurations through time	Varying

3.2 Figure 3. Geometrical classification of joint networks (Adopted from Chernyshev, 1983).

activity it is common to see thrust faulting originate in one slipped bedding plane, pass upward through a few beds as a gentle curvilinear surface, and terminate in a second slipped plane.

Large faults are rarely represented by single fracture surfaces. Usually lesser faults feather off them at acute angles as a consequence of the wrenching of the rock. Shear fractures (discussed above) are often oriented parallel to or close to the feathering breaks.

Fracture traces (or *linears, lineaments*) are narrow linear trends detectable on high altitude and satellite photography. Most karst landscapes display them. On the ground they are zones of closely spaced high-angle faults of minor displacement, plus their feathering fractures and associated shear fractures.

The hydrogeologic and speleogenetic role of faults varies with their type, size, and the diagenetic record since they were formed. At one extreme they constitute the predominant fissure in a groundwater basin, like a trunk river channel on the surface. Large normal and reversed faults often have quite low permeability, however, due to the presence of cataclasite or mylonite (clayey crush fillings) or calcite precipitates in them. Instead, there is higher permeability in the fracture trace zone of feathering fissures, which can be up to several hundred meters wide.

For example, Dublyansky and Kiknadze (1984) summarized hydrogeologic data from tunnel construction works and drilling in carbonate mountain massifs showing that most water intakes occur in the feathering fissure zones, particularly (up to 90%) from the downthrown blocks. In many caves the fault planes themselves only control local passage segments, although the overall trend of cave development may follow the fault-trace zone. In intermediate cases cave systems extend between fault zones and beyond them, sometimes utilizing them locally and at other places being barred or deflected by them. The North of Lake Thun System in Switzerland (Chapter 5.3.3) provides a very good example.

At the other extreme, the fault and its zone function as an underground dam, prohibiting part of or all groundwater flow along or across it and dividing the aquifer into separate compartments that are more or less watertight. Where there has been great displacement on a fault, karst and nonkarst rocks can be juxtaposed across it, which also limits or prohibits groundwater recharge or discharge, and the cave development associated with it.

In artesian basins and disrupted basins, faults and fracture-trace zones often play an important role in achieving vertical hydraulic communication between different aquifers that are separated by aquiclude formations.

Lithology, Fissuring, Speleogenesis, and the Diagenetic Stages in Karst Rocks

The principle of *inheritance* is important in geology. It implies that new features or conditions developing in a rock may be guided by previous conditions to a significant extent. For speleology we can distinguish:

- *Posthumous inheritance*, when the new features inherit and adapt the structure of quite different previously fossilized features.
- *Direct inheritance*, when the development of a particular feature may be suspended for long periods of time but continues in the same style each time that it is reactivated.

The porosity and effective porosity of karst aquifers can be highly dynamic over geologic time. All categories of rocks, not merely the soluble ones, experience significant alterations of porosity during their lifespans. What sets the karstifiable rocks apart from others are the dynamics and magnitude of the alterations. Their nature and sequence (inheritance) may be fundamentally important in the development of a cave. To overview it, we follow the diagenetic stages proposed by Choquette and Pray (1970) outlined in Chapter 3.1.

Eogenetic Stage

A sediment acquires a "fabric-selective" porosity (such as spaces between settled grains or shell fragments) when it is formed that is commonly 40 to 80% in carbonates and 5 to 10% in gypsum. Early diagenetic processes such as compaction, cementation, dissolution, and replacement eventually lead to considerable overall decrease in porosity, although fabric-selective dissolution can locally enhance it.

In addition to the different depositional environments (from the lacustrine or intertidal to the deep marine), the subsequent history may vary widely, from no subaerial exposure,

3.2 Plate 1A. This is an an example of cyclic deposition on a carbonate platform. The alternation of massive and thin beds seen in the illustration is a common feature of such deposits. Fine but impenetrable bedding planes can be seen within the massive beds. These particular rocks are fully dolomitized, as is suggested by the abundance of vugs in the massive beds. Ordovician dolostones, Northwest Territories, Canada.

3.2 Plate 1B. This is cyclic thick-to-thin limestone beds exposed in a river cave passage in Ontario, Canada. Initial cave penetration was in the package of thin beds in the roof, succeeded by entrenchment and lateral undercutting. (Photos by D.C. Ford).

to intermittent exposure alternating with further deposition, to subsequent uplift and continuous exposure, or to permanent burial to various depths. All these variations may introduce different rock fabrics or other features. Ultimately, the eogenetic stage in most important karst rocks concludes with their deep burial.

An important effect of early diagenesis is the development of considerable vertical heterogeneity. This is particularly pronounced in cases of intermittent exposure and deposition on carbonate platforms, where complex sequences of indurated and more porous horizons can be created by case hardening, cementation, and selective dissolution in various zones of migrating freshwater lenses (e.g. Beach, 1995).

A wide range of syngenetic features can be formed and preserved along subaerial unconformities, to influence subsequent fluid flow and speleoinception. The contraction or subsidence fracturing due to desiccation, dehydration, or shrinkage that may occur in fresh sediment may be of particular importance. Although such fissures are commonly filled and sealed with sediment during the next depositional cycle, they can introduce significant "hidden" discontinuities into a bed that may focus interstitial flow or guide rock fracturing in the later stages of diagenesis.

Fabric-selective dissolution may also enhance porosity in deep marine depositional environments (Budd et al., 1995). Local synsedimentary tectonic movements may deform original stratification, causing changes in thickness and complications in facies pattern, thus introducing considerable heterogeneity in porosity. Sequence boundaries and unconformities in bedding can constitute important planes for conduit initiation and development (see Chapter 3.3 for more details).

Large caves can develop rapidly in young limestones, as consequences of marine mixing corrosion or mechanical washout of loose grains inside of or beneath cemented case hardening; details are given in Chapter 5.1.

Mesogenetic Stage

Mesogenesis refers to the burial of a young rock with its depositional fluid and its evolution until later uplift and erosion of overlying rocks exposes it to the circulation of meteoric water (Chapter 3.1). Depth of burial may be many kilometers and the duration of burial from millions to thousands of millions of years.

3.2 Plate 2. Thick-bedded platformal limestone dissected by rectangular joint systems. Ordovician rocks in the Fourth Chute, Bonnechere River, Ontario, Canada (Photo by D.C. Ford).

3.2 Plate 3. Complex joint patterns in Ordovician dolostone beds, Northwest Territories, Canada (Photo by D.C. Ford).

During the subsiding phase of mesogenesis, bulk porosity tends to decrease in most diagenetically immature rocks, although some clastic sediments (many sandstones for instance), do not experience as much as finer-grained deposits such as clays or carbonate mudstones. Carbonates that formed in shallow environments with intermittent exposure often undergo a considerable amount of induration before burial, so that they may preserve the bulk porosity and conduits acquired during eogenesis. Contrary to the belief of many geologists, conduits and voids of substantial size can survive burial of several kilometers. However, further decrease in matrix porosity generally occurs due to cementation processes. To a considerable extent, the heterogeneity inherited from the eogenetic stage determines the patterns of interstitial fluid flow during burial. Heterogeneity in matrix porosity, bedding planes with impurities, unconformities with changes of lithology across them, minor clastic beds, paleokarst conduits inherited from the eogenetic stage—all these sedimentologic, diagenetic, and karstic features may serve as preferential paths for fluid during mesogenesis and so contribute to later speleogenesis (see Chapters 3.3 and 5.2).

Fracturing, which usually commences during eogenesis, may continue through mesogenesis, depending on the tectonic setting and evolution of a given basin. Some late Proterozoic-early Paleozoic carbonate sequences experienced four to five different episodes of major tectonic activity while still within the mesogenetic stage. Even in rather stable cratonic regions, subsidence and uplift may differ both spatially and over time. This can impose considerable stress and cause fissuring, particularly where there is folding or the reactivation of ancient faults in the basement.

Inheritance is important; particular bedding planes either retain or develop preferential penetrability, and their separation can determine the frequency, continuity, and other parameters of joints and shear fractures developing within and across them later on. Preexisting joint sets that are favorably oriented with respect to the new tectonic stress fields may extend and deepen. Even contraction and subsidence cracking that was sealed during sedimentation may provide weaknesses and other framework for new fracturing when the new stress fields are applied.

Occurrence of clastic interbeds within soluble rocks, or alternating sulfates and carbonates, present special structural conditions to guide fluid flow and to favor speleogenesis in a deep-seated setting. Some important features have been briefly addressed above. This issue is treated in more detail in Chapters 3.4 and 5.2.

The effect of increasing depth (increasing lithostatic pressure) upon the closure of fissures is frequently overestimated. Fissure surfaces are irregular, the "closed" areas where their walls directly contact each other due to the crushing or shearing of projections is commonly less than 1% even at depths of several hundred meters, as noted. Bedding planes are more prone to closure than vertical fractures, but there is much drill-core evidence of open planes at depths up to several kilometers.

Even in very deep rocks, fissures tend to become open during the uplift phase of mesogenesis, when considerable tensile stress can develop. The likelihood of a fissure closing and becoming impenetrable also depends on its history. Fresh (young) fissures can be shut tight more effectively because the relief in their opposite walls is quite precisely mirrored. Fissures that developed due to tectonism, unloading, or weathering during uplift and exposure will tend to stay partly open during successive burials, even to great depths, because the relief of the walls has lost much of its concordance due to diagenetic (particularly dissolutional) processes.

In summary, during mesogenesis, effective

3.2 Plate 4A. This is a "master joint set" plus lesser joints in rectangular contact with it. This master set dominates all formations (Ordovician thick -to-thin-bedded limestones plus minor shales) on Anticosti Island, Quebec, which is 200 km in length and 50 km wide. The set is oriented at 110-290 degrees, parallel to the Grenville Front, a Proterozoic plate tectonic suture in eastern Canada. (Photo by D.C. Ford).

3.2 Plate 4B. This is an example of widely spaced joints in the standard rectangular intersection with bedded carbonate rock. Note scale figures standing and seated at center left on the top of the quarry face. (Photo by A.N. Palmer).

porosity and lithologic and structural characteristics can change considerably. Fissure networks may develop and become more complicated through time, particularly during changes of tectonic regime and upon uplift. Preconditioning processes occur widely, and caves can develop under certain favorable hydrodynamic and geochemical conditions (see Chapter 5.2).

Telogenetic Stage

When a rock rises to the surface due to tectonic uplift and denudational removal of the cover beds, three main sets of factors influence further development of porosity and fissuring:

- Increasing circulation of meteoric water, which flushes out basinal brine, with corresponding changes in fluid chemistry and gas composition.
- Unloading and stress release.
- Weathering.

The first set of processes is addressed in Chapter 5.2 in this volume. The effects of the latter two is discussed below.

When a rock is approaching the surface and the overburden load decreases, stress fields in it can be substantially rearranged and stresses relieved. This can change the forces applied to existing fissures and permit them to extend or new fissures to form between them. The depth at which such changes commence, and their character depend on the rate of relaxation (which, in turn, depends on the regional and local geologic structure and tectonic regime) and on the balance between the uplift and the denudational lowering of the surface.

In general, when the uplift and denudation are at a low rate balance, stress-release effects commence at greater depths, although the effects are biased toward increase in fissure width rather than the formation of additional new fissures. When high rates balance, or the balance is shifted toward uplift, stress-release effects will express at shallower depths and can produce more newly formed fissures. Chernyshev (1983) suggests that changes in fissuring characteristics become quite prominent at depths of 15 to 30 m, although they commence gradually at greater depths, probably 100 to 150 m or even more in some instances.

Due to stress release, the first and foremost changes will affect fissure sets that are subparallel to the surface. Fissure apertures open toward the surface following an exponential law, and in some circumstances, the spacing of penetrable fissures in a set decreases obeying a hyperbolic law (Chernyshev, 1983). The range of aperture widths increases considerably. Fragmented or discontinuous joint networks may become continuous. These changes are less effective in sets that are not parallel to the surface. In low-relief areas with low dip, bedding planes are most affected, along with joint sets which are close to and parallel with incised valleys (Renault, 1968).

In high relief areas, particularly in folded mountain regions with high uplift rates, almost all sets of fissures are affected, and the overall changes in fissure network parameters are most dramatic although spatially irregular, as is the relief.

Weathering strongly affects fissure networks. The most common effects include the widening of fissures and an increase in their frequency in existing sets. New fissures may form, including some with unusual orientations: for example, Pluhar and Ford (1970) showed that there is cracking normal to stylolite seams, with new karren forming along the cracks. Changes induced by weathering are more pronounced and penetrate to greater depths along existing discontinuities. Bulk porosity and permeability commonly increase in exposed carbonates, although the extent also depends on physicochemical properties of the material and on climatic conditions. The changes due to unloading and weathering are very important in the formation of epikarst (see Chapter 3.5). In contrast to most carbonates, exposed gypsum tends to form a sealed crust along the surface rather than a higher permeability zone, especially in dry climates (Klimchouk, 1996).

Conclusions

The lithologic and structural controls of speleogenesis are complex, especially in the carbonate rocks. They evolve throughout sedimentation, eogenesis, mesogenesis, and telogenesis and can change drastically between these stages.

Inheritance in the evolution of different kinds of prespeleogenetic porosity causes increasing heterogeneity in its distribution. The importance of fabric-selective porosity and stratigraphic elements tends to diminish with time in favor of fissure-network porosity. Networks are composed of complex planar and curvilinear surfaces interconnecting in three dimensions, constructed from fissures of various origins, generations, and ages.

The initial conditions for speleogenesis can thus be very varied. Conditions in deep-seated settings may tend to favor uniform speleogenetic development, whereas in shallow settings increased heterogeneity in the fissure parameters can favor selective development. Modeling of conduit initiation and early development needs to take these considerations into account.

References

Barton, N., and Stephasson, O., eds., 1990, *Rock Joints:* Rotterdam, A.A. Balkema, 994 p.

Beach, D.K., 1995, Controls and effects of subaerial exposure on cementation and development of secondary porosity in the subsurface of Great Bahama Bank, *in* Budd, A.D., Saller, A.H., and Harris, P.M., eds., Unconformities and porosity in carbonate strata: *American Association of Petroleum Geologists Memoir,* no. 63, p. 1-34.

Bogdanov, A.A., 1947, Dependence of cleavage intensity on the bed thickness: *Soviet Geology*, v. 6, p. 32-41 (in Russian).

Budd, A.D., Saller, A.H., and Harris, P.M., eds., 1995, Unconformities and porosity in carbonate strata: *American Association of Petroleum Geologists Memoir,* no. 63, 306 p.

Cander, H., 1995, Interplay of water-rock interaction efficiency, unconformities, and fluid flow in a carbonate aquifer: Floridan aquifer system, *in* Budd, A.D., Saller, A.H., and Harris, P.M., eds., Unconformities and porosity in carbonate strata: *American Association of Petroleum Geologists Memoir,* no. 63, p. 103-124.

Chernyshev, S.N., 1983, *Fissures in rocks:* Moscow, Nauka. 240 p. (in Russian).

Choquette, P.W., and Pray, L.C., 1970, Geologic nomenclature and classification of porosity in sedimentary carbonates: *American Association of Petroleum Geologists Bulletin,* no. 54, p. 207-250.

Cruden, D.M., 1976, Major rock slides in the Rockies: *Canadian Geotechnical Journal*, v. 13, p. 8-20.

Dennis, J.G., 1987, *Structural Geology:* Dubuque, Wm. C. Brown Publishers, 448 p.

Dubljansky, V.N., and Kiknadze, T.Z., 1983, *Hydrogeology of karst of the Alpine folded region of the south of the USSR:* Moscow, Nauka, 125 p.

Dunham, R.J., 1962, Classification of carbonate rocks: *American Association of Petroleum Geologists Memoir,* no. 1, p. 108-121.

Embry, A.F., and Klovan, J.E., 1971, A Late Devonian reef tract in northeastern Banks Island, Northwest Territories: *Canadian Petroleum Geology Bulletin*, v. 19, p. 730-781.

Ford, D.C., 1967. The sinking streams of Mt Tupper—a remarkable karst in Glacier National Park, B.C. Canadian Geographer, v. 11, no. 1, p. 49-52.

Ford, D.C., 1971, Geologic structure and a new explanation of limestone cavern genesis: *Cave Research Group of Great Britain Transactions,* v. 13, p. 81-94.

Ford, D.C., 1997, Principal features of evaporite karst in Canada: *Carbonates and Evaporites*, v. 12, p. 15-23.

Ford, D.C., and Williams, P.W., 1989, *Karst geomorphology and hydrology:* London, Unwin Hyman, 600 p.

Friedman, G.M., 1997, Dissolution-collapse breccias and paleokarst resulting from dissolution of evaporite rocks, especially sulfates: *Carbonates and Evaporites*, v. 12, p. 53-63.

Klimchouk, A.B., 1996, Dissolution and conversions of gypsum and anhydrite, *in* Klimchouk, A., Lowe, D., Cooper, A., and Sauro, U., eds., Gypsum karst of the world: *International Journal of Speleology*, v. 25, no. 3-4, p. 21-36.

Klimchouk, A.B., and Andrejchouk, V.N., 1996a, Sulphate rocks as an arena for karst development, *in* Klimchouk, A., Lowe, D., Cooper, A., and Sauro, U., eds., Gypsum karst of the world: *International Journal of Speleology*, v. 25, no. 3-4, p. 9-20.

Klimchouk, A.B., and Andrejchouk, V.N., 1996b, Breakdown development in cover beds, and landscape features induced by intrastratal gypsum karst, *in* Klimchouk, A., Lowe, D., Cooper, A., and Sauro, U., eds., Gypsum karst of the world: *International Journal of Speleology*, v. 25, no. 3-4, p. 127-144.

Klimchouk, A.B., Andrejchouk,V.N., and Turchinov, I.I., 1995, *Structural prerequisites of speleogenesis in gypsum in the western Ukraine:* Kiev, Ukrainian Speleological Association, 104 p.

Leighton, M.W. and Pendexter, C., 1962, Carbonate rock types. *American Association of Petroleum Geologists Memoir,* no. 1, p. 33-61.

Moore, C.H., 1989, Carbonate diagenesis and porosity: developments in Sedimentology 46. New York, Elsevier, 338 p.

Palmer, A.N., 1995, Geochemical models for the origin of macroscopic solution porosity in carbonate rocks, *in* Budd, A.D., Saller, A.H., and Harris, P.M., eds., Unconformities and porosity in carbonate strata: *American Association of Petroleum Geologists Memoir,* no. 63, p. 77-101.

Pluhar, A., and Ford, D.C., 1970, Dolomite karren of the Niagara Peninsula, Ontario, Canada: *Zeitschrift für Geomorpholgie*, v. 14, p. 392-410.

Renault, P., 1968, Contribution a l'étude des actions mechaniques et sedimentologiques dans la speleogenese: *Annales de Spéléologie,* v. 22, p. 5-21, 209-267; v. 23, 259-307, 529-596, v. 24, p. 313-337.

Scholle, P.A., Bebout, D.G., and Moore, C.H., eds., 1983, Carbonate depositional environments: *American Association of Petroleum Geologists Memoir,* no. 33, 365 p.

Twiss, R.J., and Moores, E.M., 1992, *Structural Geology:* New York, W.H. Freeman and Co. 532 p.

3.3. Role of Stratigraphic Elements in Speleogenesis: The Speleoinception Concept

David J. Lowe

Abstract

Inception, the earliest phase of cave development, may begin during diagenesis. Within sedimentary rock sequences inception is generally related to specific favorable horizons or zones within the rock mass. These relatively thin "inception horizons" tend to display atypical chemical and physical properties compared to the primary properties of the bulk of potentially cavernous rock successions. Commonly they correspond to depositional breaks or interruptions, particularly to boundaries between major depositional cycles.

Thus, according to the *Inception Horizon Hypothesis*, inception in sedimentary sequences (as typified by carbonate rocks) is inevitably related to and guided by thin relatively impure layers within thicker otherwise pure beds, or at boundaries between impure and pure lithologies. Growth of incipient voids occurs potentially across the full lateral extent of inception horizons, generally very slowly during extended time scales. Growth may progress simultaneously at more than one stratigraphic level in a sequence under deeply buried, confined, or artesian conditions. Voids along individual inception horizons can be linked hydrologically by others that form concurrently or subsequently along tectonic or lithogenetic fissures. Later, "interference" between the imprinted inception framework and evolving surface landscapes leads to structurally advantageous elements of the potential three-dimensional network being selected, linked, and enlarged to form the skeletons of developing cave systems.

The Necessity For Inception

It is yet to be clearly demonstrated where, or even if, a line can be drawn between the changes within a carbonate rock mass that are ascribed to diagenesis (the chemical and physical changes that occur as primary sediments become rocks) and changes that are customarily ascribed to speleogenesis.

Though the term diagenesis should perhaps be applied strictly to the alteration of sediment into rock, and speleogenesis might arguably be viewed as referring to the (real or apparent) removal of rock to create voids, areas of potential overlap are apparent between these views.

Diagenetic changes are many and varied. Almost inevitably in carbonate successions they include relative dewatering. This involves fluid transfer within the sediment mass, driven by chemical, physical, or even biological changes. Fluid migration effects can encompass the broad expanses of depositional basins during long time scales (Neuzil, 1986, 1994), and changes imprinted by rock-forming processes might influence later speleogenesis. If so, certain aspects of diagenesis and of speleogenesis form parts of a continuum. Within this continuum some diagenetic changes can be viewed as early elements of speleogenesis, whereas in some contexts apparently speleogenetic effects can be viewed as late diagenetic changes.

It remains convenient to consider speleogenesis as a separate set of processes and changes. To do so a limit must be imposed on when speleogenesis is deemed to begin. Among many concepts within the inception horizon hypothesis of cave origin (Lowe, 1992a) is the assumption of an early phase of speleogenesis, termed inception. In initially noncavernous sequences, inception takes place before the initiation of speleogenesis as generally accepted, and it may be envisioned broadly as encompassing changes that occur in the transition from rock without conduits to rock with conduits.

An open question remains whether inception exists in its own right as an essential part of the cave development process. Or whether, as some workers believe, the changes ascribed to inception lie outside speleogenesis and are simply aspects of diagenesis.

In the inception horizon model, inception is defined as ending when conditions capable of driving and supporting organized laminar flow (as against migrational seepage) are established locally within the carbonate rock mass. This marks the start of the gestation phase of cave development (Lowe, 1992a). During gestation, inception effects that extended across entire horizons (or nonstratigraphic fissures) become subordinate to the imposition of erosional processes that are concentrated locally in response to secondary factors such as fold structures, or to the intersection of the aquifer by the downcutting land surface. Gestation, within the inception horizon model, is defined as continuing until turbulent breakthrough within evolving conduit systems. Some authors have referred to an initiation phase, and as most commonly used this encompasses both inception and gestation (3.3 Figure 1).

Arbitrary definitions have been applied to void sizes encountered within developing karst drainage systems. For instance, one set of "working" definitions was provided by Atkinson (1986), who subdivided voids into:

- **a** Intergranular voids or pores
  10^{-3} to 10^{-1} mm (0.001 to 0.1 mm)
- **b** Fractures
  10^{-1} to 10 mm (0.1 to 10 mm)
- **c** Fissures
  10 to 10^{2} mm (10 to 100 mm)
- **d** Conduits
  10^{2} to 10^{4} mm (0.1 to 10 m)

On this basis, explorable caves belong to the conduit void type. Conduits therefore include cavities that can be observed by direct exploration and also those that lie below the arbitrary anthropocentric size limit inherent within most cave definitions. The latter voids are observable if revealed by boreholes or exposed by natural or artificial dissection, but they cannot be explored.

At the lower size limit, intergranular voids can be examined and measured directly or indirectly, but their mode of development and enlargement cannot readily be observed under natural conditions. Intergranular pores are common in many rock types and can be much larger than the table suggests in rocks such as reef limestones.

Likewise, isolated cavities commonly described as vugs, reaching sizes covered by categories **c** and **d**, occur in Paleozoic and younger carbonates, although their age and mode of origin may be difficult to confirm. In some well-cemented young rocks, however, and in older rocks that have undergone significant recrystallization due to diagenetic, tectonic, or metamorphic effects, intergranular pores and larger isolated voids can effectively be absent before speleogenesis (in its broadest sense). Such absence necessitates a view that, if speleogenesis is to commence in void-free rocks, either:

- The earliest speleogenesis must follow voids related to or generated along secondary tectonic (fault and joint), or lithogenetic (joint) fissure discontinuities.
- Or, the earliest speleogenesis must follow voids related to or generated along primary stratigraphic (bedding-related) discontinuities.
- Or, the earliest speleogenesis must follow preexisting (paleokarstic) "weaknesses," in some situations a subset of the last option.
- The earliest speleogenesis must follow locally favorable combinations of all three of the above options.

Evidence is still accumulating for the scale of the role of paleokarstic weaknesses in later speleogenesis. However, such a role doubtless exists, at least locally (Wenrich and Sutphin, 1994; Osborne, 1995; and Chapter 3.6, this volume). If such local effects continue to be recognized, a more widespread paleokarst influence may become apparent. Meanwhile, returning to the possibilities above, the most attractive assumption, based on what can be seen in explorable caves, is that the final option above generally applies.

Within this broad assumption, a spectrum of options must exist in different types of carbonate successions or geologic terrains. At one extreme, guidance of conduit development may be dominated by tectonic fractures (if few or no more favorable stratigraphic discontinuities exist in the local sequence). Many subvertical shafts could penetrate deep into (or up through) the rock without any horizon capable of supporting bedding-related development being penetrated. This may be the case in the Newby Moss area of Ingleborough, UK, or among the deep shafts of the Astraka Plateau in Greece. At the opposite extreme, development would be guided dominantly or exclusively by stratigraphic elements, where no significant tectonic or lithogenetic fissures cut the rocks, or where such options for underground drain development are less favorable than those related to bedding.

Whichever bias or combination dominates in any particular situation, a point must exist during the karstic rock's diagenetic-tectonic-speleogenetic history when transition occurs from rock without conduits to rock with conduits. This can neither be observed nor tied definitively to any critical void-size threshold, as many variables are involved in different situations. Whereas later cave development is characterized by solvent transfer under either laminar or turbulent flow conditions, the earliest speleogenesis may not involve flow at all, at least not as commonly understood. Early fluid movement might be best viewed as seepage, which could initially be oscillatory rather than directional.

Although the quantities of fluid involved during the progression of these basin-wide changes must be enormous, representing a significant part of the original sediment mass, the volume affecting any individual bedding plane at a given time is presumably relatively small. As fluids would initially be saturated with soluble minerals such as calcite or evaporites, migration through, and expulsion from, the sediment pile must be seen as providing only slow removal of minerals along the migration routes, rather than the fluid acting as an aggressive solvent.

Some authors on speleogenesis have recognized such early, diagenetic, fluid transfer, but the tendency has been to refer its effects to a "precave" phase. The tacit corollary is that cave genesis is strictly a later process, commencing only when a flow system begins to become organized in a way that allows selective enlargement of a linked subset of all precave voids or protoconduits. Almost by definition, such organized flow, with its ultimate product of an integrated underground drainage system, must "target" upon an output point. This may be either a resurgence at the landscape surface (whether subaerial or subaqueous) or through an output point or surface into adjacent rocks.

Though theoretically possible, observations of the latter option are rare. Examples of (undrained phreatic) caves being explored to an output into adjacent noncarbonate aquifers are lacking, but the principle seems valid in the context of some types of maze cave development. In contrast, earlier (diagenetic or precave) fluid migration may involve a far greater proportion of interaquifer transfer, occurring initially under conditions that predate the types of landscape, with valleys and coastlines, that will support resurgence establishment during later speleogenesis. Thus, though perhaps it is merely a question of semantics, I argue that any subdivision between the development of precaves and that of caves is artificial, as caves logically cannot form unless precave voids exist, from which they can be selected and enlarged.

The Role of Stratigraphy and Rock Type

Stratigraphy is the study of rock sequences according to various criteria, such as relative rock age (chronostratigraphy), the fossils they contain (biostratigraphy), or locally recognizable rock units (lithostratigraphy). The last of these is most

Simplified view of the life cycle of conduits/caves within a single "tier" of a typical carbonate succession, indicating:

① Terminology in common and informal use;
② Part of the terminology developed by Worthington (1991);
③ Terminology used by Lowe (1992a).

Elapsed **TIME**, increasing from left to right, is not to scale

NO CONDUITS	⇒ CONDUIT GROWTH ⇒		TURBULENT THRESHOLD	⇒ CAVE ⇒ GROWTH	OVERALL DECAY OF CAVES	NO CAVES
①	INITIATION			various phases of DEVELOPMENT	abandonment and eventual removal	①
②	INITIATION PHASE	LAMINAR FLOW NOTHEPHREATIC PHASE		SYNGENETIC AND PARAGENETIC CONDUIT GROWTH PHASES	STAGNATION AND DESTRUCTION PHASES	②
③	INCEPTION	GESTATION		DEVELOPMENT	ABANDONMENT and DESTRUCTION	③
true point of "cave" **origin**	Period of transitional change		Turbulent flow begins	Major growth phases	can be complex and multiphase	Host rock removed

3.3 Figure 1. Selected examples of terms used to describe cave development phases.

obviously important to cave development. A rock succession's lithostratigraphy is the sum of the individual beds making up the sequence. Most commonly several similar rock beds are grouped together and referred to by a particular name.

For instance, the Gordale Limestone member in Yorkshire, UK (3.3 Table 1) contains many limestone beds of similar appearance and composition. But it also contains a smaller proportion of thinner beds of different rock types. The Gordale Limestone member itself is part of a thicker unit known as the Malham Formation, which includes another member, the Cove Limestone. The two members are sufficiently different to be recognized and mapped separately in some areas but alike enough to be considered part of the same formation.

The Malham Formation and other older formations display enough differences to allow independent recognition and mapping, but have enough features in common to form part of a still larger group known as the Great Scar Limestone. This example emphasizes that a named rock unit at any scale of subdivision may be dominated by a single rock type, in this case limestone, but may also include lesser proportions of other rocks. In the context of speleogenesis, these

GROUP	FORMATION	MEMBERS
Great Scar Limestone	Malham	Gordale Limestone
		Cove Limestone
	Kilnsey	Kilnsey Limestone
		Kilnsey Limestone-with-Mudstone
	Chapel House Limestone	Not divided further

3.3 Table 1. The broad lithostratigraphic divisions of some Dinantian carbonate rocks (Carboniferous) in the

other rock types can be just as important as the surrounding limestones.

Formations and members are usually separated by more pronounced bedding planes than those between the lesser beds they contain. Such major planes are prominent for several reasons. They may originally have been occupied by noncarbonate beds that have been eroded preferentially. They may mark points where the lower rock was karstified soon after deposition, before higher beds were deposited. Some stand out because adjacent rock is more prone to dissolution than rocks above and below. Many major bedding planes represent intervals when one cycle of deposition ended and another began.

Cyclic carbonate deposition is well-studied, and several characteristics are known to mark many cycle boundaries. These include the presence of fine-grained limestones, dolomitic beds, and evaporite minerals, such as gypsum, at the tops of some cycles. Clay-rich beds, such as shale or fine-grained volcanic tuff, paleokarstic features, fossil soils, and thin coal seams may occur between the cycles. Thus, some major bedding planes appear very obvious, due to the presence of noncarbonate rocks or a well-marked erosion surface. Others are less obvious, such as where fine-grained limestone at the top of the lower-cycle passes across a bedding plane into a coarse-grained limestone at the base of the next higher cycle.

The term "inception horizon" was coined early in a doctoral research project (Lowe, 1992a). It served as a working title for beds within carbonate successions (usually associated with cycle boundaries) that display "aberrant" chemistry, such as pyrite-rich beds that can generate sulfuric acid. As the project progressed, use of the term grew to include other major bedding planes with chemical or physical characteristics that make them particularly favorable to the inception of cave development.

Though acknowledged as being generic in the extreme, this extended definition recognizes the crucial fact that only a small proportion of the bedding planes present in any carbonate sequence seem to support cave inception. Additionally, the broad definition allows inclusion of the many different types of "favorable horizons" recognized by other workers (see, for example, Ford and Williams, 1989, p. 35-36). These include horizons that seem to owe their favorability to the effects of subsequent tectonic movement rather than being related directly to an initial lithologic peculiarity (e.g., Knez, 1996).

Stratigraphy, Depositional Cycles, and Inception Horizons

Rock type is linked closely to stratigraphy. Carbonate sediments with specific characteristics are deposited under conditions that reflect particular cycle stages. Secondary circumstances can lead to similar sediment being deposited at different cycle stages. But in a depositional basin with little terrigenous debris input, most of an eventual carbonate succession comprises relatively pure limestone. Only the bottoms and tops of cycles show significant departure from the dominant rock type (3.3 Figure 2). Trace amounts of terrigenous debris occur disseminated through most carbonate masses, but significant noncarbonate beds tend to occur

WATER DEPTH	DOMINANT LITHOLOGY	THICKNESS AND PERMEABILITY	INCEPTION ROLE
	first bed of overlying cycle		
at or above sea-level	coal, seatearth, volcanic fallout and other terrigenous deposits	relatively thin; some beds with significant primary permeability	generation/transfer of sulphuric acid from *in situ* pyrite (etc)
shallow water ⥣	calcite mudstone, with dolomite and evaporite minerals such as gypsum	thin; initially impermeable; possible horizon of syngenetic **(paleo) karstification**	Sulphuric acid from evaporite; dolomite/clay mineral reactions; aqueous dissolution
⥣ **these rocks make up most of the cycle**	**mainly oolitic limestone**	**relatively thin, with potential primary permeability**	**upper surface is favored location of early dissolution**
⥣	**paler, thicker-bedded bioclastic limestone**	**thick (major part of cycle); initially impermeable**	**early dissolution restricted to tectonic or lithogenetic fissures**
⥣	**dark, thinner-bedded bioclastic limestone**	**relatively thin; initially impermeable**	**as above**
⥣	calcareous mudstone	thin; very low primary permeability	as above; possibly some sulphuric acid from pyrite oxidation
deep water	non-calcareous mudstone	thin; very low primary permeability	sulphuric acid formed from pyrite
	last bed of underlying cycle		

3.3 Figure 2. Schematic view (not to scale) of the rocks comprising an idealized major carbonate depositional cycle (shown with a thick border) and the thick relatively pure rocks (shaded) that comprise most of the cycle. Most inception activity occurs within the thinner less pure beds (unshaded).

between cycles. Some represent accumulations of airborne volcanic dust that has fallen through the water column and dominated sediment supply due to complementary failure of carbonate input. Others may be truly terrigenous, representing weathering residues and various windblown, water-borne, or proximal volcanic deposits.

Some carbonate sequences within a single major cycle (such as the Asbian Malham Formation of northwest Yorkshire, UK, referred to above) display superimposed minor cyclicity punctuated by thin argillaceous beds (the shales of Waltham, 1971a). Minor perturbations in sea level or debris supply responsible for the minor cyclicity probably reflect contemporaneous tectonic-volcanic activity.

The different speleogenetic roles of the main succession and the thinner relatively widely separated beds of impure limestone and clastic rock are easily overlooked. Many impressive phreatic tunnels and vadose canyons have walls of massive pure limestone, yet generally this reflects later stage growth, not the passage's mode of origin. Simply by looking at the walls of a 20-m-high canyon it might be assumed that pure limestone is the single prerequisite necessary for cave development, but this assumption is superficial and inadequate. The question of how such a passage began to develop must also be asked. Commonly the critical evidence is overlooked, lying just below a vadose canyon ceiling or perhaps at the extreme lateral limits of a major tube.

It has long been recognized that cave development is at least partly guided by bedding planes, sometimes termed bedding plane fissures (Ford and Ewers, 1978). Though recognized, the true significance of this link has not always been made clear. There is no evidence that bedding planes inevitably provide potential water routes before dissolution begins in a carbonate mass. Certainly, in some posttectonic situations, poorly continuous void systems may exist along a bedding plane, resulting from differential motion and concomitant poor fit across the plane.

Bedding Planes

As demonstrated by Knez (1996) at Skocjanska Jame in Slovenia, slipped bedding planes can provide potential water routes and a focus for major cave development. It is not clear, however, why such voids, among the class of "real partings," (Lowe, 1992a, Figure 4.14) should occur only along a few of the available planes. In Skocjanska Jama only three out of 62 bedding planes in the measured sequence are affected. We must at least suspect that primary (that is, lithologic) factors either influence the degree of movement or affect the quality of postmovement fit on a minority of planes.

An alternative view is that, if slippage occurs on all planes, the same primary factors that would influence inception in unslipped rocks continue to impose guidance after slippage. Whatever the actual situation, such slip-related weaknesses should not exist in pristine rock, and if absent, cannot influence pretectonic speleogenesis.

Potentially advantageous drainage routes commonly exist where a favorable joint and a bedding plane intersect. If such relationships exist in a pristine rock mass, they would not be expected to involve tectonic joints.

Lithogenetic (broadly "desiccation") joints may be involved at a relatively early stage. But this alone does not explain observations that joint intersections with some bedding planes are favorable, whereas intersections with others are not. Accepting that some joint-bedding intersections (and some slipped bedding planes) have speleogenetic significance in specific postuplift and posttectonic situations, it is apparent that development of many penetrable bedding plane voids must be unrelated to these factors. Equally, even where involvement of such factors is demonstrable, the affected bedding planes, among the many available, owe their selection to factors predating, and thus unrelated to the effects of tectonism or the impact of joint development.

Great potential exists here for the "chicken and egg" argument. Whether a few major bedding planes seem to favor later cave development due to their "opening up" by specific and stratigraphically localized processes during inception—or that speleoinception itself is only enabled by the preexistence of favorable (inception) horizons, depends upon the viewpoint adopted.

Realistically, the origin (as opposed to the later enlargement, integration, and ramification) of most caves is not related directly to the thick pure and massive limestone exposed along passage walls and with only minor interruption, in most vertical shafts. Original tiny water routes, parts of which evolve into explorable voids, are guided by the bedding planes that subdivide the massive rock, and specifically by major bedding planes, where conditions are most favorable (3.3 Figure 3, p. 70).

Generally, favorable conditions are not provided by the many planes where relatively pure carbonates are in contact. They are provided by those (fewer) planes where the carbonates are impure, where clastic or more readily soluble partings are preserved, or where a lower bed has undergone earlier karstification—in short, the so-called inception horizons.

Clastic Beds

It has been suggested that the importance to speleogenesis of clastic beds such as shales or volcanic clays lies in their imperviousness (e.g., Waltham, 1971b; Crawford, 1984). Though perhaps true locally and briefly in some situations, this explanation is generally unsatisfactory. Conditions change through time as lithogenetic and tectonic fractures develop. But in primitive successions, clastic or impure beds are no less pervious than the surrounding pure carbonates, and some of them are probably more pervious.

Even if drainage could penetrate thick beds of pristine (unfissured) carbonate, it remains difficult to imagine its progress being halted or diverted permanently by thin beds of relatively unresistant clay. Such barriers, seeming unlikely during cave origin, seem still less likely to survive if a significant conduit develops and turbulent water flows along the clastic horizon. If clay-rich horizons can function as aquicludes, there is no reason why initial conduit development in confined situations should necessarily perch above them, with the aquicludes forming "umbrellas" that divert drainage away from underlying beds. If such aquicludes exist, they could limit upward, just as readily as downward, movement (as shown schematically in 3.3 Figure 3). This possibility, remarked by earlier workers, was reemphasised by Klimchouk (1992, 1997) in the context of artesian situations.

Original Drainage Routes

Rather than being *de facto* aquicludes that support perched drainage within carbonate sequences, major bedding planes and associated lithologic peculiarities probably provide the original drainage routes within the rock mass. If so, it is the initial and continuing imperviousness of the carbonates that temporarily maintains any hydrologic perching.

This view has much in common with ideas developed by Katzer (1909) and supported by Martel (1921). During inception, almost all rock-bedding combinations presented by inception horizons are potentially more pervious than the surrounding pure limestones. In pristine carbonate sequences the impure carbonate or noncarbonate beds may provide the first viable water routes (the aquifers), while adjacent pure rocks remain initially impermeable (the aquicludes). Special cases occur at the base and top of the carbonate mass, where initially more permeable rock types, such as sandstone and shale, are commonly in contact with initially less permeable carbonates.

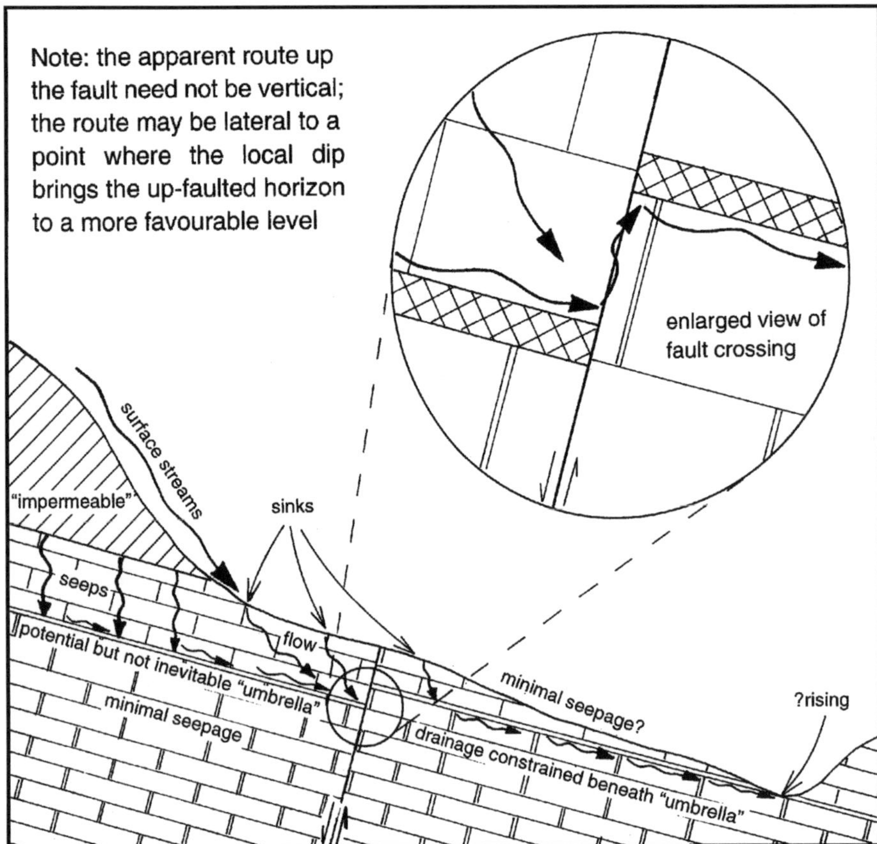

3.3 Figure 3. Schematic two-dimensional view of one potential mechanism for the guidance of passage inception below a supposedly impermeable "umbrella." Though illustrating a situation that forms part of an open, gravitational, flow system, similar guidance beneath relatively impermeable beds is also encountered in confined, commonly artesian, flow systems. Note that deeper flow could occur beneath the "umbrella" on the up-gradient side and seepage-fed flow could occur above the "umbrella" on the down-gradient side.

Inception Horizon Conditions

Thus, most potential inception horizons present conditions absent within massive pure carbonates above and below them. For instance, some are marked by primary physical properties that differ from those above and below. Chemical modification during late diagenesis can impose secondary changes upon preexisting physical properties such as crystallinity and grain size and hence porosity. Other horizons might contain gypsum (that may either dissolve or be reduced to form hydrogen sulfide) and sulfides that can be oxidized to produce sulfuric acid.

If the clay mineral kaolinite is present adjacent to dolomite rock, a transformation can occur that generates carbon dioxide, among other products, and so on. Their earliest effect may be simply to allow stratigraphically localized concentration of the movement of already saturated fluids that takes place naturally during diagenetic dewatering. If so, the subtly modified horizons represent weaknesses that can subsequently provide potential preferred routes, enabling fluid transfer through effectively impermeable carbonates. In so doing they also provide potential to turn nonaggressive water moving along them into corrosive solutions, whether of acids or of soluble salts, that can dissolve adjacent carbonates.

During inception, minuscule amounts of fluid (compared with those involved in later speleogenesis) are driven across large distances by mechanisms other than normal hydraulic flow, and limited dissolutional removal of minerals or rock may occur across the full lateral extent of favored horizons. The amount of material dissolved and transported away in a given time is also tiny, compared with the effects of later speleogenesis. But the effect is to open up laterally extensive zones of potentially higher permeability within the otherwise impermeable rock mass.

As discussed below, such zones can form simultaneously at several levels in a sequence and they may be connected in the third dimension by dissolution zones along tectonic or lithogenetic fissures. If diagenetic processes drive inception, the guidance for the bedding-related element of an eventual three-dimensional conduit network must be imprinted before imposition of the folds, joints, and faults that later add a second level of guidance to speleogenesis.

Inception Preceded Modern Landscapes

In the above context, it is vital to reiterate that according to the Inception Horizon Hypothesis (Lowe, 1992a), at the time inception commenced, modern landscapes did not exist. Hence, the earliest speleogenesis (in a broad sense) affecting the full lateral extent of inception horizons, was purely a reflection of stratigraphy. Early development predated the formation of—and was effectively unrelated to—the sinks and risings (and the hydraulic gradients between them) that typify modern local hydrologic systems. Even in areas that now possess well integrated underground drainage networks, however, conditions almost identical with those encountered during inception still persist. Within sections of inception horizons remote from main conduits, local underground and surface conditions dictate how far explorable cave passages will eventually enlarge back into the inception horizons before the rock mass is eventually removed by erosion.

Many caves explorable today, and some smaller voids that connect to them, have obviously evolved beyond this primitive stratigraphically imposed state. This evolution took place as the early rock mass passed through several interdependent stages. Tectonic deformation, whether gentle or violent, during relative uplift of the original (cave-free) rock mass, imposed the fundamental property of bedding dip. In postuplift or posttectonic situations, this property is ubiquitous. It exists at all values from effectively horizontal to vertical (whether or not overturned), and at all scales from very local to regional.

In post tectonic conditions, truly horizontal strata are arguably a rarity. Though they may exist locally, the effect of lack of dip is limited in extent and may commonly be of negligible importance within an overriding regional dip framework. Most beds perceived as "horizontal" are in fact revealed as inclined when examined regionally, although their geologic gradient may be unresolvable by eye. Gradient is confirmed only by detailed and extensive surveying.

After uplift, fluid motion that previously occurred horizon-wide, in response to basin-scale readjustments within the rock mass, became directed within the constraints of each now-dipping inception horizon, following the guidance of geologic structures. Bearing in mind that the massive carbonates above and below them may still act temporarily as effective aquicludes, inception horizons within anticlinal crests clearly can dominate flow guidance in buried situations. Those in synclinal troughs may provide the major influence when open-system conditions develop.

Fracture Development from Uplift

Fracture development, most obviously in the form of faults, thrusts, and joints, also accompanies uplift and folding. The open or incipient fissures generated have a double significance in later cave development. Most obviously they present potential pathways for fluid migration. Some fracture fissures are continuously or discontinuously open, due to the rock being pulled apart, or due to mismatch of walls after vertical or lateral movement. Others are effectively tightly closed. But even those seeming initially "tight" may have solvent fluids driven along them by tidal or seismic forces (as described, for instance by Davis, 1966, and Sibson et al., 1975).

Uplift has several other effects. It introduces a potential for broadly gravitational drainage of the uplifted rock mass down to a new base level, whether or not the output at new hydraulic "low points" is to the land surface or into adjacent rock units. Simultaneously, parts of previously buried karst rock masses can be brought to outcrop tectonically, usually due to the effects of faulting or thrusting.

Otherwise, uplift of the rock mass initiates a cycle of surface erosion. During this erosion, incising valleys intersect previously buried inception horizons and fissures. This provides potential outlets to the surface (as well as inlets from the surface in suitable situations) and encourages targeted integration of underground drainage.

In some situations uplift is sufficient to modify climatic regimes. It can raise karstic rocks to altitudes where precipitation is extreme, where freeze-thaw weathering can begin to disassemble the rock mass, or where glaciers and meltwater can begin to gouge out valleys.

Through all of this, the geologic structures imposed upon the rock mass continue to guide the development of cave frameworks. But passages also begin to extend into rock that lies beyond the constraints of the original inception horizons.

Water supply conditions permitting, passages that formed initially within the guidance of the framework imposed by inception horizons and fractures will enlarge in the most favorable direction or directions (under phreatic conditions) or will grow by downward incision and lateral erosion under vadose gravitational conditions.

Inception Horizons in the Yorkshire Dales, UK

Examples of at least three distinct variants of stratigraphic (or inception horizon) guidance of cave development have been identified within the Dinantian (Lower Carboniferous) succession of the Yorkshire Dales and adjacent areas in northern England (3.3 Figure 4). Similar situations undoubtedly exist within other Dinantian sequences of the United Kingdom and in sequences of all ages throughout the world.

The three beds considered here show local evidence of buried karstification (not necessarily synonymous with paleokarstification in this context) to a greater or lesser degree. They also demonstrate the potential influence of noncarbonate lithologies as initial aquifers, or as a source of strong acid.

The third example, though the least well studied, appears to point to the potential importance to inception of rocks derived from sediments rich in high-solubility minerals, which are precipitated under hypersaline conditions at the end of a depositional cycle.

The highest of the three example guiding beds (stratigraphically) lies at the boundary between the Asbian and Brigantian stages (3.3 Figure 4). This approximates to, but is somewhat lower than, the boundary between the D1 and D2 coral/brachiopod zones and coincides with the boundary between major depositional cycles (mesothems) D5B and D6A (Ramsbottom, 1977).

Above the boundary is the massive Lower Hawes Limestone, which is lithologically similar overall to the underlying Gordale Limestone Member or its lateral equivalents (3.3 Figure 4, p. 71). Fracture-guided shafts pass through the beds above and below the boundary, but extensive subhorizontal (bedding-parallel) passages have developed along the intervening horizon.

This horizon was clearly emergent in some areas between deposition of the two cycles. It displays a variety of lithologies across the area (3.3 Figure 4). But the effects of the pyrite-rich Notts Pot Coal Seam and its associated seat clay provide the most spectacular examples of its guiding effects in caves such as Notts Pot and Short Drop Cave on Leck Fell. These are toward the western limit of the preserved Dinantian outcrop.

In this area, ongoing effects of dissolution

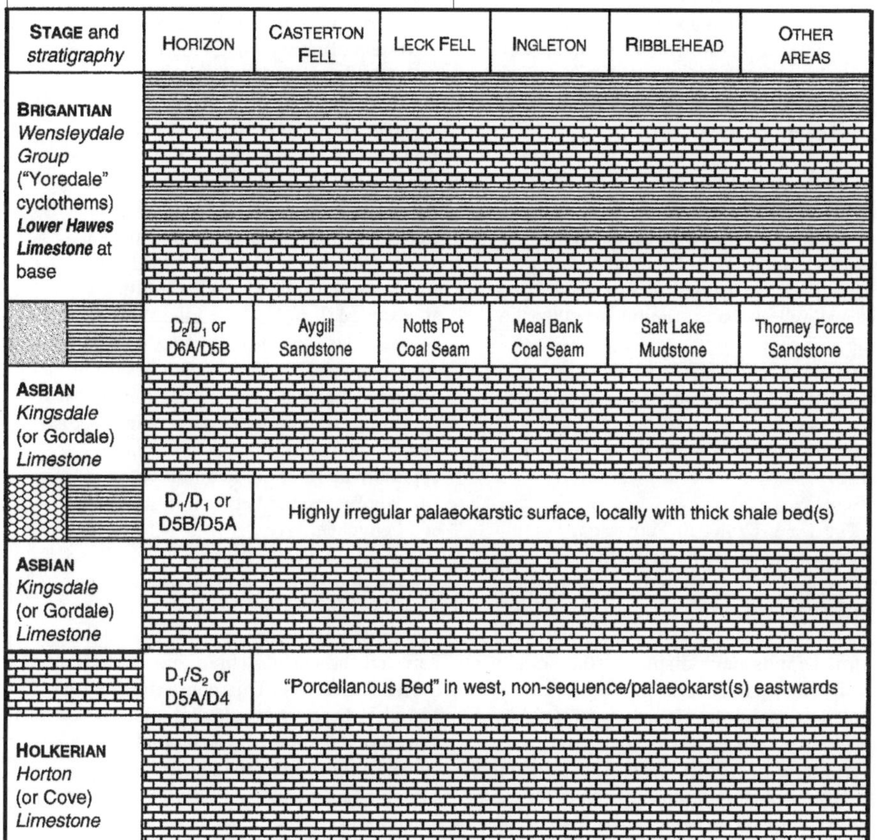

3.3 Figure 4. Schematic view of the relative positions and ages of three major "inception horizons" in the Yorkshire Dales area, UK (not to scale).

by sulfuric acid are also commonly displayed at some surface and underground locations. At underground sites the original favorable bedding plane remains visible, usually at or close to ceiling level. Significantly, the majority of later passage enlargement has occurred below rather than perched above the horizon. This is true even though ceiling doming indicates that phreatic-zone dissolutional erosion has taken place above as well as below the original level of the bed, with later incision of steep-sided vadose trenches below.

Acidified Seepage from Floods

At some points, acidified seepage emerges from the horizon during modern flood episodes. The walls beneath these seepages are marked by deep subvertical grooves and iron staining. This is in contrast to the majority of the passage walls, which are smooth, with local scalloping, and unstained.

This suggests that the bulk of the accessible passage cross sections owe their development to subsequent carbonic acid dissolution or vadose corrasion. However, the earliest development along the guiding horizons was dependent upon sulfuric acid, derived from pyrite oxidation. The ongoing sulfuric acid dissolution clearly has a minimal effect in the context of the modern passages as a whole. But when there were no major passages, its contribution to porosity development must have been significant.

Other Horizon-Guided Caves

Elsewhere, the same horizon has guided development of many caves, now left hanging, in the sides of the Ribble Valley. These have been effectively truncated by the valley's incision, though whether the truncation occurred when the passages were mature, or at an earlier stage, leaving inception voids to achieve cave proportions later, is currently unknown. Passage segments guided by the same horizon were:

- The Long Crawl in Penyghent Pot.
- The entrance crawl in Little Hull Pot.
- Several small caves in the Little Hull to Birkwith area.

Normal faults have displaced the horizon locally, and passages in the same bed at two topographic levels are separated by vertical or subvertical shafts.

In the same general area, this horizon likely guided early drainage from as far west as the Allotment on Ingleborough eastward toward or beyond Ribblesdale. This was down a primitive regional hydraulic gradient before the routes were effectively drained by relative uplift and the incision of Crummackdale, and the development of a new local, and as yet immature, resurgence (Austwick Beck Head) close to the base of the limestone sequence. Though a similar role may be assumed for the bed throughout its lateral extent, its speleogenetic role has not yet been confirmed east of Littondale.

About 60 m lower in the sequence is a major paleokarstic surface that probably marks the boundary between the late and early Asbian depositional cycles (D5B/D5A; 3.3 Figure 4). Around Leck Fell the boundary shows considerable relief, possibly displaying contemporaneous surface erosional features of the grike (Kluftkarren) type. Elsewhere the bed is less distinct. Its position is commonly marked by spring lines where small amounts of local percolation emerge at significantly higher levels than the major resurgences, particularly during periods of heavy rainfall.

Also, around Leck Fell the horizon includes a bed of shale that thickens dramatically (locally to more than 2 m) as it infills paleokarstic depressions and grikes. The horizon has had a significant effect upon development in caves such as:

- Notts Pot (the Ireby Inlet).
- Lost Johns Cave (Lyle Cavern High Levels, Shale Cavern).
- Ireby Fell Cavern (Glory Holes and NE Inlet).
- Several parts of the West Kingsdale System (Turbary Inlet, Good Friday Passage).

Nowhere does the bed show convincing evidence of having acted as an aquiclude. Yet its involvement in guiding underground drainage and cave development is clear. In this case it may be the paleokarstic or cycle-boundary features of the horizon that are significant rather than the infilling shale, as eastward from Leck Fell the shale becomes generally less obvious. Even so, extensive subconduit and conduit growth has occurred, as demonstrated by springs and minor caves at the level of the main Davidsonina septosa bed, which marks a hiatus at the same approximate level in the middle of the Gordale Limestone between Ingleborough and Penyghent.

Asbian and Holkerian Horizon

The third major stratigraphic guide (3.3 Figure 4) lies at, or close to, the boundary between the early Asbian and Holkerian cycles (D5A/D4). This approximates the boundary between coral/brachiopod zones D1 and S2 and the base of the Gordale Limestone. In some areas the most noticeable lithlogic factor associated with this boundary is a thin bed of lime mudstone, known traditionally as the Porcellanous Band (or Bed).

Some past workers (e.g., Glover, 1973, 1974) suggested that the Porcellanous Band is an aquiclude that guides the bedding-parallel vector of cave development. However, and very significantly, the bed commonly forms the roof rather than the floor of passages.

Thus, although such relationships may exist theoretically (3.3 Figure 3, discussed above), some other explanation of the bed's role may be preferable. In fact, the Porcellanous Band is just one of several lime mudstone beds that occur within a small stratigraphic interval, the whole of which marks a regressive phase at the close of the Holkerian (D4) cycle. Lime mudstone beds are just one element within this sequence, which may also have included dolomitic and evaporite beds associated with late-cycle hypersaline depositional conditions.

The horizon associated with the Asbian-Holkerian boundary has exercised a strong guiding influence on cave systems beneath Casterton Fell (at the extreme western limit of the Dinantian outcrop):

- Leck Fell.
- Both sides of Kingsdale.
- Ingleborough.

Entrenched in it are important passages such as:

- The main drains in Lost Johns Cave.
- Notts Pot.
- Parts of the West Kingsdale System.

Major modern resurgences such as Leck Beck Head and Keld Head are at the same stratigraphic level, and rather than this being coincidental, I deduce that buried inception routes that were eventually unroofed existed along the horizon prior to the incision of Kingsdale and the Leck Beck Valley. Farther east, many passages in the Gaping Ghyll System and adjacent caves beneath Ingleborough display the same initial guidance, though here the currently active phreatic drains are at a still lower level.

Fourth (Lower) Horizon

In the eastern Dales area, a fourth (lower) cycle boundary horizon has had a significant role. The boundary between the Cove Limestone and the underlying Kilnsey Limestone (Holkerian-Arundian) is deduced to guide conduits that now terminate at major valley-bottom risings such as Malham Cove and Black Keld. The speleogenetic role of this boundary zone has not yet been studied, as its conduits are accessible only by divers.

As with the stratigraphically higher situation of Leck Beck Head and Keld Head, the presence upon it of two such important resurgences at widely differing topographic levels cannot be dismissed as coincidence. If this boundary does function as an inception horizon, as suggested in the eastern Dales, it is probably also active though water-filled to the west. However, as yet there is no direct evidence, as limestones older than late Holkerian do not appear at outcrop across much of the area, and may locally be absent due to the discontinuous deposition of the earliest Carboniferous rocks on a high relief lower Paleozoic surface.

Earliest Speleogenesis

Again it must be emphasized that, as in the general situation outlined above, these beds' initial guidance is deduced to have been exercised during earliest speleogenesis. It may have begun before the sequence was affected by tectonism and uplift and before drainage integration, phreatic dissolutional enlargement, and local vadose downcutting. This view is important since, if valid, it provides a solution to arguments that have occupied local cave explorers and scientists for many years. Workers such as Glover (1973, 1974, 1976) and Dyson (1969, 1970, 1971) upheld the regional importance of such beds to cave development, though not in the context described here. Others, such as Brook (1971, 1976a, 1976b) and Latham (1976) suggested that if the beds had any role in speleogenesis it could only be at a local level.

Much of the argument against the beds exercising a regional role hinged upon the demonstrable, and sometimes large, variation in absolute elevation of the individual beds and the passages related to them. These elevations could not be tied either to each other, to any known past or present resurgence levels, or to supposed surface erosion levels, a once fashionable concept.

There is no reason to suppose that individual beds would continue to lie consistently at the same altitude following uplift, dip imposition, and faulting. Nor did there seem any reason that cave development grading to known resurgences should be affected by the same (even if favorable) horizons among the many that exist across the entire region.

Early Flow Routes Imprinted Region

The assumption that primitive potential flow routes were imprinted regionally across entire horizons, possibly before uplift and tectonism and certainly before modern landscape development, explains the seeming paradox. Only fragments of such potential regional routes, linked locally by later-stage fractures, would be enlarged and integrated into underground drainage systems, leaving the rest of the horizons' lateral extent without significant passage growth. The horizon-wide early inception activity probably took place under buried closed-system (artesian) or more traditionally accepted deep-phreatic conditions. Within these conditions, a regional hydraulic gradient carried fluids continuously across a wide area, locally moving against the imposed geologic gradient.

Development of the chosen horizon fragments into accessible caves would later proceed under open-system shallow-phreatic conditions as local outlets were imposed in response to surface downcutting. Vadose erosion would remove the active passages farther from the inception level as surface incision continued. Over time, passages cut below (or locally above) the preferred set of interlinked bedding planes and fractures would eventually become graded toward a series of valley bottom resurgences, created sequentially as the deepening valleys intersected structurally guided flow routes related to each inception horizon in turn.

Throughout the later phases, the guidance that was imprinted at the earliest stage, and the processes responsible for the imprinting, remain potentially active. Other sections of the inception horizons, originally lying along the regional hydraulic gradient and still at the same level and with the same geologic gradient (dip) as the chosen fragments would still exist.

However, they would not necessarily lie along the most favorable local hydraulic gradient at any given time. They could be left effectively abandoned, undeveloped or underdeveloped, though still with the ability to collect and carry superficial drainage to small permanent springs or ephemeral flood resurgences on a still more local level. Likewise, as conduits related to higher favorable beds were sequentially drained, phreatic (in some cases closed-system) conditions would continue to operate in other beds that were stratigraphically or tectonically and topographically lower, allowing for the existence of deep underflow (Worthington, 1991), perhaps unrelated to the shallower hydraulic systems.

Reasons Behind these Conclusions

The reasoning underlying these conclusions hinges on several interlinked observations and premises that need not be examined in depth. These may be summed up as follows, again by reference to the Yorkshire Dales example:

- Direct exploration, including diving, has confirmed the existence of an extensive and continuous set of cave passages, known collectively as the "Three Counties System" (Brook, 1968). This stretches from the western limit of the Dinantian outcrop (the Dent Fault, west of Casterton Fell) eastward at least as far as the eastern side of Kingsdale.

  Within this system many of the stratigraphically guided passages are related to the inception horizons described above, with subvertical links along fractures that connect different horizons or single horizons displaced by faults.

  Beyond East Kingsdale, cavern development exists at the same stratigraphic levels beneath Ingleborough and areas still farther east. Cave development in these areas now appears unrelated to that of the Three Counties System, being separated physically by the deep basement-floored trench of Chapel le Dale. As implied above, the latter view may be incorrect, as underground drainage, even if only via inception voids, could have been regionally continuous before Chapel le Dale was incised.

- Within the Three Counties System, later stage conduits that were conceived upon and then enlarged at structurally favorable locations on each of the known inception horizons appear to have existed concurrently. Some higher phreatic passages were clearly abandoned rapidly, without vadose incision, as flow was diverted to lower levels after uplift or surface downcutting.

- The morphology of the abandoned higher level phreatic conduits suggests that they originally followed their guiding horizon downdip or along the plunging axes of minor folds to high-level resurgences. The conduits were abandoned in sequence as the surface valleys cut through and below them, rather than ever being graded toward modern resurgences.

  Abandoned and heavily silted major phreatic tunnels (with local underfit vadose modification), originally guided by the higher inception horizons, must once have led to the protoresurgences. They are now truncated. Most are plugged by glacial drift or blocked by breakdown or scree. The full significance of surviving trunk conduit fragments high on the valley sides has commonly been overlooked.

- Elements lying west and east of the relatively recently incised surface valleys of Leck Beck and Kingsdale drain to common resurgences that lie on the same (Asbian-Holkerian) inception horizon in the two valley bottoms. Similar situations likely pertained when higher inception horizons carried the active drainage at higher levels. The active resurgences are erosional "windows" intersecting the current phreatic zone. Present-day flow to them may be updip or downdip at the same horizon, depending upon local structure.

- Still more significantly, abandoned and currently blocked links along inception horizons seem to exist across modern drainage divides, between elements of the system that drain to one valley, into areas that now drain to adjacent valleys. Arguably a similar situation exists beneath Ingleborough (now separated from the Three Counties System by Chapel le Dale).

This is where the interplay of the geologic structure with the incision of surface valleys has allowed drains to enlarge along modern local hydraulic gradients to many sites around the mountain, but where several higher level resurgences still function in flood conditions and provide tantalizing indications of former links between drainage systems that today seem discrete.

- A final more general point is not specific to the area of the Three Counties System. Throughout the dales there are significant stratigraphically biased concentrations of mainly abandoned conduits exposed in fault-line scarps. Though many entrances have been over enlarged by nonkarstic processes, the reason for passage origin and placement has been obscure. The context of each set has to be examined in detail in its own right, but a general conclusion is that it is unlikely that they developed from and carried drainage away from the fault-line scarps.

  Equally, there is no obvious reason why sets of conduits should have developed toward the fault-line scarps after their formation. Structural and stratigraphic conditions in at least some of the cases point strongly to this fact. Void systems, probably related to a regional hydraulic system, existed along specific horizons before the formation of the fault-line scarps and perhaps before the development of the fractures themselves. The general possibilities of pretectonic void development related to inception horizons were examined briefly (Lowe, 1992a), and ongoing work in Slovenia (introduced by Šušteršič, 1997) confirms the viability of "preorogenic" conduit inception.

Region-Wide Buried Conduits

As the Dinantian sequence in the Yorkshire Dales area was covered by later Carboniferous clastic rocks until geologically recently, the obvious interpretation of the pointers described is that an extensive system of buried conduits existed at several stratigraphic levels beneath the entire area. Surface lowering and incision of the main surface valleys led to sequential unroofing first of high level risings (now abandoned and lost following further incision) and then of the current risings.

In this context it is possible to postulate early continuity of drainage not only across the area covered by the Three Counties System but beyond the area, across the modern valley of Chapel le Dale and still farther eastward. If so, the original drains certainly were formed by flow along a regional hydraulic gradient unrelated to the modern landscape. This flow may have occurred under closed-system conditions, down a regional geologic gradient, moving off the essentially anticlinal (but structurally complex) tectonic high of the Pennine Hills. Passing beneath the younger Paleozoic and Mesozoic rocks in the basin to the east, at least part of such drainage eventually moves upward via fractures into younger aquifer rocks under artesian pressure.

Thus, I deduce that inception activity is related to processes that began to operate under closed-system conditions before the currently visible or provable hydraulic gradients were imposed—possibly even before the tectonic activity that was responsible for the structural features of the area.

Subsequently, following uplift and surface landscape formation, ongoing cave development under open-system conditions became concentrated along favorable elements of the early framework. These elements include those where the guiding bed was deformed into gently plunging synclinal folds. Such folds have guided successions of drains along the same structural and hydraulic gradients, such as in the abandoned and active "master caves" of the Easegill System and in the active main drains of Lost Johns and Short Drop Caves. In the latter two systems, streamways more than 100 m vertically apart still follow routes within the core of a gently plunging syncline, having been conceived within and then incised below the Brigantian-Asbian and Asbian-Holkerian boundary inception horizons.

Similar examples may be observed in Dinantian rocks throughout the United Kingdom. Elsewhere the boundaries between the Brigantian, Asbian, and Holkerian cycles are also important. But in areas south of the Derbyshire Peak District, the boundaries between older cycles that are not represented or are attenuated in the north are locally of great importance. In the Forest of Dean and South Wales areas, for example, the Arundian-Chadian and Chadian-Courceyan boundaries have both had a profound influence.

General Overview of Inception Horizons

The examples described above are based generally upon observations of the effects of particular noncarbonate or impure carbonate beds within the Dinantian successions of the United Kingdom. Similar effects have been noted in rocks of different facies and different ages. For example, again in the United Kingdom, it has been suggested that the Cretaceous Chalk is effectively noncavernous due to an ubiquitous primary porosity and related permeability, although fissure-flow of drainage of unspecified type is widely recognized by hydrologists.

On the basis of the relatively few dissolutional caves that have been observed in the British Chalk, I pointed out that here too cave-forming dissolutional activity seems to be concentrated along a limited number of aberrant horizons within an otherwise relatively homogeneous succession (Lowe, 1992b). As in the Dinantian sequences, these horizons may reflect hiatuses between depositional cycles and display atypical chemical compositions.

Another example may be provided by the work of Palmer (1989) in part of Mammoth Cave, Kentucky, USA. This revealed that guidance of dissolutional cave development has been dominated by relatively few horizons among 24 recognized within a 100m limestone sequence. Depending upon viewpoint, it might equally be argued that this does not necessarily support the hypothesized role of inception horizons, as not all bedding planes in a given sequence would be expected to support development, especially within a small area. However, such argument avoids the question of why some planes are susceptible to cave development, not only locally but across wide areas, while adjacent planes are not. Elsewhere, work that is still in its early stages in Slovenia (Šušteršič, 1994, 1997) appears to demonstrate that specific beds, accepted by the author as inception horizons, have guided the development of several caves, including Vranja Jama, Najdena Jama, and Kloka Jama.

Inception Before Later Speleogenesis

As mentioned above, inception may conveniently be considered as being complete when potential conditions for further speleogenesis are established across parts of each susceptible bedding plane—though some processes characteristic of inception will continue to operate during later phases. The timescale of the change from inception to further speleogenesis must depend upon local factors. Potentially, a carbonate sequence with only inception voids could undergo uplift and tectonic deformation followed by extremely rapid surface downcutting. This would lead to significant postinception development in a (geologically) short time.

In more stable areas, inception voids that began to open during diagenesis might lie essentially dormant at depth for many millions of years, though arguably still undergoing inception, until external changes trigger ongoing speleogenesis. The obvious limiting case is that a sequence may possess no active inception horizons. Or if inception horizons are present they may never be activated for further speleoenesis, in which case, unless there is a high density of nonstratigraphic fissuring, the carbonate mass will be removed by erosion before significant speleogenesis takes place. This is the so-called State 0 conditions of Ford and Ewers (1978).

Post External Changes

Following external changes, underground drainage lines tend to become localized at zones along each bedding plane, where conditions support establishment of true hydraulic flow. Such lines can be at the cores of flexures imposed during rock compaction or formed in response to tectonism, and these flexures may be so gentle as to be undetectable to the naked eye.

Broadly speaking, synclinal flexures will be favored in open-system conditions. But under buried closed-system (artesian) conditions, routes into and along anticlinal cores may be preferred, as long recognized in the related context of hydrocarbon migration. As laminar flow begins, a transformation takes place from bedding planes covered by seemingly random sets of distributaries (termed bedding plane anastomoses) to those where drainage targets in preferred parts of the early network, commonly termed victor tubes (Ford and Ewers, 1978).

Within the inception horizon view, as in more traditional models, as victor tubes develop they capture drainage from adjacent voids, whether these lie along the inception horizon or relate to other types of fissure. The complete inception horizon remains potentially active, as a network of interconnected microvoids, but as with dendritic surface drainage systems, individual feeders become smaller the more remote they are from the victor tube they feed.

While phreatic conditions persist, victor tubes and their tributaries continue to enlarge by dissolution of the tube walls. The rocks above and below the guiding horizon may be equally susceptible to dissolution, and a roughly circular or elliptical passage cross section develops. Otherwise, rocks above and below the horizon have differing solubilities. So an asymmetric cross section is formed, with most passage growth above or (more commonly) below the initial tube. During this phase, input of aggressive seepage continues from the original bedding on each side of the main tubes. But the importance of the chemical processes that drove the initial dissolution decreases as the tubes enlarge and turbulent water begins to be transmitted. Overall dissolution rates are also enhanced by mixing chemistry that occurs where minor seeps from the original bedding encounter the main flow in the expanding victor tube, and this localized enhancement of dissolution can affect the shape of the passage cross section.

The preexistence of subvertical links in the inception framework is demonstrated later when, after relative uplift, the circular or elliptical phreatic tubes may be abandoned as drainage diverts to lower levels. Where no subvertical links exist to allow instant diversion, the tube shape is modified. Any stream that continues to use it entrenches its floor, cutting down through the massive pure carbonates to produce a characteristic keyhole cross section.

Depending upon the rate of uplift and the erosional capacity of the stream, deep and narrow vadose canyons can develop beneath the original level. Examples of such incised canyons, which are normally relatively high and relatively narrow and are cut below guiding planes, exist in most cavernous areas. Commonly, incised canyons meander sharply, and although the meanders may be emphasized by the effects of vadose erosion, the initial bends must reflect ones that existed in the original bedding-guided phreatic conduit (perhaps partly reflecting initial anastomosis) or were incised into its floor by a meandering underfit stream.

Conclusions

This brief examination of an as yet only hypothetical role of stratigraphy in carbonate speleogenesis glosses over many important issues, or looks at issues from a viewpoint unlike that taken by other models of stratigraphic involvement.

It is emphasized that according to this model, although large tubes and deep vadose canyons have cut through high purity carbonates, they owe their origin to widespread activity involving a few major bedding planes and the favorable rock properties of generally less pure rocks at these levels. Whether these bedding/rock combinations are referred to as inception horizons or simply as favorable beds, their role and importance are the same.

In most situations they form only part of the overall framework of speleogenesis—the complementary part being provided by tectonic or lithogenetic fissures. These arguments relate primarily to speleogenesis in carbonate rock sequences. But similar types of relationships probably also influence cave origin and guidance in other karstic successions.

Gypsiferous sequences, where evaporite beds that superficially seem massive but contain bedding-related differences of crystallinity and different levels of insoluble impurities, provide obvious parallels (Klimchouk, 1996). Other subtle lithologic differences related to the lithostratigraphy may also influence early conduit development in less obviously karstic rocks, such as sandstone and quartzite.

References

Atkinson, T.C., 1986, Soluble rock terrains, *in* Fookes, P.G., and Vaughan, P.R., eds., A handbook of engineering geomorphology: *Surrey University Press,* p. 241-257.

Brook, D., 1968, The Three Counties System: *University of Leeds Speleological Association Review*, no. 3, p. 15-19.

Brook, D., 1971, Mudstone beds and cave development in Craven: *University of Leeds Speleological Association Review*, no. 3, p. 15-19.

Brook, D., 1976a, The Porcellanous Band (The Philosopher's Stone): *British Cave Research Association Bulletin*, no. 13, p. 10-11.

Brook, D., 1976b, Hydrological grooves in Kingsdale Master Cave: *British Cave Research Association Bulletin*, no. 14, p. 8-9.

Crawford, N.C., 1984, Karst landform development along the Cumberland Plateau escarpment of Tennessee, *in* LaFleur, R.G., ed., *Groundwater as a geomorphic agent:* Boston, Allen and Unwin, p. 294-339.

Davis, S.N., 1966, Initiation of ground-water flow in jointed limestone: *National Speleological Society Bulletin*, v. 28, p. 111-118.

Dyson, N., 1969, The Meal Bank Coal Seam and its influence on cave development: *Bradford Pothole Club Bulletin*, v. 5, no. 6, p. 16-18.

Dyson, N., 1970, Statistical analysis of caves influenced by the Meal Bank Coal Seam: *Bradford Pothole Club Bulletin*, v. 5, no. 7, p. 1.

Dyson, N., 1971, Statistical analysis of potholes influenced by the Meal Bank Coal Seam: *Bradford Pothole Club Bulletin*, v. 5, no. 8, p. 15-16.

Ford, D.C., and Williams, P.W., 1989, *Karst Geomorphology and Hydrology:* London, Unwin Hyman, 600 p.

Glover, R.R., 1973, Gaping Gill-some underground controls of development: *Craven Pothole Club Journal,* v. 5, p. 8-11 and 58-66.

Glover, R.R., 1974, Cave development in the Gaping Gill System, *in* Waltham, A.C., ed., *Limestones and caves of northwest England:* Newton Abbot, David and Charles, p. 343-384.

Glover, R.R., 1976, The Porcellanous Band in Yorkshire caves: *British Cave Research Association Bulletin,* no. 12, p. 11-12.

Katzer, F., 1909. Karst und Karsthydrographie. Zur Kunde der Balkanhalbinsel. Sarajevo: Kajon, v. 8, no. 3, 94 p.

Klimchouk, A.B., 1992, Large gypsum caves in the western Ukraine and their genesis: *Cave Science,* v. 19, p. 3-11.

Klimchouk, A.B., 1996, Speleogenesis in gypsum, *in* Klimchouk, A., Lowe, D., Cooper, A., and Sauro, U., eds., Gypsum Karst of the World: *International Journal of Speleology,* v. 25, no. 3-4, p. 61-82.

Klimchouk, A.B., 1997, Artesian speleogenetic setting: *12th International Congress of Speleology*, La Chaux-de-Fonds, Proceedings, v. 1, p. 157-160.

Knez, M., 1996, Vpliv lezik na razvoj kraških jam: primer Velike doline Škocjanska Jame. (The bedding-plane impact on development of karst caves; an example from Velika Doline, Škocjanska Jame): *Zbirka ZRC*, p. 14.

Latham, A.G., 1976, The Porcellanous Band in Yorkshire caves: *British Cave Research Association Bulletin,* no. 13, p. 9-10.

Lowe, D.J., 1992a, *The origin of limestone caverns: An inception horizon hypothesis:* Manchester Metropolitan University, unpublished Ph.D. thesis, 512 p.

Lowe, D.J., 1992b, Chalk caves revisited: *Cave Science,* v. 19, p. 55-58.

Martel, E.A., 1921, Nouveau traité des eaux souterraines: Paris, Delagrave, 838 p.

Neuzil, C.E., 1986, Ground water flow in low-permeability environments: *Water Resources Research,* v. 22, p. 1163-1195.

Neuzil, C.E., 1994, How permeable are clays and shales?: *Water Resources Research,* v. 30, p. 145-150.

Palmer, A.N., 1989, Stratigraphic and structural control of cave development and groundwater flow in the Mammoth Cave region, *in* White, W.B., and White, E.L., eds., *Karst Hydrology-Concepts from the Mammoth Cave Area:* New York, Van Nostrand Reinhold, p. 293-316.

Ramsbottom, W.H.C., 1977, Major cycles of regression and transgression (mesothems) in the Namurian: *Yorkshire Geological Society,* Proceedings, v. 41, p. 261-291.

Sibson, R.H., Moore, J.McM., and Rankin, A.H., 1975, Seismic pumping–A hydrothermal fluid transport mechanism: *Journal of the Geological Society of London,* v. 131, p. 653-659.

Šušteršic, F., 1994, The Kloka Cave and the speleo-inception (English summary of *Jama Kloka in Zacetje): Naše Jame,* v. 36, p. 9-30.

Šušteršic, F., 1997, Cave patterns north of the Planinsko Polje (Slovenia): *12th International Congress of Speleology,* La Chaux-de-Fonds, Proceedings, v. 1, p. 117-120.

Waltham, A.C., 1971a, Shale units in the Great Scar Limestone of the southern Askrigg Block: *Yorkshire Geological Society,* Proceedings, v. 38, p. 285-292.

Waltham, A.C., 1971b, Controlling factors in the development of caves: *Cave Research Group of Great Britain Transactions*, v. 13, p. 73-80.

Wenrich, K.J., and Sutphin, H.B., 1994, Grand Canyon caves, breccia pipes and mineral deposits: *Geology Today,* v. 110, no. 3, p. 97-104.

Worthington, S.R.H., 1991, *The karst hydrogeology of the Canadian Rocky Mountains:* McMaster University, unpublished Ph.D. thesis, 227 p.

3.4. Hydrogeologic Control of Cave Patterns

Arthur N. Palmer

Abstract

Cave patterns are controlled by a hierarchy of hydrogeologic factors. The location and overall trend of a cave depends on the distribution of recharge and discharge points within the karst aquifer. Specific cave patterns, i.e. branchwork vs. maze patterns, are controlled mainly by the nature of the groundwater recharge. Individual passage configurations are determined by the structural nature of the bedrock and by the geomorphic evolution of the aquifer.

The origin of branchwork caves is favored by point recharge sources of limited catchment area. Floodwater recharge, especially through sinking streams, tends to produce maze caves or local mazes superimposed on branchwork caves. Through floodwater activity, anastomotic mazes form in prominently bedded aquifers, network mazes in prominently fractured aquifers, and spongework mazes in highly porous or brecciated rocks. Epikarst, network caves, and spongework caves are also produced by diffuse or dispersed infiltration into the karst aquifer, and network and spongework caves can be the product of mixing of two waters of contrasting chemistry. Ramiform caves are produced most often by rising water rich in hydrogen sulfide, which oxidizes to sulfuric acid. Deep-seated processes that help to initiate cave development include the interaction between carbonates and sulfates, which can greatly increase the solubility of dolomite, gypsum, and anhydrite, while calcite precipitates.

Although tightly confined artesian conditions have long been associated with the origin of maze caves, they actually have no inherent tendency to form mazes. The slow movement of groundwater close to equilibrium with dissolved bedrock, typical of tightly confined artesian aquifers, is the least favorable setting for maze development.

Introduction

The hydrogeologic setting of a karst aquifer is the most significant factor in determining the cave patterns within it.

At the very broadest scale, the location of a cave is determined by the geographic distribution of recharge and discharge areas for its host aquifer, i.e. the position of the aquifer in relation to the overall drainage pattern.

Within individual caves, the fundamental passage pattern is controlled mainly by the nature of groundwater recharge to the karst aquifer. It is at this scale that the basic types of cave patterns are determined, such as branchworks and mazes. Specific details of passage shape and orientation are influenced mainly by the type of initial porosity, particularly the relative influences of fractures, bedding-plane partings, and intergranular pores, as well as their structural attitudes. The local erosional history helps to determine the vertical distribution of passages within a cave. Even within a single karst aquifer these hydrogeologic factors are so varied, both spatially and temporally, that each cave acquires a unique pattern.

Localization of Caves by Patterns of Bedrock Exposure and Recharge

Any geologic setting that allows the movement of aggressive groundwater through soluble rock will favor the origin of caves and related surface karst features. The most significant factors include a humid climate, widespread exposure of soluble rock at the surface, and entrenchment of major rivers to provide substantial relief of the soluble rocks above fluvial base level. The length of time of exposure is also important, because much of the time in forming a cave is consumed in the initial stages of fissure widening (see Chapters 4.2.2 and 4.2.3).

Beyond a certain minimum value, the magnitude of topographic relief is not significant. For example, widespread karst and caves have developed in the low plateaus of the central U.S.A., where the relief in places is no more than 30 m. A large catchment area is important in forming large caves, even if much of the catchment area consists of insoluble rocks that simply drain water to local areas of exposed soluble rock. A capped karst aquifer, in which the potentiometric surface lies below the base of the capping unit, may be highly favorable to cave development. This is because the caprock concentrates the recharge and protects abandoned upper levels from erosion or collapse.

Usually such an aquifer receives almost as much water as it would if it were not capped by impermeable rocks. The difference is that the recharge is simply focused into smaller areas and tends to be more aggressive. Mammoth Cave, the world's longest cave, is of this type (see Chapter 5.3.8 in this volume). Friar's Hole Cave, in West Virginia, U.S.A., is an example of a huge cave that has formed in well-drained limestone whose surface exposures are limited to a few small windows in thick overlying sandstones and shale (3.4 Figure 1). The main effect of the caprock is to produce a distribution of passages that is conspicuously less uniform than those of caves in uncapped aquifers.

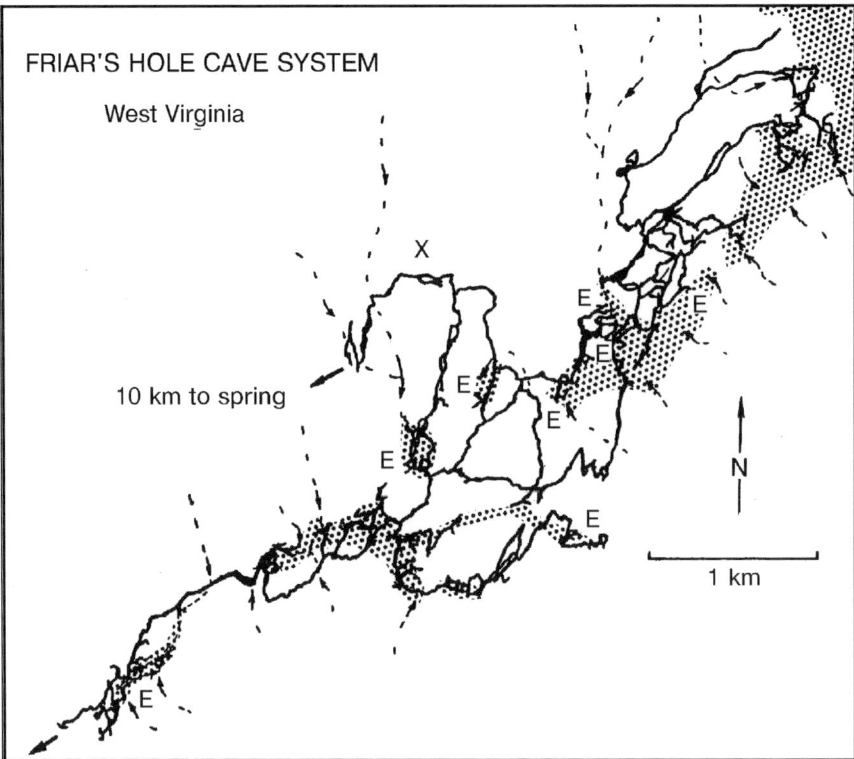

3.4 Figure 1. Generalized map of the Friar's Hole Cave System, West Virginia, U.S.A., showing restricted passage distribution caused in part by limited recharge areas. The main limestone aquifer is exposed only as inliers (shown with dot pattern) surrounded by hills capped by interbedded siliciclastic and carbonate rocks. Sinking streams are shown as dashed arrows. The few passages that extend under the caprock, such as the Rubber Chicken Streamway (X), have no local vadose infeeders. E indicates entrance, (map by D. Medville and others, geology from Worthington, 1984).

Cave origin is enhanced where surface runoff from large catchments is concentrated into a few small areas of groundwater recharge. Caves are thus most abundant in soluble rocks that lie beneath perched stream valleys or glacial cirques, or which border areas of exposed insoluble rocks. Convex topography is less favorable, because runoff is divergent and infiltration tends to be dispersed.

Caves formed by upward groundwater flow are likewise clustered around zones of favorable inflow or outflow. Most commonly, caves formed in this way are focused in areas where rising groundwater is able to escape, for example where a relatively impermeable caprock has been breached (Ford, 1995) or where there is a relative low in the potentiometric surface (see Klimchouk, Chapter 5.2). An example of deep-seated karst groundwater rising along faults through a low-permeability caprock in the Illinois Basin, U.S.A., is given later in this section. In many cases, cave origin is concentrated around the outflow points because that is where mixing of different waters and oxidation of hydrogen sulfide are most likely. Caves of the Black Hills, South Dakota, U.S.A., provide a possible example (see Chapter 5.2.2).

Points where groundwater rises into soluble aquifers are also sites of intense cave development. These are most common at the edges of tectonic basins, where fracturing allows basinal water to rise into mountain masses. Many of the large caves of the Rocky Mountains, U.S.A., are located along these zones. Caves of the Guadalupe Mountains, New Mexico, U.S.A. can be traced downward to steeply inclined fractures that fed rising H_2S-rich water into the cavernous aquifers (see Hill, Chapter 5.2.8).

Vadose vs. Phreatic Cave Development

The great majority of caves have patterns determined almost entirely by the flow of rather shallow meteoric water. In these caves the distinction between vadose and phreatic cave inception is significant, because it provides many clues about the erosional and depositional history of the surrounding region.

Vadose water moves by gravitational and capillary forces. To form caves, vadose water must have enough velocity to remain aggressive over long distances, and therefore capillary water is of negligible importance to cave origin, although it is the source for many secondary minerals. Cave-forming vadose water is gravitational, and as a result, vadose passages follow the most direct downward path to the water table. Where possible, the water follows vertical fissures and forms shafts.

Otherwise the water follows the next-steepest available paths, which are generally inclined fractures or bedding-plane partings, and canyons with entrenched floors are the most common result. Water that flows on resistant or insoluble beds may form perched tubes, which may become entrenched by later headward erosion from points where the perching beds have been breached.

Phreatic cave passages form at or below the potentiometric surface, which is highly irregular and discontinuous in karst areas. Groundwater in this zone exhibits a downward increase in hydrostatic pressure that more or less offsets the gravitational effect, and as a result there is no inherent tendency for phreatic water to flow downward by gravity. Instead, phreatic water flows along the most efficient paths to available surface outlets. During their early phases, phreatic cave passages are entirely water filled, either perennially or during high flow. They typically have an undulatory profile that contains segments that rise in the downstream direction. Tubular or fissure-like passages with elliptical or lenticular cross sections are the most common product.

Phreatic conduits preferentially form where the groundwater flow is greatest, and this is strongly controlled by the widths of the presolutional fissures. It is common for the bedrock to be highly anisotropic (i.e. the permeability varies in different directions), with the most favorable flow paths oriented along the predominant fracture patterns. As a result, the paths of maximum groundwater flow may not coincide with the direction of steepest hydraulic gradient, so that most phreatic cave passages follow irregular and sinuous paths.

The vast majority of phreatic water is fed by vadose sources, and so vadose and phreatic cave origin operate simultaneously. These two water types differ considerably in water chemistry, however, especially during the earliest phases of cave development. Water that infiltrates through soil acquires a high PCO_2 that is retained throughout the vadose zone, as long as there is no escape route for degassing. Narrow vadose fissures have PCO_2 values that are typically 0.01 to 0.05 atm. This can be demonstrated by analysis of drip waters at the points where they first enter caves. Loss of CO_2 to the cave atmosphere is well known as a means for depositing speleothems in aerated caves. Even in vadose streams that are fed by dispersed seepage, a great deal of CO_2 can be lost to the cave atmosphere, especially if the cave is well ventilated to the surface.

Many (or most) such streams are supersaturated with respect to calcite during low flow. In some caves, seasonal deposits of calcite accumulate on stream floors. But during high flow, when a larger proportion of vadose water is contributed by overland flow directly into surface depressions, these same cave streams tend to become aggressive. It is during the highest 5 to 10% of the discharge rates that most vadose cave enlargement takes place. The effects of long-term changes in climate or in the nature of groundwater recharge patterns can produce similar alternations between calcite dissolution and deposition. These fluctuations are demonstrated by cave passages floored with rimstone dams that have been partly removed by dissolution. Oxidation of organic compounds counteracts the loss of CO_2 in certain vadose passages by augmenting the local PCO_2 and aggressiveness of the water.

Phreatic water is often regarded as less favorable for cave development simply because under closed conditions it is unable to acquire CO_2 to replace that which is consumed by the dissolution of carbonate rocks. The equilibrium CO_2 drops as dissolution proceeds. However, by the time water has reached the phreatic zone, it has already dissolved a large amount of carbonate, and whatever CO_2 content it possesses is not greatly diminished by further dissolution. If this water is even slightly aggressive where it first reaches the phreatic zone, it will retain its aggressiveness for great distances, approaching saturation only slowly.

Even though dissolution rates are relatively low, the process is relentless, with no tendency for precipitation under ordinary circumstances. There is no degassing of CO_2 in the phreatic zone, and in fact the aggressiveness may be enhanced by mixing with water from deep-seated sources. In large conduits, rapid influx of floodwaters can bring highly aggressive water deep into the phreatic zone, greatly reducing the chemical disparity between vadose and phreatic water, at least on a temporary basis.

Control of Cave Patterns by the Mode of Groundwater Recharge

Whether a cave consists of conduits (usually linked as tributaries in a branchwork pattern), or a maze of interconnected passages formed more or less simultaneously, is determined mainly by the mode of groundwater recharge. This is of course controlled in turn by many aspects of the physical setting, such as topography, geology, and climate. Nevertheless the resulting mode of recharge is the focal point through which these many diverse factors affect the basic cave pattern.

To explain how the nature of groundwater recharge affects cave patterns, some ideas are borrowed from Chapter 4.2.3 (see also Palmer, 1981a, 1991). Chemical kinetics can be combined with flow hydraulics to reveal a few simple concepts about speleogenesis, particularly with regard to competitive growth rates and the origin of maze vs. branchwork patterns.

Consider:

- L to be the distance of flow from where aggressive water first enters a soluble aquifer (or from where aggressiveness is created by mixing or oxidation),
- Q to be the discharge through a given conduit,
- r to be the mean conduit radius.

Where the ratio Q/rL is small (< 0.001 cm/s), solutional enlargement is slow and highly varied between conduits. An increase in Q is required for any increase in enlargement rate. Where Q/rL is large (> 0.001 cm/s), the enlargement rate is at a maximum for the given chemical conditions, and further increase in Q has little effect. All openings that achieve that status grow at comparable rates. Where many competing passages achieve this status simultaneously, maze caves are formed.

These conditions include short flow paths or high discharge. Steep hydraulic gradients and wide initial flow paths have a similar affect because of their control over Q. Uniform Q into many fissures is an alternate way to achieve uniform growth rate. In highly soluble rocks such as gypsum, the enlargement rate does not reach a maximum, and uniformity of recharge is more important in producing maze caves than in carbonate rocks.

During the early stages of karst development, many flow paths compete, each with a low discharge and long flow length (i.e. small ratio of discharge to length, Q/L). Rates of passage enlargement differ a great deal from one conduit to another. Only a few of the competing flow routes enlarge rapidly enough to form caves, and they do so only if they are able to increase their discharge with time (therefore increasing Q/L). Other flow paths stagnate with insignificant enlargement rates. Most caves, particularly those with branchwork patterns, are formed under these conditions. As a given cave passage grows, its rate of solutional wall retreat reaches the maximum, which is 0.001 to 0.1 cm/yr, depending on the water chemistry.

In the following paragraphs these concepts are applied to the interpretation of cave patterns and other dissolution features under a variety of different modes of recharge.

Epikarst

Dispersed infiltration through a soil-covered or bare bedrock surface produces a myriad of interconnecting solutional voids in the uppermost part of the bedrock called the epikarst, or subcutaneous zone (Williams, 1985). Mechanical enlargement of fissures is widest there because the effects of unloading are greatest, allowing infiltration to pass through with little difficulty (see Klimchouk, Chapter 3.5 in this volume). With time, near-surface openings in the epikarst enlarge simultaneously at more or less similar rates.

The quantitative explanation is as follows: Flow distances *L* are so short through the soluble rock that many competing openings grow at or near the maximum possible rate. Narrow fissures, or those receiving small amounts of recharge, enlarge at this rate for much shorter distances than wide ones, or those which receive greater amounts of recharge. Differences in the rate of fissure enlargement become greater with depth. The bottom of the epikarst therefore tends to be irregular, except where it is terminated downward by a relatively insoluble bed. The epikarst is drained by only a few major outlets, which have been favored by large discharge and initial fissure widths. These conduits bring aggressive water into the heart of the karst aquifer and are the ones that are responsible for most cave origin.

The epikarst is best developed in prominently fractured carbonate rocks that are exposed directly at the surface or below an organic-rich soil cover. It is rudimentary in areas where the quantity of infiltration is limited, or where the unconsolidated overburden contains a large carbonate fraction.

Dispersed Recharge

Certain karst aquifers receive recharge that is dispersed among numerous small fissures or pores. Most of these consist of young diagenetically immature limestones. Well-known examples include the Tertiary limestones of Florida and Quaternary limestones of the Bahamas.

Even diagenetically mature carbonates receive some dispersed recharge, but mostly in the form of small trickles that deposit speleothems while having almost no effect on cave origin (Bögli, 1980; Thrailkill and Robl, 1981; Smart and Friederich, 1986). It is less common for dispersed recharge to be the primary source of aggressive recharge during cave origin. Diffuse recharge into highly porous soluble rock has such a large surface area of contact with the bedrock that it becomes nearly saturated after only a few meters (or even centimeters) of flow. A thin epikarst with many small solution pockets may be the only karstic result. This water is rarely able to form caves unless its aggressiveness is renewed by mixing with other water of different chemistry.

With time, however, even the diffuse flow may reorganize by selective enlargement of major flow paths, together with cementation of lesser ones, to form more focused inputs to the deeper parts of the aquifer. This process is much more rapid if the bedrock is prominently fissured and contains only small amounts of intergranular porosity. As surface depressions develop, large caves can form this way even in rather porous bedrock, but they are usually limited to a few major passages along favorable flow routes. An example is the Rio Encantado System in the Tertiary Lares Limestone of Puerto Rico. This consists almost entirely of a single stream passage more than 10 km long with only a few rudimentary tributaries. Most of the water within its basin enters simply as diffuse seepage that is already supersaturated with calcite.

Dispersed recharge that enters soluble rock through a porous but insoluble caprock is capable of forming network mazes or small complex shaft-and-canyon systems (see Chapter 5.3.11). Water enters the soluble rock while still retaining most or all of its aggressiveness. Cave development is concentrated where the water first enters the soluble rock, with a result rather like a subsurface version of epikarst.

One of the largest caves of this type is Anvil Cave in Alabama, U.S.A., which contains more than 20 km of tightly intersecting fissure passages (5.3.11 Figure 2). An interesting variant occurs where dispersed water rises into the soluble aquifer from below. For example, the enormous mazes cave of Ukraine—such as Optimisticeskaja with 200+ km and Ozernaja with 117 km of passages—were formed in gypsum strata by water rising at many points from an underlying carbonate aquifer (Klimchouk, 1991; see also Chapter 5.2.1 in this volume).

Seacoast Mixing Zones

Dissolution of carbonate rocks by mixing of waters of contrasting chemistry is described extensively elsewhere in this volume (see Chapters 4.1, 5.1, and 5.2.5). Most caves in porous poorly indurated seacoast carbonates are produced by the mixing of fresh infiltrating water with saline groundwater derived from seawater. The mixing curve between freshwater and seawater is not as simple as that shown in 4.1.5 Figure 4, because the tendency for dissolution is usually lost at high percentages of seawater (Berner and Morse, 1974). Besides the obvious contrast in salinity, there is considerable difference in CO_2 content. Partly because CO_2 solubility is diminished by salinity, the saline water has a lower PCO_2 than infiltrating water. Where the saline water is overlain by a discrete freshwater lens, solutional porosity tends to develop at both the water table and at the underlying halocline (Mylroie and Carew, 1990; see also Chapter 5.1). As the hydraulic conductivity increases with time, the two zones tend to merge into a single brackish-water zone of moderate aggressiveness.

Caves consisting of irregular rooms or spongework are most common in these settings for two reasons. First, dissolution is controlled by the supply rate of the two mixing waters, and since these are being fed through rock with dispersed intergranular porosity, mixing corrosion is widespread and not focused into a few conduits. In addition, the dissolution takes place at and near the site of renewed aggressiveness (i.e. at small L), so each opening tends to enlarge at comparable rates.

Point Sources of Recharge

The most common mode of recharge to karst aquifers in humid regions is through depressions of small catchment area, particularly dolines. Solutionally aggressive water can enter a karst aquifer through many such recharge points, each of which supplies a variable but limited amount of flow. At least 60% of all caves of explorable size are fed by small point sources of recharge. Caves and recharge sources develop more or less simultaneously, with each source capable of forming a cave passage, at least an incipient one (3.4 Figure 2). Doline growth is an important factor in the development of traversable cave passages, because it provides the increasing amounts of flow that are critical for a given passage to win the enlargement race against competing routes.

Passages of this type tend to combine together as branchwork caves, in which tributaries converge and becoming fewer but larger in the downstream direction. The branching cave pattern is partly the result of the largest ones developing low hydraulic heads, which attract water from surrounding parts of the aquifer. Ford and Ewers (1978) summarize the results of hardware models in gypsum and salt that simulate this process (see also Lauritzen and Ford, Chapter 4.2.1). Branching can also occur in the vadose zone, where the heads are determined mainly by passage elevation, simply by intersection or convergence of the structures that guide the vadose water.

On cave maps the branchwork pattern is often obscured by the presence of many different stages of development, in which passages cross each other at different levels, and by complex diversion routes superimposed on the basic pattern.

Examples of large caves that clearly exhibit the branching morphology in map

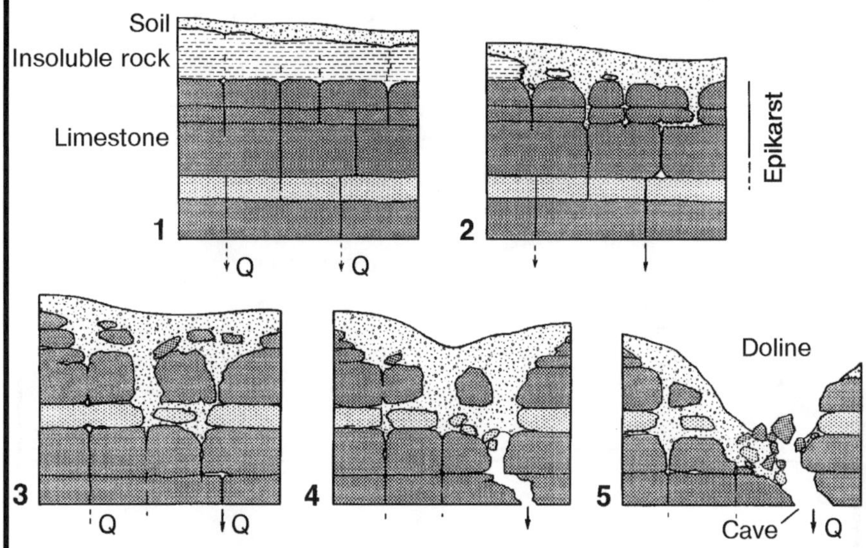

3.4 Figure 2. Evolution of the epikarst and a typical doline. The lightly shaded bed is only a visual aid for comparing stages.

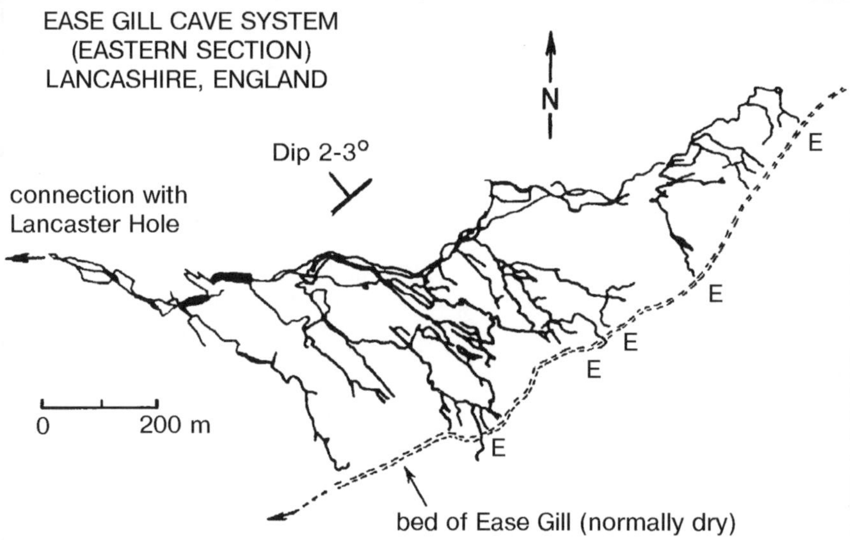

3.4 Figure 3. Map of the eastern section of the Ease Gill Cave System, England, showing a branching cave pattern on one or two levels. Tributaries are fed by small ponors in the bed of the surface stream Ease Gill. E indicates major entrance. Modified from map by P. Ashmead (see Courbon et al., 1989).

view are the Ease Gill section of the Lancaster-Ease Gill System in northern England (3.4 Figure 3) and Crevice Cave in Missouri, U.S.A. Those in which the branchwork pattern is very much present, though obscured by multiple levels, include the Mammoth Cave and Fisher Ridge Cave Systems in Kentucky, U.S.A. On the other hand, many small caves possess so few passages that the branching tendency is poorly developed.

Floodwater Recharge

Where water is able to collect over a large surface area before entering a karst aquifer, the flow conditions are considerably more dynamic than those described above. This situation is typical where runoff from large areas of insoluble rocks is able to flow directly into adjacent soluble rocks. In this way, water enters the karst aquifer through a small number of perched sinking streams, each having the highly variable flow characteristics of surface runoff. This can vary with time from mere trickles to raging torrents. Caves fed by this type of recharge have only a rudimentary branching pattern, owing to the limited number of inputs. Their overall patterns are complicated by the development of mazes and blind fissures superimposed on the main passage layout as a result of floodwater dissolution.

Breakdown, detrital sediment fill, or relatively insoluble beds in the main stream passages serve as local points of considerable energy loss for the water flow. This energy loss is proportional to the square of the discharge. The result is a formidable hydraulic gradient in the vicinity of constrictions during floods. Upstream from each constriction, floodwaters pond under high pressure, causing water to be forced into all openings in the surrounding aquifer, including those that are air filled during low flow. This water is augmented by a large amount of overland flow that is still highly aggressive and is able to enlarge the openings rapidly. The result is a maze of blind fissures, solution pockets, fissure networks, and anastomoses. In some caves these enlarged openings develop into traversable bypasses to the constriction. Bank flow from the flooding of entrenched surface rivers provides a similar example of this kind of recharge, and many riverside network mazes owe their origin to this process. Recharge from bare alpine karst can have similar characteristics, because the water is not slowed by infiltration through soil.

Caves fed by floodwater can have extremely complex patterns, and it is common for the basic branchwork pattern, if any, to be nearly obscured by secondary fissures and mazes. Examples include Sof Omar Cave in Ethiopia, many alpine caves, and certain local network or anastomotic mazes superimposed on what are otherwise branchwork caves (e.g. Blue Spring Cave, Indiana, U.S.A.).

The pattern of a floodwater maze depends on the nature of the host aquifer. Network mazes are limited to prominently fractured aquifers, while anastomotic mazes are common in prominently bedded aquifers. Rare spongework mazes occur in rocks in which fractures and partings are largely absent. Big Brush Creek Cave and Little Brush Creek Cave in Utah, U.S.A., contain floodwater passages surrounded by spongework, as well as local spongework mazes, in carbonate breccias of the Mississippian (lower Carboniferous) Humbug Formation. Some typical floodwater maze patterns are shown in 3.4 Figure 4.

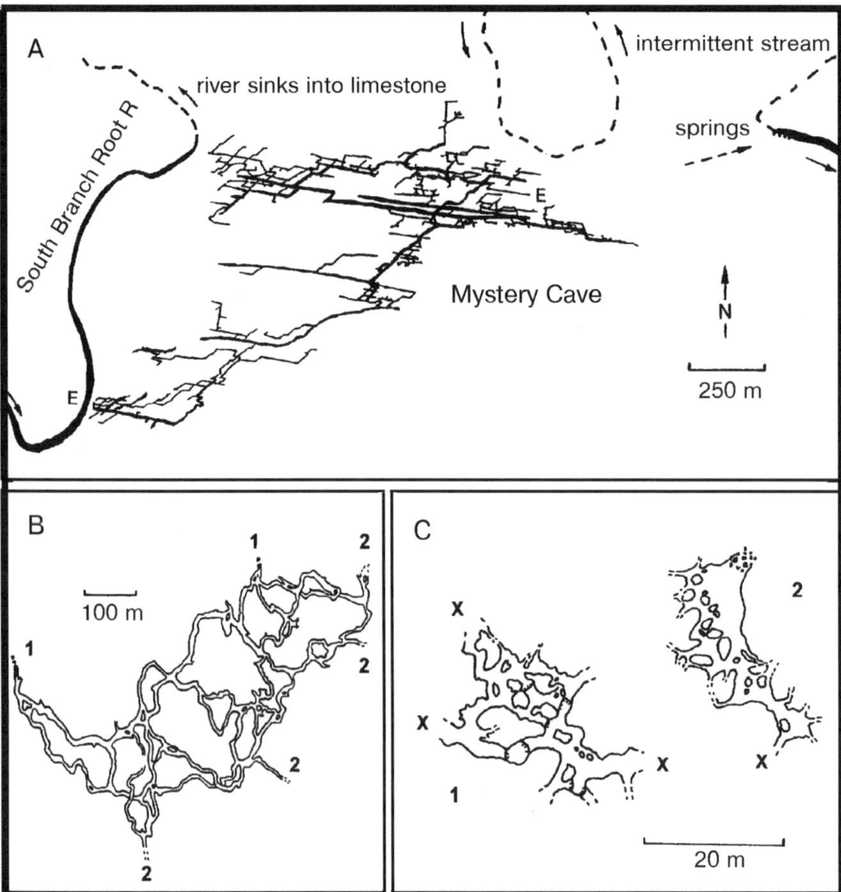

3.4 Figure 4. Floodwater maze patterns: A–Network pattern, Mystery Cave, Minnesota, U.S.A., a subterranean meander cutoff of the South Branch of Root River (courtesy of Minnesota Speleological Survey). ***E*** *indicates entrance.* ***B***–*Anastomotic pattern in the north-central part of flood-prone Hölloch, Switzerland (courtesy A. Bögli). The anastomotic pattern is superimposed on a multistory layout of conduits. Main flow through the cave is in an inaccessible phreatic conduit to the north, which periodically rises into this section of the cave through the points labeled 1. Points labeled 2 connect with the other parts of the cave.* ***C***–*Spongework patterns formed by severe flooding in Big Brush Creek Cave, Utah, U.S.A. (from Palmer, 1975). Water sinks into a large passage that degenerates into numerous distributary tubes surrounded by mazes that range from a dense three-dimensional anastomotic pattern* ***1*** *to true spongework* ***2. X*** *indicates connections to other parts of the cave.*

Hypogenic Cave Origin

Hypogenic caves are formed by water in which the aggressiveness has been produced at depth, independent of acid sources at or near the surface (see also Klimchouk, Chapter 5.2 in this volume). Caves formed by these processes show little or no genetic relationship to patterns of groundwater recharge from the overlying surface, although their spring outlets and levels of development are commonly at present or former fluvial base levels. They are most abundant in active orogenic belts but are by no means limited to those areas.

Hydrogeologic settings for this kind of cave origin are as varied as the chemical processes themselves. Common hypogenic agents of cave origin include sulfuric acid produced by the oxidation of sulfides (hydrogen sulfide, iron sulfide, and others), renewed aggressiveness caused by the cooling of ascending water, and acids generated by maturation of hydrocarbons. In humid climates all of these processes tend to be overwhelmed by the effects of epigenic carbonic acid. Hypogenic cave origin is therefore most effective in dry climates where surficial karst processes are subdued.

Hypogenic caves can also form by the mixing of contrasting water types. For example, waters that contrast in hydrogen sulfide (H_2S content can be very effective in forming solutional porosity where they mix. Water rich in H_2S frequently encounters water that contains virtually no H_2S. It is under these conditions that the mixing effect is most potent. This is shown by the strong curvature of the solubility graph at low concentrations (4.1.5 Figure 4). This process may account for deep-seated porosity that seems to be isolated from obvious sources of recharge. Wells occasionally encounter solutional openings of this type that contain stagnant sulfide-rich water (Warwick, 1968).

Mixing is also highly effective where water rises from depth and encounters near-surface meteoric water. A contrast in temperature might seem an important factor. However, it is actually of no speleogenetic benefit, because the mixing of two waters that differ only in temperature will increase the saturation ratio. Therefore, the mixing of these waters can form caves only if there is a contrast in chemistry, usually in CO_2 content. This mode of cave origin is common where the ascending water is strongly enriched with CO_2 by igneous processes or by the metamorphism of carbonate rocks. Manitou Springs in Colorado, U.S.A., are highly undersaturated as the result of this process, and it is suspected that nearby Cave of the Winds represents a now-relict byproduct (Luiszer, 1994). The caves in and around Budapest also exhibit this phenomenon (Müller and Sarvary, 1977; see also Dublyansky, Chapter 5.2.6 in this volume).

Oxidation of metallic sulfides is a potential source of cave origin (Morehouse, 1968; Jagnow, 1979), but well-documented examples are few. One example is Queen of the Guadalupes (New Mexico, U.S.A.), a vertical shaft lined by rills emanating from zones of oxidized pyrite. Pyrite oxidation along certain geologic horizons may serve as a means for localizing and accelerating the early phases of cave origin, although most of the eventual cave enlargement is performed by other types of aggressiveness (Lowe, 1992, and Chapter 3.3 in this volume).

Worthington (1991, 1994) has emphasized the importance of sulfate dissolution along deep flow paths as a mechanism for initiating cave development within large regional flow systems. Most springs that drain deep regional karst aquifers have high sulfate levels. It is likely that early cave development is enhanced by this process (Lowe, Bottrell, and Gunn, Chapter 5.2.7 in this volume). Cave development along sulfate zones is best observed in dry climates where recharge from surface sources is limited and the effects of deep-seated processes are able to survive with little modification. For example, caves in the widespread Carboniferous karst aquifer on the semiarid flanks of the Black Hills, South Dakota, U.S.A., are concentrated along former sulfate zones (Palmer and Palmer,

1989; see Chapter 5.2.2).

Deep regional flow paths can also be enlarged by the interaction between gypsum, dolomite, and calcite, even in the absence of an acid source. This is also of demonstrated importance to the origin of many caves in gypsum, regardless of depth. The incongruent dissolution of dolomite, in which dissolution of gypsum allows dolomite to dissolve while forcing calcite to precipitate, is an important but little-studied karst process (Plummer and Back, 1980; Bischoff et al., 1994). Calcite is usually the first mineral to reach saturation within a karst aquifer, while dolomite dissolves less rapidly.

When gypsum is encountered, its high solubility allows it to dissolve readily while forcing calcite to precipitate on an almost mole-for-mole basis. This process keeps the bicarbonate and pH low, allowing dolomite to continue to dissolve. In comparison with the solubilities of gypsum and dolomite in isolation, at least 1.5 times as much gypsum and 5 times as much dolomite can be dissolved by this mechanism. The presence of salinity can drive the process still further. In a simple carbonate-sulfate system, the volume of precipitated calcite is less than 30% of the combined total volume of dissolved gypsum and dolomite, resulting in considerable solutional porosity. Because the process is mediated by the relatively slow dissolution of dolomite, it provides a viable mechanism for long and deep dissolutional flow paths in gypsum. Alone, gypsum dissolves too rapidly to allow long and deep conduits to develop (see Klimchouk, Chapter 7.1 in this volume).

Oxidation of hydrogen sulfide to sulfuric acid is the most potent hypogenic cave-forming agent (3.4 Figure 5). This process is described in detail in Chapters 4.1.4, 5.2.7, and 5.2.8 (see also Davis, 1980; Egemeier, 1981; Hill, 1987). It is most common in uplifts bordering petroleum-rich basins. The best-documented example is seen in the cavernous Guadalupe Mountains of New Mexico, U.S.A., which border the Delaware Basin, one of the largest producers of oil and gas in the country. The process is now inactive in the Guadalupes, as far as is known. But active examples are known elsewhere, such as Tabasco, Mexico (Pisarowicz, 1994). Cave origin by this process is often episodic, rather than continuous, because the H_2S supply from depth tends to vary with time as the result of overpressuring of sedimentary basins or tectonic disturbances.

Large rooms and passages are formed when great concentrations of H_2S rise to the contemporary water table, as shown by their nearly horizontal attitude over large distances, despite considerable cutting across the geologic structure and lithologic boundaries. In the Guadalupe Mountains, however, these events do not seem to have coincided in time, because there is little or no correlation in the altitudes of the major rooms and passages.

Ramifying cave patterns are formed in this way, with distributary passages branching outward at various levels from the large central portions of the caves. Local areas of spongework are formed in mixing zones where aggressiveness is most extreme. This pattern agrees with the general concept that mazes are produced where the distances of flow are short from where the acidity was first generated, and relatively few isolated passages are formed by water that has flowed long distances through the soluble rock. The spongework mazes appear to have formed in zones of highest aggressiveness, at confluences between rising anoxic water and laterally

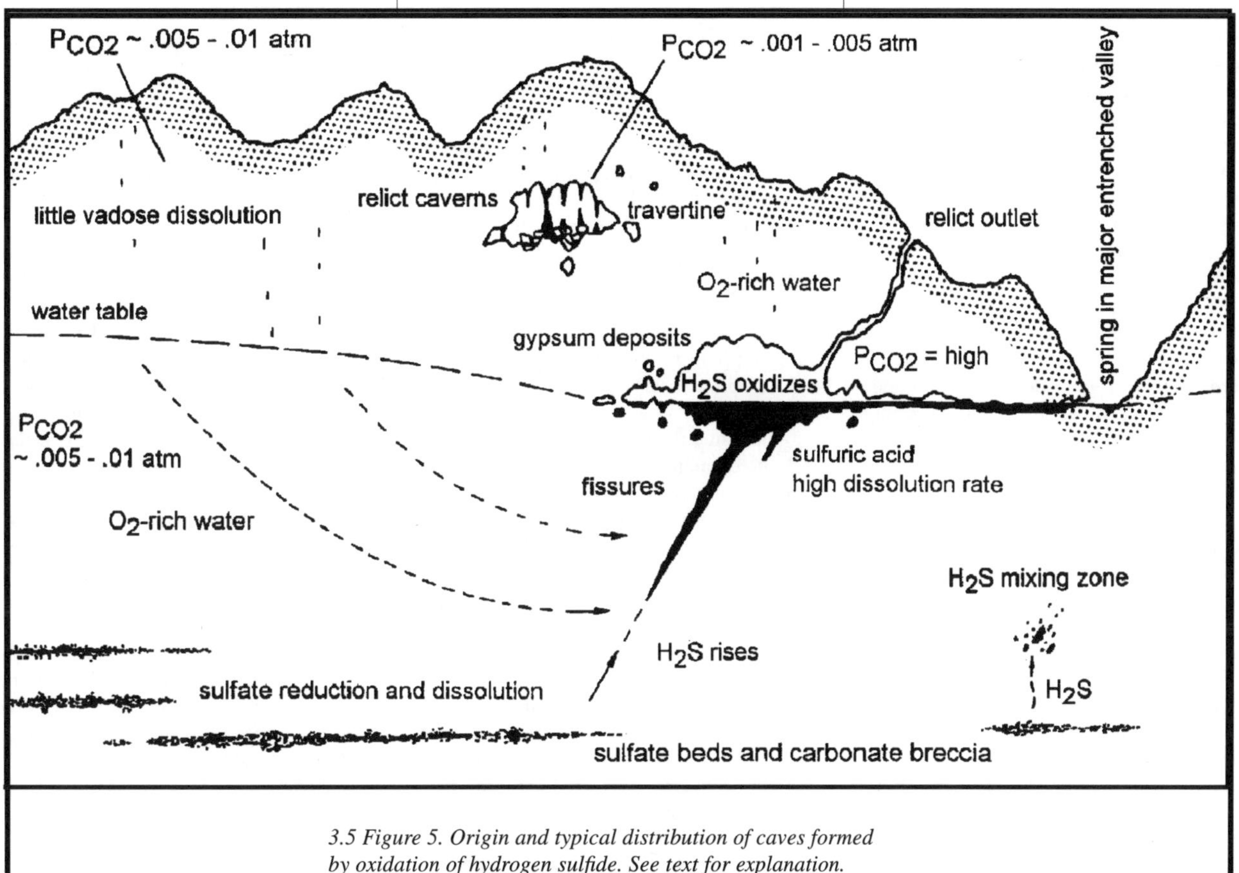

3.5 Figure 5. Origin and typical distribution of caves formed by oxidation of hydrogen sulfide. See text for explanation.

flowing oxygenated water. Some of the spongework appears to be a relic of an earlier stage of cave development that has been partly engulfed by later dissolution (Hill, 1987).

Cave origin simply by rising thermal water is theoretically feasible because of the inverse relationship between carbonate solubility and temperature (Bakalowicz et al., 1987; Dublyansky, Chapters 4.1.5 and 5.2.5 in this volume). As groundwater rises and cools it tends to become aggressive toward limestone and dolomite. Dissolution rates are very slow, however, and caves can be formed by this mechanism only where upward flow patterns persist for time periods on the order of millions of years (Palmer, 1991). The most favorable areas are at the downstream ends of deep carbonate basins where water rises along faults, or where thermal convection is induced by deep igneous sources. However, it is unlikely that many caves are formed by this process alone. Such settings are also favorable sites for mixing with shallow water and oxidation of rising hydrogen sulfide, and these processes appear to be more significant than cooling alone.

Stratigraphic and Structural Control of Cave Patterns

Geologic structure and stratigraphy control the orientation and sinuosity of individual passages. Stratigraphic variations are also important to the early phases of cave origin (see Lowe, Chapter 3.3 in this volume). Examination of cave maps often gives the false impression that caves are "out of adjustment with the surrounding topography" (Bretz, 1942). This misconception is caused by the fact that subsurface water flow is controlled to a much greater extent by local geologic structures than is surface drainage. Cave patterns reflect in great detail the geologic structures that controlled the cave-forming groundwater.

A cave's layout and character are greatly influenced by the structural nature of its host bedrock, particularly where the rock lies in the continuum between thin-bedded rocks with prominent bedding-plane partings, and massive prominently fractured rocks. An example of the former is the Ste. Genevieve Limestone of Indiana and Kentucky, U.S.A. Caves in this type of rock are strongly concordant to the strata, with sinuous curvilinear patterns. Examples of massive, prominently fractured rocks include the Hauptdolomit of the Austrian Alps, the Salem Limestone of Indiana (U.S.A.), and the massive reef carbonates of the Guadalupe Mountains, New Mexico. Caves in this type of rock are conspicuously guided by fractures and show strong discordance to the strata. Deep vadose shafts and angulate, fissure-like passages are most typical.

Because of the tendency for vadose passages to follow the steepest available routes, dip orientation is strongest in prominently bedded strata. Passage sinuosity is usually controlled by minor irregularities in local dip (Palmer, 1989a). Intervening shafts interrupt the downdip flow where steeper-dipping or vertical fractures are available, and this tendency increases with stratal thickness and tectonic deformation. Where vadose water reaches the water table, the dip tendency is replaced by a tendency for the water to follow paths that are nearly parallel to the strike of the beds (3.4 Figure 6). In this manner the incoming vadose water is able to follow essentially the same favorable beds or partings without having to flow discordantly across the strata, while at the same time maintaining shallow depths at and just below the water table, where the partings tend to be widest. The specific direction along the strike (left or right, with respect to the incoming downdip vadose flow) is determined by the direction of the nearest entrenched river valley.

The sinuosity of passages that form at or just below the water table is partly controlled by variations in strike, although fractures can also impose local irregularities in path. Again, in thick-bedded, prominently fractured bedrock the strike orientation is less probable. Probability plots of the orientation of vadose and phreatic passage trends (Palmer, 1986) show that both the vadose dip orientation and the phreatic strike orientation are most probable in well-bedded rocks, regardless of dip, and are least probable in massive, fractured rocks. Discordance of vadose passages to the strata is greatest in massive, fractured rocks, less in bedded, steeply dipping rocks, and least in bedded, gently dipping rocks. In rocks of gentle dip, passages of vadose origin tend to be long and sinuous, extending many hundreds of meters or even kilometers before encountering phreatic conditions. In areas of steeper dip, these passages are steeper and shorter, feeding vadose water to the phreatic zone much more directly.

Although maze caves depend largely on particular modes of groundwater recharge, the character of their passages is dictated by the nature of the bedrock porosity. It is no surprise that network mazes are formed in strata with prominent discordant fractures (joints or faults), and anastomotic mazes are formed in prominently bedded strata, or along one or more gently dipping fractures. Spongework mazes occur in bedrock that has a high primary porosity, such as reefs, certain breccias, or diagenetically immature carbonates. Floodwater recharge can form all three types, and uniform seepage or mixing can form networks or spongework, but not anastomotic mazes. Note that the mere presence of prominent fissure porosity does not predispose the bedrock to maze caves. For example, no matter how fractured the bedrock is, network mazes will not develop unless the mode of recharge is favorable.

The structural setting also has an influence on the depth to which meteoric cave development can take place. This topic is discussed at length by Klimchouk and Ford in Chapters 3.1 and 3.2 of this volume.

Effects of Stratigraphic Confinement

Karst researchers once intuitively considered that caves formed under artesian conditions should typically be mazes (e.g. Howard, 1964; White, 1969; Ford, 1971), especially since some of the world's largest maze caves are located in aquifers that were once confined. The most frequently cited examples are the Black Hills caves of South Dakota (see Chapter 5.2.2), These are located in the presently unconfined, near-surface parts of a carbonate aquifer that continues downdip as an extensive confined aquifer. However, there is scant evidence that artesian conditions were responsible for the maze pattern (Palmer, 1975). True confinement by itself does not produce maze caves. Any association between confined groundwater flow and maze development is coincidental. In fact, slow groundwater flow near chemical equilibrium, typical of truly confined aquifers, is least likely to produce maze caves (Palmer, 1991).

True confinement by rocks of low hydraulic conductivity is only an idealized concept. Even where the confining unit has a hydraulic conductivity many orders of magnitude smaller than that of the aquifer, the enormous area of contact between the two rock types allows a great deal of leakage between them. In recharge areas the hydraulic gradient, and therefore the leakage, is generally downward across the boundary. Farther in the down-flow direction the reverse is true. Klimchouk (Chapter 5.2) provides a lengthy discussion of speleogenetic processes in confined aquifers, a topic that is often overlooked or misunderstood.

Upward flow from soluble rocks into overlying insoluble strata is controlled by the thickness and hydraulic conductivity of the insoluble material. Although this governing effect is similar to that of recharge through an overlying insoluble bed (see Chapter 5.3.11), the tendency to produce mazes is greatly diminished by the fact that the groundwater becomes nearly saturated with dissolved bedrock far upgradient, near the recharge areas. As conduits grow by dissolution, the

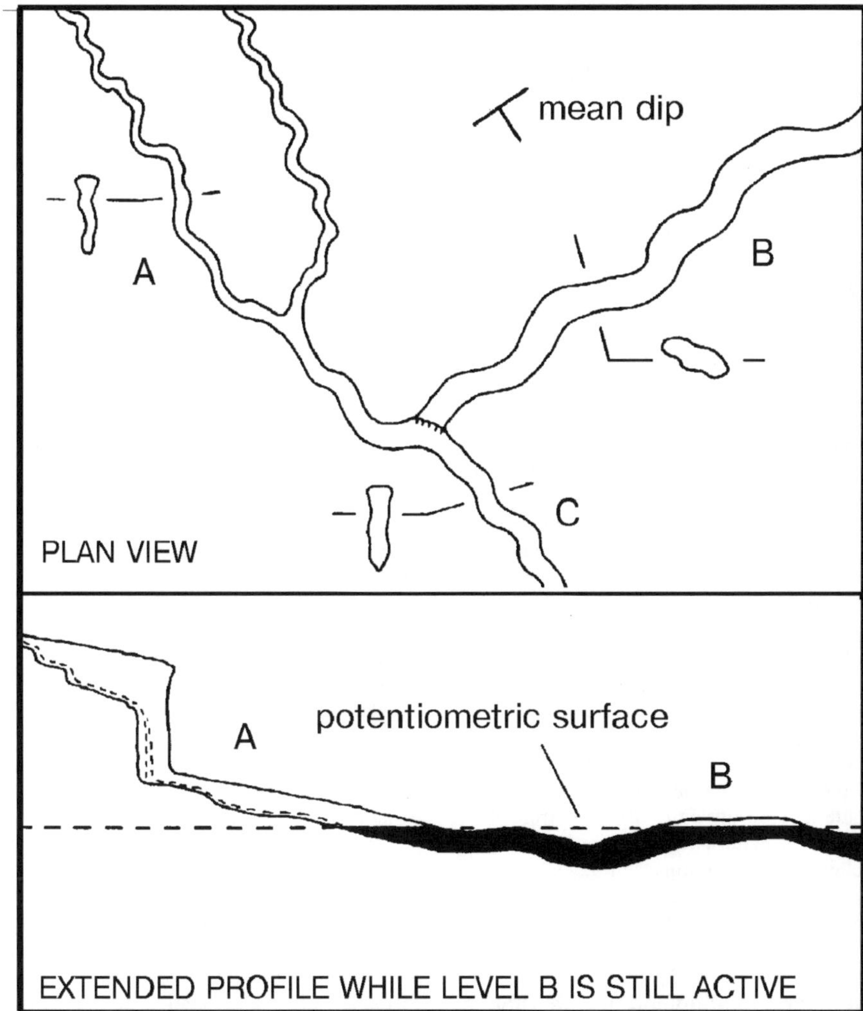

3.4 Figure 6. Typical changes in the character of a cave passage at the vadose-phreatic transition. A–Downdip canyon passages formed in the vadose zone. B–Nearly strike-oriented tubular passage formed in the phreatic zone as the downstream continuation of A. C–Downdip canyon formed by vadose water, bypassing B after a drop in the water table.

insoluble beds become increasingly effective in governing the flow rate through the system. By the time this governing effect becomes significant, however, the conduits within the soluble rock have already grown selectively at a variety of different rates—because of their low Q/L ratios—producing only a few major conduits and tributaries.

An exception to this scenario is where the flow distances through the soluble rock are so short that the water in all openings retains a large part of its original aggressiveness all the way to the discharge zone into the overlying strata. In this case it is the nature and rate of recharge that controls the maze pattern, rather than the low permeability of the overlying insoluble rock. Nevertheless, several authors (e.g. Palmer, 1981b; see also Chapter 5.3.11) have suggested the governing action of insoluble but permeable capping beds on the upward flow from a soluble unit as a cause for maze development. Whether this is a viable mechanism depends entirely on the local setting. It is most valid for short flow paths through soluble units (e.g. the gypsum mazes of Ukraine, as described in Chapter 5.2.1). But the greater the distance of flow before the water rises through the permeable caprock, the less tendency there is for the governing effect of the caprock to create maze caves in the underlying soluble rock.

Influence of Local Erosional History on Cave Patterns

Although local geologic structures can influence cave profiles, the main control over the vertical distribution of passages is the history of fluvial dissection, which in turn is determined by the tectonic and climatic history. Branchwork caves are especially well adjusted to the geomorphic history of the valleys into which they drain. During fluvial entrenchment, water in base-level cave streams tends to divert to successively lower levels (3.4 Figure 7). In a multilevel cave, the largest passages are those that have remained active for the longest time. They generally correlate with periods of nearly static base level when downcutting has slowed, come to a complete halt, or even reversed during a period of aggradation (Powell, 1970).

Thus the traditional view of cave "levels" is that they form at the same altitudes as static fluvial base levels and therefore should correlate with fluvial terraces, bedrock benches, and past or present floodplains (Davies, 1960; Droppa, 1966). This idea is valid but oversimplified. It assumes that large phreatic passages should be concentrated at the level of the outlet rivers, but it does not take into account the possibility of phreatic loops below the water table. Because few of the relevant passages are actually level, many authors have rebelled against the term "cave level," preferring instead a more generic term such as "tier" or "story."

This issue can be clarified by focusing attention on the altitude at which the passage character changes downstream from vadose to phreatic. That is, from dip-oriented canyons, which descend continu-

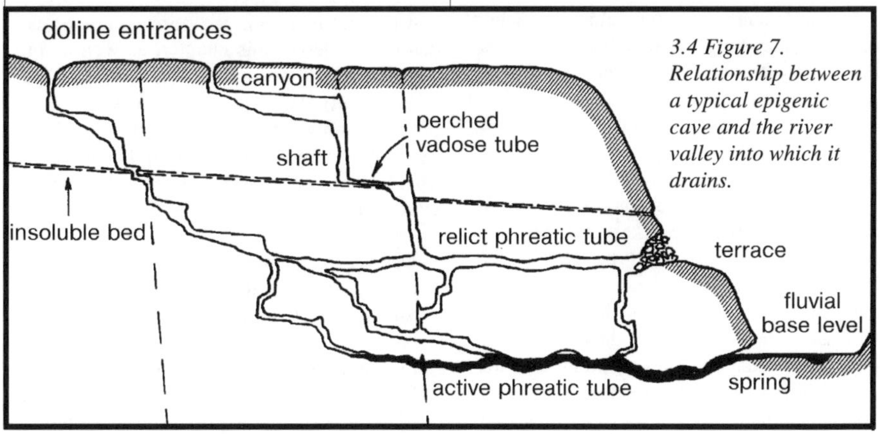

3.4 Figure 7. Relationship between a typical epigenic cave and the river valley into which it drains.

ously along their entire lengths, to phreatic tubes with strike orientation (or with no relation to the dip), and which may have undulatory profiles containing segments that rise in the downstream direction. The elevation of this transition—which is easily discerned even in a relict passage by mapping its profile—is called the "piezometric limit" (Palmer, 1987). It is at this point where the passage profile ceases to be controlled strictly by gravity and is instead influenced by the downward increase in pressure in the phreatic zone. It does not simply refer to a former water table, but instead that specific position of the water table that was stable enough to influence the passage character.

In this sense, the term "cave level" actually has a valid meaning, because the piezometric limits for various passages in a cave tend to cluster at essentially the same altitudes. Mammoth Cave (Kentucky, U.S.A.) contains the best-documented examples of this kind of cave level (see Chapter 5.3.8). Perched bodies of phreatic water can produce temporary piezometric limits, but they are usually short lived, and their influence on overall passage configurations is usually subtle.

Some caves show poor correlation with the history of fluvial entrenchment and aggradation. The vertical distribution of predominantly vadose caves, including many deep alpine caves, is determined by the perching effect of insoluble or resistant strata, or by favorable partings. Certain phreatic caves, such as those formed by hypogenic processes, exhibit a tiered pattern dictated by favorable strata. The caves of the Black Hills, South Dakota, U.S.A., are of this type (see Chapter 5.2.2). Rising water can also acquire an apparent dip orientation by following the bases of confining beds upward along the dip, as in the Guadalupe Mountains of New Mexico, U.S.A.

Worthington (1991) has presented an additional view of cave tiers that is only indirectly related to local fluvial entrenchment. The viscosity of groundwater decreases with depth because of the downward increase in temperature, a function of the geothermal gradient. His hypothesis is that the decreasing viscosity increases the flow rate along certain deep routes so that they enlarge more rapidly than those at shallower depth. As a result, cave tiers develop at rather fixed intervals as the region is dissected by erosion. Supporting this view is the fact that many large karst systems contain rather uniform ratios between the lengths of active phreatic passages and their depths below base level. Potential difficulties with this hypothesis include the tendency for the presolutional openings to diminish in size and number with depth, offsetting the viscosity effect, and the fact that higher temperatures increase the time required for cave inception (see Chapter 4.2.3). However, it is clear that the interpretation of cave tiers is not a simple matter of equating subhorizontal passages with pauses in river entrenchment.

Glaciation can have a strong impact on cave development and patterns. Meltwater from glacial ice caps and alpine glaciers has produced, or at least enlarged, many significant caves. One of the best examples is Castleguard Cave in the Canadian Rockies, which extends beneath the Columbia Ice Field (Ford, 1987; Smart, 1983). Glacial erosion has truncated the headwaters of some alpine caves. For example, certain caves in the Silvertip System of Montana, U.S.A., extend headward to areas that are no longer catchments for water, such as the middle of cirque headwalls.

Detrital sediment deposited by continental glaciers tended to block or redirect the infiltration routes and outlets of preglacial caves (Palmer, 1976; Mylroie, 1977; Rubin, 1991). In parts of New York State, for example, glacial sediment has filled surface river valleys to at least half their original depth, so that former subsurface drainage patterns are now quite different from the preglacial patterns (3.4 Figure 8). Many relict passages terminate in glacial till that was lodged directly into sinkholes and fissures by glacial action.

Aquifer Evolution and the Inheritance of Earlier Karst Features

Most aquifers evolve with time from one type to another, and in karst aquifers, solution features may be inherited from previous phases. Immediately after deposition, soluble strata are diagenetically immature and undergo rapid chemical changes (for example, the conversion of aragonite to calcite), which cause considerable redistribution of porosity and pore filling. Karst features formed during this phase are termed syngenetic (Jennings, 1968). Mixing-zone caves with spongework and ramiform morphology are common, as described later in this chapter, as well as in Chapter 5.1. Diagenetically mature strata are generally exposed to surface karst processes long after they were deposited, by gradual removal of nonsoluble rocks. These karst aquifers undergo the following evolution:

- Deep-seated, where the soluble rock is not yet exposed at the surface.
- Subjacent, where the soluble rock has been breached by erosion, but without the entire thickness exposed.
- Entrenched, where the entire thickness of soluble rock is exposed, but with the insoluble cap remaining.
- Denuded, where the insoluble cap has been removed by erosion (Klimchouk, 1996; see also Chapter 3.1 in this volume).

Caves and karst features usually accelerate in rate of development during the first two stages and reach a maximum during the last two. Karst may also be buried by other strata (forming paleokarst) and exhumed at a later time.

The karst of the gently dipping strata of southern Indiana, U.S.A., offers a clear example of the evolution of a karst aquifer from deep-seated to denuded, because all stages of development can be observed simultaneously within the same aquifer (3.4 Figure 9). The Mississippian (lower Carboniferous) strata include a variety of limestones that dip westward into the Illinois Structural Basin at less than one degree. One of these units (St. Louis Limestone), near the middle of the carbonate section, contains interbedded gypsum at depth. Downdip from the outcrop area are sulfate-rich springs fed by discharge that rises along faults in clastic beds that overlie the carbonate rocks (Ash, 1984). Hydrogen sulfide is present in the spring water during high flow.

The combined presence of sulfates and faulting allows dissolution to take place at considerably greater depths than would otherwise be possible. Dedolomitization, accompanied by precipitation of calcite, is almost certainly present, allowing both gypsum and dolomite to dissolve to greater extents than would be possible for either one in isolation (see discussion earlier in this Chapter). Farther updip toward the east, karst springs rise from depth where the overlying clastic cover has been locally breached. Diving has shown that some of these springs are fault-guided phreatic lifts as much as 45 meters deep. From their stratigraphic positions, it is probable (although not verified) that these openings are taking advantage of deep flow routes that originated along former sulfate zones. This idea is compatible with the concept of inception horizons, which are stratigraphic zones particularly favorable to dissolution (Lowe, 1992; see also Chapter 3.3 in this volume). Still farther updip, large thicknesses of the carbonates are exposed, in places with residual caps of insoluble rock. Extensive branchwork caves have developed in these areas.

In the presently accessible caves, however, there seems to be no inheritance of the deep-seated features described earlier. For example, there is no record of relict passages with phreatic loops out of adjustment with the present hydrology, despite considerable geologic mapping.

Outcropping beds that correlate with sulfate zones deeper in the basin show several significant features, including siliceous nodules that have replaced gypsum or anhydrite, local zones of contorted and brecciated bedding less than one meter thick, and minor patches of boxwork. These features all relate to the former presence of sulfates, but they have little apparent effect on the accessible caves. Of the thousands of caves in the region, including two that are more than 30 km long, nearly all are located in underlying or overlying parts of the limestone section. The brecciated and contorted beds of the St. Louis are commonly observed in the bedrock walls of dolines and may be a factor in

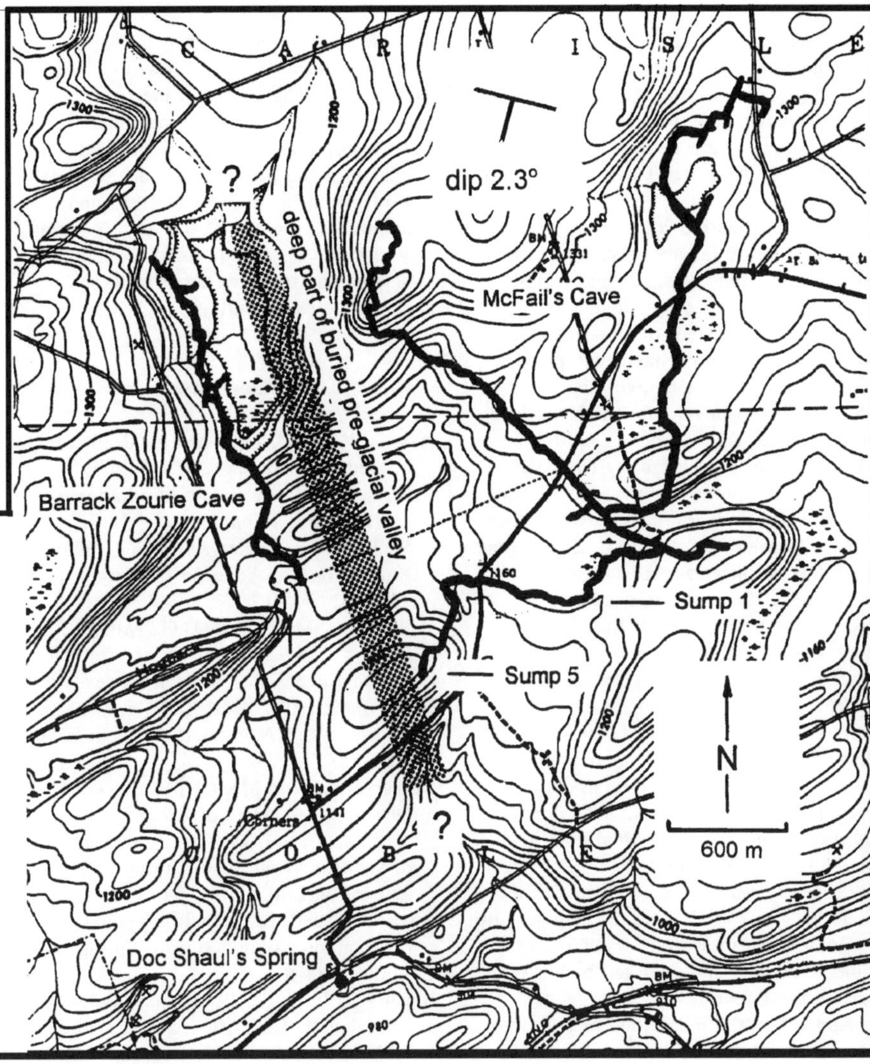

3.4 Figure 8. Relation of preglacial caves to a valley, also preglacial but now filled with glacial till (New York State, U.S.A.). The main passages of McFail's Cave consist of a vadose canyon oriented down the dip over most of its length, with nearly strike-oriented side passages that are phreatic tubes. Barrack Zourie Cave ignores the usual dip-strike orientation and instead follows a fracture-controlled trend parallel to the buried valley and at about the same level as the valley floor. If it extended down the dip, as does McFail's Cave, it would have to pass beneath a thick sequence of unkarsted rock. Both caves drain to Doc Shaul's Spring. (Valley profile based on geophysical surveys by M. Palmer and K. Milunich.)

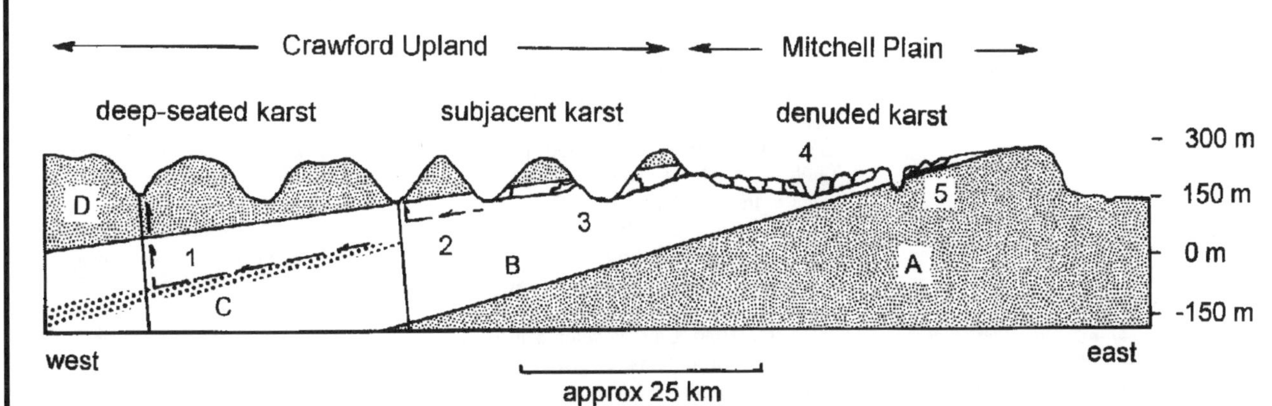

3.4 Figure 9. Geologic cross section through the karst of southern Indiana, U.S.A., showing stages of karst evolution. The variations observed today from west to east (i.e., in the updip direction) are the stages that the landscape undergoes with time at any given place. ***A****–Lower Mississippian (early Carboniferous) siltstones.* ***B****–Middle Mississippian limestones.* ***C****–intrastratal gypsum.* ***D****–Upper Mississippian rocks, mainly clastic.* ***1****–Deep-seated flow within and along boundaries of gypsum, feeding artesian sulfate springs that rise along faults in overlying clastic rocks.* ***2****–Conduits feeding artesian bicarbonate springs that rise along faults where clastic rocks have been breached.* ***3****–Shallow caves fed by recharge along edges of ridges capped by clastic rocks.* ***4****–Shallow caves fed by dolines in karst surface.* ***5****–Shallow caves perched on less soluble strata.*

localizing doline development. This is an example of how earlier deep-seated karst processes may have very little impact on later meteoric caves, even in zones where deep-seated processes are known. The same is true for the Mammoth Cave area (Chapter 5.3.8).

In contrast, earlier karst processes have had an immense impact on caves in the Black Hills (Chapter 5.2.2) and neighboring uplifts of the Rocky Mountains. The evolution of the Madison karst aquifer (early Carboniferous) has been episodic, rather than sequential, involving several phases of karst development, all superimposed on the same regions of uplift. Inheritance of prior porosity has been important in controlling the sites of later groundwater flow and cave development, especially since the climate is semiarid and the influx of meteoric water is small.

During early diagenesis, sulfate-rich beds in the upper third of the Madison were brecciated and numerous solution cavities were produced by oxidation of hydrogen sulfide released by reduction of sulfates at greater depth. Later karst development by meteoric water, probably accompanied by deeper mixing dissolution, produced dolines, fissures, and caves that followed many of the original diagenetic features. After lengthy burial, tectonic uplift in the early Tertiary exposed these same beds to further cave development by a combination of various water sources, which exhumed and enlarged both sets of earlier karst features.

Apparently the magnitude of the effect of prior karst processes depends on the scale and continuity of the early features, as well as the intensity of aggressive meteoric karst recharge. The impact of inheritance is described further by Ford (1995), Klimchouk and Ford (Chapter 3.1), Lowe (Chapter 3.3), Osborne (Chapter 3.7) nd and mchouk (Chapter 5.2).

Conclusions

Typical cave patterns are summarized in 3.4 Figure 10. A summary diagram such as this must of course include many simplifications. Note that the exact patterns of cave passages and their levels are controlled by factors not included in the diagram, because they merely modify the patterns shown. The many types of hypogenic cave origin are combined here, as their resulting cave patterns tend to be similar. Also, the patterns shown are often obscured by multiple phases of cave development.

Cave patterns are clearly a product of their hydrogeologic setting, and it is possible to interpret former hydrologic conditions through the analysis of caves. This is especially useful in the reconstruction of environments from paleokarst. Cave geometries also provide helpful insight into the prediction and interpretation of groundwater flow patterns and contaminant migration, and any opportunity to relate cave patterns to the nature of their host aquifers will assist in these applied studies as well.

References

Ash, D.W., 1984, Evidence for deep-seated groundwater movement in Middle Mississippian carbonate lithologies of south-central Indiana (abs.): National Speleological Society, Sheridan, Wyo., Proceedings of Annual Convention, p. 28.

Bakalowicz, M.J., Ford, D.C., Miller, T.E., Palmer, A.N., and Palmer, M.V., 1987, Thermal genesis of dissolution caves in the Black Hills, South Dakota: Geological Society of America Bulletin, v. 99, p. 729-738.

Berner, R.A., and Morse, J.W., 1974, Dissolution kinetics of calcium carbonate in sea water, IV: Theory of calcite dissolution: American Journal of Science, v. 274, p. 108-134.

Bögli, A., 1980, Karst hydrology and physical speleology: Berlin, Springer-Verlag, 284 p.

Bischoff, J.L., Juliá, R., Shanks, W.C., and Rosenbauer, R.J., 1994, Karstification without carbonic acid: Bedrock dissolution by gypsum-driven dedolomitization: Geology, v. 22, p. 995-998.

Bretz, J H., 1942, Vadose and phreatic features of limestone caves: Journal of Geology, v. 50, p. 675-811.

Courbon, P., Chabert, C., Bosted, P., and Lindsley, K., 1989, Atlas of the great caves of the world: St. Louis, Cave Books, 369 p.

Davies, W.E., 1960, Origin of caves in folded limestone: National Speleological Society Bulletin, v. 22, p. 5-18.

Davis, D.G., 1980, Cave development in the Guadalupe Mountains: A critical review of recent hypotheses: National Speleological Society Bulletin, v. 42, p. 42-48.

Droppa, A., 1966, Untersuchungen der Parallelität von Flussterrassen mit horizontalen Höhlen: 3rd International Congress of Speleology, Vienna, Proceedings, v. 5, p. 79-81.

Egemeier, S.J., 1981, Cavern development by thermal waters: National Speleological Society Bulletin, v. 43, p. 31-51.

Ford, D.C., 1971, Geologic structure and a new explanation of cavern genesis: Cave Research Group of Great Britain Transactions, no. 13, p. 81-94.

Ford, D.C., 1987, Effect of glaciations and permafrost upon the development of karst in Canada: Earth Surface Processes and Landforms, v. 12, p. 507-521.

Ford, D.C., 1995, Paleokarst as a target for modern karstification: Carbonates and Evaporites, v. 10, p. 138-147.

Ford, D.C., and Ewers, R.O., 1978, The development of limestone cave systems in the dimensions of length and breadth: Canadian Journal of Earth Science, v. 15, p. 1783-1798.

Hill, C.A., 1987, Geology of Carlsbad Cavern and other caves in the Guadalupe Mountains, New Mexico and Texas: New Mexico Bureau of Mines and Mineral Resources Bulletin, no. 117, 150 p.

Howard, A.D., 1964, Model for cavern development under artesian ground water flow, with special reference to the Black Hills: National Speleological Society Bulletin, v. 26, p. 7-16.

Jagnow, D.H., 1979, Cavern development in the Guadalupe Mountains: Columbus, Ohio, Cave Research Foundation, 55 p.

Jennings, J.N., 1968, Syngenetic karst in Australia: Contributions to the study of karst: Australian National University, Dept. of Geography Publication G/5, p. 41-110.

Klimchouk, A.B., 1991, Large maze caves in gypsum in the western Ukraine: speleogenesis under artesian conditions: National Speleological Society Bulletin, v. 53, p. 71-82.

Klimchouk, A.B., 1996, The typology of gypsum karst, in Klimchouk, A., Lowe, D., Cooper, A., and Sauro, U., eds., Gypsum karst of the world: International Journal of Speleology, v. 25, no. 3/4, p. 49-60.

Lowe, D.J., 1992, The origin of limestone caverns: An inception horizon hypothesis: Manchester Polytechnic University, Ph.D. dissertation, 512 p.

Luiszer, F.G., 1994, Speleogenesis of Cave of the Winds, Manitou Springs, Colorado, in Sasowsky, I.D., and Palmer, M.V., eds., Breakthroughs in karst geomicrobiology and redox geochemistry: abs. and field-trip guide, Karst Waters Institute Special Publication, no. 1, p. 91-109.

Morehouse, D.F., 1968, Cave development via the sulfuric acid reaction: National Speleological Society Bulletin, v. 30, p. 1-10.

Müller, P., and Sarvary, I., 1977, Some aspects of developments in Hungarian speleology theories during the last 10 years: Karszt es Barlang, special issue, p. 53-60.

Mylroie, J.E., 1977, Speleogenesis and karst geomorphology of the Helderberg Plateau, Schoharie County, New York: Rensselaer Polytechnic Institute, Ph.D. dissertation, 336 p.

Mylroie, J.E., and Carew, J.L., 1990, The flank margin model for dissolution cave development in carbonate platforms: Earth Surface Processes and Landforms, v. 15, p. 413-424.

Palmer, A.N., 1975, The origin of maze caves: National Speleological Society Bulletin, v. 37, p. 56-76.

Palmer, A.N., 1981a, Hydrochemical controls in the origin of limestone caves: 8th International Congress of Speleology, Bowling Green, Ky., Proceedings, p. 120-122.

Palmer, A.N., 1981b, The geology of Wind Cave: Hot Springs, S. Dakota, Wind Cave Natural History Association, 44 p.

Palmer, A.N., 1986, Prediction of contaminant paths in karst aquifers: Association of Ground Water Scientists and Engineers, 1st Conference on Environmental Problems in Karst Terranes, Proceedings, Dublin, Ohio, p. 32-53.

Palmer, A.N., 1987, Cave levels and their interpretation: National Speleological Society Bulletin, v. 49, p. 50-66.

Palmer, A.N., 1989, Stratigraphic and structural control of cave development and groundwater flow in the Mammoth Cave region, in White, W.B., and White, E.L., eds., Karst Hydrology-Concepts from the Mammoth Cave area: New York, Van Nostrand Reinhold, p. 293-316.

Palmer, A.N., 1991, Origin and morphology of limestone caves: Geological Society of America Bulletin, v. 103, p. 1-21.

Palmer, A.N., and Palmer, M.V., 1989, Geologic history of the Black Hills caves, South Dakota: National Speleological Society Bulletin, v. 51, p. 72-99.

Palmer, M.V., 1976, Groundwater flow paths in limestone solution conduits: State University of New York, Oneonta, NY, M.A. thesis, 150 p.

Pisarowicz, J.A., 1994, Cueva de Villa Luz-An active case of H_2S speleogenesis, in Sasowsky, I.D., and Palmer, M.V., eds., Breakthroughs in karst geomicrobiology and redox geochemistry: Karst Waters Institute, Special Publication, no. 1, p. 60-62.

Plummer, L.N., and Back, W., 1980, The mass balance approach: Application to interpreting the chemical evolution of hydrogeologic systems: American Journal of Science, v. 280, p. 130-142.

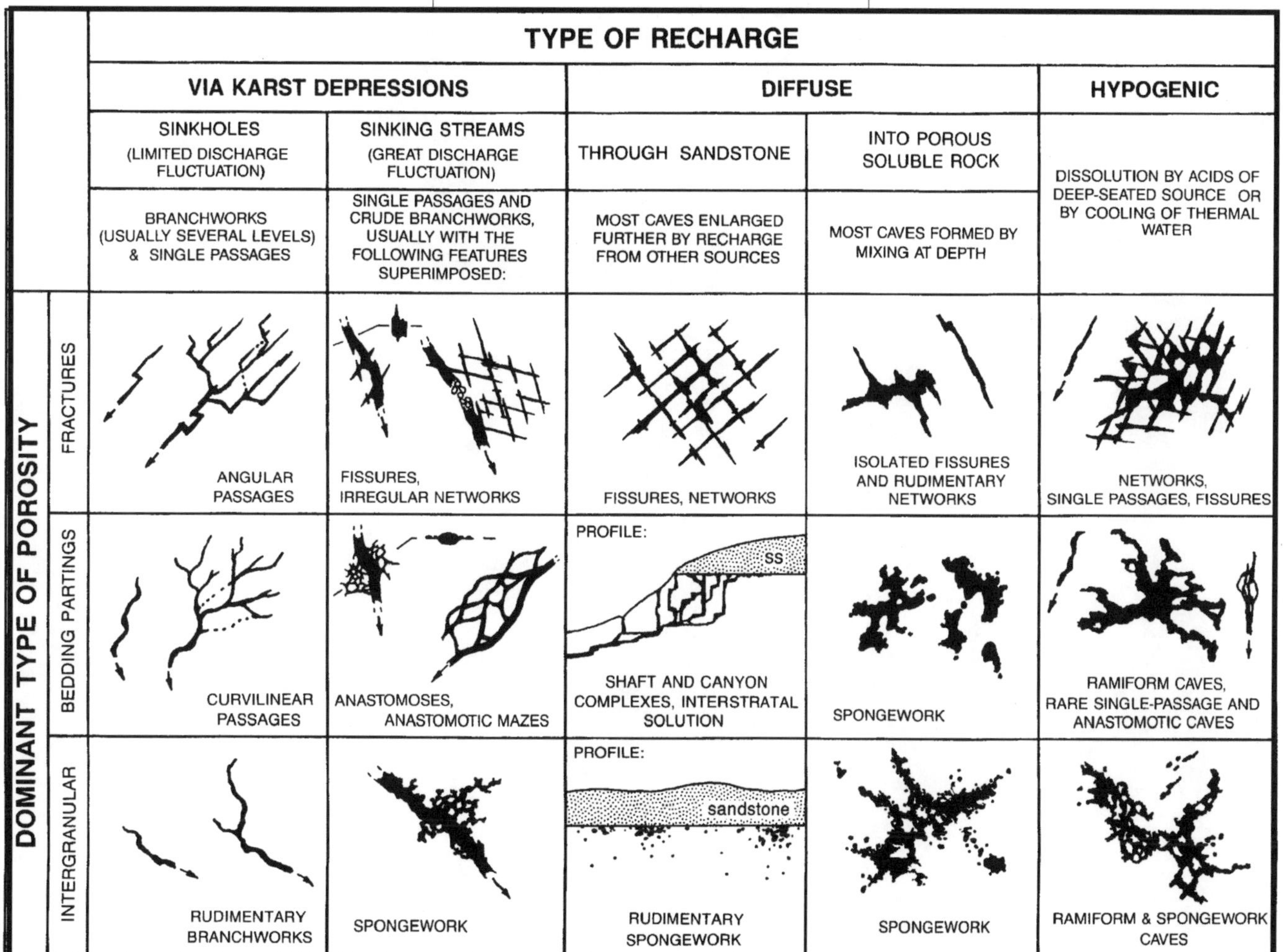

3.4 Figure 10. Cave patterns and their relationship to types of recharge and porosity. Maps are plan views unless otherwise noted. They are generalized to illustrate the dominant patterns. Many caves exhibit rudimentary forms, multiple stages of development, or combinations of more than one type (from Palmer, 1991).

Powell, R.L., 1970, Base level, lithologic, and climatic controls of karst groundwater zones in south-central Indiana: Indiana Academy of Sciences Proceedings, v. 79, p. 281-291.

Rubin, P.A., 1991, Flow characteristics and scallop-forming hydraulics within the Mill Pond Karst Basin, east-central New York, in Kastning, E.H., and Kastning, K.M., eds., Appalachian Karst: Pro-ceedings of Appalachian Karst Symposium, Radford University, p. 75-99.

Sasowsky, I.D., and White, W.B. 1994, The role of stress release fracturing in the development of cavernous porosity in carbonate aquifers: Water Resources Research, v. 30, p. 3523-3530.

Smart, C.C., 1983, The hydrology of the Castleguard Karst, Columbia Icefields, Alberta, Canada: Arctic and Alpine Research, v. 15, p. 471-486.

Smart, P.L., and Friederich, H., 1986, Water movement and storage in the unsaturated zone of a maturely karstified carbonate aquifer, Mendip Hills, England: Association of Ground Water Scientists and Engineers, 1st Conference on Environmental Problems in Karst Terranes, Proceedings, Dublin, Ohio, p. 59-87.

Thrailkill, J., and Robl, T.L., 1981, Carbonate geochemistry of vadose water recharging limestone aquifers: Journal of Hydrology, v. 54, p. 195-208.

Warwick, G.T., 1968, Some primitive features in British caves: 4th International Congress of Speleology, Ljubljana, Proceedings, p. 239-252.

White, W.B., 1969, Conceptual models for carbonate aquifers: *Groundwater,* v. 7, p. 15-21.

Williams, P.W., 1985, Subcutaneous hydrology and the development of doline and cockpit karst: Zeitschrift für Geomorphologie, v. 29, p. 463-482.

Worthington, S.R.H., 1984, The paleodrainage of an Appalachian fluviokarst: McMaster University, M.S. thesis, 218 p.

Worthington, S.R.H., 1991, Karst hydrogeology of the Canadian Rocky Mountains: McMaster University, Ph.D. dissertation, 227 p.

Worthington, S.R.H., 1994, The possible importance of sulfur minerals in initiating epigenic caves, in Sasowsky, I.D., and Palmer, M.V., eds., Breakthroughs in karst geomicrobiology and redox geochemistry: Karst Waters Institute Special Publication, no. 1, p. 80-82.

3.5. The Formation of Epikarst and Its Role in Vadose Speleogenesis

Alexander Klimchouk

Abstract

The epikarstic zone differs structurally from the underlying bulk rock mass, reflecting the higher degree of fissuring and diffuse karstification due to unloading, weathering, and dissolution processes that encompass this uppermost layer of exposed rocks. An initial distinction in the permeability between the epikarstic zone and the bulk rock mass below is caused largely by nondissolutional processes. Contrast in hydraulic conductivity allows some groundwater storage in the epikarstic zone and flow concentration at its base. Effective vertical leakage paths from the epikarstic zone (major tectonic fissures) commonly need no speleogenic initiation, as they are wide enough to support undersaturated flow through them. Shafts fed by epikarstic recharge represent headwaters of a conduit drainage system developed in the deeper parts of a massif.

A large majority of single shafts, several tens of meters deep (the most common feature among explored vertical caves), belong to this type of epikarst-fed shafts. The opening of these shafts to the surface through collapse and subsidence, with subsequent edge smoothing, is probably the main mechanism of doline formation in exposed karsts, rather than gradual preferential lowering of the surface. Therefore, focused karst landforms reflect, through specific mechanisms of epikarst morphogenesis, the permeability pattern of the upper part of the vadose zone.

Introduction

During the past two decades the enormous importance of the *epikarstic zone* (also referred to as the *subcutaneous zone*) in karst hydrology and morphogenesis has been widely recognized by karst researchers (Bakalowicz et al., 1974; Bonacci, 1987; Drogue, 1980, 1992; Ford and Williams, 1989; Gunn, 1978, 1981, 1983; Klimchouk, 1987, 1989, 1995; Mangin, 1973, 1975; Williams, 1983, 1985).

The epikarstic zone can be defined as the uppermost zone of exposed karstified rocks, in which permeability due to fissuring and diffuse karstification is substantially greater and more uniformly distributed in area, as compared to the bulk rock mass below (Klimchouk, 1997). Not only do the specific properties of this zone determine its peculiar function and role in karstification and speleogenesis, but also the contrast in permeability structure between the epikarstic zone and the underlying main vadose zone. (The bulk rock mass below the epikarstic zone, marked by intact blocks bounded by widely separated fissures, will be referred to here as the *block zone,* when the contrast in fissure permeability structure is emphasized, or as the *main vadose zone,* when the existence of an unsaturated zone with prevailing vertical percolation below the perched epikarstic aquifer is emphasized.) The ultimate hydrologic functions of the epikarstic zone are the storage and concentration of flow while it passes to the vadose zone.

At the base of the epikarstic zone, flow is focused along a few major fissures, and so is the speleogenetic development. Major tectonic fissures provide leakage paths from the epikarstic zone, and their preferential enlargement occurs from the base of the zone downward, resulting in the formation of "hidden" vertical shafts. Such shafts are headwaters draining the epikarstic zone above to conduit drainage systems developed in the deeper parts of a massif, although they are not necessarily genetically related to or contemporaneous with these systems. Opening of the shafts to the surface through collapse and subsidence, with subsequent edge smoothing, is probably the main mechanism of doline formation in exposed karst, rather than gradual preferential lowering of the surface (Klimchouk et al., 1979, 1981, 1996; Klimchouk, 1989, 1995). This illustrates well the point that the localized appearance of a karst landscape is a reflection of the pattern of speleogenesis in the upper part of the vadose zone rather than a result of surface hydrology itself.

The Nature of Epikarst

Most researchers emphasize the role of dissolution in the formation of epikarst, explaining the enhanced karst porosity at the uppermost layer of exposed carbonates by the high dissolutional potential of infiltrating water. Although this certainly is partly true, the matter is not so straightforward. The formation of the *initial contrast* in permeability between the top layer of an exposed rock and its bulk mass at depth is, in most cases, due to a combined effect of stress release during exposure and a set of weathering processes (3.5 Figure 1).

That dissolution is not the single factor generating these properties of the epikarstic zone is illustrated by the fact that some kind of stress release and weathering profiles—commonly referred to as weathering crusts or mantles—develop on any types of rocks exposed to the surface, not only on soluble rocks. On noncarbonate rocks, however, the generation of fine weathering products and lack of possibilities for suffosional removal of fine clasts via underground drainage routes limits the permeability of the weathering crust, in spite of enhanced fissure abundance. Limited internal outflow of weathered material favors deep chemical and mineralogical transformation of a rock, which further decreases permeability.

In contrast, on carbonates, dissolution predominates over other weathering mechanisms. And not only dissolved matter but also fine clastic weathering products can be removed internally via progressively enlarging shaft drainage systems below. *In situ* chemical and mineralogical transformations of the rock in epikarst are not as deep as these in weathering crusts on noncarbonate rocks. The common result is that permeability of the uppermost zone on carbonates progressively increases as compared to the underlying block zone. Thus, the formation of a weathering crust on nonsoluble rocks, and the formation of the epikarstic zone in carbonates, should be regarded as different types of hypergenesis. Based on similar reasoning, Tsykin (1976) suggested that a distinction should be made between weathering and karst processes in general, although a wider approach is more common, which treats karst as being one of components of weathering.

As a rock approaches the surface after burial and the weight of overlying rocks decreases, stress fields in it substantially rearrange, and stresses are relieved. This alters existing fissures and allows the formation of new ones. When the uplift and denudation are at a low-rate balance, stress-release effects commence at greater depths, although these effects are biased toward increase in fissure widths rather than the formation of additional new fissures. At a high-rate denudation balance, or at a balance shifted toward uplift, stress-release effects are expressed at shallower depths and produce more new fissures. The resulting changes in fissure character commonly become quite prominent at depths of 15 to 30 m. Fissure widths increase exponentially toward the surface. Frequency of fissures increases geometrically (Chernyshev, 1983). Also, the variation in widths increases considerably. These changes are less pronounced in fissures that are not parallel to the surface.

Weathering of an exposed rock (physical, chemical, and biogenic) strongly affects the fissure network. Most common effects include further widening of fissures, increase in abundance of the existing sets, and formation of new fissures, frequently of unusual orientation. Changes induced by weathering are more pronounced, and penetrate to a greater depth along the existing discontinuities. Dispersion of fissure parameters increases as a result. However, the overall effect on permeability depends on climate and on the properties of the rock. If the generation of fine-grained products exceeds the ability of a system to remove the residue via the underground drainage network, then permeability will decrease. The opposite situation is common for carbonates and produces the epikarstic zone. Therefore, the inheritance of older characteristics during karst evolution plays an important role (see Klimchouk and Ford, Chapter 3.2).

The above discussion demonstrates that the structural prerequisites for the formation of the epikarstic zone evolve as a rock is emerging to the shallow subsurface and, eventually, to the surface. Drastic changes in fissure characteristics—increase in fissure widths in particular—make inapplicable the low-order dissolution kinetics and the breakthrough concept to the interpretation of the formation of epikarst as suggested by Dreybrodt and Siemers (Chapter 4.2.4). The initial contrast in the permeability structure between the epikarstic zone and the block zone below is caused largely by nondissolutional processes.

The evolution of epikarst depends strongly on whether the vadose zone exists below the unloading-weathering zone. In most cases a vadose zone does exist below it because of the high hydraulic capacity of prominent tectonic fissures and preexisting conduit systems draining the epikarst. Vertical leakage paths from the epikarstic zone need no speleogenic initiation, as they are commonly wide enough to support undersaturated flow through them. After quick additional enlargement by dissolution or even without it, such vertical paths permit free flow of water films or sheets on the walls to form shafts of virtually unlimited hydraulic capacity. The epikarstic aquifer is perched above the main vadose zone because of the large contrast in porosity structure with the bulk rock mass below and paucity of major drains in it, not because of low hydraulic capacity of these major drains.

It is less common that epikarst evolves without a vadose zone beneath it, the situation assumed in the model of Dreybrodt and Siemers (Chapter 4.2.4). This situation occurs in coastal and oceanic settings when young carbonates emerge without substantial burial (see Mylroie and Carew, Chapter 5.1). However, structural prerequisites for the formation of epikarst in the case of young diagenetically immature carbonates with high fabric-selective porosity are quite different from the more common situation in well-indurated older rocks exposed after burial.

Major Features of Epikarstic Hydrology

Epikarstic Storage

Dispersed downward infiltration of water within the epikarstic zone becomes more difficult with depth. This is due to diminishing fissure abundance and width, so that flow acquires a considerable lateral component, with convergence toward major fissures penetrating deeper into the vadose zone and possessing high hydraulic conductivity (3.5 Figure 2). The storage capacity of epikarst depends on its maturity and on the contrast in hydraulic conductivity between the epikarstic zone and the underlying vadose zone, so that this capacity changes with karst evolution. Storage in the epikarstic zone results in significant delay in the throughput of percolating water. Hydrochemical and isotope studies from various regions demonstrate that such delay can be from several days to a few months (Gunn, 1983; Klimchouk and Jablokova, 1989; Williams, 1983).

Retention in the epikarst causes leakage to the vadose zone to be rather uniformly distributed through time. Epikarstic storage can support a base flow to shafts through quite prolonged dry periods. Williams (1983) has developed an approach to hydrograph analysis that allows the epikarstic storage component to be identified. Hydraulic response of shaft flow below the epikarstic zone to precipitation events depends on the storage volume available (maturity of epikarst, and storage accumulated during the preceding period). If the epikarstic storage is small, hydraulic response of shafts can be considerably "smoothed" as most precipita-

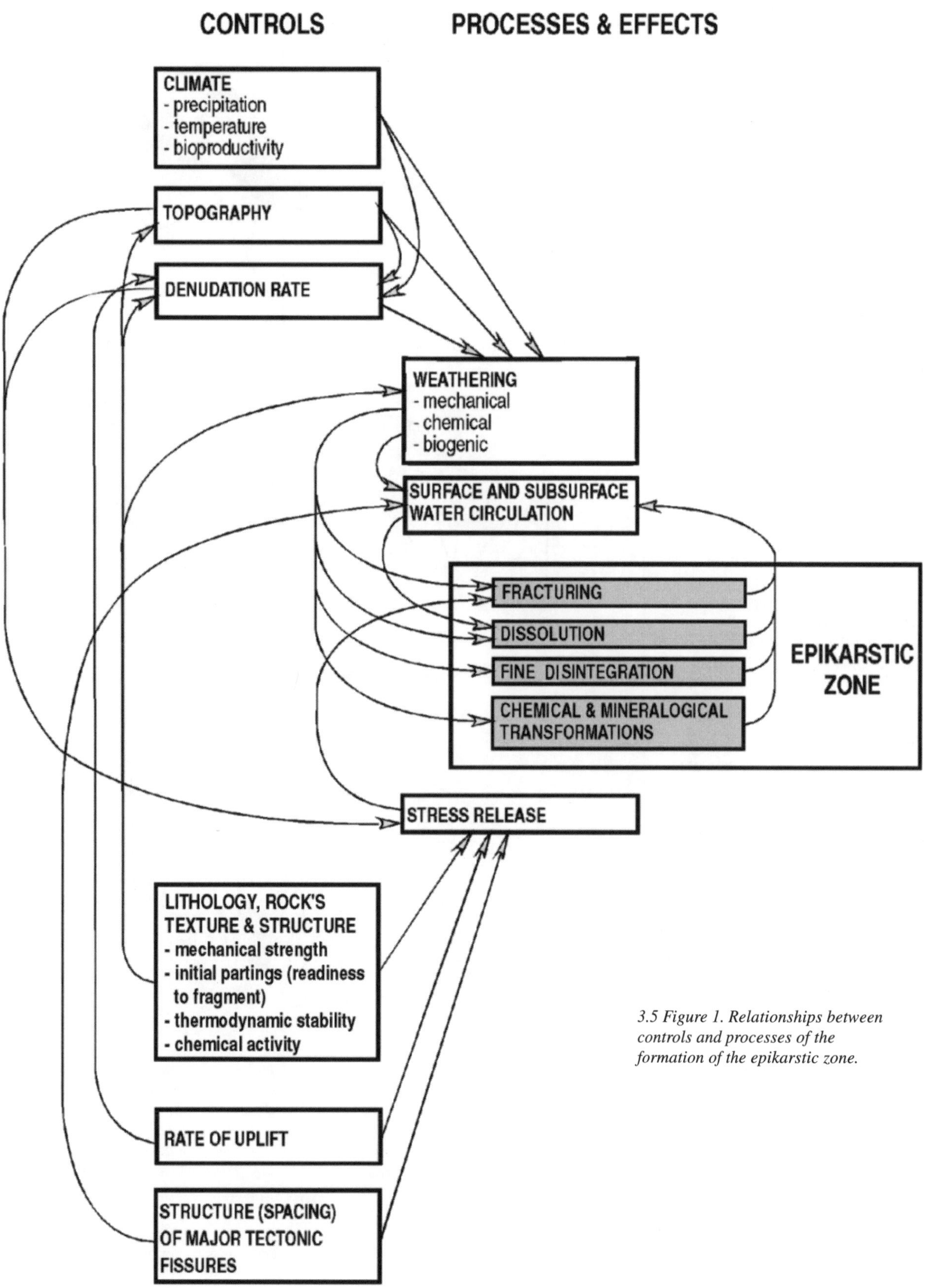

3.5 Figure 1. Relationships between controls and processes of the formation of the epikarstic zone.

tion will be retained in the epikarst. At high storage the more pronounced response of shaft flow to a precipitation event is caused mainly by a piston effect where inflow at the top causes water to be released at the bottom.

Chemical responses are commonly obscured due to a tendency for homogenization of water within the epikarstic aquifer. However, actual responses strongly depend on the local configuration of epikarstic permeability and deeper drains.

Epikarstic aquifers are commonly irregular in area as is the epikarstic zone. The continuity of an epikarstic aquifer in an area is strongly dependent on the distribution and hydraulic capacity of vertical leakage paths (major tectonic fissures) draining water down to the main vadose zone (see 3.5 Figure 2B). As the frequency of leakage points from the epikarstic zone and their hydraulic capacity increases with time, the ability of epikarst to retain water diminishes despite an increasing porosity of the epikarstic zone itself.

Flow Concentration

At the exposed limestone surface, possibilities for concentration of overland flow are commonly limited. A high abundance of wide fissures near the surface causes diffuse recharge to overwhelmingly predominate. Even in areas with well-developed doline relief, overland flow still is generally unimportant, and the dominant concentration is through the epikarst (Gunn, 1983). Yet within the epikarstic zone and at its base, the permeability pattern is organized to converge flow into major tectonic fissures that act as leakage paths to the underlying vadose zone. It is a percolation threshold at the base of the epikarstic zone that causes flow to concentrate while passing from the epikarstic zone to the vadose zone (3.5 Figure 3, A).

This feature is of great importance for speleogenesis, and more generally, for karst morphogenesis. It means that, despite diffuse recharge on the exposed karst surface, flow and hence speleogenetic development become localized at certain depths below the surface, and that input to the main vadose zone is predominantly focused.

This resolves an apparent contradiction between the dispersed mode of recharge on exposed karst surfaces and the localized appearance of most karst features, especially shafts in the vadose zone. Dolines at the surface seem to be the result of epikarstic flow concentration and speleogenesis at the base of the epikarstic zone, rather than the primary causes of flow concentration.

Many karst researchers emphasize the distinction between the two main types of authigenic waters in the vadose zone. The above mechanism of flow concentration generates a kind of input to the vadose zone that is referred to in various works as shaft flow, vadose flow, and rapid percolation. Another type of flow occurs where the water leaves the epikarstic zone through a series of narrow fissures as slow seepage. This is also referred to as vadose seepage and slow percolation. These types differ remarkably in their chemistry and capacity to perform dissolution in the vadose zone. When collected below the epikarstic zone, vadose seepage is commonly saturated to highly supersaturated with respect to calcite, with large spatial variations in chemistry. Shaft flow remains undersaturated in most cases. Studies by Gunn (1983) suggest that shaft

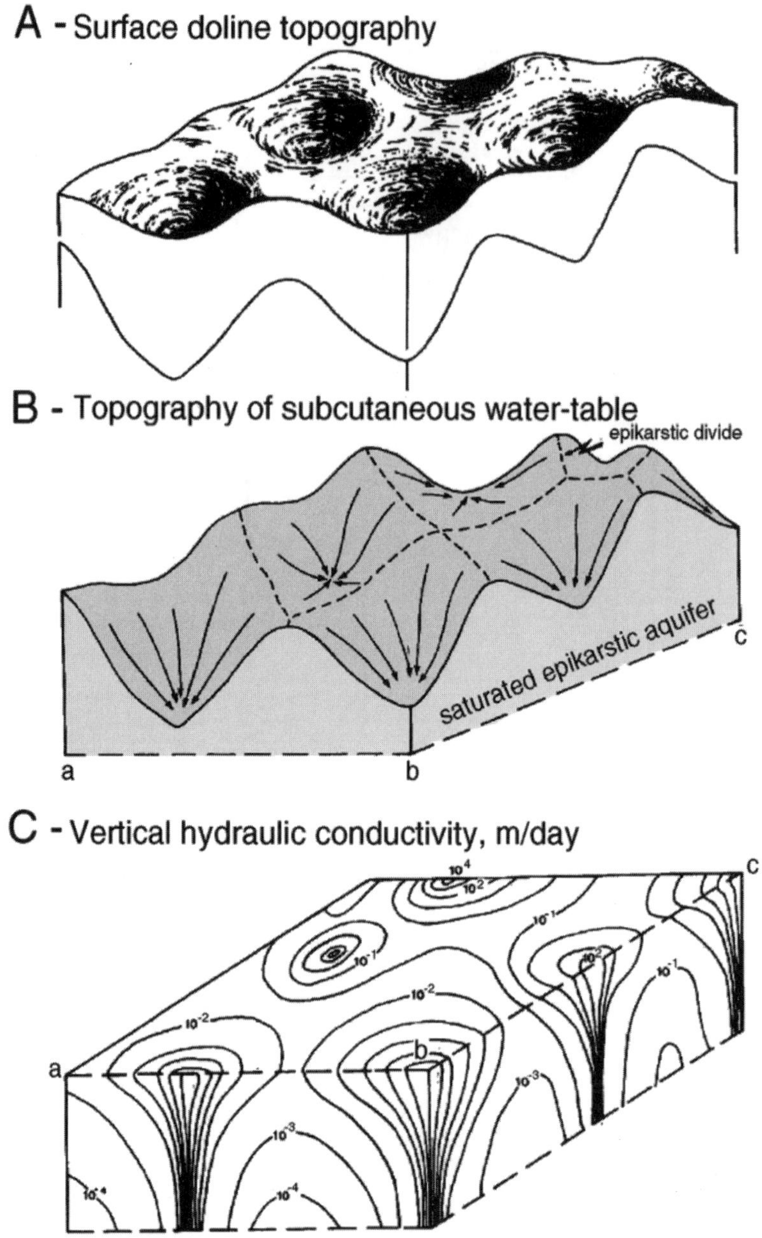

*3.5 Figure 2. The relationship between **A** surface doline topography, **B** underlying relief on the epikarstic water table, and **C** vertical hydraulic conductivity near the base of the epikarstic zone (from Williams, 1985). Note that the surface doline topography depicted in **A** corresponds to a mature karst landscape. Dolines evolve in the course of epikarstic morphogenesis in response to specific hydrologic patterns **B** and **C**, which are present even during earlier stages, before dolines have appeared.*

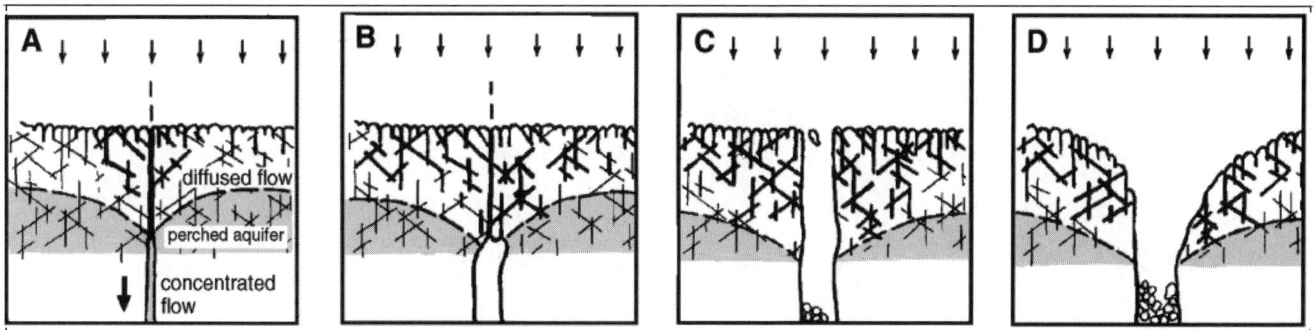

3.5 Figure 3. Model of the development of shafts at the base of the epikarstic zone and associated landforms at the surface (modified from Klimchouk et al., 1979). ***A*** *is the initial stage,* ***B*** *is the "hidden" shaft stage,* ***C*** *is the collapse stage,* ***D*** *is the doline stage.*

flow is the predominant input to an aquifer, whereas vadose seepage accounts only for small proportion of the total input.

The Formation of Shafts and Related Forms of Epikarst

The Conceptual Model

Flow concentration at the base of the epikarst along major tectonic fissures causes the formation of vertical shafts in the vadose zone (3.5 Figure 3). Preferential enlargement of deeply penetrating fissures draining the epikarst extends from the base of the epikarstic zone downward. Shafts initially develop as "hidden" features (stages A and B). Enlargement of large open fissures is due to dissolution by free-flowing sheets or films along the vertical fissure walls. Leakage of water from the epikarstic aquifer causes a drawdown cone to form around the shaft inlet. Because water that infiltrates from the surface through the epikarstic zone *within* a drawdown cone has shorter time and distance of dissolution than water infiltrating from outside the drawdown limits, enlargement of fissures in the epikarst above the drawdown cone is faster than elsewhere in the area. Moreover fine soil grains are washed out through the epikarst more readily within the drawdown cone. This results in enhanced karstification of the rock mass above the shaft, and the formation of a karren field at the surface with a diameter roughly equal to that of the drawdown cone (3.5 Figure 3B; see also 3.5 Figure 4 and 3.5 Plate 1).

The recharge-drainage confguration and growth mechanism of the shafts at the base of the epikarstic zone are somewhat similar to those of shafts formed due to leakage from overlying insoluble aquifers, which are abundant through much of the eastern United States (see Chapter 5.3.9). The shafts in both cases enlarge by retreat of vertical walls due to the dissolving action of sheets or films of water streaming down the walls, as described by Brucker, Hess and White (1972). White (1990) generalized that shaft waters range from slightly to highly undersaturated with respect to calcite. Palmer (Chapter 5.3.8) considers that water renews its aggressiveness when passing from essentially closed conditions to aerated space in a shaft.

That water leaving the epikarstic zone along major fissures is aggressive and capable of further dissolution is evidenced by the very existence of the shafts and by solution morphology of fluted walls. However, variations in dissolution chemistry, and hence in wall retreat rate, can be substantial depending on climate, lithology, and thickness of the epikarst. Temporal variations in chemical and isotope composition of shaft flow are usually low due to the homogenizing effect of the epikarstic aquifer (Gunn, 1983; Klimchouk and Yablokova, 1989).

An important difference between the above two types of shafts is that those beneath insoluble rocks, most commonly sandstones, are not associated with any substantial changes in the structure of the caprock. In contrast, the epikarstic roof above a growing shaft increasingly breaks apart into boulders due to an elaborate network of widening fissures and bedding planes, eventually resulting in collapse (3.5 Figure 3C). The shaft acquires an open entrance at this stage. Further enlargement rapidly leads to the formation of a wider mouth of the shaft, as the rocks around it are already unstable (3.5 Figure 3C). If the shaft is choked by boulders and slope wash, it can assume the bowl shape of a typical solution doline. Quite possibly, most dolines in mature exposed karsts, which are conventionally interpreted as solution dolines, are actually formed by the above mechanism, which includes collapse of the shaft roof and reworking of the shaft mouth. Considering their distinct genetic and morphologic characteristics (the latter are discussed elsewhere below), shafts formed by focused recharge at the base of the epikarstic zone can be distinguished as a specific type of vadose shaft termed *epikarst-fed shafts*.

The Role of Condensation Processes

With growing karst porosity in a massif, air exchange increases between the outside atmosphere and caves—hence condensation processes play an increasing role. Studies have shown that condensation recharge can be significant in the water budget of karst massifs. According to calculations by V. Dublyansky, condensation contributes from 0.7 to 2.6 L/s/km^2 to the average annual flow in the low-altitude karst massifs in Crimea and West Caucasus, and from 3.8 to 4.7 L/s/km^2 in the high mountainous karsts of West Caucasus (Arabika and Bzybsky Ranges). Because condensation recharge is active only during the warm season, flow produced by condensation is much higher then: from 1.4 to 6.1 L/s/km^2 in low-altitude mountains, and from 7.9 to 9.7 L/s/km^2 in high mountains (Dublyansky and Kiknadze, 1984; Dublyansky et al., 1983; see also Chapter 3.6 in this volume).

During warm periods in mountainous karst massifs, air flows into upper entrances on the plateau surface, and condensation recharge takes place, contributing to karst flow. The epikarstic zone undergoes the most active condensation, as this zone is transitional between the outside and inside atmospheres (3.5 Figure 5A). During cold periods, air outflow prevails from the interior of the karst massif through upper entrances, but little moisture outflow occurs, because the epikarstic zone is the zone of cooling, so that moisture from ascending cave air condenses there (3.5 Figure 5B). Although it provides little or no contribution to karst flow from a massif as a whole, condensation during the cold period can contribute considerably to the corrosional enlargement of epikarst-fed shafts, particularly at their tops.

The above consideration is largely speculative—although it shows that besides infiltration, condensation can also be of great significance in karst morphogenesis in the epikarstic zone and, in particular, in speleogenesis at its base. This supposed role should be further assessed by analytical and experimental studies.

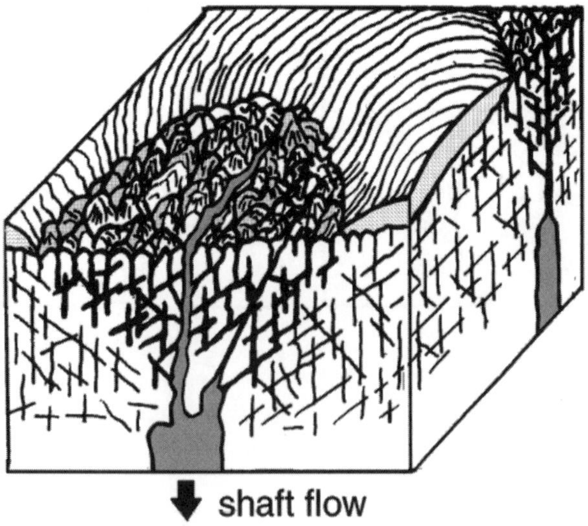

3.5 Figure 4. Block diagram showing the internal structure of the epikarst above a shaft, and the relationship between the development of shafts and karren fields (from Klimchouk, 1995).

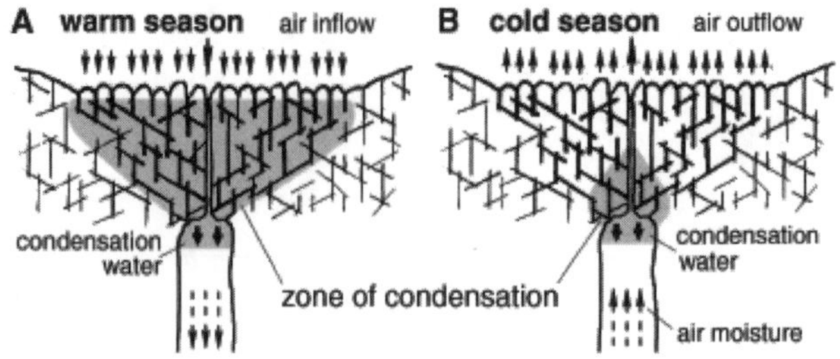

*3.5 Figure 5. Hypothetical localization of condensation in the epikarstic zone during warm **A** and cold **B** seasons.*

3.5 Plate 1. Karren fields with underlying shafts on the Kyrktau Plateau, Zerafshan Range, Tjan'-Shan', Uzbekistan. A major tectonic fissure is well recognizable that drains the epikarstic zone and guides the shafts (photo by the author).

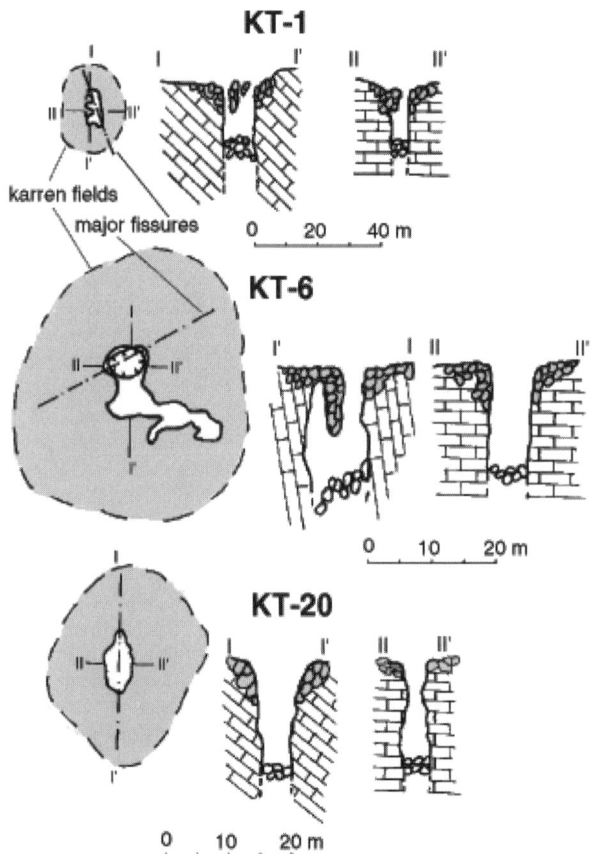

3.5 Figure 6. Typical examples of vertical shafts without associated dolines around entrances, located within karren fields, Kyrktau Plateau, Zerafshansky Range, Central Asia (From Klimchouk et al., 1981).

Field Evidence and Discussion

The above model was suggested initially to explain the formation of "hidden" shafts and shafts without entrance catchments in the Kyrktau Plateau of central Asia (Klimchouk et al., 1979, 1981). Later exploration has shown that such features are typical for many limestone massifs of this region. Most shafts are clearly associated with distinct karren fields (3.5 Figure 6). In many cases it is possible to reach well-developed hidden shafts via enlarged cracks within a karren field (see 3.5 Figure 4).

Speleologic explorations in many regions of the world provide many examples of shafts in the upper part of the vadose zone, accessible from the bottom or from side "windows" that extend upward and have blind terminations some meters below the surface. Maucci (1975) and Rossi and Sauro (1977) made early attempts to attribute their genesis to bedding-plane circulation and local perching of water in the upper part of the vadose zone. Shafts without entrance catchments are more common (stage C in 3.5 Figure 3), and are positioned in landscape in such a way that they receive no surface runoff at all. For such cases the assumption was commonly made that catchments once existed in the past, then were removed by various surface processes.

Recent study of epikarst in limestone quarries on the Sette Communi Plateau (Venetian Pre-Alps, Italy) has provided much excellent evidence for the validity of the above model (Klimchouk et al., 1996). Inspection of numerous quarries has shown that "hidden" shafts are very common features. They are cut by quarry levels or walls (3.5 Figure 7 and 3.5 Plate 2, p. 98). In the latter case, blind heads can be observed at some depth below the surface, depending on location and stratigraphic setting. The cross sections range from rounded to remarkably elongated along a guiding fissure, widths vary between 1 to 3 meters, and depths are typically about 15 to 20 m.

In any given area of uniform structure and stratigraphy, a set of vertical tectonic fissures that intersects the epikarstic zone and penetrates deeper into the vadose zone has a characteristic vertical range. Such fissures terminate downward at prominent bedding planes. Drains from shaft bottoms are commonly bedding-plane conduits or meandering conduits following vertical fissures confined to a single bed. Drains are small compared to the shaft cross sections, as has been noted for shafts fed by an overlying aquifer (Brucker et al., 1972). These drains are commonly of unexplorable size, unless the shaft is superimposed on a larger older conduit.

When epikarst-fed shafts open up to the surface after collapse of the epikarstic roof, their bottom parts are normally choked by breakdown boulders. This explains the remarkable bias toward 10 to 40-m-deep single "blind" shafts in the distribution of explored vertical caves. For instance, among 620 vertical caves with entrances on the plateaus in the Crimea Mountains, Ukraine, 366 (59%) are single shafts with depths up to 80 m, which have been interpreted as being dissolved by snowmelt (Dublyansky, 1977). Of them 67% are shafts 5 to 20 m deep, and 84% are shafts 5 to 30 m deep. It seems that most of these shafts are of the epikarst-fed type rather than related to snow melt. The fact that they commonly accommodate snow deposits is a consequence, rather than the cause, of their formation.

Only a few epikarst-fed shafts provide explorable access to deeper parts of a drainage system. Small drains from different headwater shafts converge downstream, increasing in discharge, and eventually drop down where intercepted by an underlying series of master fissures. Internal shafts in the deeper parts of a thick vadose zone differ from the epikarst-fed shafts by their greater flow. Decreasing solutional capacity of water is compensated for by the increasing erosional capability of the flow. Each next-lower meandering horizon between vertical drops serves to converge subsystems of increasing order, so that flow concentration progressively increases with depth. This is the typical structure of complex mountainous cave systems. Interception of fossil phreatic levels may further complicate the structure.

Some Implications for Karst Morphogenesis

It follows from the model described above that karren fields, hidden shafts, epikarst collapses, and dolines are genetically associated forms representing successive stages of the epikarstic process (Klimchouk, 1995). In immature karst landscapes, dolines are relatively scarce, while prevailing forms are karren fields (grikes or Kluftkarren) and shafts with fissure-like entrances in overlying epikarst or with open entrances (collapsed roofs) but without associated catchments. Hidden shafts are abundant, but most of them are inaccessible. Mature karst landscapes demonstrate a prevalence of dolines formed due to choking of original shafts by breakdown and washed-in material.

These relationships should not be taken as a strict rule, however, as many regional and local factors such as climatic conditions, tectonic structure and regime, and exposure of different stratigraphic units, may cause local deviations. Moreover, in a given mature karst landscape, different stages of epikarstic morphogenesis may coexist, as denudation will cause the epikarst to be intercepted by large fissures that previously were not hydraulically connected to the epikarstic aquifer. Hence, the development of new epikarst-fed shafts and associated early-stage forms may commence within an otherwise mature karst landscape.

Another complication arises where high-energy geomorphic processes, such as glaciation, have scoured previous epikarst and exposed largely unweathered rocks. Afterward, a new epikarstic zone starts to form, adjusting not only to prominent deeply penetrating fissures, but also to former internal shafts now open to the new surface or positioned just below the surface. The cycle of epikarstic morphogenesis is renewed, with peculiar features caused by the remarkable role of inheritance of previously formed karst permeability. Field examples, abundant in many high-alpine karsts that have experienced Pleistocene glaciations, demonstrate that the time since the last glaciation seems to be great enough to form moderately mature epikarst with young shafts beneath, although this time has not been sufficient to reach the stage of collapse and doline formation. The present dolines in such areas are commonly scarce, formed mainly due to reworking of remnants of older shafts cut by glacial scouring.

Conclusions

- **1** The epikarstic zone is drained mainly via major tectonic fissures conducting water into the underlying vadose zone. Shafts fed by epikarstic recharge form at the base of the epikarstic zone and represent the headwaters of conduit drainage systems developed in the deeper parts of a massif. It is likely that a large majority of single 10 to 40 m shafts, "blind" at the bottom, are epikarst-fed shafts. These are the most common features among explored vertical caves.

- **2** Structural prerequisites for the formation of the epikarstic zone evolve as a rock emerges into the shallow subsurface and eventually to the surface. An initial distinction between the permeability structure of the epikarstic zone and that of the bulk rock mass below is largely the result of nondissolutional processes. Vertical leakage paths from the epikarstic zone (major tectonic fissures) commonly need no speleogenic initiation, as they are wide enough to transmit undersaturated flow. After quick additional enlargement by dissolution, or even without it, such vertical paths permit free flow of water films or sheets on the walls to form shafts of virtually unlimited hydraulic capacity.

- **3** Karren fields forming at the surface above drawdown cones in the epikarstic aquifer, hidden shafts serving as leakage paths from this aquifer, epikarst collapses, and dolines are genetically related forms representing successive stages of the

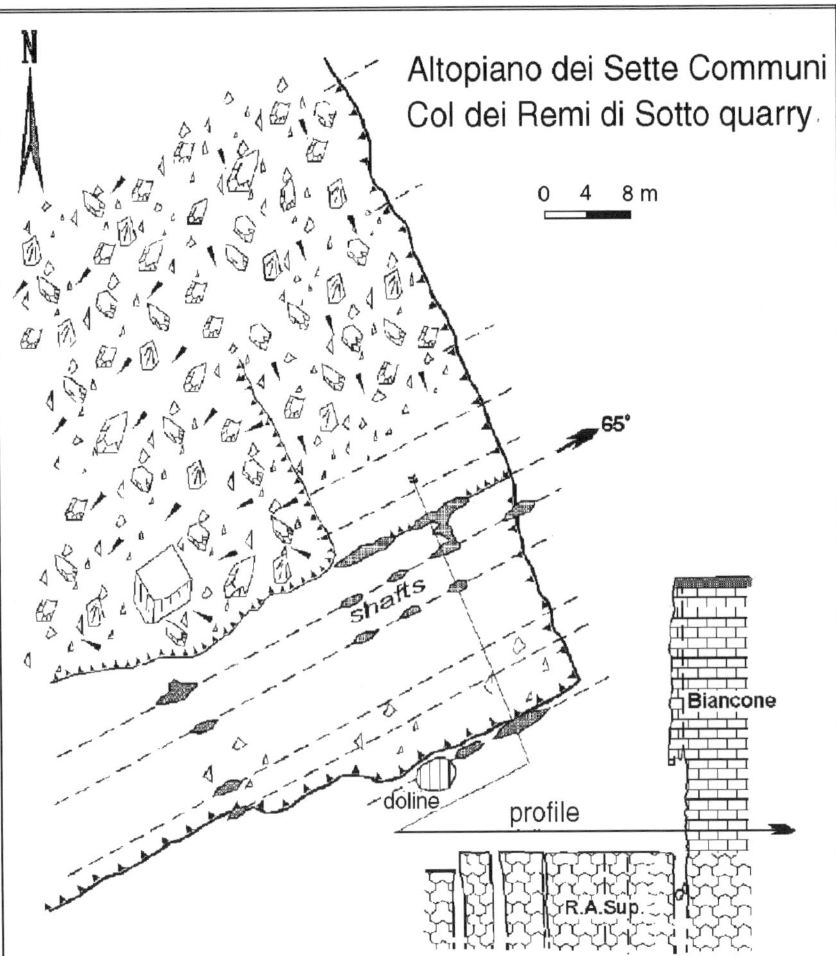

3.5 Figure 7. Plan and profile of shafts at the base of the epikarstic zone cut by quarry levels and walls in the Sette Commini Plateau, Venetian Pre-Alps, Italy (from Klimchouk et al., 1996). The shaft on the right of the profile is cut by the wall and shows the position of a blind head beneath the surface.

3.5 Plate 2. Shafts at the base of the epikarstic zone cut by quarry levels and walls on the Sette Commini Plateau, Venetian Pre-Alps, Italy (photo by the author).

epikarstic morphogenetic process. Opening of the shafts to the surface through collapse and subsidence, with subsequent edge smoothing, is probably the main mechanism of doline formation in exposed karst, rather than gradual preferential lowering of the surface.

- **4** Surface karst landforms reflect, through specific epikarstic processes, the permeability structure of the upper part of the vadose zone.

Acknowledgements

I thank Derek Ford and Arthur Palmer for their review of this paper and valuable comments. I am indebted to Arthur and Margaret Palmer for their careful and professional editing of my English, which helped to make the paper more clear and consistent.

References

Bakalowicz, M., Blavoux, B., and Mangin, A., 1974, Apports du tracage isotopique naturel a la connaissance du fonctionnement d'un systeme karstique-teneurs en oxygene-18 de trois systemes des Pyrenees, France: *Journal of Hydrology,* v. 23, p. 141-158.

Bonacci, O., 1987, *Karst hydrology, with special reference to the Dinaric Karst:* Berlin-Heidelberg, Springer-Verlag, 184 p.

Brucker, R.W., Hess, J.W., and White, B., 1972, Role of vertical shafts in the movement of ground water in carbonate aquifers: *Ground Water,* v. 10, no. 6, p. 5-13.

Chernyshev, S.N., 1983, *Fissures of rocks:* Moscow, Nauka, 240 p. (in Russian).

Drogue, C., 1980, Essai d'identification d'un type de structure de magasins carbonates fissures: Application a l'interpretation de certains aspects du fonctionnement hydrogeologique: *Mémoires de Société Géologie France,* v. 11, p. 101-108.

Drogue, C., 1992, Hydrodynamics of karstic aquifers: Experimental sites in the Mediterranean karst, southern France, *in* Back, W., ed., *Hydrogeology of Selected Karst Regions:* Hannover, Heise, p. 133-149.

Dublyansky, V.N., 1977, *Karst caves and shafts of the Crimea Mountains:* Leningrad, Nauka, 18 p. (in Russian).

Dublyansky, V.N., Kiknadze, T.Z., and Kolomietz, A.L., 1983, Vertical zonation of condensation in the fissure-karst collectors of western Georgia: *Bull. of the Acad. of Sci. of the Georgian SSR,* v. 110, p. 333-336. (in Russian).

Dublyansky, V.N., and Kiknadze, T.Z., 1984, *Hydrogeology of karst of the alpine folded region of the south of the USSR:* Moscow, Nauka, 126 p. (in Russian).

Ford, D.C., and Williams, P.W., 1989, *Karst Geomorphology and Hydrology:* London, Unwin Hyman, 601 p.

Gunn, J., 1978, *Karst hydrology and solution in the Waitomo District, New Zealand:* University of Auckland, Auckland, Ph.D. thesis, 315 p.

Gunn, J., 1981, Hydrological processes in karst depressions: *Zeitschrift für Geomorphologie,* v. 25, p. 313-331.

Gunn, J., 1983, Point recharge of limestone aquifers—A model from New Zealand karst: *Journal of Hydrology,* v. 61, p. 19-29.

Klimchouk, A.B., 1987, Conditions and peculiarities of karstification in the near-surface zone of carbonaceous massives, *in Caves of Georgia* 11: Tbilisi, Metzniereba, p. 55-65 (in Russian, res. English).

Klimchouk, A.B., 1989, *The significance of the subsurface zone in karst hydrogeology and morphogenesis:* Kiev, Institute of Geologic Science, 43 p. (in Russian, res. English).

Klimchouk, A.B., 1995, Karst morphogenesis in the epikarstic zone: *Cave and Karst Science,* v. 21, p. 45-50.

Klimchouk, A.B., 1997, The nature and principal characteristics of epikarst: *12th International Congress of Speleology,* La Chaux-de-Fonds, Proceedings, v. 1, p. 306.

Klimchouk, A.B., Stotland, A.B., and Lomaev, A.A., 1979, Karst and speleological investigations on the Kyrktau Plateau (Zeravshansky Range, Tjan-Shan): *Izvestija VGO,* v. 3, no. 5, p. 442-448 (in Russian).

Klimchouk, A.B., Rogozhnikov, V.J., and Lomaev, A.A., 1981, *Karst of the Kyrktau Massif (Zeravshansky Range, Tjan-Shan):* Kiev, Institute of Geologic Science, 54 p. (in Russian).

Klimchouk, A.B., Sauro, U., and Lazzarotto, M., 1996, "Hidden" shafts at the base of the epikarstic zone: A case study from the Sette Communi Plateau, Venetian Pre-Alps, Italy: *Cave and Karst Science,* v. 23, no. 3, p. 101-107.

Klimchouk, A.B., and Jablokova, N.L., 1989, Evidence of hydrological significance of the epikarstic zone from study of oxygen isotope composition of water, Arabika Massif, western Caucasus: *10th International Congress of Speleology,* Budapest, Proceedings, v. 3, p. 798-799.

Mangin, A., 1973, Sur la dynamique des transferts en aquifer karstique: *6th International Congress of Speleology,* Olomouc, Proceedings, v. 4, p. 157-162.

Mangin, A., 1975, *Contribution à l'étude hydrodinamique des aquiferes karstiques:* University of Dijon, DES thesis, Ann. Spéléol., Paris, v. 29, no. 3, p. 282-332; no. 4, p. 495-601; v. 30 no. 1, p. 21-124

Maucci, W., 1975, Il fenomeno dell'erosione inversa come contributo allo studio della speleogenesi: *Le Grotte d'Italia,* ser. 4, v. 4, p. 235-285.

Rossi, G., and Sauro, U., 1977, L'Abisso di Lesi: Analisi morfologica e ipotesi genetiche: *Le Grotte d'Italia,* ser. 4, v. 6, p. 73-100.

Tsykin, R.A., 1976, Weathering crust and karst, *in* Weathering crust, Moscow, Nauka, v. 25, p. 186-193.

White, W.B., 1990, Surface and near-surface karst landforms: *Geological Society of America Special Paper,* no. 252, p. 157-175.

Williams, P.W., 1983, The role of the subcutaneous zone in karst hydrology: *Journal of Hydrology,* v. 61, p. 45-67.

Williams, P.W., 1985, Subcutaneous hydrology and the development of doline and cockpit karst: *Zeitschrift für Geomorphologie,* v. 29, p. 463-82.

3.6. The Role of Condensation in Karst Hydrogeology and Speleogenesis

Victor N. Dublyansky and Yuri V. Dublyansky

Abstract

Condensation in karst occurs over a wide range of natural settings at latitudes from 25° to 70° and altitudes from sea level to 2600m. In summer (April through September), condensation introduces a significant amount of water into the karst massifs (from 0.1% to as much as 20% of the total dry-season runoff). Contrary to common belief, in winter evaporation does not withdraw appreciable amounts of water from the massifs. Evaporating at depth, the water condenses near the surface within the epikarstic zone or on the snow cover and flows back. Condensation can sustain springs during prolonged dry periods (such as summer and winter) when there is no recharge by liquid precipitation.

Condensation can play a significant role in speleogenesis, and many forms of cave macro-, meso-, and micromorphologies are attributable to condensation corrosion. It can be particularly efficient in the latter stages of hydrothermal cave development (during partial dewatering) when the temperature and the humidity gradients are highest. Coupled with evaporation, air convection, and aerosol mass transfer, condensation can play a crucial role in the formation of several speleothems, and can create peculiar patterns of cave microclimate.

Introduction

Early discussions advocating the condensation of moisture in caves can be found in texts of the ancient Greek philosophers Thales of Miletus, Empedocles, and Aristotle (6th to 4th centuries B.C.). Marcus Vitruvius Pollio (1 A.D.) stated his belief that moisture can condense from the vapors of hot water rising from depth in the Earth (Fedoseev, 1975). During the following 15 centuries, natural phenomena were interpreted in accordance with theologic dogmas.

It was only in the 16th and 17th centuries that the European philosophers, Georgius Agricola, Rene Descartes, and Athanasius Kircher brought the ideas of ancient natural philosophy back into scientific use. In 1877, Volger offered a more or less comprehensive hypothesis about formation of groundwater by the condensation process. However, it came under severe criticism (Hann, 1880), and after a relatively short debate was abandoned. Lebedev revived the ideas of Volger in a series of publications between 1908 and 1926 that went largely unnoticed outside of Russia due to the unfavorable historic background, World War 1, and the Russian revolutions.

According to the concept of Lebedev (1936), the vapor moves from areas with elevated partial vapor pressure and air temperature to areas with lower values of these parameters. For more than 70 years, this concept provided the basis for studies of natural condensation and only recently have several papers suggesting reconsideration been published (Tkachenko, 1978; Kuldzhaev, 1989). In karst, the concept of Lebedev implies that condensation should occur in caves during warm seasons, and that evaporation must prevail there during cold seasons (3.6 Figure 1).

Shestakov (1989) compiled a bibliography of more than 1000 papers on condensation. About 10% of the researchers seem to believe that only evaporation of soil and groundwater occurs and that condensation of atmospheric moisture underground is not possible. About 40% think that the input of water due to summer condensation is balanced by water withdrawal due to winter evaporation. About 50% consider the condensate to be an important component of the water balance, but abstain from numerical estimations.

Difficulties involved in quantifying the amount of condensation are reflected in many general references, for example, "due to the fact that quantitative determination of condensation is difficult and time consuming, it is not expedient to take it into account in water balance studies" (Borevskii et al., 1976, p. 120), or "due to practical difficulties in determining the condensation, the latter is conventionally lumped with the precipitation and the evaporation" (Pinneker, 1980, p. 89).

There is no agreement among karst researchers regarding the role of condensation either. For example, Martel (1894) rejected the very possibility of it. Trombe (1952) and Gvozdetsky (1954) believed that it plays a significant role in speleogenesis. Gergedava (1970) assigned it the leading role in speleogenesis!

By and large, the importance of condensation in karst and other hydrogeologic systems seems to be underestimated. For instance, a fundamental review on karst geomorphology and hydrology by Ford and Williams (1989) only mentions seasonal condensation (p. 469) and condensation corrosion, mostly in hypogene settings (p. 309). Meanwhile, the long-standing karst studies in the Crimea, Caucasus, and other regions of the former Soviet Union reveal an important role of condensation in both the formation of karst waters and speleogenesis. Being published mostly in Russian, these studies remain largely unavailable to non-Russian researchers.

Different Scales of Condensation Investigations

The differing opinions about condensation, quite common in karst literature, have often resulted from differences in methods and approaches. The latter strongly depend on the scale of the studies. For the sake of simplicity, we will divide them into four scales: global, regional, local, and individual features. It should be borne in mind that the data and results obtained at one scale cannot always be directly translated to another.

Study of Condensation on the Global Level
Condensation at the global scale can be assessed in terms of dependence on latitude and altitude. Dublyanskaya and Dublyansky (1989) used data on the latitudinal distribution of the absolute humidity of air (Khromov and Mamontova, 1963) along with data on humidity in caves at different latitudes and altitudes (Moore and Sullivan, 1978) (with corrections from later findings) to produce results shown in 3.6 Figure 2. During the warm season (July in the northern hemisphere) condensation can occur between latitudes 25 to 70; at lower latitudes evaporation is likely to prevail. During the winter (January), evaporation may prevail in caves at all latitudes. The rates of condensation and evaporation change with altitude.

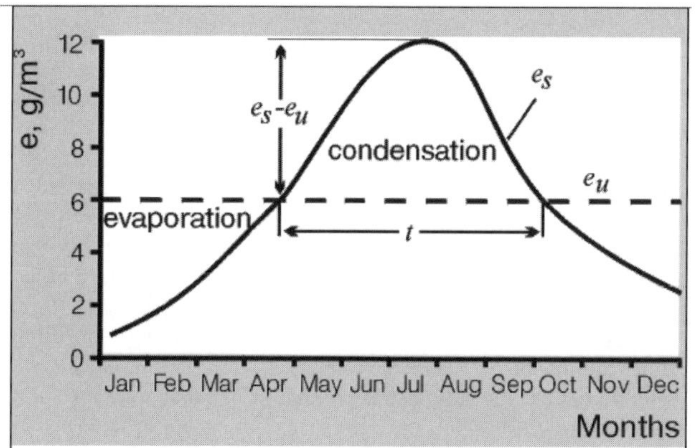

3.6 Figure 1. Potential of condensation and evaporation in the karst massifs of the Crimea Mountains (after Dublyansky and Lomaev, 1980, generalized). Many other karst massifs of the former Soviet Union exhibit a similar pattern, although the concrete values of all parameters may vary. e_s is the absolute humidity on the surface. e_U is the absolute humidity underground. (e_s-e_U) is the gradient of the absolute humidity of the air entering the cave. t is the duration of the condensation period. Condensation occurs when $e_s > e_U$.

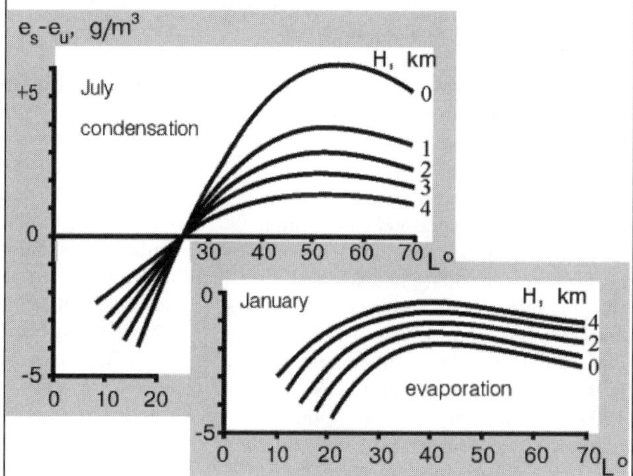

3.6 Figure 2. Possibility of condensation in karst during warm (July) and cold (January) seasons at different latitudes L and altitudes above sea level H. Condensation occurs when (e_s - e_u) > 0 (after Dublyanskaya and Dublyansky, 1989).

Shown in 3.6 Figure 2 is a broad generalization that depicts only the *possibility* of condensation. Processes occurring in real caves depend on the climatic zone in which the cave is located, the position of the cave within the karst massif, the passage of weather fronts, and many other factors.

Also, 3.6 Figure 2 implies that the input of water from summer condensation is counterbalanced by its withdrawal during winter evaporation. However, local studies of condensation suggest that this does not happen in natural settings (see section "Winter condensation" below).

Study of Condensation on the Regional Level
Regional studies deal with condensation within individual mountain ranges (e.g., the Crimea) or karst massifs (e.g., Chatyr-dag, one of nine karst massifs within the Crimea). At this level, the following features may suggest that condensation is a significant process.

Low-Discharge Springs. Such springs exist within isolated karstic or structural remnants such as pinnacles located close to mountain summits, crests, and passes. In such settings, the recharge that they can obtain from precipitation infiltration is insignificant, which suggests that condensation is the source of the waters (Slavianov, 1955; Protasov, 1959; Dublyansky, 1977; Dublyansky et al., 1985).

Constant Summer Discharges of Karst Springs. During summer dry seasons, karst springs often display hydrographs that may be approximated by smooth recession curves. This type of hydrograph is normally interpreted by a karst hydrologic model in which water drains progressively from storage in caves, fissures, and pores. Several expressions approximate the observed recession curves; the one most frequently used is

$$Q_t = Q_0 e^{-at} \tag{1}$$

where Q is the discharge (m³/s) at times t and 0, t is the time elapsed (days), and a is the recession coefficient (T^{-1}) (Ford and Williams, 1989, p. 196). Quite often, however, after reaching a certain minimum at n·10⁻²–n·100 L/s, the falling hydrograph limb levels off and remains flat (3.6 Figure 3) for a long period of time (1 to

4 months). This was reported by Jenko (1959) as a "precise constant of dry-season discharges" in springs of the Dinaric Karst (Slovenia).

Similar patterns were found for karst springs in the Crimea, the Caucasus, and some other karst areas of the former Soviet Union (Dublyansky et al., 1984, 1985, 1989) Ed. note: This phenomena can also be interpreted as an evidence of retarded discharge from the epikarstic storage. However, this interpretation does not deny the role of condensation, as the latter can contribute to the epikarstic storage.

The analysis of daily hydrographs of the Krasnopeshchernaya River in the Crimea (fed by karst springs withdrawing water from the Dolgorukov Karst Massif, with its 13.7-km-long Krasnaya Cave) for the period 1963 through 1988 reveals the fine structure of dry-season discharges. Having a mean daily value of 6 L/s, the discharge exhibits regular diurnal cycles, increasing from 4 to 8 L/s and then falling again (3.6 Figure 3, inset).

Thus, the drainage of stored groundwater is being overprinted by a certain dynamic process that stabilizes discharge of the springs at a given level specific to each spring and governs the systematic daily variations that occur around this level. Similar patterns were observed in springs at other elevations. They do not depend on the presence and composition of vegetation, which could induce an evapotranspirational cyclicity in discharge.
Also, the relationship between the discharge of springs and the variations of the gravitational field caused by the interaction of the Earth, Moon, and Sun, reported by Kinzikeev (1993) has not been confirmed in the Krasnaya Cave area. We believe that condensation provides the most appropriate explanation for the observed patterns of dry-season discharges from these springs.

Meteorological Data. For a number of years the senior author monitored the discharges of several springs suspected of being fed by condensation, plus meteorologic parameters (Dublyansky, 1977). A strong correlation was found between dry-season discharge (no precipitation for 1 to 4 months), and meteorologic parameters for 15 springs in the Crimea and 6 springs in the western Caucasus located at altitudes from 30 to 1800 m and having discharges from 0.012 to 8.660 L/s. Systematic diurnal variations of discharge Q, and water temperature T_W ,were correlated with changes in the dynamic parameters of the air that control condensation, i.e. the temperature of the air T_A, and its absolute humidity e_S. Observations during the warm seasons (June-August) over the period, 1965 to 1994, yielded statistically significant values of the coefficient of correlation (0.77 to 0.99 ± 0.05 –0.21). The response of Q and T_W to the changes of T_A and s is normally delayed.

Hence, a travel-time correction (retardation t) must be applied. The latter ranges from 1 to 15 hours for different springs (3.6 Figure 4 and 3.6 Table 1). T_W changes during a day almost synchronously with Q and does not show appreciable correlation with T_A. One may assume that it is water that controls temperature during condensation, as the specific heat capacity of water (1.0 cal/g/degree) is 4 to 6 times as large as the specific heat capacity of karst rocks (e.g., 0.16 –0.24 cal/g/degree for limestone and 0.20 to 0.25 cal/g/degree for gypsum).

Geological Engineering Data. Much about condensation associated with construction works within karst terrains (residential and industrial buildings, asphalt and concrete covers, and embankments and airport runways) has been published (Vedernikov and Larina, 1985). Condensation is viewed as a major factor in the rise of the groundwater (underflooding) within urban areas (Riazonova, 1987).

At the regional level, three groups of methods are used for quantitative determination of condensation:

- Water balance studies.
- Analytical methods.
- Microclimatic methods.

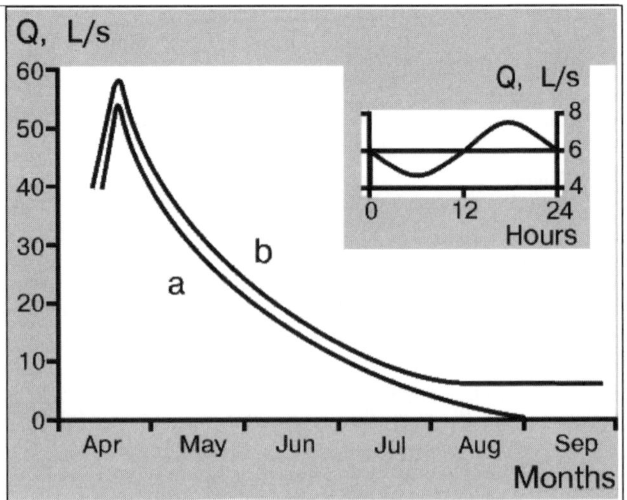

*3.6 Figure 3. Discharge hydrographs (schematized). **a** is according to the model of the drainage of stored water. **b** is according to the condensation model. Inset shows the daily oscillations of the hydrograph during a period of condensation recharge (dry season; August-September on this graph). The values shown in the inset correspond to actual values measured for the Krasnopeshchernaya River in the Crimea.*

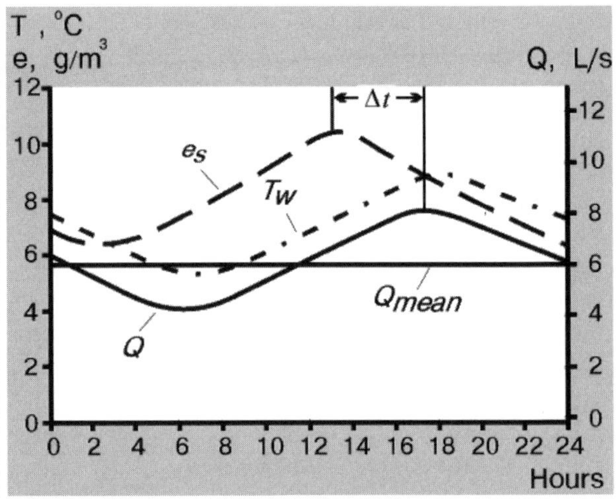

3.6 Figure 4. Daily variations of the discharge Q and the temperature T_w of a condensation-fed spring as compared with variation of absolute humidity of the atmospheric air e_s. Δt - is the lag time (travel time).

Water Balance Studies. Several attempts have been undertaken to estimate condensation in the Crimea using the water balance equation: condensation = precipitation – (evapotranspiration + runoff), (Golovkinsky, 1896; Vasilievski and Zheltov, 1932; Glukhov, 1963). The calculated values ranged from 0.7 to 25.0% of the rainfall. This approach is considered to be inappropriate, because the calculated values are of the same magnitude as are estimated errors in the other components of the water balance (15 to 20%).

Analytical methods. Between 1938 and 1986 several researchers suggested methods to calculate condensation within karst massifs. Most of the methods have important shortcomings. Some of them contain terms that are very difficult to determine Others are suitable for too general calculations or, vice versa, for too particular calculations. By way

Spring name	Altitude (m)	Period of observations	Mean discharge Q		Mean T_w		Lag time (hours)	Coefficient of correlation between Q and e
			L s^{-1}	C_V	°C	C_V		
			Crimea					
Ay-Petri	1100	July 1-2, 1959	0.012	0.03	7.2	0.04	2	0.95±0.09
Beshtekne	1000	June 26-27, 1989	6.100	0.07	6.5	0.01	3	0.78±0.17
Beshtekne	980	July 8-9, 1989	0.770	0.05	7.7	0.03	3	0.72±0.11
Pahkal-Kaya	960	June 27-28, 1994	0.014	0.07	10.1	0.04	6	0.80±0.12
Chatirdag	920	June 20-21, 1995	0.008	0.20	12.4	0.26	4	0.57±0.16
Alushta	920	July 20-21, 1991	0.054	0.04	6.5	0.03	3	0.86±0.09
Alushta	730	July 20-21, 1991	0.181	0.02	8.5	0.03	3	0.85±0.13
Mangup	575	July 25-26, 1971	0.370	0.06	10.6	0.03	4	0.92±0.12
Kizil-Koba	550	August 24-25, 1965	8.560	0.22	9.8	0.04	15	0.81±0.07
Aleshina Voda	550	August 25-26, 1965	2.350	0.17	9.6	0.03	6	0.85±0.14
Red Cave #33	510	June 26-27, 1984	1.460	0.02	9.6	0.02	8	0.81±0.16
Petrovski-1	420	June 18-19, 1986	0.110	0.05	10.9	0.04	2	0.74±0.16
Preduschelnoye	380	August 2-3, 1989	0.700	0.01	10.5	0.01	4	0.83±0.21
Petrovski-2	330	July 9-10, 1991	0.104	0.02	11.2	0.02	14	0.75±0.18
Yanishar	160	May 1-2, 1988	0.020	0.02	-	-	18	0.96±0.02
Opuka	30	July 2, 1991	0.105	0.04	13.3	0.02	10	0.73±0.18
			Caucasus					
Bagia	1800	July 12-13, 1983	0.220	0.01	8.5	0.11	1	0.99±0.05
Gelgeluk	1780	July 19-20, 1984	1.400	0.03	4.1	0.02	5	0.92±0.09
Alek	960	August 5-6, 1971	0.220	0.03	9.2	0.06	5	0.88±0.07
Alek	920	August 6-7, 1971	0.170	0.04	9.6	0.05	5	0.92±0.09
Proval	600	August 21-22, 1975	0.830	0.04	30.0	0.03	16	0.72±0.19
Akhun	210	July 27-28, 1976	0.030	0.05	8.9	0.02	7	0.87±0.14
Mean	-	-	1.132	0.05	10.1	0.03	6	0.84±0.12

3.6 Table 1. Discharge and water temperature in the condensation-fed springs in limestone massifs of the Crimea and Caucasus, and correlation between the discharge Q and absolute air humidity e.

of illustration, Sitnikov (1986) studied the dynamics of moisture and salt transfer in the aeration zone and suggested the following for estimating condensation

$$q(0,t) = -\frac{\beta\sigma(t)}{\gamma_L}\{P_W(0,t) - [P_L(0,t) - P_{AT}(0,t)]\} \quad (2)$$

where q is the condensation onto or evaporation from a surface (m/day). β is a coefficient accounting for the state of water onto or from the surface of which condensation or evaporation occurs (m/day). δ is a coefficient between 0 and 1 to account for the effective surface available for condensation or evaporation (parts of unity). γ_L is the density of water (kg/m^3). P_L and P_W are the pressure of water (Pa); and P_{AT} is the atmospheric pressure (Pa). This equation can be used for covered karst but is not applicable to directly exposed karst or to individual caves.

The microclimate method is based on the following equation introduced by Obolensky (1944) and modified by Dublyansky (1969)

$$A = V\,\varepsilon\,(e_S - e_U)\,t\,J \quad (3)$$

where A is the amount of condensate (g). V is the volume of the active part of the karst massif (m^3) (estimated from topographic and geologic maps and considering the depth of caves). ε is the fissuration and karst porosity (0 to 1) estimated by geologic, geophysical, or hydrochemical methods by density of fissuration and the volume of caves. $(e_S - e_U)$ is the difference between absolute humidities on the surface and underground (g/m³) determined by meteorological and climate data (see 3.6 Figure 1). t is the duration of the condensation period (days) (see 3.6 Figure 1). J is the frequency of air exchange (times/day) estimated by air flow velocities measured in caves and scaled to the average fissure or cave dimensions (Dublyansky, 1969).

This method has several apparent shortcomings. It is difficult to determine ε and J accurately. Also, the diffusion of water vapor, which does not depend on air movement, is not considered. In actual practice, nevertheless, it gives satisfactory results. It has been calibrated in the closely monitored massif of Krasnaya Cave (Crimea) and gave fair agreement (±10%) with the values obtained by the independent method of measuring of dry-season spring discharges.

This microclimate method was used to determine the role of condensation in recharge of karst waters within 16 karst massifs in the former Soviet Union (3.6 Table 2). The data imply that condensation accounts for 0.1 to as much as 9.1% of the total annual precipitation (mean 3.5%, coefficient of variation Cv = 0.73). The average yearly runoff from condensation is 1.86 L/s/km² (Cv = 0.79). Because condensation occurs only during the warm season, it is worthwhile to calculate seasonal (warm-season) runoff. Its mean value for the massifs studied is 4.54 L/s/km². The calculated values show good correlations with the dry-season discharges of springs within the massifs (coefficient of correlation r = 0.7 to 0.8).

Microclimatic calculations also reveal some interesting peculiarities of condensation in karst massifs located at different altitudes (3.6 Figure 5). Equation (3) contains both static (V, e) and dynamic (t, e_S, e_U, and J) terms. The values of (e_S - e_U) decrease with increasing altitude, whereas those of t increase. The combined effect is the appearance of a condensation minimum between 800 to 1600 m above sea level and two maxima, below 800 and above 1600 m.

Naturally, the results presented in this section do not give an exhaustive account of all aspects of regional condensation studies. For instance, of particular interest are the annual pattern of:

- Condensation in karst massifs occupying different altitudinal zones displaying different vegetation cover.
- Condensation in high altitude karst massifs.
- The change of the condensation pattern due to the passage of weather fronts.

These problems have not been addressed yet.

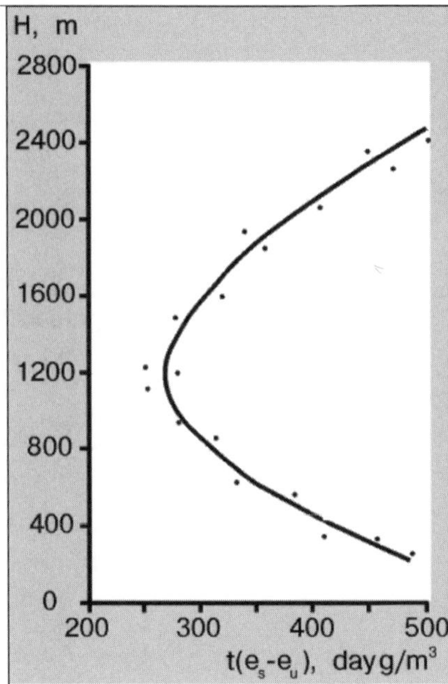

3.6 Figure 5. The dynamic parameter of condensation t (e_S-e_U) as related to altitude above sea level H. Data for the Bzib Massif, western Caucasus.

Condensation at the Local Scale

The local scale deals with individual caves and with small remnant karst hills feeding springs.

Karst massif	Surface area (km²)	The amount of condensed water			
		(mm)	% of yearly total precipitation	Contribution to runoff L/s/km²	
				Annual	Per warm season
		Western Carpathians			
Ugolsk	12.0	1	0.1	0.02	0.11
		Crimea			
Aipetri	97.7	77	6.4	2.46	5.00
Dolgorukov	79.5	25	3.0	0.81	2.94
Karabi	217.4	27	3.2	0.86	3.09
Chatirdag	23.4	69	7.2	2.38	5.30
Inner Range	293.0	11	2.0	1.85	9.60
Opuka	2.7	36	9.1	1.15	1.70
		Western Caucasus			
Kavminvody	495.0	8	1.2	0.25	0.68
Alek-Akhtsu	28.8	82	3.4	2.60	6.12
Dzikhra-Vorontsov	38.0	41	1.9	1.85	4.20
Akhshtir-Akhun	19.0	22	1.1	0.69	1.40
Arabika	517.8	134	5.6	4.27	9.36
Bzib	297.8	121	4.8	3.85	8.40
Khipsa	186.8	149	5.8	4.72	10.13
Gumishkha	263.5	61	3.1	1.95	4.37
Duripsh	40.9	4	0.1	0.12	0.28
Mean		54	3.5	1.86	4.54
C_v		0.87	0.73	0.79	0.75

3.6 Table 2. Condensation as related to recharge of some karst massifs of Ukraine, Russia, and Georgia (after Dublyansky, 1977; Dublyansky and Kiknadze, 1984; Dublyansky et al., 1985; and Vakhrushev, 1993). Calculations are by equation no. 3.

Historical and Archeological Data. Water that condensed on naturally collapsed rocks or within man-made piles of gravel was used for a water supply in ancient and medieval settlements in southern Europe and central Asia (Zibold, 1904; Tugarinov, 1955; Klimochkin, 1973; Firsov, 1990; Vakhrushev, 1993). Destruction of such structures, as well as destruction of small isolated limestone remnants (e.g., their quarrying for gravel during road construction) has led to a decrease in discharge or even cessation of condensation-fed springs (e.g., the Gaspra Isar and Morcheka Springs in the Crimea).

Observations and Microclimatic Studies in Caves. Studies during the past 30 years have revealed the extent of condensation in caves with different locations, morphology, and microclimate, as well as in underground mines (Ustinova, 1956; Protasov, 1959; Prokofiev, 1964; Lukin, 1969; Andrieux, 1970; Eremenko and Kolpashnikov, 1974; Dublyansky, 1977; Ginet, 1977; Racovitca and Viehmann, 1984).

The possibility that condensation occurs in a cave may be estimated by measuring the difference between absolute humidity at the surface and underground (e_S - e_U). Microclimatic observations from 290 caves of the Crimea and western Caucasus (more than 10,000 measurements and 5,000 continuous daily records of air pressure, temperature, and humidity) demonstrated the existence of condensation and revealed its daily, weekly, and monthly variations (Dubljanski and Sotskova, 1977; Tsikarishvili, 1981; Dublyansky et al., 1989). Condensation was observed in caves during the warm season (maximum in July and August). Evaporation prevails during the cold season, although in some particular settings condensation also can occur. The diurnal pattern of condensation is controlled by the change in temperature and humidity at the surface (maximum between 10 a.m. and 4 p.m. and minimum between 10 p.m. and 2 a.m.).

Condensation and chemosorption in caves and underground mines in hydroscopic rocks (massive halite and potash deposits) were reported by Maksimovich (1963), Eremenko and Kolpashnikov (1974), and Beltiukov (1989). Korotkevich (1970) demonstrated that at relative humidities close to 100%, the rate of condensation there may be as high as 180 to 225 mm/yr.

Detailed microclimatic studies carried out between 1960 and 1970 in 157 caves of the Crimea (Dublyansky et al., 1989) yielded an average value of condensation of 19.9 g/m^3 (*Cv* = 1.45). Condensation reaches a maximum of 75.8 g/m^3 in July in vertical shafts; values of 0.7 to 4.0 g/m^3 were measured in open caves located on the plateau, which agrees with earlier results by Ustinova (1956). Jameson and Alexander (1989) reported condensation rates of 30 to 90 g/m^3 on vertical surfaces and 45 to 200 g/m^3/day on horizontal surfaces in Snedegar's Cave in West Virginia. Similar studies have been carried out in caves of Romania, France, and some other countries (Andrieux, 1970; Racovitca and Viehmann, 1984).

Calculations of Condensation. Several methods of calculating condensation in caves have been suggested, each based on different theoretical premises. Mucke, Völker and Wadewitz (1983) suggested the empirical equation

$$m = (25 + 20W)(X_S - X_L) \quad (4)$$

where m is the condensation rate (g/m^3/hour). W is the velocity of air flow (m/s). X_s is the saturation humidity in the boundary layer between the air and cave wall (g/kg). And X_L is the humidity of the bulk of the air at the same temperature (g/kg).

Eraso (1969) pointed out that when warm air enters a cave, it cools down and a part of its moisture can condense. The decrease of the temperature is defined as

$$\Delta T = \frac{K - I}{1000\ C_p \gamma} \quad (5)$$

where ΔT is the decrease of temperature (°C), K and I are the heat content of the moist and dry air masses (kcal/m^3). Cp is the specific heat capacity of the air (cal/g/degree). And γ is the air density (kg/m^3). A similar approach was developed independently by Dublyansky and Iliukhin (1981).

Golod (1981) proposed a mathematical model of coupled aero-, thermo-, and hydrodynamic processes in caves. He assumed that condensation—being a low-intensity process—occurs at an equilibrium (along the curve of saturation). This allows calculation of condensation by:

$$P_H = P_0\ exp\ [\mu_n\ L_K/R]\ [(T_B - T_0)/(T_B\ T_0)] \quad (6)$$

and

$$dQ_K = r_L\ L_K\ dV_K \quad (7)$$

where P_H is the pressure of saturated vapor. P_0 and T_0 are the pressure and the temperature at the triple point. μ_n is the molecular weight of the vapor. L_K is the specific heat of condensation. R is the universal gas constant. T_B is the absolute temperature. dQ_K is the heat released in the process of condensation. dV_K is the amount of condensed water. And ρ_L is the density of the liquid phase.

Comprehensive thermodynamic calculations of condensation processes were performed by Bruent and Vidal (1981) in order to develop methods to protect the celebrated Paleolithic paintings in Lascaux Cave, France. The air conditioning system in the cave was constructed on the basis of these calculations.

Winter Condensation. Local studies of condensation have fostered some fundamentally new ideas about the formation of groundwaters. One problem of special interest is that of winter condensation (distillation). The warm-season condensation in caves is accompanied by an input of moisture from the outside atmosphere. The cold-season evaporation, however, does not remove all moisture from the rocks: the water evaporated deep within a karst massif and transported outward may condense within the epikarstic zone (as discussed by Klimchouk, 1995; see also Chapter 3.5) or within the snow pack (3.6 Figure 6). So there is an "internal circulation of moisture," which does not increase the total amount of water stored in the massif but does significantly increase the residence time of some portion of it and thus enhances condensation-related karstification. Winter condensation sustains the runoff from high-mountain karst massifs during periods when they are devoid of recharge by rain or snowmelt.

Winter condensation may be the mechanism that prevents the large scale removal of moisture from karst massifs by evaporation during cold seasons. The surface area of the open cave entrances—through which such removal can take place—is negligible, when compared to the total underground surface areas exposed within karst massifs.

Existence of winter condensation was confirmed by specific observations at the karst massifs of the western Caucasus (Dublyansky et al., 1985). At some of them, the snow cover lasts for 3 to 5 months and its thickness reaches 6 to 12 m (e.g., Bzib Massif). There is only insignificant runoff in the caves, and it is well correlated with the changes of negative air temperature at the surface. Detailed microclimatic surveys have shown that this runoff cannot be accounted for by melting of snow at its base due to the geothermal flux. This is because the karst massifs are deeply cooled fragments of the Earth's crust, where the "normal" geothermal gradient begins to control the rock temperatures only at great depths.

Winter condensation was also discovered in the Pinego-Kuloy Karst Region of northern Russia, where Malkov and Frantz (1981) reported the formation of condensation droplets during an 11 day period when outside air temperature ranged from –7 to –42° C.

Hydrothermal Condensation. A special case is that of hydrothermal condensation

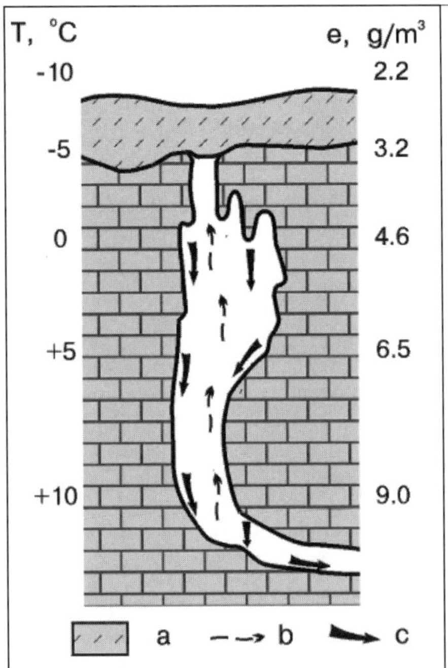

3.6 Figure 6. The pattern of air and moisture circulation characteristic of a cold-season condensation setting. ***a*** *is snow.* ***b*** *is movement of water in form of vapor.* ***c*** *is movement of water under the influence of gravity.* ***T*** *is the temperature of the cave air.* ***e*** *is absolute humidity.*

occurring above underground thermal waters during the partial dewatering stage of hydrothermal caves (3.6 Figure 7). Such a mechanism is believed to be responsible for carving the peculiar spherical niches or pockets found in certain caves of Hungary (Müller, 1974), Italy (Cigna and Forti, 1986), and elsewhere. Theoretical considerations (Szunyogh, 1982, 1990; Dublyansky, 1987) as well as laboratory experiments (W. Dreybrodt, 1994, pers. comm.) have shown that condensation in such a system is a naturally attenuating process. It slows down due to the low rate of heat transfer through the bedrock. However, in natural settings, hydrothermal condensation may become a powerful factor in karst formation if cooling of the bedrock can be maintained by ventilation. The latter is quite probable because such caves develop not far beneath the surface and within the zone of aeration. A prominent example of such convection-condensation-corrosion is known in Grotta Giusti, Italy. The cave contains a lake with a temperature of 32 to 34°C. The rate of condensation in this cave was estimated to be 8,640 L/day. The estimated amount of bedrock $CaCO_3$ dissolved was 630 g/day (Cigna and Forti, 1986).

Study of condensation at the local scale reveals many interesting problems that call for further study. Examples of such particular problems are:

- Formation of condensation moisture in "mixing fogs."
- Condensation in narrow passages due to an increase of velocity of the air flow (accompanied by a decrease of the pressure and temperature).
- Influence of cave ice on condensation (sublimation).
- Evaporation from the surface of inflowing water and further redistribution of the moisture in the cave.

Condensation Effects on Individual Features in Caves

These are investigations concerned with particular parts or features in caves (e.g., isolated rooms, near-entrance zones). They use condensation devices of various designs and have specific aims, such as studies of mineral formation, preservation of ancient graffiti and paintings, and so forth.

When it is studied at this scale, condensation can seem to be quite a heterogenous process, and the patterns of condensation and evaporation within individual caves can be extremely complex. In some caves, the regional or local seasonal patterns of condensation and evaporation outlined above can be interrupted by relatively short (from days to weeks) reversals (e.g., evaporation in summer due to passage of frontal systems, and change of the air circulation pattern).

Condensation typically occurs on the bedrock surfaces, on cave sediment (where it may be accompanied by different kinds of sorption), or on water bodies. In addition, it can occur in the cave air itself, as warm relatively moist surface air enters and cools to yield condensation fog. This can be classed as an aerosol formation process. The fog particles must grow to greater than 5 micrometers in size to become visible. Although it can occur by homogenous nucleation, more often it occurs heterogeneously—condensation of moisture onto solid aerosol particles such as dust particles, which serve as condensation nuclei. If the water is condensed on aerosol particles, its further migration inside the cave will be governed by aerosol mechanics, rather than by gas dynamics.

Dripping and splashing water (from waterfalls and flowstone) can evaporate into cave air and be transported to other locations where it can condense. Splashing produces relatively large (10 µm) hydroaerosol particles, which increases the specific surface of the liquid and fosters evaporation. Also, water for condensation and evaporation processes can be influenced by the air-flow structure of the cave (see, for example, data on the air flow directions inferred from the orientation of cave popcorn in Carlsbad Cavern; Queen, 1981).

Even though the data concerning condensation on individual features are abundant, they are mostly descriptive. Quantitative studies are quite scarce, but some examples are given below.

Prokofiev (1964) was the first to attempt direct measurement of the amount of condensate that can accumulate. He investigated several different rooms in Vorontsovskaya Cave, western Caucasus, using traps designed to collect the condensed (and partly infiltered) water. The traps yielded up to 1300 g/m^3 of water at an air exchange rate of 161.3·106 m^3/day. Studying the archeology of Medieval cave dwellings in the Crimea, Firsov (1990) noted intense condensation occurring in some crypts in the summer. He estimated that an average crypt carved in the massive limestone and having a volume of 5.5 m^3 generates 0.25 L/m^2 of condensate each day. Similar observations have been made in several other caves in Europe and Asia.

Condensation has also been studied by means of lysimeters and condensers:

- Of differing design—metallic cones, cylinders, boxes, casing tubes, polyethylene canisters.
- Of varying sizes (from 0.01 to 1120 m^3).
- And in various types of cave fill (sand,

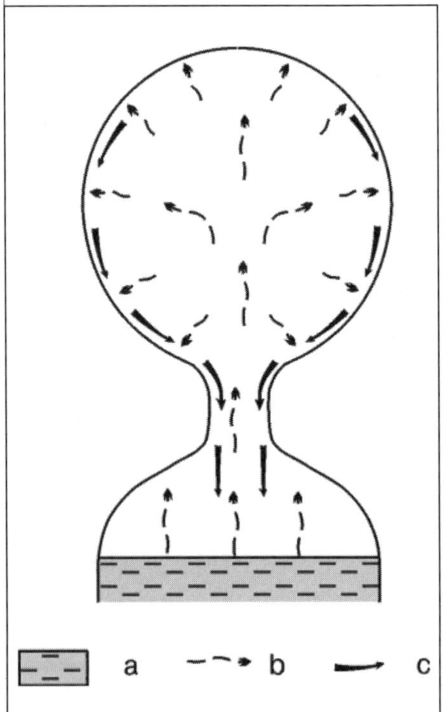

3.6 Figure 7. Condensation in a hydrothermal cave: ***a*** *is a thermal lake.* ***b*** *is the movement of water vapor.* ***c*** *is the movement of water as a liquid under the influence of gravity (after Szunyogh, 1982).*

gravel, pebbles, limestone boulders). They have been installed or constructed at altitudes ranging from 190 to 1100 m in the Crimea, Caucasus, Central Russia, Kola Peninsula, and southern and eastern Siberia. Yields of water in these experiments ranged from 0.01 to 2.3 L/day per cubic meter of the fill (3.6 Table 3). Attempts to use these values directly to calculate condensation in adjacent karst massifs have failed (Glukhov, 1963; Klimochkin, 1973).

The Role of Condensation in Karst

Increase of the Amount of Water Available for Speleogenesis

Our data suggest that the share of condensation in the water balance does not normally exceed 9% of the annual sum of precipitation in any given region (3.6 Table 2). However, condensation occurs during the warm season when there may not be much precipitation. It thus can play an important role in the dry-season discharge of karst springs and rivers, as we have demonstrated above. Winter condensation does not increase the total amount of underground water but may reduce the proportion of the moisture that escapes from the subsurface and also creates specific local patterns of water circulation. These general conclusions need to be checked and assessed quantitatively by regional and local studies.

Destructive Processes

Many workers believe that condensed water initially has exceptionally high aggressiveness with respect to limestone (Trombe, 1952; Gvozdetsky, 1954; Bernasconi, 1966; Andrieux, 1970; Gergedava, 1970; Pasquini, 1973; Ginet, 1977; Mucke and Völker, 1978). Data on the composition of condensate from different karst massifs of Eurasia are compiled in 3.6 Table 4.

Analyses were mostly performed on composite samples (up to 10,000 individual water drops) collected from cave walls. This implies that these waters had already had some opportunity to dissolve the bedrock. The condensates have quite variable amounts of mineralization (total dissolved solids Cv up to 1.15), differing according to lithology. Condensate water collected after known lengths of contact with the rock (30 to 100 m, flowing as films or rivulets down walls) was more mineralized and had lower coefficients of variation (Cv of 0.25 to 0.60). This suggests intensive corrosion of the bedrock.

Data on samples collected on cleansed cool artificial surfaces such as ice-filled containers or refrigerators are scarce. In some cases the measured total dissolved solids are low (e.g., 8.27 to 8.31 mg/L in Ochtinskaya Aragonite Cave, Slovakia; Homza, Rajman, and Roda, 1970). However, Tarhule-Lips and Ford (in press) have found that such induced condensates in caves on Cayman Brac (an island in the Caribbean) can be slightly undersaturated to slightly supersaturated with respect to calcite. These caves are short and somewhat exposed to sea breezes that carry salt and other aerosols. This problem warrants further study.

The quantitative contribution of condensation to the total dissolution budget has been attempted only for the Alek Karst Massif in the western Caucasus, (Dublyansky et al., 1985). The total chemical denudation attributed to condensation water in the warm season was estimated to be 41.2 tons (16.2 m^3) of limestone, which represents only 3.7% of the gross annual amount of karst denudation. Condensation corrosion is most intense in July (19%) and August (16%). At present, the data are insufficient to estimate the speleogenetic role of winter condensation.

Many forms of cave macro-, meso-, and micromorphologies are attributable to condensation corrosion, as reported in Chapter 6.1. These include the:

- Very distinctive spherical chambers 1 to 3 m in diameter found in some hydrothermal caves (Szunyogh, 1982, 1990).
- Cupolas and ceiling pockets or bells (Andrieux, 1970; Mucke and Völker, 1978).
- Solutionally widened fissures, niches, cells, vertical gutters, rills, edge patches, drop dents, and indentations on cave walls (Bernasconi, 1966; Eraso, 1969; Mais, 1973; Cigna and Forti, 1986; Davis and Mosch, 1988; Beltiukov, 1989; Jameson, 1989, 1995) (3.6 Figure 8).

Condensation corrosion often cuts through not only the bedrock, but also cave fill (calcite, gypsum, salt and ores) and speleothems (Hill and Forti, 1986; Tarhule-Lips 1994; Tarhule-Lips and Ford, in press).

Condensation may be particularly

Author	Years	Location	Altitude (m)	Characteristics of the setup: Volume (m^3)	Characteristics of the setup: Type of infilling	Yield L/day per each m^3 of in-filling
Golovkinsky, N.A. and Peddakas, I.K	1864-1905	Crimea, Southern Coast	300	0.04-0.14	clay, sand	0.028
Zibold, F.I.	1912	Eastern Crimea	190	1117.0	pebble	0.320
Khudiaev I.E.	1931	Crimea, Southern Coast	480	0.03	gravel	0.980
Beliavsky, A.Y.	1940	Kiev	100	0.01	sand, loess	2.324
Tugarinov, V.V.	1951	Moscow area	280	300.0	gravel	0.620
Reiniuk, I.G.	1951-1965	Kolima Region	200	from 0.007 to 0.016	sand gravel	0.022 0.370
Protasov, V.A.	1955	Crimea, Southern Coast	400	0.25	gravel	0.018
Klimochkin, V.V.	1958-1970	Siberia and Kola Peninsula	?	0.01-0.05	sand pebble gravel	0.050 0.060 0.100
Dublyansky, V.N.	1960	Western Crimea	1100	900.0	gravel	0.016
Pribluda, V.D.	1963-1976	Western Crimea	900	22.0	gravel, cobbles	0.012
Mean						0.386
C_v						1.71

3.6 Table 3. The output of different types of condensation setups constructed on the surface.

Parameters	Moisture condensed on clean surface	Stationary drops on the celling	Trails and films on the walls	Source
		Carbonate karst		
Number of samples	7	55	66	Dublyansky, 1977
Mean TDS mg/L	106	22.0	140.0	Dublyansky et al., 1985
Mean TDS C_v	0.58	1.15	0.25	Homza, Rajman, and Roda, 1970
Hydrochemical type	HCO_3-Ca(Mg)	HCO_3-Ca	SO_4-HCO_3-Ca	Klimochkin, 1973
Chemical denudation µm/year				Nemeriuk and Paltsev, 1969, Mais and Pavuza, 1994
calculated at 10 °C	-	-	2-19	Mucke and Völker, 1978
actual	-	-	0.5-4.0	Prokofiev, 1964
		Carbonate hydrothermal karst		
Chemical denudation µm/year (*)				Szunyogh, 1990
at 60 °C	-	-	200-50	
at 20 °C	-	-	30-4	
		Sulfate karst		
Number of samples	-	5	44	Dublyansky *et al.*, 1984
Mean TDS mg/L	-	2000	2100	Malkov and Frantz, 1981
Mean TDS C_v	-	-	0.27	Mucke and Völker, 1978
Hydrochemical type	-	SO_4-Ca	SO_4-Ca	
Chemical denudation µm/year				
calculated at 10 °C	-	-	92-730	
actual	-	-	90-121	
		Rock salt karst		
Number of samples	-	-	17	Beltiukov, 1989
Mean TDS mg/L	-	80000	322,000	Eremenko and Kolpashnikov 1974
Mean TDS C_v	-	-	0.8	
Hydrochemical type	-	Cl-Na(K)	Cl-Na(K)	
Chemical denudation µm/year	-	-	21,000	

All samples are composite (many drops are mixed). They were typically collected at the end of a long period of nil precipitation or when the massif was covered by snow. (*) Growth of a spherical niche; slows down with increase of its diameter. TDS is total dissolved solids

3.6 Table 4. Chemistry of condensation waters.

important in hydrothermal karst. During the dewatering stage, when free water surfaces are in air-filled caves, the heated water may evaporate, rise to the upper (cooler) parts of the cave by convection, and condense on the walls there (3.6 Figure 7). The air mass above a thermal lake is thermodynamically unstable. This is because warmer (less dense) air lies below cooler (denser) air—thus, convection readily develops. The aggressivity of the condensate is determined by the content of CO_2 in the cave air and by the thickness and flow pattern of the films of condensate.

Another situation—often particular to hydrothermal caves—is corrosion due to the presence of H_2S in the cave air. Worthington and Ford (1995) have shown that a high content of sulfate is characteristic of the thermal waters that contribute to dissolution of hypogene caves. During the stage of dewatering, the hydrogen sulfide in thermal water escapes into the cave air. Some of it redissolves in films or droplets of condensate, where it is oxidized. This produces sulfuric acid which attacks the limestone. It may convert it to gypsum, forming fragile gypsum crusts on the cave walls and ceilings. Such a mechanism of "replacement-corrosion" was first suggested by Egemeier (1981). It is important to note that replacement-corrosion must be coupled with condensation (the gypsum crust does not form on dry cave walls).

Many fossil hydrothermal caves exhibit morphological features that can best be explained by this mechanism. Examples are known in Hungary (the Buda Hills), Kirghizstan (Tyuya-Muyun Cave), Algeria (the Bibans), and elsewhere. The process of replacement corrosion is still active in caves of the Pryor Mountains, Wyoming (Egemeier, 1981) and Grotta Grande del Vento, Italy (Cigna and Forti, 1986).

Depositional Processes

Condensation coupled with evaporation and aerosol mass transfer may play a significant role in the formation of a variety of speleothems. Hypotheses suggesting a condensation origin of stalactites (though incorrect) were advanced by H. Jacob, E. de Clave, and J. Beaumont in the 17th Century (Shaw, 1992). Holland et al., (1964) included a category for condensation-related speleothems in their classification of cave formations. Cser and Maucha (1968) suggested an aerosol origin for one type of helictite. They supported their hypothesis by means of an elegant experiment in which fragments of calcite crystals were set in a plastic frame with 90 volts DC applied to the base and exposed to the cave atmosphere. Appreciable growth of crystals was observed after an

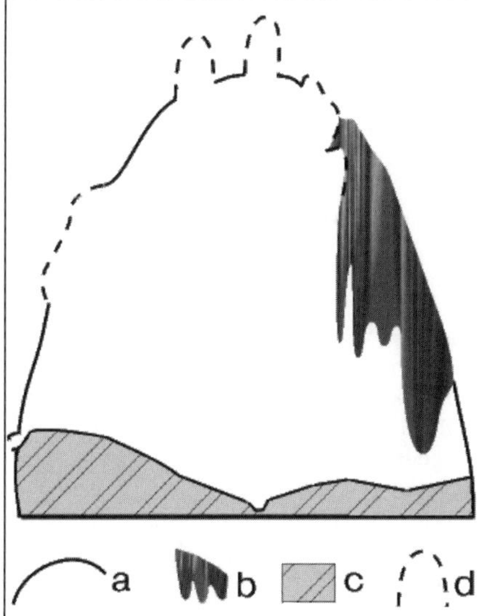

*3.6 Figure 8. Destructive action of condensed waters. **a** is a primary karst cavity. **b** is speleothems. **c** is cave sediments. **d** is morphologies caused by condensation corrosion (cupolas, niches, facets, grooves, pits).*

exposure of several months.

In *Cave Minerals of the World*, Hill and Forti (1986) cite many examples of speleothems supposedly related to condensation, including:

- Coatings and crusts (p. 29).
- Coralloids (popcorn, coral, grape, or knobstone, p. 33).
- Rims (p. 54).
- Ice frost crystals (p. 83).
- Moonmilk (p. 116).
- Clay vermiculations (p. 161).

These speleothems are composed of calcite, gypsum, halite, carnallite, nitrates, and other minerals. The list of specific depositional forms of cave ice attributed to condensation is even larger (Mavliudov, 1989). More and more often, scientists resort to aerosol condensation to explain the origin of some peculiar gypsum and carbonate speleothems (Calaforra and Forti, 1994; Turchinov, 1994; Klimchouk, Nasedkin, and Cunningham, 1995; Dublyansky and Pashenko, 1997).

Cave Microclimate

The pattern of condensation in caves is controlled by the microclimate, which it in turn alters because the process of condensation is accompanied by significant release of heat (585 kcal/kg). This affects the cave microclimate by changing the temperature and relative humidity (Andrieux, 1970; Molerio, 1981; Racoviţcă and Viehmann, 1984).

One prominent example of such an impact are the short-periodic (seconds to tens of seconds) autooscillations of air movement that may overprint longer period "cave breathing" of a barometric nature (Finnie and Curl, 1963; Dublyansky, 1977). Such autooscillations were reported from Krasnaya Cave in the Crimea (Dubljansky and Sotskova, 1989).

Conclusion

Condensation in karst is quite a complex phenomenon. Data about condensation in caves and karst are scattered in publications by karst researchers, physicists, chemists, meteorologists, hydrogeologists, mining geologists, soil scientists, geographers, glaciologists, biospeleologists, archeologists, and construction engineers. These different specialists consider particular aspects of condensation and use research methods specific to their sciences. There is no consensus in the literature on the role of condensation in the formation of karst waters or its role in the formation of caves.

Karst condensation can be studied by either a hydrogeologic or a speleogenetic approach. The two are seldom combined.

The Hydrogeologic Approach deals with the role of condensation at the scale of a karst massif. Examples of typical problems are:

- The contribution of condensation to the water balance.
- The role of condensation water in karst denudation.
- The chemistry of condensed water.
- Patterns of moisture and condensate migration within karst massifs during different seasons.

Such a large-scale approach makes it necessary to generalize many characteristics, such as the fissuration of the rock or the coefficients of air-exchange at the scale of the karst massif, which necessarily decreases the precision of the results. In the hydrogeologic approach, atmospheric moisture is considered to be a major source of water involved in condensation (and thus karst) processes.

The Speleogenetic Approach deals with the role of condensation:

- In the development of individual caves.
- The carving of smaller solutional forms.
- The creation of speleothems.

The smaller scale makes it necessary to perform more precise measurements of pertinent parameters (temperature of the air and condensate, humidity, and chemistry of condensate). These parameters can be used in laboratory and theoretical modeling of cave-forming processes. The ultimate purpose of these studies is to explore the possible role of condensation in the creation of specific cave morphologies and speleothems, and to find the mechanisms involved.

When using the speleogenetic approach, one should consider more than just the moisture carried into the cave with the outside air. Where caves contain lakes and streams, the water may evaporate from these sources, migrate, and then condense in other parts of the cave (Hill, 1987). Condensation of moisture from cave air and from evaporation of standing water is also possible due to a change of thermodynamic characteristics inside a cave (Dublyansky and Lomaev, 1980; Badino, 1995).

Condensation and Aerosols. One important problem not yet studied adequately is the relationship between condensation and cave aerosols. The moisture in the cave air may exist as more than just a form of vapor. It may also exist as all-water aerosol particles (hydroaerosols; Gadoros and Cser, 1986), as well as shells on solid aerosol particles (Pashenko et al., 1993, 1996).

The laws of gas dynamics and aerosol mechanics are different. Thus the migration of water in these two forms may display significantly different patterns. Elevated concentrations of radon, which are typical of the air in some caves, may foster radiochemical reactions between hydroaerosol particles and gaseous components of the cave atmosphere.

For instance, there is the convertion of H_2S in the air (even in very small amounts) into H_2SO_4. This dissolves in water of hydroaerosols and may drastically increase their aggressivity. It is becoming apparent that aerosols can play a significant role in mass transfer of moisture and speleothem material inside caves, and can be particularly important in the formation of speleothems (Dublyansky and Pashenko, 1997).

Although awareness of the importance of these processes is growing, neither condensation nor cave aerosols have been adequately (quantitatively) studied to date. Future studies should aim to develop a comprehensive theory of gas and aerosol mass transfer in caves, involving evaporation, convection, aerosol and gas mass transfer, and condensation. This problem is an interdisciplinary one; it warrants the combined efforts of hydrogeologists, meteorologists, physicists, and karst researchers.

Acknowledgements

The paper has greatly benefited from thorough review and comments of Dr. R.A. Jameson and an anonymous reviewer of the *Journal of Cave and Karst Studies.*

References

Andrieux, C., 1970, Evapo-condensation souterraine: *Annales de Spéléologie*, v. 25, p. 531-559.

Badino, G., 1995, Fisica del clima sotterraneo: *Memoire dell'Istituto Italiano di Speleologia,* ser. 2, v. 7, 137 p.

Beltiukov, G.V., 1989, On the formation of karst forms due to condensation, *in Problemi kompleksnogo izuchenija karsta gornikh stran:* Tbilisi-Tskhaltubo, p. 91-94 (in Russian).

Bernasconi, R., 1966, La condensation interne du karst profond: *Cavernes,* v. 10, no. 2, p. 41-46.

Borevskii, B.V., Khordikainen, M.A., and Yazvin, L.S., 1976, *Exploration and assessment of water reserves in fissure-karst reservoirs:* Moscow, Nedra, 246 p. (in Russian).

Bruent, J., and Vidal, P., 1981, Études climatiques pour la conservation des Grott Ornees deux intervention: *8th International Congress of Speleology,* Bowling Green, Proceedings, p. 659-662.

Calaforra, J.M., and Forti, P., 1994, Two new types of gypsum speleothems from New Mexico: Gypsum trays and gypsum dust: *National Speleological Society Bulletin,* v. 56, p. 32-37.

Cigna, A., and Forti, P., 1986, The speleogenetic role of air flow caused by convection: *International Journal of Speleology,* v. 15, p. 41-52.

Cser, F., and Macuha, L., 1968, Contribution to the origin of "excentric" concretions: *Karszt és Barlangkutatás*, v. 5, p. 83-100.

Davis, D.G., and Mosch, C., 1988, Pebble indentations: A new speleogen from a Colorado cave: *National Speleological Society Bulletin,* v. 50, p. 17-20.

Dubljansky, V.N., and Sotskova, L.M., 1977, Microclimate of karst cavities of the Crimea Mountains: *7th International Congress of Speleology,* Sheffield, Proceedings, p. 158-160.

Dublyanskaya, G.N., and Dublyansky, V.N., 1989, The role of condensation in the development of mountain karst, *in Problemi kompleksnogo izuchenija karsta gornikh stran:*, Tbilisi-Tskhaltubo, p. 107-108 (in Russian).

Dublyansky, V.N., 1969, Methods of calculation of moisture condensation in fissure-karst reservoirs: *Bulletin NTI,* Series Hydrogeol. and Eng.-Geol., no. 6, p. 13-17, (in Russian).

Dublyansky, V.N., 1977, *Karst caves and shafts of the Crimea Mountains:* Leningrad, Nauka, 181 p. (in Russian).

Dublyansky, V.N., Dorofeev, E.P., and Borodaeva, L.A., 1984, *Hydrochemistry of Kungur Ice Cave:* Simferopol, Ukr. NIINTI, N603Uk 84, 37 p. (in Russian).

Dublyansky, V.N., and Iliukhin, V.V., 1981, *Travels underground:* Moscow, Physical Culture and Sport, 192 p. (in Russian).

Dublyansky, V.N., and Kiknadze, T.Z., 1984, *Hydrogeology of karst in the Alpine folded zone of the southern USSR:* Moscow, Nauka, 125 p. (in Russian).

Dublyansky, V.N., Klimenko, V.I., Vakhrushev, B.A., and Iliukhin, V.V., 1985, *Karst and groundwaters of mountain massifs of the western Caucasus:* Leningrad, Nauka, 150 p. (in Russian).

Dublyansky, V.N., and Lomaev, A.A., 1980, *Karst caves of the Ukraine:* Kiev, Naukova Dumka, 179 p. (in Russian).

Dublyansky, V.N., Sotskova, L.M., and Ferbei, G.G., 1989, *Microclimate of karst caves in the Crimea Mountains:* Simferopol, Ukr. NIINTI, N2495Uk 89, 132 p. (in Russian).

Dublyansky, Y.V., 1987, Theoretical modeling of the dynamics of formation of hydrothermal caves, *in* Metodi izuchenija i modelirovanija geologicheskikh yavlenij: Novosibirsk, Institute of Geology and Geophysics, p. 97-111 (in Russian).

Dublyansky, Y.V., and Pashenko, A.E., 1997, Cave popcorn—an aerosol speleothem?: *12th International Congress of Speleology,* Chaux-de-Fonds, Proceedings.

Egemeier, J.S., 1981, Cavern development by thermal waters: *National Speleological Society Bulletin,* v. 43, p. 31-51.

Eraso, A., 1969, La corrosión climática en las cavernas: *Bol. Geol. y Minero,* v. 80, p. 564-581.

Eremenko, Y.P., and Kolpashnikov, G.A., 1974, On the methods of calculations of condensation runoff from spoils of potash mines, *in* Regim, balans i resoursi podzemnikh vod: Minsk, p. 134-141 (in Russian).

Fedoseev, N.A., 1975, *The history of study of the major problem of the hydrosphere:* Moscow, Nauka, 206 p. (in Russian).

Finnie, J., and Curl, R., 1963, On the functioning of a familiar nonlinear thermodynamic oscillator, *in International Symposium on nonlinear oscillation,* Moscow, v. 3, p. 121-123.

Firsov, L.V., 1990, *Isars:* Novosibirsk, Nauka, 470 p. (in Russian).

Ford, D., and Williams, P., 1989, *Karst geomorphology and hydrology:* London, Unwin Hyman, 601 p.

Gádoros, M., and Cser, F., 1986, Aerosols in caves–Theoretical considerations: *9th International Congress of Speleology,* Barcelona, Proceedings, v. 2, p. 90-92.

Gergedava, B.A., 1970, The role of condensation and infiltration waters in cave origin: *Trudy Vses. Geograficheskogo Obshchestva,* v. 102, no. 5, p. 196-198.

Ginet, R., 1977, Étude de la condensation atmospherique saisonneire dans la grotte de Hautcourt: *Pr. Acad. Sci.*, v. 18, p. 1615-1618.

Glukhov, I.G., 1963, The role of condensation in the water balance of mountains (on the example of the Crimea Mountains): *Izvestija VUZov*, Geol. Series, v. 3, p. 26-31 (in Russian).

Golod, V.M., 1981, Mathematical model of aerohydrodynamic processes in the aeration zone, *in* Akkumulatsija zimnego kholoda v porodakh i ego ispolzovanije: Perm, p. 6-8 (in Russian).

Golovkinskii, N.A., 1896, *Observations on precipitation in soil:* Attachment to the report of the hydrogeologist for 1896, Simferopol, 6 p. (in Russian).

Gvozdetsky, N.A., 1954, *Karst:* Moscow, Geografgiz, 351 p. (in Russian).

Hann, J., 1880, Über eine neue Quellentheorie auf meteorologicher Basis: *Ztochr. Osterreichichen Gesellsch. für Meteorol.*, v. 15, p. 482-486.

Hill, C., 1987, Geology of Carlsbad Cavern and other caves in the Guadalupe Mountains, New Mexico and Texas: *New Mexico Bureau of Mines and Mineral Resources Bulletin,* no. 117, 150 p.

Hill, C., and Forti, P., 1986, *Cave Minerals of the World:* Huntsville, Alabama, 238 p.

Holland, H.D., Kirsipu, T.-W., Huebner, I.I., and Oxburgh, U.M., 1964, On some aspects of the chemical evolution of cave waters: *Journal of Geology*, v. 72, p. 36-67.

Homza, S., Rajman, L., and Roda, S., 1970, Vznik a vyvoj krasoveho fenomenu Ochtinskej aragonitovej jaskine: *Slovensky kras,* v. 8, p. 21-68 (in Slovak).

Jameson, R.A., 1989, Features of condensation corrosion in caves of the Greenbrier Karst, West Virginia (abs.): *National Speleological Society*, Proceedings.

Jameson, R.A., 1995, Condensation, condensation corrosion, and associated features in Snedegar's and Greenville Saltpeter Caves, *in* Underground the Appalachians: *National Speleological Society,* Blacksburg, Virginia, Convention Guidebook, p. 122-125.

Jameson, R.A., and Alexander, C. E., Jr., 1989, Hydrology and chemistry of condensation waters in Snedegar's and Greenville Saltpeter Caves, West Virginia (abs.): *National Speleological Society,* Proceedings.

Jenko, F., 1959, *Hidrogeologija i vodno gospodarstvo Krasa:* Ljubljana, 236 p.

Khromov, S.P., and Mamontova, L.I., 1963, *Meteorological dictionary:* Leningrad, Gidrometeoizdat, 250 p. (in Russian).

Kinzikeev, A.R., 1993, Geodynamics and karst process, *in* Ingenernaja geologija karsta (Perm): v. 2, p. 16-20 (in Russian).

Klimchouk, A., 1995, Karst morphogenesis in the epikarstic zone: *Cave and Karst Science,* v. 21, p. 45-50.

Klimchouk, A., Nasedkin, V., and Cunningham, K., 1995, Speleothems of aerosol origin: *National Speleological Society Bulletin,* v. 57, p. 31-42.

Klimochkin, V.V., 1973, The necessity of consideration of condensation processes when determining the balance of groundwaters, *in* Vlagooborot v prirode i ego rol v formirovanii resursov podzemnikh vod: Moscow, Stroiizdat, p. 288-290 (in Russian).

Korotkevich, G.V., 1970, *Salt karst:* Leningrad, Nedra, 256 p. (in Russian).

Kuldzhaev, N.K., 1989, On the theory of a condensation origin for ground waters by A.F. Lebedev: *Izv. of Academy of Sci. of Tadzhik. SSR,* Series Phys-Techn., Chem., and Geol., v. 1, p. 103-106 (in Russian).

Lebedev, A.F., 1936, *Soil- and groundwaters*: Moscow-Leningrad, 312 p. (in Russian).

Lukin, V.S., 1969, Quantitative expression of evaporation and condensation of water in gypsum-anhydrite massifs of the Ufa Plateau: *Zemlevedenije,* new ser., v. 8, no. 48, p. 213-218 (in Russian).

Mais, K., 1973, Vorläufige Beobachtungen über Kondenswasserkorrosion in Schlenkendurhgangshöle: *6th International Congress of Speleology,* Praha, Proceedings, v. 3, p. 203-207.

Maksimovich, G.A., 1963, *Fundamentals of karst studies,* v. 1: Perm University, 444 p. (in Russian).

Malkov, V.N., and Frantz, N.A., 1981, On winter condensation, *in* Akkumulatsija zimnego kholoda v porodakh i ego ispolzovanije: Perm, p. 97-99 (in Russian).

Martel, E.A., 1894, *Les Abimes:* Paris, 578 p.

Mavliudov, B., 1989, Snow and ice formations in caves and their regime: *10th International Congress of Speleology,* Budapest, Proceedings, v. 1, p. 295-297.

Molerio, L., 1981, Hidrogeologia y climatologia de la cueva La Mariana: *Volun. hidraul.,* v. 18, no. 57, p. 2-9.

Moore, G.W., and Sullivan, G.N., 1978, *Speleology:* Teaneck, Zephyrus Press, ed. 2, 150 p.

Morozov, A.T., 1938, On the methods of study of moving vapor moisture in soil-ground: *Transact. of the Inst. of Hydrology and Melioration,* v. 22, p. 18-26 (in Russian).

Mucke, D., and Völker, R., 1978, Kondenswasserkorrosion: *Jahrb. des Höhlenforschers,* p. 7-16.

Mucke, D., Völker, R., and Wadewitz, S., 1983, Cupola formation in occasionally inundated cave roofs, *in Europ. Conf. on Speleology*, Sofia, v. 2, p. 129-132.

Müller, P., 1974, A melegforrás-barlangok és gömbfülkék lebetheziséröl: *Karszt és Barlang* (Budapest), v. 1, p. 7-11 (in Hungarian).

Nemeriuk, G.E., and Paltsev, V.P., 1969, On the methods of study of the ionic composition of vapor condensed from the air: *23rd Hydrochem. Symposium.*, Novocherkassk, Proceedings, p. 101-102 (in Russian).

Obolensky, V.N., 1944, *Course of meteorology:* Moscow-Sverdlovsk, 120 p. (in Russian).

Pashchenko, S.E., Dublyansky, Y.V., and Andreichuk, V.N., 1993, Aerosol studies in Kungur Ice Cave: *11th International Congress of Speleology*, Beijing, Proceedings, p. 190-192.

Pashenko, S.E., Dublyansky, Y.V., Anderichuk, V.N., and Pashenko, E.F., 1996, Transformation of fractal atmospheric aerosol moving through a natural cave: *Journal of Aerosol Sci.,* v. 27, suppl. 1, p. S127-S128.

Pasquini, G., 1973, Aggressive condensation: *6th International Congress of Speleology*, Olomouc, Proceedings, v. 8, p. 315-318.

Pinneker, E.V., ed., 1980, *Fundamentals of hydrogeology; General hydrogeology:* Novosibirsk, Nauka, 230 p. (in Russian).

Prokofiev, S.S., 1964, The role of condensation moisture in creation of karst caves: *Peshchery* (Perm), v. 4, no. 5, p. 35-38 (in Russian).

Protasov, V.A., 1959, Condensation waters of Crimea Mountains and their role in underground runoff: *3rd Hydrol. Congress*, Leningrad, Proceedings, v. 9, p. 98-103 (in Russian).

Queen, J.M., 1981, *A discussion and field guide to the geology of Carlsbad Caverns:* Unpublished report to Carlsbad Caverns National Park, 64 p.

Racoviţcă, G., and Viemann, J., 1984, Sur le rôle de la condensation souterraine dans la genèse des stalagmites de glace: *Trav. Inst. Spéol. "É. Racoviţză,"* v. 23, p. 89-97.

Riazanova, E.A., 1987, Classification of factors and sources of groundwater level rise, *in* Voprosi ingenernogeologicheskikh issledovanii na zastraivajemikh territorijakh: Moscow, Stroiizdat, p. 54-59 (in Russian).

Shaw, T.R., 1992, *History of Cave Science; The exploration and study of limestone caves, to 1900:* Sydney Speleological Society, 338 p.

Shestakov, F.V., 1989, *Condensation of water vapor in soil-ground and in the near-surface layer; Bibliography from 1877 to 1987:* Alma-Ata, Nauka, 77 p. (in Russian).

Slavianov, V.N., 1955, About natural condensors on mountain slopes and about the possibility of their usage for small-scale water supply, *in* Voprosi izuchenija podzemnikh vod i ingenerno-geologicheskikh processov: Moscow, AN SSSR, p. 21-32 (in Russian).

Szunyogh, G., 1982, Theoretical problems of dissolution related to spherical niches formation: *Karszt és Barlang* (Buda-pest), v. 2, p. 83-88 (in Hungarian).

Szunyogh, G., 1990, Theoretical investigation of the development of spheroidal niches of thermal water origin-Second approximation: *10th International Congress of Speleology,* Budapest, Proceedings, p. 766-768.

Tkachenko, K.D., 1978, Importance of condensation moisture and dew in the water balance of the aeration zone: *Geol. Journ.,* v. 38, no. 3, p. 27-34 (in Russian).

Trombe, F., 1952, *Traitè de spéléologie:* Paris, 385 p.

Tsikarishvili, K.D., 1981, Climatic particularities of karst caves in Georgia: *European Conf. on Speleology*, Sofia, Proceedings, v. 2, p. 386-390 (in Russian).

Tugarinov, V.V., 1955, Some results of study of water vapor condensation from the air, *in* Voprosi izuchenija podzemßnikh vod i ingenerno-geologicheskikh processov: Moscow, AN SSSR, p. 60-78 (in Russian).

Turchinov, I.I., 1994, Aerosol mineral formation in gypsum caves of western Ukraine, *in* Voprosi fizicheskoi speleologii: Moscow, MFTI, p. 51-63 (in Russian).

Ustinova, T.I., 1956, Conditions of condensation of atmospheric moisture in caves of the western part of the Crimea Mountains: *Soveshchanije po izucheniju karsta*, Moscow, v. 8, p. 3-4 (in Russian).

Vakhrushev, B.A., 1993, Usage of condensation groundwater in ancient and medieval times and at the present time, *in* Dvizhenije v noosphere; Theoreticheskie i regionalnye problemy: Simferopol, p. 92-96 (in Russian).

Vasilievski, P.M., and Zheltov, P.I., 1932, Hydrogeologic studies of Chatirdag Mountain in 1927: *Trudy Vserossiys-kogo geologo-razvedochnogo ob'edinenija,* v. 142, p. 1-99 (in Russian).

Vedernikov, V.V., and Larina, L.A., 1985, Calculation of water regime within constructional territories, *in* Raschet vodnogo rezhima zastroennikh territoriy: Moscow, Stroiizdat, p. 52-56 (in Russian).

Volger, O., 1877, Über eine neue Quellentheorie auf meteorologicher Basis: *Meteorol. Ztsechz.*

Worthington, S.R.H., and Ford, D.C., 1995, High sulfate concentrations in limestone springs: An important factor in conduit initiation?: *Environmental Geology,* v. 25, no. 1, p. 9-15.

Zibold, F., 1904, The role of underground dew in water supply for Feodosia: *Pochvovedenije,* v. 4, p. 21-29 (in Russian).

3.7. Paleokarst and its Significance for Speleogenesis

R.A.L. Osborne

Abstract

Paleokarst is difficult to define, it is often extremely difficult to recognize in caves, and its significance for speleogenesis is somewhat controversial. Paleokarst features intersected by caves include a variety of deposits, exhumed voids, and voids that have survived as open features for geologically significant periods of time. Paleokarst features may act as both aquicludes and aquifers.

In some circumstances they may play a significant role in guiding later cave development, and in other cases they may have negligible influence. Paleokarst deposits often influence the form and mineralogy of speleothems developed in recent caves. Investigating the relationships between paleokarst and modern caves requires persistent detailed field observation, backed by laboratory investigations and some skill at solving four dimensional puzzles with many missing pieces.

Introduction

It has long been recognized that features which we would today describe as paleokarst are found on and below unconformity surfaces in limestone (e.g. Lewis-Abbott, 1894). An extensive literature (see James and Choquette, 1988; Bosák, Ford, Glazek, and Horácek, 1989) describes paleokarsts and paleokarst-hosted ore bodies of various ages. In many cases this has involved the recognition of filled cave systems of considerable extent and complexity (see Rhodes et al.; 1984; Olson, 1984; D.C. Ford, 1986). Other types of ore bodies—whose karst origin is not firmly established—are developed in limestone. Ore veins are noted to act both as aquifers and as aquicludes in massive limestone, and to guide cave development T.D. Ford, 1976).

Many authors have suggested that elements of ancient karst may be incorporated in modern caves (i.e. currently open cavities) and karst drainage systems. Avias (1972) considered the Languedocian karst of southern France to be polygenetic incorporating:

> ...rejuvenated or active, dead, drowned or buried, perched fossil karst, and present-day karst.

T.D.Ford (1976) using the Permian and Lower Carboniferous limestone of the United Kingdom as examples remarked that:

> The possibility of the control of present-day drainage by fossilised systems should not be overlooked.

There has been much less study, however, of paleokarst features exposed in caves, or of the relationship between paleokarst and modern caves and karst drainage systems. Paleokarst features exposed in modern caves are not described in any of the papers in James & Choquette (1988) (although Kahle, 1988 describes caves that are interpreted as being of Silurian age, see below) and only described in two of the papers in Bosák, Ford, Glazek & Horácek (1989).

D.C. Ford (1995) proposed that paleokarst has a limited role in the development of most modern caves which have been formed by descending meteoric water. But it has a more important role in the development of those caves formed by ascending artesian or hydrothermal water.

Published reports of paleokarst deposits exposed and exhumed in caves include:

- Fluorite-bearing paleokarst at Moorfurlong Mine, Bradwell, Derbyshire (Worley and Beck, 1976);
- In the Jugholes Caves, Matlock, Derbyshire, U.K. (Worley and Nash, 1977).
- Paleokarst deposits in many eastern Australian caves (Osborne 1984, 1986, 1991, 1993b).
- Pennsylvanian paleokarst deposits in the walls of cave passages of Tertiary age in the Black Hills of South Dakota, U.S.A. (Palmer and Palmer, 1989).

Hill's work (1987) on Carlsbad Cavern and other caves of the Guadalupe Mountains, New Mexico, U.S.A., gives numerous examples of paleokarst deposits and cavities that have been truncated by more recent caves produced by strong acid solution. Recent work by Polyak et al. (1998) has used the alteration of paleokarst sediment in Carlsbad Cavern as a means of dating the strong acid solution event responsible for forming the present cave.

Proctor (1988) reported neptunian dikes exposed in and guiding the development of caves in Devon, U.K., while Jones and Smith (1988) described the exposure of cross sections through filled cave passages in the smooth walls of caves on Cayman Brac in the Cayman Islands. In later papers, Jones (1992a, 1992b, 1994) described speleothem and other deposits produced by several phases of karstification exposed in caves and surface outcrops in the Tertiary carbonates of the Cayman Islands. Wrenrich and Sutphin (1994) recognized paleokarst breccia exposures in the walls of caves in the Grand Canyon, U.S.A.

When the hydrological significance of paleokarst has been discussed, it is often unclear whether the interactions are with paleokarst sensu stricto (see below) or with relict karst conduits and those filled with relatively recent cave deposits.

Coincidence of Modern Caves with Unconformities and Paleokarst

Some of the most cavernous karst areas, which include large and well-decorated caves, occur in close proximity to unconformities and in limestone containing paleokarst and ore bodies, suggesting that there may be a genetic relationship between paleokarst, ore deposits, and modern karst.

This is well illustrated in several cave regions in the United Kingdom. In the Derbyshire area, where Walkden (1974) recognized paleokarst surfaces, extensive caves occur in close proximity to and in places intersect fluorite-bearing and galena-bearing paleokarst deposits and associated lead mines (see T.D. Ford, 1989; Marsden, 1991).

Modern caves and iron-ore bearing Triassic paleokarst occur together in the Forest of Dean (Simms, 1990; Lowe, 1989, 1993). Triassic paleokarst features, including fissure fills, caves filled with Triassic fossil-bearing sediment (Halstead and Nicoll, 1971), and paleokarst-hosted ore bodies (Simms, 1990), are reported from the highly cavernous Mendip Hills, Somerset. Drew (1975) considered this to have been: "a true karst area possessing some degree of subterranean drainage in Triassic times."

D.C. Ford (1995) gave a more complete account of the situation in the Mendips in which both Permian-Triassic and Late Triassic-Jurassic paleokarst is recognized. But he stressed that there appears to be little relationship between these paleokarsts and the modern karst system. Paleokarst-hosted hematite and interstratal karst is reported in the Carboniferous limestone of the South Wales Coalfield (Thomas, 1974; Simms, 1990), which contains some of the longest caves in Britain. The paleokarst iron-ore deposits of the Austrian Alps are intimately related to Dachstein-Mammuthöhle, one of the longest caves in Austria (Seemann, 1979).

My original search for paleokarst in New South Wales, Australia (Osborne, 1984), was partly initiated by the recognition that major cavernous karsts (Jenolan, Bungonia) are close to the grand unconformity at the base of the Sydney Basin and the regional unconformity at the base of the Late Devonian (Wellington, Wyanbene). Later work showed that major caves in Tasmania (Osborne, 1994, 1995), which include the most extensive and deepest caves developed in Palaeozoic limestone in Australia, were developed in close proximity to the Permian-Carboniferous unconformity at the base of the Tasmania Basin.

A Clear Definition of Paleokarst

Any attempt to understand the relationship between "modern" caves and paleokarst requires a clear definition of what is meant by "paleokarst." James and Choquette (1988, p. 2) defined paleokarst as:

> Ancient karst, which is commonly buried by younger sediments or sedimentary rocks and thus includes both relict paleokarst (present landscapes formed in the past) and buried paleokarst (karst landscapes buried by sediments).

Whereas this is a useful definition at the macro scale—and for geologists working on drill core—it does not assist in distinguishing between paleokarst deposits exposed in caves and modern cave sediments. Nor does it help distinguish between relict karst features formed at an early stage of the current period of karstification and truly paleokarst features formed during an ancient period of karstification and remaining as inherited or relict features in the landscape or rock mass.

Jennings (1982) considered that the term "palaeokarst" was "problematic" and proposed that paleokarst should only be used to refer to karst features that are older than the last major orogeny in the region. This definition is clearly too restrictive and would rule out most interstratal paleokarsts, most of the paleokarst features recognized by petroleum geologists (see Wright, 1991), many paleokarsts recognized in karst bauxite deposits (see Combes, 1978), many paleokarsts related to subaerial diagenesis and dolomitization (see Henrich, 1984), and many Mississippi Valley ore deposits (see Sangster, 1988) that are formed on or below disconformable erosional surfaces in relatively undeformed sedimentary strata.

Bosák, Ford, and Glazek (1989) discussed the definitions of "paleokarst" and "buried karst," and recommended that:

> Paleokarst refers to karst developed largely or entirely during past geologic periods.

In their diagram (Figure 1.1, p. 31) they indicate that relict karst and buried karst are types of paleokarst.

Use of the terms *period* and *phase* can be a cause of considerable confusion when used in relation to paleokarst. Many writers, including myself, have used these two terms either without proper definition or as synonyms (see Multiple Karstification, below). It is also often unclear whether period should be taken to mean a Period of the geologic time scale or some other length of time. Bosák, Ford, and Glazek (1989, p. 32) introduced the term "karst period," which they define as:

> Long-lasting times of continental weathering and groundwater circulation which were terminated by a marine transgression.

They go on to state that:

> Karst periods will be recognized by upper unconformities or disconformities.

Karst periods are divided into karst phases by Bosák, Ford and Glazek (1989, p. 32) in which "several generations" of karst forms may be created. Karst phases:

> ...are caused by geodynamic or major climatic change, e.g. uplift or down-warping, a phase of premafreezing, etc. (Bosák, Ford and Glazek, 1989, p. 32).

Karst produced during previous geologic Periods (or even Eras) would include relict caves and their sediment developed in landscapes where the present period of landscape development and karstification was initiated long ago. In localities where the present period of landscape development began in the Mesozoic (e.g. eastern Australia, where there has been no orogeny and possibly even no uplift in the past 90 to 100 million years), the oldest caves produced by the present period of karstification may be Late Cretaceous (Osborne, 1993a). Such caves are relict (in the sense of Jennings, 1982) but were produced during the current period of karstification, so can hardly be described as paleokarst.

The problem of definition becomes particularly important when dealing with cavities that have a long (extending over millions of years) and complex history of deposition and exhumation during what might be considered to be the current period of landscape development. Examples are the hominid-bearing cave deposits of the Transvaal, South Africa (Kent, 1980), or the Cenozoic sequence at Wellington Caves, New South Wales, Australia (Osborne, 1983).

Should, for instance, buried and partly-buried caves be considered to be paleokarst, when sedimentation and exhumation has occurred within the same cave walls that enclose a presently open "modern" cavity? One approach that might help to distinguish "modern" cave deposits from paleokarst deposits exposed in caves, would be to consider as paleokarst deposits those that are enclosed by boundaries (cave walls) older than the wall enclosing the present cave. If "modern" caves simply intersected caves formed during earlier periods of karstification, this approach would work. But "modern" caves can clearly be exhumed, rejuvenated, or reactivated paleokarst features of considerable antiquity (e.g. Paleozoic) (see below and Osborne 1993b).

It is thus clear that the definition of paleokarst proposed by James and Choquette (1988) is too broad, as it does not allow for an easy distinction between modern cave sediments and paleokarst deposits or between old relict features and paleokarst. The definition of Jennings (1982) is too narrow, as it would exclude much of what clearly is paleokarst.

A useful definition would seem to be best derived by modifying that of Bosák, Ford, and Glazek (1989):

> Paleokarst refers to karst developed largely or entirely during past **geologic** periods, *to*
>
> Paleokarst refers to karst developed largely or entirely during past **karst** periods.

Even this definition has problems. Zötl (1989), for instance, considered that karst cavities flooded by sea level rises in the Mediterranean were examples of paleokarst. Clearly these cavities and similar features elsewhere such as the Blue Holes of the Bahamas were produced during a past karst period. But are they paleokarst? Similarly Zötl (1989) described caves in desert regions that formed under wetter climatic conditions in the past as paleokarst.

These two examples beg the question as to whether the concept of a karst period should be defined in a "geologic" way that would recognize the period as beginning with the initial exposure of the karst rock by erosion (or its penetration by aggressive groundwater) and would end with its reburial, or in a more "geomorphic" way in which karst periods are seen as being terminated by significant climate changes, partial burial, or inundation by the sea.

Ford and Williams (1989) proposed a simplified definition:

> Relict karsts are those removed from the situation in which they were developed.

And:

> Paleokarsts or buried karsts are completely decoupled from the present hydrogeochemical system.

This would seem to be generally compatible with the "geologic" definition.

For the purposes of this discussion I have chosen the "geologic" definition and will thus examine only the relationships between the formation of "modern" caves and paleokarst features and deposits that are of geologically significant age or were formed during a previous period of subaerial exposure.

Multiple Karstification (Polycyclic Karst)

The idea that limestones have undergone multiple periods or phases of karstification is now well established in the international literature. Instances have been found on many continents.

Examples have been reported from Europe of karst areas having undergone four (Avias, 1972) and six (Milikhiker, 1962) periods of karstification. Five phases of karstification have been recognized in central and northern Israel by Buchbinder et al. (1983). Four "episodes" of karstification have been reported from North America (Olson, 1984) and three periods from the Cayman Islands (Jones, 1992a). Two Paleozoic and a modern "phase" of karsti-fication were recognized in eastern Australia (Osborne, 1995). In most cases these authors' use of the terms "period," "episode," and "phase" to indicate events and spans of time that would be defined as a karst period as by Bosák, Ford, and Glazek (1989). In his review Komatina (1975, p. 25) noted that:

> Karst must be considered in most cases as a complex phenomenon grossly dependent on the stratigraphic and tectonic evolution of the region, i.e. as a complex formation of stratigraphically different paleokarsts and recent karsts.

Many karst caves may therefore be the product of a complex overprinting of numerous (usually more than two) distinct periods (and possibly many phases) of karstification and speleogenesis. If this is the case it is important to ask:

- Where is the evidence for paleokarst features and deposits in modern caves?
- What has been the influence of paleokarst on speleogenesis?

Recognition of Paleokarst Exposed in Caves

The first and often most difficult step in understanding the relationship between modern caves and paleokarst processes is the recognition of paleokarst features and deposits that are exposed in caves today. The classic papers on the recognition of paleokarst (e.g. Quinlan, 1972; Wright, 1991) are largely concerned with alerting geologists to the nature of karst features so that they can be recognized in natural exposures and drill cores.

Jones and Smith (1988), some of whose observations were made in caves, emphasized how difficult it can be to recognize paleokarst features. They discussed the preservation potential and implications for the recognition of paleokarst and concluded that:

> Although an area may undergo multiple generations of karstification, it may be difficult to precisely date the different generations.
>
> Surficial karst features have a low preservation potential, whereas interface and subsurface karst features have a high preservation potential providing that:
>
> - Later surface erosion does not remove evidence of interface karst features.
> - Speleothems and other distinctive sediments are emplaced in the subsurface karst features prior to collapse.
>
> Karst features such as sinkholes, solution-widened joints, and caves can be filled with a wide variety of speleothems and sediments prior to the burial of the karst during subsequent marine transgressions. The development of speleothems and (or) the deposition of clastic sediment in any karst feature is highly variable even on a local scale; thus, some features remain open while others are partly or totally filled with speleothems and (or) sediments. (Jones and Smith, 1988, p. 1289-1290).

Jones and Smith's concerns were somewhat influenced by the environment in which they worked: young, mechanically quite weak limestones, close to sea level, where collapse and transgression are common events. Their general conclusion is that subsurface paleokarst has a high preservation potential.

There is, however, limited reporting of subsurface paleokarst. This could mean that:

- Little paleokarst is intersected by modern caves.
- Or: Paleokarst is difficult to recognize in caves.
- Or: Paleokarst has not been looked for well enough in caves.

My experience suggests that identification of

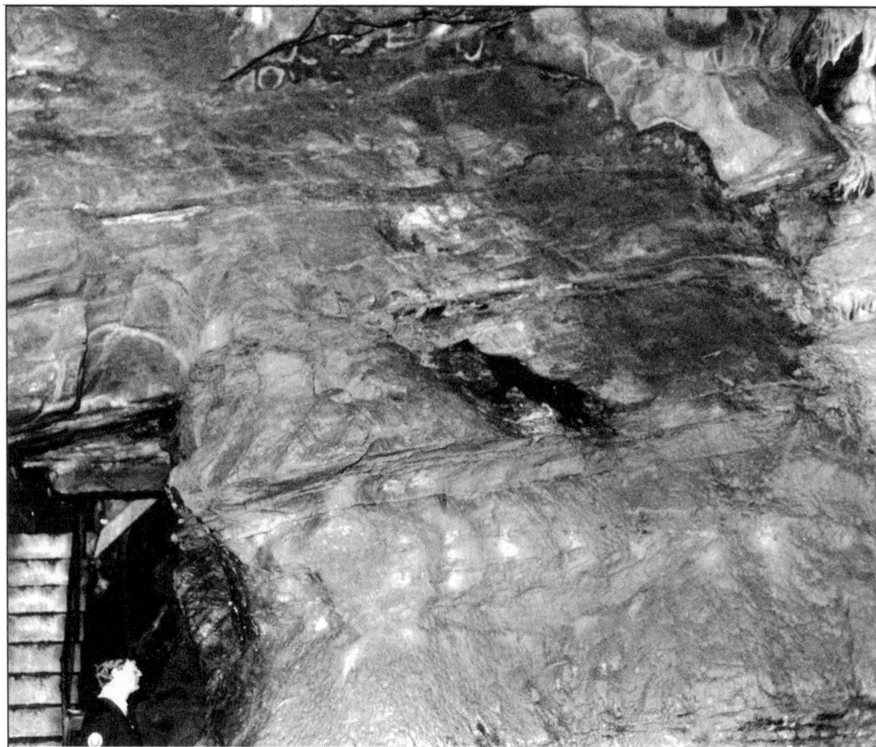

3.7 Plate 1. Exposure of unconformable, lithified paleokarst deposit, River Cave, Jenolan Caves, New South Wales, Australia. Cave guide is standing in newer passage excavated through paleokarst deposit. Boundary and part of wall of initial cavity is exposed in upper right of frame. Silurian bedrock is smooth and lighter in color than paleokarst deposit.

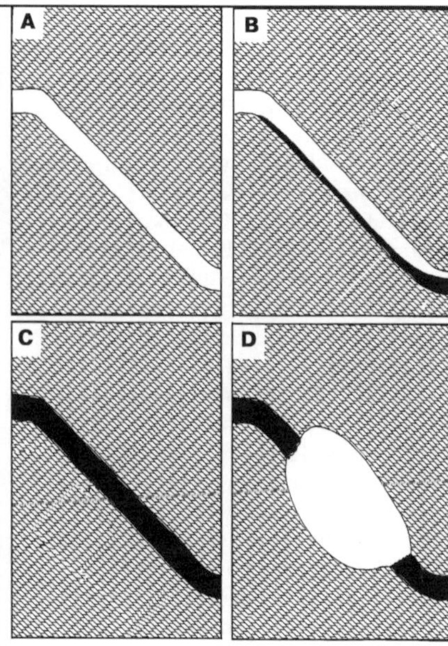

3.7 Figure 1. Formation and exposure of disconformable lithified paleokarst deposits in dipping limestone. ***A****–Dip tube forms in limestone with floor and roof parallel to bedding.* ***B****–Fine laminated clays deposited in dip tube with laminations parallel to depositional surface (i.e. at angle of initial dip close to that of limestone bedrock).* ***C****–Cavity fills completely and clay becomes lithified to shale.* ***D****–Subsequent cave development exposes shale paleokarst deposit that is now extremely difficult to distinguish from a shale bed or lens in the bedrock.*

paleokarst structures and deposits in modern karst caves can be both controversial and difficult because it is commonly difficult to distinguish between

- Paleokarst structures and deposits (including speleothem) and structures and deposits that formed during early stages of the present karst period (relict features in the sense of Ford and Williams, 1989)
- Paleokarst deposits and structures and some bedrock structural and sedimentary features of a nonkarst origin, particularly when they are exposed in caves.

This is made particularly difficult and often inconclusive because exposures of geologic structures in cave wall surfaces commonly:

- Are small and incomplete.
- Have a complex geometry.
- Are highly weathered.
- Are obscured by wall coatings and more recent deposits.

Unconformable Lithified Paleokarst Deposits

By far the most easily recognized paleokarst deposits exposed in caves are those that are strongly lithified, clearly unconformable with the bedrock, and have boundaries shaped like cave walls (3.7 Plate 1). Where these deposits are siliciclastic (e.g. quartz sand) recognition is usually straightforward.

Lithified carbonate unconformable fills can be problematic. If the lateral and vertical boundaries are not well exposed, and the bedrock is significantly folded, it may be difficult to distinguish between an unconformable paleokarst deposit and a change in bedding orientation produced by tight folding.

Disconformable Lithified Paleokarst Deposits

When exposed in cliffs or quarries, disconformable lithified paleokarst deposits are generally not difficult to recognize with some degree of certainty. In these large exposures, boundaries and shapes of filled karst cavities can usually be easily determined, despite the fill having much the same bedding orientation as the bedrock.

In caves, the boundaries may not be well exposed and are commonly only partly visible, making interpretation difficult. Even in cases of good exposures in quarries, disputes have arisen as to whether a feature is a shale lens in the original limestone or a paleokarst fill (for example, see Branagan, 1984, 1986; Webb, 1986).

The deposition of fine mud and flowstone with high initial dips parallel to the sides of bedding-controlled rift passages can also result in paleokarst features in steeply dipping limestone being misidentified as being part of the bedrock and vice versa (3.7 Figure 1). Petrographic study can often but not always resolve this dilemma.

Disconformable Poorly Lithified Paleokarst Deposits

Where disconformable paleokarst deposits are poorly lithified, it may be extremely difficult to distinguish them from "modern" cave sediments, unless they contain distinctive fossils or have a distinctly different provenance from the "modern" sediments. Given that the degree of cementation or lithification of cave sediment is not necessarily a good indicator of age, and that some authors (e.g. Kahle, 1988) have implied that unconsolidated paleokarst deposits may be as old as Silurian, it is clearly dangerous to assume (without confirming evidence) that unconsolidated sediment found in caves was necessarily deposited during the most recent period of karstification.

A good example of this situation is reported by Palmer and Palmer (1989) from Jewel and Wind Caves, South Dakota.Here many instances of unconsolidated clastic fillings are almost certainly Early Pennsylvanian in age. In some instances these sediments are spilling onto the floors of the Tertiary caves that intersect them.

Paleokarst Deposits of Volcanic or Volcaniclastic Origin

When lava flows enter karst features such as grikes and caves, where shallow intrusives (dikes and sills) break through into existing karst cavities (3.7 Figure 2B), or where flows fill karst cavities and overlie older paleokarst deposits (3.7 Figure 2C and 3.7 Plate 2) paleokarst deposits of volcanic origin are formed. These can be particularly significant as they are commonly amenable to potassium-argon dating.

It can be extremely difficult to distinguish between paleokarst deposits of volcanic origin and intrusive igneous structures such as dikes and sills, particularly where the exposure occurs in a cave wall or ceiling (3.7 Figure 2).

Where a planar volcanic rock body exposed at a cave wall has undergone spheroidal weathering, three interpretations present themselves:

- The feature is a weathered dike or sill that antedates the cave or other karst.
- The feature is a flow that has entered a preexisting karst cavity and has either been altered by contact with water or weathered (i.e. is a paleokarst feature) (3.7 Figure 2D).
- The feature is a void, younger than the volcanic rock, that has been filled with spheroidally weathered core stones derived from the surface (3.7 Figure 2E).

Paleokarst deposits of volcaniclastic origin form when tephra fills caves, grikes, and karst depressions. Identification of volcaniclastic paleokarst presents many of the same problems as occur in volcanic paleokarst, with the added complication that there must be no doubt about the petrographic identification of the rock as being of volcaniclastic and not shallow-intrusive origin.

Recognition of Exhumed Voids and Cave Passages

Exhumed voids occur where underground paleokarst features have become reactivated (and perhaps partially or completely emptied) during a more recent period of karstification. Recognition of these features is more difficult than in the cases described above where the modern karst feature intersects and exposes a paleokarst feature, because in this case the modern cave passage is to all intents and purposes the ancient one.

One might suspect that a modern cave passage is an exhumed or relict passage if it

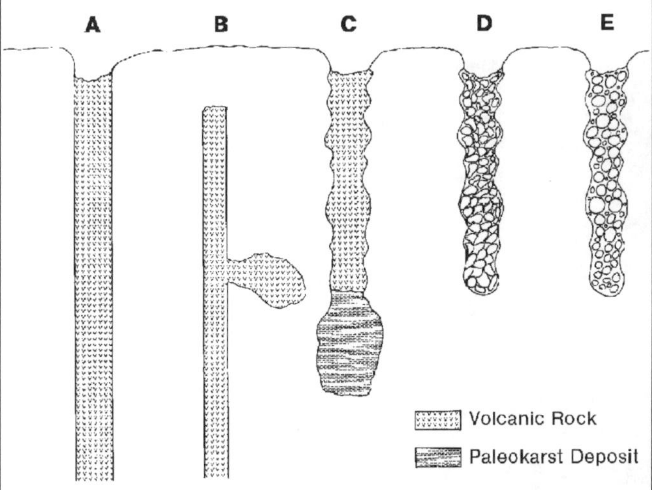

*3.7 Figure 2. Planar volcanic features that may be exposed in limestone caves: **A**–A simple dike, exposed at surface, that may act as an aquiclude in the phreatic zone and may control seepage and breakdown as it weathers in the vadose zone. **B**–A blind dike with no surface exposure that has intersected and filled a preexisting karst cavity, and the filled cavity may be exhumed as the dike weathers or is exposed by later surface erosion or cave development. **C**–A volcanic flow filling a vadose slot in which the flow overlies an older paleokarst deposit (see 3.7 Plate 2). **D**–A vadose slot filled with basalt that has either weathered in situ or has reacted with water to form pillows from which core stones are shaped to fit against the cave wall and the cave is older than extrusion of the basalt. **E**–A vadose slot filled with basalt corestones that have been transported into the cave, packed against the cave walls, but not shaped to fit against the walls, and the cave may either be older or younger than extrusion of the basalt.*

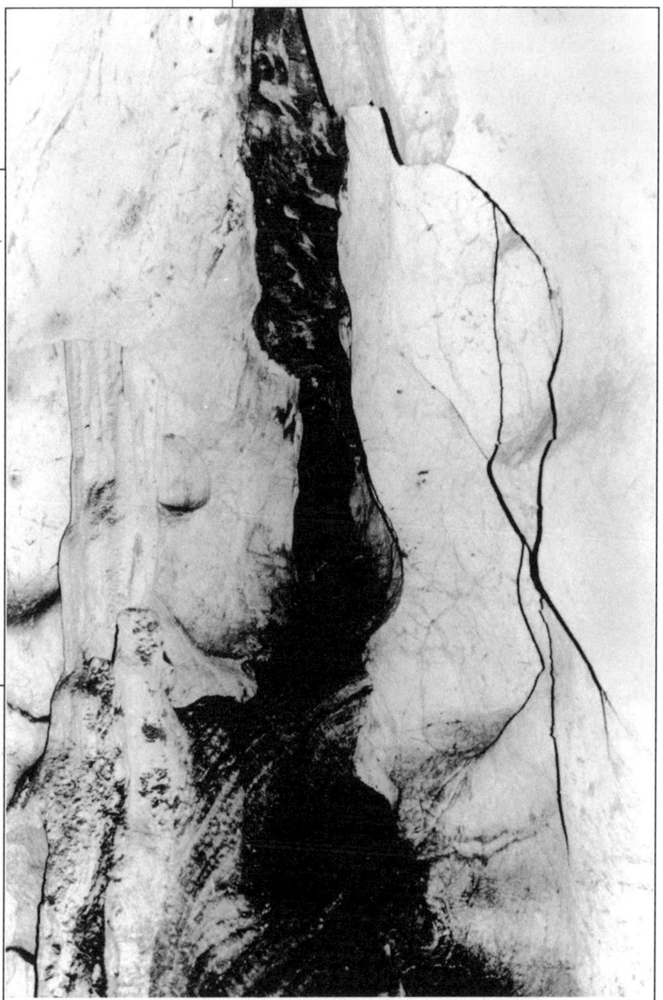

3.7 Plate 2. Basalt flow (upper, dark) overlying flowstone paleokarst deposit (lower, showing bedding) filling canyon-like passage, exposed in wall of more recent phreatic cave, Main Cave, Timor Caves, New South Wales, Australia, (Osborne, 1986).

is out of scale with the rest of the system or has a distinctly different orientation or morphology (see below). Exhumed paleokarst passages commonly contain underfit streams or are of much greater size than the passages actively forming in the caves today. Convincing evidence of exhumation takes the form of remnant (truly paleokarstic) fills whose boundaries are contiguous with the modern cave wall or can be observed in the process of being removed from a void that is clearly contiguous with the modern cave passage.

Since paleokarstic fills can be removed with little trace by modern speleogenic processes (particularly if they are chemically unstable in the vadose zone), there will be many cases where, although it might be suspected that the modern cave is an exhumed paleokarst feature, convincing evidence is lacking.

When limestone caves are filled for significant periods of time, the limestone wall rock frequently becomes altered as a result of chemical reactions between the cave walls and the sediment. Long periods of filling can thus result in the formation of alteration zones in the wall rock of the cave. This may result in silicification of the limestone, giving the cave walls a weathered appearance. Osborne (1993a) described a zone of silicification 150 mm thick at the boundary between Silurian limestone bedrock and laminated clay of presumed early Tertiary age at Bungonia Caves, New South Wales, Australia (3.7 Plate 3).

The limestone walls of the cave where the sediment had slumped away have a crumbly "weathered" appearance, and smooth clean limestone is only exposed where recent ephemeral vadose streams have eroded away the coating. Cave walls that are not made of smooth clean limestone and have a crumbly weathered appearance, particularly where the cave is developed in dense, massive, high purity limestone, should be investigated to determine if they are a result of a reaction between ancient sediment and the bedrock, and thus evidence for exhumation. This type of evidence requires very careful investigation, as many paleocave walls show little reaction with fill, and weathered walls commonly occur in young caves as a result of organic activity or condensation corrosion.

In my work at the Jenolan Caves, New South Wales, Australia, recognition of intersected paleokarst features proved to be fairly easy. It was only with the recognition of one particular exposure (Osborne, 1993c) shown here in 3.7 Plate 4, that it became clear that many of the significant passages at Jenolan are exhumed paleokarst features.

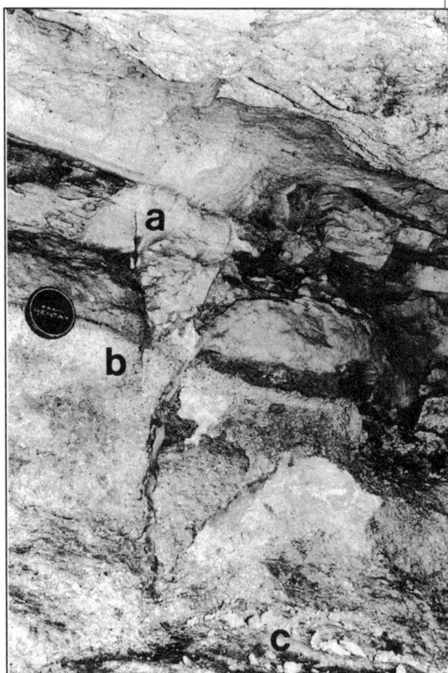

3.7 Plate 3. Laminated clay and silicified limestone "reaction zone" in Grill Cave, Bungonia Caves, New SouthWales, Australia. ***a****–Laminated clay.* ***b****–Silicified crumbly limestone.* ***c****–Unaltered limestone bedrock. Lens cap 55 mm in diameter marks the lower boundary between the clay and altered limestone (Osborne, 1993a).*

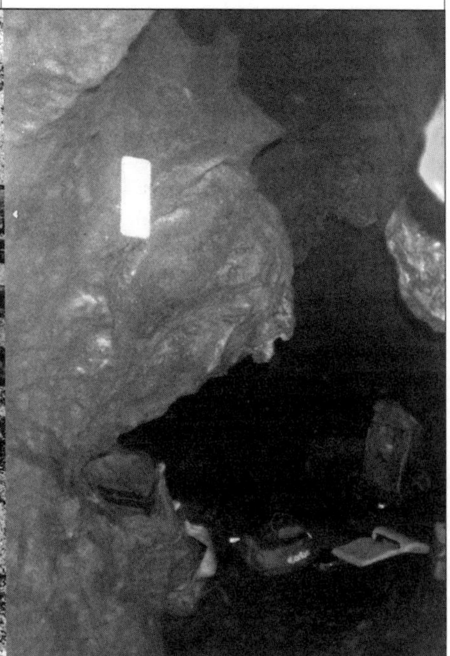

3.7 Plate 4. Exhumed cave passage in Barrelong Cave, Jenolan Caves, N.S.W., Australia (Osborne 1993c). Scale bar on left side is resting on paleokarst deposit (dark color) Note boundary to upper left with Silurian bedrock (lighter color). Paleokarst remnant on right hand side (dark color) abuts Silurian bedrock (light color).

Long-term Survival and Recognition of Open Paleokarst Voids

Open paleokarst voids are caves and other openings that were formed during a previous karst period and have survived in a relatively intact state, neither substantially collapsed nor filled. Some workers doubted that such survival would be possible at all. Davies (1951) suggested that that all caves in a vadose state must collapse within 500,000 years.

Open paleokarst voids should be distinguished from relict caves formed during the early stages of the present period of karstification, such as those described in the Sierra de El Abra, Mexico, in Chapter 5.3.1 of this volume. Open paleokarst voids may be intersected by modern caves, exposed in quarries, intersected by drill holes (either as lost section or "bit drops") or may have survived as relict features in the landscape.

One might initially imagine that the time span over which voids would remain open would be quite short, given the tendency for modern caves to rapidly fill with both clastic and nonclastic deposits. But cave sediment commonly seals entrance zones of caves. This prevents the entry of further sediment and leaves the inner portions to remain as open unfilled voids.

While remote from the surface, open voids are filled with formation water and are likely to develop wall coatings, crystal linings, or undergo significant wall-rock alteration. The origin of such crystal-lined caves, in parts of the U.S.A. where we would today recognize paleokarst, was discussed at some length by Davis (1930).

One problem with the recognition of long-surviving voids in artificial excavations is that it is often unclear whether the voids have survived in a substantially unfilled state or were filled with unconsolidated sediment or breakdown that emptied between the time of intersection and inspection. Kahle (1988, p. 241), for instance, described caves in the Silurian Lockport Dolomite of western Ohio, U.S.A., which are interpreted as having formed under subaerial exposure during the Silurian. The caves are:

> ...partially to completely filled with a variety of sizes of rock debris.

But it is unclear from both the descriptions and photographs whether open voids had survived in the caves since the Silurian or whether they were completely filled and then partly emptied as a consequence of exposure in a quarry face. One interesting aspect of the voids described by Kahle (1988) was that speleothems had become silicified.

There are, however, clear accounts of open voids surviving for geologically significant lengths of time. Craig (1988) reported open caves of Late Permian age up to 6.4 meters high intersected by drill holes in the San Andreas Dolomite of west Texas, U.S.A. Dublyansky, Chapter 5.2.9 in this book, describes cavities which may be of paleokarst origin intersected by drilling operations in Bulgaria. These resulted in bit drops of more than 500 m. Simms (1990) and Lowe (1989, 1993) described open and partly-filled caves in the Forest of Dean, U.K. that contain Triassic sediment and may date, in part, back to the Carboniferous.

Features that Should Arouse Suspicion that Caves Might Expose or be Guided by Paleokarst

Not all karst areas are likely to have their hydrology guided by paleokarst or contain caves whose development has been substantially or partially influenced by paleokarst. There are, however, regional geologic features and specific characteristics of caves that make it possible from geologic literature and cave descriptions to identify karst areas and caves where it is likely that paleokarst might either be exposed in caves or have guided at least some proportion of the cave development (3.7 Table 1).

The presence of any or all of these features in a karst area does not necessarily indicate that paleokarst guidance has played a role in cave development. They are simply pointers to areas where a detailed search for the underground effects of paleokarst is likely to be worthwhile.

One karst area where many of these features have been reported in the literature, but where paleokarst in caves has not been reported, is the South Wales Coalfield in the U.K. Both interstratal karst (Thomas, 1974) and paleokarst (Wright, 1988) are reported in the Dinantian limestone of the region. These are unconformably overlain by coarse clastic rocks at the base of the Namurian Millstone Grit.

Coase and Judson (1977, p. 271) described several features in the Dan yr Ogof cave system that meet the criteria for suspect underground paleokarst features. They describe a mineralized "fault/vein" containing galena, consisting of limestone fragments in a carbonate matrix with iron oxide. Displacements on their "faults" are "either minimal, ranging from a few centimeters to about a meter, *or else impossible to establish*" (emphasis mine).

They also report (p. 291) that:

> Many of the faults are well "decorated" with calcite formations in which the normally quite rare helictites are amongst the most prolific.

Where features such as those described above are found in caves and karst areas, a systematic search for paleokarst features in caves and evidence for paleokarst guidance over speleogenesis should be undertaken.

Regional Features	The cavernous rock occurs below an unconformity. The cavernous rock is host to ore deposits. Paleokarst features have been reported from surface exposures, quarry faces, mines, or drill hole data.
Cave Features	The cave contains large out-of-scale chambers or passages. The cave contains significant breakdown chambers. Major cave passages contain underfit streams.
The cave intersects or is guided by	"Faults" with no apparent displacement. Breccia zones said to be caused by faulting with little displacement. Mineral veins. "Shale beds" that are unconformable with bedrock. Irregular zones of dolomitization.
Some cave walls are	Lined with crystals or atypical coating. Made of non-limestone rock. Composed almost entirely of sections through speleothem. Have pieces of rock (that is clearly not bedrock) adhering to them. Made of limestone with a "weathered" or "rusty" appearance.

3.7 Table 1. Features indicating paleokarst guidance.

Paleokarst Guidance of Karst Drainage and Cave Development

Paleokarst structures and deposits can both enhance and inhibit the flow of groundwater and the hydrocarbon reservoir capacity of soluble rocks. The classic work on the influence of paleokarst on hydrology by Zötl (1989), however, is largely concerned with features that would not be considered to be paleokarst by the definition used here. Zötl's examples from the Austrian Alps are *relict* caves and those from the Mediterranean are flooded karsts (see above).

Little has been published on the influence of paleokarst on cave development. Proctor (1988) described several small caves in south Devon, U.K. in which Permian-Triassic neptunian dikes exercise strong guidance over development. In some cases the dikes acted as aquicludes, restricting cave development perpendicular to them, while in others cave development occurred preferentially along the strike of the dike. Resistant dikes also formed the hanging walls of chambers where they had acted as a barrier to dissolution. Lowe (1993) discussed the influence of paleokarst over cave development and its role in the inception of modern caves in the Forest of Dean, U.K.

D.C. Ford (1995) provided the most detailed discussion yet of the role of paleokarst in modern karstification. Ford observed that paleokarst has little on no influence on the development of caves by "per descensum" (normal meteoric processes), but is significant in the development of those caves produced by "per ascensum" (hydrothermal and artesian processes).

Exhumation of paleokarst cavities by "per ascensum" waters is described in Jewel and Wind Caves, South Dakota, guidance of cave development by paleokarst is described in Lechuguilla Cave and Carlsbad Cavern, New Mexico, and direction of cave-forming thermal waters by paleokarst is described in the Buda Hills of Budapest, Hungary.

Ford proposed that a lack of correspondence between modern and ancient drainage patterns may account for this observed lack of influence of paleokarst over "per descensum" cave forming processes, while by "per ascensum" waters are likely to intersect and be guided by paleokarst features.

Paleokarst as Aquifers

Where ancient cavities remain open, or are filled with permeable sediment such as sand, gravel, and coarse carbonates (e.g. calcarenites), they can act as aquifers and control groundwater flow and cave development when the limestone is subjected to uplift and reexposure. This simple rejuvenation of ancient caves is not widely reported in the literature and is probably quite difficult to recognize.

Exhumation, but not by fluvial action, however, is reported. Ford (1996) described exhumation as a result of hydrothermal development in Wind Cave, and I (1996) described exhumation as a result of vadose weathering of chemically unstable (?hydrothermal) paleokarst fills (Osborne, 1996).

Since both caves and cave fills are commonly modified in the time between karst periods, more complex relationships between paleokarst and modern caves are more likely to occur.

Paleokarst fills are usually more permeable than the bedrock that consists of massive lime mudstone or limestone that has already undergone significant diagenesis. Mineralizing fluids of various origins are thus more likely to affect paleokarst deposits and preexisting pores than the bulk of the limestone mass. This process is not only responsible for the formation of economic ore bodies, but also for the emplacement of noneconomic deposits of pyrite and dolomite that may be derived from basinal fluids in overlying sediments.

Permeable paleokarst deposits are potential "inception horizons" (Lowe, 1992, 1993) for cave development in subsequent karst phases or periods. The presence of sulfides within permeable paleokarst deposits can result in strong acid dissolution in the earliest stages of subsequent speleogenesis.

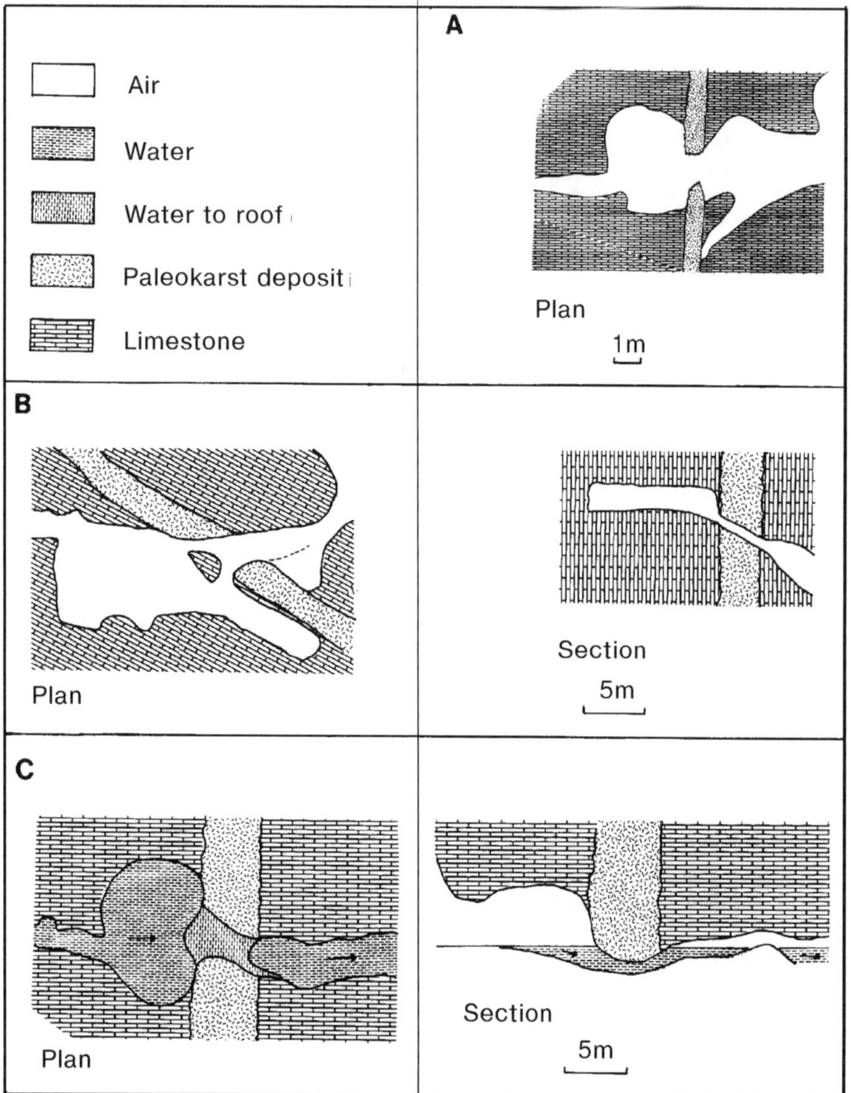

3.7 Figure 3. Paleokarst bodies acting as aquicludes: **A**–Acting as dams, based on observations in Sigma Cave, Wombeyan Caves, New South Wales, Australia (Osborne, 1993b). **B**–Forming squeezes, based on observations in Argyle Hole, Bungonia Caves, N.S.W, Australia. **C**–Forming "duck unders," based on observations in the Jenolan Underground River, Jenolan Caves, New South Wales, Australia.

Paleokarst as Aquicludes

While coarse-grained paleokarst fills and open pores are likely to act as aquifers, fine-grained fills, particularly clays, which are very common cave sediments both today and in the past, are likely to act as aquicludes. Saller et al. (1994) quote numerous cases where paleokarst has produced impermeable zones due to fine-grained sediment filling karst pores.

When coarse-grained paleokarst sediments become mineralized or dolomitized they are commonly impermeable or less soluble than the surrounding limestone, forming aquicludes in the phreatic zone (3.7 Figure 3). Similarly, unweathered volcanic and volcaniclastic paleokarst bodies are likely to act as aquicludes. Paleokarst deposits with dike-like geometry in places act as dams in phreatic pressure tubes, causing the cave passage to enlarge or bifurcate on the "upstream" side before the deposit is finally breached (3.7 Figure 3A). Elsewhere these bodies will "throttle" the pressure tube forming constrictions, "squeezes" in passages of otherwise constant cross section (3.7 Figure 3B). Paleokarst deposits are commonly associated with "sumps" in large stream passages, with water descending and dissolving a path through the limestone below the paleokarst body and then rising on the downstream side (3.7 Figure 3C).

Sulfide-bearing Paleokarst and Exhumation, Breakdown, and Chamber Formation

Paleokarst mineral deposits and paleokarst sediments that have become mineralized with sulfides are chemically unstable in the vadose and shallow-phreatic zones where they are exposed to oxygenated percolation water and the action of oxidizing bacteria. Paleokarst bodies that have acted as strong dense aquicludes in the phreatic zone are progressively transformed to weak rusty masses that are easily eroded and frequently fail by parting from the limestone walls of the cavities they fill. This is an important process in the exhumation of paleokarst cave passages. Since this process is likely to be most nearly complete in the upper sections of caves that have been exposed to vadose conditions for the longest, it is generally useful to begin the search for this type of process in the lower sections of a cave, within 10 m of the water level.

Vadose weathering of large bodies of sulfide-bearing paleokarst and associated minor veins can result in the formation of large out of scale breakdown chambers and avens in caves (3.7 Figure 4). Weathering of small sulfide-bearing veins and tabular paleokarst deposits can result in rapid

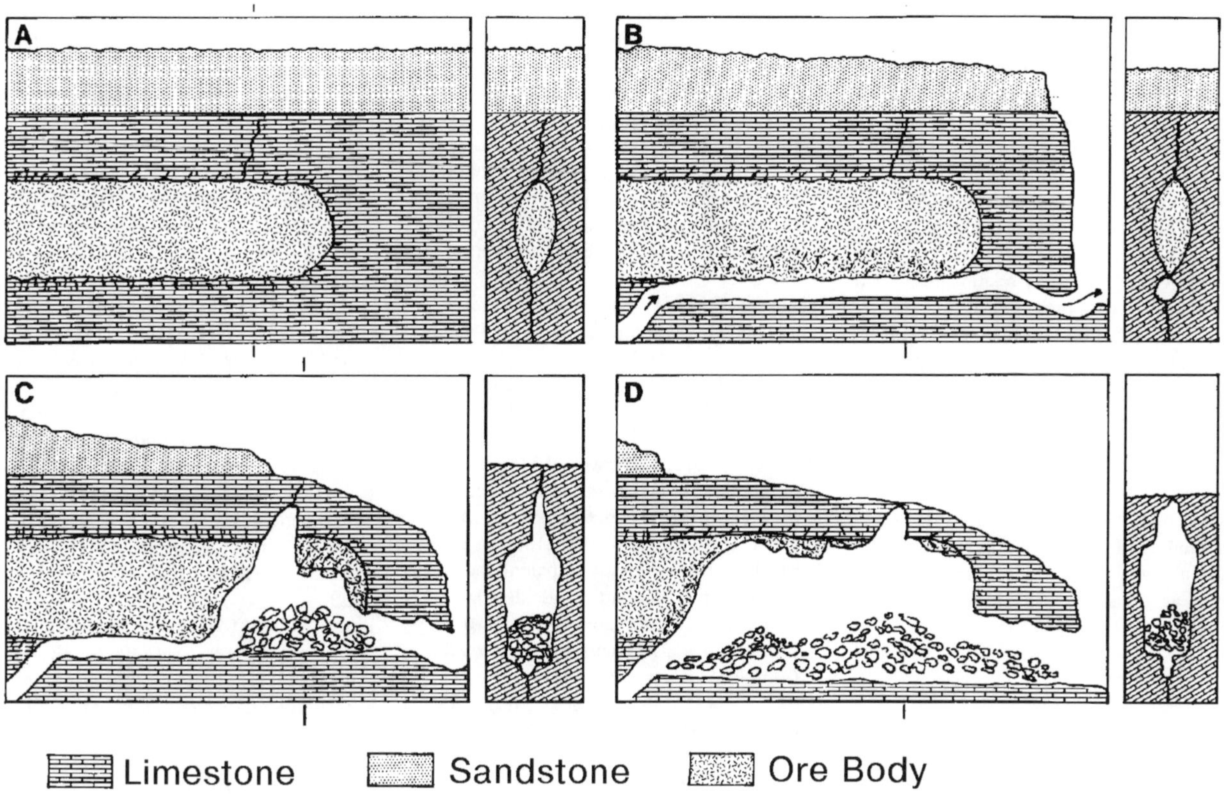

3.7 Figure 4. Vadose excavation of a paleokarst-hosted ore body, resulting in chamber and aven formation, based on observations in Wyanbene Cave, N.S.W., Australia (Osborne, 1966). **A**–Mineralized paleokarst deposit and paleokarst-hosted ore body in limestone below unconformity. **B**–Ore body acts as aquiclude, rising phreatic water forms pressure tube below body , surface erosion lowers base level and exposes body to vadose conditions, and the body begins to weather. **C**–Sedimentary cover begins to retreat, further weathering of the ore body forms a chamber, vadose water entering via an exposed joint oxidizes sulfides that results in the aven forming by vadose solution, and weak and dissolved material is removed by a modern (underfit) stream. **D**–Out-of-scale chamber forms, sulfides in small veins oxidize to form gypsum that causes breakdown of limestone adjacent to ore body, and paleokarst remnants form substrate for aragonite and gypsum speleothems while major upstream development of chamber is delineated by end of covering sediment that inhibits entry of oxygen-rich vadose water into the limestone.

shallow-phreatic dissolution, rapid vadose dissolution forming avens, and the growth of gypsum, leading to extensive breakdown of limestone and the production of large breakdown chambers.

Paleokarst and Speleothem Deposition

Because they are commonly mineralized, are chemically distinct from the bedrock, and in the case of siliciclastic paleokarst, may be more porous than the bedrock, paleokarst deposits are in places the substrate (or chemical source) for the deposition of speleothems with anomalous form and mineralogy. Wrenrich and Sutphin (1994) described speleothems composed of uranium (carnotite), copper (conichalcite), arsenic (talmessite and hornesite), and molybdenum (powellite) in Redwall Caves in the Grand Canyon, U.S.A., associated with mineralized paleokarst breccias exposed in the cave walls.

Much of the paleokarst sediment in eastern Australia contains secondary pyrite and dolomite (Osborne, 1994). Apart from where they are associated with dikes and other igneous rocks, aragonite and gypsum speleothems in eastern Australian caves are almost only found deposited on or in close proximity to dolomitic and pyrite-bearing paleokarst sediments.

Finding the Relationships Between Modern Caves and Paleokarst

Cave systems in limestones that have undergone multiple periods of karstification are a complex combination and overprinting of

- Modern cave forms.
- Modern caves whose development has been guided by paleokarst.
- Paleokarst deposits.
- Open paleokarst cavities.
- Exhumed paleokarst cavities.
- Filled paleokarst cavities that are in the process of being exhumed.

Unravelling the history of such systems and the relationships between their various components is a complex and difficult task. It is greatly assisted if the bedrock geology of the area is well understood and if detailed cave maps and Chapters are available. The crucial requirements for this task are

- Detailed field observation, documentation, and sampling.
- Petrographic analysis.
- Dating techniques where applicable.
- An ability to solve complex stratigraphic relationships.
- An understanding of regional geologic and

geomorphic history.

Weathered rocks exposed in caves are generally very difficult to identify in the field, and only when a thin section is available is the true lithology of a specimen apparent (rusty pyrite-bearing sandstone and weathered basalt can look remarkably similar in underground exposure and in hand specimen).

Exposures and boundary relationships need to be sketched and photographed. It may require many visits to the site, lots of wall washing and scraping of mud, a great deal of thinking, and some second opinions before the true nature of boundaries and relationships is established.

The recognition of paleokarst guidance of and interaction with recent speleogenic processes makes possible not only a better understanding of caves as landscape features, but also allows caves to realize their true potential as repositories of information about the past.

References

Avais, J., 1972, Karst of France, *in* Herak, M., and Springfield, V.T., eds., *Karst-Important karst regions of the northern hemisphere:* Amsterdam, Elsevier, p. 129-188.

Bosák, P., Ford, D.C., and Glazek, J., 1989, Terminology, *in* Bosák, P., Ford, D.C., Glazek, J., and Horácek, I., eds., *Paleokarst-A Systematic and Regional Review:* Amsterdam and Praha, Elsevier and Academia, p. 25-32.

Bosák, P., Ford, D.C., Glazek, J., and Horácek, I., eds., 1989, *Paleokarst-A Systematic and Regional Review:* Amsterdam and Praha, Elsevier and Academia, 725 p.

Branagan, D.F., 1984, Proterozoic palaeokarst at Torrowangee, western New South Wales?: *Australian Journal of Earth Sciences,* v. 31, p. 155-158.

Branagan, D.F., 1986, Reply to discussion: Proterozoic palaeokarst at Torrowangee, western New South Wales?: *Australian Journal of Earth Sciences,* v. 33, p. 366-367.

Buchbinder, B., Magaritz, M., and Buchbinder, L.G., 1983, Turtonian to Neogene palaeokarst in Israel: *Palaeogeography, Palaeoclimatology, Palaeoecology,* v. 43, p. 329-350.

Coase, A.C., and Judson, D.M., 1977, Dan yr Ogof and its associated caves: *Transactions of the British Cave Research Association,* v. 4, p. 247-344.

Combes, P.-J., 1978, Karst precoce et karst secondaire du trioseme horizon de bauxite dans la Zone du Parnasse (Grece): *4th International Congress I.C.S.O.B.A.*, Athens, Proceedings, p. 101-113.

Craig, D.H., 1988, Caves and other features of Permian karst in San Andreas Dolomite, Yates Field Reservoir, west Texas, *in* James, N.P., and Choquette, P.W., eds., *Paleokarst:* New York, Springer-Verlag, p. 342-363.

Davies, W.E., 1951, Mechanics of cavern breakdown: *National Speleological Society Bulletin,* v. 13, p. 36-43.

Davis, W.M., 1930, Origin of limestone caverns: *Geological Society of America Bulletin,* v. 41, p. 475-628.

Drew, D.P., 1975, The caves of Mendip, *in* Smith, D.I., ed., *Limestones and Caves of the Mendip Hills:* Newton Abbot (U.K.), David and Charles, p. 171-213.

Ford, D.C., 1986, Genesis of paleokarst and strata-bound zinc-lead sulfide deposits in a Proterozoic dolostone, northern Baffin Island, Canada-A discussion: *Economic Geology,* v. 81, p. 1562-1566.

Ford, D.C., 1995, Paleokarst as a target for modern karstification: *Carbonates and Evaporites,* v. 10, p. 138-147.

Ford, T.D., 1976, The geology of caves, *in* Ford, T.D., and Cullingford, C.H.D., eds., *The Science of Speleology:* London, Academic Press, p. 11-60.

Ford, T.D., 1989, Paleokarsts of Britain, *in* Bosák, P., Ford, D.C., Glazek, J., and Horácek, I., eds., *Paleokarst-A Systematic and Regional Review:* Amsterdam and Praha, Elsevier and Academia, p. 51-70.

Halstead, B., and Nicoll, P.G., 1971, Fossilized caves of Mendip: *Studies in Speleology,* v. 2, no. 3-4, p. 93-102.

Henrich, R., 1984, Facies, dolomitization and karstification of lagoonal carbonates: Triassic of the northern Alps: *Facies,* v. 11, p. 109-156.

Hill, C.A., 1987, Geology of Carlsbad Cavern and other caves in the Guadalupe Mountains, New Mexico and Texas: *New Mexico Bureau of Mines and Mineral Resources Bulletin,* no. 117, 150 p.

James, N.P., and Choquette, P.W., eds., 1988, *Paleokarst:* New York, Springer-Verlag, 416 p.

Jennings, J.N., 1982, Principles and problems in reconstructing karst history: *Helictite,* v. 20, no. 2, p. 37-52.

Jones, B., and Smith, D.S., 1988, Open and filled features on the Cayman Islands: Implications for the recognition of paleokarst: *Canadian Journal of Earth Sciences,* v. 25, p. 1277-1291.

Jones, B., 1992a, Void-filling deposits in karst terrains of isolated oceanic islands: A case study from Tertiary carbonates of the Cayman Islands: *Sedimentology,* v. 39, p. 857-876.

Jones, B., 1992b, Caymanite, a cavity-filling deposit in the Oligocene-Miocene Bluff Formation of the Cayman Islands: *Canadian Journal of Earth Science,* v. 29, p. 720-735.

Kahle, C.F., 1988, Surface and subsurface paleokarst, Silurian Lockport, and Peebles Dolomites, western Ohio, *in* James, N.P., and Choquette, P.W., eds., *Paleokarst:* New York, Springer-Verlag, p. 229-255.

Kent, L.E., 1980, Stratigraphy of South Africa: *Republic of South Africa Geological Survey Handbook 8*, no. 1, p. 603-628.

Komatina, M., 1975, Development conditions and regionalization of karst: *International Union of Geological Sciences, Series B, Hydrology of Karstic Terrains,* v. 3, p. 21-29.

Lewis-Abbott, J.W., 1894, The ossiferous fissures in the valley of the Shode, near Ightham, Kent: *Quarterly Journal of the Geological Society of London,* v. 50, p. 171-187.

Lowe, D.J., 1989, Limestones and caves of the Forest of Dean, *in* Ford, T.D., ed., *Limestones and Caves of Wales:* Cambridge University Press, p. 106-116.

Lowe, D.J., 1992,The origin of limestone caverns: An inception horizon hypothesis: Manchester Polytechnic University, Ph.D. thesis, 512 p.

Lowe, D.J., 1993, The Forest of Dean caves and karst: Inception horizons and iron-ore deposits: *Cave Science,* v. 20, no. 2, p. 31-43.

Marsden, A.W., 1991, Aspects of the Peak-Speedwell Cave System 1650-1990: *Cave Science,* v. 18, no. 1, p. 3-1.

Milikhiker, S.H., 1962, Issledovaniya karsta v rayone srednego techeniya reki Chusovoy svyazi s gidrotekhnicheskim stroitel'stvom, *in* Spetsial'nye Voprosy Karstovednija: Moscow, Akad. Nauk. S.S.S.R.

Olson, R.A., 1984, Genesis of paleokarst and strata-bound zinc-lead sulfide deposits in a Proterozoic dolostone, northern Baffin Island, Canada: *Economic Geology,* v. 70, p. 1056-1103.

Osborne, R.A.L., 1983, Cainozoic stratigraphy at Wellington Caves, N.S.W.: *Proceedings of the Linnean Society of New South Wales,* v. 107, p. 129-145.

Osborne, R.A.L., 1984, Multiple karstification in the Lachlan Fold Belt in New South Wales: Reconnaissance evidence: *Journal and Proceedings of the Royal Society of New South Wales,* v. 117, p. 15-34.

Osborne, R.A.L., 1986, Cave and landscape chronology at Timor Caves, New South Wales: *Journal and Proceedings of the Royal Society of New South Wales,* v. 119, p. 55-76.

Osborne, R.A.L., 1991, Palaeokarst deposits at Jenolan Caves, New South Wales: *Journal and Proceedings of the Royal Society of New South Wales,* v. 123, p. 59-73.

Osborne, R.A.L., 1993a, A new history of cave development at Bungonia, N.S.W.: *Australian Geographer,* v. 24, p. 62-74.

Osborne, R.A.L., 1993b, The history of karstification at Wombeyan Caves, New South Wales, Australia, as revealed by palaeokarst deposits: *Cave Science,* v. 20, no. 1, p. 1-8.

Osborne, R.A.L., 1993c, Geological Note: Cave formation by exhumation of Palaeozoic palaeokarst deposits at Jenolan Caves, New South Wales: *Australian Journal of Earth Sciences,* v. 40, p. 591-593.

Osborne, R.A.L., 1994, Caves, dolomite, pyrite, aragonite and gypsum: The karst legacy of the Sydney and Tasmania Basins: *28th Newcastle Symposium on Advances in the Study of the Sydney Basin,* University of Newcastle, Department of Geology Publication, no. 606, p. 322-324.

Osborne, R.A.L., 1995, Evidence for two phases of late Palaeozoic karstification, cave development and sediment filling in south-eastern Australia: *Cave and Karst Science,* v. 22, no. 1, p. 39-44.

Osborne, R.A.L., 1996, Vadose weathering of sulfides and limestone cave development-Evidence from eastern Australia: *Helictite,* v. 34, no. 1, p. 5-15.

Palmer, M.V., and Palmer, A.N., 1989, Paleokarst of the U.S.A., *in* Bosák, P., Ford, D.C., Glazek, J., and Horácek, I., eds., *Paleokarst-A Systematic and Regional Review:* Amsterdam and Praha, Elsevier and Academia, p. 337-363.

Polyak, V.J., McIntosh, W.C., Guven, N., and Provencio, P., 1998, Age and origin of Carlsbad Cavern and related caves from 40Ar/39Ar of alunite: *Science,* v. 279, p. 1919-1922.

Proctor, C.J., 1988, Sea-level related caves on Berry Head, South Devon: *Cave Science,* v. 15, no. 2, p. 39-49.

Quinlan, J.F., 1972, Karst-related mineral deposits and possible criteria for the recognition of paleokarsts: A review of preservable characteristics of Holocene and older karst terranes: *24th International Geological Congress,* sec. 6, p. 156-168.

Rhodes, D., Lantos, E.A., Lantos, J.A., Webb, R.J., and Owens, D.C., 1984, Pine Point ore bodies and their relationship to the stratigraphy, structure, dolomitization, and karstification of the Middle Devonian Barrier Complex: *Economic Geology,* v. 79, p. 991-1055.

Saller, A.H., Budd, D.A., and Harris, P.M., 1994, Unconformities and porosity development in carbonate strata: Ideas from a Hedberg Conference: *American Association of Petroleum Geologists Bulletin,* v. 78, p. 857-872.

Sangster, D.F., 1988, Breccia-hosted lead-zinc deposits in carbonate rocks, *in* James, N.P., and Choquette, P.W., eds., *Paleokarst:* New York, Springer-Verlag, p. 102-116.

Seemann, R., 1979, Die sedimentaren Eisenvererzugne der Karstgebiete der Nordlichen Kalkalpen: *Annals Naturhistorisches Museum Wein,* v. 82, p. 209-289.

Simms, M.J., 1990, Triassic palaeokarst in Britain: *Cave Science,* v. 17, no. 3, p. 93-101.

Thomas, T.M., 1974, The South Wales interstratal karst: *Transactions of the British Cave Research Association,* v. 1, no. 3, p. 131-152.

Walkden, G.M., 1974, Paleokarstic surfaces in upper Visean (Carboniferous) limestones of the Derbyshire Block, England: *Journal of Sedimentary Petrology,* v. 44, p. 1232-1247.

Webb, J.A., 1986, Discussion: Proterozoic palaeokarst at Torrowangee, western New South Wales?: *Australian Journal of Earth Sciences,* v. 33, p. 365-366.

Worley, N.E., and Beck, J.S., 1976, Moorfurlong Mine, Bradwell, Derbyshire and its geological evolution: *Transactions of the British Cave Research Association,* v. 3, p. 49-53.

Worley, N.E., and Nash, D.A., 1977, The geological evolution of the Jugholes Caves, Matlock, Derbyshire: *Transactions of the British Cave Research Association,* v. 4, p. 389-401.

Wrenrich, K.J., and Sutphin, H.B., 1994, Grand Canyon caves, breccia pipes and mineral deposits: *Geology Today,* May-June 1994, p. 97-104.

Wright, V.P., 1988, Paleokarsts and paleosols as indicators of paleoclimate and porosity evolution: A case study from the Carboniferous of South Wales, *in* James, N.P., and Choquette, P.W., eds., *Paleokarst:* New York, Springer-Verlag, p. 329-341.

Wright, V.P., 1991, Palaeokarst: Types, recognition, controls and associations, *in* Wright, V.P., ed., *Palaeokarsts and Palaeokarstic Reservoirs:* Postgraduate Research Institute for Sedimentology, University of Reading, Occasional Publication, ser. 2, p. 65-88.

Zötl, J., 1989, Paleokarst as an important hydrogeological factor, *in* Bosák, P., Ford, D.C., Glazek, J., and Horácek, I., eds., *Paleokarst-A Systematic and Regional Review:* Amsterdam and Praha, Elsevier and Academia, p. 483-509.

Part 4: Theoretical Fundamentals of Speleogenetic Processes

4.0 Overview: Implication of Experimental and Modeling Results into Speleogenetic Studies

Wolfgang Dreybrodt

The formation of caves and karst aquifers is governed by natural laws of physics and chemistry. Any kind of conceptional models must not contradict them.

Mathematical and numerical models of concepts stemming from geological field experience are important tools in unraveling the complexity of what occurs in nature. They help to uncover the underlying processes active in karst. To achieve this, any attempt at modeling will start with simplifications which are necessary to render the problem tractable and clear-cut. Their correctness, however, decides whether the model will be successful and is therefore of utmost importance.

Mathematical models are quantitative. Unlike many geological studies, they need a secure knowledge of which parameters are needed and which numbers have to be attributed to them. Therefore a good model based upon the physics and chemistry of the underlying processes can give new insights and reveal what one must look for in the field to validate them. This mutual interaction of theoretical design and experience from field work offers a good chance to advance our knowledge of karst.

The first step in designing a model should be as simple as possible but as complex as necessary. Simple models are more suitable for understanding basic processes. On the basis of this understanding, more complex steps can be added that reveal deeper insight. Beginning with highly complex models from the start may obscure the meaning of their results.

The formation of karst is governed by the equilibrium chemistry of the system H_2O-CO_2-$CaCO_3$. Up until the 1960's this was regarded as sufficient to explain cave genesis. Bögli's (1964) concept of mixing corrosion is a typical example.

Weyl's pioneering step in studying the dissolution kinetics of calcite (1958) revealed that dissolution kinetics plays an indispens-able role in widening narrow fractures into mature karst conduits, although it confused researchers when they found that caves should not exist at all (White & Longyear, 1962). Later research revealed what White (1977) called the kinetic trigger and was quantified by Palmer (1984) as nonlinear dissolution kinetics. These were the necessary ingredients for understanding early karst evolution. Karst from its earliest state is formed by water flowing in fissures and crac ks in the rock. Therefore fluid dynamics must be considered.

To understand what happens in all stages of karst evolution one has to combine all three ingredients:

- Equilibrium chemistry.
- Dissolution kinetics.
- Fluid mechanics.

Combining these three ingredients is the purpose of Part 4 of this book—which because of its many mathematical equations, might look intimidating at first glance. Since the system H_2O-CO_2-$CaCO_3$ is well understood both in equilibrium chemistry and dissolution kinetics, it clarifies limestone speleogenesis by CO_2-bearing waters from the surface. But it also gives insight into other cave-forming processes. Chapter 4.1

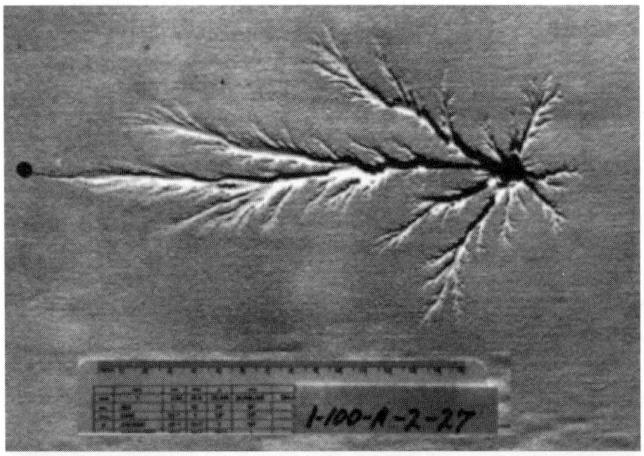

4.0 Plate A. Hardware modeling of early cave development. (Photo by R. Ewers)

4.0 Plate B. Main stream passage of Marengo Cave, Indiana, U.S.A. (Photo by A. Palmer)

summarizes the equilibrium chemistry and dissolution rates for a variety of soluble rocks. Chapter 4.2 focuses on the modeling of karst aquifers and conduits in limestone.

Chapter 4.1.1 outlines the present knowledge of the equilibrium chemistry of karst waters. Besides introducing the basic mass-reaction laws, it focuses on the chemical evolution of water entering a karst system. Water can flow with a surface open to the atmosphere (open system with respect to CO_2) allowing CO_2 exchange, or it flows in the phreatic zone completely confined by rock, such that no CO_2 exchange occurs (closed system). Intermediate conditions between these two extremes can exist,when a finite volume of water exchanges CO_2 with a finite volume of CO_2 containing atmosphere. This can happen in the soil or epikarst on the top of a karst aquifer. To understand the chemistry of waters emerging in the vadose zone one has to regard that complicated pathways can be taken when water seeps down to the cave.

Chapter 4.1.2 covers dissolution rates of limestone under various geological settings, e.g. in narrow fissures where flow is laminar, or large conduits with turbulent flow. Dissolution rates depend on the fissure widths, due to the interplay of three processes:

- Surface reaction.
- Molecular diffusion of dissolved Ca^{2+} and HCO_3^- away from the surface of the rock.
- Conversion of $CO_2 + H_2O \rightarrow H^+ + HCO_3^-$.

Enlargement rates can therefore vary one order of magnitude, even though the chemical composition of the bulk water remains unchanged. Dissolution rates also depend on the type of flow and rise significantly when the flow becomes turbulent. This has to be considered in all models. This chapter outlines the surface reactions on natural carbonates far from and close to equilibrium, on the basis of experimental data. It also shows that lithology has a considerable influence on the kinetics, which later on in Chapter 4.2.4 can be considered as a reason for the complexity of cave systems.

Equilibrium chemistry is of basic importance in estimating the overall removal of carbonate rock from karstic drainage systems. This is discussed in Chapter 4.1.3, which gives an overview on karst denudation.

Some caves are the result of dissolution by sulfuric acid with or without microbial mediation. The role of sulfuric acid dissolution in early speleogenesis is believed to be important by some authors. Chapters 4.1.4 and 4.1.5 outline the various sources of the origin of H_2SO_4 and its possible effects to speleogenesis. These chapters are more descriptive, since up to now little research has been done on the dissolution kinetics of this system. This is also the case for gypsum and anhydrite.

Chapter 4.1.6 gives an overview of what is known so far with regard to speleogenesis in these rocks. Caves can also exist in rock salt, as is reviewed in Chapter 4.1.7. Quartz-rich and silicate rocks can also exhibit karst-like features. Chapter 4.1.7 gives an overview of cave-forming processes in these rocks.

Chapter 4.2 combines flow dynamics, equilibrium chemistry, and dissolution kinetics to help understand cave genesis in limestone.

Chapter 4.2.1 reports on hardware cave modeling by forcing water with constant hydraulic head through scale models of bedding-plane partings constructed in plaster of Paris. These experiments, first performed by Ralph Ewers, show nicely how an increasing amount of flow enhances dissolutional widening, and vice versa.

Although this is not chemically similar to limestone dissolution, it represents impressively the principle of a feedback mechanism which is also characteristic for limestone. The plaster experiments are then simulated digitally by a 2D network of fractures.

Chapter 4.2.3 simplifies the two-dimensional model to a one-D plane-parallel fracture with constant hydraulic head. Widening of the fracture is first restricted to the entrance, while the exit at first widens slowly. When the exit width enlarges to several times its original size, flow increases rapidly, and breakthrough occurs. Prior to breakthrough the concentration of the solution is close to equilibrium along the entire fracture. After breakthrough, however, due to the rapidly increasing flow, the concentration becomes low and dissolution rates become large and rather uniform.

This digital model has been further simplified in Chapter 4.2.2 by assuming that the rate at the exit acts along the entire fracture, such that it remains plane parallel. In that case a strict mathematical analysis is possible, which leads to an analytical expression for breakthrough times.

In this way all the following parameters are revealed to be important:

- Fracture widths.
- Hydraulic head.
- Length of the tube.
- Viscosity of water.
- Constants determining equilibrium concentration with respect to calcite.
- Constants determining the dissolution rates.

Finally cave genesis with exclusively linear dissolution kinetics is examined, with the conclusion that large caves cannot evolve within geologically reasonable times except where there are unnaturally steep hydraulic gradients and short distances between input and output. Since these conditions may exist at dam sites, extensive karstification can endanger these structures, which suggests a possible important contribution of speleogenetic studies to civil engineering.

Two-dimensional fracture networks are examined in Chapter 4.2.4. Although the basic principles revealed in the previous chapter are still valid, extension to two dimensions allows a variety of geological settings to be modeled which better resemble natural systems. Simulation of the evolution of epikarst shows that it is highly connected to the fissured structure of the bedrock below. Varying lithologies can exert significant influence and create caves without entrances. The models show also that the complexity of caves is related to their breakthrough times. Caves needing long times for evolution tend to be less braided. This chapter shows that designing more complexity in the model leads to a variety of results, which give a better understanding of the basics of the simpler 1-dimensional models, and shows the advantage of increasing the complexity of models step by step.

Chapter 4.2.5 presents a further step. Once initial small conduits of about 1 mm width have developed, karst systems can be described by a pipe network of these conduits embedded into a continuum of the rock matrix with many smaller yet unwidened fractures. In this later stage of karstification a constant hydraulic head can no longer be supported, as assumed in all previous chapters. Instead, the recharge rate becomes the independent variable. The evolution of the conduit system is treated in a similar way as in the previous chapters. But the changing boundary conditions lead to a complicated interaction which drains flow from the continuum of the small fissures into the conduits. Although the model is incomplete—since no dissolutional widening in the continuum is assumed—there are many different scenarios of the evolution of mature karst that show the consequences of increasing model complexity. This is a pioneering step toward more realistic future models to account for the various scales of porosity and hydraulic conductivity in karst aquifers.

The question whether a plane parallel fracture is sufficiently realistic to model karst evolution is addressed in Chapter 4.2.6. Simple approximations of roughness have no dramatic effect on the behavior of one-dimensional models. It is an open question whether this is still valid in more realistic fractures with statistically distributed fracture widths.

In summary, Part 4 gives an overview of the present state of art in modeling cave genesis from the basics of the underlying physics and chemistry. Even if not all of the approaches are strictly realistic, they open a large field of research for the future.

4.1. Chemistry of Speleogenetic Processes

4.1.1 Equilibrium Chemistry of Karst Water in Limestone Terranes

Wolfgang Dreybrodt

Abstract

This chapter summarizes the equilibria of the chemical reactions occurring in $CaCO_3 - H_2O - CO_2$ solutions, as they are typical for karst water. The evolution of the chemical composition of such solutions during their interaction with limestone depends on specific geologic conditions, such as dissolution proceeding under the conditions of an open or closed system with respect to CO_2.

Since the CO_2 concentration in karst water determines its further chemical evolution and the equilibrium concentration of Ca^{2+} with respect to calcite in all cases of interest, equations are derived that describe the evolution of the chemical composition in the most general case, i.e. when a limited volume of gaseous CO_2 is in contact with a solution dissolving limestone. This includes the cases of open and closed systems with respect to CO_2 as end members. Finally I discuss the influence of foreign ions common in karst, such as Mg^{2+}, SO_4^{2-}, Na^+ and Cl^-, to the equilibrium concentration of dissolved Ca^{2+} ions with respect to calcite.

Introduction

Karst processes in carbonate rocks depend on the dissolution of limestone by water containing carbon dioxide. Carbonic acid derived from aqueous carbon dioxide acts as an aggressive agent to dissolve $CaCO_3$, which then is transported out of the karst terrain as calcium and hydrogen carbonate ions dissolved in the moving water. The maximum amount of $CaCO_3$, that can be dissolved, expressed by the maximum calcium concentration in the aggressive waters, is determined by the equilibrium chemistry of the system H_2O–CO_2–$CaCO_3$.

This equilibrium concentration, for example, establishes the maximum denudation rates in limestone terranes (White, 1984). These will occur when all the water entering a limestone area by precipitation or by allogenic rivers has achieved saturation with respect to calcite by the time it leaves the karst terrain.

To understand karstification as a dynamic process, however, one also has to know the rates of removal of limestone in mass per surface area and time by water of a given chemical composition. These rates establish the time that is necessary for the aggressive water to attain equilibrium with respect to calcite. They play a dominant role in the dynamics of karst evolution (Dreybrodt 1988, 1990, 1996; Palmer 1984, 1991), and also in this volume:

- Dreybrodt and Gabrovšek, Chapter 4.2.2,
- White, Chapter 4.1.3
- Palmer, Section 4.2.3

The time needed for a cave system to develop from the onset of karstification into a mature state is strongly influenced by these rates, which depend heavily on the under-saturation of the water. This is expressed by the difference between the equilibrium concentration of calcium with respect to calcite and its actual concentration (see Dreybrodt and Eisenlohr, Chapter 4.1.2 in this volume).

To derive these rates one must know the chemical composition of the water with respect to H_2CO_3, H^+, HCO_3^- and CO_3^{2-}. This is determined by its concentration of aqueous CO_2 and the actual calcium concentration. This chapter summarizes the equilibrium chemistry of the system H_2O–CO_2–$CaCO_3$ under relevant geologic conditions, that occur in karst processes. In this context the reader is also referred to Garrels and Christ (1965), Stumm and Morgan (1981), and Morse and Mackenzie (1990).

Reactions and Equilibria

Three different chemical reactions of water containing CO_2 attack the calcite surface

(Plummer et al., 1978). They are:

$$CaCO_3 + H^+ \rightleftarrows Ca^{2+} + HCO_3^-$$

$$CaCO_3 + H_2CO_3 \rightleftarrows Ca^{2+} + 2HCO_3^-$$

$$CaCO_3 + H_2O \rightleftarrows Ca^{2+} + CO_3^{2-} + H_2O$$

These can be combined into a global reaction:

$$CaCO_3 + CO_2 + H_2O \rightarrow Ca^{2+} + 2HCO_3^-$$

from which an important rule of stoichiometry can be seen. For each Ca^2 ion dissolved, one molecule of CO_2 is needed, which finally is converted into HCO_3^-. From these equations it is evident that the reaction of the system $H_2O - CO_2$ with $CaCO_3$ is governed by the concentrations of the species H^+, HCO_3^-, H_2CO_3, and Ca^{2+} at the calcite surface.

To elucidate the chemical equilibria between these species, it is appropriate to consider a pure carbonic acid system $H_2O–CO_2$ prior to any dissolution of $CaCO_3$. The chemical reactions in this system are:

$$CO_2^g + H_2O \rightleftarrows CO_2^{aq} + H_2O.$$

This process describes transfer of CO_2 from the gas phase into the solution, where the aqueous species CO_2^{aq} is formed. The equilibrium between CO_2^g and CO_2^{aq} is given by Henry's law. This relates the partial pressure p_{CO_2} of CO_2 in the surrounding atmosphere to the activity of the dissolved CO_2.

$$(CO_2^{aq}) = K_H \cdot p_{CO_2} \qquad (1)$$

K_H is Henry's constant in mol/(l atm) and p_{CO_2} is measured in atm. The parentheses indicate activities in *mol/l*, which are related to the concentrations by activity coefficients γ as will be discussed below. Concentrations are expressed by square brackets. In some cases it is more convenient to express p_{CO_2} by the activity (CO_2^g) of CO_2 in the atmosphere.

In this case Henry's law is given by

$$\left(CO_2^{aq}\right) = K_H RT\left(CO_2^g\right)$$

R is the gas law constant (R =2 0.0826) $\frac{\text{atm l}}{\text{mol K}}$) and T is the temperature in K.

CO_2 reacts with water to form carbonic acid

$$H_2O + CO_2 \rightleftarrows H_2CO_3$$

with the mass action equation

$$\left(CO_2^{aq}\right) = K_0 \cdot (H_2CO_3) \qquad (2)$$

Often one defines

$$\left[H_2CO_3^*\right] = \left[CO_2^{aq}\right] + \left[H_2CO_3\right]$$

Then equation 2 reads

$$\frac{\left(CO_2^{aq}\right)}{\left(H_2CO_3^*\right)} = \left(1 + \frac{1}{K_0}\right)^{-1} \qquad (2a)$$

Note that $\left(CO_2^{aq}\right) \approx (H_2CO_3^*)$, since $1/K_0 \approx 0.003$. Only 0.3 % of $\left[H_2CO_3^*\right]$ is due to free carbonic acid H_2CO_3.

H_2CO_3 dissociates into $H^+ + HCO_3^-$,

$$H_2CO_3 \rightleftarrows H^+ + HCO_3^-,$$

with the mass-action law

$$\frac{\left(H^+\right)\left(HCO_3^-\right)}{\left(H_2CO_3^*\right)} = K_1 \left[\frac{\text{mol}}{\text{l}}\right]$$

or (3)

$$\frac{\left(H^+\right)\left(HCO_3^-\right)}{\left(H_2CO_3\right)} = K_1 \cdot (1 + K_0) = K_{H_2CO_3} \left[\frac{\text{mol}}{\text{l}}\right]$$

The next dissociation step is

$$HCO_3^- \rightleftarrows H^+ + CO_3^{2-}$$

with

$$\frac{\left(H^+\right)\left(CO_3^{2-}\right)}{\left(HCO_3^-\right)} = K_2 \left[\frac{\text{mol}}{\text{l}}\right] \qquad (4)$$

Finally, we have the dissociation of water with

$$\left(H^+\right)\left(OH^-\right) = K_W \left[\frac{\text{mol}^2}{\text{l}^2}\right] \qquad (5)$$

In Equations 1 to 5 the mass-action equations have been written using activities. These are related to the concentrations by activity coefficients that depend on the ionic strength I of the solution, defined by

$$I = \frac{1}{2} \sum_i Z_i^2 \cdot [i] \qquad (6a)$$

where Z_i is the charge of ion i and [i] is its concentration. Activities (i) are related to concentrations [i] by the activity coefficients γ_i

$$(i) = \gamma_i \cdot [i] \qquad (6b)$$

The individual ion activity coefficients are usually calculated by the extended Debye-Hückel equation as

$$\log \gamma_i = -AZ_i^2 \cdot \frac{\sqrt{I}}{1 + Ba_i\sqrt{I}} \qquad (7)$$

The values A and B are dependent on temperature:

$A = 0.4883 + 8.074 \times 10^{-4}t$,
$B = 0.3241 + 1.6 \times 10^{-4}t$
t is the temperature in °C.

The values a_i represent effective ionic radii.

For various ionic strengths and ions encountered in karst water, 4.1.1 Table 1 lists ionic radii and gives activity coefficients calculated from equation. no. 7

Ion	Ionic strength I 0.001	0.005	0.01	0.05	0.1	Effective Ionic Radius a [Å]
H^+	0.967	0.935	0.915	0.856	0.82	9
Mg^{2+}	0.874	0.760	0.694	0.522	0.450	8
Ca^{2+}	0.872	0.751	0.680	0.489	0.407	6
CO_3^{2-}	0.870	0.745	0.669	0.461	0.370	4.5
SO_4^{2-}	0.870	0.743	0.667	0.456	0.363	4
HCO_3^-, Na^+	0.965	0.929	0.904	0.822	0.776	4
OH^-, Cl^-	0.965	0.927	0.902	0.815	0.765	3

4.1.1. Table 1. Activity coefficient γ for various ions encountered in karst water at $T = 15°C$ calculated by the extended Debye-Hückel expression (intermediate values can be estimated by linear interpolation).

for various ionic strengths and ions encountered in karst water.

As can be seen from equation 7 and 4.1.1 Table 1, $\gamma_i \rightarrow 1$ with I approaching zero. In natural karst water the ionic strength is below 0.05 and the extended Debye-Hückel theory can be used with confidence. It should be noted, however, that even at $I \leq 0.05$ the activity coefficients can change by up to a factor of 2. Therefore it is mandatory to use activities, even when considering relatively dilute karst water. For uncharged species such as CO_2 and H_2CO_3 the activity coefficents are given (Plummer and Mackenzie, 1974) by

$$\gamma_i = 10^{0.1 \cdot I} \approx 1, \text{ if } I < 0.1 \quad (8)$$

Equilibrium Between $H_2CO_3^*$, HCO_3^-, CO_3^{2-}, and H^+

The dissociation of carbonic acid into bicarbonate HCO_3^- and carbonate CO_3^{2-} is governed by pH as a master variable. This follows from the mass action equations given above. The total concentration of carbon in a $CO_2 - H_2O$ system is given by

$$C_T = [H_2CO_3^*] + [HCO_3^-] + [CO_3^{2-}] \quad (9)$$

The molar fractions are defined as

$$\alpha_0 = \frac{[H_2CO_3^*]}{C_T}, \quad \alpha_1 = \frac{[HCO_3^-]}{C_T}, \quad \alpha_2 = \frac{[CO_3^{2-}]}{C_T} \quad (10)$$

These ionic fractions are visualized in 4.1.1 Figure 1 as a function of pH for the case of a very dilute solution, where all the γ_i are equal to 1. In the region pH < 4 virtually no HCO_3^- and CO_3^{2-} are present and only $H_2CO_3^*$ exists. With increasing pH, carbonic acid dissociates, forming HCO_3^-. Thus, below pH = 8.3 practically no CO_3^{2-} is present. Above pH = 8.3, HCO_3^- starts to dissociate until at pH 12 only CO_3^{2-} ions exist.

It should be noted here that this behavior depends only very weakly on temperature and ionic strength and is also valid if the water contains other ions such as Ca^{2+}, Mg^{2+}, and SO_4^{2-}, which are common in karst water (Loewenthal and Marais, 1978). The data in 4.1.1 Figure 1 therefore are representative, with sufficient accuracy, for most karst waters. One important conclusion from this diagram is that in most cases of karst waters the species CO_3^{2-} can be neglected, since karst waters very rarely exhibit pH values higher than 8.3.

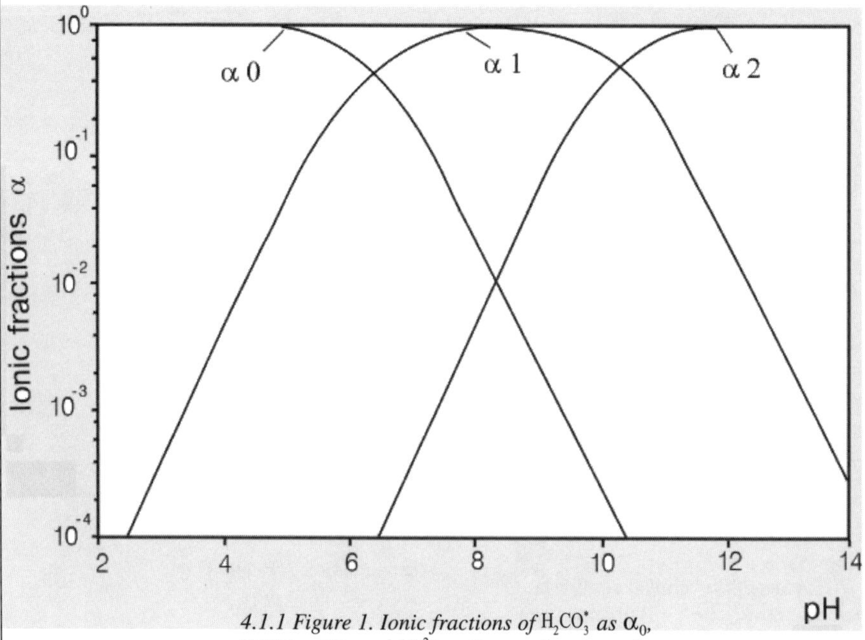

4.1.1 Figure 1. Ionic fractions of $H_2CO_3^*$ as α_0, HCO_3^- as α_1, and CO_3^{2-} as α_2 as a function of pH in the solution. Note that the scale of the ordinate is logarithmic.

Equilibrium with Respect to $CaCO_3$

If $CaCO_3$ is dissolved, the equilibrium between dissolved species and the solid is given by the solubility product K_c

$$(Ca^{2+})_{eq}(CO_3^{2-})_{eq} = K_c \left[\frac{mol^2}{l^2}\right], \quad (11)$$

where $(Ca^{2+})_{eq}$ and $(CO_3^{2-})_{eq}$ are the activities at equilibrium. To describe the degree of saturation of a solution, one uses the saturation state

$$\Omega = \frac{(Ca^{2+})(CO_3^{2-})}{K_c} \quad (12)$$

The solution is undersaturated for Ω>1, saturated for Ω=1, and supersaturated for Ω<1. Ca^{2+} reacts with HCO_3^- and to CO_3^{2-} form ion pairs. The corresponding ion pair activities are related to mass action constants according to the equations

$$K_3 \cdot (Ca^{2+}) \cdot (HCO_3^-) = (CaHCO_3^+);$$
$$K_4 \cdot (Ca^{2+}) \cdot (CO_3^{2-}) = (CaCO_3^0) \quad (13)$$

In the case of natural karst water these species can be neglected safely in all calculations and will therefore not be included in our further considerations.

The most recent data for mass-action constants and their temperature dependence for the system $H_2O - CO_2 - CaCO_3$ are given by Plummer and Busenberg (1982). It should be noted that very often the values of mass-action constant K is given as pK, related to K by

$$pK = -\log K, \quad K = 10^{-pK} \quad (14)$$

The empirical temperature dependence of all mass action constants defined above is summarized in 4.1.1 Table 2.

To calculate equilibria in an electrolytic system one has to observe the neutrality of electrical charges in the solution. The equation of electroneutrality states that the sum of all positive ionic charges is balanced by that of all negative ones, such that the solution remains uncharged. In the case of a pure $H_2O - CO_2 - CaCO_3$ system and neglecting ion pairs, this is formulated as

$$2[Ca^{2+}] + [H^+] = [HCO_3^-] + 2[CO_3^{2-}] + [OH^-] \quad (15)$$

In the region between pH = 4 to pH = 8.4 with an accuracy of 2%, this equation can be simplified to

$$2 \cdot [Ca^{2+}] = [HCO_3^-] \quad (15a)$$

This can be seen from 4.1.1 Figure 1, which shows $[CO_3^{2-}] \ll [HCO_3^-]$ for pH < 8. From this an important conclusion on ionic strengths in karst waters can be drawn. In aqueous solutions of pure $CaCO_3$ the ionic strength (cf. equation 6) is given by a sufficiently accurate approximation as

$$I = 3 \cdot [Ca^{2+}] \quad (16)$$

In the presence of Mg^{2+} derived from dolomite or magnesian calcites

$$I = 3\cdot\left([Ca^{2+}] + [Mg^{2+}]\right) \qquad (17)$$

This is of high practical interest as it allows one to obtain the ionic strength I directly from the concentration of the metal ions. In equilibrium calculations, therefore, once the concentrations of the metal ions are known, the activity coefficients no longer depend on the concentration of the carbonic ion species, and iterative procedures to calculate I and the related activity coefficients are not necessary.

Boundary Conditions for Achieving Equilibrium

Dissolution of calcite by CO_2-containing water in natural systems proceeds under a variety of different boundary conditions:

- In the saturated zone of karst aquifers or porous calcareous soils there is no interface between an atmosphere and the solution and consequently no mass transport of CO_2 into the solution. Therefore, during dissolution of calcite, CO_2 is consumed from the solution. This defines a closed system with respect to CO_2.
- In all situations where water flows in contact with limestone rocks and an unconfined CO_2-gas phase, e.g. in rivers, mass transport of CO_2 is effective between the liquid-gas interface. Thus, the CO_2 consumed by dissolution of calcite is replaced by transfer of CO_2 through the liquid-gas boundary. In this case the system is defined as open with respect to CO_2.

There is a general intermediate condition, however, between the extreme limits of open and closed systems.

- In the unsaturated zone of an aquifer, where a limited volume V_s of the solution is in contact with calcite and also forms a surface to a limited volume V_g of a CO_2-containing atmosphere, in this case, in contrast to the open system, where V_g is assumed to be extremely large, any mass exchange between this atmosphere and the solution changes the CO_2 composition of both the atmosphere and the solution. Thus, open-system conditions are the limiting case with $V_g \to \infty$, and the closed system is achieved with $V_g \to 0$.

The General Case
The purpose of this section is to formulate the chemical composition of the $H_2O - CO_2 - CaCO_3$ system during the process of dissolution for the general intermediate condition. This includes the two extreme cases of open and closed systems.

We assume that dissolution of $CaCO_3$ is sufficiently slow, such that all carbonic species are in equilibrium to each other, and also that the aqueous CO_2 is in equilibrium with the atmospheric CO_2. This assumption, however, is not generally valid. CO_2 and H_2CO_3 are only in equilibrium when flow is turbulent and when the ratio $V_s / A > 1$cm, where A is the surface area of the dissolving limestone (see Dreybrodt and Eisenlohr, Chapter 4.1.2 in this volume). In an idealized pure system H_2O-$CaCO_3$-CO_2 the chemical composition of the solution can be calculated once the initial CO_2 pressure of the atmosphere and the Ca^{2+} concentration of the solution are known from the following considerations.

Each mole of $CaCO_3$ dissolved releases 1 mol Ca^{2+} ions and 1 mol carbonate ions. The total amount of carbon in the system is therefore given by conservation of carbonate species, i.e. C-atoms, as

$$\begin{aligned} M_T &= \left([Ca^{2+}] + [HCO_3^-] + [H_2CO_3^*] + [CO_3^{2-}]\right)V_s + [CO_2^g]V_g \\ &= \left([HCO_3^-] + [H_2CO_3^*] + [CO_3^{2-}]\right)V_s + [CO_2^g]\cdot V_g \end{aligned} \qquad (18)$$

The right-hand side of equation 18 gives the amount of carbon atoms as calculated from the concentrations of all carbon-containing species in the liquid and in the gas phase. This is equal to the initial amount of carbon atoms present in the pure H_2O-CO_2 solution prior to dissolution of calcite, augmented by the amount of Ca^{2+} ions dissolved (left hand side), in as much as each $CaCO_3$ released from the solid also adds one carbon atom to the solution. According to Henry's law

$$[CO_2^g] = \frac{[H_2CO_3^*]}{K_H RT}\cdot\left(1 + \frac{1}{K_0}\right)^{-1} \qquad (19)$$

$H_2O - CO_2 - CaCO_3$

	Unit	Reference
$\log K_H = 108.3865 + 0.01985076T - 6919.53/T - 40.45154 \log T + 669365/T^2$	mol/l atm	(1)
$\log K_1 = -356.3094 - 0.06091964T + 21834.37/T - 126.8339 \log T - 1684915/T^2$	mol/l	(1)
$\log K_2 = -107.8871 - 0.03252849T + 5151.79/T + 38.92561 \log T - 56371.9/T^2$	mol/l	(1)
$K_0 = 1.7 \times 10^{-4}/K_1$	-	(2)
$\log K_W = 22.801 - 0.010365T - 4787.3/T - 7.1321 \log T$	mol^2/l^2	(3)
$\log K_3 = 1209.12 + 0.31294T - 34765.05/T - 478.782 \log T$	mol/l	(1)
$\log K_4 = -1228.732 - 0.299444T + 35512.75/T + 485.818 \log T$	mol/l	(1)
$\log K_C = -171.9065 - 0.077993T + 2839.319/T + 71.595 \log T$	mol^2/l^2	(1)

(1) Plummer and Busenberg (1982). (2) Wissbrun et al. (1954). (3) Harned and Hamer (1933). Temperature T in K.

4.1.1 Table 2. Temperature dependence of mass-action constants.

The initial values $[\,]_i$ of the carbonic species are those of the initial H_2O-CO_2 system prior to dissolution, i.e., with $[Ca^{2+}]=0$. They are given by

$$[H_2CO_3^*]_i = K_H p^i_{CO_2}(1+1/K_0) \qquad (20a)$$

$$[HCO_3^-]_i = \left([H_2CO_3^*]_i \cdot K_1/\gamma^i_H\gamma^i_{HCO_3}\right)^{1/2} \qquad (20b)$$

$p^i_{CO_2}$ is the CO_2-pressure that would be in equilibrium with the initial concentration $[H_2CO_3]_i$ prior to dissolution of $CaCO_3$. Furthermore, since in any case the initial pH is below 7, the initial concentration of CO_3^{2-} is so low (see 4.1.1 Figure 1) that it can be safely neglected.

$$[CO_3^{2-}]_i = 0. \qquad (20c)$$

A second master equation is the equation of electroneutrality, which reads

$$2[Ca^{2+}]+[H^+]=[HCO_3^-]+2[CO_3^{2-}]+[OH^-] \qquad (21)$$

Combining these equationsand the mass action laws (for details see Dreybrodt, 1988) one obtains a fourth-order polynomial for the hydrogen ion concentration.

$$[Ca^{2+}]+[H^+]-v-K_w'/[H^+]+\left(w[H^+]-K_2'/[H^+]\right)\left(2v-[H^+]+K_w'/[H^+]\right)\cdot\left(1+2w\cdot[H^+]\right)^{-1}=0 \qquad (22)$$

which can be solved numerically.

The abbreviations are

$$v = f\cdot[H_2CO_3^*]_i+[HCO_3^-]_i$$

$$f = 1+\frac{(1+1/K_0)^{-1}}{K_H RT}\chi$$

$$\chi = V_g/V_s;\quad K_w' = K_w/\gamma_H\gamma_{OH};$$

$$K_2' = K_2\gamma_{HCO_3}/\gamma_H\gamma_{CO_3}$$

$$w = f\cdot\frac{\gamma_H\gamma_{HCO_3}}{K_1} \qquad (23)$$

The activities involved in equation 23 can be calculated with sufficient accuracy by approximating the ionic strength

$$I = 3\cdot[Ca^{2+}] \qquad (24)$$

Open and closed system conditions are included in this general formulation by using $f=1$ in the closed system and $f=\infty$ in the open system. In most cases of practical interest in a natural system, pH is below 8 and the carbonate and OH^- species can be neglected in equations18 and 21. Then equation 24 is reduced to a quadratic equation, which can be employed much more conveniently.

Its solution is

$$[H^+]=-\frac{1}{2}\left(2[Ca^{2+}]+\frac{1}{w}\right)+\frac{1}{2}\sqrt{4[Ca^{2+}]^2+\frac{4v}{w}+\frac{1}{w^2}} \qquad (25)$$

Once $[H^+]$ is known, the concentrations of $H_2CO_3^*$, HCO_3^-, CO_3^{2-}, and OH^- can be obtained by use of the mass action laws and equations 21 and 24 (for details see Dreybrodt, 1988).

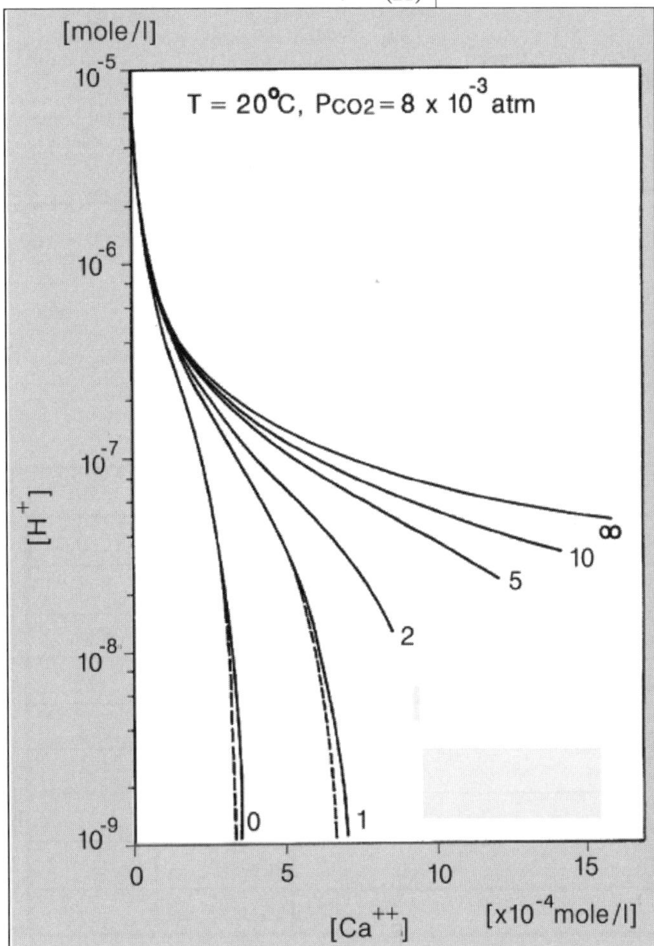

4.1.1 Figure 2. Hydrogen ion concentration $[H^+]$ as a function of $[Ca^{2+}]$in the solution as the system approaches equilibrium with calcite. The numbers give the values of $V_g/V_S=\chi$. The dashed lines show the approximation according to equation 25. The initial CO_2 pressure is $8x10^{-3}$ atm.

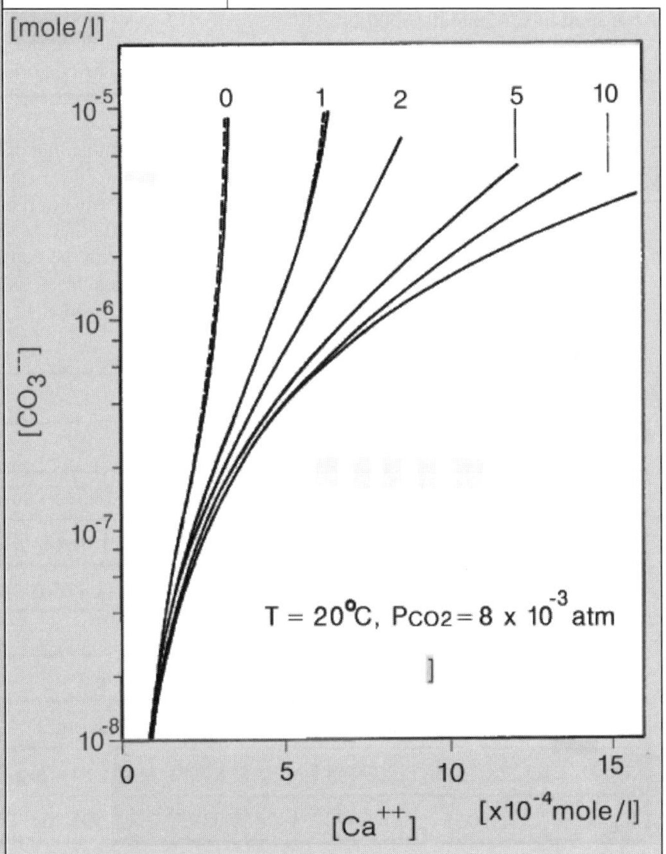

4.1.1 Figure 3. Carbonate ion concentration $[CO_3^{2-}]$as a function of $[Ca^{2+}]$ in the solution as the system approaches equilibrium. The numbers give the values of $V_g/V_s=\chi$. The dashed lines show the approximation according to equation 25. Initial conditions as in 4.1.1 Figure 2.

In 4.1.1 Figures 2 to 5 the evolution of $[H^+]$ $[CO_3^{2-}]$, Ω, and $[H_2CO_3^*]$ in an intermediate system, when Ca-dissolution proceeds freely until it attains equilibrium is shown. Thus these figures demonstrate the chemical evolution of karst water during its interaction with the limestone.

In 4.1.1 Figure 2 the hydrogen ion concentration $[H^+]$ as a function of dissolved $[Ca^{2+}]$is shown. The results of the exact calculation employing equations 22, 23, 24, and the approximation of equation 25 are compared for the evolution of the closed system, i.e. $\chi = 0$. Deviations in the values of $[H^+]$are only significant for pH > 8. Below this value the approximation can be used with confidence in *all* cases. In the closed system $[H^+]$drops very rapidly with increasing Ca^{2+} concentration and is very sensitive to changes in $[Ca^{2+}]$.

The evolution of $[CO_3^{2-}]$ for a variety of intermediate cases is illustrated in 4.1.1 Figure 3. The carbonate concentration is shown in a mirror like curve compared to $[H^+]$. A steep increase of $[CO_3^{2-}]$ is observed as $[H^+]$ decreases. For the open system $\chi = \infty$ the decrease in $[H^+]$ is much less steep as the system approaches equilibrium, and so is the increase in $[CO_3^{2-}]$. The intermediate cases show a behavior that is between the two extremes, tending to be more like a closed system for $\chi \leq 2$ and like an open system for $\chi \geq 10$.

In 4.1.1 Figure 4 the saturation state Ω (cf. equation 12) is shown for the same conditions as in 4.1.1 Figure 3.

In 4.1.1 Figure 5 the results are depicted for the $H_2CO_3^*$ concentration in the solution. Only in the open system does this value remain constant. In all other cases this value decreases linearly as $[Ca^{2+}]$ increases. This behavior results from the stoichiometry of calcite dissolution, which consumes one CO_2 molecule for each Ca^{2+} ion released. It can also be derived as an approximation of equation 18.

In all cases where $P_{CO_2} \geq 3\times10^{-4}$ atm, $[HCO_3^-]$ is small compared to $[H_2CO_3^*]$ and can be neglected. For $pH \leq 8.2$ we also can neglect $[CO_3^{2-}]$ and $[OH^-]$ in equations18 and 21. Combining these simplified equations, we obtain

$$[H_2CO_3^*] = [H_2CO_3^*] - [Ca^{2+}]\frac{1}{f} \qquad (26)$$

Thus, the slope of the lines in 4.1.1 Figure 5 is given by 1/f. There are slight deviations between the exact solution and the approximation of equation 23, as is indicated by the dotted lines for $\chi = 0$ and $\chi = 1$. For $\chi > 2$ these deviations are negligible.

Change of Boundary Conditions

In karst systems, water that flows on the surface before sinking into the ground dissolves limestone under open-system conditions. It also may percolate through unsaturated soil, thus dissolving limestone in the intermediate case, mostly with $\chi \approx 1$. When it sinks into the ground its chemical composition is given by the concentrations $[Ca^{2+}]^0$, $[HCO_3^-]^0$, $[CO_3^{2-}]^0$, $[H_2CO_3^*]^0$, and $[H^+]^0$.

From then on, dissolution proceeds under closed conditions with $\chi = 0$. Conservation of carbon atoms from then on dictates that

$$[Ca^{2+}] - [Ca^{2+}]^0 + [HCO_3^-]^0 + [H_2CO_3^*]^0 + [CO_3^{2-}]^0$$
$$= [H_2CO_3^*] + [HCO_3^-] + [CO_3^{2-}] \qquad (27)$$

Each $[\ ]^0$ refers to the chemical composition attained when the dissolution conditions change. The [ ] refers to dissolution

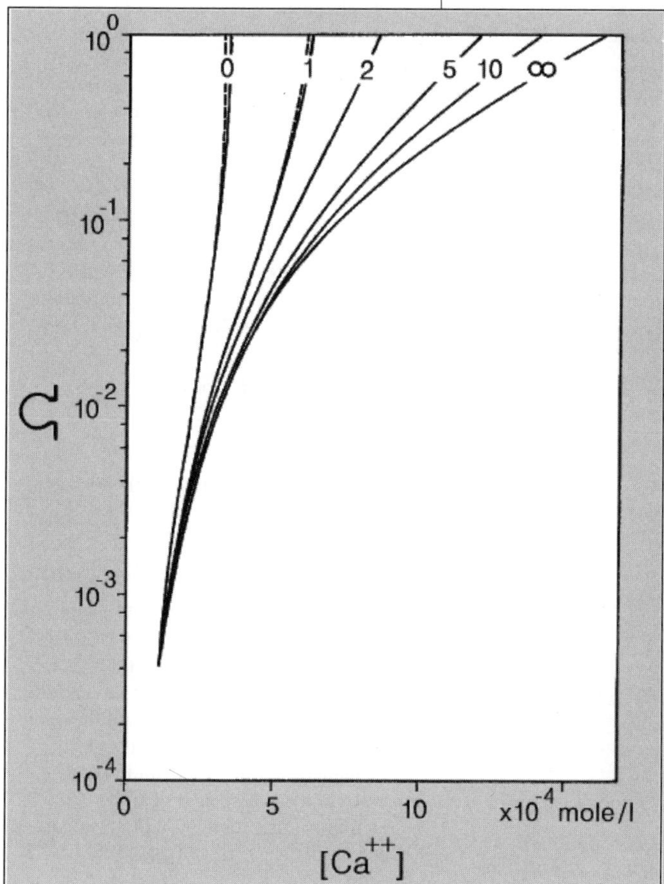

4.1.1 Figure 4. Saturation state Ω as a function of $[Ca^{2+}]$ in the solution as the system approaches equilibrium. The numbers give the values of $V_g/V_s = \chi$. The dashed lines show the approximation according to equation 25. Initial conditions as in 4.1.1 Figure 2.

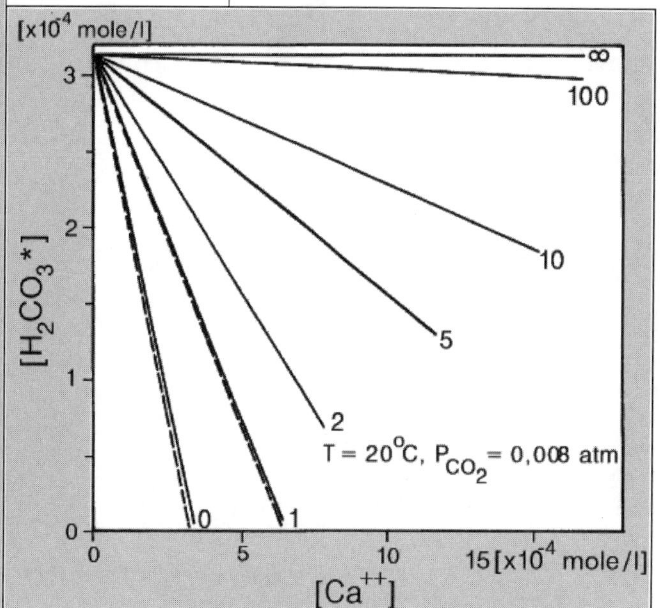

4.1.1 Figure 5. $[H_2CO_3^*]$ as a function of $[Ca^{2+}]$ in the solution as the system approaches equilibrium. The numbers give the values of $V_g/V_s = \chi$. The dashed lines show the approximation according to equation 26. Initial conditions as in 4.1.1 Figure 2.

thereafter. Proceeding as in Chapter 5.1, we obtain similarly equations analogous to equations 22 to 25 provided v is replaced by

$$v^c = f \cdot \left[H_2CO_3^*\right]^0 + \left[HCO_3^-\right]^0 + \left[CO_3^{2-}\right]^0 - \left[Ca^{2+}\right]^0 \quad (28)$$

To calculate the chemical evolution under these circumstances, we first calculate the evolution for the initial boundary conditions as in Chapter 5.1 up to the value $\left[Ca^{2+}\right]^0$. From this we obtain $\left[H_2CO_3^*\right]^0, \left[HCO_3^-\right]^0, \left[CO_3^{2-}\right]^0$, and v^c. Then we calculate the further evolution under the conditions of a closed system.

In 4.1.1 Figure 6 the evolution of $\left[H_2CO_3^*\right]$ as the system approaches equilibrium as a function of the calcium concentration is illustrated. Dissolution starts under open system conditions ($\chi = \infty$) or under intermediate conditions with ($\chi = 1$). When $\left[Ca^{2+}\right]^0$ is established, the boundary conditions change and dissolution proceeds under closed-system conditions ($\chi' = 0$) along the straight lines until equilibrium is achieved. This equilibrium concentration of calcite in its dependence on the $H_2CO_3^*$ concentration is represented by the lower limiting curve. If one extrapolates the lines representing the $H_2CO_3^*$ concentration to $[Ca^{2+}] = 0$ (dashed lines), one can read from the interception with the $\left[H_2CO_3^*\right]$-axis the value of $\left[H_2CO_3^*\right]_i$. The dashed line then represents the chemical evolution of the system as it would have developed if its initial $H_2CO_3^*$ concentration were $\left[H_2CO_3^*\right]_i$ and dissolution had started at $[Ca^{2+}]=0$ with closed-system conditions.

Thus dissolution under changed conditions is equivalent to dissolution under closed conditions with elevated initial $p^i_{CO_2}$. This explains the high calcium concentrations of about 2 mmol/l observed in karst waters derived from springs fed by considerable diffuse flow, where dissolution proceeds in narrow joints under the condition of a closed system. At a soil $p_{CO_2} = 0.05$ atm, dissolution starts at $[Ca^{2+}]=0$ under closed-system conditions and attains an equilibrium concentration of about 2 mmol/l.

At $p_{CO_2} = 0.01$ atm the equilibrium concentration drops to about 0.6 mmol/l. In most soils in temperate climates, p_{CO_2} does not exceed 0.05 atm. Thus the elevated Ca^{2+} concentrations observed indicate that soil waters that enter joints must have already dissolved a significant amount of limestone when percolating through unsaturated soil.

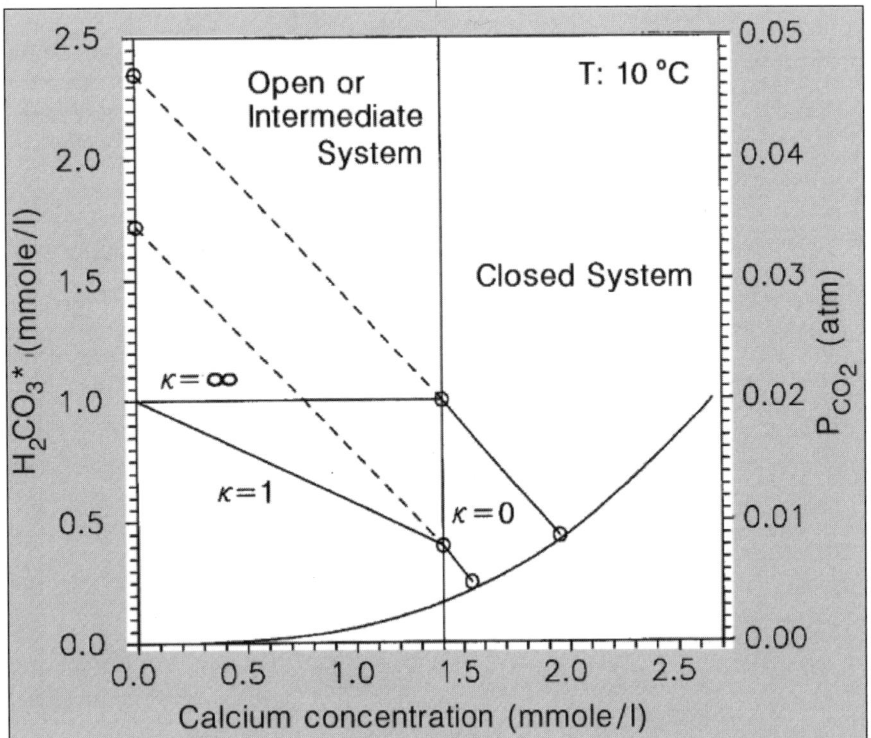

4.1.1 Figure 6. Evolution of $[H_2CO_3^*]_i$ for changing boundary conditions as a function of the calcium concentration generated by dissolution of $CaCO_3$. First the system evolves under open-system ($\chi = 0$) or intermediate conditions ($\chi = 1$). When it has reached the concentration $[Ca^{2+}]^0$ it enters narrow joints, and dissolution proceeds under closed-system conditions. The p_{CO_2} on the right scale is the pressure p_{CO_2} of the gaseous phase in equilibrium with $H_2CO_3^*$. The dashed lines represent the evolution of the system under closed-system conditions, as it would have evolved with initial $[H_2CO_3^*]_i$ as read from the intercept with the $[H_2CO_3^*]$ axis.

Saturated Calcite Solutions

In principle the solubility of calcite can be calculated from the above expressions by employing the solubility product of equation 11 and determining numerically the corresponding saturation concentration $\left[Ca^{2+}\right]_{eq}$ with $\Omega = 1$. In all cases of practical interest, however, approximations can be used that are sufficiently accurate.

Using concentrations in the mass action equations in place of activities, and employing the simplified equation of charge balance, i.e. $2\left[Ca^{2+}\right] = \left[HCO_3^-\right]$, one finds the equilibrium concentrations as a function of the p_{CO_2} in equilibrium with the solution (mathematical details are given in Dreybrodt, 1988).

$$\left[H^+\right]^3_{eq} = \frac{K_1^2 K_2 K_H^2 \gamma_{Ca}}{2K_c \gamma_{HCO_3} \gamma_H^3} \cdot \left(p_{CO_2}\right)^2 \quad (29)$$

$$\left[Ca^{2+}\right]^3_{eq} = p_{CO_2} \cdot \frac{K_1 K_c K_H}{4K_2 \gamma_{Ca} \gamma_{HCO_3}^2} \quad (30)$$

From these, $\left[CO_3^{2-}\right]_{eq}$ and $\left[HCO_3^-\right]_{eq}$ can be easily calculated with equations 15 and 4.

Equations 29 and 30 can be used directly for the open system, in as much as p_{CO_2} remains constant during the dissolution process. To obtain the saturation values for all the other cases as a function of the initial $p^i_{CO_2}$, one has to remember that p_{CO_2} drops during dissolution, because each Ca^{2+} released into solution consumes one CO_2 molecule.

The actual p_{CO_2} at equilibrium with respect to calcite is then given by expressing $\left[H_2CO_3^*\right]$ in equation 26 by p_{CO_2} using Henry's law (equation 2a)

$$p_{CO_2} = p^i_{CO_2} - \frac{\left[Ca^{2+}\right]_{eq}}{f K_H (1 + 1/K_0)} \quad (31)$$

Introducing this into equation 30 yields a cubic equation from which $\left[Ca^{2+}\right]_{eq}$ can be obtained.

In 4.1.1 Figure 7 the solubility of $CaCO_3$ is plotted expressed by $\left[Ca^{2+}\right]$ as a function of the initial $p^i_{CO_2}$ in the atmosphere of the system prior to any dissolution, for a variety of ratios V_g / V_s. With increasing volume of the gas, the curves approach that of the open system (equation 30). All the curves have been calculated by an exact equilibrium calculation. For the open system the curve is redrawn on a logarithmic plot (4.1.1 Figure 8). The dashed curve is calculated from equation 31, for T = 20°C. For $p_{CO_2} > 10^{-3}$ atm the agreement between the exact approximate calculation is almost perfect. A slight deviation becomes noticeable for $p_{CO_2} < 10^{-3}$ atm.

The nonlinearity of $\left[Ca^{2+}\right]_{eq}$ in dependence on p_{CO_2} has an important consequence, which was pointed out independently by Laptev (1939) and Bögli (1964). But the importance of the concept of mixing corrosion to subsurface karstification was recognized in the international karst literature mainly from Bögli's work.

In 4.1.1 Figure 9 (p. 134) sequation (30) as a function of $\left[H_2CO_3^*\right]$ is replotted. The area below the curve denotes undersaturated solutions. The area above corresponds to supersaturated solutions. From the nonlinearity of the saturation curve a very important conclusion can be drawn.

Consider an equilibrated solution (point A) under closed-system conditions. If this solution mixes with a saturated solution of a different composition (point B)—also under closed-system conditions—then the composition of the resulting solution is represented by point C. Depending on the ratio of the volumes of the two mixed solutions, this point lies at different positions on the straight line connecting A and B, in the region of undersaturation.

Therefore, mixing of two differently saturated waters leads to renewed calcite aggressivity of the solution in a pure $H_2O - CO_2 - CaCO_3$ system. In general, for natural waters containing foreign ions, mixing can also result in supersaturation, depending on the particular compositions of the waters involved (Wigley and Plummer, 1976).

Influence of Foreign Ions on Calcite Equilibrium

Karst waters are rarely derived from the pure system H_2O-$CaCO_3$-CO_2. Mostly due to dolomite or magnesian calcites, and due to gypsum and anhydrite—which usually are also encountered in karst areas—one usually finds Mg^{2+} and SO_4^{2-} in considerable amounts. In many cases one also observes Na^+, K^+, and Cl^-. All these ions change the ionic strength of the solution, thus affecting the ionic equilibria.

Furthermore, if minerals are dissolved which have one ion in common with $CaCO_3$, such as $MgCO_3$ or $CaSO_4$, equilibria are changed by the common-ion effect. Therefore, not only does the chemical evolution of the system change as dissolution proceeds, but also the solubility of calcite is altered by the presence of foreign ions. To discuss this in more detail, we assume that in addition to the species resulting from the pure system, the following substances might have been added:

- a substance Me_nCO_3, with the CO_3^{2-} ion common with $CaCO_3$, eg. $MgCO_3$.
- a substance $Ca\,An_m$, with the Ca^{2+} ions common with $CaCO_3$, eg $CaSO_4$.
- a substance, which dissociates into ion X^{+Z_x} and Y^{-Z_y}, which do not react with Ca^{2+}, H^+, or the carbonate species.

These ions, however, increase the ionic strength of the solution.

Using the conservation of carbonic species, which is modified by the presence of Me_nCO_3 in compliance with the charge-balance equation, one proceeds similarly as in Chapter 5 to obtain a fourth-order polynomial for $\left[H^+\right]$. This can be approximated by a quadratic equation, if one neglects $\left[CO_3^{2-}\right]$, $\left[H^+\right]$ and $\left[OH^-\right]$ in the charge-balance equation (for details see Dreybrodt, 1988). The equilibrium concentration $[Ca^{2+}]_{eq}$ with respect to calcite can be derived with the simplified charge-balance equation

$$2\left[Ca^{2+}\right] + Z_{Me}\left[Me\right] = 2\left[HCO_3^-\right] + Z_{An}\cdot\left[An\right] \qquad (32)$$

Z refers to the charge of the corresponding ion and denotes only its amount, not its sign. Note that the ions X and Y do not occur in this equation, because their charges cancel. The result is

$$\left[Ca^{2+}\right]_{eq}\left(\left[Ca^{2+}\right]_{eq} + \frac{[F]}{2}\right)^2 = \left(\left[Ca^{2+}\right]_{eq}^{p}\right)^3 \cdot \frac{\gamma_{Ca}^{p}\left(\gamma_{HCO_3}^{p}\right)^2}{\gamma_{Ca}\left(\gamma_{HCO_3}\right)^2} \qquad (33)$$

where $[F] = Z_{Me}\cdot[Me] - Z_{An}[An]$

The index p refers to the corresponding equilibrium value of $[Ca^{2+}]$ and activity coefficients in the pure $H_2O - CO_2 - CaCO_3$

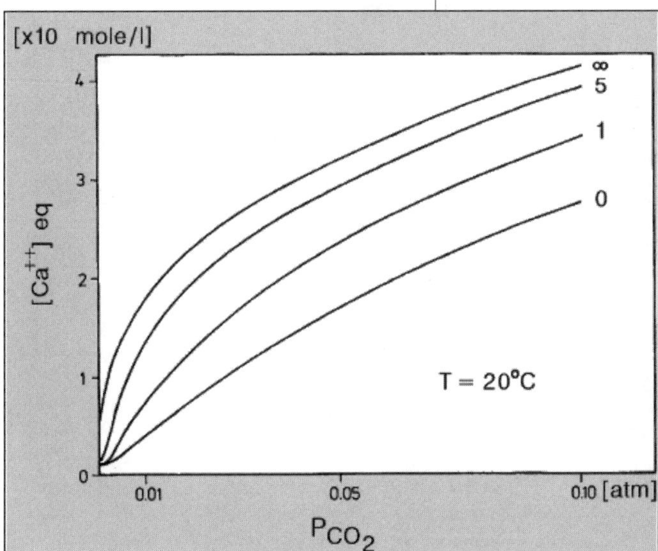

4.1.1 Figure 7. Equilibrium concentrations of $[Ca^{2+}]_{eq}$ as a function of initial $p^i_{CO_2}$ prior to dissolution for various values of $V_g/V_s = \chi$, given by the numbers of the curves.

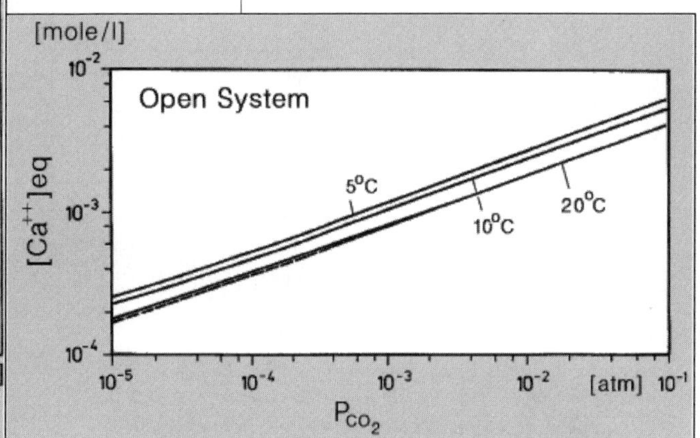

4.1.1 Figure 8. Equilibrium values of $\left[Ca^{2+}\right]_{eq}$ as a function of p_{CO_2} in equilibrium with the solution at different temperatures. The dashed line gives the approximation according to equation 30.

system. If Ca^{2+} is added as a common ion, [F] < 0. The corresponding saturation curves are shown in 4.1.1 Figure 10. They are located above the saturation curve of the pure system with $[F] = 0$. If the addition of Ca^{2+} as a common ion did not change the solubility of $CaCO_3$, then the new equilibrium curve would just be the sum of the two concentrations.

This is shown for $[F] = -5 \times 10^{-3}$ as a dashed curve. The real saturation curve lies below this dashed line, thus showing that the calcite solubility is decreased. For positive $[F]$ the common ion is CO_3^{2-}. In this case the concentration of this species is increased and the corresponding increase of the ion activity product (Ca^{2+}) (CO_3^{2-}) reduces $CaCO_3$ solubility.

In summary, the influence of foreign ions on calcite solubility can be classified into several independent mechanisms.

• **Ionic-Strength Effect***:* If indifferent substances such as NaCl are added, [F] = 0 because of the charge balance. Ionic strength is changed, however, and therefore the activity coefficients become smaller. Thus the increase in ionic strength decreases the solubility product, and calcite solubility increases. This effect is always operative but is small in karst waters.

• **Common-Ion Effect***:* Addition of common ions to a pure calcite-saturated solution increases the ion activity product, and therefore the solution becomes supersaturated. Thus, the common-ion effect always decreases calcite solubility. This can be easily seen from 4.1.1 Figure 10. This is evident for [F] > 0, i.e. with CO_3^{2-} added as a common ion. For [F] < 0, i.e. when Ca^{2+} is added as common ion, this can be seen by comparing the dashed upper curve with the curve resulting from $[F] = -5 \cdot 10^{-3}$ mol/L. The dashed curve results by adding 2.5 mmol/L of Ca^{2+} as a common ion to the pure system $CaCO_3 - H_2O - CO_2$ in equilibrium, and assuming no common-ion effect. The curve with $[F] = -5 \cdot 10^{-3}$ mol/L results also from adding 2.5 mmol/L of Ca^{2+} as a common ion but with the common-ion effect present. The reduced calcite solubility is seen from the fact that it is situated below the dashed curve.

• **Ion Pair Effect:** In the presence of sulfate ion pairs $CaSO_4^0$ is formed. Thus the concentration of $[Ca^{2+}]$ is reduced and therefore calcite solubility increases.

The total dissolved calcium $[Ca^{2+}]^t_{eq} = [Ca^{2+}]_{eq} + [CaSO_4^0]_{eq}$ at equilibrium is then given (Dreybrodt, 1988) by

$$[Ca]^t_{eq} = [Ca^{2+}]_{eq} \cdot \left(1 + \frac{[SO_4^{2-}]\gamma_{Ca}\gamma_{SO_4}}{K_{SO_4}}\right)^{1/3} \cdot \left(\frac{\gamma^p_{Ca}(\gamma^p_{HCO_3})^2}{(\gamma_{Ca}\gamma_{HCO_3})^2}\right)^{1/3} \quad (34)$$

with

$$(Ca^{2+})(SO_4^{2-}) = K_{CaSO_4}(SO_4^0)K_{CaSO_4} \approx 5 \cdot 10^{-3}\ mol/l$$

(Garrels and Christ, 1965). Thus at a sulfate concentration of 10^{-3} mol/l an increase in $[Ca]^{eq}_t$ by about 3% is expected.

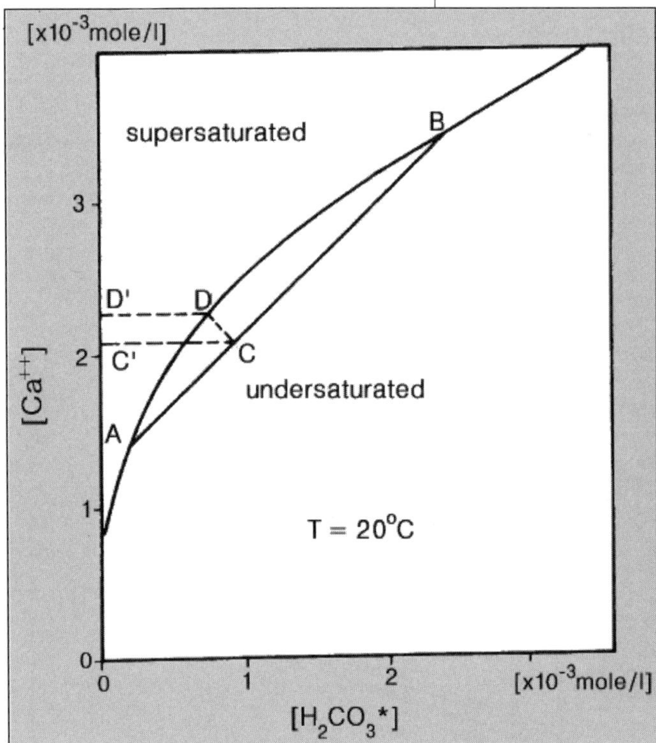

4.1.1 Figure 9. The equilibrium curve divides the $[H_2CO_3^*] - [Ca^{2+}]$ diagram into two parts. Above the curve, solutions are supersaturated. Below the curve, solutions are undersaturated. Mixing of two saturated solutions A and B leads to an under-saturated solution, for example, C. The additional amount of Ca^{2+} that can be dissolved after mixing is given by C′D′.

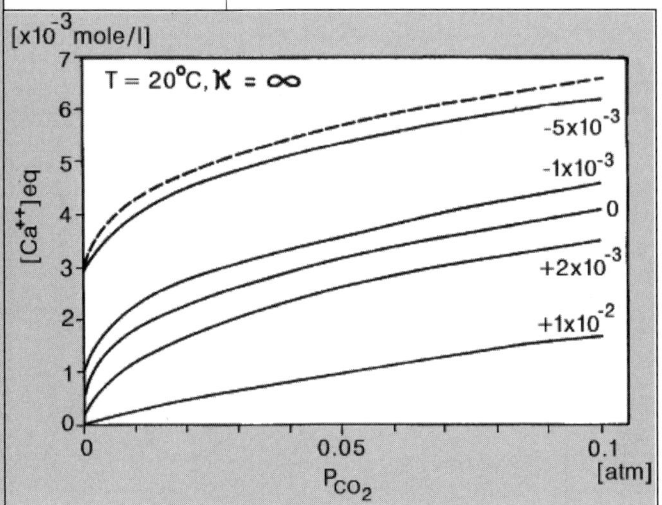

4.1.1 Figure 10. Equilibrium values of $[Ca^{2+}]$ in the presence of common ions, i.e. F≠O, as a function of P_{CO_2} in the solution. Numbers on the curves designate F in mol/l. The dashed curve results by shifting the equilibrium curve of the pure system (F = 0) vertically by 2.5×10^{-3} mol/l and designates the equilibrium curve upon addition of Ca^{2+} as a common ion with $F = -5 \times 10^{-3}$ mol/l without the common-ion effect acting. The corresponding curve with the common-ion effect present is below the dashed curve, $F = -5 \times 10^{-3}$ mol/l and shows that calcite solubility decreases.

Conclusion

The chemical reactions in karst waters have been summarized during their dissolutional interaction with $CaCO_3$. From these it was possible to derive the evolution of their chemical composition when water flows through limestone aquifers under various boundary conditions, such as under conditions open or closed with respect to CO_2. The intermediate case has also been considered, where only a limited volume of CO_2 is present. The knowledge of the chemical composition is of utmost importance to calculating dissolution rates (see next Chapter 4.1.2).

Finally, expressions have been given for the Ca concentration at equilibrium with respect to calcite for pure $HCO_3^- - Ca^{2+}$ solutions and also for more realistic karst water containing foreign ions such as Mg^{2+}, SO_4^{2-}, Na^+, K^+, and Cl^-.

References

Bögli, A., 1964, Mischungskorrosion: Ein Beitrag zum Verkarstungsproblem: Erdkunde, v. 18, p. 83-92.

Bögli, A., 1980, Karst hydrology and physical speleology: Berlin, Heidelberg, New York, Springer, 284 p.

Dreybrodt, W., 1988, Processes in karst systems–Physics, Chemistry and Geology: Berlin and New York, Springer Series in Physical Environments, no. 5, 288 p.

Dreybrodt, W., 1990, The role of dissolution kinetics in the development of karst aquifers in limestone: A model simulation of karst evolution: Journal of Geology, v. 98, p. 639-655.

Dreybrodt, W., 1996, Principles of early development of karst conduits under natural and man-made conditions revealed by mathematical analysis of numerical models: Water Resources Research, v. 32, p. 2923-2935.

Garrels, R.M. and Christ, C.L., 1965, Solutions, minerals, and equilibria: New York, Harper and Row, 450 p.

Harned, H.S., and Hamer, W.J., 1933, The ionization constant of water: J. Am. Chem. Soc., v. 55, p. 693-695.

Laptev, F.F., 1939, Aggressive action of water on carbonate rocks, gypsum and concrete, in Trudy SPETSGEO, v.1. Moscow and Leningrad: GONTI. In Russian.

Loewenthal, R.E., and Marais, G.V.R., 1978, Carbonate chemistry of aquatic systems: Theory and applications 1: Ann Arbor, Mich., Ann Arbor Science, 431 p.

Morse, J.W., and Mackenzie, F.T., 1990, Geochemistry of Sedimentary Carbonates: Amsterdam and New York, Elsevier, Developments in sedimentology no. 48, 707 p.

Palmer, A.N., 1984, Recent trends in karst geomorphology: J. Geol. Educ., v. 32., p. 247-253.

Palmer, A.N., 1991, The origin and morphology of limestone caves: Geological Society of America Bull., v. 103, p. 1-21.

Plummer, L.N., and Busenberg, E., 1982, The solubilities of calcite, aragonite, and vaterite in CO2-H2O-solutions between 0 and 90°C and an evaluation of the aqueous model for the system CaCO3-CO2-H2O: Geochim. Cosmochim. Acta, v. 46, p. 1011-1040.

Plummer, L.N., and Mackenzie, F.T., 1974, Predicting mineral solubility from rate data: Application to the dissolution of magnesian calcites: American Journal of Science, v. 274, p. 61-83.

Plummer, L.N., Parkhurst, D.L., and Wigley, T.M.L., 1978, The kinetics of calcite dissolution in CO2 systems at 25oC to 60oC and 0.0 to 1.0 atm CO2: American Journal of Science 278. 179-216.

Stumm, W., and Morgan, J.J., 1981, Aquatic chemistry: New York, Wiley, 780 p.

White, W.B., 1984, Rate processes: Chemical kinetics and karst landform development, in La Fleur, R. G., ed., Groundwater as a geomorphic agent: London, Allen and Unwin, p. 227-248.

Wigley, T.M.L., and Plummer, L.N., 1976, Mixing of carbonate waters, Geochim. Cosmochim. Acta, v. 40, p. 989-995.

Wissbrun, K.F., French, D.M., and Patterson, A., 1954, The true ionisation constant of carbonic acid in aqueous solution from 5 to 45°C: Journal Physical Chemistry, v. 58, p. 693-695.

4.1.2. Limestone Dissolution Rates in Karst Environments

Wolfgang Dreybrodt and Laurent Eisenlohr

Abstract

The removal of limestone from the bedrock at the surface and below ground by aqueous solutions containing CO_2 sculptures karst landscapes and complex karst aquifers. To understand the evolution of such karstic features requires the knowledge of dissolution rates under various hydrogeologic conditions. These rates are controlled by several complex mechanisms:

- The rate equations of Plummer et al. (1978), from which surface reaction rates can be obtained when the concentrations of reacting species at the surface are known.
- The slow reaction of CO_2 to H^+ and HCO_3^-, which provides the H^+ ion for converting carbonate to bicarbonate ions.
- Mass transport by diffusion, either in laminar or turbulent flow.
- Inhibition of surface reaction rates by the presence of impurities in natural carbonate minerals.
- Open- or closed-system conditions with respect to CO_2, under which dissolutional removal of limestone is active.

Depending on the actual conditions, each of these processes can greatly affect dissolution rates. This paper addresses these problems and provides data, which can be used to obtain realistic dissolution rates, when solutions flow laminarly in narrow fractures, but also turbulently in large conduits, and during a variety of hydrogeologic conditions. These data are also necessary as input for modeling the evolution of karst.

Introduction

In the evolution of karst in limestone terrains the dissolution kinetics in the system H_2O-CO_2-$CaCO_3$ plays a dominant role. It is widely accepted nowadays that to explain karstification it is not sufficient to consider the equilibrium chemistry only, but one also has to regard the dissolution rates by which limestone is removed in response to the chemical composition of the karst water.

Dissolution of limestone occurs under various hydraulic conditions e.g. laminar or turbulent flow and also with respect to the interaction of karst water with CO_2, which delivers the chemical energy driving karst processes. In the deep-phreatic zone of karst aquifers, all empty space, such as narrow joints and also larger conduits, are filled completely by groundwater, such that no water surface is exposed to the atmosphere. Therefore, no carbon dioxide can enter or leave.

These conditions are called closed-system with respect to CO_2. Closed-system conditions rule the early evolution of karst systems, where the water table is high and dissolution of limestone occurs in narrow joints or bedding-plane partings with apertures of about 100 μm (Dreybrodt, 1988, 1990, 1996; Ford and Williams, 1989; White, 1988). Under such conditions flow is generally laminar. Later on, in the mature state of a karst aquifer, large conduits have developed, and also a variety of fractures have been solutionally widened to several millimeters. Under these conditions flow becomes turbulent.

There are also a variety of cases where water flows on limestone rock with a free surface exposed to the atmosphere, such that exchange of carbon dioxide drives the water to equilibrium with respect to the partial pressure p_{CO_2} of carbon dioxide contained in the atmosphere. These are open-system conditions with respect to CO_2. The depth of the water body ranges from a few tenths of a millimeter, when thin sheets of rainwater run down bare rock, up to several meters in rivers. Open-system conditions are dominant on the surface of karst terrains, and in the vadose zone, and at the water table. Therefore, to understand the evolution of karst aquifers and karst landscapes in their dimensions in space and time, we must predict dissolution rates under this wide spectrum of conditions.

In this paper we discuss the basic physical and chemical mechanisms, that determine the dissolution rates in synthetic and natural carbonate minerals. We also report related experimental data. From these data, at least to a reasonable approximation, kinetic rate laws are extracted, which can be used in modeling karst processes.

Basic Reactions During Dissolution

The chemical reactions in the solution and at the surface of the mineral are illustrated schematically in 4.1.2 Figure 1. In the solution the basic reactions are the production of H^+ ions from CO_2.

$$\text{(I)}\quad H_2O + CO_2 \rightarrow H_2CO_3 \rightarrow H^+ + HCO_3^-$$

This process delivers a proton, which reacts with the carbonate ions released from the mineral

$$\text{(II)}\quad CO_3^{2-} + H^+ \rightarrow HCO_3^-$$

By this way the ion-activity product $(Ca^{2+})(CO_3^{2-})$ is kept sufficiently low, such that the equilibrium concentration of Ca^{2+} in the solution depends on the cube root of the CO_2-concentration at equilibrium (Dreybrodt, 1988) (see also Chapter 4.1.1 in this volume). It is now extremely important to note that the first step in reaction (I) is slow (Usdowski, 1982; Dreybrodt, 1988), whereas the further steps are so fast that H_2CO_3, H^+, HCO_3^-, and CO_3^{2-} are always in equilibrium. Usdowski (1982) has reported reaction times in buffered calcareous solutions with a pH between 7 and 9 at 25° C.

In such solutions, typical for karst water, it takes about 30 seconds until 50% of equilibrium is attained with respect to CO_2, H^+, and HCO_3^-. To approach 99% of equilibrium, 100 seconds are needed. The slowness of this reaction has important consequences to the dissolution rates. Until now this has not been sufficiently recognized by many geologists.

The reactions at the surface that release Ca^{2+}, CO_3^{2-} and HCO_3^- from the mineral have been investigated by the pioneering work of Plummer, Wigley, and Parkhurst (1978). Three elementary reactions, are active simultaneously.

$$\text{(IIIa)}\quad CaCO_3 + H^+ \rightleftarrows Ca^{2+} + HCO_3^-$$

$$\text{(IIIb)}\quad CaCO_3 + H_2CO_3 \rightleftarrows Ca^{2+} + 2HCO_3^-$$

$$\text{(IIIc)}\quad CaCO_3 + H_2O \rightleftarrows Ca^{2+} + CO_3^{2-} + H_2O \rightleftarrows Ca^{2+} + HCO_3^- + OH^-$$

Reactions (III) can be represented by the overall reaction

$$CaCO_3 + H_2O + CO_2 \rightleftarrows Ca^{2+} + 2HCO_3^-$$

Thus stoichiometry requires that for each CO_3^{2-} ion released from the mineral, and therefore also for each Ca^{2+} released, one molecule of CO_2 is consumed from the solution.

From measuring dissolution rates of Iceland spar in large volumes of vigorously stirred turbulent solutions, Plummer, Wigley, and Parkhurst, (1978) extracted the following rate law, commonly now referred to as the PWP equation, which describes the dissolution rate at the surface.

$$F_{PWP} = \kappa_1(H^+)_s + \kappa_2(H_2CO_3^*)_s + \kappa_3 - \kappa_4(Ca^{2+})_s(HCO_3^-)_s \quad (1)$$

The brackets ( ) denote the activities in mmole/cm^3 of the corresponding ion species at the surface of the mineral. The rate F is consequently given in mmole/(cm^2 s). κ_1, κ_2 and κ_3 are the rate constants of the forward reactions, whereas the negative term summarizes the three backward reactions, such that κ_4 becomes dependent on the CO_2 concentration in the solution. As we will discuss later, these rate equations are valid for pure synthetic calcite. For natural materials, modifications are necessary (Svensson and Dreybrodt, 1992; Eisenlohr et al., 1997). The basic reactions, however, remain unchanged.

From the PWP equation (1) the rates can be calculated, provided one knows the concentrations of the ions at the *surface* of the mineral. This presents the major problem in finding actual rates for a solution, where only the bulk concentrations c_B are known. The ions released from the mineral are transported away from its surface by molecular diffusion. Thus the ionic concentration c_S at the surface must be higher than that in the bulk. Diffusional flux away from the surface is given by

$$F_D = \frac{D}{\varepsilon}(c_s - c_B) = \alpha_D(c_s - c_B) \quad (2)$$

D is the coefficient of diffusion on the order of $1 \cdot 10^{-5}$ cm^{2}s^{-1}, ε is the thickness of the diffusion boundary layer separating the surface from the bulk of the turbulent solution, where the concentration c_B does not exhibit any gradient. The flux released from the surface by surface reactions is given by a function of $f(c_{eq} - c_s)$, which might be complicated. Thus

$$F_s = \alpha_s \cdot f(c_{eq} - c_s) \quad (3)$$

where α_s is constant. Under steady state $F_s = F_D$. This is illustrated in 4.1.2 Figure 2, which shows a plot of F_D (dashed line) and F_s (solid line) versus concentration c. From the point of intersection between these two curves, one finds the surface concentration c_s. Now, if the surface reaction rate is increased by increasing α_s, c_s approaches c_{eq} for $\alpha_s \rightarrow \infty$. Thus for large α_s one finds $F_s \approx \alpha_s \cdot f(c_{eq} - c_B)$, and the rates are controlled by diffusion. If, however, α_s becomes small, c_s approaches c_B and the rates are given by $F_s \approx \alpha_s \cdot f(c_{eq} - c_B)$. At this limit, the rates are surface controlled.

The problem becomes even more complex because stoichiometry requires a third condition. The rate of conversion of CO_2 to H^+ and HCO_3^-, must be equal to the rate of release of Ca^{2+} from the mineral. Since CO_2 conversion is a slow process, it also can be rate determining. The amount of CO_2 converted in the volume V of the solution is given by

$$V \cdot \frac{d[CO_2]}{dt} = A \cdot F_s \quad (4)$$

It must be equal to the amount of Ca^{2+} ions released from the surface area A of the mineral. Therefore, if V becomes sufficiently small, the rates are controlled by CO_2 conversion, and the ratio V/A is the controlling factor.

To obtain exact rates under these various controls, one has to solve the transport equations, which concern:

- Diffusional mass transport of all species involved in the reactions.
- Reaction rates of CO_2 conversion.
- The PWP-equation as a boundary condition for flux at the surface of the mineral.

Modeling Dissolution Rates in Open and Closed Systems

Taking these three processes into account, Buhmann and Dreybrodt (1985a, b), and Dreybrodt (1988) have put forward a model of dissolution for both open and closed systems. For the open system they assumed a stagnant planar layer of water with thickness δ covering a plane surface of calcite. This water sheet exhibits an open surface to a CO_2-containing atmosphere of partial pressure p_{CO_2}. In the closed system the water is confined by two parallel calcite planes separated by a distance 2δ. The CO_2 concentration in this solution is characterized by the partial pressure of CO_2 in equilibrium with the water prior to dissolution of calcite, i.e. at calcium concentration of zero.

Note that in the open system the CO_2 concentration remains constant during dissolution, because CO_2 consumed by dissolution is replenished from the atmosphere. In the closed system, however, the CO_2 concentration drops as the calcium concentration increases, because CO_2 is consumed.

In 4.1.2 Figure 3 the results for closed-system conditions with initial atm at 10° C are illustrated. Figure 3 shows the dissolution rates during a free drift reaction as a function of the increasing calcium concentration for various values of δ. Note that the ratio of the volume V of the solution to the area A of the surface $V/A=\delta$, for both the open- and the closed-system geometry (cf. equation 4). The numbers on the curves denote the value of δ in units of 10^{-3} cm. For small $\delta<2 \cdot 10^{-3}$ cm the rates increase linearly with δ. Thus they are proportional to the volume of the solution, which is typical, when the conversion of CO_2 is controlling the rates.

At larger values of δ, from about $2 \cdot 10^{-3}$ up to $5 \cdot 10^{-3}$ cm, the rates increase much slower until for values of δ from 10^{-2} cm up to about 10^{-1} cm they become independent of δ. In this region both CO_2 conversion and diffusional mass transport are rate controlling. For large values of $\delta > 5$ cm, mass transport finally controls the rates entirely (Dreybrodt, 1988), and the rates decrease with δ^{-1} (not shown here). In all these calculations, the water layer is assumed stagnant. This is equivalent to plug flow along the joint where the Ca^{2+} concentration increases along the flow path. We consider this as a reasonable approximation for laminar flow.

If flow is so highly turbulent that laminar boundary layers have values of $\varepsilon < 10^{-4}$ cm, then diffusional mass transport is enhanced by eddies at all scales. This can be simulated by increasing the coefficient of diffusion by a factor of about 10^4. Thus $D_{eff} = 10^4 \cdot D_m$. The 4.1.2 Figure 4 depicts the results of such a simulation. At low $d < 2 \cdot 10^{-3}$ cm the rates are equal to those calculated for laminar flow or a stagnant water film.

This again shows that in this region the rates are controlled entirely by CO_2 conversion, in as much as mass transport is enhanced by four orders of magnitude. For d values of up to $5 \cdot 10^{-3}$ cm, the rates increase linearly with d, in contrast to a stagnant film. Beyond this value the rates increase more slowly, until at values of $d > 0.1$ cm they become constant. Closer inspection of these results shows that the rates for $d > 0.1$ are identical to those calculated from the PWP equation, when inserting the bulk activities of all species.

Thus the rates are surface controlled. This is reasonable since for sufficiently large δ, CO_2 conversion becomes sufficiently fast (cf. equation 4) and turbulence avoids concentration gradients, such that surface concentrations are equal to those in the bulk. This can be visualized also by 4.1.2 Figure 2 in which the enhanced mass transport is shown by the dashed line $10^4 \cdot F_D$. The point of intersection gives a surface concentration $c_s = c_B$. It should be noted that this is an idealization seldom encountered in nature. Usually, laminar diffusion boundary layers exist, and the surface concentrations are higher than those in the bulk, and therefore the dissolution rates are reduced (details are given by Dreybrodt, 1988; Dreybrodt and Buhmann, 1991, and Liu and Dreybrodt, 1997).

In all calculations the rate laws can be approximated by a linear relation

$$F = \alpha \cdot (c_{eq} - c) \quad (5)$$

The transport coefficient α depends on temperature, on the P_{CO_2} of the solution, on δ, on ε, and on the type of flow, i.e. laminar or highly turbulent. The 4.1.2 Figure 5 depicts the values of α as a function of δ for the examples of 4.1.2 Figures 3 and 4. At low δ the linear increase of α illustrates the regime of pure CO_2 control (solid lines), which gives equal rates for laminar and turbulent flow. For $\delta > 2 \cdot 10^{-3}$ cm the values for laminar flow become constant due to mixed control by CO_2 conversion and diffusional mass transport (dashed line).

For turbulent flow a further linear increase is followed by a wide region of increasing α (dashed line), where both CO_2 control and the surface reactions are rate determining. For $\delta > 0.1$ cm the rates become constant and are controlled entirely by the surface reactions (long dashes). They therefore are given by the PWP equation. This is also depicted in 4.1.2 Figure 3 by the dashed line representing the dissolution rates at turbulent flow for direct comparison to the rates in laminar flow. The 4.1.2 Figure 5 illustrates convincingly that using the PWP equation by uncritically inserting concentrations as obtained from chemical analysis of the bulk can lead to rates, that are too high by about one order of magnitude. This often has not been recognized in many hydrogeologic applications and has led to wrong conclusions.

It should be noted that similar arguments also hold for precipitation of calcite from supersaturated solutions (Buhmann and Dreybrodt, 1985a, b; Dreybrodt and Buhmann, 1991; Dreybrodt et al., 1997). Values of the rate constant a for dissolution in laminar and turbulent flow are listed for various temperatures, P_{CO_2}, and d by Buhmann and Dreybrodt (1985a, b) and Dreybrodt (1988) for both open and closed systems and for ground water containing foreign ions (Buhmann and Dreybrodt, 1987; Dreybrodt, 1988). Typical values for a are in the order of several 10^{-5} cm/s (cf. 4.1.2 Figure 5). The 4.1.2 Tables 1 and 2 list these values of a for the open and closed system, respectively.

Experimental Verification

Several experimental methods have been used to measure dissolution rates to verify the theoretical predictions from the model above. In all of them the volume V and the surface area A must be specified. After establishing contact of the solution with the mineral, one observes the concentration c(t) of calcium as it develops in time in a free-drift experiment. According to mass balance

$$V \cdot \frac{dc}{dt} = A \cdot F(c) \quad (6)$$

Therefore by taking the time derivative of c(t) the rates F(c) can be obtained.

The first experiments that systematically investigated the dependence of the rates on the V/A ratio were performed by Buhmann and Dreybrodt (1985a, b). They observed c(t) in thin layers of water ranging from 0.03 cm up to 0.15 cm. These were covering a polished surface of marble with the open surface in contact with a CO_2-containing atmosphere with $P_{CO_2} = 5 \cdot 10^{-3}$ or $3 \cdot 10^{-4}$ atm. They found rate constants of α independent of δ, with values close to the predicted ones. To simulate fully turbulent flow, marble disks were mounted on the bottom of a cylindrical vessel and covered by a solution which was stirred by a glass propeller. The depth of the water ranged from 0.5 to 1.5 cm. Again satisfactory agreement to the predicted values of α was obtained. Similar experiments performed in closed systems also proved the validity of the model.

The shortcoming of these first experiments was that they did not give direct evidence of the important role of CO_2 conversion, in as much as sufficiently low values of δ could not be obtained experimentally. Low values of the V/A-ratio, however, can be obtained by observing dissolution in porous media consisting of sized calcite grains.

Recently Dreybrodt et al. (1996) performed such experiments. They filled the space between the electrodes of a WTW-conductivity cell with calcite particles of uniform size and added distilled water in

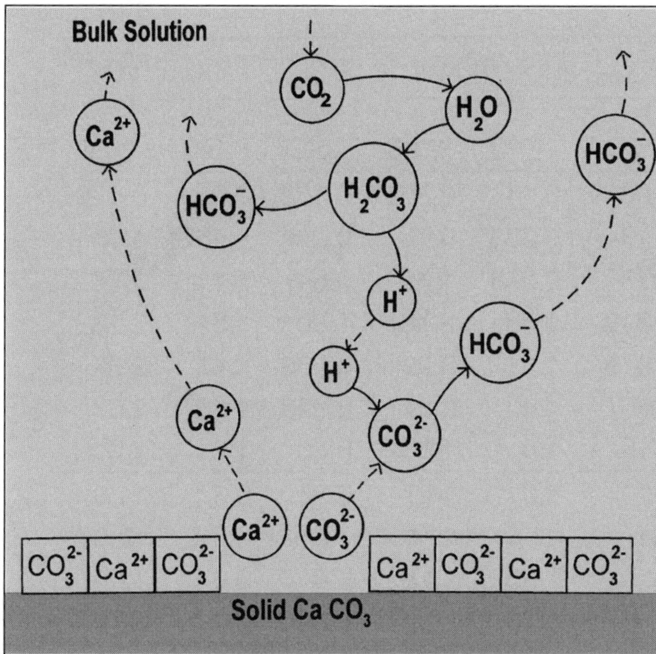

4.1.2 Figure 1. Schematic diagram of the chemical reactions during calcite dissolution. Solid arrows denote chemical reactions. Dashed arrows indicate transport of reacting species.

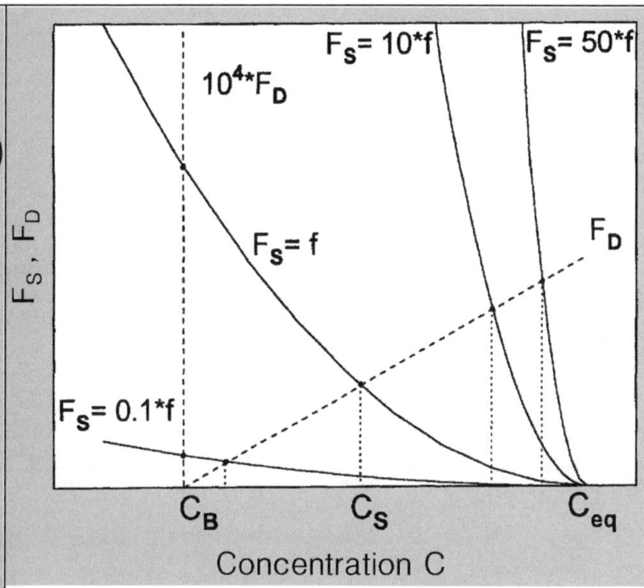

4.1.2 Figure 2. Illustration of surface-controlled or diffusion-limited dissolution rates. F_S represents the surface reaction rate, and F_D the mass transport into the bulk as a function of the concentration. The surface concentration c_S is read from the point of intersection where $F_D = F_S$. If F_S is inhibited ($F_S = 0.1 \cdot f$), c_S shifts toward the bulk concentration c_B. If F_s is enhanced ($5 \cdot f$, $50 \cdot f$), c_S approaches c_{eq}. If mass transport is enhanced ($10^4 \cdot F_D$), c_S also approaches c_B, and the rates are surface controlled.

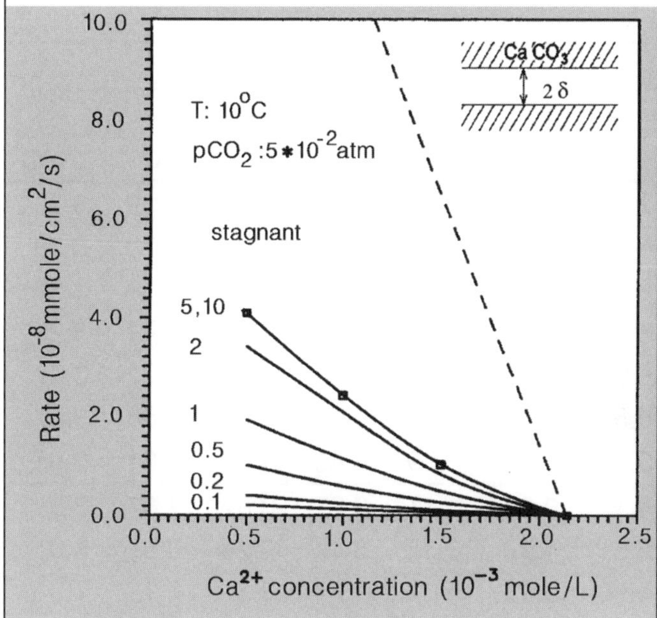

4.1.2 Figure 3. Dissolution rates from the theoretical model for a free-drift run under the conditions of laminar flow and a system closed to CO_2. The numbers on the curves denote the value of δ, i.e. half the distance between the parallel calcite surfaces in 10^{-3} cm. For $\delta = 5 \cdot 10^{-3}$ cm and $\delta = 1 \cdot 10^{-2}$ cm the curves are identical. This is shown by the open squares which correspond to $\delta = 1 \cdot 10^{-2}$. The dashed curve gives the rates for turbulent motion and δ = 1 cm. The insert in the upper right depicts the geometry of the model.

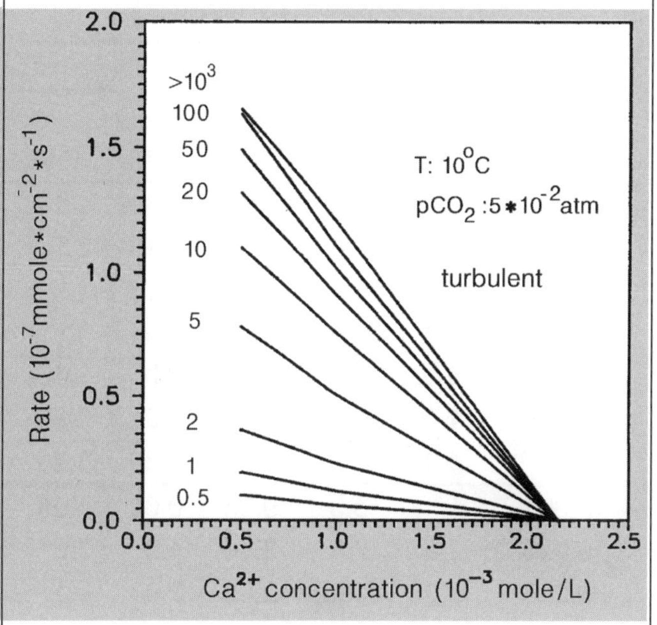

4.1.2 Figure 4. Dissolution rates as in 4.1.2 Figure 3, but now the water is in turbulent motion. This has been simulated by replacing the coefficient of molecular diffusion D_m by an effective eddy diffusion coefficient $D_{eff} = 10^{-4} \cdot D_m$. Numbers on curves denote δ in 10^{-3} cm.

	δ(cm)	$p_{CO_2} = 3 \times 10^{-4}$			$p_{CO_2} = 1 \times 10^{-3}$ atm			$p_{CO_2} = 5 \times 10^{-3}$ atm		
		5°C	10°C	20°C	5°C	10°C	20°C	5°C	10°C	20°C
(a)	0.001	0.041	0.0714	0.232	0.0619	0.11	0.358	0.153	0.25	0.679
	0.002	0.0076	0.135	0.407	0.134	0.233	0.75	0.294	0.467	1.36
	0.005	0.197	0.341	1.04	0.33	0.583	1.69	0.647	1.11	2.74
	0.01	0.37	0.635	1.84	0.619	1.08	2.76	1.00	1.54	3.13
	0.03	0.803	1.29	2.91	0.979	1.61	3.29	0.969	1.47	2.87
	0.05	0.915	1.29	2.57	0.907	1.40	2.64	0.926	1.35	2.63
	0.1	0.636	0.919	1.60	0.68	1.0	1.63	0.85	1.21	2.0
(b)	0.1	3.3	5.7	13.55	5.2	8.0	14.7	7.25	8.3	11.0
	0.2	5.9	9.5	20.0	8.4	11.5	17.5	7.9	8.8	12.0
	0.5	11.5	16.5	26.0	12.0	14.5	18.5	8.0	9.0	12.0
	1.0	16.0	20.6	28.5	13.5	15.0	19.0	8.0	9.0	12.0
	2.0	20.0	24.0	30.0	13.5	15.5	19.0	8.0	9.0	12.0
(c)	$[Ca^{2+}]_{eq}$	6.75	6.3	5.6	10.0	9.3	8.3	17.0	16.2	14.3

4.1.2 Table 1. Numerical values of $\alpha = \alpha\ (T, \delta)$ in 10^{-5} cm s^{-1} at various partial pressures of CO_2 (in atm). ***(a)*** *is for laminar flow.* ***(b)*** *is for turbulent flow.* ***(c)*** *gives the calculated Ca^{2+} concentrations at saturation, $[Ca^{2+}]_{eq}$ in 10^{-4} mmol cm^{-3}, open system.*

	δ [cm]	$p_{CO_2} = 1 \times 10^{-2}$ atm			3×10^{-2} atm			5×10^{-2} atm			1×10^{-1} atm		
		5°C	10°C	20°C	5°C	10°C	20°C	5°C	10°C	20°C	5°C	10°C	20°C
(a)	0.001	0.714	1.6	7.14	0.606	1.1	3.33	0.666	1.19	3.33	0.83	1.39	3.23
	0.002	1.3	2.8	13.3	1.18	2.1	6.25	1.25	2.25	5.26	1.05	1.73	3.33
	0.005	3.13	6.4	21.7	2.13	3.5	8.93	1.49	2.5	5.0	1.06	1.73	3.33
	0.01	5.26	9.6	21.7	2.04	3.5	8.7	1.49	2.47	4.76	1.06	1.73	3.23
	0.02	5.13	9.5	18.2	2.0	3.46	6.9	1.46	2.44	4.76	1.05	1.70	3.17
	0.05	4.76	8.4	16.1	2.0	3.3	6.5	1.43	2.43	4.67	1.04	1.68	3.07
	0.1	4.08	7.7	12.7	1.89	3.25	5.9	1.4	2.29	4.46	1.01	1.61	2.95
(b)	0.001	1.08	3.3	9.75	0.63	1.08	4.1	0.655	1.17	3.37	0.82	1.37	3.5
	0.003	2.9	7.3	22.0	1.8	3.3	9.5	1.97	3.32	8.14	2.28	3.5	6.0
	0.005	5.08	10.2	27.5	2.9	5.1	13.0	3.13	5.09	10.4	3.3	4.7	7.8
	0.01	8.6	15.1	34.0	5.4	8.75	17.0	5.2	7.3	13.0	4.54	6.05	9.0
	0.02	13.2	20.5	45.0	8.2	11.3	19.0	6.9	8.9	13.9	5.5	7.0	10.2
(c)	$[Ca^{2+}]_{eq}$	6.4	5.5	4.1	16.6	14.4	11.0	24.3	21.4	16.9	38.8	35.1	28.0

4.1.2 Table 2. Calculated values of α in 10^{-5} cm s^{-1} for various pressures of CO_2, temperatures, and film thicknesses d. (a) is for laminar flow. (b) is for turbulent flow. (c) gives the calculated Ca^{2+} concentrations at saturation, $[Ca^{2+}]_{eq}$ in 10^{-4} mmol cm^{-3}, closed system.

equilibrium with a given initial p_{CO_2}. After this, the setup was closed with respect to CO_2, and the conductivity was measured employing the same cell. The experimental setup is illustrated by 4.1.2 Figure 6. Conductivity and calcium concentration are linearly related by $c = 5.37 \cdot 10^{-3}\sigma - 0.053$ (σ in µs/cm, in mmole/L) for $CaCO_3$–H_2O–CO_2 solutions with c<3mmole/L.

Therefore the concentration can be found easily, and from the time course c(t) the rates can be determined. Theoretical considerations show that the model is also valid for such porous media if one replaces δ by the ratio V/A (Baumann et al., 1985). Measurements with values of V/A from $2.25 \cdot 10^{-4}$ cm up to $1.3 \cdot 10^{-2}$ cm revealed excellent agreement with the theoretical predictions for initial p_{CO_2} of $1 \cdot 10^{-2}$ and $5 \cdot 10^{-2}$ atm.

Furthermore, direct evidence of the important role of CO_2 conversion was obtained by adding carbonic anhydrase to the solution to obtain a concentration of 0.6 µmole/L. This enzyme enhances the conversion of CO_2 to H_2CO_3 by a factor of about 1000. Therefore one expects enhancement of the dissolution rates and correspondingly a faster increase of the concentration. The 4.1.2 Figure 7 shows the results of such an experiment with $V/A = 2.25 \cdot 10^{-4}$ and $p_{CO_2} = 5 \cdot 10^{-2}$ atm. The lower curve illustrates the rise of concentration, when no enzyme is added. Because the dissolution kinetics are linear, this rise is represented as an exponential function

$$c(t) = c_{eq}(1 - \exp(-t/\tau)); \; \tau = V/(\alpha \cdot A) \quad (7)$$

The measured time constant τ is found to be 100 s. The upper two curves represent the time course of c(t) of identical experiments, but upon addition of carbonic anhydrase (0.6 µmolar) there is a dramatic enhancement, and τ is reduced by a factor of 50. Thus a value of $\alpha = 10^{-4}$ cms^{-1} results. This value is also found when the surface reactions are rate controlling (cf. 4.1.2 Figures 3 and 4). This is what one expects. For such small values as V/A = 10^{-4} cm diffusion is fast (cf. equation 2), and since now CO_2 conversion also is fast, the rates are controlled entirely by the surface reactions.

In summary we conclude that for all geologic situations, where water flows laminarly in joints with widths less than 1 mm, the dissolution rates exhibit linear kinetics. This is a consequence of the slowness of the reaction $H_2O + CO_2 \rightarrow H_2CO_3$.

This is also true for the open system, where sheets of water flow on bare rock. In all these cases the transport coefficients a can be obtained reliably from the theoretical model.

Dissolution Kinetics Close to Equilibrium

All experiments reported so far were performed for calcium concentrations less than about 90% of the equilibrium concentration c_{eq} with respect to calcite. However, early experiments reviewed by Dreybrodt (1988), revealed that close to equilibrium natural carbonate minerals exhibit much lower rates than those predicted by the PWP rate equation.

Palmer (1991) analyzed the experimental data of Plummer et al. (1978). These were obtained from batch experiments where small calcite particles were stirred turbulently in a CO_2-containing solution with V/A>0.1 cm. By this way, the rates are controlled by surface reactions, which thus can be investigated in such an experiment. By this analysis Palmer (1991) showed that these surface-controlled rates could be described by an empirical rate law

$$F = k_{n1}(1 - c/c_{eq})^{n1} \quad \text{for } c < c_{sw} \quad (8a)$$

$$F = k_{n2}(1 - c/c_{eq})^{n2} \quad \text{for } c \geq c_{sw} \quad (8b)$$

Thus above a switch concentration $c_{sw} \approx 0.8\, c_{eq}$ the kinetics switch from a nonlinear rate law with $n_1 \approx 2$ to a higher order $n_2 \approx 4$. Such rate laws have been verified by Svensson and Dreybrodt (1992) for natural carbonate minerals in batch experiments with open system conditions for $p_{CO_2} = 5 \cdot 10^{-3}$ atm and $1 \cdot 10^{-2}$ atm at 20°C. Recently Eisenlohr et al. (1997) have verified equation 8 also for closed-system conditions in similar batch experiments.

The experimental results for the open system are illustrated in 4.1.2 Figure 8. Here we have plotted logF versus log($1-c/c_{eq}$). In this way all rate laws following the nonlinear behavior of equation 8 plot as straight lines. This is nicely exhibited in 4.1.2 Figure 8 for a limestone sample (L20) and a marble sample (M4). The uppermost curve in 4.1.2 Figure 8 depicts the dissolution rates which have been obtained from ultrapure synthetic NBS calcite. It is interesting to see that pure calcite exhibits linear rate laws quite different from natural samples.

To explain these data Svensson and Dreybrodt (1992) assumed that in contrast to pure calcite, natural materials are poisoned by impurities such as phosphate, which is present in most carbonate minerals. These impurities, also present at the phase boundary, can absorb Ca^{2+} ions from the solution and thus block these particular sites at the surface for further dissolution. Therefore a fraction Θ of the surface accessible to dissolution becomes inactive. The dissolution rate therefore reads

$$F = (1 - \Theta) \cdot F_{pwp} \quad (9)$$

Θ then represents an absorption isotherm and therefore must be a function $\Theta(c)$ of the calcium concentration. Detailed analysis of the ratio F/F_{pwp} shows that ($1-\Theta$) could be well described by a Fowler-Frumkin isotherm. This isotherm differs from the well-known Langmuir isotherm by the fact that the absorption enthalpy U_0 of the Ca^{2+} ions increases linearly with the degree of coverage Θ with sites inactive for dissolution.

This analysis showed that the switch from a lower to a higher order kinetics does not result from a change in reaction mechanism, as was often invoked by many researchers. Instead it is a consequence of inhibition, which could be explained with a well known physicochemical adsorption process, valid generally. Indeed, it turns out that all experimental data on dissolution rates thus far known in the literature can be described by such isotherms.

There is one problem, however, for all these data. Very close to equilibrium, the rates for natural minerals become extremely low, such that they cannot be detected experimentally. The limit of detection, when monitoring conductivity, is about $5 \cdot 10^{-11}$ mmole/(cm$^2 \cdot$ s), which corresponds to the removal of one molecular layer in about 10000 seconds. This means however, that for natural limestone measurements of rates for concentrations above $c > 0.99c_{eq}$ are not possible. Therefore the problem remains unsolved, whether equation 8b can be extrapolated to c_{eq}.

Inhibition by Intrinsic Impurities

Quite recently Eisenlohr et al. (1997) have performed similar batch experiments to measure the surface-controlled dissolution rates of two limestones taken from highly karstified regions and one marble from Carrara for a system closed with respect to CO_2. They used an initial $p_{CO_2} = 5 \cdot 10^{-2}$ atm and T=10°C. Their experiments revealed rate laws consistent with equation 8. The rate constants k_{n_1}, k_{n_2}, n_1, n_2 and c_{sw} however, seemed to be dependent on the V/A ratio used in otherwise identical experiments. For surface-controlled dissolution experiments, CO_2 control is not relevant, and therefore this is an unexpected behavior.

By extremely thorough experiments, using limestone and marble particles or plates of different sizes to obtain V/A ratios between 0.15 cm up to 9 cm, Eisenlohr et al. (1997) demonstrated convincingly that this dependence on the V/A value was systematic and reproducible.

The 4.1.2 Figure 9 depicts the experimentally observed dissolution rates under otherwise identical conditions for V/A=9 cm

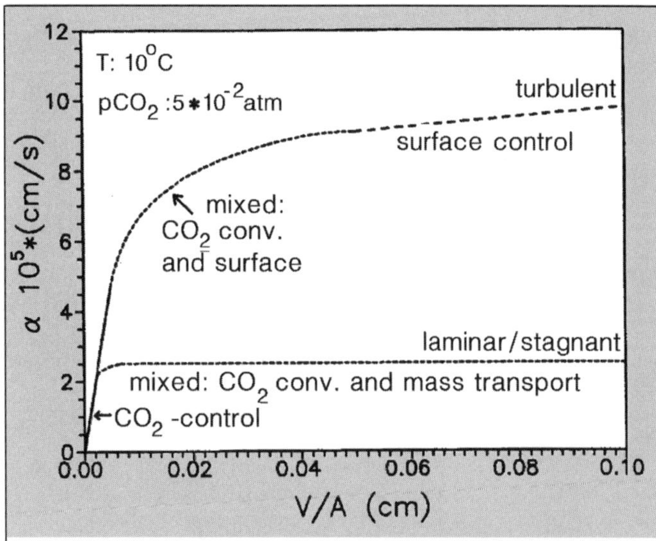

4.1.2 Figure 5. a as a function of δ = V/A for laminar flow (lower curve) and turbulent flow (upper curve). The region of pure CO_2 control is indicated by solid lines. The dashed line in the lower curve indicates mixed control by both CO_2 conversion and molecular diffusion. The dashed line in the upper curve illustrates the region where both CO_2 conversion and surface reactions are rate limiting, and the long dashed line represents the region controlled predominantly by surface reactions.

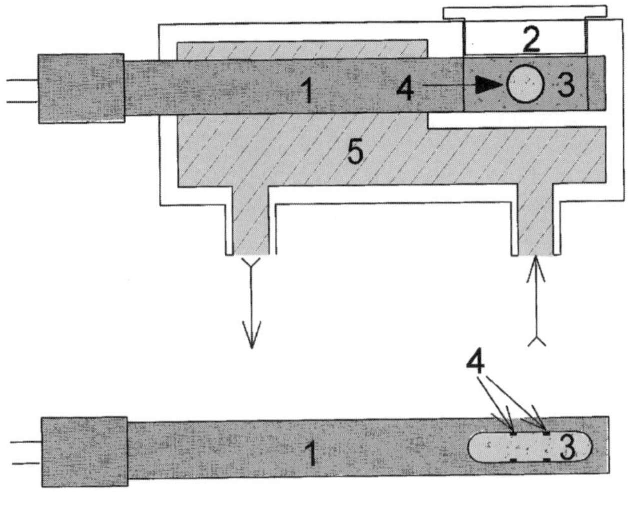

*4.1.2 Figure 6. Experimental setup. Upper drawing: The conductivity cell **1**, is surrounded by a cooling vessel **5**, using thermostated water. The space **3**, between the electrodes **4**, is filled with calcite grains of uniform size. After this space is filled with a preequilibrated solution of CO_2, the setup is sealed by the lid, **2**. Lower drawing: The WTW cell, **1**, showing the space **3**, between the electrodes **4**, in top view.*

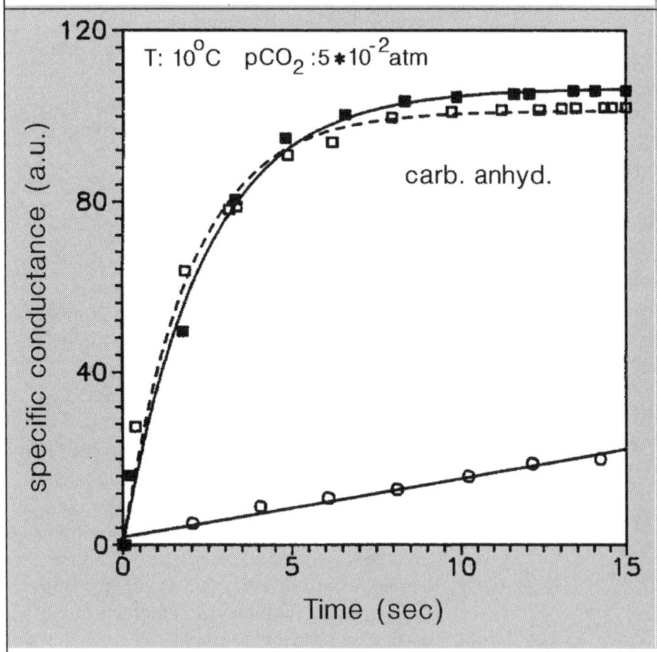

4.1.2 Figure 7. Time course of the specific conductance: Baker calcite with carbonic anhydrase (0.6 µmole/L). The upper two curves show two different experiments (t=2.2 s). The lower curve gives the time course for the same experiment without carbonic anhydrase (t=100 s).

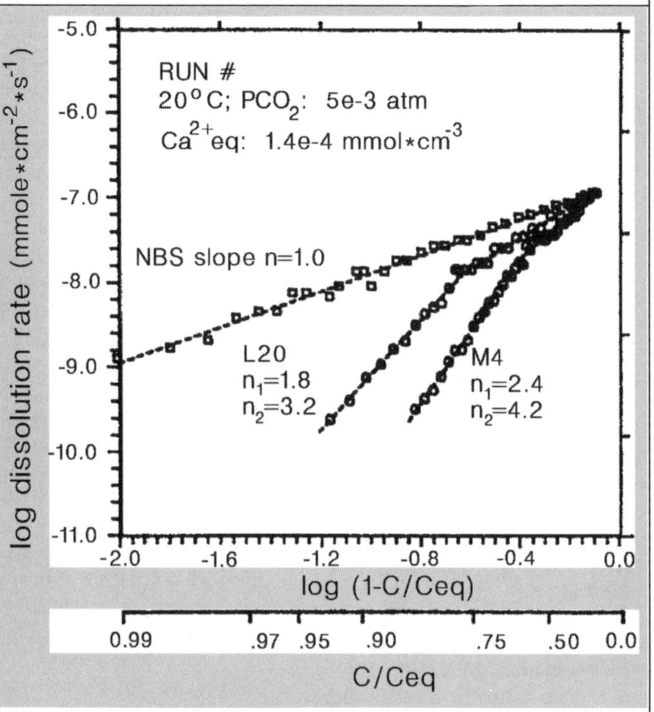

4.1.2 Figure 8. Logarithm of dissolution rates obtained from a batch experiment under open-system conditions versus log (1-c/ceq). NBS calcite exhibits a linear behavior, which is described in good approximation by the PWP equation (equation 1). Both limestone (L20) and marble (M4) exhibit inhibition and follow nonlinear rate equations (equation 8). The values of n_1 and n_2 are listed in the figure. The lower abscissa displays c/ceq.

(lower curve) and V/A=0.3 cm (upper curve). The lower curve exhibits an increase in n_2 and k_{n2}, and the rates close to equilibrium become orders of magnitude lower compared to the fresh material represented by the upper curve. The 4.1.2 Figures 10 and 11 demonstrate the dependence of n_2 and k_{n2} on the thickness d of the layer of natural carbonates, which by dissolution has been removed from the particles or the plates during the experimental run.

Values for d depend on the V/A ratio by $d=(V/A)\cdot c_{end}\cdot 0.037$, where d is in cm and c_{end} is the concentration at the end of the experiment in mole/L = mmole/cm³. In 4.1.2 Figures 10 and 11 each point represents an average of 2 to 5 identical experiments. The dashed lines simply display the observed trend of the experimental points. As can be visualized from these figures, n_2 and k_{n2} show a similar variation with the thickness d removed from the mineral. After a steep increase of n_2 and k_{n2}, within a few 10^{-4} cm the curves bend over and take constant values for $d \geq 3\cdot 10^{-4}$ cm.

Different calcium carbonate minerals used in the experiments exhibit different values for these parameters. The higher values of n_2 near 11 and $k_{n2} \approx 1\cdot 10^{-2}$ mmole/(cm² s) show that inhibition is more effective for the Jura limestone (Swiss part of the Jura Mountains) compared to the Ste. Geneviève Limestone (Mammoth Cave System, USA) and Carrara Marble (Italy), which both exhibit a simular dependence of these constants on d. These results indicate that the trends observed for the kinetic parameters apply generally for all natural calcium carbonate minerals. A similar behavior far from equilibrium is also exhibited by n_1 and k_{n1} (not shown here).

Recent Experiments
Recent dissolution experiments under open-system conditions and turbulent motion have been performed in our laboratory. The value of d was increased by repeating runs on the same limestone sample (see Svensson and Dreybrodt, 1992, for details about the setup). The results, not shown here, are in good agreement with the trends observed in 4.1.2 Figures 10 and 11. This inhibition effect in open systems with $P_{CO_2} = 1$ atm was also observed by others, but so far remained unexplained.

In a batch experiment Plummer and Wigley (1976) observed nonlinear kinetics with reaction order n_2=8. Using a rotating disk technique with V/A=60 cm Herman (1982) observed n_2=10. More recently, Svensson (1992) has carried out runs on marble and limestone particles, and plates under open-system conditions . He also observed values of n_2 up to 8.

Typical Behavior. This typical behavior of natural carbonates can be explained by the following model (4.1.2 Figure 12). One assumes that inhibitors such as Sc^{3+}, Cu^{2+}, PO_4^{3-} or others (e.g. Terjesen et al., 1961; Berner and Morse, 1974)—which are evenly distributed in the bulk of the natural carbonates—are adsorbed to the surface when they become mobile due to disintegration of the crystal lattice during dissolution. Hence the surface concentration of these inhibitors must be dependent on the thickness of the layer removed from the limestone, and therefore also on the V/A ratio used in the experiment.

This is displayed in 4.1.2 Figure 12, which shows schematically a $CaCO_3$ block with a given amount of impurities or inhibitors directly incorporated in the bulk of the mineral (small dots). By dissolution, a thin layer d is removed subsequently from the sample, and the impurities (dots) are adsorbed at the surface. Hence the surface concentration of the inhibitors increases. This is the region of accumulation, that is visualized in 4.1.2 Figure 12, by the accumulation of small dots at the surface of the $CaCO_3$ blocks.

At low d—below $3\cdot 10^{-4}$ cm—an increase in n_2 indicates the effect of this accumulation. The surface concentration of inhibitors increases with the thickness of the layer d removed by dissolution. When all possible sites for adsorption are occupied, a saturation region is reached, and the kinetic parameters show constant values. It should be noted that the saturation is reached at a depth d of about $3\cdot 10^{-4}$ cm, corresponding to about 6000 atomic layers (see 4.1.2 Figures 10 and 11). With a concentration of 10^{-4} moles of inhibiting species per mole of natural carbonate, and with all of those adsorbed at the surface, one expects a coverage of about 60%.

Auger Spectroscopy
In order to support our interpretation in 4.1.2 Figure 12, Auger spectroscopy has been employed. This method allows the determination of the concentration of elements within a few atomic layers down from the surface of the carbonate sample. For further information on Auger spectroscopy on carbonates, the reader is referred to Mucci et al. (1985).

With this method we have measured the surface chemical composition of freshly broken limestone and of limestone from which a layer with $d \approx 10^{-3}$ cm was removed by dissolution. We have found a tenfold accumulation of silicate ions at the surface of the dissolved limestone compared to the surface of the fresh sample. A weak phosphate line, however, did not exhibit any change. Through etching by Argon-ion sputtering, a depth profile of the silicate concentration reveals that it lessens until it reaches the value of the fresh sample at a depth of 10 to 14 atomic layers. We therefore infer that silicate complexes adsorbed from the interior of calcite act as inhibitors.

An important consequence follows—The rate laws observed depend on the history of dissolution. From 4.1.2 Figure 12, it should be stressed that the dissolution rates obtained from freshly broken $CaCO_3$ samples under laboratory conditions do not reflect the dissolution rate expected under natural conditions, i.e. field situations where layers of $d>10^{-3}$ cm have already been removed. This might be one reason, why so many conflicting experimental data have been observed. Therefore the dissolution kinetics in the inhibited region with $6 \leq n_2 \leq 11$ is more appropriate in modeling the early evolution of karst conduits (see Chapters 4.2.2 and 4.2.4). Up to now, dissolution rates with n_2=4 have been used in all these modeling efforts.

We therefore have used the model of Dreybrodt (1996) to calculate breakthrough times for an early karst channel with a length L of 1 km consisting of a planar parallel fracture with an initial aperture width a_0 from $5\cdot 10^{-3}$ cm up to $3\cdot 10^{-2}$ cm, subject to a hydraulic head h=50 m. The breakthrough times mark the end of early conduit evolution by a dramatic increase in the flow rates and are a measure of the intensity of karstification (Dreybrodt, 1988, 1990, 1996; Dreybrodt and Gabrovšek, Chapter 4.2.2 in this volume). In 4.1.2 Figure 13 the short dashed curves give the breakthrough times for the inhibited kinetics with n_2 =11, the long dashed curves for n_2 = 6 taken from our experiments, for various values of a_0 and (L^2/h). The full lines illustrate the breakthrough times for the kinetics with n_2 =4 taken from the experimental data obtained on freshly broken material.

From this figure one can read that when karstification times are extremely large for kinetics with n_2 =4 (e.g. $a_0 = 5\cdot 10^{-3}$ cm, l=10 km, and h=50 m), the breakthrough times are reduced by about one order of magnitude upon the action of the inhibition effect (n_2 =11).

Thus the evolution of karst aquifers over large distances of several km in joints with small aperture widths of about 50 μm becomes feasible at time scales of less than million years. This shows that karstification is much less restricted than believed up to now, where aperture widths of at least 0.01 cm were supposed as a necessary condition for the evolution of large karst aquifers in geologically reasonable times.

Summarizing—The surface-controlled dissolution rates of limestone, marble, and other natural calcium carbonate minerals are given by the empirical rate laws in equation 8. These result from inhibition of the elementary reactions (III) at the surface of the mineral. Experimental data are not available for concentrations above about $0.9c_{eq}$. Therefore it remains questionable whether equation 8b can be extrapolated to c_{eq}. During dissolution, impurities incorpo-

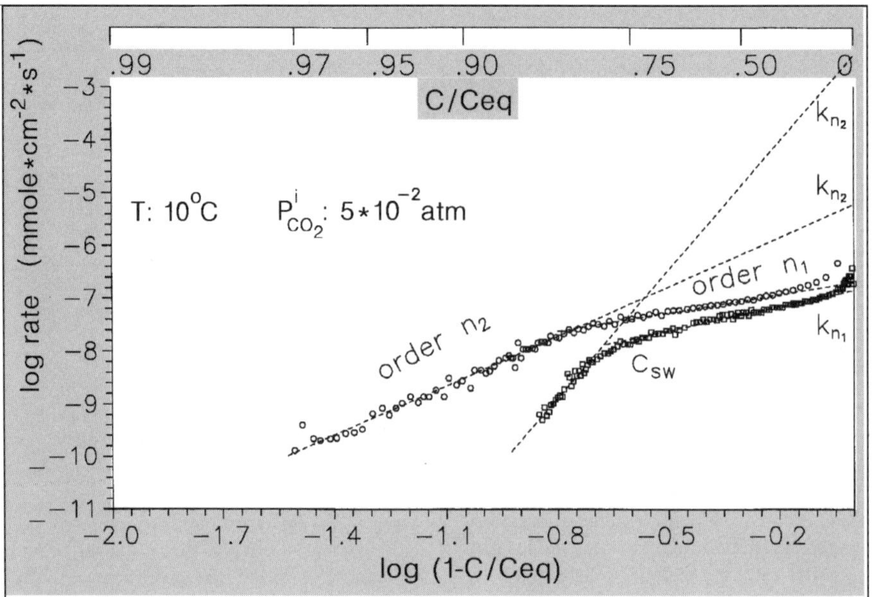

4.1.2 Figure 9. Logarithm of dissolution rates obtained from a batch experiment under closed-system conditions on Jura Limestone (Canton of Neuchâtel, Switzerland) versus log $(1\text{-}c/c_{eq})$. The upper curve (open circles) displays the dissolution rate for V/A =0.3 cm ($d \approx 0.3 \cdot 10^{-4}$ cm), n_1=1.1, and n_2=3.1. The lower curve (open squares) shows the dissolution rate for V/A=9 cm ($d \approx 6 \cdot 10^{-4}$ cm), n_1=1.6 and n_2=8.5). The upper abscissa displays c/c_{eq}.

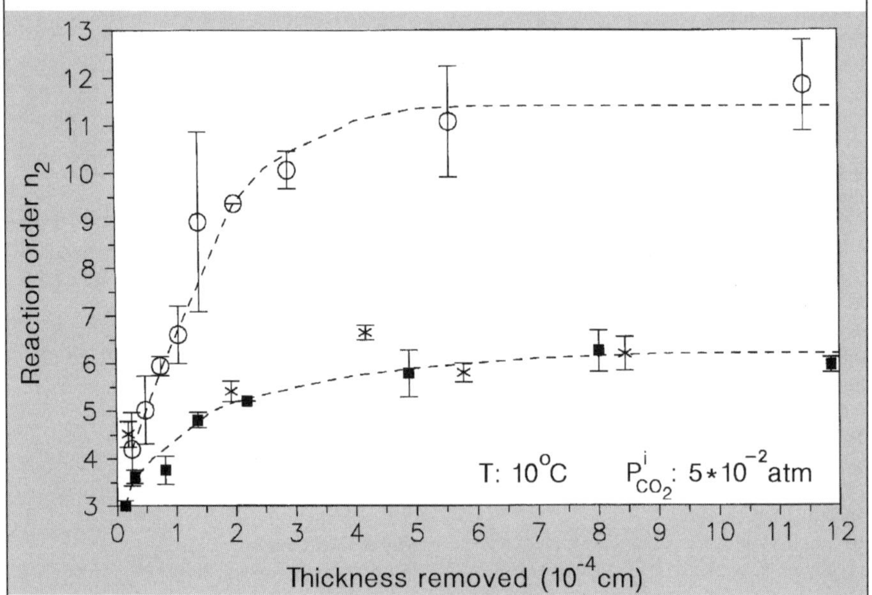

4.1.2 Figure 10. Reaction order n_2 as a function of the thickness d removed by dissolution from the surface of natural calcium carbonates. The open circles refer to Jura Limestone (Canton of Neuchâtel, Switzerland), solid squares to Ste. Geneviève Limestone (Kentucky, USA), and the stars to Carrara Marble (Toscana region, Italy).

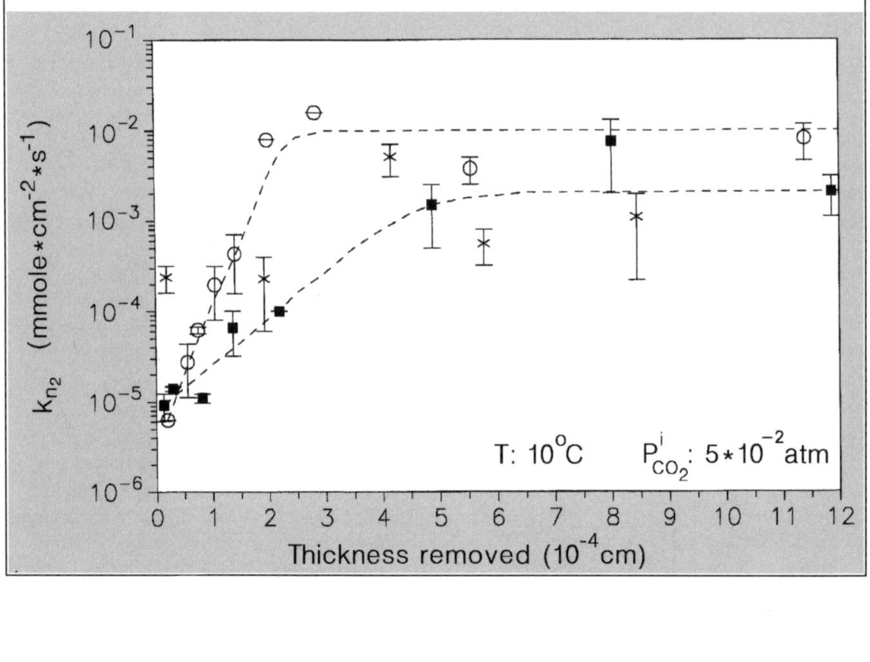

4.1.2 Figure 11. Reaction order k_{n_2} as a function of the thickness d removed by dissolution from natural calcium carbonates. The open circles refer to Jura Limestone (Canton of Neuchâtel, Switzerland), solid squares to Ste. Geneviève Limestone (Kentucky, USA), and the stars to Carrara Marble (Toscana region, Italy).

rated into the lattice of the mineral are concentrated on the surface acting there as inhibitors. Therefore the rate constants and exponents in equation 8 can change as dissolution proceeds. In any case, estimating dissolution rates by use of the PWP equation will give wrong results.

Dissolution Kinetics of Limestone in the Presence of SO_4^{2-} and Mg^{2+}

Karst water most commonly contains SO_4^{2-} and Mg^{2+} ions derived from gypsum, magnesian calcite or dolomite. These ions have been considered to act as inhibitors to calcite dissolution (Sjöberg, 1978; Mucci et al., 1989; Berner, 1967). However, the experiments providing this evidence were performed on synthetic or pure calcite. The question up to now is still open whether these ions do also influence the dissolution kinetics of natural limestone. We therefore have performed experiments to measure the dissolution rates of limestone using initial solutions saturated with a $p_{CO_2} = 5 \cdot 10^{-2}$ atm which additionally contained a fixed concentration of $CaSO_4$ or $MgCO_3$. The experiments were performed as batch experiments under closed-system conditions.

In 4.1.2 Figure 14A (p. 146), the dissolution rates of Jura Limestone for an initial solution are depicted:

- **a**—Without $CaSO_4$.
- **b**—0.5 mmole/L $CaSO_4$.
- **c**—1 mmole/L $CaSO_4$.

Within the limits of accuracy, there is a parallel shift of the curves to the new equilibrium concentrations, but no evident change in the rates. By plotting these data in a double logarithmic plot versus $(1\text{-}c/c_{eq})$ this becomes even more clear (4.1.2 Figure 14B). The three curves are very similar. This demonstrates that equation 8 is valid also in the presence of SO_4^{2-} in concentrations typical for karst water. To calculate c_{eq} in the presence of SO_4^{2-}, we have employed the methods as described in Chapter 4.1.1.

We have also performed such experiments in the presence of 0.5 mmole/L and 1 mmole/L of $MgCO_3$ and have obtained analogous results. In summary, we state that the empirical relation 8 of the pure system applies also to solutions containing SO_4^{2-} and Mg^{2+} and is therefore generally valid for real karst water.

Revisiting the Dissolution Model for Natural Carbonate Minerals

As we have seen above in the section on Modeling Dissolution Rates, the dissolution rates for a solution of a given chemical composition in its bulk depend heavily on external parameters, such as the depth of the water layer covering the mineral and the type of flow, i.e. laminar or turbulent. Only with fully developed turbulent flow and sufficiently large V/A>0.1 cm are the surface reactions rate controlling and the PWP equation applicable.

In all other cases, where CO_2 conversion or diffusional mass transport are rate limiting, one has to rely on a sound theoretical model, because experimental data are not easily available. Furthermore, for natural minerals, the model of Buhmann and Dreybrodt (1985a,b) has to be modified to account for the modified surface reactions (cf. equation 9). To obtain (1-Θ) one may use empirical rate laws F, either obtained experimentally or by means of equation 8. Then (1-Θ) is given by equation 9. The model is therefore modified by multiplying the surface reaction rate of equation 1 with this factor (1-Θ). This has the following consequences:

- Far from equilibrium, (1-Θ) is on the order of 1, and the dissolution rates will not be modified if CO_2 conversion or diffusional mass transport are rate limiting.
- Close to equilibrium, however, (1-Θ) becomes much less than 1, and in this case the rates become predominantly controlled by the surface reactions (cf. 4.1.2 Figure 2).

This is illustrated in 4.1.2 Figure 15. It shows the theoretical dissolution rates (full line) of Ste. Geneviève Limestone from the area of the Mammoth Cave System, Kentucky, U.S.A, for laminar flow in a joint with a width of 0.2 mm and inhibition of surface rates is taken into account. Initial $p_{CO_2} = 5 \cdot 10^{-2}$ atm and T=10° C.

To obtain (1-Θ) we have measured the surface reaction rates in a batch experiment. These data could be fitted to equations 8:

$k_{n1} = 1.58 \cdot 10^{-7}$ mmole/(cm^2 s),
$k_{n2} = 1.12 \cdot 10^{-5}$ mmole/(cm^2 s),
$n_1 = 1.04$, $n_2 = 3.3$, and
$c_{sw} = 0.83$)

The solid line demonstrates the theoretical dissolution rates as they occur in the joint, if inhibition is also taken into account by introducing (1-Θ) as obtained from the data above.

The surface rates obtained from the batch experiment are depicted by data points. At $c/c_{eq} \leq 0.83$ the theoretical rates exhibit a linear rate law and are lower than the experimental data. In this region, CO_2 conversion and diffusional mass transport are rate limiting, such that lithologic properties of the rock, which may determine surface reaction rates, are suppressed. This is at variance with observations of Buhmann and Dreybrodt (1985a). They found that stagnant water layers with thicknesses between 0.02 cm up to 0.15 cm open to an atmosphere with $p_{CO_2} = 5 \cdot 10^{-3}$ did not exhibit any difference in dissolution rates for various natural limestone and also for a single crystal of pure Iceland spar. Close to equilibrium, dissolution in the joint becomes controlled by surface reactions, and therefore the theoretical rates are equal to those obtained from the batch experiment.

For a further illustration of the influence of inhibition, the dashed line in 4.1.2 Figure 15 displays the results of the model of Buhmann and Dreybrodt (1985a,b), which does not take into account surface inhibition. Far from equilibrium, the rates are close to that of the model incorporating inhibition. Close to equilibrium, however, the deviations are very significant. Therefore, as a satisfactory approximation, for $c \leq c_{sw}$ one may use the results of this model and for $c > c_{sw}$ the rates obtained from batch experiments are valid.

Conclusion

In modeling the evolution of karst, dissolution rates play a crucial role. Therefore care has to be taken to consider the dependence of dissolution rates on the specific geologic situation. The early evolution of karst aquifers is characterized by laminar water flow in narrow fractures with aperture widths of about 0.1 mm or less. Therefore—as one can read from 4.1.2 Figure 15—one has to use linear kinetics, i.e. $n_1 = 1$, far from equilibrium. Close to equilibrium, it is appropriate to use the data obtained from batch experiments.

Such an approach has been taken by Dreybrodt (1988, 1990, 1996) when modeling the initial evolution of a karst conduit. Palmer (1991) in similar models, however, used nonlinear kinetics as obtained from batch experiments also far from equilibrium with $n_1 \approx 2$. In spite of this, both models yielded similar results. The reason is that natural karstification is determined by the initial dissolution rates at the exit of the fracture developing into a karst conduit. A long gestation time, during which the flow rate through the fracture increases only slowly, is terminated by a rapid increase of flow.

This breakthrough is caused by an exploding enlargement of the fracture at its exit. In a thorough mathematical analysis Dreybrodt (1996) found an analytical expression for the time at which breakthrough happens. He showed that this time depends critically on n_2, c_{eq}, and k_{n2}, but not on n_1 or k_{n1} (see also Dreybrodt and Gabrovšek Chapter 4.2.2 in this volume). Therefore, when modeling natural karst systems, it is advisable to obtain n_2, k_{n2}, and

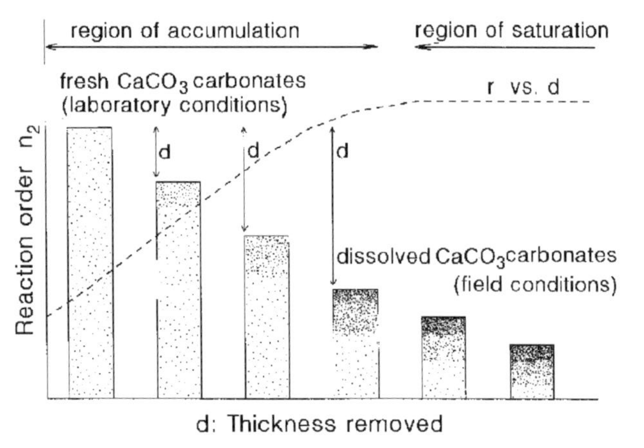

4.1.2 Figure 12. Schematic representation of surface accumulation of inhibitors from the bulk of the mineral during dissolution of natural $CaCO_3$.

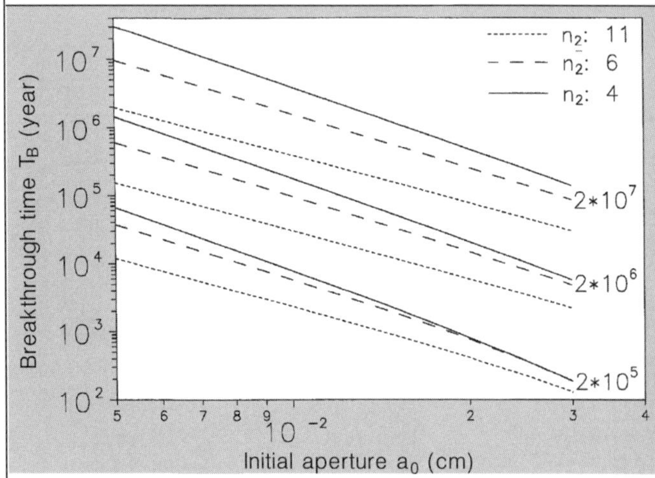

4.1.2 Figure 13. Breakthrough time as a function of initial aperture a_0 widths for kinetics with n_2=4 (dissolution for fresh Jura Limestone: solid lines); n_2=6 (nonfresh Ste. Geneviève Limestone: long dashed lines) and n_2=11 (nonfresh Jura Limestone: short dashed lines). The numbers on the curves refer to various values of l^2/h in cm.

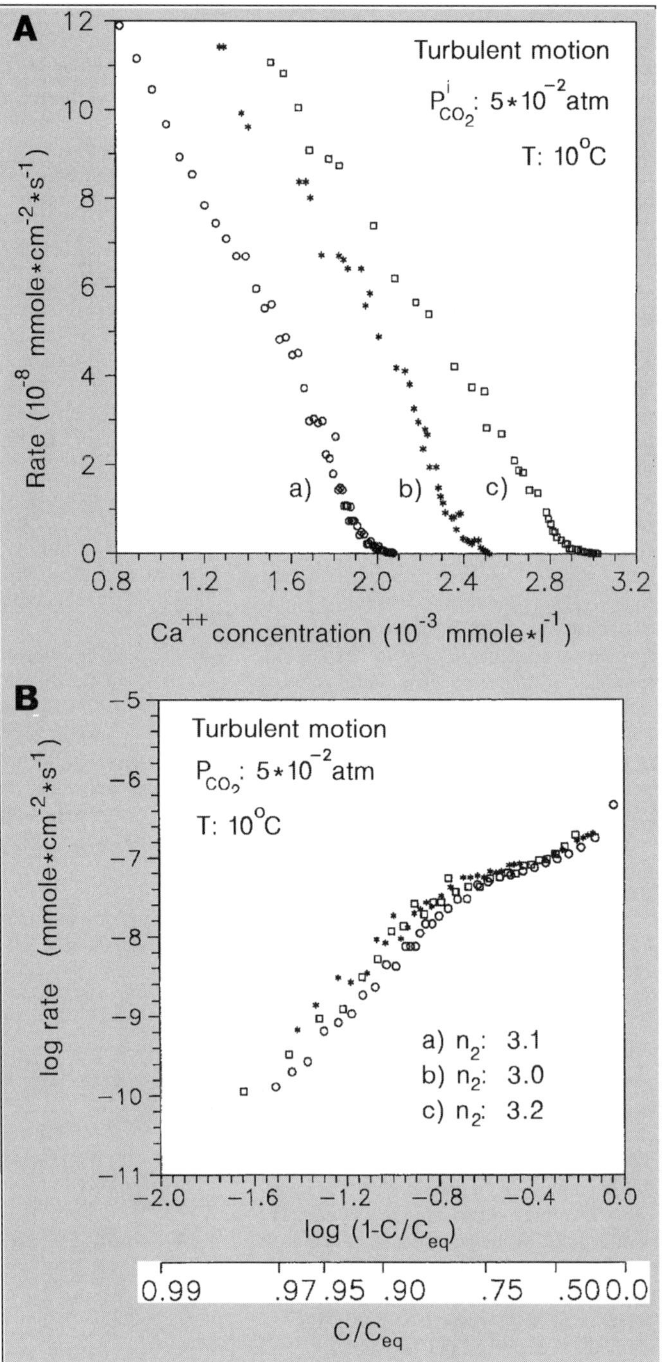

*4.1.2 Figure 14. **A:** Dissolution rates obtained from a batch experiment under closed-system conditions for Jura Limestone in the pure H_2O-CO_2-$CaCO_3$ system (a, open circles) compared to the rates when the initial solution contains 0.5 mmole/L of $CaSO_4$ (b, stars) and 1.0 mmole/L $CaSO_4$ (c, open squares), respectively.*
***B:** Logarithm of dissolution rates calculated from the plot A versus log $(1-c/c_{eq})$. The lower abscissa displays c/c_{eq}. The values of n_2 are listed in the figure. (Note that the n_2 values are similar.)*

c_{eq} from the local limestone by batch experiments.

The situation is different when modeling karst evolution under man-made conditions, where the lengths of the developing conduits are comparably short, and hydraulic gradients are extremely steep. Such a situation arises in the vicinity of dam sites. In this case the dissolution kinetics far from equilibrium also determines the evolution time of karst conduits below the dam, which might lead to water losses (Dreybrodt, 1992, 1996; Palmer, 1988). Therefore in this case first-order kinetics has to be used with rate constants derived from the theoretical model of Buhmann and Dreybrodt(1985a).

After breakthrough, flow becomes turbulent and the further evolution of conduits is determined by uniform widening, because the calcium concentration in such conduits does not differ significantly along their lengths. So far we have simulated turbulent flow by increasing the constants of molecular diffusion by several orders of magnitude. In this approximation the dissolution rates are determined entirely by surface reactions, and well-devised batch experiments will give appropriate data.

This simulation of turbulence, however, is not realistic in most natural cases, because a laminar diffusion boundary layer separates the mineral surface from the turbulent bulk of the solution. The resulting resistance to mass transport lowers the actual rates, which can decrease by almost one order of magnitude depending on the width of the boundary layer (Dreybrodt and Buhmann, 1991; Liu and Dreybrodt, 1997).

These are just a few examples. They show that care has to be taken when estimating dissolution rates in karst systems to analyze the exact conditions with respect to flow, to the ratio of the surface area A of the mineral and the volume V of solution to the temperature, and to the system conditions (open or closed with respect to CO_2). It is not only the surface kinetics provided by the PWP equation from the mechanistic model of Plummer et al. (1978) that determines the rates. Mass transport, the slow reaction $H_2O + CO_2 \rightarrow HCO_3^- + H^+$, and inhibitors in natural limestone exert substantial influence and must therefore be taken into account.

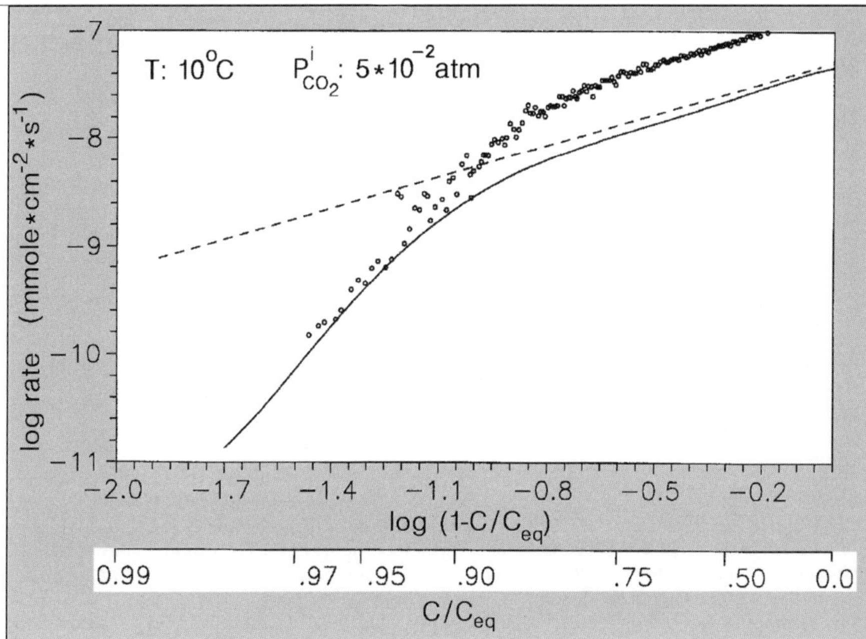

4.1.2 Figure 15. Dissolution rates (solid line) of limestone by a solution flowing in a narrow joint (δ=0.01 cm) calculated by the model employing modified surface reaction (equation 9). The data points show the rates obtained from a batch experiment in the closed system (V/A=0.27 cm, T=10° C, $p^i_{CO_2} = 5 \cdot 10^{-2}$ atm). The dashed line depicts dissolution in the joint, when the surface rates are given by the PWP rate law, i.e. when no inhibition is present.

Acknowledgement

This work was supported by grants of the Deutsche Forschungsgemeinschaft.
L.E. thanks the University of Bremen for a postdoctoral grant.

References

Baumann, J., Buhmann, D., Dreybrodt, W., and Schulz, W., 1985, Calcite dissolution kinetics in porous media: *Chem. Geol.*, v. 53, p. 219-228.

Berner, R.A., 1967, Comparative dissolution characteristics of carbonate minerals in the presence and absence of aqueous magnesium ions: *Amer. Journal Sci.*, v. 265, p. 45-70.

Berner, R.A., and Morse, J.W., 1974, Dissolution kinetics of calcium carbonate in sea water, IV: Theory of calcite dissolution: *Amer. Journal Sci.*, v. 274, p. 108-134.

Buhmann, D., and Dreybrodt, W., 1985a, The kinetics of calcite dissolution and precipitation in geologically relevant situations of karst areas: 1. Open system: *Chem. Geol.*, v. 48, p. 189-211.

Buhmann, D., and Dreybrodt, W., 1985b, The kinetics of calcite dissolution and precipitation in geologically relevant situations of karst areas: 2. Closed system: *Chem. Geol.*, v. 53, p. 109-124.

Buhmann, D., and Dreybrodt, W., 1987, Calcite dissolution kinetics in the system with participation of foreign ions: *Chem. Geol.*, v. 64, p. 89-102.

Dreybrodt, W., 1988, *Processes in karst systems–Physics, Chemistry and Geology*. Berlin and New York, Springer Series in Physical Environments, no. 4, 288 p.

Dreybrodt, W., 1990, The role of dissolution kinetics in the development of karstification in limestone: A model simulation of karst evolution: *Jour. Geology,* v. 98, p. 639-655.

Dreybrodt, W., 1992, Dynamics of Karstification: A model applied to hydraulic structures in karst terranes: *Applied Hydrogeol.*, v. 1, p. 20-32.

Dreybrodt, W., 1996, Principles of early development of karst conduits under natural and man-made conditions revealed by mathematical analysis of numerical models: *Water Resour. Res.*, v. 32, p. 2923-2935.

Dreybrodt, W., and Buhmann, D., 1991, A mass transfer model for dissolution and precipitation of calcite from solutions in turbulent motion: *Chem. Geol.*, v. 90, p. 107-122.

Dreybrodt, W., Eisenlohr, L., Madry, B., and Ringer, S., 1997, Precipitation kinetics of calcite in the system $CaCO_3 - H_2O - CO_2$: The conversion to CO_2 by the slow process $H^+ + HCO_3^- \rightarrow CO_2 + H_2O$ as a rate limiting step: *Geochim. Cosmochim. Acta* (in press).

Dreybrodt, W., Lauckner, J., Zaihua, L., Svensson, U., and Buhmann, D., 1996, The kinetics of the reaction $CO_2 + H_2O \rightarrow H^+ + HCO_3^-$ as one of the rate limiting steps for the dissolution of calcite in the system $H_2O–CO_2–CaCO_3$: *Geochim. Cosmochim. Acta*, v. 60, p. 3375-3381.

Eisenlohr, L., Madry, B., and Dreybrodt, W., 1997, Changes in the dissolution kinetics of limestone by intrinsic inhibitors adsorbing to the surface: *12th International Congress of Speleology and 6th Conference on limestone hydrology and fissured media*, La Chaux-de-Fonds, Proceedings, v. 2, p. 81-84.

Ford, D.C., and Williams, P.W., 1989, *Karst geomorphology and hydrology:* London, Unwin Hyman, 601 p.

Herman, J.S., 1982, *The dissolution kinetics of calcite, dolomite and dolomite rocks in carbon dioxide water system:* Pennsylvania State University, Ph.D. thesis, 218 p.

Liu, Z., and Dreybrodt, W., 1997, Dissolution kinetics of calcium carbonate minerals in H_2O-CO_2 solutions in turbulent flow: The role of the diffusion boundary layer and the slow reaction $H_2O + CO_2 \leftrightarrow H^+ + HCO_3^-$: *Geochim. Cosmochim. Acta*, v. 61, p. 2879-2889.

Mucci, A., Morse, J.W., and Kaminsky, M.S., 1985, Auger spectroscopy analysis of magnesian calcite overgrowths precipitated from seawater and solutions of similar composition: *American Jounal Science*, v. 285, p. 289-305.

Mucci, A., Canuel, R., and Zhong, S., 1989, The solubility of calcite and aragonite in sulfate-free seawater and the seeded growth kinetics and composition of precipitates at 25°C.: *Chem. Geol.*, v. 74, p. 309-329.

Palmer, A.N., 1988, Solutional enlargement of opening in the vicinity of hydraulic structures in karst regions: *Assoc. of Ground Water Scientists and Engineers, 2nd Conference on Environmental Problems in Karst Terranes and their Solutions*, Proceedings, p. 3-15.

Palmer, A.N., 1991, The origin and morphology of limestone caves: *Geological Society America Bulletin*, v. 103, p. 1-21.

Plummer, L.N., and Wigley, T.M.L., 1976, The dissolution of calcite in CO_2-saturated solutions at 25°C and 1 atmosphere total pressure: *Geochim. Cosmochim. Acta*, v. 40, p. 191-202.

Plummer, L.N., Wigley, T.M.L., and Parkhurst, D.L., 1978, The kinetics of calcite dissolution in CO_2-water systems at 5°C to 60°C and 0.0 to 1.0 atm. CO_2: *American Journal Science*, v. 278, p. 537-573.

Sjöberg, E.L., 1978, Kinetics and mechanism of calcite dissolution in aqueous solutions at low temperatures: *Acta Univ. Stockholm Contrib. Geol.*, v. 32, no. 1, p. 1-92.

Svensson, U., 1992, *Die Lösungsund Abscheidungskinetik natürlicher Calcitminerale in wässrigen CO_2-Lösungen nahe dem Gleichgewicht:* Bremen University, Germany, Ph.D. thesis, 123 p.

Svensson, U., and Dreybrodt, W., 1992, Dissolution kinetics of natural calcite minerals in CO_2-water systems approaching calcite equilibrium: *Chem. Geol.*, v. 100, p. 129-145.

Terjesen, S.G., Erga, O., Thorsen, G., and Ve, A., 1961, Phase boundary processes as rate determining steps in reaction between solids and liquids—The inhibitory effect of metal ions on the formation of calcium bicarbonate by the reaction of calcite with aqueous carbon dioxide: *Chem. Eng. Sci.*, v. 14, p. 267-276.

Usdowski, E., 1982, Reactions and equilibria in the systems $CO_2 - H_2O$ and $CaCO_3 - CO_2 - H_2O$ — A review: *Neues Jahrb. Mineral. Abh.*, v. 144, p. 148-171.

White, W.B., 1988, *Geomorphology and hydrology of karst terrains:* New York and Oxford, Oxford Univ. Press, 464 p.

4.1.3 Dissolution of Limestone from Field Observations

William B. White

Abstract

The overall removal of carbonate rock from karstic drainage basins can be evaluated either by direct measurement of rock surface retreat or by mass balance within known drainage basins. The dissolution rates, expressed as a uniform retreat of the land surface in mm/ka, are generally higher when measured by total rock loss from drainage basins compared with direct measurements on rock surfaces. Denudation rates can be modeled in terms of runoff, CO_2 partial pressure, and temperature, but do not discriminate well between climatic regimes.

Introduction

The Earth's surface is continuously being lowered by forces of erosion. For any given location on the surface there is a rate at which the surface altitude decreases with time. The lowering rate is, of course, offset by tectonic forces which raise the landmass. Otherwise the Earth's entire land surface would be lowered to sea level.

The normal agents of erosion—surface streams—dissect the surface into a complex network of valleys. The downcutting of valleys is episodic. When base levels are lowering, or when the landmass is rising, the streams cut relatively narrow V-shaped valleys. When base levels are relatively stable, valleys widen and develop flood plains. Renewed downcutting due to a new period of lowering base levels or rising landmass, causes new valleys to be cut and remnants of the previous valley may be left behind as upland terraces or straths.

This is the classic concept for noncarbonate rocks and dates back to the pioneering work of William Morris Davis in the late 19th century. The surface lowering of karst is rather different (4.1.3 Figure 1). A karst surface dotted with sinkholes is lowered primarily by dissolution of the bedrock and the rock mass is carried away in solution by internal runoff into the sinkholes. As a result no marker indicates the past history of the karst surface. One sinkhole plain looks much like any other sinkhole plain. Old sinkholes are eroded away and new sinkholes are developed. The overall perspective of the landscape remains constant except for a gradual decrease in altitude.

The lowering of the karst landscape is referred to as "karst denudation" and has been of interest to a great many investigators in many parts of the world. Denudation rates are a composite measure of many individual processes that take place within karst drainage basins, and there has been some disagreement as to how these should be sorted out. Further, the evolution of karst drainage basins takes place over very long periods of time, but most measurements span only a few years. There is a problem with extrapolating these short time scale measurements to the time scale of the drainage basin.

A note on units: The rate of landscape lowering has customarily been expressed in units of millimeters per thousand years, mm/ka. Direct measurements of lowering rates are given in micrometers per year, µm/a. Loss of rock as dissolved load from a drainage basin is given in cubic meters of rock per square kilometer of basin area per year, $m^3/km^2/a$. These choices of units are appropriate to the time scales of the measurements and have the additional advantage that the numerical values for denudation rates are the same in all three units.

Denudation Rate by Direct Observation

Direct Measurement of Limestone Pedestals

The most dramatic illustration of the slow retreat of limestone surfaces over time scales of tens of thousands of years is provided by limestone pedestals capped with glacial erratics of noncarbonate rock (4.1.3 Plate 1). In England, Ireland, and other regions, the retreat of the last ice sheet left a variety of boulders scattered over the underlying limestone surface. As the millennia passed, the limestone was slowly removed except for the bits protected from rainfall by the boulders. Today, the boulders appear suspended on residual pedestals of limestone. The height of the pedestal is a direct measure of the solutional retreat of the land surface (Sweeting, 1966).

Some pedestal denudation rates are shown in 4.1.3 Table 1, drawn mainly from a compilation by Jennings (1985). Pedestal measurements have limited accuracy because of the difficulty in determining pedestal height. Many pedestals are irregular, some dissolution between the top of the pedestal and the protective erratic is possible, and the reference altitude of the present-day land surface is difficult to define. Pedestal height measurements do have the overriding advantage that they span the entire Holocene and represent a time average of fluctuations in denudation rate over that period.

Direct Observation of Rock Retreat

When the karst landscape appears as a directly exposed bedrock surface, the rate of removal by rainfall, stream action, or wave action can be measured directly. The amount of rock removed on a time scale of months or years is very small but is within the range of precise micrometric measurements. The standard procedure is straightforward (High and Hanna, 1970).

Three stainless steel pins are embedded into the bedrock surface. The measurement consists of mounting a triangular steel frame with precision milled steel balls that fit exactly and very reproducibly onto the pins. In the center of the triangular frame is mounted a micrometer with a small stylus that can be lowered until it just touches the rock surface. Once the pins are installed, the observer comes by from time to time over as many years as possible and measures with the micrometer the position of the exposed bedrock surface with respect to the support pins. The retreat of the surface is on the order of micrometers per year.

Microerosion measurements are largely confined to bare surfaces on competent rock. Sources of error in the measurements include expansion and contraction of the instrument due to temperature fluctuations from one measurement to the next and damage to the rock surface by the probe tip (Spate et al., 1985). Repeated measurements at the same site show a good deal of scatter. Some data are summarized in 4.1.3 Table 2. Denudation rates are low on bare rock surfaces and considerably higher in stream channels as would be expected. There is a pronounced influence of lithology as demonstrated by Trudgill's (1976) measurements on Aldabra Atoll.

The erosion of active stream channels depends strongly on the saturation of the water. For an aggressive cave stream in County Clare, Ireland, High and Hanna (1970) found an order of magnitude decrease in the erosion rate between the swallet and the rising (4.1.3 Table 3).

Bedrock Tablets Dissolved Beneath Soil Mantle

Most karst processes of interest take place beneath a soil mantle. The background atmospheric carbon dioxide partial pressure of $10^{-3.5}$ atm produces slow denudation rates on bare bedrock surfaces. The much higher carbon dioxide pressures at the base of the soil should greatly increase the rate of bedrock removal beneath the soil mantle. Dissolution rates can be observed directly from weight loss measurements on buried rock tablets. The chosen limestone formation is cored into cylinders typically 40 mm in diameter and the cylinders are sawed into tablets 2.5 to 3.5 mm thick that are then weighed accurately and buried in a recoverable fashion in various soil environments. After a period of time, preferably a few years, the tablets are exhumed, carefully cleaned, and weighed again. The dissolution rate is then determined directly from the weight loss, the time of burial, and the surface area of the tablets.

An international project was initiated by Ivan Gams, who prepared hundreds of tablets from a single limestone in Slovenia and distributed them to interested karst researchers worldwide. At the time of a preliminary report (Gams, 1981), 101 results had been obtained. One set of these, Gams' own measurements on the Logatec Doline in Slovenia, are given in 4.1.3 Table 4.

According to these measurements, dissolution rates in the soil are not dramatically higher than those observed on bare rock surfaces.

Trudgill and his coworkers (Crabtree and Trudgill, 1985; Trudgill et al., 1994) performed the most comprehensive evaluation of the buried tablet technique. The study involved three rows of pits spaced at 10-m intervals up a slope. At each pit location, five tablets were emplaced in pits one meter apart. With eight sites along each transect, the total study involved 240 tablets, more than twice Gams' entire data base, and on one small hillside. There was a significant increase in dissolution rate upslope but the overall scatter was too large to justify calculating an absolute denudation rate.

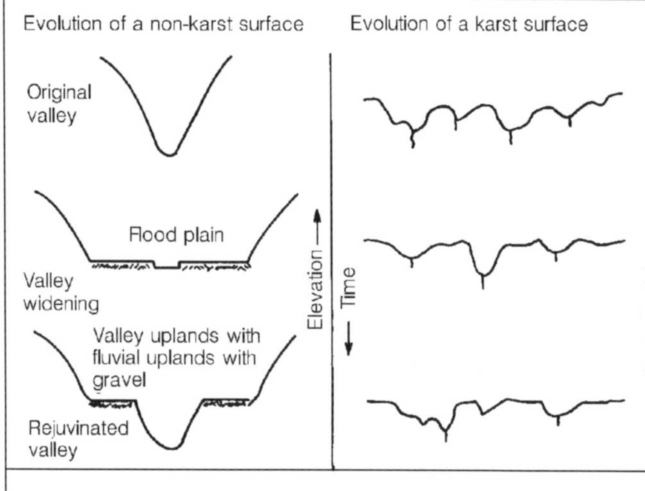

4.1.3 Figure 1. Sketch showing a denuding karst plain compared with fluvial valley downcutting.

4.1.3 Plate 1. Limestone pedestal with capping glacial erratic, County Leitrim, Ireland (photo by the author).

Location	Mean height (cm)	Elapsed time (1000 years)	D_n mm/ka
Mären Mountains., Switzerland	15	10	15
Craven, England	50	12	42
Clare-Galway, Ireland	15	12	12
County Leitrim, Ireland	51	12	42
Mt. Jaya, Irian Jaya	30	9.5	32

All data adapted from Jennings (1985), Table 4, except County Leitrim which was taken from Williams (1966).

4.1.3 Table 1. Denudation rates from bedrock pedestal measurements.

Location	Local environment	D_n (μ m/a)
Yarrangobilly and Cooleman Plain	Average bare bedrock	11
	Subsoil	12
	Cave stream	22
	Cave stream	51
	Resurgence	74
	Stream sink	200
Aldabra Atoll	Coral fragments	9
	Calcarenite	39
	Algal limestone	10
	Mean of all measurements	26

Yarrangobilly and Cooleman Plain data from Smith et al. (1995). Aldabra Atoll data from Trudgill (1975).

4.1.3 Table 2. Denudation rates by microerosion meter.

Location	Aggressivity	Meters from stream beginning	D_n (μm/a)
Stream at swallet	Very aggressive	365	500
Stream in cave	Aggressive	820	400
Stream at rising	Nearly saturated	2560	50

Data from High and Hanna (1970)

4.1.3 Table 3. Microerosion meter rates on stream bed.

Location	Depth in soil (cm)	D_n (μm/a)
Upper sunny side	3	9.5
	13	13.7
Lower sunny side	2	13.3
	13	13.7
Upper shady side	3	16.2
	13	24.6
Lower shady side	3	16.8
	13	16.5
Bottom of doline	3	33.3
	14	26.7
	50	14.8

Data from Gams (1981) from measurements on Logatec Doline, Slovenia.

4.1.3 Table 4. Denudation rates from buried limestone tablets.

Denudation Rates by Mass Balance

If one's area of interest can be delineated as a drainage basin, the average rate of land surface lowering can be determined by mass balance. The amount of water leaving the basin can be determined by a stream gage downstream from the carbonate rock outcrop area or at the spring discharging the groundwater basin. The concentration of dissolved carbonate bedrock can be determined by chemical analysis.

Knowing the amount of water leaving the basin and the concentration of dissolved carbonates, it is a straightforward procedure to calculate the amount of rock mass leaving the basin per unit time per unit area. The water leaving the surface or groundwater basin is expressed by a hydrograph, Q(t). The concentration of dissolved carbonates is expressed as a chemograph, H(t). The loss of rock from the basin is the convolution of these two functions

$$D_n = \frac{1}{N_L A} \frac{K}{\rho t_R} \int_{t_1}^{t_2} Q(t)\, H(t)\, dt \quad (1)$$

where D_n is the denudation rate in $m^3/km^2/a$, A is basin area in km^2, N_L is the fraction of the basin underlain by karstic rocks, ρ is the density of carbonate rock in g/cm^3, t_R is the period of record in years, Q(t) is the instantaneous discharge in m^3/s, and H(t) is the instantaneous (Ca + Mg) hardness in g/cm^3. The constant K contains unit conversions and has the value 10^{-12} for the units given. The time unit for integration is in seconds. Although equation 1 is exact in principle, its application to basins of complex geology and well developed karstic drainage presents many problems.

The discharge hydrographs for groundwater basins in karst tend to be "flashy," with many individual storm peaks. Likewise, the concentration of dissolved solids varies with discharge mostly because of stormwater dilution and varying inputs from sinking streams. To express the denudation rates accurately by mass balance, one must have closely spaced data or continuous records on both discharge and concentration of dissolved carbonates. The net result is illustrated in 4.1.3 Figure 2 showing dissolved load for two small karstic basins in Centre County, Pennsylvania.

Rock Spring drains a karstic Ordovician limestone aquifer along the edge of a carbonate valley bounded by a mountain ridge underlain by shale and sandstone. Of the total basin area of 14.2 km^2, 8.9 km^2 are on the mountain flanks drained by allogenic streams that sink into the carbonate. The 11.1 km^2 area of Thompson Spring lies entirely on carbonate rocks, mostly Ordovician dolomite.

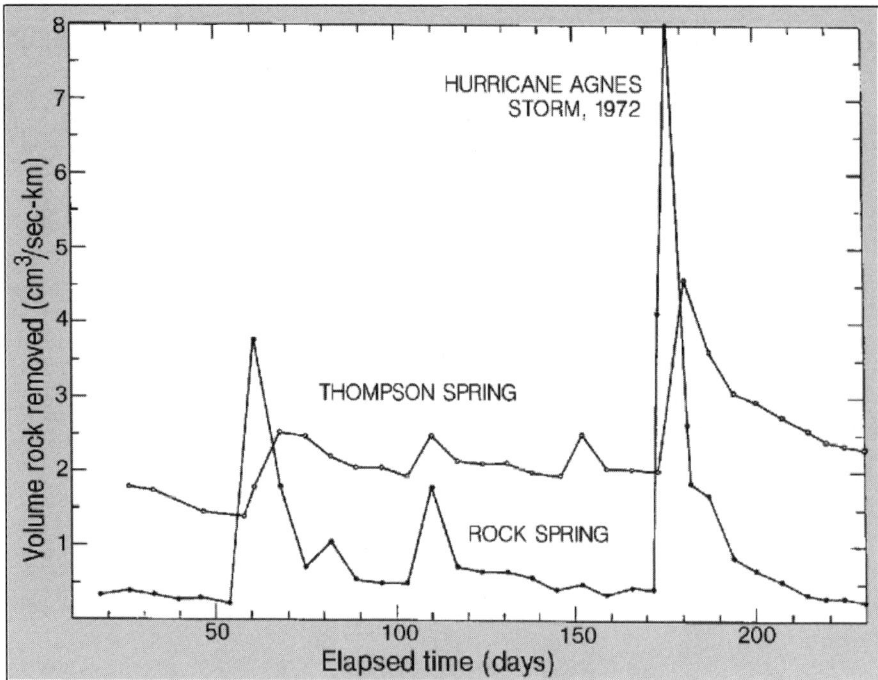

4.1.3 Figure 2. Denudation data for the Rock Spring and Thompson Spring Drainage Basins, Centre County, Pennsylvania, U.S.A. (from White, 1990).

Thompson Spring is fed mostly by fracture flow in the dolomite and is much less flashy than Rock Spring which is the outlet for a conduit.

Measurements of dissolved load at various discharges for springs in the Mendip Hills, England (Smith and Newson, 1974), the Waitomo District, New Zealand (Gunn, 1981a), and the Cooleman Plain, Australia (Jennings, 1983) show that to a good approximation the dissolved load is proportional to discharge. Specific fitting parameters for the regressions vary from basin to basin. As a practical matter, the dissolved load/discharge relationships could be worked out from chemical analysis made at various discharge stages and the resulting regression equation applied to the entire hydrograph.

Because detailed records are rarely available, many approximations have been attempted. The oldest, most simplistic, and widely used of these is Corbel's (1959) equation

$$D_n = 4\ ET/100 \quad (2)$$

where E is precipitation in decimeters and T is mean hardness in milligrams per liter. The shortcomings of the Corbel equation include the following:

- The carbonate rock density was assumed to be 2.5 g/cm^3 instead of including it as a separate parameter, but the range of densities of carbonate rocks is small and this variation is likely to be lost in the other uncertainties.
- Instead of runoff, total precipitation was taken as a measure of water flux through the basin. This simplification allows the investigator to make use of widely available rainfall records, but introduces huge errors because in most regions a substantial fraction of the precipitation is lost through evaporation and transpiration. Evapotranspiration losses do not contribute to the transport of dissolved load.
- In practice "mean hardness" has often been based on a few spot measurements of streams or springs without taking account of the dependence of hardness on discharge.
- There is no accounting for how much of any given drainage basin is karstic or how large a contribution of dissolved load is contributed by allogenic runoff from the nonkarstic portion of the basin.

The last point causes the most trouble. Of the terms in equation 1, specification of N_L is most difficult. Naively, it might be assumed that one could draw the basin drainage divide on a geologic map, scale off the total area of the basin, and proportion this between the fraction of exposed carbonate rock outcrop and the fraction of exposed noncarbonate rock outcrop. In practice there are the following complications:

- Nonkarstic rocks may contain carbonate minerals and thus contribute a dissolved carbonate load by way of sinking streams.
- The carbonate rock fraction of the basin may contain both limestone and dolomite, which have quite different weathering characteristics.
- Karstic rocks may be overlain by a caprock so that the surface outcrop does not accurately delineate the fraction of the basin that is karstic.
- As pointed out by Gunn (1981b), groundwater moving beneath a protective caprock may function as a chemically closed system and thus have a lower saturation concentration of dissolved carbonates.

In view of these difficulties, researchers have modified the Corbel equation, mostly as follows:

- By including the density of limestone as a parameter to be specified.
- By including a correction for the fraction of the drainage basin occupied by carbonate rocks.
- By attempting a correction for additional dissolved load brought in by sinking streams.
- By using either measured runoff or at least precipitation minus evapotranspiration as the measure of water flux through the basin.

Published denudation rates do not all contain the same corrections so that intercomparisons are difficult.

Smith and Atkinson (1976) produced a comprehensive review of the denudation measurements reported up to that time. They separated the measurements into an alpine-arctic set, a temperate-climate set, and a tropical-climate set, plotted these against the runoff, and arrived at the following set of regression equations.

- Arctic-alpine areas not adjusted for proportion of limestone

$$D_n = 0.0356R + 7.41 \quad r^2 = 0.767 \quad N = 18$$

- Arctic-alpine areas adjusted for proportion of limestone

$$D_n = 0.1122R + 14.33 \quad r^2 = 0.956 \quad N = 4$$

- Temperate areas not adjusted for proportion of limestone

$$D_n = 0.0552R + 7.93 \quad r^2 = 0.732 \quad N = 51$$

Temperate areas adjusted for proportion of limestone

$$D_n = 0.1604R - 22.44 \quad r^2 = 0.505 \quad N = 14$$

- Tropical areas not adjusted for proportion of limestone
$$D_n = 0.0633R + 5.69 \quad r^2 = 0.720 \quad N = 14$$

where D_n is denudation rate in $m^3/km^2/a$ and R is mean annual runoff in mm. The effect of not adjusting basin areas for noncarbonate rock is apparent in the regression coefficients where denudation rates are about a factor of two larger when proportion of limestone is taken into account.

If runoff is taken equal to precipitation minus evapotranspiration, a maximum value for the denudation rate can be calculated from the saturation hardness (White, 1984).

$$D_{max} = 10^{-3} \frac{H_s}{\rho} (P - E) \qquad (3)$$

The saturation hardness is a constant for a given CO_2 pressure and (P - E) is the total runoff expressed in mm/a. The overall chemical reaction for the dissolution of limestone is

$$CaCO_3 + CO_2(gas) + H_2O \Rightarrow Ca^{2+} + 2HCO_3^-$$

The equilibrium for the overall reaction can be written

$$\frac{a^2_{HCO_3^-}\, a_{Ca^{2+}}}{P_{CO_2}} = \frac{K_C\, K_1\, K_{CO_2}}{K_2} \qquad (4)$$

where K_C, K_{CO_2}, K_1, and K_2 are the solubility product constant for calcite, the Henry's law constant for CO_2, and the first and second ionization constants for carbonic acid respectively. These constants are known quite precisely (Plummer and Busenberg, 1982). For the pure carbonate system electroneutrality demands that $[HCO_3] = 2\,[Ca^{2+}]$. With the electroneutrality condition and converting activities to concentrations, equation 4 can be solved for $[Ca^{2+}]$.

$$[Ca^{2+}] = \left(\frac{K_C\, K_1\, K_{CO_2}}{4\, K_2\, \gamma^2_{HCO_3^-}\, \gamma_{Ca}}\right)^{\frac{1}{3}} P^{\frac{1}{3}}_{CO_2} \qquad (5)$$

The hardness of limestone-saturated water H_s is directly proportional to the calcium ion concentration (using a rounded 100 as the molecular weight of $CaCO_3$)

$$H_s = 10^5 [Ca^{2+}] \qquad (6)$$

Substituting equation 6 into equation 3 gives a general expression for the maximum denudation rate.

$$D_{max} = \frac{100}{\rho}\left(\frac{K_C\, K_1\, K_{CO_2}}{4\, K_2\, \gamma^2_{HCO_3^-}\, \gamma_{Ca}}\right)^{\frac{1}{3}} P^{\frac{1}{3}}_{CO_2}\,(P - E) \qquad (7)$$

Equation 7 assumes that the water discharging from the karst is at equilibrium. An exact calculation requires the activity coefficients for Ca^{2+} and HCO_3^-, which in turn requires knowing the ionic strength of the water. Because the activity coefficients enter as a cube root, simply setting them equal to unity (as was done by White, 1984) will decrease the calculated D_{max} by an amount on the order of 5 to 20 percent, which offsets the typical undersaturation of karst springs.

Equation 7 has no empirical parameters. It combines in a single equation the climatic variables of temperature, CO_2 partial pressure, and precipitation. It gives the following predictions for denudation rates:

- Denudation is predicted to vary linearly with precipitation corrected for evapotranspiration or with runoff as indicated by Smith and Atkinson's (1976) linear regression lines.
- Denudation rate varies with the cube root of the CO_2 partial pressure. CO_2 pressures range from 0.00035 atm in the open atmosphere to 0.03 to 0.1 atm in organic-rich soils, a range of about a factor of 100. With the cube root dependence, this amounts to only a factor of about five in denudation rate. This may account for the lack of strong discrimination between temperate and tropical sites in Smith and Atkinson's regression lines.
- The temperature dependence of the denudation rate is contained in the temperature dependence of the equilibrium constants. All other parameters being equal, denudation rate would indeed be higher at low temperature, but as the temperature rises from 5 to 25°C, the denudation rate decreases by only about 30%, an effect that is easily lost in the variations of CO_2 pressure and precipitation.

Plots of karst denudation rates determined from mass balance calculations have been published by Atkinson and Smith (1976) and by Jennings (1983). Some data are summarized in 4.1.3 Figure 3 along with Smith and Atkinson's regression lines for three climatic zones and the theoretical line for temperate-climate karst (T = 12°C; $P_{CO_2} = 10^{-2.3}$) (0.5 volume percent CO_2) from equation 7.

The theoretical curve parallels the temperate-climate regression curve. It nearly intersects a measurement of denudation in Gunong Mulu National Park, Malaysia (Sweeting, 1979), of 180 mm/ka at a runoff of 3460 mm/a. In spite of the assumptions that went into the calculation, the simple theoretical model given by equation 7 seems to be a good representation of observed denudation rates. The regressions presented by Smith and Atkinson for denudation rates corrected for the fraction of limestone in the drainage basin do not fit as well to the measurements, especially those obtained in

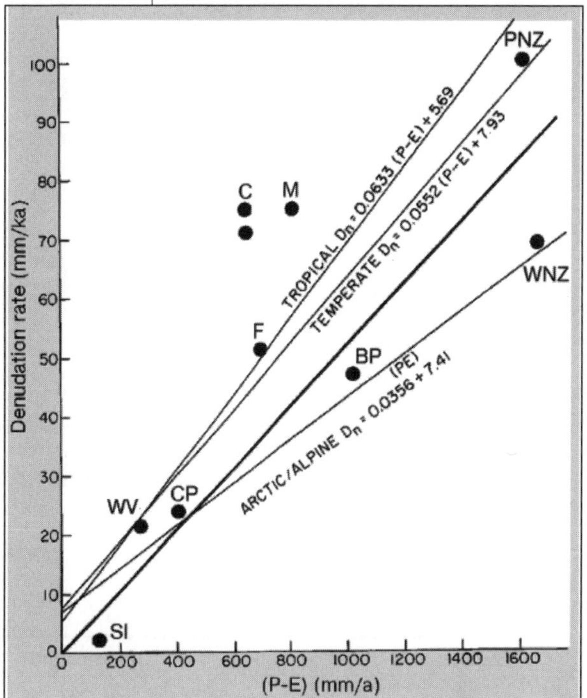

*4.1.3 Figure 3. Theoretical denudation rate (equation 7) compared with the regression lines of Smith and Atkinson (1976). Key for plotted data: **SI** is Somerset Island, Canada (Smith, 1972). **WV** is the means of three springs in Monroe County, West Virginia, U.S.A. (Ogden, 1982). **CP** is the Cooleman Plain, Australia (Jennings, 1978). **F** is the Fergus River Basin, Ireland (Williams, 1968). **BP** is the Pirin region, Bulgaria (Markowicz et al., 1972). **C** is the Thompson Spring (upper point) and Rock Spring, Centre County, Pennsylvania, U.S.A. (White, 1990). **M** is the Mendip region, England (Smith and Newson, 1974). **WNZ** is the Waitomo region, New Zealand (Gunn, 1981). **PNZ** is the Pikikiruna Range, New Zealand (Williams and Dowling, 1979).*

areas with high runoff rates.

Equation 7 allows the coefficients of Smith and Atkinson's regression equations to be interpreted in terms of temperature and carbon dioxide pressure. Assuming reasonable temperature values of "arctic-alpine," "temperate," and "tropical" climates, the CO_2 pressures were calculated (4.1.3 Table 5). The results seem in reasonable agreement with what would be expected for these environments.

Distribution of Dissolution within Karst Drainage Basins

It is of interest to know how the removal of carbonate rock is distributed through karst drainage basins. Some will be removed at the soil-bedrock contact, some deeper in the epikarst, and some in the still deeper fracture and conduit system. Several authors' attempts to partition the dissolutional activity are summarized in 4.1.3 Table 6. In all cases, the bulk of the dissolution takes place at the soil-bedrock contact, including the epikarst. Slow percolation through the soil, combined with days to weeks of storage within the epikarst, gives the infiltrating water sufficient time to approach equilibrium with respect to the carbonate minerals. Probably this is the reason the geochemical model works as well as it does. At the exit of a drainage basin, the largest part of the dissolved carbonate load contributed to the surface lowering of the karst landscape. Relatively little of it was derived from the enlargement of cave passages.

Summary and Conclusions

For more than 40 years researchers have been trying to define a single number that describes the overall rate of dissolution of complete drainage basins or entire karst regions. Denudation rates can be approached by detailed measurements on individual rock surfaces or by mass balance across drainage basins. All of these approaches make sense and give roughly accurate results that are consistent with theoretical expectations. Attempts to obtain an order of magnitude improvement have been unsuccessful. The most productive approach seems to be detailed dissolved-load and runoff measurements of small conservative basins.

	Arctic/Alpine	Temperate	Tropical
Assumed mean temperature	0 °C	10 °C	20 °C
Regression coefficient*	0.036	0.055	0.063
Calculated P_{CO_2} (atm)	$10^{-3.07}$	$10^{-2.31}$	$10^{-1.93}$

Data from White (1990).

*Regression coefficients derived by Smith and Atkinson (1976).

4.1.3 Table 5. Carbon dioxide pressures calculated from denudation rates.

Location	D_n (μm/a)	Distribution	Reference
Cooleman Plain, Australia	24	75% surface; 20% conduit; 5% covered karst	Jennings (1972)
River Fergus, Ireland	51	80% in top 8 m	Williams (1968)
Waitomo Region, New Zealand	69	37% in soil profile; 85% in soil and epikarst	Gunn (1981)
Mendips, England	50–102	1.3% in surface streams; 0.05% in cave passages; 3.6% in soil above rock; 60.1% at soil–rock interface; 35.0% in rock interstices	Smith and Newson (1974)
Pikikiruna Range, New Zealand	100	80% in top 10-30 m of marble outcrop	Williams and Dowling (1979)

4.1.3 Table 6. Distribution of dissolution within basin.

References

Atkinson, T.C., and Smith, D.I., 1976, The erosion of limestones, *in* Ford, T.D., and Cullingford, C.H.D., eds., *The Science of Speleology:* London, Academic Press, p. 151-177.

Crabtree, R.W., and Trudgill, S.T., 1985, Chemical denudation of a magnesian limestone hillslope, field evidence and implications for modeling: *Earth Surface Processes and Landforms*, v. 10, p. 331-341.

Corbel, J., 1959, Vitesse de l'erosion: *Zeitschrift für Geomorphologie*, v. 3, p. 1-28.

Gams, I., 1981, Comparative research of limestone solution by means of standard tablets: *8th International Congress of Speleology*, Bowling Green, Proceedings, p. 273-275.

Gunn, J., 1981a, Limestone solution rates and processes in the Waitomo District, New Zealand: *Earth Surface Processes and Landforms*, v. 6, p. 427-445.

Gunn, J., 1981b, Prediction of limestone solution rates from rainfall and runoff data: Some comments: *Earth Surface Processes and Landforms*, v. 6, p. 595-597.

High, C., and Hanna, F.K., 1970, A method for the direct measurement of erosion on rock surfaces: *British Geomorphological Research Group Technical Bulletin*, no. 5, 24 p.

Jennings, J.N., 1972, Observations at the Blue Waterholes, March 1965-April 1969, and limestone solution on Cooleman Plain, N.S.W.: *Helictite*, v. 10, p. 3-46.

Jennings, J.N., 1983, Karst landforms: *American Scientist*, v. 71, p. 578-586.

Jennings, J.N., 1985, *Karst Geomorphology:* Oxford, Basil Blackwell, 293 p.

Markowicz, M., Popov, V., and Pulina, M., 1972, Comments on karst denudation in Bulgaria: *Geographia Polonica*, v. 23, p. 111-139.

Ogden, A.E., 1982, Karst denudation rates for selected spring basins in West Virginia: *National Speleological Society Bulletin*, v. 44, p. 6-10.

Plummer, L.N., and Busenberg, E., 1982, The solubilities of calcite, aragonite, and vaterite in CO_2-H_2O solutions between 0 and 90°C, and an evaluation of the aqueous model for the system $CaCO_3$-CO_2-H_2O: *Geochimica Cosmochimica. Acta*, v. 46, p. 1011-1040.

Smith, D.I., 1972, The solution of limestone in an Arctic environment: *Institute of British Geographers Special Publication*, no. 4, p. 187-200.

Smith, D.I., and Newson, M.D., 1974, The dynamics of solutional and mechanical erosion in limestone catchments on the Mendip Hills, Somerset, *in* Gregory, K.J., and Walling, D.E., eds., *Fluvial Processes in Instrumented Watersheds:* Institute of

British Geographers Special Publication, no. 6, p. 155-167.

Smith, D.I., and Atkinson, T.C., 1976, Process, landforms and climate in limestone regions, *in* Derbyshire, E., ed., *Geomorphology and Climate:* London, John Wiley, p. 367-409.

Smith, D.I., Greenaway, M.A., Moses, C., and Spate, A.P., 1995, Limestone weathering in eastern Australia, Part 1: Erosion rates: *Earth Surface Processes and Landforms*, v. 20, p. 451-463.

Spate, A.P., Jennings, J.N., Smith, D.I., and Greenaway, M.A., 1985, The micro-erosion meter: Use and limitations: *Earth Surface Processes and Landforms*, v. 10, p. 427-440.

Sweeting, M.M., 1966, The weathering of limestone, *in* Dury, G.H., ed., *Essays in Geomorphology:* London, Heinemann, p. 177-210.

Sweeting, M.M., 1979, Weathering and solution of the Melinau Limestones in the Gunong Mulu National Park, Sarawak, Malaysia: *Annales de la Société Géologique de Belgique*, v. 102, p. 53-57.

Trudgill, S.T., 1976, The subaerial and subsoil erosion of limestones on Aldabra Atoll, Indian Ocean: *Zeitschrift für Geomorphologie*, suppl. 26, p. 201-210.

Trudgill, S.T., Crabtree, R.W., Ferguson, R.I., Ball, J., and Gent, R., 1994, Ten year remeasurement of chemical denudation on a magnesian limestone hillslope: *Earth Surface Processes and Landforms*, v. 19, p. 109-114.

White, W.B., 1984, Rate processes: Chemical kinetics and karst landform development, *in* LaFleur, R.G., ed., *Groundwater as a Geomorphic Agent:* Boston, Allen and Unwin, p. 227-248.

White, W.B., 1990, Surface and near-surface landforms, *in* Higgins, C.G., and Coates, D.R., eds., *Groundwater Geomorphology: The Role of Subsurface Water in Earth-Surface Processes and Landforms:* Geological Society of America Special Paper, no. 252, p. 157-175.

Williams, P.W., 1966, Limestone pavements with special reference to western Ireland: *Institute of British Geographers Transactions and Papers*, no. 40, p. 155-172.

Williams, P.W., 1968, An evaluation of the rate and distribution of limestone solution and deposition in the River Fergus Basin, western Ireland, *in* Williams, P.W., and Jennings, J.N., eds., *Contributions to the Study of Karst:* Canberra, Australian National University, p. 1-40.

Williams, P.W., and Dowling, R.K., 1979, Solution of marble in the karst of the Pikikiruna Range, Northwest Nelson, New Zealand: *Earth Surface Processes*, v. 4, p. 15-36.

4.1.4. Calcite Dissolution by Sulfuric Acid

S.H. Bottrell, J. Gunn and D.J. Lowe

Abstract

Dissolution of carbonates, particularly calcite, due to the effects of sulfuric acid is potentially important as a speleogenetic mechanism in a variety of geologic situations, including coastal mixing zones.

Introduction

The chemical processes by which limestone is dissolved by water containing dissolved carbon dioxide are well understood and are discussed in detail elsewhere in this volume (Chapters 4.1.1 and 4.1.2). Although the potential role of carbonate dissolution due to sulfuric acid is less well appreciated and has been little studied in the context of speleogenesis, it is a more important and widely occurring speleogenetic mechanism than is commonly assumed (see relevant discussion and case studies in Chapters 5.2, 5.2.7, and 5.2.8 of this volume). Here we provide a brief examination of the physical chemistry of calcite dissolution by sulfuric acid.

Sulfide Oxidation Reactions Generating Acidity

In the shallower regions of the Earth's crust, oxic groundwaters are commonly brought into contact with rocks that contain metal sulfide minerals. Most metal sulfides contain sulfur as monosulfide, in the 2-minus oxidation state, and oxidize by a reaction typified by

$$MS + 2O_2 \Rightarrow M^{2+} + SO_4^{2-} \quad (1)$$

This reaction releases no acidity and therefore does not figure directly in the dissolution of calcite. Notable exceptions are the iron sulfide minerals pyrite and marcasite, in which the sulfur is present as disulfide groups and is consequently in the 1-minus oxidation state. The oxidation of these minerals is described by the reaction

$$FeS_2 + 3.5O_2 + H_2O \Rightarrow Fe^{2+} + 2SO_4^{2-} + 2H^+ \quad (2)$$

This reaction does generate H^+ and can therefore contribute acidity to effect carbonate dissolution. Under oxygenated conditions Fe^{2+} is not stable and undergoes oxidation and hydrolysis to form ferric hydroxide

$$Fe^{2+} + 0.25O_2 + 2.5H_2O \Rightarrow Fe(OH)_3 + 2H^+ \quad (3)$$

This reaction also generates H^+. Under conditions where calcite is present in excess, the final stable end product of calcite dissolution is HCO_3^-, so the overall mass balance of calcite dissolution by acidity developed by pyrite oxidation is

$$FeS_2 + 4CaCO_3 + 3.75O_2 + 3.5H_2O \Rightarrow Fe(OH)_3 + 4Ca^{2+} + 4HCO_3^- + 2SO_4^{2-} \quad (4)$$

Thus, the oxidation of one mole of pyrite will potentially dissolve four moles of calcite, leading to the suggestion that such a mechanism may be important in porosity development in carbonate rocks that either themselves contain pyrite, or contain adjacent or intercalated beds of pyritic lithologies. Pyrite can be oxidized by other electron acceptors (oxidizing agents), the most likely of which in the natural environment is ferric iron (Fe^{3+}). This proceeds via the reaction

$$FeS_2 + 14Fe^{3+} + 8H_2O \Rightarrow 15Fe^{2+} + 2SO_4^{2-} + 16H^+ \quad (5)$$

This is known to be the mechanism by which consortia of iron- and sulfur-oxidizing bacteria (e.g. *Thiobacillus ferrooxidans, T. thiooxidans,* and *Leptospirillium ferrooxidans*) oxidize sulfide minerals. Indeed, Taylor et al. (1984, 1985) used oxygen isotopic studies to show that reaction (5) is the dominant mechanism producing sulfate in acid mine waters. This reaction generates far more acidity and would dissolve more calcite per mole of pyrite than reaction (4).

However, it must be stressed that Fe^{3+} is not an abundant species, and would be exhausted rapidly by reaction (5). In most situations Fe^{3+} is regenerated by reaction of Fe^{2+} and O_2 (Singer and Stumm, 1970)

$$Fe^{2+} + 0.25O_2 + H^+ \Rightarrow Fe^{3+} + 0.5H_2O \quad (6)$$

Although reaction (5) may account for the mechanism by which pyrite is oxidized, the availability of O_2 remains the controlling factor in pyrite oxidation. Thus, reaction (4) provides the best description of the overall stoichiometry and mass balance of pyrite oxidation and coupled calcite dissolution.

Pyrite is a very common mineral in marine sedimentary rocks, forming as the end product of bacterial sulfate reduction during sediment diagenesis (e.g. Berner 1970, 1984; Bottrell and Raiswell, 2000). Hence, pyrite oxidation is the most likely source of sulfuric acid for calcite dissolution where secondary porosity and ultimately caves are formed in sedimentary successions. However, two other reactions are of potential interest in other geologic environments. First, in metamorphosed limestone or marble the dominant iron sulfide may be the iron monosulfide pyrrhotite (FeS), rather than pyrite. This will oxidize by reaction (1) above, with no consequent generation of acidity, but the subsequent oxidation and hydrolysis of the Fe^{2+} produced will yield H^+. In this situation the overall mass balance of calcite dissolution would be given by

$$FeS + 2CaCO_3 + 2.25O_2 + 2.5H_2O \Rightarrow Fe(OH)_3 + 2Ca^{2+} + 2HCO_3^- + SO_4^{2-} \quad (7)$$

The other reaction of interest is the oxidation of H_2S dissolved in thermal waters, which also generates acidity

$$H_2S + 2O_2 \Rightarrow SO_4^{2-} + 2H^+ \quad (8)$$

Many limestone sequences are potentially exposed to thermal solutions during their burial and uplift history and thus this reaction may be important in calcite dissolution at depth. However, from a geochemical standpoint the lack of availability of O_2 or another suitable oxidizing agent in the deep subsurface (where conditions are generally anoxic) may exert fundamental limitations on the scope for calcite dissolution by this mechanism.

Kinetics

The rate at which calcite may be dissolved by sulfuric acid generated by sulfide oxidation may be governed by a number of factors. Rates of pyrite oxidation are generally very high, especially where bacterial catalysis is involved (Singer and Stumm, 1970 and references therein). But in some situations the availability of O_2 to drive the reaction may provide the limiting factor.

In the opposite instance, the rate of proton generation may not be a limiting factor. But the rate may be determined by the dissolution of calcite and neutralization of acidity. In this case the pH will fall as protons are produced faster than they are consumed. At low pH the dissolution rate of calcite increases markedly as dissolution mechanisms other than those in the circumneutral range associated with dissolution by carbonic acid come into play (Plummer et al., 1978). Thus, as pH falls the rate of calcite dissolution will increase until it balances the rate of production of acidity.

In a study of a karst aquifer affected by pyritic ore-processing wastes, Bottrell (1993) found that pyrite oxidation was balanced by calcite dissolution at a pH of ~4, where the calculated rate of calcite dissolution was 1.8 mg/m^2/s—within error of the value of 4 mg/m^2/s at pH = 4 quoted by Plummer et al. (1978).

Thus, if the O_2 supply does not provide a limitation, rates of calcite dissolution can be far higher via the sulfuric acid mechanism than via carbonic acid, simply because far lower solution pH levels can be achieved and the calcite dissolution mechanism is changed.

Sulfuric Acid Generation in the Mixing Zone

Bottrell et al. (1990) produced convincing evidence that acidity is generated by bacterial reduction of sulfates near the halocline at the base of the freshwater-saltwater mixing zone. The sulfur species are reoxidized at shallower depth, providing sulfuric acid whose contribution to wall-rock dissolution is at least as great as that of freshwater-saltwater mixing, itself a highly effective process with a clear and fundamental importance to speleogenesis.

Bosch et al. (1989), provided further insight of activity within coastal carbonate aquifers. Here ion exchange was shown to be potentially important in carbonate dissolution below the vadose zone. All water samples were depleted in sulfate following reduction to hydrogen sulfide. This gives further evidence for the potential involvement of reduced and reoxidized sulfur compounds in early speleogenesis.

These results provide a valuable link between conventional views of carbonate dissolution, the more recently recognized process of mixing-zone corrosion, and sulfuric acid dissolution.

References

Berner, R.A., 1970, Sedimentary pyrite formation: *American Journal of Science,* v. 290, p. 882-911.

Berner, R.A., 1984, Sedimentary pyrite formation: An update: *Geochimica et Cosmochimica Acta,* v. 48, p. 605-615.

Bosch, X., Custodio, E., and Pascual, M., 1989, Geochemical reactions in carbonate coastal aquifers, Catalonia, Spain, *in* Simpson, E.S., and Sharp, J.M., eds., *Selected papers on hydrogeology from the 28th International Geological Congress,* Washington, U.S.A.: p. 147-159.

Bottrell, S.H., 1993, Water chemistry in the Xingwen caves, China: *Cave Science,* v. 20, p. 87-92.

Bottrell, S.H., and Raiswell, R., 2000, Sulfur isotopes and sulfur cycling in sediments, *in* Awramik, S.M., and Riding, R., eds., *Microbial Sediments:* Springer-Verlag (in press).

Bottrell, S.H., Smart, P.L., Whitaker, F., and Raiswell, R., 1990, Geochemistry and isotope systematics of sulfur in the mixing zone of Bahamian blue holes: *Applied Geochemistry,* v. 6, p. 97-103.

Plummer, L.N., Wigley, T.M.L., and Parkhurst, D.L., 1978, The kinetics of calcite dissolution in CO_2-water systems at 5° to 60°C and 0.0 to 1.0 Atm CO_2: *American Journal of Science,* v. 278, p. 179-216.

Singer, P.C., and Stumm, W., 1970, Acid mine drainage: The rate determining step: *Science,* v. 167, p. 1121-1123.

Taylor, B.E., Wheeler, M.C., and Nordstrom, D.K., 1984, Oxygen and sulfur isotope compositions of sulfate in acid mine drainage: Evidence for oxidation mechanisms: *Nature,* v. 308, p. 538-540.

Taylor, B.E., Wheeler, M.C., and Nordstrom, D.K., 1985, Stable isotope geochemistry of acid mine drainage: Experimental oxidation of pyrite: *Geochimica. et Cosmochimica Acta,* v. 48, p. 2669-2678.

4.1.5. Dissolution of Carbonates by Geothermal Waters

Yuri V. Dublyansky

Abstract

Three major mechanisms of carbonate dissolution in hydrothermal karst systems are briefly reviewed: dissolution by rising carbonic thermal water, dissolution due to oxidation of H_2S, and dissolution due to mixing of waters of contrasting chemistry. Only the first of the above mechanisms is specific to hydrothermal karst systems.

Introduction

Significant dissolution of carbonate rocks requires acidic water. Hydrothermal, and more generally, hypogene caves are formed by acids of deep-seated origin or epigenic acids rejuvenated by deep-seated processes (Palmer, 1991). The major hypogene acids are carbonic acid, originating from igneous activity and metamorphism of buried carbonates, as well as hydrosulfuric acid (aqueous H_2S) originating from redox reactions. Thus, hydrothermal caves associated with CO_2 and caves formed by waters containing H_2S are considered to represent two major classes of hypogene caves (Ford and Williams, 1989). (Editor's note: In chapters 3.4 and 5.2, additional aspects of hypogene caves are discussed.)

Theoretical generalizations concerning the patterns of dissolution within the endokarstic zone are lacking (see Chapters 5.2 and 5.2.5 in this volume for discussions of the endokarst concept). Empirically, from deep drilling, it is known (Ezhov et al., 1992) that the mineralization of water is comparatively low (*ca.* 5 to 10 g/L). Due to high pressures, the total amount of dissolved gases may be as great as 60 to 70 vol. %. The gases are dominated by CO_2, and to a lesser extent, by H_2S. For the exokarstic zone, the major mechanisms of cave excavation are

- Dissolution by rising carbonated thermal water.
- Dissolution due to oxidation of H_2S.
- Dissolution due to mixing of waters of contrasting chemistry.

These mechanisms are reviewed below.

For the discussion of settings and specific features of hydrothermal speleogenesis see Chapter 5.2.5 in this volume.

Dissolution by Rising Thermal Water (CO_2)

Thermal waters rising from significant depth are often saturated with CO_2, which may originate from metamorphism of carbonate rocks or from magmatic and volcanic activity. The solubility of CO_2 in water depends on temperature and pressure. The typical behavior of CO_2 saturation concentration in hydrothermal environments was calculated by Malinin (1979). This is shown in 4.1.5 Figure 1a. The figure implies that water saturated with respect to CO_2 at depth becomes supersaturated as it rises toward the surface. Hence, CO_2 must exsolve in the gaseous phase and leave the system.

Rising carbonic thermal waters also cool down. Due to the inverse relationship between carbonate solubility and temperature, they may acquire and maintain solutional aggressiveness–even at decreasing CO_2 levels. The 4.1.5 Figure 1b shows that the solubility of $CaCO_3$ increases evenly along the fluid path. But near the land surface (or water table) it drops drastically. Such a nonlinear behavior leads to the appearance of two geochemical zones: a zone of carbonate dissolution, and a zone of carbonate precipitation.

The rates of fissure widening by ascending thermal water at different discharge rates were studied by Palmer (1991). The 4.1.5 Figure 2 shows that solutional widening rates increase in the direction of flow, leveling off to constant values only at very small discharge. Doubling either the thermal gradient or CO_2 content allows a potential increase in growth rate of approximately ten times, with no change in the shape of the curves. No rate increase can be fully realized without a proportional increase of discharge.

Most significantly, increasing discharge causes the enlargement rate to reach a maximum, beyond which any further increase in discharge has no effect. This implies that regardless of their size or flow, if many fissures receive thermal water of similar character, they will all enlarge at the same rate, forming a network maze.

Dissolution Due to Oxidation of Sulfides (H_2S)

Caves formed in association with H_2S bearing waters are classified as hypogene, although they are not necessarily thermal. Examples of active caves developing due to oxidation of hydrogen sulfide are known in the U.S.A. (caves of the Bighorn Basin, Wyoming;

Egemeier, 1981). For more case studies relevant to dissolution due to sulfuric acid, see Lowe, Bottrell, and Gunn (Chapter 5.2.7 in this volume). On the basis of an analysis of spring chemical data, Worthington and Ford (1995) argued that elevated sulfate content is an inherent feature of thermal springs in carbonate terrain. Prominent examples of active hydrothermal karst are caves of the Bandar-e-Khamir area in southern Iran, with waters of 42°C and high contents of H_2S in their cave atmospheres (Bosák, 1993).

Sulfuric waters become aggressive when their dissolved H_2S oxidizes in oxygen-rich waters or on contact with the air to form sulfuric acid. The 4.1.5 Figure 3 shows that conversion of H_2S to H_2SO_4 produces a sharp increase of the equilibrium concentration, and therefore dissolution is enhanced. The effect is attenuated if CO_2 generated by the H_2SO_4-$CaCO_3$ reaction is degassed.

In two settings hydrogen sulfide oxidation is an important speleogenetic process. The first is subaqueous dissolution of carbonates near the water table. Large rooms of Carlsbad Cavern in New Mexico, U.S.A. are believed to be formed this way. The second setting is dissolution and subsequent replacement of calcite by gypsum and its consequent removal above the water table. This mechanism, termed *replacement corrosion*, was suggested by Egemeier (1981) for caves of the Bighorn Basin, U.S.A.

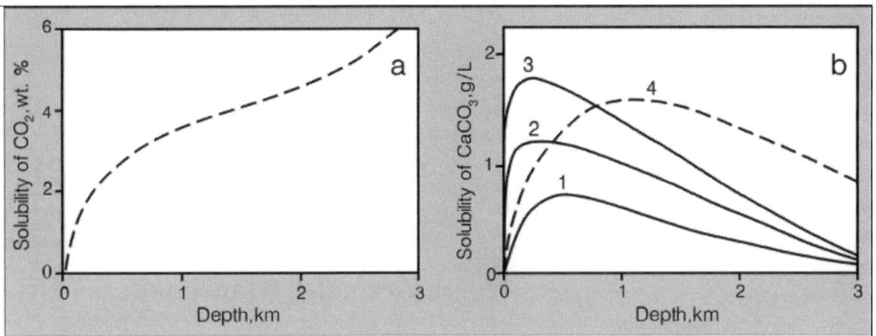

4.1.5 Figure 1. Solubility of CO_2 and $CaCO_3$ in ascending hydrothermal water (after Malinin, 1979). Temperature changes from depth toward the surface: (a) from 200 to 25°C; (b) line 1 from 225 to 100°C, line 2 from 225 to 50°C, line 3 from 225 to 25°C, line 4 depicts a solution with ionic strength I = 1, from 200 to 100°C.

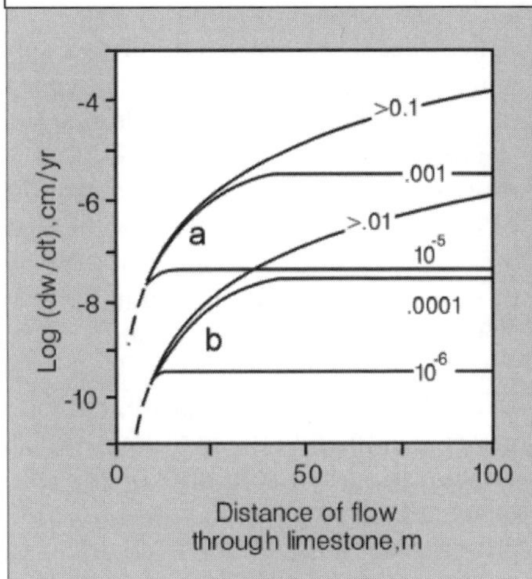

4.1.5 Figure 2. Dissolution by rising carbonic thermal water. Rate of fissure widening: (a) cooling from 50 to 45°C and (b) from 25 to 20°C. Initial CO_2 concentration = 0.005 M. Numbers near curves indicate linear discharge (velocity x width) q, cm^2/s (from Palmer, 1991).

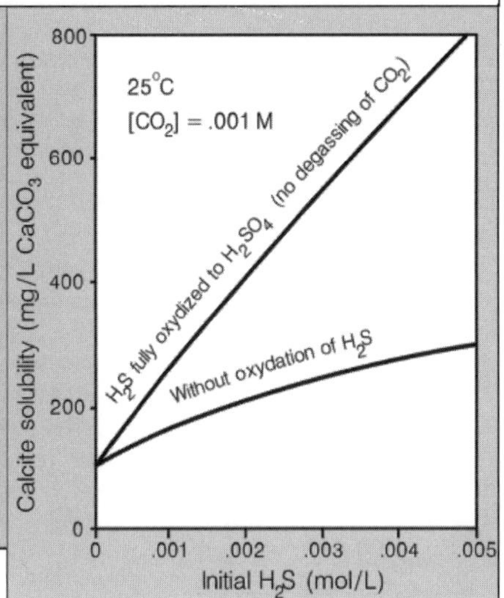

4.1.5 Figure 3. Dissolution due to oxidation of sulfides. Solubility of calcite in an H_2S- H_2SO_4 system at an initial CO_2 concentration = 0.001 M (from Palmer, 1991; simplified).

Dissolution Due to Mixing of Waters (CO_2 and H_2S)

Solutional aggressiveness can be renewed or enhanced by mixing of waters of contrasting chemistry, particularly those differing in CO_2 and H_2S content or salinity (Bögli, 1964; Palmer, 1991). Mixing of waters having different temperatures produces similar effects, due to contrasts in CO_2 contents.

The effect known as "mixing corrosion" is depicted in 4.1.5 Figure 4. The figure shows that if two saturated solutions mix, the resulting solution is undersaturated with respect to calcite or dolomite. This is responsible for the development of network maze cave systems in Budapest, e.g. Pal-völgyi, Szemlo-hegy, Ferenc-hegy, and others (Müller and Sárváry, 1977; Takács-Bolner and Kraus, 1989).

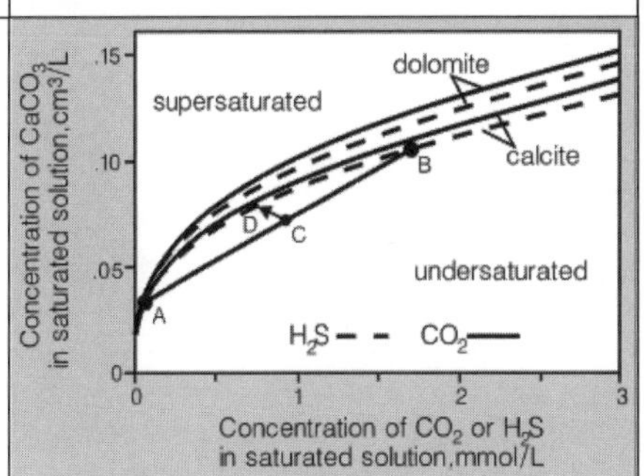

4.1.5 Figure 4. Saturation concentration of calcite and dolomite as a function of CO_2 or H_2S concentration (mmol/L). The saturation concentration is given as volume of solid dissolved per liter (cm^3/L) (Palmer, 1991). Mixing of two solutions (A, B) produces a solution that is undersaturated (C).

References

Bögli, A., 1964, Mischungskorrosion, ein Beitrag zur Verkarstungsprobleme: *Erdkunde,* v. 18, p. 83-92.

Bosák, P., 1993, Karst phenomena in the Bandar-e-Khamir area, SE Iran, *in Krasove Sedimenty*, Knihovna Ceske Speleologicke spolechnosti, v. 21, p. 74-78.

Egemeier, S.J., 1981, Cavern development by thermal waters: *National Speleological Society Bulletin,* v. 43, p. 31-51.

Ezhov, Y.A., Lysenin, G.P., Andreychuk, V.N., and Dublyansky, Y.V., 1992, *Karst in the Earth's crust: Distribution and the main types*: Novosibirsk, United Inst. of Geology, Geophysics, and Mineralogy, 76 p. (in Russian).

Ford, D.C., and Williams, P.W., 1989, *Karst geomorphology and hydrology:* London, Unwin Hyman, 601 p.

Malinin, S.D., 1979, *Physical chemistry of hydrothermal systems with carbon dioxide:* Moscow, Nauka, 111 p. (in Russian).

Müller, P., and Sárváry, J., 1977, Some aspects of development in Hungarian speleology theories during the past 10 years: *Karszt és Barlang* (Hungary), Special Issue, p. 53-60.

Palmer, A.N., 1991, Origin and morphology of limestone caves: *Geological Society of America Bulletin,* v. 103, p. 1-21.

Takács-Bolner, K., and Kraus, S., 1989, The results of research into caves of thermal water origin: *Karszt és Barlang* (Hungary), Special Issue, p. 31-38.

Worthington, S.R.H., and Ford, D.C., 1995, High sulfate concentrations in limestone springs: An important factor in conduit initiation?: *Environmental Geology,* v. 25, p. 9-15.

4.1.6. Dissolution and Conversions of Gypsum and Anhydrite

Alexander Klimchouk

Abstract

Dissolution of sulfate rocks proceeds by different mechanisms and at different rates compared to those associated with the dissolution of carbonate rocks. Gypsum dissolves by a simple two-phase dissociation. Anhydrite, when dissolved, forms a solution of calcium sulfate, which at common temperatures and pressures is in equilibrium with the solid phase of gypsum, but not with anhydrite.

The solubility of gypsum at 20°C is 2.531 g/L or 14 mM/L, which is roughly 140 times lower than the solubility of common salt, but two orders of magnitude greater than the solubility of $CaCO_3$ in pure water. The difference in solubility between gypsum and calcite decreases to 10 to 30 times if the latter is dissolved in the presence of CO_2.

The dependence of the solubility of gypsum on temperature is nonlinear, reaching a maximum at 43°C. Hydrostatic pressure does not substantially affect the solubility of gypsum, but the solubility increases sharply with pressure applied to the gypsum rock. The solubility differs for grains of varying sizes. The solubility of gypsum is boosted with the presence of other salts (foreign ions) in solution, up to three times, for example, in the presence of NaCl, and up to six times in the presence of $Mg(NO_3)_2$.

These effects are important for speleogenesis, as gypsum is commonly associated with other salts within evaporitic formations. The presence of common ions does not decrease the gypsum solubility considerably. The bacterial or thermal reduction of sulfates rejuvenates the dissolution capacity of water by consuming sulfate ions. Dedolomitization has a similar effect. Both processes have very important consequences for speleogenesis in deep-seated settings.

The dissolution of gypsum follows a first-order rate law, whereas the dissolution rate of anhydrite obeys a second-order law. Rates of gypsum dissolution are very high in the region far from saturation, but they decrease abruptly when the solution approaches high saturation levels. Because of transport-controlled dissolution kinetics, rates are strongly dependent on boundary-layer conditions within the flowing solution. They are affected by velocity, the ionic strength of the solution, and its temperature.

Conversion of gypsum to anhydrite and back to gypsum is common due to the thermodynamic instability of these minerals within the physico-chemical range of common geologic environments. The mechanisms and rates of conversion of anhydrite to gypsum are still poorly understood. They display complex dependencies on the tectonic regime, on the water-bearing properties of surrounding formations, and on both the regional and local flow regimes. Contrary to a common view, little or no expansion of volume occurs during hydration of anhydrite in most underground environments.

Introduction

The development of karst is driven by the dissolution of a host rock and the subsequent removal of dissolved matter by moving water. This process, at various stages, initiates or triggers associated processes including erosion, collapse, and subsidence. The dissolution of sulfate rocks proceeds by different mechanisms and at different rates than those of carbonate rocks. For each rock type, different factors influence the process.

This chapter attempts to summarize the present knowledge of the dissolution chemistry and kinetics of gypsum and anhydrite. These are important to the genetic interpretation of karst features in these rocks. Gypsum-anhydrite-gypsum conversions and recrystallization processes are also addressed because of their importance to karst development.

Many studies have been undertaken on the solubility and dissolution of sulfate minerals in the context of construction engineering and karst processes. Important works include those of Laptev (1939), Kuznetzov (1947), Shternina (1949), Zdanovsky (1956), Sokolov (1962), Zverev (1967), Liu and Nancollas (1971), Blount and Dickson (1973), Mel'nikova and Moshkina (1973), Wigley (1973), Gorbunova (1977), James and Lupton (1978), and Kushnir (1988). The most comprehensive recent account is that of James (1992).

Chemical Equilibria

Gypsum dissolves by a simple two phase dissociation (solid and solvent)

$$CaSO_4 \cdot 2H_2O \Leftrightarrow Ca^{2+} + SO_4^{2-} + 2H_2O \qquad (1)$$

Gypsum, like $CaCO_3$ and NaCl, dissolves reversibly, but anhydrite does not. When anhydrite dissolves it forms a solution of calcium sulfate, which at common temperatures and pressures is in equilibrium with the solid phase of gypsum, but not with anhydrite. If disequilibrium of the solid-solvent system produces undersaturation, gypsum precipitates. This is due to the instability of anhydrite under normal surface and shallow subsurface conditions of temperature and pressure (4.1.6 Figure 1).

The solubility of gypsum in pure water at 20°C is 2.531 g/L or 14.7 mM/L. It is roughly 140 times lower than the solubility of common salt (360 g/L) but two orders of magnitude greater than the solubility of $CaCO_3$ in pure water (15 mg/L). However, in the presence of CO_2 the dissolution of calcite is enhanced and the difference in solubility between calcite and gypsum decreases to 10 to 30 times.

The dependence of gypsum solubility on temperature is reported by many authors (Blount and Dickson, 1973; James, 1992; Liley et al., 1963; see 4.1.6 Figure 2). Between 0° and 30°C, the range encompassing most natural waters, the solubility of gypsum increases by 20 percent, reaching a maximum (about 2.66 g/L) at 43°C. Cigna (1985) examined the possible effects on gypsum solubility caused by mixing waters at different temperatures. He found that when mixing equal amounts of two saturated waters (one at 10°C, and the other at temperatures ranging from 40 to 100°C) the solubility in the mixture increased by between 2 and 13 percent. This effect may play some role in the karstification of areas with geothermal waters.

Anhydrite may be considered to have no characteristic solubility (James, 1992). This is because of its chemical instability in commonly encountered shallow subsurface conditions. Some values given in the literature are misleading. The true solubility of anhydrite under normal temperatures is equivalent to that of gypsum. When dissolved in water, anhydrite produces a solution of $CaSO_4$ that ultimately attains the same equilibrium concentration as the gypsum-H_2O system in pure water, 2.00 g/L at 20°C. James (1992) pointed out that anhydrite in contact with water tends toward a metastable state characterized by supersaturated solutions. These probably account for some of the high solubilities quoted for anhydrite, which range up to 3.5 g/L. The subject of gypsum-anhydrite conversion is considered in detail elsewhere in this chapter.

The 4.1.6 Figure 2B shows the solubility data for anhydrite and gypsum in their stability regions. The solubility of anhydrite is lower than that of gypsum under these pressure conditions and decreases with increasing temperature.

Hydrostatic pressure does not substantially affect the solubility of gypsum within common geologic environments. In contrast, the $CaCO_3$-CO_2-H_2O system is influenced by the presence of the CO_2 gas phase which makes it sensitive to pressure. The solubility of gypsum increases slightly at pressures exceeding 100 bars (Manikhin, 1966), but at depths of less than 1000 meters or so, the influence is negligible. The effect of pressure applied to the solid mineral is discussed later on.

Equilibrium Constants

Different values of the equilibrium constant *Kg* for gypsum are reported by various authors, reflecting varied experimental conditions and the use of different thermodynamic data in the calculations. The constants are usually given for 25°C and higher temperatures, but in many karst environments the water temperature is normally between 5 and 15°C.

Aksem and Klimchouk (1991) provided thermodynamic calculations of Gibbs free-energy values and equilibrium constants for the dissolution of gypsum in water at temperatures of 0 to 50°C (4.1.6 Table 1). The results agree closely with the values previously provided by Wigley (1973). The data in 4.1.6 Table 1 and 4.1.6 Figure 3 give the following function of temperature

$$pKg = 4.667 - 5.197 \times 10^{-9} \times t + 1.133 \times 10^{-4} \times t^2 \qquad (2)$$

Saturation Index

Karst waters in perfect equilibrium with a solid phase are rare. When a solution is undersaturated with respect to the soluble mineral, dissolution proceeds; when the solution is supersaturated, no dissolution occurs, or there may be precipitation.

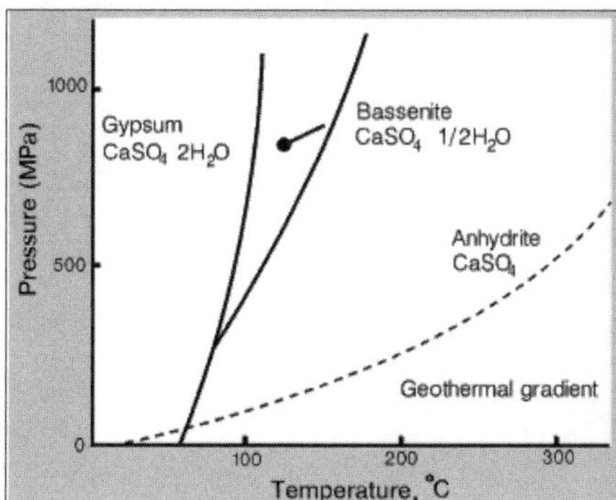

4.1.6 Figure 1. Equilibrium diagram for the system $CaSO_4$-H_2O (after Zanbak and Arthur, 1986).

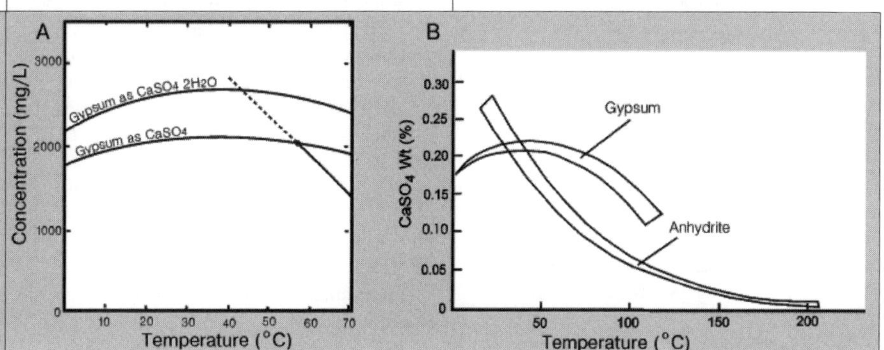

4.1.6 Figure 2. Solubility of gypsum and anhydrite: (A) = Curves for gypsum and anhydrite based on the experimental data of Blount and Dickson (1973). The upper curve shows the mass loss of gypsum rock in solution. The lower curve is calculated as $CaSO_4$ and displays the invariant point at 58°C, where gypsum, anhydrite, and liquid coexist (after White, 1988). (B) = As summarized by Zanbak and Arthur (1986).

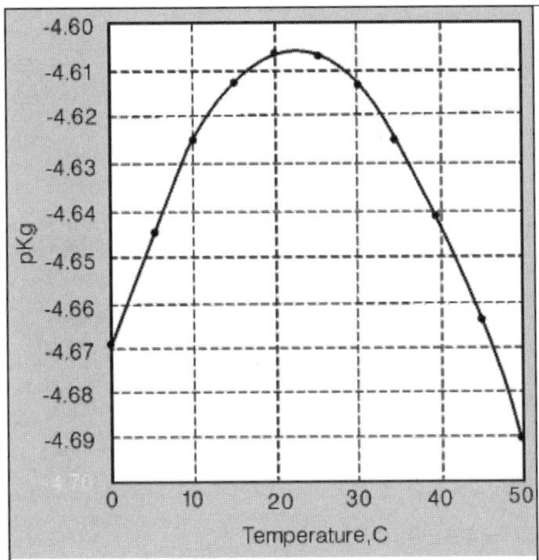

4.1.6 Figure 3. Equilibrium constant Kg as a function of temperature (after Aksem and Klimchouk, 1991).

t°C	GT, kal	pKg	Kg x 10^5
0	5839	4.671	2.131
5	5912	4.645	2.266
10	5993	4.625	2.370
15	6082	4.613	2.439
20	6183	4.607	2.472
25	6286	4.607	2.471
30	6400	4.613	2.436
35	6522	4.625	2.370
40	6653	4.642	2.278
45	6791	4.665	2.165
50	6938	4.692	2.034

4.1.6 Table 1. Free-energy values and equilibrium constants for gypsum dissolution at temperatures of 0 to 50°C (after Aksem and Klimchouk, 1991).

Precipitation does not always occur in supersaturated solutions, because its triggering and progress depend on many additional factors.

The deviation of a solution from equilibrium is measured by the saturation index SI, introduced by Langmuir (1971) and used widely by karst researchers (see White, 1988; Ford and Williams, 1989; for a description of the general concept). The saturation index is defined as the log of the ratio between the ion activity product for the mineral dissociation to the equilibrium constant *Kg* of the reaction. For gypsum the saturation index is

$$SI_{gyp} = \log([Ca^{2+}] [SO_4^{2-}] \gamma_{Ca} \cdot \gamma_{SO4})/Kg \quad (3)$$

where the [ ] denotes concentration in mole/L and the g is the activity coefficient. SI is zero if water is in equilibrium with the mineral. It has negative values for undersaturated (aggressive) solutions, and positive values for supersaturated solutions.

In natural conditions, equilibrium is rarely attained, or it is disrupted by changes in conditions that affect solubility. The dependence of the solubility on various properties of a solvent and solid are not clearly and unambiguously described, either theoretically or experimentally. The main factors affecting the solubility of gypsum are outlined below.

Main Factors Affecting the Solubility of Gypsum

Pressure Applied to the Rock

Korzhinsky (1953) showed that the solubility of minerals increases when the rock fabric experiences pressures higher than that of the groundwater. Experimental data by Manikhin (1966) suggest that the solubility of anhydrite increases sharply with increasing pressure. Each 0.01 Pa increase in pressure enhances the solubility by 3 to 5 times. Gypsum's solubility is reported to increase 4 times with each additional 0.1 Pa. Consequently, under applied stress, the solubility of anhydrite becomes higher than that of gypsum.

Pechorkin (1986) discussed the stress field in the Polazna gypsum-anhydrite massif of the pre-Urals, and referring to Manikhin's data, determined that the solubility of anhydrite in the high stress zone should be 2 to 5 times higher than in the low-stress zone. This factor is believed to play a significant role in the differentiation of dissolution-recrystallization and hydration processes at the scale of massifs. The effect is likely to be important to karst development in all environments, not just in deep-seated ones.

Grain Size

G. Hewlett (as cited by Sokolov, 1962) reported that saturation with respect to gypsum for grains of 2 mm in size is reached at a concentration of 15.3 mM/L. However, for grains of 0.3 mm the solution becomes saturated at 18.2 mM/L, an increase by 20 percent. Sonnenfeld (1984) indicated that the solubility of gypsum reaches a maximum for crystals in the size range of 0.2 to 0.5 mm, whereas the solubility of anhydrite is highest for crystals of about 2.8 mm.

Differential solubility for grains of different sizes makes it possible for interstitial (pore) waters to be undersaturated with respect to small grains but supersaturated with respect to large ones. This plays an important role in recrystallization and hydration processes (see below) and perhaps in the development of irregular small-scale porosity. Selective dissolution within differently grained rock may facilitate surface retreat by water, because of the preferential removal of small grains that initially provide a cement between the larger ones. The differential solubility of crystals of various sizes is well illustrated by observations made in the gypsum caves of the western Ukraine, where single giant crystals of selenite within the rock mass commonly protrude from the walls and ceilings as pendants. They are apparently less soluble than the surrounding finer grained matrix.

Solubility in Various Salt Solutions

All natural waters contain some dissolved salts, and it is well known that these can affect the solubility of other minerals. Ion pairing and the increase in ionic strength reduce the activities of Ca^{2+} and SO_4^{2-} ions and result in increased solubility. Ford and Williams (1989) noted that an increase of up to 10 percent in gypsum solubility is possible in typical karst waters. However, they stressed the far greater importance of the effect on the calculated saturation indexes. If pairing is not taken into account, the SI values are overestimated. Many reported cases of supersaturated water in gypsum karst are likely a result of overlooking this effect.

Ions foreign to the solid phase considerably increase the solubility of gypsum by enhancing the ionic strength of the solution. The 4.1.6 Figure 4A shows the effect of NaCl

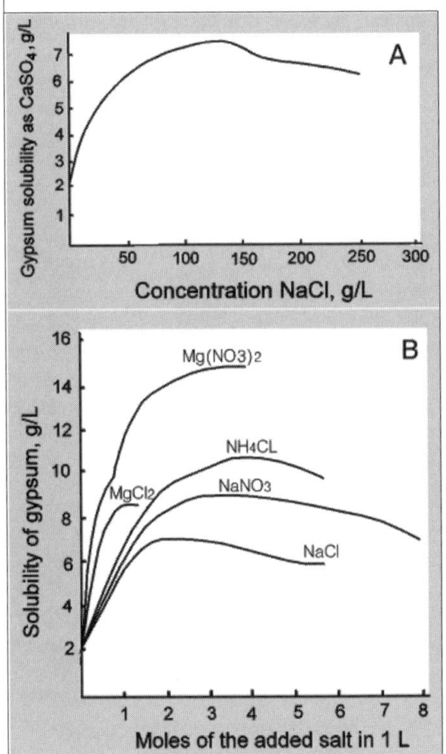

4.1.6 Figure 4. The solubility of $CaSO_4$ in salt-water solutions at 25°C (after Shternina, 1949).

(after Shternina, 1949). With increasing concentrations of sodium chloride, the solubility of gypsum increases. After quickly reaching a maximum of 7.33 g/L at 139 g/L of NaCl, it then decreases slowly but remains much higher than the solubility in pure water. The solubility of gypsum in solutions containing other salts is still higher.

The presence of $Mg(NO_3)_2$ can boost the solubility of gypsum by almost 6 times in comparison with the value for pure water. The 4.1.6 Figure 4B, also taken from the work of Shternina, shows similar curve shapes. The study of complex systems common in nature (Mel'nikova and Moshkina, 1973) indicates gypsum solubilities of 5.9 to 6.3 g/L in solutions containing high concentrations of $MgSO_4$ (5.6 to 18.2 percent) and NaCl (0.2 to 14.1 percent).

James (1992), referring to Paine et al. (1982), quoted a good example from the Poechos Dam in Peru, where direct measurements of gypsum solubility in groundwater samples from wells gave $CaSO_4$ values as high as 6.2 g/L, three times the solubility in pure water and 35 percent more than the maximum solubility in sea water. These water samples also contained Na, K, Mg, HCO_3, Cl, SO_4, and NO_3 ions.

The effect of foreign ions is very important for gypsum karst development. Other salts commonly accompany gypsum in evaporate formations, and the groundwater of many aquifers, particularly deep-seated ones, may contain high levels of dissolved salts.

The presence of common ions in solution (ones that occur in the dissolving mineral, but are introduced from another mineral) decreases the solubility of each contributing mineral. Ca^{2+} is the common ion for gypsum and calcite, and the effect occurs in many karst areas containing intercalated or adjacent sulfate and carbonate layers. The effect is more pronounced on the solubility of calcite and less on gypsum.

Study of the system Ca^{2+} - HCO_3^- - SO_4^{2-} - H_2O by Wigley (1973) allows the assessment of the relative contributions to the total concentration of calcium derived from gypsum and calcite respectively. It also allows the equilibrium-disequilibrium for each mineral to be evaluated (4.1.6 Figure 5). The partial pressure of CO_2 is an independent variable influencing the solubility of calcite, but it has a negligible effect on gypsum solubility (Sokolov, 1962). Zdanovsky (1956) suggested that the solubility of some salts, including gypsum, decreases slightly with increasing CO_2. Where only gypsum dissolves, but CO_2 is supplied to the water from soil cover or from other sources, net deposition of calcite may occur as saturation with respect to $CaCO_3$ is quickly reached.

The relationship between gypsum dissolution and calcite deposition in the presence of CO_2 in the shallow subsurface was studied by Forti and Rabbi (1981). They calculated the equilibrium pattern for the CO_2 - H_2O - calcite - gypsum system with respect to CO_2 and pH (4.1.6 Figure 6). The effect is responsible for calcite deposition in many gypsum caves that are close to the surface, but it is also responsible for the replacement of gypsum with calcite in the reducing environment of some confined aquifers. Because the effect has a small influence on gypsum solubility, however, and because much gypsum dissolution occurs without any CO_2 involvement it seems unreasonable to claim that gypsum karst is a three-component system (Forti and Rabbi, 1981).

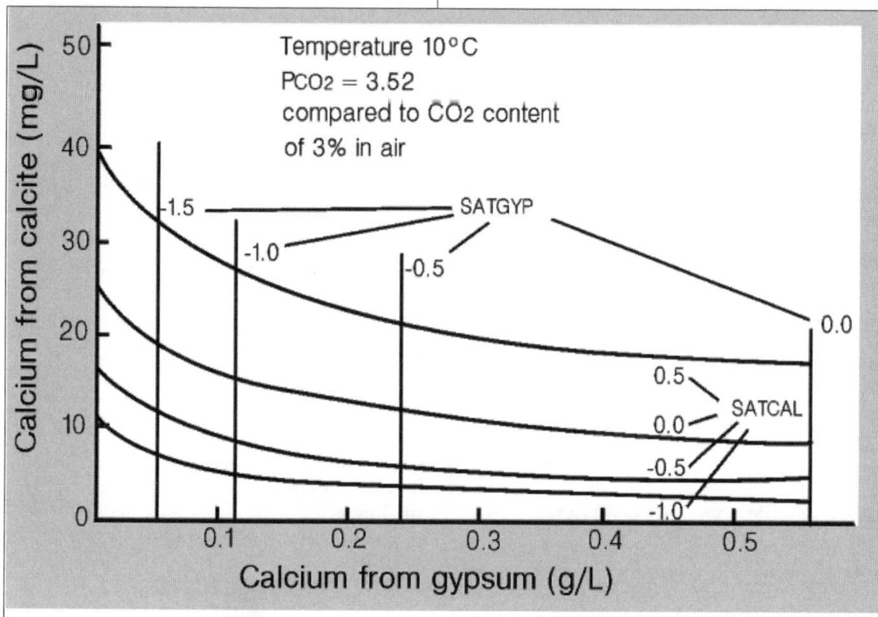

4.1.6 Figure 5. Saturation index within calcite-gypsum solutions (after Wigley, 1973).

Factors that Maintain the Dissolution Potential of Sulfates

Sulfate Reduction

The reduction of dissolved sulfates by microbes (including heterogeneous assemblages of *Desulfo-x*) is a common process in confined aquifers where sulfate rocks and dispersed organic matter are present. The process is described by the following simplified reaction induced by anaerobic bacteria:

$$SO_4^{2-} + 2CH_2O \longrightarrow H_2S + 2HCO_3^- \quad (4)$$

During sulfate reduction, sulfate ions are consumed and removed from the solution. Therefore, additional sulfate can be dissolved. Calcium and bicarbonate commonly react to precipitate $CaCO_3$, utilizing the HCO_3^- generated by the above reaction. Epigenetic calcite masses can form as a result. Calcium cations may also be exchanged with sodium derived from interbedded or surrounding rocks.

Sulfate reduction seems to be a very important mechanism in maintaining the dissolutional potential of groundwater with respect to gypsum in confined aquifers, especially with vertical cross-formational hydraulic communication. In hydrochemistry, the effect has been known for a long time, and its possible relevance for karst development was outlined by Kaveev (1963), Turyshev (1965), and other workers. Recently, I have emphasized its importance

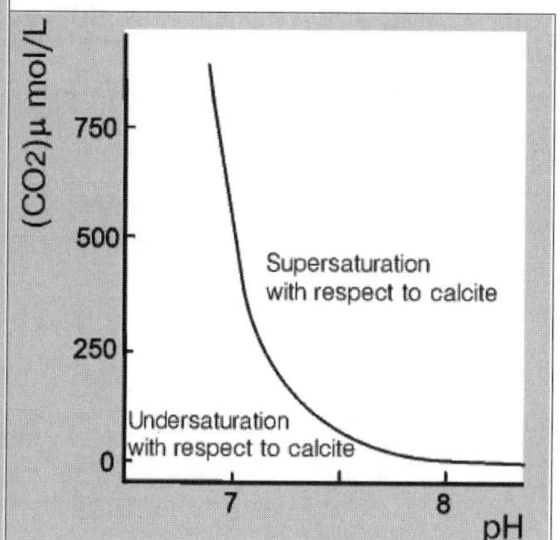

4.1.6 Figure 6. Equilibrium within the system CO_2 - H_2O - calcite - gypsum with respect to CO_2 and pH at 10°C (after Forti and Rabbi, 1981).

for speleogenesis in gypsum (Klimchouk, 1994, 1997).

Dedolomitization

Dolomite is commonly associated with or interbedded with gypsum. Stankevich (1970) pointed out that the process of dedolomitization generates further dissolutional capacity with respect to gypsum, because Ca^{2+} is removed from solution and the sulfate ions react with Mg

$$CaSO_4(aq) + CaMg(CO_3)_2(solid) \rightarrow$$

$$2CaCO_3(solid) + MgSO_4(aq) \qquad (5)$$

Palmer (Chapter 3.4 in this volume) specifies that at least 1.5 times as much gypsum can be dissolved by this mechanism, in comparison with its individual solubility. Raines and Dewers (1997) show that dedolomitization can occur only when gypsum approaches equilibrium under slow-flow conditions, when time scales of the various mineral reaction rates approach each other. This condition is readily met in deep-seated confined aquifers where the process seems to be important in supporting speleogenesis.

Suspended Crystals

Pechorkin (1986) reported experimental results suggesting that when a solution approaches gypsum saturation, small crystals originate in the presence of the solid phase. These can then be carried in suspension by flowing water. Such crystals begin to form at $CaSO_4$ concentrations of 1.1 to 1.5 g/L and reach a maximum of 10 to 15 percent of the total dissolved $CaSO_4$ at concentrations of 2.2 g/L. Thus, an additional 0.28 to 0.42 grams of gypsum can be dissolved in each liter of water. Pechorkin did not explain what causes crystallization in undersaturated solutions.

Dissolution Kinetics of Gypsum and Anhydrite

Dissolution is a heterogeneous reaction occurring at the boundary between two phases. Molecular dissociation of gypsum occurs almost instantaneously, so that dissolution rates are controlled solely by diffusion across the boundary layer. Dissolution rates thus depend on boundary-layer conditions and the concentration gradients across it. They are described by the following equation

$$dC/dt = (KA/V)\,(C_s - C)^n \qquad (6)$$

where dC/dt is a rate of change of concentration with time in a volume V of solution with a bulk concentration C, C_s is the solubility of the dissolved substance, A is the surface area of the dissolving mineral, K is a rate constant varying with boundary-layer conditions, mineral properties, and surface roughness, and n is reaction order.

Theoretical and experimental studies of the dissolution kinetics of gypsum and anhydrite are numerous, although many of the results are conflicting. The most comprehensive treatment of the topic is given in James and Lupton (1978) and James (1992). The brief summary below is based largely on these works.

The main difference in the dissolution kinetics between gypsum and anhydrite lies in the exponent n. It was shown by Zdanovsky (1956), Liu and Nancollas (1971) and James and Lupton (1978) that gypsum dissolution follows a first-order equation whereas the dissolution rate of anhydrite obeys a second-order equation. The latter reflects partial control of the surface reaction rate, which is assumed to be a hydration. This difference is shown in 4.1.6 Figure 7 by plotting concentration against time, with an overlay of theoretical curves. For gypsum, the flow time (or distance) for the solution to reach 90 percent of saturation is very short; the dissolution rate decreases by several orders of magnitude beyond this point. Similar behavior of gypsum dissolution rates was reported by Laptev (1939), Kuznetzov (1947), and Pechorkin (1986). This fact has important speleogenetic consequences.

The second-order equation for the dissolution of anhydrite, in contrast to that of gypsum, requires much longer times to achieve saturation. The travel distance for water flowing through fissures in anhydrite could be rather long before sufficient $CaSO_4$ is dissolved to precipitate gypsum. The conditions required for gypsum to be precipitated from solutions that have dissolved anhydrite are reached gradually, due to the second-order dissolution kinetics, but when they are achieved the precipitated gypsum may seal the seepage paths.

The main concern of dissolution-kinetics studies are variations in K, which is not a true constant but one that varies with changing boundary-layer conditions. These conditions affect the thickness of the layer, which varies with the flow velocity over the dissolving surface, the ionic strength of the solution, and its temperature. The appropriate values of K that encompass these variables are considered briefly below, along with such other parameters as the diffusion coefficient, which reflects ion mobility (values for the common inorganic ions are rather similar). Theoretical calculations of rate constants for transport-controlled dissolution are rarely adequate, and experimental data are used in most cases (Frank-Kamenetsky, 1987).

Gypsum and anhydrite (which are strongly dipolar molecules) tend to form thick boundary layers, which are thus easily subject to thinning (stripping) by flowing water. This explains why K values and dissolution rates are strongly dependent upon flow velocity. The 4.1.6 Figure 8 shows linear dependence on flow velocity for dissolution within laminar flow; for each doubling of flow velocity over gypsum, K doubles. But for anhydrite it only increases by one and a half times.

Note that K has small positive values even in stationary water. As turbulence sets in, K is expected to increase abruptly, but there are no experimental data for gypsum. In the case of

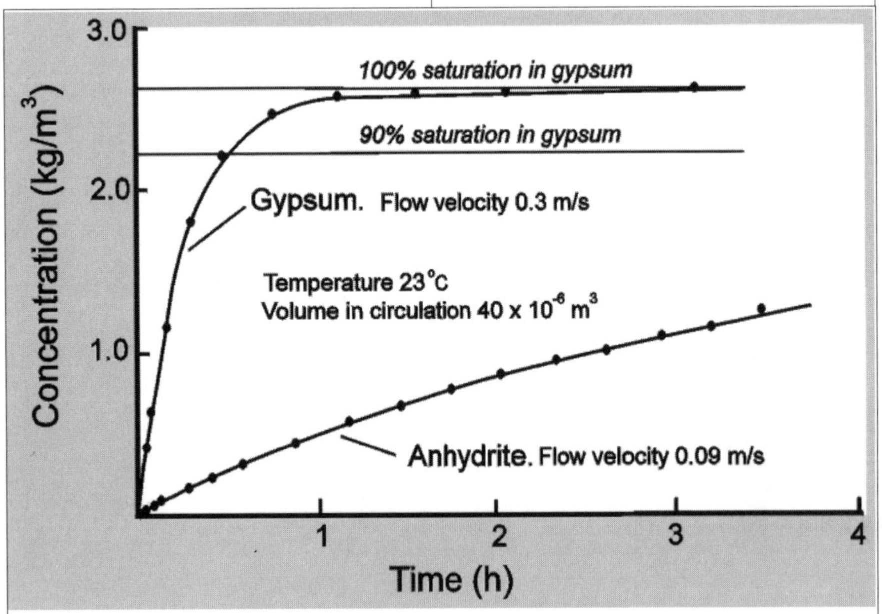

4.1.6 Figure 7. Dissolution rates of gypsum and anhydrite (after James and Lupton, 1978).

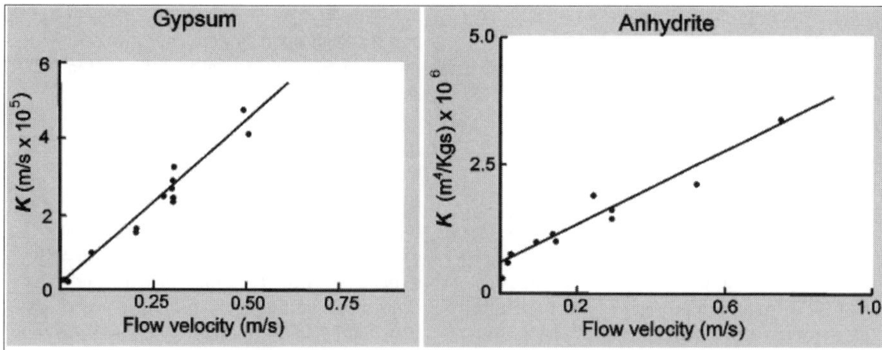

4.1.6 Figure 8. Dependence of the rate constants for gypsum and anhydrite upon flow velocity (after James and Lupton, 1978).

calcite, an increase by a factor of up to ten is reported (Buhmann and Dreybrodt, 1985a,b; Dreybrodt and Buhmann, 1991, Liu and Dreybrodt, 1997). James (1992) postulates that gypsum and anhydrite should exhibit a similar behavior. The strong dependence of gypsum dissolution rates upon flow velocity has important speleogenetic implications (see Chapter 7.1 in this volume). It also accounts for a variety of dissolutional features that form readily on gypsum surfaces.

The presence of other dissolved salts increases the ionic strength of a solution, causing diminishing of the thickness of the diffusion boundary layer and hence rising K values. This is illustrated by 4.1.6 Table 2, which summarizes data presented by James and Lupton (1978). The rate constant almost doubles for gypsum, but it increases by a factor of 9 for anhydrite, as the salt concentration rises from 0 to 10 g/L. Apparently, this effect needs to be considered when interpreting karst development in deep-seated settings, where the content of sodium and chloride ions is high. This is especially true for anhydrite rocks.

Data on the temperature dependence of K for gypsum are given in 4.1.6 Table 3. James (1992) suggests that a proportional relationship of log K to 1/T should be used to adjust K values from one temperature to another.

Gypsum-Anhydrite-Gypsum Conversions

The stability and solubility of gypsum and anhydrite are greatly affected by changes in the physical and chemical parameters that occur within common geologic environments. The conversion of gypsum to anhydrite and back to gypsum are common processes.

Geologic data suggest that in evaporitic environments at shallow depths, sulfates occur mainly in the form of gypsum, but that anhydrite predominates at depths exceeding 450 m. There are numerous exceptions to this usual situation, however with gypsum occurring at greater depths, and localized or dispersed anhydrite being found at shallow depths. Theoretical and experimental data on the stability of sulfate minerals and the mechanisms of conversion from one mineral to the other also are controversial. Some misleading views are common in the literature. This Chapter discusses the current understanding of the problem, which is important for the interpretation of karst processes and associated phenomena in gypsum and anhydrite.

The stability fields for gypsum and anhydrite are depicted in 4.1.6 Figure 1. The presence of other salts, such as sodium chloride, also affects their stability and solubility. In evaporitic basins, calcium sulfates primarily precipitate in the form of gypsum (Strakhov, 1962; Sonnenfeld, 1984). Anhydrite is believed to originate mainly by the dehydration of gypsum due to the effects of high pressure and temperature during burial. However, Sonnenfeld (1984) suggested that high pressure and temperature alone are not sufficient to explain the transition of gypsum to anhydrite. He showed that gypsum dehydration occurs widely during early diagenesis, where it takes place at shallow burial depth by interaction with hygroscopic brines of Na, Mg, or Ca chloride.

James (1992) noted that in very hot climates gypsum can dehydrate to anhydrite when it is exposed at the surface with temperatures above 42°C or where highly saline water is present. These changes are slow and mainly unaffected by diurnal cycles, but over longer periods they can be affected by seasonal changes. It can be assumed that in such conditions the conversion occurs by the dissolution of gypsum and subsequent precipitation of anhydrite, not by alteration within the solid phase.

Regardless of how the anhydrite formed, most mature gypsum rocks appear to be secondary, having formed by hydration of anhydrite to gypsum after uplift to shallow subsurface levels. Consequently, the conversion of anhydrite to gypsum is a significant process for karst development. It also has important implications for engineering and construction practices.

Rock Expansion

The common view is that the conversion of anhydrite to gypsum is accompanied by an overall increase in rock volume. Kushnir (1988) quoted an increase in rock volume of 18.25 percent, Pettijohn (1975) 30 to 50 percent, Gorbunova (1977) 64.9 percent, and Ford and Williams (1989) 30 to 67 percent. Sonnenfeld (1984) quoted an increase of 61 percent but stressed that a pressure of about 2 x 10^6 Pa, corresponding to a 60 to 75 m thickness of overlying rocks, would effectively balance the pressure generated by hydration and thus prohibit expansion. The effect is referred to widely in texts about karst (e.g. Jakucs, 1977), with the argument that such expansion would seal most of the fissures in the gypsum-anhydrite rock, preventing water circulation and karst development.

When expressed in this generalized form, such views are misleading. Closer examination of the problem reveals that expansion need not necessarily occur, and that a variety of mechanisms may be involved in the conversion processes. The problem is not

Concentration of NaCl, g/L	Gypsum: K x 10^5 (m/s)	Anhydrite: K(m^3 kg^{-1} s^{-1})
0	1.5	0.45
10	2.9	0.77
30	3.2	1.7
100	5.8	5.8

4.1.6 Table 2. Effect of dissolved sodium chloride on the dissolution rates of gypsum and anhydrite (after James and Lupton, 1978)

Temperature, C°	K x 10^5
5	0.8
15	1.7
23	2.6

4.1.6 Table 3. Effect of temperature on the rate constant of gypsum dissolution with temperature; flow velocity 0.25 m/s (after James and Lupton, 1978).

clear theoretically, especially when rate processes are concerned, and the field data are controversial. Reported observations of heave and swelling, claimed to have resulted from the hydration of anhydrite, may represent specific local conditions. Geologic observations of folded structures in gypsum and the deformation of adjacent layers (assumed to prove expansion by anhydrite-gypsum conversion) may well be explained by other mechanisms (see Klimchouk et al., 1995, for an example).

Furthermore, other field data show that in some underground and open-cast anhydrite mines, no heave has occurred (e.g. Kaiser, 1976; James, 1992). Experimental data and interpretation also conflict, suggesting that expansion during the conversion from anhydrite to gypsum is not always the rule.

Nekrasov (1945) derived an expression describing the limit of compression Δ_{lim} in a system caused by full hydration

$$\Delta_{lim} = (A/d_a + B) - C/d_c \qquad (7)$$

where A is the mass of the original substance of density d_a, B is the quantity of added water (density = 1 g/cm^3), and C is the mass of hydration product of density d_c. A system will compress in proportion to the volume of water involved in the reaction. This means that the exact changes depend on whether the process proceeds in an open or a closed system.

Theoretical calculations (Zanbak and Arthur, 1986; Pechorkin, 1986; Kushnir, 1988; James, 1992) suggest that when anhydrite converts completely to gypsum, the molar volume of the solid phase increases by factor of 1.626, but the overall volume of the system reduces by 8.7 percent. Pechorkin (1986) reported experimental data for a closed system. He used 18 to 22 gram samples of anhydrite placed respectively in distilled water and in a saturated solution of $CaSO_4$. These were hermetically sealed for 1.5 years under normal pressure conditions. Complete conversion to gypsum occurred, resulting in a reduction in the overall system volume of 3 percent in the case of the distilled water and 2.8 percent for the saturated solution.

Simultaneously, the solid volumes increased respectively by 3.1 percent and 4.1 percent. However, the short time reported for the complete conversion to gypsum apparently conflicts with the results of another experiment performed by James (1992). He used a small disk of anhydrite immersed in water for 12 years. This displayed the growth of gypsum crystals on it, but it was not fully converted to gypsum.

Mechanisms of Hydration

In nature the mechanisms and rate of hydration of anhydrite to gypsum depend on many factors including:

- Texture and structure of the rock.
- Form and chemical composition of the water.
- Temperature and pressure conditions.

Most authors believe that hydration proceeds through the dissolution phase, so that anhydrite dissolves to provide a solution of $CaSO_4$, which then precipitates from solution as gypsum (e.g. Kuznetsov, 1947; Mossop and Shearman, 1973; Quinlan, 1978; Kushnir 1988; James, 1992). However, Pechorkin (1986) argued that hydration through dissolution-precipitation accounts for only a minor proportion of rehydrated rocks. He considered that the main process proceeds through the diffusion of water molecules (or hydroxyl ions) into the anhydrite crystal lattice. Crystal-lattice defects are said to favor this process. This is also supported by data suggesting that the crystal-lattice defects in gypsum are inherited from anhydrite (Pechorkin, 1986). In reality, it is likely that the mechanisms of dissolution-precipitation and diffusion are closely interrelated.

Two main types of water contact anhydrite rocks:

- Interstitial water, which is retained in pores within the rock.
- Water that circulates freely through joints and other partings.

The former is disseminated throughout the rock mass, while the latter contacts only the surfaces of large rock blocks. I believe that interstitial water plays the more important role in the hydration of anhydrite rocks, even though its volume is relatively small due to the low porosity of anhydrite (note that not only the effective porosity, which is negligible in anhydrite, but total porosity should be considered).

If fissuring within a deep-seated anhydrite is rare, then such a system can be viewed as closed, with no additional water entering or leaving the system. At temperatures and pressures in which anhydrite is stable, the associated water saturated with $CaSO_4$ is in dynamic equilibrium with the mineral. When the rock becomes less buried and moves out of the stability field of anhydrite, the equilibrium is disturbed and the interstitial solutions precipitate gypsum. In closed or semiclosed conditions only partial conversion may be achieved, resulting in mixed anhydrite-gypsum rock, apparently with no expansion of the solid phase.

Conversely, some shrinkage of the overall solvent-solid system may cause some water to be sucked from adjacent beds into the hydration zone. With continuing emergence of the rock to progressively shallower depths, imposed fissuring and free-water circulation can result in open-system conditions, allowing water to partially recharge the remaining pore spaces. In this situation, localized hydration along flow paths becomes increasingly important. Water circulation through open fissures in anhydrite and gypsum at shallow depths may be fast enough to ensure that dissolution will remove any excess gypsum. In this situation, no overall expansion of the rock may be expected. The importance of the dissolutional removal of material is supported by the fact that the porosity of secondary gypsum is higher when compared with that of anhydrite.

This explanation combines several possible hydration mechanisms and encompasses most of the known geologic characteristics of gypsum-anhydrite formations. It suggests that in natural conditions the mechanisms and rates of conversion of anhydrite to gypsum depend

- On the tectonic setting.
- The water-bearing properties of surrounding sediments.
- Both the regional and local flow regimes.

It also suggests that in most cases no expansion in volume occurs during hydration. Expansion resulting in heave can be expected where thin layers of anhydrite are suddenly (in a geologic sense) released from their confining pressure and exposed to water. Perhaps a specific mode and rate of water ingress is required for expansion to occur. This view is in agreement with the fact that heaves definitely identified as due to hydration of anhydrite to gypsum have been reported in tunnels and mines (James, 1992).

Recrystallization

Sulfate rocks undergo recrystallization throughout their diagenetic and catagenetic history. Evaporites precipitated from aqueous solutions contain connate pore water preserved from their original deposition. Some of these connate brines are expelled from the pores by compaction during burial, but some remain. When meteoric water begins to circulate through open partings, it can replace part of these interstitial connate brines and induce recrystallization. The gypsum-anhydrite-gypsum conversions discussed above further complicate the water-rock interaction. All these processes continuously disturb the water-rock equilibrium and are accompanied by recrystallization of the deposits.

Recrystallization considerably affects the various properties of gypsum and anhydrite by altering, among other things, a rock's texture, structure, porosity, and strength. Consequently, it may influence karst development in many ways. Accretionary recrystallization is an important factor, because generally the solubility of gypsum is higher for the smaller crystals. The different solubilities and dissolution rates for crystals of mixed size are the main cause of recrystallization and directly influence the karst process (discussed above). However, the most important effect of recrystallization on karst is the alteration of the rock permeability. Two extreme examples are cited below to illustrate the possible effects.

In the western Ukraine, recrystallization has caused severe textural and structural differentiation of the buried gypsum sequence, with the formation of three distinct layers (Klimchouk et al, 1995). This differentiation has also caused the formation of largely independent superimposed networks of lithogenetic fissures confined to each layer (see Chapter 3.2 for a discussion of lithogenetic fissures).

These fissure networks have served as primary paths for meteoric waters, which have entered the sequence from the underlying aquifer and circulated upward under artesian conditions (Klimchouk, 1992; see also Chapter 5.2.1 in this volume). The structure of the lithogenetic fissuring was exploited by dissolution to generate the structure of huge maze-cave systems. Thus, textural-structural differentiation of the gypsum by recrystallization has been a primary guiding factor of this speleogenetic effect.

In Sicily, where gypsum massifs are exposed at the surface, a distinct crust up to 1 meter thick is formed in which all the open fissures tend to be sealed. This is probably the result of gypsum recrystallization caused by the loss of interstitial water from the exposed rock, and by dissolution-precipitation processes related to local climatic conditions. The exact mechanisms are not yet clear and need to be studied, but the effect upon karst development is obvious. The crust prevents dispersed recharge to the gypsum massifs from the surface, and water is thus allowed to penetrate deeper into the gypsum only along selected major fissures and faults.

Another morphogenetic effect of recrystallization of the uppermost exposed layer is the formation of small ridges and blisters (or tumuli). These occur where the crust coincides with the sedimentary bedding. These forms clearly result from the deformation of the geomechanically independent outer layer (detached from the substrate) by compressive stress, presumably caused by recrystallization. For expansional recrystallization to occur, however, certain conditions are required

- Open bedding planes subconcordant with the surface.
- Pathways for meteoric water to reach the bottom of the outer layer.
- Appropriate climatic conditions.

Contraction and fracturing of the outer layer precede the expansional recrystallization, having first provided conditions 1 and 2 above. Meteoric waters that escape surface evaporation and drain to the shallow subsurface, are drawn continuously upward back to the surface by capillary action through the pores in the outer layer, and this leads to accretionary recrystallization. The stresses generated by the volume expansion are released through swelling of the outer layer and are manifest as ridges and blisters.

References

Aksem, S.D., and Klimchouk, A.B., 1991, *Study of equilibria in a rock-solution system and some other problems of sulfate karst hydrochemistry*: Kiev Karst and Speleol. Center, no. 1, 25 p. (in Russian, res. English).

Blount, C.W., and Dickson, F.W., 1973, Gypsum-anhydrite equilibria in systems $CaSO_4$-H_2O and $CaCO_3$-NaCl-H_2O: *American Mineralogy,* v. 58, p. 323-331.

Buhmann, D., and Dreybrodt, W., 1985a, The kinetics of calcite dissolution and precipitation in geologically relevant situations of karst areas; 1. Open system: *Chemical Geology,* v. 48, p. 189-221.

Buhmann, D., and Dreybrodt, W., 1985b, The kinetics of calcite dissolution and precipitation in geologically relevant situations of karst areas; 2. Closed system: *Chemical Geology,* v. 53, p. 109-124.

Cigna, A., 1985, Some remarks on phase equilibria of evaporites and other karstifiable rocks: *Le Grotte d'Italia,* ser. 4, v. 22, p. 201-208.

Dreybrodt, W., and Buhmann, D., 1991, A mass transfer model for dissolution and precipitation of calcite from solutions in turbulent motion: *Chemical Geology,* v. 90, p. 107-122.

Ford, D.C., and Williams, P.W., 1989, *Karst Geomorphology and Hydrology:* London, Unwin Hyman, 601 p.

Forti, P., and Rabbi, E., 1981, The role of CO_2 in gypsum speleogenesis: *International Journal of Speleology* v. 11, p. 207-218.

Frank-Kamenetsky, D.A., 1987, *Diffusion and heat transfer in chemical kinetics:* Moscow, Nauka, 502 p. (in Russian).

Gorbunova, K.A., 1977, *Karst in gypsum of the USSR:* Perm University, 83 p. (in Russian).

Jakucs, L., 1977, *Morphogenetics of karst regions: Variants of karst evolution:* Budapest, Akademiai Kiado, 284 p.

James, A.N., 1992, *Soluble materials in civil engineering:* Chichester, Ellis Horwood, 435 p.

James, A.N., and Lupton, A.R.R., 1978, Gypsum and anhydrite in foundations of hydraulic structures: *Geotechnique*, v. 28, p. 249-272.

Kaiser, W., 1976, Behaviour of anhydrite after addition of water: *Society of the International Association of Engineeering Geologists Bulletin*, v. 13, p. 68-69.

Kaveev, M.C., 1963, About influence of carbon dioxide, originated during destruction of oil deposits, on development of karst processes: *Doklady AN SSSR,* v. 152, p. 3 (in Russian).

Klimchouk, A.B., 1994, Speleogenesis in gypsum and geomicrobiological processes in the Miocene sequence of the Pre-Carpatian region, *in* Sasowsky, I.D., and Palmer, M.V., eds., Breakthroughs in Karst Geomicrobiology and Redox Geochemistry: Abstract and Field Trip Guide, *Karst Water Institute, Special Publication,* no. 1, p. 40-42.

Klimchouk, A.B., 1997, The role of karst in the genesis of sulfur deposits, PreCarpathian region, Ukraine: *Environmental Geology,* v. 31, p. 1-20.

Klimchouk, A.B., Andrejchouk, V.N., and Turchinov, I.I., 1995, *Structural prerequisites of speleogenesis in gypsum in the western Ukraine:* Kiev, Ukrainian Speleological Assocociation, 104 p.

Korzhinsky, D.S., 1953, Essay on metasomatic processes, *in* Osnovnye problemy v uchenii o magmaticheskikh rudnykh mestorozhdenij: Moscow, AN SSSR Publ. (in Russian).

Kushnir, S.V., 1988, *Hydrogeochemistry of sulfur deposits of the Pre-Carpathians:* Kiev, Naukova dumka, 179 p. (in Russian).

Kuznetsov, A.M., 1947, On dissolution of gypsum and anhydrite: *Trudy Estestvenno-nauchnogo instituta*, Perm, v. 12, no. 4, p. 127-133 (in Russian).

Langmuir, D., 1971, The geochemistry of some carbonate ground waters in central Pennsylvania: *Geochimica et Cosmochimica Acta,* v. 35, p. 1023-1045.

Laptev, F.F., 1939, *Aggressive action of water on carbonate rocks, gypsum, and concrete:* Moscow-Leningrad, 120 p. (in Russian).

Liley, P.E., Touloukian, Y.S., and Gambill, W.R., 1963, Physical and chemical data, *in* Perry, J.H., ed., Chemical Engineer Handbook: New York, McGraw-Hill Book Co., 4th edition, p. 30-75.

Liu, S.T., and Nancollas, G.H., 1971, The kinetics of dissolution of calcium sulfate dihydrate: *Journal of Inorganic Nuclear Chemistry,* v. 33, p. 2295-2311.

Liu, Z., and Dreybrodt, W., 1997, Dissolution kinetics of calcium carbonate minerals in H_2O-CO_2 solutions in turbulent flow: The role of the diffusion boundary layer and the slow reaction $H_2O + CO_2 \Leftrightarrow H^+ + HCO_3^-$: *Geochimica et Cosmochimica Acta,* v. 61, p. 2879-2889.

Manikhin, V.I., 1966, On the question of solubility of calcium sulfate under high pressures: *Geokhimicheskie Materialy*, v. 13, p. 193-196 (in Russian).

Mel'nikova, Z.M., and Moshkina, I.A., 1973, Solubility of anhydrite and gypsum in the system Na-Mg-Ca-Cl-SO_4-H_2O: *Izvestija Sib. Otdel. AN SSSR*, ser. khim. nauk, v. 4, no. 21, p. 176-182 (in Russian).

Mossop, G.D., and Shearman, D.J., 1973, Origins of secondary gypsum rocks: *Institute of Mining and Metall. Bulletin*, p. B147-B154.

Nekrasov, V.V., 1945, Change of systems volume during solidification of hydraulic viscous materials: *Isvestiya AN SSSR*, otd. tekhn. nauk, no. 6 (in Russian).

Paine, N., Escobar, E., Hallowes, G.R., Sodha, V.C., and Anagnosei, G., 1982, Surveillance and reevaluation of the Poechos Dam, right wing embankment, Peru: *114th Cong ICOLD*, Rio de Janeiro, Proceedings, Qn 52, R. 19, p. 333-343.

Pechorkin, A.I., 1986, *Geodynamics of sulfate karst:* Irkutsk University Publ., 172 p. (in Russian).

Pettijohn, F.J., 1975, *Sedimentary rocks:* New York, Harper and Row, 3rd ed.

Quinlan, J.F., 1978, *Types of karst, with emphasis on cover beds in their classification and development:* University of Texas at Austin, Ph.D. thesis.

Raines, M.A., and Dewers, T.A., 1997, Dedolomitization as a mechanism for karst generation in Permian Blane Formation, southwestern Oklahoma, USA: *Carbonates and Evaporites,* v. 12, p. 24-31.

Shternina, E.B., 1949, Solubility of gypsum in water solutions of salts: *Izvestija sectora fiz.-him. analiza IONH AN SSSR,* v. 17, p. 203-206 (in Russian).

Sokolov, D.S., 1962, *Principal conditions of karst development:* Moscow, Gosgeolizdat, 321 p. (in Russian).

Sonnenfeld, P., 1984, *Brines and evaporates:* London, Academic Press.

Stankevich, E.F., 1970, On a possibility of development of deep-seated karst: *Voprosy Karstovedeniya*, no. 2, Perm University, p. 43-47 (in Russian).

Strakhov, N.M. 1962. *Principles of the theory of lithogenesis. Vol.III Regularities of composition and distribution of aridic sediments:* Moscow; Academy of Sciences Publ. 550 p., (in Russian).

Turyshev, A.V., 1965, About a possible way of the formation of karst cavities at great depths, *in Gidrogeol. Sbornic,* no. 4 (Trudy Instituta Geologii UFAN SSSR, no. 76), Sverdlovsk (in Russian).

White, W.B., 1988, *Geomorphology and hydrology of karst terrains:* New York, Oxford University Press, 464 p.

Wigley, T.M.L., 1973, Chemical evolution of the system calcite-gypsum-water: *Canadian Journal of Earth Sciences,* v. 10, p. 306-315.

Zanbak, C., and Arthur, R.C., 1986, Geochemical and engineering aspects of anhydrite-gypsum phase transitions: *Association of Engineering Geologists Bulletin.*, v. 23, p. 419-433.

Zdanovsky, A.B., 1956, *Kinetics of dissolution of natural salts in conditions of forced convection:* Leningrad, Goskhimizdat, 219 p. (in Russian).

Zverev, V.P., 1967, *Hydrogeochemical investigations of the system gypsum-groundwater:* Moscow, Nauka, 99 p. (in Russian).

4.1.7. Dissolution of Salt

Amos Frumkin

Abstract

Halite, the mineral of sodium chloride, dissolves by simply dissociating to Na^+ and Cl^-. Special features such as density stratification within water may arise from the high solubility, 360 g/L, while the rapid kinetics promotes intense karstification within a short time. Salt water enhances limestone dissolution by the ionic-strength effect. Other soluble chlorides reduce halite solubility through the common-ion effect.

Discussion

Salt is the most soluble karst rock. Rock salt is commonly composed of halite (NaCl) with minor amounts of more soluble minerals, such as sylvite (KCl) and carnallite ($Kcl \cdot MgCl_2 \cdot 6H_2O$).

Halite dissolution consists of dissociation to Na^+ and Cl^-. Being independent of CO_2 partial pressure, this process is simple compared with limestone dissolution. Mixing corrosion is not effective in salt. Slow-moving aggressive water may reach saturation shortly after initial contact with rock salt. The equilibrium constant for halite, $Kh=(Na^+)(Cl^-)$, is 101.51 at 0°C, rising to 101.58 at 25°C. Halite content (represented by Na^+ concentration) of saturated brines is inversely proportional to the concentration of other soluble chlorides (4.1.7 Figure 1), which reduce halite solubility through the common-ion effect (Lerman, 1970).

This explains why Ca-Cl brines in springs of the Mount Sedom salt diapir (see Chapter 7.2.1) contain less Na^+ than do saturated floodwaters (for example, Frumkin, 1994). Pure halite dissolution results in similar equivalent ratios of Na^+ and Cl^-, as expected (4.1.7 Figure 2).

Mixing of saturated salt waters of different temperatures would precipitate some salt (Cigna, 1986). The solution rate depends principally on the mode of flow and of contact between rock and water. Standing water in small rock salt pools can become saturated a few hours after flooding. Dissolution rate is controlled by diffusion in this case. Laminar fast-moving films on salt shaft walls were found to reach saturation in a few minutes. Turbulent stream flow remains undersaturated during high flood stages after flowing for a few minutes along several hundred meters of salt-cave channels. Solute concentration is inversely proportional to the discharge of vadose streams, as in limestone karst (4.1.7 Figure 3). The physical impact of aggressive water on the salt surface is important in controlling dissolution rates. Rapid incision of a salt-channel bed occurs at the foot of small waterfalls, where a hydraulic jump probably adds to turbulence and water-rock contact (Frumkin and Ford, 1995). Impacting raindrops form directional karren on subaerial outcrops, influenced by the prevalent local wind direction.

The regional karst denudation rate in Mount Sedom rock salt karst is estimated to be about 0.5 to 0.75 mm/year (Frumkin 1994), which is 1 to 2 orders of magnitude higher than limestone denudation rates in moister terrains. The maximum instantaneous downcutting rate calculated in a Mount Sedom cave is 0.2 mm/s (Frumkin and Ford 1995), which is about eight orders of magnitude higher than the rates in limestone cave streams.

The effect of pressure changes on salt solubility in normal speleogenetic environments is small and may be ignored.

With a solubility of 360 g/L at 20°C, the specific weight of salt karst water rises rapidly with an increase of solute content, giving rise to gravitational layering of standing water within caves and aquifers. The speleogenetic effect of increasing viscosity with salt content is still to be studied.

References

Cigna, A.A., 1986, Some remarks on phase equilibria of evaporites and other karstifiable rocks, in Atti simposio internazionale sul carsismo nelle evaporiti: *Le Grotte d'Italia*, v. 12, p. 201-208.

Frumkin, A., 1994, Hydrology and denudation rates of halite karst: *Journal of Hydrology*, v. 162, p. 171-189.

Frumkin, A., and Ford, D.C., 1995, Rapid entrenchment of stream profiles in the salt caves of Mount Sedom, Israel: *Earth Surface Processes and Landforms*, v. 20, p. 139-152.

Lerman, A., 1970, Chemical equilibria and evolution of chloride brines: *Mineralogical Society of America Special Papers*, no. 3, p. 291-306.

...End...

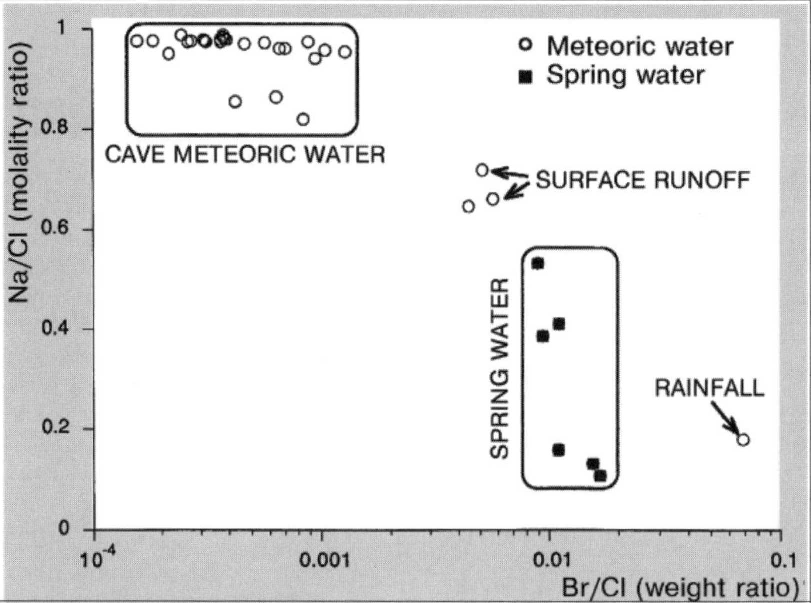

4.1.7 Figure 1. The relationship between Na/Cl and Br/Cl ratios for water samples in the Mount Sedom salt karst. Br^- content (together with K^+, Mg^{+2}) limits halite solubility and maximum Na^+ content due to the common-ion effect.

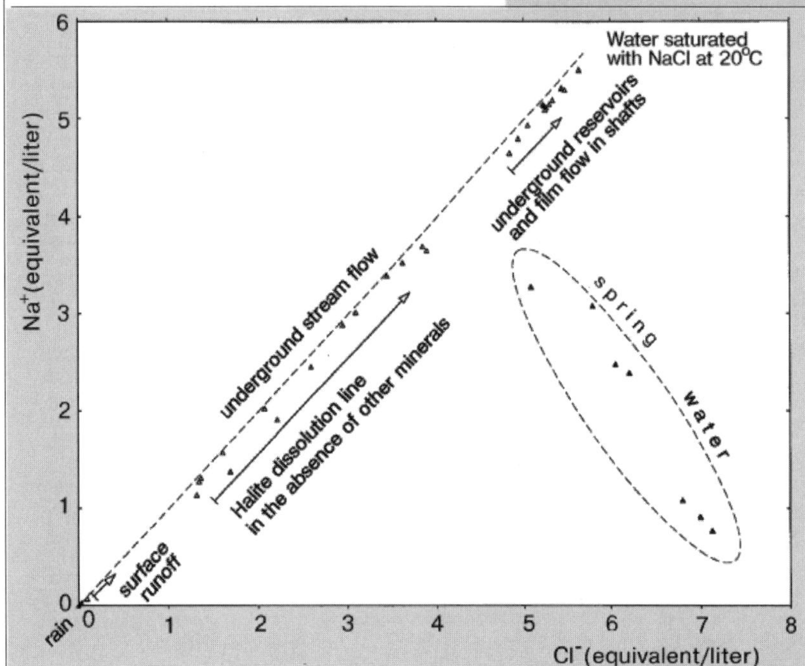

4.1.7 Figure 2. Salt karst water can be classified according to its Na^+ and Cl^- content, as exemplified by Mount Sedom water. Dashed line shows halite dissolution in pure water. Meteoric water falls close to this line because halite is the main mineral dissolved. Meteoric water evolves from rain (lower left) through cave streams to saturated reservoirs and film flow (upper right). Some water emerges from the salt karst without reaching saturation. Aquifer spring water (ellipse) includes a higher concentration of Ca^{+2}, Mg^{+2}, K^+, Br^-, and develops differently.

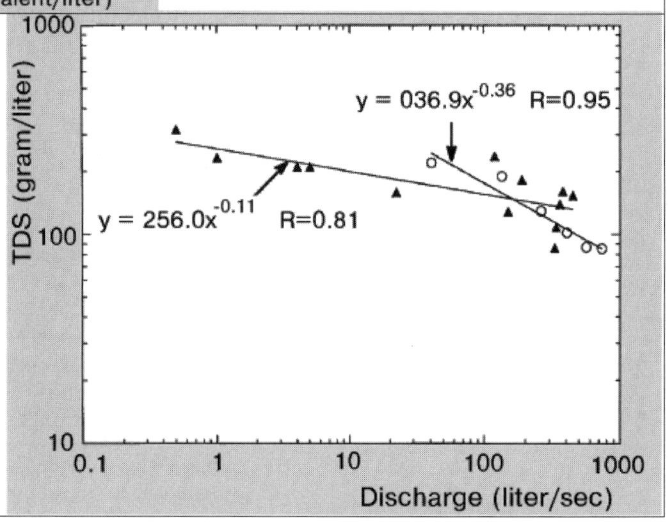

4.1.7 Figure 3. Relationship of solute concentration (TDS) to discharge during two flood events in Sedom Cave. TDS decreases with discharge, as it does in carbonate karst and surface rivers.

4.1.8. Dissolution of Quartz and Silicate Minerals

J.E.J. Martini

Abstract

In silica and silicate rocks, the majority of karst caves are developed in quartzite and therefore the dissolution of quartz is the most important process to consider. In this surficial environment, quartz dissolves by hydration of silica, producing silicic acid, which is not ionically dissociated. The solubility is not significantly influenced by the ionic and organic matter content of the water and is relatively constant at about 10 mg/L.

In contrast, the solubility of silicate minerals, like feldspar, is more complex, higher than quartz, and strongly controlled by the ionic properties of the solution, such as the pH. Generally the dissolution is incongruent, producing microcrystalline phyllosilicates and oxide minerals. Due to this last property, plus their higher solubilities, which lead to silica oversaturation relative to quartz, the presence of silicates is a negative factor in the development of karst in quartzite, when the latter contains a significant amount of silicates as impurities.

The other important factor considered is the extremely low rate of quartz dissolution. This allows water circulating along joints in quartzite to remain undersaturated in this mineral for very long distances. This slow rate permits the dissolution of quartz along crystal boundaries and leads to the weathering of quartzite into soft "neosandstone," and eventually into sand. The development of karst channels is not derived directly from the dissolution of quartz but by the weathering process outlined above, developed along joints and bedding planes, followed by piping forming the cavities when the material is sufficiently friable.

Introduction

Although fields of rock pinnacles and tower crags, evocative of karst morphology, are common on quartzite and to a lesser extent also on granite, silica and silicate rocks are not regarded as favorable for the development of cave systems. Nevertheless the caves exist. But in contrast to the classic karsts in limestone, dolomite, gypsum, and salt, they are localized and developed only on very restricted surfaces of the rock massifs.

The largest sinkholes and stream passages occur in quartzite, mainly in Venezuela, Brazil, and South Africa (see Chapters 7.3.1 and 7.3.2 in this volume). It is essential to briefly review some of the aspects of quartz dissolution, which are relevant to understanding the speleogenesis in these rocks. The dissolution of silicate minerals also has to be taken into account, although it is of secondary importance.

Chemistry of Quartz Dissolution

The dissolution of quartz in water is essentially a simple process of hydration forming silicic acid (see review by Dove and Rimstidt, 1994)

$$SiO_2(s) + 2H_2O \Leftrightarrow H_4SiO_4\,(aq)$$

and the equilibrium constant at 25°C is

$$K \cong a_{H4SiO4\,(aq)} = 1.1\ x10^{-4}$$

The silicic acid dehydrates partially forming polymers such as

$$2H_4SiO_4\,(aq) \Leftrightarrow H_6Si_2O_7\,(aq) + H_2O$$

The formation of this polymer, however, is negligible in the domain of quartz undersaturation. Only in strongly supersaturated solutions does it become the predominant species. The monomer dissociates ionically

$$H_4SiO_4\ (aq) \Leftrightarrow H_3SiO_4^- + H^+$$

This reaction increases the solubility of quartz, but only above pH 9. Below this value the solubility is not significantly affected by pH changes and by the ionic content of the solution. It remains relatively constant around 10 mg/L. Complexing of silicic acid with certain organic acids, like citric acid, may increase the quartz solubility (Bennett, 1991).

At room temperature the rate of quartz dissolution is extremely slow and is controlled by the rate of bond breaking and hydration of silica at the surface of the mineral (review by Dove and Rimstidt, 1994). In laboratory experiments this very slow rate is illustrated by the time necessary to approach saturation, several years in pure water. At 25°C this rate is 10^{-17} moles cm^2/s (Bennett, 1991) but can be increased by adding cations and organic compounds.

For instance, in concentrated NaCl solutions at high pH values, it can reach 10^{-14} moles cm^2/s (Dove and Rimstidt, 1994). In the presence of citrate, the rate increases up to $10^{-15.5}$ (Bennett, 1991).

Chemistry of Silicate Mineral Dissolution

At room temperature in pure water, silicates dissolve and form the same complex silicate species as in the case of quartz dissolution–species which are poorly ionically dissociated. Cations balancing silicate in the minerals, such as K^+, Na^+, Ca^{2+}, Mg^{2+}, Fe^{2+}, Fe^{3+}, and Al^{3+}, form more ionically dissociated species, especially the monovalent and divalent ions. In order to balance the absorption of H^+ into silica species, OH^- is produced from the dissociation of water, resulting in alkaline solutions. Since OH^- is involved in the solubility products of the silica minerals, the pH influences their solubility, which increases with the acidity.

This is in contrast with quartz solubility, which is approximately constant at pH values below 9. The dissolution of most rock-forming silicates is further complicated by the simultaneous deposition of new mineral phases. These secondary minerals commonly contain trivalent cations and consist of clay minerals and iron and aluminium hydroxides. They generally form at moderate pH values, and are excluded in extremely acidic and alkaline domains, where the dissolution is practically congruent.

The dissolution of silicates can be illustrated by the case of orthoclase, which has been extensively investigated owing to its importance in the formation of clay minerals. The process can be written as

$$K\,Al\,Si_3O_8(s) + 8H_2O \Leftrightarrow K^+ + Al^{3+} + 3H_4\,SiO_4\ (aq) + 4OH^-$$

The dissolution is complex and involves transitional phases. The minimum solubility is of a few mg/L near neutrality, to several hundred mg/L at pH values below 4 (Wollast, 1967; Berner, 1978; Holdren and Speyer, 1986). Below this last pH figure the dissolution is apparently congruent, whereas above it kaolinite ($Al_2\,SiO_5\,(OH)_4$) or gibbsite ($Al(OH)_3$) form, depending on leaching conditions during the experiments. As in the case of quartz, the dissolution rate of orthoclase is controlled by the reaction at the water-mineral interface. At 25°C, this rate is about $10^{-15.5}$ moles cm^2/s, a figure which is constant between pH 3 and 8, but increases at both lower and higher values (Helgeson et al., 1984).

Dissolution of Quartz and Silicates from Field Observations

The solubility of quartz can also be assessed from surface-water and groundwater analyses. Quartzite waters are usually very pure, with silica as the dominant solute. For instance in ground water from the Table Mountain Formation (South Africa)—a fairly pure quartzite—the silica content in 17 samples ranged from 1 to 30 mg/L, averaging 6.3 mg/L (Bond, 1946). This value suggests a slight undersaturation in quartz. The single high figure of 30 mg/L indicates oversaturation. This might be due to the presence of silicate impurities, like mica, feldspar, or chlorite, or to seasonal evaporation, which can concentrate silica well above the solubility limit of quartz. The latter does not precipitate, however, due to slow kinetics.

In quartzite cave systems of Mpumalanga Province, South Africa, comparable values were obtained from stream water, ranging from great undersaturation to saturation in quartz (Martini, 1979). In the high-rainfall tropical region of southern Venezuela, in water drained from the Roraima Quartzite, the silica values ranged from less than 1 mg/L in surface water on tabular mountain tops to 5-7 mg/L in the valley streams (Chalcraft and Pye, 1984). Here too, quartz undersaturation to saturation is suggested.

Due to a high content of organic acids, low pH values are recorded from most quartzite waters, for instance, 3.5 to 4.7 in southern Venezuela (Briceño et al., 1991). This suggested to these authors that organic acids leached from soil do not appreciably enhance the solubility of quartz, inasmuch as the values of silica in solution remain low. This is in agreement with experiments showing that complexing of silica with organics is effective only at high pH values (Iler, 1977; Ingri, 1977). Although the solubility of quartz is not affected, it remains possible, however, that the rate of dissolution is somewhat increased by organics. In conclusion, the chemistry of water in quartzite areas suggests that of the reactions reviewed above only the simple hydration of silica plays a significant role in the dissolution of quartz.

Higher silica content is measured in water draining terrains containing silicate minerals. For instance, this is the case for most major rivers in the world. They show a general oversaturation in silica with respect to quartz (Livingstone, 1963). The high-rainfall area of Kwazulu Natal, South Africa, can be given as an example: groundwaters of granitic terrains contain an average of 35 mg/L SiO_2, ranging from 20 to 53 mg/L (Bond, 1946). These high values are in agreement with a solubility and solution rate higher than that of quartz for many common silicates, such as feldspar, as measured from laboratory experiments (Wollast, 1967; Berner, 1978).

A highly variable dissolution rate of the silicate minerals is evidenced by their resistance to weathering. Talc is an example of a mineral with an extremely slow rate, as it remains practically unaltered after prolonged leaching in lateritic soil (Akpanika et al., 1987). At the other extreme of the range, wollastonite is relatively rapidly dissolved, as on outcrops it forms smooth surfaces and fluting comparable to the morphology of limestone (Martini, 1986). Although some silicates have relatively fast dissolution rates, they remain slow compared to carbonate minerals. It is common knowledge among field geologists that where outcrops of marble contain inclusions of silicate minerals, the latter stand in relief and do not show much evidence of solution, whereas the carbonate is deeply corroded by rain. This apparent resistance to dissolution is applicable to most common rock-forming minerals such as feldspar, mica, chlorite, and amphiboles.

Morphology and Texture Resulting from the Subsurface Dissolution of Quartz and Silicates

In order to explain speleogenesis and the formation of pinnacle karst in quartzite, several models have been proposed. The resistance of quartz to dissolution was the main problem most authors had to face. It was suggested that in a first phase, quartz was hydrated to opal, which is more soluble and therefore amenable to karst development (White et al., 1966). The transformation of quartz into opal, however, is not possible thermodynamically at surface conditions (Martini, 1979; Chalcraft and Pye, 1984).

It was also proposed that the quartzite had been initially softened by hydrothermal alteration along fissures (Szczerban et al., 1977). This process might have taken place locally where there is evidence of past hydrothermal activity, but it is unlikely in the majority of the cases of quartzite alteration, which obviously developed from downward movement of rain water. Weathering in a high-rainfall climate under alkaline conditions, with pH values above 9, has been suggested (Marker, 1976), but this too is unrealistic (Martini, 1979).

A model commonly proposed by geomorphologists involves dissolution of a more soluble matrix cementing detrital quartz grains (Marker, 1976; Briceño and Schubert, 1990). This matrix could be siliceous or of another nature. Genuine quartzite, however, has no matrix and is a practically monomineralic rock resulting from the recrystallization of quartz sand into a compact mosaic of crystals. The model of a more soluble matrix can only be advocated for impure quartzite and sandstone.

Microscopic observation shows that the weathering of true quartzite is penetrative through the mass of the rock and consists of quartz dissolution along crystal junctions (Martini, 1979; Chalcraft and Pye, 1984). These junctions are progressively enlarged until the quartzite loses its cohesion and is reduced to sand. In the subsurface, this weathering progresses preferentially along joints and bedding planes, because meteoric water is able to penetrate these initial planar fissures. They are enlarged hardly at all by dissolution, however, which affects essentially the mass of the quartzite on both sides of the joints. This is in contrast with the classic karst rocks, such as limestone in which the dissolution is confined to the wall of the joints, which are enlarged into cave passages (see Chapters 3.2.1 and 3.2.2 in this volume).

Nevertheless, there are exceptions to the general rule that no purely solutional cavity can develop directly from joint enlargement in quartzite. Sizeable solution cavities have been reported in quartzite associated with Pb-Ag mineralization in Colorado (Lovering et al., 1978). They formed far below the water table by circulation of saline solutions at temperatures of several hundred degrees centigrade. Under these conditions, the solubility and the rate of quartz solution is much higher than in the cool surficial environment.

In quartzite, the restriction of dissolution to crystal contacts has been attributed to the very slow rate of quartz dissolution (Martini, 1979, 1981). As the solution can remain undersaturated for a very long time, dissolution can occur in these very thin planar spaces, where the flow can only be extremely slow. In the case of carbonates, due to the much faster solution rate, dissolution along crystal contacts is usually greatly impeded. Another explanation could be that in quartzite the bonds between crystals are looser than in the case of calcite, thus also explaining the nature of quartzite weathering. This hypothesis, however, remains to be demonstrated.

In the silicate rocks such as granite, schist, and basalt, the weathering is penetrative as in the case of quartzite and also leaves a soft residue called saprolite. The process, however, is more complex because the dissolution is incongruent, and secondary minerals are formed, most commonly phyllosilicates (clay). In granite, feldspars are transformed into kaolinite, whereas quartz remains intact at first. This is not only due to the slower solution rate of quartz but to the fact that the dissolution of feldspar oversaturates groundwater with respect to quartz. This impedes the dissolution of quartz.

As discussed previously, in the case of quartzite weathering it was assumed that the rock was monomineralic. If it contains minor accessory silicates, their dissolution would interfere with the dissolution of quartz, as in the case of granite weathering. The distance a solution can travel through the rock before becoming saturated in silica with respect to quartz is shortened, and the dissolution does not proceed deeper as long as these silicates are not completely leached.

When considered from both the theoretical aspect and from field observations, the dissolution of quartz and silicates ought not to produce solution passages in the near-surface environment. Nevertheless, important stream caves have been reported. These are most often in quartzite of high-rainfall areas (see Chapter 7.3 in this volume) but to a lesser extent also in granite (Finlayson, 1982). They are not directly formed by dissolution, but by piping or suffosion, after preliminary weathering of the rocks along joints and bedding planes.

The process is mechanical and consists of a grain-by-grain removal of particles by fast-moving groundwater. These channels develop at a faster rate than in usual solution passages. They start to form at the emergence of aquifers and then propagate upstream. Cave passages in quartzite are generally vadose. This speleogenetic model is accepted by the majority of the authors (Ollier, 1969; Colvée, 1973; Ötvös, 1976; Martini, 1979; Briceño and Schubert, 1990).

References

Akpanika, O.I., Ukpong, E.E., and Olade, M.A., 1987, Mineralogy and geochemical dispersion in tropical soil overlying a talc deposit in south western Nigeria: *Chemical Geology,* v. 63, p. 109-119.

Bennett, P.C., 1991, Quartz dissolution in organic-rich aqueous systems: *Geochimica et Cosmochimica Acta,* v. 55, p. 1781-1797.

Berner, R.A., 1978, Rate control of mineral dissolution under Earth's surface conditions: *American Journal of Science,* v. 278, p. 1235-1252.

Bond, G.W., 1946, A geochemical survey of the underground water supplies of the Union of South Africa: *South Africa. Geological Survey Memoir,* v. 41, 208 p.

Briceño, H.O., and Schubert, C., 1990, Geomorphology of the Gran Sabana, Guyana Shield, south eastern Venezuela: *Geomorphology,* v. 3, p. 125-141.

Briceño, H.O., Schubert, C., and Paolini, J., 1991, Table-mountain geology and surficial geochemistry: Chimantá Massif, Venezuelan Guyana Shield: *Journal of South American Earth Sciences*, v. 3, p. 179-194.

Chalcraft, D., and Pye, K., 1984, Humid tropical weathering of quartzite in south eastern Venezuela: *Zeitschrift für Geomorphologie*, v. 28, p. 321-332.

Colvée, P., 1973, Cueva en cuarcitas en el cerro Autana, Territorio Federal Amazonas: *Boletin Sociedad Venezolana Espeleologia*, v. 4, p. 5-13.

Dove, P.M., and Rimstidt, J.D., 1994, Silica-water interactions, *in* Heaney, P.J., Prewitt, C.T., and Gibbs, G.V., eds., *Silica: Reviews in Minerology*, v. 29, p. 259-308.

Finlayson, B., 1982, Granite caves in Girraween Park, south-east Queensland: *Helictite,* v. 20, p. 53-59.

Helgeson, H.C., Murphy, W.M., and Aagaard, P., 1984, Thermodynamic and kinetic constraints on reaction rates among minerals and aqueous solutions. II Rate constants, effective surface area, and the hydrolysis of feldspar: *Geochimica Cosmochimica Acta,* v. 48, p. 2405-2432.

Holdren, G.R., and Speyer, P.M., 1986, Stoichiometry of alkali feldspar dissolution at room temperature and various pH values, *in* Colman, S.M., and Dethier, P.P., eds., *Rates and chemical weathering of rocks and minerals:* London, Academic Press, 603 p.

Iler, R.K., 1977, Hydrogen-bonded complexes of silica with organic compounds, *in* Bentz, G., and Lindquist, I., eds., *Biochemistry of silicon and related problems:* New York, Plenum Press, p. 53-76.

Ingri, N., 1977, Aqueous silicic acid, silicates and silicate complexes, *in* Bentz, G., and Lindquist, I., eds., *Biochemistry of silicon and related problems:* New York, Plenum Press, p. 3-52.

Livingstone, D.A., 1963, Chemical composition of rivers and lakes: *U.S. Geological Survey Professional. Paper 440-G*, 64 p.

Lovering, T.S., Tweto, O., and Lovering, T.G., 1978, Ore deposits of the Gilman District, Eagle Country, Colorado: *U.S. Geological Survey Professional Paper 1017*, 90 p.

Marker, M.E., 1976, Note on some South African pseudokarst: *Boletin Sociedad Venezolana Espelologia,* v. 7, p. 5-12.

Martini, J.E.J., 1979, Karst in Black Reef Quartzite near Kaapsehoop, Eastern Transvaal: *Annals Geological Survey of South Africa,* v. 13, p. 115-128.

Martini, J.E.J., 1981, The control of karst development with reference to the formation of caves in poorly soluble rocks in the eastern Transvaal, South Africa: *8th International Congress of Speleology,* Bowling Green, Proceedings, p. 4-5.

Martini, J.E.J., 1986, Can caves form in wollastonite?: *Bulletin of South African Speleological Association,* v. 27, p. 22-30.

Ollier, C.D., 1969, *Weathering:* Edinburgh, Oliver and Boyd, 304 p.

Ötvös, E.G., 1976, "Pseudokarst" and "Pseudokarst terrains": problems of terminology: *Geological Society of America Bulletin,* v. 87, p. 1021-1027.

Szczerban, E., Urbani, F., and Colvée, P., 1977, Cuevas y simas en cuarcitas y metalimolitas del Grupo Roraima, meseta de Guaiquinima, estado Bolivar: *Boletin Sociedad Venezolana Espeleologica*, v. 8, p. 127-154.

White, W.B., Jefferson, G.L., and Haman, J.F., 1966, Quartzite karst in south-eastern Venezuela: *International Journal of Speleology,* v. 2, p. 309-314.

Wollast, R., 1967, Kinetics of the alteration of K-feldspar in buffered solutions at low temperature: *Geochimica et Cosmochimica Acta,* v. 31, p. 635-648.

4.2 Modeling of Initiation and Propagation of Single Conduits and Networks

4.2.1 Hardware and Software Modeling of Initial Conduit Development in Karst Rocks

Derek Ford, Stein-Erik Lauritzen, and Ralph Ewers

Abstract

Most dissolutional conduits develop along fissures (bedding planes, joints, or faults) in which the initial opening differs considerably from place to place, i.e. they are anisotropic. The propagation of conduits between input and output points in an anisotropic fissure was studied systematically with three types of hardware models:

- Electric-current analogs, and
- Sandbox analogs—both to investigate flowfield geometry.
- Artificial fissures cast in plaster of paris, with random anisotropic elements on their surfaces.

A first series of experiments studied the case of a single input to the fissure. The earliest protoconduits were observed to extend in a radial (Darcian) array, but one principal and several secondary tubes quickly extended down the hydraulic gradient, robbing competitors of their flow.

This mode of propagation was tested with a resistor-network computer model using 100 by 100 nodes, Weyl's (1958) equation for dissolution and transport with parameters appropriate for plaster, and three choices of opening–isotropic, anisotropic but homogeneous, and heterogeneous. The computer model results closely matched those of the hardware models.

In a second and third series of plaster hardware experiments, these cases were explored:

- Multiple inputs to the fissure in one rank.
- Multiple inputs in multiple ranks.

In the competition between inputs, some principal tubes in near ranks first "breakthrough" to the output boundary. This reorients the flowfields of failed nearby competitors, which then extend to join the principal via their closest secondaries. The process extends outward and to the rear, linking up all inputs in a "cascading system."

Introduction

The great majority of dissolutional cave passages in limestone, dolostone, and gypsum develop initially along natural fissures such as bedding planes, joints, and faults, and then become enlarged into the matrix of the rock on either side of them. In this Chapter we consider the initiation and propagation of passages through such fissures, from points of water input to output. In hydrologic terms, such exploited fissures can be considered to be very thin aquifers bounded by an aquitard or aquiclude that is the matrix.

At the present time, standard hydrogeologic models outside of karst research do not allow for the fact that dissolution may enlarge selected routes through selected fissures, radically altering the local hydrologic conductivity of a volume of rock. Instead, the individual fissures are considered to remain unmodified by groundwater flow through them, and to further simplify the analysis, they are treated as simple spaces of constant width between perfectly parallel walls (the "parallel-plate model").

Initial parallel plate conditions plus uni form dissolutional enlargement from them are illustrated in 4.2.1 Figure 1A. Passages like this can be found in caves, particularly where just one or two joint sets intersect within just one or two rock beds that are of uniform composition. It is common to find them forming mazes of limited extent in river cave entrances, for example, where they are created by floodwater backing up in an earlier

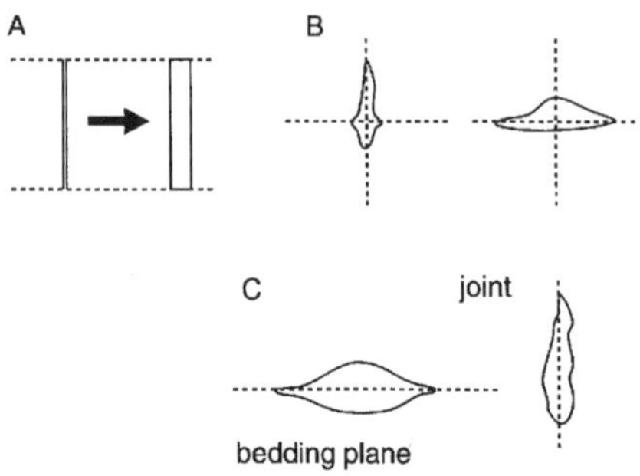

*4.2.1 Figure 1. (**A**) The dissolutional enlargement of an ideal (isotropic) parallel plate joint opening in a rock bed bounded by two bedding planes of much smaller aperture. (**B**) Development of linear passages at the intersections of a joint and a bedding plane, both of which are anisotropic. (**C**) Passage development in anisotropic bedding planes and joints.*

principal passage (Palmer, 1975; Groves and Howard, 1994).

In our experience, however, true parallel plate conditions are quite rare in the fissures that groundwater can penetrate in karst rocks. Rather, some patches of the fissure have a much wider opening than others adjoining them. In technical terms, conditions in the fissure are "anisotropic" because water or other fluids may flow more readily in some directions (along open patches) than in others (the more tightly closed areas). This compels the flow to be somewhat sinuous, threading routes from one wider patch to the next. A dissolutional enlargement that remains nearly as simple and linear as the parallel plate may occur, however, where it follows the junction of two such anisotropic fissures. This is shown in 4.2.1 Figure 1B. Here the vertical passage represents the case where the vertical fissure (usually a joint) was slightly more open on the average than the horizontal fissure (usually a bedding plane), and the horizontal passage (bedding-plane type) represents the converse.

The most fundamental natural case is illustrated in 4.2.1 Figure 1C. The fissure is anisotropic and there is no sustained guidance along intersections with other openings. This case applies in many bedding-plane- and joint- or fault-guided cave passages. Bedding planes are the more important hydrogeologically because they tend to be continuous over much greater areas than do individual joint or fault planes. They may also display greater anisotropy because of depositional effects, differential slippage, or small-scale dissolution preceding main phases of cave genesis ("nothephreatic" solution of Jennings (1985); "inception horizons" of Lowe (1992); see also Chapter 3.3 in this volume).

Mammoth Cave, Kentucky, and Hölloch, Switzerland, are two very great limestone cave systems (the longest and second longest caves in the world) where bedding-plane guidance is predominant. It is also common in well-bedded gypsum and gypsum-anhydrite sequences. When crawling flattened along such passages, knowing that the plane extends at impenetrable dimensions for hundreds or thousands of meters to either side, one poses the question "Why did the water enlarge this particular route?" The purpose of this chapter is to present an answer.

Hardware Models of Conduit Propagation in an Anisotropic Fissure

In the natural sciences, complex development is often approached by devising simplified models. There are three principal categories:

- Conceptual models (or "thought experiments," e.g. White, 1969).
- Mathematical models, including computer modeling as exemplified in Chapters 4.2.2 to 4.2.6.
- Physical or "hardware" models.

The latter may include scaled-down versions of the real thing, such as tiny river channels in sand in a laboratory flume, and analog models such as electric current substituting for water flow, (e.g. Bedinger, 1966).

In a comprehensive series of experiments Ewers (1966, 1972, 1982) and Waterman (1975) used three types of hardware models progressively to investigate the patterns of groundwater flow, dissolutional conduit propagation, and linkup along bedding planes. Studies began with the case of a single input of water and expanded to consider multiple inputs in one rank or file, and then multiple ranks. The first type of model was an analog, using the flow of electric current through copper sulfate solutions in a shallow pan to define the strength of fields of flow from the inputs. The second, also an analog, used dyed water injected into a thin layer of sand packed between parallel glass plates to determine the geometry of the flow fields.

In the third, most important, type of model, actual dissolutional caves were created at small scale in plaster of paris (artificial gypsum) and in salt. Gypsum and salt are to be preferred to natural limestone because their much greater solubility allows individual experiments to be completed in a reasonable time span. The chief experimental apparatus is shown in 4.2.1 Figure 2A.

Plaster blocks measuring 40 by 28 by 3 cm were cast carefully and sealed watertight and airtight around their edges and backs to expose a single soluble working surface for each experiment. There were two different types of surface:

- A low-anisotropy surface created by casting the plaster onto #100 grit sandpaper (4.2.1 Figure 2B).
- A high-anisotropy surface on #100 grit paper with blobs of polyester resin added at random to create lower permeability patches.

In some late work, narrow discontinuous linear grooves were also scribed onto the second type surfaces to simulate the linear intersections of 4.2.1 Figure 1B. In each experiment the working surface was clamped against an inflated transparent vinyl pillow. The transparency permitted all stages of the progress of dissolutional conduits between input and output to be observed and photographed. Flow and diffusion could be studied by injecting colored dyes.

To be truly comparable to the real thing, hydraulic models such as those in 4.2.1 Figure 2 must meet three tests of similarity (Yalin, 1971).

- First, lengths, areas, and volumes in the model should be proportional to those in the prototype—"geometric similarity."
- Second, the geometrically similar forms should undergo similar changes of shape or position (e.g. extension) as they evolve—"kinematic similarity."
- Third, values for the dimensionless parameters of fluid systems such as Froude number or Mach number should be identical.

As the model caves of 4.2.1 Figures 2B and Figure 3 illustrate, the gypsum models satisfy the geometric and kinematic criteria. Only the Reynolds number (inertial forces/viscous

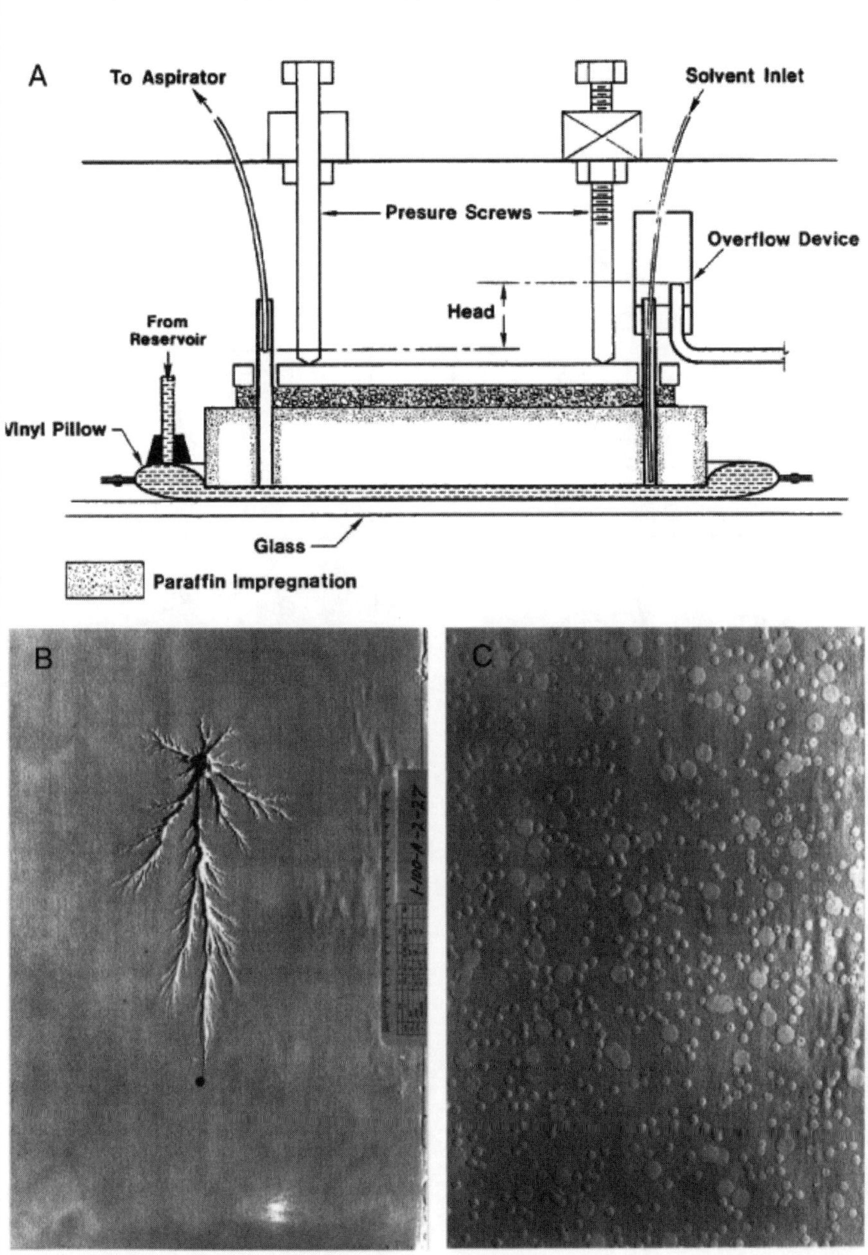

4.2.1 Figure 2. (A) The model apparatus for the study of dissolution along a fissure surface in a plaster of paris block (Ewers, 1982). (B) An example of the low-anisotropy type plaster working surface with completed experiment (1-100-A-2-27), in it. (C) A high-anisotropy type surface freshly prepared for an experiment.

forces) is important during early cave development, because groundwater velocities are very low; calculated Reynolds number for the models is around 2.0, within the range of 1.0 to 4.0 that is expected to apply during initial (Darcy) flow in nature.

Where groundwater flows slowly through a very narrow fissure in limestone, dolostone, or gypsum, it is expected to become chemically saturated with the dissolved mineral (or nearly so) within very short distances, as discussed in various sections of Chapter 4.1. Realistic dissolution models therefore should begin with the solution becoming nearly saturated before it has progressed more than a small proportion of the distance between input and output points. Under fixed boundary conditions of pressure and temperature, it is difficult to add the last few ions required to obtain 100% saturation in either natural or model systems. However Weyl (1958), considered the distance (relaxation length) needed to achieve 90% saturation to be an appropriate scaling parameter and derived the relation

$$L_9 = 0.304 \cdot v \cdot d^2 / D \qquad (1)$$

where v = velocity of flow, d = diameter of a circular tube and; D = the diffusion constant of the mineral (cm^2/s). For typical plaster models, L_9 = 0.1 cm or less during the early stages (e.g. 4.2.1 Figures 3A to 3A), more than two orders of magnitude less than the straight-line distance between the inputs and the outputs on the working surfaces.

At the beginning of each experiment, a head of water was applied to a point near the right edge of the simulated bedding plane as depicted in 4.21 Figure 3A. Discharge occurred along the left edge. The initial pressure field within the bedding plane is depicted in 4.21 Figure 3C1 and profiled along the central axis in 4.21 Figure 3B. Regions of slightly higher permeability near the input quickly developed larger openings due to their higher flow rate and the increased penetration distance they provided before the solvent became effectively saturated. Thus, the solvent flow within the bedding plane fissure quickly became channeled into a set of radial distributary tubes (4.21 Figure 3C2).

Undersaturated water could then be delivered by these tubes to points farther downstream where other regions of slightly higher permeability maintained channelized flow at all times during the experiments. The solvent moving along the virgin parting a short distance beyond the tube extremities was assumed, for all practical purposes, to be saturated, because no detectable dissolution occurred there.

The tube leading by the shortest available path in the direction of discharge, the principal tube (P), tended to grow most rapidly and at a slowly increasing rate (4.2.1 Figure 3C3). Such a path led in the direction of steepest potentiometric slope. This path was seldom a straight course because of the irregular distribution of permeability within the plane (the anisotropy). The greatest solvent flow followed this course.

The increasing rate of elongation along this tube was due to two factors related to the length change.

- Its increasing length reduced the resistance of the entire flow path from input to resurgence, because the proportion of virgin parting became progressively less.
- The tube's increasing length steepened the gradient in the remaining virgin bedding plane.

Both of these factors increased the flow rate at the downstream terminus of the evolving tube, which increased its rate of propagation. The succession of piezometric gradients in

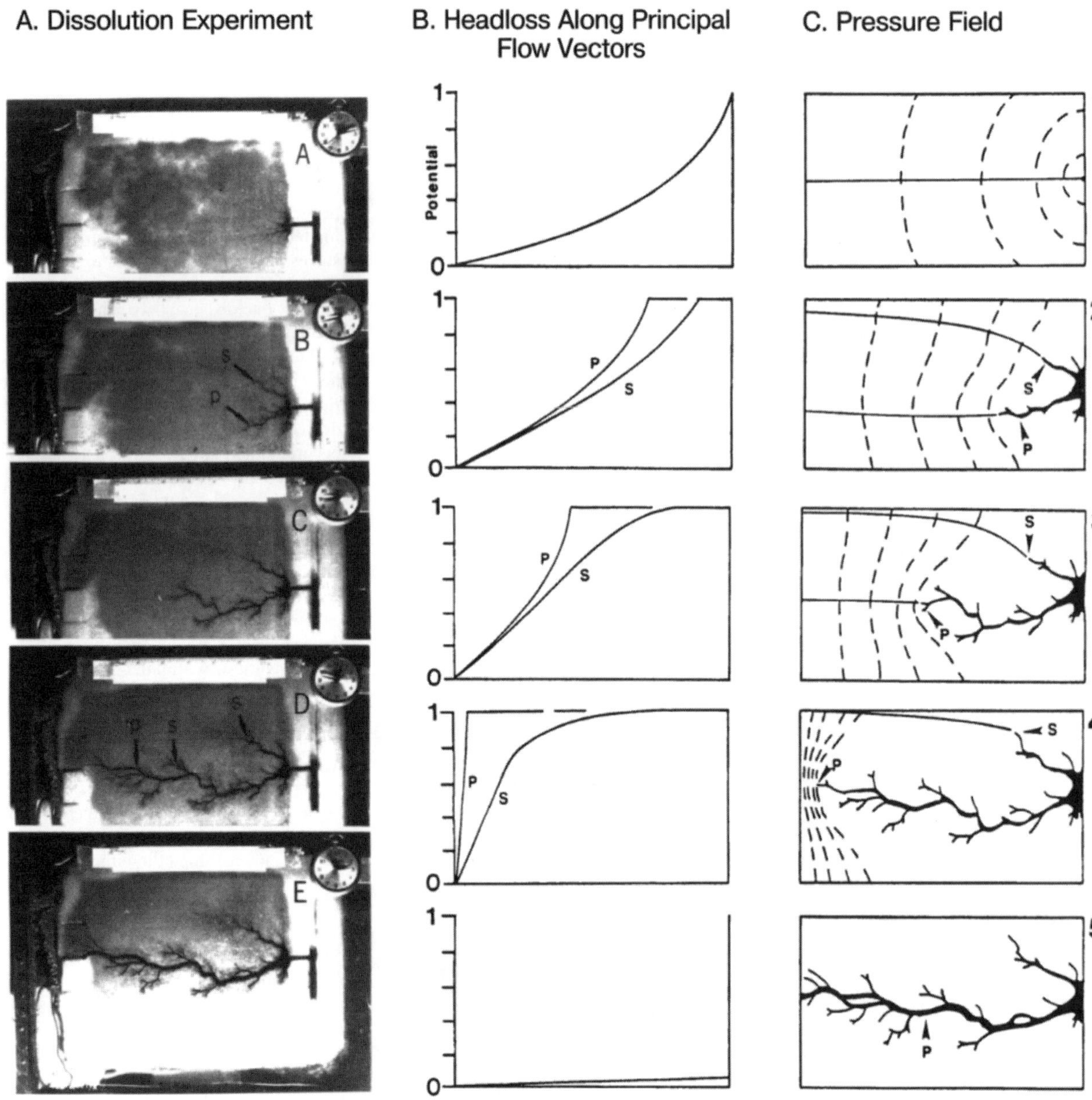

4.2.1 Figure 3. An example of a single-input dissolution experiment in plaster of paris: A–The progression of the dissolutional conduits; p is the principal tube, s are the secondary tubes. B–The change of the distribution of headloss along the principal and secondary tubes with the progress of the experiment. C–Plan representation of the headloss (distortion of the pressure field).

4.2.1 Figure 3C illustrate this phenomenon.

The remaining or secondary tubes (S) and the tube branches diverging from the principal tube (4.2.1 Figure 3C2, 3, and 4) showed lower initial rates of development, due to their longer discharge routes with their proportionally higher resistance to flow. In addition, the growth of the principal tube distorted the flow field in such a way as to further decrease the gradient and increase the flow-path length for the secondary tubes. These grew with decreasing vigor as a consequence, while the principal tube tunneled down-gradient at an ever accelerating pace. The flow-field distortion was due to the contrast in head loss across the high-resistance parting compared to the head loss along the very low-resistance tubes. The tubes transmitted the gravitational potential energy of the head at the input virtually undiminished to their extremeties. At the same time, the principal tube delivered solutionally aggressive water to the margins of the developing network.

Interconnections between the inactive secondary tubes—which provided paths parallel to the principal path—enlarged as their role shifted from one of lateral development of the network to one of parallel supply routes for the developing frontiers of the network downstream (4.21 Figure 3C5). This change in function modified the dendritic aspects of the network and increased its anastomotic character. The enlargement of the principal tube, and the tubes parallel to it, was the beginning of a process of gradual engulfment of the secondaries.

In this way, the tube network was propagated across the bedding plane in a narrow nearly straight band, closely following the original potentiometric slope. Deviations from a straight course occurred only in response to factors affecting the pressure field. The final result was a network closely resembling the "anastomotic bands" described by Ford (1971).

In nature as in the models, these early passages are very small, probably no more than a few centimeters in diameter at the input ends when breakthrough first occurs at the outputs, even in big cave systems. American readers can view excellent examples of these tiny passage patterns in many caves formed in the well-bedded gently dipping limestone of Indiana, Kentucky, Tennessee, and Missouri. Commonly the lower bed has fallen away locally from the inception bedding plane, which is equivalent to removing an opaque pillow in the model. Many spectacular examples are exposed in the ceiling in Kentucky Avenue, visited on the Half Day Tour in Mammoth Cave.

Software Modeling of Anastomosis Development in Plaster of Paris Models

Computer modeling experiments on dissolutional anastomosis development provide many opportunities to test the kinetic theory of microchannel growth. Computer models are particularly useful for sensitivity analysis. That is, one may vary the various input parameters systematically to investigate the conditions for realistic behavior of the model. The following is based on the work by Lauritzen et al. (1992) that aimed to investigate the Ewers model for a single input.

The advantage of modeling dissolution of gypsum is that the process does not involve any rate-limiting chemical reaction steps, as dissolution of limestone does. Dissolution of gypsum is a process of dissociation and diffusion of an ionic solid. The surface processes are almost instantaneous, hence this is an overall diffusion-controlled process (Liu and Nancollas, 1971). We have therefore adapted the diffusion-controlled approach suggested for calcite by Weyl (1958) to our model.

Resistor Network Computer Model

Fluid flow through porous media is modeled using a numerical "resistor" network method, which is a type of finite-difference technique (Odling and Webman, 1991). In the model, the porous medium (here a "porous" two-dimensional surface, the "parallel plate" bedding plane-parting) is represented by a square grid. Each element of the grid represents a volume of porous medium with an individual permeability. Because the permeability of each grid element can be set independently, the model allows a wide variety of permeability fields to be modeled.

At each grid node, the sum of the flows is zero through the four grid elements to which it is connected, i.e. there is a local conservation of flow (4.2.1 Figure 4). This results in a set of linear finite-difference equations for all grid nodes

$$(p_0 - p_1)k_{01} + (p_0 - p_2)k_{02} + (p_0 - p_3)k_{03} + (p_0 - p_4)k_{04} = 0 \quad (2)$$

where p_1 to p_4 are the pressures at the grid nodes and k_{01} to k_{04} are the permeabilities assigned to grid elements connected to the node. In order to simulate the pressure gradient during a hardware experiment, high pressure was assigned to nodes along the left margin and low pressure along the right margin of the grid. In order to solve the large matrix of equations (here for 100 x 100 nodes) for pressure and flow fields, a conjugate gradient algorithm was used (J. Petersen, unpublished; see Lauritzen et al., 1992) was applied. This algorithm was specially designed to be highly vectorizable, thus allowing fast solution of the system of equations.

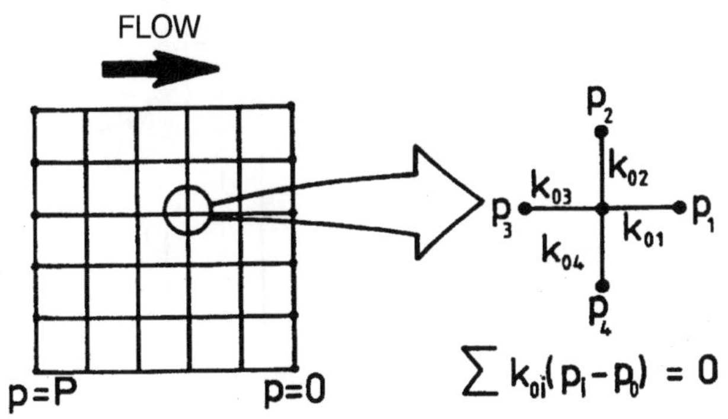

4.2.1 Figure 4. The resistor network method in which the medium is discretized onto a grid. Each grid element is assigned a permeability value, then pressure boundary conditions are imposed, and assuming mass conservation, the pressure at each node in the grid is determined (from Lauritzen et al., 1992).

Modeling Dissolution Effects

Weyl's (1958) model for the wall-fluid dissolution effect and solute transport converts to the following partial differential equation

$$\frac{\partial C}{\partial t} = D\nabla^2 C - v_z \frac{\partial C}{\partial z} \quad (3)$$

where C is the concentration, D is the diffusion constant, and z is the flow velocity that we will consider to be in the positive v_z direction, i.e. along the channel. Furthermore, assuming laminar flow demands that the product of the average flow velocity and the diameter of the tube be less than 64 cm^2/s (Batchelor, 1967), which is satisfied in our experiment. This leads to the differential equation (Lauritzen et al., 1992)

$$D\left[\frac{\partial^2 C'}{\partial r^2} + \frac{1}{r}\frac{\partial C'}{\partial r} + \frac{\partial^2 C'}{\partial z^2}\right] = 2\overline{v}(1 - \left(\frac{r}{a}\right)^2)\frac{\partial C'}{\partial z} \quad (4)$$

where $C' = 1 - C/Cm$ (Cm is the maximum, or saturation concentration), a is the radius of the capillary, and r is a variable radius between 0 and a, used for calculating the velocity and concentration profile across each channel. The boundary conditions are zero concentration at the beginning of the tube and a maximum concentration Cm, as z goes to infinity. The concentration is always Cm at the tube wall, where r = a.

Equation (4) was solved by assuming a series solution, which is nontrivial due to the large range in the values involved. Quadruple precision was necessary in order to obtain a sufficient number of terms in the series solution (see Lauritzen et al., 1992, for details).

In a network, four channels are connected at each node, leading solutes into and out of it. The neighboring channels receiving liquid are assumed to have no influence on the node sending it, thus simplifying concentration calculations. For each time step, an amount of wall rock was removed according to the increase in concentration between successive nodes, thus increasing the permeability. The final algorithm goes as follows:

- Solve the flow problem to obtain potentials where hydraulic conductivities k are given.
- Solve for the concentrations C at the end of each time step.
- Increase the hydraulic conductivity k.
- Return to the first step to start a new time step.

Model Parameters
Because the initial conditions of the hardware models used by Ewers (1982) are difficult to ascertain (e.g. initial channel diameter fields, effective diffusion constants, head values) a series of experiments was first conducted to reproduce the conditions of the laboratory experiments as closely as possible. Moreover, it was necessary to explore the influence of the various parameters on the nature of the dissolution patterns.

For the numerical experiments, an area of 0.5 by 0.5 m was discretized on a square grid of 100 x 100 nodes, so that each grid element represented a block of parallel wall material of 0.5 by 0.5 cm. Zero permeability was assigned to the top and bottom margins of the grid, providing no-flow boundary conditions. To simulate discrete point inputs, i.e. 4.21 Figure 3, all horizontal grid elements along the left margin were set to zero permeability (no-flow), except the central four, which were assigned a channel diameter of 2 mm each. The pressure at the inlet nodes at the left margin was set to a positive value and all those on the right margin to zero (linear output boundary), giving a range of possible pressure gradients. Each grid element was given a diameter corresponding to the initial experimental conditions and was assigned a permeability reflecting this initial channel radius.

For plaster of paris ($CaSO_4 \cdot 2H_2O$), the density was set to 1.43 g/cm^3, and a saturation concentration in water Cm of 2.4 g/L. Another unknown factor that controls the rate of wall retreat is the effective diffusion constant, which describes the hydrodynamic dispersion rate of $Ca^{2+}_{(aq)}$ and $SO_4^{2-}{}_{(aq)}$. This is related to the molecular diffusion constant and the Peclet number, which itself depends on the grain size and average particle velocity (Dreybrodt, 1988).

Effective diffusion constants are up to an order of magnitude greater than the molecular diffusion constants. In the first set of experiments, the initial permeability field, pressure gradient, and effective diffusion constant were varied systematically (all other parameters kept constant) to test the outcome on channel geometry.

Initial Permeability Field. In 4.2.1 Figure 5 are depicted the effects of using (a) a constant field of 100 μm, roughly equal to the #100 grit sandpaper size used by Ewers; (b) uniformly distributed over the range 60 to 140 μm; and (c) log-normal with a mean ln(2r) of -9.8 (i.e. 112 μm) and a standard deviation of 6.0. The fields were all spatially random and uncorrelated.

The constant-permeability field produced an asymmetrical tree of enlarged channels around each inlet (similar to the failed plaster-of-plaster bedding-plane experiments noted above) and two dominant straight channels that propagated straight across the model (4.2.1 Figure 5a).

On the other hand, the field composed of uniformly distributed apertures produced a single dominant channel with some tendency to branch into secondary conduits (4.2.1 Figure 5b). The initial permeability field composed of log-normally distributed apertures (4.2.1 Figure 5c) produced a dendritic pattern resembling those obtained in the vinyl pillow hardware models. Log-normally distributed initial apertures were therefore used in the succeeding experiments, although a different random-number-generator seed was used in each series of experiments.

Variable Pressure Gradient. Variation of pressure head (0.4, 0.2, and 0.1), produced a pattern that increased in growth rate and complexity (branching) with increasing pressure gradient (4.2.1 Figure 6). The pattern in 4.2.1 Figure 6c closely resembles the anastomotic bands of Ford (1971). Increasing the pressure gradient allows fluid to travel greater distances before attaining saturation, which results in longer micro-channels and a larger incidence of branching.

Effective Diffusion Constant. The results are shown in 4.2.1 Figure 7. The effective

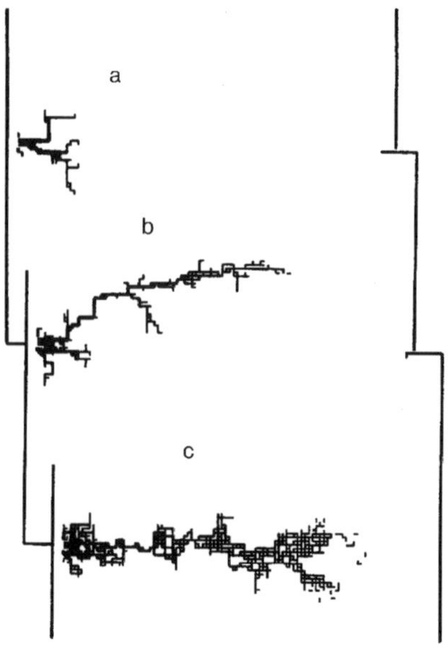

*4.2.1 Figure 6. Numerical model results for different pressure gradients. Initial channel width field as for 4.2.1 Figure 5c. Pressure gradients of (**a**) 0.1 (700 time steps), (**b**) 0.2 (400 time steps), and (**c**) 0.4 (150 time steps) (from Lauritzen et al., 1992).*

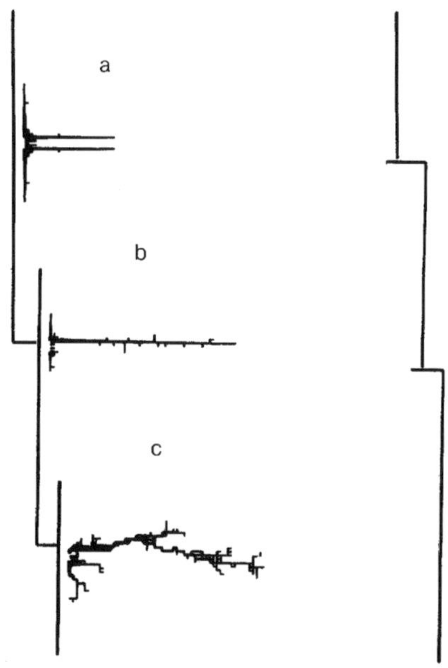

*4.2.1 Figure 5. Numerical model results for different initial aperture width fields. (**a**) constant width at 100 μm (400 time steps), (**b**) uniformly distributed over the range 60 μm to 140 μm (250 time steps), and (**c**) log-normally distributed width mode at 112 μm and standard deviation of 6.0 (350 time steps) (from Lauritzen et al., 1992).*

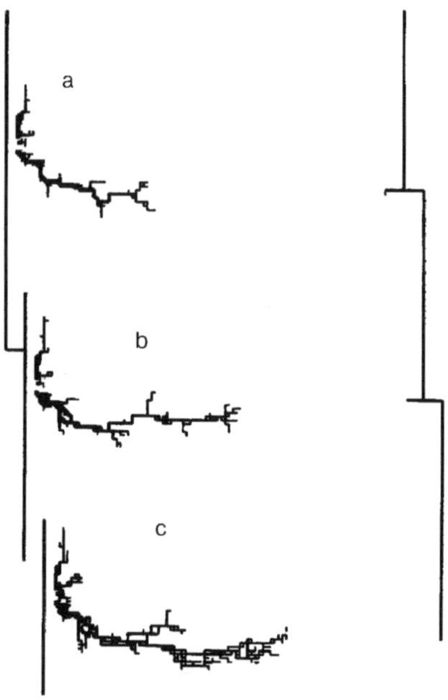

*4.2.1 Figure 7. Numerical model results for different effective diffusion constants. Initial aperture width field and pressure gradient as for 4.2.1 Figure 6b. Effective diffusion constants (**a**) 1x10^{-5} cm^2/s (300 time steps), (**b**) 2x10^{-5} cm^2/s (400 time steps), (**c**) 4x10^{-5} cm^2/s (600 time steps) (from Lauritzen et al., 1992).*

diffusion was varied ($1x10^{-5}$, $2x10^{-5}$, $4x10^{-5}$), with other conditions as in 4.2.1 Figure 6b. A lower effective diffusion constant allows faster growth rate of individual chanrnels and produces more ramified patterns (4.2.1 Figure 7a). A highly effective diffusion constant produces a more slow-growing anastomosing pattern. Smaller effective diffusion constants also produce conduits that narrow more rapidly farther from the inlet.

Time Dependence of Concentrations and Pressure Gradients
In order to test the model's ability to closely simulate the hardware experiments, several runs with optimized parameters were performed. The 4.2.1 Figure 8 shows the results of one such run. In the first sequence (a-d), concentration within the microchannels is coded as a gray scale so that the most aggressive fluid (black) becomes lighter in color until saturation (white). The final outcome of the run is chemical and hydraulic breakthrough, where aggressive water is conveyed all the way through the model. Also, the competition between principal and secondary conduits is evident as the network evolves.

In frames e-f of 4.2.1 Figure 8, the corresponding piezometric surface (pressure field) is mapped for two stages in the development of the network, clearly demonstrating the close correlation between the tip of each microchannel and the steepest areas of the pressure field.

Hardware Models of Conduit Propagation with Multiple Input Points

The simple single input plaster experiments described above were expanded progressively to explore all of the more complex groundwater input situations that we could imagine occurring on fissured karst rocks (limestone, dolostone, and gypsum) that are fully exposed in hydrologically unconfined settings, i.e. as bare rock or mantled only with soil. Full details are set out in Ewers (1982, p. 221-286); see also Ford and Williams (1989, p. 253-261).

Multiple Inputs In a Single Rank. These occur in many karst regions, especially where streams can collect in channels at the surface on insoluble rocks and reach the limestone or other soluble rock, at a sharp geologic contact. Model results are illustrated in 4.2.1 Figure 9.

At an anisotropic bedding plane inputs compete and one of the inputs will get ahead. Upon breakthrough there is a steep headloss above it (4.2.1 Figure 3B), reorienting the local hydraulic gradients toward it. Nearest-neighbor inputs then connect (breakthrough) to it in steps 2, 3, and 4. The 4.2.1 Figure 9 (p. 192) is drawn from an actual model result. Readers will appreciate that it is difficult to forecast in advance of the first breakthrough precisely which branching tubes will be involved in the later connections 2, 3, and 4. The same is true in natural caves, where only the connected tubes are likely to be enlarged to more than a few cm in diameter. This accounts for much of the sinuous twisting and turning encountered in enterable bedding-plane passages.

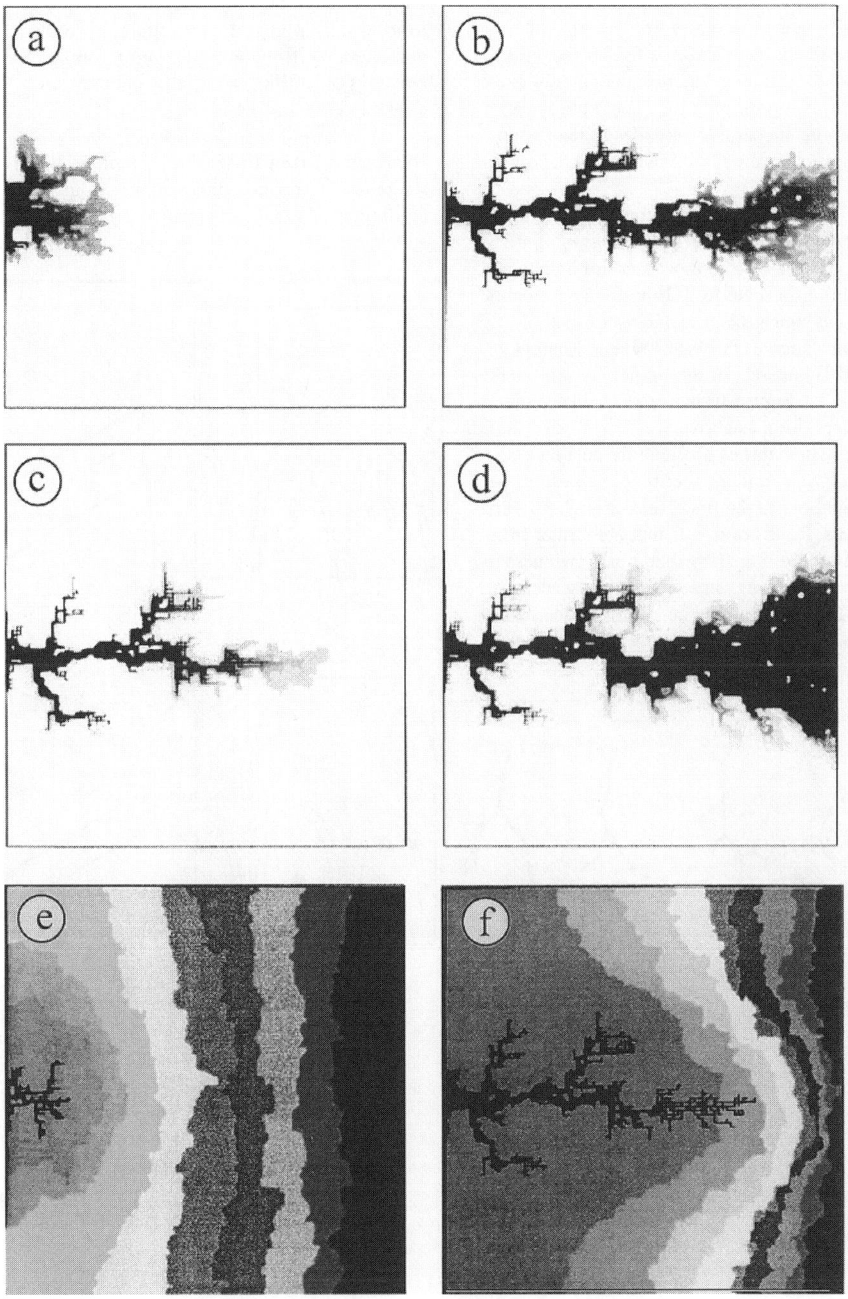

*4.2.1 Figure 8. Concluding experiment. In **a-c** Anastomosis propagation and distribution of chemical aggressiveness (concentration) in the channel network, where undersaturated water (C=0, black) enters from left and attains saturation concentration (Cm, white). **(d)** Chemical and hydraulic breakthrough is attained (note that individual competition between principal and secondary conduits also behaves as predicted from the hardware model experiments). **(e-f)** Two situations of anastomosis development with the piezometric surface (pressure field) mapped, and the apex of each microchannel in the anastomosis frame-work is closely correlated with areas of highest gradient of the pressure field (from Lauritzen et al., 1992, unpublished poster, EUROCK92').*

In *general systems* terminology, these models and natural caves are evolving as *cascading systems* (von Bertalanffy, 1968), one breakthrough (or energy cascade) reorienting the local water table and so tripping off the next cascade. The main system of Das Hölloch, Switzerland (Bögli, 1970) is a particularly fine example of cascade linkage across a fissure (i.e. strike-oriented) in what has been essentially a one-rank-of-inputs situation. There have been four or more successive cascades as the system adjusted to allogenic lowering of the outlet springs.

The most complex patterns arise where there are *many successive ranks of inputs.* This is the case in most areas of holokarst, ranging in scale from limestone pavements (clints-and-grikes) to extensive sinkhole plains such as those of Indiana, Kentucky, and Tennessee, or the tropical cockpit karsts. In 4.2.1 Figure 10 the same sequence of competition as above are shown—first among the distributaries of the individual inputs, next by lateral connection of nearest neighbors to the first-breakthrough principal tubes. Ranks of inputs that are farther from the output boundary then link to winners in the nearer ranks in successive cascades.

This pattern may be broadly discerned in the organization of passages in Mammoth Cave and nearby systems such as Procter, Roppel, Parker, and Whigpistle, that drain to springs along the Green River, Kentucky, U.S.A. This is the lengthiest regional aggregation of known and mapped conduits and shafts (>600 km). However, these cave patterns are greatly complicated by effects of lithology, stripping of cover strata, lowering and lateral shifting of spring points, and episodes of infilling with sand and clay, as explained in Chapter 5.3.8.

Restricted Input Case. This completes the range of significantly different input-output configurations that can occur (4.2.1 Figure 11).

The karst rock is exposed only along a narrow valley floor. Groundwater entering it drains to springs at the valley mouth. In this case, the sinkpoints tend to line up behind one another in a single file, with breakthrough commencing at that which is closest to the spring and proceeding headward beneath the valley.

This is a common type of cave. The example shown is complicated because, once again, the network displays more than a single phase of development and is incompletely explored. The Friar's Hole System, West Virginia (Worthington, 1984) is another spectacular example in North America.

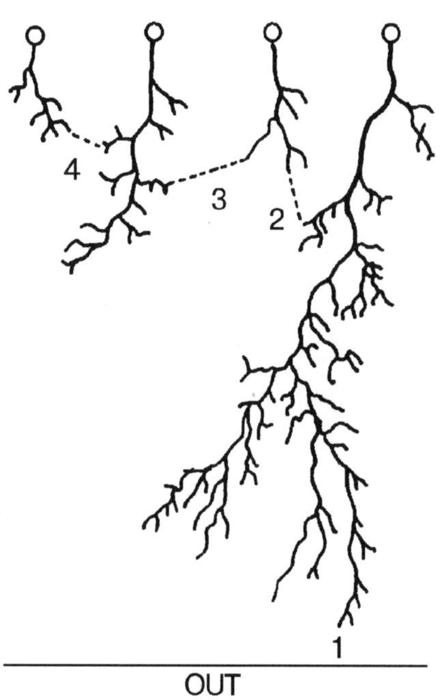

4.2.1 Figure 9. Competitive extension with multiple inputs in a rank. numbers and dashed lines indicate the predicted sequence of breakthrough connections that will occur and their location (both figures based on hardware simulations by R.O. Ewers).

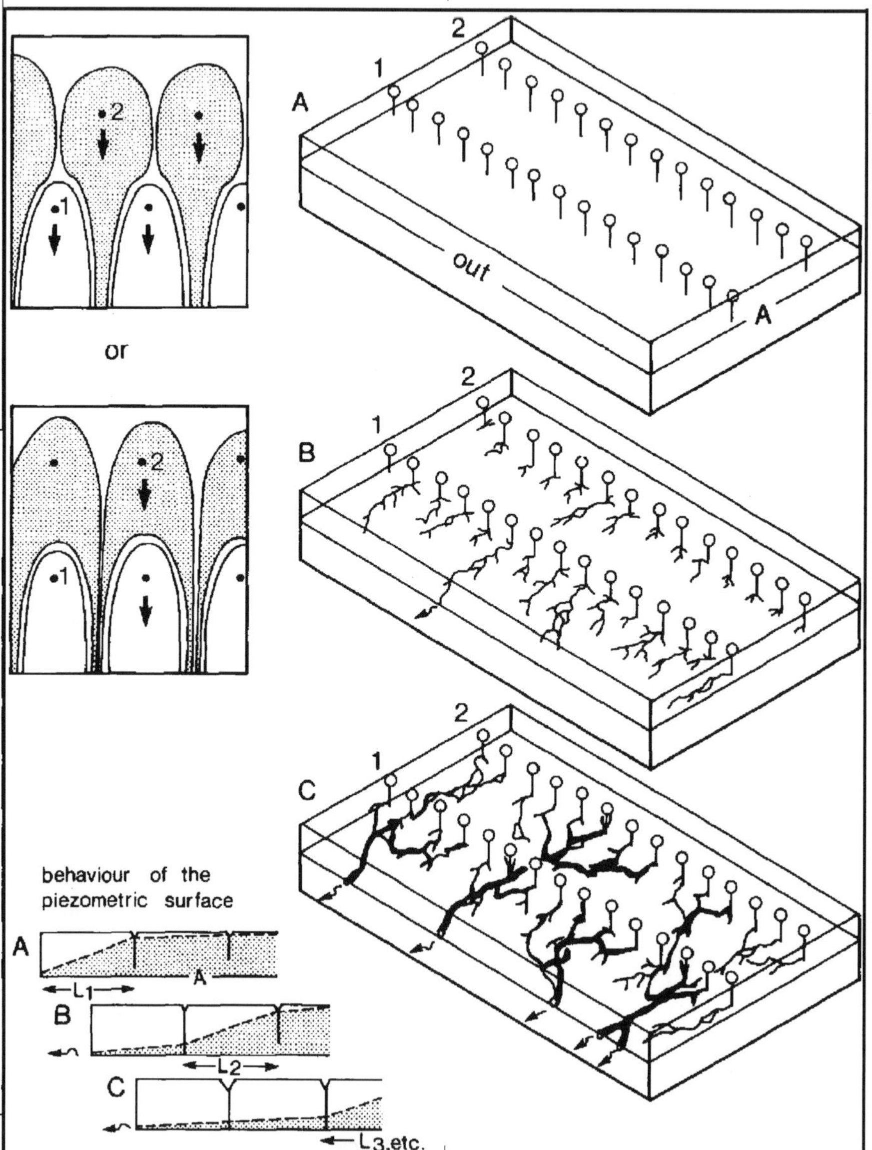

4.2.1 Figure 10. The initiation of cave systems with multiple ranks of inputs, drawn for the case of two ranks. Flowfield configurations limit rear-rank access to an output boundary. The conduit network is connected up in a headward sequence, with recession of the zone of steep hydraulic gradient (from Ford and Williams, 1989, p. 258).

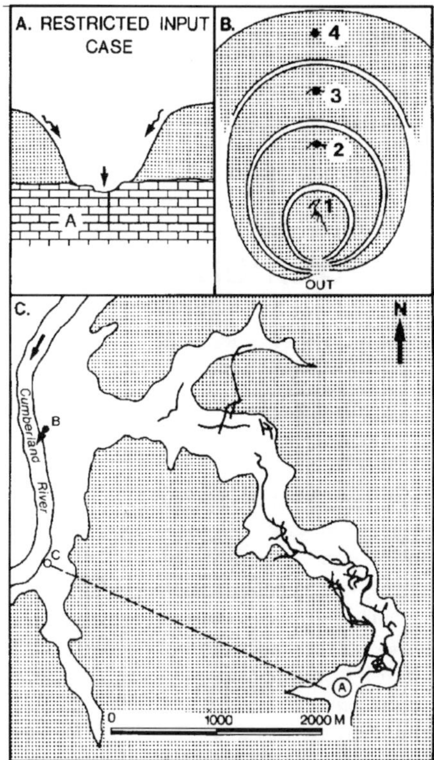

4.2.1 Figure 11. Cave system initiation where inputs are restricted to karst rocks becoming exposed in a river valley floor. The example shown is Cave Creek, Kentucky (from Ewers, 1982). The cover strata here are thick shales. The known caves underlie the valley floor and drain to springs in the Cumberland River at point B. The dashed line signifies a new capture route that passes underneath the remaining cover rocks to a younger spring at a lower altitude.

Conclusions

Hardware modeling has yielded significant insights into the propagation and linkup of dissolutional conduits in fissured soluble rocks. However, it must be emphasized that in real caves the construction of these comparatively simple patterns of conduits is vulnerable to major distortions:

- By local lithologic and structural factors such as perching on a chert bed or following the axis of a syncline.
- By great differences in the quantities of water supplied to different inputs in a group, for example, when allogenic flow to a stream sink is compared to seepage into nearby holokarst dolines
- By multiphase effects. The pattern building is analogous to that of the development of dendritic channel networks on the Earth's surface but with one very significant difference.

The ideal river pattern is wholly stochastic, the individual channel being initiated at some random point on a surface, from which it can extend upslope and downslope until it chances to encounter another channel segment and so amalgamates. In a certain sense, the location of initial input points in a karst branchwork can also be considered to be stochastic. But the successive linkups of the principal tubes are deterministic, because they are directed by local reorientation of the water table following breakthrough. Despite the strong deterministic factors in cave pattern building (lithological and structural as well as hydraulic), however, it is rarely feasible to predict the exact locations where principal conduits will be found underground, for the reasons made evident in the figures presented here.

References

Batchelor, G.K., 1967, *Introduction to fluid dynamics:* Cambridge University Press, 186 p.

Bedinger, M.S., 1966, Electric-analog study of cave formation: *National Speleological Society Bulletin,* v. 28, p. 127-136.

Bögli, A., 1970, Le Hölloch et son karst: Neuchâtel, Editions la Baconnière, 100 p.

Dreybrodt, W., 1988, *Processes in karst systems–Physics, chemistry and geology*: Berlin, Springer Verlag, 288 p.

Ewers, R.O., 1966, Bedding-plane anastomoses and their relation to cavern passages: *National Speleological Society Bulletin,* v. 28, p. 133-140.

Ewers, R.O., 1972, *A model for the development of subsurface drainage routes along bedding planes:* University of Cincinnati, M.Sc. thesis, 84 p.

Ewers, R.O., 1982, *Cavern development in the dimensions of length and breadth:* McMaster University, Ph.D. dissertation, 398 p.

Ford, D.C., 1971, Geologic structure and a new explanation of limestone cavern genesis: *Transactions of the Cave Research Group of Great Britain,* v. 13, p. 81-94.

Ford, D.C., and Williams, P.W., 1989, *Karst geomorphology and hydrology:* London, Unwin Hyman, 601 p.

Groves, C.G., and Howard, A.N., 1994, Early development of karst systems, 1.: Preferential flow path enlargement under laminar flow: *Water Resources Research*, v. 30, p. 2837-2846.

Jennings, J.N., 1985, *Karst Geomorphology:* Oxford, Basil Blackwell, 293 p.

Lauritzen, S-.E., Odling, N., and Petersen, J., 1992, Modeling the evolution of channel networks in carbonate rocks, *in* Hudson, J.A., ed., *ISRM Symposium*, Eurock '92: London, Thomas Thelford, p. 57-62.

Liu, S.T., and Nancollas, G.H., 1971, The kinetics of dissolution of calcium sulfate dihydrate: *Journal of Inorganic and Nuclear Chemistry,* v. 33, p. 2311-2316.

Lowe, D.J., 1992, *The origin of limestone caverns–An inception horizon hypothesis:* Manchester Polytechnic University, Ph.D. dissertation, 512 p.

Odling, N., and Webman, I., 1991, A "conductance" mesh approach to the permeability of natural and simulated fracture patterns: *Water Resources Research,* v. 27, p. 2633-2643.

Palmer, A.N., 1975, The origin of maze caves: *National Speleological Society Bulletin,* v. 37, p. 57-76.

von Bertalanffy, L., 1968, *General system theory–Formations, development, applications:* New York, G. Braziller, 289 p.

Waterman, S.E., 1975, *Simulation of conduit network development on bedding planes:* McMaster University, M.Sc. thesis, 116 p.

Weyl, P.K., 1958, The solution kinetics of calcite: *Journal of Geology,* v. 66, p. 163-176.

White, W.B., 1969, Conceptual models for carbonate aquifers: *Ground Water,* v. 7, p. 15-21.

Worthington, S.R.H., 1984, The palaeodrainage of an Appalachian Fluviokarst: Friar's Hole, West Virginia: McMaster University, M.Sc. thesis, 218 p.

Yalin, M., 1971, *Theory of hydraulic models:* London, McMillan.

4.2.2. Dynamics of the Evolution of Single Karst Conduits

Wolfgang Dreybrodt and Franci Gabrovšek

Abstract

The evolution of karst conduits by calcite-aggressive water flowing in initially narrow fractures requires a nonlinear rate law $F_n(c) = k_n(1 - c / c_{eq})^n$ for limestone dissolution close to equilibrium with respect to calcite. A mathematical analysis of the evolution of limestone dissolution rates of water in such early narrow fractures as a function of the distance from the input reveals an exponential decrease of the dissolution rate for linear dissolution rate laws *(n=1)*, such that subsurface karstification is prevented.

For nonlinear kinetics *(n>2)*, however, the decrease of rates proceeds by a hyperbolic relation, such that dissolution rates at the exit of the fracture are still sufficiently high to create a feedback mechanism by which after a long gestation time a dramatic increase in the widths of the conduits is established. After this breakthrough event, uniform widening along the entire channel determines the further evolution. The time to achieve breakthrough is given by $T = a_0 / 2\gamma F(L,0)$, where $2\gamma F(L,0)$ is the initial widening in cm/year at the exit of the conduit.

This, however, is only true when the inflow solution is at less than 99% of saturation. Otherwise the positive feedback loop is switched off, and the conduit widens uniformly along its entire length with rates of 10^{-9} cm/year to enlarge extremely narrow fractures with initial widths of several 10ths of micrometers over distances of kilometers to sizes of about 0.1 mm within several 10ths of millions of years. This provides a general explanation for the concept of inception horizons, where other mechanisms usually have been assumed.

Introduction

The dissolution rate of limestone by CO_2-containing calcite-aggressive water flowing through a narrow fracture with a width of several 10ths of a millimeter is given (Dreybrodt and Eisenlohr, Chapter 4.1.2 in this volume; Dreybrodt et al., 1996; Eisenlohr et al., 1997; Svensson and Dreybrodt, 1992) by

$$F_1(c) = k_1(1 - c / c_{eq}) \text{ for } c \leq \text{c} \quad (1a)$$

$$F_n(c) = k_n(1 - c / c_{eq})^n \text{ for } c > \text{c} \quad (1b)$$

where c is the concentration of dissolved calcite in the solution and c_{eq} the equilibrium concentration with respect to calcite in mole/cm^3. The rate constants k_1 and k_n are in [mole cm^{-2} s^{-1}] and are related by

$$k_n = k_1(1 - c_s / c_{eq})^{(1-n)} \quad (2)$$

where c_S is the switch concentration, where the reaction order changes from 1 to n. Typically $0.75 < c_S / c_{eq} < 0.9$ for natural limestone.

To answer the question, how does the profile of an initially uniform fracture develop in space and time, numerical models have been used assuming a constant head driving the water through the fracture. These models show (Dreybrodt, 1990, 1996; Palmer, 1991; Groves and Howard, 1994; Palmer, Chapter 4.2.3 in this volume) that a positive feedback loop is operative.

Widening of the fracture causes increasing flow through it, and therefore the dissolution rate increases along the entire fracture, until finally a dramatic increase of flow rate causes a dramatic enhancement of the widening of the fracture. This breakthrough event terminates the initial state of conduit evolu-tion. To gain more insight into this positive feedback mechanism, we present a simple mathematical model, which avoids numerical modeling by finite differences.

Dissolution of Limestone in a Plane-Parallel Fracture

We replace a natural irregular fracture by a uniform fracture with initial aperture width a_0 and a breadth b_0. We furthermore assume that $a_0 << b_0$ (typically $a_0 \approx 0.01$ cm, $b_0 \approx 100$ cm). The length of the fracture is L and the hydraulic head h is acting on it.

A part of this fracture is illustrated by 4.2.2 Figure 1. At a distance x from the input, the concentration of calcite is *(cx)*. It increases by the material dissolved from the walls between x and $x + dx$ to a value $c(x+dx) = c(x) + dc$. The flow velocity of the water is $v(x)$, $A(x)$ is the cross sectional area of the fracture, and $P(x)$ its perimeter at position x. According to the law of continuity, the flow rate through the conduit does not depend on the distance x and is given by

$$Q = v(x) \cdot A(x) \tag{3}$$

Now, conservation of mass requires that the amount of calcite dissolved per time from the walls between x and $x + dx$ must be equal to the difference of calcite transported by the flow Q out of the fracture at $x + dx$ to that entering it at x. Thus

$$Q \cdot dc = F[c(x)] \cdot P(x) \cdot dx \tag{4a}$$

Integration yields

$$\int_{c_o}^{c} \frac{dc}{F(c)} = \int_{0}^{x} P(x)dx / Q \tag{4b}$$

where c_0 is the concentration at the entrance $x = 0$ of the conduit. During the initial state of speleogenesis, flow is laminar and the flow rate $Q(t)$ at time t through the fracture is given by

$$Q(t) = \frac{h}{R}, \quad R = \frac{12\eta}{\rho g} \int_0^L a^{-3}(x,t)\, b^{-1}(x,t) \cdot M^{-1}(x,t)\, dx \tag{5}$$

where $a(x,t)$ and $b(x,t)$ are its dimensions, which now vary in space and time, η is the viscosity of the solution, ρ its density, and $M(x,t)$ is a geometric shape factor with values between 0.3 to 1 (Beek and Mutzall, 1975), g is the Earth's gravitational acceleration, and L is the length of the conduit. Units are [g], [cm], [s], and [mole/cm^3]. Equations 2 to 5 specify the parameters that determine karstification in its initial state.

The first group of parameters represents the chemical driving forces. They are n_1, k_n, n, c_S, c_0, and the equilibrium concentration c_{eq}. The kinetic constants n_1 and k_n depend on temperature and on the lithology of the bedrock (Dreybrodt and Eisenlohr, Chapter 4.1.2 in this volume; Eisenlohr et al., 1997). The kinetic exponent n takes values, $2 < n < 11$. The equilibrium concentration c_{eq} is related to temperature and to the amount of carbon dioxide available on the surface of the karst terrain. Finally c_0 is the calcite concentration of the solution entering the fracture.

The physical driving forces are mainly related to the hydraulic head h, i.e. the relief of the fracture system. If meteoric precipitation is high, such that an ample amount of water is available at the surface, this value remains constant in time. In the early state of karstification, this can be assumed as realistic, because the narrow fractures cannot transport large quantities of water and therefore the water table remains close to the surface. In the later state of mature karst, the amount of water available might not suffice to keep this high water table. Therefore, the amount of meteoric precipitation determines the further development of the karst aquifer (see also Sauter and Liedl, Chapter 4.2.5 in this volume).

The hydrogeologic setting is characterized by the geometry of the fracture, which determines its resistance R (equation 5). Thus the corresponding parameters are the aperture width a_0, the breadth b_0, and the length L of the fracture. Furthermore, the viscosity η depends on temperature ($\eta = 1.8 \cdot 10^{-2}$ at 0°C, $\eta = 1.33 \cdot 10^{-2}$ at 10°C, and $\eta = 0.90 \cdot 10^{-2}$ dynes/cm^2 at 25°C). Therefore, the amount of water transported through a system of fractures almost doubles if the temperature is raised from 2°C (alpine karst) to 25°C (tropical karst).

As a first step we calculate the initial dissolution rates along the initial fracture, which at the onset of karstification has not yet been widened. By use of equations 1 to 5, one finds

$$\int_{c_o}^{c} \frac{dc}{k_1(1 - c/c_{eq})} = 2(a_0 + b_0) \cdot x / Q_0 \tag{6}$$

when linear dissolution kinetics acts to distance x (c.f. equation 1a). As the water flows along the fissure, its dissolved calcite concentration increases until at location x_s it attains the value c_s, where the rates switch to a nonlinear rate law (c.f. equation 1b). In this case one finds

$$\int_{c_s}^{c} \frac{dc}{k_n(1 - c/c_{eq})^n} = 2\,(a_0 + b_0) \cdot (x - x_s)/Q_0 \tag{7}$$

where Q_0 is the initial flow rate, and is found from equation 5 as

$$Q_0 = \frac{\rho g a_0^3 \cdot b_0 \cdot M_0}{12\eta} \cdot \frac{h}{L} \tag{8}$$

Integration of equation 6 and equation 7 and employing equation 2 yields

$$F(x) = k_1 \exp(-x/L_1)(1 - c_0/c_{eq})$$

for $x \le x_s$ (9)

and for $n \neq 1$

$$F(x) = \frac{k_n(1 - c_s/c_{eq})^n}{\left(\frac{x - x_s}{L_n} + 1\right)^{n/(n-1)}}$$

for $x > x_s$ (10)

where

$$L_1 = \frac{Q_0 c_{eq}}{2(a_0 + b_0) \cdot k_1} = \frac{\rho g h}{12\eta L} \cdot \frac{a_0^3 \cdot b_0 \cdot M_0}{2(a_0 + b_0)} \cdot \frac{c_{eq}}{k_1} \tag{11}$$

and

$$L_n = L_1/(n - 1);\ n \neq 1 \tag{12}$$

The following scenario illustrates this in 4.2.2 Figure 2: Water with concentration $c_0 = 0$ enters into the fracture with $b_0 >> a_0$. The dissolution rate at the entrance is $F(0)$. We have plotted $F(x)$ in units of $F(0)$ versus x/L_1. At the entrance, the dissolution rates are determined by linear kinetics and drop exponentially until at $x = 2L$, the concentration has reached the switch point at $c_S = 0.86 \cdot c_{eq}$ (dotted line). From then on, dissolution proceeds by a nonlinear rate law, and the rates drop by a hyperbolic relation. This is shown by the solid lines for $n = 2$, 4, and 10.

However, if no switch occurred, the rates would continue to drop exponentially as shown by the dotted line, and within a short distance of $x = 30L$ they would be reduced by 13 orders of magnitude. In contrast to that, the rates are reduced only moderately by about 3 orders of magnitude, when nonlinear kinetics are active.

To translate the data of 4.2.2 Figure 2 into real situations, one has to know the values of L_1 and F(0). Using $c_{eq} = 2 \cdot 10^{-6}$ mole/cm^{-3} and $k_1 = 5 \cdot 10^{-11}$ mole cm^{-2} s^{-1} one finds $L_1 \approx 1.6 \cdot 10^8 \, a_0^3 h / L$. In natural karst systems $a_0 \approx 2 \cdot 10^{-2}$ cm and $h/L \approx 10^{-2}$. Therefore $L_1 \approx 10$ cm. Initial dissolution rates are converted from mole cm^{-2} s^{-1} into cm/year by the factor of $1.17 \cdot 10^9$. Thus the initial rates at the entrance are about $5 \cdot 10^{-2}$ cm/year. If only first order kinetics were operative, the dissolution rates would drop to about 10^{-45} cm/year at a distance of 10 m away from the entrance.

Therefore, caves never could develop under the conditions assumed here. The key to the evolution of limestone caves is the nonlinear kinetics, which allows the water to penetrate deep into the rock without losing its solutional power. This was first pointed out by Palmer (1984). From 4.2.2 Figure 2 one reads that at a distance of about 1 km, dissolutional widening is about $5 \cdot 10^{-9}$ cm/year for $n \geq 4$.

The initial dissolution rate at the exit of a yet unwidened initial fracture of length L and width a_0 can now be obtained from equations 10, 11, and 12 by considering that $L >> x_S$, $L >> L_n$, and $a_0 >> b_0$

$$F(L,0) = k_n \left(\frac{\rho g h}{24 \eta L^2} \cdot a_0^3 \cdot \frac{c_{eq}}{k_n (n-1)} \right)^{\frac{n}{n-1}} \quad (13)$$

Note that the dissolution rates are independent of c_0 and k_1 and that this equation represents an approximation, which is valid as long as $L >> L_0$. In this case the rates are entirely determined by the higher order kinetics. The reason is that first order kinetics act only close to the entrance at a distance up to about $x_s \approx 10$ cm, which is orders of magnitudes smaller than L. Therefore, the nonlinear kinetics is active along practically the entire fracture.

To illustrate the evolution of a cave under such conditions, 4.2.2 Figure 3 shows the results of a numerical model described elsewhere (Dreybrodt 1988, 1990, 1996). In this special case, the parameters used are:

- $a_0 = 0.02$ cm
- $b_0 = 100$ cm
- $L = 10^5$ cm
- $c_{eq} = 2 \cdot 10^{-6}$ mole cm^{-3}
- $k_1 = 4 \cdot 10^{-11}$ mole cm^{-2}s^{-1}
- $k_n = 4 \cdot 10^{-8}$ mole cm^{-2}s^{-1}
- $n = 4$
- $c_S = 0.9 c_{eq}$

The flow rate Q(t) as a function of time is shown in 4.2.2 Figure 3a. After a long gestation time of about 32,000 years, where Q(t) increases by no more than one order of magnitude, there is a dramatically steep increase within a very short time span. We call this event breakthrough, and the time at which it occurs breakthrough time T_B.

The 4.2.2 Figure 3B depicts the profiles of the initially uniform fracture as they evolve in time. Note that the scale of width is logarithmic. At the input of the fracture, dissolution rates are high, and a funnel-like shape is created. Since the rates drop rapidly, this shape is restricted to close to the input. The rest of the fracture opens up more evenly with a slow widening at the exit which finally accelerates dramatically, and within a short time span of approximately 100 years opens up and consequently breakthrough occurs. During this evolution the kinetic length L_n increases in a similar way as Q(t) (cf. equations 11 and 12).

As a consequence, the concentration of the solution at the exit decreases slowly until at breakthrough it drops rapidly to values close to the concentration c_0 at the input. From then on the fast first-order kinetics governs dissolution rates along the entire conduit, and during this new state of maturity an even and comparatively quick widening of the conduit results. This is illustrated in 4.2.2 Figure 3c, which shows the concentration profile $c(x)$ for various times until breakthrough. Note that in 4.2.2 Figure 3c the length scale is logarithmic.

After the breakthrough event, turbulent flow sets in and after that the amount of water needed to sustain the hydraulic head may be no longer available. Therefore, the hydraulic head drops to a new value, which is determined by the amount of meteoric precipitation. Consequently the upper entrance parts become vadose, and further phreatic evolution of the conduits proceeds below the new water table.

Estimation of Breakthrough Time

Nonlinear Kinetics

To obtain a deeper understanding of the processes illustrated in 4.2.2 Figure 3, we give a derivation of the breakthrough time that does not require numerical models.

From the knowledge of the dissolution rate at the exit of a plane parallel fracture, it is possible to give an estimation of how the width of the exit develops with time. To this end we assume that the fracture is widened evenly along its entire length with the dissolution rate acting on its exit. In reality, this is not the case, because the dissolution rate at the exit is lower than everywhere else in the fracture. Therefore the fracture, instead of developing a bell-shaped profile, would maintain an even width $a(t)$ after time t. This is illustrated by 4.2.2 Figure 4. The initial tube has widened to a tapered form. The width at the exit is $a(t)$. The dashed profile would have evolved if the dissolution rate at the exit had operated along the entire length.

We now by use of equation 13 calculate the dissolution rate $F(L,t)$ at the exit of the idealized uniform fracture with uniform width $a(t)$

$$F(L,t) = F(L,0) \cdot \left(\frac{a^3(t)}{a_0^3} \right)^{n/(n-1)} \quad (14)$$

This rate is smaller than the real rate, because the flow rate through the uniform fracture is less than that through the tapered one.
The time to widen the initial fracture evenly by Δa is given by

$$T_1 = \frac{\Delta a}{2\gamma F(L,0)} = \frac{a_0}{2\gamma F(L,0)} \cdot \frac{\Delta a}{a_0} = T \cdot \frac{\Delta a}{a_0} \quad (15)$$

where T is the time needed for doubling the width of the fracture

$$T = \frac{a_0}{2\gamma \, F(L,0)} \quad (15a)$$

$2\gamma F(L,O)$ is the retreat of bedrock in cm/year, where the factor $\gamma = 1.17 \cdot 10^9$ transforms the rates from mole cm^{-2} s^{-1} to cm/year.

After this time, the dissolution rates have increased, and a further uniform widening by Δa needs the time

$$T_2 = \frac{\Delta a}{2\gamma \, F(L,T_1)} = T \left(\frac{\Delta a}{a_0} \right) \cdot \left(\frac{a_0}{a_0 + \Delta a} \right)^{3n/(n-1)} \quad (16)$$

After m time steps, the width has increased to $(a_0 + m \cdot \Delta a)$ and the time T_{m+1} for further widening by Δa is given by

$$T_{m+1} = \frac{\Delta a}{2\gamma \, F(L,T_M)} = T \left(\frac{\Delta a}{a_0} \right) \left(\frac{a_0}{a_0 + m\Delta a} \right)^{3n/(n-1)} \quad (17)$$

Therefore, the total time is T_M necessary for widening the fracture to $(a_0 + M\Delta a)$ is

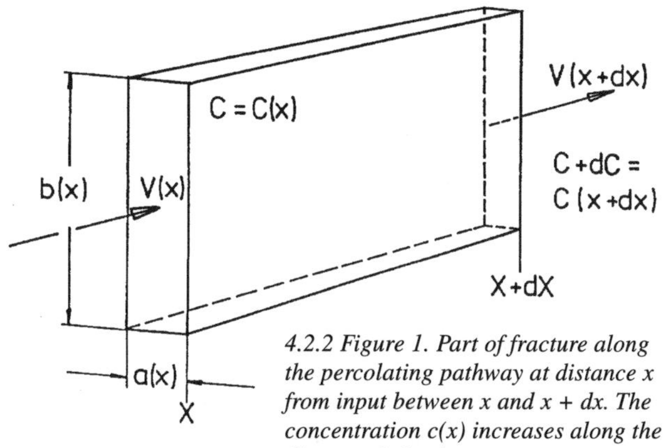

4.2.2 Figure 1. *Part of fracture along the percolating pathway at distance x from input between x and x + dx. The concentration c(x) increases along the distance dx by dc.*

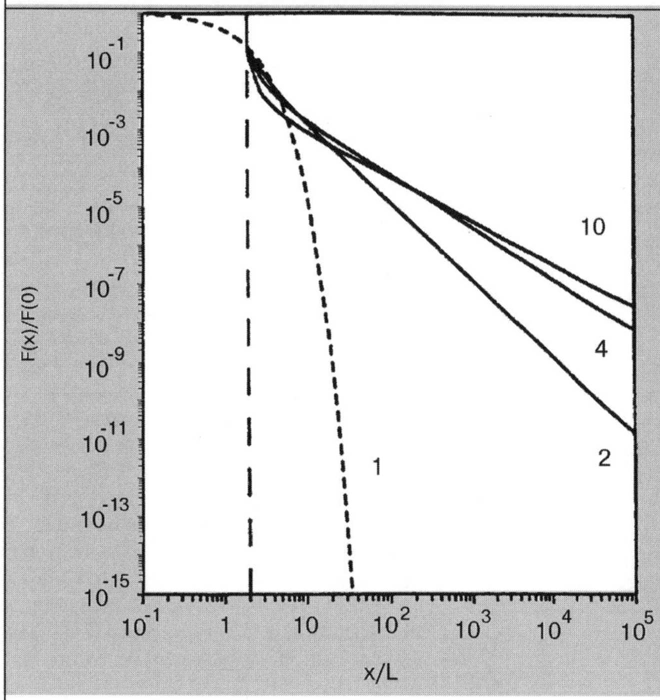

4.2.2 Figure 2. *Dissolution rates in units of the initial rate F(0) as a function of distance measured in units of L_1. The water flows to a distance 2L1and dissolves calcite with linear kinetics; then it switches to kinetics with n_2 = 2, 4, or 10 (numbers on full lines). The dashed line shows dissolution with linear kinetics, when no switch occurs.*

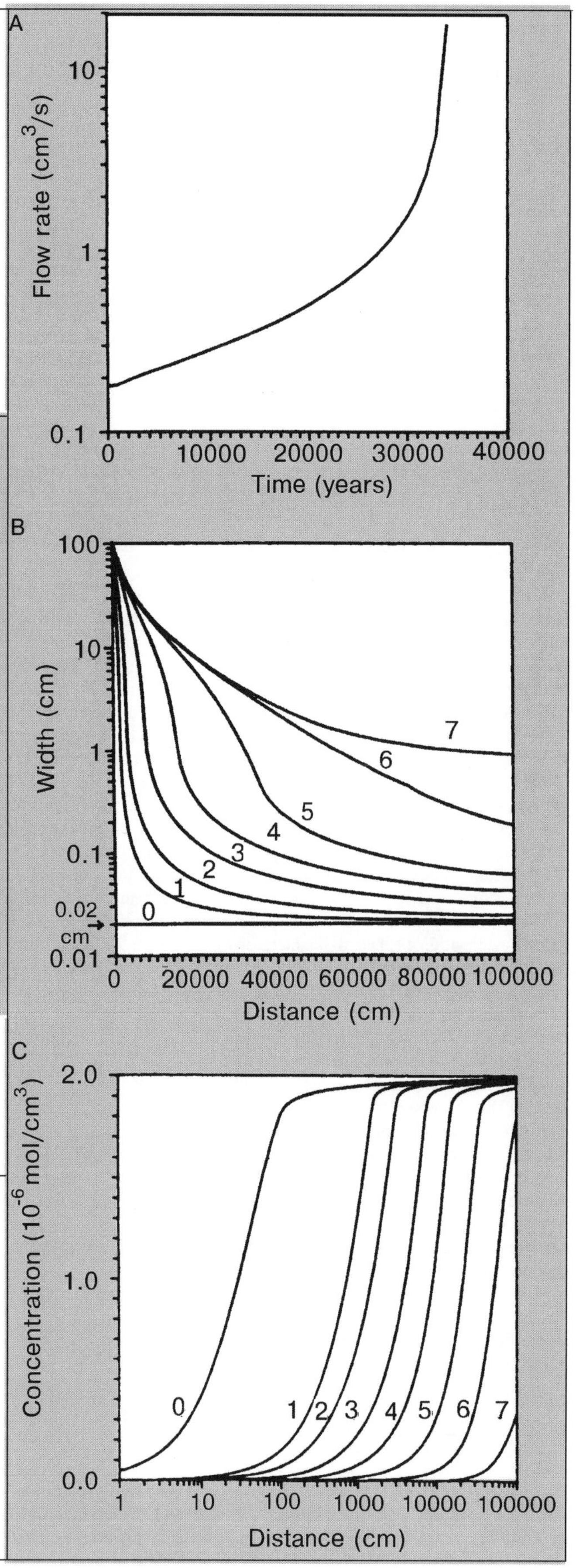

4.2.2 Figure 3. ***A*** *is flow rate Q(t) as a function of time for a conduit with a_0 = 0.02 cm, b_0 = 100 cm, L = 10^5 cm, i = 0.05, $c_{eq} = 2 \cdot 10^{-6}$ mole cm^{-3}, $c_s = 0.9 \cdot c_{eq}$, $k_1 = 4 \cdot 10^{-11}$ mole cm^{-2} s^{-1}, n =4, and $k_n = 4 \cdot 10^{-8}$ mole cm^{-2} s^{-1}. The steep increase in flow rate marks the time of breakthrough. Note the logarithmic scale of Q.* ***B*** *is profiles of the widths along the distance x for increasing times. From bottom to top T_0 = 0 years, T_1 = 10,000 years, T_2 = 20,000 years, T_3 = 30,000 years, T_4 = 33,000 years, T_5 = 34,000 years, T_6 = 34,100 years, T_7 = 34,110 years. Note the dramatic increase in diameter in the last 100 years. Note also the logarithmic scale of width.* ***C*** *is concentration profiles along the distance x for the times T_0 to T_7 listed from left to right. The concentration front, i.e. the steep increase of the curves, moves slowly toward the exit as the conduit enlarges slowly. Note that the scale of distance is now logarithmic.*

$$T_M = \sum_{m=1}^{M} T_m = T \cdot \sum_{m=1}^{M} (\Delta a / a_0) \cdot (1 + (m-1)\Delta a / a_0)^{-3n/(n-1)} \quad (18)$$

The 4.2.2 Figure 5 shows widths of the fracture as it develops in time for several values of η and $\Delta a / a_0 = 0.001$. The curves depict plot of T_M versus $a_0 + M\Delta a$.

After an initially slow widening, the widths increase rapidly and diverge at break-through time T_B. This time is given by the limit $M \to \infty$, and one finds by numerical calculation that

$$T_B = \frac{n-1}{2n+1} \cdot T \quad (19)$$

This becomes numerically almost independent of the ratio $\Delta a / a_0$ for $\Delta a / a_0 < 10^{-3}$. It should be noted that this value of T_B represents an upper limit, owing to the approximation of uniform widening along the entire fracture (cf. 4.2.2 Figure 4). In reality the fracture is wider and therefore it carries a larger flow. Therefore, according to equation 13 the dissolution rates are larger, and correspondingly T_B becomes smaller.

After breakthrough, the flow-rate Q increases with $a(t)^3$. Therefore the length $L_1(t)$ increases correspondingly such that the concentration approaches zero at the exit and the dissolution rates are uniform along the entire fracture.

The positive feedback is nicely demonstrated by equation 17. Due to the increase of the dissolution rate at the exit by a power law of $a(t)$ with exponent $3n/(n-1)$, the times for widening the exit by Δa decrease correspondingly. Thus, with increasing fracture width the dissolution rates are enhanced and this accelerates widening until breakthrough occurs. This, however, is only possible if the hydraulic head remains constant. It is the increase of discharge under constant hydraulic head due to the increasing width of the fracture, that causes the positive feedback. After breakthrough, the flow rate eventually becomes limited, and the dissolution rate becomes constant in time.

Using equations 13 and 15, the breakthrough time can be written as a function of all parameters, that determine the evaluation of the conduit on a fracture

$$T_B = \frac{1}{2\gamma} \cdot \frac{n-1}{2n+1} \left(\frac{1}{a_0}\right)^{\frac{2n+1}{n-1}} \left(\frac{24\eta L^2 (n-1)}{\rho g h c_{eq}}\right)^{\frac{n}{n-1}} (k_n)^{\frac{1}{n-1}} \quad \text{[years]} \quad (20)$$

Units in equation 17 are cm, mole, g, and s. $\gamma = 1.17 \cdot 10^9$ converts the units of T_B into years.

So far we have assumed that the length L_n in equation 10 is several orders of magnitude smaller than the length of the conduit, such that $L >> L_n$ and the +1 in the denominator of equation 10 can be safely neglected. In this case the dissolution rate F_{in} at the input is orders of magnitude higher than the initial rate $F(L,O)$ at the output. The feedback mechanism expressed by equation 14 is valid only under these conditions.

However, $F(L,O)$ cannot increase beyond F_{in}. The general valid expression for $F(L, t)$ can be derived from equation 10 as

$$F(L,t) = F_{in} \left(\frac{(L - x_s) \cdot a_0^3}{L_n \cdot a^3(t)} + 1\right)^{-n/(n-1)} \quad (21)$$

Therefore, when $a(t)$ has increased to a sufficiently large value, the growth of the exit rate becomes smaller until finally it reaches the constant value F_{in}. From then on the conduit increases linearly in time and uniformly along its entire length. In most cases flow becomes restricted to a constant value before this happens. But also, then the feedback mechanism is no longer active, and the conduit increases linearly in time.

If, however, the solution enters into the conduit with a concentration c_o sufficiently close to equilibrium with respect to calcite, and $c_o > c_s$, equation 7 has to be changed by replacing $c_s = c_o$ and $x_s = 0$. By integration of the modified equation 7 one finds

$$\tilde{L}_n = \frac{\rho g a_0^3 b_o h \cdot M_0 \cdot c_{eq}}{24\eta(a_o + b_0) L \cdot k_n (1 - c_o / c_{eq})^{n-1} \cdot (n-1)} \quad (22)$$

The dissolution rates in this case are given by equation 21, considering $x_s = 0$ and $L_n = \tilde{L}_n$.

In any case, the feedback mechanism expressed by equation 14 is only operative as long as $La_0^3 / (\tilde{L}_n a^3(t)) >> 1$, and it is effectively deactivated when this ratio takes values of about 0.1 (cf. equation 10).

Therefore, when the width $a(t)$ has reached the value

$$a_s(t_s) \approx 2a_0 \left(\frac{L}{L_n}\right)^{1/3} \quad (23)$$

the exit rate takes the value $F_s(L, ts) \approx 0.9 \cdot F_{in}$, and beyond that time t_s the feedback mechanism is no longer active.

Consequently the width of the conduit increases linearly in time and uniformly along the entire conduit.

This is illustrated by 4.2.2 Figure 6A. It shows the evolution of the idealized conduit for several values of c_o. In a double logarithmic plot, 4.2.2 Figure 6A depicts the width $a(t) / a_0$ versus t / T_B. Curves A and B represent the case where the input concentrations $c_0 < c_s$, $c_A = 0.75\ c_{eq}$, and $c_B = c_s = 0.9\ c_{eq}$. In both cases breakthrough is independent of c_o.

This is generally true as long as $c_0 \leq c_s$. As soon as $a(t)$ reaches $a_s(t_s)$, the feedback mechanism is deactivated, and growth continues with a constant rate $2\gamma \cdot k_1 (1 - c_0 / c_{eq})$ given by equation 1. This is shown by curve B. The open circle marks the deactivation point (a_s, t_s). A similar behavior is also shown by curves C ($c_0 = 0.096\ c_{eq}$) and D ($c_o = 0.9825\ c_{eq}$). In this case $c_0 > c_s$ and therefore the higher-order kinetics acts along the entire conduit from the very beginning of its evolution. The deactivation points are marked by the open circles on these curves in 4.2.2 Figures 6A and 6B.

For curve C, $L / \tilde{L}_n \approx 300$, and for curve D, $L / \tilde{L}_n \approx 25$

For curves E, F, and G these values are 4.6, 2, and 0.6 respectively. This is because the input concentrations are sufficiently close to equilibrium (E:$c_o = 0.99\ c_{eq}$; F:$c_o = 0.9925\ c_{eq}$; G:$c_o = 0.9925\ c_{eq}$). Therefore, the initial dissolution rates at the exit (cf. equation 21) are $0.1F_{in}$, $0.23F_{in}$, and $0.53F_{in}$ respectively. Consequently, the feedback mechanism is reduced so efficiently that breakthrough does not occur. It should be noted that the logarithmic plot distorts the linear growth beyond c_s. To illustrate this region we have replotted curves C–G in a linear scale (4.2.2 Figure 6B). Clearly curves C, D, E, F, and G exhibit maximal constant growth rates beyond c_s and comparatively lower rates before this time.

Therefore, when hydrogeologic conditions favor concentrations sufficiently close to equilibrium, $\tilde{L}_n$ takes values of several kilometers, and even widening in the order of 10^{-9} cm/year may enlarge narrow fractures with aperture widths of several 10's of micrometers over distances of kilometers in time scales of several 10ths of million years. This might be a general and ubiquitous mechanism creating sufficiently wide fractures as a prerequisite for later karstification, which then can occur under changing hydrogeologic conditions that deliver solutions of suffi-ciently low input concentrations to the thus widened fractures.

Linear Kinetics

In the beginning of the chapter, we stressed the importance of higher order kinetics for conduit development. All our calculations were made for the case of higher order kinetics ($n = 4$). For completeness, we present with slightly different treatment, the case where only first order kinetics acts along

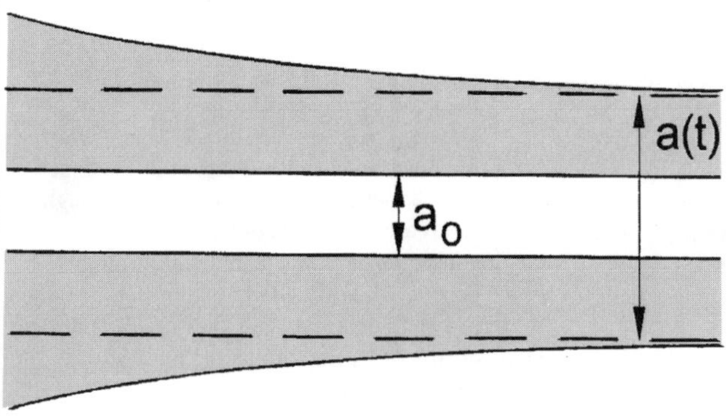

4.2.2 Figure 4. Profile of a fracture with initial width a_0 after onset of karst evolution (solid line). The dashed line illustrates the profile that would have developed if the dissolution rate at the exit had been acting along the entire length of the fracture.

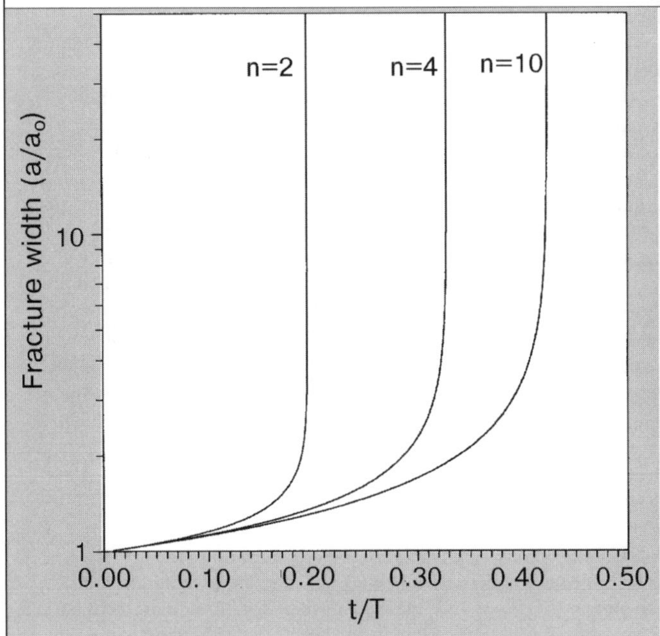

4.2.2 Figure 5. Evolution of the width of the idealized fracture as a function of time. The width is in units of a_0, and the time in units of T (cf. equation 15a.) The numbers on the curves give the order of the kinetics acting at the exit. Note that T depends on n. Therefore the time scales are different for n = 10, 4, and 2. The evolution of the fracture width is calculated from equation 18. ($M = 10^4$ and $\Delta a / a_0 = 10^{-2}$).

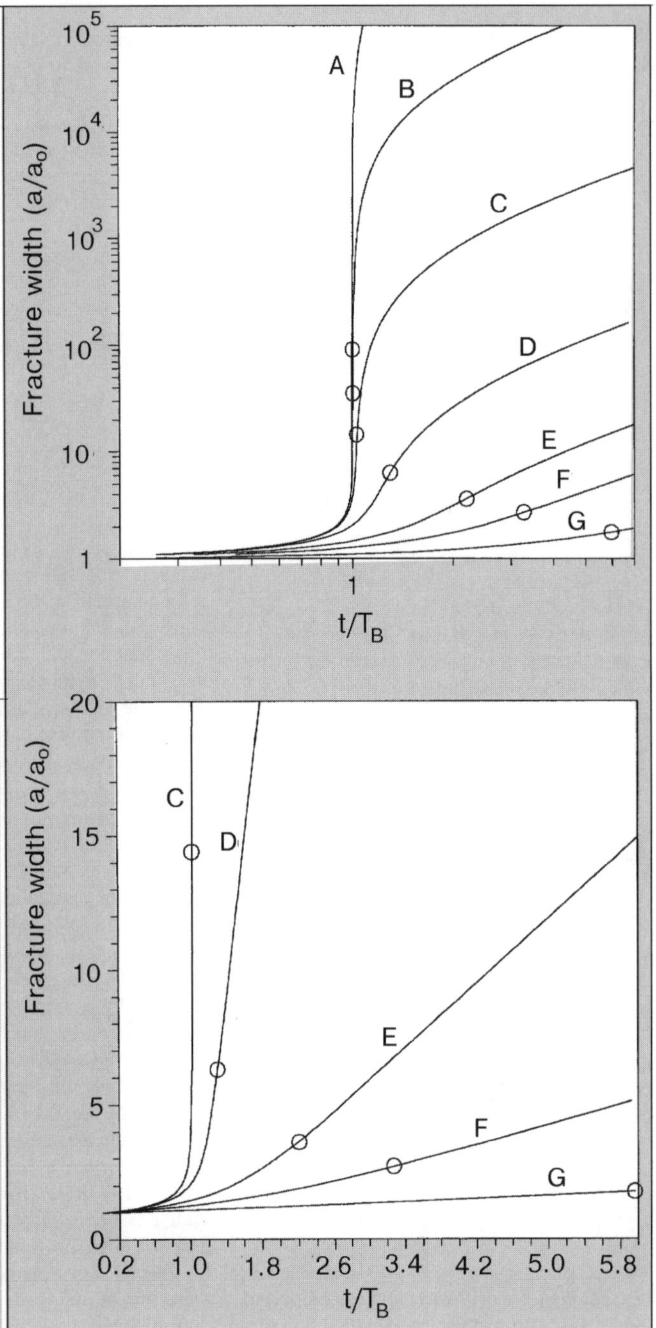

4.2.2 Figure 6. Diagram A is the evolution of the idealized fracture for various input concentrations c_o in a double logarithmic plot.
A: $c_o = 0.75\ c_{eq}$ B: $c_o = 0.9\ c_{eg}$ C: $c_o = 0.96\ c_{eq}$
D: $c_o = 0.9825\ c_{eq}$ E: $c_o = 0.99\ c_{eq}$ F: $c_o = 0.9925\ c_{eq}$
G: $c_o = 0.995\ c_{eq}$
The open circles mark the deactivation points, beyond which the positive feedback loop is no longer operative. Diagram B is Diagram A in a linear plot to illustrate the linear growth during the time after deactivation of the feedback mechanism.

the entire fracture and the solution enters with $c_o = 0$.

Equations 9 and 12 give the rate dependence on x for the case of first order kinetics. Again we assume the rates to be constant and equal to that at the exit. Then the width of the fracture stays constant along the entire length as in 4.2.2 Figure 4, and from equation 9 we find

$$F(x,t) = F(L,t) = k_1 \exp(-L/L_1(t)) \quad (24)$$

where

$$L_1(t) = a(t)^3 \frac{\rho g h c_{eq}}{24\eta L k_1} = L_1^0 \left(\frac{a_0}{a(t)}\right)^3$$

for b>>a (25)

where $L_1^0 = L_1(t=0)$.

Similarly, as in equation 17 we get the expression for the time T_m, needed for further widening by Δa starting at the time

$$T_{M-1} = \sum_{m=1}^{M-1} T_m$$

$$T_m = \frac{\Delta a}{2\gamma F(L, T_{M-1})} \quad (26)$$

Using equations 24 to 26 we find the total time T_m necessary for widening the fracture from a_0 to $(a_0 + m\Delta a)$ as

$$T_M = \frac{\Delta a}{2\gamma k_1} \sum_{m=1}^{M} \exp \left(\frac{L}{L_1^0} \left(\frac{a_0}{a_0 + (m-1)\Delta a} \right)^3 \right) \quad (27)$$

In 4.2.2 Figure 7 is illustrated the evolution of the fracture width at the exit as a function of time in units of $T = \Delta a / 2\gamma k_1$. Note that this is the time to open the fracture by Δa at its entrance. The initial width is $a_0 = 0.04$ cm, the length is $L = 10^4$ cm, the hydraulic gradient $h/L = 0.07$ and $k_1 = 4 \cdot 10^{-11}$ mole $cm^{-2}\,s^{-1}$ and $c_{eq} = 2 \cdot 10^{-6}$ mole/cm^{-3} is a typical value for first order kinetics of limestone (cf. Chapter 4.1.2 in this volume).

The typical breakthrough event is caused by the positive feedback, which reduces each timestep $\Delta T = T_M = T_{M-1}$ due to the decreasing value of $(a_0 / (a_0 + (m-1)\,\Delta a)^3$. This feedback is strong as long as the exponent in the exponential is large compared to 1. Therefore a breakthrough behavior occurs if $L / L_1^0 >> 1$. If this is not the case and $L \approx L_1^0$, the rates at the entrance and at the exit become comparable, and feedback becomes weak or is even completely deactivated. This is also the case at later times T_M, if $L \approx L_1(T_M)$. In both cases further widening is fast and uniform along the entire fracture. In the following, therefore, we define breakthrough time as the time when the condition $L = L_1(T_B)$ is valid.

To further elucidate the behavior under the regime of linear kinetics, we have plotted in 4.2.2 Figure 8 the dependence of breakthrough time on the hydraulic gradient for a fracture with initial widths.

a is $a_0 = 10^{-2}$ cm.
b is $a_0 = 2 \cdot 10^{-2}$ cm.
c is $a_0 = 3 \cdot 10^{-2}$ cm.
d is $a_0 = 4 \cdot 10^{-2}$ cm.

All other parameters are those from the example in 4.2.2 Figure 7.

In comparison, we have also plotted the corresponding breakthrough times for nonlinear kinetics with $n = 4$ and $k_4 = 4 \cdot 10^{-8}$ mol $cm^{-2}\,s^{-1}$ acting along the entire fracture. These are exhibited as straight lines in the double logarithmic plot. The arrows at the upper scale denote the corresponding values where $L_1^0 / L = 1$ ($L_1^0 = 2.04 \cdot 10^8 a_0^3 \cdot h / L$). As can be visualized, breakthrough for linear kinetics is practically instantaneous for $L_1^0 \geq 0.1L$, and rises steeply for lower values of the hydraulic gradient until it reaches 10^9 years. Most limestone is younger than this age, so we can take it as the upper limit for karstification.

Hence there exists a limiting gradient ($h/L = 2$ for curve **a**) below which karstification cannot proceed. On the other hand, for hydraulic gradients such that $L_1^0 \geq 0.1L$, breakthrough occurs practically immediately $h/L \geq 10$ for curve **a** and $h/L > 0.2$ for curve **d**. Therefore, if L is of the order of 100 m, and hydraulic gradients are high, a fast development of an aquifer draining large amounts of water becomes feasible (cf. also Chapter 4.2.4) solely under the action of first order kinetics.

Similar conclusions from numerical models were given previously (Palmer, 1991; Dreybrodt, 1992, 1996). Therefore, in the case of large hydraulic gradients and relatively short fracture lengths in the order of several hundred meters, as they exist at dam sites, an intensive karstification may endanger those structures (Dreybrodt, 1992, 1996). For low hydraulic gradients less than 0.1, however, breakthrough times increase so rapidly (even if the fracture length is only 100 m), that karstification in natural systems ($L \geq 100$ m, $h/L < 0.1$) is not possible if only first order kinetics is active.

As we have stated above, the breakthrough times obtained by the application of equations 18 and 27 represent an upper limit, owing to the assumption of uniform widening with the dissolution rate acting at the exit.

To obtain more information, we have used a finite difference model (Dreybrodt, 1996) to calculate accurate breakthrough times for the case d ($a_0 = 0.04$ cm). The results are depicted by the dashed lines in 4.2.2 Figure 8. For nonlinear kinetics the breakthrough times are lower by a factor of about 2. For first order kinetics, however, the breakthrough times are reduced significantly. As a consequence, immediate breakthrough occurs, when $L_1^0 \geq 0.1L$, stressing the danger of karstification at dam sites under the realm of first order kinetics. A breakthrough at 200 years would even result for $L_1^0 > 0.05L$.

Of course karstification below dam sites is also active under nonlinear kinetics. Breakthrough times within the lifetime of the structure, typically 200 years, can be read from 4.2.2 Figure 8. Thus for an initial fracture width of 0.02 cm, hydraulic gradients larger than 0.3 are sufficient to create a fast developing karst channel within 200 years.

To illustrate the evolution of the fracture width as a function of time under the action of linear kinetics, we have depicted in 4.2.2 Figure 9a the profiles obtained from the finite difference calculations at various times. Each profile was plotted when the flow through the fracture had increased by a factor of 1.5 with respect to the previous plot. The 4.2.2 Figure 9b presents the flow rate as a function of time. The open circles represent the respective times for the profiles in 4.2.2 Figure 9a. As can be easily seen, the first four profiles have evolved before breakthrough and dissolution is restricted close to the entrance. At breakthrough, the exit is still restricted, but then it opens up rapidly. Thus, as long as $L >> L_1(t)$, dissolution is restricted within a depth of several L_1^0, the result being denudation close to the surface.

Discussion

One may consider the breakthrough time as a measure of the intensity of karstification in a limestone terrain. The shorter these times are, the higher is the degree of karstification. Thus, equation 20 specifies the intensity of karstification dependent on the various parameters already specified.

The hydrogeologic forces are reflected by the geometric parameters a_0, b_0, and L. But they are tied to the physical and chemical driving forces by the parameter group $[L^2\eta / (hc_{eq})]$. In regions of low temperature, such as alpine or arctic karst, one expects low levels of CO_2 with p_{CO_2} from 0.1% up to 1% and consequently low equilibrium concentrations. In tropical karst, p_{CO_2} ranges from 0.2% up to 10% (Brook et al., 1983; Ford and Williams, 1989; White, 1988; Yuan Daoxian et al., 1991). Consequently calcite equilibrium concentrations are higher by at

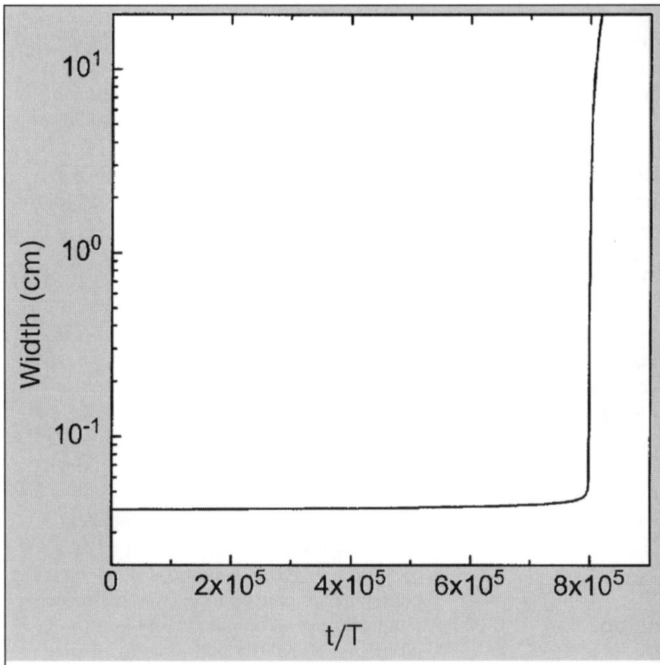

4.2.2 Figure 7. The evolution of the widths of a fracture under the action of linear kinetics, when $L_1(0) \ll L$, $a_0 = 0.04$ cm, $h/L = 0.07$, $L_1^0 / L = 0.076$, and $L = 1 \cdot 10^4$ cm.

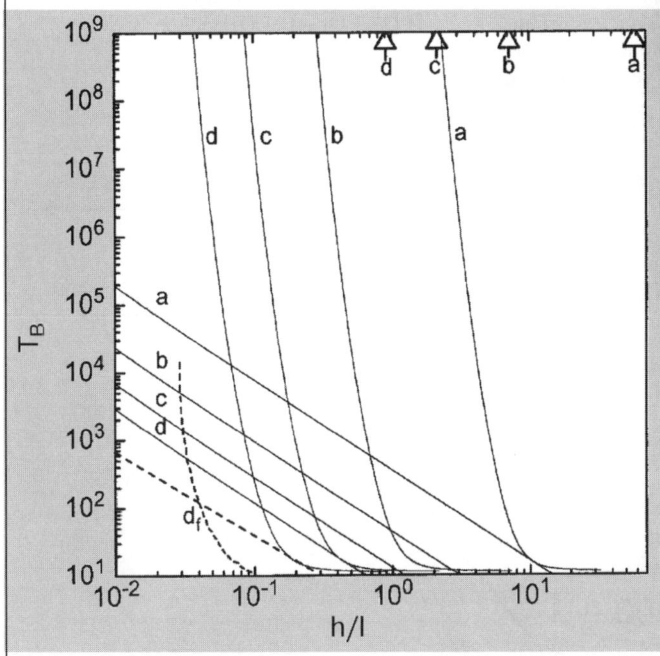

4.2.2 Figure 8. Breakthrough times given in years, for linear kinetics for various initial values of a_0. All other parameters are as in 4.2.2 Figure 7. The straight lines represent the corresponding breakthrough times under the action of nonlinear kinetics. The arrows indicate the h/L-value, where $L_1^0 / L = 1$. The dashed lines show the results of a finite differences calculation for case d.

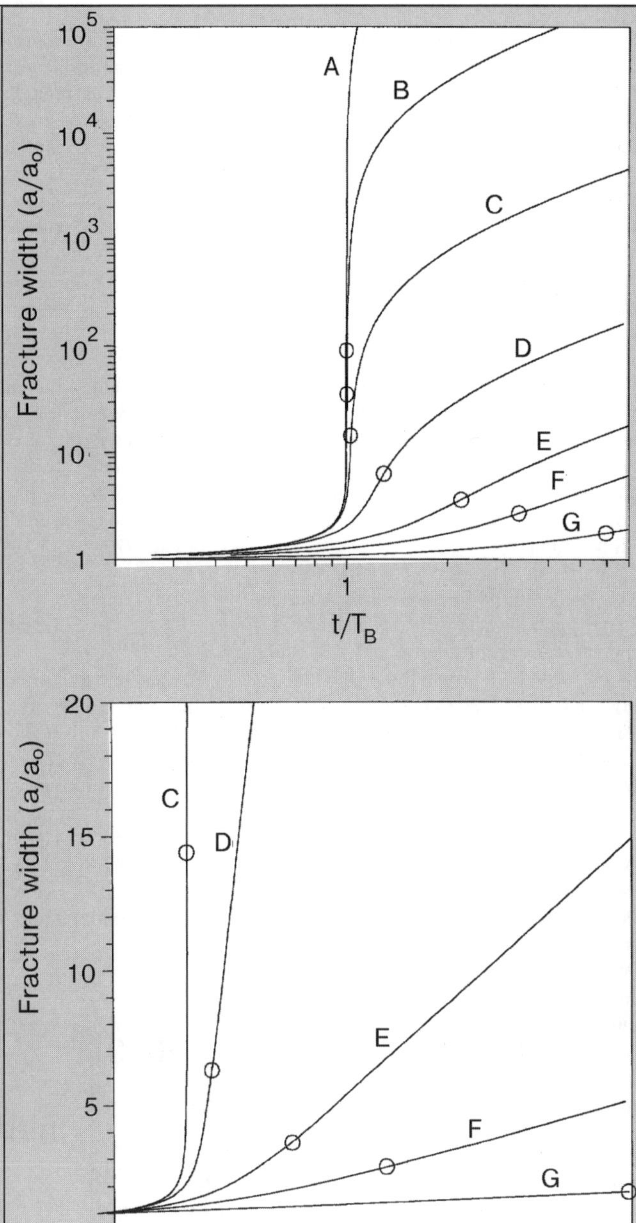

4.2.2 Figure 9. Upper panel A is evolution of the fracture widths under the action of linear kinetics for case d in 4.2.2 Figure 8, h/L = 0.03. The times for the profiles are indicated in lower panel B, which depicts the flow rate as a function of time. Breakthrough time T_B = 2500 years for this example.

least a factor of two in tropical karst.

Considering the temperature dependence of the viscosity—which is lowered by a factor of 2 from 5°C to 25°C—and the temperature dependence of k_n—which increases by a factor of two in this temperature range (Palmer, 1991)—breakthrough times for otherwise identical settings, are lowered by a factor of two.

One interesting feature is the fact that breakthrough times depend critically on n. For $n = 4$ an increase of L by a factor of only two extends the breakthrough time by a factor of six. Therefore, in areas of comparable hydraulic head h, one expects intensive karstification with many small caves where the geologic setting allows comparatively short percolating pathways. By contrast, in regions of large L, only few but long caves will be found.

Karstification times depend most sensitively on the initial fracture width. A variety of investigations using various methods report fracture widths between $5 \cdot 10^{-3}$ and 10^{-1} cm, with high probabilities of widths between 0.01 and 0.03 cm (Motyka and Wilk, 1984; Wilk et al., 1984; Böcker, 1969; Pailliet et al., 1987; Bandis et al., 1985; Johns et al., 1993).

For wide fractures with $b_0 >> a_0$ and $a_0 = 2 \cdot 10^{-2}$ cm, breakthrough times are several tens of thousand years for karst areas, where i is high (0.1) and L is of the order of 100 m. Such times are realistic and have been reported by Ford (1988), Mylroie and Carew (1986), and Bakalowicz (1982) as minimum times required for the evolution of a karst aquifer to its maturity. For large systems with $L \approx 10$ km and $i = 0.01$, correspondingly larger times in the order of 10 million years are needed.

A further interesting point should be addressed. As we have shown, the breakthrough times are controlled by the dissolution rates at the exit of the fracture and therefore depend only on k_n and n. This leads to a most important conclusion. Because the kinetic length of the first order kinetics at the entrance is extremely short, the concentration drops quickly to c_s in most realistic cases after a distance of about $10^{-3} \cdot L$. From there on, higher order kinetics take over (cf. 4.2.2 Figure 3C).

Therefore, higher-order kinetics act along practically the entire tube. Consequently it does not matter whether the water entering into the fracture has already dissolved limestone from the overlying soil. As long as the concentration c_o remains below c_s, breakthrough times will not be influenced considerably. This is also shown by numerical calculations, which in comparison to $c_o = 0$ show no more than an increase of about 10% for $c_o = c_s$, provided c_{eq} is not altered. For these reasons, in contrast to current opinions, the initiation of karstification is not enhanced when allogenic water with low hardness enters a limestone terrain.

After breakthrough, however, turbulent flow sets in, and the flow rates become restricted by the amount of water available at the surface. Due to the large flow velocities, the concentration of dissolved calcite changes little along the entire flow path and stays close to its value at the input. Therefore, dissolution rates are uniform along the conduit.

These rates, however, depend on the concentration of the inflowing water, and cave development in the mature state is enhanced when allogenic water of low calcium concentration enters into the cave (Dreybrodt, 1996). Maximal rates are about 10-2 cm/year, but rates lower by one order of magnitude should be realistic (Dreybrodt and Buhmann, 1991; Liu and Dreybrodt,1997).

Conclusion

To conclude this work, 4.2.2 Figure 10 illustrates the history of the formation of a mature conduit. It shows the conduit width at the exit as a function of time. (Because during the initial phase the width is of the order of 10^{-2} cm, its size is exaggerated by a factor of 5000 to be visible.) During the gestation or initial phase, the exit width varies extremely slowly until it has reached twice its initial width. Then breakthrough soon sets in, and turbulent flow is established. The concentration of Ca^{2+} drops to low values, such that a uniform dissolution rate is active along the entire channel length. The initiation time depends critically on the length and the initial width of the fracture, and for realistic cases it covers the range from about 1000 to 10 million years.

After breakthrough, enlargement of the conduit proceeds under phreatic conditions at a rate of about 200 mm/ky. In contrast to the initiation phase the enlargement phase is similar for all phreatic conduits and independent of their former development.

Eventually the water table may drop, and consequently the conduit becomes vadose with a stream flowing in it. Then entrenchment of canyons with similar rates will occur. Finally the passage becomes completely vadose, and growth stops until at a later time by varying external conditions it may become phreatic again and resume growth. The active phreatic phase is comparatively brief, since it takes only about 100,000 years to create a passage of about 30 m diameter, which is not often encountered in nature.

Finally, this work shows that to understand the evolution of cave systems one has to consider the dissolution kinetics of limestone. An understanding by looking only for influences of the calcite equilibrium of the system $H_2O - CO_2 - CaCO_3$ is not possible.

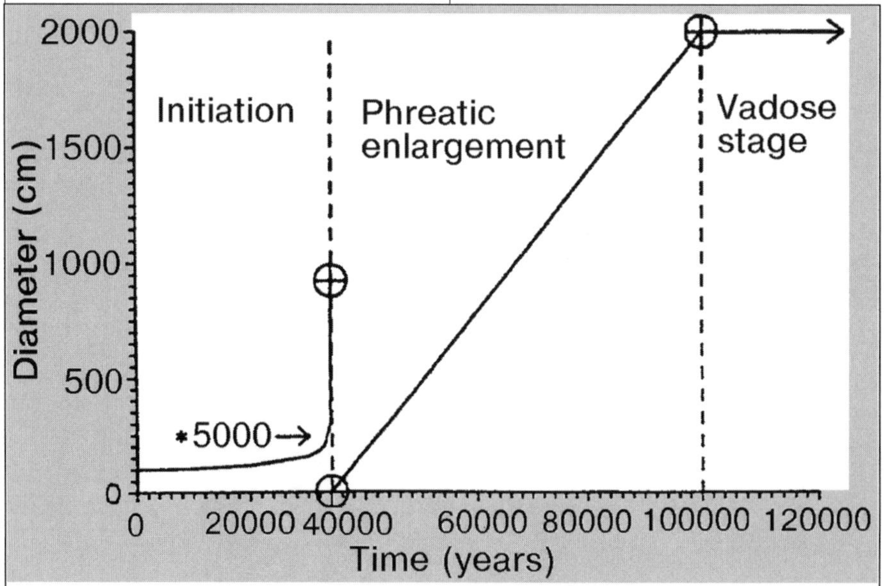

*4.2.2 Figure 10. Schematic drawing of the evolution of a conduit from initiation to maturity. The x symbols mark the transitions from initiation to the phreatic stage and from there to the vadose stage. The diameter of the fracture younger than 40,000 years (at the *5000 label) is exaggerated 5000 times.*

Acknowledgement

Franci Gabrovšek thanks the University of Bremen for financial support.

References

Bakalowicz, M., 1982, La genese de l'aquifer karstique vue par un geochemiste, *in Karst-Larra 82:* Pomplona, Fural Navarra, p. 159-174.

Bandis, S.C., Makurat, A., and Vik, G., 1985, *Predicted and measured hydraulic conductivity of rock joints:* Oslo, Norges Geotekniske Institutt, 16 p.

Beek, W. J., and Mutzall, K.M.K., 1975, *Transport phenomena*: New York, Wiley, 298 p.

Brook, G.A., Folkoff, M.E., and Box, E.O., 1983, A world model of soil carbon dioxide: *Earth Surface Processes and Landforms,* v. 8, p. 79-88.

Böcker, T., 1969, Karstic water research in Hungary: *Bull. International Association of Scientific Hydrology,* v. 14, no. 4, p. 7-20.

Dreybrodt, W., 1988, *Processes in karst systems–Physics, chemistry and geology*: Berlin, Springer Verlag, 288 p.

Dreybrodt, W., 1990, The role of dissolution kinetics in the development of karstification in limestone: A model simulation of karst evolution: *Journal of Geology,* v. 98, p. 639-655.

Dreybrodt, W., 1992, Dynamics of karstification: A model of karstification in the vicinity of hydraulic structures in karst terranes. *Applied Hydrogeology,* v. 1. p. 20-32.

Dreybrodt, W., 1996, Principles of early development of karst conduits under natural and man-made conditions revealed by mathematical analysis of numerical models: *Water Resources Research,* v. 32, p. 2923-2935.

Dreybrodt, W., and Buhmann, D., 1991, A mass transfer model for dissolution and precipitation of calcite from solutions in turbulent motion: *Chemical Geology,* v. 90, p. 107-122.

Dreybrodt, W., Lauckner, J., Zaihua, L.,Svensson, U., and Buhmann, D., 1996, The kinetics of the reaction $CO_2 + H_2O \rightarrow H^+ + HCO_3^-$ as one of the rate limiting steps for the dissolution of calcite in the system $H_2O - CO_2 - CaCO_3$: *Geochimica Cosmochimica Acta,* v. 60, p. 3375-3381.

Eisenlohr, L., Madry, B., and Dreybrodt, W., 1997, Changes in the dissolution kinetics of limestone by intrinsic inhibitors adsorbing to the surface: La Chaux-de-Fonds, Switzerland: *12th International Congress of Speleology and 6th Conference on Limestone Hydrology and Fissured Media,* La Chaux-de-Fonds, Proceedings, v. 2, p. 81-84.

Ford, D.C., 1988, Characteristics of dissolution cave systems in carbonate rocks, *in* James, N.P., and Choquette, P.W., eds., *Paleokarst:* New York, Springer Verlag, 601 p.

Ford, D.C., and Williams, P.W., 1989, *Karst geomorphology and hydrology:* London, Unwin Hyman, 601 p.

Groves, C.G., and Howard, A.D., 1994, Early development of karst systems. 1. Preferential flow path enlargement under laminar flow: *Water Resources Research,* v. 30, p. 2837-2846.

Johns, R.A., Steude, J.S., Castanier, L.M., and Robert, P.V., 1993, Nondestructive measurements of fracture aperture in crystalline rock cores, using X-ray computer tomography: *Journal of Geophysical Research,* v. 98, p. 1889 - 1900.

Liu, Z., and Dreybrodt, W., 1997, Dissolution kinetics of calcium carbonate minerals in H2O-CO2 solutions in turbulent flow: The role of the diffusion boundary layer and the slow reaction $H_2O + CO_2 \Leftrightarrow H^+ + HCO_3^-$: *Geochimica Cosmochimica Acta,* v. 61, p. 2879-2889.

Motyka, I., and Wilk, Z., 1984, Hydraulic structure of karst-fissured Triassic rocks in the vicinity of Olkusz (Poland): *Kras i Speleologia,* v. 14, no. 5, p. 11-24.

Mylroie, J.E., and Carew, J.L., 1986, Minimum duration of speleogenesis: *9th International Congress of Speleology,* Barcelona, Proceedings, v. 1, p. 249-251.

Pailliet, F.L., Hess, A.E., Cheng, C.H., and Harding, E., 1987, Characterization of fracture permeability with high-resolution vertical flow measurements during borehole pumping: *Ground Water,* v. 25, p. 28-40.

Palmer, A.N., 1984, Recent trends in karst geomorphology: *Journal of Geological Education,* v. 32, p. 247-253.

Palmer, A.N., 1991, The origin and morphology of limestone caves: *Geological Society of America Bulletin,* v. 103, p. 1-21.

Svensson, U., and Dreybrodt, W., 1992, Dissolution kinetics of natural calcite minerals in CO2-water systems approaching calcite equilibrium: *Chemical Geology,* v. 100, p. 129-145.

White, W.B., 1988, *Geomorphology and hydrology of karst terrains:* New York, Oxford University Press, 464 p.

Wilk, Z., Motyka, J., and Józefko, I., 1984, Investigations of some hydraulic properties of karst solution openings and fractures: *Annales Societatis Geologorum Poloniae Rocznik Polskiegotowarystwa Gelogicznego,* v. 43, no. 1-2, p. 15-43.

Yuan, D., 1991, Karst of China: Beijing, Geological Publishing House, 224 p.

4.2.3. Digital Modeling of Individual Solution Conduits

Arthur N. Palmer

Abstract

The genesis and evolution of solution conduits is well suited to digital modeling, in which idealized fissures or tubes are divided into many small length increments and allowed to grow in small time steps under specified hydraulic and geochemical conditions. Modeling of single conduits reveals clearly the functional relationships between conduit growth rate and other variables. In such models the hydraulic gradient is usually (but not necessarily) held constant, so that the discharge can increase with time as the openings enlarge.

In a typical karst aquifer, initial growth rates are very slow throughout all but the upstream ends of the conduits, but as the discharge increases the enlargement rate also increases, reaching a maximum rate of wall retreat beyond which further gains in discharge have little effect on solutional growth rates. The maximum rate of wall retreat is about 0.001 to 0.1 cm/yr in limestone, depending on the chemical conditions. The time required for a conduit to reach this rate is inversely dependent on the initial fissure width, hydraulic gradient, and P_{CO2}, and directly proportional to the fissure length. Initial fissure width (or tube radius) is by far the most sensitive factor.

With these functional relationships known, it is then possible to extrapolate conceptually from single fissures to multiple-passage systems. In a typical karst aquifer there is a great diversity of enlargement rates among the various competing flow routes, and only a few conduits grow to cave size. If the ratio of discharge to length is great enough in all openings, however, they will all enlarge at comparable rates, forming maze caves.

Introduction

The chemical and hydrologic processes that form dissolution caves are complex enough that they are difficult to evaluate by purely analytical or intuitive methods. However, the topic is well suited to finite-difference modeling, in which the initial presolutional openings are divided into a large number of small segments whose growth is calculated over small time increments as dissolution proceeds. This method is similar to calculus, but with the use of finite increments instead of infinitesimal ones. The length and time increments must be small enough that they, and the changes that take place within them, are insignificantly small compared to the size and time scale of the overall system.

Modeling the growth of a single conduit is a logical first step in understanding the evolution of caves. From there, the way is clear to interpret the development of complex systems, as is discussed in Chapters 4.2.4 and 4.2.5. Although finite-difference modeling of a single conduit is straightforward and computationally simple, the number of repetitive calculations is so formidable that computer modeling is almost essential.

Model Design

Numerical models of cave development must solve several equations simultaneously:

- The chemical mass balance, in which the mass of material removed from the walls of the cave must equal the mass carried away in solution.
- The hydrologic mass balance, in which the mass of water entering any cave segment during a given time increment must equal the mass that exits, plus any change in storage that occurs within the segment.
- An equation for either laminar or turbulent flow of water.

- A kinetic equation that describes the dissolution rate.

The solutional growth of a single planar fissure is considered here as the most appropriate standard. Other geometries (e.g. tubes of circular cross section) are just as easily modeled and have essentially the same functional relationships. In the field, any initial "fissure" usually consists of a series of interconnected segments of varied length, width, and roughness, and the fissure width may also increase through nonsolutional means. Although it would not be difficult to model such complex fissures, the goal here is simply to evaluate the relationships involved in their evolution, and not to examine the details of any specific system. Therefore a single long fissure (e.g. fracture or parting) of initially uniform width is considered (see 4.2.3 Figure 1).

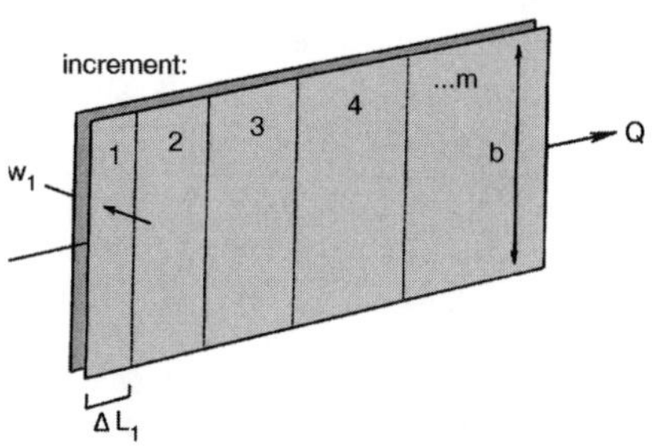

4.2.3 Figure 1. Idealized fissure divided into small increments of increasing length in the direction of flow, where Q = discharge, w = fissure width, b = long dimension of fissure cross section, ΔL = increment length, m = total number of increments (see equations 1 to 3).

The model is designed to simulate the flow of solutionally aggressive water through the fissure and to track the history of fissure enlargement. There are several choices as to the type of system. Dissolution of limestone (calcite) by the carbonic acid reaction is the most appropriate, although provision can be made for other reactions.

Closed-conduit flow (vs. open-channel flow) is also the most appropriate choice. This involves dissolution that is isolated from the initial CO_2 source, so that the equilibrium P_{CO2} and saturation concentration diminish with flow distance. The most problematic design step is how to treat the discharge through the system.

One choice is to consider a constant discharge and to allow the hydraulic gradient to decrease with time as the fissure enlarges and becomes more efficient in transmitting the flow. This does not work well, however, because the tiny amounts of water admitted by the initial narrow fissure are too small to enlarge it rapidly enough during its later stages to form a cave within a geologically feasible time. The other extreme is to hold the hydraulic gradient constant and allow the discharge to increase as the fissure enlarges.

Neither scenario is entirely realistic, because in a natural system both the discharge and gradient vary with time. Discharge usually increases because the catchment area expands, but in an unspecified and variable way. Meanwhile the gradient usually diminishes as the passage grows, despite the increase in discharge.

Furthermore, the fissure width may enlarge mechanically by tectonic stress, erosional unloading, or other forces while the solutional development is taking place. Any discharge or fissure-widening history can be modeled, but as a first attempt it is best to hold the gradient and nonsolutional fissure width constant, because this approach reveals most clearly the functional relationships among the many variables.

Laminar flow in a fissure is described by the Hagen-Poiseuille equation. Even if the initial width of the fissure is uniform, it does not remain so, because it enlarges more rapidly at the upstream end. Therefore the equation needs to be modified to consider a large number of fissure segments of varied width (w_x) and length (L_x). The subscript x refers to the individual passage segment under consideration. The finite-difference equation for the head loss (Δh_x) in a single segment is

$$\Delta h_x = \frac{12Q\mu\Delta L_x}{w_x^3 b\gamma} \qquad (1)$$

where Q = discharge, μ = viscosity of water (0.013 dyne-s/cm^2 at 10°C), b = fissure height (long dimension of the fissure cross section), and γ = specific weight of water (980 dynes/cm^3). Summing the head losses throughout the entire fissure gives

$$Q = \frac{b\gamma\Delta h_{total}}{12\mu\sum_{x=1}^{m}\left(\frac{\Delta L_x}{w_x^3}\right)} \qquad (2)$$

where Δh_{total} is the head loss over the entire fissure length, and m = total number of segments. When turbulent flow develops, the discharge is governed by the Darcy-Weisbach equation, which can be stated in finite-difference form as

$$Q = 2b\left[\frac{g\Delta h_{total}}{f\sum_{x=1}^{m}\left(\frac{\Delta L_x}{w_x^3}\right)}\right]^{1/2} \qquad (3)$$

where g = gravitational field strength (980 dynes/gram) and f = friction factor, which varies with fissure width and the height of the roughness projections (f is roughly 0.1 in natural fissures; see Castillo et al., 1972). Many semiempirical equations are available for the dissolution of calcite. The most widely used, although not the most accurate fit to experimental data, is that of Plummer et al. (1978). Other rate equations of at least equal applicability have been devised by Curl (1968), Sjöberg and Rickard (1984), Buhmann and Dreybrodt (1985a,b), Dreybrodt (1987, 1988, 1990), Dreybrodt et al. (1996), and Dreybrodt and Eisenlohr (Chapter 4.1.2 in this volume). In an attempt to simplify these relationships, Palmer (1984, 1988, 1991) used regression analysis to evaluate the original laboratory measurements of Plummer and Wigley (1976), Rauch and White (1977), Plummer et al. (1978), and Morse (1978) to fit a generic rate equation. This equation masks the specific chemical relationships but more accurately fits the data and is easily applied to numerical calculations:

$$\Delta C_x = \frac{2k}{w_x}(1 - C/C_s)^n \Delta t_x \qquad (4)$$

where k = reaction coefficient, n = reaction order, and Δt_x = residence time for water within the passage segment (= ΔL_x/velocity), C = concentration of dissolved bedrock and C_s = saturation concentration, and therefore C/C_s = mean saturation ratio within the passage segment. C_x = increase in C within a given passage segment during Δt_x. The term ($2/w_x$) is equivalent to the surface area of water-rock contact divided by the water volume, because A/V = ($2b\Delta L_x$)/($w_x b\Delta L_x$) in any fissure segment. For most limestones at C/C_s less than approximately 0.7, k is typically 0.01 mg-cm/L-s and n is roughly 2. At greater C/C_s, k (= k_2) is roughly 0.1 and n (= n_2) is roughly 4. Although recent experiments show that n_2 may be as high as 11 (see Dreybrodt and Eisenlohr, Chapter 4.1.2), the qualitative interpretation of the modeling is not affected.

Values of k vary with temperature and lithology, and both n and k vary with P_{CO2}. The C/C_s value at which the change in reaction order takes place is also a function of temperature. Ranges of values for these variables are given by Palmer (1991) and Dreybrodt (1996). These rate laws have been verified by recent experimental work in open systems (Svensson and Dreybrodt, 1992) and closed systems (Dreybrodt et al., 1996; Eisenlohr et al., 1997; Dreybrodt and Eisenlohr, Chapter 4.1.2 in this volume).

The hydrologic mass balance is trivial in a single closed conduit, because there are no significant changes in storage or alternate flow routes to consider. However, the solutional enlargement is governed by the chemical mass balance

$$\rho_r\left(\frac{\Delta A_x}{\Delta t}\right) = Q\left(\frac{\Delta C}{\Delta L}\right)_x \quad (5)$$

where ρ_r = bulk bedrock density (e.g. 2.7 g/cm^3 for calcite, which must be adjusted for intergranular porosity), ΔA_x = increase of cross-sectional area within a given segment, Δt = time increment (seconds), and $(\Delta C/\Delta L)_x$ = increase in solute concentration (g/cm^3) within a given segment. This reduces to:

$$\Delta w_x = \frac{Q}{b\rho_r}\left(\frac{\Delta C}{\Delta L}\right)_x \Delta t \quad (6)$$

where Δw_x = increase in fissure width in a given segment during the time increment Δt.

Computational Steps

The first step in modeling an idealized fissure is to define the initial fissure width and length, the overall head difference, temperature (to find viscosity), friction factor (for turbulent flow), k, n, initial C, and C_s (for both open and closed systems). C_s can be determined with reaction-path geochemical software, or from published graphs of calcite saturation. In a closed system, every mole of calcite dissolved is balanced by the consumption of one mole of CO_2. The P_{CO2} and therefore the C_s of the solution decrease with distance in the direction of flow.

As C approaches C_s, the effective C_s decreases toward the closed-system value. With very little error, C_s can be assumed to decrease linearly from C_s (open) to C_s (closed) as C increases. For example, at 10°C and initial P_{CO2} = 0.01 atm, the original C_s = 212 mg/L, but after closed-system dissolution the final C_s = 56 mg/L. The values for n and k decrease as well, according to the decrease in effective P_{CO2}. At any given C/C_s, the values for C_s, n, and k must be interpolated between the initial and final values.

The fissure must be divided into a large number of small length increments. Because the dissolution rate is most rapid at the entrance and diminishes with distance, it is important to use tiny length increments at the upstream end to avoid instability in the calculations. Their length can be increased in the downstream direction to reduce the number of calculations needed.

An appropriate way to determine the segment lengths (ΔL_x) is to perform a preliminary run in which water is passed through the initial fissure, so that the same small amount of dissolution occurs within each segment. Equation 4 can be used, with Δt_x replaced by ΔL_x/velocity. Subdivision of the conduit into segments that each achieve 1% increments of C/C_s is recommended, up to C/C_s = 0.99%. But the remaining conduit should be divided into at least 100 additional segments.

The downstream end of the system—which grows at very slow rates just below saturation—exerts the most important control over cave inception (see also Dreybrodt and Gabrovšek, Chapter 4.2.2, equation 15). Segments defined in this way will not retain their initial increase in C/C_s throughout the entire simulation.

But this preliminary step achieves the goal of preventing sharp discontinuities in dissolution rate from one segment to the next.

The dissolutional history must be divided into a great number of small time increments (for example, 100 years each). The time step can begin larger (e.g. 1000 years) but should diminish as the growth accelerates. The dissolution rate and amount of solutional widening are calculated within each passage segment, from upstream to downstream, and then the calculations are repeated for successive time increments.

Evolution of the conduit is modeled as follows:

A Calculate Q within the fissure, using equation 2. Calculate the Reynolds number (N_R) to determine whether the flow is in fact laminar. In a fissure, $N_R = \rho_w Q/b\mu$, where ρ_w = water density and μ = viscosity. N_R is uniform throughout the length of the fissure, because it is independent of the width variations between individual segments. $N_R < 1000$ indicates laminar flow, in which case equation 2 is valid. $N_R > 1000$ indicates turbulent flow, which means that Q must be recalculated with equation 3, which will be valid for all subsequent time steps.

For each segment, perform the following steps:

B Calculate ΔC with an appropriate rate equation (for example, equation 4) to obtain C at the end of the passage segment. Determine the new C_s (as CO_2 is depleted in the closed system). Calculate C/C_s, k, and n at the end of the segment. Note that there are many choices of rate equation (see list of cited references), and a versatile program might allow several choices for comparison.

C Find the mean C/C_s, k, and n within the segment by averaging the beginning and end values. Each length increment is short enough that the changes within it can be considered linear.

D Calculate Δw_x with equation 6, determine the new width within the segment, and replace the old value. (Q could be recalculated at this point, but its increase will be negligible.)

E Repeat steps **B** through **D** as many times as necessary until the new width converges on a stable value. I recommend using an integral form of equation 4, which makes this repetitive loop unnecessary (Palmer, 1991).

F Repeat steps **A** through **E** for each successive fissure segment in the downstream direction, using the C/C_s at the end of the previous segment as the initial value in the new segment. Segment lengths could be recalculated, as in the preliminary run described previously, but this adjustment has an insignificant effect.

G When steps **A** through **F** have been performed in each segment of the fissure, from upstream to downstream, the process is then repeated for the next Δt increment. The simulation can be terminated at any arbitrary time, such as when the C/C_s at the downstream end has reached an arbitrary low value. The change from high-order to low-order kinetics is appropriate). This is usually called the "breakthrough time."

At the end of each time step, the fissure growth can be displayed on the monitor and saved in a file as a table of time vs. width in each segment. Additional items to be displayed or saved include the C/C_s at the end of each segment, the discharge through the fissure, and the Reynolds number. The point at which the final C/C_s has dropped to the range of low-order kinetics (breakthrough time) is also recorded. The program can be terminated at any time thereafter, as its solutional enlargement will be fairly uniform from then on. Beyond that point, the enlargement rate by dissolution is nearly independent of discharge and can be calculated as follows (see Palmer, 1991)

$$S = \frac{31.56k(1 - C/C_s)^n}{\rho_r} \quad (7)$$

where S = rate of bedrock retreat, and the constant 31.56 adjusts the units to cm/yr. This equation is valid for all types of flow and conduit geometry and compares well with field measurements of bedrock retreat (e.g. High, 1970).

Modeling Results

Results of a typical finite-difference model are shown in 4.2.3 Figure 2 for a fissure 1 km long, initial width = 0.01 cm, initial P_{CO2} ($P_{CO2}{}^{i}$) = 0.01 atm, hydraulic gradient (i) = 0.01, and temperature of 10°C. In a natural system the rapidly enlarged upstream part of the system is usually affected in a complicated way by vadose flow conditions and soil fill, and it is simply incorporated into the irregular karst surface. After the breakthrough time, the enlargement rate is essentially uniform throughout the entire fissure.

On the basis of the modeling results, it can be seen that the length of the fissure (L) and the hydraulic gradient (i) have essentially equal, though opposite, effects on the solutional history and can be combined into a single term, i/L, with units of length^{-1} (Palmer, 1981; Dreybrodt, 1996). Modeling by Dreybrodt and Gabrovšek (Chapter 4.2.6) shows that initial wall roughness has little effect on breakthrough times. In models that assume surface-reaction control of dissolution rate (e.g. equation 4), the mean enlargement rate within the conduit increases with discharge, but only up to a certain maximum value described by equation 7, where C/C_s = initial saturation ratio where the water first enters the conduit. For highly unsaturated water this results in a maximum enlargement rate of about 0.1 cm/yr.

However, it is more typical for water to enter a cave only after it has already reached a substantial saturation ratio, so the actual maximum rate is usually at least 10 to 100 times less.

Without an increase in discharge, however, there can be no increase in solutional enlargement rate. Therefore, only those favorable conduits that are able to acquire increasing discharges are able to grow into caves. An increases in catchment area (e.g. by growth of dolines) is the most obvious way for a conduit to gain discharge. Because there is only a given amount of available water, however, an increase in flow to certain conduits may require that the others must diminish in discharge and grow more slowly. In typical karst aquifers, the result is a few discrete conduits that grow into caves, with numerous surrounding openings that exhibit little if any enlargement.

Total fissure length is also an important variable. The mean rate of conduit growth depends specifically on the ratio of discharge to flow distance (Q/L). In 4.2.3 Figure 3 the increase in fissure enlargement rate as a function of Q for various flow distances is shown. The maximum enlargement rate is achieved at Q/bL > 0.001 cm/s (or Q/rL > 0.001 cm/s in tubes), and where many competing flow routes exceed this threshold simultaneously, maze caves are formed.

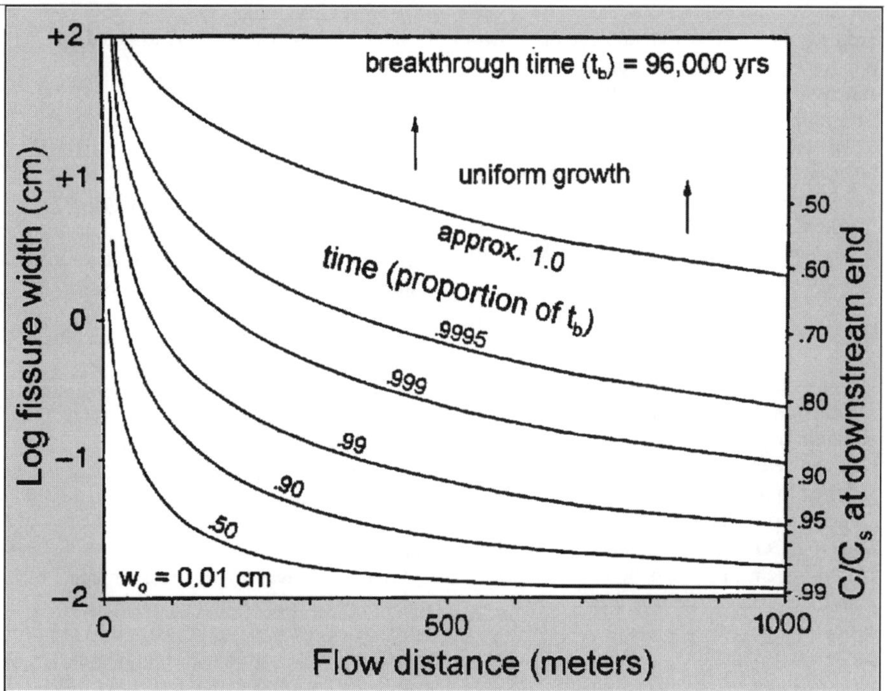

4.2.3 Figure 2. Early growth history of an idealized fissure 1000 m long at 10°C, with w_o = 0.01 cm, i= 0.1 (dimensionless), $P_{CO2}{}^{i}$ = 0.01 atm, and C_s = 212 mg/liter (from Palmer, 1991).

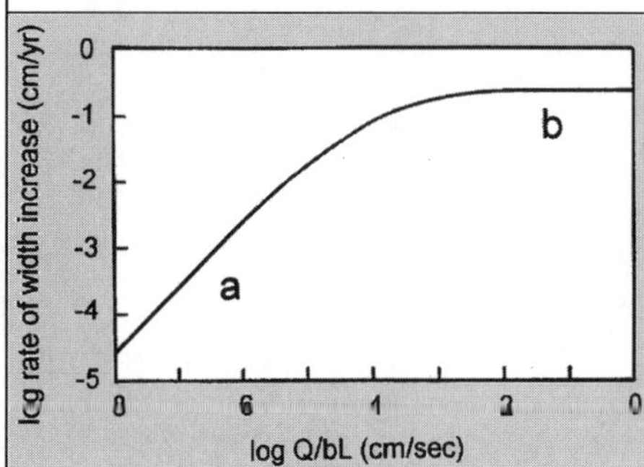

*4.2.3 Figure 3. Rate of increase in fissure width vs. discharge during the early stages of dissolution. Growth rate cannot accelerate without an increase in discharge, because b and L are constant. In **a**, early growth rates at low discharge depend only on the mass balance, equations 5 and 6. In **b** maximum growth rates at high discharge depend on the dissolution kinetics, equation 7. The rate of width increase shown here is twice the rate of wall retreat (S) in equation 7.*

The flow distance (L) is usually measured from the point at which recharge first enters the soluble rock. However, it can also refer to the point at which aggressiveness is produced, as in the case of mixing within a soluble rock. For example, maze caves are very common within short distances of where mixing has occurred. Part of the reason for the maze pattern is the multiplicity of flow routes in a mixing environment, but the ability of many openings to enlarge at comparable rates is still governed by the proximity to, or short distances of flow from, the site where aggressiveness is produced.

Estimating the time required to form a cave is one of the primary goals of cave modeling, but only in conceptual or relative terms, e.g. how the many controlling variables affect that time. A real cave involves too many uncertainties and changing conditions for a model to reveal reliable data about its developmental history.

The evolution of a typical cave passage can be divided into at least four stages:

- Inception
- Rapid growth
- Maturity
- Decay.

During the first stage, the incipient conduit transmits a small but growing discharge. Enlargement rates are very slow, and water exits at the downstream end very close to saturation.

The second stage begins when the water is able to pass through the entire conduit while maintaining enough undersaturation to preserve the low-order kinetics, so the growth rate is very rapid, at least from a geologic standpoint—usually about 0.001 to 0.1 cm/yr. This represents the "breakthrough time."

The later stages involve abandonment and ultimate collapse, which are of no concern here. Stage 1 is the most critical in determining the time required to form a cave passage.

My own modeling shows the breakthrough time for fissures in limestone to be

$$t \approx \alpha w_0^{-3.1} (i/L)^{-1.4} (P^{i}_{CO_2})^{-1.0} \text{ yr} \quad (8)$$

where w_o = initial fissure width, with the assumption that dissolution is performed by carbonic acid. Other sources of aggressiveness can be expressed as the equivalent P_{CO2} required to produce the same C_s. In cgs units, values for α are about 1×10^{-12} for systems open to a CO_2 source and about 5×10^{-12} for closed systems, in which the P_{CO2} decreases with flow distance (Palmer, 1988, 1991). From that time onward, the growth rate is governed by equation 7. A graph of breakthrough time, according to equation 8, is shown in 4.2.3 Figure 4. The breakthrough time coincides roughly with the change from laminar to turbulent flow (equation 3). Models of passage growth rarely need to consider turbulent flow, because the dissolution rate is usually almost uniform over the entire length of the conduit under those conditions.

Independent modeling by Dreybrodt (1996) gives similar functional relationships, as shown below with terms rearranged for comparison with Equation 8:

$$t \approx \alpha w_0^{-3} (i/L)^{-1.3} k_2^{0.33} C_s^{-1.3} \text{ yrs} \quad (9)$$

in a fissure at a constant C_s in mole/cm^3 (therefore an open system) and 10°C, $k_2 = 1.5 \times 10^{-9}$ mole/cm^2-s (reaction coefficient for the higher-order kinetics). Other dimensions are in centimeters. The constant α is 9×10^{-15} for fissures (b >> w) and 9×10^{-14} for conduits of circular cross section. These results are compatible with earlier models of James and Kirkpatrick (1980) and Dreybrodt (1990). Dreybrodt (1996) also derived equation 9 analytically, arriving at exactly the same functional relationships (see also Dreybrodt and Gabrovšek, Chapter 4.2.2 in this volume).

Equation 8 gives longer breakthrough times than equation 9. For example, in an open system with $i/L = 10^{-6}$, $w_o = 0.01$ cm, and $C_s = 2 \times 10^{-6}$ moles/cm^3 (equivalent to $P_{CO2} = 0.01$ atm), equation 8 (Palmer, 1991) gives 29,000 years, and equation 9 (Dreybrodt, 1996) gives 20,000 years. Although the discrepancy between the two estimates might seem large, it has little geomorphic significance. Inasmuch as the functional relationships are almost identical, the discrepancy is due mainly to the choice of k values. Both equations are designed for typical limestone bedrock, whose k values vary with composition and texture. Ironically, larger k values (more rapid dissolution rates) *increase* the breakthrough time, because the aggressiveness of the water is used up more rapidly in the upstream parts of the system, retarding the dissolution rate in the downstream parts.

Equations 7 to 9 suggest that under optimum conditions (aggressive water at high gradients and short flow distances), a traversable cave passage can form in about 10,000 years.

However, many (probably most) cave passages require times on the order of 10^5 years to reach their full size, as shown by radiometric and paleomagnetic dating (e.g. Schmidt, 1982). In rather static geologic settings, such as in bathyphreatic systems, times on the order of 10^6 years are likely required.

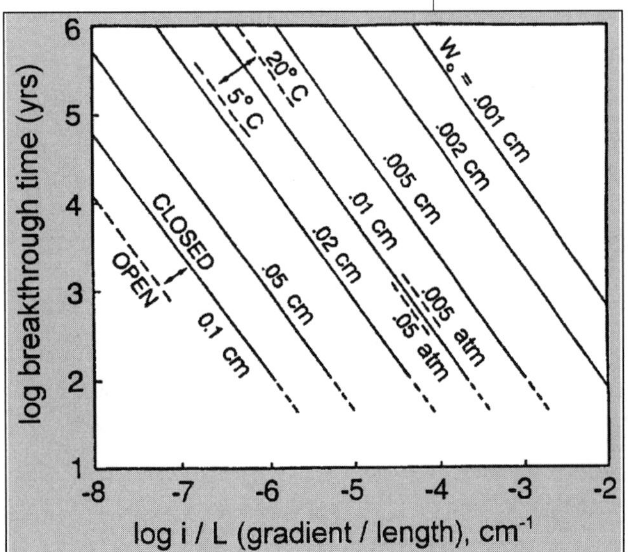

4.2.3 Figure 4. Breakthrough times in idealized fissures (times required to reach low-order dissolution kinetics over their entire length), shown vs. hydraulic gradient (i), fissure length (L), initial fissure width (w_O), temperature, initial P_{CO2} (atm), and closed vs. open conditions (from Palmer, 1991; cf. Dreybrodt, 1996).

Chemical Variants

The modeling results are so systematic and quantifiable that it is easy to draw false conclusions about rates of cave origin. The "standard" model of conduit development described above does not apply everywhere, and it is much simplified in comparison with most natural systems.

For example, it is unrealistic to assume that the entire flow system consists of closed-conduit flow that is closed to CO_2 uptake. Much or even most of the dissolution takes place in the epikarst at and near the bedrock surface in the vadose zone. This water enters the conduits with a substantial C/C_s value, far greater than zero, and the initial high P_{CO2} does not diminish nearly as much with flow distance as it would if the entire dissolution took place under closed conditions. In modeling cave passages it is reasonable to dispense entirely with the low-order (rapid) dissolution, and to allow the water to enter at the transition C/C_s. This strategy has essentially no effect on breakthrough time.

Mixing of flow between interconnecting fissures may be important in maintaining aggressiveness (e.g. Bögli, 1964). This process is obviously not included in models of single conduits, although it can be simulated by slightly decreasing C/C_s at selected intervals along the conduit. This process can be important in accelerating the enlargement rate in the critical downstream part of the conduit, whose growth is ordinarily limited by near-equilibrium conditions. A similar result is produced by generating acid within the conduit, for example through the production of CO_2 by oxidation of organic carbon or sulfides.

These situations are easily modeled, but there are so many uncertain variables that it is difficult to choose a representative scenario. In any case, if renewed aggressiveness is modeled by revising the C_s and saturation ratio at the end of each passage segment, breakthrough times can be greatly reduced. For example, consider a fissure with $P_{CO2}{}^{i}$ = 0.01 atm at 10°C, $i/L = 10^{-6}$ cm^{-1}, and initial width = 0.01 cm, in which C_s is boosted a small amount at the end of each meter of flow. Breakthrough times are reduced by the following percentages of the original:

C_s Increase (mg/liter/meter)	Breakthrough Time (% of standard)
0 (standard)	100%
0.001	98%
0.01	87%
0.1	44%

Even a small increase in aggressiveness can have an important effect in diminishing the breakthrough time. But the significance of this process can be exaggerated by inappropriate model design. Renewed aggressiveness is usually achieved by oxidation, which is severely limited during the early stages of fissure enlargement by the very small discharges. Furthermore, the oxygen content of phreatic groundwater is rather low, especially if the water has infiltrated through organic-rich soil. Oxygen is consumed in the downstream direction, and so a uniform production of aggressiveness as shown above is unrealistic. Although substantial increases in CO_2 have been measured within aerated conduits fed by surface water (Bray, 1972; Wood and Petraitis, 1984), they hardly represent the tiny closed-system fissures of the model.

Extrapolation to Multiple Dimensions

Single-conduit modeling has the virtue of revealing how the cave-forming variables relate to each other in the simplest possible way. Although it is more difficult to extend this approach to two dimensions, many have done so (e.g. Groves and Howard, 1994; Howard and Groves, 1995; see also Lauritzen and Ford, Chapter 4.2.1, Dreybrodt and Siemers, Chapter 4.2.4, and Sauter and Liedl, Chapter 4.2.5). In general, these approaches involve passing water between two boundaries of fixed head, in a manner similar to the single-conduit model.

My own approach to 2-D modeling (Palmer 1990) allowed recharge to enter vertically at all nodes at a variety of rates, for example, at rates that were uniform, random, or controlled by the mean fissure widths at each node. Heads were allowed to adjust to the catchment area allotted to each node and to the hydraulic character of the fissure network. This work was terminated prematurely, partly because of the primitive nature of computers at the time. After the model started, its progress needed to be checked only every few days. Although it could be made to simulate one's impression of how a real aquifer should behave, the results were highly dependent on arbitrary choices of initial variables. Its main function was to validate the concepts extrapolated from the single-conduit models.

In fact, two-dimensional models have done this quite well. For example, those in the literature cited above show that branching and passage complexity increase at small flow distances or breakthrough times, large hydraulic gradients or initial widths, and large or uniform discharge. These are identical to the conclusions extrapolated conceptually from single-conduit modeling.

Conclusions

Digital modeling of karst conduits helps to reveal the quantitative relationships among the many variables involved in cave origin. Virtually any hydrochemical scenario can be modeled. Unless the exact conditions are known, however, any model-based estimate of the time required for cave origin—especially breakthrough time—is only a crude estimate.

The concepts described here are discussed in relation to field examples in several other chapters, particularly Chapters 3.1, 3.4, 5.2, and 5.3.11.

References

Bögli, A., 1964, Mischungskorrosion: Ein Beitrag zum Verkarstungsproblem: *Erdkunde,* v. 18, p. 83-92.

Bray, L.G., 1972, Preliminary oxidation studies on some cave waters from south Wales: *Transactions, Cave Research Group of Great Britain,* v. 14, p. 59-66.

Buhmann, D., and Dreybrodt, W., 1985a, The kinetics of calcite solution and precipitation in geologically relevant situations of karst areas. 1: Open system: *Chemical Geology,* v. 48, p. 189-211.

Buhmann, D., and Dreybrodt, W., 1985b, The kinetics of calcite solution and precipitation in geologically relevant situations of karst areas. 2: Closed system: *Chemical Geology,* v. 53, p. 109-124.

Castillo, E., Karadi, G.M., and Krizek, R.J., 1972, Unconfined flow through jointed rock: *Water Resources Bulletin,* v. 8, p. 266-281.

Curl, R.L., 1968, Solution kinetics of calcite: *4th International Congress of Speleology,* Ljubljana, Proceedings, p. 61-66.

Dreybrodt, W., 1987, The kinetics of calcite dissolution and its consequences to karst evolution from the initial to the mature state: *National Speleological Society Bulletin,* v. 49, p. 31-49.

Dreybrodt, W., 1988, *Processes in karst systems: Physics, chemistry and geology:* Berlin, Springer-Verlag, 288 p.

Dreybrodt, W., 1990, The role of dissolution kinetics in the development of karst aquifers in limestone: A model simulation of karst evolution: *Journal of Geology,* v. 98, p. 639-655.

Dreybrodt, W., 1992, Dynamics of karstification: A model applied to hydraulic structures in karst terranes: *Applied Hydrogeology,* v. 1, p. 20-32.

Dreybrodt, W., 1996, Principles of early development of karst conduits under natural and man-made conditions revealed by mathematical analysis of numerical models: *Water Resources Research,* v. 32, p. 2923-2935.

Dreybrodt, W., Lauckner, J., Zaihua, L., Svensson, U., and Buhmann, D., 1996, The kinetics of the reaction $CO_2 + H_2O^+ H^+ + HCO_3^-$ as one of the rate limiting steps for the dissolution of calcite in the system H_2O - CO_2 - $CaCO_3$: *Geochimica et Cosmo-chimica Acta,* v. 60, p. 3375-3381.

Eisenlohr, L., Madry, B., and Dreybrodt, W., 1997, Changes in the dissolution kinetics of limestone by intrinsic inhibitors adsorbing to the surface: *12th International Congress of Speleology and 6th Conference on Limestone Hydrology and Fissured Media,* La Chaux-de-Fonds, Proceedings, v. 2, p. 81-84.

Groves, C.G., and Howard, A.D., 1994, Early development of karst systems, 1. Preferential flow path enlargement under laminar flow: *Water Resources Research,* v. 30, p. 2837-2846.

High, C. J., 1970, *Aspects of the solutional erosion of limestone, with special consideration of lithological factors:* University of Bristol, Ph.D. dissertation, 228 p.

Howard, A.D., and Groves, C.G., 1995, Early development of karst systems, 2. Turbulent flow: *Water Resources Research,* v. 31, p. 19-26.

James, A.N., and Kirkpatrick, I.M., 1980, Design of foundations of dams containing soluble rock and soils: *Quarterly Journal of Engineering Geology,* v. 13, p. 189-198.

Morse, J.W., 1978, Dissolution kinetics of calcium carbonate in sea water; VI. The near-equilibrium dissolution kinetics of calcium carbonate rich deep sea sediments: *American Journal of Science,* v. 278, p. 344-353.

Palmer, A.N., 1981, Hydrochemical controls in the origin of limestone caves: *8th International Congress of Speleology,* Bowling Green, Proceedings, p. 120-122.

Palmer, A.N., 1984, Recent trends in karst geomorphology: *Journal of Geological Education,* v. 32, p. 247-253.

Palmer, A.N., 1988, Solutional enlargement of openings in the vicinity of hydraulic structures in karst regions: *2nd Conference on Environmental Problems in Karst Terranes,* Dublin, Ohio, Association of Ground Water Scientists and Engineers, Proceedings, p. 3-13.

Palmer, A.N., 1990, Numerical modeling of karst processes [abs.]: Occasional Paper 4, Dept. of Geography, University of Wisconsin at LaCrosse; reprinted as Geo2 v. 17, nos. 2,3, p. 80.

Palmer, A.N., 1991, Origin and morphology of limestone caves: *Geological Society of America Bulletin,* v. 103, p. 1-21.

Plummer, L.N., and Wigley, T.M.L., 1976, The dissolution of calcite in CO_2-saturated solutions at 25°C and 1 atmosphere total pressure: *Geochimica et Cosmochimica Acta,* v. 40, p. 191-202.

Plummer, L.N., Wigley, T.M.L., and Parkhurst, D.L., 1978, The kinetics of calcite dissolution in CO_2-water systems at 5° to 60°C and 0.0 to 1.0 atm CO_2: *American Journal of Science,* v. 278, p. 179-216.

Rauch, H.W., and White, W.B., 1977, Dissolution kinetics of carbonate rocks. 1: Effects of lithology on dissolution rate: *Water Resources Research,* v. 13, p. 381-394.

Schmidt, V.A., 1982, Magnetostratigraphy of sediments in Mammoth Cave, Kentucky: *Science*, v. 217, p. 827-829.

Siemers, J., and Dreybrodt, W., 1998, Early development of karst aquifers on percolation networks of fractures in limestone: *Water Resources Research,* (in press).

Sjöberg, E.L., and Rickard, D.T., 1984, Temperature dependence of calcite dissolution kinetics between 1 and 62°C at pH 2.7 to 8.4 in aqueous solution: *Geochimica et Cosmochimica Acta,* v. 48, p. 271-288.

Svensson, U., and Dreybrodt, W., 1992, Dissolution kinetics of natural calcite minerals in CO_2-water systems approaching calcite equilibrium: *Chemical Geology,* v. 100, p. 129-145.

Wood, W.W., and Petraitis, M.J., 1984, Origin and distribution of carbon dioxide in the unsaturated zone of the southern High Plains: *Water Resources Research,* v. 20, p. 1193-1208.

4.2.4. Cave Evolution on Two-Dimensional Networks of Primary Fractures in Limestone

Wolfgang Dreybrodt and Jörg Siemers

Abstract

Two-dimensional nets of initial fractures are constructed on a square lattice by occupying the lines between nearest neighbor sites by a water leading fissure of width a_0 and length l with an occupation probability p. For $p > 0.5$ percolating nets occur which conduct water.

To simulate cave genesis we calculate the water flow rates driven by the hydraulic head h through all fissures. By employing nonlinear dissolution rates of the type $F_n(c) = k_n(1 - c / c_{eq})^n$ the widening of the fractures is obtained. At the onset of karstification, flow is evenly distributed along all fractures. As the system develops, solutional widening creates preferred pathways that attract more and more flow until at breakthrough both widening and flow increase dramatically.

A numerical analysis of breakthrough times dependent on the parameters a_0 (width of the initial fractures), L (horizontal dimension of the aquifer), h (hydraulic head acting on it), and the chemical parameters of the dissolution kinetics reveals that the relation valid for one-dimensional conduits (Dreybrodt and Gabrovšek, Chapter 4.2.2 in this volume) is also valid for two dimensions. We present the evolution of karst aquifers for a variety of geologic conditions including control by differing lithologies composing the aquifers. The model is also applied to simulate the evolution of epikarst. We finally discuss the evolution of karst conduits beneath dam sites, where unnaturally steep hydraulic gradients may generate conduits responsible for high water loss within the lifetime of the structure.

Introduction

To understand processes of the early state of karst genesis, one-dimensional models have been developed that couple widening of primary fractures in limestone by dissolution, with flow rates of the water driven from the input to the output (Dreybrodt, 1988, 1990, 1996; Palmer, 1991; Groves and Howard, 1994a). These models show that the width of the fracture at its exit increases slowly by dissolution until it has been doubled. From then on, a positive feedback loop is enhanced, such that within an extremely short time, width and flow rates increase dramatically.

The time required for this to happen is termed the breakthrough time. It terminates the early evolution. This time can be estimated to be:

- $T_B \approx a_0 / F(l,0)$ [years], where
- a_0 [cm] is the initial width of the fracture.
- l its length.
- $F(l,0)$ is the limestone dissolution rate in cm/year at the end of the developing channel at the onset of its evolution.

From this an expression has been derived that quantifies breakthrough time as a function of the parameters that determine karstification (Dreybrodt, 1996; see also Dreybrodt and Gabrovšek, Chapter 4.2.2 in this volume). These are the initial width a_0, the breadth b_0, the length l of the fracture, and the hydraulic head h between the input and the output, which is assumed as constant until breakthrough occurs.

Furthermore, the kinetics of dissolution plays a role of utmost importance. It is quantified by the equilibrium concentration c_{eq} with respect to calcite. This can be attained by the aggressive solution, and by the constants determining the dissolution rates as a function of the calcium concentration in solution.

For narrow initial fractures with widths of $5x10^{-3}$ cm up to $1x10^{-1}$ cm, these rates are given by a linear relation $F_1(c) = k_1(1 - c/c_{eq})$ (Buhmann and Dreybrodt, 1985; Dreybrodt et al., 1996). Close to equilibrium at $c \approx 0.8\ c_{eq}$ the rates, however, switch to a higher order law $F_n(c) = k_n(1 - c/c_{eq})^n$, where n ranges from 4 to 11 for karstifying limestone (Svensson and Dreybrodt, 1992; Eisenlohr et al., 1997). Without this switch to a higher order, karstification for natural systems, i.e. evolution of conduit aquifers, might not be possible at all.

The work described so far enables a first understanding of the early evolution of one-dimensional karst conduits in dimensions of space and time. More realistic models, however, are required. Real karst systems are three-dimensional or at least two-dimensional. Aquifers conducting water from an input to an output can develop only if percolation pathways exist. To give an answer whether such pathways do exist, we employ some basic results of percolation theory (Stauffer, 1985). Imagine a set of joints to be represented by a square lattice as depicted in 4.2.4 Figure 1. Each space between two neighboring points represents a possible location of a fracture–but fractures are not evenly distributed. Therefore we assume that at each of these spaces is occupied by a real fracture with an occupation probability $p \leq 1$.

As an example, 4.2.4 Figure 1a displays a small net of 30*x*30 with a probability $p = 0.4$, which means that, on average, 40% of the possible locations contain a real fracture. We further assume that water enters on the left side of the grid and leaves at the right side, whereas the top and the bottom of the grid are impermeable. In 4.2.4 Figure 1 we have plotted the real fractures by thin and thick lines. In drawing (**a**) we see only fractures (thin lines) that are not connected from the input to the output, cannot conduct water, and an aquifer cannot develop. Drawing (**b**) illustrates the situation for $p = 0.5$ where the clusters have merged into each other (thick lines) and percolating pathways arise conducting water from the input to the output. From the shape of these percolating pathways we could expect a dendritic cave system. The remaining fractures indicated by thin lines cannot conduct water, because they are either located in isolated clusters or are dead ends.

Drawing (**c**) gives the situation with $p = 0.7$ where we obtain a conductive network, and the resulting cave system can possibly be maze like. Percolating pathways only can exist if $p \geq p_c$, where p_c is called the critical percolation threshold (Stauffer, 1985; Balberg, 1986; Sahimi, 1995). Therefore in each setting of fractures with $p < p_c$, karst development would not be possible. For values of $p \geq p_c$, but still close to p_c, the water is lead through a few pathways that may evolve to large conduits until the water table drops. If p is close to 1, however, all the many fractures are employed to carry the water, and the water table may drop even when the diameters of the evolving conduits are still small. Therefore in highly fractured rock karstification may be intensive but major cave systems are missing. In 4.2.4 Figure 1 the boundary conditions, i.e. the regions of input and output, which are simple, may be restricted to a few input and output points.

Correspondingly, even though a cluster extending over the entire area may exist, it may not lead water to the output. Therefore, it is obvious that the hydrologic setting is of utmost importance to the evolution of a system of conduits.

Furthermore, the fractures in a real network have statistically distributed aperture widths a_0, breadths b_0, and lengths l. From all the percolating pathways, the ones that will exhibit a short breakthrough time, on average have a small flow resistance, i.e. a small average length and large average width, and are also subject to the most favorable dissolution conditions. After breakthrough of one of these pathways, the boundary conditions change due to a dropping water table, and only those pathways connecting the new water table to the output will further develop, whereas others will cease to grow.

The first crude models using brickwork-patterns of fractures on small 7x6 nets were presented by Groves and Howard (1994b). Their networks, however, were too small to provide sufficient spatial resolution. But they showed that breakthrough does occur, just as in one-dimensional fractures, and therefore they provided the basis for more extended investigations such as those reported in this work.

From this information, one may expect that the concept of breakthrough as derived from one-dimensional models (Dreybrodt, 1996) is also valid for two-dimensional fracture networks. Correspondingly, one has to ask the question, whether the equation for breakthrough time derived for single one-dimensional fractures (Dreybrodt, 1996) is still meaningful (see also Dreybrodt and Gabrovšek, Chapter 4.2.2 in this volume). This is by no means trivial, because depending on the occupation probability, the total length of a percolating pathway can vary quite considerably.

To answer this question, we present a model of two-dimensional karstification within sufficiently large networks to reveal details of cave evolution. Then we use numerical sensitivity analysis to investigate the dependence of breakthrough times on the various physical, hydrogeologic, and chemical parameters involved. Finally, we present model simulations for a variety of geologic situations, that show the potential of our improved two-dimensional model of percolation networks.

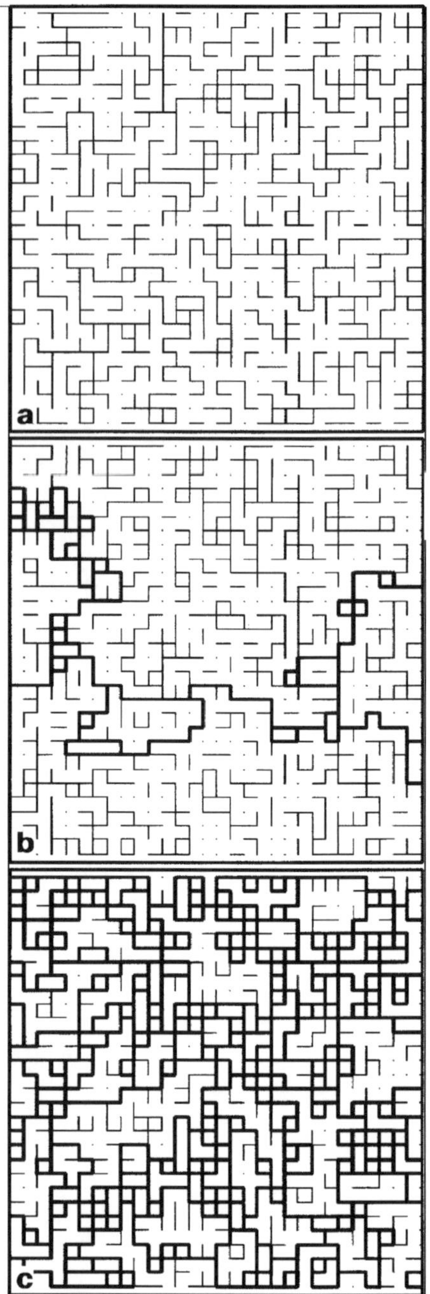

*4.2.4 Figure 1. Percolation networks on a square lattice with occupation probability: **a**—p = 0.4. **b**— p = 0.5. **c**—p = 0.7. The thin lines represent fractures, that cannot carry flow. The thick lines are fractures within clusters that carry water from the left side to the right side of the square lattice.*

Model Structure

As a first step, a percolation network with given p is created. For simplicity, in contrast to nature, we assume that the length, initial aperture width, and breadth are the same for all fractures. Next the boundary conditions for input and output are defined, and all fractures that cannot transport flow are omitted. The remaining percolation backbone represents all percolating pathways. To calculate the flow rates in each fracture, a yet unknown hydraulic head h_i is assigned to each node where three or more fractures are connected. The series of single fractures connecting such a point i to the next possible intersection point j constitutes an isolated flow resistance R_{ij}. For laminar flow, this is given (Beek and Mutzall, 1975) by

$$R_{ij} = \frac{12\eta}{\rho g}\int_i^j \frac{dx}{a^3(x)b(x)\,M(x)} \qquad (1)$$

x represents the distance from node i along the series of fractures connected to node j, $a(x)$ and $b(x)$ are the profiles of the fracture diameters, ρ is the density of water, η its dynamic viscosity, and g the Earth's gravitational acceleration. Flow rates within these elements of the network are calculated by a mass conservation equation for each of the nodes.

$$\sum Q_{in} - \sum Q_{out} = \sum_j Q_{ij} = 0;$$

$$\frac{h_i - h_j}{R_{ij}} = Q_{ij} \qquad (2)$$

Q_{in} is the flow rate toward i, whereas Q_{out} is the flow rate away from it. Q_{ij} is the flow rate in the pathway connecting nodes i and j. This set of linear equations is solved by standard methods, using a special CG-iteration procedure for sparse matrices (Jacobs, 1977). Once the heads h_i and h_j are known, the flow rates Q_{ij} are given (equation 2). Then we calculate dissolution widening for all tubes in the following way.

A First, at time t = 0 we select an input point at the boundary of the network. We specify the concentration c_{in} of the solution that flows into the input tube. Then we apply the one-dimensional transport-dissolution model (Dreybrodt, 1996) to calculate the concentration profile along this tube, including the concentration of the solution leaving it at the next node where this tube is connected to another one. Next we calculate the new profile of the tube after a time step Δt. We repeat this procedure for all input points at the boundary. In the dissolution model, the following dissolution kinetics are used (Dreybrodt, 1996)

$$F_1 = k_1\left(1 - \frac{c}{c_{eq}}\right), \text{ for } c < c_s$$

$$F_n = k_n\left(1 - \frac{c}{c_{eq}}\right)^n, \text{ for } c > c_s \qquad (3)$$

where c is the concentration of dissolved calcite in the solution, k_1 and k_n are rate constants, and c_{eq} is the equilibrium concentration with respect to calcite (see also Dreybrodt and Gabrovšek, Chapter 4.2.2 in this volume).

B In the next step we select all nodes where the concentrations of all the inflowing solutions are known. We assume complete mixing of these solutions before they are transferred into conduits transporting the flow away. For these, as described above, the one-dimensional transport-dissolution model is used to obtain the concentration of the solution at their exits, and also their new profiles after the time step Δt. Then we repeat procedure B until the new profiles of all conduits have been obtained. We have also performed simulations without complete mixing, and these give similar results.

For the new network, we calculate the new flow rates and the new profiles at time $2\Delta t$. By repeating procedures A and B, the profiles of all conduits and their flow rates can be obtained as a function of time. It should be noted that the boundary conditions and the input concentrations can be changed after each time step, thus taking account of external changes by climate or geologic changes (e.g. uplifting, downcutting of base level). It is also possible, instead of specifying hydraulic heads at each arbitrary mode, to use fixed-input flow rates. Thus a variety of geologic and climatic conditions can be simulated by the model.

Numerical Sensitivity Analysis of Breakthrough Time

In 4.2.4 Figure 2, an example of the breakthrough behavior typical for all simulations is given. The network with $p = 0.7$ simulates the initial hydrogeologic setting. The length L of the square is 3 km. The left side contains a variety of input points at an hydraulic head of $h = 150$ m. The right side contains output points, all at a hydraulic head of 0. The upper and lower sides of the square are assumed to be impervious. The width of the fractures is $a_0 = 0.03$ cm and their breadth $b_0 = 100$ cm. The concentration of the inflowing solution is $c_0 = 0$. The equilibrium concentration with respect to calcite is $c_{eq} = 2x10^{-6}$ mole/cm^3.

The rate constants of reaction kinetics are:

- $k_1 = 4x10^{-11}$ mole cm^{-2} s^{-1}
- $k_4 = 4x10^{-8}$ mole cm^{-2} s^{-1}
- $n = 4$
- $c_s = 0.9$

In 4.2.4 Figure 2, drawing (**a**) shows the average widths of the fractures after 9960 years. These average widths are those that a uniform conduit with even spacing would exhibit if it had the same resistance as the real conduit with a profile $a(x)$. Drawings (**b**) and (**c**) illustrate the further evolution as the conduits penetrate into the system. Finally, after 17,762 years, the first channel has reached an exit point. The distance of penetration $L_p(t)$ can now be defined by the largest distance from the input, where channels have been widened to 0.1 cm. Breakthrough at time T_B is achieved when $L_p(T_B) = L$.

As a function of time in years, 4.2.4 Figure 3 shows:

- The penetration length L_p/L.
- The width $a(t)/a(T_B)$.
- The water flow $Q(t)/Q(T_B)$.
- The concentration $c(t)/c_{eq}$ of the solution at the breakthrough exit point.

All these quantities show a typical breakthrough behavior. We define the breakthrough time as the time at which the concentration of the water leaving the major exit point drops below $10^{-3} \cdot c_{eq}$. In all cases, flow was laminar as checked by calculating the Reynolds number to be below 1000.

Using such a network, but changing only one of its parameters at a time, we calculate the dependence of the breakthrough time as a function of these parameters. The result is summarized by the following expression

$$T_B = (a_0)^{\frac{2n+1}{n-1}} \cdot \left(\frac{L^2\eta}{hc_{eq}}\right)^{\frac{n}{n-1}} \cdot k_n^{\frac{1}{n-1}} \cdot c(p,s) \qquad (4)$$

The constant $c(p,s)$ is dependent only on the properties of the initial fracture system. It depends on the occupation probability p and the geologic setting s. This equation closely resembles that of the one-dimensional case, provided the length l of the conduit is replaced by the spatial dimension L of the fracture system.

See 4.2.4 Figure 4 for a verification of equation 4. Here we have plotted the breakthrough times in years calculated for various a_0 as a function of a_0 in a double logarithmic plot, whereby we have used results from the fracture system shown in 4.2.4 Figure 2. The straight line plotted through the calculated points shows the validity of the power law, and the exponent can be calculated from its slope. Similar results are obtained from plots of T_B against (L^2/h), c_{eq}, and k_n.

The exponents obtained are within ±10% of those predicted from equation 4. We have also calculated the breakthrough times with their dependence on the various parameters as defined above for a variety of different values of n between $n = 3$ up to $n = 11$ and have found equation 4 to be valid in this range. We have also performed such calculations on a variety of different initial fracture systems with identical boundary conditions and different occupation probabilities. These also have verified the predictions of equation 4 (Siemers and Dreybrodt, 1998).

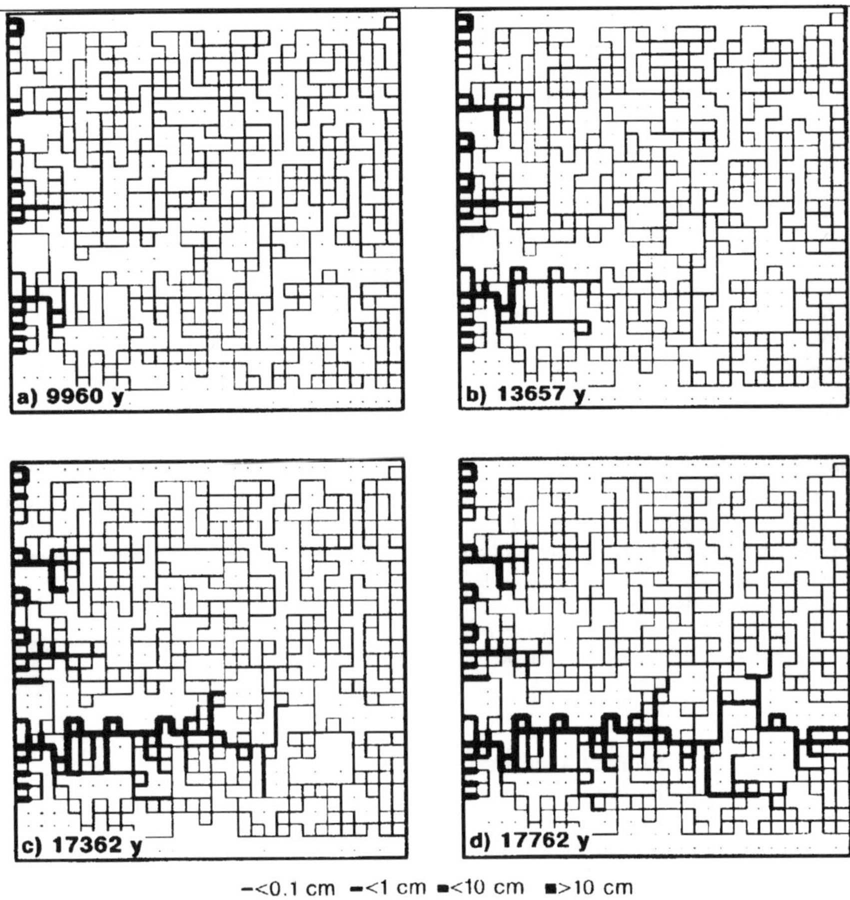

*4.2.4 Figure 2. Evolution of a karst aquifer with p = 0.7 at 9960 years in **a**. 13,657 years in **b**. 17,362 years in **c**. 17,762 years in **d**. The upper and lower boundaries are impervious. The left boundary is at constant head h = 150 m, whereas the right boundary is h = 0 m. The thickness of the lines represents the average aperture of the fractures, following the explanation below the figure. Parameters* $a_0 = 0.03$ *cm,* $b_0 = 100$ *cm,* $h = 15{,}000$ *cm,* $L = 3x10^5$ *cm,* $n = 4$, $\hat{c}_s = 0.9$, $c_{eq} = 2x10^{-6}$ *mole/cm*^{3}*, and* $k_n = 4x10^{-8}$ *mole cm*$^{-2}$*s*$^{-1}$.

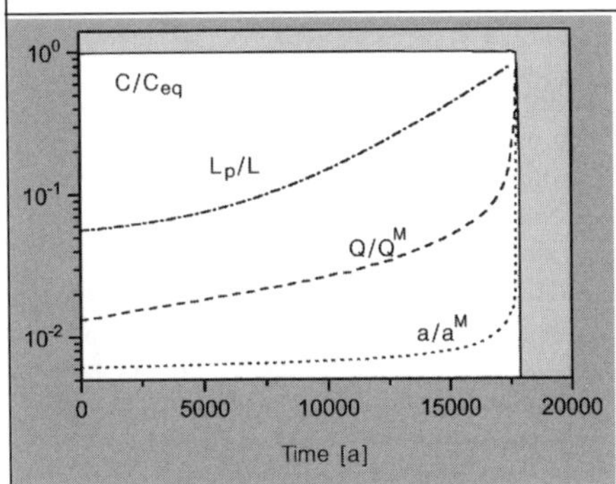

4.2.4 Figure 3. Evolution of penetration length L_p/L*, flowrate* Q/Q_M*, fracture width* a/a_M *and the concentration at the exit of the winning pathway.* Q_M *and* a_M *are the values achieved at breakthrough,* $Q_M = 518$ *cm*^{3}*/s,* $a_M = 4.86$ *cm. At breakthrough the concentration switches to values close to zero.*

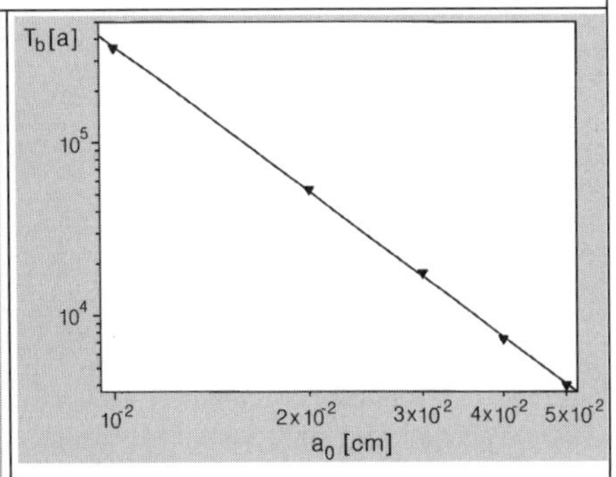

4.2.4 Figure 4. Double logarithmic plot of breakthrough time as a function of the initial aperture width a_0 *of the fracture, as obtained from computer runs on the net of 4.2.4 Figure 2. Only* a_0 *is changed. Everything else remains constant.*

Evolution of Cave Systems Under Various Geologic Conditions

To demonstrate the potential of the two-dimensional model, in the following we present the results of simulations for a variety of geologic conditions.

Ewers High-Dip Model

The 4.2.4 Figure 5 shows the hydrogeologic setting for cave evolution on a steeply dipping bedding plane parting. Water input is provided at the upper outcrop of the bedding plane. Only one spring located below is provided for output. This model was first proposed by Ewers (1982). It is also discussed in the textbooks of Ford and Williams (1989) and Dreybrodt (1988).

We have simulated cave evolution in such a setting (4.2.4 Figure 5). The length of the square is 3 km, $p = 0.7$. We assume all sides of the square to be impervious with the following exceptions. At the left side we have defined three input points at a hydraulic head of $h = 150$ m. There is one output point at the upper right side, located at base level, $h = 0$. The following parameters have been used in this model:

- $p = 0.7$
- $a_0 = 10^{-2}$ cm
- $b_0 = 100$ cm
- $c_0 = 0$
- $c_{eq} = 2x10^{-6}$ mole/cm^3
- $c_s = 1.5x10^{-6}$ mole/cm^3
- $k_I = 4x10^{-11}$ mole cm^{-2} s^{-1}
- $k_{II} = 4.2x10^{-5}$ mole cm^{-2} s^{-1}
- $n = 11$.

Eleventh order kinetics was used as it seems to be appropriate from recent experimental data (Eisenlohr et al., 1997). See also Dreybrodt and Eisenlohr, Chapter 4.1.2 in this volume.

In 4.2.4 Figure 5, drawing (**a**) shows the fracture widths that have evolved after 280,285 years where a karst channel has developed from the upper input toward the output, penetrating half the distance. The other two inputs located at a greater effective distance from the output have less successfully created only short channels.

At (**b**) only 1300 years later, however, the upper channel has reached the output, causing a dramatic increase in the flow rate, as plotted in 4.2.4 Figure 6. Because these high flow rates cannot be delivered from the storage of the system, the boundary conditions at the upper input are changed to a constant inflow of 500 cm^3/s. At such high flow, the concentration at the output, which was close to saturation prior to breakthrough, drops to low values (see 4.2.4 Figure 6) and the upper channel widens at a rate of several cm per hundred years. Flow still remains laminar as checked by the Reynolds number.

Drawing (**c**) of 4.2.4 Figure 5 shows the system after another 10,500 years. Due to the low hydraulic head established in the upper channel by its restricted input, the channel originating from the second input point by integrating into the first channel has created a second breakthrough event. Therefore, at this input point also, the inflow is limited (500 cm^3/s), whereas the lowest input remains at a constant head.

In 4.2.4 Figure 5, drawing (**d**) shows the situation only 800 years later at 292,823 years. The lowest channel has succeeded in reaching the upper network. From this moment on, there is almost equal widening everywhere in the conduits, and the structure of the conduit system will remain unchanged if the boundary conditions remain unchanged. It should be noted that the evolution of the first channel in 4.2.4 Figure 5 took 280,000 years, whereas the further development needed only 12,500 years. The 4.2.4 Figure 6 shows the evolution of flow rates Q leaving the exit, the width a of the exit conduit, and the concentration of the water at the exit.

From this one sees that the first breakthrough event takes the system into a state of adolescence, going to maturity a short time later. Prior to the first breakthrough, the width and the flow rates increase slowly. The concentration at the exit is high, and therefore dissolution rates are low. After breakthrough, the concentration c drops, dissolution rates become high, and $a(t)$ increases quickly.

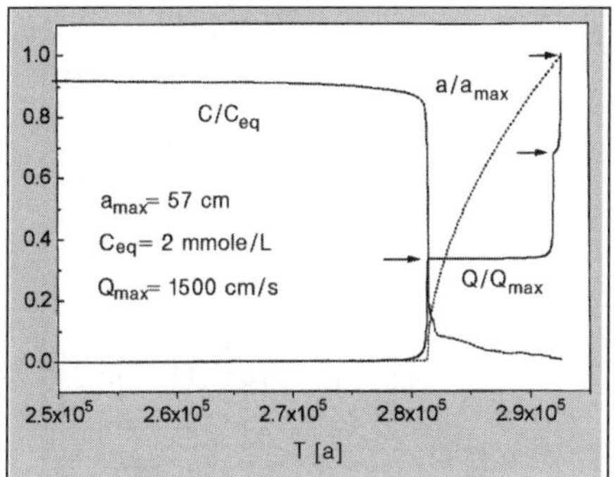

4.2.4 Figure 6. Time dependence in years of flow rate $Q(t)$, fracture width $a(t)$ at the exit, and concentration $c(t)$ of the water at the exit. All values are normalized to the maximal values at the end of the run. The arrows designate the breakthrough events in 4.2.4 Figure. 5 ***b*****,** ***c*****, and** ***d*****.**

Ewers Low-Dip Model

As a further example, 4.2.4 Figure 7 illustrates the evolution of caves in Ewers low dip model (Ewers and Quinlan, 1981). See also Ford and Williams (1989) and Dreybrodt (1988). In this case, several ranks of multiple inputs exist with equal heads. To obtain such a case we have changed the boundary conditions of the network depicted in 4.2.4 Figure 2. The upper and lower sides of the square are still impervious. The left side is at a hydraulic head of $h = 150$ m and has various inputs. The right side at base level ($h = 0$) exhibits many output points. Finally, we have introduced inside the net a row of 4 input points with fixed head $h = 150$ m, depicted by circles. We have used $a_0 = 0.02$ cm. All the other parameters are those of 4.2.4 Figure 5. This is a situation often encountered in karst. The plane square represents an extended horizontal bedding plane, where water descends to the input points from an upper plateau through open joints. Low-relief karst plains such as in Missouri, U.S.A., are representative examples for this case.

In 4.2.4 Figure 7, (**a**) shows the system after the first breakthrough from one of the inner inputs to base level. Channels have also developed from the other inner inputs but have not yet succeeded in reaching base level. This is achieved (**b**) 500 years later at 26,363 years. Due to the restricted input, the hydraulic head is lowered at all inner input points. Therefore karst channels grow downstream from the inputs at the left side. They reach (**c**) those conduits already created in another 19,000 years. If there are further arrays of input points, this scenario is repeated. Thus, dissolution creates channels growing downstream. Integration of the cave system, however, proceeds upstream. The reduction of breakthrough times in comparison to the case presented by 4.2.4 Figure 5 results from the fact that breakthrough times are related to the hydraulic head and the effective length l_{eff} between the input and the output by equation 4, where n =11.

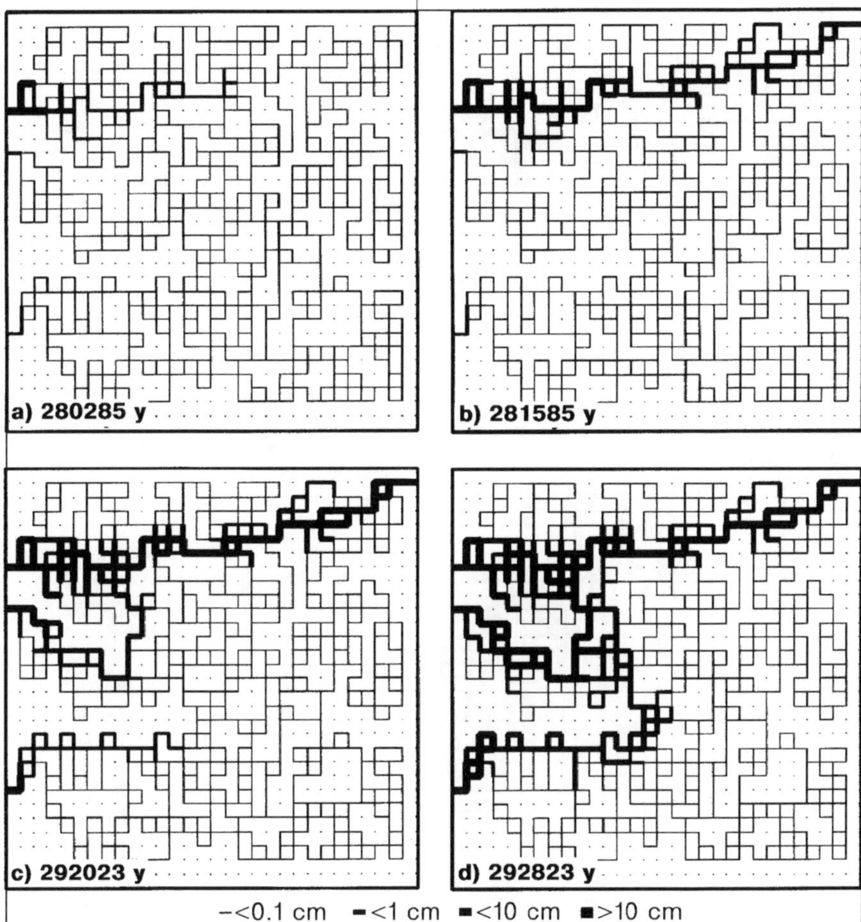

4.2.4 Figure 5. Illustration of Ewers' (1982) high-dip model: Evolution of a karst aquifer of 3x3 km^2, with 3 inputs (left side) at h = 150 m and one output at base level (h = 0). In ***a*** *first karst channels develop until* ***b*** *where a first breakthrough event (the upper channel) has reached base level. In* ***c*** *the next channel is directed toward the upper channel, and the next breakthrough event occurs. In* ***d*** *the lowest channel is finally integrated into the network.*
Parameters: $a_0 = 0.01$ cm, $b_0 = 100$ cm, $h = 15{,}000$ cm, $L = 3 \times 10^5$ cm, $n = 11$, $c_s = 0.75c_{eq}$, $c_{eq} = 2 \times 10^{-6}$ mole/cm^3, and $k_n = 4.2 \times 10^{-5}$ mole cm^{-2}s^{-1}.

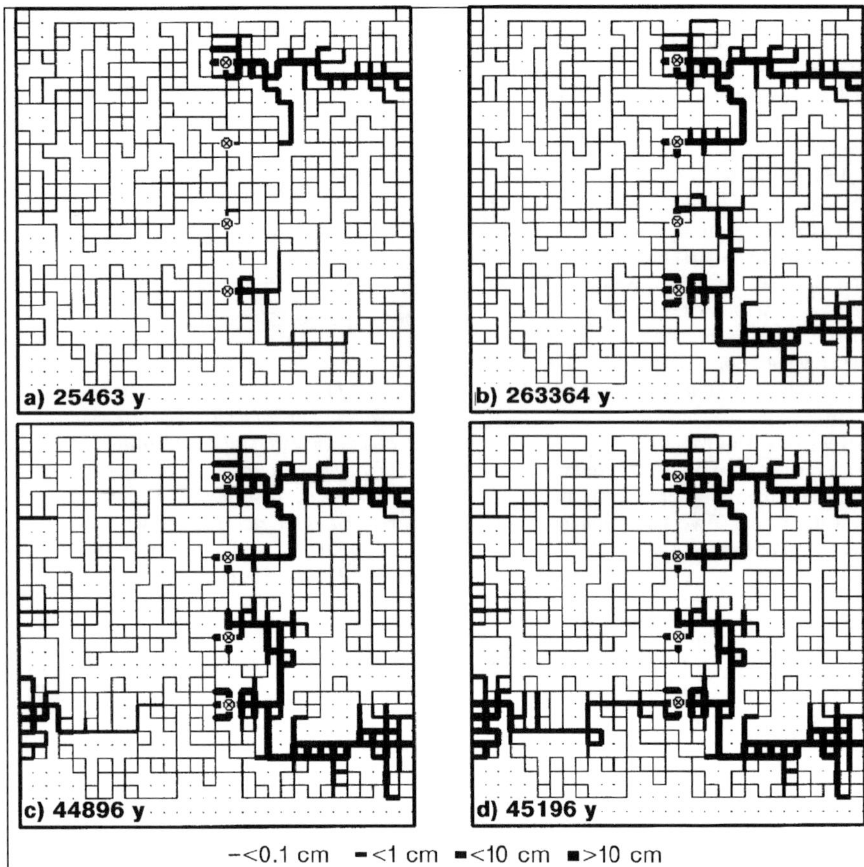

*4.2.4 Figure 7. Evolution of a karst aquifer (low-dip model). The input points at the left side and inside the network (crossed circles) are at equal head. Therefore, in a first step, karst channels are created only from the inner input points. After breakthrough of the upper channel in **a**, the lowest reaches base level in **b**, and the channels evolving from the two inner inputs are quickly integrated. From then on, in **c-d**, the channels grow from the left side and are integrated into the first conduit system. Parameters $a_0 = 0.01$ cm, $b_0 = 100$ cm, $h = 15{,}000$ cm, $L = 3x10^5$ cm, $n = 11$, $\hat{c}_s = 0.75c_{eq}$, $c_{eq} = 2x10^{-6}$ mole/cm^3, and $k_n = 4.2x10^{-5}$ mole cm^{-2}s^{-1}.*

Vertical Caves

As a further example, 4.2.4 Figure 8 shows a vertical cave in its final state after 5700 years. The boundary conditions are as follows: The left side and the lower side of the aquifer are impermeable. The upper side represents a plateau 3 km wide at an altitude of 300 m with various input points. The right side illustrates a vertical cliff with various output points at different heights, the lowest at base level, e.g. a valley. All the other parameters are those of 4.2.4 Figure 5. Note that the vertical dimension now is only 0.3 km.

It should be noted at this point that the output points are at various heights. Therefore, water could flow from a higher output point to a lower one via the fracture network. In reality, however, there is only flow out of all these points. This has to be taken into account in the boundary conditions.

Therefore, in each time step of the computation we checked whether the flow rate at all output channels is directed toward the exit points. If this was not the case, we set the flow at this point to zero and recalculated the pressure distributions until the required flow conditions were obtained. The system evolves as follows: First the input nearest to the edge creates breakthrough at the uppermost outlet after 3496 years. Its input is then restricted. Other inputs, still at high hydraulic head, develop channels integrating with the first channel system, and vertical shafts are created simultaneously.

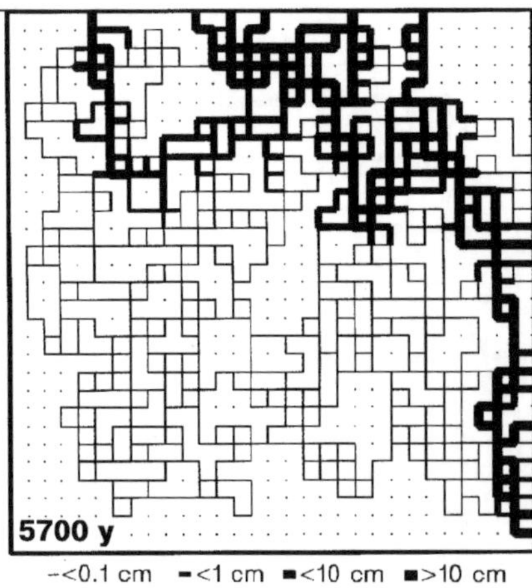

4.2.4 Figure 8. Final state of a vertical cave system: The upper side of the square represents a plateau at 300 m altitude. The left side represents a cliff of 300 m relief. Lowest point of the cave system is at base level. The percolation network utilized for this example was constructed with $p = 0.7$, $a_0 = 0.01$ cm, $b_0 = 100$ cm, $L = 3x10^5$ cm, $n = 11$, $c_s = 0.75c_{eq}$, $c_{eq} = 2x10^{-6}$ mole/cm^3, and $k_n = 4.2x10^{-5}$ mole cm^{-2}s^{-1}.

Artificial Situations

Close to hydraulic structures such as dam sites, unnaturally steep hydraulic gradients are created, and a supply of water driven by these high gradients through fractures beneath the foundation and across the abutment rock is provided. Therefore the question arises, whether within the lifetime of the dam, dissolutional widening might create channels sufficiently wide to cause serious loss of water. We therefore have applied our model to a situation illustrated by 4.2.4 Figure 9.

We have constructed a percolation net with an occupation probability $p = 0.7$ and imposed the following boundary conditions. A grouting curtain below the dam creates an impervious layer 10 m wide and 200 m deep and thus blocks the flow. This is depicted by the shaded area in (**a**) of 4.2.4 Figure 9. The basal width of the dam is 50 m and impervious (shaded). To the left of the dam, water is impounded at a height of 200 m and several input points are available. At the right side $h = 0$ m. The remaining sides of the 300-m-wide square for simplicity are assumed to be impervious. The width of the initial fractures is $a_0 = 0.015$ cm. All the other parameters are the same as in 4.2.4 Figure 5.

In 4.2.4 Figure 9:

- **a** Represents the channels that have evolved at breakthrough after 343 years,
- **b** Illustrates this situation if grouting is reduced to a depth of 100 m.

After breakthrough, turbulent flow sets in and dissolutional widening is on the order of 0.2 cm/year. From this, by use of the Darcy-Weisbach equation, water losses on the order of several m^3/s are estimated at channel widths of 5 cm. Thus after breakthrough, only 25 years are necessary to create serious water loss. If the depth of the grouting curtain is reduced to 150, 100, or 50 m, the corresponding breakthrough times are 255, 99, and 27 years, respectively. Thus sufficiently deep grouting curtains are not only necessary to prevent flow through wide fractures already existing, but also to avoid the formation of new ones.

It should be noted at this point, that similar calculations with one-dimensional models (Dreybrodt, 1992, 1996; Palmer, 1988) yield breakthrough times several times longer than those obtained with two-dimensional nets.

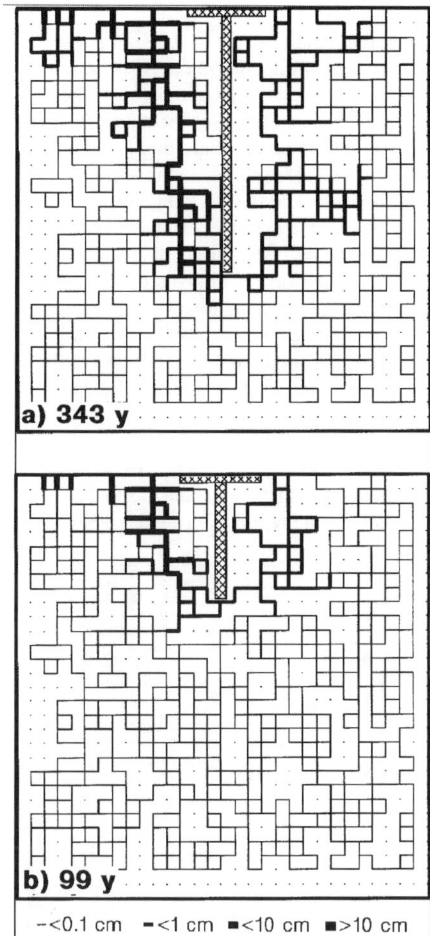

*4.2.4 Figure 9. Evolution of conduits below a dam. In **a** with a grouting curtain 200 m deep, breakthrough occurs after 343 years. In **b** reducing the depth of grouting to 100 m results in breakthrough after 99 years. In both cases the channel widths at the exit have increased by a factor of 10 to 0.16 cm, where* $a_0 = 0.015$ *cm,* $b_0 = 100$ *cm,* $h = 20{,}000$ *cm,* $L = 3x10^4$ *cm,* $n = 11$, $\hat{c}_s = 0.75\ c_{eq}$, $c_{eq} = 2x10^{-6}$ *mole/cm*^{3}, *and* $k_n = 4.2x10^{-5}$ *mole cm*$^{-2}s^{-1}$.

This is because we have used higher-order kinetics with $n = 11$, which seems to be more appropriate from recent experimental data (Eisenlohr et al., 1997) than kinetics with $n = 4$ used in the one-dimensional model. See also Dreybrodt and Eisenlohr, Chapter 4.1.2 in this volume.

Evolution of Epikarst

The uppermost layer of the karst rock is highly fractured due to stress release and weathering. This upper zone extends downward several meters. Beneath this zone, fracturing of the rock mass is not so intense, and the average distance between fractures is much larger than in the epikarstic zone (Williams, 1983; Ford and Williams, 1989). To model the evolution of epikarst, we have generated the network shown in 4.2.4 Figure 10.

The upper layer (not drawn to scale) is 10 m thick and is highly fractured with an occupation probability of $p = 0.8$. Below, a less fractured ($p = 0.6$) rock mass extends toward an output at 1500 m distance. The whole structure is subjected to a hydraulic gradient of 0.05. The initial average aperture width is $a_0 = 0.02$ cm. The right and left side are assumed to be impervious. Input at $h = 75$ m is provided at the upper boundary, whereas the lower side of the square exhibits exit points at $h = 0$. The width of the whole structure is 50 m. The chemical parameters are the same as in 4.2.4 Figure 5.

In 4.2.4 Figure 10, drawings (**a–c**) illustrate the evolution of the system. At the beginning, dissolution is concentrated at the upper highly fractured epikarst layer where the widths of the fractures are enlarged rapidly. A network of wide fractures evolves along all routes of flow that are directed toward the main inputs into the lower less fractured rock mass. The hydraulic head at all these inputs is almost even during the initiation phase. Consequently passages evolve along the percolating pathways toward base level. Finally, breakthrough occurs in the most favorable one. At that time the upper layer has become highly karstified with maximum fissure widths of 43 cm. The structure of karstification is closely connected to the structure in the rock beneath. Our model suggests that the evolution of epikarst takes place as the first step in the process of initial karstification, and a perched subcutaneous aquifer may exist a long time before large karst conduits have evolved to drain the terrain effectively.

Lithologic Control of Karst Evolution

So far we have assumed that the kinetics of limestone dissolution is the same everywhere in the rock. In other words, the rock mass is homogeneous with respect to its dissolution kinetics. This, however, need not necessarily be so. To this end we have modeled a situation where the first third of the limestone terrain consists of a rock with higher-order dissolution kinetics with $n = 11$, and $k_n = 4.2x10^{-5}$ mole cm^{-2} s^{-1}. This is shown as a shaded area in 4.2.4 Figure 11.

The remaining area exhibits a different rock type with $n = 4$ and $k_4 = 4x10^{-8}$ mole cm^{-2} s^{-1}. The boundary conditions and the parameters otherwise are those as in 4.2.4 Figure 2. There is an abrupt increase in the dissolution rates when the solution crosses the boundary between these different types of rock. At the onset of karstification, small

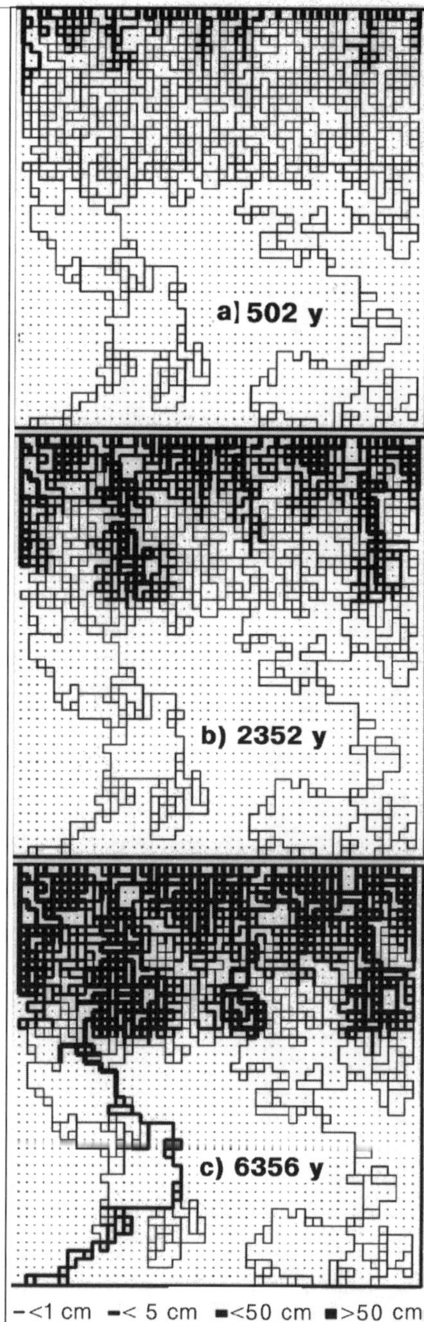

4.2.4 Figure 10. Evolution of epikarst: The upper layer is 10 m deep (not drawn to scale) with p = 0.8. It is underlain by bedrock extending to 1500 m, which is less intensely fractured (p = 0.6). The average aperture width is $a_0 = 0.02$ *cm. The width of the terrain is 50 m. The left and right sides are impermeable. In **a**, T = 500 years. In **b**, 2350 years. In **c**, 6350 years. At breakthrough, the highly karstified upper layer exhibits maximal widths of 43 cm at about 1 m spacing where* $a_0 = 0.02$ *cm,* $b_0 = 100$ *cm,* $h = 7500$ *cm,* $L = 5x10^3$ *cm,* $n = 11$, $\hat{c}_s = 0.75\ c_{eq}$, $c_{eq} = 2x10^{-6}$ *mole/cm*^{3}, *and* $k_n = 4.2x10^{-5}$ *mole cm*$^{-2}s^{-1}$.

channels develop at the input. The abrupt increase in dissolution rates at the rock boundary, however, is sufficiently large that channels also grow from the rock discontinuity toward the exit from the very beginning. Some of these channels are connected to those growing from the input, whereas others remain isolated.

Finally, breakthrough occurs along the integrated channel network. Breakthrough time is in the same order of magnitude as in the case of 4.2.4 Figure 2. This is to be expected, because the initial dissolution rates at the exit points are of similar magnitude in both cases, and as has been shown by Dreybrodt (1996) and Dreybrodt and Gabrovšek (Chapter 4.2.2 in this volume) the initial rates at the exit determine the breakthrough time.

The behavior of karst evolution changes completely if the first third of the terrain is occupied by limestone with 4th-order kinetics and the remainder by $n = 11$. In this case, there is an abrupt drop in the dissolution rates by many orders of magnitude when the aggressive water crosses the border between the two lithologies. Therefore, conduit evolution first proceeds in the region with $n = 4$ as depicted by 4.2.4 Figure 12, (a and b). Due to the extremely slow widening in the region of 11th order no increase of permeability occurs in that area. Therefore, the flow rates remain restricted, even when large conduits have evolved in the 4th-order region. No breakthrough occurs even after 50 million years. This example shows that varying lithology has a significant influence on karst evolution.

To further explore this matter, we have changed the 11th-order kinetics to 6th-order in the shaded region. During the first million

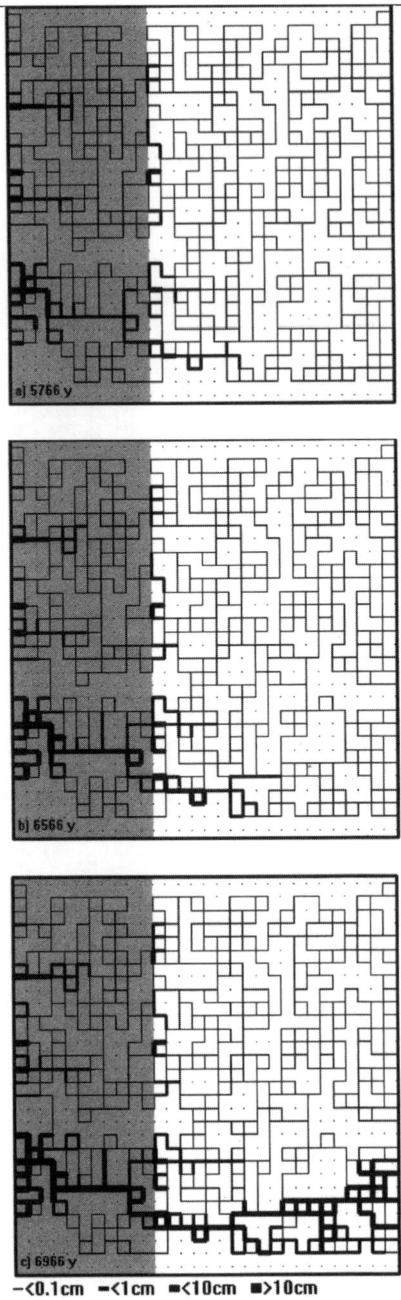

*4.2.4 Figure 11. Influence of lithology: The shaded area is occupied by rock with 11th order kinetics ($k_{11} = 4.2 \cdot 10^{-5}$ mole cm^{-2}s^{-1}, $\hat{c}_s = 0.75\ c_{eq}$). The unshaded area exhibits 4th order kinetics ($k_4 = 4 \cdot 10^{-6}$ mole cm^{-2}s^{-1}, $c_s = 0.9$). Everything else as in 4.2.4 Figure 2, **a** is breakthrough at 5766 years. **b** is 6566 years. **c** is 6966 years—where $a_0 = 0.03$ cm, $b_0 = 100$ cm, $h = 15{,}000$ cm, $L = 3x10^5$ cm.*

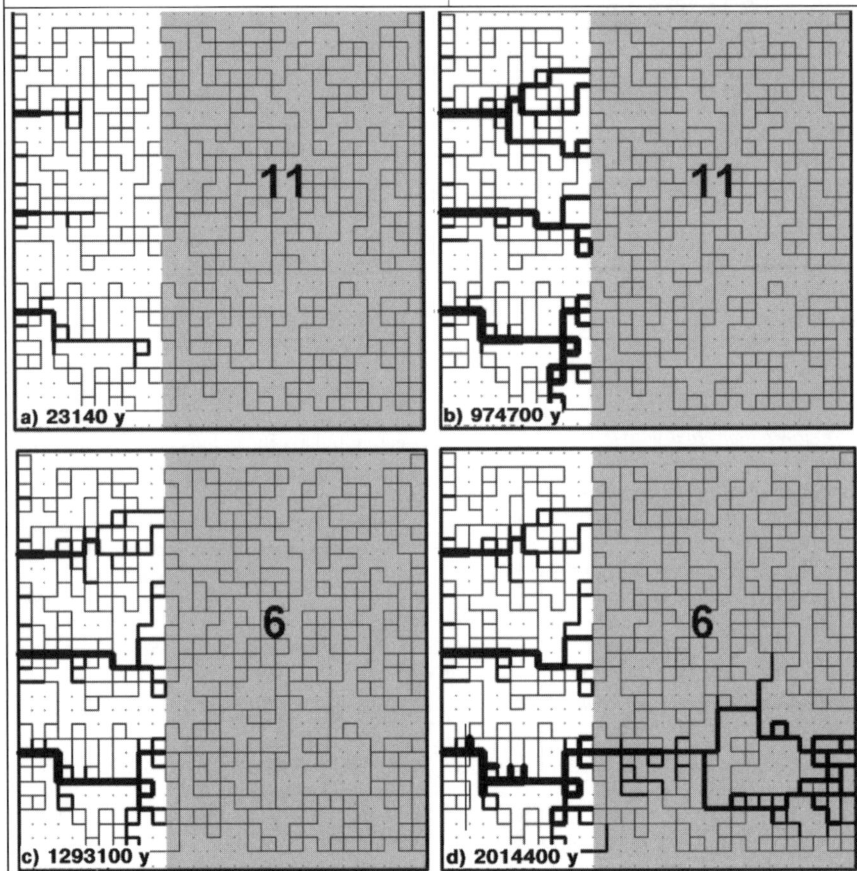

*4.2.4 Figure 12. Influence of lithology: All parameters as in Figure 11, but now the water flows first through rock with 4th-order kinetics (unshaded area, $k_4 = 4x10^{-8}$ mole cm^{-2}s^{-1}, $c_s = 0.9$) and then enters into a region of 11th-order kinetics ($k_{11} = 4.2\ x\ 10^{-5}$mole cm^{-2}s^{-1}, $\hat{c}_s = 0.75\ c_{eq}$, shaded area). After 23,140 years a system has evolved that continues to grow exclusively in the 4th-order region. Breakthrough is not achieved even after 50 million years. In **a** 23,140 years. In **b** 974,700. Drawings **c** and **d** show the evolution of the same setting, but the dissolution kinetics now has $n_6 = 6$, $k_6 = 1x10^{-6}$ mole cm^{-2}s^{-1}, in the shaded region. In this case a delayed breakthrough occurs. In **c**, 1,293,100 years. In **d**, 2,014,400 years.*

years, dissolution is restricted to the 4th-order terrain. But since dissolution rates of 6th order are less inhibited, a slow change of permeability occurs in the region of 6th order, which causes feedback and therefore breakthrough after 2 million years.

This example shows that regions of slower dissolution kinetics can delay breakthrough significantly.

A more complex situation is displayed by 4.2.4 Figure 13.

The first 20 percent of the section consists of a rock type with $n = 11$ dissolution kinetics, then with $n = 4$, after which the pattern is repeated, and finally is terminated with $n = 11$. The shaded areas in 4.2.4 Figure 13 show the regions with $n = 11$. Otherwise all parameters and boundary conditions are as in 4.2.4 Figure 11.

In the beginning the system evolves as in 4.2.4 Figure 11, inasmuch as the kinetics change from 11th to 4th order. At the boundaries from 4th to 11th order, the rates drop, such that no increase in fracture widths occurs. At the second boundary from $n = 11$ to $n = 4$, rates increase again sufficiently to produce significant widening of the fractures in the second region with $n = 4$. Therefore, an isolated cave system grows. The computer simulation was terminated after 50 million model years. No breakthrough occurred.

These examples show that lithologic control can be of utmost importance. All model simulations with homogeneous rock inevitably lead to a breakthrough behavior creating through caves. Isolated cave systems of sufficient size and also large systems without exits of transversable size can arise, however, in areas of varying lithologies.

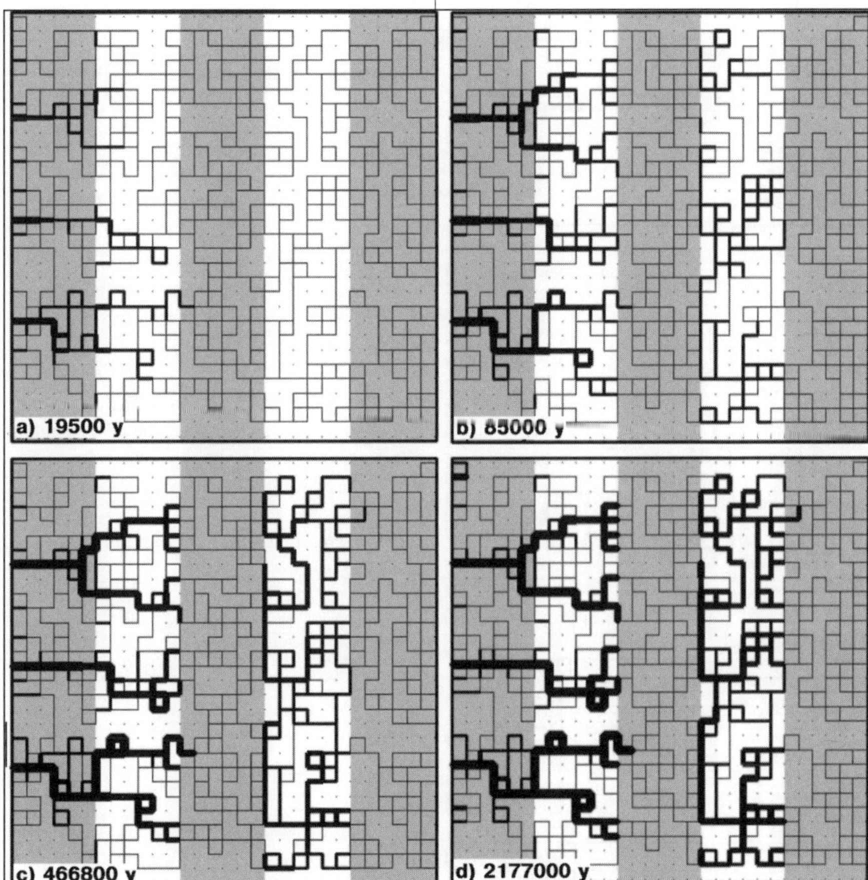

4.2.4 Figure 13. Influence of lithology: The shaded areas denote 11th-order kinetics, the unshaded 4th-order. Otherwise all parameters are as in 4.2.4 Figure 11. ***a*** *is 19,500 years.* ***b*** *is 85,000 years.* ***c*** *is 466,800. And* ***d*** *is 2,177,000 years.*

Dependence of Cave Patterns on Breakthrough Times

Finally we want to address a very unexpected result of the model calculations. During the runs described in the section on sensitivity analysis it turned out that the cave patterns depend on breakthrough times in such a way that simple mainly linear patterns arise for large breakthrough times. But if one changes only one of the many parameters on an identical percolation network, such that breakthrough time decreases, the patterns become branched and more complex. This is illustrated by 4.2.4 Figure 14, where we have used the same network as in 4.2.4 Figure 2 and also all the same parameters, except a_0.

In 4.2.4 Figure 14:

- **a** Depicts the pattern for $a_0 = 0.01$ cm at breakthrough (T_B = 355,308 years). Compared to 4.2.4 Figure 2 ($a_0 = 0.03$ cm) a simple almost unbranched linear cave pattern is formed.

- **b** Shows the pattern at $a_0 = 0.02$ cm (T_B = 54,000 years) which is more complex. In both cases the pathway of breakthrough differs from that in 4.2.4 Figure 2.

Drawings (**c** and **d**) display the patterns at breakthrough with:

- $a_0 = 0.04$ cm (T_B = 7,800 years) (**c**)
- $a_0 = 0.05$ cm (T_B = 4,000 years) (**b**)

The patterns follow different pathways and become more complex with increasing a_0. We have performed similar model runs changing the parameters h, k_n, and c_{eq}. In all cases we found similar more complex patterns when breakthrough times became smaller.

This shows that the selection of the many competing pathways that start to grow at the onset of karstification is not determined only by their flow resistance. It is not necessarily the pathway with the lowest flow resistance that finally achieves breakthrough. The dynamics of dissolution rates, which is complex, is also of importance. Otherwise, one would expect that changes of the chemical parameters should not affect the shape of the breakthrough patterns.

They do, however, as illustrated by 4.2.4 Figure 15, where all parameters are as in 4.2.4 Figure 2 except c_{eq}, which is:

- **a** $c_{eq} = 1x10^{-6}$ mole/cm^3
- **b** $c_{eq} = 1.5x10^{-6}$ mole/cm^3
- **c** $c_{eq} = 2x10^{-6}$ mole/cm^3
- **d** $c_{eq} = 2.5x10^{-6}$ mole/cm^3

Drawing **c** is identical to **d** in 4.2.4 Figure 2 shown here for comparison. As in 4.2.4 Figure 14, more complex patterns and shifting of pathways occur as the breakthrough time decreases.

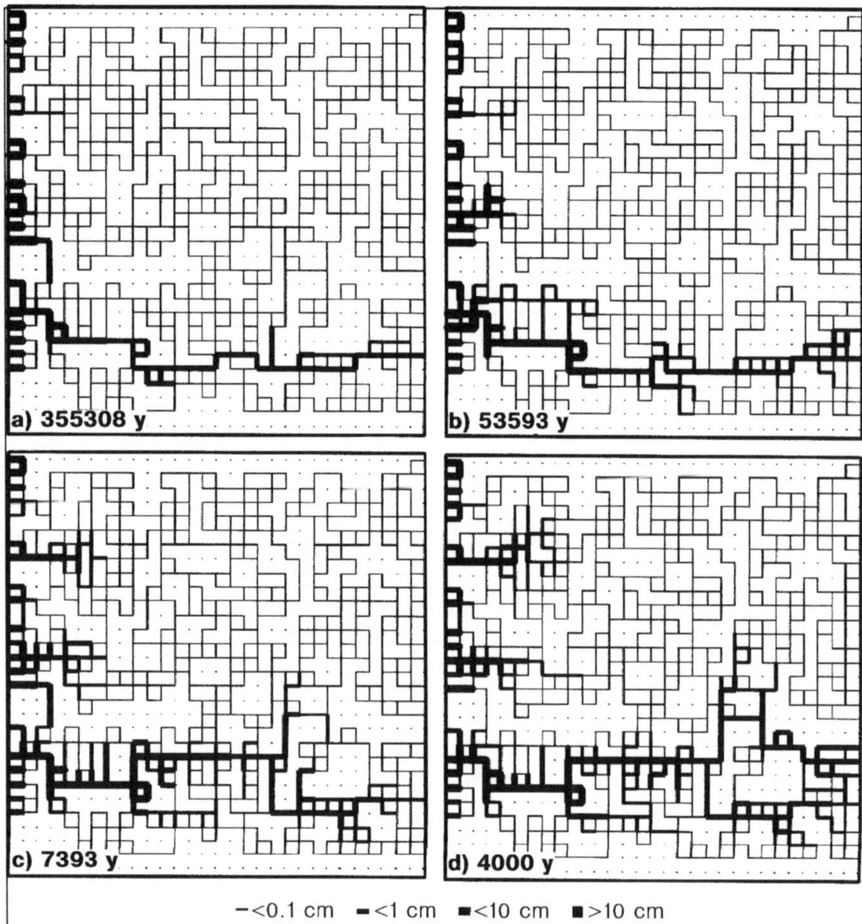

4.2.4 Figure 14. Conduit patterns at breakthrough for the fracture system as in 4.2.4 Figure 2 where all parameters are the same except for a_0:

a *a_0 = 0.01 cm, T_B = 355,308 years.*
b *a_0 = 0.02 cm, T_B = 53,593 years.*
c *a_0 = 0.04 cm, T_B = 7393 years.*
d *a_0 = 0.05 cm, T_B = 4000 years.*

These breakthrough times are also depicted in 4.2.4 Figure 4.

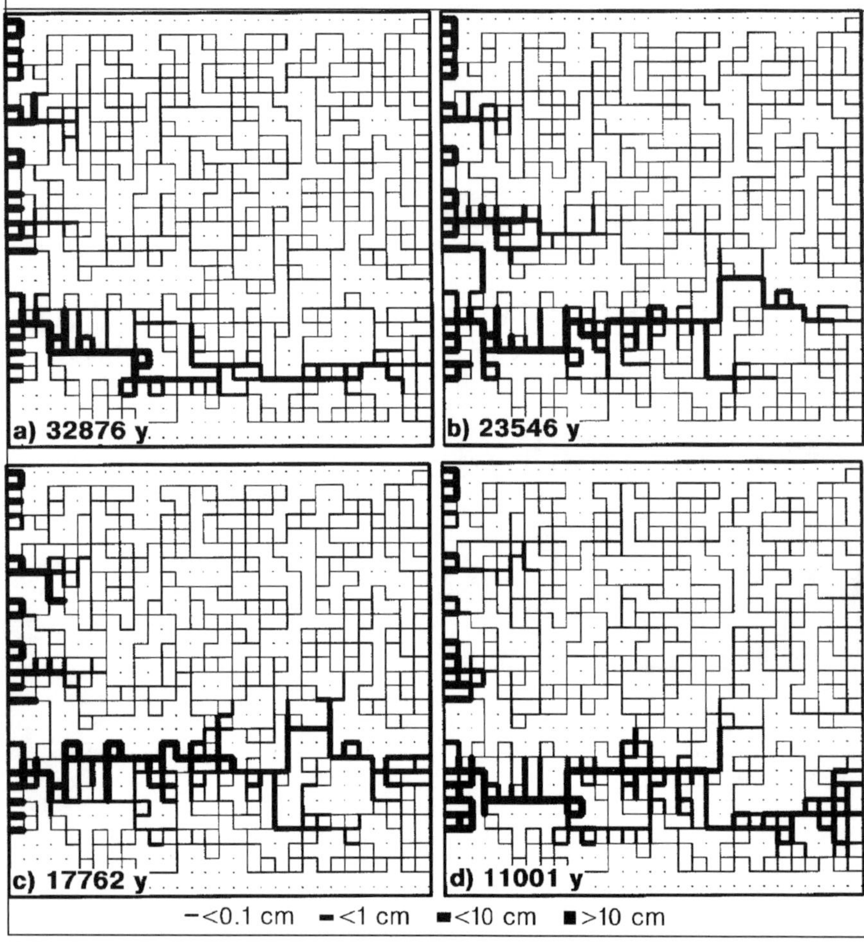

4.2.4 Figure 15. Conduit patterns at breakthrough for the fracture system as in 4.2.4 Figure 2 where all parameters are the same except for c_{eq}:

a *$c_{eq} = 1x10^{-6}$ mole/cm^3, T_B = 32,876 years.*
b *$c_{eq} = 1.5x10^{-6}$ mole/cm^3, T_B = 23,546 years.*
c *$c_{eq} = 2x10^{-6}$ mole/cm^3, T_B = 17,762 years.*
d *$c_{eq} = 2.5x10^{-6}$ mole/cm^3, T_B = 11,001 years.*

*Illustration **c** is identical to that of d in 4.2.4 Figure 2.*

Conclusion

We have presented a two-dimensional model of cave evolution in systems of statistically distributed fractures in limestone. The model shows that the main principles of breakthrough as derived from one-dimensional models remain valid. The evolution of cave patterns, however, is not only determined by the initial geometric setting—such as the fracture widths and lengths and the initial boundary conditions—but it also depends in a complex manner on the chemical parameters of the inflowing solutions and on the dissolution kinetics of the rock.

Therefore, a change in climate affecting only the equilibrium concentration can influence the pattern of the evolving conduit systems. This is a rather unexpected result that has been verified by a large number of computer runs on different geologic settings. To understand these complex mechanisms, more detailed mathematical analysis of the model is required.

We have demonstrated the potential of the new model on a two-dimensional percolation network for a variety of geologic settings, including variation in lithology with respect to the dissolution kinetics. Whereas in lithologically homogeneous rock breakthrough inevitably creates through caves, that connect input and output points, spatial variations of lithology give rise to isolated cave systems that are not connected to input or output points.

Acknowledgment

This work was supported by a grant of the Deutsche Forschungsgemeinschaft.

References

Balberg, I., 1986, Connectivity and conductivity in 2d and 3d fracture systems: *Annales of Israel Physical Society,* v. 8, p. 89-101.

Beek, W.J., and Mutzall, K.M.K., 1975, *Transport phenomena:* New York, Wiley, 298 p.

Buhmann, D., and Dreybrodt, W., 1985, The kinetics of calcite dissolution and precipitation in geologically relevant situations of karst areas, 1. Closed system: *Chemical Geology,* v. 53, p. 109-124.

Dreybrodt, W., 1988, *Processes in karst systems–Physics, chemistry, and geology:* Berlin, Springer Verlag, 288 p.

Dreybrodt, W., 1990, The role of dissolution kinetics in the development of karst aquifers in limestone: A model simulation of karst evolution: *Journal of Geology,* v. 98, p. 639-655.

Dreybrodt, W., 1992, Dynamics of karstification: A model applied to hydraulic structures in karst terranes: *Applied Hydrogeology,* v. 1, p. 20-32.

Dreybrodt, W., 1996, Principles of early development of karst conduits under natural and man-made conditions revealed by mathematical analysis of numerical models: *Water Resources Research,* v. 32, p. 2923-2935.

Dreybrodt, W., Lauckner, J., Zaihua, L., Svensson, U., and Buhmann, D., 1996, The kinetics of the reaction $CO_2 + H_2O \rightarrow H^+ + HCO_3^-$ as one of the rate-limiting steps for the dissolution of calcite in the system H_2O–CO_2–$CaCO_3$: *Geochimica et Cosmochimica Acta,* v. 60, p. 3375-3381.

Eisenlohr, L., Madry, B., and Dreybrodt, W., 1997, Changes in the dissolution kinetics of limestone by intrinsic inhibitors adsorbing to the surface: *12th International Congress of Speleology and 6th Conference on Limestone Hydrology and Fissured Media,* La Chaux-de-Fonds, Proceedings, v. 2, p. 81-84.

Ewers, R.O., 1982, *Cavern development in the dimensions of length and breadth*: McMaster University, Ph.D. dissertation, 398 p.

Ewers, R.O., and Quinlan, J.F., 1981, Cavern porosity development in limestone: Low dip model from Mammoth Cave, Ky.: *8th International Congress of Speleology,* Bowling Green, p. 721-731.

Ford, D.C., and Williams, P.W. , 1989, *Karst geomorphology and hydrology:* London, Unwin and Hyman, 601 p.

Groves, C.G., and Howard, A.D., 1994a, Minimum hydrochemical conditions allowing limestone cave development: *Water Resources Research,* v. 30, p. 607-615.

Groves, C.G., and Howard, A.D., 1994b, Early development of karst systems. 1. Preferential flow path enlargement under laminar flow: *Water Resources Research,* v. 30, p. 2837-2846.

Jacobs, D., 1977, *The state of the art in numerical analysis:* London, Academic Press, 978 p.

Palmer, A.N., 1988, Solutional enlargement of opening in the vicinity of hydraulic structures in karst regions: 2nd Conference on Environmental Problems in Karst Terrains and Their Solutions, Dublin, Ohio, Proceedings, *Association of Groundwater Science and Engineering,* p. 3-13.

Palmer, A.N., 1991, The origin and morphology of limestone caves: *Geological Society of America Bulletin,* v. 103, p. 1-21.

Sahimi, M., 1995, *Flow and transport in porous media and fractured rock:* Weinheim, VCH Verlagsges, 482 p.

Siemers, J., and Dreybrodt, W., 1998, Early development of karst aquifers on percolation networks of fractures in limestone: *Water Resources Research,* v. 34, p. 409-419.

Stauffer, D., 1985, *Introduction to percolation theory*: London, Taylor and Francis Ltd, 124 p.

Svensson, U., and Dreybrodt, W., 1992, Dissolution kinetics of natural calcite minerals in CO_2-water systems approaching calcite equilibrium: *Chemical Geology,* v. 100, p. 129-145.

Williams, P.W., 1983, The role of the subcutaneous zone in karst hydrology: *Journal of Hydrology,* v. 61, p. 45-67.

4.2.5 Modeling Karst Aquifer Genesis Using a Coupled Continuum-Pipe Flow Model

Martin Sauter and Rudolf Liedl

Abstract

This contribution presents a modeling approach for the simulation of karst aquifer genesis during geologic time periods. The approach used is based upon a hydraulic coupling of a pipe network to a continuum in order to represent the well-known duality of karst aquifer flow systems. In addition, the growth of the karst conduits due to carbonate dissolution can be simulated so that a feedback mechanism with respect to the hydraulic properties of the pipe network is established. The model is applied to some typical scenarios of karst aquifer genesis, i.e. the evolution of a dendritic tube system and of a maze cave network. It is found that the model results can successfully reproduce the basic features of karst development, e.g. the development of maze caves in a carbonate aquifer that is covered by a sandstone caprock.

Introduction

The understanding of the genesis of karst aquifer systems in space and time has been the objective of research for some time. For a long period this field of research had been approached from a more descriptive point of view. With the understanding and quantification of the carbonate dissolution kinetics (Plummer and Wigley, 1976) it was possible to investigate the main factors involved in karst development (e.g. Palmer, 1981, 1991; Dreybrodt, 1990, 1996). The application of these principles, however, has up to now only applied to the boundary conditions of constant head, e.g. near hydraulic structures (Palmer, 1988; Dreybrodt, 1992), and early development of karst channels (Palmer, 1991; Dreybrodt, 1990, 1996; Groves and Howard, 1994).

The understanding of the karstification processes at a regional, (i.e. catchment) scale is believed to be most important for the quantification of the geometry of karst aquifers as well as their parameter distribution in space. This information is most useful for the prognosis of flow and transport in karst systems. Generally, geologic, geomorphologic and paleohydrologic information is available and could be used for the specification of boundary conditions in space and time.

In karst aquifers, very characteristic flow regimes can be observed. Following recharge events, large fluctuations in groundwater levels and at the same time high flow velocities are recorded. Despite the high flow velocities measured with tracer tests, the postevent recession of spring discharge is relatively slow. Hydraulic tests show a typical increase in hydraulic conductivities with an increase in the scale of investigation. For example, flow velocities calculated from hydraulic measurements or derived from the shape of the piezometric surface do not coincide with those determined from tracer tests. These observations explain why the flow and transport processes cannot be analyzed nor modeled using a simple continuum porous medium approach (Bear, 1979).

Today, karst aquifers are generally considered as groundwater flow systems consisting of a so-called fissured system. This represents the mass of the permeable fractured limestone, and a conduit system, which represents the karst tube system enlarged by the action of calcite dissolution. In the literature, the fissured system is commonly termed a "diffuse system." For consistency we prefer to use the term "fissured system." The conduit system, which only represents less than 0.1 percent of the bulk volume of the aquifer, is characterized by high hydraulic conductivities and high transport velocities. The hydraulic conductivity of the fissured system is several orders of magnitude lower, although its storage is on the order of several percent and therefore dominates long-term flow processes. The two flow systems are hydraulically coupled, and the exchange of water is controlled by the potential difference between them.

Quinlan and Ewers (1985) differentiated karst aquifers according to their hydraulic characteristics into:

- Mature systems, dominated by the highly conductive conduit system (e.g. alpine karst systems).
- "Diffuse flow" systems (fissured systems), where continuum flow conditions prevails.
- Mixed systems, which are a combination of both, where conduit and fissured system flow are both important.

Most karst aquifers can be classified as mixed systems, i.e. they show the typical karst features with high transport velocities and they also provide considerable storage, important for water resources purposes. Therefore, any attempt to realistically model the development of karst *aquifers* has to take into account the characteristic duality of the karst flow system.

Here a new modeling approach is presented. It integrates the chemical dissolution kinetics of carbonate rocks as well as the special hydraulic characteristics of karst systems, thereby allowing the simulation of the development of karst systems at a catchment scale.

It applies to karst systems that in their state of evolution have reached sufficiently high hydraulic conductivities (K m/s) within the fissured system, such that all groundwater recharge flows through the carbonate system, i.e. no surface runoff occurs.

Furthermore, the conduit system is in an early state of evolution with channel widths of up to about 1 mm. With this model, the sensitivity of the development of karst aquifers to a change in various parameters is demonstrated. It is also shown how the understanding of the genesis of karst aquifers and the modeling of their development can assist in the characterization of the conduit system that dominates flow and transport in karst aquifers.

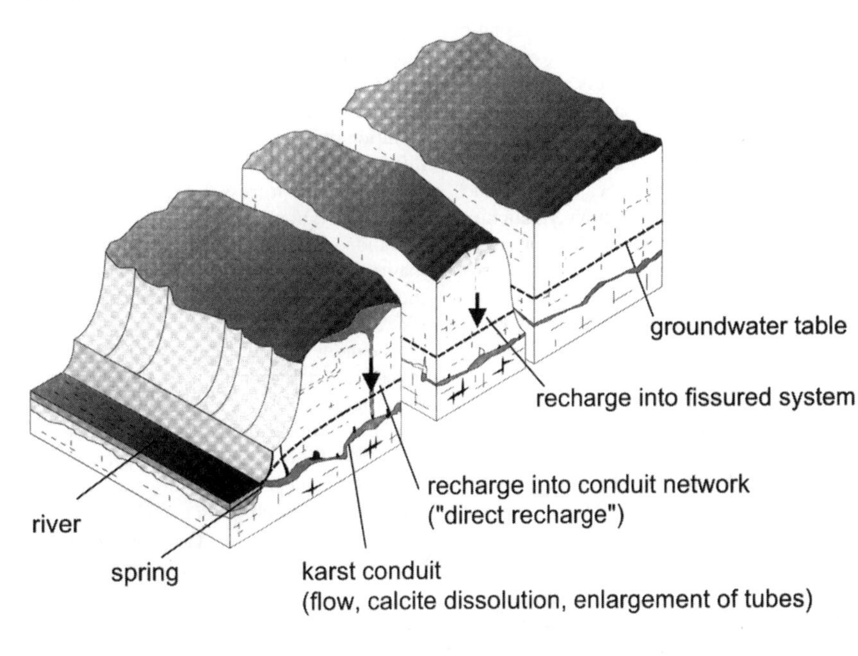

4.2.5 Figure 1. Schematic diagram of a karst aquifer with the main processes identified as the controlling factors for karst aquifer genesis.

Processes

The 4.2.5 Figure 1 summarizes the most important processes for karst aquifer genesis. Groundwater flow is represented by a dualistic flow system, i.e. a highly conductive low storativity (about 0.0001) conduit system and a much less permeable but high storativity (about 0.02) fissured system. The conduit system is characterized by high flow velocities (up to several 100 m/h). Depending on the flow velocities, laminar or turbulent flow conditions can occur. Flow between the two systems is controlled by the difference in potential. Flow from the conduit system into the fissured system usually occurs immediately after recharge events, when rapid infiltration in the phreatic conduit system causes the potential to rise. A few days after the event, the flow direction is reversed and groundwater is slowly released from storage in the fissured blocks over a period of several weeks to months, depending on the maturity of the karst system.

There are two different modes of infiltration of groundwater recharge:

- Direct recharge, which reaches the phreatic zone within a few days via highly conductive vertical shafts and dolines.
- Indirect recharge, arriving at the water table considerably delayed.

This delay can be explained either by the effect of an epikarst horizon, which acts as a buffer for the infiltrating water (Williams, 1983), or by slow flow through less permeable vadose blocks. It is important to distinguish between these two types of recharge components, because of their different infiltration rates, their different quantities, and different degree of undersaturation with respect to calcite. Direct recharge is highly undersaturated due to the rapid transfer of this recharge component through the vadose zone, whereas indirect recharge lies close to equilibrium due to the long residence time of the dissolved calcite concentrations in the epikarstic or vadose zone.

Dissolution of carbonate occurs at the conduit surface. The dissolution rate depends on the flow regime, temperature, P_{CO2} (partial pressure of carbon dioxide), and the equilibrium concentration of dissolved calcite. Different dissolution rates apply for dissolved calcite concentrations far from and close to equilibrium (Buhmann and Dreybrodt, 1985; Dreybrodt et al., 1996). The removal of calcite leads to an increase in the diameter of the tubes and consequently to an increase in hydraulic conductivity of the conduit network. The increase in flow leads to enhanced calcite dissolution, i.e. a positive feedback mechanism develops between flow and dissolution processes.

Modeling Approach

Brief Outline of Mathematical and Numerical Models

A mathematical model for karst genesis has been established taking into account the relevant processes governing flow, solute transport, calcite dissolution, and enlargement of tubes in karst aquifers. In the following the basic features of this approach are briefly summarized. More details are given by Hückinghaus et al. (1998).

Groundwater flow in the fissured system is modeled by a continuum approach using the Boussinesq equation (Berkowitz et al., 1988). Flow in the conduit system, which is assumed to consist of cylindrical tubes, is governed by Kirchhoff's rule stating that total inflow and total outflow balance for each node of the conduit network (Horlacher and Lüdecke, 1992). The relationship between discharge and hydraulic head difference characterizing pipe flow is adapted to the flow conditions prevailing within each tube, i.e. the model is able to distinguish between laminar and turbulent flow. Exchange of groundwater between the conduits and the fissured system is assumed to be proportional to the hydraulic head difference between these flow systems (Sauter, 1992). Additionally, it is possible to distribute groundwater recharge to both flow systems according to spatially varying weighting coefficients (4.2.5 Figure 2).

Transport of ions in the tubes is described by the one-dimensional advection equation with an additional source term, accounting for the increase in ionic concentration due to calcite dissolution. At the nodes of the conduit network, additional inflow of ions from the fissured system and from groundwater recharge is considered by assuming instantaneous complete mixing.

The modeling of carbonate dissolution is based on experimental findings (e.g. Buhmann and Dreybrodt, 1985; Dreybrodt et al., 1996; and Eisenlohr et al., 1997) that lead to the conclusion that the dissolution kinetics is "slow" close to equilibrium and relatively "fast" away from equilibrium (see also Chapter 4.1.2; Berner and Morse, 1974). Additionally, the model accounts for ionic diffusion, which acts as a limiting factor for the dissolution process under laminar flow conditions (Dreybrodt, 1990). Finally, conservation of carbonate mass leads to mathematical expressions that determine the increase of the tube diameters with time (4.2.5 Figure 3).

The model equations are solved by a numerical approach that is implemented in the program CAVE (Carbonate Aquifer Voids Evolution) program presented by Clemens et al. (1996). Flow in the fissured system is simulated by central finite differences using the MODFLOW program (McDonald and Harbaugh, 1988). For modeling flow in the conduits, the iterative Newton-Raphson method is applied in order to deal with the non-linearities occurring under turbulent conditions. It is important to note that flow conditions may change for each tube within each time step. Advective transport of ions is modeled by employing an upstream finite-difference scheme, where the evaluation of the reaction kinetics has been integrated for each tube. According to the amount of carbonate dissolved, the increase of the tube diameter is calculated.

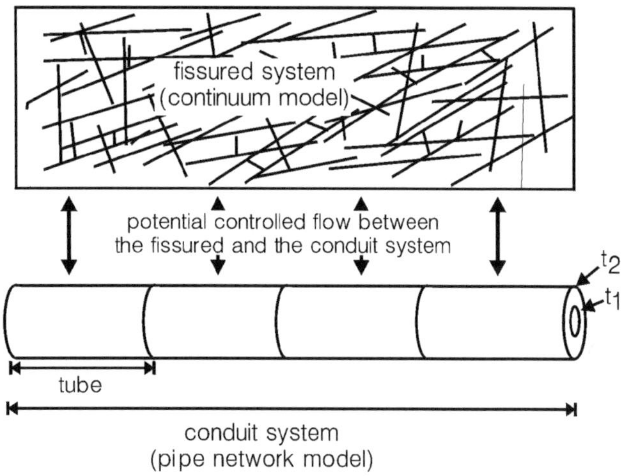

4.2.5 Figure 2. Conceptual modeling approach.

Model Verification

The CAVE program has been verified by choosing program options that allow the comparison of model results with those obtained by Dreybrodt (1996), Groves (1993), and Groves and Howard (1994) for single tubes that are not coupled to a fissured system. These investigators employ the dissolution rates

$$F_1(c) = \alpha\left(c_{eq} - c\right) \quad \text{if } c<c_{sw} \qquad (1)$$

and

$$F_4(c) = \beta\left(c_{eq} - c\right)^4 \quad \text{if } c>c_{sw} \qquad (2)$$

with c = concentration, c_{eq} = equilibrium concentration, a and b = kinetic rate constants, and $c_{sw} = 0.9c_{eq}$ relative concentration where the kinetics changes from order 1 (fast) to order 4 (slow).

In the simulations of Groves (1993) and Groves and Howard (1994), c_{eq} varies due to the consumption of CO_2 during calcite dissolution. This has been considered in CAVE by mimicking Groves' approach to calculate the equilibrium concentration (Clemens et al., 1998). The comparison of the

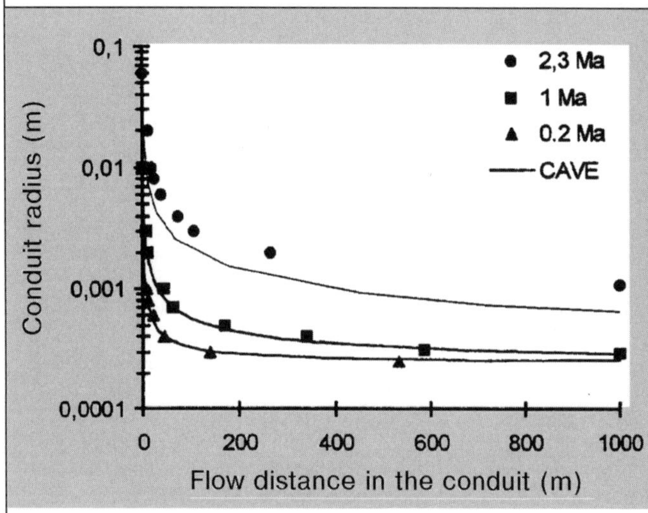

4.2.5 Figure 3. Tube radii obtained by employing the CAVE program, compared to the results from the model of Groves (1993). (Ma is millions of years.)

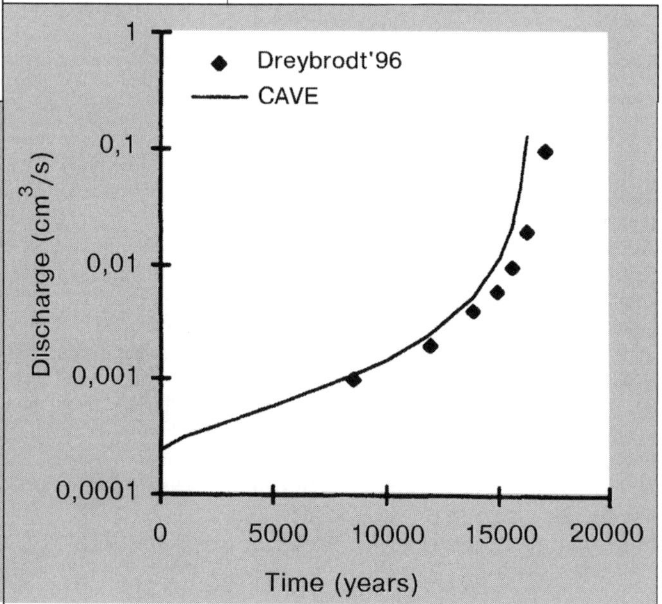

4.2.5 Figure 4. Discharge for a single tube comparing the results of the CAVE program to the model of Dreybrodt (1996).

model results is shown in 4.2.5 Figure 3.

In 4.2.5 Figure 4 the results of CAVE are compared with the data published by Dreybrodt (1996) who studied karstification at large scales. It is found (Clemens et al., 1998) that the discharge predicted by CAVE increases faster than the values obtained by Dreybrodt (1996). so that breakthrough is achieved after 16,500 years instead of 17,300 years. This small difference of 4.6 percent does not decrease even if the model grid is refined. Further detailed investigations will be required in order to explain this deviation.

Summarizing this section, it should be mentioned that CAVE has been shown to simulate different carbonate dissolution approaches with high reliability. This statement is supported in more detail by the examples presented by Clemens et al. (1998).

Sensitivity Analyses

For analyzing the sensitivity of the duration of karst development, the magnitudes of the relevant process parameters have been varied independently (Sauter et al., 1997). To this end a rectangular model domain (1350 by 1250 m) with three no-flow boundaries has been studied. On the fourth side, a fixed-head condition represents a river. The conduit system is connected to the river and forms a central series of 25 tubes. The initial diameter of the tubes is 0.4 mm, and 1 percent of total groundwater recharge (400 mm/year) is assumed to enter the pipe network directly with a dissolved calcite concentration of 0. The data set of chemical parameters has been taken from Dreybrodt, 1988, 1990 and field investigations in southwestern Germany (Sauter, 1992). Based upon these "standard values" the parameters are varied within ranges that are believed to occur in nature (4.2.5 Table 1).

The main results of these sensitivity analyses are summarized in 4.2.5 Figure 5. On the vertical axis, the change in the karstification period is shown. This is defined as the time when turbulent flow is observed in all tubes. In this study the equilibrium concentration of dissolved calcite and the initial tube diameter have been identified as the most important controlling factors of karstification. If the equilibrium concentration or the initial tube diameter are independently reduced by 50 percent, the duration of karstification increases by about 18,500 years or 10,000 years, respectively. A considerably smaller effect can be attributed to changes of the kinetic rate constants, total groundwater recharge, and the concentration threshold c_{sw} (not shown), while the water viscosity and the distribution of groundwater recharge between the conduits and the fissured system seem to play a negligible role for the duration of karst development.

It should be added, however, that there may be an additional reduction of the karstification period due to an increase of the percentage of direct groundwater recharge into the conduits during the course of aquifer development (Sauter et al., 1997). Therefore, it seems to be advisable to investigate the impact of temporally increasing direct recharge, that may even be related to additional effects such as changes of concentrations of direct as well as indirect recharge.

Applications

Two main types of cave geometries are generally distinguished: dendritic and maze-cave patterns. Dendritic caves are characterized by a tree-like structure, converging in the downgradient direction. Maze caves typically consist of networks of more or less regularly intersecting passages. Several different conceptual models describe dendritic and maze-cave development. According to White (1969) dendritic caves develop in areas with concentrated recharge. Palmer (1975, 1991) believes that "diffuse" infiltration through an insoluble caprock, i.e. equally distributed groundwater recharge, is the main factor for initiating certain maze caves. In fact, a large proportion of the investigated maze caves are or have been capped by a thin fractured sandstone formation. Klimchouk (1992; see also Chapter 5.2.1 in this volume) attributed the development of maze caves to distributed artesian flow from beneath a soluble formation.

CAVE has been applied to several scenarios to study the genesis of dendritic or maze caves and to investigate the controlling factors determining which type of cave morphology is most likely to develop

Parameter	Minimum	Standard	Maximum
Ca^{2+} equilibrium concentration (mmole/l)	1.0	2.0	3.0
Initial tube diameter (mm)	0.2	0.4	1.0
Dissolution rate constant close to equilibrium ($cm^{10}/mole^3/s$)	$2.5 \cdot 10^{12}$	$12 \cdot 10^{13}$	$5 \cdot 10^{13}$
Dissolution rate constant far from equilibrium (cm/s)	$2.5 \cdot 10^{-6}$	$2 \cdot 10^{-5}$	$15 \cdot 10^{-4}$
Groundwater recharge (mm/a, or year)	160	400	1829
Kinematic viscosity (m^2/s)	$1.8 \cdot 10^{-6}$	$1.3 \cdot 10^{-6}$	$1 \cdot 10^{-6}$
Direct recharge (%)	0.5	1.0	2.0

4.2.5 Table 1. Data sets for sensitivity analyses.

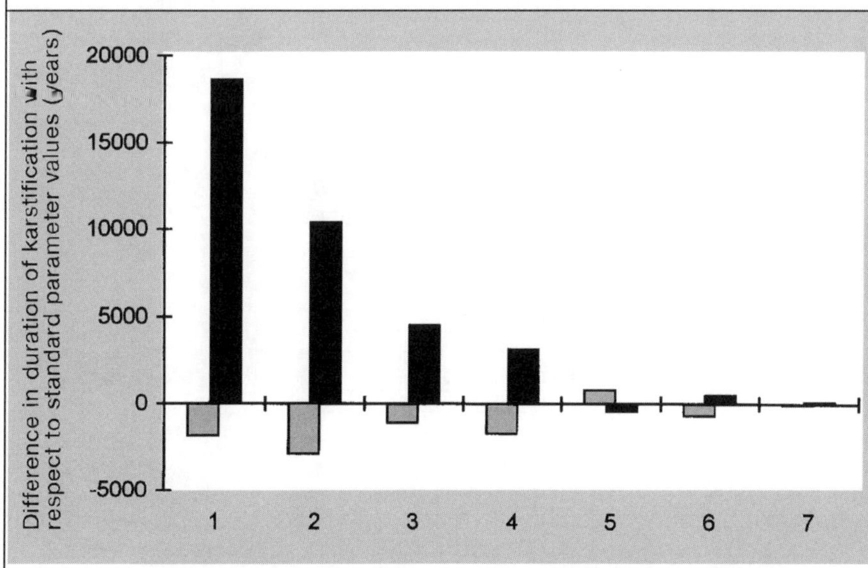

4.2.5 Figure 5. Results of sensitivity analyses. Parameters: No. 1 is the equilibrium concentration of dissolved calcite. No. 2 is the initial conduit diameter. No. 3 is the rate constant for dissolution close to equilibrium. No. 4 is the groundwater recharge. No. 5 is the rate constant for dissolution far from equilibrium. No. 6 is the kinematic viscosity. No. 7 is the fraction of direct groundwater recharge. Gray bars are maximum values of parameter. Black bars are minimum values of parameter.

(Clemens et al., 1997; Hückinghaus et al., 1998). Two examples are presented below.

Dendritic System

The dendritic system investigated in this example is shown in 4.2.5 Figure 6. The tube diameters are initially uniform. Three sides of the rectangular model domain are no-flow boundaries. At the fourth side, a constant head value is specified representing a river. It is assumed that the aquifer is unconfined, and the water within the fissured system shows a certain degree of undersaturation ($c=0.9c_{eq}$). A constant total recharge is applied uniformly over the catchment area and 1 percent is injected evenly distributed into the conduit system (direct recharge). The fissured system and the tubes are hydraulically coupled via differences in hydraulic head only, so no fixed-head boundaries lie at the upstream ends of the conduit system.

At the beginning of the simulation, the equipotential lines for the fissured system are parallel to the fixed-head boundary. This is because the tube diameters are not large enough to affect the flow pattern. Due to the specification of direct recharge into all nodes of the conduit network (4.2.5 Figure 6) each tube is enlarged under fast first-order dissolution kinetics only near its inlet.

However, the remaining part of each tube increases under fourth-order dissolution kinetics so that this slow process is dominant during the initial stage of conduit development. The penetration length of each tube, i.e. the reach where fast kinetics is active, increases with time, but it is important to note that the penetration lengths grow at different rates.

To explain this behavior, one has to recall that in model domains with uniform recharge—such as the one displayed in 4.2.5 Figure 6—the hydraulic gradient increases as the fixed-head boundary is approached. This leads to higher flow velocities in the tubes close to the spring, thus reducing the residence time of the water and increasing the calcite dissolution. As a consequence, tube number 1 is the first to experience a breakthrough, i.e. fast dissolution kinetics prevails throughout the tube.

After this breakthrough, the flow rapidly increases and changes from laminar to turbulent so that there is an additional increase of the dissolution rate constant that accelerates the growth in diameter even more. Therefore, the hydraulic head gradient in tube number 1 decreases and the position of the greatest gradient, which is associated with the highest dissolution rate, propagates away from the spring in an upgradient direction (Ewers, 1982, Rhoades and Sinacori, 1941; Clemens et al., 1996).

As a consequence, the enlargement of the conduit system moves farther upgradient. It also considerably affects the flow pattern of the fissured system that is shown in 4.2.5 Figure 7. Obviously, the growing conduits are draining the fissured system.

At the end of this section, we should mention that flow patterns such as the one presented in 4.2.5 Figure 7 are the result of an interplay between several processes within the karst system. Hückinghaus et al. (1998) demonstrate that the dendritic system of 4.2.5 Figure 6 would develop in a different manner or over different time periods if:

- Fixed-head boundaries were specified at the upstream nodes of the conduit system.
- There were no water exchange between the tubes and the fissured system
- Direct recharge were neglected.

CAVE is believed to provide an efficient modeling tool in order to identify the relative importance of the proccsses mentioned above.

Maze Cave Development

CAVE has also been used to study some of the phenomena controlling the development of maze caves. Here we present the results obtained from integrating an insoluble caprock into the hydraulic flow system, described in detail by Clemens et al. (1997). A rectangular finite-difference model grid (4.2.5 Figure 8) was subdivided into two layers. The bottom layer represents the fissured system of the limestone formation. The top layer represents the less permeable sandstone caprock. Flow between the two layers is potential controlled. The bottom layer contains a regular network of conduits (4.2.5 Figure 9), with one conduit having a slightly larger initial diameter in order to test whether this difference in diameter is a

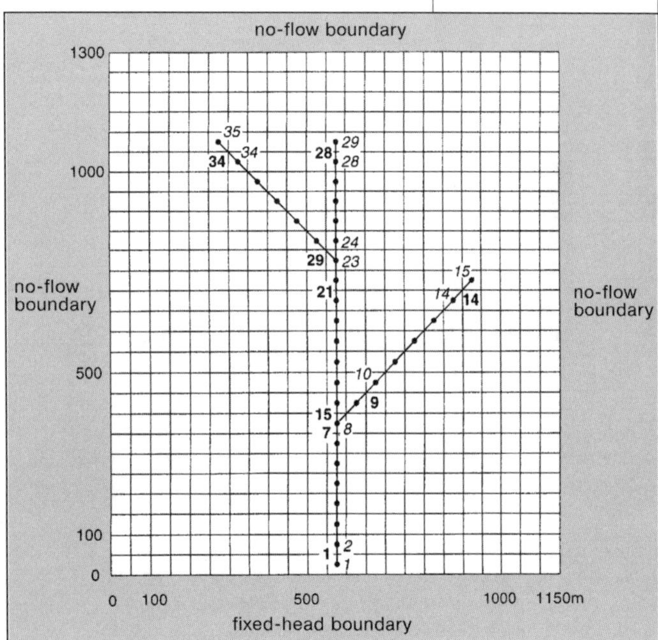

4.2.5 Figure 6. Model domain and dendritic conduit system.

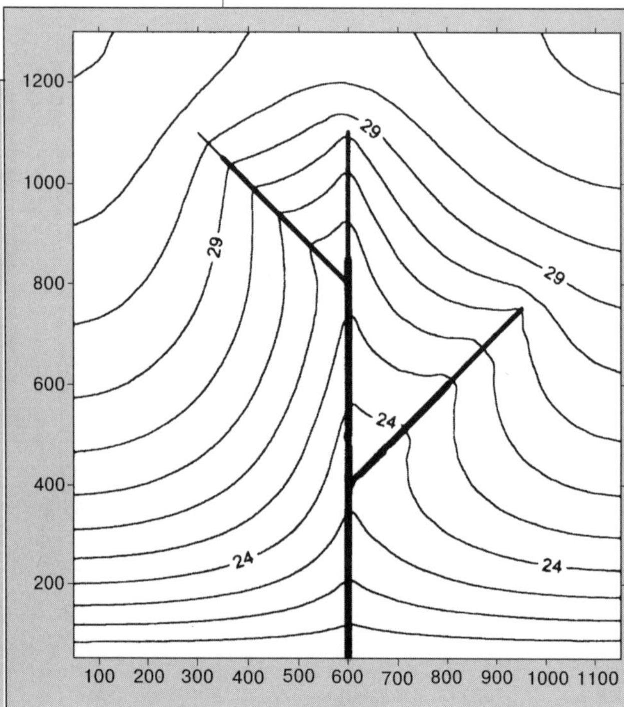

4.2.5 Figure 7. Enlargement of tubes and equipotential lines within the fissured system.

critical factor in the development of dendritic patterns. Recharge infiltrates uniformly into the sandstone. The dissolved calcite concentration of water entering the conduit system is 0 mole/L, while water flowing through the fissured system is close to equilibrium ($0.9c_{eq}$) with respect to calcite. The resulting composition (i.e. "dissolution capacity") depends on the relative proportions of direct recharge and fissured system crossflow components and their respective concentrations. The only discharge boundary is at the left side in the sandstone layer, i.e. groundwater discharge from the limestone aquifer is only possible via the less permeable sandstone.

The simulation results of the conduit enlargement are displayed in 4.2.5 Figures 10 and 11 for two types of model setups:

- Case **a** with the less permeable sandstone caprock.
- Case **b** without the less permeable sandstone layer.

Maze cave patterns develop in the case where flow is inhibited by the sandstone layer (a). Dendritic structures develop when inflow and outflow are both directly through the limestone (b). In case (a) the characteristic

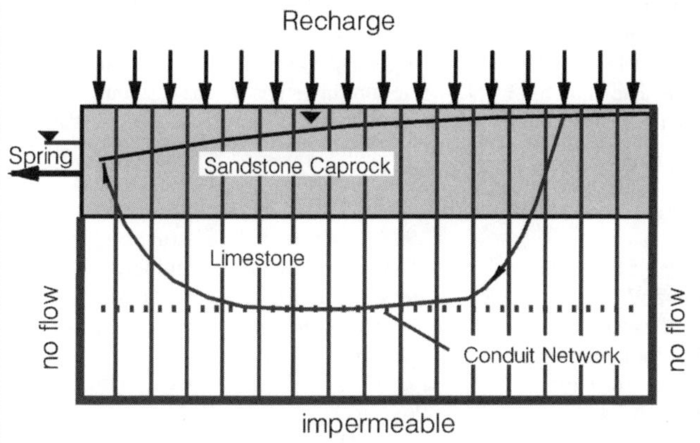

4.2.5 Figure 8. Model setup for the simulation of the genesis of maze caves.

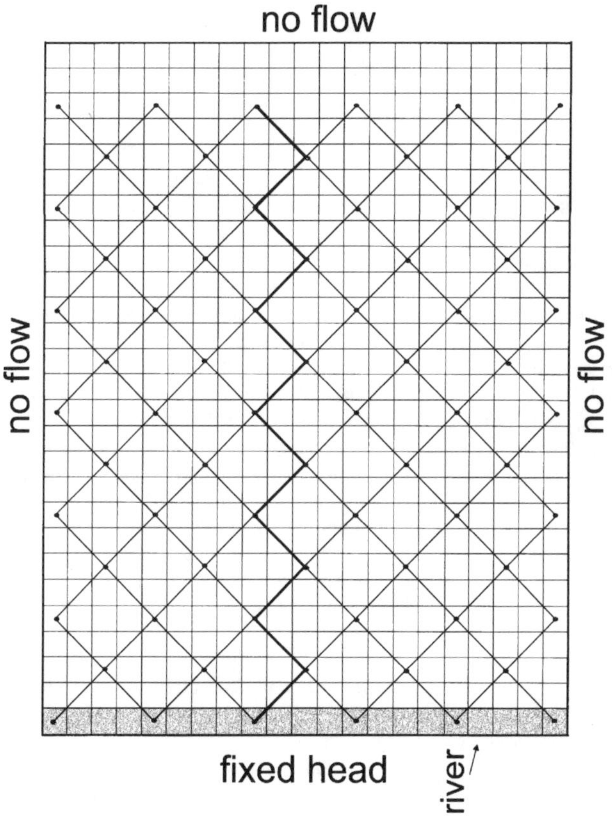

4.2.5 Figure 9. Plan view of conduit network in the limestone formation. Parameters: c_{eq} = mole/L. k = m/s. $c_{sw} = 0.9c_{eq}$. Direct recharge component = 1 percent of total recharge. Initial tube diameter is 0.4 mm, except for center tubes which start development with 1 mm diameter.

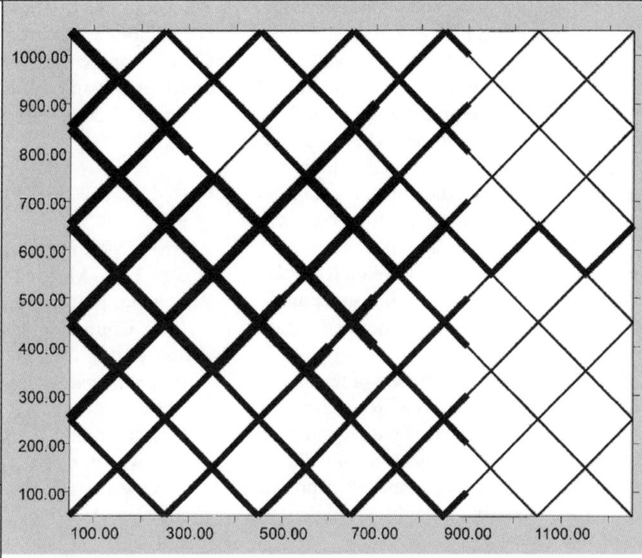

4.2.5 Figure 10. Conduit diameters for a flow system with caprock after approximately 6000 years (Clemens et al., 1997). Due to the flow inhibition by the less-permeable caprock, the feedback mechanism between flow and dissolution rates cannot develop. This type of hydraulic configuration leads to the formation of maze cave patterns.

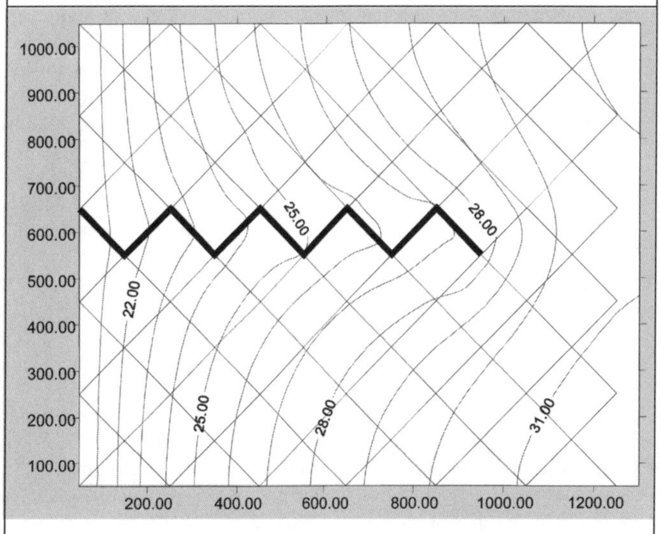

4.2.5 Figure 11. Potential distribution and conduit diameters for a flow system without a caprock after approximately 2500 years (Clemens et al., 1997). Discharge is allowed directly from the limestone formation. The initiated feedback mechanism between flow and dissolution rates selectively enlarges single conduits and therefore leads to dendritic cave morphologies. (Parameters as in 4.2.5 Figure 9).

feedback mechanism (i.e. higher flow rates induce higher dissolution rates and vice versa) cannot develop because the less conductive sandstone layer controls the flow through the whole system. Therefore the conduits are enlarged more uniformly. If the feedback between flow and calcite dissolution is not inhibited (case b), characteristic dendritic patterns evolve, i.e. the preferential development of a few single conduits is observed.

Within the flow field, the highest dissolution rates occur in the area of highest hydraulic gradient, that is next to the regional outflow at the beginning of the simulation. In the growing tube, the potential drops and therefore the location of the highest gradient. That is, the position of highest dissolution rates shifts in an upgradient direction similar to the development of the dendritic network described in the previous section. This shows how karst conduit systems develop from the spring upgradient, and with time more and more differentiated regional aquifer drainage patterns develop.

While a uniform distribution of recharge via an insoluble caprock is a prerequisite for the evolution of maze cave conduit structure (Palmer, 1975), the dominant factor is the flow inhibition by a less permeable caprock, that could be clearly shown by the CAVE modeling results.

Summary and Conclusions

Karst aquifer genesis is the result of a complex interplay between hydraulic and chemical processes. A coupled continuum-pipe-flow modeling approach can successfully be used to understand the main processes and factors controlling the development of karst systems. The dominant factors are:

- Initial and boundary conditions.
- The equilibrium concentration of dissolved calcite.
- The calcite dissolution kinetics.

Assuming that the equilibrium concentration varies only slightly with time, the factors and parameters controlling the flow through the system can be regarded as the most important ones for the enlargement of karst conduits. The modeling approach can be applied to show how different conduit structures develop under different boundary conditions. This is illustrated by modeling the development of the main types of karst networks, i.e. maze caves and dendritic patterns.

Our numerical model CAVE has proved to be a very useful tool for the investigation of karst aquifer genesis, particularly for the examination of various existing conceptual models and hypotheses. It is presently being extended to include three-dimensional conduit networks, time-variant boundary conditions, and direct groundwater recharge that depends on the maturity of the conduit system. Further, the influence of a short-term higher degree of calcite undersaturation on karst aquifer development is being investigated.

Acknowledgments

We would like to acknowledge the essential contributions of our Ph.D. students Torsten Clemens and Dirk Hückinghaus in model development and application. The valuable advice by Georg Teutsch is most appreciated.

References

Bear, J., 1979, *Hydraulics of groundwater:* New York, McGraw-Hill, 567 p.

Berkowitz, B., Bear, J., and Braester, C., 1988, Continuum models for contaminant transport in fractured porous formations: *Water Resources Research,* v. 24, p. 1225-1236.

Berner, R.A., and Morse, J.W., 1974, Dissolution kinetics of calcium carbonate in seawater, IV: Theory of calcite dissolution: *American Journal of Science,* v. 274, p. 108-134.

Buhmann, D., and Dreybrodt, W., 1985, The kinetics of calcite dissolution and precipitation in geologically relevant situations of karst areas, 2: Closed systems: *Chemical Geology,* v. 53, p. 109-124.

Clemens, T., Hückinghaus, D., Sauter, M., Liedl, R., and Teutsch, G., 1996, A combined continuum and discrete network reactive transport model for the simulation of karst development, *in* Kovar, K. and van der Heijde, P., eds., Calibration and reliability in groundwater modeling: *IAHS Publication,* v. 237, p. 309-318.

Clemens, T., Hückinghaus, D., Sauter, M., Liedl, R., and Teutsch, G., 1997, Simulation of the evolution of maze caves: *12th International Congress of Speleology and 6th Conference on Limestone Hydrology and Fissured Media,* La Chaux-de-Fonds, Proceedings, v. 2, p. 65-68.

Clemens, T., Hückinghaus, D., Sauter, M., Liedl, R., and Teutsch, G., 1998, Simulation of the development of karst aquifers by using a pipe flow model coupled to a continuum model, 2: Model verification and sensitivity analysis: *Water Resources Research,* in preparation.

Dreybrodt, W., 1988, *Processes in karst systems–Physics, chemistry and geology*: Berlin, Springer-Verlag, 288 p.

Dreybrodt, W., 1990, The role of dissolution kinetics in the development of karst aquifers in limestone: A simulation of karst evolution: *Journal of Geology,* v. 98, p. 639-655.

Dreybrodt, W., 1992, Dynamics of Karstification: A model applied to hydraulic structures in karst terranes: *Applied Hydrogeology,* v. 1, p. 20-32.

Dreybrodt, W., 1996, Principles of early development of karst conduits under natural and man-made conditions revealed by mathematical analysis of numerical models: *Water Resources Research,* v. 32, p. 2923-2935.

Dreybrodt, W., Lauckner, J., Zaihua, L., Svensson, U., and Buhmann, D., 1996, The kinetics of the reaction $CO_2 + H_2O \Rightarrow H^+ + HCO3^-$ as one of the rate limiting steps for the dissolution of calcite in the system $H_2O - CO_2 - CaCO_3$: *Geochimica et Cosmochimica Acta,* v. 60, p. 3375-3381.

Eisenlohr, L., Madry, B., and Dreybrodt, W., 1997, Changes in the dissolution kinetics of limestone by intrinsic inhibitors adsorbing to the surface: *12th International Congress of Speleology and 6th Conference on Limestone Hydrology and Fissured Media,* La Chaux-de-Fonds, Proceedings, v. 2, p. 81-84.

Ewers, R.O., 1982, *Cavern development in the dimensions of length and breadth*: McMaster University, Ph.D. Thesis, 398 p.

Ford, D.C., and Williams, P.W., 1989, *Karst geomorphology and hydrology:* London, Unwin Hyman, 601 p.

Groves, C.G., 1993, *Early development of karst systems*: University of Virginia, Ph.D. Thesis, 253 p.

Groves, C.G., and Howard, A.D., 1994, Early development of karst systems, 1: Preferential flow path enlargement under laminar flow: *Water Resources Research,* v. 30, p. 2837-2846.

Horlacher, H.B., and Lüdecke, H.J., 1992, *Strömungsberechnung für Rohrsysteme:* Expert Verlag, 218 p.

Hückinghaus, D., Clemens, T., Liedl, R., Sauter, M., and Teutsch, G., 1998, Simulation of the development of karst aquifers by using a pipe flow model coupled to a continuum model, 1: Theory: *Water Resources Research,* in preparation.

Klimchouk, A.B., 1992, Genesis of the maze caves in gypsum in the western Ukraine: *Cave Science,* v. 19, no. 1, p. 3-11.

McDonald, M.G., and Harbaugh, A.W., 1988, A modular three-dimensional finite-difference ground-water flow model: *United States Geological Survey Techniques of Water Resources Investigations,* v. 6, ch. A1.

Palmer, A.N., 1975, The origin of maze caves: *National Speleological Society Bulletin,* v. 37, p. 56-76.

Palmer, A.N., 1981, Hydrochemical controls in the origin of limestone caves: *8th International Congress of Speleology,* Bowling Green, Proceedings, p. 120-122.

Palmer, A.N., 1988, Solutional enlargement of opening in the vicinity of hydraulic structures in karst regions: *2nd Conference on Environmental Problems in Karst Terranes and Their Solutions,* Nashville, Association of Groundwater Scientists and Engineers, p. 16-18.

Palmer, A.N., 1991, Origin and morphology of limestone caves: *Geological Society of America Bulletin,* v. 103, p. 1-21.

Plummer, L.N., and Wigley, T.M.L., 1976, The dissolution of calcite in CO_2-saturated solutions at 25°C and 1 atmosphere total pressure: *Geochimica et Cosmochimica Acta,* v. 40, p. 191-202.

Quinlan, J.F., and Ewers, R.O., 1985, Groundwater flow in limestone terranes: Strategy rationale and procedure for reliable and efficient monitoring of groundwater quality in karst areas: *5th National Symposium and Exp. on Aquifer Restoration and Groundwater Monitoring,* Columbus, p. 197-234.

Rhoades, R., and Sinacori, N.M., 1941, Patterns of groundwater flow and solution: *Journal of Geology,* v. 49, p. 785-794.

Sauter, M., 1992, Quantification and forecasting of regional groundwater flow and transport in a karst aquifer (Gallusquelle, Malm, SW Germany): *Tübinger Geowissenschaftliche Arbeiten*, ser. C, v. 13, p. 1-150.

Sauter, M., Liedl, R., Clemens, T., and Hückinghaus, D., 1997, Karst aquifer genesis–Field evidence and modeling approaches: *12th International Congress of Speleology and 6th Conference on Limestone Hydrology and Fissured Media,* La Chaux-de-Fonds, Proceedings, no. 2, p. 107-110.

White, W.B., 1969, Conceptual models for carbonate aquifers: *Groundwater,* v. 7, p. 15-21.

Williams, P.W., 1983, The role of the subcutaneous zone in karst hydrology: *Journal of Hydrology,* v. 61, p. 45-67.

4.2.6. Influence of Fracture Roughness on Karstification Times

Wolfgang Dreybrodt and Franci Gabrovšek

Abstract

Karst aquifers develop along water-transmitting fractures that are rough by nature. Most numerical models on the evolution of the widths of those fractures by limestone dissolution, however, approximate such rough fractures by two parallel smooth planes.

Here we investigate the influence of fracture roughness on the breakthrough time of conduits. All results show that even when the flow through a fracture is reduced by its roughness by a factor of ten in comparison to a corresponding smooth fracture, breakthrough time is increased only by a factor of 4. This gives a first answer to the influence of natural fracture roughness on the evaluation of karst and shows that extreme roughness, unlikely in nature, is necessary to exert significant influence on karstification time.

Introduction

Most models of cave evolution in its early state approximate the initially rough fractures by two equidistant ideal planes (Dreybrodt, 1990; Palmer, 1991; Groves and Howard, 1994a; Groves and Howard, 1994b; Lauritzen et al., 1992; Dreybrodt, 1996; Siemers and Dreybrodt, 1998; see also Chapters 4.2.2. to 4.2.5). The first estimations on the influence of roughness upon cave evolution were performed by Groves and Howard (1994a). They replaced the uniform aperture widths of the initial pathways by statistically distributed apertures and found similar behavior of cave evolution, i.e. a slow increase of aperture width over long time, followed by an abrupt opening termed as breakthrough.

The times to achieve breakthrough for a single one-dimensional idealized initial smooth fracture depends on the initial fracture width a_0 by (see equation 20, Chapter 4.2.2)

$$T_s^B \propto \left(\frac{1}{a_0}\right)^{\frac{2n+1}{n-1}} \qquad (1)$$

where n is the order of the dissolution kinetics acting at its exit. The question arises as to how this time changes with increasing roughness of the fracture with an average width $\overline{a}_0$, provided all the other parameters (see Equation 20 in Chapter 4.2.2) remain constant.

A first approach to an answer can be taken by using statistically distributed apertures along the fracture. We have used such an approach in the finite-difference model and have found that the breakthrough times do not change significantly provided that the standard deviation s did not exceed $0.3a_0$. If s becomes close to a_0, however, it is highly probable that the fracture becomes blocked and its flow resistance becomes high enough to increase the breakthrough time dramatically. Therefore this approach needs many calculations with different configurations of the distribution of initial aperture widths, and is thus inconvenient and not well suited to yielding clear-cut results on the influence of roughness on breakthrough times.

A more convenient approximation can be taken by replacing an irregular roughness as it occurs in nature by a regular one, which allows calculation of the hydrodynamic resistance analytically. This was done recently by Ge (1997). He investigated the validity of the cubic law, i.e. the dependence of flow rate on the third power of the average fracture width $\overline{a}$, for a general rough fracture. He also gives analytical results for a fracture consisting of one confining ideal plane and the other plane shaped by regularly spaced triangles, as illustrated in 4.2.6 Figure 1. In this case the flow rate Q_r is changed simply by a correcting roughness factor $f_r \leq 1$, such that

$$Q_r = f_r \cdot Q_s \qquad (2)$$

where Q_s is the flow rate of the smooth fracture with aperture a_0 equal to the average width $\overline{a}_0$ of the rough fracture, and f_r depends by an analytical expression on the geometry of the idealized rough plane.

This can be easily incorporated into the

finite-difference program by correcting the flow rates calculated for the ideal smooth fractures with the roughness factor. As a consequence we find that the breakthrough times also follow a relation

$$T_r^B = \tau_r \cdot T_s^B \tag{3}$$

where the factor $\tau_r \geq 1$ is given by an analytical expression that depends only on the ratio of the roughness amplitude to the average initial width a_0.

Breakthrough Times in Rough Fractures

The cubic law for laminar flow in an ideally smooth plane-parallel fracture is stated as

$$Q_s = \frac{a^3 \rho g}{12\eta} \cdot \frac{dh}{dx} \tag{4}$$

where Q_s is the flow rate per unit breadth of the fracture, ρ is the density of the water, g is the earth's gravitational acceleration, dh/dx is the hydraulic gradient, and a is the aperture width. In a rough fracture it has been shown by Ge (1997) that the flow Q_r is given by

$$Q_r = Q_s \cdot f_r(\frac{r}{2\overline{a}}) \tag{5}$$

where $f_r(r/2\overline{a})$ is a function of the relative roughness $r/2\overline{a}$. In the following, we restrict our discussion to an ideally rough surface as depicted in 4.2.6 Figure 1. For this regularly shaped roughness the roughness factor f_r is given (Ge, 1997) by

$$f_r(\frac{r}{2\overline{a}}) = \left(\frac{2\frac{r}{2\overline{a}}}{1-\frac{r}{2\overline{a}}} + 1 \right)^2 \cdot \left(\frac{\frac{r}{2\overline{a}}}{1-\frac{r}{2\overline{a}}} \right)^{-4} \cdot \Phi(\alpha) \tag{6}$$

where a is the angle of inclination as depicted in 4.2.6 Figure 1 and $\Phi(\alpha)$ is the function of this angle. Note that $\alpha \to 0$ when $r \to 0$. In this case $\Phi(\alpha) \to 1$. For most of the natural rough fractures we assume that $30° < \alpha < 60°$. Then $\Phi(\alpha)$ ranges between 0.7 and 0.9 (Ge, 1997). In our calculations we used $\alpha = 35°$ where $\Phi(\alpha) \approx 0.84$; f_r is a measure of the reduction of flow induced by the roughness of the fracture. Therefore, for a completely rough surface, $r/2\overline{a} = 1$, and f_r becomes zero. This is because in that case the fracture is blocked to flow. The 4.2.6 Figure 2 depicts f_r as a function of $r/2\overline{a}_0$

We have incorporated this roughness factor into the calculation of the flow rates in the finite-difference model and have calculated the evolution of rough fractures for various values of initial relative roughness $r/2\overline{a}_0$. The 4.2.6 Figure 3 shows the evolution of flow rates through such fractures with initial average width $\overline{a}_0$ for various values of $r/2\overline{a}_0$ as a function of time. A typical breakthrough behavior is observed in all cases, but breakthrough time increases with roughness. It should be noted that in these calculations the following assumptions are made:

- The roughness amplitude r is maintained during the dissolutional widening.
- Only the widths $a(x)$ at all positions (x distance from the entrance of the fracture) change by dissolutional widening.

Thus during the evolution of the conduit the relative roughness $r/2\overline{a}$ decreases continuously and the correction factor f_r loses influence. Once the relative roughness $r/2\overline{a}$ has decreased to 0.2, which still is significant, the fracture develops almost like an ideally smooth one. Therefore the evolution of the fracture is delayed only in its very early state. Since the relative roughness decreases with increasing $\overline{a}(x)$ the reduction of flow rate also decreases and the feedback mechanism leading to breakthrough remains operative (see Chapter 4.2.2). As a first approximation using equation 19 in Chapter 4.2.2 and equation 5 here we find for the breakthrough time T_r^B of the rough fracture

$$T_r^B = T_s^B \cdot f(r/2\overline{a}_0)^{\frac{-n}{n-1}} \tag{7}$$

This is depicted by the dotted line in 4.2.6 Figure 4.

We have performed a large number of finite-difference calculations with various $\overline{a}_0$, and also with different fracture lengths. The results are illustrated by 4.2.6 Figure 4. Here we plotted the ratio T_r^B / T_s^B versus $r/2\overline{a}_0$. T_s^B is the breakthrough time in a smooth fracture ($r = 0$, $\overline{a}_0 = a_0$) with initial width a_0. All other parameters are identical. It is interesting to note that all points are close to the dotted line. To confirm this first crude approximation, we take a similar approach as in Chapter 4.2.2 of this book. We assume that the dissolution rate $F(l,t)$ at the exit acts along the entire fracture and that during the time Δt the average width changes by Δa, but r remains unchanged. Thus during the widening of the fracture, the relative roughness $r/2\overline{a}_0$ decreases with time as

$$\frac{r}{2a(m\Delta t)} = \frac{r}{2(\overline{a}_0 + m\Delta a)} \tag{8}$$

where m denotes the mth time step. After time T_M, when the average width has increased to $\overline{a}(T_M) = \overline{a}_0 + m\Delta a$, the dissolution rate at the exit is

$$F(l,T_M) = F(l,0) \cdot \left(\frac{\overline{a}_0 + m\Delta a}{\overline{a}_0} \right)^{\frac{3n}{n-1}} \cdot \left(\frac{f\left(\frac{r}{2\overline{a}_0}\right)}{f\left(\frac{r}{2(\overline{a}_0 + m\Delta a)}\right)} \right)^{\frac{n}{n-1}} \tag{9}$$

This results from the fact that $F(l,T_M)$ is related to the flow rate Q by equations 10 to 14 in Chapter 4.2.2 and by use of equation 2. The time ΔT_{m+1} to widen the fracture after the time T_M by Δa is therefore

$$\Delta T_{m+1} = T \cdot \left(\frac{\Delta a}{\overline{a}_0} \right) \cdot \left(\frac{\overline{a}_0}{\overline{a}_0 + m\Delta a} \right)^{\frac{3n}{n-1}} \cdot \left(\frac{f\left(\frac{r}{2\overline{a}_0}\right)}{f\left(\frac{r}{2(\overline{a}_0 + m\Delta a)}\right)} \right)^{\frac{-n}{n-1}} \tag{10}$$

with $T = \dfrac{\overline{a}_0}{2\gamma F(l,0)}$

Breakthrough time is given by the sum (see Chapter 4.2.2)

$$T_r^B = T \cdot \sum_{m=0}^{\infty} \frac{\Delta a}{\overline{a}_0} \left(1 + \frac{m\Delta a}{\overline{a}_0} \right)^{\frac{-3n}{n-1}} \cdot \left(\frac{f\left(\frac{r}{2\overline{a}_0}\right)}{f\left(\frac{r}{2(\overline{a}_0 + m\Delta a)}\right)} \right)^{\frac{-n}{n-1}} \tag{11}$$

For a smooth fracture, breakthrough time is (see equations 18 and 19 in Chapter 4.2.2)

$$T_s^B = T \cdot \sum_{m=0}^{\infty} \frac{\Delta a}{a_0} \left(1 + \frac{m\Delta a}{a_0} \right)^{\frac{-3n}{n-1}} = \frac{n-1}{2n-1} \cdot T \tag{12}$$

Dividing equation 10 by equation 9 yields

$$\frac{T_r^B}{T_s^B} = \tau(r / 2\bar{a}_0)$$

$$\tau = f\left(\frac{r}{2\bar{a}_0}\right)^{\frac{-n}{n-1}}.$$

$$\frac{\sum_{m=0}^{\infty}\left(1+\frac{m\Delta a}{\bar{a}_0}\right)^{\frac{-3n}{n-1}} \cdot \left(f\left(\frac{r/2\bar{a}_0}{1+m\Delta a/\bar{a}_0}\right)\right)^{\frac{n}{n-1}}}{\sum_{m=0}^{\infty}\left(1+\frac{m\Delta a}{a_0}\right)^{\frac{-3n}{n-1}}} \qquad (13)$$

The full line in 4.2.6 Figure 4 illustrates this function. The perfect agreement to the values calculated by finite differences shows its general validity and also the high quality of the so-called crude approximation.

Discussion

As can be seen from 4.2.6 Figure 4, extreme relative roughness ($r / 2\bar{a}_0 \geq 0.8$) is necessary to exert significant influence on the breakthrough times. For a relative roughness of 0.6, which is twice as much as depicted by 4.2.6 Figure 1, the breakthrough time increases only by a factor of 2. Even if the initial roughness reduces the initial flow rate by a factor of ten (see 4.2.6 Figure 2) breakthrough time increases only by a factor of 5. This will also hold for more general types of roughness, as long as the cubic law can be modified by a roughness factor f_r.

A further example where such factors can be expressed analytically has been given by Ge (1997). In this case the fracture has constant *a*, but both planes are sinusoidal with a fixed wavelength and amplitude, which increases the tortuosity of the flow path.

Of course it is difficult to estimate the roughness factor for naturally rough fractures that exhibit roughness on both confining sides. Nevertheless the idealized rough fracture as depicted by 4.2.6 Figure 1 can be regarded as a first approximation, and the results can be reasonably generalized as follows. As long as the ratio $Q_r/Q_s \geq 0.3$ for the initially rough fracture, roughness influences breakthrough times by no more than a factor of two. To increase the breakthrough time by one order of magnitude, a $Q_r/Q_s = 0.02$ is needed.

Thus extreme roughness is necessary for a drastic change, and it seems that in most natural cases its influence is not very significant. Although this result might seem qualitative, it gives a first estimation of the reliability of smooth-fracture models of conduit evolution.

References

Dreybrodt, W., 1990, The role of dissolution kinetics in the development of karstification in limestone–A model simulation of karst evolution: *Journal of Geology* v. 98. p. 639-655.

Dreybrodt, W., 1996, Principles of early development of karst conduits under natural and man-made conditions revealed by mathematical analysis of numerical models: *Water Resources Research,* v. 32, p. 2923-2935.

Ge, S., 1997, A governing equation for fluid flow in rough fractures: *Water Resources Research,* v. 33, p. 53-61.

Groves, C. G. and Howard, A. D. 1994a, Early development of karst systems. 1: Preferential flow path enlargement under laminar flow. *Water Resources Research* v. 30, p. 2837-2846.

Groves, C. G. and Howard, A. D., 1994b, Minimum hydrochemical conditions allowing limestone cave development, *Water Resources Research* v. 30, p. 607-615.

Lauritzen, S.-E., Odling. N., and Peterson, P., 1992, Modeling the evolution of channel networks in rocks, in Hudson, J.A., ed., *ISRM Symposium, Eurock '92*: Thomas Thelford, London, p. 57-62.

Palmer, A. N., 1988, Solutional enlargement of opening in the vicinity of hydraulic structures in karst regions: *2nd Conference on Environmental Problems in Karst Terraines and Their Solutions,* Association of Groundwater Scientists and Engineers, Nashville, Tenn., p. 16-18.

Palmer, A. N., 1991, The origin and morphology of limestone caves. *Geological Society of America Bulletin* v. 103, p. 1-21.

Siemers, J. and Dreybrodt, W., 1998, Early development of karst aquifers on percolation networks of fractures in limestone. *Water Resources Research* v. 34. p. 409-419.

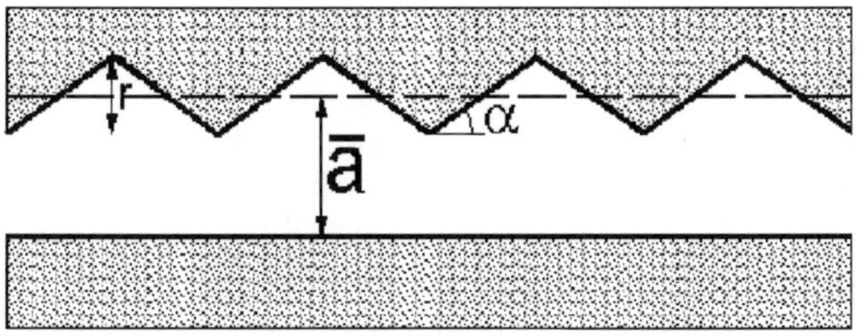

4.2.6 Figure 1. Geometry of an ideally rough fracture with a saw-toothed upper plane. Initial relative roughness $r / 2\overline{a}_0 = 0.3$.

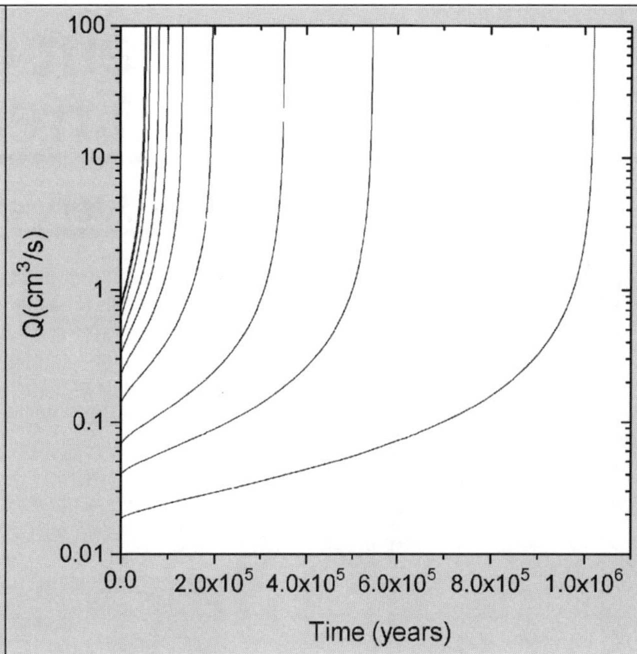

4.2.6 Figure 3. Breakthrough curves calculated by the finite-difference model for a single fracture with varying initial relative roughness. Parameters are: $l = 4x10^5 cm$ $b_0 = 100$ cm $n = 4$ $k_4 = 4x10^{-8}$ mol cm^{-2}/s^{-1} $\overline{a}_0 = 0.03$ cm $dh/dx = 0.05, 0, 0.166, 0.333, 0.5, 0.583, 0.66, 0.833, 0.875$, and 0.916 from left to right respectively.

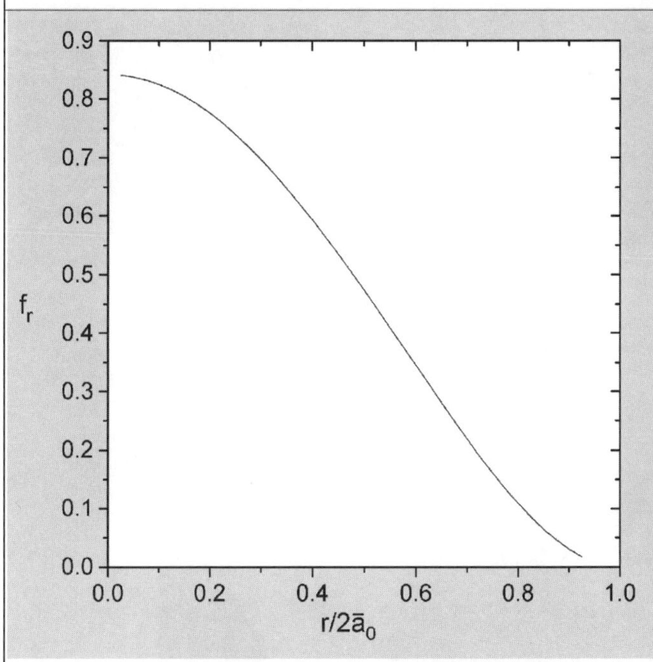

4.2.6 Figure 2. Roughness factor f_r as a function of relative initial roughness $r / 2\overline{a}_0$ for $a = 35°$.

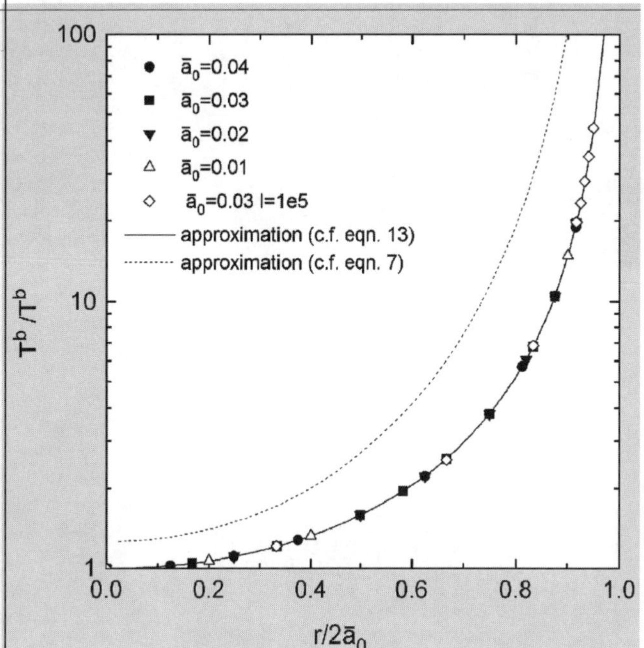

4.2.6 Figure 4. Dependence of the ratio between breakthrough times for initially smooth and rough fractures with initial relative roughness $r / 2\overline{a}_0$. The various symbols represent results of the finite-difference model for various initial widths and lengths. Other parameters are as shown in 4.2.6 Figure 3. The dotted line represents the first crude approximation given by equation 7. The solid line depicts equation 13.

Part 5: Development of Cavities and Cave Systems in Different Settings

5.0 Overview

An important point that has emerged from cave studies of the three decades is that no single speleogenetic model applies to all geologic and hydrologic settings. Given that settings may also change systematically during the evolutionary geologic cycles outlined above (Chapter 3.1), an evolutionary approach is called for.

This is attempted in Part 5, which is organized to give extended accounts of speleogenesis in the three most important settings that we recognize:

5.0 Plate 1. Coeymans Dome, a 25 m waterfall shaft in McFails Cave, New York, U.S.A. (photo by A.N. Palmer).

5.0 Plate 2. Oficho Cave, Western Caucasus, Georgia (photo by V. Kisseljov).

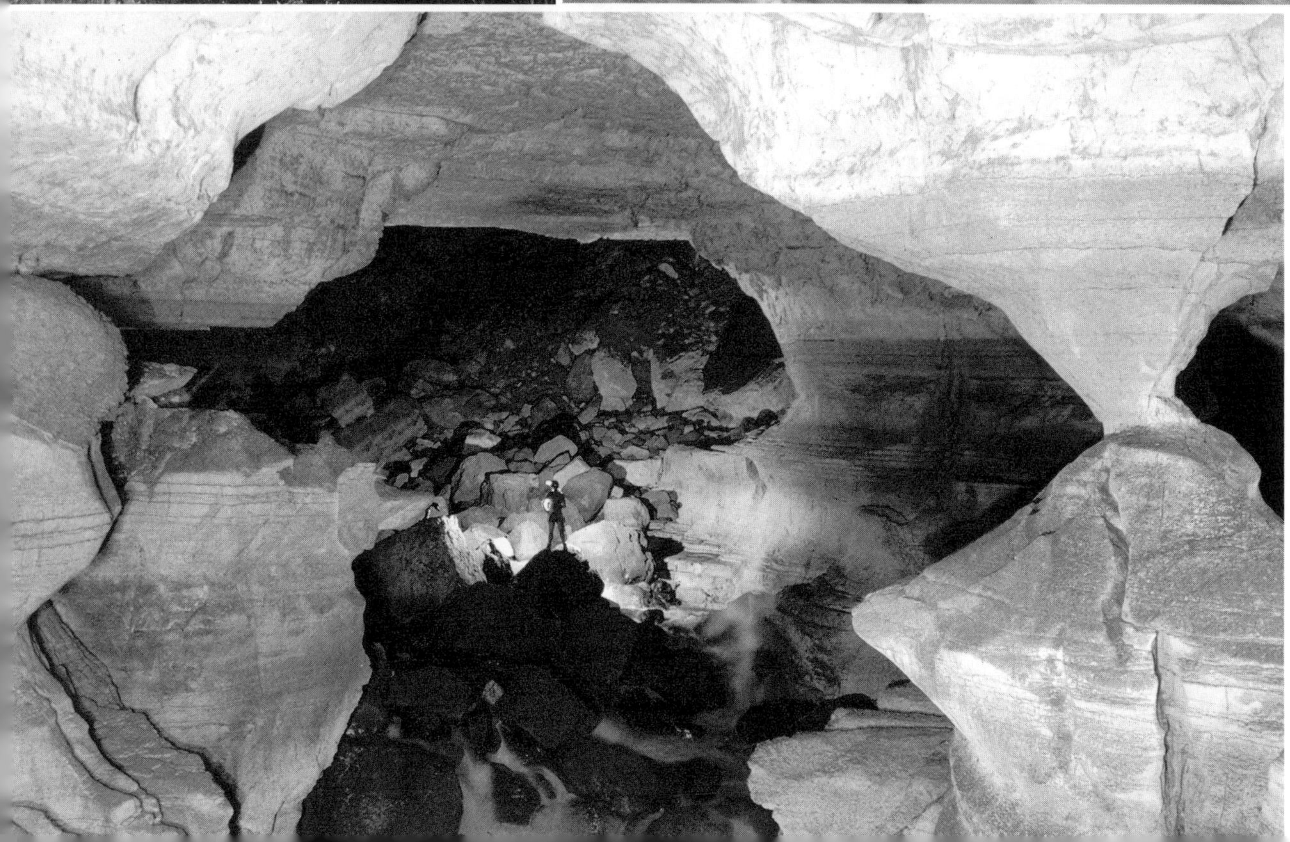

5.0 Plate 3. Big Rapids and Balcony, Sof Omar Cave, Ethiopia (photo by A. Klimchouk).

- Coastal and oceanic (Chapter 5.1)
- Deep-seated and confined (Chapter 5.2)
- Unconfined (Chapter 5.3)

Each chapter begins with a review of modern ideas on cave development in the setting, followed by representative case studies. The latter include new accounts of some "classic" caves as well as descriptions of other little-known cave systems and areas. Readers may determine for themselves how well the real field examples fit the general models presented in the introductory sections.

Mylroie and Carew in Chapter 5.1 summarize specific features of cave and karst development in young rocks in coastal and island settings that result from the chemical interactions between fresh and salt waters, and the effects of fluctuating sea level during the Quaternary. The case studies include a review of syngenetic karst in coastal dune limestones, Australia (S. White, Chapter 5.1.1) and an example of speleogenesis on tectonically active carbonate islands (Gunn and Lowe, Chapter 5.1.2).

Klimchouk in Chapter 5.2 reviews conditions and mechanisms of speleogenesis in deep-seated and confined settings, one of the most controversial but exciting topics in modern cave research. Conventional karst and speleogenetic theories are concerned chiefly with shallow unconfined geologic settings. These suppose that the karstification found there is intimately related to surface conditions of input and output, with the dissolution being driven by downward meteoric water recharge.

The possibility of hypogenic karstification in deeper environments has been neglected for a long time, and the quite numerous instances of karst features found at significant depths have usually been interpreted as buried paleokarst.

However, the past decade has seen a growing recognition of the variety and importance of hypogene dissolution process, and of speleogenesis under confined settings that often precedes unconfined development. These include Hill, 1987, 1995; Klimchouk, 1994, 1996, 1997; Lowe, 1992; Lowe and Gunn, 1995; Mazzullo and Harris, 1991, 1992; Palmer, 1991, 1995; Smart and Whitaker, 1991; Worthington, 1991, 1994; and Worthington and Ford, 1995.

Confined (artesian) settings were commonly ignored as sites for cave origin. This is because the classic concept of artesian flow implies long lateral travel distances for groundwater within a soluble unit, resulting in a low capacity to generate caves in the confined area.

However, the recognition of nonclassical features in artesian flow—namely the occurrence of cross-formation hydraulic communication within artesian basins, the concepts of transverse speleogenesis, and of the inversion of hydrogeologic function of beds in a sequence—calls for a revision of the theory of artesian speleogenesis and of views on the origin of many caves.

It is proposed that artesian speleogenesis is immensely important to speleoinception and also accounts for the development of some of the largest known caves in the world. Typical conditions of recharge, the flow pattern through the soluble rocks, and groundwater aggressiveness favor uniform, rather than competing, development of conduits, resulting in maze caves where the structural prerequisites exist. Cross-formational flow favors a variety of dissolution mechanisms that commonly involve mixing. Hydrogeochemical mechanisms of speleogenesis are particularly diverse and potent where carbonate and sulfate beds alternate and within or adjacent to hydrocarbon-bearing sedimentary basins.

Hypogene speleogenesis occurs in rocks of varied lithology and can involve a variety of dissolution mechanisms that operate under different physical constraints but create similar cave features. Case studies include:

- The great gypsum mazes of the western Ukraine (Klimchouk, Chapter 5.2.1).
- Great maze caves in limestones in the Black Hills, South Dakota (Palmer, Chapter 5.2.2), and Siberia (Filippov, Chapter 5.2.3).
- Karstification in the Redwall Aquifer, Arizona (Huntoon, Chapter 5.2.4).
- A summary of the peculiar features of hydrothermal speleogenesis (Y. Dublyansky, Chapter 5.2.5).
- Hydrothermal caves in Hungary (Y. Dublyansky, Chapter 5.2.6).
- Sulfuric acid speleogenesis (Lowe, Bottrell, and Gunn, Chapter 5.2.7, and Hill, Chapter 5.2.8).
- An outstanding example of a hydrothermal cavity—in fact the largest ever recorded by volume—in the Rhodope Mountains (V. Dublyansky, Chapter 5.2.9).

Recognition of the scale and importance of deep-seated speleogenesis and of the hydraulic continuity and cross-formational communications between aquifers in artesian basins is indispensable to correctly interpret the evolution of karst aquifers, speleogenetic processes and associated phenomena, regional karst water-resource evaluation, and the genesis of certain karst-related mineral deposits.

These and other theoretical and practical implications still have to be developed and evaluated. This offers a wide field for further research.

Ford in Chapter 5.3 reviews the theory of speleogenesis that occurs where normal meteoric water sinks underground through the epikarst or dolines and streamsinks, and circulates in the limestone or other soluble rocks without any major artesian confinement. These are termed *common caves* (Ford and Williams, 1989) because they probably account for 90% or more of the explored and mapped dissolutional caves that are longer than a few hundred meters. This estimate reflects the bias in exploration. Caves formed in unconfined settings and genetically related to surface recharge are the most readily accessible and hence form the bulk of documented caves.

Common caves chiefly display branchwork forms where the dissolutional conduits occupy only a tiny proportion of the total length or area of penetrable fissures that is available to the groundwater. The rules that govern the selection of the successful linkages that will be enlarged into the branchwork pattern are supported in the models presented in Chapter 4.2.

In longitudinal section, caves may be divided into deep phreatic, multiloop, mixed loop and water table, and ideal water table types, with drawdown vadose caves or invasion vadose caves above them. Many large systems display a mixture of the types. The concepts of plan-pattern construction, phreatic, water table, or vadose state, and multiphase development of common caves are illustrated in the case studies that follow the introduction.

They are organized broadly to begin with examples of comparatively simple deep phreatic and multiloop systems: El Abra, Mexico, by Ford, Chapter 5.3.1 and Castleguard Cave, Canada by Ford, Lauritzen, and Worthington, Chapter 5.3.2.

Examples then are given of large and complex multiphase systems. These are North of Lake Thun System, Switzerland, by Jeannin, Bitterli, and Häuselmann, Chapter 5.3.3, and Mammoth Cave, Kentucky, U.S.A. by Palmer, Chapter 5.3.8.

Next are representatives of mixed vadose and phreatic development in mountainous regions—The Alps, by Audra, Chapter 5.3.4, The Pyrenees, by Fernandez, Calaforra, and Rossi, Chapter 5.3.5, and Mexico, by Hose, Chapter 5.3.6.

Finally, cases are described where there is strong lithologic or structural control—folded appalachians, by W. White, Chapter 5.3.7, and gypsum caves in southern Spain, by Calaforra and Pulido-Bosch, Chapter 5.3.10.

Two special topics are considered: Speleogenesis of vertical shafts in the eastern United States (W. White, Chapter 5.3.9), and maze origin by diffuse recharge through overlying formation (Palmer, Chapter 5.3.11).

The set concludes with two instances of nearly ideal water table cave development. these are in Belize and Hungary (Ford, Chapter 5.3.12). And finally, a review of the latest models of speleogenesis from the region where modern karst studies began, the Classical Karst of Slovenia and Trieste, by Šušteršič, Chapter 5.3.13..

5.1. Speleogenesis in Coastal and Oceanic Settings

John E. Mylroie and James L. Carew

Abstract

Carbonate islands and coasts are directly influenced by the chemistry of mixing marine and fresh water, and by the migration of sea level in response to Quaternary glacioeustacy. Carbonate islands are unique in particular because they are limited in recharge to their own areal extent, whereas carbonate coasts can obtain recharge from adjacent non-carbonate areas.

Carbonate islands fall into three categories:

- Simple carbonate islands that are entirely carbonate from the land surface to a depth exceeding the lowest Quaternary glacioeustatic sea-level lowstand (approximately –125 m).
- Carbonate-cover islands which are entirely carbonate, but a carbonate-noncarbonate contact exists within the range of Quaternary glacioeustacy.
- Carbonate-rimmed islands where noncarbonate rocks are present at the surface.

Simple carbonate islands have only autogenic recharge and develop pit caves in the epikarst, banana holes at the top of the freshwater lens, and flank-mar in caves at the margin of the freshwater lens. Carbonate-cover islands develop additional caves at the carbonate-noncarbonate contact, and carbonate-rimmed islands form stream caves from noncarbonate allogenic recharge.

Blue holes and large conduit caves exist on carbonate islands and coasts. They are overprinted by Quaternary glacioeustacy, and their development is linked to processes not operating at their locations today.

Introduction

Unique cave and karst features develop in soluble rocks in coastal and oceanic settings as a result of conditions associated with proximity to the marine environment. For the purposes of the discussion presented here, only carbonate rocks (limestone and dolomite) are considered. Conditions that are of great significance in karst development in coastal and island settings result from the chemical interactions between fresh and salt waters, and from the effects of fluctuating sea level during the Quaternary (Mylroie and Carew, 1995; Mylroie et al., 1995a).

These conditions have affected all carbonate coasts and islands in the world, but, carbonate coastlines and islands are found preferentially in the tropics and subtropics, because of the influence of climate on the deposition of carbonates. In fact, subaerial Quaternary limestone is found only in subtropical and tropical settings, and there only in oceanic or coastal environments, because plate tectonic processes have not yet placed youthful carbonates into continental-interior or high-latitude settings.

It is also useful to separate carbonate islands from carbonate coasts on continents (or very large islands that hydrologically behave like continents). The major difference between these two situations is that coastal carbonates commonly receive surface or subsurface recharge derived from adjacent noncarbonate land areas, whereas carbonate islands do not. Carbonate island recharge is limited by the island's areal extent, and the chemical properties of that recharge water is controlled by the local island environment. There is, of course, a continuum between conditions on carbonate islands, continental carbonate coasts, and classic continental karst that is free of marine influence. Oceanic carbonate islands may be regarded as the simplest examples of the influence of the marine environment on cave and karst development, and they will be the initial focus of this presentation. The examples used here to illustrate cave and karst development on carbonate islands and coasts are from the Atlantic region.

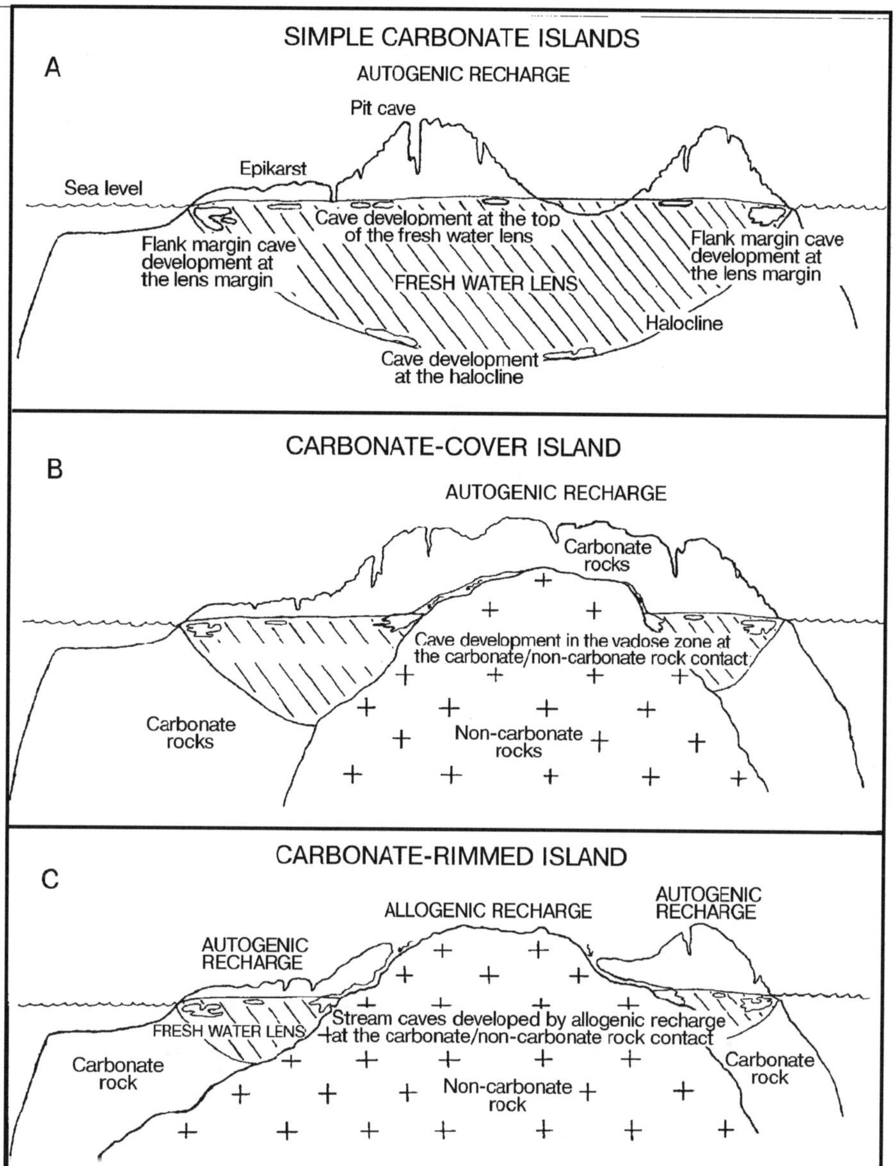

*5.1 Figure 1. Features of the three main types of carbonate islands. Section **A** is a simple carbonate island. Section **B** is a carbonate-cover island. Section **C** is a carbonate-rimmed island*

Carbonate Islands

Carbonate islands can be divided into the following three categories:

- Simple carbonate islands.
- Carbonate-cover islands.
- Carbonate-rimmed islands (5.1 Figure 1).

Simple carbonate islands are entirely carbonate from the land surface to a depth exceeding the lowest Quaternary glacio-eustatic sea-level lowstand (approximately 125 m). The land surface of carbonate-cover islands is entirely carbonate, but a carbonate–non-carbonate contact exists within the range of Quaternary glacioeustacy. Carbonate-rimmed islands comprise those where noncarbonate rocks lie at the surface in the interior.

In the Atlantic region, the Bahamas, the Caymans, and Isla de Mona (Puerto Rico), are examples of carbonate islands; Bermuda is a carbonate-cover island; and Jamaica, Puerto Rico, and Barbados are carbonate-rimmed islands. Given islands of the same size and climatic setting (i.e. same amount of recharge), conditions on the three different island types listed above result in quite different cave and karst development.

Simple Carbonate Islands

Today, simple carbonate islands are primarily late Tertiary (i.e. Miocene-Pliocene) or Quaternary in age. Their recharge is entirely autogenic and diffuse at large scale. Surface streams are generally absent, and the karst landforms associated with streams (e.g. streamsinks and blind valleys) are also absent. The fresh groundwater typically floats on underlying denser marine groundwater and forms a freshwater lens (5.1 Figure 1A) that approximates the Dupuit-Ghyben-Herzberg model (Vacher, 1988).

Although the idealized lens shape is often distorted by differences in rock lithology, porosity, permeability, transmissivity, and climatic variables such as water budget (e.g. Vacher and Wallis, 1992), it provides a useful approximation for discussion. These young carbonate rocks usually have good permeability. As a result, the position of the freshwater lens is controlled by sea level, such that the top of the lens is at or slightly above sea level, the margin of the lens is at sea level, and the base of the lens increases in depth away from its margin (5.1 Figure 1A). Given a density difference between fresh water and sea water of about 1 part in 40 (1.000 g/cm^3 versus 1.025 g/cm^3), the base of the freshwater lens extends below sea level about 40 times the amount that the top of the lens is above sea level. Small carbonate islands rarely have a lens that is more than 40 m thick, so the top of the lens is generally less than 1 m above sea level. Quaternary glacio-eustatic sea level fluctuations have caused sea level to migrate over approximately a 130 m range (-125 m, +6 m). Therefore, the freshwater lenses in carbonate islands have similarly migrated.

On simple carbonate islands, cave development can be expected in the following three environments:

- The land surface that receives meteoric recharge.
- The top of the freshwater lens that receives and mixes with recharge water that has traversed the vadose zone.
- The bottom of the freshwater lens where sea water and fresh water mix (the halocline or mixing zone).

Meteoric recharge to the land surface results in dissolution of the surface and produces an etched and weathered zone referred to as epikarst. The epikarst is drained by diffuse flow and by local concentrations of descending water that produce vadose shafts called *pit caves* (Pace, et al., 1993). As a result of multiple piracies of source areas within the epikarst, a region may contain numerous abandoned pit caves. Pit caves are found on simple carbonate islands composed of late Quaternary limestone, such as the Bahamas (Pace, et al., 1993), and on islands composed of Miocene-Pliocene carbonates such as Isla de Mona, Puerto Rico, U.S.A. (Mylroie, et al., 1995b).

Because the autogenic recharge that produces these pit caves has limited dissolutional potential, most pit caves end before reaching the top of the fresh-water lens, and the water continues downward to the lens as diffuse flow. Pit caves are generally less than 10 m deep, and as a result of the piracies mentioned above, they commonly connect to adjacent pit caves to form complexes with passage lengths exceeding 50 m (5.1 Figure 2). These features can be locally abundant, and they may pose significant land-use problems.

The top of the fresh-water lens is also a region where cave development can be prolific, but most such caves are relatively small (Mylroie and Carew, 1988, 1995). The phreatic water of the fresh-water lens and the vadose water descending from the epikarst may both be saturated with respect to $CaCO_3$. But, if their $CaCO_3$ in solution differs, they are theoretically capable of further dissolution when they mix (Bögli, 1980).

The top of the fresh-water lens also represents a density interface at which organic matter may collect. Oxidation of the organic matter can produce CO_2 that further drives dissolution in that environment. If organic loading is sufficient, reducing conditions may develop, and those conditions allow still greater dissolution, in a manner similar to that described for haloclines in blue holes (Bottrell, et al., 1991).

Field evidence from the Bahamas (Harris, et al., 1995) shows that phreatic voids with lateral dimensions of 1 to 15 m developed at the top of the fresh-water lens during the last interglacial sea-level highstand (oxygen isotope substage 5e, approximately 15,000 years in duration). These voids are vertically restricted such that their depth-to-width ratio is less than one (e.g. 1 to 3 m deep and 3 to 10 m wide). This is a consequence of their genesis at the top of the lens.

These voids commonly collapse to produce sinkholes called *banana holes*, and they are known to occur in numbers that can exceed 3,000/km^2 (Harris, et al., 1995). The expression of these phreatic voids as banana holes in the Bahamas is a result of the development of the voids where the land surface was only a few meters above the lens, so the void roofs were thin and prone to collapse.

Isla de Mona, Puerto Rico, has fewer banana holes than the Bahamas. There the fresh-water lens was many meters below the subsequent land surface, and only the largest voids that formed at the top of that fresh-water lens have collapsed to the surface (Mylroie, et al., 1995b). Banana holes can be completely or partially collapsed (5.1 Figure 3), and where only partial collapse has occurred, chambers of appreciable size may remain. These features are an extreme land-use hazard in the Bahamas (Wilson, et al., 1995).

The third area of substantial cave development is in the mixing zone, or halocline, at the base of the fresh-water lens (The term *halocline* is used when the fresh-water to salt-water boundary is sharp. The term *mixing zone* is used when a broad

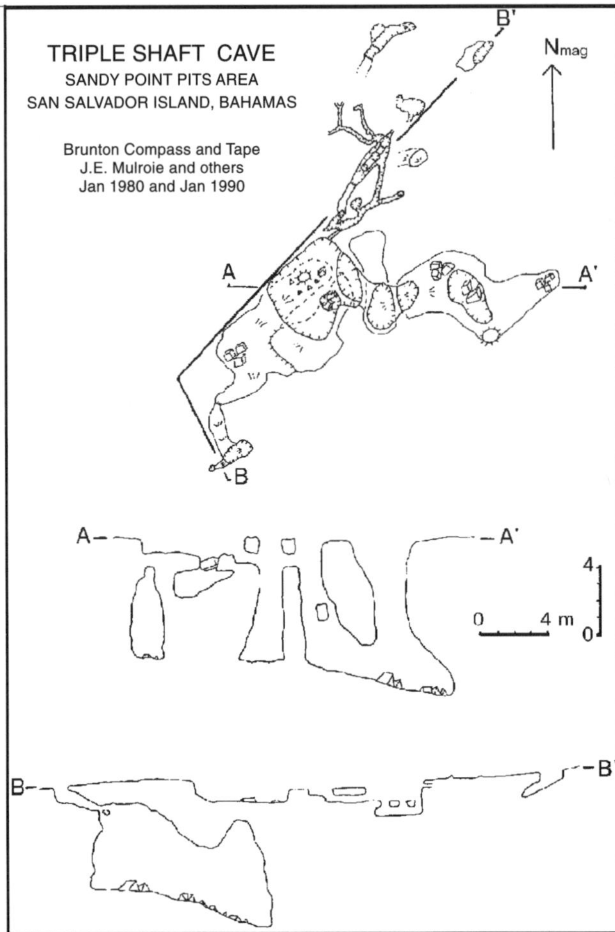

5.1 Figure 2. Pit cave formed from drainage collected in the epikarst, showing complex development from self-piracy, Triple Shaft Cave, San Salvador Island.

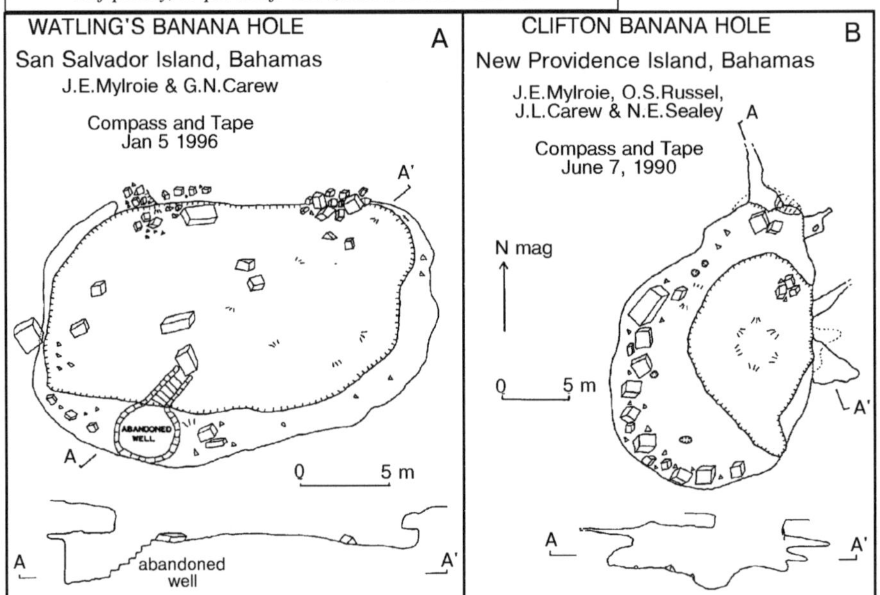

5.1 Figure 3. Banana holes formed by roof collapse of shallow phreatic voids formed at the top of a past fresh-water lens. Panel ***A*** *is Watling's Banana Hole, San Salvador Island, Bahamas, a totally collapsed banana hole with an abandoned well on one side. Panel* ***B*** *is Clifton Banana Hole, New Providence Island, Bahamas, showing a mostly intact roof (from Mylroie, et al., 1991).*

zone of variable salinity separates the two water types.) In a manner analogous to the mixing of fresh water from the vadose zone with fresh water at the top of the lens, the mixing of fresh water at the base of the lens with underlying marine water results in a mixing zone with enhanced dissolutional potential (Plummer, 1975).

Also, like the top of the lens, the base of the lens is a site of density contrast, and organic material collects at that horizon. Enhanced dissolution as a result of oxidation-reduction reactions in the mixing zone has been documented in the blue holes of the Bahamas (Bottrell, et al., 1991). Therefore, the mixing zone is a location where large-scale dissolution is to be expected.

Within the mixing zone, the distal margin of the lens is a region of enhanced cave development, because thinning of the lens results in a decrease in cross-sectional area, with a consequent increase in flow velocity (Raeisi and Mylroie, 1995). In addition, at such locations, the favorable geochemical environments for dissolution (mixing of dissimilar waters and oxidation-reduction reactions) at the top and bottom of the lens are superimposed. The caves formed at the lens margins are the largest found in simple carbonate islands. They are called *flank margin caves* as they develop at the margin of the fresh-water lens, under the flank of the enclosing high ground (Mylroie and Carew, 1990). These caves are not true conduits.

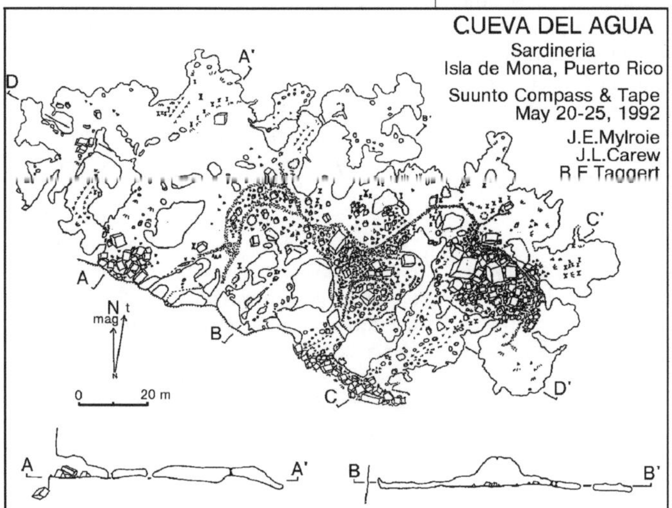

5.1 Figure 4. Flank margin cave formed by dissolution in the discharging margin of a fresh-water lens, Cueva del Agua Sardinera, Isla de Mona, Puerto Rico (from Mylroie, et al., 1995b).

Instead they represent former mixing chambers and as such did not transmit groundwater any faster than it was delivered to them by the diffuse flow of the carbonate aquifer. Evidence from analyses of stable sulfur isotopes in gypsum crusts indicates that oxidation-reduction reactions were involved in the development of at least some of these caves in the Bahamas (Bottrell, et al., 1993).

Flank margin caves have a consistent morphology of large oval chambers, (5.1 Figure 4) with numerous remnant dissolution pillars and thin-wall partitions between chambers. Commonly, the chambers continue in toward the island interior as branching voids or as tubes that terminate abruptly.

These abrupt terminations represent the location of the mixing front at the time the caves were drained. Because these caves developed at the lens margin, they are an excellent indicator of sea-level position at the time the caves developed. Flank margin caves, like banana holes, do not form with a macroscopic opening to the surface. Only after surface erosion breaches a flank margin cave can the void be entered by humans.

For tectonically stable islands such as the Bahamas, the changes in sea level that formed and subsequently drained the caves were caused by glacioeustacy (Carew and Mylroie, 1995). The banana holes and flank margin caves that are subaerial today in the Bahamas must have been produced by a past higher interglacial sea-level highstand. These caves are found at altitudes of 1-6 m above modern sea level and must have developed in a fresh-water lens at that position in the past. Given a 1-2 m per 100,000 yr isostatic subsidence rate of the Bahamas, only the last interglacial sea-level highstand (Sangamon, or oxygen isotope substage 5e, circa 125,000 yr) was both high enough and recent enough to place the fresh-water lens at the position necessary for development of these caves (Carew and Mylroie, 1995).

Earlier sea-level highstands were either not high enough (oxygen isotope stage 7, circa 220,000 yr), or were too long ago (oxygen isotope stage 9, circa 330,000 yr) to have left banana holes or flank margin caves above modern sea level.

On the other hand, tectonically uplifted Isla de Mona, Puerto Rico, has flank margin caves that formed at a variety of past sea-level stillstand positions. Some of these caves have been dated by paleomagnetic analyses of internal deposits. They are more than 1,000,000 years old (Mylroie, et al., 1995b). These caves also show evidence of dissolutional overprinting; that is, they contain features which indicate that they have undergone repeated cycles of vadose and phreatic conditions. In many places, massive vadose speleothems and the wall rock to which they are attached are cut by later phreatic dissolutional surfaces.

It has been suggested that these younger dissolutional surfaces could be a result of condensation corrosion (Tarhule-Lips and Ford, 1995). But the preponderance of evidence suggests reoccupation of existing caves by dissolutionally aggressive phreatic water (Mylroie, et al., 1995b).

Carbonate-Cover Islands

Generally, cave and karst development on carbonate-cover islands is very similar to that of simple carbonate islands. However, because these islands have a relatively thin carbonate veneer on top of noncarbonates, and the carbonate–noncarbonate contact is within the range of Quaternary glacioeustacy, that contact influences the nature of the fresh-water lens during high sea levels and collects and deflects vadose flow during low sea levels (5.1 Figure 1B).

Bermuda is the classic example. There, Quaternary limestone, primarily eolianite, thinly covers a basaltic pedestal (Mylroie, et al., 1995a). The epikarst on Bermuda is similar to that seen in the Bahamas, but the caves are markedly different. Most caves on Bermuda are large collapse caves that developed by vadose dissolution at the limestone–basalt contact during previous sea-level lowstands with subsequent progradational collapse of that void to the near surface (Mylroie, 1984; Mylroie et al., 1995a).

The perching of vadose flow at the limestone–basalt contact resulted in wide undercuts at the contact, promoting collapse. The ability of small vadose streams to produce such undercuts allows a small amount of recharge to create a wide unsupported ceiling, as has been noted in similar situations on Kangaroo Island, Australia (Jennings, 1968). As a result of collapse prograding upward from the contact, these caves as entered today generally lack dissolutional surfaces.

The walls and ceilings are fracture surfaces, and the floors are piles of breakdown or sediment, which commonly extends below modern sea level (5.1 Figure 5). Flank margin caves also exist on Bermuda, but because of Bermuda's moist climate and greater surficial erosion, most of them have been destroyed by retreat of the hillsides in which they were contained (Vacher and Mylroie, 1991).

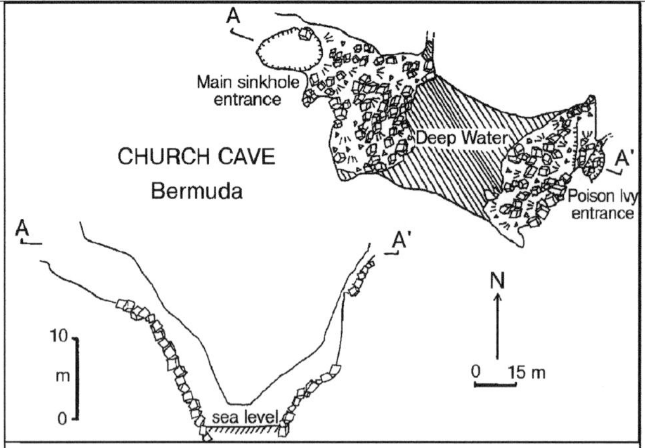

5.1 Figure 5. Church Cave, Bermuda, a collapse cave formed by vadose dissolution along the carbonate–basalt contact during a glacioeustatic sea-level lowstand, followed by progradational collapse upward toward the surface, partially flooded at modern sea level conditions (from Mylroie and Carew, 1995).

Carbonate-Rimmed Islands
Carbonate-rimmed islands are those where significant noncarbonate outcrops are exposed at the surface (5.1 Figure 1C). Commonly these islands have a noncarbonate central region rimmed by carbonates, but a variety of carbonate–noncarbonate configurations are possible. The carbonate part of these islands behaves much like a simple carbonate island or a carbonate-cover island (the noncarbonate lithologies commonly underlie the carbonates in these situations) in that the carbonates receive autogenic meteoric recharge. However, the carbonates also receive allogenic recharge from the noncarbonate region.

In this respect the carbonate-rimmed islands behave similar to any continental–interior setting where allogenic water from noncarbonates encounters carbonates.

The unique aspect of the carbonate-rimmed islands is that this allogenic water eventually interacts with sea water in the halocline of the fresh-water lens. So, carbonate-rimmed islands have karst features such as sinking streams and blind valleys, and stream caves are common. For example, Puerto Rico—a carbonate-rimmed island—has large stream caves such as Rio Camuy (Gurnee and Gurnee, 1974) (5.1 Figure 6). Only in the coastal areas of Puerto Rico do simple carbonate-island flank margin caves exist, such as Cueva del Indio, a flank margin cave on the north coast (Frank, 1993).

Blue Holes
Blue holes are deep water-filled shafts found in the Bahamas and many other coastal and island carbonate settings around the world. The term had fallen into diverse and casual usage, and recently the term *blue holes* was reviewed and redefined as follows:

> Blue holes are subsurface voids that are developed in carbonate banks and islands; are open to the Earth's surface; contain tidally-influenced waters of fresh, marine, or mixed chemistry; extend below sea level for a majority of their depth; and may provide access to cave passages (Mylroie, et al., 1995c, p. 231).

Blue holes may be further subdivided into *ocean holes* and *inland blue holes* as follows:

- An ocean hole is a blue hole that opens directly into the present marine environment. It usually contains marine water with tidal flow.
- An inland blue hole is a blue hole isolated by present topography from surface marine conditions. It opens directly onto the land surface or into an isolated pond or lake. It contains tidally influenced water of a variety of chemistries from fresh to marine (Mylroie, et al., 1995c, p. 231).

Blue holes are created by several geologic processes:

- Flooding of pit caves developed during glacioeustatic sea-level lowstands by a return to high sea level.
- Progradational collapse to the surface of deep-seated voids within the carbonate platform.
- Bank-margin failure (usually parallel to the island margin) to produce deep flooded fractures (Mylroie, et al., 1995c).

Many complex cave systems open at depth into blue holes. These features represent a difficult explorational challenge, and several works have been written about their exploration (e.g. Farr, 1991; Palmer, 1989).

The genesis of blue holes clearly is not necessarily tied to Quaternary glacioeustacy. For example, Deans Blue Hole, on Long Island, Bahamas, is over 200 m deep. This is deeper than any Quaternary lowstand of sea level, even allowing for considerable isostatic subsidence (Wilson, 1994). Deans Blue Hole most likely represents progradational collapse upward from a deep-seated void.

As a result of Quaternary glacioeustacy, blue holes have been overprinted by numerous transitions from vadose to fresh-water-phreatic to marine-phreatic conditions. Not surprisingly, their geology is complex, and deciphering the various factors that played a role in their development is difficult.

Regardless of specific origin, blue holes have become part of the water-flow system underlying many carbonate platforms (Whitaker and Smart, 1990), and they provide direct access into an unique submerged environment.

Conduit Caves in Carbonate Islands and Banks

Cave divers have reported long conduit caves underlying large carbonate platforms.These include Conch Sound Cave off Andros Island on Great Bahama Bank (Farr, 1991) or the Lucayan Cave System of Little Grand Bahama Bank (Palmer, 1989). But drained conduit caves are not known from the subaerial Bahamas.

A speculative reason for the absence of abandoned conduit caves may be that during current and earlier sea-level highstands, the existing islands are too small in aerial extent to develop a large lens. In the Bahamas, when sea level is below -10 m, the carbonate banks become entirely exposed to form very large islands. Total island area today is 11,400 km^2, but that would increase to 136,000 km^2 during glaciations (Meyerhoff and Hatten, 1974).

These large islands would probably have had very large fresh-water lenses. Under those conditions the discharge may have been ample enough that preferred flow pathways developed and long conduit caves formed. On small islands with small lenses, however, the available recharge relative to the island perimeter is such that preferred flow paths do not develop and there is no conduit flow. Hence very large carbonate islands with very large lenses may be analogous to continental carbonate coasts, which also have conduit caves.

Carbonate Coasts
Carbonate coasts in continental settings behave differently than do carbonate islands. The primary reason for this difference is that carbonate coasts usually have large-scale recharge from the continental landmass. This large landmass is commonly noncarbonate. It provides large amounts of allogenic recharge with an aggressive dissolutional chemistry, which is not available on carbonate islands with limited autogenic recharge. A classic example of a carbonate coast in a continental setting is the Yucatan Peninsula of Mexico.This is where pioneering work on

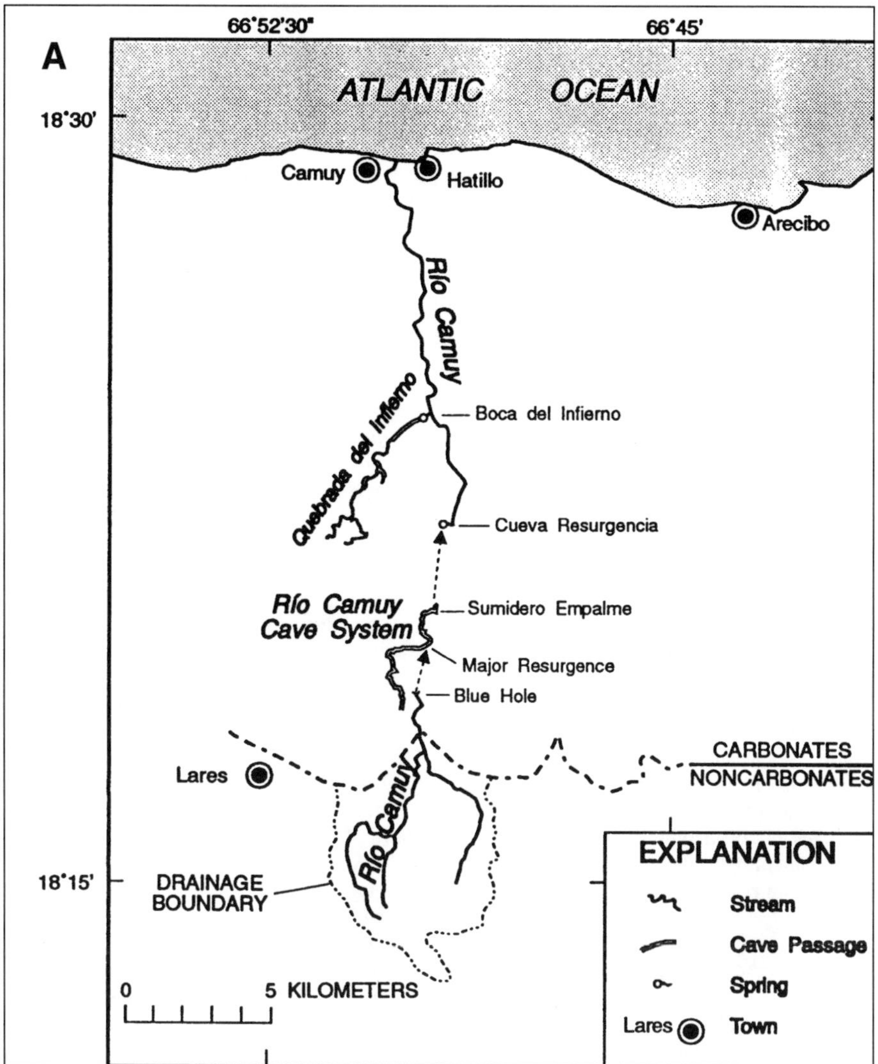

5.1 Figure 6. *The Rio Camuy Cave System, Puerto Rico. Panel A is the plan view of the Rio Camuy area, showing allogenic catchment on non-carbonate rock that forms a stream cave upon crossing onto carbonate rock; dotted lines with arrowheads show dye-tested but unexplored flow routes. Panel B is the cross section of Rio Camuy Cave, showing the cave migrating upsection in the limestone in response to the base level formed by current sea level (reproduced from Troester, 1994, with permission).*

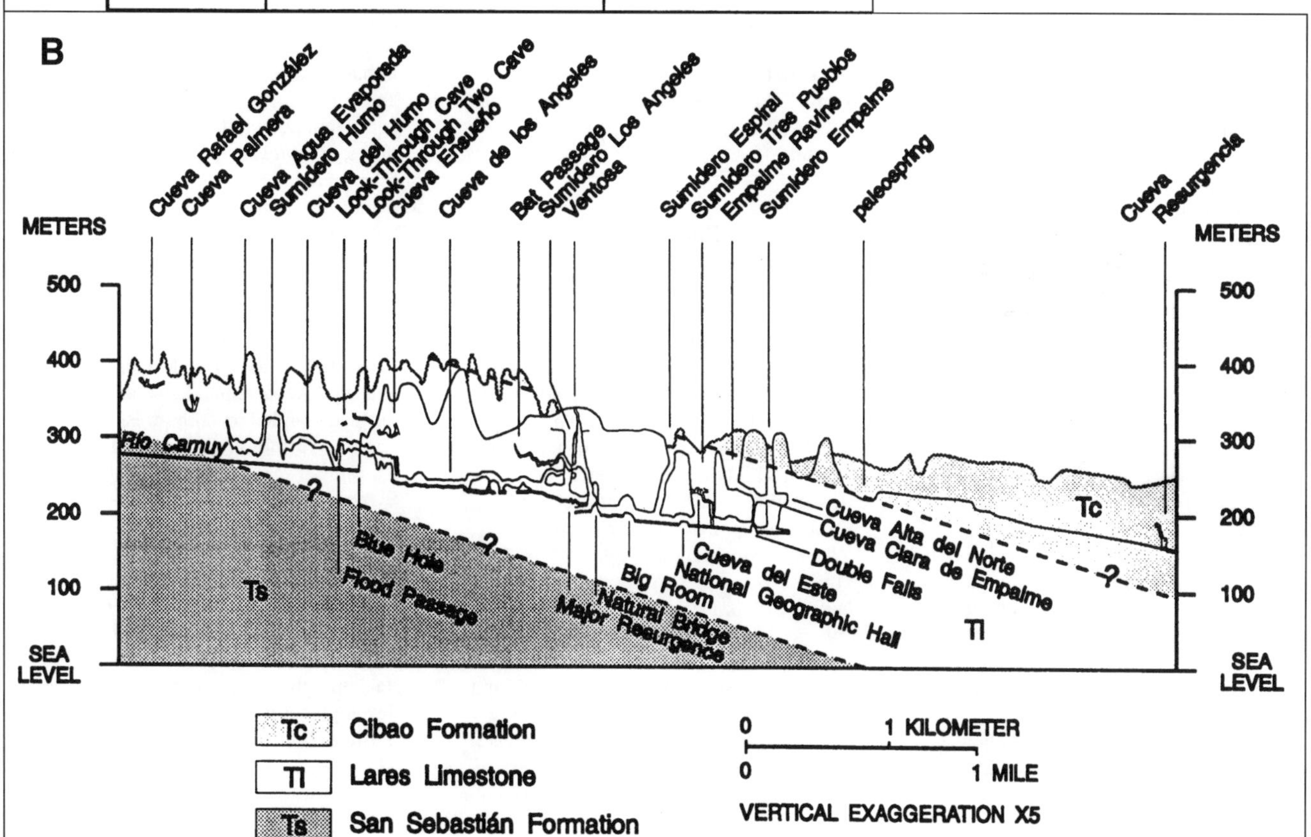

limestone dissolution in the mixed-water environment was done (e.g. Back, et al., 1984, 1986). The area also served as "ground truth" for mathematical models of carbonate dissolution by mixed waters (e.g. Sanford and Konikow, 1989).

Some of the largest explored and mapped underwater caves in the world exist in the Yucatan (e.g. Exley, 1994). These caves show conduit morphology and they currently discharge fresh water to the Caribbean Sea. In many of these caves, a sharp halocline can be observed, even in conduits with noticable flow. Where these caves reach the coast, headward collapse of the conduits produces small bays and inlets called *caletas* (Back, et al., 1984).

The caves contain abundant vadose speleothems. Their morphology reflects the overprinting of vadose and phreatic conditions resulting from glacioeustacy in the Quaternary. With the exception of the caletas, it is not clear to what degree fresh/salt-water mixing has controlled the development of these conduit systems. The large available recharge provided by the land mass of the Yucatan Peninsula (and perhaps areas to the southwest) may be the major governing factor in the production of these extensive caves.

Summary

The development of caves and karst in coastal and island carbonates was controlled primarily by the position of sea level, which in turn was controlled by glacioeustacy during the Quaternary. The mixing of fresh and salt water in these areas provides enhanced geochemical dissolution that is not found in continental situations, except for hypogenic caves (Palmer, 1991), and where mixing of dissimilar groundwaters enhances dissolution. The caves in the Guadalupe Mountains region of New Mexico were formed by mixing of H_2S-rich basinal water with O_2-rich shallow phreatic water. These caves are an excellent large-scale analogue to the flank margin caves of simple carbonate islands (Mylroie, 1991).

The allogenic recharge of the carbonate coasts of continents and carbonate-rimmed islands create caves most like those of continental interiors, with some modification by fresh/salt-water mixing. Because of the geochemical restrictions from limited autogenic recharge, the caves found on simple carbonate islands and carbonate-cover islands develop primarily by mixing of dissimilar waters. In addition, perching of vadose water at the carbonate–noncarbonate contact of carbonate-cover islands results in the development of large voids at depth, by producing undercut chambers that are larger than available recharge would seem to allow. Progradational collapse of these voids yields large cave chambers without dissolutional surfaces.

On tectonically stable carbonate islands, such as the Bahamas, only those caves formed during the last interglacial (oxygen isotope substage 5e, circa 125 ka) sea-level highstand have avoided the dissolutional overprinting effects from Quaternary glacioeustacy. That highstand of sea level was relatively short, so these caves provide severe constraints on the conditions for cave development. Flank margin caves, which develop in the distal margin of a fresh-water lens, may be the result of the fastest carbonate dissolutional cave development mechanism known. These caves formed where two separate regions of mixing and oxidation/reduction reactions were superimposed, and that may account for this rapid cave development. While the flank margin caves of the Bahamas show that caves can develop in an extremely short time (ca. 15,000 years), the flank margin caves of Isla de Mona, Puerto Rico, show that they can persist for over a million years.

Within continental interiors, caves with configurations similar to flank margin caves may represent ancient caves formed on a former carbonate coast or island, or they may be caves that formed in their current setting by mixing of basinal brine with fresh shallow phreatic water (Mylroie, 1991).

Acknowledgments

We wish to thank the Bahamian Field Station on San Salvador Island, Mississippi State University, the University of Charleston, and the U.S. Geological Survey for support of the field work that made much of our research possible. We thank our many colleagues and students for their assistance, comments, and criticisms.

References

Back, W., Hanshaw, B.B., and Van Driel, J.N., 1984, Role of groundwater in shaping the eastern coastline of the Yucatan Peninsula, Mexico, *in* LaFluer, R.G., ed., *Groundwater as a geomorphic agent:* Boston, Allen and Unwin, 390 p.

Back, W., Hanshaw, B.B., Herman, J.S., and Van Driel, J.N., 1986, Differential dissolution of a Pleistocene reef in the ground-water mixing zone of coastal Yucatan, Mexico: *Geology*, v. 14, p. 137-140.

Bögli, A., 1980, *Karst hydrology and physical speleology:* New York, Springer-Verlag, 284 p.

Bottrell, S.H., Smart, P.L., Whitaker, F., and Raiswell, R., 1991, Geochemistry and isotope systematics of sulphur in the mixing zone of Bahamian blue holes: *Applied Geochemistry*, v. 6, p. 99-103.

Bottrell, S.H., Carew, J.L., and Mylroie, J.E., 1993, Bacterial sulphate reduction in flank margin environments: Evidence from sulphur isotopes, *in* White, B., ed., 6th *Symposium on the Geology of the Bahamas*, Port Charlotte, Florida, Proceedings, Bahamian Field Station, p. 17-21.

Carew, J.L., and Mylroie, J.E., 1995, Quaternary tectonic stability of the Bahamian Archipelago: Evidence from fossil coral reefs and flank margin caves: *Quaternary Science Reviews*, v. 14, p. 144-153.

Exley, S., 1994, *Caverns measureless to man:* St. Louis, Cave Books, 325 p.

Farr, M., 1991, *The darkness beckons:* St. Louis, Cave Books, 280 p.

Frank, E.F., 1993, *Aspects of karst development and speleogenesis, Isla de Mona, Puerto Rico: An analogue for Pleistocene speleogenesis in the Bahamas*: Unpublished MSc. Thesis, Mississippi State University, 349 p.

Gurnee, R., and Gurnee, J., 1974, *Discovery at the Rio Camuy:* New York, Crown Publishers, 183 p.

Harris, J.G., Mylroie, J.E., and Carew, J.L., 1995, Banana holes: Unique karst features of the Bahamas: *Carbonates and Evaporites*, v. 10, p. 215-224.

Jennings, J.N., 1968, Syngenetic karst in Australia, *in* Williams, P.W., and Jennings, J.N., eds., Contributions to the study of karst: Canberra, Australian National University Research School of Pacific Studies, Publication G5, p. 41-110.

Meyerhoff, A.A., and Hatten, C.W., 1974, Bahamas salient of North America: Tectonic framework, stratigraphy, and petroleum potential.: *American Association of Petroleum Geologists Bulletin*, v. 58, p. 1201-1239.

Mylroie, J.E., 1984, Speleogenetic contrast between the Bermuda and Bahama Islands, *in* Teeter, J.W., ed., *2nd Symposium on the Geology of the Bahamas,* Fort Lauderdale, Proceedings, CCFL Bahamian Field Station, p. 113-128.

Mylroie, J.E., 1991, Cave development in the glaciated Appalachian karst of New York: Surface-coupled or saline-freshwater mixing hydrology?, *in* Kastning, E.H., and Kastning, K.M., eds., Appalachian Karst Symposium Proceedings, Huntsville, National Speleological Society, p. 85-90.

Mylroie, J.E., and Carew, J.L., 1988, Solution conduits as indicators of late Quaternary sea level position: *Quaternary Science Reviews*, v. 7, p. 55-64.

Mylroie, J.E., and Carew, J.L., 1990, The flank margin model for dissolution cave development in carbonate platforms: *Earth Surface Processes and Landforms,* v. 15, p. 413-424.

Mylroie, J.E., and Carew, J.L., 1991, Erosional notches in Bahamian carbonates: Bioerosion or groundwater dissolution?, *in* Bain, R.J., ed., *5th Symposium on the Geology of the Bahamas*, Port Charlotte, Florida, Proceedings, Bahamian Field Station, p. 185-191.

Mylroie, J.E., and Carew, J.L., 1995, Karst development on carbonate islands, Chapter 3, *in* Budd, D.A., Harris, P.M., and Saller, A., eds., Unconformities and porosity in carbonate strata: American Association of Petroleum Geologists Memoir 63, p. 55-76.

Mylroie, J.E., Carew, J.L., Sealey, N.E., and Mylroie, J.R., 1991, Cave development on New Providence Island and Long Island, Bahamas: *Cave Science,* v. 18, p. 139-151.

Mylroie, J.E., Carew, J.L., and Vacher, H.L., 1995a, Karst development in the Bahamas and Bermuda, *in* Curran, H.A., and White, B., eds., Terrestrial and shallow marine geology of the Bahamas and Bermuda: *Geological Society of America Special Paper 300*, p. 251-267.

Mylroie, J.E., Carew, J.L., Frank, E.F., Panuska, B.C., Taggart, B.E., Troester, J.W., and Carrasquillo, R., 1995b, Comparison of flank margin cave development: San Salvador Island, Bahamas and Isla de Mona, Puerto Rico, *in* Boardman, M.R., ed., *7th Symposium on the Geology of the Bahamas,* San Salvador Island, Proceedings, Bahamian Field Station, p. 49-81.

Mylroie, J.E., Carew, J.L., and Moore, A.I., 1995c, Blue holes: Definition and genesis: *Carbonates and Evaporites*, v. 10, p. 225-233.

Pace, M.C., Mylroie, J.E., and Carew, J.L., 1993, Petrographic analysis of vertical dissolution features on San Salvador Island, Bahamas, *in* White, B., ed., 6th Symposium on the Geology of the Bahamas, Port Charlotte, Florida, Proceedings, Bahamian Field Station, p. 109-123.

Palmer, R.J., 1989, *Deep into blue holes:* London, Unwin Hyman, Ltd., 164 p.

Palmer, A.N., 1991, Origin and morphology of limestone caves: *Geological Society of America Bulletin,* v. 103, p. 1-25.

Plummer, L.N., 1975, Mixing of sea water with calcium carbonate ground water, *in* Whitten, E.H.T., ed., Quantitative studies in geological sciences, *Geological Society of America Memoir 142*, p. 219-236.

Raeisi, E., and Mylroie, J.E., 1995, Hydrodynamic behavior of caves formed in the fresh-water lens of carbonate islands: *Carbonates and Evaporites,* v. 10, p. 207-214.

Sanford, W.E., and Konikow, L.F., 1989, Porosity development in coastal carbonate aquifers: *Geology,* v. 17, p. 249-252.

Smart, P.L., Dawans, J.M., and Whitaker, F., 1988, Carbonate dissolution in a modern mixing zone: *Nature,* v. 335, p. 811-813.

Tarhule-Lips, R.F., and Ford, D.C., 1995, Speleothem dissolution on Cayman Brac and Isla de Mona: *Geological Society of America Abstracts with Programs,* v. 27, no. 2, p. 346.

Troester, J.W., 1994, *The geochemistry, hydrogeology, and geomorphology of the Rio Camuy drainage basin, Puerto Rico: A humid tropical karst:* Pennsylvania State University Ph.D. dissertation, 345 p.

Vacher, H.L., 1988, Dupuit-Ghyben-Herzberg analysis of strip-island lenses: *Geological Society of America Bulletin*, v. 100, p. 580-591.

Vacher, H.L., and Mylroie, J.E., 1991, Geomorphic evolution of topographic lows in Bermudian and Bahamian Islands–Effect of climate: *in* Bain, R.J., ed., *5th Symposium on the Geology of the Bahamas*, Port Charlotte, Florida, Proceedings, Bahamian Field Station, p. 221-234.

Vacher, H.L., and Wallis, T.N., 1992, Comparative hydrology of Bermuda and Great Exuma Island, Bahamas: *Groundwater,* v. 30, p. 15-20.

Vogel, P.N., Mylroie, J.E., and Carew, J.L., 1990, Limestone petrology and cave morphology on San Salvador Island, Bahamas: *Cave Science*, v. 17, no. 1, p. 19-30.

Whitaker, F.F., and Smart, P.L., 1990, Circulation of saline groundwaters through carbonate platforms; evidence from the Great Bahama Bank: *Geology*, v. 18, p. 200-204.

Wilson, W.L., 1994, Morphology and hydrology of the deepest known cave in the Bahamas: Dean's Blue Hole, Long Island: in Boardman, M.R., *7th Symposium on the Geology of the Bahamas*, San Salvador Island, Proceedings, Bahamian Field Station,p. 21.

Wilson, W.L., Mylroie, J.E., and Carew, J.L., 1995, Caves as a geologic hazard: A quantitative analysis from San Salvador Island, Bahamas: *in* Beck, B.F., ed., *Karst geohazards,* Brookfield, Illinois, A.A. Balkema, p. 487-495.

5.1.1. Syngenetic Karst in Coastal Dune Limestone: A Review

Susan White

Abstract

Studies in coastal eolianite systems show that rapid speleogenesis is possible under such relatively poorly lithified conditions. Simultaneous lithification of Pleistocene carbonate dunes and the development of solutional karst features can be explained in terms of the precipitation and solution of carbonate grains within the dunes, and the characteristics and position of the water table over time. An area of such rapid subaerial syngenetic karst in southwestern Victoria, Australia, is described in terms of its caves and karst features and their speleogenetic controls.

Introduction

Most studies of karst landscapes and their processes have been concerned with competent, consolidated, and generally well-jointed limestone in areas of sufficient rainfall for solution to occur. However, limestone is an extremely variable lithology and substantial karst features, including extensive caves, can be found in less consolidated and poorly jointed rocks. Particular problems are involved in the study of karst processes in softer less compact limestones, such as chalk, coral reefs, and eolian calcarenite. Studies in coastal eolian calcarenite in Australia and the Bahamas show that speleogenesis can be more rapid than previously perceived from classical karst ideas (Jennings, 1968; Carew and Mylroie, 1987-1988; White, 1994). Schemes of rapid speleogenesis in young porous limestones such as eolianites include the simultaneous lithification of the sediment and the formation of karst landforms. Jennings (1968) discussed these problems in relation to the dune limestone of southern Australia, and he developed the syngenetic scheme of karst evolution for this relatively unconsolidated limestone. This scheme has been further developed by White (1994). Carew and Mylroie (1987-1988) have developed a similar scheme involving the Pleistocene limestone of the Bahamas.

Karst development on coastal eolianite ridges is dependent on several interrelated conditions. The dune ridge must have a sufficiently high proportion of soluble material that is pure enough for the solution processes not to be impeded by insoluble residue left after solution has occurred. The higher the proportion of carbonate, the more likely the karst features will develop. Adequate moisture must be available for karst processes to be operative. As the dunes have a high primary porosity and permeability, the percolation of water is not dependent on the presence of joints, although the dune bedding planes may also be important as percolation routes for water. The percolating water is made more aggressive by CO_2 enrichment from soil air, and this is enhanced by the vegetation covering the dunes. The development of underground karst forms is dependent on the ability of the limestone to mechanically support a cavity. Insufficient strength in the limestone will result in solutional cavities collapsing before they are very large.

The lithification of the calcarenite is an important factor in the development of karst landforms. As such dune ridges were deposited during the middle Pleistocene, and the caves have been excavated since then under conditions of higher water tables, the dates of the dunes combined with the limited lithification in the dunes indicate that diagenesis of the calcarenite is occurring now and must have been occurring since the middle Pleistocene (Carew and Mylroie, 1987-1988; White, 1994). The simultaneous nature of the lithification processes of the carbonate dunes into eolian calcarenite rock and the development of solutional karst features in the dunes is the characteristic feature of such limestone and can be explained in terms of precipitation of carbonate within the dunes.

Cementation of the limestone is due to solution of the sand-sized and finer carbonate fraction, and the redeposition of this as a secondary calcite meniscus cement. Such cementation is controlled by the rate of percolation and runoff of the water, the availability of CO_2 and the gaseous partial pressures of CO_2 in relation to the solubility equilibria. The dissolution of aragonite in the sediments, also seems important in this rapid diagenetic process. The precipitation of carbonate in dunes was described by Jennings (1968). However, his mechanism implied the drying of carbonate-rich films and capillaries of water in balance with CO_2 partial pressures. More recent chemical work (Bathurst, 1975; Bögli, 1980; Gardner and McLaren, 1993) has shown that evaporation is not necessarily important. Precipitation of calcite may occur due to variations in CO_2 partial pressure and other factors relating to the $CaCO_3$-H_2O-CO_2 chemical equilibria. The meniscus and rim cements are the earliest forms of cementation. Very little pore filling cement—supplied by the reprecipitation of dissolved aragonite—is present (Gardner and McLaren, 1993). However, it has been shown that no single cement type is diagnostic of the vadose zone (McLaren, 1993).

The lithification process involving solution by CO_2-enriched percolating water and redeposition of calcium carbonate has resulted in the formation of a hardened kankar (indurated layer or cap rock) in the dunes. This feature has been termed cap rock. although it does not necessarily occur only as a capping on the dunes. (Jennings, 1968; White, 1994). It is necessary for the origin of many karst forms, especially caves, as it gives the otherwise relatively unconsolidated calcarenite structural strength.

Syngenetic Karst in Southeastern Australia: A Case Study at Bats Ridges

Bats Ridges is an area of calcareous dune ridges in southwestern Victoria (5.1.1 Figure 1). Karst landforms are found in similar Tertiary to Pleistocene calcareous ridge systems of the Otway Basin in southeastern Australia, but the ridges at Bats Ridges show the most intensive karst development (White, 1989). The area is an excellent site for showing rapid subaerial syngenetic speleogenesis.

The caves are concentrated in a strandline calcareous dune deposited during the high sea level of the middle Pleistocene (5.1.1 Figure 2). The dune has been dated using thermoluminescence techniques on the quartz fraction at 290±37 ka (personal data). The source of the calcareous material for these dunes was the shelly Tertiary sea-floor sediment. This was exposed and subjected to vigorous wave attack (Kenley, 1976) and remobilized into dunes during low Pleistocene sea levels. In the late Pleistocene, the sea receded farther south. The dunes are now about 90-100 m above present sea level at their crest.

The main ridge has an ENE alignment. This is most likely parallel to the main shoreline of middle to late Pleistocene time. The caves are found, in the main, away from the central ridge on the side spurs of the

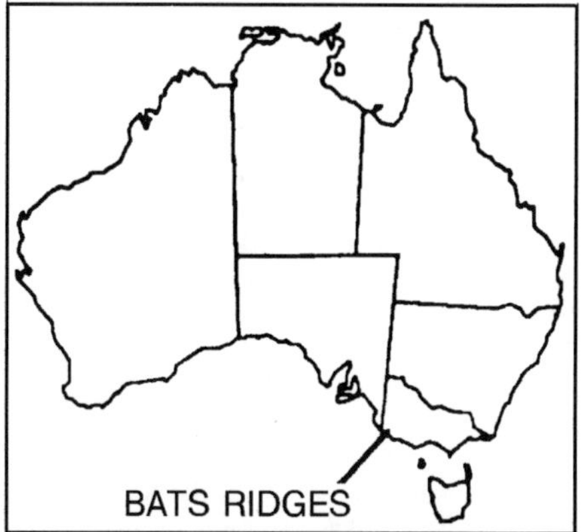

5.1.1 Figure 1. Location of Bats Ridges, Victoria, Australia

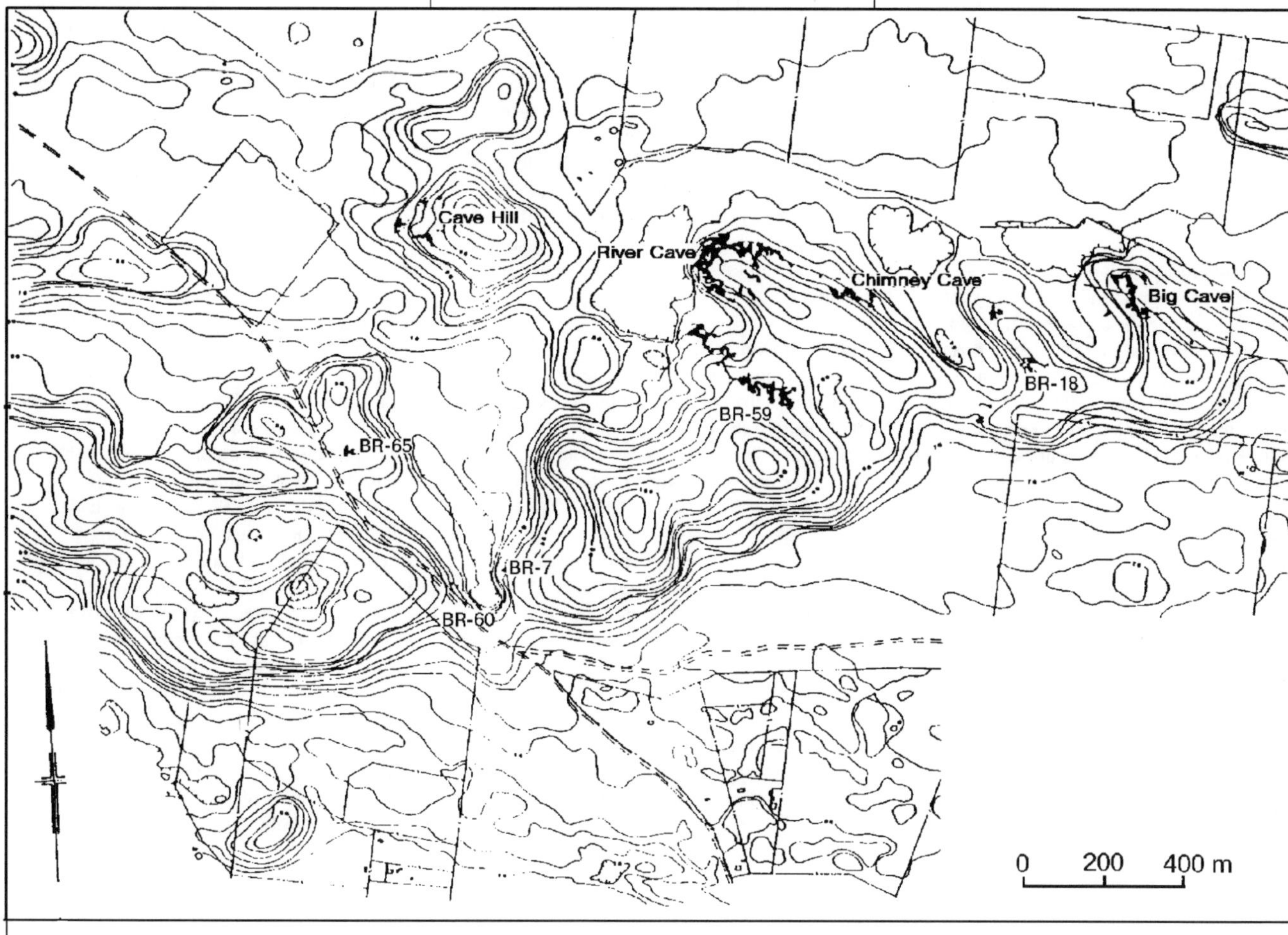

5.1.1 Figure 2. Dune ridge and cave passages, Bats Ridges.

ridge system. These dunes have developed a soil cover—derived primarily from the calcareous dune sands—and are covered by a dominant eucalyptus woodland.

Bridgewater Formation

On the basis of lithology, the Bats Ridges dunes are considered to be part of the Bridgewater Formation (Boutakoff, 1963). The rock is a well-sorted, fine- to medium-graine, bioclastic carbonate sand which is commonly laminated and variably cemented by calcite. Cross-bedding is a prominent feature. Fossil soil horizons are an integral part of the formation, but at Bats Ridges the fossil soil horizons are absent.

An important lithologic aspect is the development of an indurated layer in the original calcareous dunes. It is not clear from the nature of the feature whether this can be classed as a true calcrete. There is no clear evidence of either biological activity or pedogenic processes as described by Tucker and Wright (1990).

This kankar is necessary for the development of any major karst features such as caves, since the original sand dunes have insufficient strength to resist collapse after even relatively small amounts of material have been removed by solution. The cap rock is present near the surface of the dunes and can be seen in cave entrances where a cross section of part of a dune is obtainable. But its full extent and exact position in the dune is unknown. The thickness of the cap rock as observed in these exposures at cave entrances is about 1 m.

Chemical Composition

Thin-section and mineralogical analyses show that the chemical composition of the Bridgewater Formation is typically variable (Kenley, 1976) but generally contains 72 to 84 percent $CaCO_3$ (Coulson, 1940). At Bats Ridges the $CaCO_3$ content ranges between 73.8 percent and 92.7 percent, with between 1.5 percent and 17 percent insoluble residues.

Lithologic analyses show that the rock is largely composed of calcite bioclastic grains with a low magnesium content, and the pore spaces are partially filled with low-magnesium calcite cement. The insoluble residue consists of organic material, fine clay, and fine to medium sand-sized fragments of quartz. These fragments are subrounded to rounded and are scratched, typical of quartz grains that have been subjected to eolian processes (Krinsley and Doornkamp, 1973).

It is these quartz grains that have been dated using thermoluninescence techniques at 290+37 Ka, placing the host rock at mid Pleistocene and revealing the time frame available for speleogenesis.

Cementation

The cementation is variable and is predominantly a meniscus cement at grain contacts. In a few cases the cement fills the small pores and spaces within fossil fragments, e.g. the pores of echinodermal structures. Under subaerial conditions the presence of meniscus cement implies cementation in the vadose zone as diagenesis of the rock proceeds, especially as overprinting with later cement has not occurred (Dunham, 1971; McLaren, 1993).

The cementation has resulted in a limestone cap rock of sufficient strength to support the roof of caves formed in the dunes. However, it would be erroneous to assume that the cap rock is a hard, dense limestone. It is harder and has more compressive strength than the less cemented layer. but it is still a relatively porous rock, particularly when compared to dense limestone such as the Paleozoic limestone of East Gippsland. This primary porosity can be also be observed in thin section (White, 1994).

Generally, where the cap rock exists, caves form beneath it. However, the fresh collapses, which are a characteristic modifying feature in the caves, are not usually composed of indurated limestone. The caves retain their linear pattern despite modification by collapse. The collapsed areas often show domed roof structures which do not expose the cap rock. The indurated layer within the dune can withstand such collapse. On the other hand, cave entrances where the cap-rock exposures can be clearly seen are just below the ridges of the dunes. The cap rock then should be seen as a layer of dune limestone, formed during diagenesis which has sufficient strength to withstand complete collapse when caves and hollows are dissolved beneath it (White, 1994).

Solution Conditions

The development of caves is dependent on conditions promoting solution. These include the purity of the limestone and the availability of sufficiently aggressive groundwater under the cap rock. Much of the groundwater in the Otway Basin is saturated (or nearly so) in $CaCO_3$.

But the acidifying action of the many swampy dune swales has probably been important in the creation of conditions of undersaturation with calcite in the water. This in turn has resulted in cave formation in these dune ridges.

Surface Drainage

The Bats Ridges area has limited surface drainage and is part of a catchment draining eastward via a system of artificial drains which flow into Wattle Hill Creek. This surface water is intermittent and seasonal. In the wetter months (July to October–November) a small lake in the interdune swales generally holds water (see 5.1.1 Figure 2). This lake connects into River Cave (BR-4) and the outflow points flow into the main lake.

Prior to the diversion of water from the Bats Ridges Drain (circa 1970), surface water was rare and often only present after two successive wet winters as in 1968 and 1969 (F. Davies 1974, pers. comm.). The chemical analyses confirm that the water that is present in River Cave is surface water deflected into the cave via the northern lake from Bats Ridges Drain. Such surface water tends to be undersaturated in calcite. But it is only present for relatively short periods and is therefore not a major factor in current cave development. On the other hand, percolation water is relatively saturated in calcite and results in the deposition of calcite speleothems in the caves.

Water in the Caves

River Cave has extensive areas of passage rendered impassable by water in many wet seasons. However, this water is from the diversion from the Bats Ridges Drain into the small lake north of the BR-4 entrance mentioned above. Evidence of water having at least partially filled other cave passages can be seen in mud marks on walls. These indicate that water levels in such caves have been higher in the past. In some caves the mud has a deposit of calcite (usually moonmilk) over it, indicating that some time has elapsed since the last mud deposition. This also implies that aqueous flow through the rock is not impeded greatly by the mud.

With the exception of River Cave (BR-4), only two caves—Hut Cave and Big Caves, (BR-6 and BR-5)—have water present, even in extremely wet years. These two caves contain water either in years with exceptionally long and wet winters, or when there have been two or more wet years in succession. Both these caves are shown in 5.1.1 Figure 3 to be lower in altitude than other caves in the area (60 m above sea level).

So this observed water is probably the result of the water table rising in wet conditions. This implies that the water table under the dunes is less than 53 m above present sea level but can rise to about 60 m in wet conditions. This is comparable to water levels reported in Mines Department bores (Department of Minerals and Energy (Ground water) unpublished data). The Bridgewater Formation is a groundwater aquifer in the area with seaward water flow. The water table is between 0.4 m and 1.9 m below the lower interdune swales, still well below most areas of cave passages and therefore unlikely to be forming the observed caves.

The conclusion therefore, is that the caves show evidence of higher groundwater levels in the past than at present. The exception of River Cave (BR-4) is due to surface-water diversion through the cave and not to

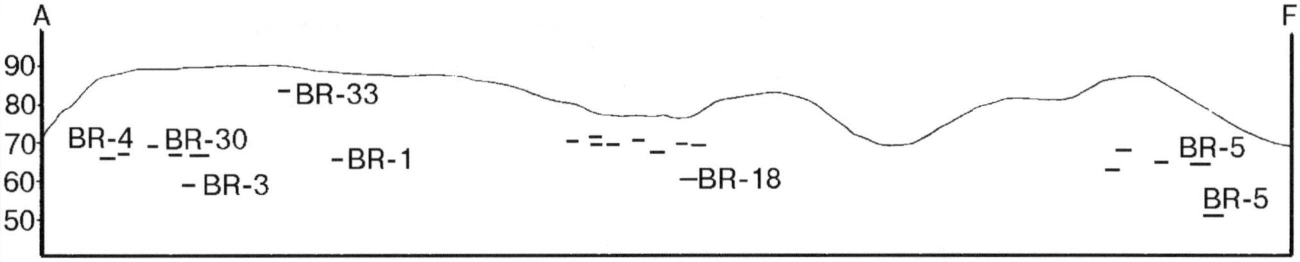

5.1.1 Figure 3. Relative cave base levels

seasonal rises in the water table. The passages of River Cave are modified by this vadose water, but the major cave formation cannot be attributed to it. The results indicate that the period of the Pleistocene relevant to the formation of the caves was probably a period of wetter conditions with higher water tables than at present.

Horizontal Development

Many caves have low, flat, wide chambers (flatteners). These show extensive horizontal development, which is in conjunction with the level of the water table. The horizontal development is unlikely to be due to preferential dissolution along the bedding planes, because of the dune cross-bedding. Caves formed under such conditions are typically linear. Successful cave development remains dependent on the strength of the overlying rock.

The presence of both solution pipes and roof avens is characteristic of caves at Bats Ridges. Solution pipes are most likely to be the result of the enlargement of a water percolation pathway. Once the pipe is enlarged enough, it becomes subjected to the processes of cementation and develops a secondary calcite cement coating. This in turn results in the solution pipe having a hard smooth surface which does not develop solution runnels or pitting. Soil-filled pipes of similar structure are present in many areas of the Bridgewater Formation (Boutakoff, 1963; Blackburn et al., 1965; Kenley, 1976). But they are absent from the Bats Ridges calcarenite. The absence of soil-filled pipes may be due to the purity of the limestone, in that there are relatively low amounts of insoluble residue present.

The Bats Ridges dunes are more soluble than some other similar dunes, and this is expressed in the formation of the caves. Similar cave development on other equivalent dunes is highly variable and seems to relate to the proportion of soluble carbonate sand present in each dune system.

Subsequent to cave formation, caves have been modified by clastic sediment deposition primarily from rock residue after solution and speleothem formation, especially moonmilk. A single speleothem has been dated by U/Th techniques and indicates that the caves were present and contained conditions conducive to the formation of calcite speleothems in the caves between 26 and 6 Ka (Paul Holgate, pers. comm.).

Caves have formed at the top of a previous water table and the karst features present at Bats Ridges are primarily the result of previous conditions favorable to cave formation. The present conditions are favorable to the modification of these features by processes present under subaerial conditions. This simultaneous lithification of the carbonate dunes into eolian calcarenite rock and the development of solutional karst features in the dunes is the characteristic feature of the speleogenesis in this area.

References

Bathurst, R.G.C., 1975, *Carbonate Sediments and their Diagenesis, 2nd ed.:* New York, Elsevier, 658 p.

Blackburn, G., Bond, R.D., and Clarke, A.R.P., 1965, Soil development associated with stranded beach ridges in south east S.A: *C.S.I.R.O. Soil Publication,* v. 22.

Bögli, A., 1980, *Karst hydrology and physical speleology:* Berlin, Springer-Verlag, 284 p.

Boutakoff, N., 1963, The geology and geomorphology of the Portland areas: *Geological Survey of Victoria Memoir,* no. 22, 171 p.

Carew, J., and Mylroie, J., 1987-1988, Rapid karst denudation, San Salvador, Bahamas: *Geo²,* v. 15, p. 14-15.

Coulson, A.L., 1940, Caves in the Portland District: *Victorian Naturalist,* v. 57, p. 37-39.

Dunham, R.J., 1971, Meniscus cement, *in* Bricker, O.P., ed., *Carbonate cements:* Baltimore, John Hopkins Press, p. 297-300.

Gardner, R.A.M., and McLaren, S.J., 1993, Progressive vadose diagenesis in late Quaternary eolianite deposits?, *in* Pye, K., ed., The dynamics and environmental context of eolian sedimentary systems: *Geological Society of London Special Publication,* no. 72, p. 219-234.

Jennings, J.N., 1968, Syngenetic karst in Australia, *in* Jennings, J.N., and Williams, P.W., eds., Contributions to the study of karst: *Australian National University Research School of Pacific Studies Publication,* no. G5, p. 41-110.

Kenley, P.R., 1976, South western Victoria Quaternary, *in* Douglas, J.G., and Ferguson, J.A., eds., Geology of Victoria: *Geological Society of Australia Special Publication,* no. 5, p. 89-153.

Krinsley, D.H., and Doornkamp, J.C., 1973, *Atlas of quartz sand and surface textures:* New York, Cambridge University Press, 91 p.

McLaren, S.J., 1993, Use of cement types in the palaeoenvironmental interpretation of coastal eolianite sequences, *in* Pye, K., ed, The dynamics and environmental context of eolian sedimentary systems: *Geological Society of London Special Publication,* no. 72, p. 235-244.

Tucker, M.E., and Wright, V.P., 1990, *Carbonate sedimentology:* Oxford, Blackwell, 482 p.

White, S., 1989, Karst features in Pleistocene dunes, Bats Ridges, western Victoria: *Helictite,* v. 27, no. 2, p. 53-7.

White, S., 1994, Speleogenesis in eolian calcarenite: A case study in western Victoria: *Environmental Geology,* v. 23, p. 248-255.

5.1.2. Speleogenesis on Tectonically Active Carbonate Islands

John Gunn and David J. Lowe

5.1.2Plate 1. Cliff face on 'Eau with cave entrance (see arrow).

Abstract

Studies of the geologically young relatively porous limestones on Tongatapu Island in the Tongan Archipelago suggest that the effect of dissolution at the interface between fresh and saline groundwater has been—and continues to be—crucial to the inception and development of underground conduits within young carbonate rock sequences. So far as it is possible to reconstruct the earliest speleogenetic events in the older preserved sequences on nearby 'Eua Island, it seems that the processes that acted upon young reefal and back-reef carbonates during the Eocene were effectively the same as the processes that have acted on subsequent deposits and are still active today.

It is commonly assumed that tectonism promotes the erosional removal of any early speleogenetic activity on carbonate islands and coasts. However, there is evidence on 'Eua to suggest that littoral cave systems, and higher level conduits that target upon them, may survive gentle uplift, or even more extreme tectonism. This raises the possibility that some of the caves that can be explored today in both tropical and extra tropical areas owe their origin to development of cavernous porosity in the littoral zone that was penecontemporaneous with rock formation.

Introduction

In Chapter 5.1, Mylroie and Carew reviewed the broad principles of cave development on carbonate islands and coasts. Their examples are from the Atlantic region of Central America. The majority are from areas that are relatively stable and in which the influences of sea-level changes in response to Quaternary glacioeustacy are clearly manifest. In contrast, we examine conduit inception and development on the Tongan Archipelago in the Pacific Ocean, where there is a long and continuing history of tectonic activity.

The Tongan Archipelago is conveniently divided into three island groups, Vava'u in the north, Ha'apai in the middle and the Tongatapu Group in the south (5.1.2 Figure 1). Although caves are known in all three groups, only those in the Tongatapu Group are discussed here. Within the island groups there is also a broad east-west division whereby the oldest and geologically most complex islands occur in the east and the youngest in the west. In the Tongatapu Group the easternmost island 'Eua has a complex geological history stretching back at least to early Eocene times. Since this time, the island's basement of igneous rocks has been submerged and uplifted several times, and a sequence of carbonate rocks of Eocene to Holocene age now encircles the basement core. During its complex history, the island has also been subjected to tectonic compression, and the rock mass is cut by faults at all scales. Most important from the speleogenesis viewpoint are several major wrench (strike slip) fractures trending between westerly and northwesterly, many related smaller wrench faults, and complementary sets of tension gashes that cut the blocks between the faults.

To the northwest, Tongatapu itself is at a stage analogous to that reached by 'Eua at the end of the Miocene or early Pliocene. Its basement is currently buried by younger rocks and lies well below sea level. A relatively subdued relief of Pleistocene rocks remains. There is no example of a still younger island northwestward within the Tongatapu Group, but only a short distance to the northwest, the gradation is completed by volcanic islands, both active and quiescent, of the Ha'apai Group. These islands represent the stage reached by 'Eua in late Eocene time. They are essentially volcanic piles, but carbonate buildups around them are assumed to be analogous to Eocene carbonates that hold most of the caves of the 'Euan uplands (Lowe and Gunn, 1986; Lowe, 1989).

If we accept the adage that "The present is the key to the past," the threefold east-west division presented by the Tongan Islands provides a rare opportunity to review the caves preserved in a relatively ancient island in the context of recently active and currently active processes in progressively younger islands. Such a full study is as yet unattainable, as juvenile carbonates surrounding volcanic islands such as Falcon, Kao, Tofua, and the recently active Metis Shoal await study by "diving geologists" to carry out a program similar to that which has produced valuable results in the Bahamas (e.g. Bottrell et al., 1990). However, examination of

Tongatapu and 'Eua alone provides an insight into the later lives of caves developed at sea level.

Cave Development in Young Porous Limestones: Tongatapu

Taylor and Bloom (1977) suggest that the bulk of modern Tongatapu developed in the lee of a small paleoisland that emerged long before the rest of the visible island formed and which now forms the highest part of the island (c. 65 m above sea level). A succession of Pliocene-Pleistocene-Holocene coral limestones that retain a relatively high primary porosity overlies a basement of Miocene volcanic rocks that in turn overlie older Tertiary rocks. It is possible that calcareous volcanic arenites on 'Eua (Lowe and Gunn, 1986) are the distal equivalent of the thick Miocene sequence preserved at depth beneath Tongatapu.

Fragments of dissolutional caves, assumed to have formed when their host rock was at sea level, are preserved in uplifted patch reefs exposed in quarries at Fu'alu Knoll in the middle of Tongatapu (written communication, D.H.R. Spennemann, Australian National University, 1988). The cave fragments (5.1.2 Figure 2) include some with a broad, arched profile similar to those preserved in more recent caves close to sea level. The fragments are related to vertical dissolutional pits, reminiscent of pits formed at the intersection of "grikes." The latter features are deduced to postdate the formation of the caves. The patch reefs themselves may be among the oldest rocks exposed on the island, and they now lie some 25 m above sea level in the Fu'alu Knoll Quarry area.

In the southeast of the island are several caves (Lowe and Gunn, 1986; Iliffe and Sarbu, 1990; 5.1.2 Figure 2) at levels at or close to 25m above sea level. Caves such as Oholei Cave, Easter Cave and Fatumu Cave exhibit a mutually similar morphology and

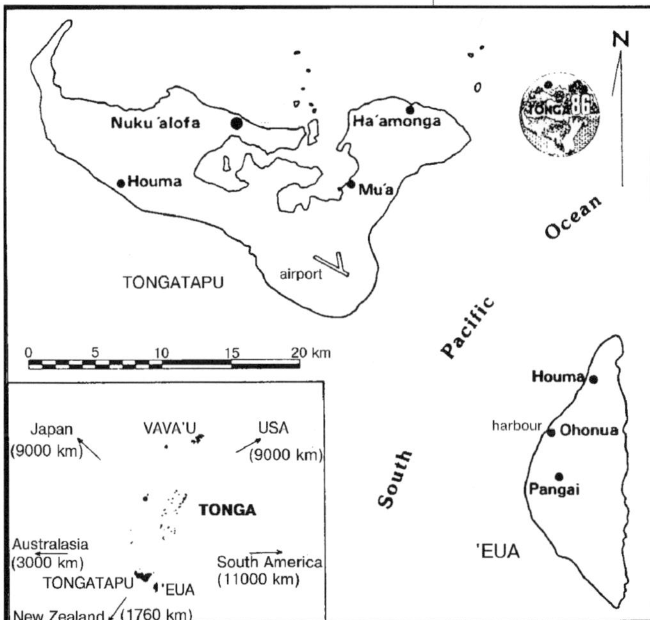

5.1.2 Figure 1. Location map of Tonga.

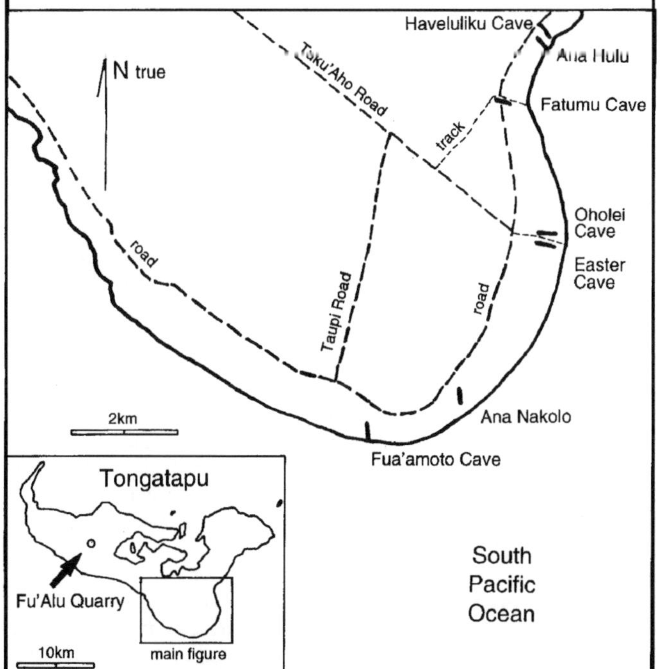

5.1.2 Figure 2. Locations of caves and other sites on Tongatapu.

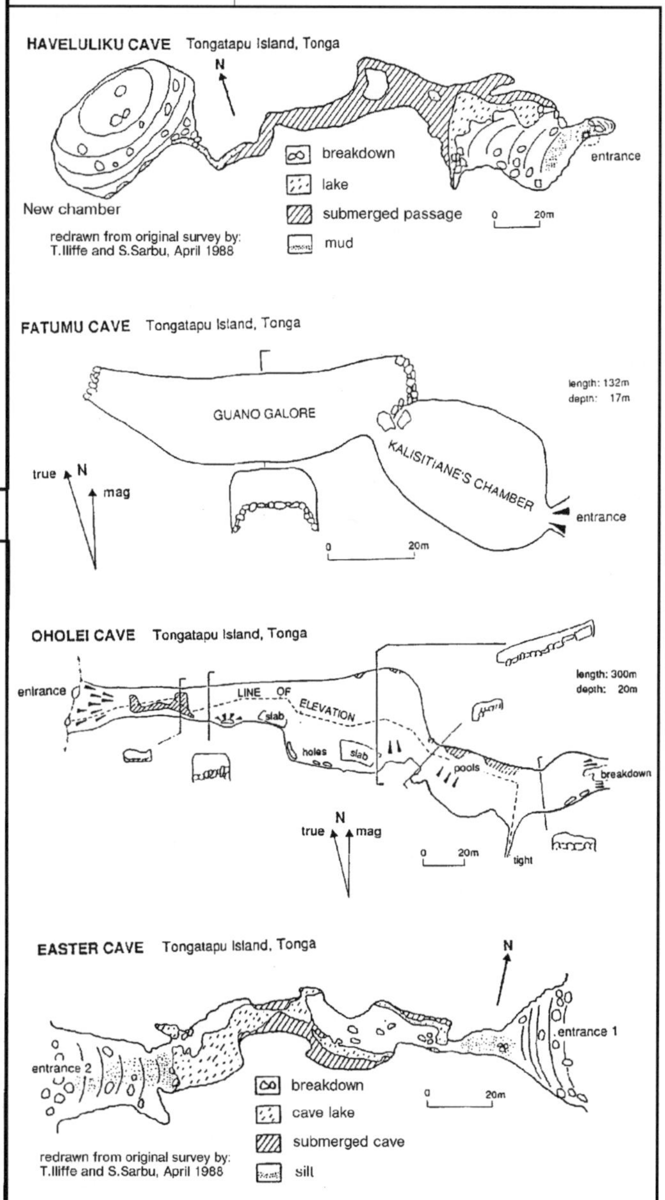

5.1.2 Figure 3. Plans of caves on Tongatapu.

locally maintain a profile similar to those exposed at Fu'alu quarry (5.1.2 Figures 3 and 4). These caves trend approximately perpendicular to the present coast and generally include a relatively deep midsection that in places includes pools of fresh water where the top of the freshwater lens is encountered. At the ends of these caves—adjacent to the coast and inland—inclined passages join the deep mid sections to the surface. Easter Cave has open entrances at both its seaward and landward ends, Fatumu Cave only at its seaward end, and Oholei Cave only at its landward end. But the lack of entrances in these latter two cases reflects blockages rather than the absence of passages.

These caves, and possibly others at this approximate elevation elsewhere on the island, probably had an origin similar to caves closer to sea level. However, they appear to have been modified to a lesser or greater degree by breakdown or foundering, in a manner similar to that discussed by Jennings (1968) on the basis of unpublished ideas presented by Hill (1957).

Caves such as Ana Hulu and Fua'amotu (Lowe and Gunn, 1986), also on the south-eastern coast, are entered close to sea level. Their morphology is similar to that of the higher-level caves. Many passage sections suggest foundering into preexisting sub-horizontal voids, but in these caves sections of uncollapsed passage survive. The caves include flooded sections that are assumed to be connected to the sea, although diving to depths of at least 20m has revealed no open routes. Another cave, with its entrance elevation between the two sets discussed above, the Haveluliku Cave (Iliffe and Sarbu, 1990), was extended by diving beyond a flooded subhorizontal section at and below sea level, to regain upward trending passage. The overall morphology is similar to that displayed by the higher level caves (5.1.2 Figures 3 and 4).

The typical long profile of most of these caves, that of an elongate and flattened "U" (5.1.2 Figure 4), is distinctive, yet its origin is problematical. It is unclear whether the steeply inclined "entrance" passages are a primary feature, whether they mirror primary slopes (though modified by breakdown), or whether their roofs represent inclined fissures due to collapse and their floors are merely of foundered debris. If they are a result of collapse, it is unclear why the stoping process should propagate through a vertical range in excess of 20 m at these localities and not elsewhere in the known passages, where smaller scale stoping is common. Hence, more probably the 'entrance' inclines are a primary or modified primary feature. They owe their origin at least in part to dissolution along the inclined dispersion zone between fresh and salt water at the base of the freshwater lens. It is tempting to suggest that the broadly symmetrical "U" profile indicates formation beneath the freshwater lens under a relatively narrow fringing reef, such as proposed by Taylor and Bloom (1977) as being an early stage in the development of Tongatapu. However, this seems unlikely because the supposed reef body would have a width of only 100 to 200 m. At this size it might not support a freshwater lens, unless under conditions far wetter than today's.

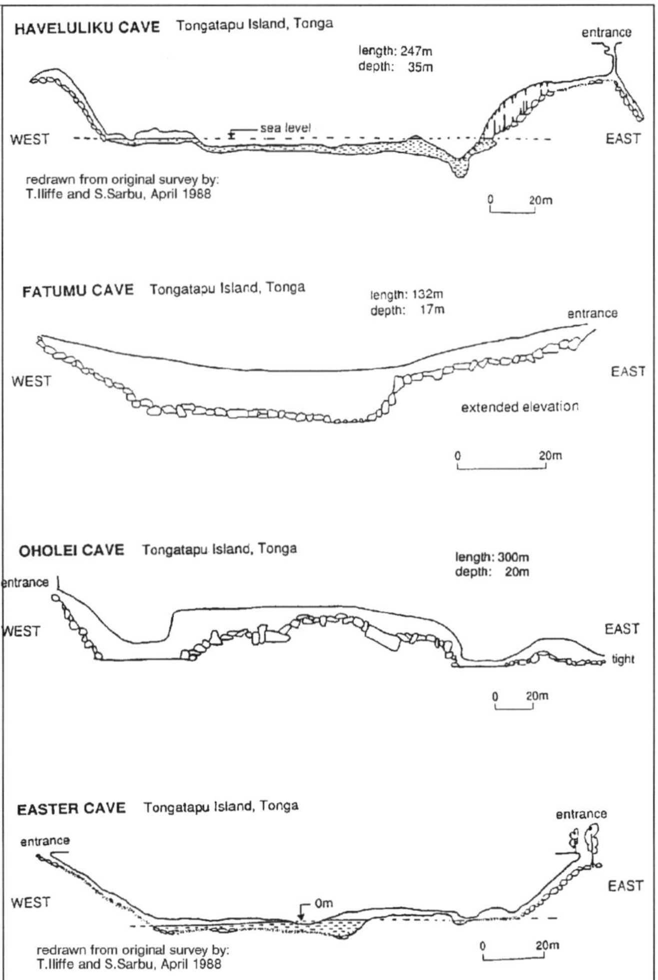

5.1.2 Figure 4. Profiles of caves on Tongatapu.

Nonetheless, the coincidence of the level of caves such as Oholei, Easter, and Fatumu with the level now occupied by the rocks attributed to this early fringing reef leaves an element of doubt.

Summing up, the caves currently known in Tongatapu represent sections on at least three "levels" that are assumed to have developed relative to separate sea-level still-stands, as described by Mylroie and Carew (Chapter 5.1 in this volume). Although isolated fragments preserved within patch-reef deposits in the interior of the island lie at a similar level to more extensive coastal caves (c. 25 m above sea level) it is uncertain if they are of the same age because the island is known to have formed and been uplifted unevenly.

Of the coastal caves, those at the highest levels have been extensively modified by breakdown and those at the lowest known level have passages at or below sea level. By comparison with areas such as the Bahamas (Mylroie et al., 1995) or the Trobriand Islands (Ollier, 1975), it would be expected that deeper dissolutional development underlies the lowest currently known passages, at the base of the freshwater lens. A lack of known resurgences on the island implies that the dispersion zone intersects the rock surface below sea level. In this context, it is tempting to wonder whether extensive "terraces" of calcite that are being deposited at sea level along part of the south coast of the island owe their origin to submarine resurgence and degassing of supersaturated water from beneath the island.

Tectonic Influences: The Caves of 'Eua

Processes similar to those currently active on Tongatapu—and the same as those also responsible for the higher cave relics—may be deduced to have been active throughout the history of 'Eua Island. Whereas the recorded history of Tongatapu is probably one of net emergence from the sea (with minor climatic oscillation), the whole of proto-'Eua was likely resubmerged due to tectonic effects at least twice since the Eocene. The evidence for resubmergence is more equivocal than the evidence for uplift, but several sets of observations add support.

Most conclusive is the paleontologic evidence for the presence of younger consolidated infill within early Eocene carbonates at the bottom of Danger Cave (Tappin and Ballance, 1991). The nature of the fill and its content of marine foraminifera with a Miocene to Holocene age range suggest that it was emplaced below sea level. Similar fills have been observed but not analyzed in

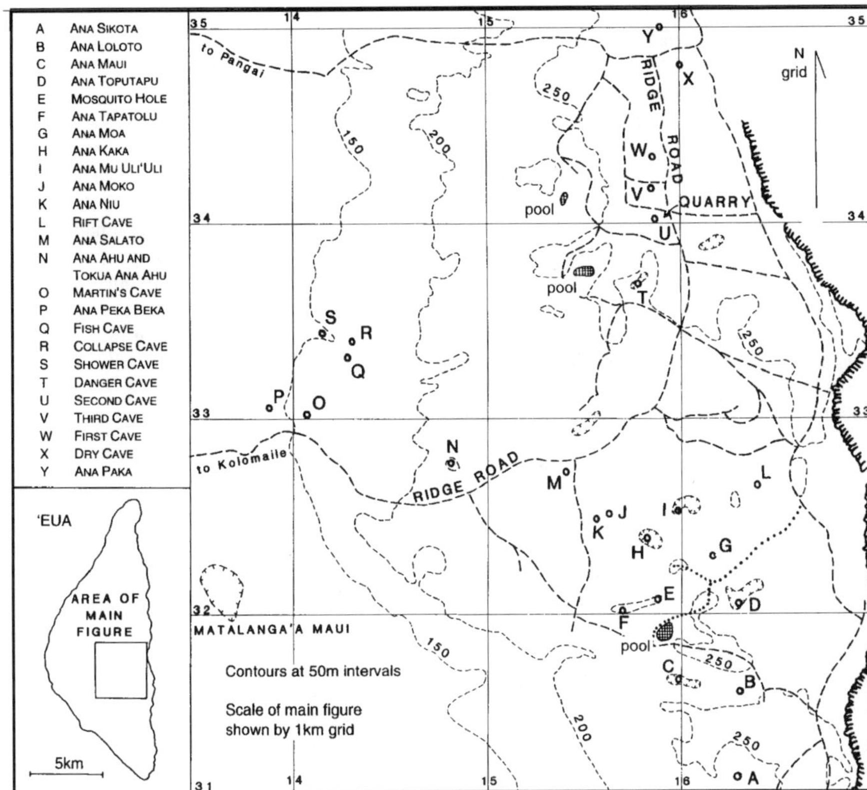

5.1.2 Figure 5. Locations of 'Euan cave sites mentioned in text.

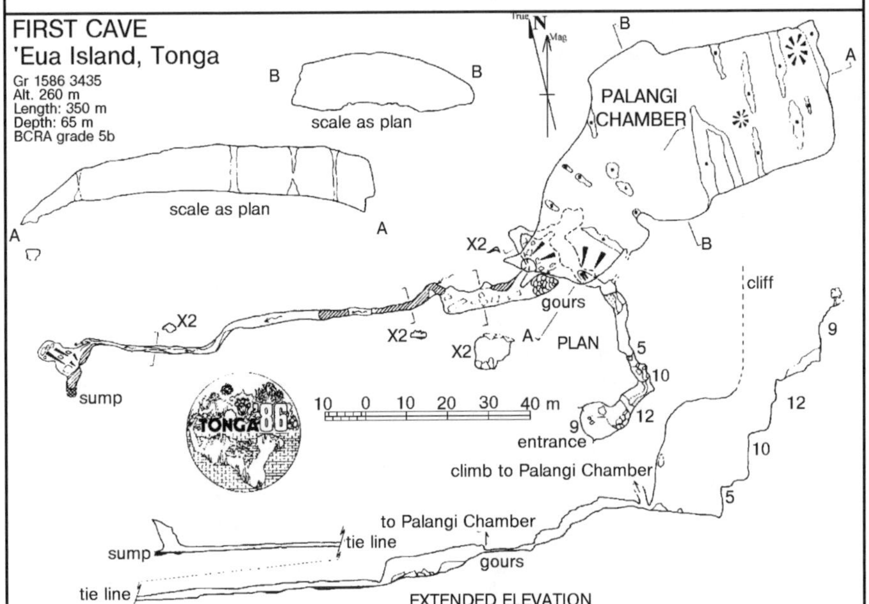

5.1.2 Figure 6. Palangi Chamber in First Cave.

several other 'Euan caves including Third Cave and First Cave (Lowe and Gunn, 1986).

In caves such as Ana Ahu (Lowe and Gunn, 1986) and Ana Maui (Lowe, 1989) similar fill has been cemented by later calcite deposits. Subsequently the fill has been reworked, and the calcite cements have in turn been partially redissolved. Whether the redissolution was achieved during later episodes of marine resubmergence or simply due to high freshwater levels caused by backup behind the clastic-calcite blockage has not yet been confirmed. Also in Ana Maui are extensive unconsolidated noncarbonate (but locally calcite-cemented) clastic deposits that range from very coarse "superconglomerate" to finely graded sands and clays. The spatial relationships of these deposits suggests that they are mass-flow deposits, akin to submarine turbidites, that were carried along existing water-filled cave passages under the influence of seismic shock waves related to uplift or subsidence of the island. Turbiditic deposits have been reported in underground situations in New South Wales by Osborne (1983).

Related to the later removal of parts of these deposits by misfit streams in some caves are local concentrations of titaniferous magnetite. These heavy mineral deposits seem to have been winnowed out of the parent deposit and redeposited in slacker water. An unproved parallel can be drawn between these magnetite sands and similar but far more extensive concentrations of "iron sand" that are found on many Pacific beaches, particularly in New Zealand.

Subhorizontal passages preserved in 'Euan caves at levels now ca. 200 m above sea level were probably conceived when their host rocks were close to sea level during the late Eocene or Oligocene. Since then they have been preserved as the island has oscillated between uplift and subsidence. During this activity the effects of tectonism have affected the pre-existing passages.

Particularly convincing are high-level fragments such as the magnificent *Palangi Chamber* in First Cave (5.1.2 Figure 6). This remnant of a huge old passage is preserved between two faults. It bears no genetic relationship to a younger, fissure-guided doline drain (First Cave itself) that has fortuitously opened access to the chamber by utilizing one of its limiting faults. To the south, a similar situation exists where the horizontal drain of Ana Maui is truncated by a major strike-slip fault, possibly continuing (but displaced laterally by ca.150 m) as a fault-bounded chamber in Ana Sikota (Lowe, 1989).

Similar isolated subhorizontal fragments are preserved at various lower levels within the main Eocene carbonate mass, but also within a variety of younger carbonates. One fragment in the main Eocene limestone is entered through a cave mouth perched c. 120 m above the sea and 20 m below the terrace surface above (see 5.1.2 Plate 1 on p. 238).

Though probably related to developments such as the main horizontal drain in Ana Maui, this particular opening is blocked by clastic debris. The altitudes of some passage fragments agree closely with those of terrace surfaces on the island's margin, particularly the well-marked *"400-foot"* terrace. However, since most of the horizontal segments are isolated due to fault activity, it is doubtful that such level-for-level correlation has any widespread validity (5.1.2 Figure 7). With or without this direct correlation there is a clear parallel between the types of cave now visible on 'Eua and the spectrum of cave types described by Ollier (1975) on the Trobriand Islands (5.1.2 Figure 8). Of these types, various subhorizontal elements may best be explained as having formed in a freshwater-lens-marginal environment similar to that deduced to have guided cave growth on Tongatapu and in the Trobriands. On 'Eua, as on Tongatapu, there is a net loss of surface water into the freshwater lens (though in the case of 'Eua the term freshwater lens may not be strictly applicable to the complex flooded zone beneath the island).

In broad terms, water sinking on the eastern side of the island is deduced to enter and remain within Eocene carbonates. No significant risings are known above sea level. Westward, water sinks within Miocene and Pliocene rocks, some rising again from the same rocks, only to be reswallowed in younger (?), topographically lower rocks. With the exception of the risings from the Miocene and Pliocene rocks, no significant resurgences are known above sea level on this side of the island (5.1.2 Figure 7).

It can thus be deduced that dissolutional activity continues beneath the island. Such processes are certainly active within relatively recent and supposedly rather porous and isotropic rocks. Whether similar recent dissolution has formed passages to submarine resurgences in the Eocene limestone on the east side of the island is unknown. All caves fully explored to date within these rocks terminate against fracture planes. We have to assume that these fractures postdate passage development and have been responsible for offsetting and isolating passage fragments, as described in the cases of First Cave, Ana Maui, and Ana Sikota above. Any water that the explored caves still transmits passes into narrow immature, and flooded fault-guided routes, perhaps to reenter former passage continuations in those situations where fault displacement has moved them to a lower relative position. Clearly such routes could be laterally extensive and could ultimately transmit the drainage back into Eocene rocks or into any younger material that is juxtaposed by faulting or by simple reef superposition. Submarine and possibly littoral resurgences have been reported on the eastern coast of the island, but they are yet to be studied geologically or hydrologically.

Finally it must be pointed out that many of the cave passages explored to date on 'Eua falls into the *Inakebu* type of Ollier (1975). These comprise vertical and subvertical fissures guided by faults and tension gashes. These may locally be linked by short sections of subhorizontal dissolutional passage, assumed to have formed within a limited number of favorable horizons. At their lowest levels most caves meet subhorizontal passages analogous to Ollier's (1975) *Bhabwatu* (upper-lens-marginal or water table) and/or *Tumwalau* (uplifted phreatic, but probably best considered as relict lower-lens-marginal) cave types. Cave fragments visible in cliff faces on the east coast of 'Eua are not true sea caves and owe their present form to erosional truncation processes identical to those discussed by Ollier of the Trobriands.

To date, however, the missing element in the overall picture of ancient freshwater lens development in the 'Euan context is the complete exploration of a subhorizontal cave segment entered by a vadose shaft system back to the surface via an incline such as is demonstrated within the caves of Tongatapu. Why such a link has not been discovered may reflect the exploratory techniques employed to date, which have concentrated upon downward exploration. In First Cave (Lowe and Gunn, 1986) and Ana Maui (Lowe, 1989) upward-trending passages have not been pursued to definite ends. In First Cave the unexplored upward passage is guided by a fault plane rather than being a simple incline parallel to an ancient lens margin as is met in the Tongatapu caves, though this structural involvement need not negate the possibility of an unexplored route back to the surface. Ana Maui was assumed to terminate in boulders in a zone of breakdown on a major fault. Cave survey data suggest that this termination is quite close to the surface, and the presence of bats in this part of the cave, remote from the known entrance, suggests that a currently undiscovered opening to the surface does exist in this area.

Conclusions

Limited evidence collected from a range of caves on two islands in the Tongan Archipelago indicates that the effect of freshwater-lens marginal mixture dissolution has been, and continues to be, crucial to the establishment of underground conduits within young,

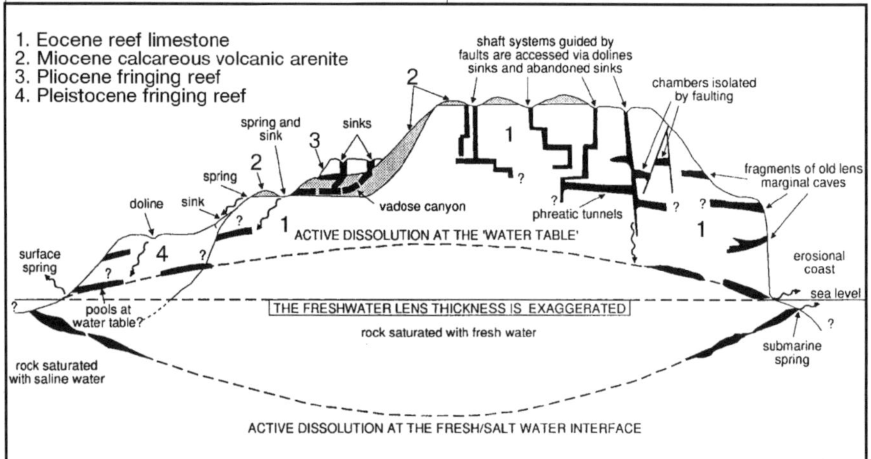

5.1.2 Figure 7. Schematic cross section showing the relationship of caves to geology and the freshwater lens.

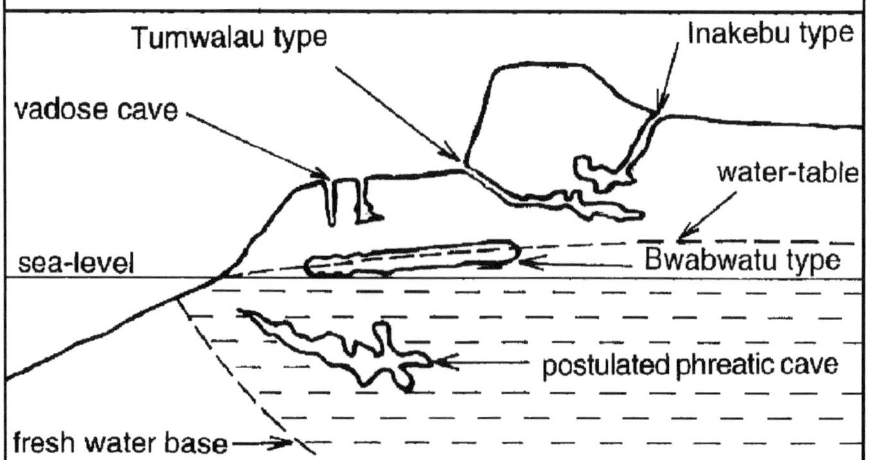

5.1.2 Figure 8. Cave passage types as recognized at the Trobriand Islands (redrawn from Ollier, 1975).

relatively porous, rock sequences. It is well documented in other areas that passages formed at the upper and lower limits of the freshwater lens within the phreatic zone may be relatively raised, due to sea-level lowering, and undergo modification within the vadose zone. Equally well documented is the common occurrence of subsequent sea-level rises during which the originally phreatic but vadose-modified passages are returned to the phreatic zone, where additional dissolution of wall rock and redissolution of speleothems occurs. In this context it is not unreasonable to consider that simple or complex oscillations of relative sea level due to the interaction of tectonic and, for example, global climatic effects can affect the history of cave development.

In Chapter 5.1 in this volume, Mylroie and Carew point out that

> ...carbonate coastlines and islands are found preferentially in the tropics and subtropics, because of the influence of climate on the deposition of carbonates... and that ...subaerial Quaternary limestone is found only in subtropical and tropical settings, and there only in oceanic or coastal environments, because plate tectonic processes have not yet placed youthful carbonates into continental interior or high latitude settings...

We do not dispute either of these points. However, we have shown from 'Eua that conduits developed at or close to sea level in young relatively porous limestone may survive tectonic uplift. By inference, they may also survive lateral migration of the rock masses, driven by plate tectonic movement. Hence, it is reasonable to assume that rock sequences, deposited and first karstified in the tropics, can "drift" out of the tropics, still carrying remnants of early speleogenesis.

Currently, little reported evidence supports the possibility that cavernous porosity developed, for instance, in the littoral zone of a subtropical Dinantian (Early Carboniferous) reef could have had any effect upon the development of cave systems that may be explored in Dinantian rocks in temperate zones today. However, there is no evidence that relationships of this kind do not exist. Closely related topics, such as the link between preexisting cavities and subsequent mineral emplacement, are gaining in credibility (Lowe, 1993; Wenrich and Sutphin, 1994; Ford, 1995; Lowe, Chapter 8.2 of this volume).

As yet, however, only a relatively few workers (e.g. Palmer and Palmer, Chapter 5.2.2 of this volume) have given serious consideration to the survival and reintegration of penecontemporaneous cave systems or inception voids into more recent cave systems. With only a few exceptions, an entire research area with potential applicability to speleoinception has been widely ignored. The question of whether sections of extant cave systems may be related to the effects of early dissolutional activity remains largely unresearched in all but a few areas.

Acknowledgments

Many of the explorations that underpinned this study were carried out by members of the Tonga '86 and Tonga '87 expeditions. We thank all expedition sponsors and team members for their invaluable support, particularly Norman Flux, whose solo reconnaissance work in late 1986 confirmed the viability of the second visit. Much of the overview presented has been built with the benefit of hindsight, and several aspects need analytical confirmation and/or rechecking in the field.

References

Bottrell, S.H., Smart, P.L., Whitaker, F., and Raiswell, R., 1990, Geochemistry and isotope systematics of sulphur in the mixing zone of Bahamian blue holes: *Applied Geochemistry,* v. 6, p. 97-103.

Ford, T.D., 1995, Some thoughts on hydrothermal caves: *Cave and Karst Science,* v. 22, p. 107-118.

Iliffe, T.M., and Sarbu, S., 1990, Anchialine caves and cave fauna of the South Pacific: *National Speleological Society News,* v. 48, p. 88-96.

Hill, A.L., 1957, *The origin of the Kelly Hill Caves, Kangaroo Island, South Australia:* Unpublished paper read to the Cave Exploration Group (South Australia), April 1957.

Jennings, J.N., 1968, Syngenetic karst in Australia, *in* Williams, P.W., and Jennings, J.N., eds., *Contributions to the study of karst*: *Australian National University Research School of Pacific Studies Publication,* no. G5, p. 41-110.

Lowe, D.J., 1989, *Tonga '87–The report of the 1987 speleological expedition to 'Eua Island, Kingdom of Tonga:* Private publication, 28 p.

Lowe, D.J., 1993, The Forest of Dean caves and karst: Inception Horizons and iron-ore deposits: *Cave Science,* v. 20, p. 31-43.

Lowe, D.J., and Gunn, J., 1986, Caves and limestones of the islands of Tongatapu and 'Eua, Kingdom of Tonga: *British Cave Research Association Transactions,* v. 13, no. 3, p. 105-130.

Mylroie, J.E., Carew, J.L., and Vacher, H.L., 1995, Karst development in the Bahamas and Bermuda, *in* Curran, H.A., and White, B., eds., Terrestrial and shallow marine geology of the Bahamas and Bermuda: *Geological Society of America Special Paper,* no. 300, p. 251-267.

Ollier, C.D., 1975, Coral island geomorphology–The Trobriand Islands: *Zeitschrift für Geomorphologie,* v. 19, p. 164-190.

Osborne, R.A.L., 1983, Cainozoic stratigraphy at Wellington Caves, New South Wales: *Linnaean Society of New South Wales,* Proceedings, v. 107, pt. 2, p. 131-147.

Tappin, D.R., and Ballance, P.A., 1991, Contributions to the sedimentary geology of 'Eua Island: Reworking of an oceanic forearc, *in* Ballance, P.A., Herza, R.H., and Stevenson, A.J., eds., *Geology and resources of island arcs–Tonga, Lau, Fiji region*: Circum-Pacific Council for Energy and Mineral Resources Earth Science Studies, p. 1-20.

Taylor, F.W., and Bloom, A.L., 1977, Coral reefs on tectonic blocks–Tonga Island Arc: *3rd International Coral Reef Symposium,* Miami, Proceedings, p. 275-281.

Wenrich, K.J., and Sutphin, H.B., 1994, Grand Canyon caves, breccia pipes and mineral deposits: *Geology Today,* v. 10, no. 3, p. 97-104.

5.2. Speleogenesis Under Deep-Seated and Confined Settings

Alexander Klimchouk

Abstract

The terms deep-seated speleogenesis, hypogenic speleogenesis, and artesian speleogenesis refer to closely related and overlapping (although not entirely equivalent) concepts. As concerns groundwater hydrodynamics, the vast majority of deep-seated and hypogenic karst develops under confined settings—or settings that are unconfined but paragenetic or subsequent to confinement. Ccrtain diagnostic features of confined groundwater circulation and deep-seated environments distinguish these conditions from those formed in unconfined settings. The last few decades have seen a growing recognition of the variety and importance of hypogenic dissolution processes and of speleogenesis under confined setting, which commonly precedes unconfined development.

Views of artesian speleogenesis are controversial and it was commonly ignored as a mechanism for cave origin. This is because the classic concept of artesian flow implies long lateral travel distances for groundwater within a soluble unit, resulting in a low capacity to generate caves within the confined area. However, recognition of aspects derived from nonclassic views of artesian flow allows a revision of the theory of artesian speleogenesis and of views on the origin of many cave types. These are:

- The role of cross-formation hydraulic communication within artesian basins.
- The concept of transverse speleogenesis.
- The inversion of the hydrogeologic function of beds in a sequence.

Under artesian speleogenesis, discharge through a cave is always hydraulically controlled. It is constrained either by the hydraulic capacity of the passages, or by the hydraulic capacity of the major confining bed or other overlying formations. In contrast to normal phreatic conditions, the discharge and enlargement rate do not increase dramatically after the kinetic breakthrough in the early evolution of conduits. Dissolution rates depend mainly on the mass balance rather than on solution kinetics during the artesian stage.

Artesian speleogenesis is immensely important to speleoinception. But it also accounts for the development of some of the largest known caves in the world and of many smaller caves. Typical conditions of recharge, the flow pattern through the soluble rocks, and groundwater aggressiveness favor uniform, rather than competing, development of conduits, resulting in maze caves where the proper structural prerequisites exist.

The most common flow pattern favoring artesian speleogenesis is upward cross-formation flow in areas of topographic and potentiometric lows. The hydrodynamic influence of prominent valleys or depressions may extend more than 1000 meters below the surface. Artesian speleogenesis and flow through soluble beds are commonly transverse, with conduit development occurring across the beds rather than laterally. Cross-formational flow favors a variety of dissolution mechanisms that commonly involve mixing. Hydrogeochemical mechanisms of speleogenesis are particularly diverse and potent where carbonate and sulfate beds alternate and within or adjacent to hydrocarbon-bearing basins.

Introduction

Traditional karst and speleogenetic theories are largely concerned with shallow and unconfined settings and ultimately imply a close hydrologic and morphogenetic relationship between the surface and sub-surface. For a long time, the common view was and still largely is that karst processes and speleogenesis commence only when soluble rocks are exposed—whether immediately after deposition or after prolonged burial—to receive recharge from the overlying surface, and that karst rapidly diminishes with depth. Features at considerable depth were commonly treated as paleokarst.

However, exposure of a soluble formation at the surface represents only a minor, although important, part of its history. According to the evolutionary approach adopted in this book, speleogenesis should be considered throughout the entire hydrogeologic cycle (see Klimchouk and Ford, Chapters 3.1 and 3.2 in this volume). It commonly—though not always—commences in a deep-seated intrastratal setting and is inherited by later shallower settings.

During the last decade there has been a rapid increase in recognizing karst development that has no apparent relation to the surface, although some works discussing hydrogeochemical mechanisms for deep karstification, and interpreting relevant features as active karst in progress, appeared in the Soviet literature in the 1950's through the 1970's (Durov, 1956; Kaveev, 1963; Maximovich, 1979; Sokolov, 1962; Stankevich, 1970; Turyshev, 1965, and others). This chapter reviews recent advances in understanding deep-seated speleogenesis. It is followed by some well-documented case studies.

Relationships Between Deep-Seated, Confined, and Hypogenic Speleogenesis

Several closely interrelated sometimes poorly defined and partially overlapping concepts and terms are relevant to deep-seated or confined speleogenesis or karst development:

- Deep-seated, intrastratal (or interstratal).
- Confined or artesian.
- Hypogenic.

Also relevant are the terms:

- Endokarst.
- Hydrothermal karst.
- Sulfuric-acid karst.

All these terms may have genetic meanings. Or they may be used simply to indicate spatial attributes. A conceptual and terminological problem was brought up during the 1991 symposium in the USSR, "Settings of karstogenesis: deep-seated karst, endokarst, hydrothermal karst" (Kungur, 1991).

To avoid confusion, we must consider at least some of these ideas and terms. Others are discussed in detail in Chapter 3.1, and are briefly reviewed here for convenience.

Deep-Seated Karst

Deep-seated karst is considered here to be developed without any exposure of the soluble rock to the overlying and closely adjacent surface and is not related to it. The term does not bear a genetic meaning.

Intrastratal karst is considered here to be developed within rocks already buried by younger strata, where karstification is later than deposition of the cover rocks. This meaning does not relate to genesis but implies stratigraphic—although not necessarily hydrogeologic—confinement of karst development by overlying nonsoluble strata (stratigraphic separation of karst from the surface) and emphasizes the evolutionary aspect (i.e. karst is later than the cover). Intrastratal karst can be deep-seated, subjacent, entrenched, or denuded (the latter represents the former intrastratal karst). This subdivision relates karst settings to the depth of erosional entrenchment and the degree of denudational exposure (see Chapter 3.1). In appropriate settings, and depending on the cover lithology, intrastratal karst can develop in confined or unconfined conditions. Deep-seated karst is commonly intrastratal. It also can be buried karst, but in such case it falls in the category of paleokarst.

Confined or Artesian Karst

Confined or artesian karst develops where groundwater is under pressure in a bed or stratum confined by a less permeable rocks or sediment above it. Head in such an aquifer lies above the bottom of the confining bed. As types of flow system and regime along with hydrogeochemistry are the major controls of speleogenetic mechanisms, the term confined karst (speleogenesis) tends to bear genetic meaning (Klimchouk, 1997b). Deep-seated karst is predominantly artesian, so this chapter focuses on examining speleogenesis in confined settings.

Hypogenic karst

Hypogenic karst is usually considered the opposite of the common epigenic karst due to carbonic acid dissolution formed by water recharged from overlying or directly adjacent surfaces. The term and concept of hypogenic (or hypogene) speleogenesis has seen increasing use during the recent decade, although still with some uncertainty as to its meaning. Two approaches appear in recent works. Ford and Williams (1989) and Worthington and Ford (1995) define hypogenic caves as those formed by hydrothermal water or by water containing hydrogen sulfide. Hill (see Chapter 5.2.8 and references therein) also tends to narrow the idea of hypogene karst and speleogenesis to H_2S-related processes and forms. Palmer (1991) defined hypogenic caves in a broader way, as those formed by acids of deep-seated origin, or epigenic acids rejuvenated by deep-seated processes. In Chapter 3.4, he presents the definition in a slightly modified and even broader form as follows:

> Hypogenic caves are formed by water in which the aggressiveness has been produced at depth beneath the surface, independent of surface or soil CO_2 or other near-surface acid sources.

This modification is important, as it allows us to include in this class features formed by still surface-independent but nonacidic sources of aggressiveness, and because it does not rely on depth as a criterion. Palmer further specifies that hypogenic caves show little or no genetic relationship to patterns of groundwater recharge from the overlying surface, although their spring outlets and levels of development may relate to present or former fluvial base levels. The above broader definition of hypogenic caves is adopted here.

A common misconception about hypogenic karst is that it is believed to be unrelated to aggressive *meteoric* circulation (e.g. Budd, Saller, and Harris, 1995). Although this can be the case in some instances—such as expulsion circulation or magmatic circulation that may also generate karst (see Chapter 3.1)—it is not necessarily true. Deep circulation of meteoric water from distant recharge areas can expose rejuvenated aggressiveness to carbonates or maintain original aggressiveness toward sulfates, generating features that fall into the hypogenic class as it is defined above.

The concept of hypogenic karst and speleogenesis clearly bears a genetic meaning, as it refers to sources of aggressiveness, as well as, indirectly, to types of circulation systems. In the form adopted here, this term nearly coincides with the concept of artesian karst. The criterion of nonrelevance to the surface recharge and sources of aggressiveness implies substantial separation of groundwater circulation from the overlying surface, i.e. some degree of confinement or rising flow.

The typical exception is upward discharge from basinal or deep-seated circulation where confining strata have locally been breached by erosion, although this can also be viewed as a special case of artesian karst development. Hydrothermal and sulfuric acid speleogenesis are variants of hypogenic speleogenesis, but are not exclusive.

Although the concepts of hypogenic karst and deep-seated karst are closely related, they are not equivalent. Hypogenic karst—though most common in deep-seated settings—can also develop in shallow exposed settings (and manifest itself at the surface) mainly where deep-seated circulation discharges upward.

Endokarst

Exokarst vs. *Endokarst.* Ezhov and Vdovin (1970), and Ezhov and Lysenin (1986), have developed some important new ideas on the global hydrodynamic zonation of underground water. These views allow a distinction to be made between *exokarst* and *endokarst* (Ezhov et al., 1991, 1992). This new interpretation remains independent of and differs drastically from the traditional meaning of these terms in the international literature. (Their conventional usage simply denotes roughly surface vs. underground karst.) The conventional usage can result in a confusion of concepts and terms. But the new views of exokarst and endokarst seem more justified, as they correlate well with broader geologic concepts.

According to Ezhov and Lysenin (1986), the principal zones distinguished within the groundwater hydrosphere are:

- The *hydrostatzone*, where fluid pressures generally exceed hydrostatic pressure by no more than 10 to 15 percent.
- The *lithostatzone*, where fluid pressures are created mainly by the lithostatic load and endogenous sources, and they approach lithostatic pressures.
- A transitional zone, the *mezozone*, lies between these two principal zones, where pressures are highly irregular spatially and temporally.

The upper part of the mezozone is marked by maximum compaction of rocks and serves as a global isolator between the two principal zones. Relaxation of extra-high pressures within the lithostatzone occurs due to upward breakthroughs of fluids from below, through the mezozone into the hydrostatzone.

The *hydrostatzone* is the realm of traditional hydrogeology, and all the well-known characteristics of meteoric and expulsion circulation, as well as of porosity and karst distribution (which generally diminish with depth) relate to it. This zone is considered the *exokarstic story,* and karst systems in it as *exokarst* (Ezhov et al., 1991, 1992).

In contrast, the *endokarstic story,* which encompasses the mezozone and lithostatzone, is characterized by decompaction of rocks due to fluid rupture and by a high degree of fissuring and solutional porosity. High temperatures and pressures in this zone cause high aggressiveness of fluids with respect not only to carbonates, but also to silicates and aluminosilicates, which have low solubilities under normal conditions.

Conditions and processes within the endokarst are as yet very poorly known, but the general concept is sufficiently well substantiated theoretically and is supported by available data. The above concept of endokarst is particularly important in the discussion of hydrothermal karst (see Dublyansky, Chapter 5.2.5 in this volume) and of karst development in quartzite. Most of this book and the present chapter, however, deal with exokarst.

Hydrothermal Karst

Hydrothermal karst and speleogenesis broadly encompasses karst circulation where the groundwater temperature exceeds 20° C (Ford, 1995; Dublyansky, Chapter 5.2.5 in this volume). However, this conventional limit includes all groundwater in deep-seated environments below a certain depth. To generate karst there, several dissolutional mechanisms can operate where temperature does not have the ultimate control. Therefore, in this broad context, the concept of hydrothermal karst does not have a genetic meaning (Ezhov et al., 1991). Most features of cave patterns, morphology, and sediments previously listed as specific to hydrothermal speleogenesis (Bakalowicz et al., 1987; Dublyanski, 1980, 1990; Ford, 1995) are in fact more specifically common for hypogenic karst—for the confined type of flow system (Klimchouk, 1997b).

However, there is a specific mechanism for dissolution by rising and cooling thermal water. Decreasing temperature in such water can maintain or create solutional aggressiveness even at decreasing CO_2 levels (Dublyansky, 1990; Palmer, 1991; see also Dublyansky, Chapter 4.1.5 in this volume). Doubling the thermal gradient allows a potential increase in fissure growth rate of approximately ten times. Steep thermal gradients also give rise to specific features of groundwater circulation, such as thermal convection. Therefore, the development of conduits in areas of steep thermal gradients represents a case of specifically hydrothermal speleogenesis that can be a basis for a narrow, yet genetic understanding of hydrothermal karst.

Concepts of Artesian Flow

The vast majority of deep-seated and hypogenic karst develops in artesian settings, or in settings unconfined but paragenetic or sequential to artesian. Specific conditions of confined groundwater circulation give rise to important features of artesian speleogenetic development that differs in many respects from that in unconfined settings.

In contrast to a water-table or unconfined aquifer, in which the water table is under atmospheric pressure, an artesian (confined) aquifer contains water at pressures greater than atmospheric. At any point, the head in such an aquifer lies above the bottom of the upper confining bed of low permeability. Any breaching of the upper confining bed, for example by a well or a fault, will cause water to flow upward to a level sufficient to balance the pressure in the aquifer (the potentiometric surface). This is mainly because water enters the aquifer at levels higher than the bottom of the confining bed through the most of the area.

Confined aquifers and artesian aquifers are commonly considered synonymous, and I adopt that usage here, although it is not strictly correct. "Artesian" was originally applied to aquifers in which the potentiometric surface lay above the ground. So the water from the aquifer would overflow naturally without being pumped. Obviously, not all the wells in a confined aquifer qualify to this at all times, especially in the downflow areas of the aquifer.

Also, in the early stages of the evolution of a confined groundwater basin, pore pressures are generated primarily due to geostatic pressure and expulsion of water from sediments. Hydrostatic conditions prevail only after considerable relief develops, and increasing meteoric circulation replaces expulsion circulation (see Klimchouk and Ford, Chapter 3.1 in this volume).

The Classic Concept

The classic concept of artesian flow implies that recharge occurs only in limited areas where aquifers crop out at the surface (usually at basin margins) and that groundwater moves laterally through separate aquifers in the confined area. This brings about a major problem in interpreting artesian speleogenesis. With a considerable distance and travel time through a soluble rock unit, water should not be capable of significant dissolution in the confined flow area. The origin of maze caves is most commonly attributed to artesian settings. But strong arguments developed from hydrodynamic (Palmer, 1975) and kinetic (Palmer, 1991) considerations question the possibility of maze-cave origin where lateral artesian flow is implied through soluble rocks . shere is a common view that "true" artesian conditions (in the sense of the classic concept) offer limited hydrodynamic and chemical potential for karstification (see Palmer, Chapter 3.4 in this volume).

The Nonclassic Concept

The nonclassic approach to hydrodynamics of artesian basins acknowledges that there are virtually no impervious rocks or sediments, just large contrasts in permeabilities. It implies hydraulic continuity in basins and close cross-formational communication between aquifers. It was introduced by the pioneering work of Mjatiev (1947) and Hantush and Jacob (1955, cited by Töth, 1995). It has been well recognized and developed in mainstream hydrogeology during the last few decades from aquifer and well data (on a local scale) and from basin hydraulics and water-resources evaluation (on a regional scale). The works by Shestopalov (1981, 1989) and Töth (1995) provide important reviews and discussion of new developments.

According to these views, recharge to and discharge from a given aquifer (vertical groundwater exchange) may take place across dividing beds throughout the whole confined flow area (5.2 Figure 1). It is controlled by the head relationships in the system. These are in turn controlled significantly by surface topography and by the presence of zones of enhanced permeability within the dividing beds (stratigraphic "windows," zones of enhanced fissuring, and the like). Adoption of the nonclassic concept has several important speleogenetic implications, that are considered elsewhere in this chapter.

Paleohydrogeology and Hydrostratigraphy of Artesian Basins

Groundwater circulation in a basin adjusts to the pattern of maximum and minimum fluid potentials. During continuous sedimentation and burial, circulation is predominantly driven by pressure heads resulting from compaction of sediments which generates an overall upward expulsion of water. With increasing anisotropy in the permeability due to late diagenesis flow acquires a considerable lateral component, although the cross-formational component still remains important.

With later uplift and expanding continental exposure, meteoric water invades the system, driven by hydraulic gradients imposed mainly by topography. In the continental regime, meteoric circulation increasingly flushes out connate and resurgent water from a basin, although compaction expulsion may still predominate in its deeper parts. In a typical basin, stratigraphically lower (older) aquifers crop out at successively higher altitudes along the basin margins, imposing greater heads. Hence the regional flow system is normally characterized by an increase in head with depth, so that ascending cross-formational flow predominates within the confined areas. However, local topographic highs can impose local flow patterns in the upper part of the sequence (5.2 Figure 2).

In the interior parts of a basin, lateral circulation in aquifers is normally slow. Communication across confining beds is even slower, commonly involving time scales of centuries or millennia. However, such movement may account for initiation of proto-conduits (e.g. Lowe, 1992). Pronounced conduit development is more likely when uplift and geomorphic differentiation substantially increase groundwater circulation rates.

The hydrostratigraphy of an artesian basin is determined mainly by the relative permeabilities of rock units. In a typical basin, aquifers are separated from each other and from any upper unconfined aquifer, by low-permeability beds. Initial permeabilities of common aquifers (e.g. many medium- to coarse-grained clastic rocks and sediments) are normally greater then those of soluble units such as massive limestones or sulfates prior to speleogenesis. Thus soluble rocks initially act as separating beds (aquitards). As late diagenesis and tectonism impose fissure permeability, soluble units increasingly transmit groundwater between "normal" (nonkarstic) aquifers in zones of sufficient hydraulic gradient. Such simplified hydrostratigraphy, where a soluble unit is sandwiched between formations with initially higher permeability, seems ideal for speleogenesis. When conduits have developed within soluble units, conventional karst wisdom views the situation as a karst aquifer sandwiched between aquitards, without recognizing that the initial conditions were quite the opposite.

Most thick soluble rock sequences include a combination of layers of varying permeability. Beds of higher initial porosity and relatively diffuse permeability (such as oolitic beds) may exist within an otherwise massive and poorly fissured carbonate sequence. They will act as aquifers, and hydraulic communication across confining beds will improve with time through speleogenesis. More complex relationships occur in thick, lithologically heterogeneous sequences composed, for instance, of intercalated carbonate, sulfate, and clastic beds. Switching of hydrogeologic functions of different beds in a sequence during the speleogenetic evolution of the soluble ones is quite common in artesian settings (Lowe, 1992; Klimchouk, 1992, 1994a, 1997b). This is because changes in karst permeability through time are much more dynamic and drastic than those in nonsoluble beds.

Hydrodynamics of Artesian Basins

The concept of hydraulic continuity in artesian basins implies considerable cross-formation communication between aquifers and hence more complex flow patterns than envisioned by the classic view of artesian aquifers. Besides marginal recharge areas and lateral flow components, this pattern includes (Shestopalov, 1981, 1989; Töth, 1995):

- Laterally alternating recharge and discharge areas (areas of correspondingly

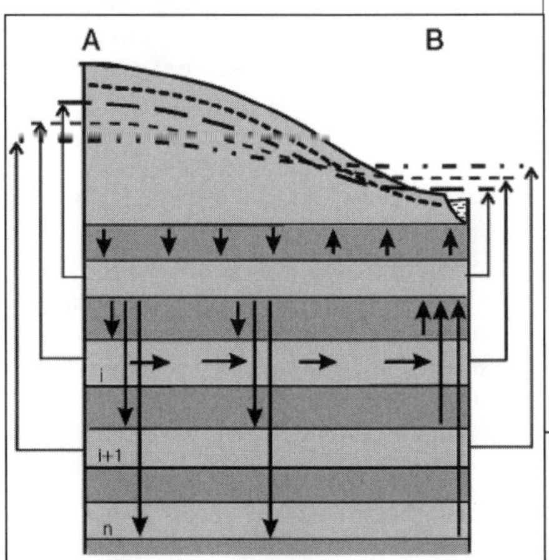

*5.2 Figure 1. Flow pattern in a multistory artesian aquifer system (from Shestopalov, 1989). **No. 1** is aquifers. **No. 2** is confining (separating) beds. **No. 3** is potentiometric surfaces of the aquifers. **No. 4** is directions of transverse flow. **No. 5** is directions of lateral flow.*

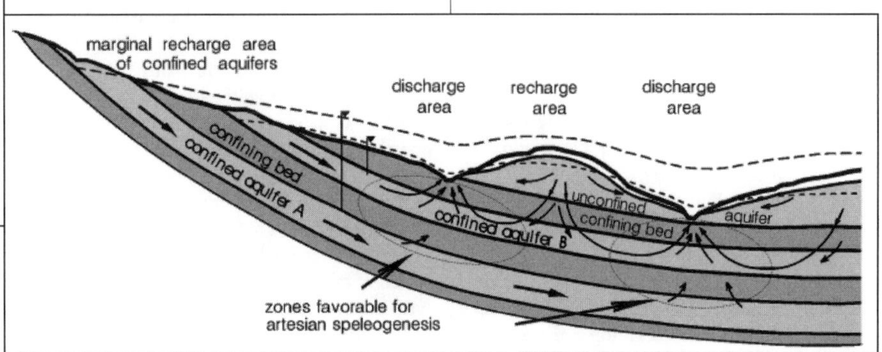

5.2 Figure 2. Flow pattern in a typical artesian basin (from Klimchouk, 1997b).

descending and ascending cross communication) that occur throughout the entire region of confined flow.
- Superposition of recharge-discharge regimes for particular aquifers in a system.
- Flow systems at different scales

The amount and direction of hydraulic communication across homogenous dividing beds of low permeability depends on the relationship of heads between adjacent aquifers (see 5.2 Figure 1). The typical patterns of recharge, discharge, and flow are that of cratonic regions with stratified sedimentary cover and low to moderate relief, where the main marginal recharge areas are not much higher than the areas of confined flow.

Mjatiev (1947) generalized that the recharge areas in an artesian aquifer are—besides the uplifted marginal outcrops—all the areas within the basin where the head is lower than in any adjacent aquifers. Discharge areas are those where the head relations between adjacent aquifers are the oposite and the combined head terms are greater than zero, and where a concave potentiometric surface is characteristic. For a given aquifer in a multistory artesian system, there is a gradual vertical transition between net recharge and discharge, which both occur simultaneously.

This is why Shestopalov (1988) termed the areas of potentiometric highs and lows respectively the areas of downward (A) and upward (B) percolation (5.2 Figure 1). Potentiometric highs correspond to topographic highs, whereas potentiometric lows coincide with topographic lows, most commonly river valleys.

In basins where stratigraphically lower aquifers crop out along marginal recharge areas and where heads are generally great (as in high-relief platforms and foreland basins), hydraulic gradients between aquifers are predominantly upward, so that upward percolation will prevail throughout the entire area of confined flow. Local topographic lows (valleys) impose increased head gradients, so that cross-formation communication and the combined discharge from the artesian system will be maximized in such areas. Basinal flow patterns of this type are represented by the Wyoming and Arizona foreland basins, as illustrated by Huntoon (1993, 1996).

In addition to hydraulic gradients, heterogeneous vertical permeabilities through the confined flow area exert a strong control over cross-formational flow between aquifers. This flow can be greatly enhanced in areas where permeabilities across a dividing bed are locally increased due to the presence of stratigraphic "windows," fault zones, and so forth. The coincidence of such areas with zones of maximum hydraulic gradient between aquifers produces the greatest effect. Such a situation is commonly represented under erosional valleys, as they normally develop along zones of weakness that also induce potentiometric lows in the underlying confined aquifers. This is why erosional valleys and other pronounced topographic lows are important indicators of zones of preferential speleogenesis, even in the deep parts of basins.

Detailed studies of regional hydraulics of artesian basins in many regions strongly support the above general views (Shestopalov, 1988, 1989; Töth, 1995). For instance, the peculiar features of the main Ukrainian basins are summarized by Shestopalov as follows:

The distribution of potentiometric highs and lows in aquifers is determined to a considerable extent by the overall surface topography. The dip of the formations is not the decisive factor in determining directions of regional confined flow. The topography, along with lithologic and structural factors, determines directions of hydraulic communication between aquifers across confining beds. Downward percolation prevails along major divides, whereas upward flow is characteristic of areas below river valleys and other topographic lows. Regional confining formations of low permeability reduce cross-formation communication within a system but do not isolate aquifers completely. The above effects can be traced to considerable depths in artesian basins. In particular, the draining influence of major rivers normally extends to depths of 1000 to 1500 m and more.

The permeabilities of confining beds generally decrease with their submergence to greater depths. But beneath river valleys, the permeabilities in confining beds increase considerably (up to 10 times) as compared to adjacent areas. This is due to the facts that the valleys are themselves developed in tectonically weakened and disturbed zones, and also due to the relaxation of rocks from the load of formations removed by valley erosion.

The rate of lateral flow decreases with the increase in depth of aquifers from the margins toward the interiors of basins. However, beneath valleys a notable increase of flow rate is characteristic even for the central parts of basins. Despite the fact that permeabilities of confining beds diminish with depth, the rate of vertical water exchange remains more stable than the rate of lateral flow, so that the relative significance of cross-formation flow increases with depth. The rate of upward percolation across confining beds under valleys is generally higher than the rate of downward percolation in the vicinity of topographic divides.

The influence of adjacent structural uplifts and massifs on the hydrodynamics of artesian basins is not only that the marginal recharge areas are located there, but that within the edges of basins there are particularly favorable conditions for enhanced vertical groundwater exchange between aquifers.

This is due to the following factors:

- Decreasing depths of aquifers and general reduction in thickness of the sedimentary cover.
- Gradual transition of clay-rich sequences to sandy-carbonate ones.
- Increase of fissuring due to weathering and unloading toward the basin edges.

The above general patterns are commonly complicated by faults, fault-cored anticlines, structural domes, and stratigraphic "windows." These features create additional, preferential paths for focused upward and downward groundwater exchange, which sometimes considerably complicate the regional hydrogeologic pattern.

Within the Dnieper-Donetzk basin (eastern Ukraine) the flow in aquifers is supported largely by vertical water exchange (up to 88 to 100 percent) rather than by lateral communication with marginal recharge areas. In the central parts of the basin, lateral inflow from the adjacent side comprises only 10 to 32 percent of the total groundwater flux (Shestopalov, 1989).

In summary: cross-formational hydraulic communication is one of the most important factors determining the resources and chemical composition of groundwater in the upper hydrogeodynamic story of artesian basins. Rates of vertical water exchange depend not only on the permeability, thickness, continuity, and number of dividing confining beds, but also on the tectonic regime of a region. The uplift trend and neotectonic activity favor cross-formational communication between aquifers.

The concept of cross-formation hydraulic communication has been largely overlooked in karst hydrogeology and speleogenetic studies. It obviously has an immense importance, and provides a broad perspective for speleogenetic implications. The characteristics of basin hydrodynamics considered above determine a location of zones particularly favorable for speleogenesis, where not only speleoinception in the meaning of Lowe (1992) is possible, but the full development of cave systems (Klimchouk, 1990, 1992, 1994a, 1997b).

Dissolution Potential of Artesian Groundwater

Sources of groundwater aggressiveness to drive speleogenesis in confined settings are mainly hypogenic and quite varied. Epigenic sources, dependent on soil CO_2 or other near-surface sources of acidity, are of limited significance, and only at close proximity to unconfined marginal recharge areas and at the top of the uppermost confined or semiconfined aquifer where percolation is downward from the overlying unconfined aquifer. Farther downflow, either laterally or verticclly, the epigenic solutional capacity of groundwater quickly ceases and needs to be rejuvenated by some process to drive dissolution again.

Hypogenic sources of aggressiveness received increasing attention from karst researchers during the last decade due to the growing appreciation of deep-seated speleogenesis (Ford and Williams, 1989; Hill, 1987, 1995; Lowe, 1992; Lowe and Gunn, 1995; Palmer, 1991, 1995; Smart and Whitaker, 1991; Worthington and Ford, 1995). The relevant geochemical processes are discussed in various chapters in this volume (Bottrell, Gunn, and Lowe, Chapter 4.1.4; Dublyansky, Chapter 4.1.5; Klimchouk, Chapter 4.1.6; Palmer, Chapter 3.4). They are briefly summarized below, particularly in light of the cross-formational flow described earlier.

Carbonic Acid Dissolution

Dissolution by carbonic acid groundwater (see Chapter 4.1.1), which is greatly dominant in epigenic environments, also operates as a hypogenic agent. However, the origin of the acidity is different: it can be related to CO_2 generated by igneous processes, by thermometamorphism of carbonates, or by thermal degradation and oxidation of deep-seated organic compounds by mineral oxidants. The latter is particularly common in oil and gas fields where waters contain high CO_2 concentrations (Kaveev, 1963). Initial P_{CO2} may reach much higher values in deep settings than that in epigenic environments, although it is less evenly distributed, being focused along local pathways of cross-communication. In view of the sluggish flow conditions and predominant mass-balance control of dissolution, creation of considerable voids by hypogenic carbonic acid depends mainly on rejuvenation of aggressiveness, e.g. by mixing or by a drop in temperature.

Hydrosulfuric Acid Dissolution

Dissolved hydrogen sulfide, originated mainly from the reduction of sulfates, forms a weak acid capable of dissolving carbonates. H_2S is a common component of basinal brines or deep meteoric groundwater. In expulsion water, both sulfates and organic matter are inherited from the original sedimentation and are dispersed, so H_2S production is also dispersed. However, generation of H_2S at considerable rates can occur only where sulfate sources are sufficient and where the produced H_2S can escape from the reducing zones. These settings are typical of the margins of sedimentary basins containing evaporate formations (e.g. Delaware Basin and Guadalupe Mountains, Texas and New Mexico, U.S.A.; Hill, 1987, 1995, 1996, Chapter 5.2.8 in this volume) or deep fault zones in tectonically active regions (e.g. Vento Cave, Frasassi, Italy). The most favorable settings for H_2S generation due to sulfate reduction are where sulfates are available for dissolution by water enriched with organic matter, the settings common within or adjacent to many oil- and gas-bearing regions containing evaporites (e.g. Delaware Basin; Western Ukraine, Klimchouk, 1994b, 1997a, Chapter 5.2.1). The above references emphasize exemplary examples of cave development.

To create large cavities, rejuvenation of the solutional capacity is needed. This occurs mainly by mixing, which can produce considerable undersaturation. Mixing of water of contrasting H_2S content in closed systems may result in even more pronounced dissolution than that from mixing of water differing in P_{CO2} because one of the water sources more commonly has a low H_2S content that provides the greatest boost in aggressiveness (Palmer, 1995; Chapter 3.4). Dissolution by hydrosulfuric acid is probably the most important speleogenetic process in deep-seated anoxic environments and may occur even at the stage of expulsion circulation. In shallower settings, within the upper hydrogeodynamic story where meteoric circulation predominates, the speleogenetic importance of H_2S is due mainly to its oxidation to sulfuric acid.

Oxidation of Sulfides to Sulfuric acid

Oxidation of H_2Sor of metallic sulfides produces sulfuric acid, a very strong dissolutional agent. The effect of sulfuric acid produced by oxidation of metallic sulfudes is generally dispersed, as is pyrite, the most common metallic sulfide. But it may be substantial enough to form caves, or to open up bedding planes (Lowe, 1992) where pyrite is localized in ore bodies or along certain horizons or bedding planes. More pronounced is the role of dissolution by sulfuric acid where H_2S-bearing water rise to interact with shallow oxygenated groundwater. Oxidation of H_2S can occur under confined conditions, where waters enriched in H_2S rise to enter a higher aquifer that conducts oxygenated water delivered laterally from the marginal recharge area. An outstanding example is the artesian belt in the gypsum karst of the western Ukraine, where H_2S generated in gypsum caves rises to flow into an overlying limestone aquifer containing oxygenated water to produce sulfuric acid and create considerable porosity, either dispersed or localized as large cavities (Klimchouk, 1997a). Perhaps some parts of morphologically complex cave systems in the Guadalupe Mountains, which are classic examples of sulfuric acid speleogenesis, were formed when groundwater in the carbonates of the Artesia Group and Capitan Formation were still locally confined although oxygenated. However, the classic interpretation attributes the main speleogenetic development to unconfined settings where H_2S oxidation and sulfuric acid dissolution culminates upward at the water table (Hill, 1987). Water table control of sulfuric acid speleogenesis is apparent where cave passages develop horizontally and discordantly to bedding and geologic boundaries.

As the sulfuric acid mechanism depends on oxygen, it is believed to be strongly limited in depth by the zone of shallow meteoric groundwater circulation. However, it can be speculated that H_2S can react with oxygen from *radiolysis*, a process occurring even at great depths (see below for details).

Cooling of Thermal Water

The inverse relationship between calcite solubility and temperature makes dissolution possible by rising and cooling thermal water even at constant CO_2 level (Dublyansky, 1980; Bakalowicz et al., 1987). The effectiveness of the process is studied in detail by Dublyansky (1990) and Palmer (1991), and the relevant speleogenetic effects are reviewed by Dublyansky (Chapter 5.2.5 in this volume). The pronounced speleogenesis attributed to this mechanism can be expected in high-gradient zones where ascending flow is localized along faults. In cases of more dispersed upward flow of thermal water this mechanism is only of secondary importance, as it is obscured by other hypogenic dissolutional mechanisms (Palmer, 1991, 1995).

Nonacidic Mechanisms

Several processes can create deep-seated solutional porosity through nonacidic dissolution or by replacement processes that produce a net increase in porosity. They may form caves by themselves or may simply initiate paths that enhance flow and trigger other cave-forming mechanisms.

Dissolution of Sulfates. Gypsum and anhydrite are the most common sulfates in intrastratal settings. They readily dissolve by simple dissociation. Gypsum solubility, quite high compared to carbonates, is further enhanced by several factors commonly operative in deep-seated environments (see Klimchouk, Chapter 4.1.6). Most notably the solubility of calcium sulfate increases by up to 3 times in the presence of various other dissolved salts, and by several times where considerable pressure is applied to the rock. More important for porosity development is that several processes common in deep-seated settings can maintain and rejuvenate groundwater aggressiveness with respect to gypsum. They are:

- Chemical reduction of sulfates, which consumes sulfate ions and enables further dissolution.
- Dedolomitization, which leads to decreasing calcium ion concentration and reaction of sulfate ions with Mg.

These processes also affect the origin of porosity in carbonates (see below) so that interbedded sulfates and carbonates are particularly favorable for speleogenesis. Situations where sulfate beds receive aggressive recharge from vertically adjacent formations to form substantial conduit systems or cavities are common in deep-seated and confined settings (see Klimchouk, Chapter 5.2.1 and Chapter 7.1).

Sulfate Reduction. The bacterial or thermal reduction of sulfates by organic compounds occurs widely in anoxic zones and is a very important process favoring deep-seated speleogenesis. For the process to occur on a large scale, considerable amounts of sulfates are needed, so that intense sulfate reduction is commonly linked with karstification in adjacent or host sulfate beds. This process rejuvenates the potential of groundwater to dissolve sulfates by removing sulfate ions, thus promoting speleogenesis in gypsum and anhydrites (Klimchouk, 1994b; 1997a). One of the byproducts is epigenetic calcite enriched in isotopically light carbon, which commonly replaces the original sulfates (e.g. the "castile buttes" in the Delaware Basin and the "Ratynsky limestone" in the western Ukraine). The increase in porosity in closed systems is ideally as much as 50 percent compared to the original gypsum and 20 percent for anhydrite, although these values are rarely achieved (Palmer, 1995). Also, hydrogen sulfide is generated, which has already been shown to be an important speleogenetic agent. Sulfate reduction is particularly effective in creating porosity within oil and gas basins where interbedded or adjacent sulfate and carbonate rocks are present and organic carbon compounds are readily available. Karstification and reduction of sulfates are dynamically related in many instances: the latter favor initiation of percolation paths and karstification in several ways while the former enhances the sulfate reduction by developing a flow pattern effective in terms of supply of reactants and outflow of reaction products (Klimchouk, 1997a).

Dedolomitization. The process of dedolomitization is believed to be an important mechanism of karstification where sulfates and carbonates occur together (Bischoff et al., 1994; Palmer, 1995; Raines and Dewers, 1997; Stankevich, 1970; see also Palmer, Chapter 3.4 in this volume). The presence of sulfate in solution forces calcite to precipitate, keeping the bicarbonate and pH low and allowing dolomite to continue to dissolve. Palmer (Chapter 3.4 in this volume) indicates that at least 1.5 times as much gypsum and 5 times as much dolomite can be dissolved by this mechanism, in comparison with the individual solubilities of these minerals. The presence of salinity, common in deep-seated environments, can drive the process still further. Raines and Dewers (1997) show that dedolomitization can only occur when gypsum approaches equilibrium under slow flow conditions, when time scales of the various mineral reaction rates approach each other. This condition is readily met in deep-seated confined aquifers.

Special Hydrogeochemical Features of the Deep Parts of Hydrocarbon-Bearing Basins

Deep-seated Water with low Total Dissolved Solids. It is commonly believed that the content of total dissolved solids (TDS) increases with depth in sedimentary basins, and that the lower hydrodynamic story is characterized by the almost universal occurrence of brine, commonly of Cl-Na-Ca or Cl-NCO_3-Na composition and with a TDS content of 200- to 40 g/L. However, studies made in recent decades reveal that in many oil and gas fields the associated deep-seated groundwater displays comparatively low TDS contents (<30 g/L, commonly 1.5-10 to 10 even <1 g/L). This inversion of geochemical zonality is observed regionally in some basins below a certain depth, although more commonly the low-TDS water form local anomalies compatible in scale with the related hydrocarbon fields. Some anomalies have column-like shapes that propagate upward from hydrocarbon bodies across different formations for many hundreds of meters (Kartzev, 1985).

According to summary publications (e.g. Kolodij et al., 1991), there can be a variety of origins for deep-seated low-TDS groundwater. Regionally such water are predominantly resurgent, generated as the result of thermal dehydration of clay-rich and hydrated minerals (Kartzev, 1985). It is shown in many works that water can be released from thick clay or evaporitic formations in great quantity and can readily account for aquifer reserves. Such water commonly have a TDS content of 1 to 6 g/L, varying chemical composition, and a high CO_2 content. Above the zones where they form such water may be localized along high-permeability pathways (Kolodij et al., 1991).

Local water of low TDS in hydrocarbon fields (within or at the bottom of productive zones) are thought to be formed by condensation and segregation as a result of fluid transport in the gas or liquid hydrocarbon phase, with later underground condensation of water, or segregation of the water-oil emulsion due to cooling (Kolodij, 1975; Kolodij et al., 1991). For such a mechanism to operate, steep temperature gradients and a large thickness of a sedimentary cover are required, as well as quick vertical migration of considerable amounts of hydrocarbons from high-temperature zones to zones of lower temperature and pressure. Waters formed by condensation and segregation, having a low TDS content and predominantly Cl-Na-HCO_3 composition, contrast strikingly with background water. Extensive theoretical, experimental, and field data support this genetic model (Kolodij, 1975; Kolodij et al., 1991).

It is also assumed that the generation of so called "organogenous" water by certain redox reactions may play a role in forming low-TDS water. In particular, such water can be produced by sulfate reduction and oxidation of hydrocarbons. However, the mass balance and hydrogeologic consideration provided by Kolodij et al. (1991) suggest that such processes are unlikely to be important, mainly because of constraints on the supply of oxidizers (sulfates).

Deep-seated low-TDS water may produce dissolutional porosity. More importantly, its mixing with background water may have a pronounced dissolutional effect. Resurgent water is forced from overpressured zones in clay or evaporite formations into overlying rocks, which may be carbonates. Mixing of resurgent water with connate water trapped within carbonates is likely to generate an effective dissolutional mechanism. When condensation and segregation of hydrocarbon fluids occur within carbonates, mixing of the resultant water with background water can be similarly effective in enhancing dissolutional porosity.

Radiolysis. Recently the geochemical role of radiolysis of groundwater has been increasingly appreciated in mainstream hydrogeochemistry but not yet in the context of karst. Many geochemical features of deep-seated water have been interpreted as caused by and related to radiolysis. This refers to decomposition of water and hydrocarbons under the impact of radioactivity (Gutsalo, 1971; Vovk, 1979a). Radiolysis of water in a closed system generates free hydrogen and highly reactive compounds such as OH, H_2O_2, and HO_2. Free oxygen is produced by further reactions between these compounds. The presence of oxygen in considerable amounts (up to 50 vol % of dissolved gasses) in water at depths of 2 to 3 km is found at many oil and gas fields. Its origin by radioactive bombardment is supported by a close relationship between the Ra and He contents. Elevated radioactivity and Ra content (1 to 3 orders of magnitude higher than that of the background) are characteristic of deep groundwater in oil and gas fields (Kartsev, 1972; Vovk, 1979b).

Vovk (1979a, b) has demonstrated that oxygen and other products of radiolysis account for several reactions involving oxidation of hydrocarbons, sulfides, and other compounds. The most pronounced effects are observed at oil-water or gas-water contacts and in close proximity to them. Radiolysis is a powerful source of acidity in deep-seated environments, and it can favor dissolution through many mechanisms. In particular, oxidation of hydrocarbons by OH and oxygen generates CO_2, and oxidation of sulfides produces sulfuric acid. Thus the dissolution of carbonates by sulfuric acid need not be limited only to shallow zones of oxygenated groundwater. Further studies are needed to assess the potential effects of radiolysis on dissolutional processes.

"Oil-field Karst." The potential effects of the above processes, combined with the effects inferred from sulfate reduction processes, which are also commonly associated with hydrocarbon deposits, give enough reasons to consider areas of gas and oil fields as particularly favorable for karstification. The concept of *sulfuric acid oil-field karst* suggested by Hill (1995) should be extended to a broader concept of *oil-field karst,* with various dissolutional mechanisms involved. The common views of karst porosity in petroleum geology as paleokarst reservoirs that simply host hydrocarbons, are gradually giving a way to a new concept of dynamically linked processes of karstification and the formation of hydrocarbon fields.

Mixing Processes and Transition Zones

It is generally accepted that solutional aggressiveness can be rejuvenated or enhanced by mixing of water of contrasting chemistry. Due to the concavity of solubility curves, mixing of water that differ in P_{CO_2}, H_2S, or salinity produces a solution of lower saturation ratio, even if the source solutions are supersaturated and incapable of further dissolution. The concept of mixing corrosion was introduced into the Western karst and speleogenetic studies by Bögli (1964), although in the Soviet literature the effect was described in detail as far back as 1939 by Laptev, who stressed its importance in the development of deep-seated karst features and solution conduits at the intersections of tectonic fissures where different water mix (Laptev, 1939). Runnels (1969), Wigley and Plummer (1976), Ford and Williams (1989), and Palmer (1991, 1995) expanded the concept by showing the effects of mixing of waters differing in salinity, H_2S, and contents of dissolved species (chemical mixing).

The role of mixing corrosion (Moore and Sullivan, 1997, use the term "mixture dissolution") under shallow unconfined continental settings is modest. It is more pronounced under bathyphreatic conditions. In common unconfined settings, most flow originates from basically the same meteoric recharge source, although it may differ in flow routes and residence time. Mixing effects are more conspicuous where waters of different natures mix, such as those having a drastic contrast in evolution and chemistry, as in coastal oceanic settings (see Mylroie and Carew, Chapter 5.1 in this volume). They commonly involve chemical mixing (the foreign-ion effect) and various redox reactions that may lead to dissolution. Similarly, or even to a greater extent, mixing corrosion is of immense importance in deep-seated confined settings.

It is useful to distinguish between *mixing-disequilibrium corrosion* (dissolution due to disruption of equilibrium by mixing) and *mixing-reaction corrosion* (dissolution due to chemical reactions between two mixing liquid phases). An example of the latter is dissolution by sulfuric acid generated where mixing between H_2S-bearing water and oxygenated water occurs. Both phenomena are common in deep-seated settings.

Under predominantly sluggish-flow conditions, groundwater within a certain rock or sediment is commonly at equilibrium with respect to a given solid phase. Groundwater commonly becomes undersaturated when it passes from one bed or formation to another of different lithology or mixes with water of differing origin and evolution.

During burial, connate water is expelled to vertically adjacent formations due to differential compaction of contrasting lithologies. At certain depths, resurgent (dehydration) waters are generated by certain sediments. These waters mix with connate water in pores and are expelled to adjacent collector formations. During continental exposure, meteoric circulation extends to increasing depth, causing mixing of meteoric water with basinal brines. Not only are different water chemistries brought into play, but contrasting geochemical environments as well. With continuing uplift and denudation, progressively deeper aquifers pass through the transition zone between the meteoric and expulsion systems. This pattern is further complicated by an interplay between regional (deeper) and local (shallower) circulation and by discordant zones of enhanced vertical conductivity, such as faults or zones of fissuring, that account for local anomalous hydraulic cross-communication and mixing.

Emphasis must be placed on the importance of cross-formational hydraulic commuication in triggering and maintaining mixing hydrogeochemical processes, and generally, in driving deep-seated karstification. Vertical circulation of groundwater (upward in deep sections, and lateral alternation between downward and upward in shallower systems) is the main cause of mixing-disequilibrium corrosion and mixing reactions. It brings in interacting water of contrasting chemistry and geo-chemical environment, and causes migration of various reactants and reaction products between horizons. The greater the contrast in the physical and chemical attributes of rocks and sediments in a geologic section, the more varied and intense are the hydrogeochemical processes and dissolutional mechanisms.

The transition zones commonly relate to prominent bedding planes or unconformities that mark sharp differences in rock type or primary porosity. Mixing corrosion can accentuate bedding planes and form adjacent layers of pervasive solutional porosity. Such transitional zones can be laterally extensive and misinterpreted as buried unconformity-related paleokarst. Mixing processes can also be imposed by the presence of stratiform or discordant bodies of unusual chemistry, such as hydrocarbon deposits, within a thick and rather homogeneous stratum. The transition zones in this case are localized around the bodies and downgradient from them. Discordant vertical hydrogeochemical anomalies related to zones of enhanced conductivity are what create the transition zones in each intercepted aquifer. The latter two types of localized transition zones may account for the formation of either pervasive porosity in local areas, or larger isolated cavities and vertical conduits (e.g. breccia pipes). In the upper hydrogeodynamic story, where confined circulation is more active, mixing dissolution may produce large caves, either dispersed or localized (see below).

Dissolution caused by mixing of contrasting water should not be confused with dissolution caused by recharge coming from an adjacent insoluble bed (commonly underlying), which is aggressive with respect to the new solid phase even without mixing. This case is rather common within the upper hydrogeodynamic story. However, the initial dissolutional capacity of water is commonly enhanced or rejuvenated by mixing effects, which may increase the differences in permeability and flow dynamics between the feeding and recipient beds.

Origin and Development of Conduits and Cavities in Deep-Seated and Confined Settings

The origin and distribution of aggressiveness described above embrace one of the three main factors that guide speleogenesis. In deep-seated confined settings, speleogenesis commences where water, aggressive with respect to a given solid phase, comes from an adjacent lithologic unit, or where groundwater becomes aggressive due to mixing or other effects such as a drop in temperature or radiolysis. Palmer (1995) noted that as sources of aggressiveness are spatially and temporally limited and flow rates are low, solution porosity tends to be localized rather than distributed over large flow distances. The karstification rate is controlled by the mass balance rather than by solution kinetics.

Other important variables guiding speleogenesis are structural factors (see Klimchouk and Ford, Chapter 3.2 in this volume) and types of flow system (see Palmer, Chapter 3.4 and sections below). An attempt is made below to integrate these factors into a model of speleogenesis under confined settings.

Confined vs. Phreatic Conditions

The term "phreatic" refers to conditions where water saturates all voids in a rock or sediment, in contrast to vadose conditions, above the water table, where voids are water filled only temporarily, if ever. Water in phreatic conduits is always confined by the host rock and possesses some hydraulic head above the conduit ceiling. This has given rise to some confusion where the terms "phreatic" and "artesian" ("confined") have been misleadingly understood as being equivalent, especially where deep-phreatic conditions are concerned.

For example, Glennie (1954) termed water rising from such deep phreatic paths "artesian." Jennings (1971, p. 97) noted that such usage is in a strict sense incorrect, but it serves as a reminder that consolidated rock can act virtually as its own aquiclude. It is necessary to limit the term "artesian (confined)" to prevailing flow conditions in a whole aquifer (or a system of aquifers where there is major geologic confinement), rather then to flow conditions within a single conduit. Use of the term "phreatic" should be restricted to the lower zone in unconfined aquifers, limited above by a water table that is free to rise and fall.

The speleogenetic importance of the distinction between phreatic and confined conditions has not been fully recognized in karst science. Within phreatic conditions, discharge through a developing flow path is governed by the resistance of the path itself, particularly in its narrowest part. Discharge increases with the growth of the conduit until the amount of available recharge begins to limit the flow (Palmer, 1984, 1991). In confined conditions, discharge through a conduit only slightly depends on its diameter, being governed mainly by resistance of the least permeable bed in the down-gradient direction (confined outflow control). This point is examined later in detail.

Transverse vs. Lateral Speleogenesis

The conventional approach to speleogenesis implies that groundwater flows laterally through an aquifer, from a recharge boundary to an output boundary. This applies either to unconfined settings (with the exception of the vadose zone) and to confined settings within the classic artesian concept. Furthermore, it is commonly implied (see, for instance, all model designs in Chapter 4.2 and elsewhere in this volume) that water flows along the long dimension of a fissure (which is commonly lateral relative to bedding), or along a pathway that combines long dimensions of several laterally connected fissures. Long flow lengths and therefore low discharge to length ratios (particularly during the early stages of speleogenesis) are inferred by such views (see Palmer, Chapter 4.2.3). Similarly, the parameter of passage length, or cave development, derived from speleologic mapping, tacitly implies the meaning of the length of flow that formed a passage. Such views represent what I call here *lateral speleogenesis*, a concept that is generally adequate when applied to unconfined settings. It is deeply rooted in the speleogenetic literature and is commonly generalized to encompass speleogenesis in confined settings, resulting in misleading implications.

It has been shown throughout this chapter that vertical cross-formation hydraulic communication is particularly important in confined settings. Where vertical commonly upward circulation occurs through a fissured soluble bed, flow actually rises along a fissure height, or along a sequence of heights of vertically connected fissures. Flow distances through a soluble rock are rather short, commonly of the order of meters or tens of meters, thus allowing rather high discharge to length ratios even under sluggish flow conditions. Where laterally continuous fissure networks are present, maps of caves formed in this way may display tens or even a few hundred kilometers of passages. The flow length is conventionally associated with the length of some laterally continuous series of passages across a cave field, which can be hundreds of meters or a few kilometers. But these figures have nothing to do with the actual flow pattern.

The term *transverse speleogenesis* is suggested here to denote solutional enlargement of fissures normal to bedding in stratified soluble rocks, driven by flow directed transversely relative to the bedding and the long dimension of fissures, between adjacent aquifer formations. The characteristic features of caves formed in such a way are numerous blind terminations of passages. Even a single fissure can enlarge to a passable size, remaining blind-terminated at both ends. If such a passage is later on encountered by a mine, it usually presents a puzzling case for a conventionally thinking karst researcher who tends to consider the soluble unit to be *a priori* the main aquifer and infer that such an apparently elongated passage must be formed by lateral flow.

Cave patterns resulting from transverse speleogenesis are strongly controlled by the pattern of permeability in a soluble bed (or a composite unit) and in the adjacent formations, as well as by the overall hydrostratigraphic arrangement. See Klimchouk and Ford (Chapter 3.2 in this volume) for a classification of fissure networks. In the cross-sectional aspect, there can be a single fissured bed of small thickness sandwiched between diffuse aquifers, in which each fissure directly connects the bottom and top boundaries. Two-dimensional (single-story) caves are formed in this situation, for example, the network maze Botovskaya Cave in Siberia (Filippov, Chapter 5.2.3). More commonly, several beds or horizons of varying character lie within a soluble unit, each bed containing a largely independent fissure network. Fissures along a given horizon are rarely coplanar with vertically adjacent fissures, but they may have vertical connections at many points. Moreover, fissure frequency may differ between horizons.

Such a discordance in permeability structure between horizons (or between the lower and upper aquifers) causes some lateral component in the generally transverse flow. Because of this lateral component, the morphology of some passages can be shaped correspondingly, giving a misleading impression of generally lateral flow through a soluble unit. Multistory (three-dimensional) maze caves with stratiform levels may form, which further favors the misleading interpretation of their development to be lateral.

The outstanding examples are the gypsum mazes in the western Ukraine. Here two to four levels of passages (fissures) are integrated into cave systems with a few 10's to more than 200 kilometers of lateral passage length, but with only few 10's of meters of actual transverse length of cave-forming flow between the lower and upper aquifers, including the lateral component within individual stories (Klimchouk, Chapter 5.2.1 in this volume).

Extensive 3-D maze caves in the 100-m-thick Pahasapa Limestone of the Black Hills, South Dakota, U.S.A. where passages are arranged into stratiform stories, also may have formed in this way (Ford, 1989). But Palmer leaves open the possibility of mixing from downward recharge through the overlying formation, rather than the opposite (see Chapter 5.2.2 in this volume). An example of more extended vertical transverse speleogenesis is given by Brod (1965; 5.2 Figure 3-II) from eastern Missouri. Here, rectilinear fissure caves and small maze clusters, developed along the bottom of a soluble unit by ascending recharge from a basal sandstone, continue upward with a succession of pits and passages. These breach the upper beds of varying lithologies and eventually provide focused discharge outlets for the artesian aquifer.

The plan aspect of caves of this kind displays the structure of fissuring even more clearly. As mentioned above, single seemingly isolated conduits can form if the inherited fissure network is fragmented. More commonly, laterally connected fissures are arranged into clusters, either small ones consisting of few elements, or large clusters that form extensive networks. Such clusters (networks) can be superimposed, as shown above. Remarkable examples of small clustered rectilinear fissure caves occur in the presently breached Neogene limestones in the Prichernomorsky artesian basin, the south Ukraine (5.1 Figure 3, IA, 1B), or recently explored caves in Tertiary gypsum beds in the Paris Artesian Basin (5.1 Figure 3 III). In both instances the caves were encountered by mines. The example from eastern Missouri also belongs to this set. Examples of extensive networks are mentioned above.

When interpreting the origin of hypogenic caves in stratified formations, the possibility of transverse speleogenesis vs. conventional lateral development should always be evaluated. The latter seems to be unlikely or rare in confined settings.

Hydraulic Conditions for Initiation and Development of Artesian Caves

The specifics of artesian speleogenesis arise from the fact that discharge through a passage is always hydraulically controlled. It is dependent on either:

- The hydraulic capacity of the passage, during the early stages of speleogenesis under internal hydraulic control.
- The hydraulic capacity of the major confining bed, during the main stage of artesian development.
- The hydraulic capacity of the overlying nonkarstic aquifer that becomes the confining bed, during the late stage of artesian development.

The two latter cases display the external hydraulic controls that can also be termed constrained-outflow control. Recharge to a cave system can be viewed as unlimited, considering that the discharge is constrained; this is by definition an inherent feature of artesian settings. Rates of conduit enlargement depend more on the mass balance than on solution kinetics (Palmer, 1995).

In contrast, in vadose and phreatic settings the hydraulic control on discharge that is characteristic of the early stages of phreatic speleogenesis, switches to catchment control at the development stage, in which the available recharge becomes the limiting factor in flow through the system (Palmer, 1984). This, along with kinetic factors, favors

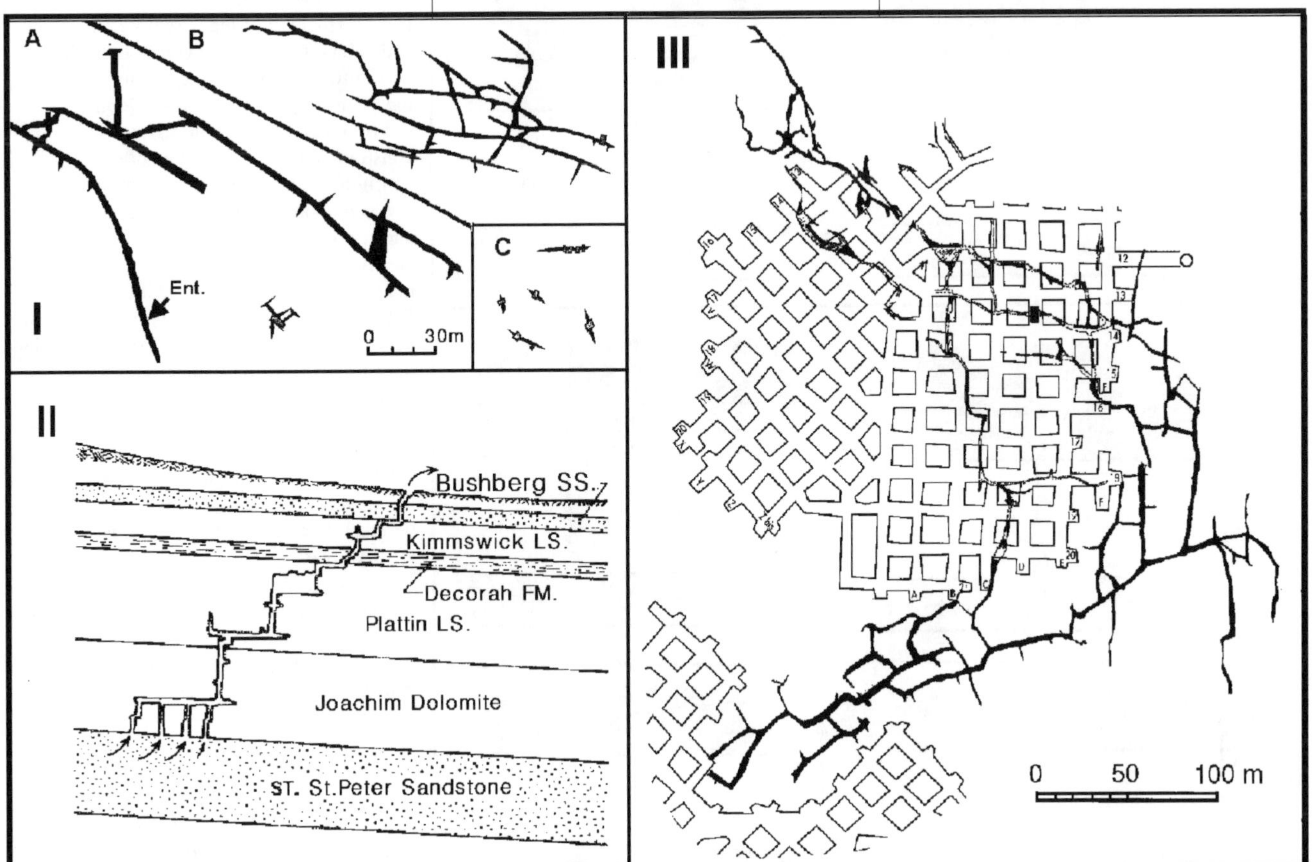

5.2 Figure 3. Typical morphologies of caves formed by transverse flow across soluble beds: ***Panel I*** *is single-fissure caves and caves guided by small clusters of connected fissures, encountered by mines in the Prichernomorsky Artesian Basin, south Ukraine (from Pronin, 1995).* ***Panel II*** *is fissure-like caves and ascending pits in eastern Missouri (from Brod, 1964).* ***Panel III*** *is the Denis Parisis gypsum cave encountered by a mine in the Paris Artesian Basin (from Beluche et al., 1996). For examples of extensive maze caves of the transverse type, see 5.2.1 Figure 6 (the western Ukrainian gypsum mazes), 5.2.2 Figures 1 and 2 (the caves of the Black Hills), and 5.2.3 Figure 3 (Botovskaya Cave in Siberia).*

competing development and selectivity of speleogenesis.

Another particular feature of artesian speleogenesis is that recharge to a soluble unit comes from the adjacent permeable but insoluble formation, not from the surface. The feeding aquifer acts as a governor to discharge through fissures in the soluble unit causing differences in their width to have little effect on discharge through fissures. As shown by Palmer (1975, 1991; see also Chapter 5.3.11) a ten-fold difference in fissure width produces only a two-fold difference in discharge. This also favors uniform growth of many alternative conduits rather than their competing development.

Consider the hydrostratigraphic architecture where two nonkarstic porous aquifers are separated by a soluble unit, with the whole system confined by an upper, insoluble aquitard (5.2 Figure 4). The soluble unit is virtually impermeable unless fissuring provides initial paths for percolation between the unit's lower and upper boundaries. Some vertically-upward head gradient exists between the aquifers, steep enough to drive slow percolation through a connected fissure path. Three cases, considered below, differ in their structural prerequisites for speleogenesis.

Basic Case: Uniform Fissuring (Jointing) in the Soluble Unit (5.2 Figure 4A). At the initiation and early development stages in unconfined phreatic conditions the width of a fissure imposes strong control over discharge and growth rate. Moreover, a positive feedback loop operates between these variables causing strong competition between alternative flow paths (Palmer, 1991; and Chapter 4.2 in this volume). In contrast, in confined conditions discharge and growth rate do not increase so readily because of the regulatory role of recharge from the adjacent formation. This makes the initiation stages generally slower than in a case of unconfined situations, suppresses hydraulic competition between alternative paths, and favors their uniform initiation.

After breakthrough, the discharge through the early conduit and the growth rate will increase to some extent, although not drastically. This causes drawdown in the lower aquifer and a depression cone to form around the fissure inlet. Discharge to the fissure will then depend on the aquifer transmissivity and storativity, the overall head gradient between the aquifers, and local conditions around the conduit inlet. But it will depend only slightly on further increases in conduit width. The head differences between the aquifers will be rapidly minimized within the depression cone. Hence, reaching a steady regime, discharge and enlargement rates increase only slightly with further growth of the conduit through the main stage of artesian development.

From then on, flow through a conduit is constrained by the resistance of the upper confining bed. Vertical conductivity of it is low and approximately constant. So the flow and rate of conduit enlargement are low and steady. Enlargement rates likely remain below the theoretical maximum calculated by Palmer for limestones (1991). However, given geologically lengthy time periods of artesian development, substantial caves can develop even in such low-dynamic conditions.

Circulation in the artesian system can be considerably enhanced in areas where vertical head gradients are large (particularly beneath incising valleys, see 5.2 Figure 2) and where permeabilities across the confining bed are locally great.

Two stages can be envisioned in the subsequent history of artesian speleogenesis, which result in a substantial increase in flow through the system and in rates of conduit enlargement (see 5.2 Figure 5).

- First, local breaching of the major confining units will cause the upper porous aquifer to become the main governor of flow. Its conductivity is much higher than that of the former confining bed, but is still much lower than the vertical conductivity of the already developed conduit system in the soluble unit. Hence the flow will increase substantially but will then stabilize at some higher value, as the overlying insoluble bed as a confining bed becomes the major constraint to flow through the conduits.
- Second, the breaching of this directly-

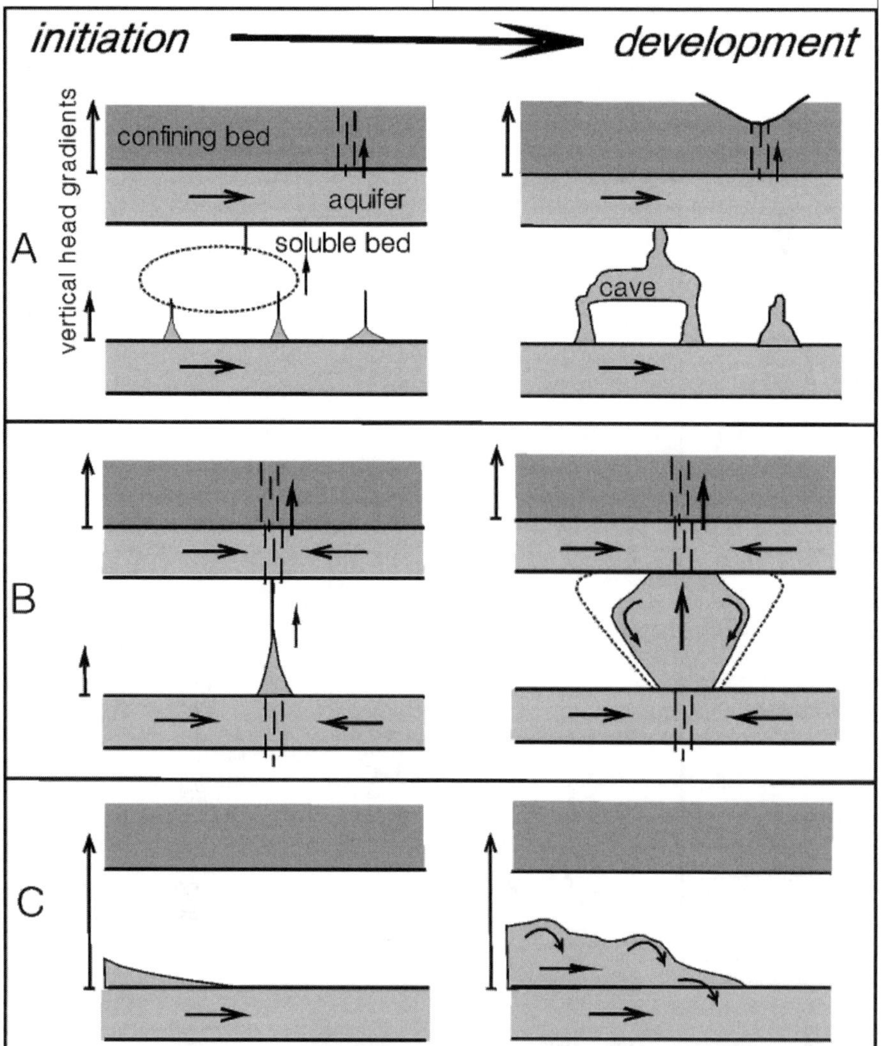

5.2 Figure 4. Initiation and development of conduits and voids under artesian conditions (from Klimchouk, 1996): ***Panel A*** *is transverse speleogenesis across uniformly distributed common fissures.* ***Panel B*** *is transverse speleogenesis across a major tectonic disruption along an otherwise low-fissured bed.* ***Panel C*** *is lateral contact speleogenesis in the case of a massive impermeable bed.*

overlying bed will finally eliminate the artesian confinement and mark a transition to common phreatic conditions. The aquifer that underlies the soluble unit will also lose its confinement, at least within the areas where conduits have evolved, because of their great vertical hydraulic conductivity.

The above two events constitute the late stage of artesian speleogenesis. The increase in flow through conduits at this stage, and in their rates of enlargement, will not lead to competing development, as many alternate conduits are already uniformly enlarged far enough to support very high discharge to length ratios and maximum enlargement rates, the main conditions of the Palmer (1991) interpretation of maze patterns. Most of the growth of artesian cave systems probably occurs during this late stage, despite its rather short duration compared to the main stage of artesian development.

Heads in the artesian system range gradually between areas of downward percolation and areas of upward percolation. During the main artesian stage, lateral flow although not substantial is focused within the adjacent insoluble aquifers, whereas it is negligible to nonexistent in the soluble unit. This is because the lateral hydraulic conductivity in porous (granular) aquifers are characteristically much greater and more homogenous than that in the still poorly karstified soluble unit. Transverse speleogenesis seems to be the ultimate mode of conduit initiation and early development under confined settings. During the late artesian stage, initial breaching of confining beds (always occurring at a local scale) increases the lateral head gradients and activates lateral flow in the whole system.

Where maturely evolved conduits in the soluble unit comprise a laterally continuous network, lateral flow focuses on that unit because its hydraulic conductivity is now much greater compared to those of adjacent aquifers. But, lateral continuity of conduit networks in transverse mazes rarely extends over a few hundred meters.

At larger scales mazes are always clustered as a result of structural, lithologic, and hydraulic conditions. Therefore, at scales of a few kilometers or more, the flow pattern in such composite aquifer systems includes lateral components that still predominate within the nonsoluble aquifers and vertical components through the soluble unit in the form of conduit clusters. (For a representative example shown by dye tracing see Klimchouk, 1997a,c). However, at smaller scales the lateral component may increase with time even within the maze clusters. If breaching of artesian confinement occurs right within the continuous cluster of conduits, the local lateral component increases drastically, and flow effectively focuses on the soluble unit.

The maze origin by diffuse recharge through overlying insoluble but permeable formations, as described by Palmer (1975; see also Chapter 5.3.11), can also be viewed as a particular example of transverse artesian speleogenesis. Two reasons can be put forward in favor of such a reinterpretation.

- First, the areally extensive exposure of insoluble cap rocks (most commonly sandstone) is in most instances, the evolutionary stage of a former artesian setting where the major confining formation has been removed by denudation. Thus the skeletal outline of maze patterns could be created according to the above scenario, with further development after the removal of a confining bed above the sandstone as Palmer suggests.
- Second, even if the hydrostratigraphy considered by Palmer is taken as a starting point, transverse speleogenesis involving high discharge to length ratios could still be the case, with downward percolation through the soluble unit within topographic highs and upward percolation beneath early valleys (compare 5.2 Figure 2 and 5.3.11 Figure 4, stage 1).

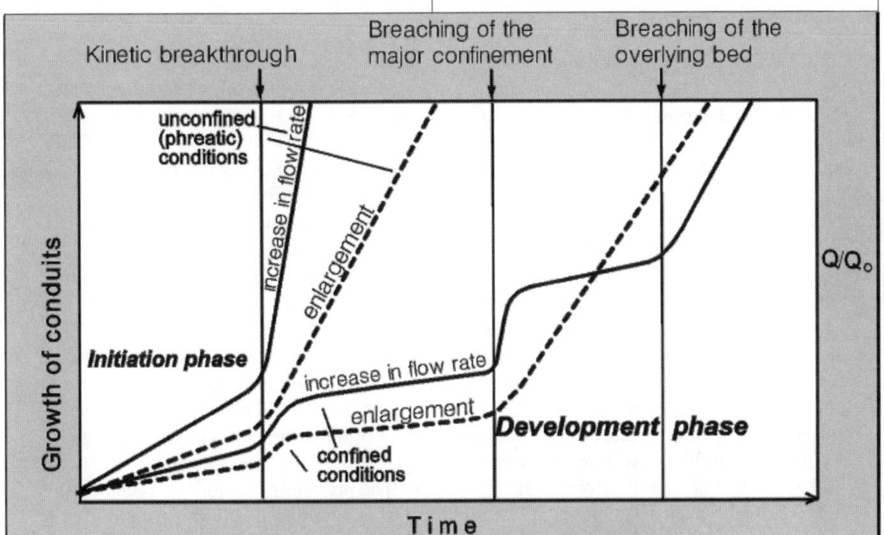

5.2 Figure 5. Schematic diagram comparing the dynamics of conduit growth through the initiation and development phases under common phreatic and confined conditions.

The possibility of transverse speleogenetic initiation in the limestone during Stage 1 (5.3.11 Figure 4) seems more realistic than lateral speleogenic initiation implied by the figure. However, the presence of an aquifer underlying the limestone becomes an important consideration in such a reinterpretation. The question under discussion is the origin of maze patterns, although Palmer rightly emphasizes that either way a substantial part of cave enlargement is performed during subsequent denuded or entrenched stages by vadose percolation or backflooding water.

Extreme Case: Tectonic Shearing of the Soluble Unit (5.2 Figure 4B). In case a fault or shear zone crosses an otherwise poorly fissured soluble bed and provides a path of high permeability, transverse flow will be focused along it. A potential distinction in comparison with the Case 1 above of uniform fissuring is that such extreme disruption commonly intersects several aquifers and confining beds below and above the given soluble unit, allowing substantial discharge through it. Breakthrough conditions can be reached fairly rapidly, perhaps even immediately, and the rate of further enlargement may achieve its maximum possible value, because a comparatively large discharge through the conduit can be maintained if abundant leakage occurs across the confining bed. Outflow from the system may be constrained on a much higher level than in the basic uniform fissuring case, although this depends on local circumstances. Either way, the development of quite large isolated voids can be envisioned, especially if a strong hydrogeochemical mechanism is involved, or the soluble bed consists of sulfates. In that case, natural convection-circulation may contribute substantially to the process (see more on this point below).

Large voids formed in the soluble bed can trigger major breakdown and the development of *vertical through structures* (VTS;

Klimchouk and Andrejchouk, 1996). These are commonly referred to as breccia pipes, collapse columns, and "geological pipe organs." These characteristic features of deep-seated karst may reach remarkable vertical extension, up to many hundreds of meters, by upward stoping across a multistory artesian system that includes soluble beds. VTS are not merely breakdown structures but complex hydrogeologic structures whose development depends on (and favors) focused cross-formational groundwater circulation and continuing dissolution of intercepted soluble beds and infallen clasts (Huntoon, 1996; Klimchouk and Andrejchouk, 1996).

Extreme Case: A Nonfissured Soluble Unit (5.2 Figure 4C). In this case the nonfissured soluble unit is virtually impermeable and does not permit transverse speleogenesis unless the unit is breached by fissuring or by dissolution due to natural convection. Instead, some lateral dissolution may occur along the contact with the adjacent aquifer. Lateral contact speleogenesis is quite common in confined settings where evaporites are underlain by an aquifer or sandwiched between aquifers. The consideration given below applies mainly to evaporites. In the case of carbonate beds, deep-seated dissolution along the contacts probably accounts only for initiation (inception) of conduits along the bedding.

The mechanisms and shapes of the initiation and development of solution conduits may vary depending on the nature of the adjacent aquifer and initial flow paths along the contact. Least likely under confined settings (in the absence of distinct lateral input and output boundaries) is the initiation of conduits along the contact between soluble and insoluble rocks.

More realistically, water comes into contact with the soluble bed from below, along the interface between it and the granular porous aquifer material, or along water-conducting fissures in the adjacent bed, which terminate at the contact with the soluble rock. In this case the dissolved load can be removed from the lithologic interface by diffusion or convection, then flowing out with the regional flow. The latter situation is most favorable for conduit initiation and development. Actual flow paths are guided by the arrangement of connected initial apertures along bedding planes, or by channels determined by intersections of fissures in the adjacent insoluble rock with the bedding contact, or by more transmissive zones in the adjacent granular material.

When an aquifer underlies an evaporite rock, and some initial space is created, further enlargement and upward growth can be driven largely by local natural convection circulation. It is likely that some large voids can develop in this way.

Speleogenetic Effects of Natural Convection

Natural convection due to density differences is an important phenomenon in deep-seated settings. Large-scale (regional) convection cells driven by density gradients (thermal or salinity) are generally recognized as playing a substantial role in past and present deep groundwater circulation in artesian basins and hydrogeologic massifs. Natural convection circulation commonly creates patterns that favor dissolution through various geochemical mechanisms. These aspects have been discussed in relation to many karst regions (e.g. Florida and the Black Hills, U.S.A.; Buda Hills, Hungary; and many coastal areas and areas of evaporite formations). The effects are particularly prominent in high-gradient zones of geothermal systems and where fresh groundwater comes in contact with evaporites.

Anderson and Kirkland (1980) generalized that the density of brines that develop when salt deposits are dissolved can cause continuous gravitational movements of brine. Quinlan, Smith, and Johnson (1987) further stressed that

> …if the source of the dissolving water is artesian or otherwise continuous, a flow cycle is developed in which the salt itself supplies the density gradient that becomes the vehicle of its own dissolution (p.83).

Similar effects, although on smaller scales, commonly operate in intrastratal gypsum dissolution.

At the local scale (the scale of a cave system or part of it), dissolution by natural convection circulation can contribute substantially to shaping the cave morphology, and even to building up a cave pattern. Curl (1966) provided a theoretical analysis of conduit enlargement by natural convection in a limestone aquifer, depicting transitional conditions that determine the prevalence of natural convection or forced flow regimes. He found that with sufficiently slow water circulation (common in artesian settings), convection caused by density differences may be the primary flow mode for limestone removal. This is made possible by even extremely small compositional differences.

Clearly the effect becomes much stronger in aquifer systems containing evaporites and low-TDS water, and where thermal gradients are involved. Particular morphological features resulting from natural convection are discussed for hydrothermal limestone caves by Bakalowicz et al. (1987), Collignon (1983), and Dublyansky (Chapter 5.2.5 in this volume), for caves in gypsum by Kempe (1972), Klimchouk (1996, 1997c, Chapter 7.1 in this volume), and in salt by Frumkin (1994; Chapter 7.2 in this volume).

They include dome-like cupolas, flat ceilings, inclined wall facets, keyhole cross-sections, and ceiling half-tubes, although each of these features can be produced by other processes. The morphologic effects are particularly prominent and diverse where low-density fresh water enters a soluble unit from the underlying aquifer (Klimchouk, 1997c; Chapter 5.2.1 in this volume). In such favorable situations the upward growth of speleoforms due to natural convection likely accounts for—or at least considerably contributes—to the early transverse speleogenesis when hydraulic gradients between aquifers are not sufficient to drive the initial flow across the soluble unit.

The Problem of Artesian Origin of Caves

The long-lasting discussion of the possibility of the artesian origin of maze caves can be resolved on the basis of a proper recognition of cross-formational hydraulic communication in artesian basins, and of transverse speleogenetic development. Works that previously accepted the artesian origin of maze caves (Howard, 1964; White, 1969; Ford, 1971) or rejected this possibility (Palmer, 1975, 1991), followed the classic concept of lateral artesian flow through a soluble unit—although Ford has adopted some elements of transverse speleogenesis by distinguishing the type of artesian caves with basal injection (1988) and by proposing a "lifting" origin for the South Dakota mazes (1989).

In contrast, Palmer (1991) showed that artesian flow in the classic sense of full confinement was unlikely to form maze caves. He examined the hydraulic-kinetic conditions within a simple loop in which water diverges into two branches that rejoin downstream, and showed that these branches will not develop at competitive rates except at very high discharge to length ratios, which would occur most likely under floodwater conditions or within short flow paths. He concluded that slow groundwater flow near chemical equilibrium, typical of confined aquifers, is least likely to produce maze caves (Chapters 3.4 and 5.2.2).

However, maze origin can be readily explained by the concept of transverse speleogenesis and nonclassic views of artesian hydrodynamics and hydrostratigraphy. Given the proper structural prerequisites for transverse speleogenesis (see above) the following hydrodynamic and chemical conditions allow and indeed favor the development of maze caves in artesian settings:

- The predominant mode of recharge is from an adjacent aquifer, commonly the underlying one. All fissures in the soluble unit receive a compatible amount of areally dispersed recharge provided the permeability structure in the feeding aquifer is at least an order of magnitude more closely spaced than that in the soluble unit (as in the case of porous siliclastic aquifers in contact with a fissured soluble unit). Variations in fissure widths have very little effect on the amount of recharge into the fissures.

- The head difference between aquifers separated by a soluble unit varies gradually through the area, also supporting uniform transverse flow through fissures.

- Water chemistry in the feeding aquifer is commonly rather uniform in the above situation. Water of uniform aggressiveness with respect to a soluble rock is delivered to each fissure open to the recharge boundary (lithologic contact). The pattern of cross-formation hydraulic communication and of transverse flow through the soluble unit generally favors dissolutional processes, as the solution is aggressive when coming into contact with the new solid phase, or because various mixing processes are triggered.

- In the evolution of a conduit, there is no dramatic increase in discharge and enlargement rate after the breakthrough. The boost in the enlargement rate after breakthrough, and consequent recharge piracy, are the major reasons why differences in breakthrough times between flow paths (caused by variations in initial widths and in the chemistry of the feeding water) determine the predominant competing conduits under unconfined phreatic conditions, resulting in branching patterns (see in this volume Dreybrodt and Gabrovšek, Chapter 4.2.2, and Palmer, Chapter 4.2.3). In contrast, the dynamics of conduit growth does not differ greatly between the initiation and development phases under artesian settings, and competing development is inhibited because of the constrained output control (Klimchouk, 1996, 1997b; Palmer, 1981, p. 14-16). As discharge is rather insensitive to conduit width, and enlargement rates are slow even after breakthrough, fissures less successful in achieving the earliest breakthrough continue to grow. Eventually, all fissures will be enlarged rather uniformly and obscure any initial differences in width.

- Transverse speleogenesis implies rather high discharge to length ratios even under sluggish flow conditions, because flow distances are short. But because of the output hydraulic control and low discharge, they can still not be sufficient to support maximum enlargement rates during the main artesian stage. When discharge increases due to local breaching of artesian confinement, all conduits readily exhibit very high discharge to length ratios and maximum enlargement rates, as they are already uniformly initiated.

All these reasons lead to this conclusion:

- Where cross-formational flow dominates, artesian conditions do characteristically produce uniform enlargement of all available fissures. This results in maze caves where proper structural prerequisites exist.

More generally, the recognition of nonclassic views of artesian flow, the role of cross-formational hydraulic communication in artesian basins, and the concept of transverse speleogenesis, brings a "second wind" to the views of speleogenesis in confined (artesian) settings. The new concept of artesian speleogenesis requires revisiting established conventional views on the origin of many caves that display uniform passage morphology and/or maze patterns. It bypasses the criticism of the artesian origin of maze patterns under truly confined conditions. It may help to resolve some still-controversial cases, such as the origin of the Black Hills caves, for which many conflicting or partially overlapping hypotheses have been produced (see overview by Palmer, Chapter 5.2.2).

Thus, for example, hydrothermal or mixing-dissolution effects, inferred for the Black Hills caves, can find their proper place within the new artesian concept. An emphasis on hydrothermal mechanisms for the development of maze caves in the Buda Hills in Hungary should not obscure the fact that these caves are attributed to a confined flow system (see Dublyansky, Chapter 5.2.6 in this volume).

The broad evolutionary approach to speleogenesis implies that some caves may inherit prior development through greatly changing settings. Hence, the problem of cave origin requires specifying the processes that are operative and the features produced during each of the main stages. The skeletal outline of a cave pattern is perhaps the most definite feature that can be attributed to certain recharge modes and flow systems (Palmer, 1991; Chapter 3.4 in this volume). As confined conditions almost inevitably pass into unconfined ones, phreatic through vadose, each subsequent setting may contribute substantially to cave development, sometimes adding the majority of a cave's volume. Unless the original pattern has altered dramatically, many caves still can be placed in the artesian class, although subsequent effects should be properly acknowledged.

The specific speleogenetic environment—paragenetic or sequential to artesian—is created by settings of marginal outflow from artesian basins to the adjacent massifs, the best examples being the Guadalupe Mountains, U.S.A. (Carlsbad Cavern and Lechuguilla Cave) and the Kugitang Mountains in Turkmenistan (Cupp-Coutunn System and other caves in the area). However, the majority of cave development in the Guadalupe Mountains is believed to be unconfined, with upward inflow of hypogenic fluids from the adjacent basin (see Hill, Chapter 5.2.8).

Even in the absence of major confinement, heterogeneity of conductivities through the thick carbonate sequence should cause some semiconfined situations, thus favoring maze patterns. Focused inputs from below and a possibility of large discharges through the systems due to low confinement explain the large volumes. The source of hypogenic fluids and of their aggressiveness in both examples was the adjacent artesian basins. Consequently, despite an apparent deviation from the "pure" artesian models, this environment is closely related genetically to the evolution of artesian basins and to the regional confined flow. Hence it can be viewed as a subclass of artesian speleogenesis.

Stages of Artesian Speleogenesis

The evolution of caves in confined settings may involve a prolonged time span and gradually changing conditions. Several major stages of artesian speleogenesis can be envisioned (Klimchouk, 1997b).

Early Artesian Stage. During continuing subsidence and burial, groundwater circulation in a basin is driven mainly by pressure heads developed by compaction and dehydration of sediments. Some speleoinception effects can operate at this stage. Substantial dissolutional porosity development is questionable, although it is likely to occur where resurgent waters are expelled from clay or evaporitic formations into adjacent soluble rocks.

Main Artesian Stage. As the tectonic regime reverses between subsidence and uplift, and geomorphic development locally exposes basinal formations, fluid flow becomes increasingly driven by hydraulic gradients. Early speleogenesis, predominantly transverse, occurs widely during this stage, even in deep-seated horizons. Under favorable conditions true conduits can develop. Flow is slow, but elapsed time spans are commonly large.

Late Artesian Stage 1. As continuing uplift brings artesian aquifers closer to the eroding surface, circulation within the entire system, and its cross-formational flow components, are increasingly activated. The full development of cave systems occurs in a pattern of upward cross-formational communication in zones of potentiometric lows, where valleys are incised into the upper confining bed. Switching of hydraulic functions occurs between the initial porous aquifers and soluble units as speleogenesis proceeds. Deposition of fine clastic sediments, redistributed from occasional cavern breakdown, may begin at this stage.

Late Artesian Stage 2. Breaching of the major artesian confinement at local discharge areas dramatically increases the flow through the system and enhances cave development within established karst zones. However, output discharge is limited until the directly overlying permeable bed is breached within a given karstified zone. This is the last stage of artesian speleogenesis.

Further entrenchment of major valleys and accompanying partitioning of the major confining bed and other capping formations leads to radical reorganization of recharge-discharge configurations and inversion of flow patterns. Artesian cave systems become relict. Several models of speleogenesis and karst morphogenesis address these open karst settings, but commonly they fail to explain relict artesian systems adequately. These models interpret the superimposed cave development in terms of phreatic, water table, and vadose flow conditions. They tend to overlook the possibility that a previous artesian history could have influenced their development.

The succession of stages outlined above applies only to a given formation. Within a given basin, different stages may occur simultaneously in different formations, at different depths, or in different parts of the basin.

Major Features of Artesian Caves

These can be summarized as follows:

- The frameworks of basic cave systems show no genetic relationship to modern landscapes. However, in the context of paleogeomorphology, active and significant cave growth is normally induced by and converges toward valleys incising into upper confining formations.
- Cave patterns display strong structural guidance, and passages within a given geologic and hydrostratigraphic position are quite uniform in size and morphology. Two- or three-dimensional (multistory) rectilinear mazes are typical, with many blind-terminated passages. However, prominent isolated cavities or extended vertical features are also characteristic of artesian settings where transverse flow is focused along rare but highly effective paths.
- Ceiling cupolas or half-tubes that originated by the action of natural convection cells or currents are common medium-scale features.
- Clastic cave sediments are represented mainly by fine silts and clays. This can be partly autochthonous (comprising insoluble residues) although it is mainly allochthonous sediment carried into artesian systems from overlying formations only during the late artesian stages. Breakdown is induced by a decrease in hydrostatic pressure, and increasing flow velocities allow some transport and redistribution of clastic material. However, hydraulic gradients in the aquifer remain much more uniform, and the energy of water flow remains much lower than in unconfined settings, so that sediments are fine grained. Whereas sediments in unconfined phreatic and especially vadose caves are characterized by great variation over short distances and by rapid facies changes, artesian caves normally contain sediment that is much more uniform, and display similar facies even on a regional scale.
- Speleothems commonly although not inevitably include "exotic" hydrothermal minerals, or minerals deposited as the products of redox reactions that are typical in the transitional zones.

Many of the above features were suggested by V. Dublyansky (1980) and Y. Dublyansky (1990). They have been widely cited (e.g. Bakalowicz et al., 1987; Ford, 1995) as being characteristic of hydrothermal speleogenesis. However, from the above considerations, they clearly are characteristic not of a particular set of physical (thermal) conditions or dissolutional mechanisms, but of a specific (confined) type of flow system and regime. Deep basinal flow is commonly heated, and its upward discharge generally creates thermal anomalies. As noted above, the hydrothermal karst concept is largely (although not completely) encompassed within the artesian karst concept.

Conclusions

The terms "deep-seated," "hypogenic," and "artesian" karst and speleogenesis refer to closely related and largely overlapping concepts, although they are not entirely equivalent. But most deep-seated and hypogenic speleogenesis occurs under confined hydrogeologic settings.

Adoption of the nonclassic concept of artesian flow and the concept of transverse speleogenesis gives a wide perspective for development of the new theory of artesian karst and speleogenesis. The theory is adequate to explain the hydraulic and chemical mechanisms involved, as well as the peculiar distribution, patterns, and morphology of resultant caves. The main specifics of artesian speleogenesis arise from the fact that recharge comes to a soluble unit from the adjacent formation and discharge through a cave is constrained by hydraulic properties of the feeding bed and major confining bed or other overlying formation. Dissolution rates depend mainly on the mass balance rather than solution kinetics. Confined conditions with transverse flow characteristically produce uniform enlargement of all available fissures, resulting in maze caves where proper structural prerequisites exist.

Artesian speleogenesis occurs in rocks of varied lithology. It involves various chemical dissolution mechanisms that operate under different physical conditions, although resulting in similar cave features. This indicates clearly that the main factor responsible for the genetic specifics of this kind of speleogenesis is the type of flow system, not a single chemical or physical characteristic of the genetic environment.

Artesian speleogenesis can have an immense importance to speleoinception. It accounts for some of the largest known caves in the world, as well as many smaller caves. The majority of presently unconfined karst rocks have experienced more or less prolonged episodes of basinal development before undergoing major uplift and exposure. Thus, the possibility that artesian karstification has preceded unconfined karst development should never be overlooked.

Recognition of the scale and importance of artesian speleogenesis and of the hydraulic continuity and cross-formational communication between aquifers in artesian basins is indispensable for a correct interpretation of speleogenetic processes (evolution of karst aquifers) and associated phenomena, regional karst water-resource evaluations, and the genesis of certain karst-related mineral deposits. It has numerous other theoretical and practical implications, and ultimately requires a modification of the traditional paradigm of karst science, which is based largely upon concepts of unconfined karst.

Acknowledgements

I thank Derek Ford, Arthur Palmer, and Margaret Palmer for their review of this paper and valuable comments. I am also indebted to Arthur and Margaret Palmer for their careful editing of my English, which helped to make the paper more clear and consistent.

References

Anderson, R.Y., and Kirkland, D.W., 1980, Dissolution of salt deposits by brine density flow: *Geology*, v. 8, p. 66-69.

Bakalowicz, M.J., et al., 1987, Thermal genesis of dissolution caves in the Black Hills, South Dakota: *Geological Society of America Bulletin*, v. 99, p. 729-738.

Beluche, F., Kens, J., and Teyssier, D., 1996, Le reseau Denis Parisis, Bethemont-la-Foret, Val-d'Oise: *Spelunca*, v. 63, p. 31-37.

Bischoff, J.L., et al., 1994, Karstification without carbonic acid: Bedrock dissolution by gypsum-driven dedolomitization: *Geology*, v. 22, p. 995-998.

Bogli, A., 1964, Mischungskorrosion-ein Beitrag zur Verkarstungsproblem: *Erdkunde*, v. 18, p. 83-92.

Brod, L.G., 1964, Artesian origin of fissure caves in Missouri: *National Speleological Society Bulletin*, v. 26, p. 83-114.

Budd, A.D., Saller, A.H., and Harris, P.M., 1995, Foreword, *in* Budd, A.D., Saller, A.H., and Harris, P.M., eds., Unconformities and porosity in carbonate strata: *American Association of Petroleum Geologists Memoir*, no. 63, p. 7-12.

Collignon, B., 1983, Speleogenese hydrothermale dans les Bibans (Atlas Tellen-Nord de l'Algerie): *Karstologie*, v. 2, p. 45-54.

Curl, R.L., 1966, Cave conduit enlargement by natural convection: *Cave Notes*, v. 8, p. 4-8.

Dublyansky, V.N., 1980, Hydrothermal karst in the Alpine folded region of the south of the USSR: *Kras i Speleologia* (Poland), v. 3, no. 12, p. 18-36.

Dublyansky, V.N., and Andrejchouk, V.N., 1991, Superficial, underground, deep-seated karst, *in* Obstanovki karstogenesa: Glubinnyj karst, endokarst, gidrotermokarst: Tesisy dokladov regional'nogo sovehchanija, Kungur, p. 55-64 (in Russian).

Dublyansky, Y.V., 1990, *Regularities of the formation and modeling of hydrothermal karst*: Novosibirsk, Nauka, 151 p. (in Russian, res. in English).

Durov, S.A., 1956, On the question about the origin of the salt composition of karst water: *Ukrainian Chemical Journal*, v. 22, p. 106-111 (in Russian).

Ezhov, Y.A., *et al.*, 1991, Deep-seated karst. Endokarst. Hydrothermal karst, in Karstogenetic settings: Deep-seated karst, endokarst, hydrothermal karst: Abstracts of papers of a regional symposium, Kungur, March 1991, p. 4-13 (in Russian).

Ezhov, Y.A., et al., 1992, *Karst in the Earth crust: Distribution and the main types*: Novosibirsk, Institute of Geology, Geophysics, and Mineralogy, 76 p. (in Russian).

Ezhov, Y.A., and Lysenin, G.P., 1986, The vertical hydrodynamic zonality of the Earth crust: *Soviet Geology*, v. 8, p. 111-120 (in Russian).

Ezhov, Y.A., and Vdovin, Y.P., 1970, On the question of vertical hydrodynamic zonality of the Earth's crust: *Soviet Geology*, v. 8, p. 66-76 (in Russian).

Ford, D.C., 1971, Geologic structure and a new explanation of limestone cavern genesis: *Cave Research Group of Great Britain Transactions*, v. 13, p. 81-94.

Ford, D.C., 1988, Characteristics of dissolutional cave systems in carbonate rocks, in James, N.P., and Choquette, P.W., eds., *Paleokarst*: New York, Springer-Verlag, p. 25-57.

Ford, D.C., 1989, Features of the genesis of Jewel Cave and Wind Cave, Black Hills, South Dakota: *National Speleological Society Bulletin*, v. 51, p. 100-110.

Ford, D.C., and Williams, P.W., 1989, *Karst geomorphology and hydrology*: London, Unwin Hyman, 601 p.

Ford, T.D., 1995, Some thoughts on hydrothermal caves: *Cave and Karst Science*, v. 22, no. 3, p. 107-118.

Frumkin, A., 1994, Morphology and development of salt caves: *National Speleological Society Bulletin*, v. 56, p. 82-95.

Glennie, E.A., 1954, Artesian flow and cave formation: *Cave Research Group of Great Britain Transactions*, v. 3, p. 55-71.

Gutsalo, L.K., 1971, Radiolysis of water as the source of free oxygen in the underground hydrosphere: *Geokhimiya*, v. 12, p. 1473-1481 (in Russian), *Geochemistry International*, v. 8, p. 897-903 (1972).

Hill, C.A., 1987, Geology of Carlsbad Cavern and other caves in the Guadalupe Mountains, New Mexico and Texas: *New Mexico Bureau of Mines and Mineral Resources Bulletin*, no. 117, 170 p.

Hill, C.A., 1995, H_2S-related porosity and sulfuric acid oil-field karst, in Budd, A.D., Saller, A.H., and Harris, P.M., eds., Unconformities and porosity in carbonate strata: *American Association of Petroleum Geologists Memoir*, no. 63, p. 301-306.

Hill, C.A., 1996, Geology of the Delaware Basin, Guadalupe, Apache, and Glass Mountains, New Mexico and West Texas: Permian Basin Section, *Society of Economic Paleontologists and Mineralogists Publication*, no. 96-39, 480 p.

Howard, A.D., 1964, Model for cavern development under artesian ground water flow, with special reference to the Black Hills: *National Speleological Society Bulletin*, v. 26, p. 7-16.

Huntoon, P.W., 1993, The influence of Laramide foreland structures on modern ground-water circulation in Wyoming artesian basins, in Snoke, A.W., Steidtmann, J.R., and Roberts, S.M., eds., Geology of Wyoming: *Geological Survey of Wyoming Memoir*, no. 5, p. 756-789.

Huntoon, P.W., 1996, Large-basin groundwater circulation and paleo-reconstruction of circulation leading to uranium mineralization in Grand Canyon breccia pipes, Arizona: *The Mountain Geologist*, v. 33, no. 3, p. 71-84.

Jennings, J.N., 1971, *Karst*: Cambridge, M.I.T. Press, 252 p.

Kartsev, A.A., 1972, *Hydrogeology of oil and gas deposits*: Moscow, Nedra, 280 p. (in Russian).

Kartsev, A.A., 1985, Low mineralized waters of oil and gas deposits, their origin and significance, in Kolodij, V.V., ed., *Low-mineralized waters of deep-seated horizons of oil- and gas-bearing provinces*: Kiev, Naukova Dumka, p. 4-6 (in Russian).

Kaveev, M.C., 1963, About the influence of carbon dioxide originated during the destruction of oil deposits on development of karst processes: Doklady Akademii Nauk SSSR, v. 152, no. 3, (in Russian), p. 254-257.

Kempe, S., 1972, Cave genesis in gypsum with particular reference to underwater conditions: *Cave Science*, v. 6, no. 49, p. 1-6.

Klimchouk, A.B., 1990, Artesian genesis of the large maze caves in the Miocene gypsum of the western Ukraine: *Doklady Akademii Nauk Ukrainskoj SSR*, ser. B, v. 7, p. 28-32 (Russian, res. in English).

Klimchouk, A.B., 1992, Large gypsum caves in the western Ukraine and their genesis: *Cave Science*, v. 19, p. 3-11.

Klimchouk, A.B., 1994a, Speleogenesis under confined conditions, with recharge from adjacent formations: Publ. Serv. Geol. Luxembourg, *Comptes Rendus du Colloque International de Karstologie a Luxembourg*, v. 27, p. 85-95.

Klimchouk, A.B., 1994b, Speleogenesis in gypsum and geomicrobiological processes in the Miocene sequence of the Pre-Carpatian region, *in* Sasowsky, I.D., and Palmer, M.V., eds., Breakthroughs in Karst Geomicrobiology and Redox Geochemistry, Abstracts and Field Trip Guide: *Karst Water Institute, Special Publications*, no. 1, p. 40-42.

Klimchouk, A.B., 1996, Speleogenesis in gypsum, in Klimchouk, A., Lowe, D., Cooper, A., and Sauro, U., eds., Gypsum karst of the world: *International Journal of Speleology*, v. 25, no. 3-4, p. 61-82.

Klimchouk, A.B., 1997a, The role of karst in the genesis of sulfur deposits, Pre-Carpathian region, Ukraine: *Environmental Geology*, v. 31, p. 1-20.

Klimchouk, A.B., 1997b, Artesian speleogenetic setting: *12th International Congress of Speleology*, La Chaux-de-Fonds, Proceedings, v. 1, p. 157-160.

Klimchouk, A.B., 1997c, The role of speleogenesis in the Miocene gypsum in the western Ukraine in groundwater circulation in the multi-story artesian system, *in* Gunay, G., and Johnson, A.I., eds., *Karst Waters and Environmental Impacts:* Rotterdam, Balkema, p. 281-291.

Klimchouk, A.B., and Andrejchouk, V.N., 1996, Breakdown development in cover beds, and landscape features induced by intrastratal gypsum karst, in Klimchouk, A., Lowe, D., Cooper, A., and Sauro, U., eds., Gypsum karst of the world, *International Journal of Speleology*, v. 25, no. 3-4, p. 127-144.

Kolodij, V.V., 1975, Underground condensational and solutional waters of oil, gas condensate, and gas deposits: Kiev, Naukova Dumka, 122 p. (in Russian).

Kolodij, V.V., et al., 1991, *Low-mineralized waters of the deep-seated horizons of oil- and gas-bearing confined basins of Ukraine*: Kiev, Naukova Dumka. 184 p. (in Russian).

Kungur Symposium, 1991, *Karstogenetic settings: Deep-seated karst, endokarst, hydrothermal karst*: Abstracts of papers of a regional symposium, Kungur, March 1991, 102 p. (in Russian).

Laptev, F.F., 1939, *Aggressive action of water on carbonate rocks, gypsum, and concrete*: *in* Trudy SPETSGEO, v. 1, Moscow-Leningrad, GONTI, 120 p. (in Russian).

Lowe, D.J., 1992, *The origin of limestone caverns: An inception horizon hypothesis*: Manchester Metropolitan University, Ph.D. thesis, 512 p.

Lowe, D.J., and Gunn, J., 1995, The role of strong acid in speleo-inception and subsequent cavern development, *in* Barany-Kevel, I., ed., Environmental effects on karst terrains: Acta Geographica (Hungary), v. 24, p. 33-60.

Maximovich, G.A., 1979, Karstosphere and the types of karst, in Karst I gidrogeologia Preduralja: Sverdlovsk, p. 16-21 (in Russian).

Mjatiev, A.N., 1947, Confined complex of underground waters and wells: *Izvestija AN SSSR*, Otd. Tekhnich. Nauk, no. 9, p. 33-47 (in Russian).

Moore, G.W., and Sullivan, N., 1997, *Speleology—Caves and the cave environment*: St. Louis, Cave Books, 3rd ed., 176 p.

Palmer, A.N., 1975, The origin of maze caves: *National Speleological Society Bulletin*, v. 37, p. 56-76.

Palmer, A.N., 1981, The geology of Wind Cave, Wind Cave National Park, South Dakota, U.S.A.: Hot Springs, South Dakota, *Wind Cave Natural History Association*, 44 p.

Palmer, A.N., 1984, Geomorphic interpretation of karst features, *in* LaFleur, R.G., ed., Groundwater as a geomorphic agent: Boston, Allen and Unwin, p. 173-209.

Palmer, A.N., 1991, Origin and morphology of limestone caves: *Geological Society of America Bulletin*, no. 103, p. 1-21.

Palmer, A.N., 1995, Geochemical models for the origin of macroscopic solution porosity in carbonate rocks, in Budd, A.D., Saller, A.H., and Harris, P.M., eds., Unconformities and porosity in carbonate strata: *American Association of Petroleum Geologists Memoir*, no. 63, p. 77-101.

Palmer, M.V., and Palmer, A.N., 1989, Paleokarst of the United States, in Bosák, P., Ford, D., Glazek, J., and Horácek, I., eds., *Paleokarst: A systematic and regional review*: Praha, Academia, p. 337-365.

Pronin, K.K., 1995, Karst caves in the Odessa region: Dep. Paper No. 953 UK95 (Ukraine), 42 p. (in Russian).

Quinlan, J.F., 1978, *Types of karst, with emphasis on cover beds in their classification and development*: University of Texas at Austin, Ph.D. Thesis, 325 p.

Quinlan, J.F., Smith, R.A., and Johnson, K.S., 1987, Gypsum karst and salt karst of the United States of America, *in* Atti symposio international sul carsismo nelle evaporiti: *Le Grotte d'Italia*, ser. 4, v. 13, p. 73-92.

Raines, M.A., and Dewers, T.A., 1997, Dedolomitization as a mechanism for karst generation in Permian Blane Formation, southwestern Oklahoma, U.S.A.: *Carbonates and Evaporites*, v. 12, p. 24-31.

Runnels, D.D., 1969, Diagenesis, chemical sediments, and mixing of natural waters: *Journal of Sedimentary Petrology,* v. 39, p. 1188-1201.

Shestopalov, V.M., 1981, *Natural resources of underground water of platform artesian basins of the Ukraine*: Kiev, Naukova Dumka, 195 p. (in Russian).

Shestopalov, V.M., 1988, *Methods of study of underground water natural resources*: Moscow, Nedra, 168 p. (in Russian).

Shestopalov, V.M., ed., 1989, *Water exchange in hydrogeological structures of the Ukraine-Water exchange under natural conditions*: Kiev, Naukova Dumka, 288 p. (in Russian).

Smart, P.L., and Whitaker, F.F., 1991, Karst processes, hydrology and porosity evolution, *in* Wright, V.P., Esteban, M., and Smart, P.L., eds., Paleokarsts and paleokarstic reservoirs: *P.R.I.S. Occasional Publications Series,* University of Reading, no. 2, p. 1-55.

Sokolov, D.S., 1962, *Principal conditions of karst development*: Moscow, Gosgeolizdat, 321 p. (in Russian).

Stankevich, E.F., 1970, *On the possibility of development of deep-seated karst*: Perm University, Voprosy Karstovedeniya 2, p. 43-47 (in Russian).

Töth, J., 1995, Hydraulic continuity in large sedimentary basins: *Hydrogeology Journal*, v. 3, no. 4, p. 4-15.

Turyshev, A.V., 1965, About one possible way of formation of karst cavities at great depths, *in* Gidrogeologichesky Sbornik 4, (Trudy Instituta Geologii UFAN SSSR 76), Sverdlovsk (in Russian). p. 65-67

Vovk, I.F., 1979a, *Radiolysis of underground waters and its geochemical role*: Moscow, Nedra, 231 p. (in Russian).

Vovk, I.F., 1979b, Natural radiolysis and peculiar features of the composition of groundwaters of oil and gas deposits, *in* Regional'nyje uslovija dinamiki, okhrana I popolnenije podzemnykh vod USSR (collection of papers): Kiev, Naukova Dumka, p. 12-20.

White, W.B., 1969, Conceptual models for carbonate aquifers: *Ground Water*, v. 7, p. 15-21.

White, W.B., 1988, *Geomorphology and hydrology of karst terraines*: Oxford University Press, 464 p.

Wigley, T.M.L., and Plummer, L.N., 1976, Mixing of carbonate waters: *Geochimica et Cosmochimica Acta*, v. 40, p. 989-995.

Worthington, S.R.H., 1991, *Karst hydrogeology of the Canadian Rocky Mountains*: McMaster University, Ph.D. Thesis, 227 p.

Worthington, S.R.H., 1994, The possible importance of sulfur minerals in initiating epigenic caves, *in* Sasowsky, I.D., and Palmer, M.V., eds., Breakthroughs in karst geomicrobiology and redox geochemistry: *Karst Waters Institute Special Publication*, no. 1, p. 80-82.

Worthington, S.R.H., and Ford, D.C., 1995, High sulfate concentrations in limestone springs: An important factor in conduit initiation?: *Environmental Geology*, v. 25, p. 9-15.

5.2.1. Speleogenesis of The Great Gypsum Mazes in The Western Ukraine

Alexander Klimchouk

Abstract

Speleogenesis in the western Ukraine proceeded mainly under confined hydrogeologic conditions. While such development still continues in part of the territory, other parts exhibit entrenched karst settings. Huge relict maze cave systems have been explored here, five of which are currently the longest known gypsum caves in the world. They account for well over half of the total length of gypsum caves explored around the world.

This unique concentration of large caves is due to the local coincidence of several factors:

- Confined speleogenetic settings with dispersed upward recharge from a basal aquifer.
- Specific structural prerequisites of artesian speleogenesis (nature and frequency of fissuring).
- Favorable regional evolution (rapid uplift and fossilization of maze systems).
- Presence of an overlying limestone bed and a widespread clayey protective cover (which prevented later infilling or destruction of the caves).

Maze cave development in the western Ukraine is not specific to gypsum. It is quite characteristic, however, of confined flow and transverse speleogenesis in a multistory aquifer system—provided that proper structural conditions exist. The main advantage of speleogenetic studies in gypsum is that, due to its high solubility and fast dissolution kinetics, morphologic evidence of the flow pattern is well displayed and can be rather easily reconstructed. The great gypsum caves in the western Ukraine can serve as a model example of transverse artesian speleogenesis.

Introduction

The great gypsum karst of the western Ukraine, developed in Miocene (Badenian) gypsum, provides the world's foremost examples of speleogenesis under artesian conditions. Differential neotectonic movement has resulted in various parts of the territory displaying different types (stages) of intrastratal karst, from deep-seated, through subjacent, to entrenched.

Important previous works on the gypsum karst of the region include Ivanov and Dublyansky (1966), Dublyansky and Ivanov (1970) and Dublyansky and Smol'nikov (1969). Abundant new data were obtained, and their interpretation was developed during the 80s and 90s (Andrejchouk, 1988; Klimchouk, 1986; 1990, 1992, 1996, 1997b; Klimchouk and Andrejchouk, 1988; Klimchouk, Andrejchouk, and Turchinov, 1995; Klimchouk and Rogozhnikov, 1982).

Geologic Setting

The Miocene gypsum sequence is widespread along the southwestern edge of the Eastern European Platform, in the transition zone between the platform and the

5.2.1 Figure 1. Location map of the gypsum karst of the western Ukraine.

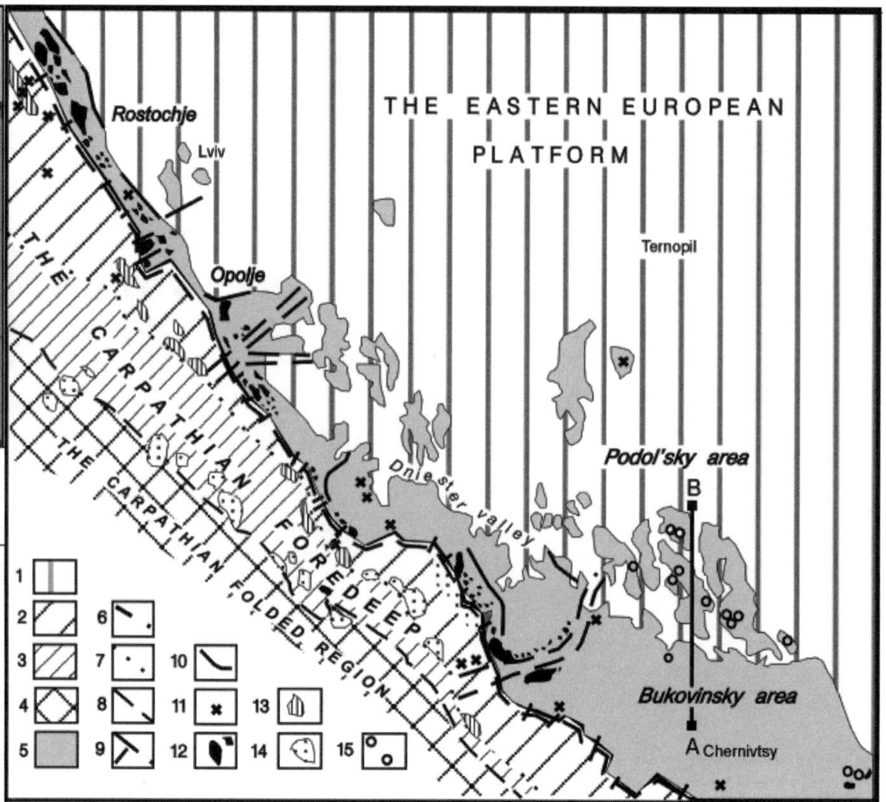

*5.2.1 Figure 2. Distribution of the gypsum strata, sulfur deposits, and large caves in the western Ukraine (modified from Polkunov, 1990). **No. 1** is the Eastern European platform fringe. **No. 2** is the Carpathian foredeep, outer zone. **No. 3** is the Carpathion foredeep inner zone. **No. 4** is the Carpathian folded region. **No. 5** is sulfate rocks on the platform. **No. 6** is the platform foredeep boundary. **No. 7** is the outer to inner zone of the foredeep. **No. 8** is the foredeep and folded region boundary. **No. 9** is other major faults. **No. 10** is flexures. **No. 11** is sulfur occurrence. **No. 12** is sulfur deposit. **No. 13** is gasfield. **No. 14** is oilfield. **No. 15** is large maze cave in gypsum. See profile **A-B** in 5.2.1 Figure 4.*

Carpathian Foredeep. Gypsum stretches northwest for about 300 km as a belt ranging from several to 40 to 80 km wide (5.2.1 Figures 1 and 2). The present extent of sulfates on the platform is about 20,000 km^2.

Most Miocene rocks along the platform margin overlie eroded Cretaceous strata, which include terrigenous and carbonate sediments, mostly marl and sandstone, together with detrital and argillaceous limestone. The Miocene succession comprises deposits of Badenian and Sarmatian age. The Lower Badenian unit, beneath the gypsum, includes mainly carbonaceous, argillaceous, and sandy beds (50 to 90 m thick) adjacent to the foredeep. These grade into rocks of calcareous bioherm and sandy facies (10 to 30 m thick) toward the platform interior (5.2.1 Figure 3).

The gypsum sequence (10 to 40 m thick) is variable in structure and texture. In the Podol'sky area, where most of the accessible caves are located, it includes three units which in ascending order contain:

- Crypto- and microcrystalline massive gypsum.
- Bedded microcrystalline gypsum.
- Mega-crystalline gypsum.

Gypsum within the upper unit displays large spherulitic structures (Klimchouk, Andrejchuk, and Turchinov, 1995). Close to the foredeep, the gypsum is more homogeneous and aphanitic in texture, and the anhydrite content of the sequence increases. There are sporadic thin interbeds of carbonate and clay.

A layer of pelitic and cryptocrystalline limestone, ranging from several 10's of centimeters to more than 25 m in thickness, overlies much of the gypsum. This limestone contains two genetic varieties that differ in carbon isotopic composition. The pelitic limestone (locally called "Ratynsky") has normal "evaporitic" δC^{13} values. The other crypto- and microcrystalline variety was formed epigenetically by replacement of the gypsum during sulfate reduction, and it is characterized by "light" carbon, with δC^{13} from -32 to -65 percent. Where the limestone thickness exceeds 1 to 2 m, it consists of mainly epigenetic calcite, which locally replaces the gypsum stratum entirely. Together this limestone and the gypsum comprise the Tyrassky Formation.

The Tyrassky Formation is overlain by the upper Badenian unit, which begins with argillaceous and marly lithothamnion limestone (the Ternopol'sky' Beds, 2 to 5 m thick). Above this is a succession of clay and marl. Its lower part is in the upper Badenian (the Kosovsky Formation). Its upper part is in the lower Sarmatian. The total thickness of clay is 40 to 50 m in the Podol'sky area, reaching 80 to 100 m toward the foredeep, and thickening to several hundred meters close to the regional faults that separate the platform edge from the foredeep.

The Miocene succession is overlain by late Pliocene and Pleistocene glaciofluvial sand and loam in the northwest section of the gypsum belt, and by sand and gravel alluvial terrace deposits left by the pre-Dniester River during the late Pliocene and early Pleistocene in the Podol'sky area. Many buried valleys, of early to middle Pleistocene age, are entrenched 30 to 50 m into the Kosovsky and Sarmatian clay and locally into the upper part of the Tyrassky Formation.

The present distribution of the Miocene formations and the levels of their denudation vary in a regular way from the platform interior toward the foredeep (Andrejchouk, 1988; Klimchouk et al. 1985; Klimchouk and Andrejchouk, 1988). The Tyrassky Formation dips 1 to 3° toward the foredeep and is disrupted by block faults in the transition zone. To the south and southwest of the Dniester Valley, large tectonic blocks drop down as a series of steps, the thickness of clay overburden increases, and the depth of erosional entrenchment decreases (5.2.1 Figure 4). This variation, the result of differential neotectonic movement, played an important role in the hydrogeologic evolution of the Miocene aquifer system. It determined the recharge-discharge and flow conditions, and hence guided the type of development of cave systems in the gypsum.

Age			Lithostratigraphy	Thickness, meters: area adjacent to the foredeep	Thickness, meters: the Podols'ky sub-region	Hydrostratigraphy
Plio-Pleistoc.			Alluvial and fluvio-glacial sands, loams and gravels	0-40	0-30	Aquifer
Neogene	Lower Sarmatian	Volynsky Formation	Clays and marls, with minor aleuritic and sandy beds N_1S_1	20-80	20-50	Confining bed
Neogene	Upper Badenian	Kosovsky Formation	Clays and marls with minor sandstone beds at the lower part. Lithothamnion marly limestones (Ternopol'sky Beds in the Podol'sky region) N_1Bd_3	20-80	2-15	
Neogene	Middle Badenian	Tyrassky Formation	Pelitic marine and epigenetic limestones, locally substitute the gypsum ("Ratynsky Beds")	5-30	1-5	Aquifer (the supra-gypsum aquifer)
Neogene	Middle Badenian	Tyrassky Formation	Gypsum, with increasing content of anhydrite towards the foredeep, with minor beds of clays N_1Bd_2	10-40	10-20	Properties changed through time and the area from aquiclude to conduit aquifer
Neogene	Lower Badenian	Opol'sky Formation	Lithothamnion limestones, sands and sandstones, locally marls N_1Bd_1	50-90	10-30	Aquifer (the sub-gypsum aqufer)
Cretaceous	Lower & Upper		Carbonaceous sands and sandstones, detritic limestones, locally marls and marly limestones K_{1+2}	10-50	10-30	Properties vary through the area from aquiclude to aquifer

5.2.1 Figure 3. Generalized litho- and hydrostratigraphy of the upper part of the sedimentary sequence at the southwest edge of the Eastern European Platform.

Hydrogeologic Setting and Karst Types

In hydrogeologic terms, the region represents the southwestern part of the Volyno-Podol'sky Artesian Basin (Shestopalov, 1989). Sarmatian and Kosovsky clay and marl provide an upper confining sequence. The sandstone interbeds in the lower part of the Kosovsky Formation, the Ternopol'sky Beds (lithothamnion limestone) and the limestone bed of the Tyrassky Formation (all above the gypsum) and the lower Badenian sandy carbonate beds (below the gypsum) are the original aquifers, the latter being the major regional one. The hydrogeologic role of the gypsiferous unit has changed with time, from initially being an aquiclude, intervening between two aquifers, to a karstified aquifer with well-developed conduit permeability. Underlying Cretaceous sediments have variable properties across the area. Regional flow is from the platform interior, where confining clay and the gypsum are largely denuded, toward the large and deep Dniester Valley and the Carpathian Foredeep.

The main factors that determine contemporary hydrogeologic conditions are:

- The differences in the depth of the gypsum.
- The level of areal denudation of the land surface.
- The depth of erosional entrenchment by the major valleys.

Three subparallel zones are distinguished, each characterized by a different type of gypsum karst (see 5.2.1 Figure 4).

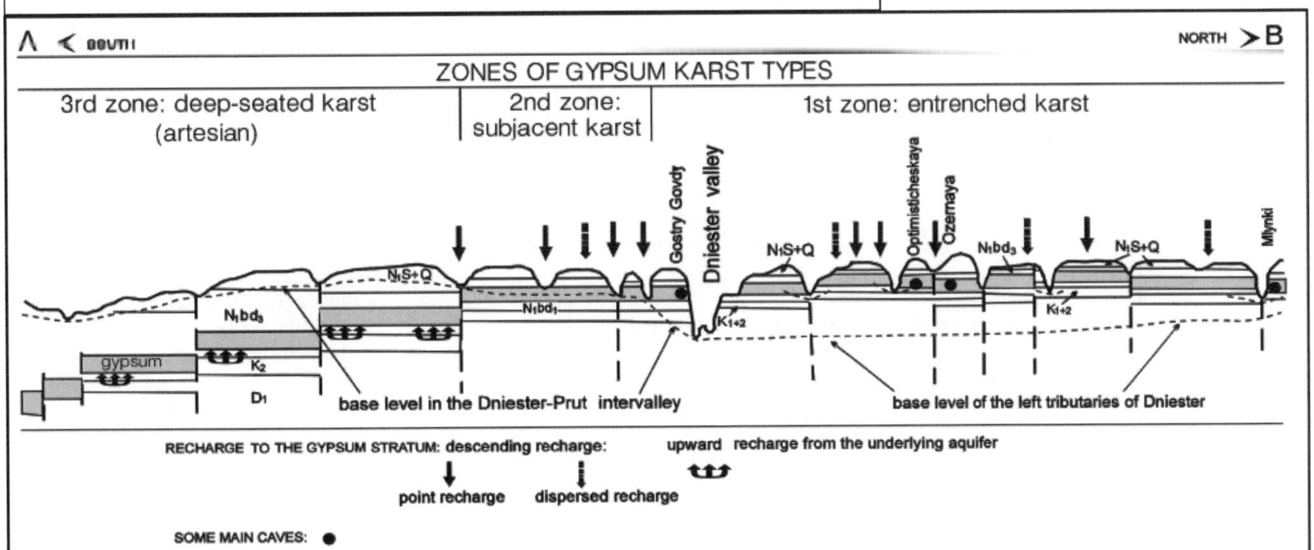

*5.2.1 Figure 4. Types of gypsum karst and hydrogeologic setting in the Podol'sky and Bukovinsky areas, the southeastern section of the gypsum belt (line **A-B** on 5.2.1 Figures 2 and 5) (modified from Klimchouk and Andrejchouk, 1988).*

First Zone—Entrenched Karst

In the first zone, the Miocene rocks and underlying formations are deeply entrenched by valleys of the major northern tributaries of the Dniester, separated by wide intervalley plateaus where the gypsum and clay overburden remain largely intact. The Miocene formations are almost entirely drained, and only in the central parts of the intervalley plateaus do the subgypsum units contain unconfined underground water. In some tectonic blocks the water table extends upward into the lower part of the gypsum (Ozernaya Cave) with multiyear fluctuations that range from 3 to 5 m. Huge maze cave systems in the gypsum are relict, having formed under former artesian conditions. Modern dissolution is restricted to the lower part of the gypsum, where the water table is present, at rare points of focused vertical percolation (where vertical dissolution pipes develop), and along linear underground streams that are fed via swallow holes that receive periodic surface flow. At present, focused point recharge is the main recharge mode of the Miocene aquifer is in this zone. Doline density increases locally where the capping clay is removed, as within old, high, river terraces. Discharge takes the form of springs flowing from the subgypsum unit, or from the underlying Cretaceous formations.

These conditions typify an intrastratal entrenched karst, according to the classification introduced in Chapter 3.1. They are characteristic of the area within the interior of the platform, to the northeast and north of the Dniester Valley, and for the Podol'sky area in particular. A narrow area on the southern bank of the Dniester Valley also lies within this zone, but this grades into the second zone.

Second Zone—Subjacent Karst

To the southwest and south of the Dniester Valley, the depth of the gypsum formation increases, while the depth of erosional entrenchment diminishes. The narrow (3-to-15-km-wide) second zone is between the Dniester and Prut Valleys, where the water table lies within the gypsum or locally within the overlying unit. Floors of erosional valleys lie above, or are entrenched into the karstified gypsum unit, causing diverse karst features to develop, with intermittent streams, swallow holes, ascending karst springs, and estavelles. Recent collapse and subsidence dolines are common, as the water table and localized streams intensify breakdown processes in the gypsum by active dissolution and erosional removal of cave fills. Cave systems inherited from the confined stage are accessible only as small fragments due to infilling and a high water table position, but drilling data and observations in the neighboring first zone where the gypsum is drained and has been extensively quarried indicate their wide distribution. Relatively small linear through caves are also common, and their origin is attributed to contemporary conditions.

In general the groundwater flow in the Miocene aquifer is directed northward to the Dniester Valley, and south-southwestward to the Prut Valley. Flow becomes confined in the latter direction, within the third zone, where the Tyrassky Formation lies at an even greater depth beneath the increasing thickness of clayey overburden. The second zone represents subjacent karst conditions, grading locally into early phases of entrenched karst.

Third Zone—Deep-Seated and Subjacent Karst

The third zone stretches along the boundary between the platform and the foredeep. Within this zone the formations comprising the Miocene aquifer lie at depths of 30 to 100 m beneath the Kosovsky clay, which is cut in part only by shallow erosional valleys, so that flow is confined.

Recharge conditions differ between a narrow northwest and a wider southeast section of the gypsum belt (see 5.2.1 Figure 2). In the northwest, recharge occurs within the adjacent unconfined area (where both the clay and the gypsum are denuded) by infiltration directly into the subgypsum beds. To the southeast of the gypsum belt, recharge occurs from the neighboring area of subjacent gypsum karst (the second zone) via karst systems.

On the opposite flank of the confined flow area, along the foredeep margin, regional faulting has brought the Miocene aquifer into lateral contact with an even thicker Kosovsky clayey sequence. Thus further flow in this direction is prevented, and upward discharge occurs locally, focused upon areas where the confining properties of the Kosovsky clay is weakened by stratigraphic or tectonic discontinuities, or by incised erosional valleys. The latter situation is common in the northwest part of the gypsum belt, where discharge is commonly focused along paleovalleys buried by Quaternary glaciofluvial sand and loam, and in the southeast section, where discharge converges toward the major modern Prut Valley.

Lower Badenian sandy carbonaceous sediment that lies directly below the gypsum is the major aquifer. Misleadingly, the gypsum unit was long considered an aquiclude separating the subgypsum and supragypsum aquifers. Numerous indications of karstification in the gypsum were not properly interpreted. However, it has recently been shown (Klimchouk, 1990, 1992, 1996) that extensive karst systems develop due to the effects of upward transverse hydraulic communication through the gypsum between the original aquifers (see below). Such systems, which are modern analogues of relict maze caves that are known in the first zone, develop preferentially in areas of potentiometric lows, where an overall discharge from the artesian aquifer system occurs. Analysis of data from the many cavities intercepted in the gypsum by exploratory drilling in different deposits suggests that their morphology and structure are similar to those of explored relict caves (Klimchouk, 1997a,b).

Gypsum karstification within the third zone is not exposed at the surface across most of those areas where the clayey overburden thickness exceeds 45 to 60 m. This represents a zone of deep-seated karst. However, in areas where the natural groundwater circulation has been affected by an anthropogenic impact (such as opencut quarrying or groundwater abstraction) and karst processes have consequently been intensified in the gypsum, collapse and subsidence dolines may develop at the surface, being induced by karstification in the gypsum at still greater depth. Moreover, local areas of shallow karst lie within the third zone, related particularly to the most uplifted tectonic blocks. Here the gypsum lies at a relatively shallow depth and major valleys have been incised, breaching artesian confinement (as in the Krivsky and Mamalyzhsky blocks in the Bukovinsky subregion; Andrejchuk, 1988). Locally this results in drainage of the upper part of the gypsum unit, and widespread development of collapse and subsidence phenomena. Such cases correspond to subjacent karst.

Caves and Their Genesis

General Characteristics

Fourteen large caves over 1 km in length are known in the region. Eleven are north of the Dniester Valley within the Podol'sky subregion, and nine of these caves lie within a narrow belt subparallel to the Dniester Valley (5.2.1 Figure 5). Two caves (Mlynki and Ugryn') are outside this belt some 15 to 20 km to the north. All of these caves are within the first zone, as is one more cave, Gostry Govdy, on the south bank of the Dniester. Two other large caves, Zoloushka and Bukovinka, are in the Bukovinsky subregion near the Prut River, in the area of artesian flow within the Miocene aquifer system (third zone) but within local especially uplifted blocks where the upper part of the gypsum has been entrenched and drained. Because of this, water table conditions prevail in the gypsum unit within this area. In the area of Zoloushka Cave, additional water table lowering has been caused by groundwater abstraction related to quarrying. So the upper story of the cave has been drained and become accessible to exploration.

All the large gypsum caves in the region are mazes developed along vertical and steeply inclined fissures and are arranged into laterally extensive networks. Aggregating

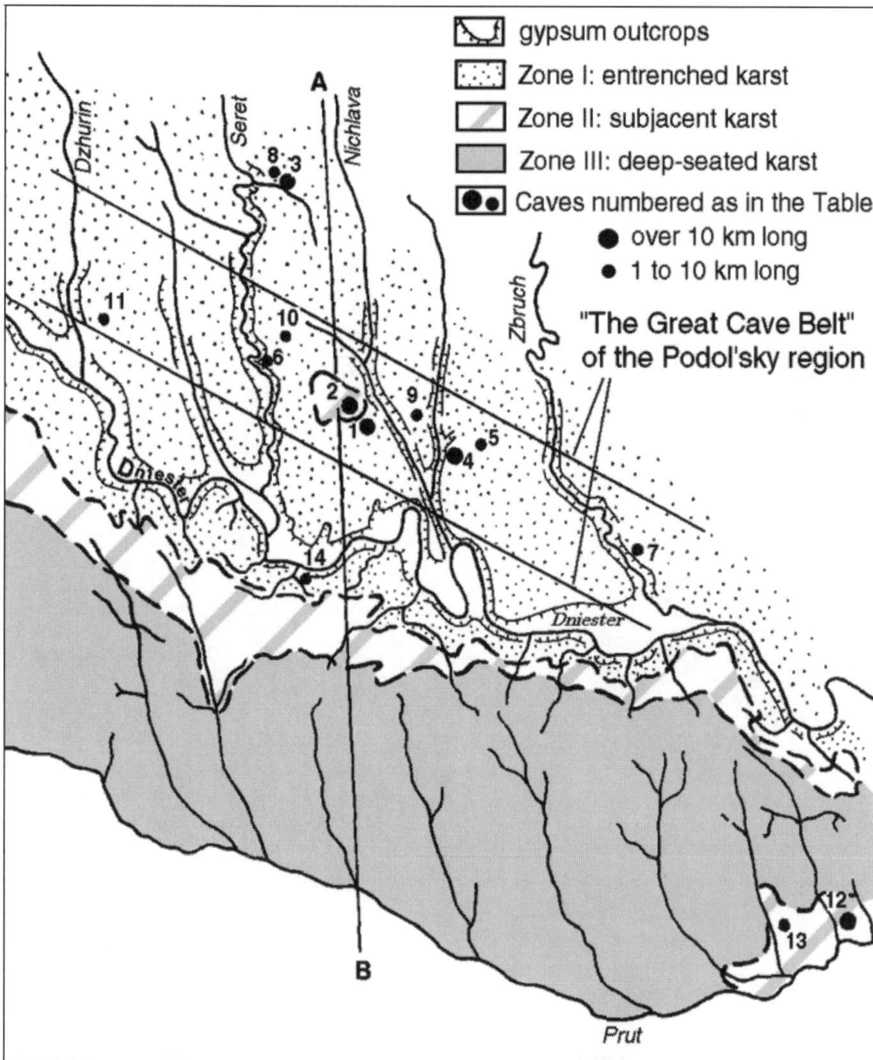

*5.2.1 Figure 5. Distribution of karst types, and the location of the main caves in the Podol'sky and Bukovinsky subregions. Numbers of the caves correspond to those in the table. Line **A-B** corresponds to the profile of 5.2.1 Figure 4 (modified from Klimchouk and Andrejchouk, 1988).*

passages form lateral two- to four-story systems that extend over areas of up to 1 km^2. Such areas, termed here cave fields, are defined by drawing an arbitrary boundary closely comprising a cave. Some typical cave patterns are illustrated in 5.2.1 Figure 6 (p. 266). Significant morphologic parameters of the caves are summarized in 5.2.1 Table 1 (p. 268).

Optimisticheskaja Cave is the longest gypsum cave and the second longest cave of any type known in the world. It has more than 200 km of surveyed passages (5.2.1 Figure 7). The region contains the five longest known gypsum caves in the world. It accounts for well over half of the total known length of gypsum caves on the Earth. By area and volume, the largest caves are Ozernaya (330,000 m^2 and 665,000 m^3) and Zoloushka (305,000 m^2 and 712,000 m^3), followed by Optimisticheskaja Cave (240,000 m^2 and 500,000 m^3).

The absolute parameters of the cave systems change as speleologic exploration progresses. Specific parameters are more informative. Specific volume (the cave volume to length ratio) characterizes "voluminousness" (or average cross-sectional area) of cave passages in a cave system. For the caves of the region, this measure ranges from 1.6 (Gostry Govdy Cave) to 8.0 (Zoloushka Cave) m^3/m. The average for the region is 3.84 m^3/m.

Passage network density is characterized conveniently by use of the ratio of cave length to the unit area of a cave field (km/km^2). This parameter varies for the region from 118 (Slavka Cave) to 321 (Bukovinka Cave) km/km^2, with an average value of 200 km/km^2.

The availability of detailed morphometrical data on caves and the host rock bodies allows the calculation of coefficients of karstification of the gypsum strata both in terms of area and volume (fractions of total area and volume of the rock within a cave field, occupied by passages). The former parameter varies from 17 percent (Komsomol'skaja Cave) to 71 percent (Zoloushka Cave). High values are also characteristic for the fields of Verteba Cave (61 percent), Bukovinka Cave (57 percent) and Ozernaya Cave (51 percent). The average value is 37 percent. The volumetrical coefficient of karstification varies from 2 to 3 percent (Dzhurinskaja, Mlynki, Slavka and Komsomol'skaja caves) to 12 percent (Verteba Cave), with an average value of 5 percent.

Cave Genesis: Hydrodynamic Pattern and Evolution

Maze caves in the region developed under confined conditions (and are still developing there in the third zone). This development was due to upward transverse groundwater circulation between the preexisting subgypsum and supragypsum aquifers (Klimchouk, 1990, 1992). Such flow patterns are characteristic of potentiometric low areas, related to topographic lows (valleys) that commonly coincide with zones of enhanced fluid conductivity created within the capping clay by tectonic or stratigraphic discontinuities (5.2.1 Figure 8, p. 268). Overall discharge from artesian aquifer systems occurs in such areas. Mechanisms of cave development in such settings are discussed in detail in Chapter 5.2 where it is shown that maze patterns will result if appropriate structural conditions exist. The western Ukrainian maze caves provide the most outstanding and unambiguous evidence for the transverse artesian speleogenetic model presented in Chapter 5.2.

Across the entire region studied, confined hydrogeologic settings (with constrained discharge and sluggish flow) prevailed during most of the Pliocene, when slow speleogenetic initiation probably occurred. By the late Pliocene to early Pleistocene the early, shallow, and wide pre-Dniester Valley had formed, as evidenced by the extent of the alluvium of the 7th and 6th terraces preserved to the northeast of the modern valley. This initial erosional entrenchment into the confining clay established better conditions for discharge and activated the transverse groundwater flow within the artesian system. The "great cave belt" of the Podol'sky region (see 5.2.1 Figure 5) lies totally within the limits of this ancient pre-Dniester Valley. Its northeastern boundary approximates the limits of the old alluvial deposits that are preserved within the modern intervalley

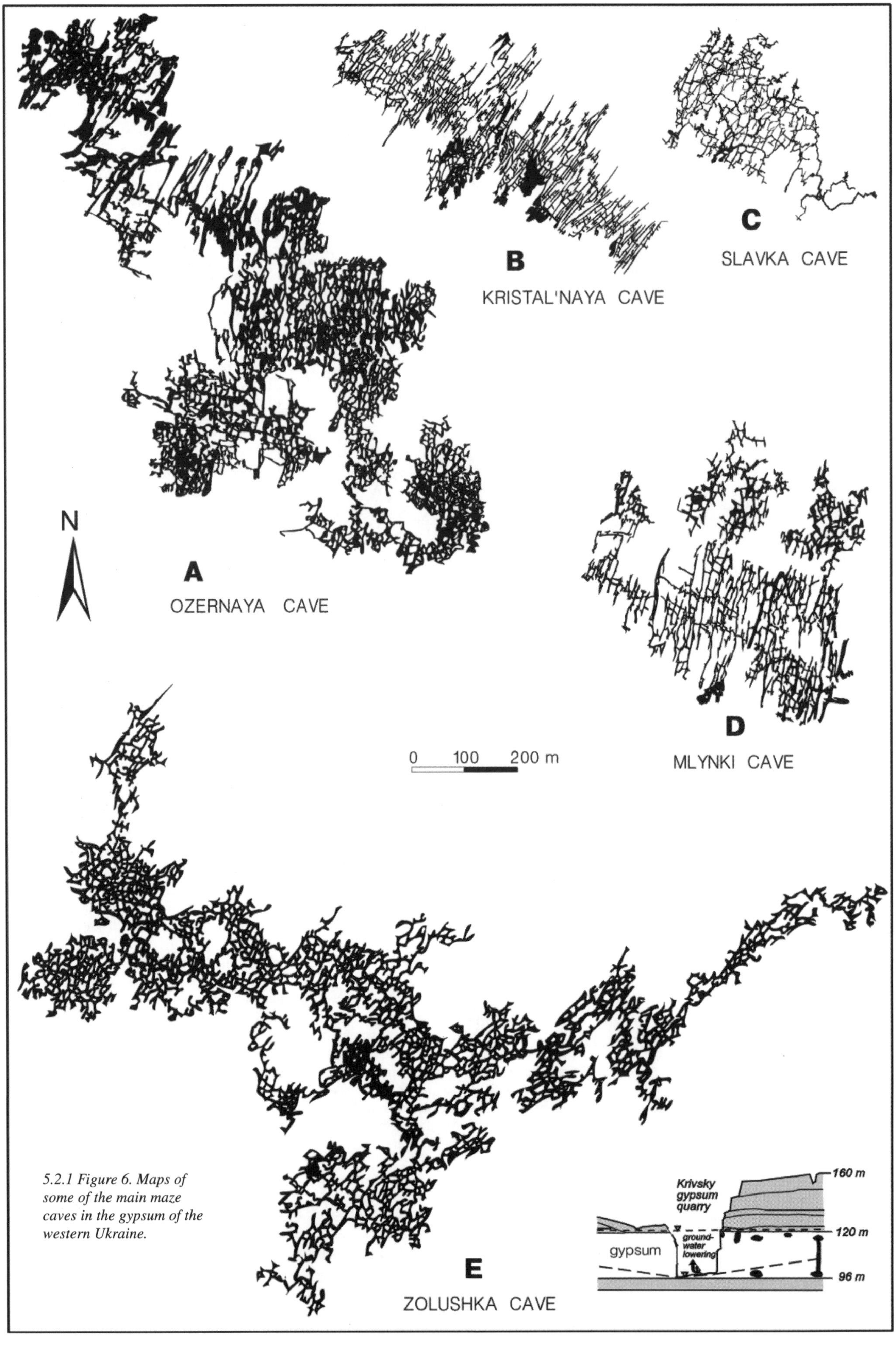

5.2.1 Figure 6. Maps of some of the main maze caves in the gypsum of the western Ukraine.

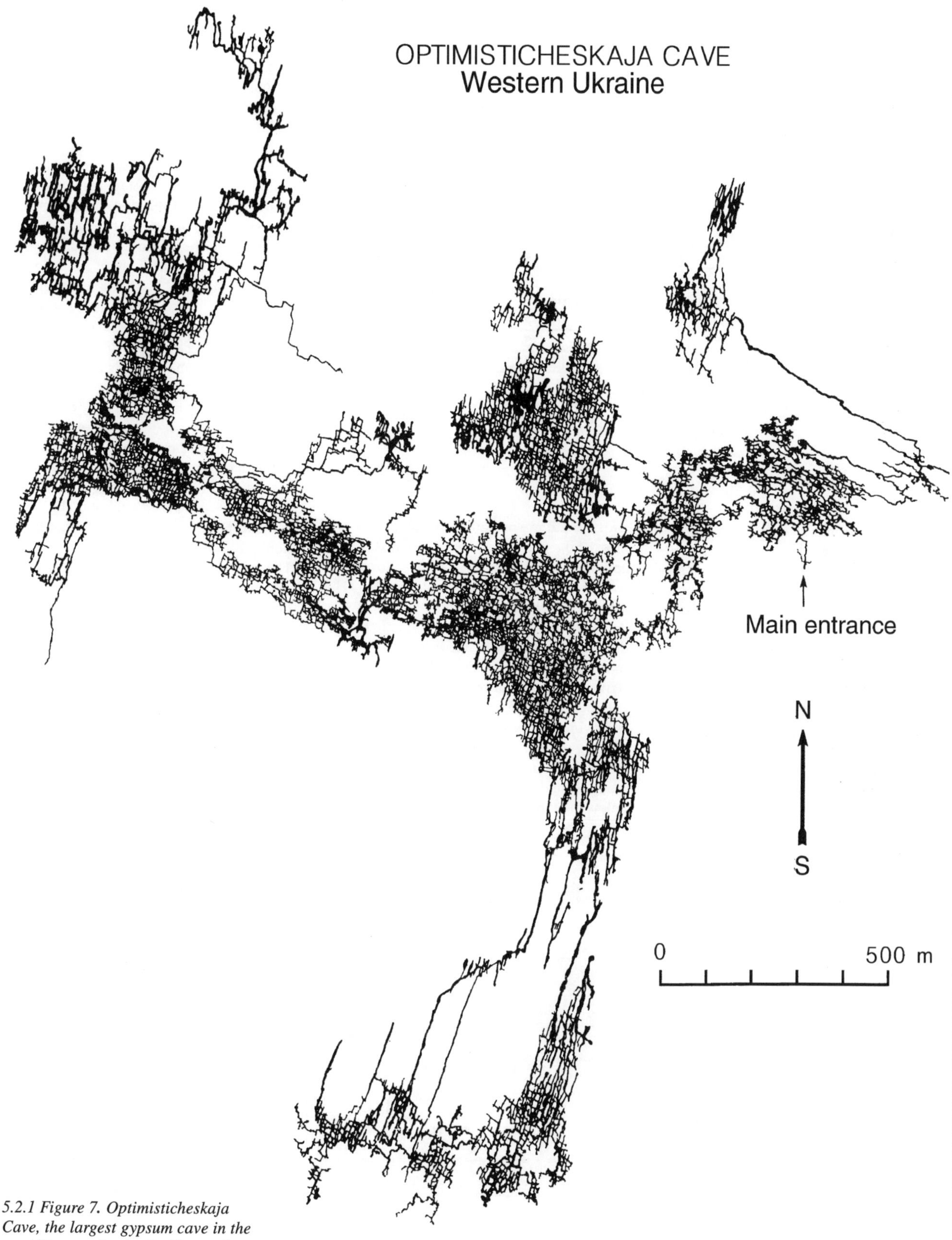

5.2.1 Figure 7. Optimisticheskaja Cave, the largest gypsum cave in the world (survey by the Tsyklop and Bat'kivshchina Speleological Clubs, Lvov).

No	Cave name	Development, m	Specific volume m^3/m	Density of passages, km/km^2	Coefficient of karstification, %	
					in area	in volume
			The Podol'sky sub-region			
1	Optimisticheskaja	208000+	2.6	298	38	4
2	Ozernaja	111000	6	173	51	6
3	Mlynki	25000	3.3	123	24	3
4	Kristalnaja	22000	5	161	28	7
5	Slavka	9100	3.8	118	24	3
6	Verteba	7800	6	206	61	12
7	Atlantida	2520	4.5	168	30	4
8	Ugryn	2120	3.8	193	36	7
9	Jubilejnaja	1500	2.3	277	37	4
10	Komsomol'skaja	1240	2.1	124	17	3
11	Dzhurinskaja	1130	2.3	126	18	2
			The Bukovinsky sub-region			
12	Zoloushka	92000	8	208	71	6
13	Bukovinka	2400	2.5	321	57	4
14	Gostry Govdy	2000	1.6	272	27	4
	Totals	487820	-	-	-	-
	Averages	-	3.84	198	37	5

5.2.1 Table 1. Parameters of large caves and cave fields (see text for definitions).

plateaus. The two maze caves that lie north of this alluvial limit are related to a separate buried valley that has been traced in that area.

Toward the end of the early Pleistocene and through the middle Pleistocene, rapid uplift in the Podol'sky subregion resulted in further incision of the Dniester, making a much narrower valley only slightly wider than the modern one. Also, the northern tributaries of the Dniester incised rapidly, dividing the area north of the Dniester into large, elongated, subparallel plateaus. This led to a substantial acceleration of groundwater circulation within the Miocene artesian system and eventual breaching of its artesian confinement marking a stage of subjacent karst. Probably the majority of cave growth occurred during these transitional periods, although wholly inheriting the skeletal outline established during the main artesian stage. This is supported by the observation that mazes located in a more favorable position relative to the local entrenchment and discharge zones display high specific volume values (e.g. Ozernaja and Verteba, 6.0 m^3/m). Whereas caves less connected to incising valleys are much less voluminous (e.g. Komsomol'skaya and Gostry Govdy Caves, 1.6 to 2.1 m^3/m), although their cave patterns are largely the same. With further deepening of the surface drainage during late Pleistocene times, conditions of entrenched karst were established. Where the water table was established in the gypsum for a prolonged time, further widening of passages occurred. Eventually, with the drop of the water table below the lower gypsum contact, cave systems became entirely fossilized.

In the area to the south-southwest of the Dniester Valley, overall uplift rates during most of the Pleistocene were much slower, and there was a relative subsidence of some tectonic blocks adjacent to the foredeep. This imposed slow rates of speleogenetic development, which became active only during the late Pleistocene in some of the more uplifted blocks, such as those of Krivsky and Mamalyzhsky that were entrenched by the Prut Valley (Zoloushka and Bukovinka Caves). In the northwestern part of the gypsum belt, karstification was intensified during the middle Pleistocene, being related to valleys that were incised below the top of the Tyrassky Formation but subsequently buried when the local neotectonic movement changed from uplift to subsidence. Karst and speleogenetic processes have been reactivated more recently in response to Holocene uplift (which has encompassed the whole region), and also as a reflection of the increasing effect of anthropogenetic impacts (opencut quarrying and groundwater abstraction).

The hydrodynamic and hydrochemical conditions of cave development during the mature and active stage of artesian speleogenesis can be illustrated by examples from northwest of the gypsum belt (Jazovsky and Nikolaevsky deposits) where confined conditions still prevail. Groundwater entering the gypsum from below (from the subgypsum sandy-carbonate aquifer) have a total dissolved solids (TDS) content ranging from 0.4 to 0.6 g/L, and are very aggressive with respect to gypsum, being able to dissolve it at rates ranging from 2.48 to 25.57 mm/yr. Water chemistry and dissolution rates in cavities within the gypsum vary substantially, depending upon the cave system configuration (the actual flow path within the cave system) and flow velocity. The TDS content ranges from 1.3 to 2.1 g/L. Dissolution rates range from 0.16 to 3.46 mm/yr. The above enlargement rates are obtained (Klimchouk et al., 1996) from areas where the flow conditions have been influenced by quarrying activity that may approximately correspond to the late artesian stage when initial breaching of confinement occurred. The rates during the main artesian stage are perhaps of the order of the lowest reported values.

Under conditions of transverse circulation in an artesian system, all available fissures in the gypsum, which hold similar positions within analogous flow paths, will enlarge at

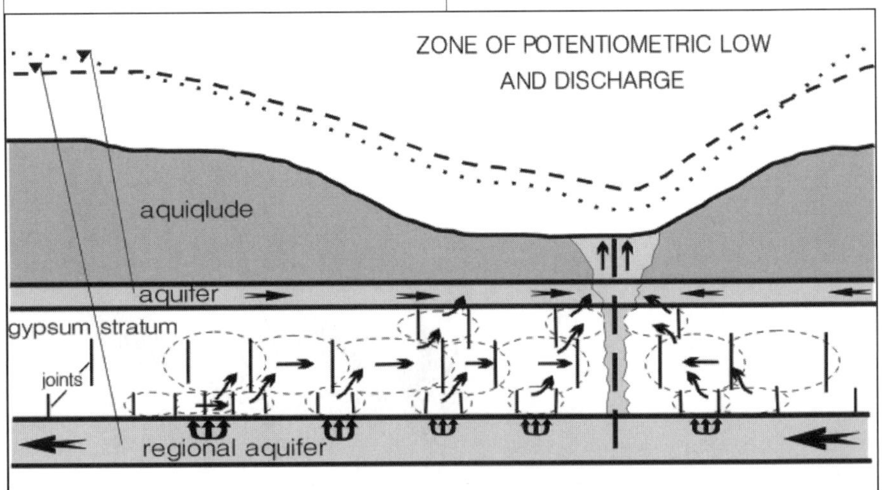

5.2.1 Figure 8. The initial stage of the "ascending" development of multistory maze caves under artesian conditions in the western Ukraine (from Klimchouk, 1996).

comparable rates because of the availability of widely dispersed aggressive recharge from below and suppressed hydraulic competition (Klimchouk, 1977c; see Chapter 5.2). This behavior generally favors the development of maze cave structures, but the actual conduit arrangement in any situation will depend upon the initial fissure pattern in a given locality.

Structural Prerequisites of Speleogenesis
The maze and multistory structure of caves in the region is preconditioned by the structure of the fissure patterns in the gypsum. Fissures are arranged in largely independent laterally extensive networks restricted to individual horizons of the gypsiferous unit that differ in texture and structure (5.2.1 Figure 9). This feature, together with some topologic peculiarities of the fissure patterns, indicates the lithogenetic rather than tectonic origin of the fissures (Klimchouk, Andrejchouk and Turchinov, 1995). Fissure patterns at each level have their own characteristic frequency, orientation, and degree of lateral connectivity, and elements of these are inherited by the passages that comprise each level of a cave system. This is the fundamental reason for different structures being displayed by passage networks at different levels (5.2.1 Figure 9). The extent and the mode of the vertical connectivity between stories of fissures (passages) ranges substantially between tectonic blocks. Besides the overall transverse groundwater circulation pattern in these artesian systems, there is a considerable lateral component in flow and in the cave development. This is caused by

- Good lateral connectivity of fissure networks in some horizons.
- The discordance in the permeability structures between the subgypsum and supragypsum aquifers, and between different horizons in the gypsum unit.

All of the caves include some morphologic elements within the lower part of the gypsum bed, which provided upward recharge to developing cave systems from the underlying aquifer. In most cases the fissures in the lowermost horizon do not form extensive laterally connected networks, so that recharge to a continuous network at the next higher story ("master story") occurred through the separate fissures or localized fissure clusters (Ozernaya, Slavka, Dzhurinskaja, Zoloushka and some areas of Optimisticheskaja Caves).

Such "feeder" conduits (5.2.1 Plate 2A and B, p. 270) are commonly numerous and uniformly distributed, providing relatively dispersed inflows of aggressive water to the conduit network of the master story and supporting uniform dissolutional widening of all passages. In some other cases (such as Atlantida Cave) the lateral flow component and the development of master passages occurred chiefly along the base of the gypsum unit, with vertical "ascending" domepits and small networks formed locally at the upper story (5.2.1 Plate 3A and B, p. 271). Outflow paths to the overlying aquifer may also vary. In places, all fissures (passages) in the uppermost network directly contact the base of the overlying bed, each fissure providing out-flow upward. More commonly, separate ascending domepits develop that are terminated at the upper boundary. In all cases, the multistory conduit systems developed with the ultimate function of transmitting groundwater upward between the subgypsum and supragypsum aquifers (see 5.2.1 Figure 8).

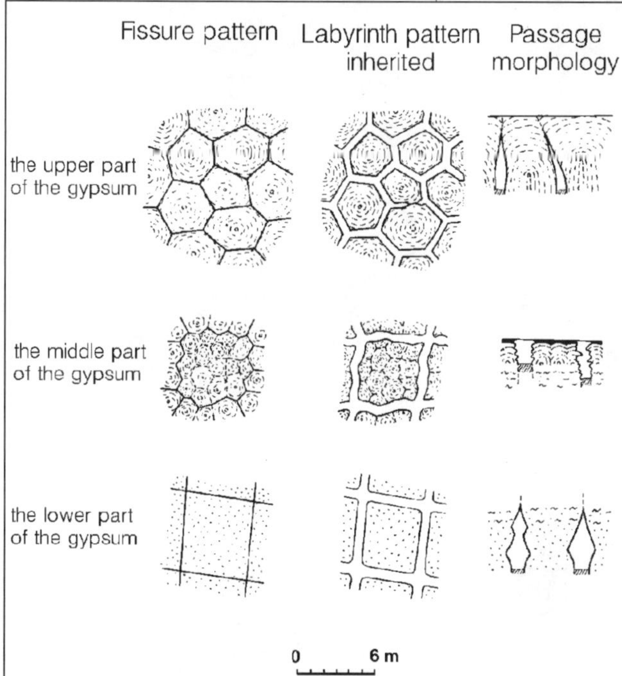

5.2.1 Figure 9. Fissure and passage patterns and passage morphology in the different horizons of the gypsum unit in Optimisticheskaja Cave (after Klimchouk, Andrejchouk and Turchinov, 1995).

Meso-Morphology of Caves
The shapes and sizes of the passages, and of the smaller-scale forms produced within the passages, depend upon geologic and hydrodynamic factors, the most important of which are the following:

- Shape and size of initial fissures.
- Distinctive structural and textural features of the gypsum within a given horizon.
- Position and function of a given morphological element in a karst circulation system.
- Distinctive effect of the local hydrogeologic evolution on the late stages.

Because of the locally varying role of these factors, passage morphology can vary substantially between caves, different parts of a single cave, and particularly between the cave levels.

Vertical structural and textural differentiation of the gypsum unit (into three distinct subhorizons across most of the Podol'sky area) is believed to have resulted largely from late diagenetic recrystallization of the rock (Klimchouk, Andrejchouk, and Turchinov, 1995). Such differentiation is the main cause of the development of distinct lithogenetic fissure (and hence passage) patterns, each confined to a specific layer (see 5.2.1 Figure 9). Also, the varying position and function of passage levels in transverse cave systems contribute further to the morphologic distinction between levels. Thus, at least the first three of the factors listed above combine to influence the development of different morphologies of passages at different levels.

The converse is even more consistently true: passages developed at the same level within a particular cave area (passages occupying the same geologic position and providing the same hydrologic function in a system) are characterized by uniform morphology, as all four of the factors listed above act uniformly.

Passages typically have a cleft-shaped cross section, with a flat or gothic-arched ceiling. Also common are relatively wide (2 to 4 m) passages with symmetric horizontal notches and inwardly inclined facets in the walls, or with two or more levels of notching (5.2.1 Plate 1, p. 270) and 5.2.1 Figure 10B, C, D, p. 271). The ceilings of these passages can also be flat or gothic-arched. Such morphologic features are the result of dissolution driven by natural convection, under either artesian or water table conditions (Kempe, 1972; Klimchouk, 1997d; Chapter 6.1).

Dissolution cupolas and domepits are widespread, having developed from a master passage up to 12 m above ceiling level. They can be "blind" (5.2.1 Plate 3A), or open upward into a passage at the next higher level. Some terminate at the base of the overlying limestone bed. Such domepits may be extended laterally in their upper parts, if a higher level fissure was intercepted (5.2.1

5.2.1 Plate 1. Typical cross sections of master passages in gypsum maze caves of the western Ukraine: ***A, B*** *and* ***C*** *are Ozernaja Cave.* ***D*** *is Zoloushka Cave. The guiding fissure is normally recognizable in passage ceilings, except where a fissure ends abruptly at the bottom of the next upper bed* ***(A)****. The widened part of the cross section continues downwards in a rift-like profile* ***(B)****, although this is obscured in most instances by clay fill (A, C and D).*

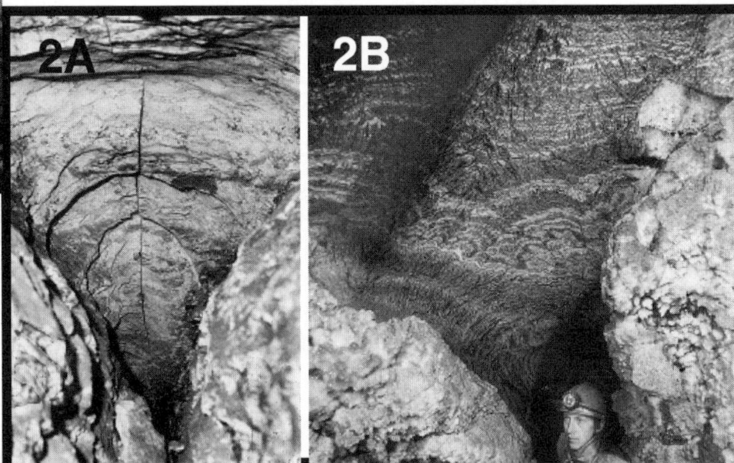

5.2.1 Plate 2. Related morphologic elements playing different hydrologic functions in a cave system and illustrating the effects of buoyant density currents. ***A*** *and* ***B*** *are feeder conduits where they connect to master passages (Ozernaja and Dzhurinskaja Caves).* ***C*** *and* ***D*** *are master passages (Dzhurinskaja Cave) with ceiling half-tubes that normally begin from feeder conduits and end up at a domepit, as illustrated by* ***E*** *(Optimisticheskaja Cave).*

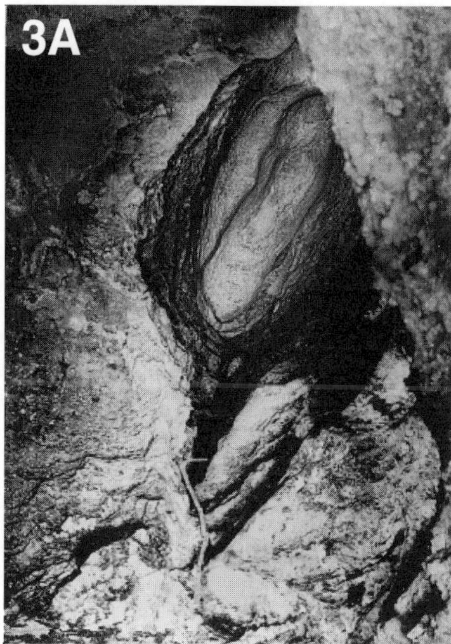

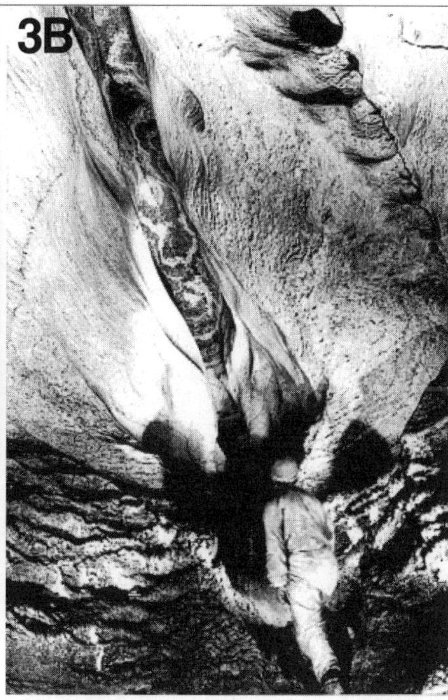

5.2.1 Plate 3. Atlantida Cave: ***A*** *is an "Ascending" elongated domepit formed by dissolution due to natural convection along a higher level fissure that was intercepted by a growing master passage (viewed upward from a master passage).* ***B*** *is a similar feature, but greatly elongated at the upper part along the maximum lateral extent of the upper fissure, having the form of a passage but terminating at both ends. In Atlantida Cave such features, developed from adjacent master passages along laterally connected upper fissures, often join at the upper level to form small networks and provide "bypasses" between the master passages.*

Plate 3B). Cupolas and domepits, as well as another common feature, ceiling half-tubes (5.2.1 Plate 2C, D), are also formed largely by dissolution driven by natural convection. The development of ceiling half-tubes is accomplished by buoyant currents of relatively fresh water that will always tend to occupy the highest available position in a passage containing more highly saturated bulk water (Klimchouk, 1997d). Such half-tubes can commonly be traced continuously from the "feeder" conduits at a lower level (5.2.1 Plate 2A, B), through master passages (5.2.1 Plate 2C, D), to domepits that open to the next higher level or reach the upper boundary of the gypsum (5.2.1 Plate 2E; 5.2.1 Figure 11).

During the final stage of artesian speleogenetic development, when the upper confining bed is breached locally by an incising valley or areas of collapse, groundwater flow through a cave system increases dramatically, due to unconstrained discharge. Accelerated growth of the passages along preferred flow paths results in development of the large trunk passages that are recognized in most of the region's caves (5.2.1 Plate 1D). In some cases, passages within a particular part of a maze may be enlarged substantially relative to average passage size in adjacent areas. This is exemplified by the Chamber of Chernovtsy Speleologists in Zoloushka Cave, where only small pillars remain within a large space that was formed by the coalescing of enormously enlarged passages. Another mechanism responsible for comparatively rapid passage enlargement, which operates more uniformly within specific areas at a particular level, is horizontal notching in response to dissolution at the water table.

Typical features formed under modern entrenched karst conditions are vertical dissolution pipes, which grow due to a focused descending percolation from overlying formations. They are 1 to 3 m wide, extend downward through the full thickness of the gypsum from its top, and are commonly superimposed upon relict artesian passages. Recent linear "through" caves carrying active streams are relatively rare in the first zone but more common in the second zone. Normally they develop along the base of the gypsum, originating from doline ponors and extending towards the nearest entrenched valley. Some such caves may enter areas of relict mazes, where the active streams dissipate, and water filters downward into the underlying aquifer (as is the case in the entrance passage of Optimisticheskaja Cave). Such areas of the caves are prone to short-term flooding after heavy rainfall or active snowmelt.

Cave Sediments and Formations

By far the predominant clastic sediment in the maze caves of the region are successions of fine clay, with minor beds of silty clay (Klimchouk, 1984). These fill passages to a variable extent and can reach 5 to 7 m in thickness. The clay material is derived largely from overlying clayey formations. It has been introduced into the caves mainly during the later stages of artesian speleogenesis, when breakdown processes became active and flow velocities increased, allowing redistribution of suspended clastic material. However, gradient fields within the aquifer remained relatively uniform, so that the sediment is very fine grained, exceptionally well sorted, and maintains a uniform facies over considerable distances. The number of silty beds increases in the upper part of the clay sediment sequence, reflecting the more variable hydraulic environment that was transitional from confined to unconfined, during the subjacent and early entrenched karst stages. Repeated transitional cycles from a reducing to an oxidizing geochemical environment are marked by the presence of Fe and Mn hydroxide layers (Klimchouk and Rogozhnikov, 1982). Massive deposition of Fe and Mn compounds has occurred in Zoloushka

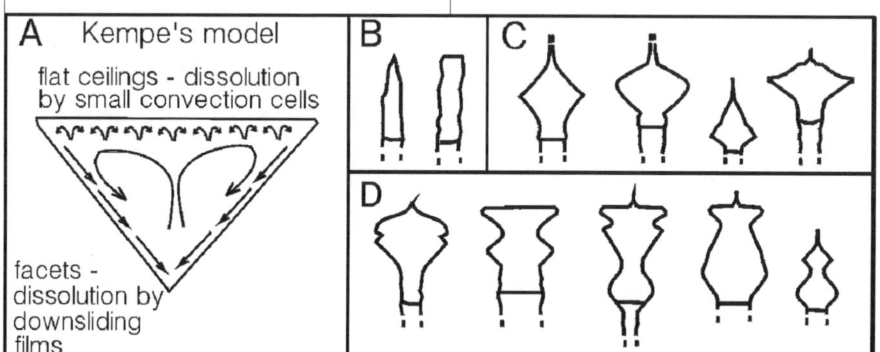

5.2.1 Figure 10. Wall notching and wall facet formation by dissolution caused by natural convection. Panel ***A*** *is Kempe's model (1972) of the morphologic effects of dissolution due to natural convection circulation. Panels* ***B, C, D*** *are typical passage cross sections from the gypsum caves of the western Ukraine.*

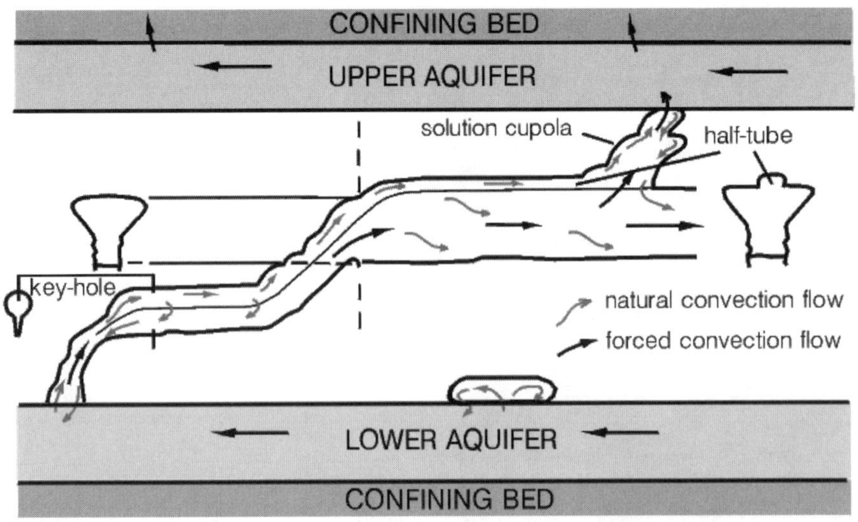

5.2.1 Figure 11. The formation of upward solution forms by buoyant currents. The diagram shows schematically the relationship between natural convection flow and forced flow on the mature stage of artesian speleogenesis, when a connection has already been established through the gypsum, but the forced flow is still slow due to the major discharge constraint of the upper confining bed (from Klimchouk, 1997d).

Cave, where a rapid transition of the type described has been caused by groundwater abstraction during the past 50 years (Volkov, Andrejchouk, and Janchuk, 1987). Sandy gravel sediment occurs in the upper part of the cave fill, but it is limited to areas that surround active ponors. It has been reworked and redeposited by streams from the old (late Pliocene to early Pleistocene) alluvial sediment that commonly overlies Sarmatian clay.

Breakdown deposits are also quite common in the caves. They include chip, slab, and block breakdown material from the gypsum, as well as more massive breakdown from the overlying formations. Locally the plastic Sarmatian clay is injected into caves in breakdown zones, due to the pressure of the overburden. The distribution of the breakdown deposits is governed by passage size, by local textures and structures within the gypsum of a given area or layer, and by tectonic faults.

Various chemical deposits, speleothems, and formations are quite common in the gypsum caves of the western Ukraine (Rogozhnikov, 1984; Turchinov, 1993). Calcite speleothems (stalactites, stalagmites, flowstone, and helictites) occur locally in zones of water percolation from overlying formations. Among other carbonate minerals, rhodochrosite occurs within a flowstone in Optimisticheskaja Cave (Turchinov, 1993). Gypsum crystals of different habit, size, and genesis are the most common cave formations. Those covering the walls in many parts of the caves have a largely subaerial origin, having formed by evaporation of a saturated solution moving as a thin film on the walls.

Also common are gypsum formations deposited by evaporation of interstitial solutions seeping from the host rock (local crust, and "flowers" of fibrous gypsum). "Gypsum snow," like hoar frost on passage floors, and gypsum rims are assumed to be precipitated from aerosols (Klimchouk, Nasedkin, and Cunningham, 1995). Silicates are represented by lutecite, a variety of cristobalite, found in Optimisticheskaja Cave (Turchinov, 1993). Oxides and hydroxides (Fe and Mn compounds) are common within the clay fill of many caves, and also occur as crusts, stalactites, and stalagmites (Zoloushka Cave). Manganese minerals are represented by birnessite and asbolite-buserite in Zoloushka Cave (Volkov, Andrejchouk and Janchuk, 1987).

Gypsum Maze Caves and the Origin of Native Sulfur Deposits

Large bioepigenetic deposits of native sulfur in the pre-Carpathian region are genetically associated with the same gypsiferous unit, and are all located within the third zone described above (see 5.2.1 Figure 2). Sulfur is embedded in epigenetic calcite that partially (at the top) or wholly replaces the gypsum. The origin of these sulfur deposits is quite controversial. Recently a new interpretation, which implies a fundamental role of gypsum speleogenesis in the sulfur origin, has been suggested (Klimchouk, 1997b). It is evident that, in general, bioepigenetic sulfur formation is related to karst, because dissolved sulfates are needed to fuel the large-scale sulfate reduction that is a key process in the sulfur cycle. But in the case of the pre-Carpathian sulfur deposits, the role of speleogenesis as a governor of groundwater circulation in the artesian aquifer system was decisive (Klimchouk, 1997a). The proposed model for the origin of the sulfur deposits hinges upon the transverse groundwater communication between the subgypsum and supragypsum aquifers, through maze cave systems in the gypsum. Such a flow pattern and speleogenetic evolution within the gypsum provided the spatial and temporal framework within which the sulfur cycle processes took place, as well as controlled the geochemical environment and the migration of reactants and reaction products between them. The model explains the well-accepted relationship of sulfur deposits to buried valleys and karst zones, and resolves many contradictions in earlier interpretations of the geologic features of the Tyrassky Formation and the hydrogeologic features of sulfur deposits (Klimchouk, 1997b).

Conclusions

- Large maze caves in the Miocene gypsum in the western Ukraine were formed under confined hydrogeologic conditions and became relict when breached by the main entrenched valleys in the interior part of the platform. In other parts of the region, which lie directly adjacent to the foredeep, speleogenesis under confined settings is still in progress.
- The main factors responsible for the formation of the huge multistory maze caves in the region are:
  - The confined setting, with ascending transverse groundwater communication through the gypsum between the lower and upper adjacent aquifers, hence with dispersed aggressive recharge to the gypsum.
  - The quasi-uniform (in plan) and multistory (in profile) occurrence of fissures in the gypsum. Transverse hydraulic communication and speleogenetic development are favored in areas of discharge (potentiometric lows) where confining properties of the capping clay are weakened or locally breached due to stratigraphic or tectonic discontinuities and incised erosional valleys.
  - Groundwater circulation driven by natural convection played an important role in establishing the cave patterns, particularly during the early stages of speleogenesis. Upward development of cavities by dissolution driven by natural convection helped to vertically connect fissured horizons in the gypsum and, eventually, the lower and upper adjacent aquifers.
- Speleogenesis in the gypsum was the major factor that governed groundwater circula-

tion in the whole Miocene artesian aquifer system through its evolution. Before the onset of speleogenesis, the gypsum stratum behaved as an aquiclude hydraulically separating the subgypsum and supra-gypsum aquifers. Speleogenetic development has changed the hydrogeologic function of the gypsum from an aquiclude to a conduit aquifer, with cave systems providing an effective vertical and lateral hydraulic communication. At the stage of entrenched karst the presence of cave systems in the gypsum has induced sinkhole landscape development that provided point recharge of groundwater and its rapid transmission to the lower unconfined aquifers.

References

Andrejchouk, V.N., 1988, *The tectonic factor and peculiarities of the sulfate karst of Bukovina: Geology, geomorphology, and hydrogeology of karst:* Sverdlovsk, 66 p. (in Russian).

Dublyansky, V.N., and Ivanov, B.N., 1970, Karst of the Podol'sky-Bukovinsky karst region: Kiev University, *Fizicheskaja geografija I geomorfologija,* no. 4 (in Ukrainian).

Dublyansky, V.N., and Smol'nikov, B.N., 1969, *Karstological and geophysical investigations of karst cavities of the Pridnestrovskaja Podolija I Pokutje*: Kiev, Naukova Dumka. 151 p. (in Russian).

Ivanov, B.N., and Dublyansky, V.N., 1966, Superficial and underground karst of the south-west flank of the Russian Platform, *in* Voprosy izuchenija karsta Russlkoj ravniny: Moscow (in Russian).

Kempe, S., 1972, Cave genesis in gypsum with particular reference to underwater conditions: *Cave Science,* v. 6, no. 49, p. 1-6.

Klimchouk, A.B., 1984, The experience of the detailed study of water-mechanical sediments of a large cave system, *in* Peshchery (Caves): Perm University, p. 70-88 (in Russian).

Klimchouk, A.B., 1986, Genesis and development history of the large gypsum caves in the western Ukraine: *Le Grotte d'Italia,* v. 4, no. 13, p. 51-71.

Klimchouk, A.B., 1990, Artesian genesis of the large maze caves in the Miocene gypsum of the western Ukraine: *Doklady Akademii Nauk Ukrainskoj SSR,* ser. B, v. 7, p. 28-32 (in Russian).

Klimchouk, A.B., 1992, Large gypsum caves in the western Ukraine and their genesis: *Cave Science,* v. 19, p. 3-11.

Klimchouk, A.B., 1994, Speleogenesis under confined conditions, with recharge from adjacent formations: *Publ. Serv. Geol. Luxembourg* Comptes Rendus du Coll. Int. de Karstol. a Luxembourg, v. 27, p. 85-95.

Klimchouk, A.B., 1996, Gypsum karst in the western Ukraine, *in* Klimchouk, A., Lowe, D., Cooper, A., and Sauro, U., eds., Gypsum karst of the world: *International Journal of Speleology*, v. 25, no. 3-4, p. 263-278.

Klimchouk, A.B., 1997a, The role of speleogenesis in the Miocene gypsum in the western Ukraine in groundwater circulation in the multi-story artesian system, *in* Gunay, G., and Johnson, I., eds., *Karst Waters and Environmental Impacts:* Rotterdam, A.A.Balkema., p. 281-292.

Klimchouk, A.B., 1997b, The role of karst in the genesis of sulfur deposits, PreCarpathian region, Ukraine: *Environmental Geology,* v. 31, p. 1-20.

Klimchouk, A.B., 1997c, Artesian speleogenetic setting: *12th International Congress of Speleology,* La Chaux-de-Fonds, Proceedings, v. 1, p. 157-160.

Klimchouk, A.B., 1997d, Speleogenetic effects of water density differences: *12th International Congress of Speleology,* La Chaux-de-Fonds, Proceedings, v. 1, p. 161-164.

Klimchouk, A.B., and Andrejchouk, V.N., 1988, Geologic and hydrogeologic conditions of development of large gypsum caves in the western Ukraine and their genesis, *in* Peshchery (Caves), Gypsum and anhydrite caves: Perm University, p. 12-25 (in Russian).

Klimchouk, A.B., Andrejchouk, V.N., and Turchinov, I.I., 1995, *Structural prerequisites of speleogenesis in gypsum in the western Ukraine*: Kiev, Ukrainian Speleological Association, 104 p.

Klimchouk, A.B., et al., 1985, Geologic and hydrogeologic conditions of karst development of the Pridnestrovsky Podolia, *in* Sokolovskij, I.L., and Klimchouk, A.B., eds., Karst of the Ukraine: *Fizycheskaja Geographija i Geomorphologija,* Kiev, no. 32, p. 47-54 (in Russian).

Klimchouk, A.B., et al., 1996, Dissolution of gypsum from field observations, *in* Klimchouk, A., Lowe, D., Cooper, A., and Sauro, U., eds., Gypsum karst of the world: *International Journal of Speleology*, v. 25, no. 3-4, p. 37-48.

Klimchouk, A.B., Nasedkin, V.M., and Cunningham, K.I., 1995, Speleothems of aerosol origin: *National Speleological Society Bulletin,* v. 57, p. 31-42.

Klimchouk, A.B., and Rogozhnikov, V.Y., 1982, *Conjugate analysis of the development history of a large cave system (with the example of Atlantida Cave):* Kiev Institute of Geology Nauk, 58 p. (in Russian).

Polkunov, V.F., 1990, Mineral concentrations of native sulfur, *in* Geotektonika Volyno-Podolii: Kiev, Naukova Dumka, p. 229-232.

Rogozhnikov, V.Y., 1984, Water-chemical deposits in karst labyrinthic caves of the Podol'sky Pridnestrovje, *in* Peshchery (Caves): Perm University, p. 46-55 (in Russian).

Shestopalov, V.M., ed., 1989, *Water exchange in hydrogeologic structures of the Ukraine–Water exchange under natural conditions*: Kiev, Naukova Dumka, 288 p. (in Russian).

Turchinov, I.I., 1993, Secondary mineral formations of gypsum caves of the western Ukraine: *Svet (Light),* Kiev, v. 3, no. 9, p. 29-37 (in Russian, English summary).

Volkov, S.N., Andrejchouk, V.N., and Janchuk, E.A., 1987, Modern iron-manganese formations of the Zoloushka Cave: *Mineralogicheskij Sbornik Lvov,* v. 4, p. 79-83 (in Russian).

5.2.2. Speleogenesis of the Black Hills Maze Caves, South Dakota, U.S.A.

Arthur N. Palmer and Margaret V. Palmer

Abstract

Caves of the Black Hills of South Dakota, U.S.A., are located in the Madison Limestone of Mississippian (early Carboniferous) age in a zone of diagenetic breccia and late Mississippian paleokarst. Most of the caves are extremely complex networks with multiple stratigraphically controlled tiers. Today they are essentially hydrologic relics. Their history is as complex as the caves themselves:

- The earliest cave openings were formed by diagenetic processes, mainly by the dissolution and reduction of sulfates. Oxidation of hydrogen sulfide produced many small and rather isolated voids lined by brecciated bedrock.
- Late Mississippian exposure produced caves, dolines, and surface fissures, which were later filled with basal Pennsylvanian (late Carboniferous) sand and clay of the Minnelusa Formation.
- Deposition of sedimentary strata buried these early karst features to depths of at least two kilometers. During this time, voids that had not been entirely filled by Pennsylvanian sediment were lined by a thin layer of scalenohedral calcite, and later by quartz.
- Uplift of the Black Hills at the end of the Cretaceous Period and beginning of the Tertiary Period exposed the Madison Limestone once again, allowing rapid groundwater flow through it. The earlier caves and solution pockets were enlarged at this time.
- A thick layer of rhombohedral calcite precipitated on the cave walls, probably as a result of stagnation of groundwater caused by late Tertiary aggradation, which blocked spring openings.
- Both before and after the calcite wall crust was deposited, deep subaerial weathering produced boxwork, with veins of calcite that had replaced earlier sulfates, as well as thick accumulations of carbonate sediment.

The Tertiary cave enlargement probably involved mixing of at least two of the following water sources: artesian flow from recharge along the carbonate outcrop area, diffuse recharge through the overlying sandstone, and rising thermal water. There is evidence for all three sources, but the relative importance of each is still uncertain.

Introduction

The Black Hills, located on the border between South Dakota and Wyoming, U.S.A., contain some of the most complex caves in the world, both in layout and in developmental history. These include Jewel and Wind Caves, with respective surveyed lengths of 195 km and 138 km (5.2.2 Figures 1 and 2). They have the unique distinction of having been cited as classic examples of three entirely different modes of origin: artesian (Darton, 1901; Tullis and Gries, 1938; Howard, 1964), thermal (Owen, 1898; Bakalowicz et al., 1987), and rejuvenated paleokarst (Palmer and Palmer, 1989, 1995). In fact, all three origins have contributed. The caves are polygenetic, having undergone several stages of development:

- Early Carboniferous diagenetic processes, including dissolution and brecciation.
- Middle Carboniferous karst development, with later filling by detrital rocks.

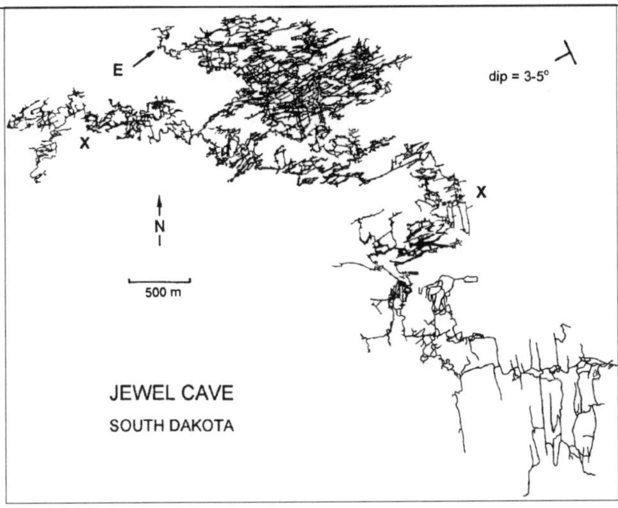

5.2.2 Figure 1. Map of Jewel Cave, Jewel Cave National Monument. In the southeastern section, explored since 1993, only the survey lines are shown and do not indicate the true passage widths. Sections beyond points labeled X have been explored since 1987. Map courtesy of the U.S. National Park Service. E is the entrance.

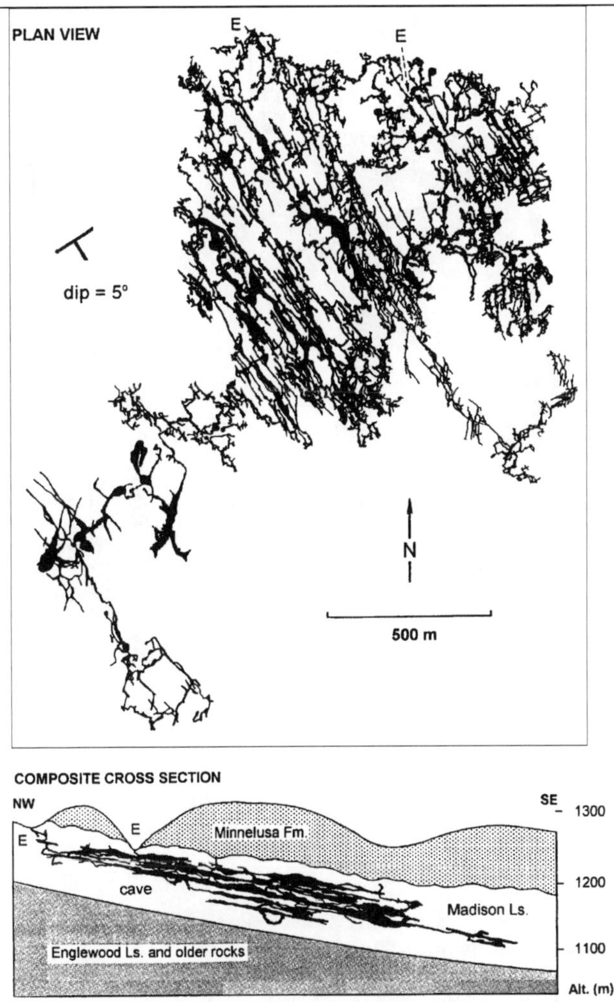

5.2.2 Figure 2. Map and profile of Wind Cave, Wind Cave National Park, South Dakota. The most recent discoveries are in the southwestern part of the cave. Note the stratal control of the cave tiers, and the almost total lack of passages in the lower and uppermost parts of the Madison Limestone. Map courtesy of U.S. National Park Service. E is the entrances.

- Deep burial and mineralization.
- Renewed cave development during the Tertiary, in response to orogenic uplift of the Black Hills.
- Modification by subaerial processes, including condensation corrosion.

Geologic Setting

The Black Hills consist of a small but prominent structural dome that was uplifted during the Laramide Orogeny (latest Cretaceous and early Tertiary). The uplift correlates in age and geologic setting with the Rocky Mountains farther west. Precambrian igneous and metamorphic rocks occupy the mountainous core of the uplift, which rises to altitudes of up to 2207 meters at Harney Peak. Paleozoic and Mesozoic sedimentary rocks dip off the flanks of the uplift in all directions, with their eroded edges forming concentric ridges and valleys that encircle the mountains (5.2.2 Figure 3).

The major caves of the region are in the Madison Limestone, locally called the Pahasapa Limestone, which consists of 100 to 190 m of limestone and dolomite of early-middle Mississippian age (early Carboniferous). The top of the Madison was exposed to intense karst development near the end of the Mississippian and then buried by Pennsylvanian (late Carboniferous) detrital rocks of the Minnelusa Formation. The Minnelusa consists mainly of quartz sandstones up to 250 m thick containing interbedded shale, carbonates, and sulfates. In the plains surrounding the Black Hills, the Minnelusa is overlain by more than 2 km of mainly detrital sedimentary rocks that range in age from Permian to Tertiary, but erosion has removed them from the cave areas. 5.2.2 Figure 4 shows the stratigraphic column for the area.

Throughout the Tertiary, uplift and erosion alternated with periodic accumulation of detrital sediment. The present land surface acquired its shape mainly during the Eocene Epoch, as shown by the fact that much of the surface is presently being exhumed from a cover of Oligocene siltstone that accumulated around the perimeter of the Black Hills. Although the mountains themselves were not covered by continental glaciers, climatic variations during the Pleistocene caused fluctuations in river levels, which left a series of terraces in the major river valleys. The caves show no apparent relation to the terraces (Ford et al., 1993).

The sedimentary rocks dip away from the center of the uplift at an average of about 5 degrees, except along the eastern flank, where the dips are 20 to 30 degrees or more. Most of the Madison Limestone is massive and competent, with prominent joints and faults. Most of these fractures are perpendicular to the bedding, although some faults are low-angle thrusts. Where the Madison Limestone is exposed at the surface, a great deal of water infiltrates and flows downdip as artesian groundwater. Folding in the southeastern Black Hills has brought the Madison relatively close to the surface 5 to 25 km downdip from the main Madison outcrop belt. Most of the artesian water emerges there as warm sulfate-rich springs concentrated along the crests of anticlines, where upward flow takes place along fractures and breccia zones in the Minnelusa Formation (Rahn and Gries, 1973).

Character of the Black Hills Caves

Nearly all the caves in the Black Hills share the same general pattern and geologic history. With few exceptions they are network mazes

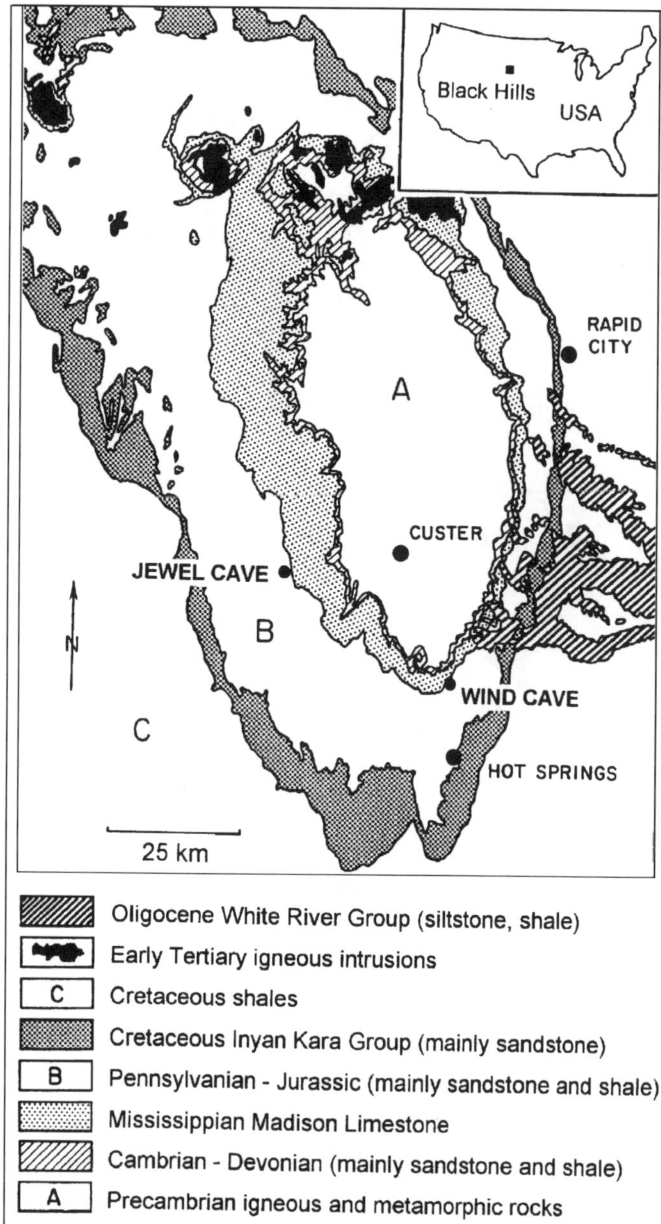

5.2.2 Figure 3. Geologic map of The Black Hills.

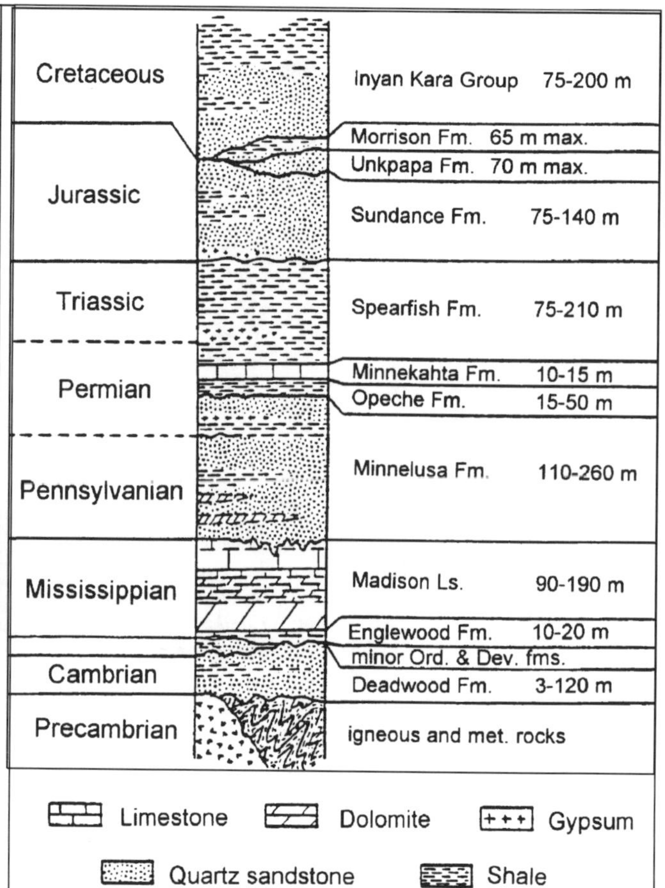

5.2.2 Figure 4. Paleozoic and Mesozoic strata exposed in the Black Hills.

arranged in several distinct tiers that correspond to favorable strata in the carbonate rocks. They dip at roughly the same angle as the strata (5.2.2 Figure 2). The lowest passages are fracture-controlled fissures in massive dolomite. The intermediate passages are mainly wide tubes and irregular rooms in prominently bedded limestone and dolomite (5.2.2 Plate 1). Some passages in these strata are bounded above or below by thin chert beds. These two lower tiers of passages are highly interconnected, but with abrupt, discordant, and ungraded passage junctions. Sharp changes in ceiling and floor levels are common. The uppermost level in the caves consists of isolated wide, vaulted rooms or irregular galleries in massive limestone (5.2.2 Plate 2). These voids are generally connected to underlying passages by narrow vertical funnels in a prominent zone of bedded chert that separates the bedded limestone and dolomite from the overlying massive limestone.

The caves are concentrated in areas where the limestone is covered by a thin cap of Minnelusa sandstone. Very few caves show a clear relation to surface topography or to the present pattern of groundwater recharge. Entrances are accidental features formed by the intersection of the caves by the erosion surface. Cave passages show little evidence either of extending entirely updip into the out-crop zone or down-dip far into the phreatic zone. These limits may be biased by explorational barriers, but they are curiously prevalent throughout the major caves of the Black Hills. Thus it is difficult to interpret the nature of groundwater sources or outlets, as is so easily done in most other karst areas.

Very few Black Hills caves contain running water, except during periods of heavy runoff from the surface. Wind Cave is the only one that extends down to the present water table. Several passages in the lowest cavernous strata terminate downdip in ponded water. Some of these lakes are fed by a small perennial trickle of water, but none of this water is aggressive. In fact, the lakes are supersaturated and are depositing calcite as rafts and wall crusts.

The present caves intersect many late Mississippian paleokarst features, whose fill of detrital sediment is easily recognized by its bright red or yellow color. The sediment spills into the more recent cave passages, exposing Mississippian paleocaves (5.2.2 Plate 3). Most of the vaulted upper-level rooms in the massive limestone contain narrow, irregular tubes extending upward from their ceilings. In many of them one may climb up to the paleokarst surface at the base of the Minnelusa. Many of these rooms and rising tubes are partly filled with Pennsylvanian sediment.

Cave patterns seem to be well adjusted to

the structure of the Black Hills. Most of the major fracture-controlled passages are oriented radially from the center of the uplift, like the spokes of a wheel. However, these fractures greatly predate the Laramide uplift of the Black Hills, as shown by the fact that many of them have guided the diagenetic and paleokarst features as well as the Cenozoic caves. As is true for many other structural features of western North America, the basic outline of the Black Hills was apparently established during the Precambrian and has remained a slightly positive feature of tectonic unrest throughout the Paleozoic Era. The fractures that so conspicuously control the major cave passages today were inherited and enlarged from pre-Laramide times.

5.2.2 Plate 1. A typical middle-level passage in Jewel Cave (part of the Scenic Tour), showing irregular fissure cross section and calcite spar coating 15 cm thick. Note where the crystal coating has spalled off the wall at the lower right.

Geologic History of The Black Hills Caves

Diagenetic Processes

When the Madison Limestone was deposited early in the Mississippian period—about 340 to 330 million years ago—it originally contained considerable gypsum and anhydrite. Soon afterward, during brief periods of subaerial exposure late in the Mississippian period, the sulfates were dissolved, altered, and replaced, resulting in considerable brecciation of the surrounding limestone and dolomite beds, as well as dissolution voids, altered bedrock, and boxwork. These processes are described briefly below (see also Palmer and Palmer, 1989, 1995). The original sulfates still survive in structural basins between mountain uplifts.

As much as 25 to 30 percent of the Madison thickness has been brecciated by sulfate-related processes. The uppermost breccia is mainly the result of sulfate dissolution, with simultaneous collapse of the overlying carbonate beds into the voids. This breccia is concordant to the strata, with subrounded fragments and a carbonate-rich detrital matrix. Sando (1974) described similar features in correlative beds of neighboring states. Stratigraphically lower breccia was produced where dissolution of original anhydrite caused supersaturation of the groundwater with gypsum, resulting from the fact that anhydrite at temperatures less than about 30° C is more soluble than gypsum. Crystallization of gypsum in fissures caused the carbonate bedrock to shatter, resulting in stratally discordant breccia bodies containing angular fragments (5.2.2 Plate 4). The matrix consists of calcite that is colored yellow-red by iron oxide. Microscopic examination of the calcite shows abundant filaments of iron-fixing bacteria. The filaments and iron oxide indicate that the deposition of the calcite matrix also involved the reduction of sulfates to hydrogen sulfide, production of metal sulfides, and later oxidation.

Reduction of sulfates to hydrogen sulfide also caused the carbonate rock to be dissolved in scattered zones where the hydrogen sulfide was oxidized to sulfuric acid. Evidence for this process is seen today as isolated voids up to a few meters in diameter in relatively fresh walls exposed by massive collapse within the later caves. These voids are lined with shattered bedrock cemented by the conspicuous yellow-red calcite, showing the effect of wedging by crystallizing gypsum. Early solutional voids of this type served as the framework for many of the later caves and karst features.

Some cave walls contain white siliceous rinds that separate the friable outer bedrock from the unaltered bedrock beneath. Precipitation of secondary silica was favored where infiltrating high-pH water from evaporative environments at the surface encountered the near-neutral or low-pH zones around sulfate bodies. The presence of microbial filaments within the silica suggests that this mineralization may have been biologically mediated. However, microbial filaments are common in redox environments and may have played no role in precipitating the silica.

These diagenetic and alteration processes have resulted in caves with unusually

5.2.2 Plate 2. A typical upper level in Wind Cave (part of the Fairgrounds Tour), a wide arched room in massive limestone. Note the floor coating of authigenic carbonate sediment from subaerial weathering, combined with detrital sand and clay from pockets of Pennsylvanian fill intersected by the cave.

5.2.2 Plate 3. *Paleokarst filling at the base of the Minnelusa Formation in Wind Cave. Large white fragments consist mainly of Mississippian limestone and sandstone surrounded by a matrix of red-brown sand, silt, and clay.*

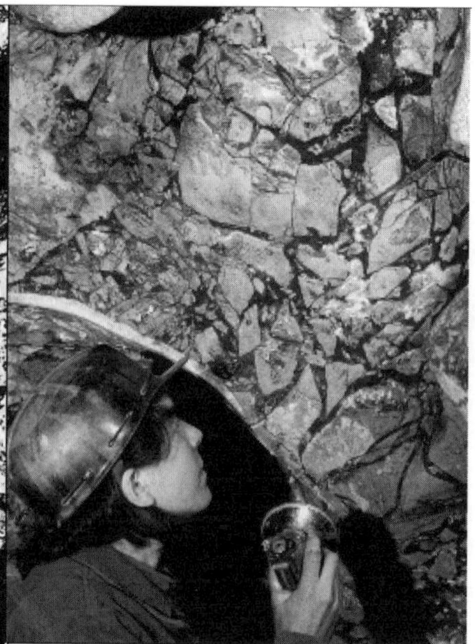

5.2.2 Plate 4. Brecciated Madison Limestone in the Madison Limestone in Jewel Cave. The angular fragments were produced by recrystallization of gypsum. The matrix consists of reddish brown calcite, which has replaced gypsum. The calcite cave lining has been almost completely removed by condensation

weathered and multicolored walls, and with a variety of features rarely seen in other caves. Most or all of these processes owe their origin to the now-vanished sulfates.

Origin of Boxwork

Many of the bedrock walls exposed in the caves exhibit prominent boxwork, which consists of paper-thin intersecting veins of calcite that project into the caves as resistant fins, and from which the intervening bedrock has been removed by weathering (5.2.2 Plate 5). Most of the calcite veins are roughly perpendicular to the bedding, but some lie along bedding planes and others show splayed patterns. Vein spacing ranges from a few centimeters to several decimeters. The veins are similar in composition and origin to the matrix of the angular breccias, but they formed along narrow fractures resulting from stresses produced when anhydrite was hydrated to gypsum. The intervening pockets, from which the bedrock has fallen out, average about 10 cm deep, although in places their depths exceed 1/2 meter.

The bedrock is less resistant than the boxwork veins not simply because it is less crystalline, but because it has been considerably altered to a friable sand consisting of calcite crystals held together at their contacts by a sparse cement of secondary quartz. The calcite crystals in the bedrock are mostly pseudomorphs after gypsum, and the quartz is the remnant of an early matrix that formed around former dolomite crystals. Much of the original bedrock was apparently removed by H_2S-H_2SO_4 dissolution processes, which left many millimeter-sized pores. These porous zones readily weathered away during cave development, as well as later when they were in contact with the moist cave atmosphere. Oxidation of metallic sulfides in the cave atmosphere has colored the weathering zones in places with a variety of bright reds and yellows.

Boxwork veins are limited almost entirely to dolomite beds within the Madison. Petrographic analysis shows that the calcite veins have replaced former gypsum, and that they predate the late Carboniferous paleokarst (Palmer and Palmer, 1989). This process has been observed in various stages of completion at the contact between gypsum and dolomite beds in the Permian San Andres Formation of New Mexico, U.S.A.

Boxwork has also been erroneously attributed to other processes. The traditional explanation is that it consists simply of more resistant calcite veins projecting from the bedrock. Palmer (1981), noted that the boxwork was concentrated only around the caves. So he originally assumed that the veins were precipitated after the cave formed, where supersaturated infiltrating water was able to lose CO_2 through the porous dolomite and precipitate calcite in the fractures. This mechanism is valid for certain calcite veins in young carbonate sediment, but it is not appropriate in the Black Hills caves.

The origin of the veins has also been attributed to calcite precipitation in fractures as the result of dedolomitization during enlargement of the caves during the Tertiary Period (Ford and Bakalowicz, 1983; Millen and Dickey, 1987). However, this is not a feasible mechanism either.

The boxwork veins are truncated by the late Carboniferous paleokarst features, and therefore greatly predate the Tertiary enlargement of the caves. In addition, it is not clear how calcite can precipitate within the fractures of a dissolving dolomite surface.

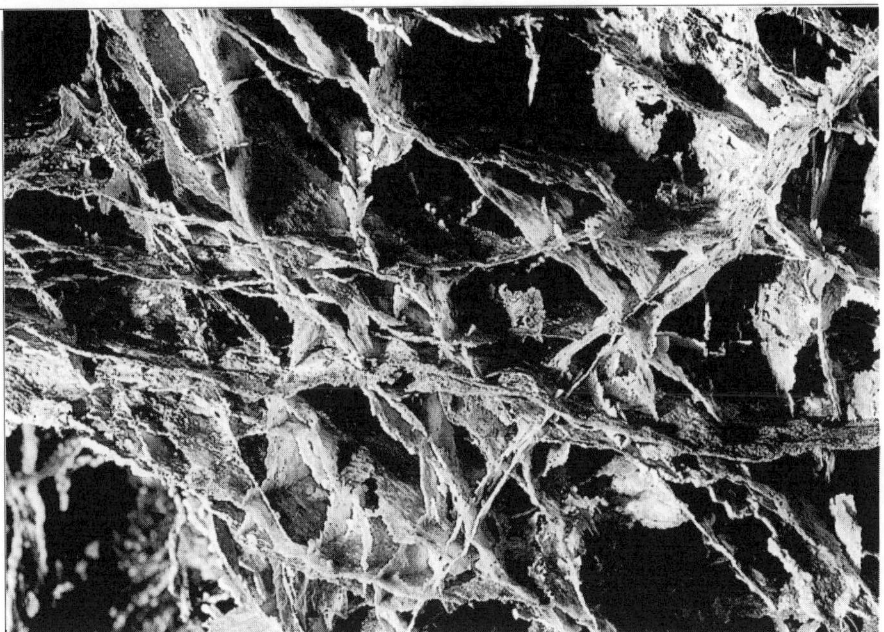

5.2.2 Plate 5. Calcite boxwork in a dolomitic bed in Wind Cave. The fractures were produced by mechanical stresses caused by hydration of anhydrite. The veins were originally anhydrite and/or gypsum. Long dimension of photo is approximately 75 cm.

Development of Middle Carboniferous Paleokarst

Exposure of the Mississippian carbonate rocks 320 to 310 million years ago produced a broad karst plain containing sinkholes and fissures, and underlain by caves. These karst features were later filled partly or completely by the late Carboniferous Minnelusa Formation about 310 million years ago. These features are now part of the most extensive paleokarst horizon of western North America, remnants of which are clearly visible throughout the Rocky Mountains and Colorado Plateau.

Relict Mississippian caves are abundant below the paleokarst surface and are concentrated about 20 to 50 m below the top of the Madison Limestone. Although most are (or were) filled with detrital sediment carried in from the surface, connections to the paleokarst surface are obscure. The caves are mainly isolated domed chambers, rather than continuous systems. They are rarely more than 10 m in height or more than 100 m in lateral extent and seem to be entirely phreatic in origin, because they lack vadose features such as canyon passages, fluted vertical shafts, and speleothem remnants. Some caves received very little surface-derived sediment and seem to have survived more or less intact without filling.

The late Carboniferous Minnelusa fill consists mostly of red and yellow sand, silt, and clay. It includes 60 percent quartz sand, 20 percent muscovite and biotite, 10 percent feldspar, 10 percent clay minerals, and traces of hornblende, and tourmaline. The bottom-most deposits are mainly laminated red and white carbonate sand derived from the local carbonate bedrock. Cloud-shaped white calcite bodies in the sediment indicate by their shape that they are replacements of original sulfate nodules.

In places, the paleokarst fill also contains fragments of limestone, sandstone, and chert (5.2.2 Plate 3). The sandstone blocks were derived from late Mississippian strata that originally overlay the limestone but which were entirely removed from the cave area during the origin of the middle Carboniferous karst surface.

Deep Burial

The paleokarst surface was eventually buried to depths of several kilometers by sedimentary rocks ranging from late Carboniferous to Tertiary age. Groundwater in the Madison Limestone became nearly static. Several distinctive deposits can be recognized from the deep-burial stage. Early voids that were not completely filled with sediment acquired a lining of clear to white scalenohedral calcite ("dogtooth spar") 1 to 2 cm thick. It is brightly cathodoluminescent, representing anoxic conditions, and has isotopic ratios ($\delta^{18}O = -17$ to -19 ‰ and $\delta^{13}C = -4$ to -5 ‰) that are compatible with the high temperatures of deep burial. It lines pockets in both the paleofill and bedrock but is truncated by later cave development, and so it is most likely of Mesozoic age. Some upper-level rooms are still lined by remnants of this spar where a later coating of sediment protected it from dissolution during Cenozoic cave enlargement.

Euhedral quartz crystals up to several millimeters long were deposited in and around faults. Most are confined to narrow fractures, but the largest are located in Carboniferous-age voids and are superimposed on the scalenohedral calcite as caps up to a millimeter thick, or as crystals up to 0.5 cm long on hematite cores. Quartz crystals are visible mainly where collapse has exposed the voids in the walls of later caves. They were probably deposited by thermal water during the first stages of the late Cretaceous Laramide Orogeny.

Cenozoic Cave Enlargement

All of the above features were intersected by Cenozoic caves when uplift of the Black Hills during the late Cretaceous and early Tertiary steepened the local hydraulic gradients enough to allow large amounts of groundwater to flow through the Madison Limestone. Cave enlargement appears to have taken place under phreatic conditions, except for some later subaerial condensation corrosion. Although a few entirely new caves formed, most were simply enlargements of earlier voids. Caves are concentrated in the upper half of the Madison Limestone within zones of paleokarst and earlier diagenetic features. Most of them seem to be of Paleocene and Eocene age, as shown by the fact that the present topography dates from the late Eocene. Since then the caves have undergone considerable modification by weathering and by mineral deposition.

The numerous upper-level rooms in the larger Black Hills caves are clearly exhumed relics of paleocaves. Red detrital fill has subsided from many of them into underlying passages, and this process continues today. Other passages at similar levels contain little or no detrital fill and yet have geometries similar to those of the exhumed paleocaves, which poses the question of whether these are enlarged Carboniferous caves that were never filled by sediment because of poor connections with the paleokarst surface. The horizon of upper-level rooms is slightly discordant to the strata, and it is feasible to imagine that it represented a former mixing zone. The present caves also intersect many linear sediment-filled fissures that connect directly with the paleokarst surface and extend as much as 50 m below it. Large ceiling expanses in certain upper-level rooms in Wind Cave consist of semiconsolidated sand and clay at the base of the Minnelusa Formation where the paleokarst surface extends relatively far into the limestone.

Deposition of Subaqueous Calcite
During the late Tertiary, and in places continuing through the Quaternary, water-filled caves acquired a palisade calcite crust with rhrombohedral terminations ("nailhead spar") up to several tens of centimeters thick (see 5.2.2 Plate 1). This calcite is distinct from the deep-burial dogtooth spar. Aside from the difference in crystal structure, the nailhead spar is not cathodoluminescent, contains many detrital impurities that form reddish bands within the crust, and has a different range of isotopic signatures ($\delta^{18}O = -12.5$ to -18 ‰ and $\delta^{13}C = -2.5$ to -7 ‰).

Bakalowicz et al. (1987) interpreted the nailhead spar to be hydrothermal, although elevated temperature may also have been caused by stagnation of groundwater after burial of the karst system by Oligocene sediment (see geologic map, 5.2.2 Figure 3). Pliocene faults have broken and displaced the crystal crust in many areas, especially in Jewel Cave, which shows the antiquity of some of the crusts. However, the crust is still being deposited in water-table pools in the lower levels of Wind Cave.

Subaerial Modification
Caves of the Black Hills have been greatly modified by subaerial processes, both before and after the wall coating of nailhead spar was deposited. These processes include weathering, in which condensation corrosion has played a major role, and deposition of a variety of speleothems.

Chemical weathering of the bedrock in the caves has produced a thick rind of multicolored carbonate sand that falls from the walls and ceiling at the slightest touch. Red, yellow, and other earthy colors are produced by oxidation and are most intense in iron-rich dolomitic rocks. Brightly colored beds on the weathered faces of breakdown blocks lose most or all of their color when traced to freshly exposed sides. Some of the colors may be caused by oxidation of sulfide minerals in the bedrock.

The intensity of weathering is due to the peculiar nature of the caves. Physical weathering is absent, and water that infiltrates from the surface is very corrosive. As infiltrating water passes downward through the Minnelusa Formation, its first encounter with carbonates is at interbedded limestone and dolomite sandwiched within the detrital strata. Because the overlying sandstone shields the water from soil CO_2, these carbonates dissolve in an environment that is essentially closed to further uptake of CO_2. Therefore the CO_2 level in the water drops nearly to zero. When the water emerges into the cave, it readily absorbs CO_2 from the cave air and becomes aggressive. In the porous bedrock surface of the caves, the water rarely forms discrete drips but instead moves down the walls as capillary moisture within the weathered zone.

Further corrosion is accomplished by condensation moisture, which accumulates where moist warm air rises from lower levels and condenses on the colder surfaces of upper-level passages (see 5.2.2 Plate 4). Evidence for this process is shown by the accumulation of discrete droplets on non-porous surfaces such as flagging tape, chert, and calcite crystals—and only in upper levels exposed to vertical air circulation. Condensation water absorbs CO_2 from the cave atmosphere, and since it originally contains little or no dissolved load, it is highly aggressive.

The smooth, rounded, and cuspate surfaces of many upper levels in the Black Hills caves have been accentuated (and in places even produced) by subaerial weathering. Weathered bedrock powder has accumulated on cave floors to depths of as great as a meter (5.2.2 Plate 2). In many passages, earlier deposits of this kind have been truncated by later subaerial corrosion. Subaerial weathering has also caused the boxwork veins to project outward from the cave walls (5.2.2 Plate 5).

Subaerial speleothems consist mainly of evaporative deposits such as aragonite frostwork and popcorn. Many are deposited by the same water that causes subaerial corrosion at higher levels—water that has seeped downward into warmer evaporative zones lower in the cave. There is rarely enough water to produce gravitationally oriented speleothems such as stalactites.

In some areas of Wind Cave, helictites have grown from the floor and have acquired a coating of the same spar that coats the nearby walls. These helictites show evidence for a subaqueous origin, although the growth mechanism is not certain (Davis, 1989; LaRock and Cunningham, 1995). A natural rise in the water table since 1995 has inundated most of them with calcite-saturated water, and so their growth, or at least their coating with calcite, is still an active process.

Similar water-level rises in the past have cemented many of the carbonate deposits produced by subaerial weathering. Weathered surfaces and evaporative crystals (e.g. aragonite) also lie beneath the crust of nailhead spar, showing that subaerial conditions pre-dated the spar, in addition to post-dating it.

Speleogenetic Controversies

The origin of the Black Hills caves can be traced all the way back to diagenetic events during the Mississippian Period, more than 300 million years ago. But regardless of how much these early events influenced the cave patterns, there is still considerable controversy as to the mechanism for the considerable enlargement of early features during the Cenozoic Era. Jewel Cave, in particular, shows evidence for a great amount of Cenozoic enlargement.

The earliest hypothesis (Owen, 1898) considered the caves to be hydrothermal. Many features, such as cones of carbonate sediment from subaerial weathering, were misinterpreted as ancient vents for thermal fluids and gases.

A similar interpretation was proposed by Bakalowicz et al. (1987) on the basis of isotopic evidence and resemblance of certain cave features to those of thermal caves in Europe. Quartz and hematite in Wind Cave attributed to elevated temperatures (White and Deike, 1962) are actually relics of the deep-stage deposition. The prevailing opinion throughout the 20th Century has been that the caves were formed by artesian water (Darton, 1901; Tullis and Gries, 1938; Deal, 1962; Howard, 1964). This was due to the fact that the Madison Limestone around the Black Hills is one of the best examples of an artesian aquifer in the country.

However, the pattern of the caves does not support a purely artesian origin (Palmer and Palmer, 1989). They are almost entirely limited to parts of the limestone overlain by a thin cover of Minnelusa Formation. Cave passages very rarely extend updip into the regions where the Madison Limestone is exposed directly at the surface, and they do not extend downdip to where the Minnelusa is overlain by other less permeable rocks (although paleokarst features are found in all areas). The largest passages are developed along prominent fractures oriented in the same direction as the stratal dip, making travel in the cave in that direction relatively easy. However, there has been little success in extending these passages either updip or downdip beyond the known limits of the caves. New discoveries (particularly in Jewel and Wind Caves) have been limited almost entirely to extensions along the strike of the beds (5.2.1 Figures 1 and 2).

On the basis of these geologic relationships it seems that Cenozoic cave enlargement was related to the surface topography, but in a very indirect way. It is important to remember that the present topography is essentially inherited from the Eocene. Yet the caves show an almost total disregard for present-day hydrology. Broad valleys of late Tertiary or early Quaternary age, now almost entirely relict, extend through the cave areas, and it is possible that they served as focused inputs for water (Palmer, 1981), although the cave patterns do not suggest inputs of that kind.

The relation of the caves to the distribution of the Minnelusa outcrop is suggestive. If the Minnelusa has acted as a confining unit, why should the even greater degree of confinement farther downdip not be just as conducive to cave development? If water rose upward through the Minnelusa from thermal sources, why should some of the water not have bypassed the Minnelusa and exited through the areas where the Minnelusa was missing? And why are the caves concentrated in only the upper 30 to 60 percent of the Madison Limestone? On the other hand, the Minnelusa may have served as a source of water to the caves. At least one modern researcher (Michael Wiles, Jewel Cave National Park, personal communication, 1997) believes this to be the major source, and that it is still active today. Wind Cave extends upward to the sandstone contact in many areas, but Jewel Cave does so in only a few places. The majority of cave development is far below the sandstone/limestone contact.

In short, we are left with very rigid boundary conditions. Nearly all caves of the Black Hills, especially the large ones, occupy only the upper part of the carbonate sequence and terminate both updip and downdip. Their very sparse entrances are purely accidental, caused by erosional intersection of upper-level passages. The only feasible conclusion is that the caves formed in a zone of mixing, where two or more sources of water were required to produce an aggressive solution. The available sources are meteoric artesian water, rising thermal water, and diffuse infiltration through the Minnelusa Formation. Cave origin by mixing must have involved any two or all three of these sources. The apparent restriction of the caves to the zone of exposed Minnelusa Formation may be a hint that diffuse infiltration was a factor in the late-stage enlargement of the caves.

But regardless of which mechanism was responsible for their Cenozoic enlargement, the Black Hills caves represent perhaps the world's best example of exhumed and modified paleocaves. The interpretation of the present-day caves is made difficult by the fact that their extent is unknown. Both Jewel Cave and Wind Cave are still being actively explored. Their present boundaries have been expanded far beyond what they were only a few years ago. In each case the expansion has been along the regional strike of the beds, with no extension either updip or downdip.

In particular, exploration in Jewel Cave is expanding the boundaries of the cave more rapidly than ever before, with no end in sight. Passages are becoming larger toward the southeast but seem to be diminishing in size toward the northwest. Until the true pattern of these caves becomes better known, a full interpretation of their origin will be elusive.

References

Bakalowicz, M.J., Ford, D.C., Miller, T.E., Palmer, A.N., and Palmer, M.V., 1987, Thermal genesis of dissolution caves in the Black Hills, South Dakota: *Geological Society of America Bulletin,* v. 99, p. 729-738.

Darton, N.H., 1901, Preliminary description of the geology and water resources of the southern half of the Black Hills and adjoining regions in South Dakota and Wyoming: *U.S. Geological Survey, 21st Annual Report,* Pt. 4, p. 489-599.

Davis, D.G., 1989, Helictite bushes–A subaqueous speleothem?: *National Speleological Society Bulletin,* v. 51, p. 120-124.

Deal, D.E., 1962, *Geology of Jewel Cave National Monument, Custer County, South Dakota, with special reference to cavern formation in the Black Hills:* University of Wyoming, M.S. thesis, 183 p.

Ford, D.C., and Bakalowicz, M.J., 1983, *Preliminary report upon McMaster University geochemical and isotopic studies in Wind and Jewel Caves:* Hot Springs, S.D., Wind Cave National Park, Open-File Report, 16 p.

Ford, D.C., Lundberg, J., Palmer, A.N., Palmer, M.V., Schwarcz, H.P., and Dreybrodt, W., 1993, Uranium-series dating of the draining of an aquifer: The example of Wind Cave, Black Hills, South Dakota: *Geological Society of America Bulletin,* v. 105, p. 241-250.

Howard, A.D., 1964, A model for cavern development under artesian groundwater flow, with special reference to the Black Hills: *National Speleological Society Bulletin,* v. 26, p. 7-16.

LaRock, E.J., and Cunningham, K.I., 1995, Helictite bush formation and aquifer cooling in Wind Cave, Wind Cave National Park, South Dakota: *National Speleological Society Bulletin,* v. 57, p. 43-51.

Millen, T.M., and Dickey, D.N., 1987, A stable isotopic investigation of waters and speleothems in Wind Cave, South Dakota: An application of isotope paleothermometry: *National Speleological Society Bulletin,* v. 49, p. 10-14.

Owen, L.A., 1898, *Cave regions of the Ozarks and Black Hills:* New York, Johnson Reprint Co., (1970 reprint), 228 p.

Palmer, A.N., 1981, *Geology of Wind Cave, Wind Cave National Park, South Dakota:* Hot Springs, S.D., Wind Cave National History Association, 44 p.

Palmer, A.N., and Palmer, M.V., 1989, Geologic history of the Black Hills caves, South Dakota. *National Speleological Society Bulletin,* v. 51, p. 72-99.

Palmer, A.N., and Palmer, M.V., 1995, The Kaskaskia paleokarst of the Northern Rocky Mountains and Black Hills, northwestern U.S.A.: *Carbonates and Evaporites,* v. 10, p. 148-160.

Rahn, P.H., and Gries, J.P., 1973, Large springs in the Black Hills, South Dakota and Wyoming: *South Dakota Geological Survey, Report of Investigations,* no. 107, 46 p.

Sando, W.J., 1974, Ancient solution phenomena in the Madison Limestone (Mississippian) of north-central Wyoming: *U.S. Geological Survey Journal of Research,* v. 4, p. 133-141.

Tullis, E.L., and Gries, J.P., 1938, Black Hills caves: *Black Hills Engineer,* v. 24, p. 233-271.

White, W.B., and Deike, G.H., 1962, Secondary mineralization in Wind Cave, South Dakota: *National Speleological Society Bulletin,* v. 24, p. 74-87.

5.2.3 Speleogenesis of Botovskaya Cave, Eastern Siberia, Russia

Andrej G. Filippov

Abstract

Botovskaya Cave is located in the Angaro-Lensky Artesian Basin on the southern Siberian Craton, Russia. It developed under confined conditions in 6- to 12-m thick Ordovician limestone strata. The cave is a subhorizontal two-dimensional maze 32 km long.

The limestone beds are confined above and below by massive marine sandstone units which contain thick silty and argillaceous layers. Cave passages are guided by an orthogonal fissure system. Embryonic passages are tube-shaped with oval and round cross sections. The mature passages are corridor-like with wide low notches at their bases between sandstone beds.

The cave system probably originated due to mixing corrosion involving meteoric artesian water flowing from a major recharge area, and ascending water migrating toward surface valleys from underlying artesian aquifers.

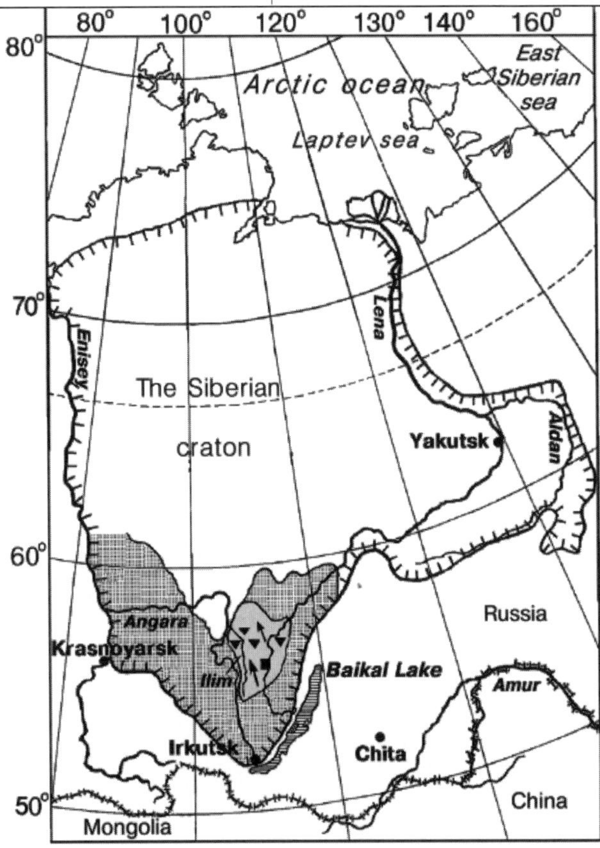

5.2.3 Figure 1. Location map of the Angaro-Lensky and Verkhnelensky Artesian Basins. No. 1 is the boundary of the Siberian Craton. No. 2 is the Angaro-Lensky Artesian Basin. No. 3 is the Verkhnelensky Artesian Basin. No. 4 is Botovskaya Cave. No. 5 is large springs of fissure-stratal and fissure-karstic groundwater from Lower Ordovician sandstone and limestone. No. 6 is direction of groundwater flow.

Introduction

Botovskaya Cave is an outstanding example of a maze cave developed under confined conditions. It is located in Russia, in the middle-Siberian tableland on the Verkholenskoje high plateau between 900 and 1100 m. Tectonically, this territory belongs to the Siberian Craton (5.2.3 Figure 1). The latitude and longitude of the cave are 55° 18' north and 105° 21' east.

Botovskaya Cave is presently inactive. Its entrances are in a cliff face 310 m above the Lena River. The carbonate and clastic Ordovician rocks form a plateau, which dips gently (2 to 10°) to the north. The plateau is dissected by steep river valleys up to 400 m deep. Marine carbonate and clastic rocks of Middle and Late Cambrian age outcrop along the bottom of the valleys (5.2.3 Figure 2).

Cave Morphology

The cave is a nearly horizontal, two-dimensional, rectilinear maze (5.2.3 Figure 3) developed in 6- to 12-m thick limestone beds. Passages follow vertical fissures, and numerous blind terminations are typical. The cave lies predominantly at one level. But in places passages are divided by a sandstone bed into two overlying channels. Passage cross sections are controlled by the thickness of the limestone and sandstone. Some small channels have oval or rounded shapes if entirely confined to a limestone bed (5.2.3 Plate 1, 5.2.3 Figure 4A). Ceilings are flat where they reach a sandstone layer (5.2.3 Figure 4B) except where weathering has

5.2.3 Figure 2. Geologic map and cross section of the Botovskaya Cave area. No. 1 to 4 are the Ordovician sequence—No. 1 is the Badaranovky Member: sandstone, siltstone (O_1bd). No. 2 is the Iya Member: sandstone, siltstone, argillite (O_1is). No. 3 is the upper member of the Ust'-Kout Formation: sandstone, siltstone, argillite (O_1us_2). No. 4 is the lower member of the Ust'-Kout Formation: sandstone, limestone, basal conglomerate (O_1uk_1). No. 5 and 6 are the Cambrian strata— No. 5 is the Ilga Member: sandstone, siltstone, argillite (C_3il). No. 6 is the Verholensk Member: sandstone, siltstone, argillite, marl, gypsum ($C_{2\text{-}3}vl$). No. 7 is the Quaternary alluvium (Q). No. 8 is the bedding orientation. No. 9 is the profile of the Botovskaya Cave System. No. 10 is the entrance to Botovskaya Cave.

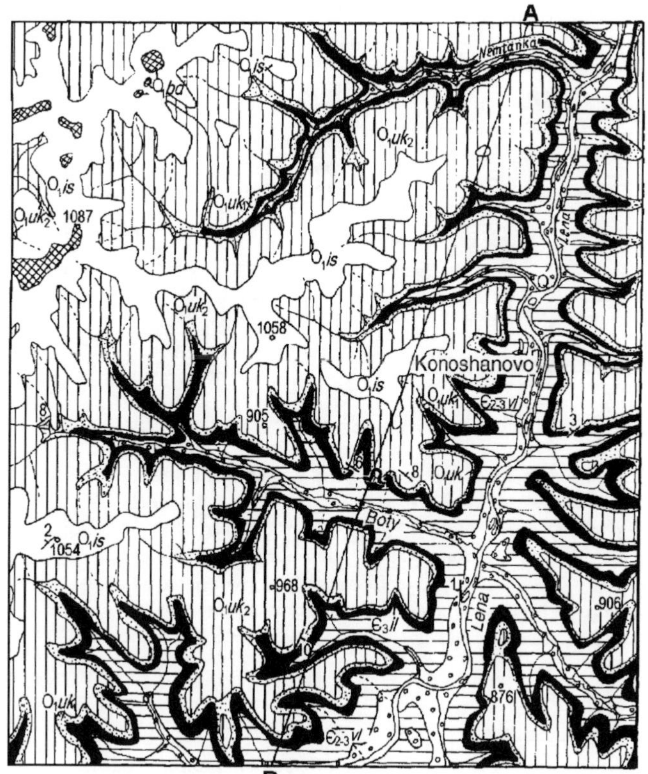

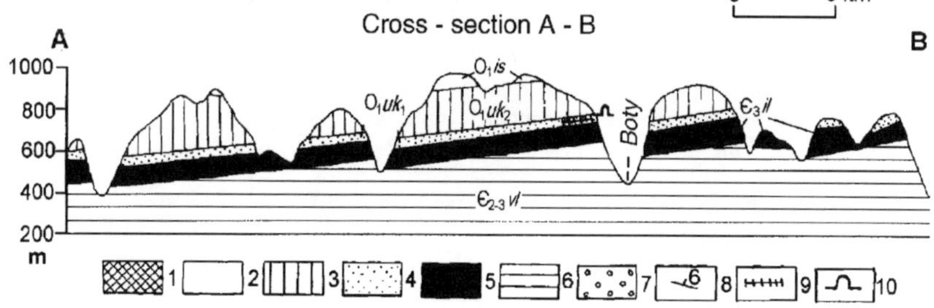

caused the sandstone to crumble, enlarging the fissures upward (5.2.3 Plate 2). On the other hand, the floor ends downward in 1 to 2 cm-wide cracks that continue downward into the underlying sandstone (5.2.3 Figure 4C). Some passages in the limestone bed are corridors with vertical, slightly undulating walls , complicated by smooth sandstone projections (5.2.3 Figure 4E –G). Many of the ceilings do not reach the overlying insoluble sandstone (5.2.3 Figure 4A, C, G). But a guiding fissure enlarged by solution is always present. Typical of many cross sections are sharp horizontal notches at floor level, which lie just above and also between sandstone layers (5.2.3 Figure 4F). In many instances, these notches were covered by clay fill, but they were exposed during paleontologic excavations.

As of May 1997, the total mapped length of the cave was 32,050 m. The average passage height and width are 2.0 m and 2.9 m, although they can reach 12.5 m and 10.0 m. The area of the cave is about 90,000 m^2, the volume is about 150,000 m^3, and the passage density is 136 km/km^2.

Geology

Botovskaya Cave is developed in algal limestone of the lower member of the Ust'-Kout Formation (Lower Ordovician). The limestone is up to 6 to 12 m thick and dips gently to the northeast at 6 to 8 degrees. It lies between marine sandstone units of the same formation. The limestone beds crop out in cliffs and occur as blocks and debris along valley slopes. The sandstone has gentler slopes covered by pine forest, whereas the cliffs and debris beneath them are either bare or covered by sparse bushes. The limestone beds are easily followed along both sides of

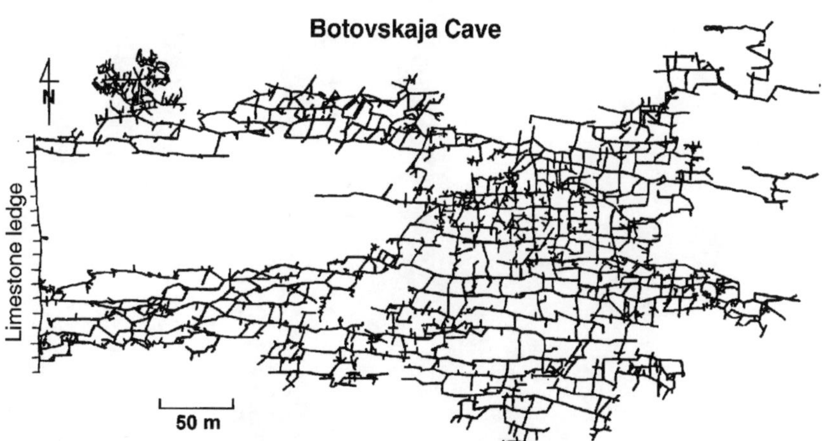

5.2.3 Figure 3. Map of Botovskaya Cave, eastern Siberia, showing 16.3 km of surveyed passages as of 1992 (presently the cave length is 32 km). The survey was conducted between 1989-1992 by the Irkutsk City Speleo-Club and the Arabika Speleo-Club. Areas explored in 1994-1996 are not shown on the map.

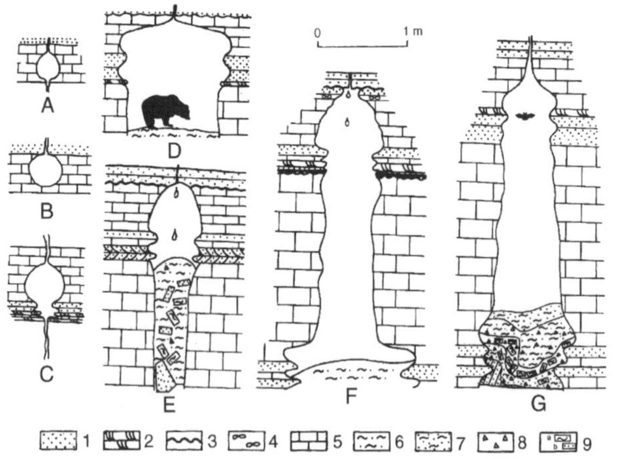

5.2.3 Figure 4. Passage cross sections of Botovskaya Cave. No. 1 is massive sandstone. No. 2 is cross-bedded sandstone. No. 3 is ripple marks. No. 4 is sandstone with contorted bedding. No. 5 is limestone. No. 6 is sandy clay. No. 7 is clayey sand. No. 8 is small clasts (2 to 10 mm). No. 9 is large clasts (>10 mm)—a is clasts of ponded clay, b is sandstone clasts.

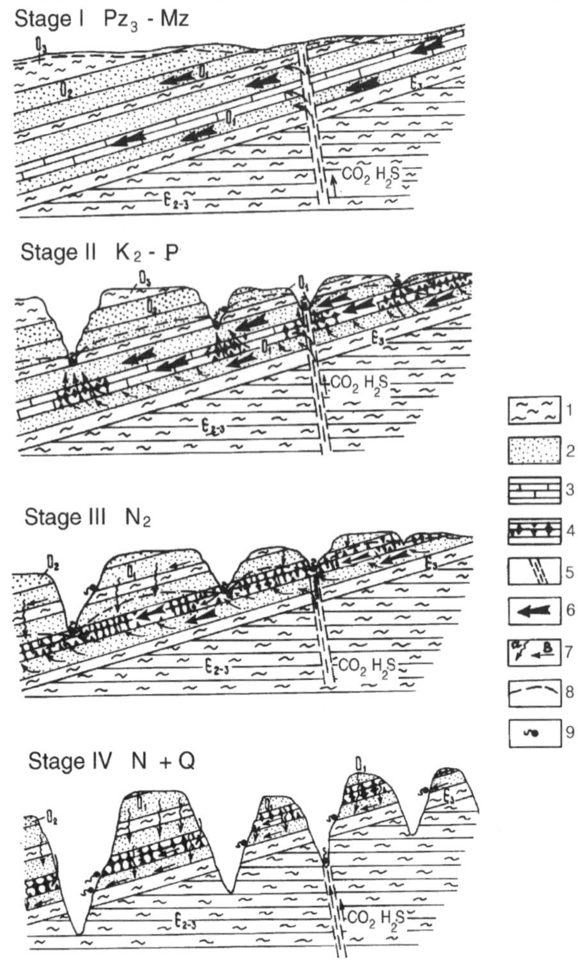

5.2.3 Figure 5. Hydrogeologic evolution of the Botovskaya Cave System. No. 1 is siltstone and argillite. No. 2 is sandstone. No. 3 is limestone. No. 4 is cave passages. No. 5 is the fault zone. No. 6 is lateral artesian circulation. No. 7 is cross-formational cir-culation: a is descending infiltration. b is ascending artesian circulation. No. 8 is the water table. No. 9 is springs.

5.2.3 Plate 1. Tube-shaped cave passage confined to the limestone bed intersected by the cliff face (photo by A.G.Dokuchaev).

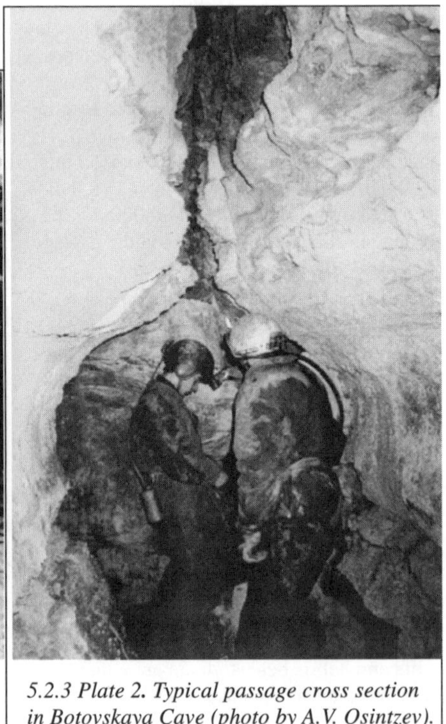

5.2.3 Plate 2. Typical passage cross section in Botovskaya Cave (photo by A.V. Osintzev).

the Boty River from its mouth to its upper reaches, as well as along numerous other tributaries of the Lena River. Relics of intersected passages lie on both sides of the Boty River.

The caves are in gray limestone, commonly sandy, algal, oolitic, or in places marly containing pebbles or cobbles of marl. In various parts of the cave, the number of sandstone layers varies. Less commonly, silt and marl layers also occur. The thickness of the limestone beds varies slightly. In cross section the sandstone layers characteristically form horizontal projections into the passages, that can be traced virtually throughout the entire cave. They form shelves up to 0.5 m wide in places. A more detailed geologic description is given in Filippov (1994).

Secondary Cave Deposits

Floor deposits consist mainly of fine polygenetic clay and breakdown. Clay covers the passage floors almost everywhere except where breakdown occurs. Cave ice, calcite and aragonite flowstone, condensation, and pool formations occur in some locations.

Sections of deposits opened by research trenches show that sedimentation in ponded water occurred in the past. However, in most of the trenches thinly layered clay occurs only as clasts enclosed in disintegrated weathered sandstone. The clay was disrupted during subsidence into underlying cavities. Breakdown or weathered debris is found at all passage junctions. In places there are large boulder chokes of sandstone slabs. Most of these boulder chokes are not covered by clay or flowstone.

Hydrogeology

Hydrogeologically the area is in the second-order Verkhnelensky Artesian Basin which is part of the larger Angaro-Lensky Artesian Basin (see 5.2.3 Figure 1). Confined hydrogeologic conditions were dominating in pre-Neogene time when the slightly permeable argillaceous and siltstone beds with limestone and sandstone interbeds were below the level of erosional entrenchment (5.2.3 Figure 5). Today, after uplift, the rocks of the lower member of the Ust'-Kout Formation in the cave area are dry.

A regional survey suggests that downdip the same sandy limestone and thick-bedded sandstone contain a fissure-conduit stratiform aquifer at depths as great as 240 m. The aquifer extends to the southwest of the caves. The groundwater discharges as springs with low flow rates ranging from 0.1 to 0.5 or 2 to 10 L/sec. The groundwater regime is constant because the aquifer receives its recharge not only from direct infiltration of meteoric precipitation but also from leakage of groundwater from the overlying aquifers. The waters are presently unconfined within the zone of active circulation. They are characterized by HCO_3-Mg-Ca and low total dissolved solids (0.25 to 0.50 g/L).

In the downdip direction, large springs discharge fissure-conduit groundwater from the Lower Ordovician sandstone and limestone within the Verkhnelensky Artesian Basin in the Lena, the Ilim, and the Kupa River valleys. The linear discharge zones stretch for hundreds of meters (Tkachuk, 1968). Lateral circulation is typical for these Lower Ordovician aquifers.

At present, ascending brines discharge at the surface in the upper Lena River Valley, including the area of Botovskaya Cave (Dzuba, 1984). The width of the joint zone—through which recent ascending discharge occurs—ranges between 3 to 5 and 10 to 15 km. Such zones extend from tens to some hundreds of kilometers (Dzuba et al., 1974).

The cave itself has no flowing water. But seepage water along fissures plus condensation moisture, occurs throughout the cave, locally forming small pools.

Cave Origin

The origin of maze caves is still a controversial topic (Ford, 1989; Klimchouk, 1991, 1994, 1997; Palmer, 1975, 1991; Palmer and Palmer, 1989; see also Palmer, Chapters 3.4 and 5.3.11, and Klimchouk, Chapter 5.2 in this volume). Of the three main existing genetic models, two can be disregarded with respect to Botovskaya Cave. A floodwater origin is inappropriate because of the lack of any relationship between the cave and past or present surface streams. The model of origin by diffuse recharge through overlying insoluble but permeable rocks does not apply as well because the thick poorly permeable Ust'-Kout Formation in the overlying rocks prevented any appreciable downward filtration into the limestone.

The characteristics of the cave and its geohydrogeologic settings are best explained by the model of cave origin under artesian conditions developed by Klimchouk (1994, 1997)—that is, by groundwater circulating upward through the limestone between the lower and upper sandstone aquifers. Generally speaking, the origin of Botovskaya Cave is similar to that of the great gypsum mazes in the western Ukraine (Klimchouk, 1991)—although the lithology, local geologic history, and settings impose different features.

Virtually all available fissures in the limestone beds are uniformly enlarged to form the dense passage pattern. This was possible under conditions where aggressive recharge entered all the fissures from the underlying sandstone aquifer. The presence below and above the limestone of the sandstone units—originally more permeable than the limestone—favored hydraulic cross-communication via fissures in the limestone beds. The minor sandstone bed within the otherwise carbonate sequence has caused some separation of passages into two levels. Where a single fissure crossed the whole sequence, the passages were preferentially enlarged within the limestone beds, forming overlying conduits. The separating sandstone bed was breached, giving rise to the single high passages (5.2.3 Figure 4D, E, F, and G) that form most of the network. Such passages are characterized by cross sections markedly widened near the base just above the underlying sandstone and within underlying limestone interbeds.

Numerous sandstone fragments are identified within clay filling in such passages where ceilings extend up to the top of the limestone or to the upper intact sandstone (5.2.3 Figure 4D–G). This suggests that the former overlying sandstone bed was breached. In rare instances, conduits at two levels remained separate, or were developed along independent (noncoplanar) fissures confined to each limestone bed (5.2.3 Figure 4A–C).

The presence of laterally connected fissure networks has caused a considerable local lateral flow component within the limestone, particularly when hydraulic conductivity had been increased in the course of speleogenesis. Lateral flow through the system increased dramatically during the latest stages of artesian speleogenesis, when the hydraulic potential field became rather unhomogeneous, and when artesian confinement was breached by surface valleys. The formation of some tube-like conduits, and most of the enlargement of large passages, probably occurred during these stages.

It can be assumed that groundwater entering the limestone from the sandstone aquifer possessed sufficient aggressiveness to enlarge the fissures. Groundwater in the presently confined Ordovician aquifers in other parts of the Angaro-Lensky Artesian Basin have considerable CO_2 and H_2S gases. These gases come from the underlying Proterozoic sequences along the fault zones and could rejuvenate dissolutional potential at certain points in the lateral flow system of the confined carbonate aquifer.

Cave morphology and sediments also show evidence of prolonged vadose and subaerial development. Remnants of erosional channels are found in the cave deposits (Filippov, 1994) and in the floor of many of the passages that are subsequently filled by sandy sediment with clasts of sandstone, siltstone, and flowstone. Apparently this occurred late in the evolution of the cave when the shape of the cave had been largely established.

Evolution of the Fissure-Conduit Aquifer and the Cave System

Four main stages can be distinguished in the evolution of the fissure-conduit aquifer and the cave system (5.2.3 Figure 5):

- Stage 1—early artesian.
- Stage 2—partial erosional breaching of the overlying clastic beds and activation of upward flow beneath the major valleys.
- Stage 3—breaching of the fissure-conduit aquifer by erosion.
- Stage 4—deep entrenchment of erosion valleys below the former aquifer.

According to the karst typology introduced by Klimchouk (1996) and Klimchouk and Ford (Chapter 3.1 in this volume):

- Stages 1 and 2 correspond to intrastratal deep-seated karst.
- Stage 3 to intrastratal subjacent karst.
- Stage 4 (present stage) to entrenched karst.

Stage 1 was the longest. It was characterized by low surface relief and confined hydrogeologic settings in the large artesian basin. During this stage, connate water was slowly replaced by meteoric water from distant recharge areas circulating from the southeast to the northwest. Discharge took place along zones of linear folds and associated faults. The stage lasted from the late Paleozoic through the Mesozoic. Some incipient speleogenetic development may have occurred along vertical fissures during this stage.

Stage 2 is characterized by uplift and subsequent erosional entrenchment. The Verkhnelensky structural high began to form at the end of the Late Cretaceous (Zamaraev et al., 1976). Valley entrenchment led to the establishment of a regional pattern of potentiometric lows in the confined aquifers, and to considerable activation of groundwater circulation in general and to upward flow beneath valleys in particular. The formation of the maze conduit pattern is attributed mainly to this stage.

Stage 3 is marked by breaching of the major confining beds, by further activation of groundwater circulation, and by the rapid transition from confined conditions, through phreatic, water table, and vadose stages. This stage was the shortest and can be dated to the Miocene, as the cave lies above the level of Pliocene river terraces.

Stage 4 is characterized by deep entrenchment of valley bottoms below the karst system and draining of the fissure-conduit aquifer. Diagenesis of cave sediment and the formation of subaerial speleothems occurred during this stage and continues today. Other processes in the cave include local dissolution by infiltration and condensation water, erosion and redistribution of clastic deposits by temporal streams with shifting courses, and breakdown of the limestone walls and minor sandstone beds separating overlying passages.

References

Dzuba, A.A., 1984, *Brines discharge in the Siberian Craton:* Novosibirsk, Nauka, 156 p. (in Russian).

Dzuba, A.A., Kustov, Y.I., Pospeev, V.I., and Trofimuk, P.I. 1974, Open and latent discharge of brines in the Angaro-Lensky Artesian Basin, *in Groundwater deposits of the Irkutsk region:* Novosibirsk, Nauka, p. 176-182 (in Russian).

Filippov, A.G., 1994, Botovskaya Cave, *in Problems of physical speleology:* Moscow, p. 142-160 (in Russian).

Ford, D.C., 1989, Features of the genesis of Jewel Cave and Wind Cave, Black Hills, South Dakota: *National Speleological Society Bulletin,* v. 51, p. 100-110.

Klimchouk, A.B., 1991, Large maze caves in the western Ukraine: Speleogenesis under artesian conditions: *National Speleological Society Bulletin,* v. 53, p. 71-82.

Klimchouk, A.B., 1994, Speleogenesis under confined conditions, with recharge from adjacent formations: Publ. Serv. Geol. Luxembourg, *Comptes Rendus du Coll. Int. de Karstol. a Luxembourg,* v. 27, p. 85-95.

Klimchouk, A.B., 1996, The typology of gypsum karst according to its geological and geomorphological evolution, *in* Klimchouk, A., Lowe, D., Cooper, A., and Sauro, U., eds., Gypsum karst of the world: *International Journal of Speleology*, v. 25, no. 3-4, p. 49-60.

Klimchouk, A.B., 1997, Artesian speleogenetic setting: *12th International Congress of Speleology,* La Chaux-de-Fonds, Proceedings, v. 1, p. 157-160.

Palmer, A.N., 1975, The origin of maze caves: *National Speleological Society Bulletin,* v. 37, p. 56-76.

Palmer, A.N., 1991, Origin and morphology of limestone caves: *Geological Society of America Bulletin*, no. 103, p. 1-21.

Palmer, A.N., and Palmer, M.V., 1989, Geologic history of the Black Hills caves, South Dakota: *National Speleological Society Bulletin,* v. 51, p. 72-99.

Tkachuk, V.G., ed., 1968, *Hydrogeology of the USSR: Irkutsk region:* Moscow, Nedra, 495 p. (in Russian).

Zamaraev, S.M., et al., 1976, *Structure and development history of the Predbaikalian Foredeep:* Moscow, Nauka, 134 p.

5.2.4. Karstification Associated with Groundwater Circulation Through the Redwall Artesian Aquifer, Grand Canyon, Arizona, U.S.A.

Peter W. Huntoon

Abstract

The karstified Redwall artesian aquifer discharges significant quantities of water to a small number of large springs in the Marble and Grand Canyons of Arizona, U.S.A. The locations of the springs are topographically controlled, being situated on the flanks of regional structural depressions at locations where the depressions have been dissected by the canyons. The springs serve as the lowest potentiometric spill points for the aquifer.

Modern caves behind the springs seem to be adjusted to the hydraulic boundary conditions governing circulation through the aquifer. These caves seem to be organized parallel to modern hydraulic gradients and are thus fairly independent of preexisting dissolution-enhanced fracture permeability. This indicates that sufficient time has elapsed since the modern circulation system boundaries became established for the flow regime to have created optimally oriented karstic permeability pathways.

Dry remnant caves occur in dewatered sections of the Redwall Aquifer that obviously predate dissection of the aquifer by the Colorado River. In contrast to the active caves, the dry caves are characterized by keyhole and slot passageways that are predominantly localized along joints and normal faults. The fractures date largely from late Tertiary extensional tectonism. These older caves are interpreted to be remnants of dissolution conduits in what was a more widespread regional Redwall artesian aquifer prior to incision of the Grand Canyon.

Recharge to the Redwall Aquifer takes place primarily as vertical circulation in normal fault zones where the faults have propagated upward through the overlying Supai confining layer. The water enters the faults directly from the land surface or as leakage from shallower aquifers that drain to the faults.

Redwall Caves

Cliffs in the lower Paleozoic carbonate section in the Marble and Grand Canyons provide profiles through zones of keyhole and slot caves that are localized along vertical fractures (5.2.4 Plate 1). These caves seem to be remnants of dissolution cave networks that developed in the lower carbonate section while it was under artesian conditions prior to and during early dissection by the Colorado River. Most of the fractures along which the caves are localized date from late Tertiary time based on the association of the fractures with Tertiary normal faults (Huntoon, 1990), thus most of the fractures predate the canyon. Many of these caves are now dry in locations where the carbonate rocks are elevated above river level and dip toward the river. Some such caves are more than 1200 meters above the Colorado River, indicating that they are of great antiquity.

A small number of large karst springs on the south side of the canyons discharge from still active parts of the Redwall artesian system (5.2.4 Figure 1). Three of the largest springs, East Fence, Blue, and Havasu collectively drain as many as 100,000 square kilometers of the surrounding plateaus to the east and south, and yield approximately 0.4, 6.5 and 1.8 m^3/s respectively (Johnson and Sanderson, 1968; Huntoon, 1981).

Most of the Redwall caves treated here are inaccessible. The active conduits are water filled behind the springs, whereas the dry remnant caves are usually too small to enter or are exposed high on inaccessible cliffs. The dry caves of this type that have been entered on the north side of the eastern Grand Canyon are localized along fractures, primarily joints, but locally along minor high-angle normal faults. They are most commonly found in association with late Tertiary extensional tectonic fractures. But in some cases they occur along Laramide or even older fractures. The most extensive of the dry

5.2.4 Plate 1. Keyhole fissure caves in the exhumed Redwall artesian aquifer that are localized along joints in the late Tertiary Eminence Graben, north wall of Marble Canyon, Arizona, U.S.A. View is of the upper 100 m of the Mississippian Redwall Limestone.

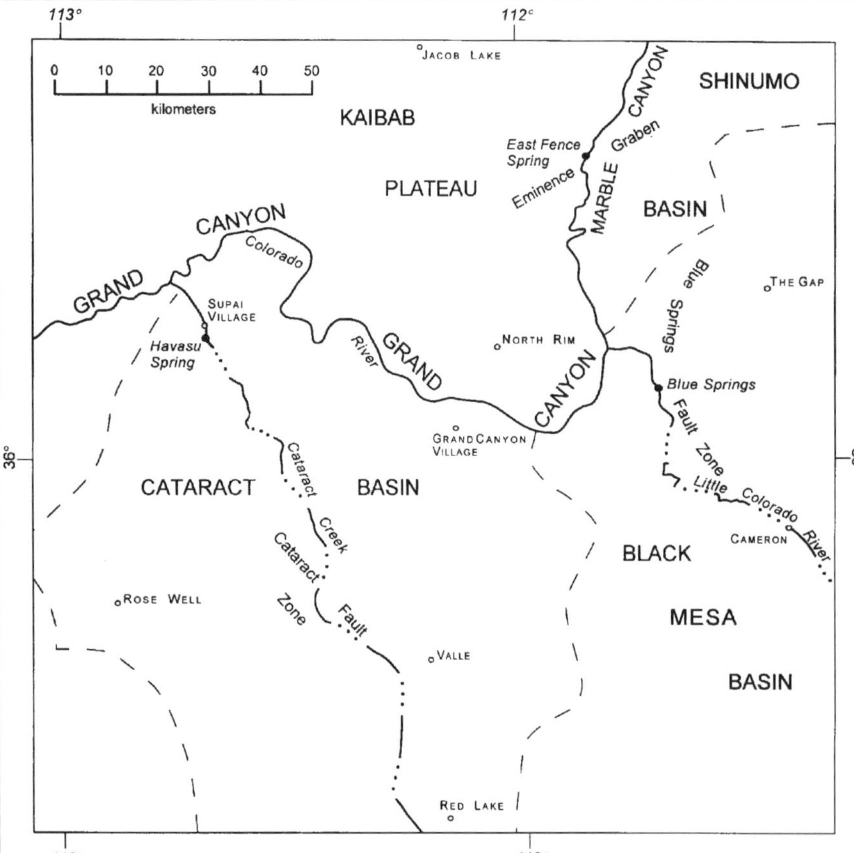

5.2.4 Figure 1. Map showing the locations of the principal Redwall artesian groundwater basins and springs in the vicinity of the Marble and Grand Canyons, Arizona, U.S.A. Dashed lines show the approximate locations of groundwater divides.

caves have been explored for distances of up to a few kilometers. They are developed along intersecting laterally extensive master joints and minor normal faults. The joints commonly parallel normal faults.

The caves occur in some locations as dense clusters that comprise mazes of parallel tubes within tectonically extended zones. The most conspicuous exposed clusters of tubes are in the Redwall Limestone within the moderately fractured Eminence Graben that trends across Marble Canyon (5.2.4 Plate 2). Here numerous horizontal fissure caves, oriented parallel to tectonic trends, are developed along joints and minor normal faults within the downdropped graben block (5.2.4 Plate 3).

The solutional enhancement of permeability along extensional fractures prior to incision of the canyons makes sense. The minute separations that tended to develop between the surfaces of joints and small-displacement normal faults provided the most permeable and laterally interconnected pathways through the otherwise nontransmissive carbonate section. Consequently, circulation and attendant dissolution preferentially occurred along such fractures. In many locations, dissolution favored joints parallel to normal faults rather than the faults themselves, indicating that the gouge that develops along faults probably serves to reduce fracture permeability. Rates of dissolution during the long geologic interval prior to dissection undoubtedly were very small because hydraulic gradients were minimal.

Physical and Temporal Setting of the Redwall Aquifer

The Grand Canyon is incised into a 1.5 km-thick section of Paleozoic rocks (5.2.4 Figure 2) that crops out over the southwestern part of the Colorado Plateau physiographic province. The province is a large Laramide-Tertiary uplift that comprises predominantly flat-lying sedimentary rocks. It is segmented into individual plateaus that are usually bounded by north-trending Laramide monoclines and superimposed late Tertiary normal faults that are spaced from 11 to 50 km apart.

The lower Paleozoic carbonate sequence that where saturated comprises the Redwall Aquifer includes the conformable Cambrian Muav Limestone, undivided Cambrian carbonates, Devonian Temple Butte Formation and Mississippian Redwall Limestone (5.2.4 Figure 2). The thickness of the carbonate section in the eastern Grand Canyon is about 400 m, but it increases to 750 m in the west. The carbonates are sandwiched between very effective confining strata, specifically the thick micaceous Cambrian Bright Angel Shale below, and the Pennsylvanian-Permian

5.2.4 Plate 2. High-altitude oblique aerial photograph looking toward the east across Eminence Graben, Marble Canyon, Arizona, U.S.A. Arrows point to the bounding faults. The down-dropped block is riven with minor normal faults and fractures. Plates 1 and 3 are photo-graphs of the canyon walls within the graben.

5.2.4 Plate 3. Parallel dissolved slots in the Thunder Springs Member of the Redwall Lime-stone from the exhumed Redwall Aquifer in Marble Canyon, Arizona, U.S.A. The permeability in the aquifer was highly anisotropic with the maximum principal per-meability tensor oriented parallel to the slots. Top to bottom is about 65 m. Some of the fissure widening here can be attributed to etching by the Colorado River as it dissected these caves.

Supai confining layer above. The Supai is largely comprised of virtually impermeable siltstone and shale.

The limestone and dolomite that constitute the Redwall Aquifer are generally dense and nontransmissive except where karstified. The upper unit in this sequence, the Redwall Limestone, is composed of microcrystalline limestone and dolomite with zones of inter-bedded cherty limestone. The Muav Lime-stone at the base is dense and commonly contains silt and clay impurities.

Development of water from wells drilled into the Redwall Aquifer is predicated on intersecting interconnected dissolution-widened fractures and solution cavities. Permeable zones in the Redwall Aquifer are heterogeneously distributed. The permeability within the karstified zones is highly anisotropic, with the maximum principal hydraulic conductivity tensor commonly oriented parallel to past or modern hydraulic gradients.

The Redwall Aquifer was an artesian aqui-fer of broad regional extent prior to excava-tion of the Grand Canyon in late Miocene and Pliocene time. Circulation rates for fluids in the aquifer were sluggish, owing to generally very small permeabilities and extremely gentle hydraulic gradients. The cutting of the 1.5-km-deep Marble and Grand Canyons first exposed and then fully penetrated this aquifer. Incision of the canyons caused hydraulic gradients in the aquifer to reorganize toward the canyons and steepen, thus greatly enhancing the cir-culation rate through the aquifer.

Dissolutional enhancement of permeability along the fractures prob-ably was maximized when the Colorado River first penetrated the Supai confining strata. Initially, the invigorated circulation within the aquifer was largely restricted to available fracture permeability paralleling regional late Tertiary normal fault zones, and possibly in a few locations along permeable fractures associated with older Laramide monoclines.

Consequently, the caves preferentially developed in the extensional tectonic zones. Dissolution continued on both sides of the canyon during the early stages of dissection. As the canyon deepened, however, the elevated parts of the Redwall Aquifer began to dewater, leaving the caves high and unsaturated on the canyon walls.

A younger generation of caves has devel-oped at the base of the Cambrian Muav Limestone along favorably oriented regional normal fault zones within the elevated Kaibab Plateau north of the Colorado River (Hun-toon, 1974). These modern caves serve as large-capacity drains that have dewatered the elevated parts of the old Redwall artesian systems in that region.

The situation to the south and east of Mar-ble and Grand Canyons where the Redwall Aquifer is saturated is vastly different. Here the strata dip away from the canyons, so the lower Paleozoic carbonates constituting the Redwall Aquifer are fully saturated. The springs issue from the Redwall Limestone through structurally depressed outcrops that serve as topographic spill points along the margins of the basins.

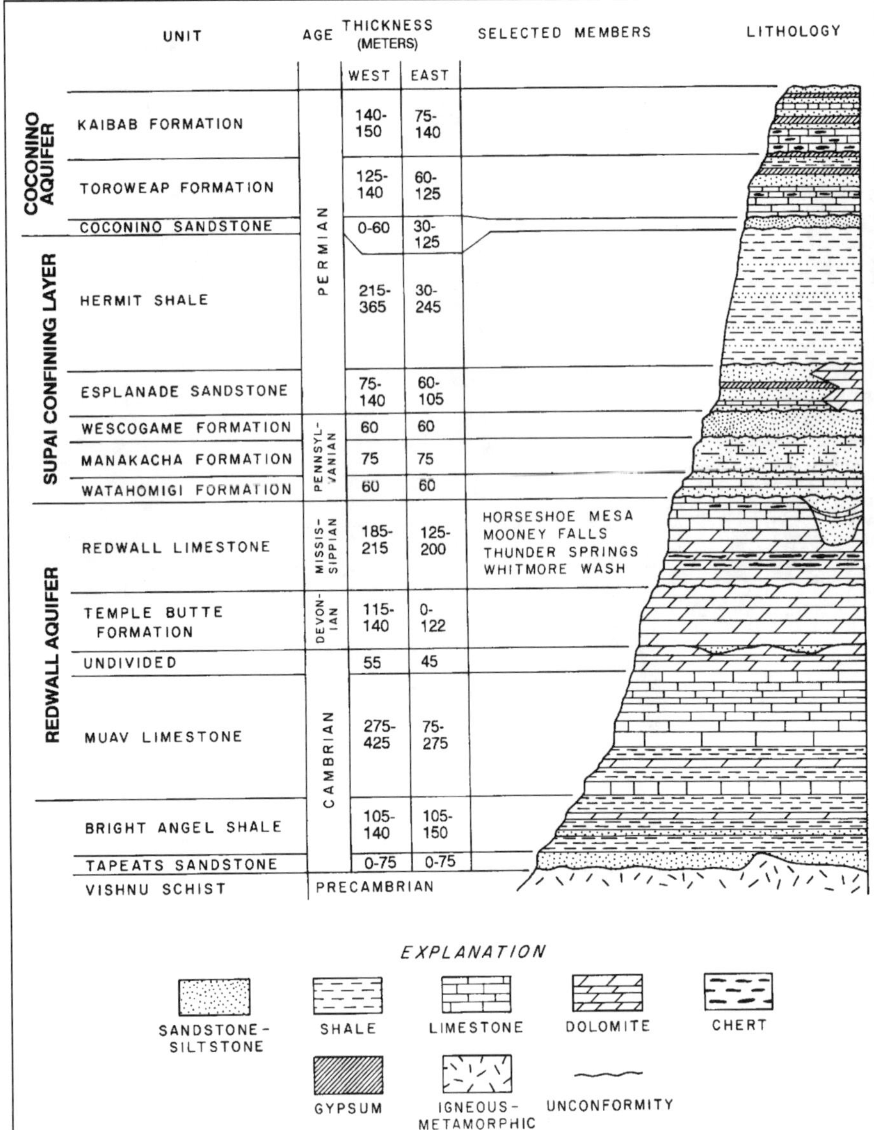

5.2.4 Figure 2. Paleozoic rocks and hydrostratigraphy in the walls of the Grand Canyon, Arizona, U.S.A. The unlabeled deep channel deposit shown at the top of the Redwall Limestone is the Mississippian Surprise Canyon Formation.

In the cases of the hydrologically significant Fence and Blue Springs, the structural lows surrounding the springs coincide with swarms of late Tertiary normal faults. Extension across these faults in part caused the tectonic subsidence of those groundwater basins. In contrast, Havasu Spring discharges from a minor normal fault that does not seem to be interconnected with the extensive normal-fault zone centered in Cataract Basin which it drains (5.2.4 Figure 1).

The discharges from East Fence, Blue, and Havasu Springs are large, stable, and warm. Total dissolved solids are large, indicating that the springs drain sizable groundwater basins.

- The Shinumo Basin drains to East Fence Spring. This basin is at least 600 km^2.
- Next in size is the Cataract Basin at approximately 10,000 km^2. It drains to Havasu Spring.
- The largest is the Black Mesa Basin at about 70,000 km^2. It drains to the Blue Springs.

These groundwater basins are bounded to the south and east by groundwater divides that separate them from adjacent basins, rather than physiographic or structural boundaries (McGavock, 1968; Cooley et al., 1969; Loughlin, 1983). The Redwall Limestone is buried throughout these basins by a minimum of 500 m of rock.

Recharge to the Redwall Aquifer

Recharge to the Redwall Aquifer occurs primarily through widely spaced extensional fault zones where the faulting has propagated upward through the entire Paleozoic section. Fracture permeability associated with the normal faults provides vertical circulation pathways through the Supai confining layer into the aquifer. Recharge to the faults originates both from the land surface and from drainage from shallower aquifers.

The hydraulic behavior of the vertical recharge pathways has been observed in the Cataract Basin. The surface of the basin comprises very gently warped erosionally resistant Permian carbonates. A well-exposed extensional fault zone occurs at the structural center of the basin, and the ephemeral channel of Cataract Creek meanders across this zone in a slightly incised canyon. Numerous sediment-choked swallow holes and fissures open along the high-angle normal faults in the floor of the canyon. When the channel floods, these sinks accept a considerable volume of the flood flow.

Elsewhere in the Cataract Basin, topographically closed subsidence depressions on the downthrown side of normal faults and small karst depressions dissolved in the Permian carbonates, conduct other ephemeral surface flows to fault zones where those waters are similarly lost. Losses to the subsurface through internally drained areas is significant. This is revealed by the fact that of the 7,822 km^2 surface catchment area for Cataract Creek, 541 km^2 are drained internally (Melis and others, 1996).

Certainly there are other sources for recharge to the Redwall Aquifer including

- Downward circulation within fractured volcanic intrusives that penetrate through the Paleozoic section.
- Recharge in highly localized areas where the rocks constituting the Redwall Aquifer crop out or subcrop in lacoliths.
- Plus a trivial amount of vertical leakage through the Supai confining strata or through breccia pipes that penetrate the Supai confining layer.

The majority of the recharge, however, occurs as loss of surface flood flows to permeable normal fault zones, and as lateral leakage of water from the shallow Coconino Aquifer (5.2.4 Figure 2) to those same fault zones. Once in the fault zones, the water readily moves downward through the Supai confining layer.

Residence Time and Implications for Modern Karstification

Residence time for water in the Shinumo, Black Mesa, and Cataract Basins is generally long as revealed by the large dissolved solids content and warm temperature of waters discharged from the springs. Representative dissolved solids from the East Fence, Blue, and Havasu Springs are respectively 1,370, 2,300, and 580 mg/L, and temperatures are 20.8°, 20.5°, and 21°C (Loughlin and Huntoon, 1983).

By contrast, the water that discharges from karst drains on the Kaibab Plateau typically has a temperature of 10° to 16°C and total dissolved solids of less than 200 mg/L.

As the following example will illustrate, these parameters are variable—a fact that can be used to deduce additional information about the character of the karst within the Redwall Aquifer.

On February 20, 1993, a major flood developed in the headwaters of Cataract Creek, washing out two small reservoirs and several stock tanks. The peak discharge in Cataract Creek was estimated to be 391 m^3/s (Melis et al., 1996). The ensuing deluge inundated the normally dry Cataract channel where it crosses the Cataract Fault Zone. A tremendous but unmeasured volume of water recharged directly into the fault zone 65 km upstream from Havasu Spring. This was the latest and largest of three floods in the early 1990s, the other lesser floods having occurred on September 3, 1990, and July 25, 1992.

On April 1, 1995, George Billingsley of the U.S. Geological Survey (written communication, December 1996) observed that the temperature of the water issuing from Havasu Spring was discernibly below 21°C, Total dissolved solids were significantly less than normal, as revealed by the lack of spontaneous precipitation of calcium carbonate to cause rapid buildup of travertine. One explanation for the improved water quality is dilution by the flood water that recharged the aquifer two years previously. The travel time for this water was well in excess of porous-media rates. The implications are that the permeability in the block of unfractured rock lying between the point of recharge in the fault zone and Havasu Spring is karstic, and that karstic permeability is independent of regional normal-fault trends.

Discussion

Emphasis in this paper has been placed on the occurrence of karstic conduits in extensional fracture zones. These observations come from the exhumed dewatered parts of the Redwall Aquifer. Thus the caves sampled represent remnants of some of the earliest circulation patterns to develop in the aquifer.

Those caves date from prior to or during the early stages of excavation of Marble and Grand Canyons. The small fragments of these old cave systems that have been entered reveal that they tend to be bedding-parallel mazes developed along intersecting sets of fractures.

Permeability within the dissolutionally enhanced zones is extremely anisotropic, with the maximum principal hydraulic conductivity tensor oriented roughly parallel to local structural trends. The intermediate principal hydraulic conductivity tensor is oriented vertically within the fault zones. And the minimum principal tensor is perpendicular to the fractures. The permeability contrast between the dissolutionally enhanced fracture zones and the vast expanse of adjacent unfaulted carbonate country rock is on the order of ten or more orders of magnitude.

However, these paleocaves do not seem to be fully representative of the modern permeability structure within the Redwall Aquifer. More than one million years have elapsed since the Marble and Grand Canyons were excavated to their present depths. This in turn established the hydraulic boundary conditions for the modern artesian systems. The observations from the Cataract system indicate that the karstic permeability has already adjusted to the current hydraulic regime.

Specifically, the flow system has dissolved new conduits through the structurally barren block lying between the Cataract Fault Zone in the recharge area and Havasu Spring. The modern conduits parallel the current hydraulic gradient, so the maximum principal permeability tensor has reoriented parallel to the gradient. Consequently, the geologically young karstification has functionally supplanted the previous permeability enhancements along less favorably oriented fractures.

The extremely large karstic permeability found in the Redwall Aquifer ensures that hydraulic gradients in the permeable zones are very small. Gentle gradients facilitate the development of dissolution mazes that in turn maximizes storage within the aquifer. The presence of large storage is revealed in part by the constancy of discharges from the major springs.

In contrast, the partly unsaturated karstic drains dissolved in the Muav Limestone under the Kaibab Plateau are characterized by simpler, linear, large-capacity caves that carry underground rivers. Hydraulic gradients within those systems are steeper, being more characteristic of vadose caves, and groundwater storage is rather minimal. Also, flow rate varies dramatically, both seasonally and in response to precipitation events.

References

Cooley, M.E., Harshbarger, J.W., Akers, J.P., and Hardt, W.F., 1969, Regional hydrogeology of the Navajo and Hopi Indian Reservations, Arizona, New Mexico, and Utah: *U.S. Geological Survey Professional Paper 521A,* 61 p.

Johnson, P.W., and Sanderson, R.B., 1968, Spring flow into the Colorado River, Lees Ferry to Lake Mead, Arizona: Arizona State Land Department, *Water Resources Report,* no. 34, 26 p.

Huntoon, P.W., 1974, The karstic groundwater basins of the Kaibab Plateau, Arizona: *Water Resources Research,* v. 10, p. 579-590.

Huntoon, P.W., 1981, Fault controlled ground-water circulation under the Colorado River, Marble Canyon, Arizona: *Ground Water,* v. 19, p. 20-27.

Huntoon, P.W., 1990, Phanerozoic structural geology of the Grand Canyon: *in* Beus, S.S., and Morales, M., eds., *Grand Canyon geology:* New York, Oxford University Press, p. 261-309.

Loughlin, W.D., 1983, *The hydrogeologic controls on water quality, ground water circulation, and collapse breccia pipe formation in the western part of the Black Mesa Basin, Coconino County, Arizona:* University of Wyoming, M.S. thesis, 117 p.

Loughlin, W.D., and Huntoon, P.W., 1983, Compilation of available ground water quality data for sources within the Grand Canyon of Arizona: Department of Geology and Geophysics, University of Wyoming, *Report to U.S. National Park Service,* 9 p. plus appendices.

McGavock, E.H., 1968, Basic ground-water data for southern Coconino County, Arizona: Arizona State Land Department, *Water Resources Report,* no. 33, 49 p.

Melis, T.S., Phillips, W.M., Webb, R.H., and Bills, D.J., 1996, When the blue-green waters turn red, historical flooding in Havasu Creek, Arizona: *U.S. Geological Survey Water Resources Investigations Report 96-4059,* 136 p.

5.2.5. Hydrothermal Speleogenesis–Its Settings and Peculiar Features

Yuri V. Dublyansky

Abstract

Three major settings of hydrothermal karst development are:

- Endokarst.
- Deep-seated karst.
- Shallow karst.

Endokarst develops at great depth, where the pressure exceeds the strength of the rock. Voids can exist only if they are filled with overpressured fluid that prevents them from collapse. Deep-seated hydrothermal karst develops in response to changes of pressure and temperature of upwelling fluids. Shallow karst encompasses the interface between thermal and low-temperature water or the zone near the upper surface of thermal water.

Two large zones form within hydrothermal systems:

- A zone of carbonate dissolution.
- A zone of carbonate precipitation.

Four major conditions create and enhance solutional capacity in hydrothermal systems

- Elevated temperature gradients (for carbonated waters).
- Elevated rate of discharge (for carbonated waters).
- Oxidation of hydrogen sulfide.
- Mixing waters of contrasting chemistry.

These features readily occur in the shallow hydrothermal karst setting, and the largest hydrothermal caves are formed there.

Morphologies and dimensions of hydrothermal caves range from pores, individual rooms, single-conduit caves, and large three-dimensional mazes. Cave deposits hold clues as to their origin in their mineralogy, morphology, chemistry, isotopic properties, and fluid-inclusion temperatures.

Introduction

Hydrothermal caves are those in which hot water has played some part in their genesis and development (Ford, 1995). This definition, though quite simple, is sometimes difficult to apply, because it requires another definition of what water should be called hot or thermal. Scholler (1962) considered that thermal water at its resurgence should be at least 4°C warmer than the mean average air temperature in the region. Other researchers suggest different temperature differences, e.g., 5°C (White, 1969) or 15°F (8.3°C) (Waring, 1965). This type of definition is fairly satisfactory when it is applied to a still-active process (i.e. when the temperature at the resurgence can be measured) in areas with a moderate climate.

One example of difficulties with this definition applies to climatic extremes. At Spitsbergen or the Siberian Arctic islands, with their below-zero mean annual temperature, any spring water must be considered hydrothermal. However, for those settings, where neither resurgence temperatures nor annual climatic averages are applicable (for example, for deep-seated waters tapped by boreholes) one needs a more formal definition of thermal water. The same applies to fossil processes, for which paleoclimatic temperatures are often unknown. Conventionally, the temperature of 20°C is considered to be the lower limit of the hydrothermal environment (Ford, 1995), although some meteoric karst systems in hot arid climates may exceed this without any thermal input (for instance, the mean annual temperature in some karst areas such as Venezuela is more than 20°C at the surface, and the global mean annual temperature of the planet is 23°C and rising). The boundary of thermal waters (those with T > 20°C) lies at a depth of 20 to 50 m in areas of active volcanism, at about 500 m on platforms, and decreases to 1000 to 1500 m in permafrost areas (Frolov, 1976).

The sources of extra heat for thermal waters are many and varied. Generally, elevated water temperatures are rather uncommon for exogenous processes. Most hydrothermal systems relate to hypogenic sources of energy, which are powered by the internal heat of the Earth (radiogenic decay, cooling of magmatic bodies, and volcanic activity). In that sense, most hydrothermal caves may be classified as hypogenic. The waters taking part in the process also vary in origin. They may be juvenile (from igneous activity), metamorphic (derived from metamorphism of buried rocks), connate (sedimentary waters trapped in sediments and then expelled), deeply circulating meteoric waters, or mixtures of these types.

Development of Hydrothermal Karst in Various Settings

In the Earth's crust, the properties of fluids and rocks change with depth, along with and dependent on changes in the temperature and pressure. Temperature in the upper part of the Earth's crust increases with depth at a rate of 21 to 30°C/km. The higher geo-thermal gradients (35 to 50°C/km) are report-ed from geodynamically active regions, and still higher gradients (as much as, and higher than 100°C/km) are measured in regions of active volcanism. In terms of hydrothermal karst, this means that below a certain depth, all waters fall in the category of thermal water, and any karst occurring there must be classified as hydrothermal. It must be appreciated that, on a global scale, meteoric waters occupy only a very thin "film" within the uppermost part of the Earth's crust, whereas below that "film" lies the realm of thermal waters.

Water pressure changes with depth in a nonlinear manner, from hydrostatic in shallow settings to nearly lithostatic at great depth. The following three hydrogeodynamic zones can be distinguished (Ezhov and Lysenin, 1986):

- Zone of hydrostatic pressures.
- Transitional zone.
- Zone of lithostatic pressures.

It is convenient to subdivide hydrothermal karst settings into the following three categories (5.2.5 Figure 1):

- Shallow hydrothermal karst.
- Deep-seated hydrothermal karst.
- Endokarst.

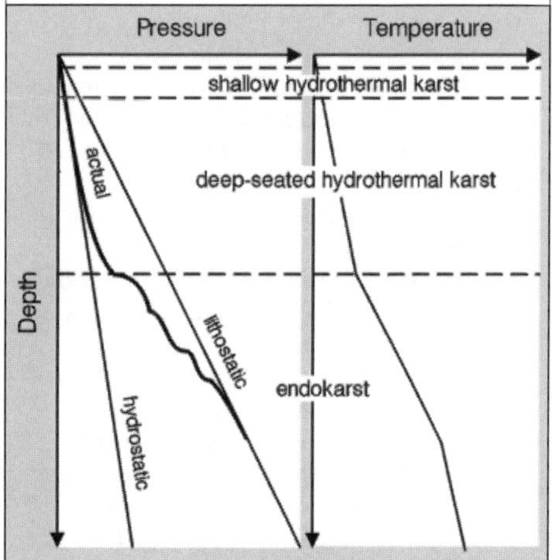

5.2.5 Figure 1. Change of pressure and temperature with depth, and settings of hydrothermal karst.

Endokarst

The notion of endokarst was introduced by Ezhov et al. (1988). It is a well-known fact that the rate of karst development decreases with increasing depth. However, deep drilling for oil and gas reveals that solutional porosity of Paleozoic and Mesozoic carbonate rocks at depths of 4 to 5 km and more may reach 18 to 28 perceent, and the porosity of aluminosilicate rocks may be as great as 25 to 30 percent (Yakutseni, 1984). The pressures of fluids at these depths are typically greater than hydrostatic and may approach lithostatic values.

At those levels, where the pressure exceeds the strength of the rock, pores and cavities may exist only if they are filled with overpressured fluid (water), which prevents them from failure. Inasmuch as the process of dissolution in such an environment is termed endokarst (Ezhov et al., 1992), the whole zone of the Earth's crust where the pressure exceeds the strength of the rock is termed the *endokarstic story*, as opposed to the exokarstic story above.

The depth where processes of endokarst may begin (the roof of the endokarstic story) vary with dependence on the physical-mechanical properties of the rocks. This depth is smaller in regions of Alpine orogenesis (e.g., 500 m in the Indol-Kuban Depression, Russia) and greater within ancient platforms (e.g., 4 km in the Caspian Depression, Russia). Also, it is smaller in basins and greater in adjacent mountains (Ezhov and Lysenin, 1988). In most of the hydrocarbon-producing regions, the roof of the endokarstic story lies at a depth of 4 to 5 km.

This zone of the Earth's crust lies well beyond the realm of "classic" hydrogeology; not much is currently known about it. Fluids taking part in endokarstic dissolution are juvenile and metamorphic, having originated in the course of defluidization of the Earth's crust. There can be only a small proportion of connate water in them, and no meteoric water. The temperature of fluids here always exceeds 80 to 100°C. The size of pores does not normally exceed several centimeters. Though this type of karst apparently does not produce traversable caves, it may play a significant role in the creation of deep-seated reservoirs for hydrocarbons because it is the zone of the "maturation window" in oil genesis and migration.

Deep-Seated Setting

This covers a wide range of depths in the exokarstic story (*ca.* 0.3 to 4.0 km) where the temperature gradients are small, pressures are close to hydrostatic or slightly elevated, and the influence of temperature changes at the Earth's surface is practically absent. Karst that develops in such a setting is genuinely hydrothermal.

Processes of cave excavation and cave infilling occur in response to the change of physicochemical parameters of fluids moving toward the Earth's surface, such as a decrease in temperature and pressure. It is thought that dissolution related to the elevated content of carbonic acid is the leading factor of initiation and enlargement of caves in the deep setting. However, Worthington and Ford (1995) and Lowe and Gunn (1995) demonstrated the important role of dissolution processes involving sulfuric acid.

At certain specific combinations of bedrock solubility and frequency of fractures, net dissolution is accompanied by collapse. Characteristic collapse breccias are known in many hydrothermal karst areas related to Pb-Zn deposits in dolomitic rocks (Dzulynsky, 1976; Dzulynsky and Sass-Gustkiewicz, 1985). Waters taking part in the process may be juvenile, connate, meteoric, or a mixture of these types. Since the residence time for all waters of deep-seated hydrothermal karst is very long, it is often difficult or impossible to determine the actual proportions of these waters in hydrothermal fluids.

Shallow Hydrothermal Setting

The shallow setting describes the interface between thermal and low-temperature waters or near the free surface of the thermal waters. In this zone the pressures are low (down to atmospheric) and temperatures rarely exceed 100° C. This is an environment of transition between hydrothermal and "common" cold karst (Dublyansky, 1997). Temperature gradients may be significant, which lead to the appearance of some specific and powerful processes, like thermal convection. Also, this is a zone where upwelling thermal waters meet colder oxidized meteoric ones, which may induce specific reactions and processes, such as H_2S oxidation, mixing corrosion, and cooling corrosion.

Shallow hydrothermal karst may occur both above and below the thermal water table. All enterable active hydrothermal caves (those containing hot water) are examples of shallow hydrothermal karst. Examples of such caves are known in Turkmenistan (Bakharden Cave, with a lake with temperatures of 35 to 37°C), Italy (Grotta Giusti, 32 to 34°C), Hungary (Molnár János Cave, 18 to 24°C), and many other places.

The processes involved in shallow hydrothermal karst are many and varied. Thus, the caves formed in such a setting often exhibit extremely diverse macro-, meso-, and micromorphologies. Characteristic dimensions of individual caves are appreciably larger than those of deep-seated hydrothermal karst. Also, speleothems of shallow hydrothermal karst are much more varied than those of deep-seated hydrothermal karst.

Chemistry of Fluids and Processes of Cave Excavation

Significant dissolution of carbonate rocks requires acidic water. Hydrothermal, and more generally, hypogenic caves are formed by acids of deep-seated origin, or epigenic acids rejuvenated by a deep-seated process (Palmer, 1991). (Editor's note: In Chapters 3.4 and 5.2 the concept of hypogenic caves is expanded further.)

The major hypogenic acids are carbonic acid, originated from igneous activity and metamorphism of buried carbonates, as well as hydrosulfuric acid (aqueous H_2S) from redox reactions. Thus hydrothermal caves associated with CO_2 and caves formed by water containing H_2S are considered to represent two major classes of hypogenic caves (Ford and Williams, 1989).

Theoretical generalizations concerning the patterns of dissolution within the endokarstic story are lacking. Empirically, from deep drilling it is known (Ezhov et al., 1992) that dissolved solids are comparatively low (*ca.* 5 to 10 g/L) (these values are actually very high compared with normal karst water). Due to high pressures, the total amount of dissolved gases may be as great as 60 to 70 vol percent. The gases are dominated by CO_2 and to a lesser extent by H_2S.

For the exokarstic story, the major mechanisms of cave excavation are:

- Dissolution by rising CO_2-rich thermal water.
- Dissolution due to oxidation of H_2S.
- Dissolution due to mixing of waters of contrasting chemistry.

These are reviewed in Chapter 4.1.5.

Morphology of Hydrothermal Caves

Hydrothermal karst, like common "cold water" karst, produces a large variety of cave sizes and morphologies. Hydrothermal solutional features range in size from solutionally enlarged pores to extensive cave systems with total mapped passage lengths exceeding 100 kilometers.

A comprehensive morphologic classification for hydrothermal caves has not been developed yet. Below, I shall discuss several morphologic types of hydrothermal caves reported in the literature.

Being formed at depth due to the action of rising waters, most hydrothermal caves exhibit a lack of genetic relationships to recharge from overlying or directly adjacent surfaces (Dublyansky, 1989, 1990; Palmer, 1991). Caves must be uplifted with the rock and be intersected by surface erosion, common karst, or by drilling, mining, or quarrying to be studied. This also holds true for many shallow hydrothermal caves (though the latter can show a relationship with thermal spring resurgences or travertine bodies deposited on the land surface). The caves of the Buda Hills, Hungary, where virtually all the hydrothermal caves were discovered due to human activities (quarrying, construction work, and others) are examples.

Solutional Porosity

Solutionally enlarged pores, small anastomotic cavities, and layers of rock with solutionally increased transmissivity are commonly observed in borehole cores from geothermal deposits and oil fields. These small voids may form extensive horizons, and in places become parts of oil and gas reservoirs. For instance, in the West Siberian oil-bearing structures, solutional voids may account for as much as 5 to 15 percent of the entire rock volume (Belkin and Medvedsky, 1989). Such horizons are commonly formed at depths as much as or even exceeding 4 km, where the temperature may exceed 100°C (Ford, 1995). In places these small cavities become lined or entirely filled with secondary hydrothermal minerals (calcite, quartz, rhodochrosite, dolomite, and pyrite), as was reported for some oil fields in West Siberia (Belkin and Medvedsky, 1989).

Layers of solutionally enhanced transmissivity appear where the movement of the fluids is slow, of the order of a few millimeters per year or less, and where hydraulic structures which could concentrate the flow (faults and fissures) are absent. Such zones of enhanced porosity may become inception horizons for future caves (Ford, 1995).

Isometric Rooms

Roughly spherical pockets or rooms with diameters typically ranging from 0.5 to 8.0 m were reported from Khod Konyom Cave in the Crimea. The cavities, lined with scalenohedral Iceland spar crystals and filled with red clay, were truncated by a later vertical vadose cave at a depth of 80 to 200 m. Another example from the Crimea is in Foross Cave, a nearly spherical room some 7 m in diameter with smaller solutional channels 30 to 50 cm across extending 1 to 2 m in various directions and lined with Iceland spar crystals (Dublyansky, 1990).

Similar caves occur in Kirghizstan, where Medieval excavations in the Birksu mercury mine (Turkestan Range) uncovered a series of spherical rooms 3 to 8 m in diameter. The rooms coalesce into two- and three-dimensional multicuspate clusters following the bedding planes and minor faults. Originally these rooms were entirely filled with massive hydrothermal calcite containing veinlets of cinnabar, but the ore was removed by the Medieval miners.

A nearly spherical cave some 15 m in diameter was discovered by divers in Heviz Lake in Hungary. The cave is developed in sandstone, and its bottom is 45 m below the water table. Two springs of 17 and 40°C emerge through the cave (Plózer, 1977) and are examples of little or no mixing of flow between two discrete conduits until they reach the surface.

Individual Chambers

Caves of this type are composed of one or several large individual chambers. The chambers commonly have a length of 100 to 200 m, width of 30 to 60 m, and height of 80 m or more. A characteristic of such caves is that their specific volume (volume to length or depth ratio) always exceeds 100 m^3/m (Dublyansky, 1979, 1980).

Examples of caves belonging to this type are Bakharden Cave in Turkmenistan (300 m^3/m), Novoafonskaya Cave in Abkhasia (306 to 500 m^3/m), Karani Cave in the Crimea (200 m^3/m), and Baritovaja and Petrova Caves in Kirghizstan (112 and 230 m^3/m). An outstanding example of a single hydrothermal chamber is the Madan Cavern in Rhodopy, Bulgaria (see Chapter 5.2.9.

Single-Conduit Caves

Single-conduit caves are beautifully exemplified by the hydrothermal Hellespont, Spence, and Kane Caves in Wyoming, U.S.A. (Egemeier, 1981). The caves are nearly horizontal tube-shaped conduits 60 to 600 m long. They seem to have developed in a vadose setting, where ascending H_2S-bearing fluid comes in contact with the air, which induces replacement corrosion. The gentle floor gradients are supposedly controlled by oxygenation and solutional downcutting, rather than by the water table. The caves contain peculiar gypsum crusts and mounds. Thermal springs discharge through the three above-mentioned caves, and several inactive caves having a similar pattern are known in the region.

Bush-like Caves or Caves with Cupolas
Such caves typically consist of a basal chamber from which a branching pattern of rising passages develops. The branches are composed of coalesced spherical cupolas whose typical size is 0.5 to 1.5 m, and cupolas terminate most of the branches. Such caves are known in Hungary (Müller and Sárváry, 1977), Sátorköpuszta Cave in the Pilis Mountains being the prototype.

Another Hungarian example is Bátori Cave in the Buda Hills. This type of cave is believed to be created by delivery of hot water to a single point at the base of carbonate rocks of low fissure density (Ford and Williams, 1989). It is taken as an example of monogenic hydrothermal karst, although the exact mode of formation for such caves is not determined yet.

Specifically, one hypothesis ascribes their formation to the convective movement of hot water in the phreatic zone (Rudnicky, 1978, 1989). Another model invokes the convective movement of moist air above the hot water table, coupled with condensation corrosion (Szunyogh, 1982, 1989). D. Ford (pers. com.) believes that such caves are formed by a combination of the two mechanisms, with paleokarst caves serving as initial water conduits. He envisages that rising heated water invades a paleokarst target, opening up a relatively small route through it. Later exterior erosion lowers the spring position. Once that is stabilized, a basal chamber forms at the new water table, and spherical chambers develop upward by condensation corrosion, probing weaknesses in the paleokarst. The gas is ventilated via small exit routes.

Another location where cupola-shaped niches occur in abundance is the Azérous in Northern Algeria (Collignon, 1983). Hydrothermal caves there have various patterns—from a single passage (Rhar Sidi Bacou Cave) to three-dimensional network and ram-iform-maze patterns (Rhar Medjraba and Rhar Amalou Caves).

Most caves have at least two patterns superposed on one another, one of them always represented by multiple cupolas. The cupolas have rather regular diameters of about 1 m. They commonly coalesce, developing in massive Cretaceous limestone without any apparent relation to fissures.

In most cases cupolas are restricted to the uppermost parts of the caves. Only in Rhar Kab Shrir Cave, in the section called Gruyere, which is composed of a multitude of spherical niches, is there development below the network maze. These two sections of the cave are believed to represent two consecutive stages of cave development (Collignon, 1990).

Phreatic Maze Caves
Maze caves are the most common type of hydrothermal cave systems. Maze caves were typified by Palmer (1991) who distinguished network, anastomotic, spongework, and ramiform caves. The following discussion is based on this classification.

Network Caves. These are angular grids of intersecting passages formed by the widening of nearly all major fractures within favorable areas of soluble rocks. Closed loops and relatively high and narrow passages are common. Two-dimensional rectilinear maze systems are created where rising water is trapped in densely jointed carbonate rock below a relatively impervious bed (Ford and Williams, 1989). Prominent examples of such a pattern are Cserszegtomaji-kut Cave and Acheron-kut Cave in Hungary, developed in Triassic dolomite under the cover of Miocene sandstone (Kárpát, 1982, 1983).

Multistory rectilinear network caves occur more frequently. Examples of such a pattern are caves of the Buda Hills, Hungary (Pálvölgy, Szemlo-hegy, Ferenc-hegy), caves of the Black Hills, South Dakota, U.S.A. (Wind Cave, Jewel Cave). Many carbonate-hosted zinc-lead ore deposits (the so-called Mississippi Valley type) exhibit a network pattern of solutionally widened fractures later filled–partly or entirely–with sulfides (Jefferson City Mine, Tennessee, U.S.A.; Devil's Hole Mine, England).

Anastomotic Caves. These caves consist of curvilinear tubes that intersect in a braided pattern with many closed loops. They usually form a two-dimensional array along a single favorable parting or low-angle fracture. Rare three-dimensional variants follow more than one geologic structure. The most prominent example of a supposedly hydrothermal 3-D anastomotic cave is Pobednaya Cave in Kirghizstan, with its 1.5 km of very narrow tube-shaped crawlways (Mikhailev, 1989) and unusual barite mineralization.

Spongework Caves. These consist of interconnected solution cavities of varied size in a seemingly random three-dimensional pattern. Such caves appear to form by coalescing of intergranular pores and minor interstices.

Ramiform Caves. These consist of irregular rooms and galleries wandering three-dimensionally with branches extending outward from the main areas of development. Passage interconnections are common, producing a continuous gradation with spongework and network caves. Jósef-hegi Cave in the Buda Hills, Hungary, is probably a good example of such a pattern.

Cave Deposits

Speleothems may give further insight into the past cave environment. The most reliable data can be obtained from subaqueous deposits (those formed below the water table). Much more uncertainty is involved in interpreting subaerial cave deposits from those formed at and above the water table, because they are often formed far from thermodynamic equilibrium. The following sections compare some properties of speleothems found in caves of deep-seated and shallow (transitional) hydrothermal karst. Generalization is based on empirical data obtained from five hydrothermal karst regions as follows: Northern Algeria, Hungary, southern Kirghizstan; Crimea (Ukraine), and South Dakota (U.S.A.).

Mineralogy
Calcite is the most common mineral in hydrothermal caves developed in carbonate rocks. In addition, such minerals as quartz, barite, fluorite, and sulfides are often reported. (Ore-related hydrothermal karst, where the list of minerals can be quite large, is not considered here.) Shallow hydrothermal karst caves do not normally contain any "exotic" minerals.

Character of Cave Deposits
Large individual calcite crystals, palisade aggregates, thick crusts, and sediment reflecting very low (essentially stagnant) hydrodynamic conditions are common in deep-seated hydrothermal caves. The size of individual calcite crystals can be as large as 10 to 30 and even 100 cm. The crystal morphology is normally simple, dominated by scalenohedra ("dogtooth spar").

By contrast, deposits of shallow hydrothermal karst often reflect an unstable environment. Crusts and cave wall linings are typically less thick (thickness normally increases from the water table downward). Individual crystals are rare. The size of crystals in aggregates typically ranges from several millimeters to a few centimeters. The dominant crystal morphology is a combination of scalenohedra and prisms with the crystal tip blunted by flat rhombohedra ("nailhead spar"). For an example of the association between dogtooth and nailhead spar, see Palmer and Palmer on the Black Hills caves (Chapter 5.2.2).

In places minerals are contaminated by significant amounts of clay (e.g., caves of

Nagykömázsa Quarry, Hungary), which indicate that the paleowaters were dynamic enough to carry the particles. Typically, crystals have many defects. In addition to subaqueous euhedral crystals and crusts, two more types of speleothems occur in shallow hydrothermal karst: waterline deposits (rafts, folia, cave cones), and subaerial deposits. However these two types are also common in cold karst. Waterline and subaerial deposits of shallow hydrothermal karst have not been studied in sufficient detail to develop criteria diagnostic of their origin.

Luminescence

A simple and efficient way of distinguishing high-temperature vs. moderate temperature calcite was developed by Y. Shopov (pers. com.). On being excited by a powerful photographic flash, calcite crystals glow with different colors. Specifically, most calcites found in deep-seated hydrothermal caves exhibit a short duration (<0.3 s) orange-red glow, sometimes accompanied by a longer (1 to 20 s) blue-white afterglow. Calcites from shallow hydrothermal karst typically yield only a lengthy (1 to 20 s) blue-white glow. Our fluid-inclusion studies demonstrated that in one of the Hungarian samples calcite which formed at >60°C had a red (+blue) luminescence, whereas the younger part of the same crystal formed at < 60°C yielded only a blue glow (Dublyansky, 1995).

Isotopic Properties

Calcite crystals from deep-seated hydrothermal caves typically have $\delta^{18}O$ varying from -20 to -6 percent PDB and $\delta^{13}C$ from -3 to +3 PDB. The values of $\delta^{13}C$ in isotopic profiles taken along the growth axes display little or no variation (2.5.2 Figure 2a), which reflects deposition from fluids well equilibrated with the bedrock (very long residence time). The typical isotopic values for transitional karst calcites are $\delta^{18}O$ from -15 to -8 percent PDB and $\delta^{13}C$ from -9 to -2 percent PDB.

The $\delta^{13}C$ may vary significantly in the growth direction (2.5.2 Figure 2b). This reflects possible occasional input of water charged with isotopically light soil-derived organic carbon.

Content of Trace Elements

Hydrothermal cave calcite commonly has elevated amounts of trace elements. For instance, calcite from five hydrothermal karst areas contain the following:

- Mn 12 to 120 ppm.
- Fe 80 to 900 ppm.
- Pb 0.5 to 19 ppm.
- Sr 40 to 770 ppm.

The contents of these elements in shallow hydrothermal karst calcite are normally lower as follows:

- Mn 3 to 20 ppm.
- Fe 70 to 230 ppm.
- Pb 0 to 9 ppm.
- Sr 5 to 45 ppm.

Temperature and Salinity of Paleowaters from Fluid Inclusions

Temperatures of hydrothermal karst minerals determined by fluid inclusions range from 250 to 55°C. Mineral-forming water trapped in inclusions contains dissolved salts. The concentrations of these salts in some inclusions reaches tens of wt percent (they are brines). In contrast, most calcite from transitional karst caves contains all-liquid inclusions, which indicates temperatures less than approximately 40°C. In those rare cases when two-phase gas-liquid inclusions are present, they have homogenization temperatures of 55 to 30°C. The water trapped in such inclusions invariably has low salinities.

Conclusions

As shown above, four major features create and enhance solutional capacity in hydrothermal systems. These are:

- Elevated temperature gradients (carbon dioxide waters).
- Elevated rate of discharge (carbon dioxide waters).
- Oxidation of hydrogen sulfide.
- Mixing of waters of contrasting chemistry.

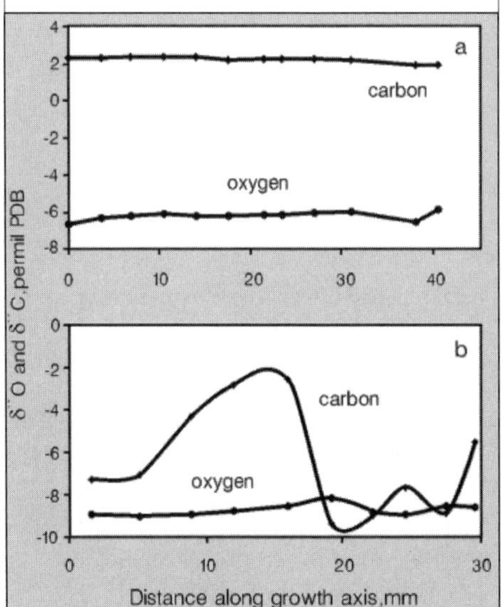

*5.2.5 Figure 2. Stable-isotopic profiles along the axis of growth of individual crystals. **a** is hydrothermal calcite (~80°C) from spherical solutional rooms (deep-seated hydrothermal karst) truncated by younger passages, Khod Konyom, Crimea. **b** is low-temperature hydrothermal calcite (35°C and less) from solutional cavities (shallow hydrothermal karst) in Esztramos Quarry, Hungary. Temperatures were determined from fluid inclusions.*

Places where these features have the best development may be considered the most favorable for finding hydrothermal caves.

The shallow zone seems to provide the most favorable environment for the development of large hydrothermal caves. All four features can readily occur there.

In the deep-seated zone, the largest caves should be expected along the routes of concentrated hydrodynamic flow.

Examples of such settings are:

- Zones of large deep-seated faults.
- Marginal parts of deep sedimentary basins, where water is expelled from the center of the basin,

References

Belkin, V.I., and Medvedsky, R.I., 1989, Deep-seated karst and the problem of interconnected ore- and oil-forming processes, *in* Ore-bearing karst of Siberia: Novosibirsk, Siberian Institute of Geology, Geophysics, and Mineral Raw Material, p. 93-103.

Collignon, B., 1983, Spéléogenése hydrothermale dans les Bibans (Atlas Tellien–Nord de l'Algérie): *Karstologia* (France), v. 2, p. 45-54.

Collignon, B., 1990, Les karst hydrothermaux d'Algérie: *10th International Congress of Speleology,* Budapest, Proceedings, p. 758-760.

Dublyansky, V.N., 1974, The giant hydrothermal karst cave in Rhodopy: *Peshchery* (Perm), v. 14-15, p. 233-237 (in Russian).

Dublyansky, V.N., 1979, Hydrothermal karst in the Alpine folded region of the south of the USSR: *Physical Geography and Geomorphology* (Kiev), v. 21, p. 61-64 (in Russian).

Dublyansky, V.N., 1980, Hydrothermal karst in the Alpine folded belt of the southern part of USSR: *Kras i Speleologia* (Poland), v. 3, no. 12, p. 18-36.

Dublyansky, Y.V., 1989, The main principles of development and diagnostic criteria of a carbonate hydrothermal karst: *10th International Congress of Speleology,* Budapest, Proceedings, p. 77-79.

Dublyansky, Y.V., 1990, *Characteristics of formation and modeling of hydrothermal karst:* Novosibirsk, Nauka, 151 p. (in Russian, English res.).

Dublyansky, Y.V., 1995, Speleogenetic history of the Hungarian hydrothermal karst: *Environmental Geology,* v. 25, p. 24-35.

Dublyansky, Y.V., 1997, Transition between hydrothermal and cold-water karst: *12th International Congress of Speleology,* La Chaux-de-Fonds, Proceedings, v. 2, p. 267-270.

Dzulynsky, S., 1976, Hydrothermal karst and Zn-Pb sulfide ores: *Roczn. Pol. Tow. Geol.,* v. 46, p. 217-230.

Dzulynsky, S., and Sass-Guskiewicz, M., 1985, Hydrothermal karst phenomena as a factor in the formation of Mississippi Valley-type deposits, *in* Handbook of strata-bound and stratiform ore deposits: pt. 4, v. 13, chap. 5, p. 391-439.

Egemeier, S.J., 1981, Cavern development by thermal waters: *National Speleological Society Bulletin,* v. 43, p. 31-51.

Ezhov, Y.A., and Lysenin, G.P., 1986, Vertical hydrodynamic zoning of the Earth's crust: *Soviet Geology,* v. 8, p. 111-120 (in Russian) .

Ezhov, Y.A., and Lysenin, G.P., 1988, Significance of the zone of transitive pressures in the underground hydrosphere: *Soviet Geology ,* v. 8, p. 107-114 (in Russian).

Ezhov, Y.A., Lysenin, G.P., and Andreichuk, V.N., 1988, Endokarst, *in* Collection of Papers–1987: *Sverdlovsk Institute of Geology and Geochemistry,* p. 36-40 (in Russian).

Ezhov, Y.A., Lysenin, G.P., Andreychuk, V.N., and Dublyansky, Y.V., 1992, *Karst in the Earth's crust: Distribution and the main types:* Novosibirsk, United Institute of Geology, Geophysics and Mineralogy, 76 p. (in Russian).

Ford, D.C., and Williams, P.W., 1989, *Karst Geomorphology and Hydrology:* London, Unwin Hyman, 601 p.

Ford, T.D., 1995, Some thoughts on hydrothermal caves: *Cave and Karst Science,* v. 22, p. 107-118.

Frolov, N.M., 1976, *Hydrogeothermal studies:* Moscow, Nauka, 280 p. (in Russian).

Kárpát, J., 1982, A Cserszegtomaji-k'tbarlang (Chersegtomay Pit Cave): *Karszt és Barlang* (Hungary), no. 1, p. 35-40 (in Hungarian).

Kárpát, J., 1983, Az Acheron-Kútbarlang: *Karszt és Barlang* (Hungary), no. 1-2, p. 25-28 (in Hungarian).

Lowe, D.J., and Gunn, J., 1995, The role of strong acid in speleo-inception and subsequent cavern development, *in* Barani-Kevel, I., ed., Environmental effects on karst terrains: *Acta Geographica,* (Szeged), v. 34, p. 33-60.

Mikhailev, V.N., 1989, *Karst of the Kirghizia:* Frunze, Ilim, 146 p. (in Russian).

Müller, P., and Sárváry, J., 1977, Some aspects of development in Hungarian speleology theories during the last 10 years: *Karszt és Barlang* (Hungary), Special Issue, p. 53-60.

Palmer, A.N., 1991, Origin and morphology of limestone caves: *Geological Society of America Bulletin,* v. 103, p. 1-21.

Plózer, I., 1977, Situation of Hungarian cave diving in 1976: *Karszt és Barlang* (Hungary), Special Issue, p. 67-69.

Rudincky, J., 1978, Role of convection in shaping subterranean karst forms: *Pr. nauk. Usl,* Katowice, Poland, v. 280, p. 92-101.

Rudincky, J., 1989, Relation between natural convection and cave formation in hydrothermal karst: *10th International Congress of Speleology,* Budapest, Proceedings, p. 14-16.

Scholler, H., 1962, Les eaux souterraines: Paris, Masson and Cie.

Szunyogh, G., 1982, A hévizes eredet¸ gömbfölkék kioldódásanak elméleti vizsgálata (Theory of the dissolving of spherical cavities formed by thermal water): *Karszt és Barlang* (Hungary), no. 2, p. 83-88 (in Hungarian).

Szunyogh, G., 1989, Theoretical investigation of the development of spheroidal niches of thermal water origin–second approximation: *10th International Congress of Speleology,* Budapest, Proceedings, p. 766-768.

Takács-Bolner, K., and Kraus, S., 1989, The results of research into caves of thermal water origin: *Karszt és Barlang* (Hungary), Special Issue, p. 31-38.

Waring, G.A., 1965, Thermal springs of the United States and other countries of the world: *U.S. Geological Survey Professional Paper,* no. 492, 383 p.

White, D.E., 1969, Thermal and mineral waters of the United States–brief review of possible origins: *23rd International Geological Congress,* Prague, Proceedings, v. 19, p. 269-286.

Worthington, S.R.H., and Ford, D.C., 1995, High sulfate concentrations in limestone springs–An important factor in conduit initiation?: *Environmental Geology,* v. 25, p. 9-15.

Yakutseni, V.P., 1984, *Intensive gas accumulation in the Earth:* Leningrad, Nauka, 122 p. (in Russian).

5.2.6 Hydrothermal Speleogenesis in the Hungarian Karst

Yuri V. Dublyansky

Abstract

Hydrothermal karst caves in Hungary were formed in a variety of speleogenetic settings. The oldest small solutional cavities (vugs) were formed in the deep-seated low-gradient zone at temperatures in excess of 90°C. The temperature decreased with time, as inferred from fluid-inclusion studies of cave calcite. The caves were formed by "normal" carbonate dissolution. Large two-dimensional mazes of the Buda Hills were formed in a shallow low-temperature (but still hydrothermal) setting. The leading speleogenetic processes were normal carbonate dissolution (including mixing and cooling corrosion) and sulfuric acid corrosion. Temperature at this stage was lower, about 50°C and less. Above the thermal water table in the subaerial zone, some specific and powerful speleogenetic processes occurred, namely, condensation corrosion (by CO_2-bearing water) and replacement corrosion involving sulfuric acid reactions. Three-dimensional bush-like caves composed of connected spherical niches were formed as a result of this subaerial karstification.

Introduction

Hydrothermal karst has long been recognized in Hungary (Pávai-Vájna, 1931; Jakucs, 1977), and a hydrothermal origin has been proposed for many caves in virtually all of the karst areas of the country. The caves have unusual morphology lacking in vadose features, uncommon types of speleothems, and no connection with the surface (many caves were discovered by quarrying), all of which imply that the speleogenetic processes differ from those in common meteoric karst.

The recognized hydrothermal origin of earlier veins, crystal-lined vugs, and zones of altered rock strongly hinted at a hydrothermal origin for the caves. Today, thermal springs are often found at the margins of the karst massifs. In places they emerge through caves, such as at Héviz and Miskolc-Tapolca.

Various aspects of the cave and speleothem formation have been discussed in numerous publications (Kovács and Müller, 1980; Takács-Bolner and Kraus, 1989, Ford and Takácsné Bolner, 1991; Nádor, 1992; Dublyansky, 1995). This text summarizes the most important speleogenetic features of the Hungarian hydrothermal karst.

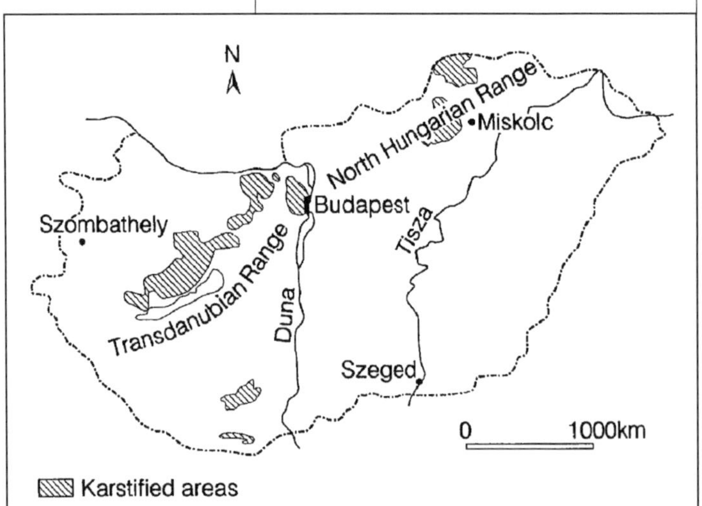

5.2.6 Figure 1. Hungary, showing the Transdanubian and North Hungarian Ranges.

Geologic Setting

The karst massifs containing the best known examples of hydrothermal karst belong to the Transdanubian Range (Buda Hills and Gerecse Mountains) and the North Hungarian Range (Bükk and Aggtelek-Rudabánya Mountains; 5.2.6 Figure 1). The northeastern segment of the Transdanubian Central Range—the Buda Hills—is the best studied of all Hungarian karst terrains and serves as a model area for the hydrothermal karst development in the country.

The oldest exposed strata are Triassic limestone and dolomite. The upper part of the sequence is cut by a paleokarst surface of Late Cretaceous to middle Eocene age (Wein, 1977). The shallow marine Szépvölgy Limestone and shelf Buda Marl of middle to upper Eocene age were deposited unconformably upon the Triassic strata. Karstifiable Triassic-Eocene rocks were overlain in the Oligocene by deep water marine Kiscell Clay up to 600 m thick.

In the early Miocene there was block faulting, the central part of the Buda Hills was uplifted, and Oligocene clay was partly or entirely eroded away. Along the foothills step-faulted blocks of carbonate rocks subsided to depths of 1000 to 3000 m, creating a basement for surrounding young sedimentary basins. Most of the Buda Hill caves occur within the Eocene formations. The other hydrothermal karst areas have similar geologic and tectonic patterns, except that the caves are mostly developed in the Triassic rocks.

Effects of Hydrothermal Activity

Hydrothermal Minerals

Numerous veins of milky-white calcite are encountered in quarries or intercepted by cave passages. Scalenohedral calcite crystals ("dogtooth spar") up to 5 to 7 cm long are often found in veins and in dissolutional vugs. In places the calcite entirely fills solutional voids, creating bodies several meters in size. Veinlets and crusts consisting of small (several mm to 1.5 cm) tabular barite crystals are in places associated with the calcite, or may form independently. Minor quartz, fluorite, pyrite, metacinnabarite, cinnabar, and hematite have also been reported from the Buda Hills (Nagy and Pelikán, 1973; Dublyansky, 1995). The minerals listed above are typical of a low-temperature hydrothermal assemblage, although they were first interpreted as high-temperature volcanic derivatives (Schafarzik, 1928).

Altered Carbonate Rocks

Zones of deeply altered and disintegrated rocks 1 to 3 meters wide occur along tectonic fissures in carbonate rocks in the Buda Hills. The products of alteration consist of dispersed quartz, kaolinite, illite, montmorillonite, and calcite (Nádor, 1992). In some karst areas, Triassic dolomite has been altered to a state of dolomitic sand (analagous to grus), which can be extensive enough for industrial mining.

Caves consisting of numerous coalesced spherical niches creating characteristic bush-like patterns were discovered in some of the quarries after removal of the grus. This type of alteration, quite common in Mississippi-Valley-Type Pb-Zn deposits elsewhere, has been attributed to hydrothermal fluids carrying sulfuric acid (Jakucs, 1977).

Another type of hydrothermal alteration forms "silicate towers. These are fossilized spring orifices created along the channels of ancient thermal springs where the carbonate host rock has been replaced by silica. Being less soluble, such metasomatic bodies stand out as towers in the later erosional relief (Jakucs, 1977).

Travertine

Small patches of travertine—carbonate precipitates deposited by ancient springs—are found throughout the Buda Hills. They range in age from late Miocene-Pliocene to late Pleistocene (Scheurer and Schweitzer, 1980; Takács-Bolner and Kraus, 1989). Their distribution corresponds to the location of hydrothermal spring caves.

Hydrothermal Caves

Small dissolutional cavities—vugs, fissures, and small chambers lined or entirely filled with crystalline calcite—are often encountered in the Hungarian karst. In many instances they are intercepted by younger supposedly hydrothermal caves. Many of the large (enterable) caves in Hungary are believed to be of hydrothermal origin, as first suggested by Pávai-Vajna (1931). The caves are characteristically mazes with equidimensional passages (5.2.6 Figure 2) or bush-like upward-branching groups of spherical niches (5.2.6 Figure 3). These caves have very few vadose erosional features. They are not rich in "normal" vadose speleothems such as stalactites and stalagmites, but are often decorated with various unusual formations (Takács-Bolner and Kraus, 1989).

Thermal Springs

Thermal springs are quite abundant at the margins of the Hungarian karst massifs. Dozens are known within Budapest alone. The water temperature ranges between 20 and 65°C, and the total dissolved solids content ranges from 600 to 2000 mg/L, being composed of bicarbonates, chlorides, and sulfates of sodium, magnesium, and calcium. The higher the temperature, the more CO_2 and respectively HCO^{3-}, Ca^{2+} and Mg^{2+} are dissolved in the water. Thermal water of these springs is interpreted to be meteoric water that infiltrated deep into the carbonate rocks, became heated, and then rose beneath the

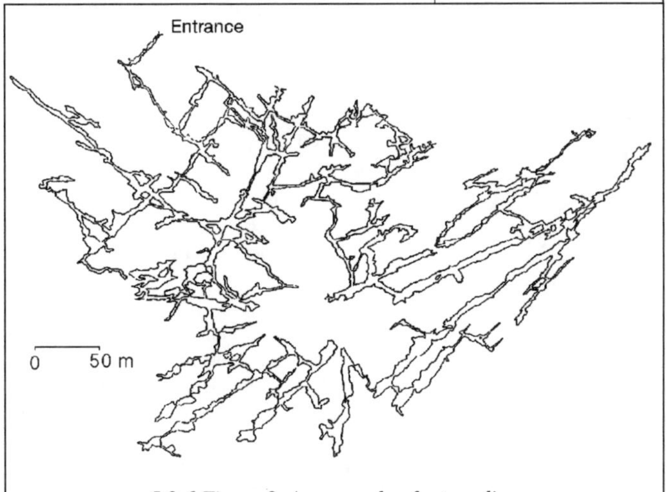

5.2.6 Figure 2. An example of a two-dimensional rectilinear maze, Pal-völgy Cave (from J. Kárpát and K. Takácsné-Bolner).

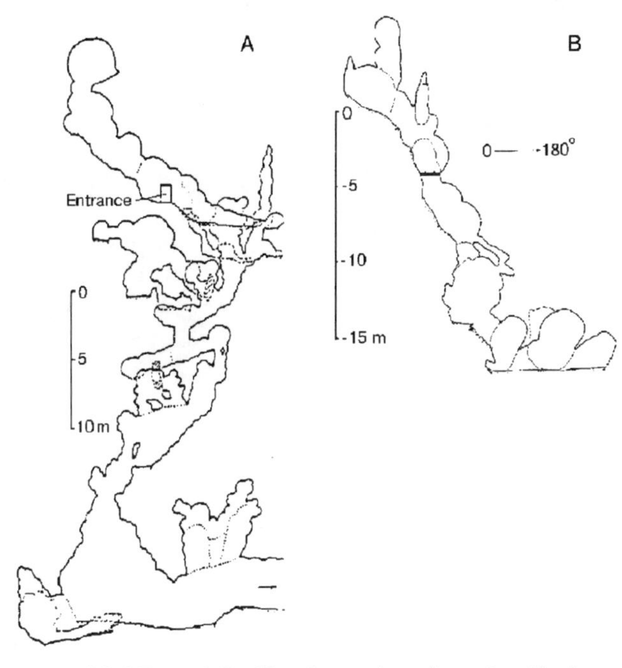

*5.2.6 Figure 3. Profiles of some three-dimensional bush-like caves: **A** is Sátorkö-puszta Cave (by M. Juhász). **B** is Bátori Cave (from P. Borka and J.Kárpát).*

confining impermeable strata of adjacent sedimentary basins (5.2.6 Figure 4; Alföldi, 1979; Kovács and Müller, 1980). In some cases the thermal springs discharge directly through caves (e.g., Molnár János Cave in the Buda Hills, and Miskolc-Tapolca Spa in the Bükk Mountains).

Geologic Model of the Hungarian Hydrothermal Karst

A geologic model of hydrothermal speleogenesis in the Buda Hills was developed by Kovács and Müller (1980) on the basis of the ideas of Schafarzik (1928), Vendel and Kisházi (1964), and Liebe and Lorberer (1978). In it, the first stage of hydrothermal activity in the area was triggered by andesite volcanism in the Miocene. Groundwater flowed in a circular path (a closed convective cell pattern) in the carbonate rocks beneath a 500- to 1000-m thick cover of impermeable Kiscell Clay. Dissolution of small hydrothermal cavities and bedrock alteration occurred at that stage. Calcite, barite, and other hydrothermal minerals were also deposited at temperatures as high as 100°C (Müller, 1989).

The second stage of hydrothermal activity was caused by an increase of the heat flux in the area during the Pliocene, when the Buda Hills were uplifted and the Kiscell Clay was breached by erosion. Meteoric water descended to significant depth, became heated and then upwelled to the surface along the contacts with the impermeable rocks of adjacent sedimentary basins (open convective cell pattern).

At the deepest parts of its path the water became charged with CO_2 released as a result of metamorphism of deep-seated carbonate. The high content of dissolved CO_2 caused high aggressiveness on the ascending branch of the path. It was enhanced significantly where upwelling thermal water met descending meteoric water (mixing corrosion).

The model of Kovács and Müller (1980) has been revised by Dublyansky (1995). Most problematic is their first stage. Both the volume and the velocity of water movement in massive carbonate rocks buried under some 1000 m of impermeable clay would have been very low, and the development of large-scale convective cells in such a setting is hardly possible. Besides, the thick cover of impermeable rock would not have allowed CO_2 to escape from the system. Hence the physical mechanism permitting vug formation and subsequent calcite deposition is not clear. Also, the fluid inclusions in minerals formed under such conditions should display a high CO_2 content, which was not observed in the Hungarian hydrothermal calcite and barite.

It has been suggested (Dublyansky, 1995) that the first stage, as presented by Kovács and Müller, may have been responsible only for alteration of the bedrock limestone and dolomite. All hydrothermal karstic features, such as:

- smaller solutional forms containing hydrothermal minerals,
- larger forms containing phreatic calcite crusts, and
- large hydrothermal caves,

are the product of a single stage of hydrothermal activity (the second stage of Kovács and Müller), which began in the Pliocene and continues today.

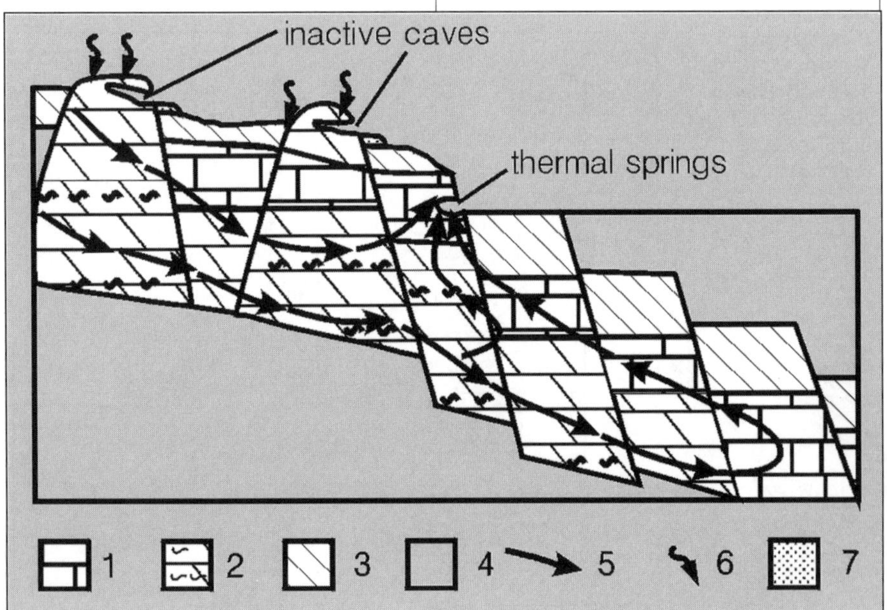

*5.2.1 Figure 4. Hydrothermal flow in Buda Hills (by Alföldi, 1979; Kovács and Müller, 1980). **1** is karstifiable rocks (Triassic and Eocene carbonates). **2** is local aquitards (marls, **3** is clay (Oligocene). **4** is younger sediments (Neogene). **5** is thermal waters. **6** is meteoric waters. **7** is travertines.*

Hydrothermal Activity in Hungary

Current Hydrothermal System

The evolution of the interpretation of thermal water systems in Hungary has been summarized by Maucha (1989). In the 1950's, the thermal waters of Hungary were considered to be "juvenile" or related to postvolcanic activity. Vendel and Kisházi (1964) were the first to demonstrate that these waters represent "normal" meteoric waters (atmospheric precipitation) that infiltrate in recharge areas, descend to a significant depth where they acquire high temperatures, upwell toward the Earth's surface where they mix with local cold water and discharge at the boundaries of karst massifs. This concept, confirmed by detailed studies by Hungarian hydrogeologists, is universally accepted today.

Temperatures and chemical composition of water. Thermal springs in the Buda Hills have temperatures up to 65°C, with higher temperatures (87°C) found in waters tapped by drill holes (Liebe, 1989). The spring water (in mg/L) contains dissolved:

- Ca^{2+} (76-125),
- Mg^{2+} (29-54)
- HCO_3^- (360-510)
- Cl^- (20-45)
- SO_4^{2-} (10-100).

The waters of thermal wells (after Maucha, 1989) contain (in mg/L):

- Ca^{2+} (90-250)
- Mg^{2+} (25-70)
- HCO_3^- (490-800)
- Cl^- (90-150)
- SO_4^{2-} (100-290).

Thermal waters also contain sodium (Liebe, 1989).

Isotopic Pattern and Age of the Water. An isotopic study of the current hydrothermal systems in Hungary was made by Deák (1979, 1989). The relationship of stable oxygen and deuterium concentrations follows the "meteoric water line" $\delta D = 8\ \delta^{18}O + 10$, which suggests their surficial origin. In different karst areas, radiocarbon dating revealed a strong correlation between the apparent age and the temperature of the thermal waters. For instance in the Eger area, spring water at 30°C was shown to be about 6,000 yr old, whereas well water with T = 73°C yielded an age of 15,000 yrs.

Although the absolute ages may not be accurate, due to problems associated with this dating method (mixing of different waters, isotope fractionation, and enrichment by limestone-derived "dead" carbon), the data may suggest a general trend of age versus temperature. Waters will have passed through deeper zones to be heated to higher temperatures, and their residence times should be longer. Similarly, the $\delta^{13}C$ value of water increases from -16.4% for "young" cold karstic water in springs (T = 9°C) to –3.4 percent for the putative 15,000-year-old thermal water (T = 73°C) encountered in a well at a depth of 460 to 480 m. This might be accounted for by the same general model—meteoric water that descended to a significant depth, with a long filtration path and residence time, exchanged its "light" soil-derived carbon for the "heavier" carbon of the bedrock limestone or metamorphic CO_2.

Ancient Hydrothermal System

Features of the past hydrothermal systems under the Hungarian karst massifs are recorded by minerals deposited from thermal waters. Calcite is the dominant mineral associated with the Hungarian hydrothermal caves. The total amount of barite is hundreds of times smaller, and all other minerals occur in almost negligible amounts. Our studies have shown that the calcite from the Hungarian hydrothermal karst can be subdivided into two types, which record changes in depositional environment. Vein-vug calcite is deposited as vein fillings and vug linings. The crystals have a scalenohedral ("dogtooth") habit. This type of calcite is commonly associated with barite and is most abundant in the Buda Hills. Phreatic cave crust calcite lines larger solutional cavities as palisade crusts or entirely fills them. The crystals have rhombohedral terminations. This type of calcite is abundant in the Gerecse, Bükk, and Aggtelek-Rudabánya Karsts.

Temperature and Chemical Composition of Paleofluids. Vein-vug calcite yields homogenization temperatures ranging from 95 to 40°C (Dublyansky, 1995). Consistent results (60 to 40°C) are obtained from barite. Two large calcite crystals reveal a decrease in temperature during calcite growth from about 80 to 50°C. The fluids trapped in the inclusions of vein-vug calcite contain chlorides of Na, Mn, K, and Ca. Concentrations of solutes in some inclusions reach 20 to 23 wt %. Most inclusions, however, contain dilute solutions (ca. 0 wt %). Phreatic cave crust calcite contains far fewer inclusions useful for thermometry. In three occurrences (Nagy kömázsa Quarry and Fecke Cave in the Bükk Mountains, and Sárkány-lyuk Cave in the Gerecse Mountains) homogenization temperatures range from 55 to 30°C. Calcite from two more sites (Bajót Cave in the Gerecse Mountains and Esztramos Quarry in the Aggtelek-Rudabánya area) contained only one-phase liquid inclusions, which implies formation temperatures lower than 35 to 30°C. Freezing experiments on inclusions from phreatic cave crust calcite gave no measurable numerical data, which suggests fluid of low salinity.

Isotopic Pattern. Studies of ^{13}C and ^{18}O in calcite from the Buda Hills have been reported by Ford and Takácsné Bolner (1991) and Nádor (1992). Strong depletion trends were found for ^{18}O (-5.1 to -9.7%) (relative to PDB) in upper Eocene limestone and $\delta^{13}C$ remained at or close to the PDB standard (-1.4 to +2.0%). These imply that significant alteration of the bedrock has occurred, probably during Oligocene burial.

Hydrothermal vein-vug calcite is more strongly depleted in $\delta^{18}O$ (-11.2 to -17.6%), which is typical of hydrothermal calcite in general (Bakalowicz et al., 1987). The values of $\delta^{18}O$ increase in individual crystals along the growth direction (that is, with decreasing temperature; 5.2.6 Figure 5A). This type of calcite is slightly enriched in ^{13}C (+1.2 to +2.4%), which was first interpreted as favoring deposition during Oligocene burial in an environment essentially "closed" with respect to "light," soil-derived carbon (Ford and Takácsné Bolner, 1991).

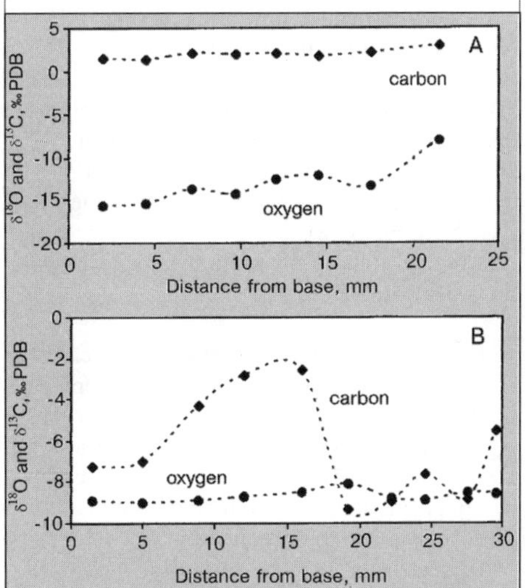

5.2.6 Figure 5. Stable isotopic patterns: A is vug-vein calcite. B is phreatic cave crust calcite. The profiles along the growth axe of the two individual calcite crystals, with bedrock to the left.

Later, it has been argued (Dublyansky, 1995) that the positive values of $\delta^{13}C$ might just as well be characteristic of calcite formed in deep parts of a hydrologic system involving deep circulation of meteoric water, similar to the one operating in Hungary today. Unvarying carbon isotope values in deep-seated calcite suggest that the calcite was deposited from fluids equilibrated with the environment.

Low-temperature phreatic cave crust calcite from Esztramos Quarry has shown wide variations in $\delta^{13}C$ (from -2.6 to -9.3%), while maintaining a quite stable $\delta^{18}O$ (-8.1 to -9.0%; Dublyansky, 1995; 5.2.6 Figure 5B). This calcite was probably deposited in a backwater reservoir setting, close to the Earth's surface. The low values and large variations for carbon might be indicative of varying input of meteoric water enriched in light soil-derived carbon. The less variable oxygen values might be interpreted as reflecting a fairly stable temperature during calcite deposition, lower than that characteristic of vein-vug calcite deposition, which is consistent with the fluid inclusion data.

Summary

The temperature of the paleohydrothermal system that operated in the Hungarian karst massifs was comparatively low. Fluid inclusion analysis gives estimates of about 90°C from the calcite and barite samples. The temperature of the system gradually decreased with time. The latest generations of calcite were formed at "lukewarm" temperatures (55 to 30°C and probably less). The highest temperatures occurred in the Buda Hills area, where almost all vein and vug-filling calcite began to form at 80 to 90°C. The phreatic cave crust calcite is not typical of the Buda Hills.

Conversely, in the karst areas east and west of the hills such crusts of rhombohedral calcite crystals lining cave walls are dominant. The fluids responsible for formation of these crusts had a temperature of 55°C and less. At the highest temperature stages the fluids contained chlorides of Na, Ca, and Mg. High concentrations of NaCl measured in some inclusions may be explained as trapped formation water during the early stages of hydrothermal activity. Subsequently, the saline water was flushed away and the system became an essentially "fresh-water" one in the later stages of mineral formation.

The temperature and chemistry of the paleofluids and the present hydrothermal system in Hungary are similar and did not differ significantly. Although detailed features such as the position of the recharge and discharge areas and the direction of flow, have been changing with time, the general behavior of the system has remained the same.

Hydrothermal Speleogenesis in Hungary—The Processes Involved

Hydrothermal karst is characterized by two major hydro- and thermodynamic zones:

- A deep zone with a low thermal gradient.
- A shallow zone with a high thermal gradient (Dublyansky, 1991b).

The deep zone embraces the bulk of any hydrothermal system and may extend to a depth of several km. The temperature of fluids there is not affected appreciably by surface temperatures. Although the water in this zone may be meteoric in origin, its chemical properties are "buffered" by the bedrock due to the long circulation path and the long residence time.

This is essentially a two-phase liquid+solid (water+rock) system. In contrast, the shallow zone is comparatively thin, extending to several tens of meters below and above the thermal water table. Thermal gradients in this zone may be quite high intensifying many processes. Some specific speleogenetic processes including evaporation, air convection, condensation-corrosion mineral replacement, and aerosol and gas mass-transfer develop in caves above the thermal water table.

Deep-seated Hydrothermal Setting
The solubility of calcite ($S_{calcite}$) in hydrothermal environments is controlled by three major factors (Holland and Malinin, 1979):

- Partial pressure of CO_2 ($\sim pCO_2^{1/3}$).
- Temperature ($\sim T^{-1}$).
- Ionic strength of a solution ($\sim I$).

In turn, the amount of dissolved CO_2 and its phase state depend on temperature and pressure. It was shown by Malinin (1979) that a decrease of temperature and pressure in upwelling water leads to continuous loss of dissolved CO_2 into a gas phase (effervescence) that in turn decreases the $S_{calcite}$.

At the same time, however, a decrease in temperature enhances $S_{calcite}$. The final result of these two opposite trends is shown in 5.2.6 Figure 6. Upwelling thermal water saturated with CO_2 remains undersaturated with respect to $CaCO_3$ along most of its path and dissolves any carbonate bedrock that occurs. Close to the surface the solubility of $CaCO_3$ drops sharply, which leads to deposition of calcite.

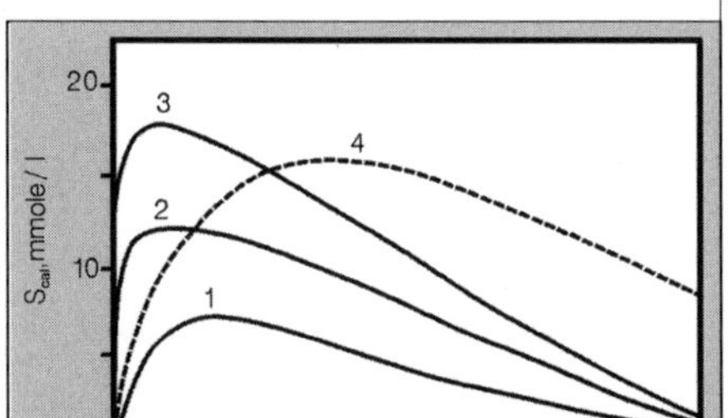

5.2.6 Figure 6. Solubility of $CaCO_3$ in ascending CO_2-bearing thermal waters (from Malinin, 1979). Solid lines represent pure water. Dashed line represents NaCl solution with the ionic strength = 1. Temperature changes toward surface as 1 is 225° to 100°C. 2 is 225 to 50°C. 3 is 225 to 25°C. 4 is 225° to 100°C.

The actual depth of calcite deposition depends on many parameters including: temperature, partial pressure of CO_2, ionic strength, and the velocity of upwelling water. The general pattern, however, remains essentially the same. This model explains the spatial relationships between solutional forms and calcite deposits in hydrothermal karst (Dublyansky and Dublyansky, 1984) and is applicable to the hydrothermal karst of Hungary (Dublyansky, 1991a).

Shallow Hydrothermal Setting
Subaqueous Karstification: Mixing-Cooling Corrosion. An increase of aggressiveness due to mixing of two karstic waters with different solute concentrations has long been recognized (Bögli, 1980). In the Hungarian karst this type of corrosion may have occurred where upwelling thermal waters met descending cold karstic ones (Kovács and Müller, 1980; Müller 1989). This process is believed to be responsible for excavation of large essentially two-dimensional rectilinear maze caves in the Buda Hills, developed at the thermal water table (see 5.2.6 Figure 2).

Cooling-induced dissolution may develop in two situations:

- Cooling of hot mineralized water increases its aggressiveness due to the inverse temperature solubility of carbonate.
- Mixing of two waters of similar mineralization but different temperatures causes an effect analogous to common mixing corrosion due to differences in CO_2 content.

The first situation could play a role at spring orifices. The mixing of waters with different temperatures is common in Hungarian karst (e.g., 18 and 26°C in Molnár János Cave; 17 and 40°C in Héviz Spring Cave; Takács-Bolner and Kraus, 1989).

Subaerial Karstification: Condensation Corrosion. The development of caves above thermal water tables as a consequence of condensation corrosion was suggested by Müller (1974) in an attempt to explain the origin of the spherical niches that are common in Hungarian caves. In fact, upward-branching, three-dimensional caves such as Sátorkö-puszta and Bátori (5.2.6 Figure 3) consist entirely of interlocked sequences of such niches in which individuals are 0.5 to 10 m in diameter. Mathematic modeling suggests that this mechanism enlarges caves very slowly (5 to 20 µm/yr) due to the low thermal conductivity of the bedrock limestone (Szunyogh, 1982, 1989). The calculations, however, assume that the physicochemical system was exclusively a carbonate one. The presence of sulfuric acid originated from the oxidation of hydrogen sulfide would probably dramatically change the overall picture.

The Role of Sulfur. Many features in the Hungarian caves indicate the presence of different species of sulfur in the thermal paleowaters. They deposited barite and smaller amounts of sulfides. Pulverization of the Triassic dolomite has been attributed to the presence of free sulfuric acid in ascending thermal water (Jakucs, 1977).

Many Hungarian hydrothermal caves contain gypsum speleothems (e.g., Sátorkö-puszta, Bátori, Jósef-hegy, Solymári-ördöglyuk, and others), although gypsum strata are not known from the geologic record. Gypsum speleothems are more common in three-dimensional caves where condensation-corrosion could have been the leading cave-forming mechanism. Finally, modern thermal karst waters in Hungary contain from 10 to 290 mg/L SO_4^{2-} (Maucha, 1989).

Dissolution of caves by sulfuric acid in a subaqueous setting has been described by Hill (1987; see Chapter 5.2.8 in this volume). In it, hydrogen sulfide is oxidized at or near the water table to sulfuric acid, which attacks carbonate bedrock. A model of hydrothermal karst development in a subaerial setting (replacement-corrosion mechanism) has been developed by Egemeier (1981). Hydrogen sulfide dissolved in water may escape into the cave air and be incorporated into condensation water films and droplets on the cave walls, where it is readily oxidized.

The sulfuric acid converts limestone to gypsum. Similarly, dolomite is converted to gypsum and epsomite. Sedimentary dolomite normally contains an excess of $CaCO_3$, which may be preferentially removed, leaving loose dolomitic grus. This process may be responsible for the "soft walls" (layers of loose sandy material up to 3 to 4 cm thick) reported from some Hungarian hydrothermal caves (e.g., Sátorkö-puszta in the Pilis Mountains and Beremend Crystal Cave in the Villány Mountains). Being fragile, most gypsum crusts eventually fall to the cave floor. The fallen gypsum accumulations can be recrystallized or dissolved and removed from the cave.

The presence of sulfuric acid in condensation water may significantly enhance the rate of cupola enlargement. Moreover, the conversion of limestone to gypsum produces additional CO_2 and that increases carbonate aggressiveness. All the processes outlined in this section may be operating simultaneously, although their relationships may have changed in time.

Conclusion

Hungary has excellent examples of hydrothermal karst developing in various stages, from deep-seated ancient to modern active thermal spring caves. No place in the world has such a variety of hydrothermal karst within such a small area. Hungary may serve as a perfect "test site" for speleogenetic studies of hydrothermal karst.

Acknowledgments

The work was commissioned by the National Authority for Nature Conservation and Hungarian Speleological Institute, Budapest. I am thankful to K. Takács-Bolner, K. Székely, A. Nádor, J. Tardy, S. Kraus, G. Szunyogh, and many other Hungarian speleologists for field assistance and fruitful discussions both underground and on the surface. I appreciate the help of V. Reutsky in conducting fluid-inclusion analyses.

Part of the field and laboratory work was supported by a Natural Sciences and Engineering Research Council of Canada Research Grant to D.C. Ford. The work was completed during an International Visiting Scientific Fellowship of the Council that was taken up at McMaster University. Dr. D.C. Ford is thanked for his assistance at all stages of this study.

References

Alföldi, L., 1979, Budapesti hévizek (Thermal waters of Budapest): *VITUKI Közl.*, no. 20, 102 p.

Bakalowicz, M.J., Ford, D.C., Miller, T.E., Palmer, A.N., and Palmer, M.V., 1987, Thermal genesis of dissolutional caves in the Black Hills, South Dakota: *Geological Society of America Bulletin,* v. 99, p. 729-738.

Bögli, A., 1980, *Karst hydrology and physical speleology:* Berlin, Springer-Verlag, 284 p.

Deák, J., 1979, Environmental isotopes and water chemical studies for groundwater research in Hungary, *in Isotope Hydrology–1978,* Vienna, International Atomic Energy Agency, p. 221-249.

Deák, J., 1989, Dating the karst thermal waters in the Eger area: *10th International Congress of Speleology,* Budapest, Proceedings, v. 2, p. 570-572.

Dublyansky, Y., 1991a, A Budai-hegység hidrotermális paleokarsztja–A folyadékzárvány-vizsgálatok elsö eredményei (Paleohydrothermal karst of the Buda Hills–The preliminary results of study by fluid inclusion method): *Karszt és Barlang,* no. 1-2, p. 19-24.

Dublyansky, Y.V., 1991b, Zonalnost i mekhanism gidrotermokarsta (Zonation and mechanisms of hydrothermal karst), *in* Environments of karstogenesis: Kungur, p. 32-36.

Dublyansky, Y.V., 1995, Speleogenetic history of the Hungarian hydrothermal karst: *Environmental Geology ,* v. 25, p. 24-35.

Dublyansky, Y.V., and Dublyansky, V.N., 1984, Obrazovanie gidrotermokarstovikh polostei (Formation of hydrothermal karst caves): *Peshchery,* v. 19, p. 37-42.

Egemeier, S.J., 1981, Cavern development by thermal waters: *National Speleological Society Bulletin,* v. 43, p. 31-51.

Ford, D.C., and Takácsné Bolner, K., 1991, Abszolút kormeghatározás és stabil izotóp vizsgealatok budai barlangi kalcitmint·kon (U-series dating and stable isotope analyses on calcite precipitates from Buda caves): *Karszt és Barlang,* no. 1-2, p. 11-18.

Hill, C., 1987, Geology of Carlsbad Cavern and other caves in the Guadalupe Mountains, New Mexico and Texas: *New Mexico Bureau of Mines and Mineral Resources Bulletin,* no. 117, 151 p.

Holland, H.D., and Malinin, S.D., 1979, The solubility and occurrence of non-ore minerals, *in* Barnes, H.L., ed., *Geochemistry of hydrothermal ore deposits:* New York, John Wiley and Sons, p. 461-508.

Jakucs, L., 1977, Morfologicheskie i evolutsionnie typi karstov Vengrii (Morphological and evolutionary types of karst in Hungary): *Acta Geographica,* Szeged, v. 17, p. 65-103.

Kovács, J., and Müller, P., 1980, A Budai-hegyek hévizes tevékenységének kialakulása és nyomai (Origin and traces of hydrothermal activities in the Buda Range): *Karszt és Barlang,* no. 2, p. 93-98.

Liebe, P., and Lorberer, Á. 1978, A karsztos héviztárolók áramlási és hömérsékleti viszonyainak vizsg·lata (Study of flow and temperature conditions in a karstic thermal water aquifer): *International Symposium on Karst Hydrology,* Proceedings, v. 1, p. 79-110.

Liebe, P., 1989, Thermalkarst-systems in Hungary–Exploration and problems thereof: *10th International Congress of Speleology,* Budapest, Proceedings, v. 2, p. 566-569.

Malinin, S.D., 1979, *Fizicheskaja khimija gidrotermalnikh sistem s uglekislotoi* (Physical chemistry of hydrothermal systems with carbon dioxide): Moscow, Nauka, 111 p.

Maucha, L., 1989, Karst water resources research in Hungary and its significance: *Karszt és Barlang,* spec. issue, p. 39-50.

Müller, P., 1974, A melegforrás-barlangok és gömbfülkék lebetheziséröl: *Karszt és Barlang* (Budapest), no. 1, p. 7-11 (in Hungarian).

Müller, P., 1989, Hydrothermal paleokarst of Hungary, *in* Bosák, P., Ford, D., Glazek, J., and Horácek, I., eds., Paleokarst–A systematic and regional review: Praha, Academia, p. 155-163.

Nádor, A., 1992, *Paleokarsts and long term karst evolution of the Buda Mountains, Hungary:* Eötvös Lóránd University, Ph.D. thesis, 171 p.

Nagy, B., and Pelikán, P., 1973, Metacinnabarit és cinnabarit a Csillaghegyi Rókahegyen (Metacinnabarite and cinnabar in the Roka Hill, Csillaghegy): *MÁFI Évi Jelt.,* p. 51-55.

Pávai-Vájna, F., 1931, Über die Rolle heisser Lösungen, Dämpfe und Gase bei der Höhlenbildung: *Hidr. Közl.,* no. 10-11, p. 63-69.

Schafarzik, F., 1928, Visszapillantás a budai hévforrások fejlödéstörténetére (The overview of thermal springs of Budapest): *Hidr. Közl.,* no. 1, p. 9-14.

Scheurer, G., and Schweitzer, F., 1980, A budai hévizforrások fejlödéstörténete a felsö-pannontól napjainkig (History of development of the Budda thermal springs from the Uupper Pannonian, Miocene-Pliocene, to now): *Hidr. Közl.,* no. 5, p. 391-395.

Szunyogh, G., 1982, A hévizes eredetü gömbfülkék kioldódásanak elméleti vizsgálata (Theory of the dissolving of spherical cavities formed by thermal water): *Karszt és Barlang,* no. 2, p. 83-88.

Szunyogh, G., 1989, Theoretical investigation of the development of spheroidal niches of thermal water origin– Second approximation: *10th International Congress of Speleology,* Budapest, Proceedings, v. 3, p. 766-768.

Takács-Bolner, K., and Kraus, S., 1989, The results of research into caves of thermal water origin: *Karszt és Barlang,* spec. issue, p. 31-38.

Vendel, M., and Kisházi, P., 1964, Összefüggések melegforrások és karsztvizek között a Dunántúli Középhegységben (Relationships between thermal springs and karst waters in the Transdanubian Range): *MTA Müszaki–Tudományos Osztály Közl.*

Wein, G., 1977, *A Budai-hegység tektonikája* (Tectonics of the Buda Hills): Budapest, MAFI Alk.Kiadv., 76 p.

5.2.7. Some Case Studies of Speleogenesis by Sulfuric Acid

D.J. Lowe, S.H. Bottrell, and J. Gunn

Abstract

Minerals that can weather to produce sulfuric acid directly or indirectly, with or without microbial mediation, occur as trace components in most carbonate sequences, but they are more concentrated at specific horizons. These horizons comprise beds of nontypical lithology, together termed inception horizons, and are commonly associated with breaks between major depositional cycles. Some cycle boundaries are marked by concentrations of sulfide minerals, particularly pyrite, that are readily oxidized to generate sulfuric acid. Cycle boundaries may also be marked by the presence of primary evaporite minerals such as gypsum, and their removal by direct dissolution or by their reduction to hydrogen sulfide may be implicated in early porosity development. Although few caves in carbonate sequences are largely, or entirely, the product of calcite dissolution by sulfuric acid or of evaporite removal, such processes may play an important role in cave inception. This chapter examines several situations where processes other than carbonic acid dissolution have played an important role in secondary porosity generation and have influenced subsequent speleogenesis.

Introduction

The chemical processes by which limestone is dissolved in water containing dissolved carbon dioxide are well understood and are discussed in detail elsewhere in this volume (Chapters 4.1.1 and 4.1.2). However, it is widely assumed that dissolution due to sulfuric acid (see Chapter 4.1.4 for a discussion of its physical chemistry) is unimportant in most carbonate sequences. This assumption is made primarily because analyses of water from active underground drains rarely include significant amounts of sulfate. An alternative explanation of the lack of sulfate in many spring waters is that whereas sulfuric acid may play an important part in calcite dissolution under restricted inception flow conditions, its significance will decrease as conduits enlarge.

In particular, carbonic acid byproduct levels are likely to increase significantly relative to those of sulfuric acid after turbulent breakthrough, and a still more significant jump will occur on transition from phreatic to vadose conditions. In parallel, the usually relatively small thicknesses of primary sulfates preserved in some carbonate successions may be removed by direct dissolution during early stages of underground water movement, or they may be reduced to hydrogen sulfide. This in turn may react directly with carbonates or may be oxidized to produce sulfuric acid (Egemeier, 1981; Van Everingden and others, 1985; Sando, 1988). Broad support for these ideas is provided by the work of Worthington (1991) and Worthington and Ford (1995). They found that flows from artesian or deep phreatic elements in multilevel systems contain significantly higher sulfate concentrations and relatively less bicarbonate, whereas sulfate (or other strong acid related anion) values in water from vadose and shallow phreatic systems are swamped by bicarbonate ions.

Case Studies of Speleogenesis by Sulfuric Acid

This section briefly reexamines case studies covering porosity-permeability enhancement (conduit inception) by sulfuric acid dissolution of calcite (Lincolnshire Limestone, England) and cave development by the same process (Guadalupe Mountains, New Mexico, and Level Crevice Cave, Iowa, U.S.A.). Porosity-permeability enhancement and assumed cave development related to removal or chemical changes of gypsum within carbonate-dominated sequences are also considered, with particular reference to the Mississippian rocks of southern Indiana.

Secondary Porosity Development and Conduit Inception in the Lincolnshire Limestone, England

The Lincolnshire Limestone (Ashton, 1980), which appears at outcrop in eastern England, is a predominantly calcitic marine limestone. It was originally rich in pyrite and organic matter. Primary evaporite minerals such as gypsum are absent. Strike is northerly, with an eastward dip of about 1 degree. Its thickness ranges from 25 to 35 m. The limestone is confined by estuarine shale and clay of the Rutland Formation (formerly the "Upper Estuarine Series") above and by similar beds of the Grantham Formation below. Within the aquifer, the water chemistry and the distribution of ^{14}C ages reveal the presence of two rather separate bodies of water. Beneath the areas of outcrop—and for a few kilometers downdip into the confined zone—water flow is rapid with surface discharge to springs and rivers under natural conditions. This flow has been considerably augmented during the past 50 years by private and public abstractions.

This "surface skim" is underlain by older water that infiltrated into the aquifer before and after the last glaciation, the first post-glacial recharge being about 10,000 years ago. This older water is restricted by the confining beds and may be viewed, simplistically perhaps, as being stagnant. The natural flow pattern in the aquifers is from easterly, *i.e.* down the dip of the aquifer, but the pattern is disrupted by the heavy impact of modern abstractions on this natural flow.

Within the Lincolnshire Limestone, this simple pattern of flow from west to the east is further complicated by the separation of the aquifer into two subaquifers by the Kirton Shale Member. This is a regionally developed and prominent marl that occurs stratigraphically approximately in the middle of the Lincolnshire Limestone. However, normal faults, striking north-south, throw these subaquifers together at many points and also act as conduits for strong regional flows of water in a north-south direction.

Water in this aquifer displays the chemical character typical of water in carbonate aquifers worldwide. In the outcrop region, waters are dilute, fully oxygenated, and of the Ca-HCO_3 type. As the waters become confined, oxygen is utilized, and sulfate concentration increases. The sulfate added is isotopically light, with a composition close to that in diagenetic pyrite in the limestones and marls of the aquifer and its confining beds (Moncaster et al., 1992).

Deeper in the aquifer, where ^{14}C content indicates an age of 5,300 years (Bishop and Lloyd, 1990), the water chemistry indicates a long period of water-rock interaction, with a characteristic Na-Ca-HCO_3 composition resulting from Ca-for-Na exchange on clay minerals. Sulfate is present at around 100 mg/L. In these preanthropogenic waters, it is derived entirely from pyrite oxidation (Moncaster et al., 1992).

Interestingly, these concentrations of sulfate are three times that which could be produced by oxidation of pyrite in water equilibrated with the atmosphere, so an additional oxidant source is required.

Oxidation of pyrite takes place largely at the margins of the confined zone where annual variation in groundwater heads repeatedly saturates and dewaters the limestone and overlying clay. This process may also be responsible for drawing air into the unsaturated zone as groundwater heads fall and probably provides the additional oxidant source.

Clearly, pyrite oxidation is a major process, and 100 mg/L sulfate corresponds to 200 mg/L (~2 mmol/L) of calcite dissolution, a significant enhancement of porosity generation in this zone of the aquifer. This corresponds closely to the 2 mmol/L excess Ca concentration of water affected by pyrite oxidation in the Xingwen Karst of China (Bottrell, 1993), which may indicate that in both cases availability of oxygen to drive pyrite oxidation limits the reaction. Although the Lincolnshire Limestone does not contain enterable caves, it has a high permeability due to well-developed dissolutionally enlarged fractures, the development of which must be influenced by pyrite oxidation reactions.

Role of Sulfuric Acid in Cave Development in the Guadalupe Mountains, New Mexico, U.S.A.

A major publication (Hill, 1987) describing cave development in the Capitan Reef of the Guadalupe Mountains, New Mexico, U.S.A., consolidated ideas from earlier short papers (Hill, 1981, 1985, 1986). A multistage history of speleogenesis was described—including a major phase of dissolution due to sulfuric acid—by a mechanism previously proposed by D.G. Davis (1979, 1980) and resembling that described by S.J. Egemeier (1973, 1981). Within a history that included three stages of dissolution, Hill considered that only Stage III was dominated by sulfuric acid, formed from hydrogen sulfide derived biogenically from hydrocarbon deposits in the adjacent Delaware Basin. The suggested reactions are as follows

$$CaSO_4 + CH_4 \Rightarrow H_2S + CaCO_3 + H_2O \text{ (Delaware Basin)} \quad (1)$$

$$2H^+ + SO_4^{2-} + CH_4 \Rightarrow H_2S + CO_2 + 2H_2O \text{ (Delaware Basin)} \quad (2)$$

$$2H_2S_{(aq)} + O_{2(aq)} \Rightarrow 2S_{(s)} + 2H_2O_{(l)} \quad (3)$$

$$2H_2O_{(l)} + 2S_{(s)} + 3O_{2(aq)} \Rightarrow 2HSO_4^-{}_{(aq)} + 2H^+_{(aq)} \quad (4)$$

$$2H^+ + SO_4^{2-} + CaCO_3 + 2H_2O \Rightarrow CaSO_4 \cdot 2H_2O + H_2O + CO_2 \quad (5)$$

Hill's basic model for these processes is well supported by analytical results and by observations in the caves. It has been further developed in subsequent papers (e.g. Hill 1990, 1995; Chapter 5.2.8). As Hill stated (1995, p. 16): "The Delaware Basin...is probably the best example of sulfuric acid karst anywhere in the world." However, the Hill model for sulfuric acid speleogenesis during Stage III relies to some degree upon earlier speleogenetic activity in Stages I and II. This was deduced by Hill (1987) to have commenced soon after deposition and diagenesis of the Capitan Reef and associated fore- and backreef deposits during the Permian. In contrast to the "special" case of the Stage III sulfuric acid dissolution, Hill's explanation of the earlier stages relies upon more traditional ideas of speleogenesis by carbonic acid.

We have no first-hand knowledge of the Guadalupe caves. But from this stance of being uninvolved in the earlier debate, we would raise several points relating to the cave development models proposed for the area.

First, we suggest that in a speleogenetic history that began soon after the deposition of the host rocks in a Permian reefal environment, the probability of development and preservation of conduit permeability by dissolution adjacent to a contemporary freshwater lens—as reported in other late Permian rocks in the region (Craig, 1988)—cannot be ignored. A possibility must also exist that a proportion of open or loosely filled conduits would survive subsequent uplift. These would be "stacked" at several levels, as described in the Trobriand Islands (Ollier, 1975) and Tonga (Lowe and Gunn, 1986; Lowe, 1989; and Chapter 5.1.2 in this volume). If preserved, such conduits would provide preferred sites for concentration of later underground drainage and dissolution.

Second, if the presumed presence and preservation of stacked lens-marginal conduits is a reality, then much of the argument for the later bathyphreatic inception and development of passages cutting across the bedding may no longer be necessary. Relationships between conduits in the forereef area and those in the bedded backreef deposits are potentially analogous to those observed in the reef hinterland of 'Eua Island, Tonga.

Following from this, a possibility emerges that secondary conduit enlargement (Hill's Stage II) could be partly due to dissolution by sulfuric acid derived from the backreef Yates Formation. This mechanism, suggested by Jagnow (1977), was justifiably considered by

Hill as inadequate to have caused the Stage III passage growth. However, if preexisting lens-marginal conduit permeability was present, its involvement at Stage II may be a realistic possibility.

Role of Sulfuric Acid in the Development of Level Crevice Cave, Iowa, U.S.A.

Level Crevice Cave, Iowa, was opened to the surface by mining activity (Morehouse, 1968). It is unclear whether the absence of a surface connection was due to an impermeable cover or whether connections smaller than the anthropocentric limit (the limit of human accessibility) preexisted. If the former, the point is significant, as few earlier authors had acknowledged that caves (other than isolated dissolutional chambers) could form without surface input and output points.

The cave is formed in the Middle Ordovician Galena Formation and has three bedding-parallel levels or tiers. The bedrock is dolomite, and calcite, pyrite, marcasite, galena, and limonite occur in the cave. Morehouse described the cave as "joint controlled," but stated (1968, p. 2):

> The cave passages also appear to have been controlled by development at favorable stratigraphic levels...

This stratigraphic guidance, previously noted by Howard (1960), was linked to differential bedrock fracturing or brecciation, and uneven solubility due to varied porosity and chemical composition.

Water analyses by Morehouse (1968) showed that:

> ...sulfate ion is present in the cave waters in fairly large concentrations."

He concluded that sulfuric acid dissolution was dominant in the cave. The reaction chain deduced by Morehouse is

$$4FeS_2 + 15O_2 + 2H_2O \Rightarrow 2Fe_2(SO_4)_3 + 2H_2SO_4 \quad \text{(A)}$$

$$2Fe_2(SO_4)_3 + 12H_2O \Rightarrow 4Fe(OH)_3 + 6H_2SO_4 \quad \text{(B)}$$

$$4Fe(OH)_3 \Rightarrow 2Fe_2O_3{\cdot}6H_2O \quad \text{(C)}$$

$$8H_2SO_4 + 8CaMg(CO_3)_2 \Rightarrow 8Ca^{2+} + 8Mg^{2+} + 8SO_4^{2-} + 16HCO_3^- \quad \text{(D)}$$

Combining equations A through D

$$4FeS_2 + 15O_2 + 14H_2O + 8Ca.Mg(CO_3)_2 \Rightarrow 2Fe_2O_3{\cdot}6H_2O + 8Ca^{2+} + 8Mg^{2+} + 8SO_4^{2-} + 16HCO_3^-$$

Morehouse pointed out that the stoichiometry of the sulfuric acid reaction should produce a sulfate ion concentration half that of the bicarbonate ion. His analytical results agreed with this ideal ratio within 4 percent. Other observations supported the view that sulfuric acid dissolution dominates in the cave.

Massive and disseminated pyrite and marcasite occur in the wall rocks. Morehouse (1968, p. 10) stated:

> ...it does not take much sulfuric acid (or, therefore, much pyrite or much marcasite) to significantly affect cave development.

The cave waters were saturated with dissolved oxygen at all times. Limonite ($Fe_2O_3{\cdot}nH_2O$) was present within the dolomite wall rocks in amounts between 0.9 and 2.3 percent. No sulfate concentrations exist near the cave to account for the high sulfate values in the analyses. Lack of seasonal fluctuation in analytical data suggests that dissolution is independent of carbon dioxide levels in overlying soils. Finally, *"valley bars,"* which exist where passages narrow beneath surface valleys, are reciprocal to the expectation if carbonic acid dissolution were dominant.

In the latter case, dissolution should be greater where percolation passes through less bedrock and has less opportunity to achieve saturation. Larger passages would be expected at these points. In the case of the sulfuric acid reaction, downward percolating water is acidified in passing through the overlying rock, being theoretically less acidic where the rock is thinnest, beneath valley floors. The validity of this argument is uncertain, as laterally moving water must contribute to development, and vertically descending water must constantly be reacting with carbonate in penetrating the bedrock.

As an alternative to equation A, Morehouse suggested that iron bacteria *(Crenothrix* and *Gallionella)*, common in the cave, gain energy by oxidizing ferrous to ferric iron.

$$4FeS_2 + 15O_2 + 14H_2O \Rightarrow 4Fe(OH)_3 + 8H_2SO_4 \quad \text{(Alternative A)}$$

Ferric hydroxide excreted by the bacteria rearranges as in equation C of the inorganic mechanism. Morehouse noted that equal numbers of moles of pyrite-marcasite, oxygen, and water are used in the inorganic and bacterial reactions and that the same number of moles of sulfuric acid are produced. Whichever mechanism operates, the result is the same. Morehouse concluded that the primary development (inception) of the cave was achieved by the sulfuric acid reaction under artesian conditions, as outlined by Howard (1964). He reasoned that artesian conditions developed because the Galena Formation was sealed beneath the Maquoketa Shale, except where relatively few steep valleys, such as those cut by glacial meltwater, had incised the carbonate sequence.

This supports the proposed chemistry, as percolation reaching the cave level would have reached equilibrium with respect to the carbonic acid reaction, but still be saturated with oxygen. The general reasoning and proposed reactions remain basically sound. But it is no longer certain that the inferred linking of inception to Quaternary time (by reference to downcutting by glacial streams) is realistic (Lowe, 1992). The cave was drained by mining, so at least part of the system had not reacted to surface downcutting by that time. Inception by sulfuric acid dissolution of calcite could have been active long before breakthrough to turbulent flow conditions and before the cave acquired its present dimensions.

Conduit Inception Related to Evaporite Minerals

Hydrochemical and isotopic studies of spring waters by Krothe and Libra (1983) provided unequivocal quantitative pointers to the involvement of primary evaporite minerals in conduit inception in southern Indiana. Here Pennsylvanian rocks overlie a mainly carbonate Mississippian sequence. The uppermost Mississippian formations are more varied lithologically than those below, resembling cyclothemic sequences of similar age in northern England.

Underlying beds include the St. Louis Formation, which contains economic quantities of evaporite, the aspect of which passes from being interbedded, where preserved at depth, to occurring only as nodules closer to outcrop. Horizons that are highly evaporitic at depth are represented by brecciated rock at outcrop. Krothe and Libra deduced the presence of two flow systems. A shallow system carries meteoric water entering the system through open fractures and passing through traditional and anthropocentrically defined cave passages. Other waters utilize a deep regional system recharged mainly by diffuse flow.

Ion concentrations in the shallow system are highly variable, and carbonic acid dissolution of calcite is the dominant process. Deep flow is characterized by limited ion concentration variation and high sulfate levels.

On the basis of Krothe and Libra's (1983) discussion, it is assumed that initial void creation in the sequence is dominated by a dissolution-precipitation interaction involving calcite, dolomite, and gypsum. However, descriptions of other "special" cases of speleogenesis in other geologic situations (Egemeier, 1981; Van Everingden et al., 1985; Hubbard et al., 1986) raise possibilities that sulfate minerals may undergo more drastic chemical changes that can contribute to enhanced porosity-permeability development. Although the geologic situations described, and the proposed reaction chains differ in detail, they can be considered together and simply as:

- a Reduction of primary (marine) evaporites to produce hydrogen sulfide.
- b Oxidation and hydrolysis of hydrogen sulfide to form acidic sulfur dioxide and, more importantly, sulfuric acid.
- c Dissolution of carbonates remote from the primary evaporite source by the acids generated during (b).

The dissolutional mechanisms described in different areas by the various cited authors relate to sulfate reduction in deeply buried anaerobic flow situations followed by oxidation and hydrolysis on meeting an oxygen-rich cave atmosphere near the downstream limit of underground flow. Such relationships have been observed and are readily acceptable.

However, it should not be overlooked that oxygenated conditions may also exist at shallow depths within the rock mass, above the anaerobic reduction zone. Situations must exist where water saturated with hydrogen sulfide generated at depth will encounter well-oxygenated groundwater, or even air-filled voids, close to the vadose zone/phreatic zone interface. In such situations chemical or microbially mediated oxidation, comparable to that recorded in Bahamian blue holes (Bottrell et al., 1990), could result. Production of strong acids in such a situation would be expected to contribute significantly to porosity-permeability enhancement and, hence to cave development.

Conclusion

Speleogenesis due to sulfuric acid dissolution, as demonstrated in the Guadalupe Mountains and Level Crevice Cave, has often been regarded as a "special case" and processes based exclusively upon carbonic acid as the norm. However, evidence from the Lincolnshire Limestone demonstrates that sulfuric acid may play an important part in conduit inception. Production of sulfuric acid in southern Indiana seems not to be confirmed but may be suspected to occur on the basis of analogous changes undergone by primary evaporitic sulfates elsewhere. In the cases discussed, the conduits are now at some depth. Thus consideration must be given to the possibility that currently observable "shallow" systems elsewhere were conceived under conditions similar to those that prevail in the Lincolnshire Limestone.

Deep speleoinception can begin before host carbonates are uplifted and unroofed and before surface evolution provides viable conditions for "normal" hydraulic flow. Evaporites occur within many deeply buried carbonate successions. After exhumation, the highest elements of incipient deep flow would transform into shallow flow systems, under conditions more normally associated with underground drainage. Early in the transformation, residual evaporites would be removed from the rock mass by speleogenesis *(sensu lato)* and by near-surface weathering. Such changes explain the evaporite distribution in buried and outcrop elements of the St. Louis Formation (Indiana) and may also explain a lack of sulfate in near-outcrop shallow flow near recharge areas around the Big Horn Basin of Wyoming (Egemeier, 1981).

In contrast to the idea that sulfuric acid dissolution is a special case, we suggest that in many areas it affects all stages of speleogenesis. But its greatest importance, instrumentally and volumetrically, is during inception and the subsequent phase of laminar flow (which we term *gestation*) prior to turbulent breakthrough. We cannot yet show conclusively that speleogenesis would not proceed without these "exotic" mechanisms. But we would point out that many unanswered questions in preexisting cave development theories cease to present problems if the alternative processes are accepted. Processes appearing quantitatively insignificant in post-gestational speleogenesis may have relatively great importance before establishment of turbulent conduit flow.

Recent geochemical models of calcite dissolution by carbonic acid (Groves and Howard, 1994a, 1994b; Howard and Groves; 1995, Dreybrodt, 1990, 1996; Palmer, 1991; see also Chapters 4.1.2, 4.2.2, and 4.2.3 in this volume) suggest that strong acid dissolution may not be *essential* for conduit inception. However, it is likely to be important under most inception conditions, although the amount of carbonate dissolved is small. As gestation and turbulent flow development progress, the proportion of total conduit enlargement by strong acid generally decreases, but strong acid dissolution continues throughout the life of the conduit.

Analyses of seepages from incipient inception horizons, and discharges from deep phreatic and artesian systems, commonly confirm a dominance of sulfur-bearing ions as compared to carbonate and bicarbonate. Equally, as integrated drainage systems develop, and hydraulic conductivity improves, relatively constant yields of sulfur-bearing ions are progressively overwhelmed by increasing production of carbonic acid reaction products. From this viewpoint, the effect of "exotic" acids is perceived as negligible—from the speleoinception viewpoint it is considerable and fundamental.

Acknowledgements

The concepts in the paper are based partly upon work undertaken by other explorers and scientists. We thank them for their efforts and conclusions, and apologize if we have taken any of their ideas out of context.

References

Ashton, M., 1980, The stratigraphy of the Lincolnshire Limestone Formation (Bajocian) in Lincolnshire and Rutland (Leicestershire): *Proceedings of the Geological Association,* v. 91, p. 203-223.

Bishop, P.K., and Lloyd, J.W., 1990, Chemical and isotopic evidence for hydrogeochemical processes occurring in the Lincolnshire Limestone: *Journal of Hydrology,* v. 121, p. 293-320.

Bottrell, S.H., 1993, Water chemistry in the Xingwen caves, China: *Cave Science,* v. 20, p. 87-92.

Bottrell, S.H., Smart, P.L., Whitaker, F., and Raiswell, R., 1990, Geochemistry and isotope systematics of sulphur in the mixing zone of Bahamian blue holes: *Applied Geochemistry,* v. 6, p. 97-103.

Craig, D.H., 1988, Permian karst in San Andreas Dolomite, Yates Field, West Texas, *in* James, N.P., and Choquette, P.W., eds., *Paleokarst:* New York, Springer-Verlag, p. 342-363.

Davis, D.G., 1979, Cave development in the Guadalupe Mountains, New Mexico and Texas: *Caving International,* v. 3, p. 43.

Davis, D.G., 1980, Cave development in the Guadalupe Mountains: A critical review of recent hypotheses: *National Speleological Society Bulletin,* v. 42, p. 42-48.

Dreybrodt, W., 1990, The role of dissolution kinetics in the development of karst aquifers in limestone: A model simulation of karst evolution: *Journal of Geology,* v. 98, p. 639-655.

Dreybrodt, W., 1996, Principles of early development of karst conduits under natural and man-made conditions revealed by mathematical analysis of numerical models: *Water Resources Research,* v. 32, p. 2923-2935.

Egemeier, S.J., 1973, *Cavern development by thermal waters with a possible bearing on ore deposition:* Stanford University, Ph.D. dissertation, 88 p.

Egemeier, S.J., 1981, Cavern development by thermal waters: *National Speleological Society Bulletin,* v. 43, p. 31-51.

Groves, C.G., and Howard, A.D., 1994a, Minimum hydrochemical conditions allowing limestone cave development: *Water Resources Research,* v. 30, p. 607-615.

Groves, C.G., and Howard, A.D., 1994b, Early development of karst systems. 1.: Preferential flow path enlargement under laminar flow: *Water Resources Research,* v. 30, p. 2837-2846.

Hill, C.A., 1981, Speleogenesis of Carlsbad Cavern and other caves of the Guadalupe Mountains: *8th International Congress of Speleology,* Bowling Green, Proceedings, p. 143-144.

Hill, C.A., 1985, Speleogenesis of Carlsbad Cavern and other caves of the Guadalupe Mountains, New Mexico: *GEO², Newsletter of the National Speleological Society Section on Cave Geology and Geography,* v. 12, no. 2, p. 30-31.

Hill, C.A., 1986, Carlsbad Cavern and other caves in the Guadalupe Mountains, New Mexico and Texas: A sulfuric acid genesis related to the oil and gas fields of the Delaware basin: *9th International Congress of Speleology*, Barcelona, Proceedings, v. 1, p. 267-269.

Hill, C.A., 1987, Geology of Carlsbad Cavern and other caves in the Guadalupe Mountains, New Mexico and Texas.: *New Mexico Bureau of Mines and Mineral Resources Bulletin,* no. 117, 150 p.

Hill, C.A., 1990, Sulfuric acid speleogenesis of Carlsbad Cavern and its relationship to hydrocarbons, Delaware Basin, New Mexico and Texas: *American Association of Petroleum Geologists Bulletin,* v. 74, p. 1685-1694.

Hill, C.A., 1995, Sulfur redox reactions: Hydrocarbons, native sulfur, Mississippi Valley-type deposits, and sulfuric acid karst in the Delaware Basin, New Mexico and Texas: *Environmental Geology,* v. 25, p. 16-23.

Howard, A.D., 1960, Geology and origin of the crevice caves of the Iowa, Illinois, and Wisconsin lead-zinc district: *Journal of the Yale Speleological Society,* v. 2, no. 4, pt. 1, p. 61-95.

Howard, A.D., 1964, Processes of limestone cave development: *International Journal of Speleology,* v. 1, p. 47-60.

Howard, A.D., and Groves, C.G., 1995, Early development of karst systems. 2.: Turbulent flow: *Water Resources Research,* v. 31, p. 19-26.

Hubbard, D.A., Jr., Herman, J.S., and Bell, P.E., 1986, The role of sulphide oxidation in the genesis of Cesspool Cave, Virginia, U.S.A.: *9th International Congress of Speleology,* Barcelona, Proceedings, v. 1, p. 255-258.

Jagnow, D.H., 1977, *Geologic factors influencing speleogenesis in the Capitan Reef Complex, New Mexico and Texas:* University of New Mexico, M.Sc. thesis, 197 p.

Krothe, N.C., and Libra, R.D., 1983, Sulfur isotopes and hydrochemical variations in spring waters of southern Indiana, U.S.A.: *Journal of Hydrology,* v. 61, p. 267-283.

Lowe, D.J., 1989, *Tonga '87–The Report of the 1987 speleological expedition to 'Eua Island, Kingdom of Tonga:* Private publication, 28 p.

Lowe, D.J., 1992, *The origin of limestone caverns: An inception horizon hypothesis:* Manchester Metropolitan University, Ph.D. thesis, 512 p.

Lowe, D.J., and Gunn, J., 1986, Caves and limestones of the islands of Tongatapu and 'Eua, Kingdom of Tonga: *Transactions British Cave Research Association,* v. 13, p. 105-130.

Moncaster, S.J., Bottrell, S.H., Tellam, J.H., and Lloyd, J.W., 1992, Sulfur isotope ratios as tracers of natural and anthropogenic sulfur in the Lincolnshire Limestone aquifer, eastern England, *in* Kharaka, Y., and Maest, A.S., eds., *Water-Rock Interaction:* Rotterdam, Balkema, p. 813-816.

Morehouse, D.F., 1968, Cave development via the sulfuric acid reaction: *National Speleological Society Bulletin,* v. 30, p. 1-10.

Ollier, C.D., 1975, Coral island geomorphology–The Trobriand Islands: *Zeitschrift für Geomorphologie,* v. 19, p. 164-190.

Palmer, A.N., 1991, The origin and morphology of limestone caves: *Geological Society of America Bulletin,* no. 103, p. 1-21.

Sando, W.J., 1988, Madison Limestone (Mississippian) Paleokarst: A Geologic Synthesis, *in* James, N.P., and Choquette, P.W., eds., *Paleokarst:* New York, Springer-Verlag, p. 256-277.

Van Everingden, R.O., Shakur, M.A., and Krouse, H.R., 1985, Role of corrosion by H_2SO_4 fallout in cave development in a travertine deposit–Evidence from sulfur and oxygen isotopes: *Chemical Geology,* v. 49, p. 205-211.

Worthington, S.R.H., 1991, *Karst hydrogeology of the Canadian Rocky Mountains:* McMaster University, Ph.D. thesis, 360 p.

Worthington, S.R.H., and Ford, D.C., 1995, High sulfate concentrations in limestone springs: An important factor in conduit initiation?: *Environmental Geology,* v. 25, p. 9-15.

5.2.8. Sulfuric Acid, Hypogene Karst in the Guadalupe Mountains of New Mexico and West Texas, U.S.A.

Carol A. Hill

Abstract

Carlsbad Cavern, Lechuguilla Cave, and other caves in the Guadalupe Mountains are probably the world's best examples of karst formed by sulfuric acid in a hypogene setting. Four episodes of karstification have occurred in these mountains from Late Permian time to the present, the sulfuric acid episode being the last of these four.

Sulfuric acid karst can be recognized by its:

- Large passage size.
- Ramiform-spongework pattern.
- Horizontal passages connected by deep pits and fissures.
- Location beneath structural and stratigraphic traps.
- Gypsum and native sulfur deposits.
- Sulfuric acid H_2S indicator minerals—endellite, alunite, natroalunite, and tyuyamunite.

Guadalupe caves formed in a diffuse-flow aquifer regime where caves may have acted as "mixing chambers" for hypogene-derived H_2S and meteoric-derived fresh water. How cave hydrology has been related to regional hydrology during the late Tertiary to present is poorly understood. Sulfuric acid karst is an integral part of H_2S-degassing hydrocarbon basins, which also can contain economic sulfur and Mississippi- alley-type ore deposits.

Introduction

Over 300 caves are known to exist in the Guadalupe Mountains of New Mexico and west Texas, U.S.A. Carlsbad Cavern is perhaps the most famous of these caves, but it is neither the deepest nor the longest. Lechuguilla Cave, also located in Carlsbad Caverns National Park, is the deepest (486 m) and longest (>150 km) cave in the Guadalupe Mountains, the deepest and third longest in the United States, and the fifth longest in the world.

Over the past 20 years, empirical evidence for a sulfuric acid speleogenesis for these caves has been mounting, so that now this theory has been accepted by most speleologists. However, the recognition of this new type of speleogenesis was a slow process taking over two decades (5.2.8 Table 1).

A sulfuric acid model of speleogenesis has been an important conceptual breakthrough in the study of karst systems. Not only can caves form by "epigene" processes from descending, CO_2-rich water, but they can also form by "hypogene" processes from deep-sourced H_2S-rich water (Palmer, 1991).

Specifically for the Guadalupe Mountain area, this model explains the peculiarities of Guadalupe caves that have confused geologists for decades. Such peculiarities include cave passages that seem unrelated to surface topography or to groundwater flow routes and passages that are huge but which terminate abruptly. Especially enigmatic have been the massive gypsum blocks, native sulfur, and waxy endellite clay in these caves. A sulfuric acid origin can explain all of these features (Hill, 1987).

Hill (1996) defined four episodes of karst development in the Guadalupe Mountains:

- Stage 1, fissure karst of Late Permian age.
- Stage 2, spongework karst of Mesozoic age.
- Stage 3, thermal karst of Miocene age.
- Stage 4, sulfuric acid karst of late Miocene to Pleistocene age.

Each of these four episodes has had an effect

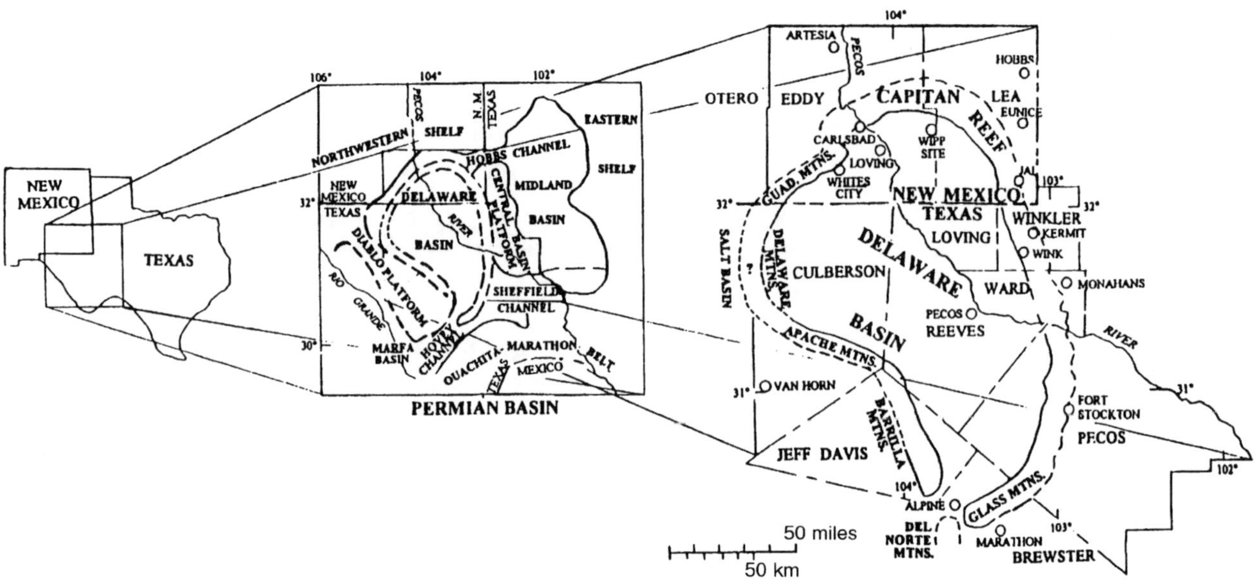

5.2.8 Figure 1. Location map of Permian Basin, southeastern New Mexico and west Texas, U.S.A. The caves of the Guadalupe Mountains are in the northeast section of the Delaware Basin where the Capitan reef is exposed. The Capitan reef is also exposed in the Apache and Glass Mountains to the south. But it is in the subsurface on the east and north sides of the Delaware Basin. It is unknown to the west in the Salt Basin section of the Delaware Basin (from Hill, 1996).

Bretz (1949)	Carbonic acid dissolution in zone of saturation.
Egemeier (1971, 1973)	Replacement-solution sulfuric acid theory for Kane caves, Wyoming. Suggests possibility Carlsbad Cavern formed by a similar mechanism.
Jagnow (1977)	Suggests sulfuric acid dissolution for Guadalupe caves, with pyrite as source of acid, similar to Dubuque, Iowa, U.S.A., caves.
Davis (1979)	Suggests a possible hydrocarbon source for H_2S/sulfuric acid.
Hill, 1979	Performs first sulfur isotope determination on gypsum block, Big Room, Carlsbad Cavern. First empirical evidence supporting a hydrocarbon-cave connection.
Hill (1981)	Performs 11 sulfur isotope analyses on gypsum and sulfur to establish the hydrocarbon-cave connection. Suggests that endellite is a sulfuric acid indicator mineral.
Hill (1987)	Gives detailed account of Guadalupe cave deposits and how they relate to sulfuric acid model. First mention of basinal degassing and connection of Mississippi Valley-type deposits with caves.
Hill (1990)	Expands on the role of hydrocarbons in sulfuric acid cave genesis.
Hill (1996)	Ties together sulfuric acid caves (in both Guadalupe and Glass Mountains) with evolution and degassing of entire Delaware Basin
Polyak and Güven (1996)	Sulfur isotope values on alunite and natroalunite show these to also be sulfuric acid indicator minerals.
Polyak et al. (1997)	Dates on alunite show origin of caves to be between 12 to 4 Ma and relative to their altitude and position in the Capitan reef.

5.2.8 Table 1. History of sulfuric acid theory of speleogenesis, Guadalupe Mountain caves (after Jagnow, 1996).

on the overall morphology and hydrology of the caves, but only the last, sulfuric-acid, episode of karstification will be emphasized in this chapter.

Geologic Setting

The Guadalupe Mountains are in the northwest part of the Delaware Basin, an intracratonic basin of Late Permian age delineated by the Capitan Reef Complex (5.2.8 Figure 1). The stratigraphic, structural, and hydrologic setting of the Delaware Basin and Guadalupe Mountains has been described in detail by Hill (1996), and this chapter is a summary of that work.

Stratigraphic Setting

The Capitan Reef Complex includes backreef, reef, and forereef beds of Late Permian (Guadalupian) age (5.2.8 Figure 2). Backreef rock consists of the dolomitic Seven Rivers, Yates, and Tansill Formations. Reef rock consists of the massive (non-bedded) Capitan Limestone. Forereef rock consists of dolomitized reef talus (breccia). The Capitan is the only limestone section of rock in the reef complex. Caves are developed within this limestone between less-soluble, dolomitized, backreef and forereef rock.

Four main zones of preferential cave

development are controlled by stratigraphy:

- Below the Yates transition into the massive Capitan (Jagnow, 1977; 5.2.8 Figure 2).
- At the contact between the massive reef facies and breccia forereef facies of the Capitan.
- At the transition of the Capitan with backreef facies.
- Beneath the Yates contact with the Seven Rivers.

Structural Setting
The Delaware Basin area lies between the structurally active Basin and Range Province on the west and the nonstructurally active Great Plains Province on the east. The Guadalupe Mountains are a Basin and Range fault block that has been uplifted on its western side (Western Escarpment) relative to the down-dropped Salt Basin Graben. The fault block has also been exhumed by differential erosion and solution-subsidence of evaporite rock in the basin, creating the Reef Escarpment.

Major folds in the area consist of two anticlines (Reef Anticline and Guadalupe Ridge Anticline) and one syncline (Walnut Syncline). Major joint sets consist of a conspicuous system of near-vertical joints that trend parallel to the Reef Escarpment, and a less-conspicuous system that trends perpendicular to the Reef Escarpment. Most of the cave passages in the Guadalupe Mountains are controlled by these two major joint sets.

Hydrologic Setting
The Capitan Aquifer includes the following, all of which act as a connected hydrologic unit:

- Capitan Limestone.
- The underlying Goat Seep Dolomite.
- Their contemporaneous backreef units: Grayburg, Queen, Seven Rivers, Yates, and Tansill Formations.

In the Guadalupe Mountains the Capitan Aquifer parallels the Reef Escarpment. It has a basinward edge that is abruptly and sharply defined, and a shelfward edge that is gradational. Groundwater movement in the aquifer is controlled by a basinal evaporite facies (Castile Formation). This has a very low permeability that confines water in the aquifer on its basinward side, by a reef zone of very high permeability, and by a backreef zone of moderate (near the reef) to low permeability. The Capitan Aquifer is recharged by precipitation in the Guadalupe Mountains, where backreef beds slope downdip into the reef (5.2.8 Figure 2). The discharge point for the Capitan Aquifer in the Guadalupe Mountains is Carlsbad Springs at Carlsbad, New Mexico (5.2.8 Figure 1).

Sulfuric Acid Caves

Sulfuric acid caves usually form either as a result of:

- Oxidation of sulfide ore masses (e.g., Dubuque, Iowa caves, Morehouse, 1968).
- Oxidation of H_2S supplied by hydrocarbon basins (e.g., Guadalupe Mountains, New Mexico caves; Hill, 1990).

The number of known sulfuric acid caves related to hydrocarbon basins (5.2.8 Table 2) is small when compared to the vast number of

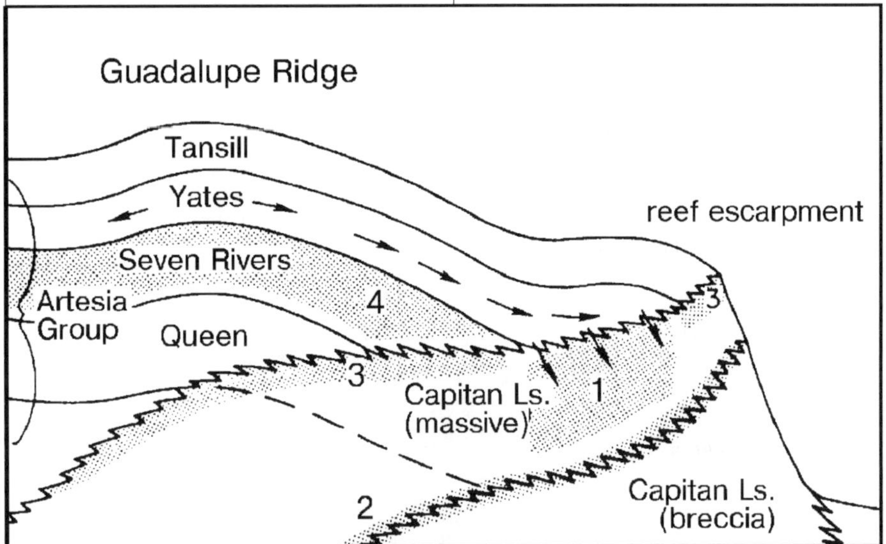

5.2.8 Figure 2. Cross section of the Capitan Reef Complex showing backreef, reef, and forereef stratigraphic units, and also showing four main zones of preferential cave solution controlled by stratigraphy (after Jagnow, 1977).

Location	Characteristics
Kane Caves, Wyoming. Egemeier (1973)	Gypsum rinds and mounds. Native sulfur H_2S in air and thermal springs. Active replacement-solution. Associated with hydrocarbons in Big Horn Basin.
Carlsbad Cavern, Lechuguilla Cave, other Guadalupe caves Hill (1981, 1987, 1990), Polyak & Mosch (1994), Polyak and Güven (1996)	Gypsum blocks and rinds. Native sulfur. Sulfuric acid indicator minerals: endellite, alunite, natroalunite, and tyuyamunite. Large passages; ramiform-spongework pattern. Associated with hydrocarbons in Delaware Basin.
Fiume-Vento Cave, Italy. Cucchi & Forti (1988)	Gypsum blocks and rinds. Endellite. Large passages; ramiform-spongework pattern. Associated with hydrocarbons in Adriatic Sea Basin.
La Cueva de Villa Luz, Tobasco, Mexico. Pisarowicz et al. (1992)	Actively forming today by sulfuric acid process. H_2S measured in air @ 40 ppm. pH of drip water = 1. Gypsum rinds and native sulfur replacing bedrock. Milky-white river (containing gypsum and sulfur). Associated with hydrocarbons in Campeche Basin.
Las Brujas Cave, Malargue, Argentina. Forti et al. (1993)	Gypsum rinds and blocks. Associated with hydrocarbons of Andean overthrust belt.

5.2.8 Table 2. List of known sulfuric acid caves associated with hydrocarbon basins.

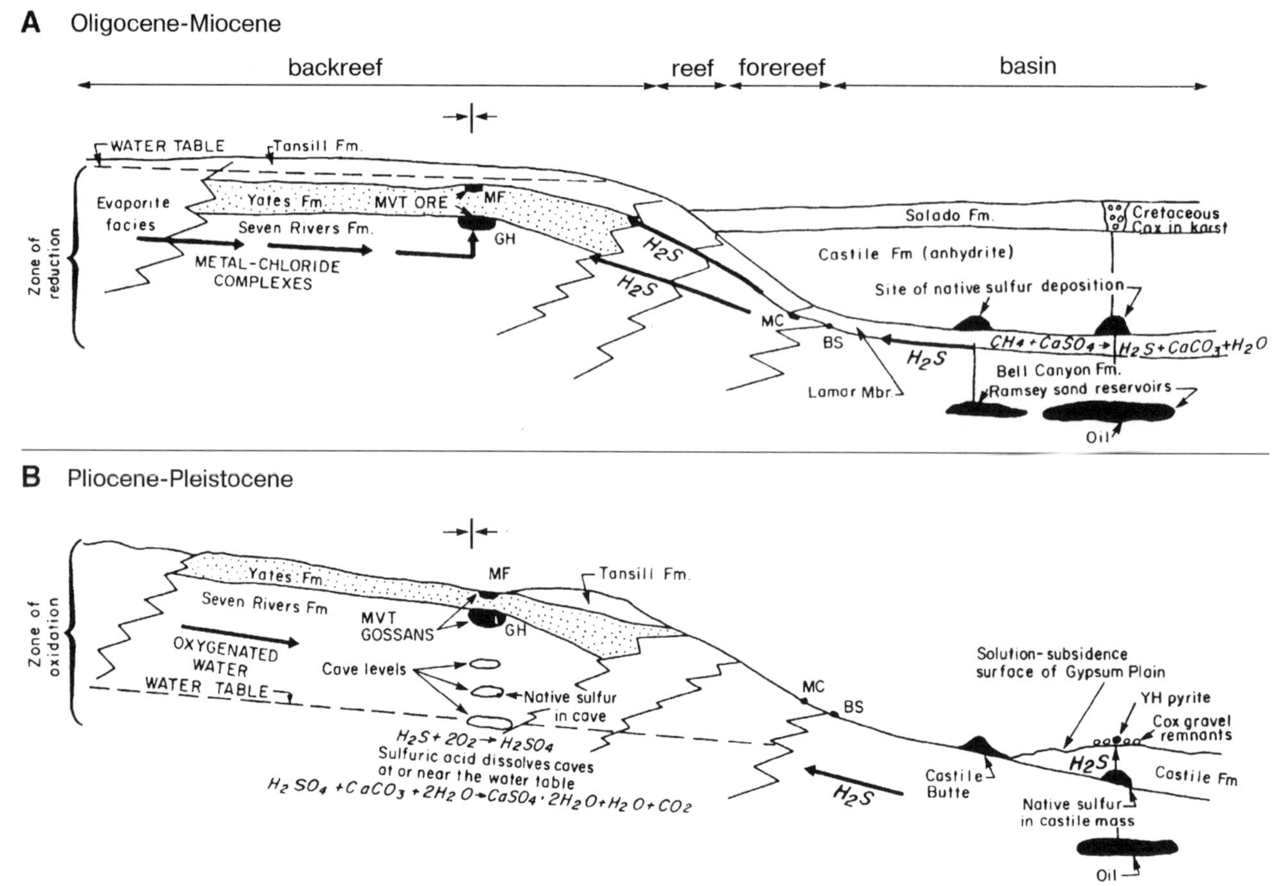

5.2.8 Figure 3. Location of caves and Mississippi Valley-type deposits in the Guadalupe Mountains with respect to structural and stratigraphic traps for H_2S (from Hill, 1996).

carbonic acid caves (most caves in the world). But this number may increase as sulfuric acid cave processes become more universally recognized by speleologists. The characteristics of the sulfuric acid caves in the Guadalupe Mountains are summarized below.

Passage Size

The size of Guadalupe cave passages can be huge. The Big Room of Carlsbad is one of the largest cave rooms in the world. Many other passages exceed 15 m in height and width. Sulfuric acid can readily dissolve limestone and create large cave passages.

Passage Morphology

Sulfuric acid caves display a ramiform and spongework pattern compared to branched, network, and anastomosic patterns more characteristic of carbonic acid caves (Palmer, 1991; see 3.4 Figure 10). Cave passages are large and deep, but end abruptly both in the horizontal and vertical direction. Major horizontal levels are connected by deep pits and fissures.

Boneyard, solution pockets and domes, and chambers that are globular and irregular are also characteristic of this type of cave, where irregular rooms and galleries wander three-dimensionally, with branches extending outward from main areas of development.

Such a ramifying pattern is typical of caves formed by the oxidation of rising H_2S, where the main passages and large domed rooms correspond to favored paths (injection points) of H_2S (Palmer, 1996). The larger the H_2S input, or the longer it is injected into the system, the bigger the cave rooms or passages become.

Structural and Stratigraphic Traps

Caves in the Guadalupe Mountains are located along structural and stratigraphic traps because it is there that H_2S can accumulate. For example, Carlsbad Cavern is on the Reef Anticline and Lechuguilla Cave and Cottonwood Cave are on the Guadalupe Ridge Anticline—structural traps for gas. Lechuguilla and Cottonwood are also located directly beneath the siliciclastic Yates Formation, a stratigraphic trap for H_2S. These structural-stratigraphic traps are also where Mississippi Valley-type (MVT) sulfide mineralization is located in the Guadalupe Mountains (Hill, 1993b; 5.2.8 Figure 3).

Gypsum and Sulfur

Massive gypsum blocks (up to 10 m high) and native sulfur (up to thousands of kilograms) are distinguishing features of Guadalupe Mountains caves. These deposits are one of the most important criteria for establishing a hypogene sulfuric acid origin for these caves. Epigenic carbonic-acid caves do not contain these types of deposits.

Importantly, the gypsum and sulfur are significantly enriched in the light isotope of sulfur–values as low as $\delta^{34}S = -25.6$ ‰ for gypsum and –25.8 ‰ for sulfur have been measured (5.2.8 Figure 4). These sulfur isotope results are crucial to understanding the process of speleogenesis that produced the large Stage 4 cave passages in the Guadalupe Mountains.

The pronounced enrichment of light sulfur implies that hydrocarbons and sulfur bacteria were involved in speleogenesis, as only biologically aided reactions could have produced the large isotopic fractionations that characterize the cave gypsum and sulfur.

Endellite, Alunite, Natroalunite, and Tyuyamunite

Other distinctive deposits in Guadalupe Mountain caves are (Hill, 1981; Hill, 1987; Polyak and Mosch, 1994; Polyak and Güven, 1996):

- Colorful waxy endellite [$Al_2Si_2O_5(OH)_4 \cdot 2H_2O$].
- Alunite [$K_2Al_6(SO_4)_4(OH)_{12}$].
- Natroalunite [$Na_2Al_6(SO_4)_4(OH)_{12}$].
- Tyuyamunite [$Ca(UO_2)_2(VO_4)2 \cdot 5\text{-}8H_2O$].
- Metatyuyamunite [$Ca(UO_2)_2(VO_4)_2 \cdot 3\text{-}5H_2O$].

The clay mineral endellite and the sulfate minerals alunite and natroalunite are pH-indicators; that is, they are minerals which are known to form in a low-pH environment. Sulfur isotope values of alunite and natroalunite ($\delta^{34}S$ = -9.8‰ to -28.9‰) confirm that the origin of these minerals is also related to H_2S and hydrocarbons (Polyak and Güven, 1996). The presence of high uranium (in uranium minerals and in montmorillonite clay) and high radon in these caves is also indicative of a H_2S system (Hill, 1987).

Hydrology of Guadalupe Caves

The caves of the Guadalupe Mountains are a combination of bathyphreatic and water-table: vadose water has in no way been responsible for their genesis. Bathyphreatic conditions have been responsible for the strong vertical development of these caves, and water-table conditions have been responsible for the horizontal development of cave passages along certain levels (corresponding to past regional base levels). Bathyphreatic development probably occurred earlier than water table development, perhaps much earlier.

Initial uplift and tilting of the Guadalupe Mountain fault block occurred as early as 30 to 20 Ma, with maximum uplift and tilting at about 15 to 5 Ma (Hill, 1996). This means that H_2S could have been migrating from basin to reef throughout the middle to late Tertiary, with the potential for cave formation. This is a possibility recently confirmed by Polyak et al. (1998) who $^{40}Ar/^{39}Ar$-dated alunite in the caves at 12 Ma for caves in the southwestern section of the mountains (Virgin Cave) to 4 Ma for caves in the northwestern section of the mountains (Carlsbad and Lechuguilla).

The main corridors or passages (e.g., Main Corridor, Carlsbad Cavern; Main Corridor, Cottonwood Cave) could have begun forming as conduits for the transmission of water in the early stages of bathyphreatic cave development. According to this idea, diffuse flow from the surrounding recharge area converged along these incipient corridors, ascended toward base level, and discharged at a major spring, just like Carlsbad Spring discharges at the city of Carlsbad today.

The entrance to Carlsbad Cavern, high upon the reef, may represent one such still-intact paleospring. But other paleosprings (e.g., those for Lechuguilla, Cottonwood, and Hell's Below), have not been preserved because they have been truncated or obliterated by valley downcutting and erosion. As base level progressively lowered over time, the position of these springs shifted toward the northeast along the reef front (Hill, 1987; Jagnow, 1988). Some cave passages rise in a series of arcs toward former outlets (paleosprings), and this configuration may show that the mixing of O_2-rich water and H_2S-rich water took place well below the outlet and over vertical distances as great as 250 m (Palmer, 1996).

The ramifying nature of cave passages in the Guadalupe Mountains attests to the diffuse-flow aquifer conditions that formed them. Mylroie (1991) and Mylroie et al. (1995) noted the morphologic similarity between Guadalupe Mountain caves and those in the Bahamas and Bermuda. They attributed both to a diffuse-flow and mixed-water mechanism (H_2S water–O_2-rich fresh water and sea water–fresh water, respectively).

According to this model, both sets of caves represent mixing chambers that receive water as diffuse flow and that discharge the water, after mixing, as diffuse flow. Over time, the mixing front migrates headward up the routes of incoming diffuse flow (Mylroie et al. 1995). And, in the case of the Guadalupe caves, the mixing front also migrates downward along the flanks of chambers, thereby creating "maze" areas at the bottom of large rooms (Hill, 1987).

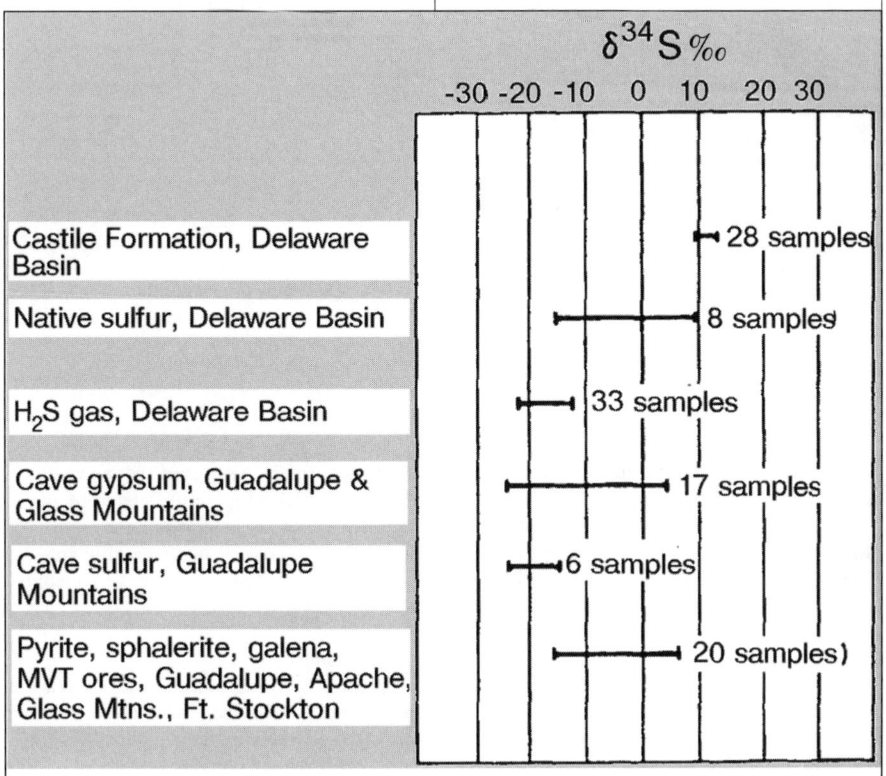

5.2.8 Figure 4. Sulfur isotope values for various types of sulfur deposits in the Delaware Basin (from Hill, 1996). In addition, alunite and natroalunite cave deposits have sulfur isotope values ranging from -9.8 to -28.9‰ (Polyak and Güven, 1996).

Relationship of Guadalupe Caves to Regional Hydrology

The relationship of cave development in the Guadalupe Mountains with respect to regional hydrologic events over time is poorly understood. It is known that the caves must have been related to past positions and levels of the ancestral Pecos River, and that the major cave levels are attributable to long periods of stability as defined by these former base levels. But further connections are speculative. A major question that needs to be

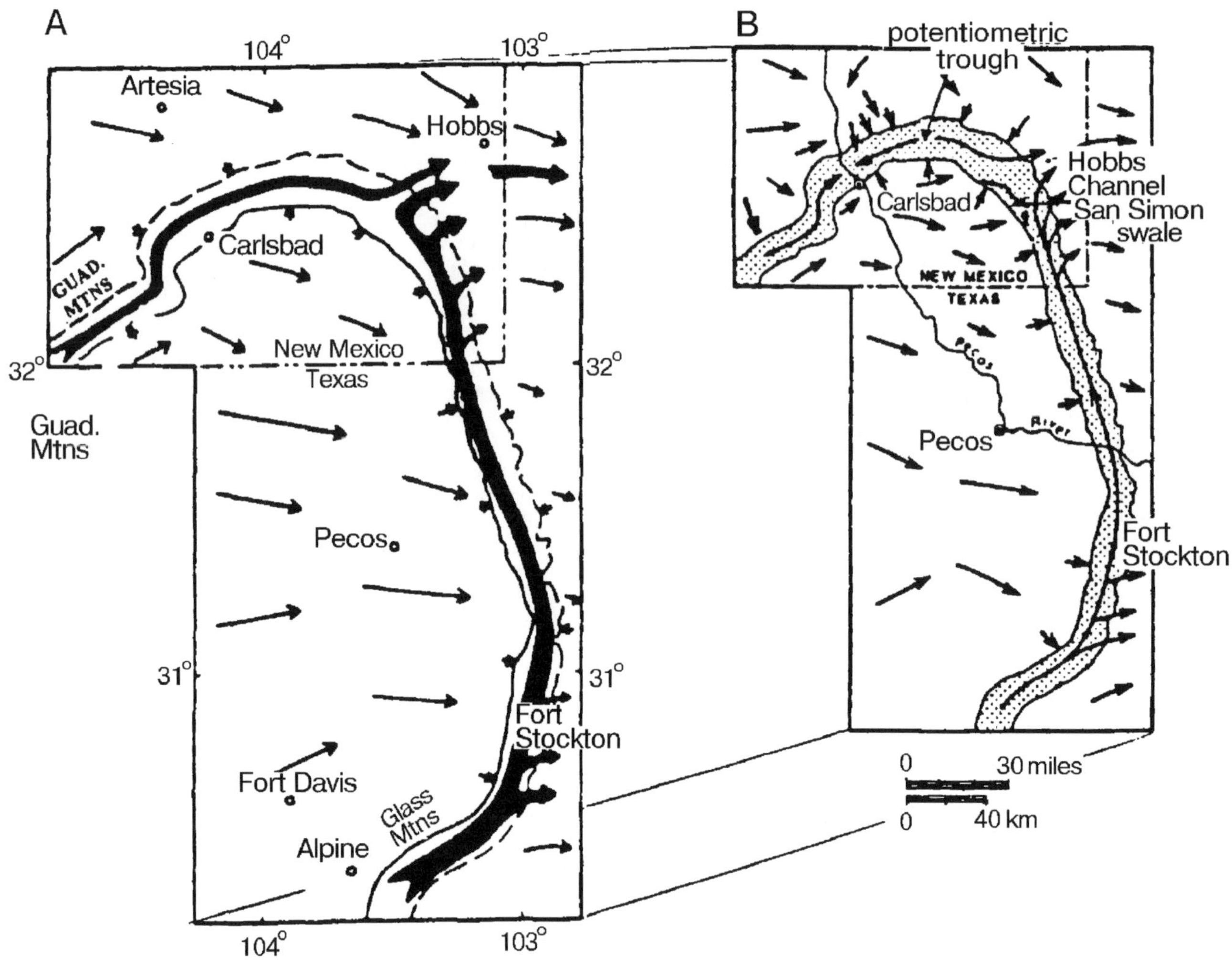

*5.2.8 Figure 5. **A** is flow of water in the Capitan Aquifer prior to entrenchment of the ancestral Pecos River at Carlsbad, New Mexico. **B** is the flow of water in the Capitan Aquifer after entrenchment, continuing to the present. The draining of the lower levels of Carlsbad Cavern (Big Room and Lower Cave) and Lechuguilla Cave may correspond to this entrenchment (after Hiss, 1980).*

answered in this regard is: How do the locations of paleosprings (e.g., the entrance to Carlsbad Cavern) correspond to the position of the ancestral Pecos River over time?

I suggested one possible intriguing connection between the lower cave levels of Carlsbad and Lechuguilla and the breaching of the Pecos River at Carlsbad (i.e., the time when water in the Capitan Aquifer began to exit at Carlsbad Springs rather than at Hobbs, New Mexico; 5.2.8 Figure 5; Hill, 1996). Bachman (1984) claimed that this incision of the Pecos River and its hydrologic connection with the Capitan Aquifer occurred at about 600,000 ybp. This time frame roughly correlates with the draining of water out of the Big Room and Lower Cave levels in Carlsbad. It raises this question: Could the breaching have been responsible for a sudden draining of these levels and other equivalent levels in other Guadalupe caves? No aquifer water is encountered in any caves in the Guadalupe Mountains. Dates on the cloud linings in the lowest passages—Lake of Clouds, Carlsbad, and Lake of White Roses, Lechuguilla—are >350,000 ybp and ~600,000 ybp by the uranium series and electron spin resonance methods, respectively (D. Ford and K. Cunningham, pers. comm.). These dates are supportive of this idea, but by no means prove it.

Why is Sulfuric Acid Karst Important?

The caves in the Guadalupe Mountains are genetically related to economic hydrocarbon, native sulfur, and Mississippi Valley-type sulfide deposits by a series of redox reactions (Hill, 1995b; 5.2.8 Figure 6). Therefore, the processes which are responsible for producing sulfuric acid caves should be of interest to mainstream geologists as well as to speleologists.

Hydrocarbons

The process by which sulfuric acid karst is produced can also be responsible for reservoir porosity (i.e., porosity that contains hydrocarbons). "H_2S-related porosity" refers to porosity created in a H_2S system where dissolution is produced by the mixing of waters of different H_2S content or by the oxidation of H_2S (Palmer, 1991). "Sulfuric acid oil-field karst" refers to a specific kind of H_2S-related porosity where carbonate reservoirs of cavernous size have been

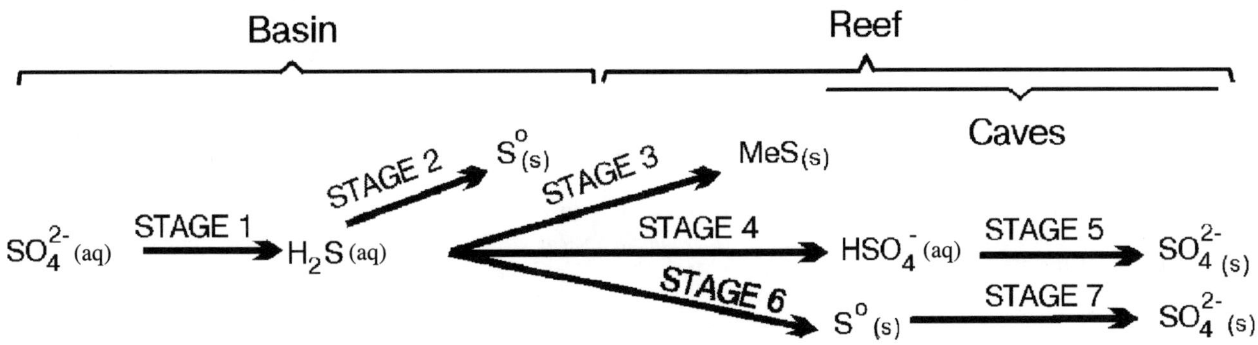

5.2.8 Figure 6. Series of redox reactions genetically connecting hydrocarbons, economic sulfur deposits, Mississippi-Valley-type deposits, and sulfuric acid caves (from Hill, 1995b).

dissolved by a sulfuric acid mechanism (Hill, 1995a). In an H_2S system, porosity can be produced entirely in a deep subsurface circulation setting and does not need to represent a paleokarst surface or dissolution in the shallow-phreatic or vadose zones.

Examples of sulfuric acid oil-field karst (i.e., karstic reservoirs dissolved by a sulfuric acid mechanism) may be the Lisburne Field, Prudhoe Bay, Alaska, and some Middle East fields such as Agha Jara in Iran and the Kirkuk Field in Iraq (Hill, 1995a).

Native Sulfur

Economic sulfur deposits in the Delaware Basin include those in the Rustler Springs and Fort Stockton sulfur districts. Freeport's Culberson sulfur mine in the Rustler Springs district is one of the largest sulfur mining operations in North America. The sulfur resource for this mine was estimated by Crawford and Wallace (1993) to be about 7.7 x 10^9 kg (76 million long tons). The H_2S needed for the formation of this economic sulfur is supplied in Stage 1 (5.2.8 Figure 6), where hydrocarbons react with evaporite rock to form H_2S, and where this H_2S is oxidized to native sulfur at the redox boundary. This sulfur is isotopically light, as is the cave sulfur (5.2.8 Figure 4), because it is derived from bacterial reactions involving hydrocarbons.

Mississippi Valley-type (MVT) Deposits

MVT sulfides in the Guadalupe Mountains are located along the same structural-stratigraphic traps as the caves (5.2.8 Figure 3). They also have the same range of isotopic compositions as native sulfur in the basin (5.2.8 Figure 4). These correlations suggest that the H_2S produced in Stage 1 (5.2.8 Figure 6) migrated from basin to reef (5.2.8 Figure 3). There it formed metal sulfides (pyrite, sphalerite, galena) in the chemically reduced saline-water zone. Later, as the water table dropped, sulfuric acid caves formed along these same H_2S traps at or near the water table in the oxygenated fresh-water zone. The MVT deposits surrounding other intracratonic basins may also be due to basinal degassing rather than to basinal compaction-dewatering, based on findings in the Delaware Basin (Hill, 1993a,b).

Concluding Remarks

Sulfuric acid karst is not only important to speleology, but it is also important to understanding the evolution of intracratonic basins. Why do some basins host native sulfur deposits while others do not? Why do some basins host enormous Pb-Zn (MVT) deposits while others display caves around their carbonate margins? Speleologists in the next millenium should consider applying their knowledge of sulfuric acid karst processes to the field of economic geology. It is my contention that the role of the migration of H_2S has been underestimated not only as a speleogenetic mechanism, but also in the formation of ore deposits and in the creation of reservoir porosity.

References

Bachman, G.O., 1984, Regional geology of Ochoan evaporites, northern part of Delaware Basin: *New Mexico Bureau of Mines and Mineral Resources Circular 184,* 22 p.

Bretz, J H., 1949, Carlsbad Caverns and other caves of the Guadalupe block, New Mexico: *Journal of Geology,* v. 57, p. 447-463.

Crawford, J.E., and Wallace, C.S., 1993, Geology and mineralization of the Culberson sulfur deposit, *in* Love, D.W., et al., eds., Carlsbad region, New Mexico and west Texas: *New Mexico Geological Society Guidebook,* 44th Annual Field Conference, p. 301-316.

Cucchi, P., and Forti, P., 1988, Speleogenetic evolution of the "Fiume-Vento" karst system (San Vittore, Genga, Marche, Italy): *International Symposium on Physical-Chemical Hydrogeology of Karst,* Kosice, p. 193-199.

Davis, D.G., 1979, Geology and speleogenesis of Ogle Cave: Discussion: *National Speleological Society Bulletin,* v. 42, p. 42-48.

Egemeier, S.J., 1971, *A comparison of two types of solution caves:* Unpublished report to Carlsbad Caverns National Park. 7 p.

Egemeier, S.J., 1973, *Cavern development by thermal waters with a possible bearing on ore deposition:* Stanford University, Ph.D. dissertation, 88 p.

Forti, P., Benedetto, C., and Costa, G., 1993, Las Brujas Cave (Malargue, Argentina): An example of the oil pools control on speleogenesis: *Theoretical and Applied Karstology,* v. 6, p. 87-93.

Hill, C.A., 1981, Speleogenesis of Carlsbad Cavern and other caves of the Guadalupe Mountains: *8th International Congress of Speleology,* Bowling Green, Proceedings, p. 143-144.

Hill, C.A., 1987, Geology of Carlsbad Cavern and other caves in the Guadalupe Mountains, New Mexico and Texas: *New Mexico Bureau of Mines and Mineral Resources Bulletin,* no. 117, 150 p.

Hill, C.A., 1990, Sulfuric acid speleogenesis of Carlsbad Cavern and its relationship to hydrocarbons, Delaware Basin, New Mexico and Texas: *American Association of Petroleum Geologists Bulletin,* v. 74, p. 1685-1694.

Hill, C.A., 1993a, A hydrogen sulfide, basinal-degassing model for the origin of Mississippi Valley-type ore deposits, *in* Shelton, K.L., and Hagni, R.D., eds.,

Geology and geochemistry of Mississippi Valley-type ore deposits: University of Missouri-Rolla Press, p. 121-132.

Hill, C.A., 1993b, Sulfide/barite/fluorite mineral deposits, Guadalupe Mountains, New Mexico and west Texas: *New Mexico Geology,* v. 15, no. 3, p. 56-65.

Hill, C.A., 1995a, H_2S-related porosity and sulfuric acid oil-field karst, *in* Budd, D.A., Saller, A.H., and Harris, P.M., eds., Unconformities in carbonate strata—Their recognition and the significance of associated porosity: *American Association of Petroleum Geologists Memoir,* no. 61, p. 301-306.

Hill, C.A., 1995b, Sulfur redox reactions, native sulfur, Mississippi Valley-type deposits, and sulfuric acid karst, Delaware Basin, New Mexico and Texas: *Environmental Geology,* v. 25, p. 16-23.

Hill, C.A., 1996, Geology of the Delaware Basin, Guadalupe, Apache, and Glass Mountains, New Mexico and West Texas: *Society of Economic Paleontologists and Mineralogists—Permian Basin Section,* Publication 96-39, 480 p.

Hiss, W.L., 1980, Movement of ground water in Permian Guadalupian aquifer systems, southeastern New Mexico and western Texas, *in* Dickerson, P.W., and Hoffer, J.M., eds., *Trans-Pecos region, New Mexico Geological Society Guidebook,* 31st Field Conference, p. 289-294.

Jagnow, D.H., 1977, *Geologic factors influencing speleogenesis in the Capitan Reef complex, New Mexico and Texas:* University of New Mexico, M.S. thesis, 197 p.

Jagnow, D.H., 1988, The geology of Lechuguilla Cave: *National Speleological Society News,* v. 46, p. 422-425.

Jagnow, D.H., 1996, History of sulfuric acid theory of speleogenesis in the Guadalupe Mountains: *National Speleological Society Convention,* Salida, Program with Abstracts, p. 50-51.

Morehouse, D.F., 1968, Cave development via the sulfuric acid reaction, *National Speleological Society Bulletin,* v. 30, p. 1-10.

Mylroie, J.E., 1991, Cave development in the glaciated Applachian karst of New York: Surface-coupled or saline-freshwater mixing hydrology: *Appalachian Karst Symposium,* Radford, Virginia, Proceedings, p. 84-90.

Mylroie, J.E., Carew, J.L., and Vacher, H.R., 1995, Karst development in the Bahamas and Bermuda: *Geological Society of America Special Paper,* v. 300, p. 251-267.

Palmer, A.N., 1991, Origin and morphology of limestone caves. *Geological Society of America Bulletin,* no. 103, p. 1-21.

Palmer, A.N., 1996, Cave patterns in the Guadalupe Mountains: *National Speleological Society Convention*, Salida, Program with Abstracts, p. 51-52.

Pisarowicz, J., 1992, Cueva de Villa Luz–H_2S parameters: *National Speleological Society, Section of Cave Geology and Geolography Newsletter. Geo²:* v. 20, p. 9.

Polyak, V.J., and Güven, N., 1996, Mineralization of alunite, natroalunite, and hydrated halloysite in Carlsbad Cavern and Lechuguilla Cave, New Mexico: *Clays and Clay Minerals*, v. 44, p. 843-850.

Polyak, V.J., and Mosch, C.J., 1994, Tyuyamunite and metatyuyamunite?: *National Speleological Society Bulletin,* v. 57, p. 80.

Polyak, V.J., McIntosh, W.C., Güven, N., and Provencio, P., 1998, Age and origin of Carlsbad Cavern and related caves from $^{40}Ar/^{39}Ar$ of alunite: *Science,* v. 279, p. 1919-1922

5.2.9. A Giant Hydrothermal Cavity in the Rhodope Mountains, Bulgaria

Victor N. Dublyansky

Abstract

A giant cavern was intercepted by drilling at the Madan tin-zinc ore deposit. The cavern is developed in Archean and Proterozoic marble, has an estimated volume of 237.6 million m^3, has a maximum intersected vertical dimension of 1341 m, and is filled with thermal water under high pressure. A brief description of the geologic and hydrogeologic setting is given.

Geology and Hydrology

A body of Archaean marble was discovered in the course of exploratory drilling at the Madan tin-zinc ore deposit in 1959. In the course of subsequent drilling in 1966-1969, several boreholes entered a giant cavern (5.2.9 Figure 1, 5.2.9 Table 1). Two boreholes did not reach the bottom of the cavern, even though the drill was extended to 316 m (borehole 15) and 1341 m (borehole SH-1).

The Madan ore district is located in southern Bulgaria, near the border with Greece in the western Central Rhodopes. The relief is mountainous and includes the Andra River Basin. The bedrock consists of metamorphic, igneous, and sedimentary rocks.

The metamorphic complex is up to 11,000 m thick and consists of two units:

- Archaean—biotite gneiss with layers of amphibole-biotite, muscovite, and feldspar gneisses, aplite muscovite gneiss, and marble of the Madan-I Horizon.
- Proterozoic—migmatized gneiss, marble of the Madan-II and III Horizons, biotite gneiss, and amphibolite.

Igneous rocks are represented by Paleozoic stocks and dikes of serpentinite, hornblendite, gabbro, and quartz rhyolite, as well as by late Oligocene rhyolite dikes. Sedimentary rocks include Oligocene conglomerate-breccia filling depressions in the roof of the metamorphic rocks.

One syncline and two anticlines striking 310 to 335°, as well as the Madan Overthrust (amplitude up to 1.2 km), are developed on the western slope of the Central Rhodopean Anticlinorium within the limits of the district. Hydrothermal mineralization follows two systems of tectonic faults striking 325° to 355° and 275° to 305°. The first zone of marble (Madan-I) was outlined by 23 boreholes within an area of 17 km^2 stretching from the Erma River Valley on the north to the border with Greece on the south. Its roof was encountered at a depth of –348 to –121 m (5.2.9 Figure 1). The total thickness of the zone is unknown, because none of the boreholes has reached the underlying beds (the thickness of drilled marble ranges from 180 to 791 m). Judging from lowering of the drill columns in boreholes 5 and SH-1, the thickness of the zone may be as great as 1800 to 2000 m. Significant differences in the position of the roof of the zone intersected by adjacent boreholes suggest block faulting. The marble of the Madan-I Horizon is white,

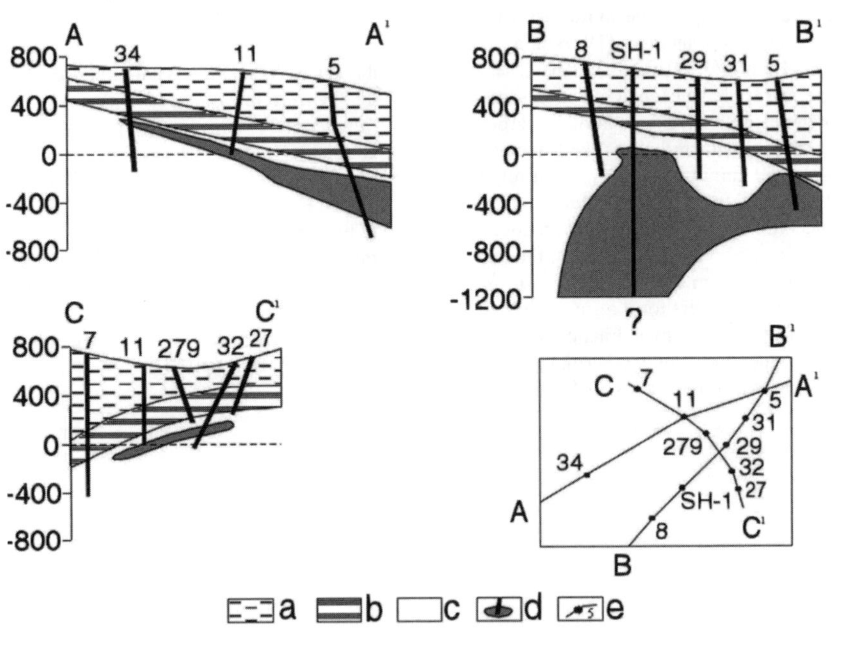

*5.2.9 Figure 1. Location of boreholes in the Madan ore district and schematic geologic cross sections along profiles A to A', B to B', and C to C'. **a** is biotite gneiss. **b** is aplite gneiss. **c** is marble (Madan-1). **d** is boreholes and encountered caverns. **e** is borehole profiles and numbers.*

Cavern	Depth, m		Height of the Cavern, m
	Roof	Bottom	
11	-644.3	-658.8	14.5
32	-562.0	-606.0	44.0
5	-797.2	-1113.0*	>315.8
SH-I	-667.9	-2009.0*	>1341.1

* bottom not reached

5.2.9 Table 1. Caverns discovered by drilling at the Madan ore deposit.

CaO	CO_2	SiO_2	Fe_2O_3	Al_2O_3	K_2O	SrO	Na_2O	P_2O_5	MgO	BaO	MnO	TiO_2	Total
55.00	43.48	1.00	0.15	0.10	0.07	0.06	0.05	0.04	0.02	0.01	0.01	0.01	100.00

5.2.9 Table 2. Chemical composition of the Madan-I marble (average of 250 analyses, %).

Aquifer	Mean Depth m	Number of Samples	pH	HCO_3	SO_4	Cl	Ca	Mg	Na	K	Al	TDS
Gneisses	360	6	6.0 to 6.4	60	640	10	235	48	20	12	55	1080
Marbles	680	2	6.6 to 8.4	327	193	56	32	6	195	20	12	841

5.2.9 Table 3. Mean chemical composition of groundwater in the Madan area (mg/L).

sugar-like, fine grained, and massive. Its mean chemical composition is given in 5.2.9 Table 2. Matrix porosity reaches 3%. The marble hosts layers of muscovite, biotite, and two-mica gneiss 1 to 80 m thick. The marble has experienced metasomatic alteration near the contact with aplite muscovite gneiss forming 2- to 180-m thick cavernous zones in the quartzose rocks. Cavities are abundant in this zone, and they range in size from 5 cm to several meters and are filled with hot water. Some of them are lined with low-temperature hydrothermal fluorite (formation temperature calculated to be 80 to 90°C from fluid inclusions).

Fissure water is tapped near faults in the gneiss overlying the Madan marble. The mean discharge of the wells is 0.004 L/s, whereas near the faults it can be as high as 0.04 L/s. Many wells yield outflowing water with heads reaching 15 m. The temperature of the upper-zone water is ambient, gradually increasing with depth from 23 to 33°C. The water belongs to the calcium-sulfate type and has total dissolved solids ranging from 1.2 to 1.8 g/L.

The Madan marble contains fissure-conduit thermal water. It is characterized by a significant head: their potentiometric levels can be as high as –478 to –486 m. The results of airlift pumping show a mean discharge for the wells of 4.3 L/s, with a maximum discharge of 9 to 10 L/s. The transmissivity of the marble ranges from 800 to 2000 m^2/day (mean 1000 m^2/day); the mean coefficient of piezometric conductivity is $3x10^5$ m^2/day. The water belongs to the sodium-silica-bicarbonate type and has TDS ranging from 1.0 to 1.2 g/L (5.2.9 Table 3). The gases N, CO_2, and He have been detected in some boreholes. Intensive deposition of carbonate material is characteristic of the flowing wells: a hole 160 mm in diameter was entirely sealed within 25 to 35 days by alternating concentric layers of calcite and aragonite.

Hydraulic connection between the two aquifers has been observed in several wells. Together the waters form a single thermal anomaly expressed by a drastic increase in rock temperature and geothermal gradient. The gradient is as great as 17.5°C/100 m in the central part of the area (borehole 279) and decreases to 4.5°C/100 m on the periphery. The temperature of the roof of the marble bed at depth of -449 m reaches 90°C. The highest measured temperature of water was 129.6°C at a depth of –1279 (–359 m below sea level). The Madan thermal anomaly is thought to be related to young magmatic bodies 28 km to the south near the Grecian town of Ksanti, or alternatively to an ancient deep-seated fault serving as an avenue for upwelling thermal water. The discharge area for the water is not known with certainty. It might be in the area of the Termia mineral spring (discharge 25 L/s, T = 45 to 57°C, TDS = 1.5 to 1.6 g/L), in Greece, 7.5 km from the Bulgarian border.

Hydrothermal processes have led to the development of the giant cavern in the Madan-I marble horizon (5.2.9 Figure 1). Its known dimensions are summarized in 5.2.9 Table 1. Sebev (1970) estimates its mean length as 800 m, width as 620 m, height as 479 m, and volume as 237,600,000 m^3. This giant system is filled with thermal water having a pressure of 37 atm near the ceiling and 170 atm near the floor. The temperature of the water increases with depth from 85.9 to 90.2°C at –110 to 120 m to 129.6°C at –1270 m.

The upper part of the cavern, which is located above sea level, will become available for direct study after the completion of a utility shaft that will lower the level of the thermal water in the marble.

References

Dublyansky, V.N., 1974, Giant hydrothermal karst cavern in Rhodopes (Bulgaria): *Peschery (Caves),* v. 14-15, p. 233-237 (in Russian).

Sebev, D.G., 1970, Giant cavern in Rhodopes: *Rodopski Pescherniak,* no. 50 (Chepelare), 34 p. (in Bulgarian).

Serafimov, S., and Milev, T., 1977, Particularities of geothermal characteristics of the Erma Reka deposit: *Fizicheskie Processi Gornogo Proizvodstva* (Moscow), v. 3, p. 113-115 (in Russian).

5.3. Speleogenesis Under Unconfined Settings

Derek Ford

Abstract

It is estimated that 90% or more of the longer caves that speleologists have explored and mapped were created by meteoric water circulating underground without important artesian confinement. These are "unconfined" or "common" caves. They are often three-dimensionally complex systems of galleries and shafts. For ease of analysis, they may be divided into

- Plan patterns (the dimensions of length and breadth).
- Long sections (length and depth).

Plan patterns are dominantly branchworks in their form. They are angular where joints and faults are dominant controls and curvilinear where bedding planes are more important. Irregular networks and anastomotic mazes are common subsidiary components. Modeling has determined the controls of the pattern building to a significant extent.

The geometry of cave *long sections* may be divided into deep phreatic, multiloop, mixed-loop water table, and ideal water table types, with drawdown vadose caves or invasion vadose caves above them. Many large systems display a mixture of the types.

Most of the longer and deeper common caves that are known are of multiphase origin, displaying two or more different levels of principal passages that are related to the past and present altitudes of the springs. The reconstruction that occurs when new passages develop below older ones is responsible for much of the complexity seen on plans of caves.

The case studies that follow this introductory chapter illustrate many of these features of cave genesis.

Introduction

Chapters 5.3.1 to 5.3.13, which follow this chapter, present studies by many different authors of cave systems that have been created by normal meteoric water sinking underground through the epikarst or dolines and streamsinks, and circulating in limestone or other soluble rocks without any major artesian confinement. These have been termed *common caves* (Ford and Williams, 1989, p. 246) because they probably account for 90% or more of the explored and mapped dissolutional caves that are longer than a few hundred meters. They do not include caves created largely or entirely by waters (thermal or otherwise) ascending into the cavernous rock from deeper strata, or marine mixing zone caves, the subjects of previous chapters of this part of the book.

Common cave systems are usually large in all three spatial dimensions. For purposes of analysis, it is conventional to divide three-dimensional phenomena like these into two-dimensional components, and this happened historically in the case of studies of cave genesis. Most early proposals focused on cave development in length and depth (i.e. in projected long sections between groundwater sinkpoints and springs) and were greatly concerned with the relationship between cave and water table. With the benefit of hindsight, I believe that it might have been more useful to have considered length and breadth first, i.e. the genesis of the plan patterns of cave systems. The focus on length and depth probably reflects the early European speleologists' attention to the deep shaft-and-drain types of caves that were beginning to be explored in the Alps, Pyrenees, and elsewhere. E.A. Martel, widely considered to be the father of scientific speleology, was certainly attracted to them (e.g. Martel, 1921).

Development of the Plan Patterns of Unconfined Caves

The maps of large unconfined cave systems are often very complicated. They look like mazes or tangles of spaghetti that seem to lack any apparent logical organization. Usually this is because these systems are multiphase, with one or more older higher levels of galleries wandering across those that actually convey the water today.

Speleologists in many nations have found that most active cave-river patterns are broadly dendritic or treelike in their form, focusing flow from several tributaries toward a single spring outlet or to a narrow band of distributary springs. Common caves display chiefly the branchwork forms in Palmer's (1991) general model for the plan patterns of all genetic cave types that is reproduced in 3.4 Figure 10 (p. 89).

Angular connections dominate branchworks where joints and faults are the principal structural guides of the conduits. Curvilinear connections dominate where the conduits develop primarily along bedding planes. Many cave systems display mixtures of the two patterns. The "irregular networks" and "anastomotic mazes" shown in 3.4 Figure 10 tend to occur as subsidiary components of the branchwork patterns, usually at their upstream ends where flash floods from surface channels can create extreme but short-lived hydraulic gradients.

Worthington (1991) analyzed 96 published cases where active river-cave passage systems have been explored and mapped along their full length from input points to output points. He found that their index of sinuosity (measured length along the passages divided by the straight line distance from input to output) fell between 1.4 and 1.7 in most instances. This is quite low, indicating that the conduit systems tend to evolve with fairly direct connections between head and mouth, like most surface river channel systems.

In a fully developed branchwork cave, the dissolutional conduits occupy only a tiny proportion of the total length or area of penetrable fissures that is available to the groundwater. The rules that govern the selection of the successful linkages that will be enlarged into the branchwork pattern are investigated in the series of models presented in Chapter 4.2 and need not be repeated at length here. A most important feature of the dynamic behavior is the "breakthrough" when a dissolutional protoconduit approximately one centimeter in diameter reaches an output boundary such as a spring or an earlier cave passage. This may abruptly drop the resistance to flow along its line by one or more orders of magnitude because nonlinear laminar or turbulent flow can commence (Ewers, 1982; White, 1988), thereby accelerating the dissolutional enlargement of the conduit. In the most simple case of propagation (a single groundwater input to a fissure), competing protoconduits form distributary patterns out from the input point by exploiting the wider pores and throats in the fissure, until breakthrough creates one winner or principal tube.

Where, instead, there are several competing inputs, one of them will get ahead, and upon breakthrough, the piezometric surface is drawn down above it, reorienting the local hydraulic gradients toward it. Nearest neighbor inputs then connect to it, reorienting the gradients again and drawing in further neighbors. In General Systems terminology (von Bertalanffy, 1967), common caves evolve as "cascading systems," one breakthrough (or energy cascade) reorienting the local water table and so triggering the next cascade. The problem for the speleologist is to decide which particular inputs will win or where exactly the breakthroughs will occur, i.e. to predict the actual patterns of cave passages that will emerge. At present this is infeasible in most real instances.

The most complex patterns arise where many potential groundwater input points are spread over all the area of the evolving karst basin. This is the case in most areas of holokarst. These range in scale from local limestone pavements (clints-and-grikes) occupying a few hundred square meters to extensive sinkhole plains such as those of Indiana, Kentucky, and Tennessee (U.S.A.).

The structure of Mammoth Cave, Kentucky, U.S.A., is analyzed as an example in Section 5.3.8. It is the lengthiest regional aggregation of known and mapped conduits and shafts (>600 km). Note, nevertheless, that the map has many gaps, indicating that information about the conduit system remains very incomplete, especially if we compare it to the information contained in maps of surface river patterns at the same scale.

It must be emphasized that the comparatively simple patterns of cave conduits constructed in all of the models are liable to major distortions caused by many different factors. These include:

- Local lithologic and structural controls such as perching on a chert bed or deflection down the axis of a syncline.
- Great differences in the quantities of water supplied to different inputs in a group, for example, when allogenic flow to a stream-sink is compared to seepage into nearby holokarst dolines.

Pattern building is analogous to the development of dendritic river channel networks on the Earth's surface. However, there is this very significant difference. The ideal river pattern is wholly stochastic, the individual channel being initiated at some random point on a surface, from which it can extend upslope and downslope until it chances to encounter another channel segment and so amalgamates. In a certain sense, the location of initial input points in a karst branchwork can also be considered to be stochastic. But the successive linkups of the principal tubes are deterministic, because they are directed by local reorientation of the water table following breakthrough.

Development of the Long Sections (Length x Depth) of Unconfined Caves

Lowe (Chapter 2.2) and White (Chapter 2.3) have reviewed the evolution of ideas up to the 1960's concerning the development of cave long profiles. The models advanced between (roughly) 1900 and 1950 are quite a striking example of the evolution of scientific thought. Most investigators were concerned to determine what is (or should be) the principal locus of cave genesis in relation to the water table. By the close of this period, almost every alternative possibility seems to have been advocated by one group or another.

Vadose proponents held that caves of enterable size develop chiefly above the water table, because the dissolutional attack is freshest there (closest to the inputs) and groundwater flow is fastest. In deep phreatic or random-depth hypotheses, caves essentially follow the streamtubes in a Darcian flow net below the water table, although adapting to local lithologic or structural controls. As a third alternative, the master caves follow the water table itself, because that is the most direct route between the more diffused flow in the vadose zone and the outlet springs. In 5.3 Figure 1, I attempt to summarize these conflicting models of what cave systems should look like.

In reality all of these types of caves (plus many variants and mixtures of them) can be found in the unconfined settings, although in any given geographic area just one or a few of them tend to be predominant. I offered a resolution of the problem with the pair of conceptual models shown in 5.3 Figures 2 and 3. They are the product of several iterations (Ford, 1968, 1971; Ford and Ewers, 1978; Ford and Williams, 1989, p. 261 and following) and are intended to apply to those aquifers in limestone, dolostone, gypsum, and anhydrite in which fissure porosity and transmission is predominant before the integrated dissolutional cave systems develop. It is important to appreciate that normally the water table in such rocks has not stabilized with a low gradient to a spring before the cave has begun to develop. At the outlet boundary, its altitude may be determined by some external control such as a river channel in the floor of a valley or an impermeable barrier of insoluble rocks, but behind this position it responds to the evolving three-dimensional geometry of the cave system, because that is the master plumbing draining the aquifer.

Determining the Development of Phreatic or Water Table Caves

The four basic types of phreatic or water table caves that (I contend) can be predicted to occur in the "Four State Model" in Ford's discussion in Chapter 5.3.3 are shown in 5.3 Figure 2.

- State 1—Bathyphreatic.
- State 2—Phreatic with multiple loops.
- State 3—Mixture of phreatic and water table.
- State 4—Water table.

Bathyphreatic. First, where the fissures that can be penetrated by flowing water under the prevailing conditions are very widely spaced, cave systems may be compelled to follow deep loops below the spring levels, because no shallower routes are open. Bathyphreatic cave systems are formed as a consequence. The maximum depths they attain in unconfined settings are not established but are known to exceed 300 meters below low stage (dry season) water tables (see Chapter 5.3.1).

Multiple-Loop Phreatic. These caves (5.3 Figure 2B) are created where there is a significantly higher frequency of penetrable fissures. Note that the altitude of the tops of the higher loops in the model fix the stable position of the water table and *not vice versa.* Hölloch, Switzerland (Bögli, 1970), is an example where the vertical amplitude of the looping is about 100 m, increasing to 180 m during high floods. It follows one principal bedding plane in the strike direction. Other caves loop down the stratal dip and back up joints or faults, as in 5.3 Figure 2B (and see Chapter 5.3.2). Rossi et al. (1997) describe examples from the Cantabrian Mountains, Spain, where the looping is up the dip and down the joints, which is comparatively unusual.

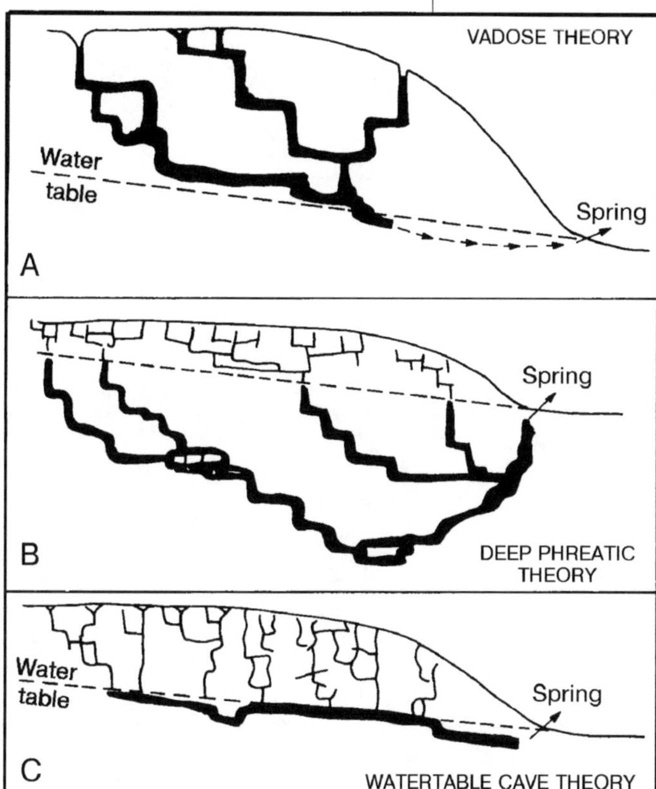

5.3 Figure 1. Schematic diagram to illustrate the vadose, deep phreatic, and water table theories of common cave genesis.

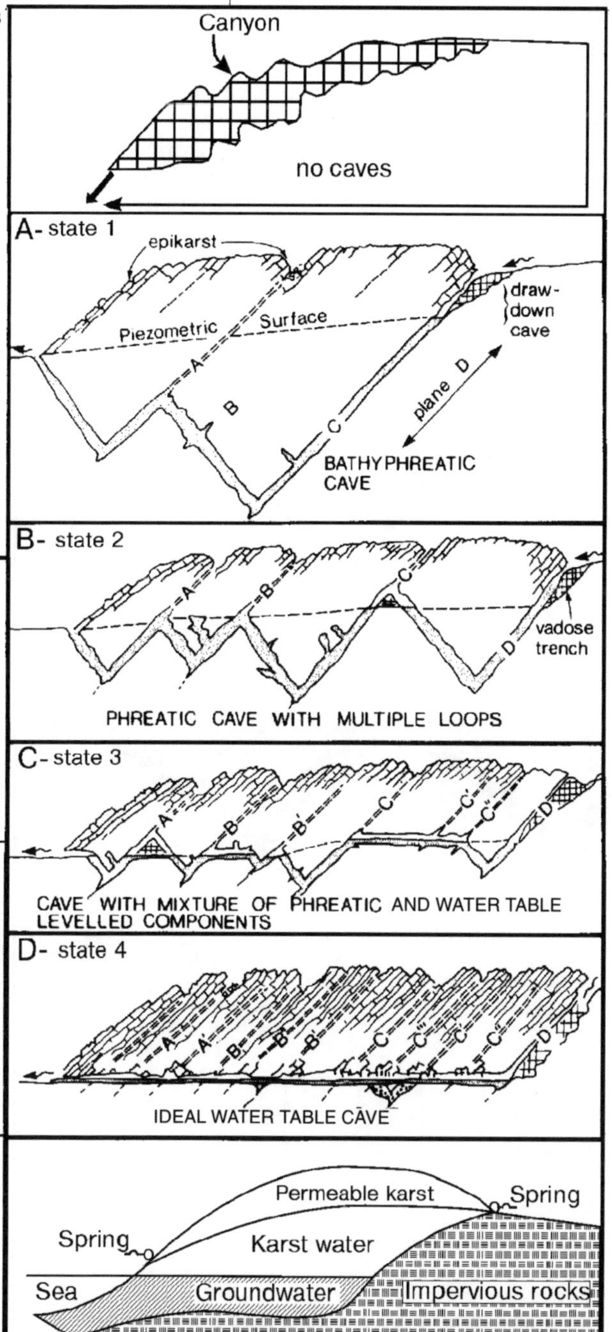

5.3 Figure 2. The "four state model" of cave system geometries (Ford, 1971) amplified to include top, *State Zero, where fissure frequency or aperture is too low to permit cavern genesis, and* bottom, *State 5, where fissure frequency or matrix porosity are too high to permit caves to develop to an enterable scale.*

Mixture of Phreatic and Water Table. Caves that are a mixture of shorter shallower loops and quasihorizontal canal (i.e. water table) passages represent a third, still higher state of fissure frequency and diminishing resistance. In steeply dipping rocks, the horizontal segments tend to follow major joints or to propagate along the strike in bedding planes. The Swildon's Hole–Wookey Hole System in the Mendip Hills, England, is an outstanding example. Exploration from its upstream end has passed eleven shallow phreatic loops separated by canal sections and is arrested at a twelfth, deeper loop. At least eight loops have been overcome from the downstream end (Wookey Hole), where exploration is currently halted at -80 m in a ninth and deeper loop.

Water Table Caves. In State 4 (5.3 Figure 2D) the frequency of penetrable fissures is so great or the resistance is so low that low gradient very direct routes to the spring outlets are constructed at shallow depths below the early water table. As they enlarge, the water surface is lowered into them, although they may continue to flood during storms or wet seasons. Such "ideal water table caves" that consist of wadeable or swimmable canal passages with low roofs are very common. Hundreds of short examples pass through residual limestone towers on alluvial plains in southern China, for example.

It must be stressed that the models in 5.3 Figure 2A-D are highly simplified. They are drawn to show caves draining down the dip in steeply dipping rocks, because that is the most simple means by which to illustrate the relationships of loops and canals to each other and to the geologic structure. Many great caves drain principally along the strike, however, or are mixtures of dip and strike sections. Some drain counter to stratal dip, as noted in the Cantabrian examples. They demonstrate the same basic range of geometries. Where strata dip as steep as 2 to 5° or more, the bedding planes tend to entrain groundwater to great depth. Their gradients are steeper than most initial hydraulic gradients, so there is a quasi-artesian trapping effect, as Glennie (1954) noted. As a consequence, the deeper State 1 and 2 types of caves are favored.

In contrast, where strata are nearly horizontal, individual bedding planes can extend to the perimeter of the karst, where they crop out to offer many potential spring outlets. In addition, such conditions are also most suitable for perching groundwater streams on aquitards such as dolomite beds and thick shale layers. As a result, State 3 and 4 caves are particularly favored in near-

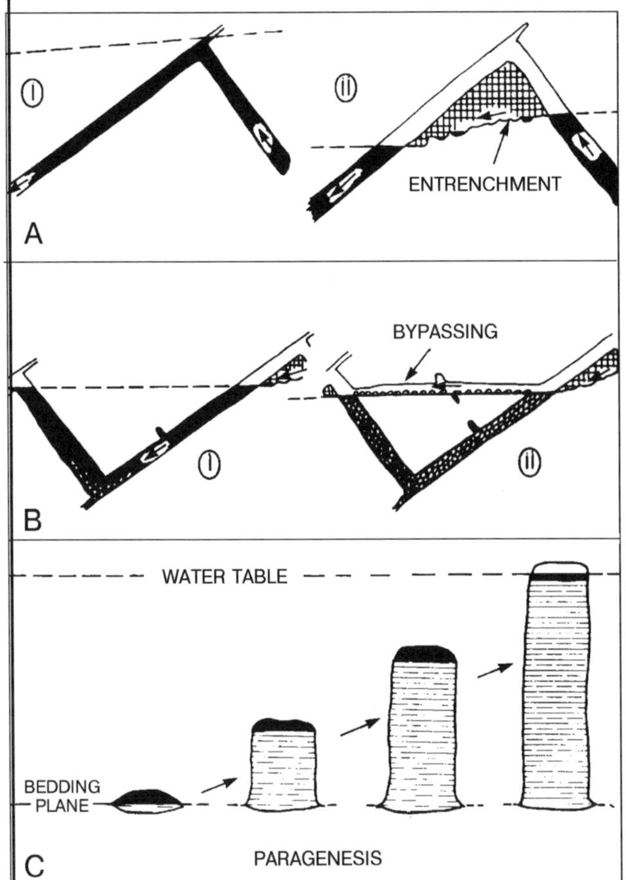

*5.3 Figure 3. Gradational features common in phreatic caves: **A** is isolated vadose entrenchment of the upward apex of a phreatic loop. **B** is development of a bypass tube above a loop clogged with detritus. **C** is development of a paragenetic gallery upward to the water table.*

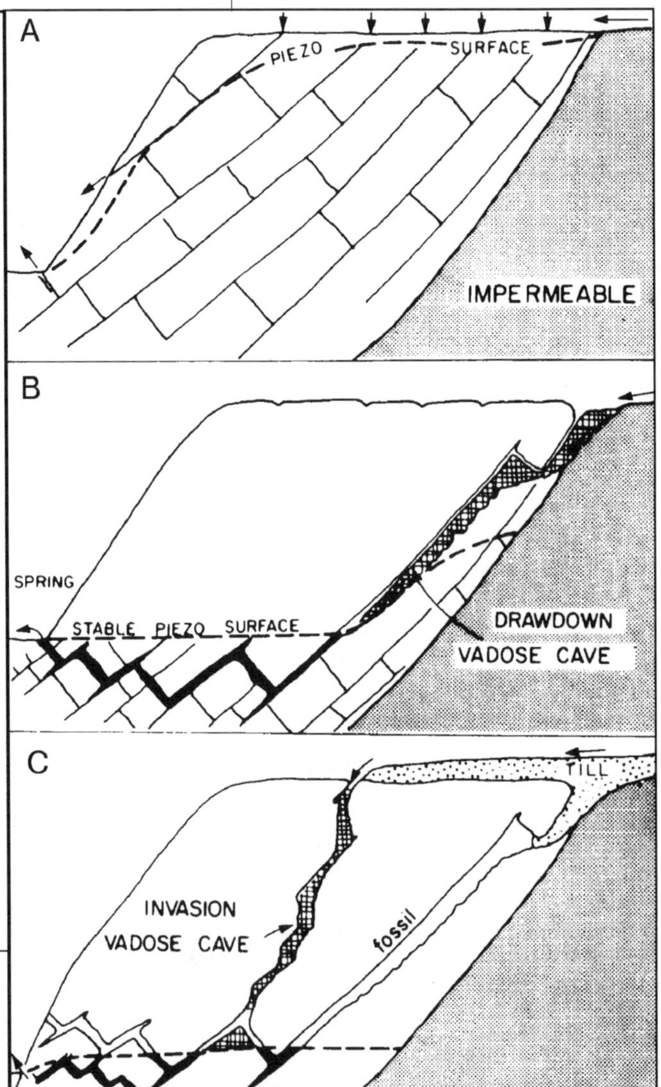

5.3 Figure 4. The development of drawdown and invasion vadose caves.

horizontal situations. In many ancient reefal rocks and some other geologic settings, bedding planes are not significant structural guides of groundwater flow, but the patterns of intersecting joints and faults produce the same range of cave types.

It is not possible to correlate these four basic phreatic and water table states to precise global values for the frequency of penetrable fissures. Even where that frequency is very low, a few fissures that are interconnected, are favorably oriented to link the potential sinkpoints to springs, and have exceptionally wide apertures, may host the ideal type of water table cave or an intermediate state. There is a State 0, however, where frequency and aperture are both too low to have permitted caves (of explorable dimensions, at least) to have developed in the available geologic time (5.3 Figure 2 top). This is often the case in arid regions (insufficient water), in very flat lowlands (insufficient hydraulic head), and in many marbles where metamorphism has sealed the bedding planes and other early fractures.

At the other extreme, the very high effective porosity that occurs in many medium to thinly bedded rocks that are densely intersected by joints (3.2 Table 2, p. 58) or in young and diagenetically immature limestones may lead to conditions of highly diffused groundwater flow. In these conditions, the dissolutional attack is too dispersed to generate caves of enterable size along any one flow route.

This condition ("State 5") is suggested in the bottom drawing in 5.3 Figure 2. This is a reproduction of one of the "classic" models of karst groundwater organization, that of Grund (1903). State 5 occurs in chalk and on coral reefs, although in the latter, channeled streams that impinge from other rocks may extend caves through them (normally of State 4).

Modifications of the Phreatic States
The four types of geometry themselves are not exclusive. Many large cave systems display mixtures of State 2 and 3 or of State 3 and 4, especially where these are also of multiphase origin (see below). In addition, the early phreatic loops can experience significant modification as a cave enlarges and matures. The principal types of modification are shown in 5.3 Figure 3.

Isolated vadose entrenchments (5.3 Figure 3A) develop where as a result of water table lowering the apices of upward loops become drained. Entrenchments more than 50 m deep and hundreds of meters long are known, e.g. the great river passage of Skocjanske Jama, Slovenia (Chapter 5.3.13) seems to be at least partially of this origin.

Bypass passages (5.3 Figure 3B) develop where sharp downward loops below a water table become obstructed by alluvial detritus. If rapid flooding occurs, a big head of water may build above the obstruction, greatly steepening the hydraulic gradient across the loop. Local fissures that are impenetrable under normal gradients may then be exploited to open one or a series of subparallel bypasses over the loop.

As originally used by Renault (1968) a *paragenetic passage* is any phreatic or water table passage where the erosional cross section is partly attributable to effects of accumulations of fluvial detritus, which shield the floor and so direct more dissolution at the ceiling (*erosione antigravitativa,* Pasini, 1975). Later authors have applied it chiefly to the situation of a steadily rising dissolutional ceiling, as in 5.3 Figure 3C. The vertical amplitude of such paragenesis can exceed 50 m, and remarkably flat roofs can be beveled across steeply dipping strata. The process terminates at the water table.

These processes act to smooth the long profiles of phreatic caves during their active lifetimes. State 2 systems may be modified to State 3 geometry, and State 3 to State 4.

Types of Vadose Caves — Drawdown and Invasion
The two fundamental types of vadose caves are illustrated in 5.3 Figure 4 (p. 322), *drawdown vadose,* and *invasion vadose*.

Drawdown Vadose Caves. When first exposed to erosion, the effective porosity in many karst areas is very low. This is especially true where strata had previously been deeply buried. As a consequence, the early water table may be close to the surface and high above the level where mature springs will later be found (5.3 Figure 4A). Early conduit development thus takes place under phreatic conditions. As the passages integrate and enlarge, they progressively drain the aquifer toward its minimum altitude, the stable water table level. The passages in between are of the *drawdown vadose* type (5.3 Figure 4B). They have a phreatic initial form (that may be preserved in the roof) and a vadose entrenchment beneath it. Although the trench can represent 90 plus percent of the cross section of the passage, the plan pattern and the skeleton of the cave are largely phreatic in origin.

Invasion Vadose Caves. These form where streams are able to enter or to collect in a rock that is already in a vadose condition (5.3 Figure 4C). The initial erosional morphology in the cave has a rough imperfect phreatic form, termed *paraphreatic* (lacking the smooth surfaces and rounding of well developed phreatic surfaces). This is usually created under flood conditions. The great systems of fluted vertical shafts connected by tortuous meandering canyons in their floors that are characteristic of high mountain settings are usually examples of this type of cave. The existence of a deep vadose zone at the start of the development of a particular cave may be due to initially high effective porosity in the rock or to the fact that it has been drained by an earlier generation of drawdown vadose caves.

Multiphase (or "Multilevel," -"Multistage") Caves

The great majority of the known large common caves seem to be multigenerational systems. They display different tiers or levels of major conduits. These have been created in response to changes in the altitude of the springs. They are separated by zones of rock containing much smaller volumes of passages or shafts that serve to connect the tiers. "Level" is the term most often used in the English literature to denote an individual tier. But it is often not strictly level (i.e. horizontal) but has a State 2 or 3 geometry. "Multiphase" or "multistage" should be used instead of "level," to indicate that the cave has two or more successive generations of passages.

Normally, the succession will be due to lowerings of the level of springs as a consequence of erosional base level lowering around the margins of the karst. Upper galleries are progressively abandoned by the streams that generated them until they become completely relict. Of course, there are also instances of the converse. That is, cave systems being reconstructed progressively above one another where base level is rising because the land is sinking or the sea is rising. Many such caves are beginning to be explored in the limestones of the Caribbean. Here sea level has repeatedly risen and fallen 100+ m during the past 1 million years or more in response to the ice ages. However, they are difficult to explore and map accurately. They are water-filled today and commonly obstructed by sediment. Multiphase caves created by the lowering of springs dominate in most regions.

In many multiphase systems, the lower younger main galleries are close to or directly beneath the older passages they replace. But in others the new galleries may follow new directions into previously noncavernous rock. This usually occurs where a major shift in the geographic location of the springs accompanies their opening at a lower altitude. Reorientation of the trend of the main drainage conduits can be 90° or more. It tends to be greatest where stratal dip is gentle (as at Mammoth Cave, Kentucky; Chapter 5.3.8 in this volume) or where tectonic movements or glacial action shift the springs from one valley to another. In the majority of instances, a new generation of main galleries is carved piecemeal by progressive captures from the previous generation that work their way upstream from the springs.

These captures, termed "diversion passages" (Ford, 1968), and "undercaptures" (Jeannin and others in Chapter 5.3.3) are responding to the cascade of local reorientations and steepenings of the hydraulic gradient in the same manner as the integration of different input passages in a first generation that is discussed above and in Chapter 4.2. Their plan patterns can become quite complicated, because only some portions of an individual diversion may be occupied and enlarged by the new generation of main galleries that is evolving, the other portions being abandoned in favor of the next diversion and left as relicts in the walls, perhaps too small to enter. An example is shown in 5.3 Figure 5 from Swildon's Hole, England. Here at least eight different undercaptures (dashed lines) from two older main passages were required to construct approximately 1 km of new, "lower level" master passage.

Conclusion

The concepts of plan pattern construction, phreatic, water table or vadose state, and multiphase development of common caves are illustrated in the case studies that follow this chapter. They are organized broadly to begin with examples of comparatively simple deep phreatic and multiloop systems (El Abra, Mexico, and Castleguard Cave, Canada), proceeding to large and complex multiphase systems such as North of Lake Thun, Switzerland, and Mammoth Cave, Kentucky), to representatives of mixed vadose and phreatic development in mountainous regions (Alps, Pyrenees, Mexico) and where there is strong lithologic or structural control (Appalachians, south Spain). The set concludes with two instances of nearly ideal water table cave development (Belize and Hungary) and a review of the latest models of speleogenesis from the region where modern karst studies began, the classic karst of Slovenia and Trieste.

References

Bitterli, T., and Jeannin, P.-Y., 1997, Entwicklungs-geschichte der Hohlen im Gebiet Hohgant-Sieben Hengste-Thunersee (Berner Oberland, Schweiz): *12th International Congress of Speleology,* La Chaux-de-Fonds, Proceedings, v. 1, p. 349-354.

Bögli, A., 1970, Le Holloch et son karst: Neuchâtel, Editions la Baconnière, 100 p.

Ewers, R.O., 1982, *Cavern development in the dimensions of length and breadth:* McMaster University, Ph.D. thesis, 398 p.

Ford, D.C., 1968, Features of cavern development in central Mendip: *Cave Research Group of Great Britain Transactions,* v. 10, p. 11-25.

Ford, D.C., 1971, Geologic structure and a new explanation of limestone cavern genesis: *Cave Research Group of Great Britain Transactions,* v. 13, p. 81-94.

Ford, D.C., and Ewers, R.O., 1978, The development of limestone cave systems in the dimensions of length and breadth: *Canadian Journal of Earth Sciences,* v. 15, p. 1783-1798.

Ford, D.C., and Williams, P.W., 1989, *Karst Geomorphology and Hydrology:* London, Unwin Hyman, 601 p.

Glennie, E.A., 1954, The origin and development of cave systems in limestone: *Cave Research Group of Great Britain Transactions,* v. 3, p. 73-83.

Grund, A., 1903, Die Karsthydrographie: *Pencks Geographische Abhandlung,* v. 7, p. 103-200.

Martel, E.A., 1921, Nouveau traité des eaux souterraines: Paris, Editions Doin.

Palmer, A.N., 1991, Origin and morphology of limestone caves: *Geological Society of America Bulletin,* v. 103, p. 1-21.

Pasini, G., 1975, Sull'importanza speleogenetica dell' "erosione antigravitativa": *Le Grotte d'Italia,* v. 4, p. 297-326.

Renault, H.P., 1968, Contribution a l'étude des actions mechaniques et sedimentologiques dans la speleogenese: *Annales de Spéléologie,* no. 22, p. 5-21, 209-267; no. 23, p. 259-307, 529-596; no. 24, p. 313-37.

Rossi, C., Munoz, A., and Cortel, A., 1997, Cave development along the water table in Cobre System (Sierra de Penalabra, Cantabrian Mountain, N. Spain: *12th International Congress of Speleology,* La Chaux-de-Fonds, Proceedings, v. 1, p. 179-182.

von Bertalanffy, L., 1967, *General Systems Theory; formulations, development, applications:* New York, G. Braziller, 289 p.

White, W.B., 1988, *Geomorphology and hydrology of karst terrains:* Oxford University Press, 464 p.

Worthington, S.R.H., 1991, *Karst hydrogeology of the Canadian Rocky Mountains:* McMaster University, Ph.D. thesis, 370 p.

Worthington, S.R.H., and Ford, D.C., 1997, Borehole tests for megascale channeling in carbonate aquifers: *12th International Congress of Speleology,* La Chaux-de-Fonds, Proceedings, v. 2, p. 195-198.

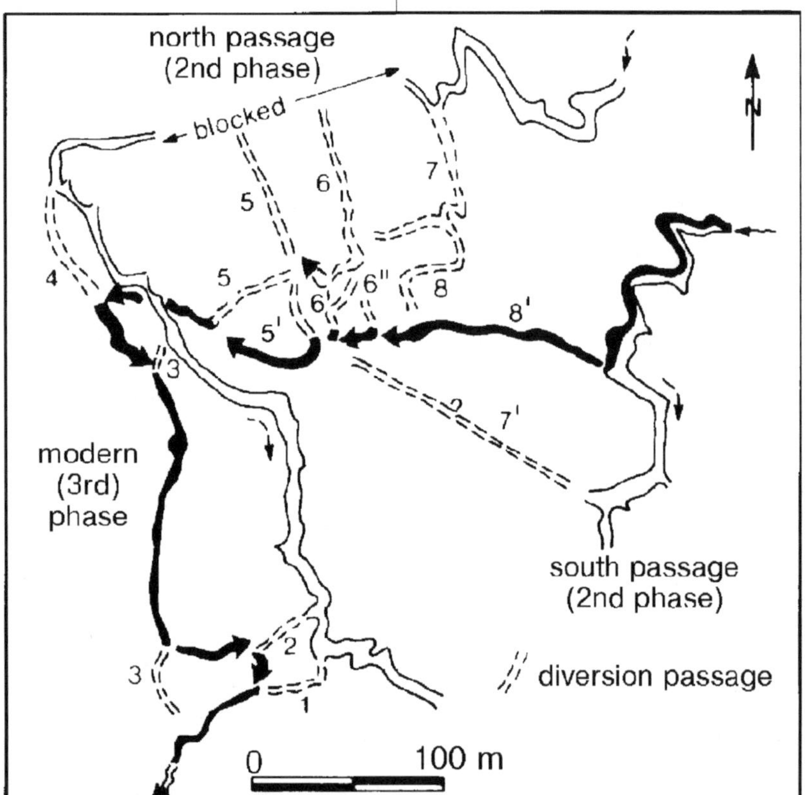

5.3 Figure 5. The central galleries of Swildon's Hole, Mendip Hills, England. Shown in solid white are the north and south passages of a second genetic phase. In black are passages of the third (modern) phase. Numbered dashed lines are a headward-working sequence of diversion passages by which the modern route was built from that of the second phase.

5.3.1. Deep Phreatic Caves and Groundwater Systems of the Sierra de El Abra, Mexico

Derek Ford

Abstract

The Sierra de El Abra, Mexico, comprise reef and back-reef strata formed during the middle of the Cretaceous. These strata were buried by clastic rocks, then uplifted, and progressively exhumed by erosion during the Cenozoic. The exposed limestone is now a mountain range about 100 km long and 2 to 6 km wide rising abruptly 400 to 650 m above the Gulf of Mexico coastal plain. West of it, the Sierra Madre Orientale contains higher limestone ranges that supply aquifers confined below the uneroded clastics. The climate is tropical wet-dry with 1000 to 2000 or more millimeters annual precipitation, depending on the altitude. Groundwater flows to a few major springs at the limestone-clastic contacts. They are mixtures of local bicarbonate water (confined and unconfined) with sulfate water estimated to circulate to depths greater than 1000 m. Base-flow discharges are about 1 m^3/s, and floods can exceed 200 m^3/s.

Explored caves in the El Abra Range are of three types:

- Swallet caves draining streams from the surviving clastic cover that display shaft-and-drain morphology with superimposed floodwater features (such as mazes) and terminate downstream at siphons or other obstructions.
- Shafts descending to water table lakes that may rise 30 m or more during rains.
- Phreatic lifting caves formed by water flowing along or up the sparse joints and bedding planes in the reefal rocks.

Relict examples have been explored to depths greater than 300 m and display both vertical and horizontal conduits more than 30 m in diameter. One example of a modern spring, Monte, has been dived to –270 m down a vertical joint fissure. Together, the known caves extend less than 10 percent of the straightline distance between major sinks and springs. In the unknown interval between, the caves are presumed to be deep phreatic loops that can collect the regional sulfate groundwater. Deep looping is a consequence of a very low frequency of penetrable joints, faults, and bedding planes in these limestones.

Introduction

The caves and groundwater systems of the Sierra de El Abra in Mexico are a good introduction to the variety of types to be presented in the series of chapters under Part 5.3, *Speleogenesis Under Unconfined Settings.* This is because the systems exhibit a mixture of unconfined speleogenesis with the confined speleogenesis illustrated in the chapters under 5.2. Important regional flows of sulfate groundwater (including thermal waters) with deep paths that are probably many tens of kilometers in length mix with local bicarbonate groundwaters as they approach springs. The topographic and geologic settings are spectacular, there are enormous discharges of floodwater through deep phreatic loops in reef and back-reef strata, and the form of the caves is attested to by a mixture of active and relict shafts and galleries at both the input and output ends of

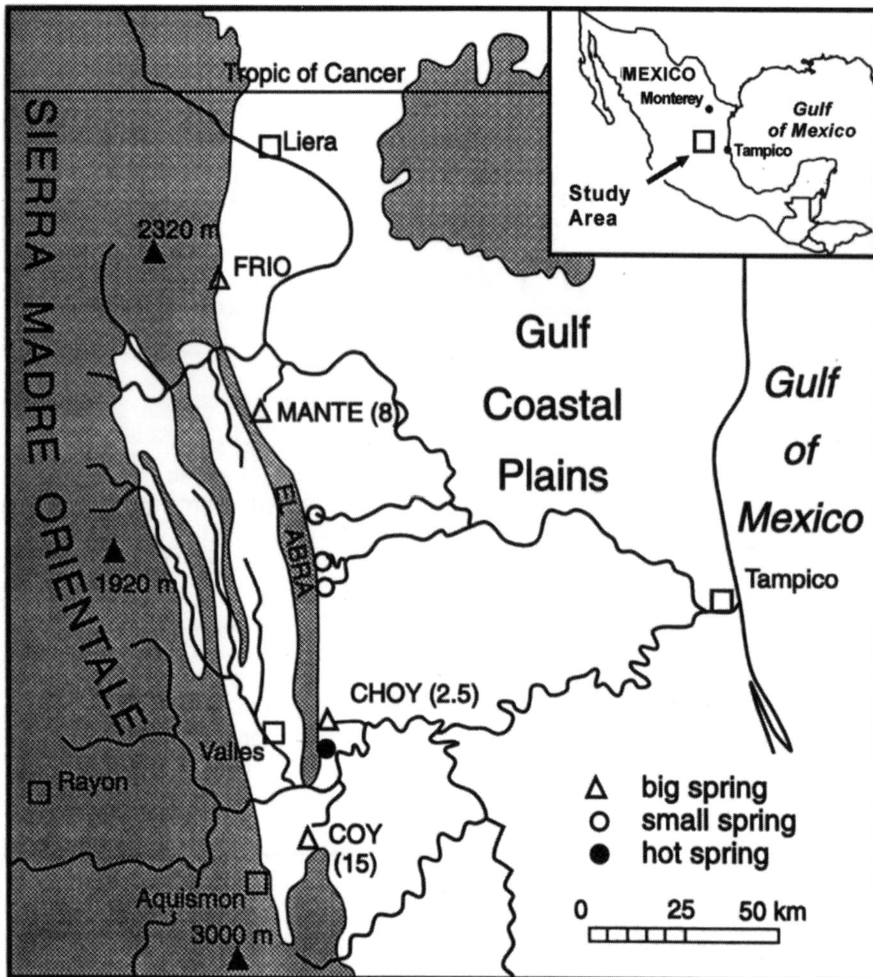

5.3.1 Figure 1. The location of the Sierra de El Abra and eastern parts of the Sierra Madre Orientale, Mexico, showing the largest regional karst springs and the hot spring at Taninul.

the systems. This report summarizes the findings of J.E. Fish, who studied the caves and karst of the El Abra for a Ph.D. thesis in the 1970s (Fish, 1977).

The regional topography is shown in 5.3.1 Figure 1. The Sierra de El Abra is the most easterly range of the highland core of Mexico, rising as a straight and steep front from the broad Gulf Coastal Plain. The range extends 100 km from a river gorge at its northern end to the town of Ciudad Valles at the south, where it dips beneath coastal plain sediment. It is 2 to 6 km in width and for most of its length has a smooth crest between 400 and 650 m above sea level. The altitude of the coastal plain along its foot is only 30 to 90 m. Along its west side the range drops away gently into a broad valley floored by shale. West of El Abra, the Sierra Madre Orientale is a mixture of mountain ranges, plateaus, and basins rising to 2000 to 3000 m. It is also composed predominantly of limestone.

The climate is tropical. Mean January temperature at Ciudad Valles is 20°C, increasing to 30°C in July. There is a pronounced wet season from June to October, associated with storms (including hurricanes) moving in from the Gulf of Mexico. Individual rainstorms can be very intense. There is a strong orographic effect; mean annual precipitation is less than 700 mm on the coastal plain, rising to an estimated 1200 to 1400 mm on El Abra and to 2000 mm or more on high ranges in the eastern Sierra Madre Orientale.

The Sierra de El Abra is covered by a dense thorn forest of drought-adapted species, including many cacti. There are only a few patches of soil of significant thickness. Elsewhere, the surface is limestone with karren pits and shallow grikes (Enjalbert, 1964).

Geology

A geologic cross section with the hypothesized regional groundwater flow added is given in 5.3.1 Figure 2 (from Fish, 1977). The Sierra de El Abra and Sierra Madre Orientale are part of a great carbonate platform of Cretaceous age, the "Valles-San Luis Potosi Platform," measuring 200 x 300 km (Carrillo Bravo, 1971). Basal red beds are overlain by the Guaxcama Formation (Early Cretaceous), which is a sequence of gypsum, anhydrite, brecciated dolostone, and limestone deposited behind barrier reefs. The formation is up to 2000 m in thickness. Above it are 1000 to 2500 m of the El Abra Formation (Early-Late Cretaceous), predominantly limestone.

In the Sierra de El Abra, the El Abra Formation comprises an eastern frontal reef 600 m in width, with back-reef beds to the west. The "reef" is a structure of rudist banks and bioherms, interreefal bioclastic debris, and carbonate mud. There are very large vugs, but replacement sparry calcite has essentially eliminated any intervug permeability. The reef is massive with just a few bedding planes that are mostly at facies contacts. There are sparse but deep vertical joints, and individuals have been traced to depths of more than 100 m and do not display any offset. The reef is at least 2000 m deep. The back reef limestone is calcarenite and calcilutite with some dolomite interbeds. Bedding is massive to indistinct near the reef, becoming more conspicuous, thinner, and flaggy to the west.

The carbonates were buried by a Late Cretaceous sequence of shale and argillaceous limestone. There was uplift with some folding and strong north-trending faulting in the early Tertiary (Laramide Orogeny). The Sierra de El Abra is believed to have been above sea level since that time and to have been progressively exposed by stripping of the weaker cover strata. The spectacular eastern reef front of the range is an erosional exhumation that may have been somewhat enhanced by folding or faulting. On the western flank of the range, river valley patterns superimposed from the cover have been partly disaggregated by capture underground but sufficient remains that storm runoff is channeled quickly into swallets, creating rapid flooding underground.

Karst Groundwater

This general region contains some of the largest karst springs that are known, including the Frio group, which discharged 700 m^3/s during a recorded flood in August 1951. The Sierra de El Abra is drained primarily by just two springs at the foot of the east front at the north and south ends of the range respectively. The north spring, Mante, is 80 m above sea level. It has a mean minimum flow of 8 m^3/s during the driest month (March), and its peak instantaneous discharge is estimated to be between 150 and 250 m^3/s. The water rises vertically to the spring from a

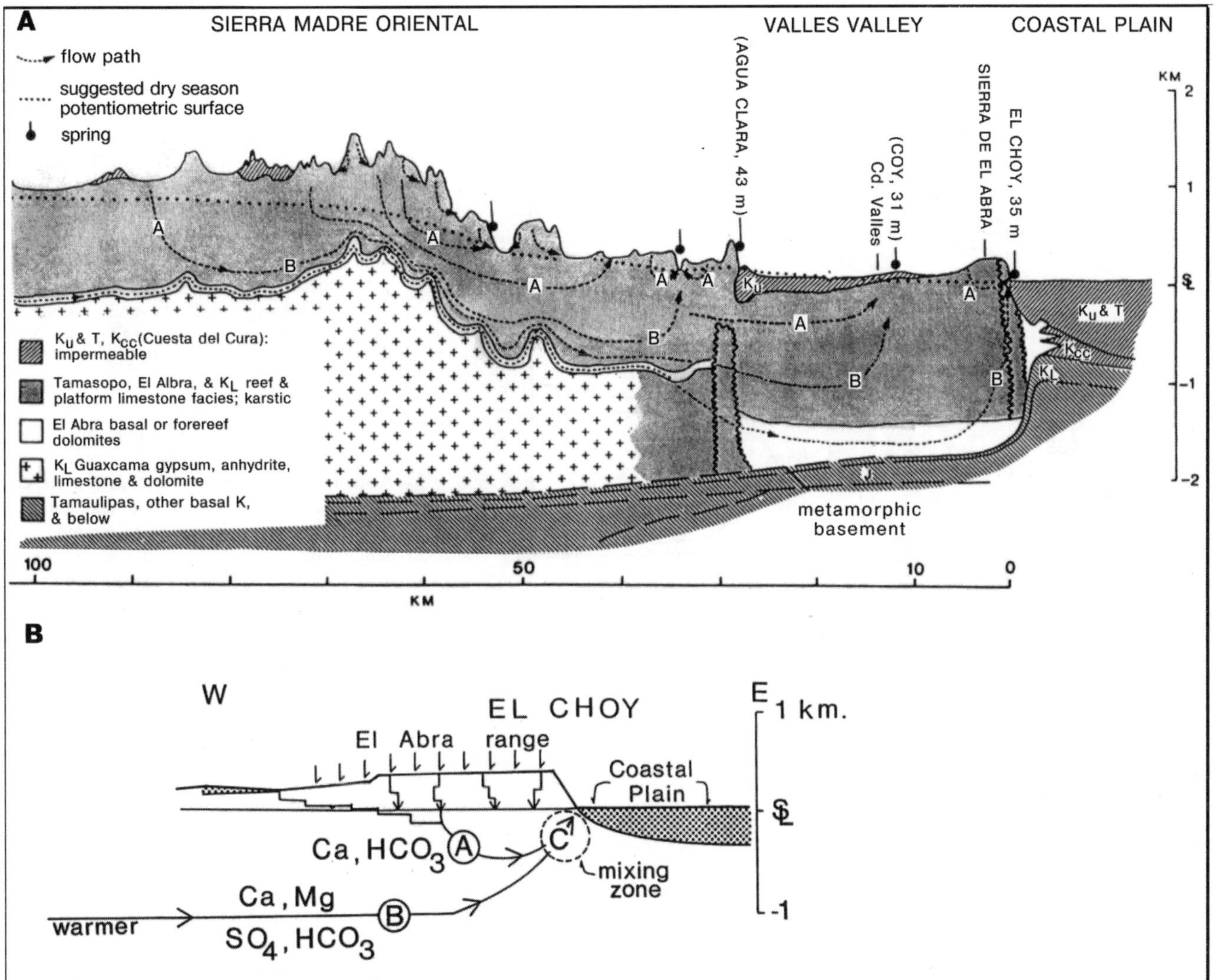

*5.3.1 Figure 2. **A** is a Geologic cross section from the Sierra Madra Orientale to the Gulf Coastal Plain showing hypothetical courses of Type A (shallow bicarbonate) and Type B (deep sulfate) karst groundwater and the El Coy and El Choy Springs. **B** is a mixing model for Type A and B groundwater beneath the Sierra de El Abra (both figures adapted from Fish, 1977).*

shaft that the late Sheck Exley dived to a depth of 270 m, noting that it continued straight down beneath him to the limit of his lights. The southern spring, Naciemiento del Rio Choy (or "El Choy") is 35 m above sea level and is obstructed by sand in its rise pool at a depth of 54 m. Its mean monthly minimum flow (March) ranged between 1.25 and 3.75 m^3/s during the period, 1961 to 1972. Its greatest recorded instantaneous discharge was 143 m^3/s (in 1967).

A small limestone hill protrudes through the coastal plain sediment 36 km south of El Choy, and from it emerges a much larger spring, El Coy (5.3.1 Figure 1). Its mean monthly minimum instantaneous discharge during 1954 to 1973 was as great as 24 m^3/s. El Coy is draining groundwater confined beneath coastal plain shale, but the magnitude of its discharge suggests that it is supplied by an efficient conduit! Despite its confinement and low altitude, it is a couple of degrees C cooler than the Mante or El Choy waters (5.3.1 Figure 3), which led Fish (1977) to propose that the bulk of its flow must derive from the very wet and cooler karst highlands of Xilitla, south of Aquismon (5.3.1 Figure 1; and see Worthington and Ford, 1995).

These three great springs each discharge a mixture of local karst groundwater with a bicarbonate composition (100 to 220 mg/L $CaCO_3$ under low-moderate flow conditions) that is termed "Type A" water in 5.3.1 Figure 2, and a regional sulfate water, "Type B." Fish (1977) developed a hydrochemical mixing model (based on Pinder and Jones' work (1969)) to show that 20 percent of the base flow at El Choy and 40 percent at El Coy is Type B water with calculated temperatures between 30 and 35°C. It is probably circulating to depths in excess of 1000 m below sea level. In floods the data indicate that its contribution is reduced to 1 percent or less. Type B water is believed to contact gypsum and anhydrite in the Guaxcama Formation beneath limestone of the El Abra Formation (5.3.1 Figure 2).

There has been one successful trace with lycopodium spores to El Choy from a swallet 23 km distant on the west flank of the El Abra Range. Flowthrough time was less than 92 hours during a period of moderate discharge, yielding a mean minimum velocity of 250 m/hour. This is in good agreement with the

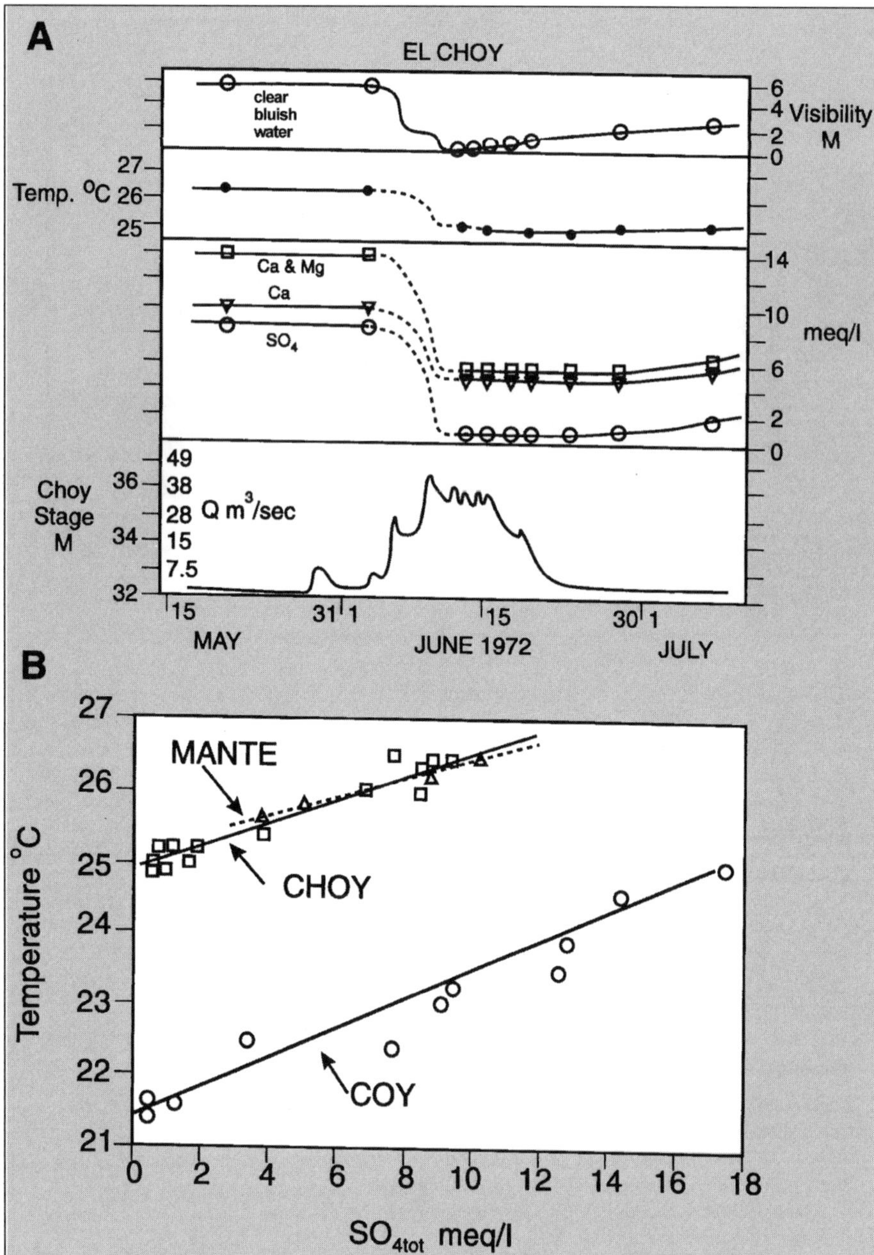

*5.3.1 Figure 3. **A** is the hydrograph record of a June 1972 flood at El Choy Spring, showing the responses of dissolved solids, water temperature, and turbidity. **B** shows the relationship between water temperature and dissolved sulfate concentration in the El Coy, El Choy, and Mante Springs (both figures from Fish, 1977).*

spring response to heavy rains. Hydrographs begin to rise within a few hours, and muddy water from the swallets appears after 4 days or so.

This rapid response is illustrated in the hydrograph and chemographs for a sequence of floods recorded at El Choy in June 1972 (5.3.1 Figure 3A). Mg^{2+} from deep dolostones and SO_4^{2-} from deep sulfates (i.e. Type B water) are diluted much more than Ca^{2+}, which the floodwaters are dissolving from the local limestone. The correlation between SO_4^{2-} concentration and water temperature at the three large springs is displayed in 5.3.1 Figure 3B. The two El Abra springs, Mante and El Choy, show identical behavior despite their considerable geographic separation. Floodwater reduces the temperature by 1.5°C and dilutes the SO_4^{2-} by a factor of ten. The El Coy water is always cooler because it originates at altitudes mostly above 2000 m in the Xilitla district. It is chilled and diluted by floods along the same trend as Mante and El Choy, but the SO_4^{2-} dilution factor approaches twenty.

As a footnote to the hydrogeology, it is interesting to consider the "Taninul Sulfur Pool." This rises at the foot of the El Abra front just 5 km south of El Choy Spring, at 39 m above sea level. It has a higher sulfate content than the large springs and a strong odor of H_2S. Its normal discharge is 0.1 m^3/s, at a temperature of 39 to 40°C, i.e. it is a "hot spring." It responds to flood events several days after they are registered at El Choy: discharge increases to 0.7 m^3/s or a little more, total dissolved solids are reduced and the temperature may fall to 30°C.

Taninul Sulfur Pool is mentioned to underscore the heterogeneity that occurs in conduit aquifers. The pool is tapping a deeper water source than the others and maintains a unique identity despite a large unconfined neighboring spring to the north and a very large confined spring to the south, both of which integrate local bicarbonate water and regional sulfate water.

Cave Systems of the Southern Sierra de El Abra

The hydrologic data demonstrate that the cave systems feeding the great springs must be large and efficient. In both geologically confined and unconfined situations, they are discharging bicarbonate floodwater rapidly and in great quantities. None of the active cave systems has been explored for as much as 10 percent of the distance between sinkpoints and springs. What is known and may be inferred will now be summarized. J.E. Fish limited his cave exploration, mapping, and monitoring to the southern section of the Sierra de El Abra because of many practical difficulties of access (Fish, 1977). All of these caves are probably within the catchment of El Choy Spring.

In 5.3.1 Figure 4 is a schematic illustration of the different types of caves and their setting within the El Abra Range.

Swallet Caves. Along the western flanks are a series of swallet caves that drain perennial streams from the shale or channeled storm water from the flaggy limestone. Their catchments are 4 to 22 km^2 in area. They are entered between 200 and 300 m above sea level, have surveyed lengths up to 8 km, and have depths to 150 m. They display much variety of form in detail due to local differences in the limestone lithology and structural controls, but all are vadose shaft-and-drain caves with many floodwater morphologic features. El Sotano de Jos (5.3.1 Figure 4) is one of the shorter examples. It terminates in a siphon that is believed to be perched. The other caves also terminate in

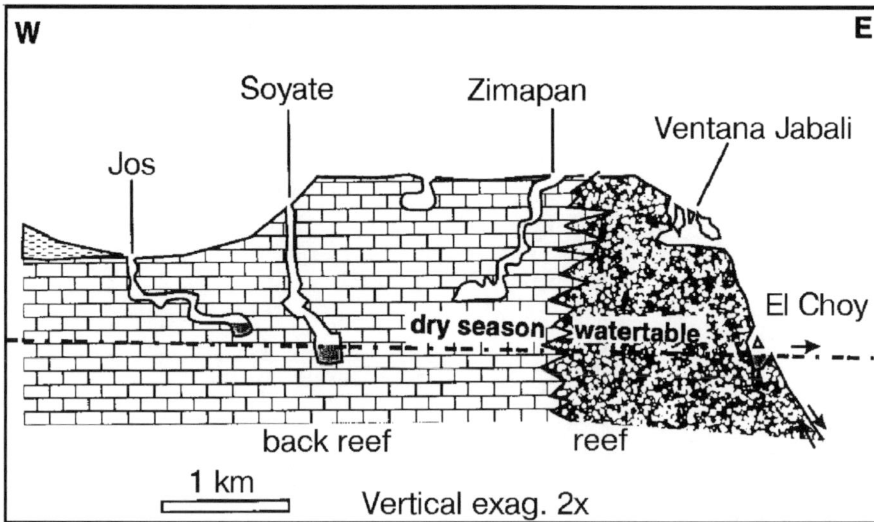

5.3.1 Figure 4. Schematic cross section of the southern Sierra de El Abra showing the relative locations of different types of caves described in the text (adapted from Fish, 1977).

perched siphons or blockages of floodwater sediment. Fish (1977) considers that the swallet caves are comparatively young and developed to drain to the rising cave at El Choy Spring.

El Sotano de Soyate (5.3.1 Figure 4) is a more enigmatic cave. The 200-m-deep entrance shaft does not take significant drainage and seems to have been truncated by the dissolutional lowering of the plateau. It displays deep pocketing in its walls that was interpreted (in 1977) as strong evidence for a phreatic origin. However, it is possible that it is a product of condensation corrosion (Chapter 3.6). A talus slope at the base of the shaft descends into a large lake chamber with blindfish (Mitchell et al., 1977). During the dry season the lake surface has been observed at 59 m above sea level, 25 m higher than El Choy Spring, which is 9 km to the southeast. It is probably at the water table. The lake has been plumbed to a maximum depth of 53 m. It may rise rapidly; one of my stage recorders fixed 11 m above the lake was inundated so quickly that differential pressure blew the thick glass window out of its tight rubber seal in the casing and into the body of the machine! Times of rise correlate with increasing discharge at El Choy, and flood markings suggest a maximum rise of about 30 m, creating a flood piezometric gradient of 6 m/km (Fish, 1977).

El Sotano de Soyate is now a hydrologic relict, a dipstick in the heart of the El Abra Range. Its paleohydrologic function is uncertain. Flow was most probably down through it, but whether this was in a phreatic, mixed, or wholly vadose mode is not clear.

La Hoya de Zimapan, 1 km west of the top of the east face, is a magnificent relict fragment of a deep phreatic cave (e.g. Ford and Williams, 1989, p. 264). The entrance passage is a gently sloping circular tube 30 m in diameter that has been intercepted by land-surface lowering. About 150 meters inside the cave it turns through 90 degrees and descends for 300 m in three large vertical shafts. At the base is a chamber measuring 90 x 225 m and up to 40 m in height.

The cave is partly filled by enormous stalagmites and has local breakdown. Otherwise, the morphology is entirely phreatic. It is presumed to continue somewhere below the floor of the great chamber, but any extensions are concealed by flowstone and some in-washed mud. There are no distinctive directional flow markings such as scallops (Chapter 6.1). This suggests that the groundwater velocity in Zimapan was low, at least during the later stages of its enlargement. In his Ph.D. thesis, Fish (1977) remained unsure whether flow was up or down the shaft. It is my opinion that it was probably upward. Zimapan was a phreatic lifting conduit discharging onto the paleocoastal plain in a manner similar to the El Coy Spring discharging upward onto the modern plain. The Zimapan recharge derived from the higher central regions of the El Abra Range to the north, which would be the first limestone areas exposed during the stripping of the clastic cover. The large horizontal chamber may mark corrosion notch dissolution at a fossil water table level during the abandonment of the high-level outlet, similar to the modern chamber in El Sotano de Soyate.

Ventana Jabali and El Choy Cave (5.3.1 Figures 4 and 5) are more straightforward examples of phreatic lifting outlet caves. They discharged water from the east front of the Sierra de El Abra onto the coastal plain at and above 350 and 35 m above sea level respectively. Both caves are within the massive impermeable reef facies and follow a few very large joints. The morphology is phreatic, with circular ascending shafts and deep wall and ceiling pockets. In Ventana Jabali the uppermost sections of the shafts have been truncated by dissolutional lowering of the plateau, about 200 m above the floor. There is very little breakdown. The bedrock floor of Ventana Jabali is buried by more than 7 meters of cave earth and guano, which was mined in the past. Groundwater probably flowed upward from a buried passage in the rear (west end) of the cave.

A profile of the modern spring cave at El Choy and a model for its development (from Fish, 1977) is shown in 5.3.1 Figure 5. It first discharged water when the coastal plain along the El Abra front was at or above 140 m above modern sea level. The plain has since been lowered more than 100 m by the differential erosion of the mechanically weak shale. As it extends for 110 km from the front to the sea, this implies a considerable volume of erosion. The upper outlets of El Choy are therefore ancient, and Ventana Jabali and La Hoya de Zimapan much older still. It is suspected that the latter are at least Miocene in age. They have retained their phreatic architecture with remarkably little damage since they were drained.

Impressive as they are, the caves that have been explored to date reveal only a tiny proportion of the active or former flow routes between the groundwater sink points and the springs. The form of the unknown systems of conduits in between remains uncertain. It is evident that many must have large cross-sectional areas and must make deep passes below the modern water table or its ancient equivalents. It is quite possible that they display State 2 multiple loop geometry (and probable that some modern upstream portions do so beyond the known limits of the swallet caves). But none of the characteristic features such as isolated vadose trenches have been found (see Chapter 5.3 in this volume).

Application of Occam's Razor (when in doubt, choose the simplest explanation) suggests that the El Abra Cave Systems be treated as single deep phreatic loops. The deep looping is a consequence of the very low frequency of penetrable bedding planes, joints, and faults in the reef and most of the back-reef facies of the El Abra Formation here.

There is little or no significant hydrodynamic difference between the technically unconfined systems discharging at El Choy and Mante on the one hand and the confined system discharging at El Coy on the other. In both situations, the deep loops also tap highly mineralized regional groundwater that contributes only a few percent of the annual discharge but may contribute more significantly to the initiation of conduits toward the springs by means of mixing corrosion.

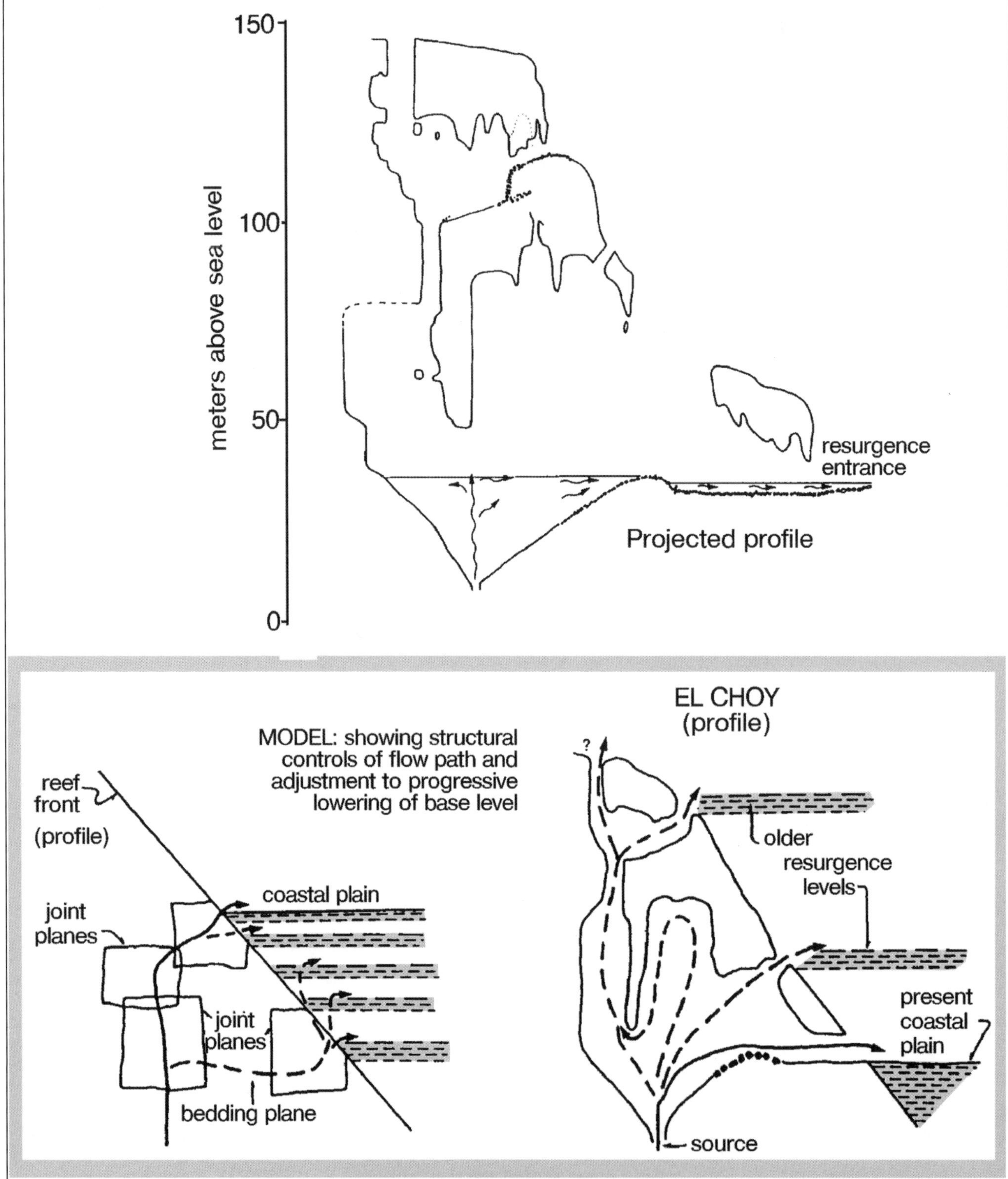

5.3.1 Figure 5. Topographic profile of El Choy Cave and Spring, with a model for its development.

Acknowledgements

Johnnie Fish's Ph.D. research in the Sierra de El Abra was funded by a Natural Sciences and Engineering Research Council of Canada grant to me.

References

Carillo Bravo, J., 1968, Exploracion geologica y posibilidades petroleros de la Plataforma Valles-San Luis Potosí (Sierra Madre Oriental-Altiplano Mexico), *in* Mesa Redonda no. 6: *Problemas de Exploracion en Areas Posiblemente Petroliferas de la Republica Mexicana,* paper no. 5, 20 p.

Carillo Bravo, J., 1971, La Plataforma Valles-San Luis Potosí: *Boletin Asociacion Mexicana Geólogos Petroleros,* v. 23, p. 1-112.

Enjalbert, H., 1964, Phénomènes karstiques Mexique et au Guatemala: *Bulletin, Association Geographie Francais,* 324-325, 30-58.

Fish, J.E., 1977, *Karst hydrogeology and geomorphology of the Sierra De El Abra and the Valles-San Luis Potosí region, Mexico:* McMaster University, Ph.D. thesis, 469 p.

Mitchell, R.W., Russell, W.H., and Elliott, W.R., 1977, Mexican eyeless Characin fishes, genus *Astyanax:* Environment, distribution, and evolution: *Texas Technical University Museum Special Publications,* no. 12, 89 p.

Pinder, G.F., and Jones, J.F., 1969, Determination of the ground-water component of peak discharge from the chemistry of total runoff: *Water Resources Research,* v. 5, p. 438-445.

Secretaría de Recursos Hidráulicos, 1969?, Datos Hidrometricos Cuenca del Río Panuco: *Boletin Hidrológico,* no. 32, sec. 1-2, (data through Dec. 1968).

Worthington, S.R.H., and Ford, D.C., 1995, High sulfate concentrations in limestone springs: An important factor in conduit initiation?: *Environmental Geology,* v. 25, p. 9-15.

5.3.2. Speleogenesis of Castleguard Cave, Rocky Mountains, Alberta, Canada

Derek Ford, Stein-Erik Lauritzen, and Stephen Worthington

Abstract

Castleguard Cave is in the Main Ranges of the Rocky Mountains of Canada. It is a relict upper level cave about 20 km in length formed in massive regularly bedded platform limestone of Middle Cambrian age. For a distance of 8 km it passes beneath Castleguard Mountain, with up to 800 m of Upper Cambrian-Ordovician cover rocks preserved above it today.

The cave is a good example of State 2 multiloop phreatic conduit development, with a vadose canyon entrenched in the upstream (higher) end of each loop. The looping in the headward and central sectors of the cave (7 km) is guided by one master bedding plane and long vertical fractures that intersect it. The master plane was slightly opened by crushing associated with differential slip of a few centimeters during tectonic uplift.

Downstream, a phreatic lift of 24 m conveyed the groundwater into a similar stratigraphically higher bedding plane that guides most of the passages there. When initiated, the cave may have been a single deep loop with a vertical amplitude of about 370 m. Once enlarged by dissolution and with stabilized springs, the greatest amplitude in the multiple loops was the 24 m required to gain the controlling bedding plane downstream.

The cave became a hydrologic relict more than 780,000 years ago but has since been invaded and modified by alpine and subglacial water on several occasions. Modern groundwater (including water sinking beneath the greatest icefield remaining in the Rocky Mountains) passes through one or more lower level cave systems that are inaccessible. The hydrologic behavior suggests that the morphology of these caves is similar to that of the known cave. The largest meltwater floods today impose a hydrostatic head of more than 300 m onto the base flow springs, temporarily rejuvenating the downstream end of the relict cave.

Introduction

Castleguard Cave is the longest cave currently known in Canada (20 km). It is also the foremost example anywhere of a cavern extending underneath a modern glacier. The cave and its local setting are shown in a schematic section in 5.3.2 Figure 1. The explored cave is now a hydrologic relict or upper level perched above one or more unknown lower levels termed "Castleguard II."

The cave displays many striking features of the interaction that can occur between glaciers and caves beneath them, plus a complex modern climatic regime, rich mineralization, and a troglobitic fauna—a fauna that has possibly survived one or more ice ages beneath deep ice cover in the heart of the Rocky Mountains. Details of these and other findings are given in a special issue of the journal *Arctic and Alpine Research* (Ford, 1983).

The focus in this chapter will be on the fundamental speleogenesis of the system. This will enable it to be compared to the other caves formed under unconfined conditions that are described here in Chapters 5.3 to 5.3.13. Two principal points are:

- Castleguard Cave is a good example of a meteoric water cave penetrating beneath a deep cover of nonkarst strata (it passes through the middle of a mountain).
- It is a very simple unequivocal instance of State 2 or multiloop phreatic cave development with accompanying drawdown vadose canyon entrenchment, of the type illustrated in 5.3 Figure 2 in Chapter 5.3 (p. 321) and 5.3 Figure 3A (p. 322).

The cave is located in the Main Ranges of the Rocky Mountains of Alberta and British Columbia, at lat 52° 10' N, long 117° 20' W. These ranges display a broad fold structure that is dominated by thick and resistant carbonate and quartzite formations of lower

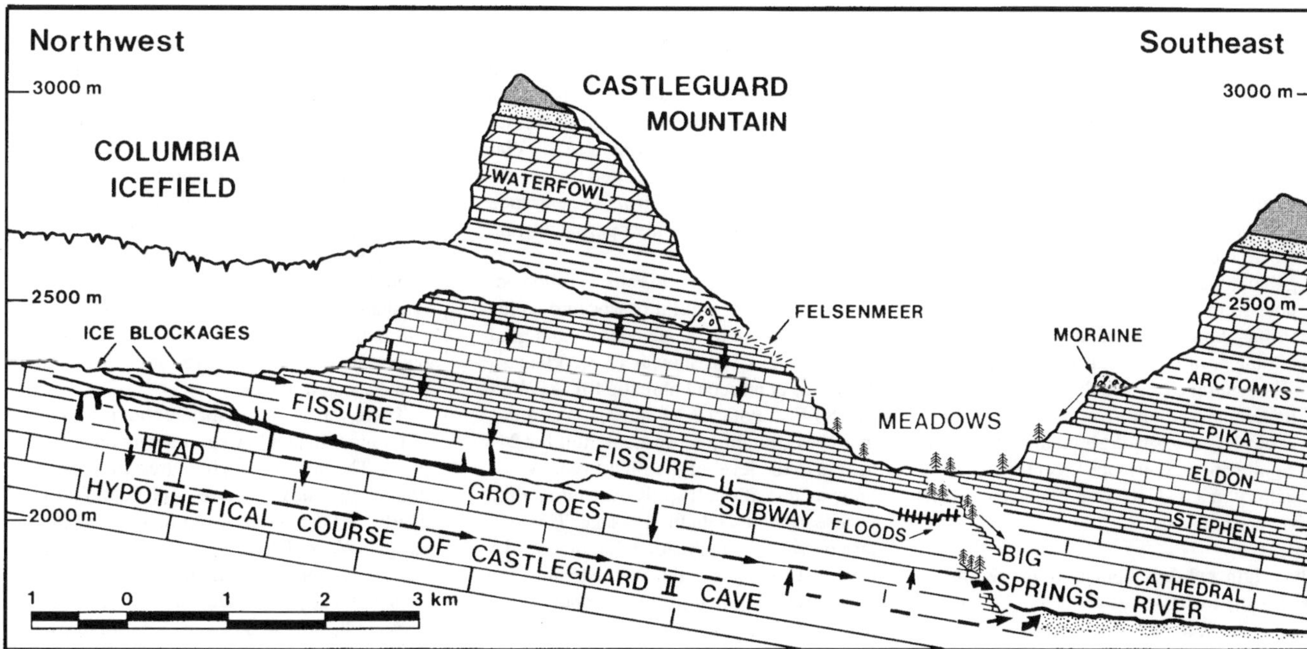

5.3.2 Figure 1. Schematic section through Castleguard Mountain, showing the geology and locations of Castleguard Cave and Castleguard II.

Paleozoic age. Altitudes extend from 1500 m (trunk valley floors) to summits at 3500 m or above. Modern mean annual temperatures are 0 to -14 degrees C across this high range, and the treeline is at about 2100 m.

These ranges support the largest remaining glaciers in the Rocky Mountains of North America. The Columbia Icefield (5.3.2 Figure 1) is the greatest of them, being 325 km^2 in area and averaging between 200 and 300 m in depth. Modern valley glaciers spill out from it for distances up to 10 km. At the peak times of the Quaternary glaciations, the icefield was much thicker, and all valleys were occupied with glaciers 400 m or more in depth. The mountains tops then stood out only as nunataks.

At Castleguard, the lithology is dominated by thick platform carbonates of Middle Cambrian age. At the base, the Cathedral Formation (>560 m) contains the cave and is composed of dark and dense crystalline limestone that is very regularly and massively bedded. At the surface, it can be seen that bedding planes that water can penetrate today are spaced 3 to 25 m apart, which is exceptionally wide by world standards. This very massive character determines that the cave passages have simple forms that are sustained for long distances.

Overlying the Cathedral Formation is the Stephen Formation, about 80 meters of calcareous shale with pyrite, shaly limestone, and some dolomite beds. The Cambrian carbonate sequence is completed by two more medium to massively bedded limestone and dolostone formations (Pika and Eldon) that are rhythmites of probable intertidal origin. They form the top of a broad strong plinth from which the summit mass of Castleguard Mountain rises (5.3.2 Figure 1) and contain some big and deep sedimentary dikes. The summit strata are weaker shale, sandstone, and dolostone of Ordovician age.

The mountain massif is a homoclinal block in which the strata dip regularly S-SE at 4-6.5 degrees. Along the southern edge, this is bounded by a sharp monoclinal flexure that trends SE and has dips up to 20 degrees S. This flexure evidently weakened the rock along its axis because the Castleguard River Valley, to which the cave and modern sub-icefield karst water drains, is entrenched into it. This flexure created the patterns of stressing and fracturing within the block that came to guide the cave, as described below.

On the surface limestone of the Cathedral Formation displays a full suite of well developed alpine karst forms such as karren, small dolines, and shafts. The Pika and Eldon strata have a much more restricted karst consisting of pitting, small grikes, and impenetrably narrow shafts. The shaly Stephen Formation does not display any surface karst features. Its rocks are essentially nonkarstic. It can be considered an aquitard ("leaky aquiclude" is perhaps a better description). Water that reaches it after sinking in the overlying Pika and Eldon karst is able to pass through it in a few places only, because the creation of Castleguard Cave in the upper Cathedral beds underneath it has placed it in the regional vadose zone, with an extreme hydraulic gradient. At the time that the cave was initiated, the Cathedral strata were in the phreatic zone, hydraulic gradients were lower, and the Stephen beds would have constituted an effective aquiclude. The early cave penetrated more than 9 km beneath this cover, broadly comparable to the penetration of Sierra Madre meteoric waters underneath coastal plain sediments to flow to the El Choy karst spring in Mexico, as described in the previous chapter.

Morphology and Genesis of the Cave

Explorers can enter Castleguard Cave only at its downstream end, which opens at an altitude of 2010 m in the north wall of the Castleguard River Valley, more than 300 m above the valley floor. From there the cave is followed northwest and upstream, ascending 386 m at generally easy gradients until it terminates at the bottom of the Columbia Icefield. In plan view (5.3.2 Figure 2), the cave has three distinct sections:

- A Headward Complex of inlet passages at the NW end, which is something of a maze because of repeated blockage and rerouting of sinking waters by glaciers in the past. The passages are abandoned today, plugged by ice or glacial detritus. Invasion vadose shafts have intersected some of them. The shafts are up to 70 m in depth, blocked by debris, narrowings, or sumps at their bases.
- A Downstream Complex at the SE end that is a warren of low tunnels along one

major bedding plane. The complex here is created by a combination of flooding and obstruction of the outlets by glacier ice in the Castleguard Valley at times in the past. One principal passage, Boon's Blunder, is ascended (as a crawl for 600 m) until it terminates abruptly in a water-filled shaft that has been dived to a depth of 20 m, where it enters a perched flooded trunk gallery. During the melt season, floods may pour from the shaft (technically, a phreatic lifting chimney), inundating the first kilometer of the cave in 1 or 2 hours.

- The central cave, which is a remarkably linear underground river passage passing beneath Castleguard Mountain. Only two significant paleotributaries enter the main gallery in a distance of 6 kilometers, which is very unusual.

The origin and development of the central cave is the focus of this chapter. As mentioned above, the passage is very straight, simple, and retains the same shape and size for a long distance. Going toward the entrance from the Headward Complex a caver traverses a typical drawdown vadose stream canyon, The Second Fissure (5.3.2 Figure 2). It has an initial phreatic ellipse partially preserved in the ceiling. This passes into a beautiful rounded phreatic tube (Holes in the Floor). It continues to descend at the same gradient and in the same direction as the canyon until there is an abrupt turn north.

The phreatic passage, "The Grottoes," maintains much the same form and dimension. But it ascends until it reaches the same level as the Second Fissure-Holes transition. Here it spills into a second lengthy vadose canyon, called The First Fissure (5.3.2 Plate 1). The First Fissure is so called because it is the first one encountered from the entrance by explorers, who can only approach from downstream.

In its turn, The First Fissure is followed gently downslope until it passes into an ideal phreatic tube on the same bearing and gradient, known as The Subway (5.3.2 Plate 2). This ends at a second phreatic lift— called The 80-Feet Pot (5.3.2 Plate 3)—a chimney that goes straight up a vertical fault for 24 meters to the level of the First Fissure-Subway transition. It spills over into a third vadose canyon, Helictite Passage. This in turn passes into a phreatic tube, one of two principal passages in the Downstream Complex. There is a final lifting chimney of 8 meters, The 20-Feet Pot, up from the Complex into the entrance passage, 100 m from daylight.

The geologic structural controls that guided this pattern of development are illustrated in 5.3.2 Figures 2, 3, and 4. Most of the known cave from the NW limits of the Headward Complex downstream to the termination of the Subway at the 24-m lifting chimney follows just one master bedding plane. The different passage segments are mainly oriented slightly aslant the true dip of the strata because they are following the intersection of particular fractures with the master plane; i.e., they are intersection-centered passages, as drawn in 5.3.2 Figure 1B. At the top of the 24-m lift, the cave enters a second master bedding plane which controls the Down-stream Complex in much the same manner except that guidance along fracture intercepts is less prominent. The 8-m chimney lifted the water into a third bedding plane system (the entrance passage) that is now almost completely destroyed by the recession of the walls of Castleguard Valley. The flooded trunk passage entered by divers from Boon's Blunder is almost certainly controlled by the master plane of the central cave.

The two master bedding planes are not especially prominent dislocations. They are by no means as prominent in the cave walls as many of the bedding planes in Mammoth Cave, Kentucky (see Chapter 5.3.8 in this volume), or the plane of the Main System in Hölloch, Switzerland, for example. Close inspection shows clear evidence of displacement, however, in the form of lensatic crushed zones with secondary riedel shears within them, slickensides and striations in places, and pinnate fractures at angles oblique to the sheared planes.

There was probably differential movement of no more than a few centimeters on either plane. The intersecting fractures are predominantly shear fractures and small faults, nearly vertical and in two principal sets that strike NW and NNE (5.3.2 Figure 2). These sets are compatible with the average stress axes that can be deduced from the various stress indicators in the cave (5.3.2 Figures 2 and 4). The main compressional axis (sigma 1) is almost due N with a dip to the south (15 degrees), while the extensional axis (sigma 2) is nearly horizontal and due E. This pattern is also compatible with the monoclinal flexure in Castleguard Valley, which presumably created it.

First Phase of Development?

Continuous phreatic passage development along the two master bedding planes can be traced from the Headward Complex to the Downstream Complex. It is possible that the early passages were expanded to crosssectional areas of 1 meter or more entirely in the phreas, which (assuming a reasonable hydraulic gradient) would imply discharge to springs above 2380 m (Worthington, 1991, p. 297-300). This probably required discharge upward through the Stephen Formation via fractures in the floor of an early Castleguard Valley. The cave then would have constituted one single phreatic loop (i.e. it was a State 1 form) with an amplitude of 370 m or more. This is quite similar to the amplitudes believed to exist for the shallow State 1 systems (or "Type A" water) in the El Abra Range, Mexico, described in the previous chapter.

Main Expansion Phase

This is illustrated in 5.3.2 Figure 3B. With entrenchment of the Castleguard Valley, the Stephen Formation was breached, and springs were established below 2020 m. With progressive enlargement of the passages, the water table in the cave fell until it was stabilized locally by the tops of the successive phreatic lifts. This was model drawdown vadose cave development. Canyons entrenched beneath the original quite-well formed phreatic ellipses along the two master planes, creating the Helictite Passage and First and Second Fissures. The canyons are deep and narrow (width = 0.3 to 3 m;

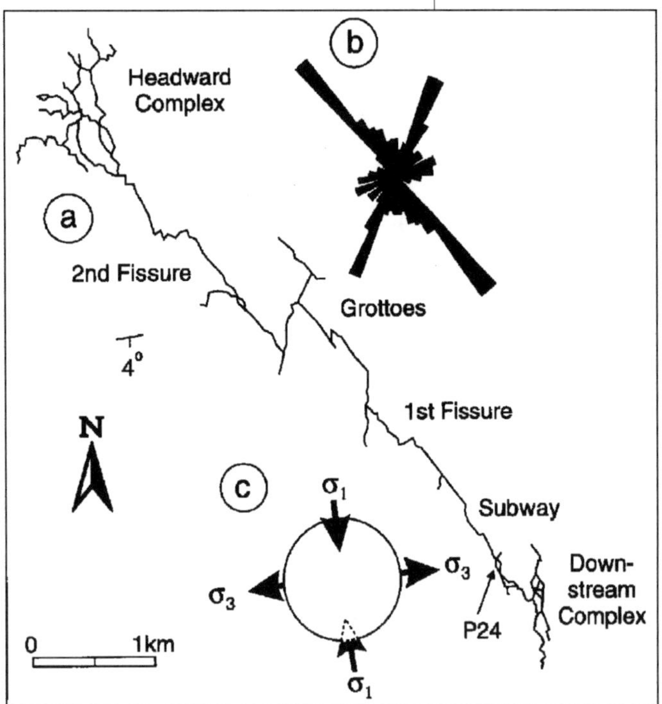

5.3.2 Figure 2. ***a*** *is a plan of Castleguard Cave.* ***b*** *is a rose diagram of fracture bearings.* ***c*** *is the orientation of principal stress axes.*

5.3.2 Plate 1. First Fissure is an ideal example of a drawdown vadose canyon. The passage was initiated along the intersection of a vertical fracture with the master bedding plane of the central cave, forming an elliptical phreatic passage about 1 m high there. Its ceiling, decorated with small aragonite and hydromagnesite encrustations, can be seen in the top center of the picture. With enlargement, the watertable was lowered into the passage and entrenchment of the narrow sinuous canyon commenced. Detritus on the ledges is the remains of silt (glacial flour) fillings (photo by D.C. Ford).

5.3.2 Plate 2. The Subway is guided by the same structure as First Fissure. It is First Fissure's downstream extension, passing beneath the water table to form this perfect example of a phreatic tube. Note that neither the master bedding plane (at left, behind the standing figure, R.O. Ewers) nor the vertical fracture that intersects it are conspicuous. But they hold the tube to a constant direction and gradient for 500 meters until it connects with the 24-meter shaft of 5.3.2 Plate 3. The channel in the floor is a good example of underfit invasion vadose entrenchment (photo by D.C. Ford).

←5.3.2 Plate 3. The 80 Feet Pot, a phreatic chimney guided by the small vertical fault seen to the left of the ladder. Water flow was straight up the chimney to spill out into the head of another vadose drawdown canyon at its top, The First Fissure- Subway transition (photo by D.C. Ford).

height to width ratio 4 to 30). The unmodified phreatic tubes and lifts—such as the Subway and 24-m chimney—average 10 m^2 in cross section.

Abandonment of the Cave and Establishment of Castleguard II

Entrenchment by glaciers and interglacial rivers in the Castleguard Valley probably lowered the potential spring positions rather steadily over time. However, the Cathedral strata have a very low effective fissure frequency, as noted above. As a consequence, the exposure of bedding planes and fractures that groundwater could penetrate to open new drainage outlets beneath the level of Castleguard Cave was small. The first major line to be exploited was as much as 267 m below the cave entrance. This site is the "Big Springs" (5.3.2 Figure 1). As a consequence of later entrenchment it now hangs 40 m above the modern valley floor, which itself consists of fluvioglacial gravel burying the true (bedrock) floor to an unknown depth.

The active cave river passages of Castleguard II have so far defied all attempts by explorers to penetrate them. Our knowledge of the structure of the modern system therefore is derived from a comprehensive study of the quantitative hydrology and hydrochemistry of the sinking waters and springs, with dye tracing and pulse analysis. This was undertaken by C.C. Smart (Smart, 1983a, 1983b, 1984, 1988; Smart and Ford, 1986). Smart's reconstruction of the complex underground architecture of the system is shown in 5.3.2 Figure 5. His sample groundwater sink points such as the Met, Midway, and Main Sinks, and Frost Pot are on top of the carbonate plinth of Castleguard Mountain, within a few meters of the edges of the modern glaciers.

Other meltwater streams can be heard flowing beneath the ice but they do not emerge. They are lost into karst sinks in the glacier beds. These all feed invasion vadose shaft and drain systems that pass through the Pika, Eldon, and Stephen strata, through or around Castleguard Cave, and down into the new active master conduits. Dyes injected into these sinks reach the springs in 3.5 to 12 hours, depending on discharge. The straight-line distance between sinks and springs is more than 4 km, with a fall of more than 750 m. These are good examples of very fast groundwaters in alpine karst systems.

In the valley below the Big Springs, many smaller springs such as Gravel, Watchman, and Artesian (5.3.2 Figure 5) rise through the fluvioglacial gravel fill. Following the summer melt season, these dry up progressively (Big Springs being the first) until only the lowest, Artesian, flows. It is a perennial spring 3 km downstream of Big Springs and 60 m lower. The aggregate peak discharge from all of these springs is about 20 m^3/s. A new "Subway" type of passage is probably evolving beneath the valley floor to handle average flows. The behavior of Big Springs indicates that it is draining a multiloop system similar to that in the cave.

When the melting is at its peak (during fine weather between mid-June and mid-September) the capacity of the Castleguard II System is exceeded. Flood water then backs up into the flooded trunk passage below Boon's Blunder and spills over into the Downstream Complex, inundating it. The total height of the hydraulic head in the integrated system (from Artesian Spring to the phreatic lift into the Blunder) is then about 370 m, which is very similar to the depth of the phreatic loop that is believed to have existed during the initiation phase.

Features Formed During the Abandonment of Castleguard Cave

Upstream of the flooding zone behind the cave entrance, modern Castleguard Cave is a hydrologic relict. The first event of the abandonment was the rejuvenation (extension) of the downstream ends of some principal tubes (see 5.3.2 Figure 4) that had been bypassed during inception by the formation of the system's main phreatic lifts upstream of them. Because they are at the bottoms of the phreatic loops, these are prime targets for the reorganization of the aquifer to drain to lower springs (Ford, 1971). The most notable rejuvenations were at the down-gradient ends of the great Holes-in-the-Floor and Subway Tubes, the principal lifts. The new passages there are much smaller than those of the main genetic phase (e.g. 5.3.2 Figure 4) and are impassable.

There are many of the more common features found in abandoned upper levels in caves everywhere, alpine and otherwise. These include the invasion shafts mentioned above. They become more common as the Headward Complex is approached, because the thickness of the aquitard strata above the cave (chiefly, the Stephen Formation) is reduced by glacial erosion. Many of the shafts pass straight down through the relict passages to terminations tens of meters below them. They are chance interceptions of the same type as the celebrated "domepits" in Mammoth Cave, Kentucky (Pohl, 1955; White, Chapter 5.3.9). Others discharge their water onto the floor of the relict passages. Here it flows as "underfit" streams that carve small vadose trenches into the wider relict canyon or phreatic channel floors (e.g. 5.3.2 Plate 3). After no more than a few hundred meters, this invasion water will sink into impenetrably small holes in the floor and pass on down into Castleguard II.

There are two major categories of clastic sediment in the cave. The first is deposits of glacial and glaciofluvial bedloads such as boulders, gravels, and sand that were swept by water or bulldozed by flowing ice into the head of the cave. All explored tubes in that area terminate at plugs of detritus or of glacier ice itself. Many deposits are partially cemented by calcite and display evidence of later erosion, indicating a long history of repeated filling and removal. The second is deposits of varved silt and clay found throughout the cave, although they are mostly swept out of the Downstream Complex by modern floods.

In the central cave, there are remains of at least three successive fillings, separated by periods of erosion when most of the previous fill was removed, and some episodes of flowstone deposition. The silt and clay range between 45 and 90% carbonate. They are deposits of glacier flour that settled out of

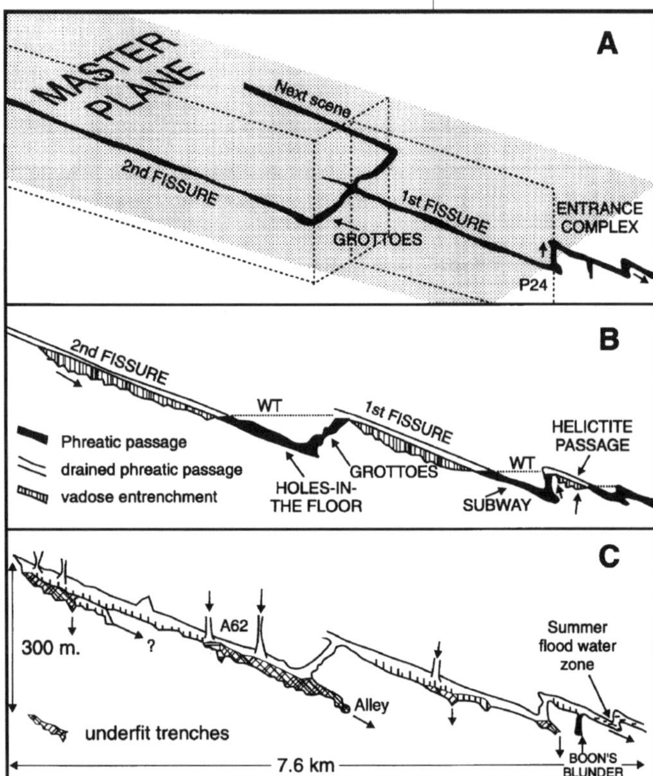

5.3.2 Figure 3. ***A*** *is the initial phreatic passages in the central cave and downstream complex, showing the master bedding planes and intersecting vertical fractures that guided them.* ***B*** *is the cave at the close of the principal enlargement phase. Drawdown vadose canyons supplied water to a succession of shallow phreatic loops.* ***C*** *is the modern cave. Small invasion vadose streams have cut shafts and underfit trenches in the drained galleries. These are lost into impenetrably small passages continuing on down into Castleguard II.*

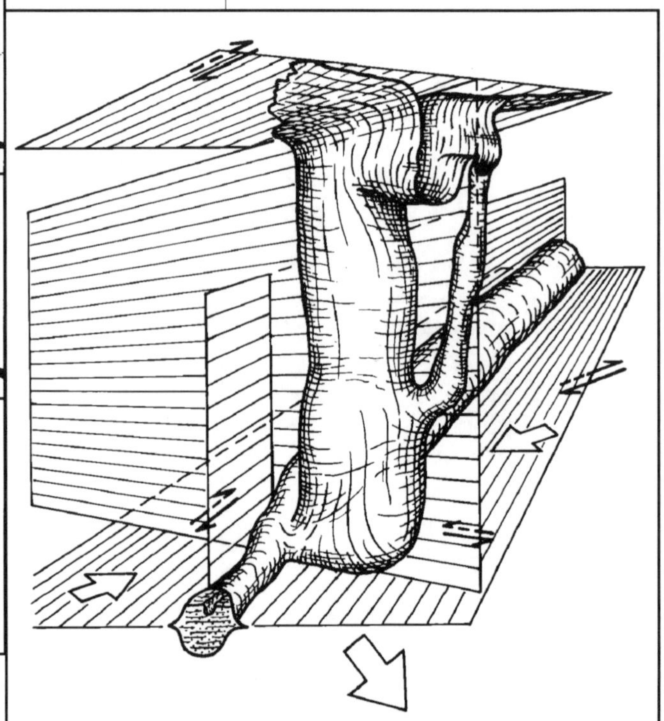

5.3.2 Figure 4. A drawing of the guiding structures and stress axes at the 24-meter phreatic shaft. The Subway on the lower master plane is in the rear and its later underfit extension that is now clogged with silt is at the front. The chimney followed a cross fault and discharged into the upper master plane, which now has a canyon entrenched in it (drawing by S.-E. Lauritzen).

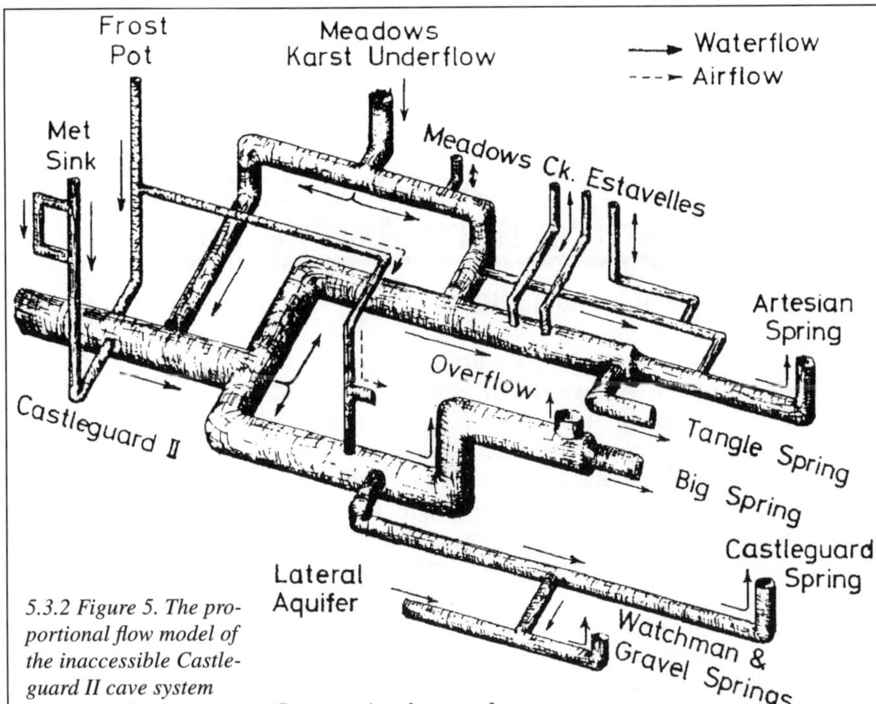

5.3.2 Figure 5. The proportional flow model of the inaccessible Castleguard II cave system and its feeder invasion vadose shafts that C.C. Smart derived from his quantitative hydrologic studies (Smart, 1983b). Cross-sectional areas of individual passages are proportional to their share of the total groundwater flow measured in the system, and their complicated pattern of interconnections are deduced from dye tracing and the flood overflow behavior.

suspension in sluggish to static water that filled the cave episodically long after Castleguard II was established (Schroeder and Ford, 1983). The likeliest interpretation is that the water was backfloods rising up from Castleguard II at times when all the valleys of the region were occupied by deep glaciers. Hydrologically the caves were subordinate to water in the glaciers and were functioning as a kind of "muddy manometer" (Smart, 1984). Varved clay and silt are common in caves in glaciated regions.

The Antiquity of the Cave

Castleguard Cave is well decorated with calcite speleothems (Harmon et al., 1983), many of which have suffered dissolution and are older than one or more of the varved silt phases. A sample of 21 broken or eroded speleothems has been dated by U-series methods, and the oldest of them was measured for paleomagnetic signals (Gascoyne et al., 1983). This work has determined that the cave was a relict and Castleguard II was well established at some time before 780,000 years ago (the date of the Brunhes-Matuyama magnetic reversal). All of the preserved varved silt filling phases are probably younger than that, but they are older than the Last Interglacial that began around 130,000 years BP.

These findings agree with those that are beginning to be found in many other mountain karst areas. Large upper cave levels that are many tens or hundreds of meters above the modern springs will almost invariably prove to be more than 1 or 2 million years in age.

Conclusions

Castleguard Cave is a good example of the State 2 multiloop cave with drawdown vadose canyons. The hydrologic behavior of the unexplored lower level or levels of the system suggests that they have a similar geometry. The looping is guided by a very few fissures (two or three slipped bedding planes plus fractures intersecting them) in massively bedded and resistant platform limestone. The cave was initiated under a deep cover of less-soluble to insoluble rocks and extended beneath them for at least 9 km between the earliest sinks and springs. It became relict more than 780,000 years ago but has since been invaded and modified by alpine and subglacial water. The greatest modern floods impose a hydrostatic head greater than 300 m on the base flow springs, temporarily rejuvenating the lowest area of the relict cave.

Acknowledgements

Research in the Castleguard Cave and Karst during the 1970s and 1980s was supported primarily by grants to the first author by the Natural Sciences and Engineering Research Council of Canada. Parks Canada is thanked for much aid in kind. All researchers at the cave are deeply indebted to their companions in the field. The cave is an unusually remote and severe place to work. Visits must be restricted to the winter season because of the flood hazard.

References

Ford, D.C., 1971, Geologic structure and a new explanation of limestone cavern genesis: *Cave Research Group of Great Britain Transactions,* v. 13, p. 81-94.

Ford, D.C., ed., 1983, Castleguard Cave and Karst, Columbia Icefields area, Rocky Mountains of Canada: A symposium: *Arctic and Alpine Research,* v. 15, p. 425-554.

Ford, D.C., Smart, P.L., and Ewers, R.O., 1983, The physiography and speleogenesis of Castleguard Cave, Columbia Icefields, Alberta, Canada: *Arctic and Alpine Research,* v. 15, p. 437-450.

Gascoyne, M., Latham, A.G., Harmon, R.S., and Ford, D.C., 1983, The antiquity of Castleguard Cave, Columbia Icefields, Alberta, Canada: *Arctic and Alpine Research,* v. 15, p. 463-470.

Harmon, R.S., Atkinson, T.C., and Atkinson, J.L., 1983. The Mineralology of Castleguard Cave, Columbia Ice Fields, Alberta, Canada: *Artic and Alpine Research,* v. 15, p. 503-516.

Pohl, E.R., 1955, Vertical shafts in limestone caves: *National Speleological Society, Occasional Paper,* no. 2, 24 p.

Schroeder, J., and Ford, D.C., 1983, Clastic sediments in Castleguard Cave, Columbia Icefields, Alberta, Canada: *Arctic and Alpine Research,* v. 15, p. 451-461.

Smart, C.C., 1983a, The hydrology of the Castleguard Karst, Columbia Icefields, Alberta, Canada: *Arctic and Alpine Research,* v. 15, p. 471-486.

Smart, C.C., 1983b, *The hydrology of a glacierised alpine karst:* McMaster University, Ph.D. thesis, 342 p.

Smart, C.C., 1984, Overflow sedimentation in an alpine cave system: *Norsk geografisk Tidsskrift,* v. 38, p. 171-175.

Smart, C.C., 1988, Artificial tracer techniques for the detemination of the structure of conduit aquifers: *Ground Water,* v. 26, p. 445-453.

Smart, C.C., and Ford, D.C., 1986, Structure and function of a conduit aquifer: *Canadian Journal of Earth Sciences,* v. 23, p. 919-929.

Worthington, S.R.H., 1991, *Karst hydrogeology of the Canadian Rocky Mountains:* McMaster University, Ph.D. thesis, 370 p.

5.3.3. Genesis of a Large Cave System: Case Study of the North of Lake Thun System (Canton Bern, Switzerland)

Pierre-Yves Jeannin, Thomas Bitterli, and Philippe Häuselmann

Abstract

The genesis of the large cave system with more than 250 km of surveyed passage in the region of Siebenhengste, Hohgant, and Lake of Thun has been reconstructed based on speleomorphologic observations (mainly where the morphology changes from vadose to phreatic). Eight flow systems (phases) and their respective conduit networks have been distinguished so far. The oldest had a phreatic level at an altitude of 1950 m. The last corresponds to today's phreatic zone at 658 m. Between each system, the water table dropped several hundred meters. This seems to be a consequence of changes in boundary conditions, mainly the springs' positions, which moved down as tectonic uplift and deepening of the nearby valleys occured.

Observations demonstrate that phreatic conduits sometimes develop close to the ancient water table, but often much deeper, down to 200 to 400 m below it. The change from one phase to the next seems to have been quick. This stepwise evolution is compatible with the results of computer models that give durations of 10,000 to 30,000 years for conduit systems to develop.

Analysis of the conduit networks of each flow system shows that their geometry is mainly influenced by the hydraulic gradients and the overall geometry of the aquifer. The orientation of discontinuity surfaces (fractures and bedding planes) or their intersections plays a subordinate role. This is also supported by numerical models in the literature.

Because we observe deep rather than shallow phreatic loops, despite a high fracture density, we assume that the heterogeneity of the discontinuity openings plays a more important role in the depth of karstification than does the spacing of the discontinuities.

Introduction

The geometry of a karstic network results from the dissolution of limestone (transport processes), which depends on the water flow conditions (hydraulic processes), which in turn depends on the distribution of voids within the limestone. Theoretical studies (Dreybrodt, 1988, 1995; Dreybrodt and Siemers, 1997; Groves and Howard, 1994a,b; Clemens et al., 1997; see also Dreybrodt and Gabrovšek, Chapter 4.2.2) show that a conduit network wide enough for human penetration can develop within a few thousand years or, within a few tens of thousands at the most. From these studies, the following parameters seem to have a significant influence on the network geometry:

- Spring locations with respect to the water catchment area.
- Initial aperture and connectivity of the voids within the limestone.
- Climatic conditions, especially the quantity of precipitation.

During the evolution of a given speleological network, some of these parameters may change. The following cases are mainly found:

- Tectonic uplift and simultaneous entrenching of the valleys, which lowers the base level and the phreatic zone within the karstic system.
- Establishment of "short cuts" across natural geologic barriers within the aquifer, leading to the lowering of the local phreatic zone, more or less independently of the regional base level.
- Tectonic displacements along faults or joints opening new flow paths.
- Infilling of adjoining valleys with detritus (aggradation) or, tectonic depression or glaciation, causing a relative rise of the

springs and phreatic zone.

- Gain or loss of a part of a catchment, leading to significant changes in the average discharge and the flow regime.
- Change from warm climatic conditions to cold (glacial) ones, leading to soil erosion with soil loss into the karst.
- Change from a humid climate to a drier one, leading to lowering of average discharges, but increasing the range of rainfall intensity and thus increasing erosion.

The reconstruction of the speleologic network in the North of Lake Thun area provides examples of the development of several successive karstic networks and of the conditions that prevailed during their development. The reconstruction is based on the fact that the position of the interface between the phreatic and vadose zones leaves obvious traces within the system, which permits us to separate the different phases of the evolution of the cave system. The spring positions are then easily extrapolated, allowing reconstruction not only of the karstic network at a given time but also the related springs. These reconstructions can then be compared to models.

The aim of this paper is to present the different phases of the genesis of the North of Lake Thun Cave System and then to compare these field observations to certain general concepts given by models.

Setting

With a length of more than 140 kilometers and a depth range of 1340 meters, the Siebenhengste-Hohgant Cave System is one of the Earth's most important. Close to this network there are also several other important caves, which so far have not been connected to it. The total length of explored conduits within the area is about 250 kilometers. This ensemble is called the North of Lake Thun Cave System in this report.

The system has a distinctive labyrinthine character, whose complex geometry can only be explained by the development of a series of superimposed cave systems that correspond to different and more or less independent times and conditions.

The North of Lake Thun Cave System is in the first Alpine range (Helvetic Nappe), directly facing the Swiss Plateau (5.3.3 Figure 1). From the edge of Lake Thun, where the principal modern spring (Bätterich) is located, it extends more than 20 km northeastward to Schrattenfluh, crossing the deep Emme Valley.

The mountain range containing the caves is the southeast-dipping limb of an anticline. Its northwest limit is now formed by high cliffs. Southeastward, the limestone disappears below thick flysch deposits of the Pennine Nappes (5.3.3 Figure 5). An important longitudinal fault, stretching from Lake Thun to Schrattenfluh (the Sundlauenen Fault), disrupts the continuity of the limestone beds toward the southeast.

The offset on the fault is 150 to 500 meters, depending on the location. West of the Siebenhengste, the frontal folds of the Helvetic Nappe crop out in the Sigriswilergrat Range, allowing us to place the original crest of the anticline between a minimum of 1 and a maximum of 3 kilometers in front of (northwest of) the Siebenhengste Cliffs.

The karst features are mainly developed within the Schrattenkalk Formation (Barremian to Aptian, Cretaceous, Urgonian facies), which is 150 to 200 m thick. This limestone lies on the Drusberg Marl (lower Barremian), which is 30 to 50 meters thick and forms an impervious base.

Most of the underground rivers flow downdip on the top of this impervious layer, more or less following the main faults. The Hohgant Sandstone (Eocene), which is locally up to 200 meters thick, overlies the Schrattenkalk. Following faults, surface water flows through the sandstone down into limestone of the Urgonian.

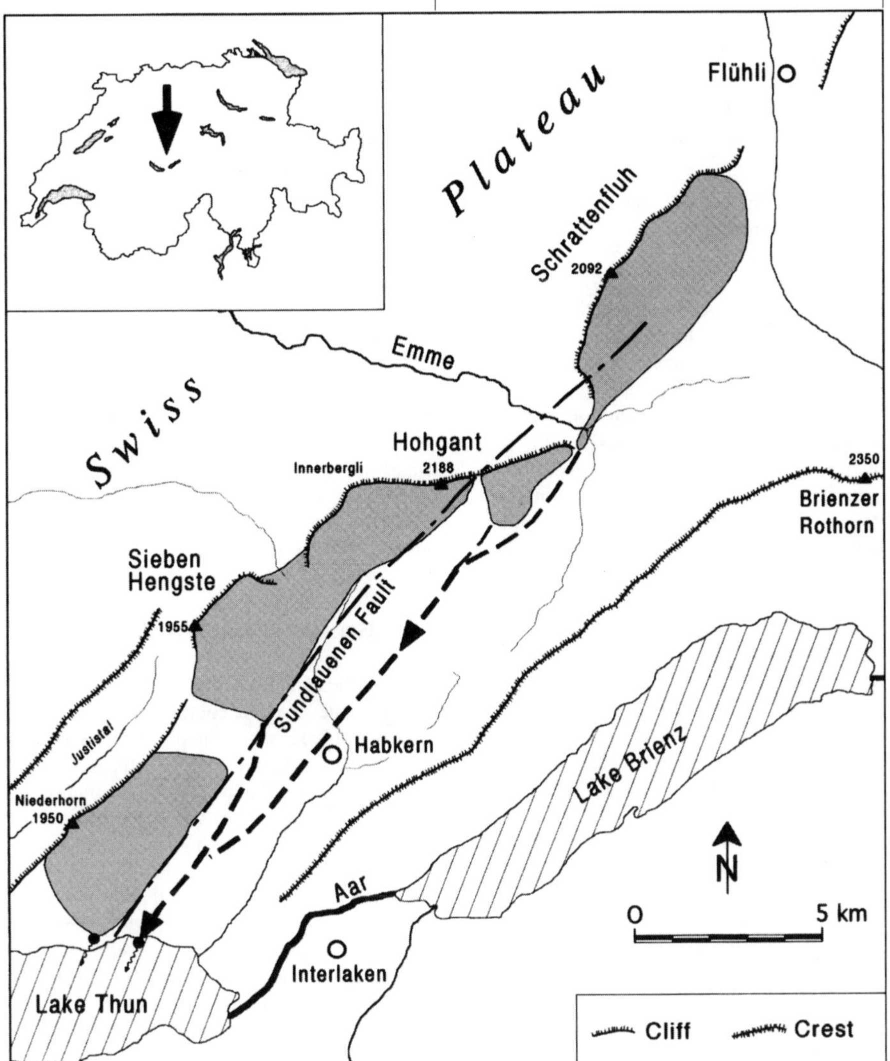

5.3.3 Figure 1. Location of the North of Lake Thun Cave System. The Helvetic Range overlooks the Swiss Plateau. The Emme River flows toward the plateau. The shaded areas indicate the various water catchment areas associated with this system.

Known Cave System

Areas northeast of Hohgant are not considered here. The remaining territory, between Lake Thun and Hohgant, can be subdivided into four sectors (5.3.3 Figure 2):

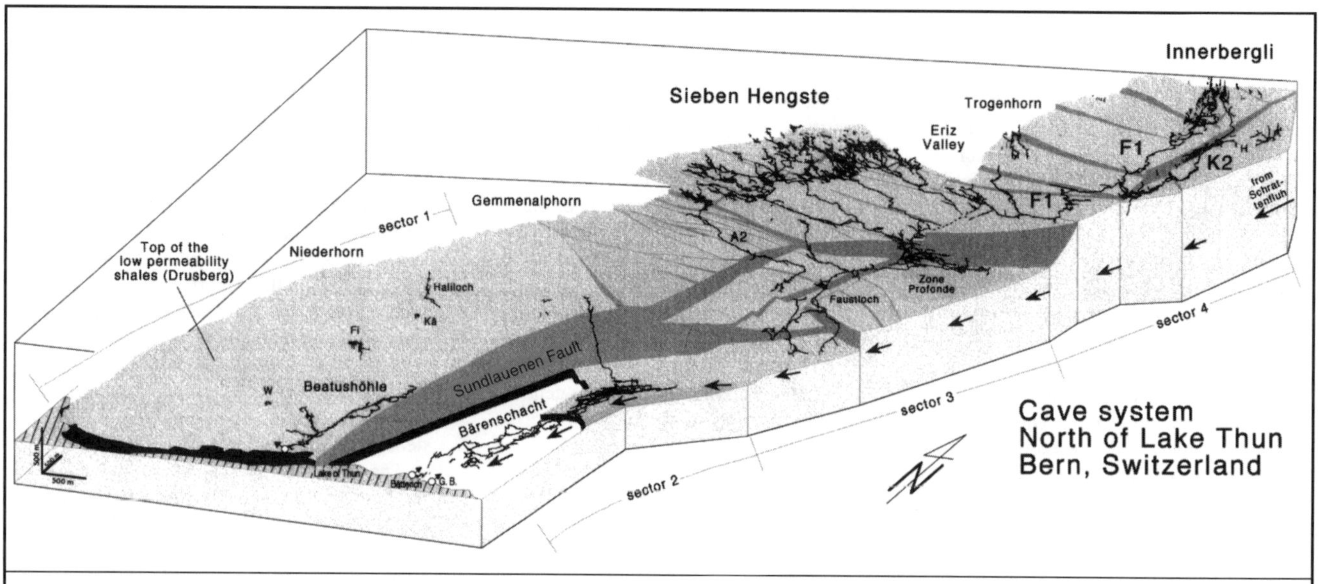

5.3.3 Figure 2. North of Lake Thun Cave System (black), top of the low permeability shale (Drusberg, light grey), and major faults (dark grey). Water from this system resurges at the Bätterich Spring and Gelber Brunnen (G.B.), and at Beatushöhle Spring.

- Sector 1—Beatushöhle area.
- Sector 2—Bärenschacht area.
- Sector 3—A2–Sieben Hengste, Zone Profonde, Faustloch area.
- Sector 4—Innerbergli area.

Sector 1 (Beatushöhle) includes five significant caves. The Beatushöhle, with about 14 kilometers, is the longest. It contains an underground river that flows out at the cave entrance. Today, most of the flow in the cave is vadose. But formerly it was entirely phreatic, despite a difference of more than 300 meters between the entrance (670 m) and the highest point of the cave (1030 m).

The four other caves display mainly phreatic fossil conduits intersected by more or less active shafts. Today there is only a limited amount of water flow in these caves. The Waldheimhöhle (W) is at about 1000 meters, the fitzlischacht (fi) between 1200 and 1400 m, the Känzlischacht (Kä) about at 1500 m, and the Haliloch between 1450 and 1750 m.

Sector 2 (Bärenschacht) corresponds to the zone directly behind the principal modern spring for the entire system. The major cave, Bärenschacht, is over 50 km in length. Because of its only recent exploration and of difficulty of access, the geologic setting of this gigantic labyrinth of phreatic galleries has not been studied in detail so far.

Some other small caves near its downstream end have also been explored. They are segments of large phreatic galleries that are now almost entirely full of sediment. The resurgences at Bätterich Spring and Gelber Brunnen (G.B.) have been dived for some hundreds of meters until obstructed by underwater breakdown.

Sector 3 (A2–Sieben Hengste–Zone Profonde–Faustloch) includes the most important caves. In the northwest, just below an area of bare limestone pavement, is the huge labyrinth of the Siebenhengste Cave System. The fossil galleries are oriented along the strike of the strata (NE), and the active galleries follow the dip of the limestone toward the SE. The combination of these two directions has produced a characteristic network pattern. From this area six active conduits (underground rivers) extend SE.

Beyond about 1.5 km, these galleries join a master conduit that is now almost entirely fossil. It changes its form between a single large conduit and labyrinths of multiple smaller conduits. This conduit possibly starts from the bottom labyrinth of K2, forms the major part of F1 and reaches the "Zone Profonde" and Faustloch. In the deep part of the Faustloch, many conduits cross the Sundlauenen Fault (which is split into two strands at this point) and then seem to descend 300 meters farther in order to reach the level of the Bärenschacht and almost the level of Lake Thun.

Sector 4 (Innerbergli) includes two major caves, F1 and K2. These are of a dendritic type, with conduits developed on the top of the impervious Drusberg Marl. F1 has many tributaries converging toward the main conduit that joins the lowest part of the Siebenhengste System. In particular, there are many tributaries below the pavement area of Innerbergli, where a large number of shafts with surface openings are known.

Downstream from sector 4, K2 reaches the Sundlauenen Fault, where there is a very important labyrinth. Today the K2 Underground River flows across the fault toward the depths of the Habkern Syncline, an area that is still being explored. Many other less important caves are perched about 100 meters above F1 and K2 and above the base of the limestone. Haglätsch Cave (H) is the most significant. The major parts of these caves are fossil phreatic galleries in which water apparently flowed westward.

Genesis of the Cave System

Most of the conduits in the system were formed under phreatic conditions. Vadose flow has only reshaped the profile of certain galleries (e.g. undercutting phreatic conduits with meandering channels) without fundamentally modifying the overall geometry. The major features resulting from vadose flow conditions are a series of shafts (invasion vadose shafts of Ford and Ewers, 1978) that pass through the limestone series vertically, connecting the modern epikarstic zone to the active conduit system.

In certain galleries it is possible to observe the exact transition point between the former phreatic and vadose zones. Such points allow us to fix the top of the "paleophreatic zone." If the levels observed at several different positions are similar, the location and altitude of the paleosprings can then be determined. Without such transition points, the altitude of phreatic conduits gives an indication of the minimum altitude of the phreatic zone.

Assuming that there was a continous lowering of the base level corresponding to the tectonic lifting during the period of time in which the caves have developed (uplift of the Alps, deepening of the valleys), it follows that the highest phreatic conduits will be the oldest. By following the morphologic continuity of the galleries, we can then approximate the shape of the paleonetwork that corresponded to a given paleospring and phreatic zone. The extrapolation of phreatic levels far away from the paleosprings is sometimes difficult, because the hydraulic gradients can be as steep as several percent under high-water conditions.

Based on these hypotheses, eight successive phases may be distinguished. The altitudes quoted in this paper are all above present sea level. Because of the uplift of the Alps, most of them do not correspond to the elevations that prevailed at the time the conduits formed.

L16-P2 Phase: Phreatic Conduits between 1800 and 1950 m

Some sections of large phreatic galleries have been found next to the Siebenhengste Ridge and in the highest parts of the Innerbergli and Hohgant areas. Although quite short, these conduits seem to indicate a flow direction toward the NE, which is parallel to the strike of the limestone (5.3.3 Figure 3). These conduits will have been developed in connection with paleosprings located more than 1300 meters above the present valleys and base levels.

Glacière–Haglätsch Phase: Phreatic Conduits between 1500 and 1720 m

This phase is recognized mainly in the Siebenhengste Cave System but can also be found in the Innerbergli–Hohgant area (5.3.3 Figures 3 and 4).

In the Siebenhengste, it is the phase when most of the galleries in the labyrinth were formed. A view of the major conduits is given in 5.3.3 Figure 4; the passage connecting the "Voie Express" (VE), via AKG, Balcon (B) to the "Glacière," was the main drain of this system during the phase. The flow was north-eastward. Its lowest known point is at 1460 m in the AKG, and the highest at 1720 m in the Glacière. It is established that this conduit is the major phreatic drain for one single phreatic zone. The contemporary spring was located in the Eriz Valley at about 1720 m. Many of the major conduits were formed quite deep below the spring altitude (more than 250 meters below the top of the phreatic zone).

In the Hohgant-Innerbergli area, this phase is indicated by the presence of some fragments of phreatic conduits between 1500 and 1720 m. Certain maze sections in Haglätsch Cave show a flow direction westward, in good agreement with a paleospring in the Eriz Valley. Some sections of phreatic conduits at the same altitude have also been found in the upper parts of the F1 and K2.

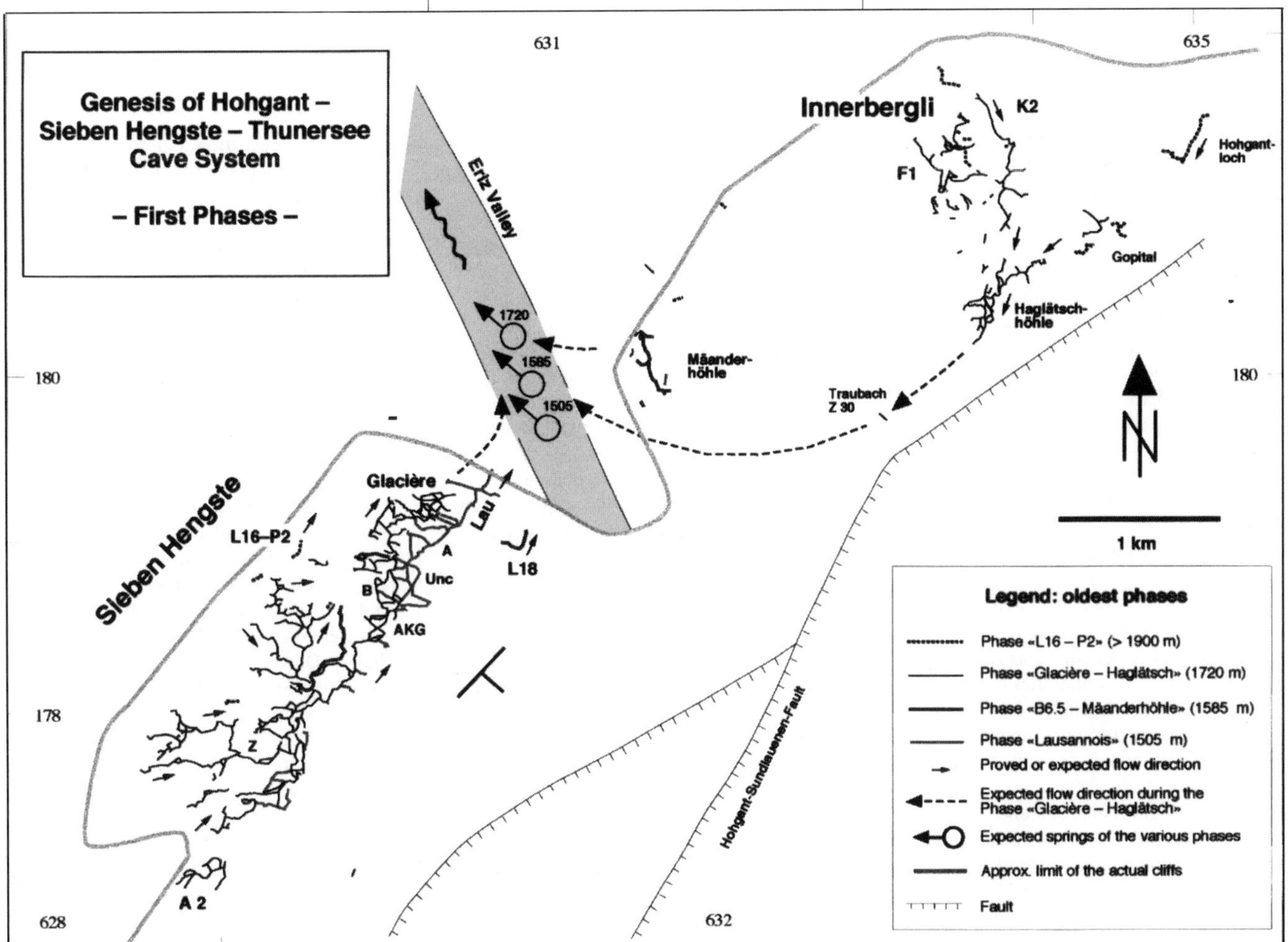

5.3.3 Figure 3. Oldest known phases of the genesis of the North of Lake Thun Cave System. Water flowed toward the Eriz Valley.

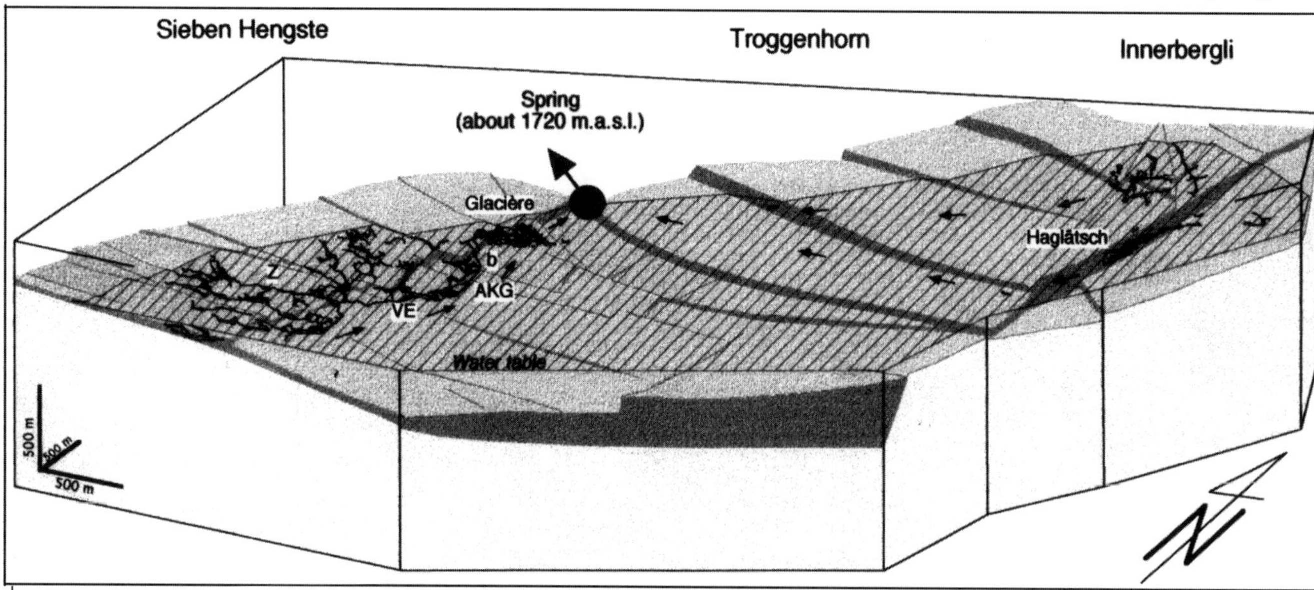

5.3.3 Figure 4. Block diagram showing the karstic system during the Glacière–Haglätsch Phase. The phreatic conduits related to the spring extended to a depth of more than 250 meters below the top of the phreatic zone.

B6.5–L18–Mäanderhöhle Phase: Phreatic Conduits Below 1585 m

The deep phreatic loop formed during the previous phase (Glacière–Haglätsch) makes the separation from this phase difficult. A transition point between vadose and phreatic conditions is discussed by Bitterli (in prep.) in B6.5 cave (B) at an altitude of 1585 m (5.3.3 Figure 3). This elevation corresponds to other phreatic galleries found in the L18 and the Mäanderhöhle, both of which are close to the expected spring position in the Eriz Valley, which is lower than the spring of the previous Glacière–Haglätsch Phase but farther upstream along the modern valley.

This level also corresponds to the highest undercapture networks found along the Balcon Gallery (Unc on 5.3.3 Figure 3). *Undercapture* is a consequence of the lowering of a spring position where water flowing through preexisting conduits is progressively captured by new discontinuities connected to the new spring. As a result, several small conduits are formed from various points along the main conduit, forming a labyrinth of small galleries that often converge toward a new lower drain. This phenomenon is typical of several parts of the North of Lake Thun Cave System (see also Ford and Williams, 1989, p. 274-278, and 5.3 Figure 4C). Farther upstream, several conduits in the Z49 region (Z on 5.3.3 Figure 3) can be related to this phase, but their distinction from conduits formed during the previous phase still needs more detailed study.

Lausannois Phase: Phreatic Conduits Below 1505 m

Once again, it is in the Siebenhengste Cave System that this phase was first observed. A detailed study of the Réseau des Lausannois (Bitterli, 1990) clearly showed the existence of a former phreatic level at 1505 m and of another later one at 1440 m. During the 1505 m phase, groundwater flow was mainly toward the NE (as in previous phases), but by the time of the 1440 m phase it was southeastward or southwestward, depending upon the location within the groundwater flow system. It is therefore possible that during floods in the 1505 m phase, the water rose up to the conduits of the 1585 m phase and the 1585 m spring behaved as an overflow.

The major conduit of the 1505 m phase was the Galerie des Lausannois (5.3.3 Figure 3), which is in fact the downstream section of the Galerie des Amours (A), which represents the convergence of the undercapture galleries of the AKG–Balcon Conduit (Unc). Farther upstream (toward the SW), there are more phreatic galleries trending NE below 1505 m. Some of them could have been formed during the previous phases but might still have been active during the 1505-m phase.

Jeannin (1990) noted that the tectonic shear stress that caused horizontal displacements on E-trending fractures can be observed only in conduits of this or earlier phases. These shifts are often of 20 to 30 cm and correspond to a NNW compression field (wrench regime). These displacements caused fracture opening, allowing water to flow along these fissures, strongly influencing the geometry of the network of the next phase (F1–Faustloch).

F1–Faustloch Phase: Phreatic Conduits between 1000 and 1440 m

The water table was lowered to 1440 m, and as mentioned before, the direction of flow changed (5.3.3 Figures 5 and 6). Downstream parts of F1 and K2 (Labyrinth) converged into a large phreatic collector (5 to 10 meters in diameter) that was between 200 and 250 meters below the water table. The flow in this conduit was directed toward the SW and collected water from the Siebenhengste area.

The Zone Profonde (ZP) and Faustloch (F) represent the continuation of this conduit, which here was some 350 to 400 meters below the top of the phreatic zone. The springs were then somewhere in the ancient Aar Valley. During this phase the Sundlauenen Fault, SE of the major conduit, clearly played the role of a dam along which the drain formed.

From the Siebenhengste catchment area down to this major conduit, water flowed along dextral faults that were activated at the beginning of this phase (see previous phase).

Under high water conditions, the water probably still rose up to the Galerie des Lausannois (L), and the 1505-m spring thus may have been the overflow spring of the system.

In the labyrinth section of Siebenhengste, conditions became vadose, and flow was down the stratal dip to the SE. Some segments of the older phreatic conduits that were oriented in that direction were entrenched by deep meandering canyons. In other places, new conduits with essentially vadose morphology have formed. The combination of the older conduits that are generally oriented NE and the canyons dipping SE has greatly increased the labyrinthine character of the cave in this area.

Downstream from Faustloch, the course of the major (trunk) paleoconduit is not established. A first possibility is that it crossed the Sundlauenen Fault and trended downdip directly toward the bottom of the modern cave or possibly to the SW part of it (Joséphine Gallery, Jo). However, proposing these two passages as continuations of the main conduit leads to two problems:

- It would imply that these conduits were formed at a depth of 650 or even 800 meters below the top of the phreatic zone.
- Because there are no limestone outcrops at this altitude SW from Faustloch, the water would have to pass through several hundreds meters of marl and flysch in order to reach the surface again.

These two problems can therefore be avoided by the following alternative. The major conduit did not cross the Sundlauenen Fault in Faustloch. Instead, it continued along the fault toward the Leimern area where the phreatic zone was continuous towards the W. The Faustloch collector ascended (via unknown conduits) toward the deep part of A2 (Keller, K). It then developed along the Topographenkluft (T) and the Sundlauenen Fault (FS) toward the SW in the direction of a spring located at about 1440 m in limestone outcrops near the Fitzlischacht (Fi) or farther west in the Justistal (Ju).

Sections of the Faustloch collector in the direction of A2 and sections of phreatic galleries farther west toward Fitzlischacht—which would confirm this hypothesis—have still to be found. However, this hypothesis leads to the postulation of a "shallow" phreatic zone that is "only" about 400 meters deep, and the problem of the spring position is solved. It seems to us that this underground course was the easiest for water and therefore the most probable.

Beatushöhle–Joséphine Phase: Phreatic Conduits below 1120 m

As soon as the spring level was lowered below 1400 m, flow through the Leimern area was no longer possible (5.3.3 Figures 5 and 6). Groundwater then had to find another route, crossing the Sundlauenen Fault to join the limestone strata located on the down-dropped SE side of the fault. At this point, the whole system became separated into two subsystems: the Beatushöhle catchment to the SW and the Siebenhengste–Hohgant catchment to the NE.

The major conduit of the Beatushöhle—a phreatic gallery—descends to about 350 meters below the paleophreatic level. The present cave entrance is probably the lowest point of this system, which might have an extension toward the west.

By this time, the Zone Profonde of the Siebenhengste Hohgant System had become partially vadose, and the rivers of F1 and K2 found a direct way through the Sundlauenen Fault. The existence of a phreatic level at 1120 meters is clearly observed in Faustloch. Farther downstream in Faustloch, the conduit crossed the Sundlauenen Fault by utilizing the Joséphine Gallery (Jo) between 800 and 900 m and then descended to the highest level of the Bärenschacht, between 700 and 800 m. In this phase the phreatic zone encompassed conduits between 700 and 1120 m, and the spring for the system was located in the Aar Valley at about 1100 m.

Dent-Tranchante–Akkordloch Phase: Phreatic Conduits below 770 m

The deepest part of Faustloch was still phreatic during this phase (5.3.3 Figure 5, p. 344), but upstream all flow became vadose. Bärenschacht has many phreatic labyrinths at this elevation. In Bärenschacht, three levels of conduits have been distinguished (Funcken, 1997), but so far it is difficult to distinguish a chronological sequence in them. The intermediate level could have been formed during this phase.

Laubloch and Akkordloch are conduits leading to the main springs of this system.

Modern Phase: Phreatic Conduits Below 558 m

The elevation of the modern water table is determined by the level of Lake Thun, at which the springs are located. The lowest level of the Bärenschacht, phreatic and scarcely explored, is still evolving today. Seasonal and other water level variations are from about 20 to a maximum of 50 meters, and no real conduit system has been developed in the high water zone.

The main collector extends toward the Schrattenfluh. It follows the Sundlauenen Fault in the SE down-dropped block (5.3.3 Figure 2). It may be joined by all the underground rivers of the Siebenhengste–Hohgant area. Upstream, it is also fed by water of the Hohlaub and especially the Schrattenfluh, some 20 km upstream from Bärenschacht. At present, our knowledge is not sufficient to determine whether this conduit is phreatic along its entire length or whether some sections are vadose.

Water tracing experiments show that there are many different "undercatchments" (Knuchel, 1973; Häuselmann and Otz, 1997):

- The Beatushöhle catchment collects water from the Beatenberg–Niederhorn area. The spring section of this system is well known because the Beatushöhle is still active today. Other parts of the catchment are "speleologically" still poorly known. Schematically, it is clear that water flows vertically downward through the limestone, reaches the top of the marl (Drusberg), and flows downdip toward the Beatushöhle, some hundreds of meters upstream of the Sundlauenen Fault. The NE limit of this catchment is in the Leimern area.
- The catchment of Gelber Brunnen and Bätterich Spring can roughly be split into two subcatchments:
  - The Siebenhengste–Hohgant catchment.
  - The Hohlaub–Schrattenfluh catchment.

The underground rivers of the Siebenhengste–Hohgant catchment are well known in their upstream (now vadose) regions. But it is only recently that the top of the modern phreatic zone has been discovered in deep parts of the Bärenschacht and Faustloch.The underground rivers of the Hohlaub–Schrattenfluh System are much less well known.

In the Schrattenfluh some important streams have been explored and look quite similar to those of the Siebenhengste. The conduit linking the water of Schrattenfluh and the Bärenschacht is not known. A tracer experiment has shown that there is very rapid flow: Between the Schrattenfluh and Bätterich Spring the dye took about 38 hours to travel the more than 21 km (550 m/h). For at least part of this distance the flow is phreatic. The phase during which this subcatchment became captured and connected to the Siebenhengste–Hohgant System is still unknown.

Dating

The absolute ages of the phases described above are unknown up to now. The only milestone has been obtained by analyzing sediment of some conduits in the Siebenhengste System (Jeannin, 1991). This study suggested that the Glacière–Haglätsch phase is older than 450,000 years. Further discussions with Swiss Quaternary specialists as well as other similar studies in Alpine areas (Maire, 1990; Audra, 1995) lead us to assume that the oldest phases (L16-P2 and Glacière–Haglätsch) are probably more than 1 million years old and could possibly be Pliocene. Absolute dating should be attempted in order to verify the relative chronologic sequence described in this paper.

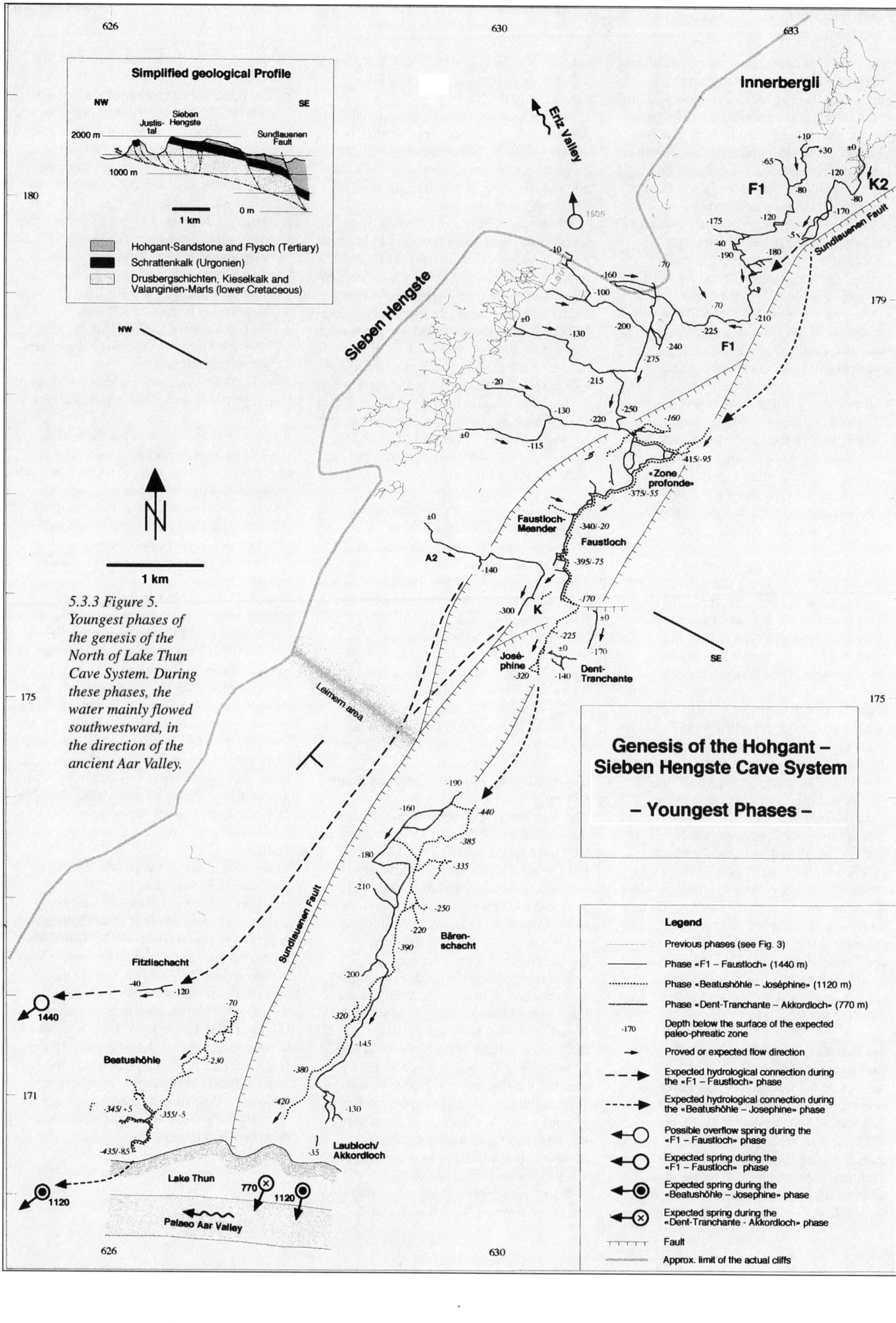

5.3.3 Figure 5. Youngest phases of the genesis of the North of Lake Thun Cave System. During these phases, the water mainly flowed southwestward, in the direction of the ancient Aar Valley.

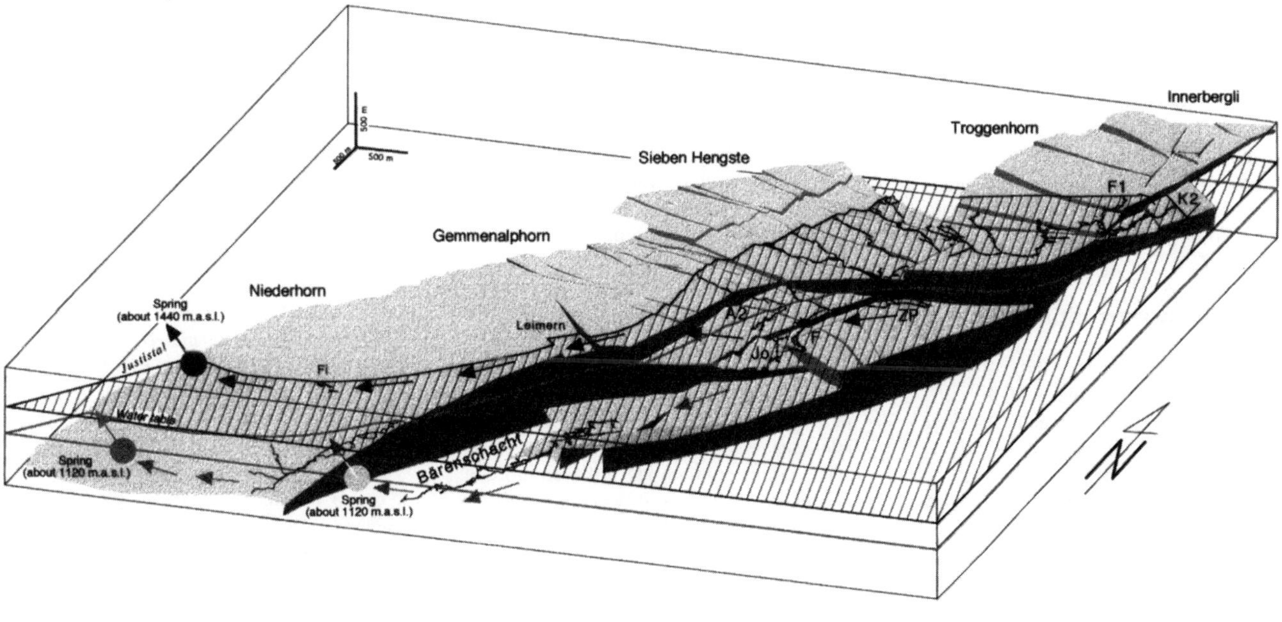

5.3.3 Figure 6. Block diagram showing the karstic systems during the F1-Faustloch and Beatushöhle-Joséphine Phases. The phreatic conduits related to the springs developed to a depth of as much as 400 meters below the top of the phreatic zone.

Discussion and Conclusions

Study of the North of Lake Thun Cave System reveals some evidence about general characteristics of karstic networks:

- Although situated in a strongly tectonised Alpine setting, conduits have usually formed 200 to 400 meters below the top of the phreatic zone, even in fairly small catchments (the Glacière–Hahglätsch and the B6.5–L18–Mäanderhöhle catchments were both smaller than 15 km^2).

- Genesis of individual flow systems seems to be quite rapid when compared to the evolution of the neighboring landscape. This is demonstrated by the fact that discrete phases of cave genesis can be distinguished, which would not be the case if the evolution of the karstic networks were continuous and slow.

Further analyses of the geometry of the karstic networks of each of the phases have been carried out. Although not presented here, some of the conclusions they lead to should be mentioned:

- The density of conduits in a given flow system that can be penetrated by a human is between 3 and 20 km per square kilometer of the phreatic zone (not of the total catchment area). Such values appear to be quite representative of "normal meteoric water" cave systems (Ford and Williams, 1989).

- Conduit networks of each of the phases have developed mainly along inclined bedding planes. This tends to make the network images two dimensional, although tilted. Only restricted areas display truly three dimensional networks of conduits.

- The overall geometry of the phreatic zone within the limestone (bottom and top of the aquifer, top of the phreatic zone) precisely delineates the possible extent of the karstic network. This statement may seem to be trivial, but any attempt to model or to assess the geometry of a karstic network (and then the hydrogeologic behavior of the aquifer) should start with a precise study of the geometry of the phreatic zone, based on geologic, speleologic, and hydrogeologic observations.

- Faults of kilometer scale seem mainly to play the role of dams (barriers) rather than being preferred drains. In the present case, this is especially due to the important displacements that disrupt the continuity of the limestone formation. Fractures of hectometer scale play a more important role in determining the flow directions (NE-trending normal faults for groundwater flow toward NE, and NW-trending strike-slip faults for flow toward the SE). Fracture orientation and distribution alone do not suffice to explain the geometry of the conduit networks. The location of the spring outlet (and then the distribution and orientation of hydraulic gradients related to it) seem to play a much more important role in this process. Tectonic movements that happened during the cave development produced an increase in the hydraulic conductivity along the reactivated faults that clearly led to the formation of new conduits along them.

Results of numerical models (Groves and Howard, 1994a, b; Howard and Groves, 1995; Clemens et al., 1997) are so far in quite good agreement with these conclusions. The major effect of the spring position and the less fundamental effect of the apertures of discontinuities is also seen in the model calculations. The effect of climatic conditions, as mentioned in the introduction, has not been presented in this paper. This information has still to be "extracted" from the abundant cave sediment in the North of Lake Thun area.

As a last point, it should be noted that the "Four State Model" of Ford and Ewers (1978) seems to contradict the present case study: In an Alpine setting where fracture density and opening is high, the model would expect to find shallow phreatic systems. In fact, the boundary conditions and the permeability field assumed in the Four State Model and those of the example presented here are not the same, so that model cannot be applied to the present case study. The same remark applies for Hölloch, although that cave is commonly quoted as an example. In fact, the Four State Model only represents a particular case corresponding to one seet of boundary conditions and permeability field among many possible cases.The model generally cannot be applied to the conditions encountered in Alpine karstic systems.

Discussion with Editor

Editors' Note

Editor Derek Ford disagrees with the assertions in the final paragraph. The Four State Model that outlines the principal phreatic and water table types of cave systems does not attempt to predict what will be the effective fissure frequency and aperture in any particular topographic or geologic setting such as thrust-plate (nappe) mountain ranges. That will be a function of the local lithology, geologic structure, diagenesis, and tectonic history. The physical boundary conditions given in basic illustrations of the model (e.g. Ford and Williams, 1989, Figure 7.14, and this volume, 5.3 Figure 2) are for the case where the drainage and stratal dip share the same direction, because that is the most simple to draw in a two-dimensional diagram.

Drainage downdip and then along the strike (as in North of Lake Thun) is probably more common at the global scale (see Ford, 1971, for discussion). The authors' meticulous account of the magnificent North of Lake Thun multiphase complex of caves suggests that it is principally of State 2. It is dominated by multiple phreatic loops, plus minor drawdown vadose and substantial later invasion vadose elements. There also may be local State 3 sections. It seems to be basically similar to the Ogof Ffynon Ddu System in Wales, although it is much larger and more complex; see Smart and Christopher, (1989, *in* T.D. Ford, ed., *Limestones and caves of Wales:* Cambridge University Press, p. 177-189) for an analysis of the Welsh cave. Their multiphase model figure for it, with undercapture passages and other common components, is reproduced as 5.3 Figure 4.

Portions of the main system of Hölloch were studied at different times in the 1970s by me (Ford, 1971) and by Ewers (1982, McMaster University Ph.D. thesis). We categorized the successive "levels" of that great cave as "irregular strike-oriented subsequent passages," placing them as State 2 or 3 components in the Four State Model's schema; see Ewers (1992, p. 320-339) for the analytical details.

Authors' Reply

We possibly misinterpreted or overinterpreted the "Four State Model" on two points:

First, for us, the model aims to link the depth of karstification to the initially "flowed through" fracture network. The higher the fracture density, the straighter the flow, and the shallower the conduit network. This mechanism is presented as the major—even the only—control parameter for karstification depth. We argue that the conduit network geometry is linked to the geometry of the flow pattern, which depends on the initial permeability field (usually anisotropic and heterogeneous). This field is linked to the spacing of discontinuities, but even more to their openings. In fact, if we consider the captions illustrating the "Four State Model" as examples of the effect of some possible permeability fields, and not as the only possible situations, then the model is acceptable. We believe that the "Four State Model" depicts the right principles, but that its usual representation seems to be too restrictive when compared to all possible cases. D. Ford admits that fracture openings end even primary porosity can also make a system be State 4.

Secondly, in our opinion, another point about the "Four State Model" is not very clear from the literature: Does the model include an evolutionary description of cave systems? In other words, does a cave system evolve from one state to another higher state through time? The case study of the Mendip Hills shown in Ford and Williams (1989, p. 266, fig. 7.16) suggests that this did happen in this case. On p. 267, the authors say that it does not happen in Hölloch. In fact, in a personnal communication, D. Ford wrote: "The model is a genetic explanation of observed patterns more than a real predicitive model." This means that the model is a classification tool of the observed cave systems and not a model for predictions. For many decades, hydrogeologists (as are we) looked for methods providing information about the "structure of flow systems" in karst aquifers. Models predicting the geometry of cave networks in a given setting would be very helpful for this purpose. This is the reason why we may have overinterpreted the Four State Model.

References

Audra, P., 1995, Karst alpins; Genèse de grands réseaux souterraines; Exemples: le Tennengebirge (Autriche), l'Ile de Crémieu, la Chartreuse et le Vercors (France): *Karstologia Mémoires,* no. 5, 280 p.

Bitterli, T. et al., 1990, Réseau des Lausannois: Höhlenforschung im Gebiet Sieben Hengste-Hohgant, no. 2, p. 2-65.

Bitterli, T., Le B6.5 (lapiaz des Sieben Hengste, BE, Suisse) : *Höhlenforschung im Region Hohgant (in prep.)*

Clemens, T., Hückinghaus, D., Sauter, M., Liedl, R., and Teutsch, G., 1997, Simulation of the evolution of maze caves: *12th International Congress of Speleology,* La Chaux-de-Fonds, Proceedings, v. 2, p. 65-68.

Dreybrodt, W., 1988, *Processes in karst systems, physics, chemistry, and geology:* Berlin, Springer Verlag, 288 p.

Dreybrodt, W., 1990, The role of dissolution kinetics in the development of karstification in limestone: A model simulation of karst evolution. *Journal of Geology,* v. 98, p. 639-655.

Dreybrodt, W., and Siemers, J., 1997, Early evolution of karst aquifers in limestone: Models on two-dimensional percolation clusters: *12th International Congress of Speleology,* La Chaux-de-Fonds, Proceedings, v. 2, p. 75-80.

Ewers, R.O., 1982, *Cavern development in the dimensions of length and breadth:* McMaster University, Ph.D. dissertation, 398 p.

Ford, D.C., 1971, Geologic structure and a new explanation of limestone cavern genesis: *Cave Research Group of Great Britain Transactions,* v. 13, p. 81-94.

Ford, D.C., and Ewers, R.O., 1978, The development of limestone cave systems in the dimensions of length and breadth: *Canadian Journal of Earth Sciences,* v. 15, p. 1783-1798.

Ford, D.C., and Williams, P.W., 1989, *Karst geomorphology and hydrology*: London, Unwin Hyman, 601 p.

Funcken, L, 1997, Bärenschacht, un exploration unique au monde. *12th International Congress of Speleology, LaChaux-de-Fonds, Switzerland,* Proceedings. v. 4, p. 9-12.

Groves, C.G., and Howard, A.D., 1994a, Minimum hydrochemical conditions allowing limestone cave development: *Water Resources Research,* v. 30, p. 607-615.

Groves, C.G., and Howard, A.D., 1994b, Early development of karst systems, 1: Preferential flow paths enlargment under laminar flow: *Water Resources Research,* v. 30, p. 2837-2846.

Häuselmann, P., and Otz, M., 1997, Wasserfärbung im Gebiet Gemmenalp-Sieben Hengste: Ziele und Resultate: *12th International Congress of Speleology,* La Chaux-de-Fonds, Proceedings, v. 2, p. 31-34.

Howard, A.D., and Groves, C.G., 1995, Early development of karst systems, 2: Turbulent flow: *Water Resources Research,* v. 31, p. 19-26.

Jeannin, P.-Y., 1990, Néotectonique dans le karst du nord du lac de Thoune (Suisse): *Eclogae Geologicae Helvetiae,* v. 83, p. 323-342.

Jeannin, P.-Y., 1991, Mise en évidence d'importantes glaciations anciennes par l'étude des remplissages karstiques du Réseau des Sieben Hengste (Chaîne bordière helvétique): *Eclogae Geologicae Helvetiae,* v. 84, p. 207-221.

Knuchel, F., 1973, Les cinq principaux essais de coloration de la région karstique situées au Nord des Lacs de Thoune et Brienz: *Cavernes,* v. 17, no. 3, p. 87-93.

Maire, R., 1990, La haute montagne calcaire; karsts, cavités, remplissages, Quaternaire, paléoclimats: *Karstologia-Mémoires,* no. 3, 73 p.

Smart, P.L. and Christopher, N.S.J., 1989, Ogof Ffynnon Ddu, *in* Ford, T.D., ed., *Limestones and caves of Wales:* Cambridge University Press, p. 177-189)

5.3.4. Pliocene and Quaternary Karst Development in the French Prealps—Speleogenesis and Significance of Cave Fills

Philippe Audra

Abstract

Three French cave systems in the Prealps in the Vercors and Dévoluy Massifs are described. It is possible to reconstruct their evolution by analyzing their morphology and by dating the karst fill using paleomagnetic and U/Th methods.

Cave development began at the end of the Miocene during uplift when inclined tubes formed in the epiphreatic zone. Later the tubes were partially blocked by weathered detritus from the surface. Uplift diverted former recharge away from the caves, and extensive calcite deposition occurred. The cave systems were reactivated during glaciations, with flooding to considerable depth. The epiphreatic zone was blocked by calcareous varves from meltwater. During interglacial periods, either calcite deposition or dissolution occurred, depending on the altitude and the density of soil cover.

Introduction

Three cave systems at the Vercors Massif—in the northern Prealps—and the Dévoluy Massif—on the border between the northern and southern Prealps—are used to characterize the lengthy speleogenetic history of the caves in the French Prealps (5.3.4 Figure 1). The systems have undergone a long evolution, starting at the end of the Miocene and extending through the Pleistocene glaciations. As a group, they are representative of the speleogenesis of the region.

5.3.4 Figure 1. Location of the cave systems.

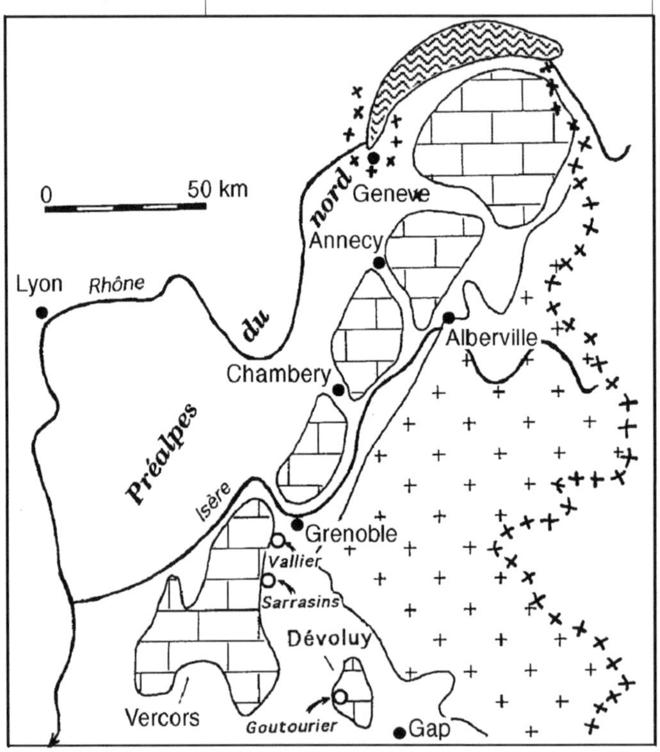

The entrance to Grotte Vallier Cave is at an altitude of 1520 m on the northeastern Vercors border, at the foot of an escarpment that rises to 1900 m and dominates the Isère Valley (200 m). The cave is developed in Urgonian (Cretaceous) limestone overlying impermeable marl. The limestone has been deformed into a recumbent fold that is inclined toward the west. There are almost 10 km of passages spread over a vertical range of 400 m (5.3.4 Figure 2i). The passage types range from rounded tubes in the limestone to large galleries that follow the marl contact and contain much breakdown. A 100-m-deep pit descends from the main cave and cuts through the saddle of the recumbent fold. The area is drained by the Bruyant Springs, which emerge at 978 m (Audra, 1994, 1995b).

Sarrasins Scialet (Shaft) Cave System is also on the eastern slope of the Vercors Massif, at 1850 m. The local top of the limestone massif is at 2000 m. (5.3.4 Figure 2ii). The system descends 440 m through shafts and canyons, and at the end continues down a large breakdown passage at the

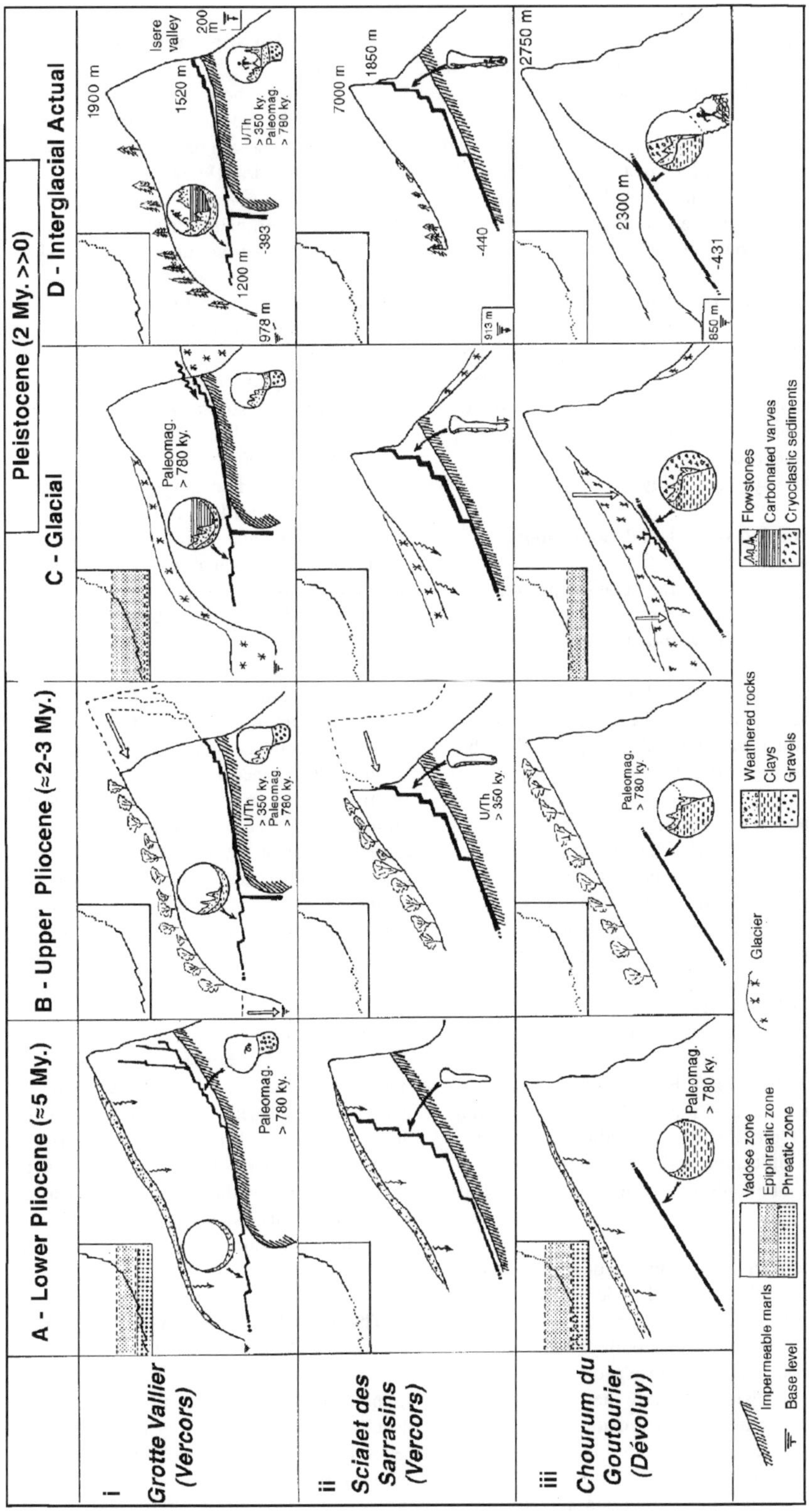

5.3.4 Figure 2. Evolution of the three Prealpine systems during the Pliocene and Quaternary.

contact with the marl, toward Goule Blanche Spring at 813 m (Audra, 1990).

The Goutourier Chourum (Shaft) Cave System is in a glaciokarstic valley on the western slope of the Dévoluy Massif at 2300 m. It is developed in the middle of thick cherty Senonian (Upper Cretaceous) limestone on the western limb of the Dévoluy Syncline, which reaches a height of about 2750 m at this point (5.3.4 Figure 2iii). The cave consists of large breakdown-modified passages that extend 431 m down the limb of the syncline. The Gillardes Springs (850 m) drain most of the Dévoluy Massif.

Comparison of the Evolution of the Three Cave Systems

Studies of the cave fill using radiometric dating (Y. Quinif, CERAK) and paleomagnetic dating (P. Rochette, CEREGE) as well as correlation with dated geomorphic features has shown that the caves began their development at the end of the Miocene (Rhodanian) during the final stage of the Alpine Orogeny (5.3.4 Figure 2).

Cave Systems Follow Post-Miocene Structures: Early Pliocene, 5 My
The late Miocene Rhodanian phase of the Alpine Orogeny created new tectonic structures and reactivated old ones (3.5.4 Figure 2A). Increased hydrologic gradients established by the uplift initiated the cave systems. A karst formed under a layer of weathered rock that then was gradually eroded and trapped in the karst fill.

Grotte Vallier. The cave has a large karstic catchment and developed on the normal limb of the recumbent fold along the marl contact, with the base level (springs) at a modern altitude of 1200 m. Upstream sections of the passages have potholes filled with quartzose and ferruginous gravel, the middle section consists of stacked floodwater tubes, and the downstream part near base level consists of U-shaped systems of tubes with elliptical cross sections (Audra, 1994, p. 231), partly filled with silt and clay.

All these sediments derive from the weathered rock cover. The sediment grain size decreases in the downstream direction due to a corresponding decrease in the velocity of the water. Some of these deposits have reversed paleomagnetic polarity, indicating an early age, greater than 780 ky (thousand years).

Sarrasins Scialet. Pits and canyons in the vadose zone lead to a large passage developed along the marl contact. Few old sediment fills have been recognized, although clastic sediment is common consisting of weathered rocks (ferruginous gravel and reddish clay) and limestone bedrock pebbles.

Goutourier Chourum. This system is developed along chert beds in the Senonian limestone, and the passages follow the dip of the chert. The passages are tubes about 2 m in diameter that descend 400 m and formed in the epiphreatic zone. They are partly filled with clay that has a reversed magnetic polarity, indicating that their age is more than 780 ky.

Continued Uplift: Pliocene, 2 to 3 My
Uplift continued during the Pliocene (3.5.4 Figure 2B), causing the valleys to be deepened by erosion, removing part of the karst recharge area. A soil cover formed and supplied CO_2 for substantial calcite deposition in the caves.

Grotte Vallier. An erosional lowering of the base level of 200 m drained the shallow phreatic caves and permitted a 100-m shaft to develop below them. The Isère River widened its valley, eroding away most of the recharge area of the cave and terminating its main discharge phase. Massive flowstone was able to accumulate in the cave, formed under a soil cover. Clastic sediment from this period, like the older underlying sediment, displays both normal and reversed polarities, suggesting great age. In the calcites, ^{234}U and ^{238}U have attained secular equilibrium, indicating that their age probably exceeds 1 to 1.25 My.

Sarrasins Scialet. This system evolved in a similar way; its shaft also contains large calcite deposits with the uranium isotopes in secular equilibrium.

Goutourier Chourum. With a change to vadose conditions part of the clay fill was removed from the cave by streams, and the first breakdown passages formed in the limestone, which has many chert beds. Calcite deposition followed this erosional stage. A flowstone floor lies on the ancient clay and has reversed magnetic polarity (>780 ky).

Reactivation by Glaciations: Pleistocene, 2 My >>0
During the Pleistocene glaciations (3.5.4 Figure 2C), the mountains were covered by ice caps, and valleys were blocked by thick tongues of glacial ice. Meltwater flowed locally.

Grotte Vallier. The glacier in the Isère Valley reached up to or above the cave entrance, permitting meltwater flow directly into the cave via a sinkhole. Because the spring was blocked by deep glacier ice, large runoff events were able to flood the cave to a depth of more than 400 m. The water carried a fine-grained clastic carbonate load. Sand was deposited in the upstream section of the cave and varved silt in the flooded section. These carbonate varves display reversed polarity (780 ky) (Audra, 1996)) and are the oldest known evidence of early glaciations in the inner Alps (Audra and Rochette, 1993). During the succeeding glaciations, after the Isère Valley was deepened, the glaciers no longer reached up to the entrance, and the cave remained dry.

Sarrasins Scialet. Because the entrance shaft opened in a cliff face just below the top of the massif, it was little affected by runoff. Small inflows of water caused slight entrenchment in the old flowstone floors.

Goutourier Chourum. Glaciokarstic valleys became entrenched into the slopes of the massif. The cave thus functioned as a sink for glacial water, but because of its high elevation, it remained in the vadose zone and did not flood. As a consequence, there are no varved deposits. To a large extent it became blocked by cherty talus.

Fluctuating Interglacial Conditions: Pleistocene, 2 My >>0
During the warm interglacial periods of the Pleistocene (3.5.4 Figure 2D), groundwater flow into the caves was moderate and carried only a small clastic load. The growth of forests was restricted by the altitude, and the deposition of flowstone was determined by the distribution of the soil cover, as it is today.

Grotte Vallier. During warm interglacial periods in the early Pleistocene, forest appears to have covered all the massif because flowstone with reversed magnetic polarity (>780 ky) was able to accumulate throughout the entire cave system. Today calcite precipitation only occurs in the downstream part of the cave, which lies below the tree line (1600 m).

Sarrasins Scialet. Because of its high altitude and position just below the mountain crest, there was little vegetation or groundwater infiltration during the interglacials, resulting in only limited flowstone deposition in the cave.

Goutourier Chourum. No deep soil cover could develop at this altitude, so calcite precipitation did not occur. On the other hand, the snowmelt supported some seasonal streams, creating new vadose canyon passages up to 10 m wide that are in good part due to mechanical erosion by the chert bed load. This is typical of high alpine caves (Maire, 1990).

In sum, there were four consecutive stages of development in these caves. Dissolution occurred during the glaciations and especially because of uplift. Glacial meltwater was also responsible for the principal sediment fills, and calcite precipitation occurred in passages beneath soil cover.

Types of Fill and Their Significance

Most types of fill, both clastic and chemical deposits, can be used to interpret the paleoclimate, and some can be used to characterize stream flow (Audra, 1995a).

Reworked Weathering Products
These cave deposits are derived from rocks that were weathered at the surface under warm wet conditions that lasted for long periods and are assigned to the Tertiary.

They consist of clay (notably smectite and kaolinite) plus rounded quartz grains and ferruginous pisoliths. Easily weathered minerals such as feldspar and biotite are absent. They are always the first sediments to be deposited and accumulated early in the cave development. Much of this sediment was carried into the karst during uplift at the end of the Miocene. It is mainly preserved in the deep parts of the alpine karst (Fernández Gibert et al., 1994; Maire, 1990).

Flowstone
Calcite deposition depends on the soil cover and the amount of infiltration (Genty, 1995). It faithfully reflects the extent of the soil-covered karst, especially in the mountains. In this sensitive environment, any change in climate is reflected by a change in the forest cover. The calcite crystals are generally transparent, columnar, and sometimes monocrystalline. Where found in high-alpine karst that is now bare and with ages exceeding the dating methods, they are evidence of warmer older periods (Maire, 1990).

Carbonate Varves
These laminated carbonate silts consist mainly of calcite grains formed by glacial abrasion of the surface karst but also contain angular quartz, feldspar, and mica grains. These deposits partly filled preexisting cave systems, such as Vallier Cave, during seasonal flooding by glacial meltwater, much like glaciolacustrine varves.

The Epiphreatic Zone—Its Role in Prealp Cave Development

In mountain karst, the phreatic zone is usually limited in extent. Because of the great relief, the vadose zone is well developed, with shafts, meandering canyons, and stream potholes. On the other hand, the less accessible epiphreatic zone has not received as much attention, although it plays an important and very different role in cave development.

Epiphreatic Zone and Cave Development in the Tertiary

There is one type of conduit whose origin has not been fully explained, and this is the inclined tube that has a continuous development over a vertical range of several hundred meters (5.3.4 Figure 2). This is a common passage type in most of the large mountain cave systems (Audra, 1993). Several theories have been suggested, such as the role of limestone resistance in the formation of large looped galleries (Ford and Williams, 1989). This theory does not contradict my own opinion.

Although the tube shape suggests a phreatic origin, the large range in altitude of the tube systems seems inconsistent with a fixed water table, which should produce subhorizontal passages developed along a base level. To understand the origin, the evolution of the rest of the system must be considered. These tubes formed following uplift at the end of the Miocene. At that time the recharge area was extensive. Moreover, the climate was warm with distinct wet seasons, as is typical of the present tropics. The cave systems received high discharge.

During this period, uplift and the relief (a few hundred meters in height) were moderate, and the caves were still forming. There were constrictions in the system that built up substantial pressure heads, up to several meters in height, and during storm events floodwater modification occurred. This seems to explain the origin of the inclined tubes. Today, inclined tubes are forming in the tropical mountain karst of China.

Therefore most of the large alpine cave systems were developed under floodwater conditions during the initial Alpine uplift. Glaciation modified these systems.

Epiphreatic Zone and Blockage of the Caves During Glaciations

During glaciations the cave systems were flooded by glacial meltwater. Varved deposits show that the flooding reached several hundred meters in height and that it was episodic. Numerous large alpine cave systems show evidence for this flooding. Slow flow velocities allowed fine silt to settle out of suspension. Throughout the epiphreatic zone the water was nonaggressive. Even delicate flowstone was only superficially corroded (Audra, 1994, 1995b).

The overall effect of the glacial meltwater was to block the epiphreatic conduits with varved deposits rather than to cause dissolutional enlargement. Well developed solutional activity was confined to the vadose zone. Filling of the inclined tube networks with dated early Pleistocene varves, as in Grotte Vallier Cave, is the major argument in favor of an early genesis for large alpine cave systems whose origin is probably late Miocene.

Conclusion

The evolution of large alpine cave systems in the French Prealps began at the end of the Miocene with the closing of the Alpine Orogeny. They developed as large inclined tubes with moderate gradients in the epiphreatic zone. These cave systems became perched after uplift and river entrenchment in the Pliocene (Audra et al., 1993), and they were rejuvenated during Pleistocene glaciations. This seems to be a general trend for Alpine cave systems elsewhere (Audra, 1994, 1995b; Jeannin, 1991; Maire, 1990; Uggeri et al., 1990).

Acknowledgements

I am grateful to D. Edwards-May, Arthur and Margaret Palmer, and Derek Ford for revising my English.

References

Audra, P., 1990, Une classique en Vercors, le scialet des Sarrasins: *Spelunca,* v. 39, p. 21-24.

Audra, P., 1993, Le creusement dans la zone temporairement noyée: Origine, processus et conséquences sur la morphologie des cavités: *3rd October Meeting,* Montpellier, p. 7-10.

Audra, P., 1994, Karsts alpins. Genèse de grands réseaux souterrains: Exemples le Tennengebirge (Autriche), l'Ile de Crémieu, la Chartreuse et le Vercors (France): *Karstologia Mémoires,* no. 5, 280 p.

Audra, P., 1995a, Signification des remplissages des karsts de montagne: *Karstologia,* v. 25, p. 13-20.

Audra, P., 1995b, Looking at alpine karst speleogenesis, case studies from France (Vercors, Chartreuse, Ile de Crémieu) and Austria (Tennengebirge). Changing karst environments: Hydrogeology, geomorphology and conservation, International Symposium, Oxford: *Cave and Karst Science,* v. 21, no. 3, p. 75-80.

Audra, P., 1996, L'apport du paléomagnétisme à la karstogenèse: le cas du chourum du Goutourier. (Massif du Dévoluy, Hautes-Alpes): Mélanges en l'honneur du Professeur Julian, Analyse Spatial, no. 38-39, p. 109-120.

Audra, P., and Rochette, P., 1993, Premières traces de glaciations du Pléistocène inférieur dans le massif des Alpes. Datation par paléomagnétisme de remplissages à la grotte Vallier (Vercors, Isère, France): *Compte-rendu à l'Académie des sciences,* sec. 2, v. 317, no. 11, p. 1403-1409.

Audra, P., Delannoy, J.-J., and Hobléa, F., 1993, Signification paléogéographique des réseaux perchés des Préalpes françaises du Nord. Exemples en Chartreuse et Vercors: *Études de géographie physique de l'URA,* v. 903, no. 22, p. 3-18.

Fernández-Gibert, E., Palomarès, M., Rossi, C., and Tortosa, A., 1994, Analisis de procedendia en arenas karsticas: Evidencia de una cobertera permotriasica erosionada en el macizo occidental de Picos de Europa: *1st Congreso Internacional sobre Picos de Europa,* Oviedo, Proceedings.

Ford, D., and Williams, P., 1989, *Karst geomorphology and hydrology:* London, Unwin Hyman, 601 p.

Genty, D., 1995, Corrélation entre l'épaisseur des lamines de deux stalagmites modernes et les données météorologiques: *Spéléochronos,* v. 6, p. 35-40.

Jeannin, P.-Y., 1991, Mise en évidence d'importantes glaciations anciennes par l'étude des remplissages karstiques du réseau des Siebenhengste (chaîne bordière helvétique): *Eclogae geologicae helvetiae,* v. 84, p. 207-221.

Maire, R., 1990, La haute montagne calcaire: *Karstologia Mémoires,* v. 3, 731 p.

Uggeri, A., Bini, A., and Quinif, Y., 1990, Datation des sédiments de la grotte Marelli (Italie, Lombardie, Varese): *Spéléochronos,* v. 2, p. 21-28.

5.3.5. Speleogenesis in the Picos de Europa Massif, Northern Spain

E. Fernández-Gibert, J.M. Calaforra, and C. Rossi

Abstract

The Picos de Europa Range in northern Spain represents one of the most important alpine karsts in the world and contains a great concentration of deep limestone caves. It comprises Carboniferous limestone with a thickness of up to 2000 m that is severely deformed by tectonic action. The structure, geologic evolution, fracturing, and Quaternary glaciation are among the main factors that influenced the evolution of the caves.

This paper proposes a speleogenetic model for the evolution, with an initial stage during which a Permian-Triassic cover provided semiconfinement for the entire carbonate series. Under these conditions, the earliest known phreatic conduits were formed. Local relict sediments in them are related to the erosion of the cover.

During a second stage, with uplift of the massif, the cover was eroded, and vadose conditions were established. These conditions were linked to glacial processes that intensified the karstification, creating an unsaturated zone almost 2000 m thick. Karstification and vadose entrenchment in this zone were especially marked along preferential flow paths coinciding with preexisting dolomitized and sulfide-mineralized fractures in the massif. These fractures may have played a precursor role in the karstic evolution of the massif as a source for more aggressive water with a higher content of weak acid due to the oxidation of the sulfides.

Introduction

The Picos de Europa Massif comprises a series of ridges running parallel to the coast of northern Spain. The massif has an area of approximately 500 km^2 and forms an entity that is both geographically and geologically distinct. It can be divided into Western, Central, and Eastern Sections that are separated from each other and from neighboring regions by deep river gorges. The highest peaks in the area are over 2500 m above sea level (5.3.5 Figure 1).

The regional climate results from the location of the massif. Despite its relatively modest dimensions, it acts as a natural barrier to the predominantly N and NW winds from the Atlantic (Puch et al., 1985). Rain frequency is about 150 days per year and is strongly affected by altitude, with over 2000 mm measured annually at 1000 m above sea level. Precipitation is most abundant between October and February (mainly in the form of snow), with a secondary maximum in April. In general, the climate is typical for a high mountain region.

Geologic Context

The Picos de Europa form a distinct region within the Cantabrian Zone of the Paleozoic Massif on the Iberian Peninsula (Julivert et al., 1971). Local strata include deposits of two important sedimentary cycles. Sediment of the first cycle appears only to the N of the region and includes Cambrian rocks (shale and sandstone of the Oville Formation) and Ordovician rocks (quartzite of the Barrios de Luna Formation), with a total thickness of nearly 650 m (Martínez García, 1981).

After a long stratigraphic hiatus, a second cycle began that was dominated by Carboniferous carbonate rocks. Some thin Upper Devonian strata underlie the limestone (sandstone of the Ermita Formation and limestone of the Baleas Formation). Above these, three formations may be distinguished:

the Genicera Formation, composed of the Griotte Limestone with a thickness of 30 to 50 m (lower Visean-Namurian), the Caliza de Montaña Formation (Serpukhovian-Bashkirian) subdivided into two members with a total thickness of 450 m, and the Picos de Europa Limestone, with a thickness exceeding 650 m. The carbonate series attains a thickness of 800 to 900 m in frontal nappes at the south part of the region and more than 1200 m in the central part.

These rocks form successively encroaching thrust sheets that strike E, and moved toward the S. The dip of the nappes increases toward the N, ranging from 40 to 90° (5.3.5 Figure 2). The structure dates from the Kasimovian (Late Carboniferous) (Marquínez, 1997).

Various formations of synorogenic character and varying lithology were deposited during nappe development. Currently they crop out in isolated and discordant form above the limestone. Subsequently, in an extensional tectonic phase, the entire region was covered by continental deposits of Permian age (comprising alluvial fans in the proximity of the Picos de Europa) and Triassic age (Bunter sandstone facies; Martínez García, 1981). The presence of the Permian-Triassic cover has played an important role in the phreatic evolution of karst in the Picos de Europa.

The modern relief of the Picos de Europa was determined by the formation of the Pyrenees Range during the late Eocene and Miocene (Vegas and Banda, 1982). Although the history of the alpine tectonics is not entirely established, various authors have emphasized the existence of important subvertical reverse fractures that trend E and

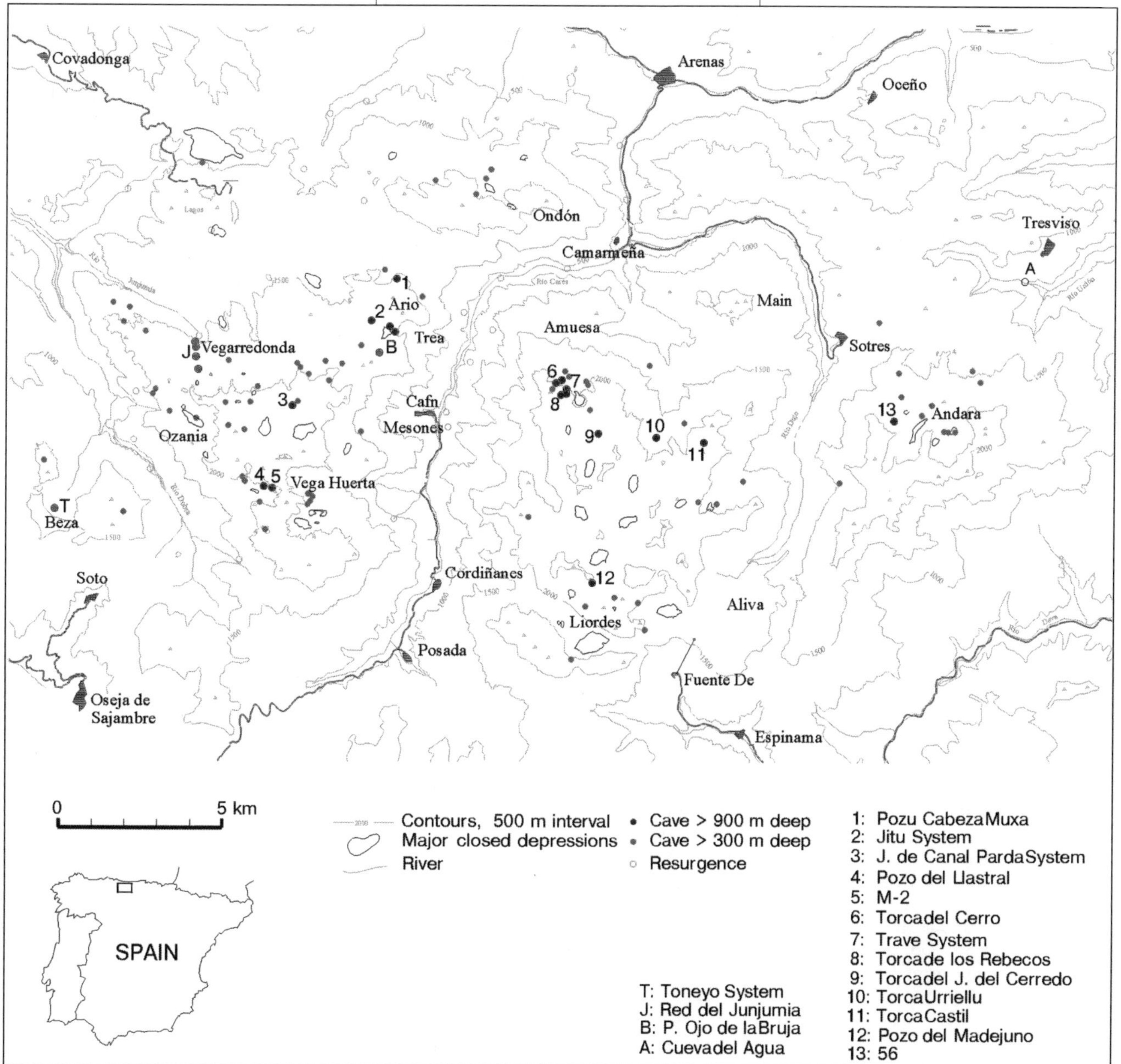

5.3.5 Figure 1. Location of the Picos de Europa Massif and its main caves.

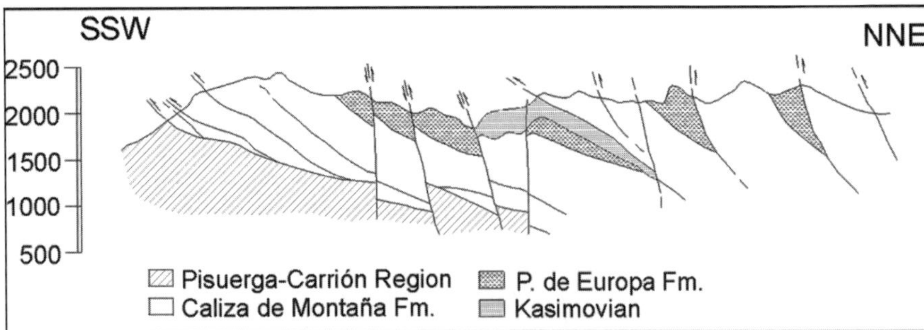

5.3.5 Figure 2. Diagrammatic cross section showing the typical geologic structure of the Picos de Europa (modified from Farias, 1982).

Cave	Depth (m)
Sistema del Trave	1441
Pozo del Madejuno	1255
Torca de los Rebecos	1255
Sima 56	1169
Sistema del Xitu	1135
Torca Urriello	1017±5
Torca Castil	1020
Sima M-2	972
Pozo del Llastral	944±5
Pozu Cabeza Muxa	939
Torca del Cerro	925
Torca del Jou del Carredo	910
Sistema del Jou de la Canal Parda	903

5.3.5 Table 1. Caves in Picos de Europa exceeding a depth of 900 m.

ESE, affecting the Hercynian (Carboniferous-Permian) structures and the Permian-Mesozoic cover.

These seem to be zones of significant movement and are commonly dolomitized and mineralized by sulfides. The origin of the minerals, which were mined in the past, is a subject of some controversy (Marquínez, 1997; Martínez García and Tejerina, 1984). The fractures are important for speleogenesis, both because they provide paths for water and because of their sulfides and dolomitization which favor karstification processes.

Some authors have proposed an uplift of the region during the Pliocene-Pleistocene (Smart, 1986; Senior, 1987). This would explain the development of the deep fluvial canyons trending perpendicular to the Hercynian structures (as in the case of the Cares and Duje Rivers) and superimposed on the Mesozoic cover.

Quaternary glaciation greatly affected the relief of the Picos de Europa. The higher areas of the three massifs were covered by extensive icefields. From these, glaciers descended to an altitude of 1400 m above sea level, sculpturing typical glacial valleys (e.g. the Duje Valley, with a length of over 7 km, and the alignment of Jou de les Boches, Jou sin Tierra, and Jou Luengo). There are many examples of moraines alongside these *Jous*, such as at Puertos de Aliva to the east of Sotres (Obermaier, 1914).

Karst in the Picos de Europa

Since the 1960s, numerous groups from Spain and abroad have carried out speleologic explorations and surveys in the Picos de Europa. As a result, the massif is considered one of the most important in the world in terms of the concentration of deep limestone caves. To date, seven of them have been explored to a depth of 1000 m, and more than 70 others are over 300 m deep (5.3.5 Table 1). The number of mapped caves increases every year. Exploration in the higher regions is limited to the summer months due to snow during the remainder of the year. In general, these caves present great technical difficulties. The average temperature inside them is approximately 4°C.

Although a large number of caves have been explored, our knowledge of the genesis and evolution of the karst is still incomplete. Nevertheless, some of the factors determining it have been identified.

Hydrogeologic Context

At altitudes above 1700 m there is practically no vegetation, and evapotranspiration is limited (200 to 300 mm). Remarkable spreads of karren promote the almost immediate infiltration of water during the intense rains. Drainage is fast, being about 55 L/s/km^2 in the catchment of Trave Cave, for example (Bigot, 1989). Total recharge in the Picos de Europa has been determined to be between 370 and 520 hm^3/yr, with a minimum discharge of 315 hm^3/yr (0.315 km^3/yr) (ITGE, 1989).

From a hydrologic point of view, the principal karst recharge occurs in the zones of greatest altitude. There is a dominant drainage in E-W directions toward the river valleys, where important springs are located, such as Farfao del Cares, with a discharge of more than 3 m^3/s, or those of Ondón, Las Pólvoras, Caín, and Reo Molin, where the flow is between 1 and 2 m^3/s.

Some authors have suggested the existence of several distinct isolated aquifers due to the thrust sheet arrangement within the massif (Hoyos and Herrero, 1989) creating a complex hydrogeologic situation. At the springs, the discharge displays large variations between maxima and minima, with ratios of up to 20 to 1 between flood and dry periods. The highest discharges occur between October and June, due to precipitation and snowmelt. The mean summer discharge is occasionally complicated by abrupt increases caused by heavy rains (Collignon, 1985).

Among the most noteworthy morphologic features of the region are large closed depressions, that sometimes retain snow throughout the year. This situation ensures a continuous input of cold aggressive water that helps to intensify karstification within the massif. In upper shafts, snow frequently accumulates and remains throughout the year, sometimes penetrating into caves to as deep as 100 m.

Prominent among the large dolines affected by glacial processes are those of the great basins of the Eastern Massif, with diameters of up to 2.5 km, or that of Jou Santo in the Western Massif, with a diameter of 800 m. During the cold periods of the Quaternary, these were sites where ice accumulated, although being of karstic origin they must have been formed previously (Smart, 1984, 1986).

Caves

The organization of the limestone in nappes produces karstifiable sequences with a thickness exceeding 2000 m. The development of a vadose zone of such magnitude has led to the formation of enormous limestone caves, which in most cases display a common vadose morphology and character.

Normally, vertical downcutting is rapid until depths very close to the piezometric surface are attained. Only occasionally do interbedded strata with insoluble content impede such entrenchment, and then caves with an extended horizontal development are formed, e.g. Sima 56 in Andara (Smart, 1985).

Large deep shafts that follow vertical or subvertical faults and thrusts (5.3.5 Figure 3) commonly alternate with narrow canyons where vadose entrenchment sometimes exceeds 50 m. In places, phreatic conduits with a diameter of up to 1 meter cut their ceilings and link them to the next series of fault-guided pitches (5.3.5 Figure 4). In the Western Massif, Senior (1987) proposed for

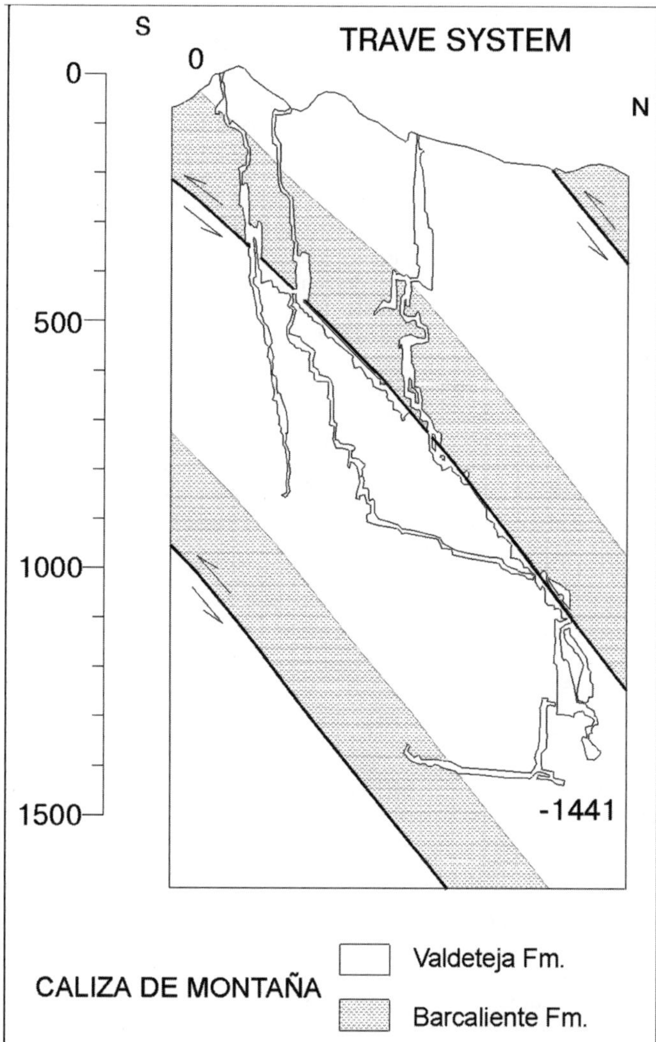

5.3.5 Figure 3. Geologic structure of the Sima del Trave, the deepest cave in the Picos de Europa (modified from Vidal, 1990).

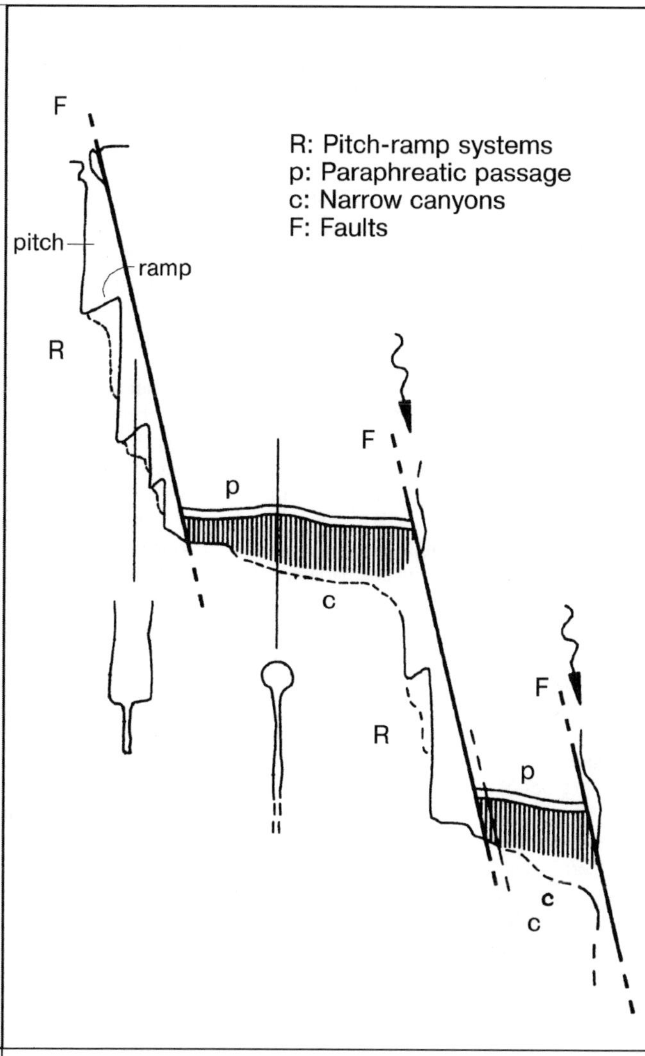

5.3.5 Figure 5. Morphology of vadose passages (modified from Senior, 1987).

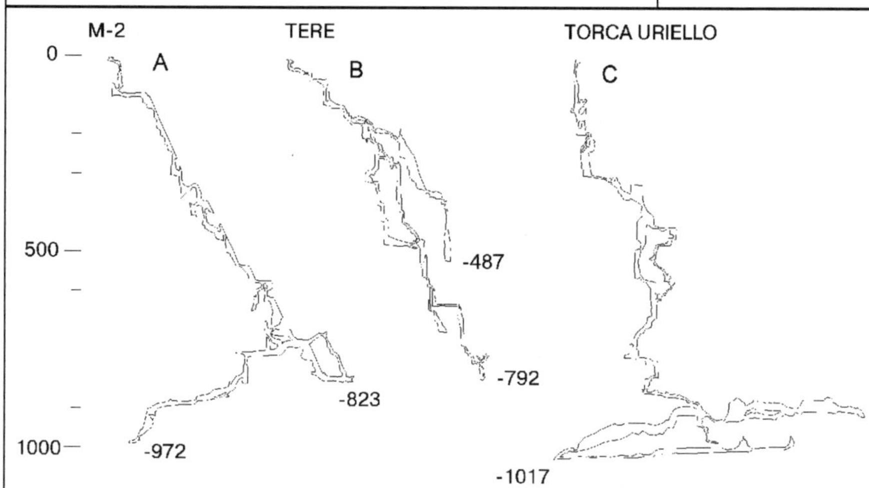

5.3.5 Figure 4. Morphology of several caves in the Picos de Europa (see text). A) Sima M-2 (modified from SEII, 1987); B) Sima Tere (modified from Sefton, 1984); C) Torca Urriello (modified from Genuite and Vidal, 1984).

these tubes a paraphreatic origin (Tratman, 1957).

The characteristic cave morphology associated with the very aggressive water typical of a glaciokarst has been termed a "pitch-ramp" system by Senior (1987). The spatial variation of the preferential water input points has often produced complexes of pitches and ramps with interconnected paths (5.3.5 Figure 5).

In the deeper parts of the cave systems, active water courses tend to occur that usually end at sumps which indicate the phreatic zone. Mud covering the walls and ceilings of neighboring conduits suggests that at periods of maximum flow the water table may rise several tens of meters. The flow response in these underground systems is extraordinarily fast. Day-to-night cycles of flow variation due to snowmelt and rises in water level of up to 8 m after a heavy storm have been described.

Some subterranean systems are perched

above the local base level due to interbeds of nonkarstic layers within the limestone (Red de Junjumia in Vegarredonda), which might be indicative of the uplift of the massif and the deepening of the karstic canyons.

At certain levels, some of the limestone caves contain conduits with completely different characteristics from those described above. These are very large and mainly horizontal phreatic conduits, up to 60 m in diameter. Good examples of such conduits include the deep galleries of the Pozo del Ojo de la Bruja in Ario, the M-2 Sima in Vega Huerta and the Torca Urriello, in addition to various sections of caves in the Ondón region (5.3.5 Figure 4A and C). In some cases, the water invading from the upper shaft systems has produced important vadose changes in them.

Cave systems with long horizontal development are relatively rare in comparison with the vertical caves. The most noteworthy among these are the Red de Toneyo, almost 19 km long (Bes, 1986) and the Cueva del Agua in Tresviso, which is more than 10 km long (Sefton, 1984).

Speleogenesis

The two types of conduits, the vadose sections and the large phreatic galleries, have very different origins. Clastic sediment in small terraces at meanders or in upper phreatic tubes are of autochthonous origin (e.g. caves in the Vega Huerta region; Sistema del Xitu in Ario; Torca Juanín in the Eastern Massif), with materials derived from the infilling of the dolomitized fractures, which are mineralized with iron oxides, pyrite, other sulfides, Fe-dolomite, and ferrous clay (Fernández-Gibert et al., 1992).

The location of some important caves demonstrates that their origin is related to the Quaternary glaciers, which concentrated water into certain sinkpoints where the recharge was especially intense due to ice melt (5.3.5 Figure 6), such as along glacier fronts and within depressions (Sima Sara in Andara, Pozo Argüelles in Vega Huerta), and in bedrock hills where the ice was highly crevassed (Pozo del Llastral, Sima 56) (Smart, 1984; Senior, 1987). It should be noted that the karst evolution must have continued during the cold periods, because the springs were located far below the lowest levels attained by the ice.

Concerning the large conduits with a phreatic morphology, analysis of their sediments has revealed that in at least two cases they have an allochthonous origin. In Sima M-2, sandy deposits at a depth of 750 m are derived from the Permian-Triassic cover that is now eroded from most parts of the Picos de Europa (Fernández-Gibert et al., in prep.). Similarly, after analyzing the heavy minerals, Smart (1986) attributes the source of sediment in the oldest conduits in Sima 56 to Permian rocks. Their position suggests that the phreatic surface has lowered over 1500 m. The rate of lowering of the base level in the Cueva del Agua at Tresviso has been estimated at between 0.3 and 0.45 m per 1000 years. Therefore, karstification could have begun between 2.5 and 4 million years ago (Smart, 1985).

Other authors have also suggested a pre-Quaternary origin for large inactive conduits elsewhere with similar phreatic morphologies but have not provided definitive evidence to justify their conclusions (Borreguero, 1986; Bes, 1986).

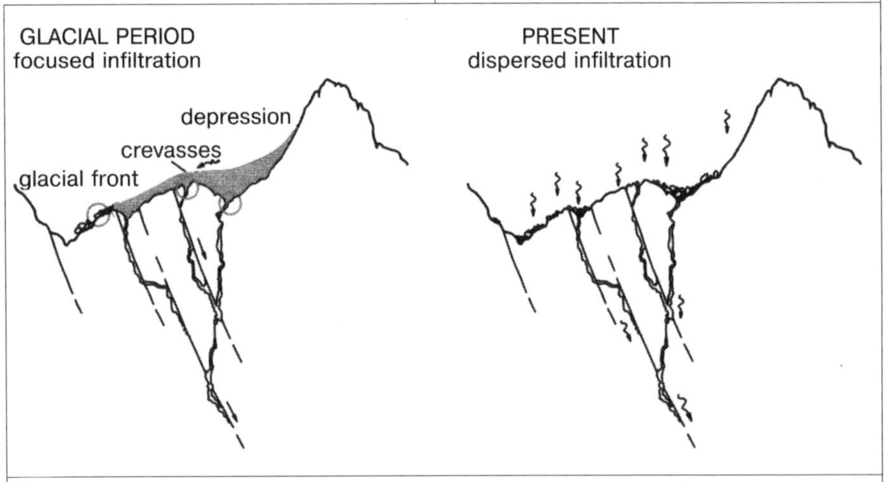

5.3.5 Figure 6. During glaciations, the karst recharge was concentrated at certain sinks.

Conclusions

The analysis of the data described above allows us to propose a speleogenetic model for the Picos de Europa region. Karstification must have begun before the Permian-Triassic cover eroded and the limestones cropped out. Under these conditions, the cover must have partially confined the massif, and thus the karst developed via phreatic conduits. The genesis of some of these large conduits is related to the oxidation of sulfides in mineralized fractures in the massif. The subsequent genesis of the deep shafts in the Picos de Europa is strongly associated with vadose hydrologic settings, and thus to the uplift of the massif, which took place while glacial processes were dominant (5.3.5 Figure 7). As the insoluble cover began to be removed by erosion, the first limestone outcrops captured the surface runoff, causing highly localized recharge into the limestone. Part of the eroded cover was carried into the conduits. Currently, this situation may be observed in some surrounding areas in the Eastern and Western Massifs.

Subsequently, after the impervious cover had disappeared, dispersed infiltration became the predominant recharge mode, a situation which continues today (Fernández-Gibert et al., in prep.). Nevertheless, during the cold periods of the Quaternary, glacial dynamics focused this inflow at certain specific sinks. During this second stage, the large dolomitized and mineralized fractures played a critical role in the development of the great limestone caves, acting as preferential planes for penetration, and in places as a source of more aggressive waters, due to the production of weak acids by the oxidation of sulfide minerals.

Acknowledgements

Thank you to Javier Lario for assistance in data processing, and to Juan Vielsa, and Juan José G. Suárez for providing useful information. This paper was partially financed by CICYT Project AMB95-0493.

References

Bes, C., 1986, Picos: Sierra de Beza 86: *Lo Bramavenc,* v. 10, p. 114-136.

Bigot, Y.V., 1989, Trave System: Third deepest cave in the world: *Caves and Caving,* v. 46, p. 10-14.

Borreguero, M., 1986, *Special Picos:* Puertos de Ondón (unpublished), 15 p.

Collignon, B., 1985, Quelques éléments de géologie et D'Hydrogéologie: *Spelunca Supplemént,* no. 19, p. 7-12.

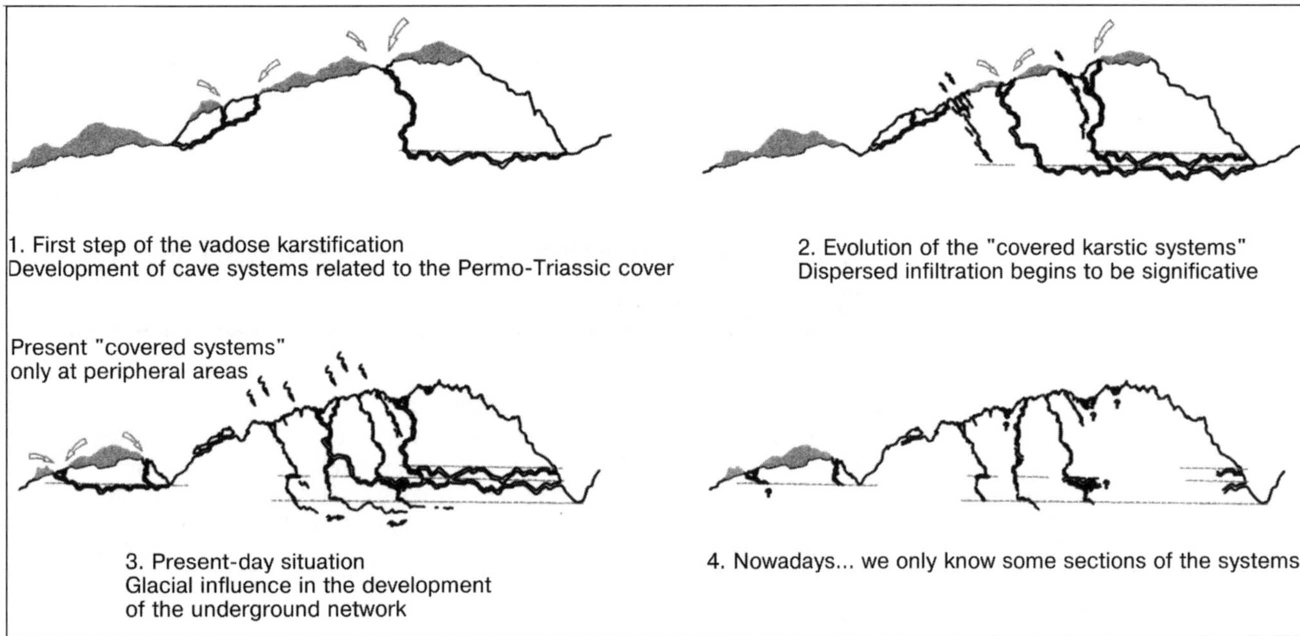

5.3.5 Figure 7. Vadose speleogenetic evolution of the Picos de Europa region (from Fernández-Gibert et al., in prep.).

Farias, P., 1982, La estructura del sector central de los Picos de Europa: *Trabajos de Geología Univ. de Oviedo,* no. 12, p. 63-72.

Fernández-Gibert, E., Rossi, C., and Ortiz, I., 1992, Espeleogénesis y procedencia de los sedimentos kársticos en el Pozo del Llastral (Unidad Frontal, Macizo del Cornión, Picos de Europa): *5th Congreso Español de Espeleología,* Actos, p. 416-425.

Genuit, P., and Vidal, B., 1984, La Torca Urriello; Massif Central Picos de Europa; Asturies, Espagne: *Spelunca,* v. 16, p. 14-18.

Hoyos, M., and Herrero, N., 1989, El karst en la Cornisa Cantábrica, El karst en España: *Monografía Soc. Esp. de Geom.,* no. 4, p. 109-120.

ITGE, 1989, Las aguas subterráneas en España: Estudio de síntesis: *Instituto Tecnológico y Geominero de España,* Estudio de síntesis.

Julivert, M., Truyols, J., and Garcia Alcalde, J., 1971, Mapa geológico de España, sheet 10 (Mieres): *Instituto Tecnológico y Geominero de España,* scale 1:200,000.

Marquínez, J.L., 1997, Mapa geológico de la Región del Cuera y Picos de Europa: *Trabajos de Geología, Univ. de Oviedo, (in press).*

Martínez García, E., 1981, El Paleozóico de la Zona Cantábrica oriental (norteoeste de España): *Trabajos de Geología, Univ. de Oviedo,* no. 11, p. 95-127.

Martínez García, E., and Tejerina, L., 1984, Fluorspar deposits associated with Carboniferous and Permian rocks in Asturias and León, northwest Spain: 9th *International Congress on Carboniferous Stratigraphy and Geology,* Champaign-Urbana, Compte Rendu, v. 9, no. 4, p. 467-478.

Obermaier, H., 1914, Estudio de los glaciares de Los Picos de Europa: *Trab. Mus. Nac. Cienc. Nat., Sr. Geol.,* no. 9, 42 p.

Puch, C., Polsson, P., and Chorvot, G., 1985, Présentation géographique et humaine: *Spelunca Supplemént,* no. 19, p. 4-6.

Sefton, M., 1984, Cave explorations around Tresviso, Picos de Europa, northern Spain, 1975-1983: *Cave Science,* v. 11, p. 199-237.

SEII, 1987, *Memória de Expedición Vega Huerta 87:* Secc. Espel. Ing. Industriales (unpublished), 22 p.

Senior, K.J., 1987, Geology and speleogenesis of the M-2 Cave System, Western Massif, Picos de Europa, northern Spain: *Cave Science,* v. 14, p. 93-103.

Smart, P.L., 1984, The geology, geomorphology, and speleogenesis of the Eastern Massifs, Picos de Europa, Spain: *Cave Science,* v. 11, p. 238-245.

Smart, P.L., 1985, Cave development in the Andara Area, with special reference to 56, *in* La Sima 56 (Picos de Europa, España): *Mem. Lancaster Univ. Spel. Soc. and Secc. Espel. Ing. Industriales,* 86 p.

Smart, P.L., 1986, Origin and development of glacio-karst closed depressions in the Picos de Europa, Spain: *Zeitschreift Geomorphologie,* v. 30, p. 423-443.

Tratman, E.K., 1957, A nameless stream: Suggested new term: *Cave Research Group Newsletter,* v. 6, p. 68-69.

Vegas, R., and Banda, E., 1982, Tectonic framework and Alpine evolution of the Iberian Peninsula: *Earth Evolution Sciences,* v. 4, p. 320-341.

Vidal, B., 1990, Picos de Europa. Sistema del Trave: -1441; T-31, -570: *L'Aven, Spéléo Club de la Seine,* v. 50, p. 63-87.

5.3.6. Speleogenesis of Sistema Cheve, Oaxaca, Mexico

Louise D. Hose

Abstract

The Cheve hydrologic complex in southern Mexico is the deepest known karst conduit system in the world, with a proved relief of 2540 m. The explored upper portion of the system, Sistema Cheve, extends to a depth of 1386 m. The water reemerges from a cave in Santo Domingo Canyon, about 18 km north of the main sink. The entire cave system developed in various carbonates along (and under) the footwall of a Laramide-age thrust fault.

Speleogenesis of Sistema Cheve began in the Neogene or Pleistocene with exposure of a carbonate block isolated by faults. Aggressive allogenic water entered the ground near the contact and formed mostly vertical passages until it arrived at the local base level, which was about 100 m higher than its present level. The stream removed ceilings and walls in the extensively fractured carbonates, resulting in local passage enlargement. Phreatic tubes formed in the middle and lower portions of the hydrologic system. Most of the conduits drained as downcutting continued in Santo Domingo Canyon, leaving mostly air-filled cave passages with a trunk stream and phreatic loops.

Introduction

The vertical relief of the explored upper portion of Sistema Cheve in southern Mexico is presently 1386 m deep (August, 1998). More remarkable is the relief of the entire hydrologic system, which is the most extensive documented in the world. Through physical exploration and a visually positive fluorescein dye trace, the proved relief of the hydrologic system is 2540 m (5.3.6 Figure 1).

The Sierra Juarez, a tropical mountain range within the Sierra Madre del Sur, contains some of the world's deepest caves. Regional hydrologic base level is represented by the Río Grande in Santo Domingo Canyon. The Sierra Mazateca, a subrange containing the 1475-m-deep Sistema Huautla (Stone and am Ende, 1997), the Zongolica-Chilchotla Caves, and the 1160-m-deep Kiajhe Xontua (Bitterli et al., 1990), lies directly north of Santo Domingo Canyon. Agua Fria de Santa Ana, the Cheve system's resurgence, is on the south side of the canyon at river level. Water from Sistema Huautla resurges along the north side of the Río Grande only 1.5 km downriver from Agua Fria.

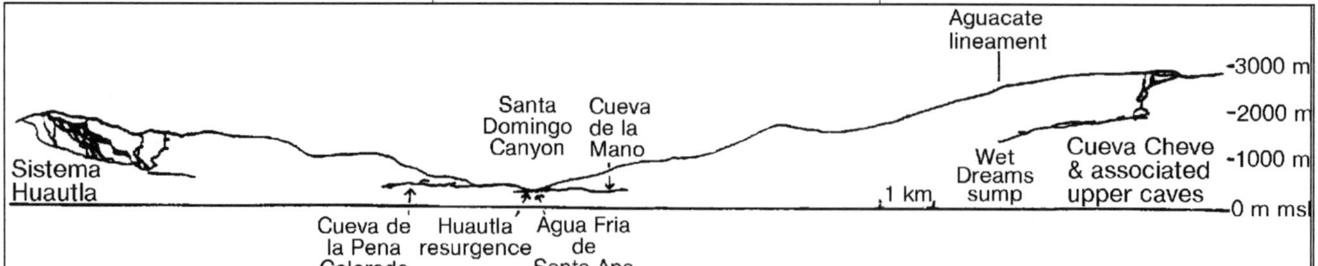

5.3.6 Figure 1. Cross section through the Sierra Juarez near Santo Domingo Canyon. View of Sistema Huautla is from 225°. Santo Domingo Canyon and adjacent caves and Cueva Cheve are viewed from 250°.

Geologic and Geomorphic Setting

Sistema Cheve is in a melange of rudist-bearing limestone, laminated micrite, forereef breccia, thick-to-massive-bedded dolomite, thin-bedded limestone and chert,

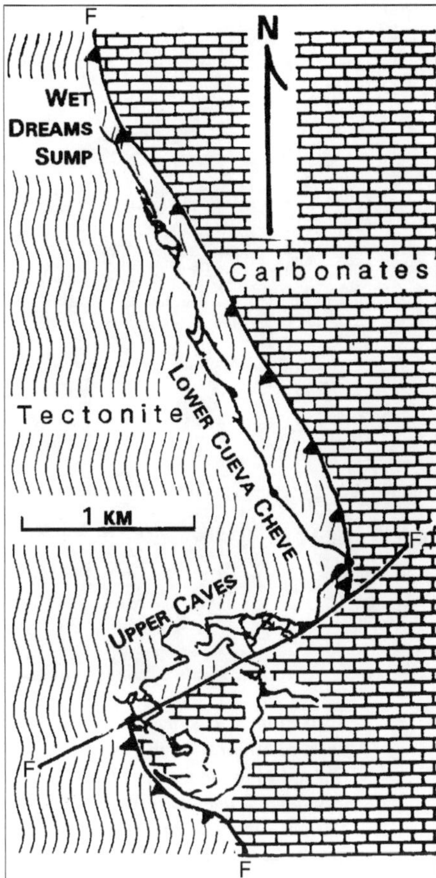

5.3.6 Figure 2. Simplified geologic map of the surface above Cueva Cheve, with a plan-view trace of the cave passages of Sistema Cheve.

thin-bedded limestone, laminated back-reef limestone, and mylonitic marble. All of these lithologies are exposed in the cave. These Cretaceous carbonates are surrounded to the west, south, and east by cataclastic rocks of the severely tectonized Cuicateco Metamorphic Complex, which include talcose and scaly serpentinite, noncarbonate mylonite, sandstone, shale, volcanic rocks, and chert-mica schist. The east-trending band of carbonates containing the cave is only 3 to 5 kilometers wide. The southernmost entrance to the system, Cueva Cheve, lies along the southern boundary of the carbonates.

The region experienced major thrust faulting during the early Tertiary as part of the Laramide Orogeny. Gradual uplift and erosion of the Sierra Juarez with a corresponding subsidence of the Veracruz Basin to the east continued into the Oligocene. Uplift and hence erosion rates accelerated during the late Oligocene and Miocene due to widespread block faulting (Helu et al., 1977; Carfantan, 1981).

The Sierra Juarez is riddled with faults, including north-northwest-trending thrust and back-thrust faults, east-northeast-trending tear faults, and approximately north-trending normal faults. The carbonate block containing Sistema Cheve is bounded by faults at the surface. Sistema Cheve is directly adjacent to and mostly beneath the hanging wall of an east-verging thrust fault that trends about N20°W and dips 60°W.

Most of the cave's known passages developed along joints that parallel the trend of the fault and probably represent the axial cleavage of an east-verging syncline and primary and auxiliary fault planes (Hose, 1992). Passages at both ends of the system (Cueva Cheve and Cueva del Mano) typically trend parallel or subparallel to the local strike of the bedding or foliation. The uppermost passages developed along the south wall of a north-northwest-dipping tear fault (5.3.6 Figure 2).

The Río Grande flows west to east across the entire Sierra Juarez through Santo Domingo Canyon. Helu et al. (1977) suggested that the drainage was established by the early Tertiary and has continuously drained a part of the Tehuacan Valley through the mountain range to the Gulf of Mexico. Some workers (Vinegra, 1966; Smith, 1994) have hypothesized that the canyon follows a major north-northwest dipping strike-slip fault. Today, the river and base level are about 300 m above mean sea level where the waters of Sistema Cheve and Sistema Huautla emerge.

Cave Description

The Cheve hydrologic system can be conveniently divided into four segments:

- Upper Cueva Cheve and associated entrances.
- Lower Cueva Cheve.
- Unexplored middle cave area.
- Cueva del Mano.

Upper Cueva Cheve and Associated Caves

The uppermost portion of the system is represented by at least four distinctly different caves that join into a single system hundreds of meters below the surface. The highest entrance is Cueva Escondida at 2798 m. The largest entrance—and the only one presently receiving a year-around stream—is Cueva Cheve. This also the lowest known entrance at 2654 m. The Cueva Cheve entrance is in a large doline approximately 1 square kilometer in area and about 150 m deep. The other major entrances to the upper system, including Escondida, are on a plateau just north of the doline. Each of the upper caves drops rapidly through a series of vertical pits up to 150 m deep (5.3.6 Plate 1) until the caves converge at various levels above 1850 m.

All passages in the upper caves seem to have formed under vadose conditions. The combination of highly fractured bedrock and active regional tectonism has resulted in massive collapse throughout the upper part of the system. Water paths are commonly lost under breakdown floors. The earlier ceilings and walls have fallen, making direct observations of earlier passage morphologies impossible.

All of the entrances to the upper caves are within 150 m of the noncarbonate terrain. Nearly all of the water flowing into the upper caves during both base flow and flood conditions is derived from the surrounding impermeable terrain. The upper passages trend overall to the east-northeast, along and under the south wall of a probable right-lateral tear fault.

Lower Cueva Cheve

A relatively simple series of passages extends from the 1850-m level and drops with an overall gradient of about 10° toward the north-northwest. The passages are mostly vadose with a few short phreatic loops that have higher air-filled bypass routes. As in the upper caves, most passages have been dramatically modified by collapse. Breakdown covers the floors, obscuring the original passage morphologies.

Lower Cueva Cheve is entirely under the hanging wall of the adjacent thrust fault, and

5.3.6 Plate 1. One of the dozens of vertical drops on the route to the bottom of Cueva Cheve (photo by Bill Stone).

5.3.6 Plate 2. *Tectonized marble walls line the Wet Dreams Passage in the lowest explored portions of Cueva Cheve (photo by Bill Stone).*

5.3.6 Plate 3. *The deepest point of exploration in Cueva Cheve lies in the waters of this sump at the end of Wet Dreams (photo by Bill Stone).*

passages are parallel to the strike in the mostly east-dipping limestone and mylonitic marble. The deepest point of exploration in the cave, 1386 m below the Escondida Entrance, is a phreatic loop formed in the mylonite at the north end of the scoured vadose canyon, Wet Dreams (5.3.6 Plates 2 and 3). Strong air flow suggests air-filled passages above the Wet Dreams Sump, but explorers have not been able to penetrate through the massive breakdown that apparently fills the bypass. A strong northeast-trending surface lineament and a weaker northwest-trending one overlie the present terminus of exploration.

Unexplored Middle Portion

Approximately 13 km separate the two explored portions of the Cheve hydrologic system. The apparent gradient of the projected passage between the Wet Dreams in Cueva Cheve and upstream Cueva del Mano is about 7°.

Cueva del Mano

Cueva del Mano, which feeds Agua Fria de Santa Ana, is a branching horizontal cave. Divers have penetrated four phreatic loops to the south, toward the Wet Dreams Sump. The tops of these loops seem to coincide with the base level of the area, and no vadose streams have been discovered in this portion of the system (Bill Stone, personal communication, 1997).

In the upstream direction, passages in Cueva del Mano trend mostly to the south-southeast and parallel to the strike of the mylonitic marble along the footwall of the same thrust fault that overlies Cueva Cheve. Air-filled passages in Cueva del Mano are branching and display evidence of phreatic or epiphreatic (floodwater) dissolution. Documented relief in Mano is less than 100 m.

Dye Trace

Fluorescein dye dumped into the stream at the Cueva Cheve entrance arrived at Wet Dreams in three days (Don Coons, personal communication, 1993). Eight days after entering the cave (five days after passing Wet Dreams), the brightly dyed water emerged from Agua Fria de Santa Ana (Nancy Pistole, personal communication, 1993).

Speleogenesis

Development of the entire Sistema Cheve seems to be a relatively recent event. The upper caves initially formed from unsaturated flow rushing east-northeastward toward a local base level that was at about 1850 m. Once the water passed the east-northeast-trending tear fault and arrived at base level, it flowed through phreatic tubes toward the Río Grande following a gradient of about 7 to 10° and emerged from phreatic conduits in Cueva del Mano. The base level at Santo Domingo Canyon was probably about 100 m higher than at present. Because of the high hydraulic gradient and steeply dipping and heavily fractured rocks in the area, conduits formed along structural discontinuities such as axial cleavage and primary and auxiliary fault planes.

Speleogenesis in Sistema Cheve probably began when the carbonate block just south of Cueva Cheve, which is now a large collapsed doline, was first exposed to the surface. Allogenic water flowing off the surrounding melange of tectonized noncarbonate rocks was then as now particularly aggressive as it entered fractures at the edge of the carbonate terrain and enlarged the mostly vertical conduits.

As the uppermost passages of early Cueva Cheve enlarged, ceilings and walls collapsed and were removed by dissolution to form the large doline south of the present cave. Downcutting by the Río Grande and the accompanying base level drop partially drained most of the conduits leaving a mainly air-filled cave with a trunk stream and phreatic loops. Erosion continues in the Sierra Juarez.

Acknowledgments

This preliminary study was supported by The Committee for Research and Exploration of the National Geographic Society Grant Number 4724-92. I also wish to thank the members of Proyecto Espeleológico Cheve, Dr. Germán Cruz, and Emily Janecek for their assistance in the field, and Bill Stone for providing his photographs.

References

Carfantan, J.C., 1981, Evolución estructural del sureste de México: Paleogeografía e historía de las zonas internas Mesozoicas: Universidad Nacional Autonómica México, *Instituto Geología Revista,* v. 5, p. 207-216.

Bitterli, T., Jeannin, P.-Y., Meyers, K., and Rouiller, P.H., 1990 : Proyecto Cerro Rabon, Mexico, 1985-1989. *Edition Speleo Projects*, Basel:German and English, 58 p.

Bitterli, T., Jeannin, P.-Y., Hapka, R., et al.,1996, Proyecto Cerro Rabón 1990-1994 (Oaxaca, Mexico). *Edition Speleo Projects*, Basel: 192 p.

Helu, P.C., Rodolfo, V.V., and Rodolfo, B.P., 1977, Origin and distribution of Tertiary conglomerates, Veracruz Basin: *American Association of Petroleum Geologists Bulletin,* v. 61, p. 207-226.

Hose, L.D., 1995, Cave in Oaxaca proven to be world's deepest known karst conduit system: *Revista Mexicana de Cíencias Geológias,* v. 12, p. 108-111.

Hose, L.D., 1992, Structural controls on the development of Cueva Cheve: A reconnaissance report: *National Speleological Society Bulletin,* v. 54, p. 89.

Smith, J., 1994, *Hydrogeology of the Sistema Huautla karst groundwater basin, Sierra Mazateca, Oaxaca, Mexico:* Western Kentucky University, M.S. thesis, 338 p.

Stone, B., and am Ende, B., 1997, The 1994 San Agustin Expedition: *Association for Mexican Cave Studies Activities Newsletter,* v. 21, p. 44-64.

Vinegra, F., 1966, Paleogeografía y tectónica del Mesozico en la provincia de la Sierra Madre y Macizo de Tezuitlan: *Boletín de la Asociación Mexicana de Geologos Petroleros Boletín,* v. 18, p. 147-177.

5.3.7. Cave Development in Burnsville Cove, Virginia, U.S.A.

William B. White

Abstract

The large cave systems of Burnsville Cove are exceptional examples of large caves formed in only 200 meters of limestone that has been strongly folded and faulted. Approximately 75 km of cave passages have been mapped. Structural and lithologic factors have controlled the synclinal shape of the Butler Cave–Sinking Creek System. The Chestnut Ridge caves are a complicated sequence of young vadose infeeder systems coupled with much older perhaps early Pleistocene trunk systems related to the Harrisburg Erosion Surface.

Introduction

Burnsville Cove is in the folded Appalachian Mountains in west-central Virginia, U.S.A., in Bath and Highland Counties. Burnsville Cove appears on the U.S. Geological Survey Burnsville and Williamsville 7.5 Minute quadrangles. It is a relatively small area consisting of 30 km^2 of exposed Silurian–Devonian carbonate rocks containing an extensively developed surface karst with large closed depressions. The master drainage for the area is the Bullpasture River, a tributary of the James River, in turn a part of the Atlantic-slope drainage of the eastern United States. The area has only intermittent surface streams, one of which, Sinking Creek, gives one cave system its name. Most of the water that collects in the cove is discharged through artesian springs into the Bullpasture River, where the river has cut a deep gorge through Bullpasture and Tower Hill Mountains (5.3.7 Figure 1).

The caves of Burnsville Cove are interesting in that they are large systems developed in highly folded, relatively thin, and discontinuous carbonate rocks. In total, roughly 83 km of cave passages have been mapped.

Stratigraphic and Structural Setting

The valley of Burnsville Cove is oriented northeast following the general trend of the folded Appalachians. The cove is bounded on the northwest by Jack Mountain and on the southeast by Tower Hill Mountain. Both consist of thick sequences of clastic rocks capped by the resistant Lower Silurian Clinch Quartzite. The valley is separated into two parts by the anticlinal Chestnut Ridge. The valleys of Sinking Creek and White Oak Draft are synclinal with carbonate rocks extending from the surface to considerable depth. The synclines plunge to the northeast taking the carbonate rocks beneath thick beds of Oriskany Sandstone and Millboro Shale. To the southwest the rising fold axes terminate the cavernous limestone at the contact with underlying clastics. In addition to the main structural features, there are many minor folds, faults, and other structural irregularities. An overview of the regional geology is given by Bick (1962).

The bedrock that crops out in the cove consists of 234 meters of Upper Silurian and Lower Devonian limestone (5.3.7 Figure 2). The lowest unit, the Tonoloway Limestone, is shaly and impure. Caves develop in it primarily through the action of vadose streams entrenching conduits formed in the overlying Keyser Limestone. The main cavernous limestone unit is the 100-meter-thick Keyser Limestone. The Keyser is divided into three segments by two tongues of Clifton Forge Sandstone. This sandstone has a profound effect on the movement of groundwater and the development of caves in the cove. At the top of the section are the Coeymans, New Scotland, and Becraft Limestones. The latter two contain a great deal of chert. Some of the alternate names and additional details of the stratigraphic sequence are shown as an expanded section of the Helderberg Limestone in 5.3.7 Figure 2.

Hydrology

Dye tracing investigations (Davis and Hess, 1982) have delineated three main drainage basins (and a fourth that is outside the present area of interest). Each basin discharges from a separate spring (5.3.7 Table 1).

The Aqua Spring Basin discharges through sumps at the back of Aqua Cave. It takes allogenic recharge from small surface tributaries on the slopes of Jack Mountain as well as diffuse infiltration from the limestone valley and internal runoff from the large closed depressions. Substantial parts of the underground route of Sinking Creek can be observed within the Butler Cave–Sinking Creek System, but more than half of the drainage lies downstream from the terminal sumps.

Cathedral Spring rises at the base of a bluff on the Bullpasture River. Diving in the spring (Simmons, 1990) reveals a fissure passage opening into a water-filled tube. This tube continues to descend, at the limit of the dive, to a depth of 45 m below river level and a linear distance of 290 meters from the entrance. The catchment basin of Cathedral Spring wraps around the Chestnut Ridge Anticline so that a part of the recharge is obtained from Jack Mountain. Recharge is obtained from Chestnut Ridge and as diffuse infiltration and internal runoff from the entire valley floor at the southwest end of the cove. Much of the northeastern portion of the Chestnut Ridge caves drains to Cathedral Spring.

Blue Spring is also an artesian discharge on the banks of the Bullpasture River. It has a long narrow catchment along the base of Tower Hill Mountain. Most of the recharge is allogenic runoff from the northwest flank of Tower Hill Mountain. Bick's (1962) geologic map shows a fault along the base of Tower Hill Mountain that may be responsible for the narrow shape of the Blue Spring catchment.

Cave Systems

The principal cave systems are listed in 5.3.7 Table 2. Outline maps of those caves with surveys that have been entered into the master data base are shown in 5.3.7 Figure 3. The most significant omission is Breathing Cave, which was one of the first caves in the cove to be investigated (Deike, 1960). The exploration prior to 1980 is described in detail by Wefer and Nicholson (1982).

Butler Cave–Sinking Creek System

The present-day Butler Cave–Sinking Creek System serves to take the allogenic recharge from the flanks of Jack Mountain along with water collected from the dry valley of Sinking Creek and deliver it to the Aqua Cave Spring. Because of the northeast-plunging syncline, much of the downstream part of the drainage

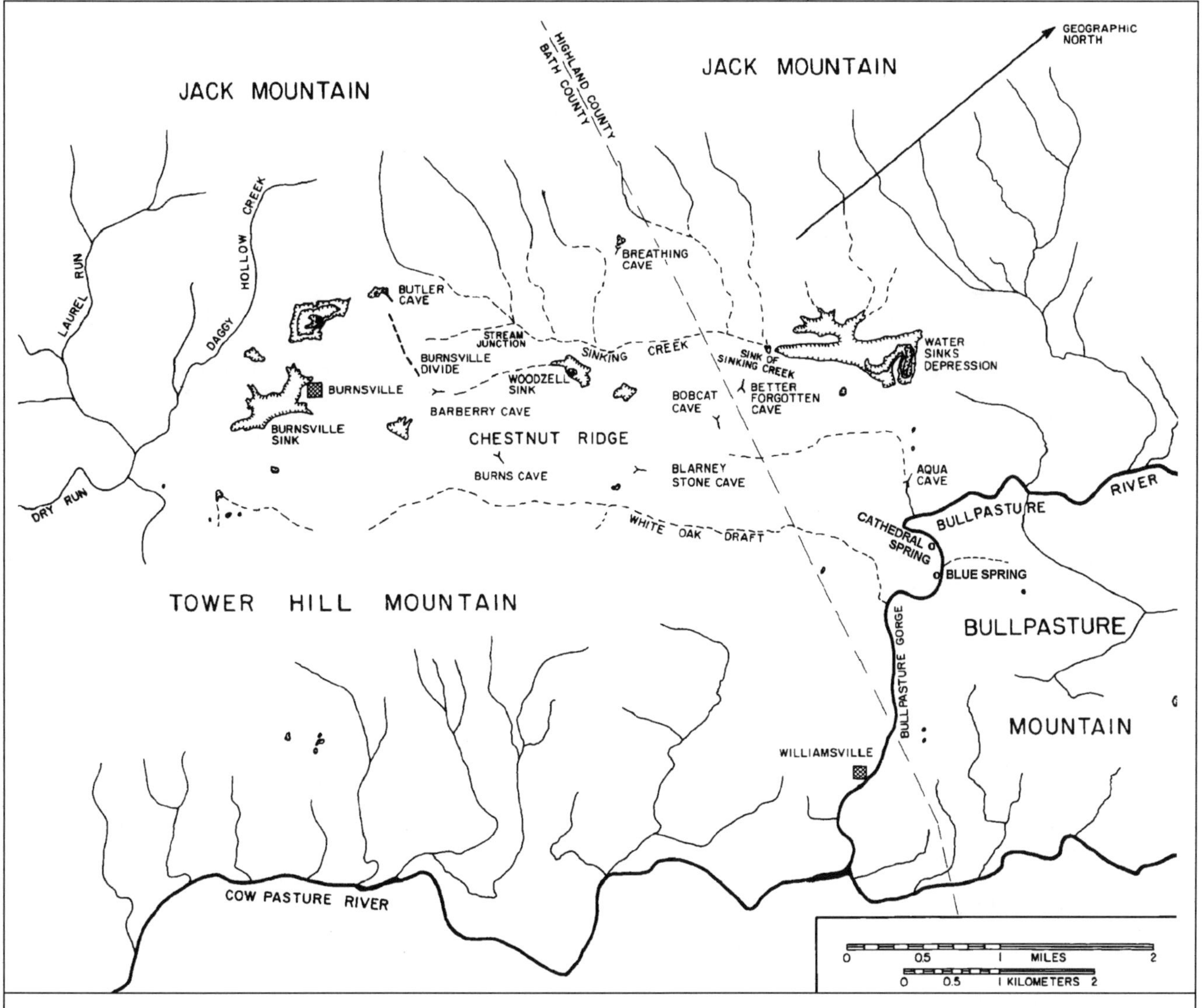

5.3.7 Figure 1. Map showing surface karst of Burnsville Cove (from White and Hess, 1982).

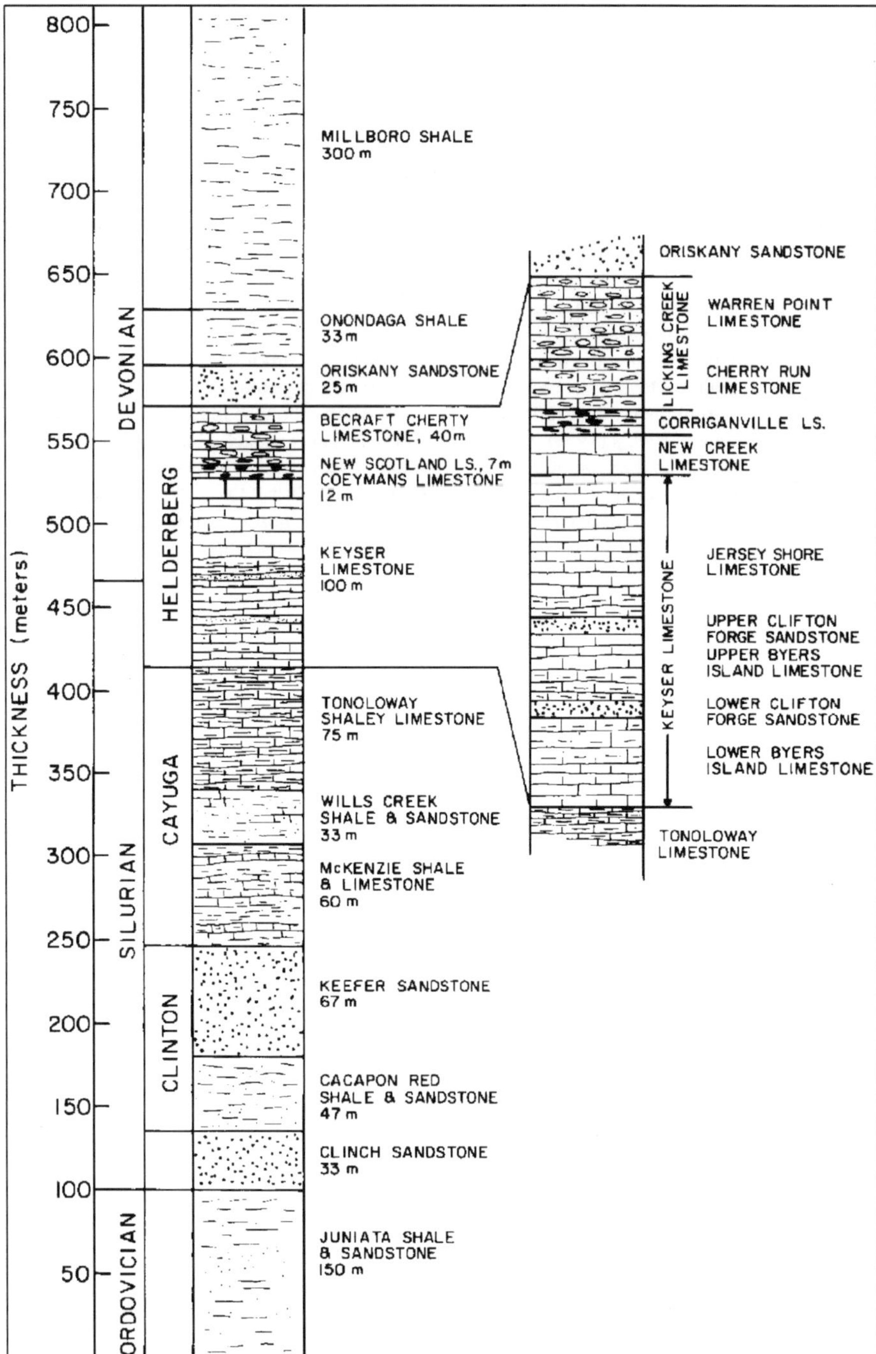

5.3.7 Figure 2. Stratigraphic column for Burnsville Cove (from White and Hess, 1982).

Spring	Catchment Area (km²)	Low Flow L/s	Low Flow/ Unit Area L/s/km²
Aqua Cave	**20.0**	**273**	**13.6**
Cathedral Spring	**10.9**	**131**	**12.0**
Blue Spring	**3.8**	**52**	**13.7**
Data from Davis and Hess (1982)			

5.3.7 Table 1. Spring discharge basins of Burnsville Cove.

Cave	Length (km)
Aqua Cave	**2.9**
Barberry Cave	**5.3**
Better Forgotten Cave	**1.2**
Bobcat Cave–Blarney Stone Cave	**22.1**
Breathing Cave	**8.0**
Burns Chestnut Ridge Cave	**3.8**
Butler Cave–Sinking Cave System	**27.3**
Helictite Cave	**9.6**

5.3.7 Table 2. Principal caves of Burnsville Cove.

system is below the water table. Very little recharge seems to come from the Chestnut Ridge side of the syncline.

The Butler Cave–Sinking Creek System consists of a central trunk passage oriented along the synclinal axis. The upstream (southwest) end of the trunk consists of a network maze of infeeders. On the northwest flank of the syncline are side caves consisting of network mazes with large sloping passages oriented in the downdip direction. There is relatively little passage development on the southeastern flank of the syncline extending up Chestnut Ridge.

The southwestern part of the trunk lies in the limestone section below the lower sandstone like the major upstream-side caves. About midway, the trunk passage breaches the lower sandstone so that the downstream portion of the main trunk passage and its terminal sumps lie between the sandstones. So does Breathing Cave, which seems to be a side cave to the system draining into it beyond the terminal sump. The lower sandstone is about 6 m thick. The trunk passage narrows where it crosses the sandstone, but there is no other evidence of the change in lithology.

Thus the main trunk of the Butler Cave–Sinking Creek System crosses a major lithologic barrier and has an overall drainage slope to the northeast that is less than the dip of the plunging synclinal axis. At the upstream end of the cave is an overlying tier of cave passages (Mbagintao Land) above the lower sandstone tongue. At the downstream end of the cave, where the main trunk lies between the sandstones, a shaft leads into a lower section of cave below the lower sandstone (Marlboro Country).

Various infeeders into the cave contribute to an active stream flowing down the plunge of the synclinal axis in the trunk passage. The upstream water sources flow into Sinking Creek, which is lost at the Sinking Creek Sump and then reappears below the lower sandstone in Marlboro Country where it again sumps.

Downstream infeeders contribute to Sneaky Creek, which is lost in perched sumps at the Rats Doom and Last Hope Sumps. Downstream from the sumps, all the streams have been traced to Aqua Cave 4 kilometers away.

Chestnut Ridge Caves

Intense exploration in the decades of the 1980s and 1990s revealed a complex of caves beneath Chestnut Ridge. These have a quite different character than the caves on the Sinking Creek side of the anticline.

The entrances to Bobcat Cave, Blarney Stone Cave, and Better Forgotten Cave are high on the ridge. The entrance series of all three caves are sequences of muddy crawls, pits, and tight canyons. They seem to be vadose infeeders linked to contemporary drainage.

In Bobcat Cave, however, a complex sequence of passages on multiple levels descends ultimately to streams with a lowest altitude 220 meters below the entrance. Many of these drain to Cathedral Spring. Within this sequence is a master trunk passage, the Burnsville Turnpike, with an average diameter of tens of meters. It is a strike-oriented passage parallel to the ridge.

In recently discovered Barberry Cave, several kilometers to the southwest, is another segment of large trunk passage at the same level and with the same orientation that may be part of the same master drain.

The Chestnut Ridge caves seem to have had a more complicated history than the Sinking Creek Caves. There are the master trunks, perhaps fed by large closed depressions, or their ancestors, near the southwestern limit of the karst. There is little allogenic recharge into the Cathedral Spring catchment, but runoff from Chestnut Ridge and the Valley of White Oak Draft contributes to contemporary active streams within the system. There is what may be a very old system on which has been superimposed the contemporary vadose caves and stream modifications in the lower levels.

Speleogenetic History of Burnsville Cove

The caves of Burnsville Cove are an excellent illustration of stratigraphic and structural control on speleogenesis. The orientation of the synclinal and anticlinal axes determines the primary orientation of cave development.

Some parts of the cave system seem to be very old. In the side caves of the Butler Cave–Sinking Creek System are thick fill sequences of poorly sorted silt, sand, and cobbles. The cobbles, many 10 to 20 cm in diameter, are made of sandstone clearly derived from the flanks of Jack Mountain. These seem to have been washed down the sloping dip passages by catastrophic floods. Paleomagnetic measurements (I. Sasowsky and V. Schmidt, unpublished data) reveal magnetically reversed sediment in Butler Cave near the trunk channel. The measurements at present are too sparse to identify the specific reversal. But the observation does establish a minimum age for sediment emplacement of 780,000 years. A fully developed cave system was already in existence when the sediment was deposited.

A clue to the age and significance of the Burnsville Cove caves is provided by the large closed depressions near the Burnsville divide shown on 5.3.7 Figure 1. They cluster near an altitude of 760 meters, also the height of Chestnut Ridge and other secondary ridges (5.3.7 Figure 4). This altitude in west-central Virginia corresponds to a valley upland and accordant ridgetop erosion surface that can be traced through much of the Appalachian Highlands and has been correlated with the Harrisburg Erosion Surface (White and White, 1991). The Harrisburg Surface may have been cut well back into the Tertiary, but examination of its karst manifestations suggests that the recognizable features of the surface are late Pliocene or early Pleistocene, only a few million years old.

The Burnsville Cove caves exhibit, therefore, particularly in the plunging synclinal shape of the Butler Cave–Sinking Creek System, an excellent illustration of stratigraphic and structural controls on cave development. At the same time, the chronological development of the caves extends from at least the early or middle Pleistocene to the present.

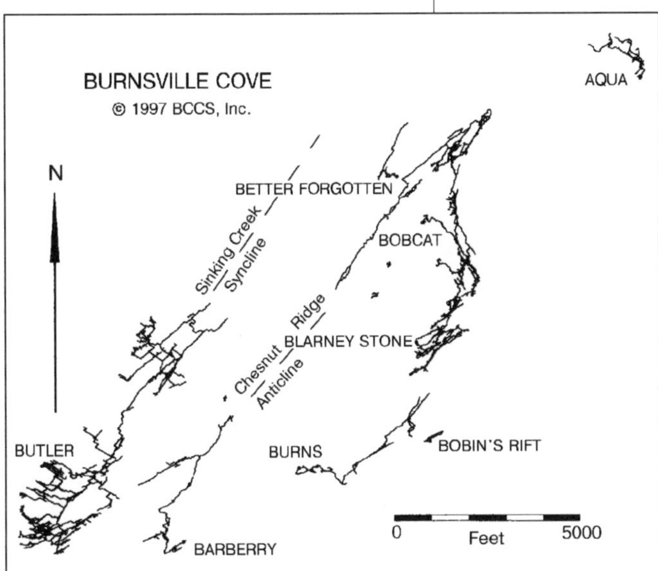

5.3.7 Figure 3. Map showing interrelationships of principal cave systems. Unpublished data provided by the Butler Cave Conservation Society. Rough positions of fold axes were extracted from the geologic map of the Williamsville Quadrangle (Bick, 1962).

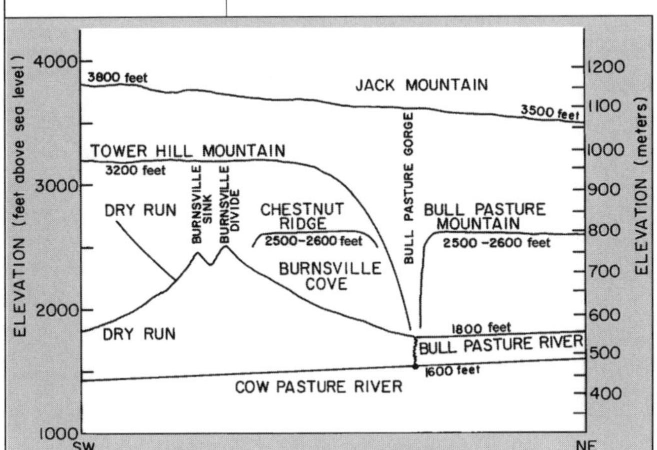

5.3.7 Figure 4. Profile sketch showing relative altitudes of ridge summits.

Acknowledgment

I am grateful to the Butler Cave Conservation Society and to Nevin W. Davis for providing information on the more recent discoveries at the Burnsville Cove caves.

This chapter is also dedicated to the memory of Fred L. Wefer, the President of the National Speleological Society, who died of cancer in the Fall of 1999. Burnsville Cove was Fred Wefer's favortite caving area. He was one of Butler Cave's earliest and most enthusistic explorers and one of the founders of the Butler Cave Conservation Society.

References

Bick, K.F., 1962, Geology of the Williamsville Quadrangle, Virginia: *Virginia Division of Mineral Resources Report of Investigations,* no. 2, 40 p.

Davis, N.W., and Hess, J.W., 1982, Hydrogeology of the drainage system, Burnsville Cove, Virginia: *National Speleological Society Bulletin,* v. 44, p. 78-83.

Deike, G.H., III, 1960, Origin and geologic relations of Breathing Cave, Virginia: *National Speleological Society Bulletin,* v. 22, p. 30-42.

Simmons, R., 1990, In search of the Cathedral System: *Butler Cave Conservation Society Newsletter,* v. 16, p. 1-8.

Wefer, F.L., and Nicholson, I.K., 1982, Exploration and mapping of the Sinking Creek System: *National Speleological Society Bulletin,* v. 44, p. 48-63.

White, W.B., and Hess, J.W., 1982, Geomorphology of Burnsville Cove and the geology of the Butler Cave–Sinking Creek System: *National Speleological Society Bulletin,* v. 44, p. 67-77.

White, W.B., and White, E.L., 1991, Karst erosion surfaces in the Appalachian Highlands, *in* Kastning, E.H., and Kastning, K.M., eds., *Appalachian Karst Symposium,* Proceedings: Huntsville, National Speleological Society, p. 1-10.

5.3.8. Speleogenesis of the Mammoth Cave System, Kentucky, U.S.A.

Arthur N. Palmer

Abstract

The Mammoth Cave System, in southwestern Kentucky, U.S.A., is in carbonate rocks of Mississippian (Early Carboniferous) age, which dip less than 1 degree toward the northwest into the Illinois Structural Basin. The cave is contained within the Chester Upland, a fluvially dissected plateau of limestone and dolomite capped by clastic rocks. However, much of the groundwater recharge to the cave today and in the past has been from the nearby karst surface of the Pennyroyal Plateau, updip from the Chester Upland, where the insoluble caprock has been removed by erosion, exposing the carbonate rocks over a broad surface.

The local bedrock is very prominently bedded, with few large fractures, and as a result the cave passages are sinuous canyons and tubes, interspersed with vertical shafts that have developed downward in sequence, bed by bed. Passages of vadose origin are oriented almost invariably down the local dip, which is complicated by many irregularities superimposed on the regional dip. Phreatic passages show no inherent relation to the dip and are mostly oriented at very shallow angles to the strike direction. Depth of cave development below the water table is limited to a few tens of meters by the small amount of structural deformation.

Dating of sediment shows that most of the earliest passages probably formed in the late Pliocene Epoch, although the very highest passages may be considerably older. These passages are few but large, and consist mainly of wide canyons and tubes that were later almost filled with silt, sand, and gravel during periods of regional aggradation.

Diversion of drainage into the Ohio River by glacial ice during the early Pleistocene caused rapid entrenchment of the Ohio and its tributaries in the Mammoth Cave region. Passages became much more numerous as the landscape was dissected and as erosional base levels fluctuated more rapidly. Several major passage levels developed during periods of relatively static base level. The Mammoth Cave System thus provides many clues to the geomorphic history of the surrounding region.

Introduction

Mammoth Cave is in southwestern Kentucky, in the heart of America's most extensive karst region. It is by far the longest cave in the world, by a factor of almost 3, with a total surveyed length of 560 km (5.3.8 Figure 1). It is the main attraction of Mammoth Cave National Park, and about 15 km of trails within the cave are open to the public. Surrounding this system are many other genetically related caves that have not yet been linked to Mammoth Cave by explored passages. Among these are Fisher Ridge Cave (146 km long), Hidden River Cave (32 km), and the Martin Ridge System (52 km), each of which is being rapidly expanded by current exploration. The astonishing length of the Mammoth Cave System has been achieved as much by the discovery of connections with neighboring caves as by exploration within the historical limits of Mammoth Cave. In a length-to-rank plot of the long caves of the world (Ford and Williams, 1989, p. 245), Mammoth Cave lies far beyond its expected length as the first-rank cave.

This cave system is probably the world's most thoroughly studied example of a cave in gently-dipping strata. Although its vertical extent is only about 120 meters, it exhibits many discrete stages of development, with as many as 8 or 9 superimposed passages in some areas. This complexity is due partly to the prominent bedding of the strata and partly to a long and intricate geomorphic history.

Geologic Setting

The Mammoth Cave System is contained within a low dissected plateau of Mississippian (Lower Carboniferous) limestone capped by insoluble clastic rocks, mainly quartz sandstone. It is at the southeastern edge of the broad Illinois Structural Basin, where the local strata dip gently toward the northwest at less than half a degree (5.3.8 Figure 2). Steep-walled stream valleys have cut through the overlying insoluble rocks, exposing 100 m of limestone in nearly flat-topped ridges with highly irregular patterns. This rugged, hilly region is the Chester Upland, within which the currently explored parts of the Mammoth Cave System are contained. Farther southeast, the insoluble caprock has been completely removed by erosion, exposing the limestone at the surface in an uninterrupted low plateau (Pennyroyal Plateau) that extends across thousands of square kilometers. The strata in the region and the landscape types developed upon them are shown in 5.3.8 Figure 3 (p. 370). The relation between the cave and the geologic setting is shown in 5.3.8 Figure 4.

The limestones of the area are of shallow marine origin and exhibit thin and prominent bedding. Three limestone formations are exposed in the cave. In ascending order they are the:

- St. Louis Limestone (60 m thick, of which only the upper 40 m is exposed in the cave).

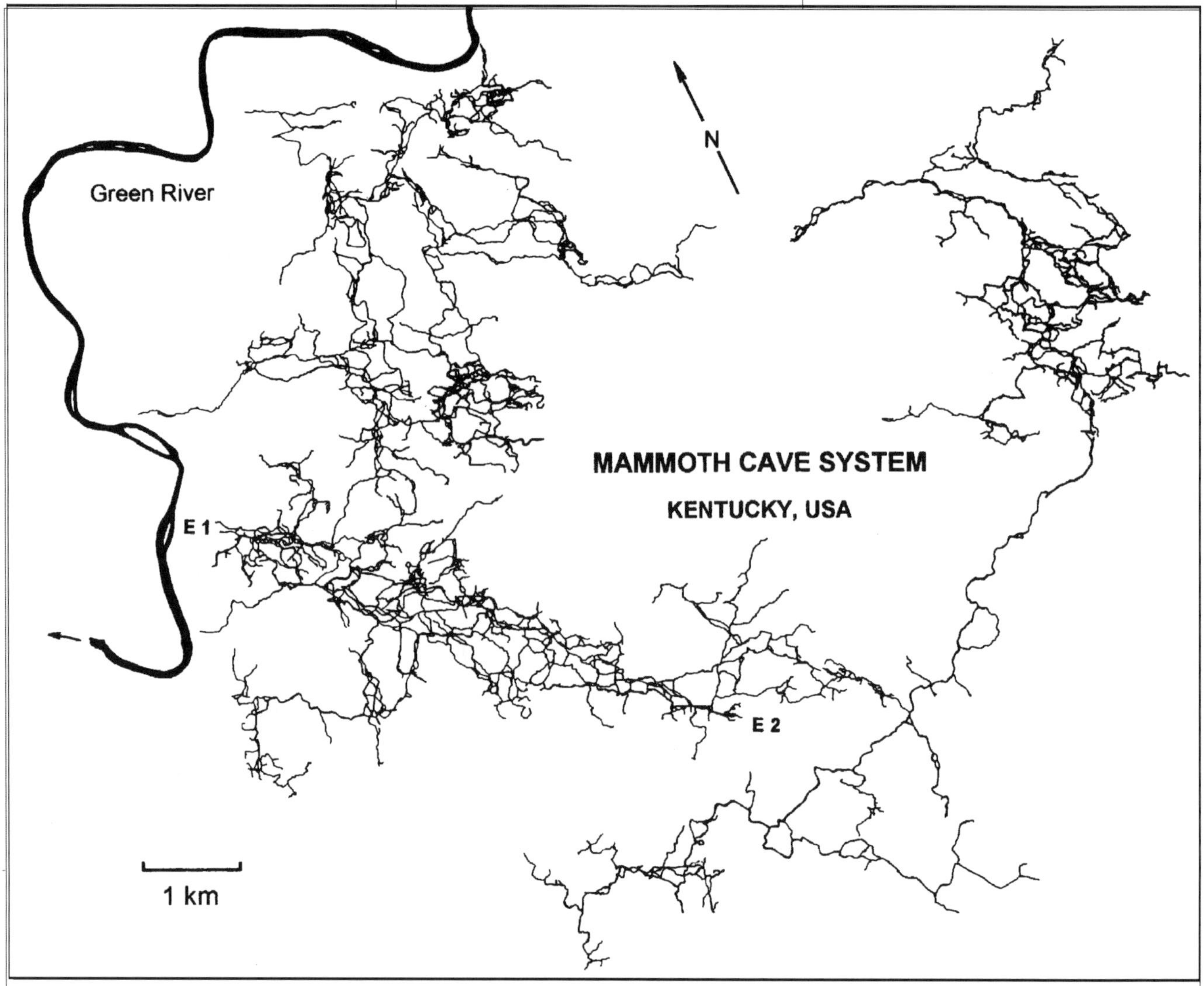

5.3.8 Figure 1. Map of the Mammoth Cave System, from surveys by the Cave Research Foundation (main part of system) and Central Kentucky Karst Coalition (Roppel Cave, in the eastern part of the system). E1 is the Historic Entrance. E2 is the Frozen Niagara Entrance. There are at least 23 other entrances to the system.

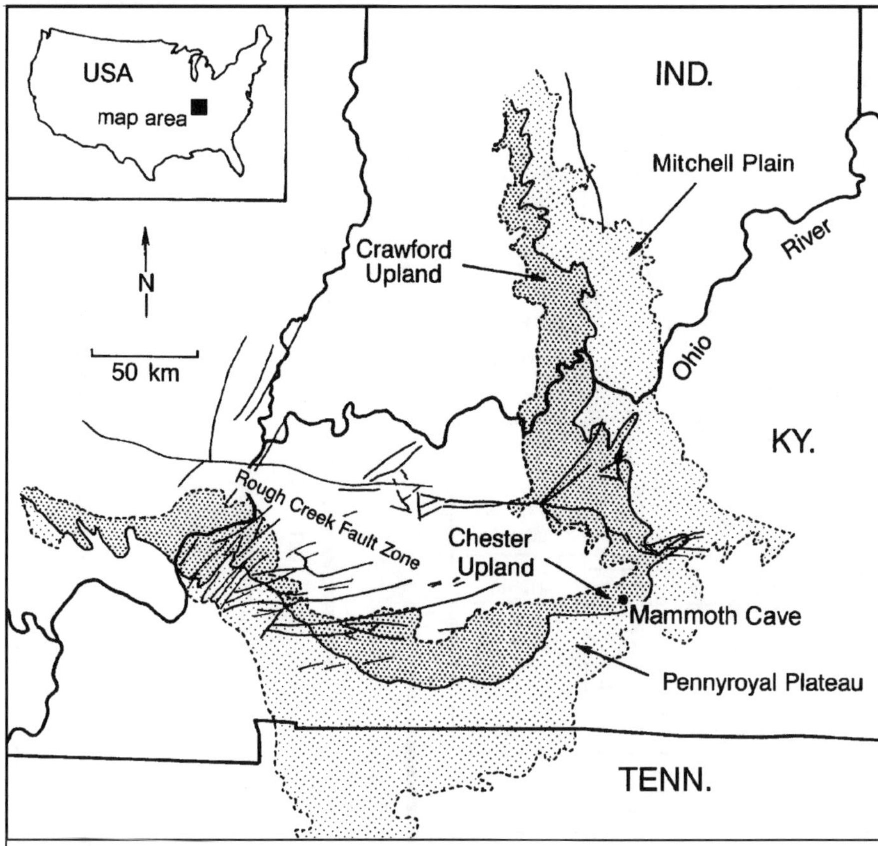

5.3.8 Figure 2. Location map for the Mammoth Cave System, showing regional geologic setting.

- Ste. Genevieve Limestone (30 to 36 m).
- Girkin Formation (30 to 43 m).

Thin beds of shale and dolomite occur throughout the section, and chert is common in the lower parts, especially in the St. Louis. Joints are narrow and of limited extent (Deike, 1967). Faults with up to 15 m of displacement occur in the region, but the largest faults observed in the cave show displacements of only a few tens of centimeters. Because prominent fractures in these limestones are so sparse, individual beds and bedding-plane partings exert the main control over the directions of passages in most caves of the region.

The Mammoth Cave System is fed by recharge from several sources, including dolines and sinking streams in adjacent parts of the Pennyroyal Plateau, karst valleys within the Chester Upland where the insoluble caprock has been breached by stream incision, and dispersed seepage along the irregular edges of the sandstone caprock (Quinlan and Rowe, 1977, 1978; Quinlan and Ray, 1981; White and White, 1989). The subsurface drainage exits at several large springs to the northwest along the banks of the Green River.

Although extensive caves are known on the Pennyroyal Plateau (e.g. Hidden River Cave to the east), and several large rivers in the Mammoth Cave System are fed by water from that same region, no passage in the cave has yet been explored beyond the limits of the Chester Upland. This is partly because most of the low-level stream passages in the cave are water filled, owing to late Pleistocene alluviation of valleys in the region.

Throughout the erosional history of the region (roughly since the Cretaceous), the karst landscape has gradually migrated northwestward toward the center of the Illinois Basin as the overlying rocks were slowly stripped away. Superimposed on this broad trend are periodic changes in stream entrenchment rates, which accelerate, decelerate, or even periodically reverse the erosional dissection of the region. To use the nomenclature of Klimchouk (1996; see also Klimchouk and Ford, Chapter 3.1 in this volume), the karst has evolved as follows:

- ***Deep-seated karst***, completely overlain by insoluble rocks, develops where a few favorable groundwater paths were enlarged, particularly along gypsum beds. This kind of flow has been demonstrated in certain zones deep within the St. Louis Limestone (Ash, 1984), with upward discharge through the caprock along major faults.
- ***Subjacent karst*** develops beneath the insoluble caprock, but with discharge into valleys that have breached the caprock. This characterizes much of the Mammoth Cave area.
- ***Entrenched karst***— Entrenched karst is the stage in which the aquifer becomes exposed by valley entrenchment, while the caprock is still retained, allowing perched flow to take place on underlying insoluble rocks. Because of the low stratal dip and small depth of entrenchment, the main karst aquifer at Mammoth Cave does not exhibit this characteristic. However, thin limestone aquifers within the otherwise insoluble caprock are of this type.
- ***Denuded karst***—Denuded karst is represented by the Pennyroyal Plateau which is the final stage of karst development, where the caprock has been entirely removed. All four stages can be seen in the Mammoth Cave area if one traverses it in the updip direction toward the southeast.

The location and prominence of karst at any given time depends on the amount of relief and the position of the exposed limestone relative to river valleys. At present the Green River is entrenched deeply into the carbonate strata (an average of 100 m at Mammoth Cave), and vast areas of exposed limestone in the region drain into the Green River and its largest southern tributary, the Barren River. Because of this deep entrenchment, karst development in the region may never have been more extensive than it is today.

Cave Pattern

Mammoth Cave consists mainly of long sinuous tubes and canyons with very low gradients, and vertical shafts up to 40 m deep. Typical passages in the cave consist of narrow canyons alternating in a stair-step manner with shafts (5.3.8 Plate 1), which feed tubular passages downstream. Most shafts are clustered around the ridge boundaries, where they are fed by diffuse infiltration through the thin eroded edge of the caprock. Many shafts and canyons intersect older abandoned passages.

The prominent bedding of the limestone has a strong control over passage morphology. Most canyons are highly sinuous, with nearly flat ceilings, and they are almost all oriented directly down the local dip. Sinuosity is controlled by local changes in dip direction caused by minor irregularities (mostly of depositional origin) along bedding-plane partings. In contrast, upper-level canyons in the cave are very large, up to

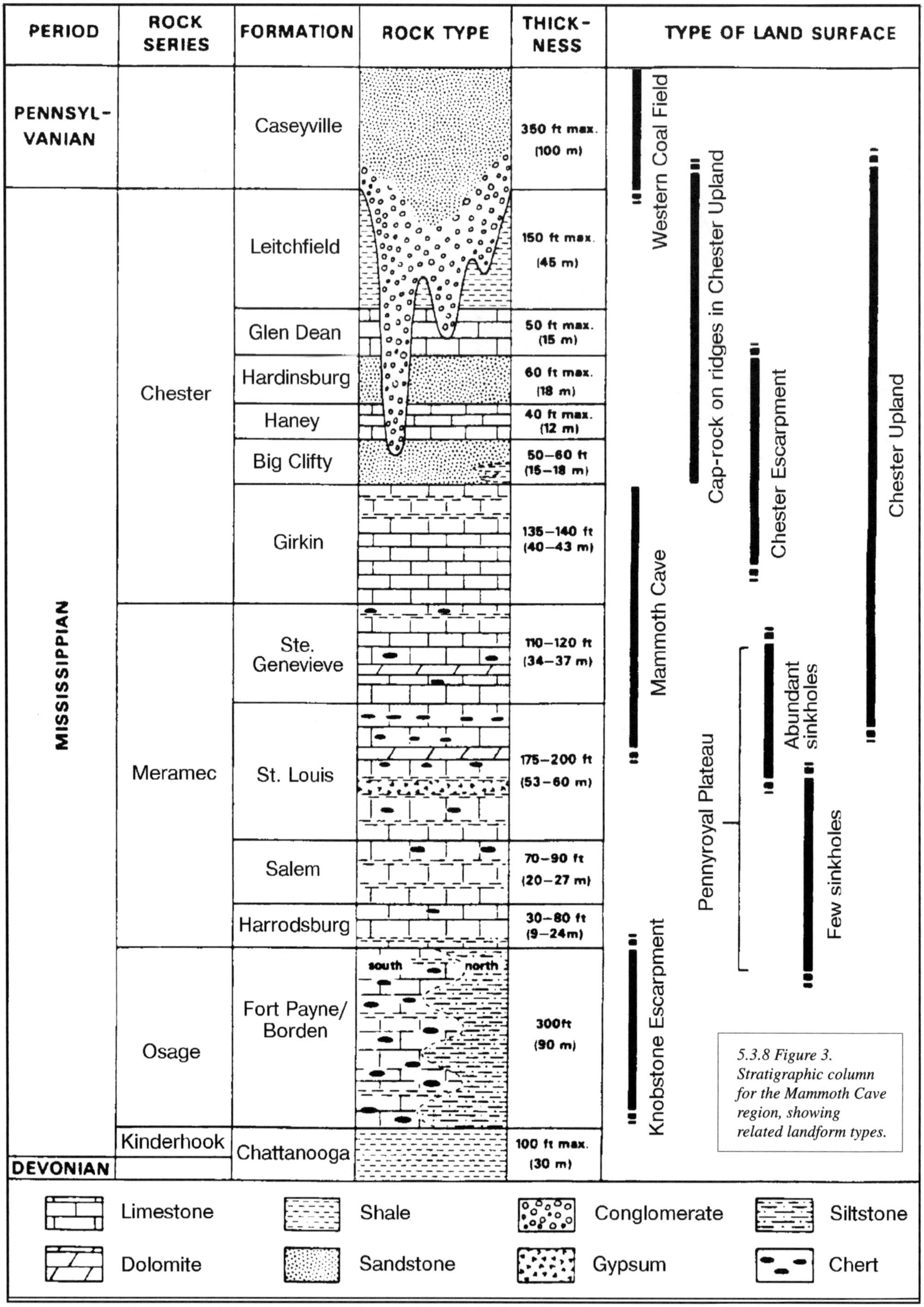

5.3.8 Figure 3. Stratigraphic column for the Mammoth Cave region, showing related landform types.

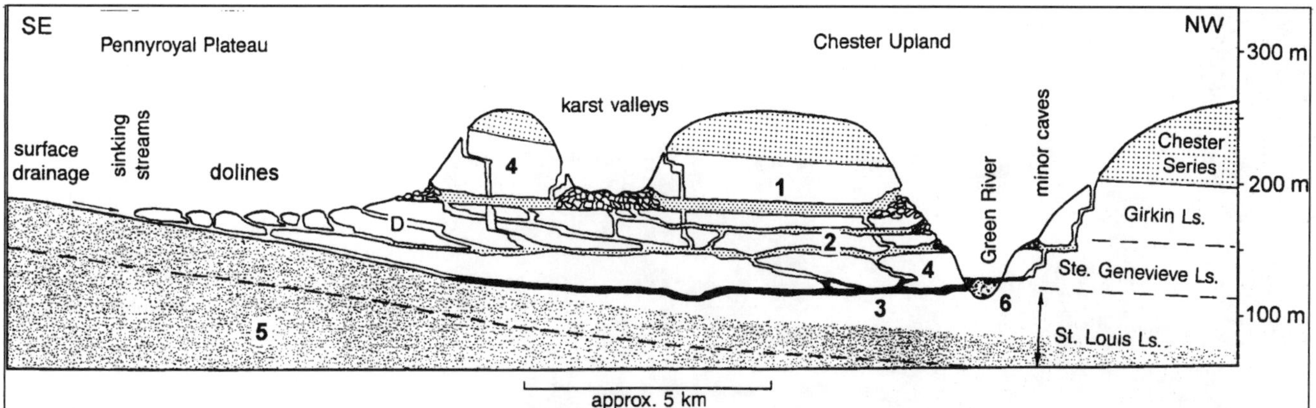

*5.3.8 Figure 4. Generalized profile through Mammoth Cave, showing the relation of cave levels to strata and to surface features. **1**—Large upper-level canyons of Tertiary age. **2**—Tubular passages of Pleistocene age. **3**—Active phreatic tubes at and below fluvial base level. **4**—Active and relict vadose canyons and shafts. **5**—Impure limestone with poor karst development. **6**—Late Pleistocene alluvial fill up to 16 m thick. None of the "minor caves" on the north side of the river are longer than 3 km.*

5.3.8 Plate 1. A typical vadose passage complex in Mammoth Cave (Waterfall Trail). The canyon feeding this waterfall is entrenched into the floor of a downdip tube perched on a resistant insoluble bed. The water has abandoned the downstream part of its canyon at this point, forming a shaft that drops 30 m to a narrow flooded shaft drain at base level, intersecting an older vadose canyon formed by water from another source.

5.3.8 Plate 2. Audubon Avenue, on the Historic Route of Mammoth Cave, a typical wide upper-level canyon at an elevation of 190 m (at its ceiling). The bedrock floor lies beneath several tens of meters of clastic sediment.

20 m wide and 30 m high, and are generally far less sinuous than their narrow counterparts (5.3.8 Plate 2).

Shafts are nearly round or oval in horizontal cross section and almost all have developed by stages through successive downward dissolution from bed to bed, as shown by sequential drains in their walls at many levels. Each drain leads to narrow canyons or perched tubes that interconnect in tangled complexes. Very few shafts have formed as single pits along major fractures. The abundance of sequential shaft drains is partly responsible for the extensive interconnectivity between various parts of the cave and accounts for the fact that many shafts can be explored from top to bottom through interconnecting passages without the use of artificial aid.

Tubular passages are curvilinear and sinuous, with lenticular cross sections elongated along the bedding. Fissure passages and linear passage segments are rare.

Most passages are strongly influenced by vadose processes. The downdip gravitational pattern is exhibited even by some of the largest passages, and many tubular passages are simply perched on strata of low solubility and show a similar downdip pattern. This dip influence is clear only from precise geologic leveling, which shows that both the magnitude and direction of the dip vary a great deal from place to place, and from bed to bed, even showing local reversals from the regional dip. The great variation in passage trends gives the false impression that the passages are simply following random paths through the limestone.

The majority of canyons in the cave formed during multiple stages. As the descending vadose water encounters successively lower bedding-plane partings, it is guided along each new path by the unique pattern of dips exhibited by each parting. In most canyons each stage follows quite a different path from the previous one, so that they coincide only at random intersections,

5.3.8 Plate 3. Great Relief Hall, on the Historic Route of Mammoth Cave, a tubular passage at the 152-meter level (at its ceiling elevation).

where it is possible to climb from one tier to another. Many vadose canyons have achieved great size, and it is important to recognize that not all the large passages in the cave are of phreatic origin.

Less common are tubular passages of distinctly phreatic origin, i.e. with evidence for upward dissolution of their ceilings, undulatory profiles, and no consistent relation to the dip direction (5.3.8 Plate 3). Many tubes are oriented nearly along the local strike of the beds. This trend is typical where groundwater passes through well-bedded rocks. Usually the most efficient outlet for phreatic water in such conditions is at shallow depth below the water table, where partings are widest. When traced upstream toward their original sources, phreatic passages formed in this way reach points where their morphologies change from tubes to canyons with persistent downdip patterns.

Within any given passage, this point represents the change from former vadose to phreatic conditions (the "piezometric limit"). This term does not apply simply to a former water table, but specifically to that part of a passage (generally an abandoned one) that demonstrates this transition. Piezometric limits are clearly displayed in Mammoth Cave by geologic leveling surveys. Passage profiles drawn from these surveys show that the deepest known extent of phreatic passage development below the piezometric limit (phreatic loops) is only about 20 m (Palmer, 1987).

The piezometric limits in the major passages cluster very tightly at several distinct altitudes that are independent of stratigraphic control. These are true "levels" in the geomorphic sense, and their elevations relate to the geomorphic history of the entire region. They are described in detail later in this chapter.

The strong vadose imprint and shallow penetration into the phreatic zone are to be expected in a clearly epigenetic cave in which descending water encounters strongly bedded rocks of highly varied lithology. They also show the strong relation between caves and the present hydrologic setting. There is little or no evidence for prior deep-phreatic conditioning of its passages. This is rather surprising when one considers that the limestone at Mammoth Cave has evolved from deep-seated conditions similar to those that now prevail farther down in the Illinois Basin. At least some inheritance of deep karst features should be expected (see Klimchouk and Ford, Chapter 3.1; Lowe, Chapter 3.3; Klimchouk, Chapter 5.2).

But careful geologic mapping and leveling of at least 50 km of passages of all kinds in the cave, as well as field examination of many more, has shown less than 100 m of passage length that cannot be clearly assigned to a shallow meteoric origin well adjusted to the present geomorphic setting. These odd segments consist of a few blind passages of irregular profile and cross section that have no clear hydrologic relation to their neighbors. Some simply may be ancient floodwater features. In any case, it is clear that Mammoth Cave has inherited virtually nothing from deep karst processes.

The strong influence of local bedding structures on the cave pattern can be illustrated in nearly every passage in the cave. For example, 5.3.8 Figure 5 shows a small segment of the longest explored river passage in the cave (Logsdon River), which is located within the cherty upper beds of the St. Louis Limestone. Structure contours on a prominent chert horizon are shown on the map on the basis of leveling surveys with a

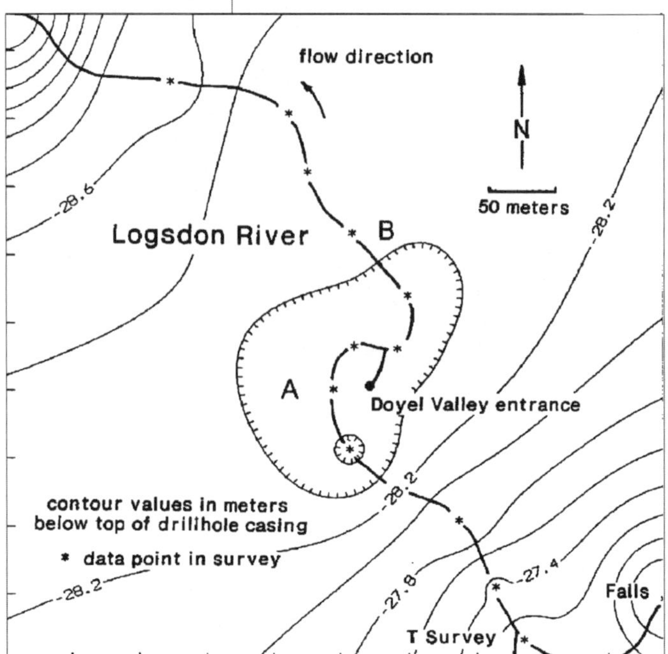

5.3.8 Figure 5. Line plot of part of Logsdon River, in the southern part of Mammoth Cave, showing structural control of a perched vadose tube. Structural contour lines are shown for a prominent chert bed in the St. Louis Limestone, with an arbitrary datum at the top of the Doyel Valley entrance (a drillhole). ***A*** *is the structural basin.* ***B*** *is the threshold of basin where vadose water was able to spill over and continue its downdip trend. Contour lines are computer-generated with a kriging routine, and they decrease in accuracy away from the passage.*

tripod-mounted automatic level. Contours were obtained with the kriging routine of the computer program SURFER (Golden Software Corp.) to avoid subjective bias. Overall, the passage has a broad curvilinear pattern that exactly follows variations in the local dip.

However, at one point the initial water flow that determined the ultimate passage pattern drained into a small closed basin (labeled A on the map), where local ponding took place. In that area the dip orientation was abandoned and the water wandered across the width of the basin, then spilled over the lowest available threshold of the basin (B on the map) to resume its downdip trend. Similar structural control of many other passages in the cave has been demonstrated (Palmer, 1977, 1981, 1989a).

The prominent bedding and its demonstrable control over passages in the Mammoth Cave System invites the interpretation that favorable bedding planes may have served as inception horizons by virtue of favorable mineralogy, as proposed for certain other regions (Lowe, 1992, Chapter 3.3; see also Šušteršič, Chapter 5.3.13). For example, abundant pyrite along certain partings could have initiated early-stage enlargement that favored later enlargement to cave size. Petrographic studies of each bed exposed in Mammoth Cave show abundant pyrite, but with no systematic relation to those partings that have guided the majority of passages.

In fact, every significant bedding plane in the entire 110-meter section of exposed carbonates has guided at least several passages in the system. The one that has guided the greatest number of passages (although not the largest ones) is the top of the Aux Vases Member of the Ste. Genevieve Limestone, a former sulfate zone exhibiting silicification and local fine-grained breccia. However, it is also shaly, and most of its passages are vadose tubes perched on these low-solubility beds. In places where these perched passages extend downstream into former phreatic conditions, they deviate upward along a bedding-plane parting in the massive overlying bed, which shows no distinctive mineralogical features conducive to cave inception. The unusually great concordance of passages to the strata is due to the equally unusual paucity of major discordant fractures.

At any time during the evolution of the cave, the pattern of active passages has been crudely dendritic. On cave maps this pattern is obscured in several ways. For example, there are many superimposed levels of old abandoned passages overlying more recent active ones. Even many active vadose passages cross over active or abandoned passages at lower levels (an impossible phenomenon in surface streams). Many passages have been reoccupied by water as the result of a late Pleistocene rise in the level of the Green River. Postglacial alluvial fill in the valley has flooded the lowest 15 meters or so of the cave, reactivating many formerly abandoned passages, so the present base-level drainage pattern adjacent to the river valley is quite complex. Finally, the geologic structure imposes a great deal of irregularity upon the cave, which disrupts the dendritic pattern.

Water Chemistry

The nature of the present water flow through the Mammoth Cave System seems to be representative of nearly the entire history of cave development. There is one exception: the earliest cave passages seem to have been fed by large sinking streams supplied by surface runoff from large areas of what is now the Pennyroyal Plateau and from valleys in the Chester Upland that are tributary to the Green River. These streams entered the limestone ridges of the Chester Upland at only a few discrete points. No present-day examples remain, although their chemistry can be inferred from measurements of similar streams in other regions. There is no reason to suspect that the water chemistry and dissolution rates in other sources of recharge to the cave differ greatly from those of the past.

In the following paragraphs, the saturation index of calcite (SI_c) is defined as log (IAP/K_{sp}), where IAP is ion activity product and K_{sp} is solubility product of calcite. Negative values represent undersaturation and positive supersaturation.

The saturation index for dolomite (SI_d) is defined as 0.5 log(IAP/K_{sp}), where the coefficient 0.5 adjusts for the fact that dolomite releases twice as many ions as calcite, so the SI values will be compatible with those for calcite.

Present water sources include:

- Sinking streams on the Pennyroyal Plateau contribute water to many of the large base-level stream passages of the cave. They have a highly variable discharge, moderately high P_{CO_2} (typically 0.001 to 0.01 atm), and widely fluctuating saturation levels ranging from highly unsaturated during peak discharge to supersaturated during low flow. Typical SI_c values range from about –1.0 to +0.4, with SI_d values slightly lower. Most dissolution in the cave takes place during high-flow events (Meiman and Groves, 1997). On the basis of measurements of large sinking streams in other karst areas, those that fed the large upper levels of the cave in the past probably had similar characteristics to those of the smaller sinking streams described here. During pre-agricultural times before deforestation, the discharge probably fluctuated much less than it does today. The general lack of floodwater features in the large upper-level passages of the cave supports this premise.
- Sinkholes on the Pennyroyal Plateau and in karst valleys in the Chester Upland supply the majority of the water to the base-level cave streams of all sizes, in addition to forming numerous minor vadose infeeders. This water acquires a moderately high P_{CO_2} from the soil (typically 0.001 to 0.01 atm), but much of its solutional potential is lost near the surface, so that it is only mildly aggressive in the caves. Typical SI_c values range from –0.5 to –0.1, and SI_d from –1.1 to –0.7. Recharge from sinkholes and sinking streams account for most of the allogenic detrital sediment carried into the cave. This consists mainly of quartz sand and gravel derived from weathering of the insoluble caprock, combined with clay, silt, and chert fragments derived from the carbonate rocks.
- Perched karst springs in the Chester Upland drain from the Haney Limestone, a thin carbonate unit sandwiched between sandstone formations within the caprock. Most of this water seeps downward through thin sandstone beds of the overlying Hardinsburg Formation. This water emerges from the Haney springs with surprisingly low SI_c (–2.0 to –0.5) and SI_d (–2.0 to –1.0), and rather low P_{CO_2} (0.002–0.008 atm). Evidently much of the limestone dissolution in the Haney takes place under essentially closed conditions relative to CO_2, because of the sandstone barrier between the soil and the limestone, and the water begins to pick up additional CO_2 only near the surface where shallow soil water is able to mix with the deeper Haney groundwater. Most of the spring water then flows on the surface over the lower sandstone (Big Clifty Formation) and enters the main cavernous aquifer through sinkholes in karst valleys. It enters the main aquifer while still very aggressive and feeds vadose passages in the cave. These passages exhibit some of the most active dissolution in the cave.
- Diffuse infiltration enters the main limestone directly through the sandstone caprock of the Chester Upland—especially where the sandstone is thin along the eroded ridge edges. Initial dissolution of the carbonate is under essentially closed conditions, as in the Haney Limestone. As a result, as the water approaches saturation with dissolved carbonates, the equilibrium P_{CO_2} drops to very low values (less than 10^{-4} atm) and the pH rises to as much as 9 or 10. These pH values cannot be measured *in situ* but are inferred from theoretical considerations, coupled with the following observations: When this seepage reaches aerated caves, it rapidly absorbs CO_2 from the cave air and the pH drops, causing substantial corrosion of the cave walls. In places tiny seeps have etched the cave walls with irregular rills as deep as several centimeters. This is surprising,

when one considers that the cave air has a very low CO_2 content (averaging about 0.00075 atm). Small discrete streams of water that enter the cave through the sandstone share some of these characteristics. This water feeds many shafts and canyons. It is aggressive but hovers just below saturation during low flow. During high flow the SI_c can drop as low as –1.0 and SI_d to about –1.5.

- Seepage directly into the limestone through overlying soil has a high P_{CO2} and enters the cave near calcite saturation, with a P_{CO2} of about 0.005–0.01 atm. Dolomite is usually undersaturated, although in places it is slightly supersaturated (SI_d up to +0.5), probably as the result of Mg enrichment during prior calcite precipitation. Loss of CO_2 to the cave air causes deposition of calcite speleothems. In the Mammoth Cave System this process is limited to relatively few areas.
- Where the sandstone caprock is thick or contains interbedded shale, diffuse seepage into the underlying carbonate is very limited. This water tends to evaporate when it reaches the caves, precipitating gypsum and other evaporite minerals. The major source of sulfate for the gypsum is oxidation of pyrite. This is shown by the presence of residual limonite (from oxidation of pyrite) in the centers of many gypsum growths. Other supporting evidence comes from sulfur isotopes: $\delta^{34}S$ in the cave gypsum ranges from –5 to –8‰, which contrasts with values that range from +14 to +19‰ for primary sulfates in the bedrock (Francis Furman, Rolla, Mo., 1998, personal communication).
- Backflooding from the Green River forces water into the caves through springs whenever the outside river level is higher than the base-level streams in the cave. It is affected by the distribution of rainfall or snowmelt, and by artificial control of the river level by dams both upstream and downstream from Mammoth Cave. During large regional floods the cave water usually rises more rapidly than the Green River, so backflooding from the river typically occurs when water levels in the cave are declining. The mean-annual flood stage for the Green River at Mammoth Cave is about 11 meters (E. White, 1989), which explains the frequency and severity of local backflooding. The chemistry of the Green River water varies with season and discharge. During low flow it is usually calcite supersaturated and during high flow (especially in winter) it is undersaturated. SI_c varies from about –0.8 to +0.4, SI_d from about –0.9 to +0.3, and P_{CO2} from about 0.001 to 0.004 atm (calculated from chemical measurements by Miotke, 1975).

Backflooding can account for a small amount of cave enlargement, enhanced by mixing of waters of contrasting CO_2 content. A considerable amount of fine-grained sediment is also deposited by backflooding (see Davies and Chao, 1959).

Cave streams that are perched on shaly beds retain their aggressiveness for long distances, with SI_c and SI_d values commonly as low as –1.0 and –2.0 respectively. Base-level streams and springs have a higher P_{CO2}, perhaps owing to oxidation of organic material, and they are closer to saturation, with SI_c values that range from about –0.5 to +0.2, SI_d about –1.0 to +0.2, and P_{CO2} about 0.002 to 0.04 atm.

As the sandstone caprock retreats by erosion, water entering the cave at any given point tends to change its character in the following sequence:

- Diffuse seepage that evaporates when it reaches the cave walls, precipitating gypsum and other evaporite minerals.
- Aggressive seeps that etch the cave walls.
- Aggressive flows through thin sandstone beds.
- Finally water that passes directly through the soil into the limestone and deposits travertine in the cave (Palmer and Palmer, 1995).

This progression is shown in many places by gypsum crusts that are peeling from moist walls, and vadose shafts that were originally fed by aggressive water, but which are now lined with calcite speleothems.

Further information about the water chemistry is given by Miotke (1975), Hess and White (1989, 1993), and Anthony et al. (1997). Although the hydrologic and chemical functions of the various water sources are fairly well known, an estimate of the cumulative effect of each source on cave enlargement, both at present and in the past, is still being investigated (see Meiman and Groves, 1997).

Geomorphic History of the Mammoth Cave System

The evolutionary history of the presently known Mammoth Cave System can be traced back as far as the late Tertiary (Miotke and Palmer, 1972; Palmer, 1989b). The sequence of passage development is fairly clear, and its relation to regional events seems to be understood, at least in general terms. Advances in dating techniques using $^{26}Al/^{10}Be$ ratios in quartz-rich sediment are beginning to clarify the ages of the uppermost passages (Granger and Palmer, 1997; see Addendum at end of this chapter). Preliminary results from this technique indicate ages up to several million years (Pliocene) for the sediment fill in upper-level passages, although the solutional

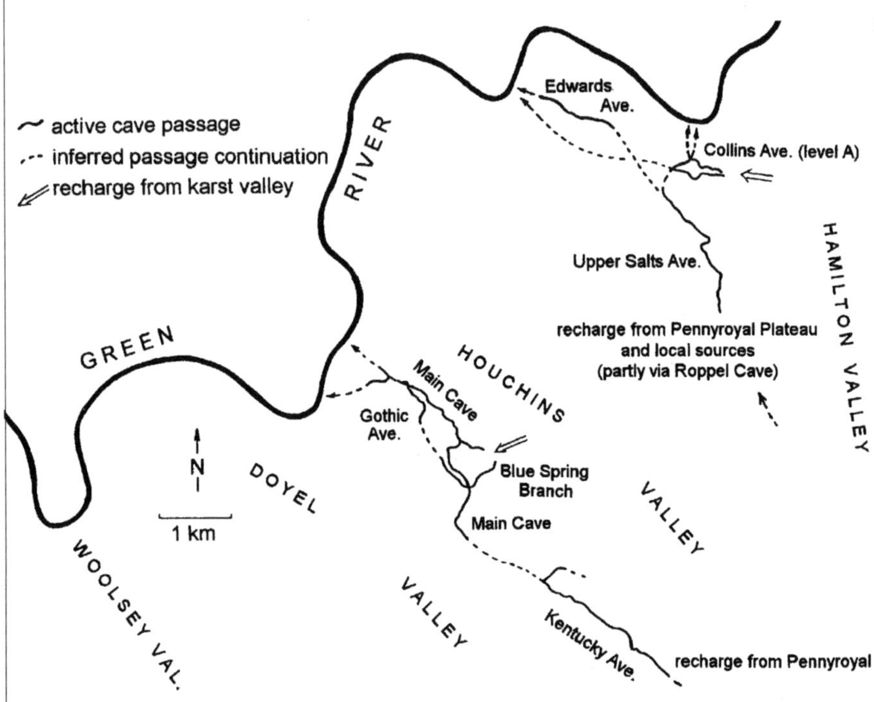

5.3.8 Figure 6. Development of major passages at levels A and B (see text), the main parts of Mammoth Cave. All of the passages shown are on level B, except Collins Avenue, which is at level A

origin of the passages may be considerably older.

The earliest passages to form in the Mammoth Cave System were large tubes and canyon passages now at least 200 m above present sea level (5.3.8 Figure 6). For convenient reference, this can be called level A. In comparison, the present altitude of the Green River at Mammoth Cave is 129 m. Of course none of the "levels" in the cave are perfectly horizontal–the term is used simply to indicate the relative position of fluvial base level while the passages were forming, as indicated by clustering of piezometric limits in the cave (see Chapter 3.4).

The largest of the passages at level A is Collins Avenue, which lies about 30 m above the Pennyroyal Plateau surface. Although it is essentially a large canyon, its irregular ceiling profile and lack of relation to the dip indicates an initial phreatic origin. Thus the Green River clearly lay at or above that level when the cave originated. The geomorphic implications are discussed at the end of this section.

Level B constitutes other large canyon passages formed at lower elevations (180 to 190 m above present sea level). In Mammoth Cave these include Main Cave and its tributaries, and in neighboring Flint Ridge the main passages of Salts and Great Onyx Caves. All of the passages at levels A and B, including Collins Avenue, were nearly filled with stratified gravel, sand, silt, and clay to depths of at least 20 m (5.3.8 Plate 2). This sediment still occupies most of the passages, and the few that are no longer filled show evidence of former filling nearly to their ceilings. More than one stage of sediment fill is evident in places. These large filled canyons can be interpreted to be the products of very gradual fluvial entrenchment of the region, with intervening periods of aggradation, probably as the result of gradual periodic climate changes during the late Tertiary.

Surface remnants of these sediments are also preserved in a few isolated interfluvial areas of the Pennyroyal Plateau (Ray, 1996). At that time the relief of the surrounding landscape was much more subdued than it is today, with the sandstone-capped ridges rising only a few tens of meters above the limestone plains. Karst was probably very sparse on the Pennyroyal Plateau, because surface drainage prevailed there with no substantial entrenchment. Surface streams drained from the Pennyroyal into the limestone of the Chester Upland, forming a few major river passages of canyon-like morphology. These include the largest passages of the system, such as Main Cave and the upper levels of Salts Cave (both at level B).

These upper levels have been interpreted by a few other researchers to be the result of paragenesis. This hypothesis considers that phreatic tubes were initiated deep beneath the water table and evolved upward by dissolution of their ceilings, because the accumulation of sediment on their floors might prohibit dissolution in any other direction. The canyons nearly filled with sediment are therefore envisioned as having formed in a single phase, rather than by sequential downcutting followed by aggradation. However, there is reason to doubt the paragenetic interpretation, including the facts that:

- Every upper-level passage shows the sediment filling, even some (such as Gothic Avenue) which are tubes, not canyons.
- Some are filled entirely to the ceiling.
- Much of the sediment contains cut-and-fill structures indicative of open-channel flow.

The oldest levels, A and B, formed in the late Tertiary prior to major glaciation of the North American continent. Pleistocene glacial advances fell short of the Mammoth Cave region. But their influence on the erosional history of the region was profound. Prior to continental glaciation, the Ohio River was not nearly so large as it is today. Drainage from most of the Appalachian Mountains flowed through what is now northern Indiana into the Mississippi River, bypassing the Ohio River entirely.

Early Pleistocene glaciers blocked this drainage and diverted it into the Ohio River, which was located approximately along the farthest advance of the ice (5.3.8 Figure 7). The Ohio's greatly increased discharge caused its entrenchment rate to accelerate. The Green River, which controls the levels in Mammoth Cave, is a tributary of the Ohio, and so it participated in the rapid entrenchment.

As a result, the regional water table dropped rapidly, and streams in the caves were diverted to lower routes. All but the largest streams on the Pennyroyal Surface were diverted underground.

Several lengthy pauses in fluvial entrenchment occurred during the Pleistocene, possibly accompanied by minor interludes of aggradation. This complex base-level history was controlled by drainage adjustments within the Ohio River Basin, changes in relative level between the continent and the sea, and probably also climatic changes.

Each lengthy pause allowed the cave streams to remain stable at the same level for a long enough period of time to enlarge the passages at that level to comparatively great size (5.3.8 Figure 8). The longest pauses resulted in tubular passage levels at the present altitudes of 168 and 152 m (levels C and D). A minor level, more of a precursor to level D, occurred at 158 m. Passages at level C are extremely well developed and clearly represent a major phase of stable base level. These include Cleaveland Avenue, Turner Avenue, and Waterfall Trail, all of which have piezometric limits within a meter of each other in altitude, despite the fact that they occupy different strata.

The lowest paleomagnetic reversal of the cave sediment occurs at this level (Schmidt, 1982), indicating an age greater than 780,000 years. Ratios of $^{26}Al/^{10}Be$ in the cave sediment are extending that date considerably.

Passages at level D include the Lost Passage, Smith Avenue, Swinnerton Avenue (158 m), and the Echo River–Great Relief

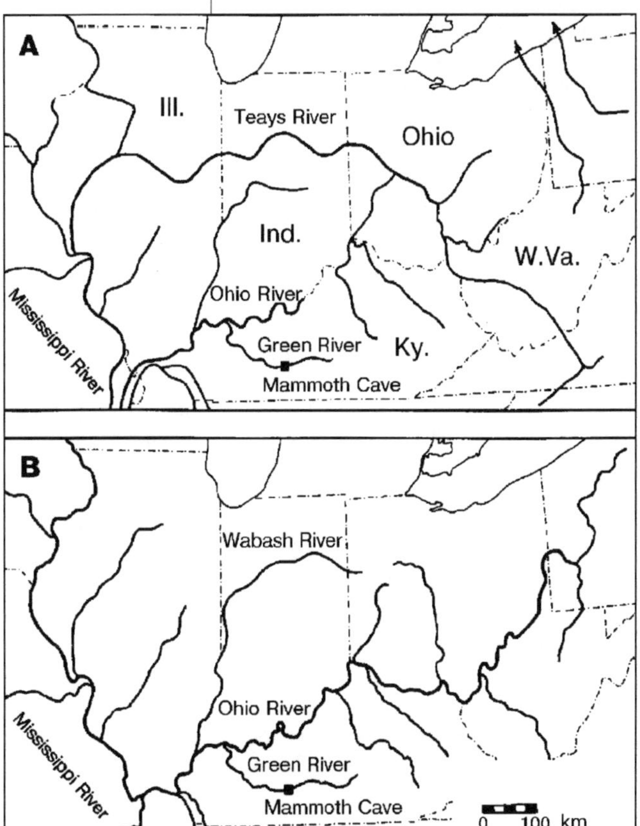

*5.3.8 Figure 7. Preglacial, **A**, and postglacial, **B**, drainage pattern in the Ohio River valley (modified from Thornbury, 1965). State boundaries are shown for reference.*

Hall–Silliman Avenue system (5.3.8 Plate 3). Again, all are in different strata. Calcite speleothems at this level have been dated by U/Th methods to more than 350,000 years (Hess and Harmon, 1981), and the passages are undoubtedly much older. Each passage listed here for levels C and D is a well-developed tube with elliptical or lenticular cross section. Level D correlates with a prominent fluvial terrace in the Green River Valley. A possible terrace at level C has been reduced to remnants, owing to subsequent valley widening. Passages lower than level D are fairly well developed, but they do not form consistent levels.

The lowest passages in the system lie below present base level and are perennially flooded as the result of late Pleistocene (probably postglacial) alluviation of river valleys to depths of 10 to 15 m. These passages now form an interconnected water-filled network (regardless of their original hydrologic function) so that dye injected at a single infiltration point tends to emerge at many different springs along the Green River.

One of the clearest examples of cave levels and the sharp transition between them is given by several of the main passages in the central part of Mammoth Cave (5.3.8 Figure 8). During the development of Level C, vadose water formed the dip-oriented canyon Boone Avenue, which joined the tube Cleaveland Avenue as a tributary.

Cleaveland Avenue is one of the best examples of a strike-oriented tubular passage, with an overall gradient of less than 20 cm per kilometer. When the water table dropped below level C, the water from Boone Avenue diverted straight down the dip, bypassing Cleaveland Avenue and following the axis of a gentle syncline to form the Pass of El Ghor, a canyon that correlates with the lower levels of Boone Avenue. The Pass of El Ghor continues as a canyon, where it joins a larger passage, Silliman Avenue.

Silliman Avenue continues as a large downdip canyon for part of its length but then changes to a tube at level D. Beyond the transition from canyon to tube, it descends as a former phreatic loop all the way down to the present base level at Echo River, then rises to level D again, where it is joined by a tubular tributary at that level, Great Relief Hall (5.3.8 Plate 3).

The amplitude of the phreatic loop was about 23 m. Thus, although the entire system of passages described here spans an altitude range of roughly 130 m to 180 m, the piezometric limits at 152 m and 168 m are sharply defined and easily recognized.

During the evolution of Mammoth Cave, its passages at successively lower levels have increased greatly in number, and in general with a corresponding decrease in passage size. This is due to erosional fragmentation of once-larger recharge areas and dissection of the Chester Upland, which has increased the number of sites for infiltration. Meanwhile the amount of recharge available to each infiltration site is reduced. The frequent changes in base level during the Pleistocene have also complicated the passage pattern.

Since the origin of the uppermost passage in the Mammoth Cave System (Collins Avenue), the Green River has entrenched at least 100 m, while the entire Pennyroyal Surface has been lowered about 30 m. Ever since the Ohio River acquired its present drainage basin, about 1 million years ago, the vigorous erosion of the Green River has achieved a net downward entrenchment of about 55 m but accompanied by very little lowering of the Pennyroyal Surface. When one considers the much slower entrenchment rates that prevailed before the present Ohio drainage was established, it is logical to assume that the age of Collins Avenue must extend back well into the Tertiary.

The clarity of the passage levels in Mammoth Cave, especially at levels C and D, give a false impression of great stability of fluvial base level during their development. The sharp transition from vadose to phreatic at former water-table levels, together with the abrupt abandonment of those phreatic levels by demonstrably vadose canyons, suggests a fluvial history consisting of relative still-stands punctuated by sudden fluvial entrenchment. The passage profiles must have been graded to base-flow conditions.

However, the sharp break between levels probably can be explained simply by an abrupt lowering of river level by headward entrenchment, rather than by uniform river downcutting. At any given point along the river, the water table would drop rather suddenly as entrenchment progressed upstream past it.

An alternative view, and one that at first might seem more realistic, is that there is simply a time delay between the lowering of base level and the abandonment of the main tubular passages. However, passage elevations at each major level are so similar that it would be necessary for all passages to be abandoned simultaneously, so that their effluent streams could stabilize lower down at the identical base level. Thus the bulk of the evidence leans toward the scenario of headward entrenchment of the Green River.

The extraordinary length of Mammoth Cave is due to the following factors:

- An unusually large drainage basin.
- Sandstone caprock that has preserved the uppermost levels from erosional destruction.
- Complex erosional embayments in the caprock that supply numerous water inputs.
- Many rather small changes in base level, allowing numerous passages to develop at successive levels.
- Thin-bedded limestone in which successive passages tend to follow bedding-plane partings independent of those that guided previous passages.

In summary, the Mammoth Cave System is an ideal example of shallow epigenetic cave development, in which the inheritance of

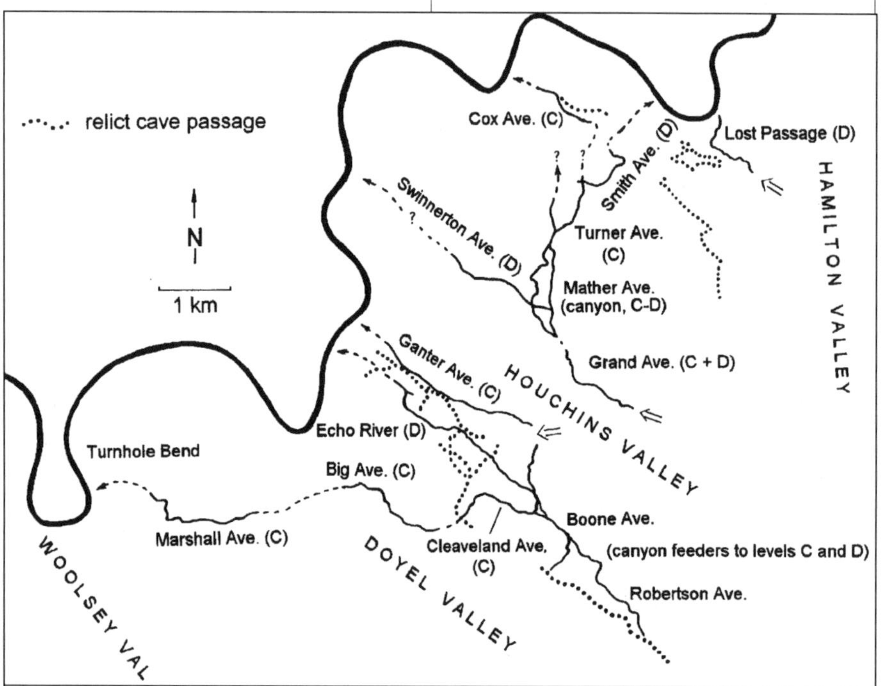

*5.3.8 Figure 8. Development of major passages at levels **C** and **D** (main parts of Mammoth Cave only).*

deep-seated karst features is virtually nonexistent. The pattern of descending vadose water is strongly controlled by the local dip of the strata, and the transition from vadose to phreatic conditions is well displayed. Baselevel control is therefore clear, even in passages that have been long abandoned by their flow. Because the Green River, into which the cave drains, is one of the major tributaries of the Ohio River, the cave provides many clues to the fluvial history of the entire east-central United States.

Addendum

Since this chapter was written, alluminum-beryllium dating of sediments in the Ohio River basin has provided a clearer correlation with the levels of Mammoth Cave (Granger et al., in review). The results are:

- Level B is older than 3.3 million years.
- A major sediment filling occurred throughout levels A and B about 2.5 million years ago, covering older local sediment. This stage appears to represent a major climatic change that caused widespread aggradation thoughout Indiana and Kentucky (see also M. Palmer and A. Palmer, 1975).
- The Teays River valley was established about 2 million years ago.
- Rapid entrenchment below level B was caused by a diversion of the Teays to form the present Ohio River.
- Sediment at level C averages about 1.4 million years old.
- Abandonment of level C was caused by rearrangement of drainage within the Ohio-Mississippi Basin.
- Sediment at level D averages about 1.1 million years old.
- Passages below level D formed during rapid incision of the Green River as the result of late Pleistocene glaciation, when the Ohio valley frequently served as a sluiceway for glacial meltwater and outwash.

References

Anthony, D.M., Groves, C., and Meiman, J., 1997, Preliminary investigations of seasonal changes in the geochemical evolution of the Logsdon River, Mammoth Cave, Kentucky: *6th Science Conference,* Mammoth Cave National Park, Proceedings, p. 15-23.

Ash, D.W., 1984, Evidence for deep-seated groundwater movement in Middle Mississippian carbonate lithologies of southcentral Indiana: *National Speleological Society,* Annual Convention, Proceedings, p. 28.

Davies, W.E., and Chao, E.C.T., 1959, Report on sediments in Mammoth Cave, Kentucky: *U.S. Geological Survey Administrative Report,* 117 p.

Deike, G.H., 1967, *The development of caverns in the Mammoth Cave region:* Pennsylvania State University, Ph.D. dissertation, 235 p.

Ford, D.C., and Williams, P.W., 1989, *Karst geomorphology and hydrology:* London, Unwin Hyman, 601 p.

Granger, D.E., and Palmer, A.N., 1997, A proposed method for determining the age and evolution of the Mammoth Cave System: *6th Science Conference,* Mammoth Cave National Park, Proceedings, p. 145-150.

Granger, D.E., Fabel, D., and Palmer, A.N., in review, Plio-Pleistocene incision of the Green River, Kentucky, from radioactive decay of ^{26}Al and ^{10}Be in Mammoth Cave sediments: *Geological Society of America Bulletin.*

Hess, J.W., and Harmon, R.S., 1981, Geochronology of speleothems from the Flint Ridge-Mammoth Cave System, Kentucky, U.S.A.: *8th International Congress of Speleology,* Bowling Green, Proceedings, v. 2, p. 433-436.

Hess, J.W., and White, W.B., 1989, Chemical hydrology, *in* White, W.B., and White, E.L., eds., *Karst hydrology–Concepts from the Mammoth Cave area:* New York, Van Nostrand Reinhold, 346 p.

Hess, J.W., and White, W.B., 1993, Groundwater geochemistry of the carbonate karst aquifer, south-central Kentucky, U.S.A.: *Applied Geochemistry,* v. 8, p. 189-204.

Klimchouk, A.B., 1996, The typology of gypsum karst, *in* Klimchouk, A., Lowe, D., Cooper, A., and Sauro, U., eds., Gypsum karst of the world: *International Journal of Speleology,* v. 25, no. 3-4, p. 49-60.

Lowe, D.J., 1992, A historical review of concepts of speleogenesis: *Cave Science,* v. 19, p. 63-90.

Meiman, J., and Groves, C., 1997, Magnitude/frequency analysis of cave passage development in the Central Kentucky Karst: *6th Science Conference,* Mammoth Cave National Park, Proceedings, p. 11-13.

Miotke, F.-D., 1975, *Der Karst im zentralen Kentucky bei Mammoth Cave:* Jahrbuch der Geographischen Gesellschaft zu Hannover, 360 p.

Miotke, F.-D., and Palmer, A.N., 1972, *Genetic relationship between caves and landforms in the Mammoth Cave National Park area:* Würtzburg, Böhler Verlag, 69 p.

Palmer, A.N., 1977, Influence of geologic structure on groundwater flow and cave development in Mammoth Cave National Park, U.S.A.: *12th International Association of Hydrogeologists,* Memoirs, p. 405-414.

Palmer, A.N., 1981, *A geological guide to Mammoth Cave National Park:* Teaneck, Zephyrus Press, 210 p.

Palmer, A.N., 1987, Cave levels and their interpretation: *National Speleological Society Bulletin,* v. 49, p. 50-66.

Palmer, A.N., 1989a, Stratigraphic and structural control of cave development and groundwater flow in the Mammoth Cave region, *in* White, W.B., and White, E.L., eds., *Karst hydrology–Concepts from the Mammoth Cave area:* New York, Van Nostrand Reinhold, p. 293-316.

Palmer, A.N., 1989b, Geomorphic history of the Mammoth Cave System, *in* White, W.B., and White, E.L., eds., *Karst hydrology–Concepts from the Mammoth Cave area:* New York, Van Nostrand Reinhold, p. 317-337.

Palmer, A.N., and Palmer, M.V., 1995, Geochemistry of capillary seepage in Mammoth Cave: *4th Science Conference,* Mammoth Cave National Park, Proceedings, p. 119-133.

Palmer, M.V., and Palmer, A.N., 1975, Landform development in the Mitchell Plain of southern Indiana: *Zeitschrift für Geomorphologie*, v. 19, p. 1-39.

Quinlan, J.F., and Ray, J.A., 1981, Groundwater basins in the Mammoth Cave region, Kentucky: *Mammoth Cave National Park Occasional Publication*, no. 1, map.

Quinlan, J.F., and Rowe, D.R., 1977, Hydrology and water quality in the central Kentucky karst; Phase I: *University of Kentucky Water Resources Institute Research Report*, no. 101, 93 p.

Quinlan, J.F., and Rowe, D.R., 1978, Hydrology and water quality in the central Kentucky karst: Phase II, Part A: Preliminary summary of the hydrogeology of the Mill Hole subbasin of the Turnhole Spring groundwater basin: *University of Kentucky Water Resources Institute Research Report*, no. 109, 42 p.

Ray, J.A., 1996, Fluvial features of the karst-plain erosion surface in the Mammoth Cave region*: 5th Science Conference,* Mammoth Cave, Proceedings, p. 137-156.

Schmidt, V.A., 1982, Magnetostratigraphy of sediments in Mammoth Cave, Kentucky: *Science,* v. 217, p. 827-829.

Thornbury, W.D., 1965, *Regional geomorphology of the United States*: New York, Wiley, 609 p.

White, E.L., 1989, Flood hydrology, *in* White, W.B., and White, E.L., eds., *Karst hydrology–Concepts from the Mammoth Cave area*: New York, Van Nostrand Reinhold, p. 127-143.

White, W.B., and White, E.L., eds., 1989, *Karst hydrology–Concepts from the Mammoth Cave area:* New York, Van Nostrand Reinhold, 346 p.

5.3.9. Speleogenesis of Vertical Shafts in The Eastern United States

William B. White

Abstract

Vertical shafts are a form of vadose cave development initiated by undersaturated water draining from overlying clastic caprock. They are related to present day topography and generally are much younger than the underlying cave passages to which they drain. Shafts are excavated by highly aggressive water streaming down the walls in supercritical flow. Shafts form very rapidly, reaching full size in a few thousand years. Development of new shafts can easily keep up with the erosional retreat of the caprock.

Introduction

Caves contain a variety of vertical passage elements in the form of solutionally widened fractures, pits, and shafts of various hydrologic function and various origins. A special category of these, termed "vertical shafts" or "domepits" in much U.S. literature, occur in the flat-lying Mississippian limestone of the eastern United States. These were first described from Mammoth Cave by E.R. Pohl (1935, 1955). But they are also recognized in many caves in Missouri, Indiana, West Virginia, Tennessee, and Alabama as well as in other caves in similar geologic settings. Within the classification of Ford and Ewers (1978) vertical shafts are a subcategory of invasion vadose caves.

5.3.9 Figure 1. Pencil sketch by James Knudsted of the Bottomless Pit vertical shaft complex, Mammoth Cave, Ky., U.S.A. (from White et al., 1970).

Vertical shafts were ascribed by J Harlan Bretz (1942) to a core-blocking mechanism involving down-cutting by bedding plane anastomosis channels. Pohl demonstrated that vertical shafts form in the vadose zone at the edges of dissected plateaus and are related to the contemporary landscape. Their geomorphic roles in the evolution of present-day topography and as a mechanism for the retreat of the Chester Escarpment in southern Kentucky were pointed out by Quinlan and Pohl (1967). Independently, Burke and Bird (1966) produced a theory of their own concerning the origin of the vertical shafts in Mendip, England, which was quite similar to concepts proposed by Pohl.

Further studies have been made of stratigraphic controls (Merrill, 1960) and interpretation of their mechanism of formation (Reams, 1965). A study of the geochemistry and hydrodynamics of vertical shaft water (Brucker et al., 1972) provided a plausible understanding of vertical shaft origin and development.

Characteristic Features of Vertical Shafts

Vertical shafts are large cylindrical voids in the vadose zone that may or may not be directly linked to underlying cave systems. They occur commonly in flat-lying carbonate rock capped by resistant shale and sandstone throughout much of the eastern and central United States. The ideal form of a vertical

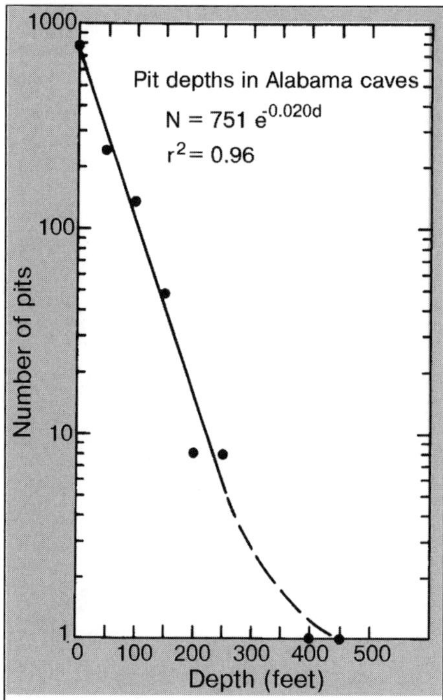

5.3.9 Figure 2. Distribution of shaft depths in Alabama. Data from the Alabama Cave Survey.

shaft is a right circular cylinder. Shaft walls usually are exactly vertical and are often fluted. The walls truncate rocks of many lithologies. Shafts are typically fed by a complicated network of small canyons at the top. Shafts often have a crosscutting relationship to cave passages. In some cases the shafts cut directly through cave passages giving rise to the alternate name "domepits" (5.3.9 Figure 1). In other cases, they simply graze the cave passage. The relationships indicate clearly that the shafts are much younger than the passages and usually have no genetic relationship with them.

Most shafts have a bedrock floor in the form of a shallow basin. Water collected in the basin is carried away by a shaft drain. Shaft drains are commonly tiny passages, in places too small to permit human exploration. One question that has not yet been satisfactorily answered is why the shaft should have been enlarged to a huge chamber while the drain is volumetrically tiny. Shaft drains connect either with the active portion of the cave system or with the land surface through small springs. Drain passages can be very long. The drain for Colossal Dome in Mammoth Cave, known as Wretched River, extends for nearly 2 km as a small crawlway before connecting into a larger drain passage.

Vertical shafts rarely span the entire thickness of the limestone. They often begin at some intermediate resistant horizon and they end on another resistant horizon. Chert beds, shale layers, interbedded sandstone, and beds of dolomite can all serve to break the continuity of vertical shafts. Shaft depths in the northern Alabama karst have an exponential distribution (5.3.9 Figure 2). Many of the shafts along the edges of the dissected Cumberland Plateau in Tennessee and Alabama have depths limited to 60 to 70 m, because of the presence of a few meters of resistant Hartselle Sandstone, which separates the cavernous limestone into two sections. A few shafts, for example in Fern Cave, Alabama, and in Ellison Cave, Georgia, have depths in the range of 120 to 160 m because the Hartselle Sandstone has been breached. Wilson (1977) has argued that these exceptionally deep caves occur along lineaments that provide pathways for water to penetrate the hydrologic barriers.

Hydrologic and Speleogenetic Role of Vertical Shafts

Vertical shafts are formed entirely in the vadose zone. Shafts are infeeders for local runoff and groundwater seepage from perched surface catchments and perched aquifers on the overlying clastic caprock. Some shafts accept input from allogenic surface streams and contain waterfalls, but many do not. They do act as the primary pathways for transport of vadose water from the overlying plateau to the main aquifer system at base level.

Shafts in the Mammoth Cave area, with few exceptions, lie along the present-day contact between the overlying clastic rocks and the underlying cavernous limestone (5.3.9 Figure 3). The caprock is gradually retreating as the plateau erodes and the valleys widen. This requires that new shafts form continuously while old shafts are abandoned along the valley sides. As the caprock retreats, the supply of aggressive water becomes concentrated on one side of the shaft, thus continuing active dissolution only on that side. The shaft then migrates, following the retreating caprock with the formation of what Watson (1966) calls "solution canyons." Cathedral Dome in Mammoth Cave, 55 m high and more than 100m long, is an example. In other locations, overlapping shafts form large chambers or shaft complexes.

Water input to shafts that have been abandoned along valley walls infiltrates through carbonate soils where it is exposed to high carbon dioxide pressure in the soil and reacts nearly to equilibrium at the soil-bedrock contact. When this water reaches the shafts, the reaction is one of flowstone deposition rather than dissolution. In large shaft complexes such as Mammoth Dome in Mammoth Cave, one side is beneath the caprock while the other side, 100 m away, is under the valley wall. On the caprock side water streams down over bare limestone, actively enlarging the shaft. On the other side are masses of active flowstone, a dramatic illustration of the shifts in vadose water chemistry over short distances.

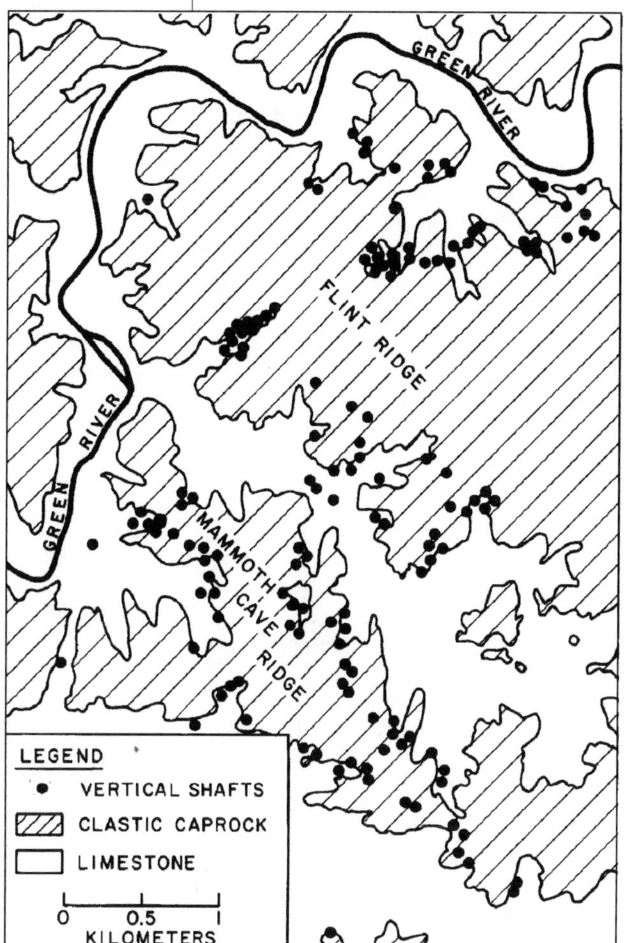

5.3.9 Figure 3. Location of vertical shafts with respect to the edge of the sandstone caprock on Flint Ridge and Mammoth Cave Ridge, Mammoth Cave National Park. Geologic base from U.S. Geological Survey Mammoth Cave Quadrangle.

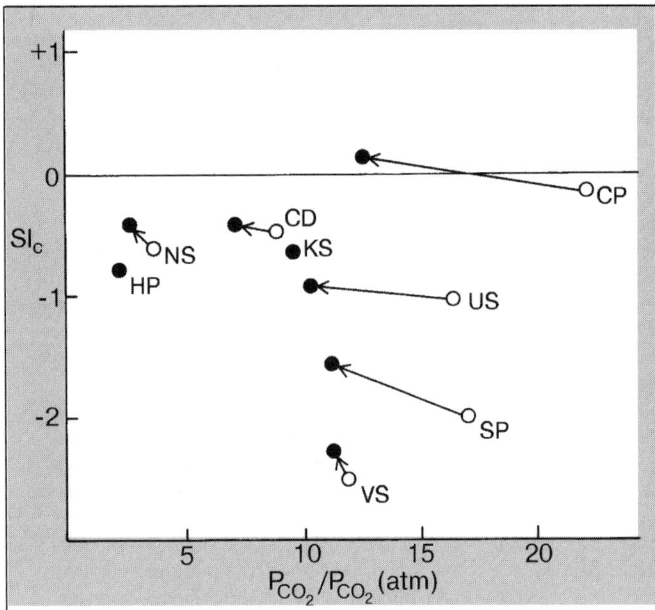

*5.3.9 Figure 4. Saturation index and carbon dioxide pressure in some vertical shaft waters in Alabama, Kentucky, and West Virginia. Carbon dioxide pressures are normalized to the atmospheric background value. Open circles are data from the top of shafts. Solid circles are from the bottom of shafts. Key to locations: **CD** is Colossal Dome, Mammoth Cave, Kentucky. **CP** is Cascade Pit, Mammoth Cave, Kentucky. **HP** is Halls Pit, Alabama. **KS** is Keller Shaft, Mammoth Cave, Kentucky. **NS** is Neversink, Alabama. **SP** is Swago Pit, West Virginia. **US** is unnamed shaft, Mammoth Cave, Kentucky. **VS** is Vahalla Shaft, Alabama.*

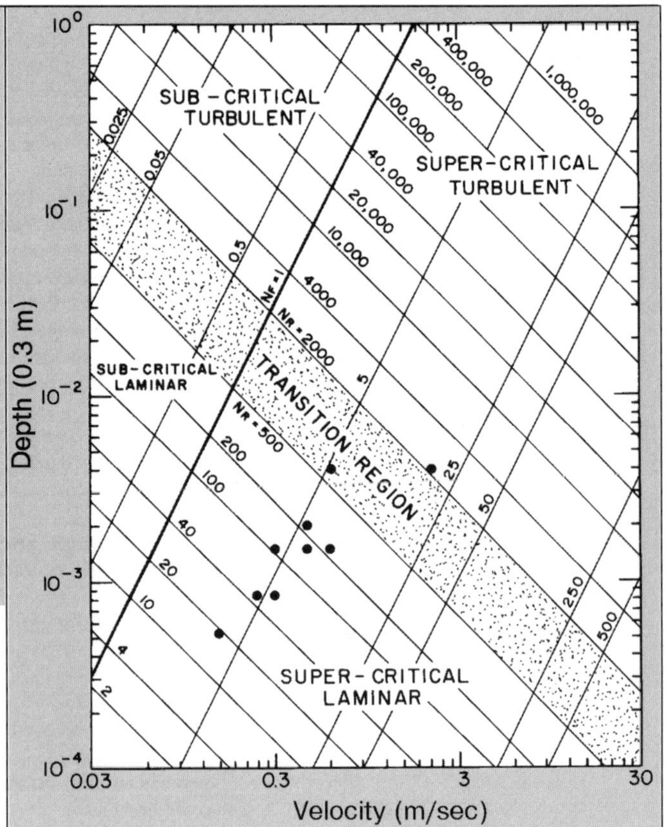

5.3.9 Figure 5. Hydraulic conditions for shaft development. Diagram assumes a wide (w >> d) rectangular channel. Froude and Reynolds numbers are then calculated and contoured, and flow regimes determined by the $N_F = 1$ boundary between supercritical and subcritical flow, and the $N_R = 500$ to 2000 transition region between laminar and turbulent flow. Solid black circles are measured water film thickness (depth) to velocity data for vertical shafts.

Vertical shafts can be accessed where they have fortuitously intersected caves passages and by exploration of shaft drains. Many shafts on the Cumberland Plateau have collapsed roofs exposing an open pit. Many such pits have been explored, and many have no human-size connections with other deeper parts of the drainage system.

Mechanisms of Vertical Shaft Development

The active agents forming vertical shafts are the thin layers of water streaming down the walls. Measurements have been made on these waters in the Swago Creek Basin in West Virginia, in the Flint Ridge portion of Mammoth Cave, and in some Cumberland Plateau shafts in northern Alabama (Brucker et al., 1972, 5.3.9 Figure 4). The waters are all highly undersaturated with respect to calcite and have carbon dioxide partial pressures about 10 times the atmospheric background. Water sampled at the top and bottom of shafts have slightly different chemistries with a trend toward lower CO_2 pressure and less undersaturation as indicated by the arrows on 5.3.9 Figure 4. Some of the shift must be due to dissolution of the shaft wall, although some may be due to degassing of carbon dioxide.

The wet walls of vertical shafts have surprising hydrodynamics. The thickness or "depth," of the water films is on the order of a few millimeters. The flow velocities, as measured by dye streamers, are in the range of 1 to 10 m per second. The velocity of thin water films on a smooth rock surface is described by the equation:

$$\bar{v} = \frac{g\delta^2}{3\upsilon} \qquad [1]$$

where $\bar{v}$ = mean flow velocity, g = acceleration due to gravity, δ = film thickness, and υ = kinematic viscosity of water.

If the circumference of the shaft is treated as the width of a wide channel, and the thickness of the moving layer of water is treated as the depth, it is possible to prepare a fence diagram (5.3.9 Figure 5) plotted as depth vs. velocity with Reynolds numbers and Froude numbers contoured on it. The measured film thickness and water flow velocity for several shafts (indicated by the points on 5.3.9 Figure 5) plot in the supercritical-laminar regime, a flow regime rarely observed in nature. Observation of disturbance waves and dye streamers confirms the calculation.

The supercritical flow may be primarily responsible for the nearly-perfect vertical walls that characterize vertical shafts. Projecting beds, ledges, and other protuberances will retard the flow of the water sheets. With lower velocity, the flow regime will go subcritical with a hydraulic jump at the transition point ($N_F = 1$). The accelerated erosion due to energy dissipation at the hydraulic jump would strip off the projections until only a streamlined vertical wall remains.

Vertical shafts are initiated along fractures or at fracture intersections. Initial dissolution rates must be comparable to those in horizontal cave passages. However, velocity will increase rapidly as the fracture intersection is opened by dissolution because of the vertical gradient. Breakthrough into a regime of fast kinetics (in the sense described by Dreybrodt, 1990) happens quickly, and the dissolution kinetics moves into a regime of maximum rate. The dissolution rate will be approximately 7 to 10 millimoles $cm^{-2} s^{-1}$. Taking into account the density of limestone, this number becomes 0.12 cm/year as the rate of wall retreat. At this rate, a vertical shaft 10 m in diameter could be dissolved out in roughly 4000 years.

References

Bretz, J H., 1942, Vadose and phreatic features of limestone caverns: *Journal of Geology,* v. 50, p. 675-811.

Brucker, R.W., Hess, J.W., and White, W.B., 1972, Role of vertical shafts in the movement of ground water in carbonate aquifers: *Ground Water,* v. 10, no. 6, p. 5-13.

Burke, A.R., and Bird, P.F., 1966, A new mechanism for the formation of vertical shafts in Carboniferous limestone: *Nature,* v. 210, p. 831-832.

Dreybrodt, W., 1990, The role of dissolution kinetics in the development of karst aquifers in limestone: A model simulation of karst evolution: *Journal of Geology,* v. 98, p. 639-655.

Ford, D.C., and Ewers, R.O., 1978, The development of limestone cave systems in the dimensions of length and depth: *Canadian Journal of the Earth Sciences,* v. 15, p. 1783-1798.

Merrill, G.K., 1960, Additional notes on vertical shafts in limestone caves: *National Speleological Society Bulletin,* v. 22, p. 101-108.

Pohl, E.R., 1935, Development of vertical shafts in limestone caves: *Geological Society of America Proceedings for 1935,* p. 96.

Pohl, E.R., 1955, Vertical shafts in limestone caves: *National Speleological Society Occasional Papers,* no. 2, 24 p.

Quinlan, J.F., and Pohl, E.R., 1967, Vertical shafts actively promote slope retreat and dissection of the solution escarpment and the Chester Cuesta in the Central Kentucky Karst: *National Speleological Society Bulletin,* v. 29, p. 109.

Reams, M.W., 1965, Laboratory and field evidence for a vadose origin of foibe (domepits): *International Journal of Speleology,* v, 1, p. 373-389.

Watson, R.A., 1966, Underground solution canyons in the Central Kentucky Karst, USA: *International Journal of Speleology,* v. 2, p. 369-376.

White, W.B., Watson, R.A., Pohl, E.R., and Brucker, R.W., 1970, The Central Kentucky Karst: *Geographical Review,* v. 60, p. 88-115.

Wilson, J.R., 1977, Lineaments and the origin of caves in the Cumberland Plateau of Alabama: *National Speleological Society Bulletin,* v. 39, p. 9-12.

5.3.10. Cave Development in Vadose Settings in a Multilayer Aquifer—The Sorbas Karst, Almería, Spain

J.M. Calaforra and A. Pulido-Bosch

Abstract

The Messinian (early Miocene) gypsum of Sorbas is a sequence of massive beds of selenitic gypsum that alternate with marl and clay attaining an aggregate thickness of up to 120 m. The strata are nearly horizontal and without any significant tectonic disturbance. Up to 12 successive gypsum units, with an average thickness of some 7 m, and an equal number of impervious intercalations can be identified. This configuration has led to the development of an interstratal karst in which large galleries have developed in the marly beds, not in the gypsum ones. The explanation for this lies in the hydrogeologic history of the area.

Initially, the gypsum karst evolved as a multilayer semiconfined aquifer under phreatic conditions, enabling the formation of small protoconduits in the individual gypsum beds, while the intervening marl and clay beds functioned as impervious barriers. During a second stage, after lowering of the piezometric surface, vadose conditions were established in which mechanical erosion processes in the intercalated marl and clay became predominant. This genetic duality means that the gypsum karst at Sorbas is an example of interstratal karstification where the current underground erosion processes and those of karstic evolution should be considered products of the hydrogeologic development of the region.

Geologic and Hydrogeologic Setting

The Sorbas Basin is one of the intramontane Neogene basins in the Betic Mountain Range where there was significant deposition of evaporites during the Messinian (early Miocene). It contains various gypsum outcrops, of which that at Sorbas is the most outstanding (5.3.10 Figure 1). It has an area of only 12 km^2 but displays a great variety of karst forms. To date, more than 800 caves have been catalogued within this small outcrop (Calaforra, 1998). Some of them have been explored for several kilometers, e.g. Cueva del Agua has a known length of 8 km, and the Covadura System is 4.5 km.

The edges and the impervious base of the gypsum aquifer are composed of silt and clay (Abad Member) deposited before the gypsum. The entire Sorbas Basin drains through the gypsum above them to a single spring, Los Molinos, which has an average flow of about 70 L/s (Pulido-Bosch and Calaforra, 1986 and 1993). This is quite remarkable for a semiarid area such as Sorbas, where the mean annual rainfall does not exceed 250 mm. Above the gypsum member are sandstone with some silty intercalations and conglomerate of a continental facies (the Sorbas Member and Góchar Formation, respectively). The semipervious nature of this overlying sequence was of great importance to speleogenetic evolution at Sorbas, as it created semiconfined conditions that affected all the gypsum strata to the western edge of the present basin.

Preevaporitic reef limestone (Cantera

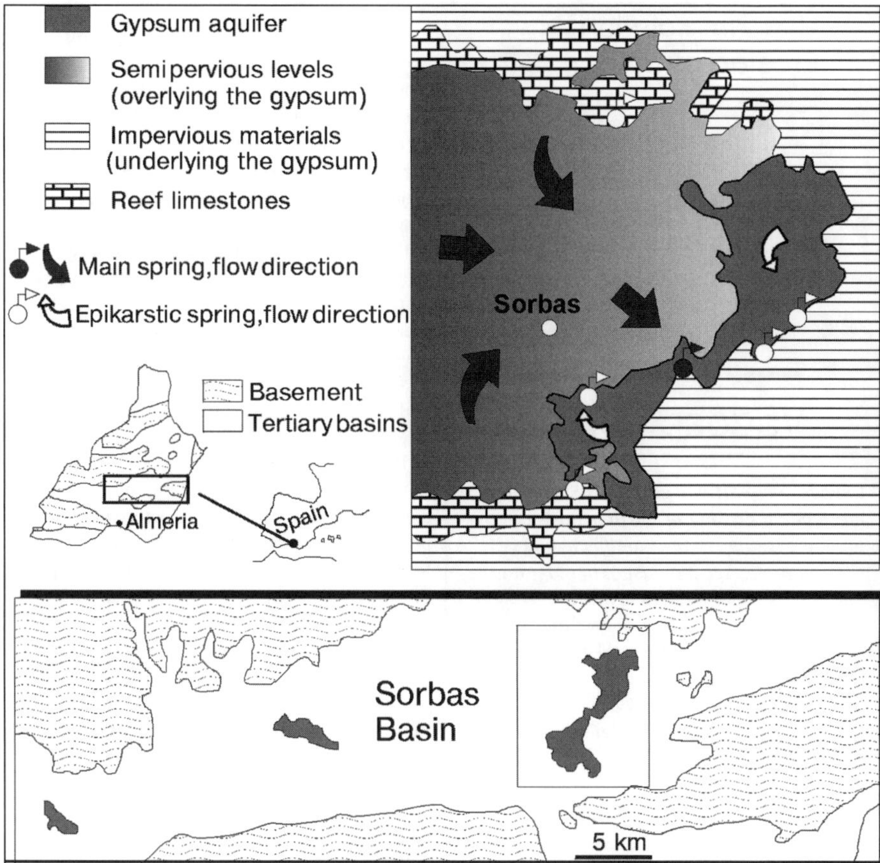

5.3.10 Figure 1. *Outcrop of gypsum karst at Sorbas, Spain.*

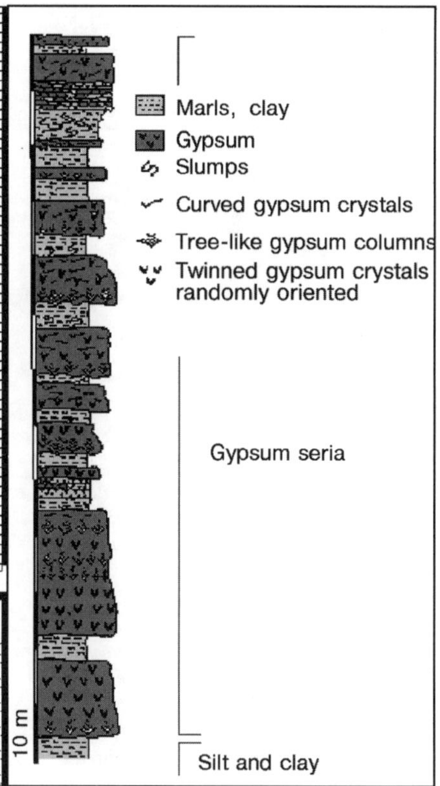

5.3.10 Figure 2. Gypsum series in the canyon of the Aguas River, showing the various gypsum-pelite intervals and the sedimentary structures that characterize the gypsum at Sorbas (modified from Dronkert, 1977).

Member) surrounds the Sorbas depression (5.3.10 Figure 1), but this has no clear hydraulic link with the gypsum, because small springs rise on the periphery of the limestone outcrops.

The Messinian gypsum (Yesares Member) comprises a cyclic sequence of well-stratified gypsum with a maximum thickness of 120 m. The individual gypsum layers may exceed 20 m in thickness and are mainly formed of selenitic gypsum with crystals up to 2 m in length. These layers are separated by marly and pelitic beds 0.5 to 3 m thick. The most complete sequence crops out in the canyon of the River Aguas (5.3.10 Figure 2). Its cross section displays twelve gypsum-pelite cycles (Dronkert, 1977, a and b; Dronkert and Pagnier, 1977). Details of typical structures can be seen in each of the massive gypsum beds, particularly the "supercones" (Dronkert, 1976), large tree-like gypsum structures, which can be observed within the caves where the base of a gypsum layer is visible on the ceiling.

The most noteworthy feature of the gypsum series, in terms of the evolution of the karst, is the alternation of gypsum and pelitic beds. This is geomorphically important, because the different levels of cave development are controlled by it. Furthermore, it allows us to assume that the gypsum aquifer functioned, at least during the early part of its hydrogeologic development, as a multilayer aquifer under semiconfined groundwater flow conditions.

Hydrogeologic Evolution of the Gypsum Karst

The hydrogeologic evolution of the gypsum karst of Sorbas passed through two well-defined stages. During the first of these, phreatic conditions predominated in the aquifer, which, as noted above sustained semiconfined flow through all the post-Messinian deposits that were not yet eroded. The gypsum series behaved as a multilayer aquifer as a consequence of the marly and clayey intercalations. Permeability within the gypsum layers was due to fissuring and some matrix porosity, with the interbedded marls constituting upper and lower impervious boundaries for each gypsum bed. As a result, the origin of the initial galleries (protoconduits) possibly took place simultaneously at the different levels within the series, depending on the location of each gypsum stratum. Subsequently, in the second stage, the piezometric surface fell, and mechanical erosion of the overlying confining rocks was accentuated. The basin passed into a vadose state, in which mechanical erosion of the soft marl intercalations came to prevail over the dissolution of the gypsum layers.

In 5.3.10 Figure 3 this dual genesis is illustrated. Within the conceptual scheme, two well-differentiated situations generally occur together in a given sector of a cave only in exceptional circumstances, such as in zones of piezometric fluctuation. First, small galleries develop within the gypsum beds under predominantly phreatic conditions (5.3.10 Figure 3A and B). Underground flow within the aquifer was slow and unable to erode the interstratal marls mechanically, so they acted as impervious barriers to any vertical flow. Under these conditions, small subcircular horizontal galleries developed along the contact between the gypsum and marl. The gypsum mass was not a true aquifer, because its transmissivity was very low, representing the first stage in the development of the karst. These passages

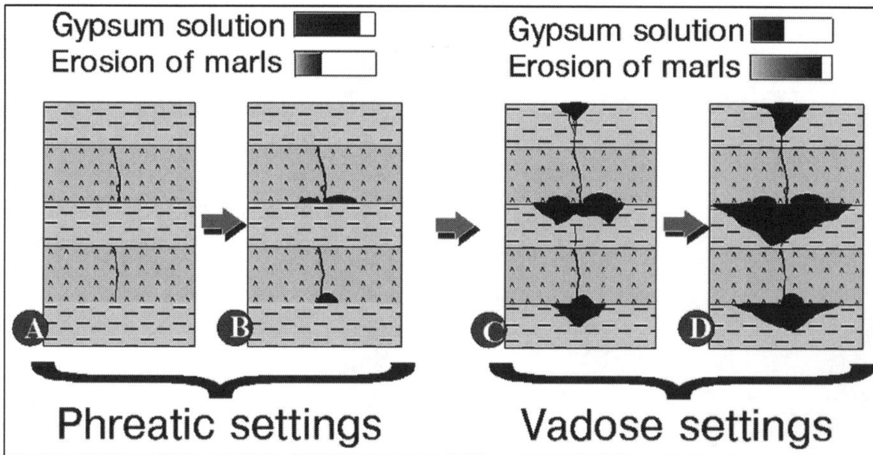

*5.3.10 Figure 3. Speleogenesis of interstratal passages in the gypsum karst at Sorbas. **A** is the formation of protoconduits under phreatic conditions as a result of heterogeneity in the alternation of gypsum-marl beds in the multilayer aquifer. **B** is passages starting to form along bedding planes as paragenetic galleries due to solution of gypsum and limited mechanical erosion of interbedded marl. **C** is a change from phreatic to vadose conditions giving rise to strong mechanical erosion of the detrital beds, but the gypsum remains unaffected. **D** is erosion processes increasing in magnitude, and the galleries developing a typical V profile due to the underground fluvial erosion of the marls. The overall process can be summarized as limited dissolution of gypsum succeeded by significant corrasion of marl (from Calaforra and Pulido-Bosch, 1997).*

developed along the planes of maximum weakness within the series, i.e. the bedding-plane contacts between the gypsum and the marl, perhaps especially where joints intersected the bedding planes. These protocoduits in places coincide with joints in the gypsum but at many other places follow a meandering course with no apparent fracture control.

Evolution of the protochannels would correspond to a paragenetic model of gallery development, although not exactly to the model proposed by Renault (1968) in which detrital sediment that accumulates during the formation of the gallery plays a key role. In the Sorbas case, it is the prior existence of the marly layer, not the accumulation of detritus to levels approaching the gallery ceilings, that initiates the paragenetic evolution of the passage. Only in the most advanced stages of the formation of these protogalleries was the deposition of allochthonous detrital material of importance, and then only locally. This also explains why it is appropriate to term these protochannels "false ceiling channels," as they predate the formation of the large modern gallery itself, whereas the genesis of authentic ceiling channels follows that of the gallery. The formation of such passages requires the influence of a slow laminar flow that is incapable of mechanically eroding the lower marly stratum, only of dissolving the gypsum above (5.3.10 Figures 3A and B).

Editors' note: Many speleologists would not consider this to be a case of paragenesis, because contemporaneous deposition of clastic sediments is not significant. In this case, the position of each inception horizon is fixed by a gypsum-marl contact, and the development is a lithologically perched type.

The second stage involved the progressive expansion of vadose conditions until the aquifer came to be completely drained. The earlier galleries became part of the unsaturated edge of the aquifer. Following a progressive fall in the piezometric surface, groundwater circulation in the unsaturated zone was restricted to preferential infiltration paths, as is evidenced by the appearance of numerous dolines at the surface. The vadose water flowed through some of the channels created during the initial phreatic stages, leading to an onset of turbulent flow and mechanical erosion of the marly levels that had previously formed impervious barriers (5.3.10 Figure 3C).

Progressively, the flow became established in the marls, as if it were a stream channel at the surface, creating the characteristic V profile of continuous vadose erosion. At roof level, closing off the inverted triangle shape of the cross section is the gypsum stratum, almost unaffected, in which the ceiling channels bear witness to their origin under quite different conditions (5.3.10 Figure 3D). In this situation, the volume of dissolution of gypsum is minimal when compared to the mechanical erosion suffered by the marl. The material most susceptible to karstification (gypsum) is not that in which the large galleries have formed. On the contrary, the galleries exist in impervious material such as marl, as a consequence of the transformation to vadose conditions. This particular gallery configuration has been termed "interstratal erosive karst," as the processes of mechanical erosion and stratal control are predominant in

5.3.10 Plate 1. Configuration of vadose galleries in the gypsum karst at Sorbas. The ceiling and floor of the gallery are formed of gypsum. The volume and shape of the gallery have developed by mechanical erosion of interbedded marl and clay. This configuration is repeated at various levels and depths.

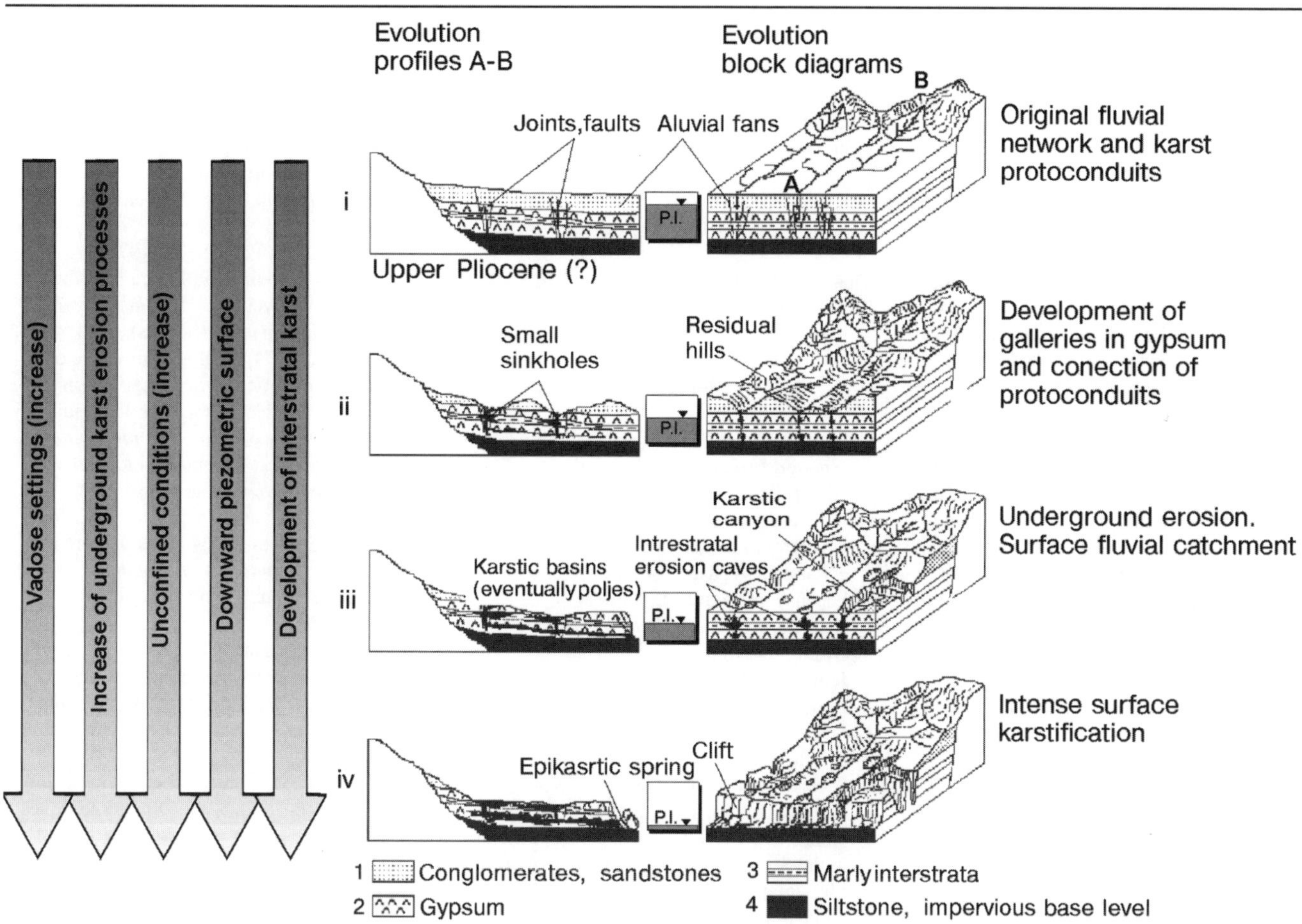

*5.3.10 Figure 4. Geomorphic evolution of the gypsum karst at Sorbas (Almería, Spain). Stratigraphy—**1** Sorbas Member and Góchar Formation. **2** Yesares Member. **3** interbedded clay in the Yesares member. **4** Abad Member. **P.L.** is the piezometric level. Stages—the first stage, **i**, is the establishment of network drainage. The second stage, **ii**, is the deep karstification (small phreatic protoconduits). The third stage, **iii**, is the capture of the surface drainage network and erosion of the interbedded marls. The fourth stage, **iv**, is the intense surface karstification. The development of the vadose setting correlates with a progressive fall in the piezometric surface and with changes from confined conditions in a multilayer aquifer to unconfined conditions with an increase in subterranean mechanical erosion.*

its formation (Calaforra and Pulido-Bosch, 1996). One of these galleries is shown in 5.3.10 Plate 1, where most of the volume excavated is within the marly layer.

Geomorphic Consequences

The evolution of an interstratal erosional karst, using the Sorbas gypsum karst as an example is shown in 5.3.10 Figure 4. In the first stage (5.3.10 Figure 4i) the rocks of the Yesares Member (gypsum and interbedded marl) are covered by semipervious material from the Sorbas Member (sandstones and marl) and the Góchar Formation (fanglomerate).

In the second stage (5.3.10 Figure 4ii) the surface drainage system becomes progressively entrenched in the semipervious materials. Such a system is not only structurally controlled but also has a subterranean equivalent in the formation of galleries along the planes of weakness and following approximately the same directions as those on the surface. As the erosion advances, small dolines begin to appear as a consequence of the subterranean dissolution. The gypsum is gradually exposed to surface weathering, such that only small scattered outcrops of the initial cover remain (5.3.10 Figure 4iii). Large karst basins are formed (e.g., that of Cueva del Agua) together with deep karstic canyons dissecting the gypsum outcrops. The surface drainage is captured at numerous points by small dolines, while at greater depths, large galleries begin to form in the marly layers. Finally, superficial karstification is accentuated, and the only remaining evidence of the original river drainage network is a multitude of dolines and sinkholes aligned in the direction of the vanished flow. A cliff is formed by differential erosion at the contact between the gypsum and the impervious base. While surface erosion continues to lower the impervious outcrops, the gypsum plateau broadly preserves its topographic form and level as well as a certain relief-inversion effect, due to the great permeability that has developed (5.3.10 Figure 4iv).

The four stages have in common one extremely significant hydrogeologic factor: the progressive lowering of the piezometric surface. This affects operations of both the geomorphic and hydrogeologic processes. In the example of the gypsum karst at Sorbas, it is a combined effect. The geomorphic consequences are evident: the fall in the piezometric surface explains numerous aspects of the gypsum karst that are currently seen, such as the formation of cliffs, capture of surface drainage, formation of deep karst canyons, and the presence of interstratal caves within the gypsum series. Such a variety cannot be separated from the speleogenetic evolution of the gypsum karst.

Conclusions

The great majority of the caves that have formed in the Messinian gypsum at Sorbas may be considered *erosional interstratal karst*. Indeed, the galleries that have formed are largely in the marly strata that are interbedded in the gypsum series. The remains of ceiling channels correspond to protochannels formed during an initial phase when groundwater within the gypsum was confined and flowed through the discontinuities in the rock and the porous matrix.

The linear erosion that entrenched the drainage network in the gypsum, together with the erosion of the overlying semipervious materials, led to the multilayer gypsum aquifer becoming unconfined. Such a process must have been rapid, with the protoconduits acting as drains in the vadose zone, leading to turbulent flow that quickly erodes the marl and widened the galleries, with the detrital materials being carried to surface canyons and cliffs that cross or border the stripped outcrop. This complete genetic hypothesis seems to be coherent and borne out by field observations.

Editors' note: Another good example of initial phreatic cave development at the base of a karstifiable stratum, succeeded by vadose entrenchment into underlying clastic beds that creates much bigger passages, is that of the "Contact Caves" of Greenbrier County, West Virginia, U.S.A. (Palmer, 1974). These caves were initiated in the basal meter or so of a platformal limestone (Mississipian) resting on shale. Galleries in the shale may be 10 m or more in depth and 30 m wide.

Acknowledgments

This work has been supported by CICYT Project AMB95-0493.

References

Calaforra, J.M., 1998, Karstología de yesos: *Monogr. Univ. (Sc. And Tecnol. Ser.),* Almería, no. 3, 384 p.

Calaforra, J.M., and Pulido-Bosch, A., 1996, Un ejemplo de karst de interstratificación: el karst en yeso de Sorbas (Almería, España), *in* Morell, I., and Fagundo, J.R., eds., *Contribuciónes a la Investigación y Gestión del agua subterranea:* Universitat Jaume I Collecció Summa Ciències Experimentals, no. 4, p. 153-160.

Calaforra, J.M., and Pulido-Bosch, A., 1997, Peculiar landforms in the gypsum karst of Sorbas (southeastern Spain): *Carbonates and Evaporites,* v. 12, p. 110-116.

Dronkert, H., 1976, Late Miocene evaporites in the Sorbas basin and adjoining areas: 1st Messinian Seminar, Erice, Proceedings, *Mem. Soc. Geol. Ital.,* v. 16, p. 341-362.

Dronkert, H., 1977a, A preliminary note on the geological setting of the gypsum in the province of Almería (SE Spain), Málaga: *Messinian Seminar,* no. 3, 6 p.

Dronkert, H., 1977b, The evaporites of the Sorbas Basin: *Instituto de Investigaciones Geológicas,* Dip. Prov., Barcelona, no. 33, p. 55-76.

Dronkert, H., and Pagnier, H., 1977, Introduction to the Mio/Pliocene of the Sorbas basin, Málaga: *Messinian Seminar,* no. 3, Field trip no. 3 (Sorbas), p. 21.

Palmer, A.N., 1974, Geologic influence on cave passage orientation in Ludington Cave, Greenbrier County, West Virginia: *4th Conference on Karst Geology and Hydrology*, West Virginia Geological and Economic Survey, Proceedings, p. 33-40.

Pulido-Bosch, A., and Calaforra, J.M., 1986, Formas kársticas en los yesos de Sorbas (Almería): *Estudios sobre Geomorfología del Sur de España,* Univ. Murcia, p. 115-119.

Pulido-Bosch, A., and Calaforra, J.M., 1993, The gypsum karstic aquifer of Sorbas (Almería), *in* Pulido-Bosch, A., ed., *Some Spanish karstic aquifers:* Univ. Granada, p. 225-241.

Renault, P., 1968, Contribution a l'étude des actions mechaniques et sédimentologiques dans la spéléogenèse: *Annales de Spéléologie,* v. 23, p. 529-593.

5.3.11. Maze Origin By Diffuse Recharge Through Overlying Formations

Arthur N. Palmer

Abstract

Where water infiltrates through a permeable but insoluble caprock into fissures in soluble rock, network caves can form in which each passage enlarges at rather comparable rates, regardless of fissure width. This process is regulated by two factors:

- The overlying rock acts as a governor to the flow, so that rather uniform amounts of water are delivered to each fissure, regardless of its size.
- The water enters the soluble rock while still highly aggressive, under which conditions each opening tends to enlarge at similar rates, regardless of discharge rate.

A typical network cave of this type develops in several stages. At first the soluble rock is less permeable than the insoluble rock, and only small amounts of water enter it. However, this water is highly aggressive and is able to enlarge the fissures in the soluble rock. With time, the soluble rock becomes more permeable than the overlying caprock, and most groundwater flow in the area passes downward into the soluble rock on its way to river outlets. Because of fluvial entrenchment the cave is eventually drained of its water, but periodic backflooding by local rivers may contribute greatly to the enlargement of the caves.

Introduction

Certain dissolutional network caves are formed by rather uniform recharge through an overlying permeable but insoluble rock formation. The process is also effective if water rises through an underlying permeable insoluble rock, as described in Chapter 5.2. In either case, water enters each fissure within a given area of soluble rock rather uniformly, with no great differences in chemical character, so that all fissures enlarge simultaneously at comparable rates (Palmer, 1975, 1991). White (1969) described maze caves in thin karst aquifers sandwiched between insoluble beds, although the origin of the maze pattern was not specified.

Field Evidence

Diffuse seepage through insoluble rock accounts for only about 5 percent of all caves, although some are quite extensive. There is abundant field evidence for network development by this process, particularly where water enters limestone through overlying quartz sandstone.

In the Appalachian Mountains of the eastern U.S.A., the Devonian Coeymans Limestone contains many extensive network caves where it is exposed beneath the permeable Oriskany Sandstone in Virginia and West Virginia. Examples include Clark's Cave and parts of Crossroads Cave in Virginia (5.3.11 Figure 1). To the north, in New York, the Coeymans is capped by relatively impermeable shaly limestone and contains no network caves of this particular origin. However, in New York the Oriskany Sandstone overlies the Becraft Limestone, which does contain many network mazes.

Network mazes are common where the sandstone cap is thin, generally less than 10 m. In the low plateaus of the east-central U.S.A.—where sandstone overlies prominently jointed limestone—local mazes are concentrated around the perimeters of ridges, where the sandstone is relatively thin. Examples include Clifty Cave in Indiana. In rare areas where thin sandstone extends over a large area, very extensive maze caves are

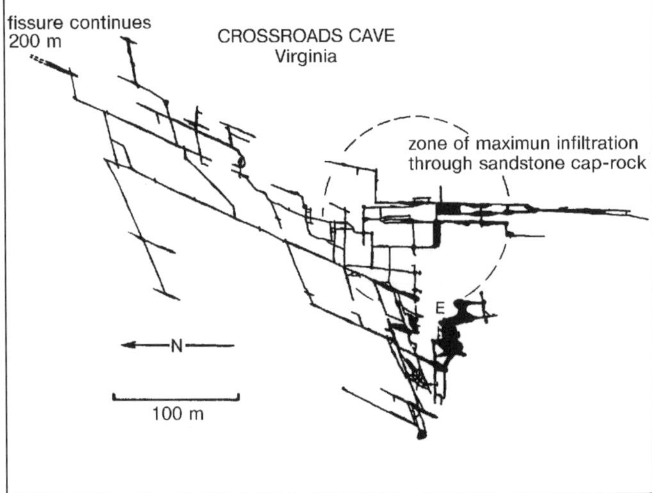

5.3.11 Figure 1. Map of Crossroads Cave, Virginia, USA (from map by D. Miller, R. Lutz, L. Slade, and D. Abbott in Douglas, 1964). This is a network maze formed partly by diffuse recharge through a thin overlying sandstone cap, and partly by floodwater activity. ***E** is the entrance.*

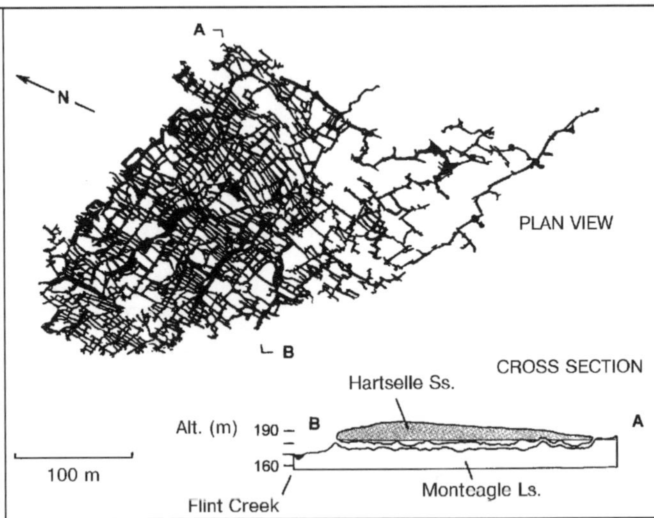

5.3.11 Figure 2. Map of Anvil Cave, Alabama, USA (from Varnedoe, 1964).

present in the limestone directly beneath. Examples are Anvil Cave in Alabama (5.3.11 Figure 2) and parts of James Cave, Kentucky. Many large network mazes capped by thin clastic rocks in Yorkshire, England, have been attributed to phreatic dissolution by Ryder (1975). But in a postscript, prompted by the appearance of a slightly earlier paper on maze caves (Palmer, 1975), he considered that infiltration through the caprock may have been responsible for at least some of the caves.

Evidence for an origin by infiltration through an insoluble caprock includes the proximity of the caves to the contact with the overlying insoluble rock, abundant fluting of walls by descending water, and termination of the maze character where the overlying insoluble rock either becomes thick, interbedded with low-permeability rocks, or was never present. Sediment in such caves is generally fine grained, much of it being deposited by backflooding from nearby surface rivers. Vadose fluting is an important criterion, because it demonstrates continued infiltration of aggressive water even after the cave has emerged above base level, but it may be absent if periodic flooding has overwhelmed the effects of infiltration from above. Dripstone and flowstone are sparse. These criteria are important in distinguishing this type of cave from those formed by other mechanisms, such as by rising water. Proximity to the sandstone cap must be assessed with care, because insoluble sediment may obscure the floors, giving the impression that cave development terminates downward at depths shallower than they actually do.

The same process operates where the overlying insoluble material is unconsolidated (for example, soil or glacial sediment), but it does not result in network caves per se. While fissures enlarge in the underlying soluble rock, the unconsolidated material subsides into them. In this way, dolines are likely to form over the largest openings and serve as catchments for increasing amounts of water. Thus the uniformity of recharge to the underlying fissures is disrupted and maze caves do not form. However, the infiltration of aggressive water into the soluble rock is quite capable of forming an epikarst in which the openings are filled with soil or sediment. This is simply another aspect of uniform growth of openings, although open caves are not the usual result.

Many sandstones are highly fissured along joints that are generally, but not invariably, coplanar with those in the underlying soluble rock. Most of the infiltration through sandstone of this type is limited to these fissures. In this case, the overlying soil serves as the collector for the infiltrating water into fractures in the sandstone.

Cave Evolution

A quantitative explanation for cave origin by diffuse recharge through an insoluble overlying or underlying formation is rather simple. It has been shown in Chapter 4.2.3 that all openings in limestone will dissolve at similar rates, forming maze caves, where flow distances from the source of aggressiveness are short, so that the ratio of discharge to flow distance (Q/L) is large. This condition is met by diffuse recharge through a thin insoluble caprock, because aggressive water is delivered directly to the soluble rock while the water still retains its full aggressiveness. Variations in enlargement rate may be produced by differences in water chemistry, but considering the localized nature of the recharge and the uniformity of the geologic setting, such differences are likely to be insignificant. Thus every fissure in the soluble rock has the potential to grow at uniform rates to produce a network maze.

Furthermore, in a network of fissures, all will enlarge uniformly if they receive similar amounts of discharge (Palmer, 1991). In the geologic setting envisioned here, the overlying or underlying permeable rock acts as a governor to the flow, so that each fissure in the soluble rock receives nearly the same discharge, regardless of fissure size. The uniformity of flow to each fissure can be explained by showing that the discharge through the caprock into fissures in the soluble rock is hardly affected by variations in fissure width. A complete analysis would require that the exact fissure spacing and thickness of the saturated zone in the insoluble rock be specified, but these complexities are unnecessary in this general discussion.

In cross-sectional view, consider that the flow in the insoluble rock is more or less radially convergent toward each fissure in the underlying soluble rock, as shown in 5.3.11 Figure 3. Darcy's Law, which applies to flow in the insoluble rock, can be integrated as follows:

$$q = \pi r K \left(\frac{dh}{dr} \right)$$

$$q \int_{r_f}^{r_0} \frac{dr}{r} = \pi K \int_{h_f}^{h_0} dh$$

$$q = \pi K \left[\frac{h_0 - h_f}{\ln\left(\frac{2r_0}{w_f} \right)} \right]$$

where q = discharge per unit length of fissure (perpendicular to the page in 5.3.11 Figure 3), K = hydraulic conductivity of the insoluble rock, h_f = hydraulic head in the fissure, r = distance into the insoluble rock from the top of the fissure, r_0 = distance to where the head value in the fissure has little effect on the head in the overlying rock, h_0 = hydraulic head in the overlying insoluble rock at r_0, r_f = half the fissure width, and w_f = fissure width (= $2r_f$).

Discharge to the fissure depends on the log of the fissure width. Thus, a ten-fold difference in fissure width produces only a two-fold difference in discharge. Heads within the fissure system will be almost uniform, because the hydraulic conductivity within it is so great compared to the flow rate. Heads within the overlying insoluble rock are likely to vary from place to place in a gradual way, such that no single fissure receives significantly more flow than any other. Minor differences in fissure width have an insignificant effect on discharge, which ensures that all fissures grow at nearly similar rates.

The validity of this interpretation has been supported by two-dimensional numerical modeling by Clemens et al. (1997). They assumed a constant head at the top of the insoluble caprock, with a uniform amount of recharge through it to an underlying limestone formation. Outflow took place at a constant-head boundary along one side of the model. A network maze was produced, even when an initially favorable route of large fissures was assigned to part of the limestone.

The evolution of a maze cave formed in this way can involve several stages with transitional boundaries (5.3.11 Figure 4).

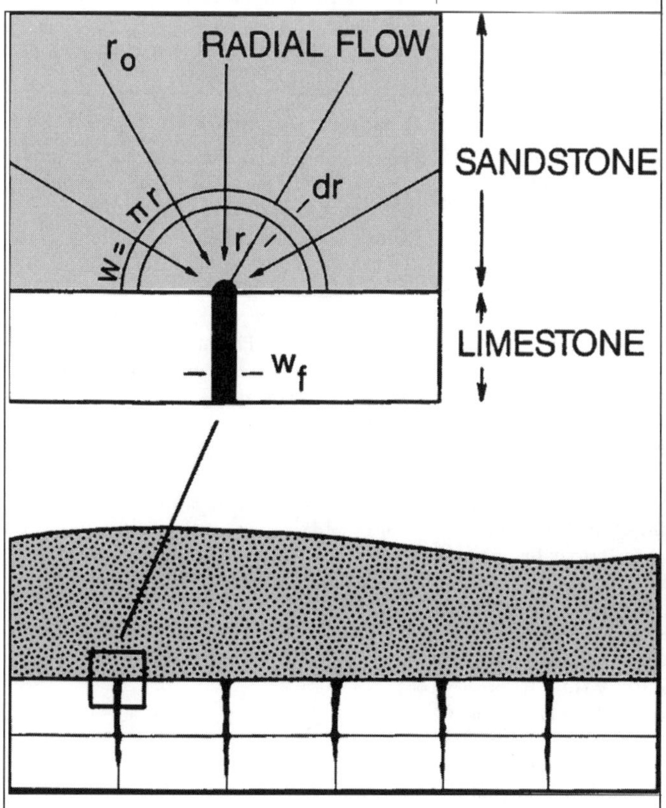

5.3.1 Figure 3. Idealized cross section of radial flow through a permeable insoluble rock into a fissure in soluble rock. See text for details.

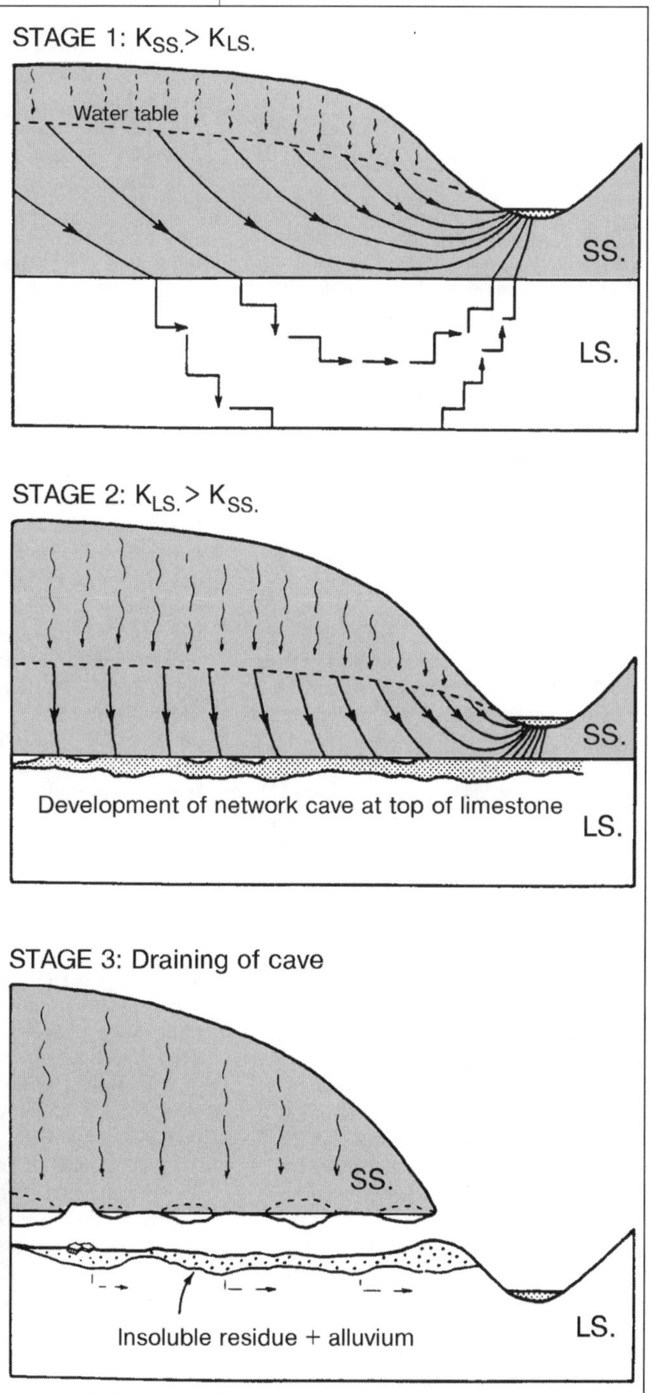

5.3.11 Figure 4. Sequence of development for a typical network cave in which at least the initial dissolution was performed by diffuse recharge through an overlying insoluble caprock (from Palmer, 1975).

For convenience, the permeable but insoluble rock will be considered to be sandstone and the soluble rock to be limestone, with the sandstone overlying the limestone.

Stages of Maze Cave Evolution

- At first the hydraulic conductivity (K) of the sandstone is greater than that of the limestone. Flow is concentrated within the sandstone, which provides the path of least head loss to nearby entrenched valleys. The small amount of water that does enter the limestone enlarges each fissure in the limestone rather uniformly just below the lithologic contact.
- As the fissures in the limestone enlarge, the hydraulic conductivity of the limestone eventually surpasses that of the sandstone, and from then on most of the water that enters the sandstone descends to the limestone, which provides a more efficient flow path. The rate of cave enlargement increases. Outflow takes place upward through the sandstone beneath stream valleys.
- Eventually the surface valleys become entrenched below the sandstone-limestone contact, and the cave becomes aerated. Water continues to descend through the sandstone, enlarging fissures in the limestone, especially at joint intersections, where local shafts with vertical flutes may develop. Backflooding from nearby rivers helps to enlarge the caves, but it also carries considerable detrital sediment into them.

Most maze caves do not owe their entire origin to the sequence outlined here. Even where infiltration through the insoluble rock has played a part in the cave origin, often other speleogenetic processes play an even greater role in cave origin. Some caves show evidence for recharge through an overlying insoluble rock—except that they only extend upward to the contact at the top of the soluble rock in a few places. The proposed origin is still valid, provided there are many individual inputs dispersed throughout the cave. In addition, many such network caves are only initiated by this type of recharge, and the majority of their growth is performed by flooding from nearby surface rivers or cave streams.

Crossroads Cave in Virginia provides an interesting contrast between maze origin beneath a sandstone cap and floodwater maze development (5.3.11 Figure 1). Parts of the cave exhibit the typical rectilinear pattern and proximity to the sandstone, with abundant vertical flutes demonstrating the aggressiveness of the infiltrating water. Joint intersections have been widened into vertical shafts by the same water source. In contrast, parts of the cave seem to have little relation to this mode of recharge, and instead show evidence of floodwater enlargement, including scallops, coarse clastic sediment, and bits of vegetal debris. These passages are also conspicuous by their acute and obtuse angles of intersection. This pattern may reflect the process of near-surface breakdown, which often produces locally unique joint sets by release of stress. Accumulation of breakdown also enhances flooding by creating constrictions within stream passages.

In soluble rocks that are not highly fractured, but which instead exhibit prominent bedding, the presence of overlying permeable but insoluble rock tends to produce interstratal dissolution. Blocks of the insoluble rock subside along the eroded edge of the caprock, as in the Mammoth Cave region of Kentucky. Caves formed by this infiltrating water are generally complexes of shafts and narrow canyons.

Water that infiltrates through an overlying insoluble formation can mix with water from more remote sources, producing maze caves in a different way from that described above. Palmer and Palmer (1989) describe a situation where diffuse recharge can mix with artesian water in the Black Hills of South Dakota, U.S.A. The highly cavernous Madison Limestone (Early Carboniferous) is overlain by the Middle and Late Carboniferous Minnelusa Formation, which consists mainly of quartz sandstone but also contains interbedded carbonates.

Water infiltrating into the sandstone encounters beds of limestone and dolomite. These dissolve in a system closed to further uptake of CO_2, since the water is isolated from soil CO_2 by the intervening sandstone. In these conditions, as the water approaches saturation with dissolved carbonates, the CO_2 content of the water drops to nearly zero.

However, most of the groundwater in the Madison is artesian, fed by recharge through the outcrop areas. This water is also largely saturated with limestone and dolomite, but at a much higher CO_2 content. Where the two waters mix, dissolution is enhanced by the traditional process of mixing corrosion, and caves can be formed or enlarged.

Most of the Black Hills caves are localized along former sulfate zones, which are much higher in permeability than surrounding parts of the limestone. It is along those zones that the artesian water flow is greatest, and where renewed aggressiveness from mixing is most likely. This process may not account for most cave enlargement in the Black Hills, because other dissolutional processes have also been effective (see Chapter 5.2.2).

However, the hydrochemical conditions that favor dissolution by this kind of mixing have prevailed throughout most of the Tertiary Period. Over this length of time, even small amounts of infiltration through the sandstone can have caused substantial cave enlargement.

Conclusions

Cave origin by diffuse infiltration through an insoluble caprock is an important process in certain karst aquifers and has resulted in some of the finest examples of network caves. These examples tend to be rather limited in areal extent. The process can also modify caves formed by other processes. However, the mere presence of an insoluble caprock does not guarantee this mode of origin. Careful mapping of a cave in relation to its geologic setting is necessary to distinguish whether the cave may instead have been formed by another mechanism, such as hypogenic processes.

References

Clemens, T., Hückinghaus, D., Sauter, M., Liedl, R., and Teutsch, G., 1997, Simulation of the evolution of maze caves: *12th International Congress of Speleology and 6th Conference on Limestone Hydrology and Fissured Media,* La Chaux-de-Fonds, Proceedings, p. 53-56.

Douglas, H.H., 1964, *Caves of Virginia:* Falls Church, National Speleological Society, Virginia Region, 761 p.

Palmer, A.N., 1975, The origin of maze caves: *National Speleological Society Bulletin,* v. 37, p. 56-76.

Palmer, A.N., 1981, Hydrochemical controls in the origin of limestone caves: *8th International Congress of Speleology,* Bowling Green, Proceedings, p. 120-122.

Palmer, A.N., 1991, The origin and morphology of limestone caves: *Geological Society of America Bulletin,* v. 103, p. 1-21.

Palmer, A.N., and Palmer, M.V., 1989, Geologic history of the Black Hills caves, South Dakota: *National Speleological Society Bulletin,* v. 51, p. 72-99.

Ryder, P.F., 1975, Phreatic network caves in the Swaledale area, Yorkshire: *British Cave Research Association Transactions,* v. 1, p. 101-124.

Varnedoe, W.W., 1964, The formation of an extensive maze cave in Alabama: *Alabama Academy of Science Journal,* v. 35, no. 4, p. 143-148.

White, W.B., 1969, Conceptual models for carbonate aquifers: *Ground Water,* v. 7, p. 15-21.

5.3.12. Caves Branch, Belize, and the Baradla-Domica System, Hungary and Slovakia

Derek Ford

Abstract

Despite their very different geographic locations, these two cave systems are quite similar in morphology and origin. They are trunk passages created by allogenic streams. They are at the water table, with channel gradients of 2 to 4 m/km for many kilometers and not interrupted by any phreatic loops in the bedrock. They are many meters in diameter and height and display prominent corrosion notching.

Caves Branch is developed in a megabreccia of Cretaceous limestone (possibly a consequence of the Chiquibul meteorite impact nearby at the Yucatan Peninsula) and represents the capture of a former allogenic river that flowed through a cockpit karst into the eastern wall of the through valley. Tributary passages draining cockpit water to the trunk are quasihorizontal and much smaller.

The Baradla-Domica System is in highly deformed Triassic carbonate rocks that were deeply karstified during the Cretaceous and then partly buried by Pannonian (Miocene) clastic sediment. Burial raised the level of the water table, and the modern caves were created along the new level by a series of allogenic streams flowing from the clastic cover. The trunk cave is now in the process of being captured by springs about 60 m lower.

Both cave systems can be considered good representatives of State 4 development, with very high effective fissure frequency being due to the brecciation at Caves Branch and to tectonic deformation succeeded by paleokarstification at Baradla-Domica. U-series dating has established that the trunk passages in both caves had achieved most or all of their modern dimension before 100,000 years BP.

Introduction

This chapter describes two examples of large caves that developed along regional water tables. The process of their enlargement probably lowered these water tables by no more than a few meters, although it regularized (smoothed) their gradients. The caves are nearly horizontal, with mean gradients between 2 and 4 meters per km. Their courses are interrupted by few siphons (permanently water-filled sections) and there are no deep phreatic loops. They are representative of State 4 cave system development as explained in Chapter 5.3. These two particular cases were selected because they have low gradients despite being located within chaotic limestone stratification, i.e. their nearly flat profiles are not consequences of predominant control by quasihorizontal bedding planes, as it is in many platform terrains with little tectonic deformation.

These two cave systems are also representative of the class of river caves that are created primarily by allogenic streams passing through quite narrow belts of the karst rocks (10 to 20 km in these instances). Such caves are to be found worldwide. They are especially common where intense compressional tectonic deformation has created belts or narrower stripes of different rocks, all of which are densely fractured.

Caves Branch and Baradla-Domica are in two very different geographic settings today. However, there are similarities in their geologic histories, and although created by allogenic streams, both also drain quite rugged karstlands with a relief of about 100 meters.

Caves Branch, Belize

Caves Branch is considered first, because in certain geologic and morphologic respects it is the simpler case. It is an instance where an allogenic river first was able to carve a valley through a karst belt and then was progressively captured into a subparallel shallow underground course in the valley wall. The capture converted the through valley into a closed polje. Caves Branch is a good example of an active trunk-river cave, displaying features that are common in many such caves elsewhere. This summary is derived from an intensive study of the karst and caves and their quantitative hydrology and water chemistry that was undertaken for a Ph.D. thesis in the late 1970s by the celebrated American speleologist, Tom Miller (Miller, 1981, 1996).

The basic topography and geology are illustrated in 5.3.12 Figure 1. The Maya Mountains are a narrow range trending E-W across the center of Belize. They are composed of Paleozoic sandstone and shale. These were metamorphosed by granitic intrusions and uplifted at some time during the Late Cretaceous to early Tertiary (Bateson anad Hall, 1971). They are in abrupt fault contact with Cretaceous limestone (Dengo and Bohnenberger, 1969). This forms a belt along the north side of the mountains 12 to 25 km in width. It descends from an altitude of 200 to 300 m along the fault to 40 to 80 m where it passes underneath a plain of Eocene and later sediment to the north (Miller, 1981). The plain functions as a base level for fluvial erosion and karst spring discharge in the limestone. It is not known how long the limestone has been exposed to subaerial denudation, but it could be since as early as the Eocene.

The limestone is now dissected into a dense cockpit karst (kegel karst) interspersed with some shallower patches of doline karst ("glades" of Miller, 1981). It is crossed by several highly alluviated valleys maintained by the principal rivers from the mountains.

The modern climatic regime is characteristic of the humid coastal tropics. January mean temperature at the capital city of Belmopan—which is on the plain 20 km from Caves Branch—is 22°C. In July it rises to 25°C. Annual precipitation is 2200 to 2600 mm, mostly falling during the period, June to December. Rainstorms are chiefly of convectional origin and may be very violent.

The cockpit karst is covered by a dense broadleaf deciduous forest (palm, cedar, and bamboo). The through valleys and poljes, being accessible, have been logged, and some are used for commercial plantations (Day, 1993). The slopes of the Maya Mountains are very steep and deeply dissected. They support a scrubby forest with open pine stands higher up. The combination of a high gradient and poor cover permits rapid runoff in the allogenic stream channels.

Caves Branch is shown in 5.3.12 Figure 2. As noted, there was once a through valley carved by the Caves Branch River flowing north from the Mountains. The bedrock valley floor is now buried by 10 m or more of alluvial gravel, sand, and silt. This aggradation can be partly attributed to flood water backing up behind clogged sink points into

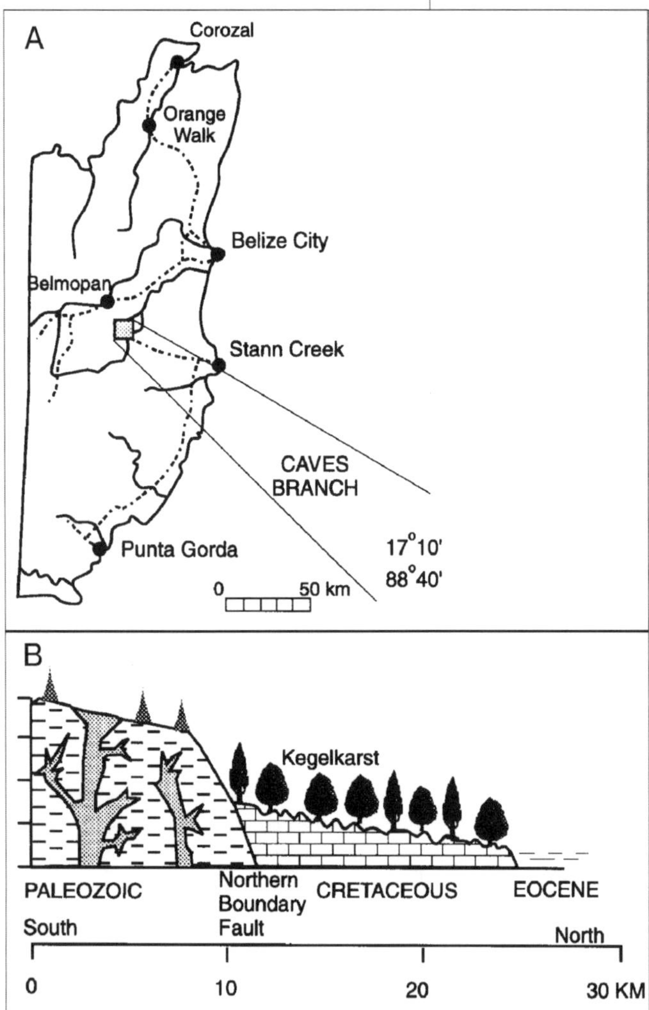

*5.3.12 Figure 1. **A** shows he location of Caves Branch in Belize. **B** is a Geologic cross-section from the Maya Mountains to the coastal plain of Belize.*

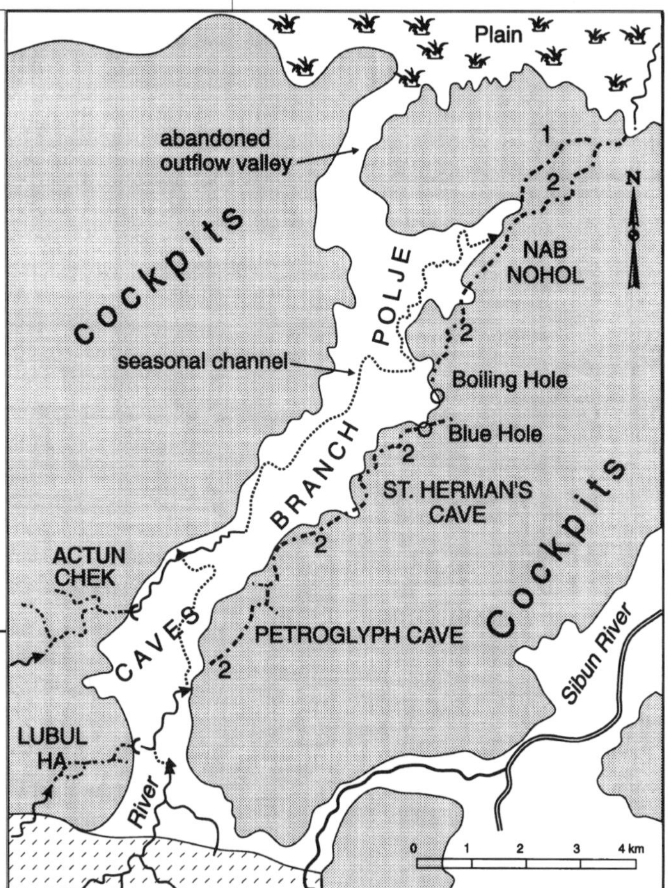

*5.3.12 Figure 2. Schematic map of the Caves Branch area showing the approximate locations and trends of the major caves. **1** is the Noloch Tata paleoriver course. **2** is a modern trunk channel, a few meters lower than Nohoch Tata.*

the caves. The river maintains a regular channel 2 to 5 m deep in the fill. At all seasons, part of its flow is lost underground a few hundred meters after leaving the metasedimentary rocks and passes into a large upstream cave, Petroglyph.

For much of the year there is also flow for long stretches farther downstream, fed by groundwater in the alluvium and by tributary allogenic streams that pass through two west-bank caves, Lubul Ha and Actun Chek. Some of this water is captured by short tributary caves draining into middle reaches of the trunk conduit system, St. Herman's Cave and upper Nab Nohol. The farthest downstream sink is now into the middle section of Nab Nohol. Wide areas of the polje around that sink are often inundated, but there is never surface discharge via the abandoned outflow valley. All water passes through lower Nab Nohol to a major spring at the northern edge of the karst.

The relationship between the allogenic catchment area of a creek and the extent to which it is able to maintain itself at the surface in a karst terrain is well illustrated by this region. The Lubul Ha tributary has a catchment of just 3.3 km^2 in the mountains and penetrates only 500 m into the karst before being swallowed entirely. The Actun Chek stream has a catchment of 20 km^2 and penetrates for 2000 m. Including these two tributaries, the allogenic catchment of Caves Branch River is 110 km^2. It once flowed through the karst entirely at the surface (16 km) but now is lost underground within 12 km. Its neighbor, the Sibun River (5.3.12 Figure 2), has a catchment of 250 km^2 and is a perennial throughflowing surface river.

Limestone Geology

The Cretaceous limestone of Caves Branch is believed to be from a back-reef and lagoonal facies because it displays a typically restricted fossil fauna (Flores, 1952). However, it is difficult to recognize any of the characteristic sedimentary features of such a setting. This is because wherever it is exposed at the surface or underground in this area, the limestone is a clast-supported breccia of disoriented rock fragments. The clasts are pebble- to boulder-sized and of gray to cream micritic limestone. These are cemented by red sparry calcite. Both the clasts and the matrix cement are more than 95 percent $CaCO_3$.

The brecciation has destroyed most or all true bedding. At a few places it is possible to trace probable individual beds or bedding planes for a few tens of meters. They display a very gentle dip to the north, but, effectively, there is no possibility of preferential flow or conduit genesis along bedding planes or in particular beds.

This is the most extensive, complete, and homogeneous brecciation of a limestone formation that I have seen. Its cause is not firmly established. Conventionally, it would be attributed to the dissolution of underlying evaporites. T.E. Miller (personal communication, 1998) offers the more exciting suggestion. He suggests it may be part of an ejecta blanket from the well-known Chicxulub meteorite impact crater at the tip of the Yucatan Peninsula, 460 km distant. The rather uniform petrologic composition of the clasts perhaps argues against this possibility. But an analysis of the dynamics of the impact by Kieffer (1998, and personal communication) shows that it is quite plausible. At this distance from the impact center, the firestorm could rip up a shallow surface layer of rock and deposit it as rubble a few kilometers farther away.

In the caves, the rock is seen to be intersected by some large vertical fractures. They are often slickensided, but the brecciation makes it difficult to determine fault offset. Air photos of the region display strong lineaments. Miller (1981) measured a mean density of 7 km/km^2 and showed that they played a significant role in determining the location and orientation of the cockpits. The smaller fractures seen in the caves are probably associated with them.

The combination of interclast matrix porosity with the permeability in the vertical fractures created a high measure of effective permeability. This favored the development of water-table caves. Miller (1981) showed that the cockpits were probably preceded by a phase of regular stream channel formation and entrenchment everywhere on the limestone, implying that the initial permeability was quite low. However, any former channel patterns are now completely disaggregated and replaced with holokarstic groundwater circulation.

Caves

The dominant caves are a system of trunk river passages in the eastern wall of the polje that channel Caves Branch River and its tributaries (5.3.12 Figure 2). They are fragmented by some breakdown obstructions and by a few short shallow siphons that Miller (1982) suggests may be caused by aggradation, i.e. where breakdown piles or banks of river gravel have created local ponds deep enough to fill the sparse low-roofed sections of a passage. There do not seem to be any phreatic loops in the bedrock, such as

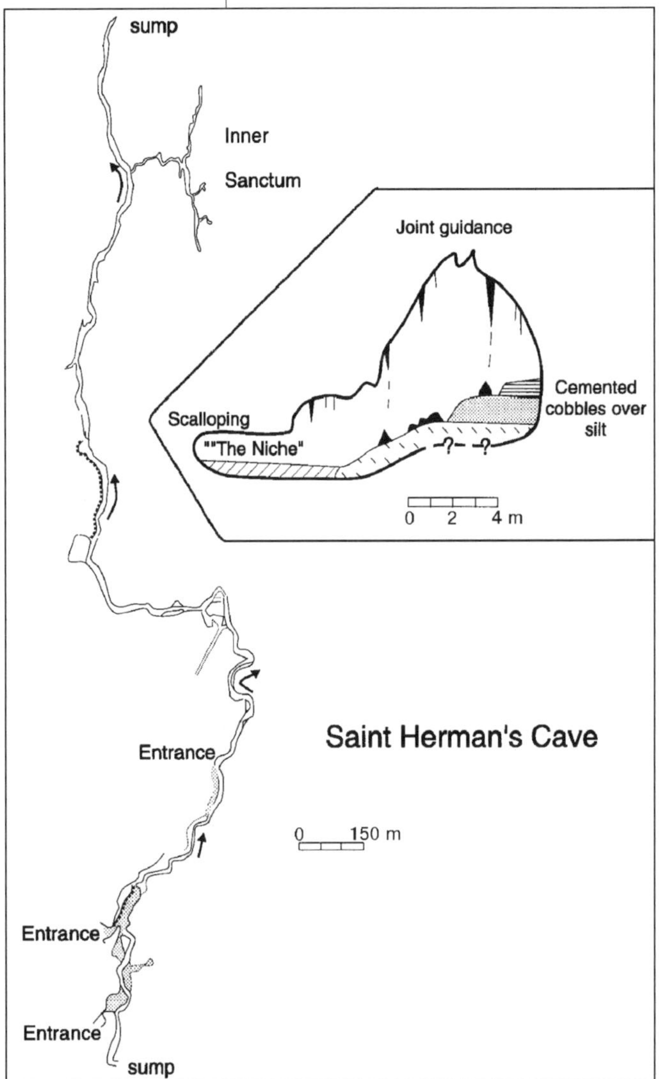

5.3.12 Figure 3. St. Herman's Cave, a sample portion of the trunk river cave between siphons. Inner Sanctum is a part of one of the smaller cave systems that drain from the cockpit karst to the trunk passage. The schematic cross section illustrates the principal features of modern cave river passages in Caves Branch.

those described in other types of caves in preceding chapters in this volume.

More than 20 km of trunk passages have been mapped. They are often guided by the vertical fractures for long distances and are quite efficient in that they are much straighter than the alluvial channel nearby in the polje. Although there are small waterfalls, rapids, and pondings in the underground river course, it maintains a mean gradient between 2 and 4 meters per km at an elevation a few meters below the alluvial surface in the polje. A plan of one section between siphons, St. Herman's Cave, is shown as an example in 5.3.12 Figure 3.

The trunk passages are on average at least 10 meters in height and 15 to 20 meters wide. A typical cross section is shown in 5.3.12 Figure 3 (from Miller, 1981). The ceilings are undulatory and display phreatic rounding and pocketing that is attributable to flooding early in the enlargement. At a few places, higher breakdown chambers may represent one or more former levels. The modern channels became entrenched and then widened laterally as they were aggraded with insoluble gravel and sand from the mountains. The bedrock floor is now rarely seen.

Throughout the trunk passages, the walls are often undercut by a deep corrosion notch (The Niche) at 1.5 m above the gravel floor of the modern channel. This suggests that the present fluvial conditions have been quite stable for a long period. U-series ages on stalagmites that have been toppled and cemented in gravel 2 to 3 meters above the modern waterline at two sites were found to range from 70,000 to 100,000 years BP. At a few places, several higher notches can be seen but they are not as widespread or regular as The Niche.

The first segment of trunk channel to develop, Nohoch Tata, is labeled 1 in 5.3.12 Figure 2. It was an initial short downstream capture of the paleoriver course, diverting it from the outflow valley. Nohoch Tata is now abandoned except in high flood and has extensive calcite flowstone and gours on its floor.

The modern trunk channel from Petroglyph Cave to Nab Nohol (labeled 2) is a few meters lower. It is not known whether this channel was assembled by a sequence of headward captures of the surface river or instead was initiated by more or less simultaneous underground diversions at several points along the polje. It is partly breached in the middle. Here surface lowering has created a karst window with two springs, Blue Hole and Boiling Hole, and some hundreds of meters of surface channel.

Miller (1982) shows that, since it was initiated, the places where water is abstracted underground have often changed as a consequence of the development of new small "pirate" caves leading from the meandering surface course. The pirate caves enter at the modern water table.

The large cockpit karst areas on either side of the polje must be drained by integrated systems of caves, but these are difficult or impossible to explore. Open vadose shafts have been discovered in the bottoms of a few cockpits but become blocked or too narrow within 30 m of the surface.

In the trunk river caves Miller (1982) recognized a distinctive category of small tributary streams (zuyhuyhas) with high total dissolved solids and no allogenic sediments. He showed that these must be the downstream termini of systems of cockpit drainage. They flow in small tubular subhorizontal passages that are 1 to 5 m above the modern trunk channels and quickly become water-filled or impassably small.

Inner Sanctum (5.3.12 Figure 3) was the longest example that could be followed. These collectors are flushed vigorously after heavy rains but drain very slowly into the allogenic trunk rivers at other times. This suggests that there is high effective porosity and a quite flat water table beneath the cockpit country. It accords with the picture of high porosity and rapid conduit development at the water table where the chemically aggressive allogenic river water was able to drain into the flanks of the polje. Both conditions are attributed to the combination of intense brecciation and dense regional fracturing of the limestone.

Baradla-Domica System, Hungary-Slovakia

The Baradla-Domica System is a trunk allogenic river cave that developed at a regional water table but is now in the process of being captured to a lower level. Where Caves Branch was an example of the lateral capture of a single river into a limestone valley wall, here there is engulfment of many separate small streams along the southern side of a small karst massif, which then amalgamate inside the massif to flow to springs at the northern flank. The geographic setting is shown in 5.3.12 Figure 4.

The Aggtelek highlands are quite rugged hills and plateaus of limestone rising to an altitude of 450 to 500 m. They are indented with dolines (Zambo and Ford, 1996) and shallow blind valleys. Along the southern side are gently rolling clay lowlands at 300 m. Streams are shallowly entrenched into the clay and sink immediately on encountering the limestone. There are some older open cave entrances a few meters above the modern swallets. The caves were occupied as early as the Neolithic (Takács-Bolner et al., 1989; Ringer, 1989), when much of the area was first cleared for agriculture. The oldest known written report of the caves dates to 1549 A.D., and the first map, of 2.2 km from the Aggtelek entrance, appeared in 1794. Imre Vass, a county engineer, greatly extended the exploration in 1825. The cave has been a significant tourist attraction since those years (5.3.12 Figure 5).

Lithology and Geologic History

The limestone is a platformal deposit of the Wetterstein Formation (Triassic, equivalent to the well-known Dachstein limestone of the Alps). It is overlain by dolostone (Guttenstein Formation), with minor sandstone and shale. This sequence was uplifted and greatly deformed by Carpathian tectonism. It was exposed to a long period of denudation beginning in the Late Cretaceous or early Tertiary. Climatic conditions were tropical or subtropical, allowing the formation of *terra rossa* soils (instead of the brown and black

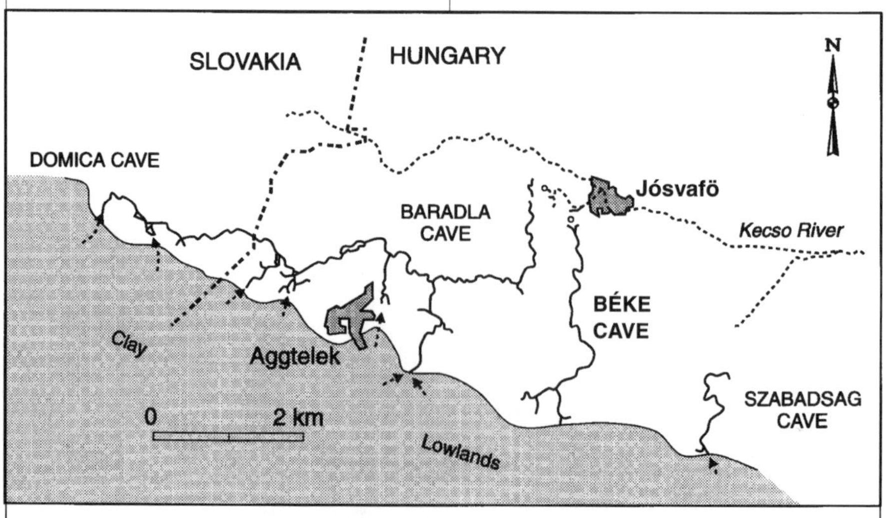

5.3.12 Figure 4. The Baradla-Domica Cave System, Hungary-Slovakia.

rendzinas of the modern era; Hevesi, 1989) and pockets of bauxite. It is important to understand that at that time the base level of erosion along the southern flank of the hills (and thus, presumably, any regional water table) was many tens of meters below the modern base level.

There is some disagreement concerning details of the subsequent history (L. Zambo, 1997, personal communication). The hills may have been completely buried and then partially exhumed by erosion before the time of the Pannonian Inland Sea (Miocene; Hevesi, 1989; Takács-Bolner et al., 1989). Pannonian sediment (gravel, sand, and clay from the Szilicky Highlands to the north in Slovakia) then filled the lowlands to or above the equivalent of 300 m today. There was regional uplift in the Pliocene, and the Neotectonic (modern) period of denudation commenced. The Kecso River (5.3.12 Figure 4) became entrenched along the northern flank of the massif, allowing early springs to develop at or above 270 m. A knick point has receded through the Kecso Valley, lowering basal spring altitudes to 220 to 230 m or below and creating steep spring-head saps that now truncate the older outlets of the caves. There is plentiful evidence of solifluction and other periglacial denudational activity in the hills during cold stages of the Quaternary, implying periodic loss of the natural forest cover.

The impacts of folding, faulting, and long exposure to denudation are very apparent in the rock walls of the caves. Strata are highly deformed, displaying steep but very irregular dips and much fracturing. Tight folds often display fills of clay or of breccia that is not cemented. There is much intrusion of *terra rossa* down steeply dipping bedding planes and in fractures that have been solutionally widened as much as 1 meter. These features are at depths ranging from 50 m to more than 100 m below the modern land surface. They indicate that, before deposition of the Pannonian sediment raised the base level and flooded the rock, the modern cave zone was part of a deeper, densely fissured vadose zone.

Baradla-Domica Cave has a mapped length of 24 km (18 km in the Hungarian part). The historic upstream entrance areas near the villages of Aggtelek and Domica are complicated by short upper levels of passages to 340 m, probably due to higher levels of the abutting Pannonian clay before its entrenchment, or to Quaternary colluvial filling. The modern sinks are at 300 m. From the Aggtelek Entrance, the modern cave river passage is joined at a breakdown chamber, the Concert Hall, at an altitude of 293 m and may be followed downstream for 7 km at a steady gradient between 3 and 4 m/km. The passage is rarely less than 10 m wide, 7 to 8 m high. It terminates at a major breakdown barrier that can be climbed into a large chamber, Oriasokterem (Giant's Hall).

5.3.12 Figure 5. *A painter's impression of the Baradla trunk river passage near the Concert Hall, by George Hering, 1835.*

An important feature of this trunk passage is that its cross-sectional area *decreases* in the downstream direction, rather than increasing in the normal manner that is to be expected as a consequence of the addition of further allogenic streams and autogenic drainage from the karst overhead. This decrease is because a sequence of headward captures of the flow is in progress, to a lower cave level draining to the modern Josva Spring at 225 m. This is illustrated (in a highly exaggerated form for the sake of clarity) in 5.3.12 Figure 6. Under low-flow conditions the river is now diverted to the lower level at sinkpoints near the Concert Hall. As discharge increases, it overflows to sinkpoints progressively farther downstream. This process is marked by general aggradation in the upstream half of the cave, but in downstream sections, which only the highest floods can reach, there are now many stretches of bare bedrock floor. The lowest sinkpoint is at the Oriasok breakdown. Flood water may reach it for a few hours every second year or so, e.g. during the heavy rains of May 1997 in central Europe. There is one recorded occasion, a flood in August 1955, when the capacity of this sink was exceeded and the overflow passed on through the breakdown and burst through an artificial seal in the wall of the Josva Valley (Jakucs, 1977).

Although the bedrock channel floor is often clean, there are both bedrock terraces and depositional terraces (sand or gravel topped by clay) throughout the downstream section of the cave. These signify one or more prolonged phases of aggradation in the past.

Beke Cave (5.3.12 Figure 4) is a smaller independent stream cave draining to a separate spring. Its general features are similar to those of Baradla-Domica (e.g. a

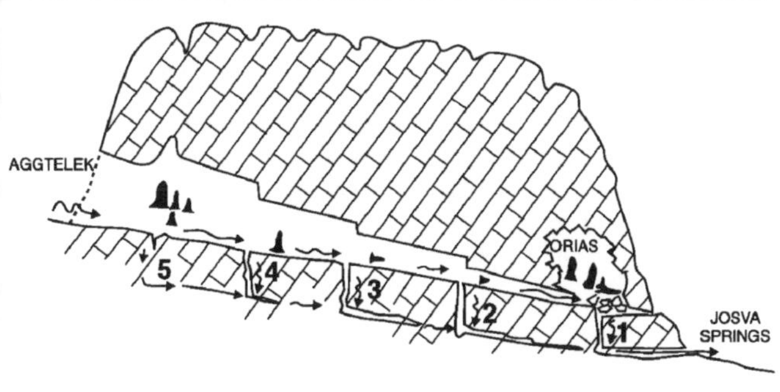

5.3.12 Figure 6. *Schematic illustration to show how stream capture to the lower cave progressively reduces the dimensions of the Baradla trunk passage in the downstream direction. This effect is exaggerated for the sake of clarity.*

sustained low-gradient river passage), but it displays stronger erosional effects of aggradation, such as multiple undulating corrosional-corrasional notches into the bedrock walls ("groove terraces" of Jakucs, 1977, p. 217 to 220). These are similar to the "niches" of the Caves Branch caves but probably developed more rapidly.

As a consequence of the high fissure frequency and a predominance of forest overhead, the cave is well decorated with calcite speleothems. These include large stalagmites and columns (2 to 3 m in diameter) growing on the cave floor, and smaller flowstone carapaces on many of the terraces. In the downstream sector of the cave (with its reduced cross section) the large stalagmites have been toppled and severely eroded by flood water except where they grow in protected recesses. The largest stalagmites in the Giants' Hall have also toppled; they are above any flood level but the breakdown floor on which they grew has been sapped by floods sinking beneath it.

Ford and Zambo (1997) investigated the pattern of terrace deposition, carapace formation, and stalagmite toppling by U-series dating of selected fallen samples ($^{234}U/^{230}Th$ method). Precise dating was difficult due to low uranium content and large amounts of contaminant thorium from the frequent floods. Nevertheless, a clear pattern emerged: the large, toppled stalagmites in the channel and Giants' Hall began to grow between 100,000 and 130,000 years ago, during the Last Interglacial (Eemian or Isotope Stage 5). During the period between 30,000 and 90,000 years, the detrital terrace deposits accumulated in the downstream cave, and flowstone began to grow on them. The big stalagmites in the channel also continued to grow, but at a lower rate than during the Interglacial. Extreme floods then swept away much of the terrace material and toppled the stalagmites during the interval of 15,000 to 30,000 years BP.

These results indicate that the river passage had achieved its modern form more than 130,000 years ago. Drainage was well established at the new lower level beneath the downstream section of the cave. The break-down event that created Giant's Hall occurred before 150,000 years BP. Throughout the Last Interglacial the runoff must have been significantly less than it is in our modern postglacial times, because the large stalagmites could begin to grow in the downstream channel; floods prohibit them from growing there today. No toppled speleothems older than the Last Interglacial were found, implying that before that time flooding was always too frequent to permit them to accumulate (i.e. the lower cave was insufficiently enlarged). These floods probably induced the breakdown.

The terrace deposits that accumulated between 30,000 and 90,000 years ago (early Würm and interstades, Isotope Stages 3 and 4) suggest that the forest cover in the catchment was much diminished, allowing more surface runoff and erosion. The phase of toppling and terrace erosion is firmly correlated with the Last Glacial Maximum cold period, which peaked about 20,000 years ago. The severe floods at that time and the cessation of all speleothem deposition even above flood level, suggest the occurrence of frozen ground and absence of forest. The fact that vigorous floods reach the Giants' Hall sink upon occasion today can be attributed to the effects of the Postglacial forest clearance and cultivation since the Neolithic (see Jakucs, 1977, p. 207 et seq.).

The modern river cave below the Baradla trunk passage remains incompletely explored. It is much smaller in dimension. Published profiles and accounts by divers suggest that it is a State 3 cave of short but steep phreatic loops (because of the steep dip) that link short horizontal canal sections with some entrenchment. The cave has a loop height range of approximately 10 meters, at an altitude between 220 and 230 m. As this passage enlarges and more clastic detritus is carried through it, its profile should become smoothed by the grading processes (entrenchment, bypassing, and paragenesis) described in Chapter 5.3. It will then adopt the form of the Baradla trunk passage 40 m above it.

Acknowledgments

T.E. Miller's research at Caves Branch, Belize, was supported by a grant to me from the Natural Sciences and Engineering Research Council of Canada. I am indebted to many Hungarian and Slovak friends for assistance in the Baradla-Domica System, especially Katalin Bolner, and Laszlo and Aniko Zambo.

References

Bateson, J.H., and Hall, I.H.S., 1971, Revised geologic nomeclature for Pre-Cretaceous rocks of British Honduras: *American Association of Petroleum Geologists Bulletin,* v. 55, p. 529-530.

Day, M., 1993, Resource use in the tropical karstlands of central Belize: *Environmental Geology,* v. 21, no. 3, p. 122-128.

Dengo, G., and Bohnenberger, O., 1969, Structural development of northern Central America: *American Association of Petroleum Geologists Memoir,* no. 11, p. 203-220.

Flores, G., 1952, Geology of northern British Honduras *American Association of Petroleum Geologists Bulletin,* v. 36, p. 404-413.

Ford, D.C., and Zambo, L., 1997, U-series dating of phases of speleothem deposition and erosion in Baradla Cave, Aggtelek National Park, Hungary: *12th International Congress of Speleology,* La Chaux-de-Fonds, Proceedings, v. 1, p. 44.

Hevesi, A., 1989, Development and evolution of karst regions in Hungary: *Karszt-es Barlang*, Special Issue, p. 3-16.

Jakucs, L., 1977, *Morphogenetics of karst regions:* Budapest, Akademiai Kiado, 284 p.

Kieffer, S.W., 1998, Dynamics of the Chicxulub Impact and implications for extinction mechanisms: *Geological Association of Canada Annual Meetings Abstract Volume*, p. A91.

Miller, T.E., 1981, *Hydrochemistry, hydrology, and morphology of the Caves Branch Karst, Belize:* McMaster University, Ph.D. thesis, 280 p.

Miller, T.E., 1996, Geologic and hydrologic controls on karst and cave development in Belize: *Journal of Cave and Karst Studies,* v. 58, p. 100-120.

Ringer, Á., 1989, The role of cave sites and their chronostratigraphy in the research of the Paleolithic of Hungary: *Karszt-es Barlang*, Special Issue, p. 51-58.

Takács-Bolner, K., Eszterhás, I., Juhász, M., and Kraus, S., 1989, The Caves: *Karszt-es Barlang,* Special Issue, p. 17-30.

Zambo, L., and Ford, D.C., 1997, Limestone dissolution processes in Beke Doline, Aggtelek National Park, Hungary: *Earth Surface Processes and Landforms,* v. 22, p. 531-543.

5.3.13. Speleogenesis in the Ljubljanica River Drainage Basin, Slovenia

France Šušteršič

Abstract

The Ljubljanica is a typical sinking river, disappearing and reappearing on the surface seven times. Data from 1,534 surveyed caves in the central part of its basin have been processed statistically. Fragments of horizontal caves are grouped in clearly expressed clusters. At least two of the clusters seem to have been separated apart along the Idria Strike-slip Fault and displaced about 12 km.

The spatial orientation of the clusters only vaguely fits the present hydrogeologic situation. The caves may be relict or reoccupied voids that formed originally in circumstances different from those of today. Most of the caves have typical phreatic shapes. These are further modified into epiphreatic channels only where there has been considerable input of mechanically transported material. The general genetic pattern is

- Initiation along bedding planes.
- Penetration into joints.
- Expansion by collapse of crushed zones and along faults.
- Filling of lower parts of the system with sediment and transformation into epiphreatic tunnels.

Introduction

The Classical Karst in Slovenia is best considered as a triangle with vertices that just encompass the cities of Ljubljana, Rijeka, and Gorizia (5.3.13 Figure 1). The present Institute for Karst Studies, in Postojna, lies close to the triangle's center. Traditionally, the area is subdivided into the Karst (Slovene: Kras) also called the Lower Karst, which lies southwest of the line from Gorizia through Postojna to Rijeka, and the High Karst, which occupies the remainder of the triangle.

Drainage of the Lower Karst and some parts of the High Karst is into the Adriatic Sea, while the rest of the area drains into the Danube Basin and the Black Sea. Within the classical triangle, the sinking Ljubljanica River is the backbone of the catchment that drains toward the Danube. It is widely known as a string of surface and underground stream segments, the latter emerging into karst poljes.

Although the surface and underground streams form one interconnected network, the river is traditionally divided into two separate branches. Within them there are six sections where the river emerges on the surface in closed basins that more or less fit the traditional interpretation of poljes. On the basis of maps and observations made predominantly by cavers, more than one third of the caves explored within the Ljubljanica catchment area seem to be of phreatic or epiphreatic origin. In the following text, these will be referred to as "horizontal" caves, although generally they are not strictly horizontal.

Prior to the 1990s, the caves in this area were thought to be epiphreatic, with their development intimately bound up with the formation of the poljes and some presumed river terraces along their margins (Gospodarič and Habič, 1979). During the second half of this century, extensive speleologic and geomorphic work was carried out in an attempt to confirm this idea. However, the great amount of information that was gathered did not support it.

Unfortunately, much of the data were collected in a way that did not readily allow them to be used to test alternative models of development. Although Gams (1963) remarked that this key probably did not fit the lock, it was not until the 1990s that studies founded upon other paradigms began to appear (Brenčič, 1992; Šebela, 1994; Šušteršič, 1992, 1994; Knez, 1996).

It is still not possible to offer a completely

satisfactory explanation of the speleogenesis within the area. Hence this paper considers its boundary conditions. The Ford and Ewers (1978) four-state model of cave genesis seems to offer a framework within which the evolution of the Ljubljanica River can be presented better than in any earlier model.

Thus their terminology is used extensively in this paper, with the more recent improvements (in my opinion) suggested by Worthington (1991) and Lowe (1992). However, this paper does not discuss these theoretical works; it just makes use of elements within them that seem to fit the observations.

Ljubljanica Catchment Area

The calculated area of the Ljubljanica drainage basin is 1779 km^2, of which 1100 km^2 is composed of karstic rocks. The location of the drainage divide is approximate, but at several segments of the boundary divides have been proved by water tracing (5.3.13 Figure 1). According to the results of comprehensive water-tracing experiments during the 1970s, the catchment of the Vrhnika Springs (5.3.13 Figure 4, no. 5513, 6185), where the main river finally leaves the karst terrain, comprises 1109 km^2 (Žibrik et al., 1976).

The mean discharge during the years 1972 to 1975 was 38.6 m^3/s, ranging from a low of 4.25 m^3/s to a high of 131.7 m^3/s, with a mean specific runoff of 34.8 L/s/km^2. Gams (1966a) established a mean denudation rate of 65 m^3/km^2/year.

The karstic rocks are generally micritic locally oolitic limestone and dominantly late-diagenetic dolomite, mostly of Mesozoic age. They were formed on the Dinaric Platform as a result of continuous sedimentation in a very uniform shallow sea. They are extremely pure, generally with less than 5 percent, but locally with as little as 0.1 percent, insoluble residue. The total thickness of the carbonate sequence is about 6850 m (Čar, 1996).

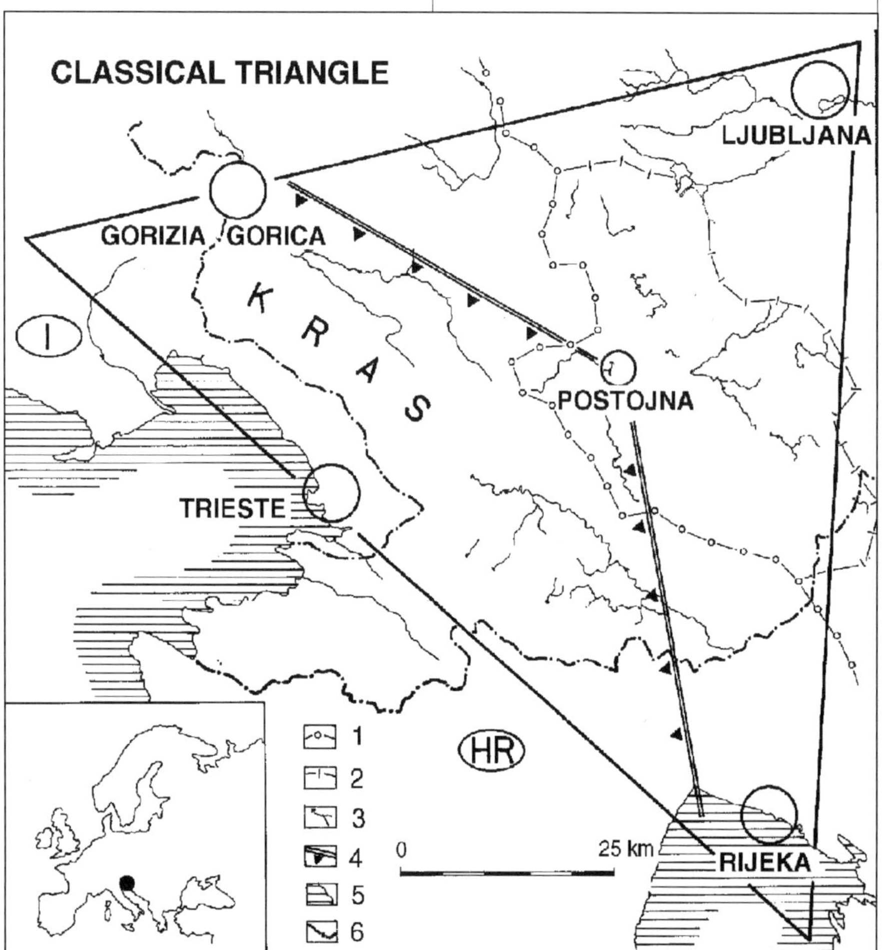

*5.3.13 Figure 1. The Classical Karst triangle. **1** is the Adriatic–Black Sea Divide. **2** is the karstic Ljubljanica basin's limits within the Black Sea Watershed. **3** is the sinking river and ponor. **4** is the southwest border of the High Karst. **5** is the Adriatic Sea coast. **6** is the international boundary.*

Among the nonkarstic rocks, only early Tertiary flysch, deposited directly on the carbonate sequence, has a significant role. Younger sediment is absent, suggesting a final emergence at the end of the Tertiary. However, due to tectonic events, the thickness of the carbonate sequence has been multiplied several times, and carbonates inevitably appear at some depth below the flysch.

Tectonically, the whole of the Ljubljanica Basin belongs to the Adriatic Subplate. The area is composed of several nappes (Placer, 1981; Premru, 1982) that were overthrust to the NE as a result of the collision of the Adriatic subplate with the European Continent. A gradual change in the direction of movement brought about the formation of the Idria Fault, a dextral strike-slip fault that runs through the area toward the NW. The total displacement along this still-active fault has never been measured systematically, and a wide range of estimates has been made. However, an apparent displacement of about 12 km is evident from the geologic map (5.3.13 Figure 3, inset map).

Except for fluvial denudation in the flysch area of the Pivka Basin, and in some areas on the Upper Triassic dolomite, most of the surface is typically karstic. Earlier workers believed that certain levels represented prekarst "erosion levels," and that some of the lower levels were remnants of prekarstic surficial drainage. Further research has proved that, with a few exceptions, this is not possible and that the relief is essentially tectonic (Habič 1981; Šušteršič, 1987a,b). The relief consists of blocks uplifted to different levels, represented by isolated hills (hums) at higher elevations and rather flat surfaces in the lower areas. The highest levels are aligned parallel with the Idria Fault, in direct contact with the southwestern margin of its main zone. These are Planinska Gora (1019 m), Javorniki (1293 m), and Snežnik (1796 m). The lowest point within the basin is the Ljubljansko Barje (Ljubljana Marsh, 294 m), where the karst water makes its final return to the surface.

Though the underground karst has been explored intensively for at least 200 years, areas with unexplored caves still exist. Most of the efforts were directed toward finding presumed underground cave connections between the main poljes, but no through connection for humans has been discovered. It has become clear that the accessible dry caves are just fragments of complex, collapsed, and in-filled systems, and the active stream caves–which are rarely observed–are just vadose or drained sections of basically drowned conduits. Among them, fewer than 20 are of speleologic significance (5.3.13 Figures 3 and 4), and only 9 are longer than 2 km. The longest one (Postojnska Jama) does not exceed 20 km.

The difference between the caves that have

Vertical	Formed exclusively in the unsaturated (vadose) zone	39.6
Horizontal	Formed at least partly in the saturated (phreatic or epiphreatic) zone	35.0
Unknown	Insufficient data to recognize the circumstances of formation	6.2
Collapse	Shape unrecognizable due to collapse	2.3
Artifical	Shape unrecognizable due to artificial alteration	0.1
Missing	Missing data	16.8

5.3.13 Table 1. Percent of each cave type in the Slovenian part of the Ljubljanica Drainage.

formed as regional drains and the vadose caves is very clear. In this paper only the former group is discussed. These caves are referred to as "horizontal caves," although, as noted, they are generally not horizontal in the true meaning of the word.

Overview of the "Horizontal" Caves
The statistics of the 1534 known caves within the Slovenian part of the Ljubljanica catchment area are given in 5.3.13 Table 1.

Generally, it is believed that the ratio between the bulk frequency of vertical and horizontal cave groups is 2:1, but this figure is based mainly upon cavers' observations. Field inspection of many features recorded as "vertical shafts" reveals that some of them, although steep to vertical, are actually phreatic loops, and it is possible that if all the caves were reexamined, the ratio might change in favor of the horizontal group. This paper is intentionally concerned with the horizontal subset and, all of the following discussion deals exclusively with this group unless stated otherwise.

It must be stressed that when ranking caves by their dimensions, only cave segments longer or deeper than 10 m generally are recorded. A first reckoning of the distribution is given in 5.3.13 Figures 2i and 2ii. From the former, it is evident that the frequency distribution is not smooth, although the data base is rather large. The first group between 0 to 9 m in horizontal length was initially identified as shafts but contains traces of phreatic passages or caves whose dimensions would normally have been too small to merit being recorded but which were registered exceptionally. Thus, the frequency of this class is not representative, and these caves are ignored in this study. Within the remainder of the population the mean length is 207 m, the median is 16 m, and the first quartile is 12 m. Thus, the idea that these caves are fragments of larger systems is well supported.

The categories just longer than 50 m and 200 m are probably enhanced by subjective judgements, as cavers may tend to hunt for and round up to "magic" numbers. The structure grouping becomes clear from 5.3.13 Figure 2iii, which includes a plot of $\log_{10}$ lengths (abscissa) against $\log_{10}$ depths (ordinate) of caves. There seem to be three major groups (A, B, C), plus one minor group (D) consisting of caves that are shorter and shallower than 10 m. Group D does not seem to alter the general pattern, but it is important to note that the frequency of caves in it is highly underestimated.

Among the others, group B that seems to consist predominantly of fragments of longer caves or systems with several tiers has the greatest frequency. If this is so, an asymptotic approach to a maximum depth of about 300 m, as predicted by Worthington (1991), could be expected. However, many such caves are accessible only via vertical vadose shafts, and thus the significance of this dimension becomes uncertain.

Most horizontal caves shorter than 160 m are seen in 5.3.13 Figure 2iii. Another drop off in frequency occurs at 320 m in length, suggesting that the apparent lengths of the caves reflect some periodic factor. These dimensions correspond to the approximate separation between faulted shattered zones that are mapped at several locations (Čar, 1982; Čar and Gospodarič, 1984; Šebela, 1994), which might trigger collapse and thus the fragmentation of what were previously continuous cave systems. Unfortunately, this idea has not been tested systematically in the field.

Zone A evidently encompasses horizontal caves that are accessible by mere chance through relatively deep vadose shafts. Most

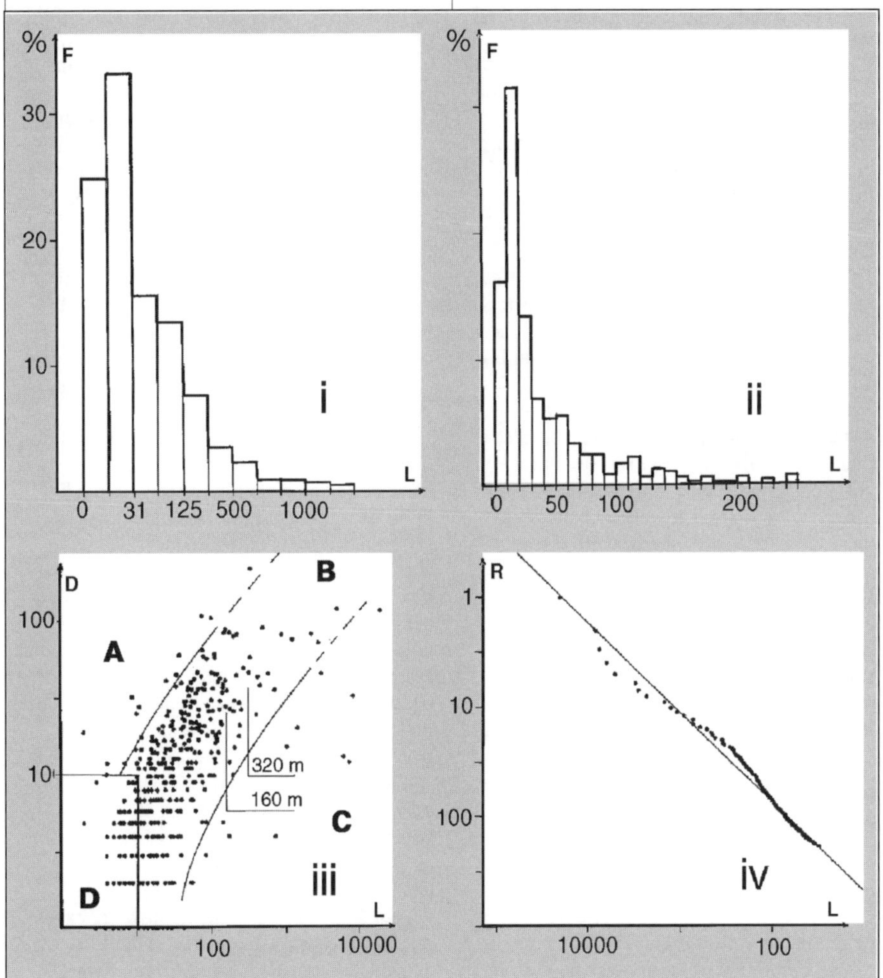

*5.3.13 Figure 2. Plot **i** is frequency F distribution of "horizontal" caves per length L of all caves. Plot **ii** is frequency F distribution of "horizontal" caves per length L of caves shorter than 250 m. Plot **iii** is a plot of $\log_{10}$ lengths L against $\log_{10}$ depths D. Plot **iv** is a Zipf diagram plot of $\log_{10}$ rank R against $\log_{10}$ lengths L.*

Cave	Q mean (m^3/s)	d_{de} (m)	Status	Q max (m^3/s)	d_{de} (m)	Status
Križna Jama (65, 6286)	1.30	4.06	Yes	16.00	14.30	Partly
Veselova Jama (905)	0.21	1.64	Yes	7.59	9.83	No
Tkalca Jama (857)	10.28	11.40	Partly	45.30	24.00	No
Postojnska Jama (747)	4.84	17.86	Yes	65.00	28.00	No
Planinska Jama (748)	14.22	13.46	No	71.96	30.30	No

5.3.13 Table 2. Likelihood (Status) that discharges (Q) in tubular passages of various caves reflect attainment of the equilibrium diameter (d_{de}).

Steam	Q mean (m^3/s)	Status	Q max (m^3/s)	Status
Lipsenjšca	1.30	Far Below	16.00	Matches
Lipsenjšca + Šerovnisca	1.51	Far Below	23.59	Surpasses

5.3.13 Table 3. Likelihood (Status) that the measured discharges (Q) at nearby streams match the calculated maximum dynamic discharge for a tube above Križna Jama.

caves in groups A and B are dry. Their walls have been heavily altered by collapse or by spalling of rock slabs, but in many cases it is clear that they were formed under Ford and Ewers' (1978) State 2 or 3. Preserved channels in the bedrock are never horizontal in the proper sense of the word. Phreatic development of the primary passages followed a relatively small number of bedding planes and joints. This holds true for both groups. If the present fissure frequency (within laminated zones there are more than 20 joints per meter) alone were to be considered, caves of State 4, or even maze caves, would be expected. In Karlovice (Velika Karlovica, 87, and Mala Karlovica, 171) and Postojnska Jama (747), only very recently formed passages are maze-like. (Cave numbering follows the central register of caves of Slovenia, maintained by the Speleological Association of Slovenia and the Karst Research Institute, Postojna.) Most passages have cross-sectional areas between 10 m^2 and 100 m^2. Scallops are generally not preserved and where preserved are barely recognizable.

Most of these caves are single passages. As a rule their ends are choked by boulders, soil, or more rarely flowstone. Based on data obtained in extensive excavations made during the construction of the motorway between Ljubljana and Postojna, I calculated that about 95 percent of the voids in the bedrock had been filled with red soil (Šušteršič, 1978a). More recent detailed mapping of the karst surface (5.3.13 Figure 5) confirms this estimate. Thus, collapse and fill control the present lengths of caves in groups A and B to a great extent.

Group C contains exclusively horizontal water caves. Their present condition matches the Ford and Ewers' (1978) States 3 or 4, but it must be stressed that none has really achieved the latter. The general pattern of these caves is a vertical zigzag of phreatic channels, partly incised by canyon-like sections. At some locations, especially in Planinska Jama (748), the mechanical stability of cave roofs has been exceeded, and extensive collapse has occurred to reestablish equilibrium, creating huge tunnel-like passages. Although some extensions end in collapse, the general rule is that even passages some kilometers long end in deep and long "sumps." In the Rak (East) branch of Planinska Jama, it is evident that the entire channel is submerged, as branching was discovered several hundred meters from the entrance of the "sump." Shorter caves belonging to groups A and B are common in the neighborhood of these caves, indicating that all of them belong to a single tier.

The first four caves, i.e. Postojnska, Križna (65, 6286), Karlovice, and Planinska, are systems that carry partly independent streams. Postojnska and Karlovice drain extensive areas with surface streams that have provided bed load, especially during the cold periods of the Pleistocene. Consequently, canyons are common, although entrenchment is modest (generally less than 10 m). It is very possible, though difficult to prove, that the passages below the present water table have been filled with fluvial sediment and that the present rivers flow close to the top of the filled tier as a consequence of this aggradation. Evidence of upward erosion (paragenesis) is common in these caves, but its true extent is difficult to establish. Križna and Planinska Jama differ somewhat, as they are closer to downstream poljes. Thus significant amounts of allogenic gravel were never transported through them.

Nevertheless, in Planinska Jama autochthonous gravel originating from the modern collapse doline of Mala Koleševka (Gospodarič, 1976) brought about local entrenchment at least some dozens of meters deep. Its main channels remained relatively high above the downstream polje, however, and the gradient between the inner part of the cave and the polje is much greater than gradients inside the system. This suggests that a relative state of equilibrium was achieved in the formation of the cave before the downstream polje had developed, at least at its present altitude.

This evidence raises the question of the general dynamic balance in these systems. Given the discharge (indirectly in Žibrik et. al., 1976, p. 49), the dynamic-equilibrium channel dimensions (Worthington, 1991, p. 73, eq. 5.9) of some of the main water caves may be calculated. In the 5.3.13 Table 2 above, d_{de} denotes the equilibrium diameter of hypothetical tubular passages, and the Status columns indicate whether observed phreatic channels have achieved these dimensions or not.

The status columns consider only phreatic tubes, however, not subsequent local enlargements. It would be interesting to relate these data to the cross sections of the caves belonging to groups A and B. Unfortunately, preserved primary profiles are only rarely accessible in them. Only a section of an elliptical tube in a quarry directly above the passages of Križna Jama seems to be a valid example. The long axis across the completely filled profile is 17 m. Although slope processes have removed the very highest part, it is fair to estimate an original short axis (height) of 12.5 m in the following calculations.

The present stream in Križna Jama flows to Šteberk Spring (4213, 5.3.13 Figure 4), and becomes Lipsenjšca Brook. Discharge can be estimated from data given by Žibrik et al. (1976) for Lipsenjšca. The flow through Križna Jama is allogenic and originates from Bloško Polje, which presently drains also to Veselova Jama (905), i.e. Žerovnišca Brook. Its hydrologic parameters are also given by Žibrik et al. (1976). It can be assumed that in some situations the entire discharge could be directed through the tube above Križna Jama. Both situations can be considered.

The cross-sectional area of the tube above Križna Jama is 166.90 m^2. According to Worthington (1991), the maximum dynamic equilibrium discharge (Q_x) should be 16.69 m^3/s. In 5.3.13 Table 3, "Status" describes whether the present discharge from recorded streams matches the calculated one.

The caves of group C, even though they are exceptionally large compared to the others in the data set, have not yet achieved dynamic-equilibrium dimensions. On the other hand, if the cross-sectional areas of all the caves in a cluster (tier) are added together, their total area is many times greater than the maximum needed to transmit the present drainage. Also, there is evidence that an inner balance once existed within Planinska Jama. It can be concluded that all these systems were formed under conditions that differ greatly from the present ones.

Spatial Distribution of Caves

It has already been noted that these horizontal cave fragments appear in clusters, implying that the concentrations might reflect the positions of abandoned or active flow corridors. The approximate mean regional density was calculated as 1.38 caves per km^2, a figure that was used to establish zones of lesser or greater density. However, neglecting caves in the surrounding region would introduce a bias due to border effects, and also would exclude the possibility of gaining insight into genetic circumstances in areas beyond the limits of the (calculated) watershed. To diminish these effects, a wider area of south-central Slovenia, totaling 5997.27 km^2, was taken into consideration, where 2489 caves have been explored and mapped in 4231.39 km^2 of karstic rocks.

Adding this area yields a value of 0.8 caves per km^2 of karstic rocks, and the statistical influence area of a cave becomes 1.25 km^2. The absolute values have changed, but the proportions, which are the main interest, remain the same. It must be noted that the predominantly Upper Triassic late-diagenetic Norian and Rhaetic dolomite plays a problematic role in karstification, and it is not possible to decide directly which sectors should be treated as karstic and which not.

Simplified spatial statistics are presented in 5.3.13 Figure 3, which is an adapted cave density map of the Ljubljanica catchment area. Four regions are defined as follows: a and b: areas where more than 4 horizontal caves and 4 vertical caves appear per influence area (see 5.3.13 Table 1), so that relatively stable proportions can be calculated.

- a is horizontal caves/vertical caves per influence area ≥ 1;
- b is horizontal caves/vertical caves per influence < 1;
- c is horizontal caves per influence area < 4, disregarding the vertical caves;
- d is no horizontal caves.

The picture is far from being chaotic, and several groupings are evident. In order to facilitate discussion, only those of special interest are numbered, and only the most important ones are discussed. The horizontal cave strips along the polje borders are considered self-evident and of no special interest.

•**1** The horizontal-cave zone contains relatively small (less than 20 m^2 in cross section) abandoned phreatic tubes about 300 m above the present water table. I interpreted them as the remains of the highest tier within the lower Ljubljanica system, between Planinsko Polje and Ljubljana Marsh, the other upper parts having been removed by erosion (Šušteršič, 1994).

•**2** The horizontal-cave zone encompasses somewhat larger much collapsed dry passages, partly in dolomite, developed within a vertical range of more than 300 m.

•**3** and **4**. The horizontal-cave zones evidently fit together and are separated by only a vertical-cave strip. They encompass some of the longest river caves in the area, i.e. Postojnska Jama (747), Planinska Jama (748), Tkalca Jama (857) and Zelške Jame (576) (5.3.13 Figures 3 and 4). Only a narrow strip of the vertical-cave zone separates them from the NW corner of the Cerkniško Polje, the site of the continuation of the Zelške Jame, i.e. Karlovice, (87, 171). Gospodarič (1976) and Šušteršič, (1978b) argued that the present direction of water flow in part of the system has been reversed with respect to the previous one.

•**5** The horizontal-cave zone is in all aspects similar to No. 2.

•**7** The main part of the horizontal-cave zone lies to the west, out of the Ljubljanica catchment area. Although some of its caves are now being reused by water flowing from the western part of the Postojna Basin southward (i.e. they belong to the Adriatic Watershed), it seems to be an abandoned tier, formed under conditions completely unlike the present ones.

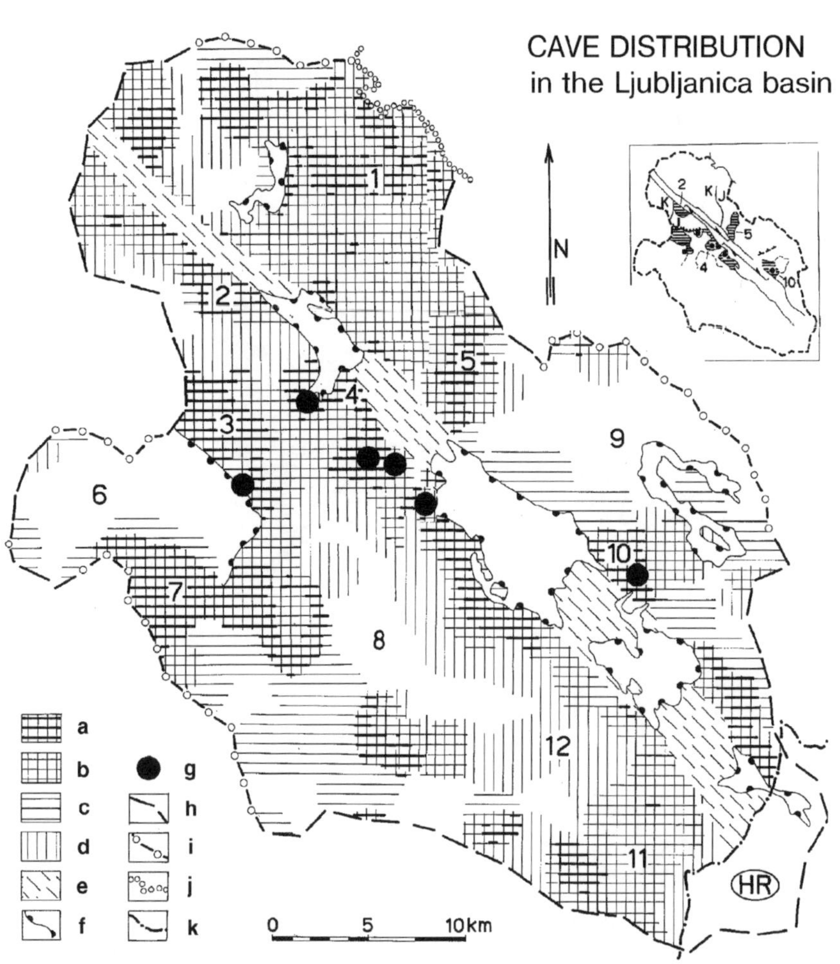

5.3.13 Figure 3. Cave distribution in the Ljubljanica Basin. ***a*** *is horizontal caves ≥ 4 and vertical caves ≥ 4 and horizontal/vertical ≥ 1.* ***b*** *is horizontal ≥ 4 and vertical ≥ 4 and horizontal/vertical < 1.* ***c*** *is 1 ≤ horizontal < 4.* ***d*** *is no horizontal caves.* ***e*** *is the Idria Fault Zone.* ***f*** *is the borders of the alluvial floors of major poljes and the lower part of the Pivka Basin.* ***g*** *is major river caves.* ***h*** *is approximate border of the karstic Ljubljanica catchment area.* ***i*** *is the well determinable border of the karstic Ljubljanica catchment area.* ***j*** *is the outflow border of the karstic Ljubljanica catchment area.* ***k*** *is the international boundary. Numbers: see text.*

- •**10** The horizontal-cave zone contains Križna Jama. Though smaller, it is similar to zones No. 3 and 4.
- •**11** The horizontal-cave zone is on the northern slopes of the Snežnik Massif. In the accessible caves, the present conditions are exclusively vadose. Relatively small predominantly reworked passages lie over a vertical range of nearly 1000 m. As neotectonic activity is lifting the core of the Snežnik Massif quite rapidly, all caves belonging to a single tier might subsequently be displaced vertically.
- •**6** This blank zone is the flysch area of the Postojna Basin.
- •**8** The blank zone is predominantly limestone, no different from the neighboring vertical-cave zone No. 12. It simply remains untouched by cave explorers.
- •**9** This blank zone is the dolomite territory of the Cerknišca catchment. It must be noted that lithologically identical dolomite appears in an area of similar extent east of zone No. 1, but only a fraction of this area is without caves.

The evident similarity of some zones, as well as the sudden termination of zone No. 11, prompts closer inspection of their relationships. Comparable zones (i.e. No. 2 and 5, No. 4 and 10) are separated by the Idria Fault Zone, and their displacements match the dextral strike slip along the fault. Approximate displacements even match the apparent horizontal displacement along the Idria Fault. Perhaps this holds true for zone No. 11 too, the counterpart of which might have been moved outside the Ljubljanica catchment area, and even out of Slovenia!

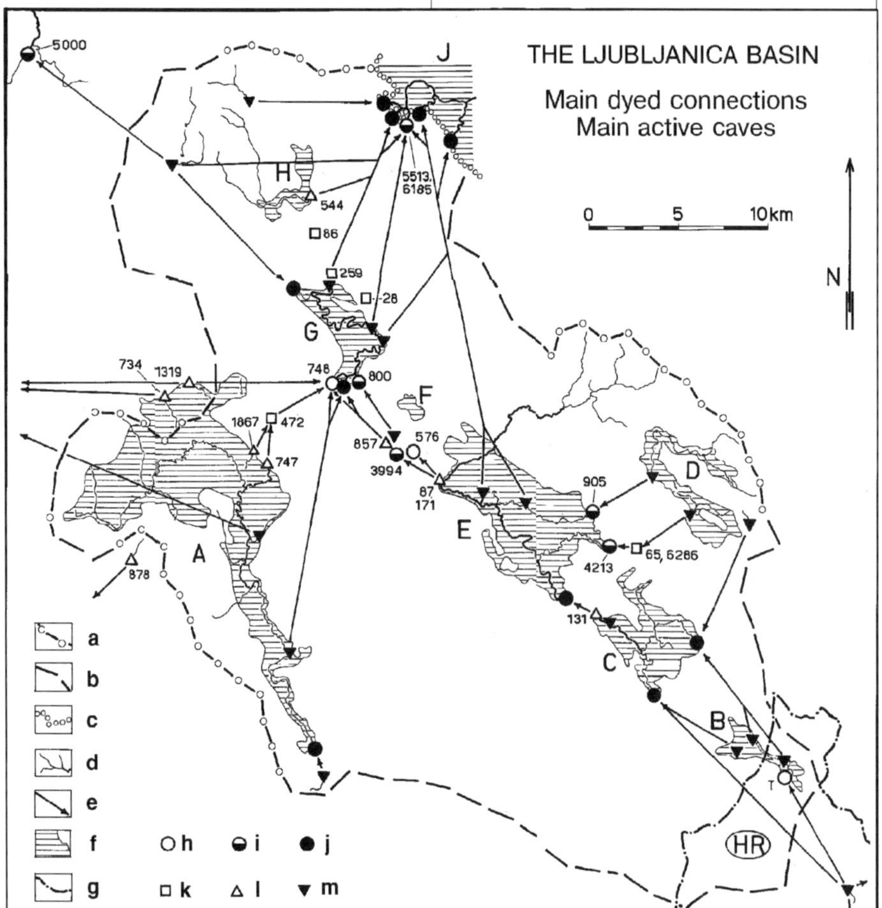

*5.3.13 Figure 4. The Ljubljanica Basin: Main dye-traced connections (simplified), Main active caves. **a** is the well determined border of the karstic Ljubljanica catchment area. **b** is the approximate border of the karstic Ljubljanica catchment area. **c** is the outflow border of the karstic Ljubljanica catchment area. **d** is the main surface streams. **e** is the traced connections. **f** is the alluvial floors of major poljes and the lower part of the Pivka Basin. **g** is the international boundary. **h** is accessible spring cave. **i** is siphon spring cave. **j** is impenetrable karst spring. **k** is major water cave accessible inside the system. **l** is accessible ponor cave. **m** is inaccessible ponor. **A** is the Pivka Basin. **B** is the Babno Polje. **C** is the Loško Polje. **D** is the Bloke (Bloško) Polje. **E** is the Cerkniško Polje. **F** is the Rakovško Polje. **G** is the Planinsko Polje. **H** is the Logaško Polje. **J** is the Ljubljansko Barje (Ljubljana Marsh). Numbers of registered caves are shown. See text.*

If returned to their presumably original positions, the paired zones fuse into single zones when the narrow strip of the Idria Fault Zone is ignored. Thus, all the major river caves, i.e. Postojnska (747, 472, 5.3.13 Figures 3 and 4), Planinska (748), Tkalca (857), Zelške (576), Karlovice (87,171), and Križna (65, 6286) form one single string. This "mega" zone is, at the same time, also the richest in shorter dry caves, i.e. fragments of conduits. Combining zones No. 2 and 5 does not produce so convincing a result, but the logic is the same. Both reconstructed "corridors" are oriented more or less westerly (however, this was not necessarily the flow direction), as are zones No. 1 and 7. Perhaps this indicates that all of them were formed in the same hydrogeologic situation, which evidently differed from the present one.

The active stream caves within the string formed by zones No. 3, 4, and 10 are the largest within the Ljubljanica Basin (the caves on the outlet side of Planinsko polje–Nos. 28, 259, 86–are modest in comparison to them). They are not situated in the region of highest discharge between Planinsko Polje and the final Vrhnika Springs. Rather, they are approximately in the middle of the system and are arranged approximately at right angles to the general flow direction. If connections confirmed by dye tests are considered, it becomes clear that, despite many bifurcations, the inaccessible connections follow quite direct lines between the inlets and outlets. On the other hand, parts of the main cave string, notably between Cerkniško and Planinsko Poljes, are situated in quite illogical positions, especially if the geologic situation at the inflow part of Planinsko Polje is taken into consideration (Šušteršič, 1978b).

The conclusion that at least part of the transmission system developed under hydrogeologic conditions quite different from those of today is the most readily acceptable. What these conditions were remains within the realm of speculation. One possibility is that water on parts of the Adriatic Plate had been draining eastward before the final collision with the European Continent.

Also, two minor but puzzling observations might be simply explained by acceptance of a significantly greater age for the majority of caves in the area. The unusual extent of silt-filled caves, when compared to the nearly negligible amount of insoluble residue in the bedrock, becomes more comprehensible if a greater time span and some possibly now-eroded sources of the material are available.

If most of the caves formed under

conditions of the Ford and Ewers' (1978) States 2 or 3, the rock was presumably not as fractured as it is at present. The subsequent strike-slip displacement of about 12 km would inevitably cause significant fracturing within the surficial zone of the crust, and caves that had not been filled with sediment would perhaps even collapse.

Some Details

Unfortunately, most previous work in the area has been directed toward picking up indications of presumed "fluvial-like" cave formation in the epiphreatic zone. That is, consciously or not, to support the well-known speleogenetic ideas first developed by Katzer (1909). Only during the past decade have some details been reinspected and reinterpreted. Two representative examples are discussed below, to provide pointers toward further more extensive research.

The Situation North of Planinsko Polje
The area about 1 km north of the main ponors of the Planinsko Poljes consists of alternations of Lower Cretaceous limestone with two intercalations of dolomite, all overlain by Upper Cretaceous limestone. The Lower Cretaceous limestone (95 to 99 percent $CaCO_3$) is micritic, rich in organic matter, and locally rich in fossil shells. The "dolomite" is microsparry dolomitized limestone (90 to 95 percent $CaCO_3$) of quite variable texture. Thus, this traditional term, dolomite, used elsewhere in this chapter, is in fact not strictly appropriate. Its strata are 6 to 8 m thick, and they contain numerous planes that are bedding-parallel, but each plane extends only some 10 m laterally (pseudo-bedding planes). Both dolomite beds are about 30 m thick. The Upper Cretaceous micritic limestone is very pure (95 to 99 percent $CaCO_3$). The general dip is 30° toward the west, but local distortions are common.

Many horizontal caves are known in the area and seem to be arranged along the lower and upper contacts of both of the dolomite packages. Typically, none of the active swallow holes can be penetrated more than some 100 m from the polje. Some hundreds of meters from the polje, floodwater appears in Najdena Jama (259; 5.3.13 Figure 4, 5008+ m long, 121 m deep), and Vranja Jama (88; 326 m long, 90 m deep). There is no direct connection between the polje border and these caves, however, as a vertical difference of about 20 m between flood levels in the polje and within the system remains more or less unchanged regardless of the absolute flood altitudes (Šušteršič, 1982).

It was concluded that a strong filter must exist between the polje and the inner part of the tier. In one of the ponor caves (Jama v Škofovem Lomu, 492, Šušteršič and Puc, 1970) a gradual diminishing of the tube cross-sectional area is observed, apparently evidence of the effect predicted by Atkinson (1986).

Due to the close proximity of the Idria Fault, the outer part of the system is much modified by collapse, and it is impossible to identify the positions of the presumed former main entrances to it, much less to detect their primary geologic structures. Inside the system (i.e. in Najdena Jama), it is evident that about 3 km of its main passages developed along the upper and lower contacts of the stratigraphically higher dolomite package, and along a third bedding plane within the Lower Cretaceous limestone, parallel with and about 5 m beneath the lower dolomite contact.

The primary tubes are lenticular to ellipsoid but their cross sections have been altered locally by entrenched canyons. Apparently the passages are orientated either downdip or along the strike, but detailed inspection reveals that this is only an approximation. Passages formed along joints are greatly subordinate, and generally they developed at breakthroughs that provided links from one major bedding plane to another. Phreatic risers, some more than 50 m upward, are common. Though some entrenchment has occurred due to local sources of bedload, most of the system is typically phreatic with no features suggestive of water table formation. Later blockfall modification in some passages is evidently related to tectonic deformation of the parent rock.

Among other abandoned dry fragments, Kloka Cave (3113, Šušteršič, 1994a) was studied in detail. Its accessible passage lies on the flank of a hill, about 100 m above the passages of Najdena Jama. Presently it is only a 10-m-long steeply inclined corridor, choked by fallen blocks, beyond which the cave undoubtedly continues. Its walls and ceiling have been modified by breakdown, but some fragments of primary phreatic tubes remain. The present corridor likely results from the collapse of the intermediate rock within a maze-like set of phreatic tubes. The channels were formed within the dolomite layer, along a pseudo-bedding plane, very close to the contact between the dolomite and the Upper Cretaceous limestone.

The observations in Kloka Cave may be summarized by the following statements:

- The cave is just a section of a larger system.
- Other than secondary developments, only traces of phreatic tubes were found.
- The cave is the highest in the area.
- The cave was formed predominantly along the pseudo-bedding planes.
- It is possible that the primary channel was formed along a true bedding plane.

Many of these statements also hold true for other caves in the area. The Vodni Rov (Water Passage) in Vranja Jama (88) is of special interest. If apparent correlations hold true over a distance of 500 m, it was formed along a single bedding plane. Its form is completely phreatic, and at first glance it appears to be formed in the dip direction. However, when measured properly, there are significant differences (Šušteršič, 1994a). Although the original plane is cut by several joints and laminated zones, they did not serve as initial structures, but water evidently penetrated into them after the main passage had been formed.

Thus, although the channel was formed along the bedding plane, it does not seem to have been influenced by any of the numerous geologic structures or orientations that now affect the bedding plane. Consequently, the passage must have begun to form before the structures were imposed. This is best explained by Lowe's (1992) theory of speleoinception.

In Najdena Jama, passages are penetrable (by scuba diving) down to a level of 388.5 m, and the highest point in Kloka Cave is at 531.8 m. They display phreatic or secondary features, and are almost always found on the upper or lower contacts of the dolomite layer. Several fragments that exist on intermediate levels generally follow the same bedding planes.

The most convincing interpretation of these observations is provided by Worthington's (1991) findings, viewing all the caves in the area as representing fragments of a single tier. Similar geologic conditions and speleologic effects are found along an older dolomite layer lying about 2 km to the east and stratigraphically 500 m lower.

Closer inspection of the development revealed three stages of relationship to geologic structures as follows:

- Initiation along bedding planes.
- Penetration into joints.
- Expansion by collapsing of brecciated zones along faults.

Although the inner parts of the tier are clearly recognizable and completely support Worthington's findings, the sites of initial phreatic descent are completely missing. More importantly, connections between the polje and the underground karst are significantly impeded. There is no trace of where the entrances to the caves behind might be. This suggests that the cave system is older than the polje, and that the present hydrologic connections are subsequent and poorly developed. Caverns revealed by drilling at depths down to 100 m below the present polje floor, at least, do not oppose this conclusion.

Situation at Laški Ravnik
Laški Ravnik lies about 3 km east of Planinsko Polje. The area is generally flat and extremely rich in solution dolines (Šušteršič,

1994b). Some main streams of the underground Ljubljanica must flow beneath it, although no active stream cave has yet been found. The known caves are short fragments evidently of phreatic origin. A bed of coarse-grained secondary dolomite with an attitude of 252°/ 32° occurs at the contact between Jurassic and Cretaceous limestones. Around this outcrop, at Javorjev Grič, an area of about 1 km^2 was mapped geologically and geomorphologically at a scale of 1:5000. Special care was taken to observe and record all detectable originally underground karst phenomena that now appear at the surface due to denudation (5.3.13 Figure 5).

During the fieldwork, it became more evident that there is not a single dolomite package, but an interfingering of wider or narrower dolomite and limestone layers (5.3.13 Figure 5). Besides the superficial karst phenomena, 4 underground types were recorded (numbers as in 5.3.13 Figure 5):

- Openings of small phreatic tubes, not larger than 0.5 m. All of them are choked within a few meters.
- Small collapsed features (<3 m) evidently continuing to choked tubes.
- Segments of (sub-)vertical tubes, formed under phreatic conditions.
- Denuded cave passages filled with silt and other sediment.

Note that no accessible cave passages were found at this location, and that true vadose shafts are omitted from the figure.

Detailed inspection of the figure reveals that despite the area being strongly affected by recent erosion, it is possible to observe that most traces of phreatic tubes lie close to either the upper or lower limestone-dolomite contact. At some places this correlation is clearly visible in the field.

It seems that around Javorjev Grič an earlier sector of an abandoned tier of caves is exposed. Again, the limestone-dolomite contact was most prone to cave-channel formation. In the northwestern quadrant, a concentration of features lies directly along the extrapolated trend of a dolomite layer that is now destroyed. The relationship between joints and the channels is less well expressed, however, and more difficult to detect.

Conclusions

Although as yet incomplete, the rough data discussed above permit some conclusions as follows:

- Horizontal caves (or fragments of them) lie in well-defined clusters, up to several kilometers in length, a few kilometers in width, and some hundreds of meters in depth. This pattern fits the notion of flow corridors within a single tier, as defined by Worthington (1991).
- At least two strings of clusters on each side of the Idria Fault can be correlated, and their displacements match the apparent displacement along the fault. This means that either the flow corridors were torn apart after formation, or similar structures, highly prone to similar karstification effects, had been established on both sides of a subsequent fault, and later activated.
- The spatial orientation of these clusters only vaguely mirrors the present hydrogeologic situation and polje locations. There is evidence that present water flow makes use–as much as possible–of older voids, inherited from different conditions. This is especially evident in the main string of active caves, which occupy unexpected positions and give an impression of being coincidental, rather than logical, parts of the modern system.
- Most of the voids in each cluster are now filled with fine-grained clastic sediment, regardless of their distance from nonkarstic source rocks, which might be quite remote. Those caves with greater lengths of accessible passage are perennially active. Diving experience indicates that rather than the classically expected short siphons, there are long submerged channels.
- Caves within a given tier were formed under phreatic conditions and reworked in vadose ones (in the hydrogeologic sense). There are no traces of epiphreatic shaping, except in the ponor caves, where abundant fluvial gravel is transported. This means either that the change from the primary to the present hydrologic conditions was relatively fast, or that activity within the

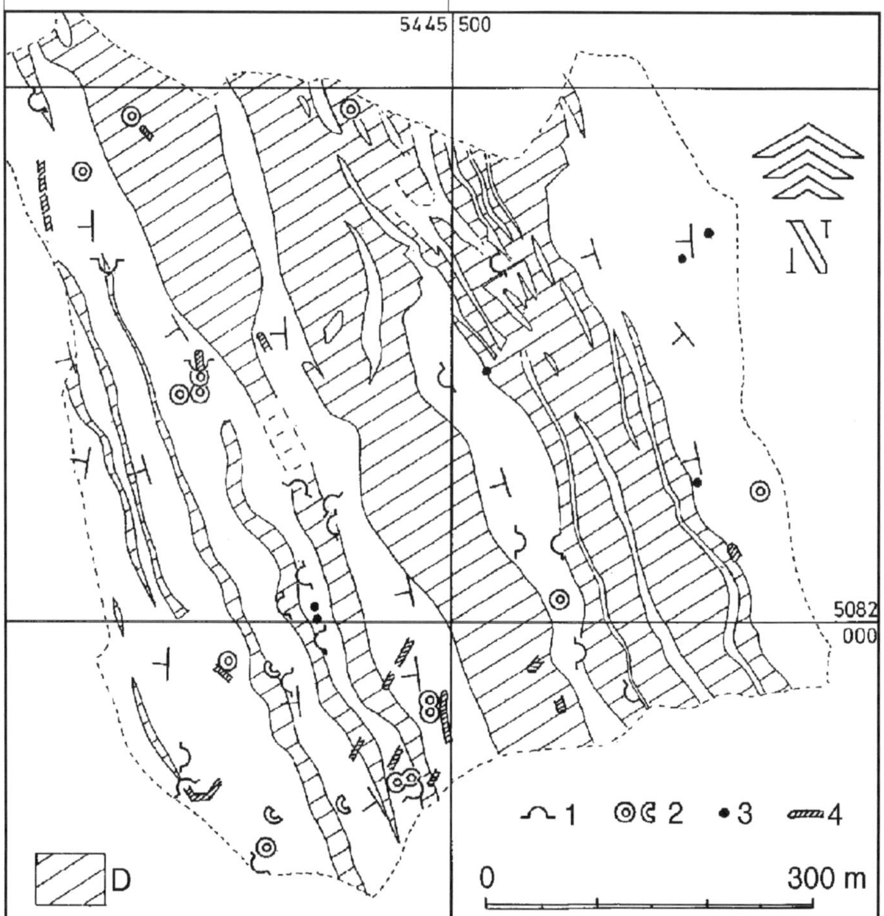

*5.3.13 Figure 5. Denuded underground karst phenomena at Javorjev Grič. **1** is openings of small phreatic tubes. **2** is small collapsed features (<3m), continuing to a choked tube. **3** is segments of (sub) vertical tubes, formed under phreatic conditions. **4** is denuded cave passages. **D** is dolomite.*

system was suspended during the transition.

- Numerous passages of relatively large dimensions in diverse positions indicate that the concentration of flow into victor tubes after breakthrough was not as fast as might be expected from theoretical considerations (White, 1988). This means either that the theoretical background is imperfectly known, or that "prekarstic" laminar flow did not appear, because voids inherited from diagenesis, or even from an earlier sedimentation phase, permitted turbulent flow from the very beginning. Inception driven by a variety of booster reactions, as described by Lowe (1992) could have a similar effect.

- Though the preserved phreatic passages are numerous, where studied in detail they are concentrated along a small number of bedding planes. In some cases, their directions evidently do not follow any current structural framework and any penetration into joints is secondary. These bedding-plane partings play the role of inception horizons, as defined by Lowe (1992).

- Joints and smaller faults are important only as master structures to guide the formation of phreatic jumps (lifts) within a tier. This probably means that they play an important role during the adaptation phase of a tier. The more tectonically disrupted zones define the areas of significant cavern collapse, or local slab spalling, during the subsequent decay of the cave. Thus, the general scheme of a cave corridor (trans)formation is:

  - Initiation along bedding planes;
  - Reorganization and penetration into joints.
  - Expansion by collapse of the brecciated zones of faults.

- Except in the direct ponor caves, where water flows on a bed of its own sediment and coarse bed-load material has made extensive adaptations (Gams, 1959), there are no traces of conventional fluvial organization of the underground karst.

- The notion of a cave system as a set of discrete tubes is being replaced by the notion of an initial maze, which may or may not evolve toward a conduit pattern. It is reasonable to speak about speleogenetic space and the degree of its utilization (Šušteršič, 1991), rather than about a system of interconnected tubes. Also, karst does not appear as the final phase after a long period of nonkarstic development. It might appear at the very moment of emergence and persist until the present, or until the karstic rocks are eventually consumed.

References

Atkinson, T.C., 1968, The earliest stages of underground drainage in limestones–a speculative discussion: *British Speleological Association Transactions,* v. 6, p. 53-70.

Brenčič, M., 1992, Košelevc: *Naše Jame,* v. 34, p. 41-51 (in Slovenian with English summary).

Čar, J., 1982, Geologic setting of the Planina Polje ponor area: *Acta Carsologica,* v. 10, p. 75-105 (in Slovenian with English summary).

Čar, J., and Gospodarič, R., 1984, About geology of karst among Postojna, Planina and Cerknica: *Acta Carsologica,* v. 12, p. 91-106 (in Slovenian with English summary).

Ford, D.C., and Ewers, R.O., 1978, The development of limestone cave systems in the dimensions of length and breadth: *Canadian Journal of Earth Science,* v. 15, p. 1783-1798.

Gams, I., 1959, About the situation and the genesis of the longest Slovene caves: *Naše Jame,* v. 1, no. 1, p. 4-10 (in Slovenian with English summary).

Gams, I., 1963, Logarček: *Acta Carsologica,* v. 3, p. 7-83 (in Slovenian with English summary).

Gams, I., 1965, On the Quaternary geomorphogenesis of the area among the karst poljes of Postojna, Planina, and Cerknica: *Geografski vestnik,* v. 37, p. 61-101 (in Slovenian with English summary).

Gams, I., 1966a, Factors and dynamics of corrosion of the carbonatic rocks in the Dinaric and Alpine karst of Slovenia: *Geografski vestnik,* v. 38, p. 11-68.

Gams, I., 1966b, On the hydrology of the territory among the poljes of Postojna, Planina, and Cerknica: *Acta Carsologica,* v. 4, p. 5-54 (in Slovenian with English summary).

Gams, I., 1970, Maximisation of the karstic underground water flows in example of the area among the karst poljes of Ceknica and Planina: *Acta Carsologica,* v. 5, p. 171-187 (in Slovenian with English summary).

Gams, I., 1994, Types of the poljes in Slovenia, their inundations and land use: *Acta Carsologica,* v. 23, p. 285-302.

Gospodarič, R., 1970, Speleological investigations of the Cerknica Cave System: *Acta Carsologica,* v. 5, p. 109-169 (in Slovenian with English summary).

Gospodarič, R., 1976, Quaternary cave development between the Pivka Basin and the Polje of Planina: *Acta Carsologica,* v. 7, p. 5-135 (in Slovenian with English summary).

Gospodarič, R., and Habič, P., eds., 1976, *Underground water tracing*: Postojna, Institute for Karst Research, 309 p.

Gospodarič, R., and Habič, P., 1979, Karst phenomena of Cerkniško Polje: *Acta Carsologica,* v. 8, p. 7-162 (in Slovenian with English summary).

Habe, F., 1976, Die morphologische, hydrographische und speleologische Entwicklung der Flyschbucht von Studeno: *Acta Carsologica,* v. 7, p. 141-213 (in Slovenian with English summary).

Habič, P., 1981, Karst relief and tectonics: *Acta Carsologica,* v. 10, p. 23-44 (in Slovenian with English summary).

Habič, P., 1987, The ponor cave Golobina at Loško Polje: *Acta Carsologica,* v. 16, p. 35-49 (in Slovenian with English summary).

Habič, P., 1989, Pivka karst bifurcation on the Adriatic-Black Sea watershed: *Acta Carsologica,* v. 18, p. 233-264 (in Slovenian with English summary).

Jenko, F., 1959, *The hydrogeology and water economy of karst:* Ljubljana, Državna založba Slovenije, 239 p. (in Slovenian with English summary).

Katzer, F., 1909, *Karst und Karsthydrographie*: Serajevo, Zur Kde. Balkanhalbinsel, no. 8, 94 p.

Knez, M., 1996, *The bedding-plane impact on the development of the karst caves:* Znanstevnoraziskovalni center SAZU, Ljubljana, Ph.D. thesis, 186 p. (in Slovenian with English summary).

Lowe, D.J., 1992, *The origin of limestone caverns: An inception horizon hypothesis:* Manchester Polytechnic University, Ph.D. thesis, 512 p.

Placer, L., 1981, Geologic structure of southwestern Slovenia: *Geologija,* v. 24, no. 1, p. 27-60 (in Slovenian with English summary).

Premru, U., 1982, Geologic structure of southern Slovenia: *Geologija,* v. 25, no. 1, p. 95-126 (in Slovenian with English summary).

Šebela, S., 1994, *The role of tectonic structures during the formation of cave channels and karst surface phenomena*: University of Ljubljana, Ph.D. thesis, 129 p. (in Slovenian with English summary).

Šikič, D., 1979, On the tectonic structure of the watershed of the River Kupa: *Geološki vesnik,* v. 31, p. 151-156 (in Slovenian with English summary).

Šušteršič, F., 1978a, Some considerations about stuffed potholes and their filling: *Naše Jame,* v. 17, p. 73-85 (in Slovenian with English summary).

Šušteršič, F., 1978b, A contribution to the knowledge of the inflow part of the Planinsko Polje and the Postojnska Vrata: *Geografski vestnik,* v. 50, p. 51-63 (in Slovenian with English summary).

Šušteršič, F., 1982, Morphology and hydrography of Najdena Jama: *Acta Carsologica,* v. 10, p. 127-155 (in Slovenian with English summary).

Šušteršič, F., 1987a, The small scale surface karst and solution dolines at the northeastern border of Planinsko Polje: *Acta Carsologica,* v. 16, p. 51-82 (in Slovenian with English summary).

Šušteršič, F., 1987b, An essay on an unconventional geomorphologic representation of Notranjska, *in* Notranjska: Zbornik 14, Postojna, Zborovanja slovenskih geografov, p. 61-72 (in Slovenian with English summary).

Šušteršič, F., 1991, On the objective of speleological science: *Naše Jame,* v. 33, p. 73-85.

Šušteršič, F., 1992a, Delovni seznam jam jugovzhodne Slovenije: *Naše Jame,* v. 34, p. 74 -108.

Šušteršič, F., 1992b, The Dolček Cave: *Naše Jame,* v. 34, p. 29-39 (in Slovenian with English summary).

Šušteršič, F., 1993, Delovni seznam jam jušnega dela osrednje Slovenije: *Naše Jame,* v. 35, p. 59-99.

Šušteršič, F., 1994a, The Kloka Cave and speleoinception: *Naše Jame,* v. 36, p. 9-30.

Šušteršič, F. 1994b, Classic dolines of a classical site: *Acta Carsologica,* v. 23, p. 123-154.

Šušteršič, F., 1994c, Delovni seznam jam jučne Slovenije–dopolnila: *Naše Jame,* v. 36, p. 112-140.

Šušteršič, F., 1995, Delovni seznam jam severne Slovenije: *Naše Jame,* v. 37, p. 98-152.

Šušteršič, F., 1996a, The pure karst model: *Cave and Karst Science,* v. 23, no. 1, p. 25-32.

Šušteršič, F., 1996b, Poljes and caves of Notranjska: *Acta Carsologica,* v. 25, p. 251-289.

Šušteršič, F., and Puc, M., 1970, The karst underground by the NE nook of the Planinsko Polje: *Acta Carsologica,* v. 5, p. 205-270 (in Slovenian with English summary).

White, W.B., 1984, Rate processes: Chemical kinetics and karst landform development, *in* La Fleur, R.G., ed., *Groundwater as a geomorphic agent:* London, Allen and Unwin, p. 227-248.

White, W.B., 1988, *Geomorphology and hydrology of karst terrains:* New York and Oxford, Oxford University Press, 464 p.

Worthington, S.R.H., 1991, *Karst hydrogeology of the Canadian Rocky Mountains:* McMaster University, Ph.D. thesis, 227 p.

Žibrik, K., Lewicki, F., and Pičinin, A., 1976, Hydrologic investigations, *in* Gospodarič, R., and Habič, P., eds., *Underground water tracing*: Postojna, Institute for Karst Research, p. 43-55.

Part 6: Meso- and Micromorphology of Caves

6.0 Overview

In Parts 2 to 5 attention is directed primarily on how the gross geometry of a cave system is established. Part 6 switches focus to the forms at meso- and microscales that can be created during enlargement of the cave. Lauritzen and Lundberg in Chapter 6.1 summarize the great variety of erosional forms (*speleogenetic facies*) that can be created by a wide range of *speleogenetic agents* operating in the phreatic or vadose zones. Some forms of cave pasages have been subject to intensive research and may be interpreted by means of simple physical and chemical principles. But many others are polygenetic and hence difficult to decipher with certainty. However, in addition to the analysis of cave patterns (see Chapter 3.4), each morphological element is a potential tool that can aid our inferences on the origin of caves and on major characteristics of respective past hydrogeologic settings.

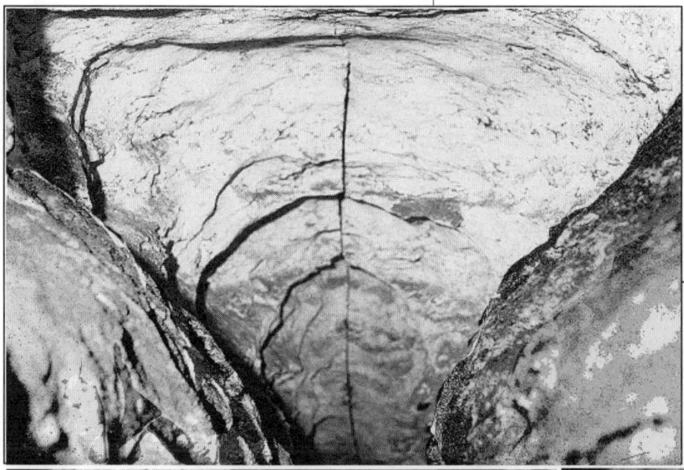

6.0 Plate 1. Feeder channel, Ozermaya Cave, western Ukraine (photo by A. Klimchouk).

In Chapter 6.2 E. White and W. White review breakdown morphology in caves. They generalize that the processes are most active during the enlargement and decay phases of cave development.

Early in the process, breakdown occurs when the flow regime shifts from pipe-full conditions to open channel conditions (that is, when the roof first loses buoyant support). Later in the process breakdown becomes part of the overall degradation of the karst system.

The chapter addresses the mechanism of breakdown formation, the geologic triggers that initiate breakdown, and the role that breakdown plays in the development of caves.

6.0 Plate 2 University Cave, North-West Caucasus (photo by V. Kisseljov).

6.0 Plate 3 Vadose canyon in Third Discovery, Blue Spring Cave, Indiana, U.S.A. (photo by A. Palmer).

6.1. Solutional and Erosional Morphology

Stein-Erik Lauritzen and Joyce Lundberg

Abstract

A cave is produced through the action of *speleogenetic agents* acting under various constraints to produce *speleogenetic facies*. These facies, expressed at the meso- and microscale, reflect the major and minor speleogenetic agents that operated on that cave. They also reflect the history of the cave, both during speleogenesis proper and during the postspeleogenetic phase, in particular the most recent history.

Geologic control is evident through the association of caves with guiding voids (the singularities that govern permeability) and of passage shape with rock chemistry (solubility). Hydrologic control guides the locus and direction of dissolution. Phreatic conditions support omnidirectional dissolution and thus hydraulically controlled tubular forms. In contrast, vadose conditions allow only unidirectional dissolution and thus gravity-controlled canyon forms and karren-like features.

Of the microforms, scallops are specific flow indicators. They yield both directional and quantitative information, such as flow rates and various hydraulic parameters specific to the cave passages. The presence of a sediment fill may further direct corrosion. In the phreatic zone this causes paragenesis. In the vadose zone, sediment causes lateral undercutting and eventually collapse.

Vadose streams display many of the forms of surface streams, such as migrating meanders, entrenchment, rock-mill potholes, and waterfalls. Vadose shafts, domepits, and condensation-corrosional forms are perhaps specific to the cave enviroment. The various vadose, phreatic, and certain water-table-specific forms are in combination powerful methods for reconstructing phases of speleogenesis as well as external base levels. Combined with speleothem dating techniques, they become important methods for determining erosion rates and landscape evolution.

Introduction

Once speleogenesis has established the gross geometry of a cave system, the cave's further development is entirely dependent on whether the conduits stay within the phreatic or vadose zones and whether they become filled with or emptied of sediment. In the strict sense (*sensu stricto*), "speleogenesis" means the creation of a cave from "scratch." That is, the process of microtube generation from an initially pristine rock mass with its tectonic features. Speleogenesis *sensu lato* means the further development into human-sized passages, i.e. the macroforms only.

These factors create a wide range of *speleogenetic agents* that in turn may give rise to an immensely variable morphology (*speleogenetic facies*). Speleogenetic agents, following Strahler's (1981, p. 366) definition of geomorphic agents, are all fluids, like water and air. These fluids have various attributes—temperature, pressure, chemical aggressiveness, kinetic energy, sediment content, aggregate state—that operate in the phreatic, water-table, or vadose zones. Each morphologic element is a potential tool for the speleologist to make inferences about the origin of caves, which is the objective of *speleogenetic analysis*: To identify the type

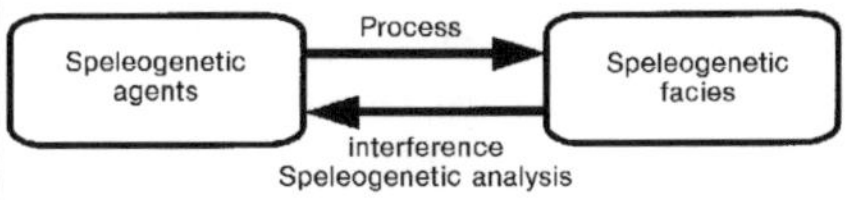

6.1 Figure 1. A set of speleogenetic agents (i.e., water in various aggregate states, in various states of movement and aggressiveness, and in various zones of the aquifer) act through certain processes on karst rocks. This results in a set of specific speleogenetic facies. Speleogenetic analysis is inference about the opposite processes. It aims at identifying specific agents and under which conditions they worked (vadose, phreatic, epiphreatic).

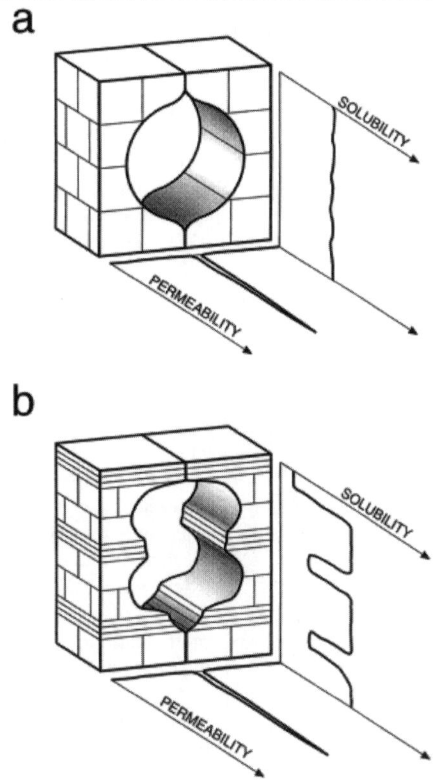

6.1 Figure 2. Lithologic structure and passage shapes. In general, cave passages form along lines of intersection between guiding voids (here a fracture) and horizons of optimal solubility. Hydrologically, the fracture is practically speaking a singularity compared to the surrounding rock mass (even more so is the resulting cave conduit). By "singularity" we refer to the virtually infinite porosity of the fracture compared to the virtually nonexistant porosity of the rock mass. ***a**—In pure, homogeneous rock, passage shape is largely controlled by hydraulic factors. Here a phreatic tube is developed along the guiding fracture. The solubility profile, or stratigraphic variation, has little importance for either passage shape or its vertical position within the fracture plane.*
***b**—Inhomogeneous rock, including less-soluble beds, exercises strong control on passage shape and also on its stratigraphic position.*

and sequence of speleogenetic agents that underlie specific forms or groups of forms (6.1 Figure 1). Because the various speleogenetic agents often operate in sequence, overprinting is common.

Some forms of cave passages have been subject to intensive research and may be interpreted by means of simple physical and chemical principles. But many of the individual forms are polygenetic and hence are difficult to decipher with certainty.

When discussing the shape of caves, it is practical to distinguish between *mesoforms*, that are of similar size to the diameter of the cave passage itself, and *microforms*, that are smaller than the passage. We may therefore say that mesoforms express the gross morphology of the passage, and in most cases they reflect how the passage was formed. Microforms are usually superimposed onto passage walls. They often indicate modifying processes that operated subsequent to the formation of the main passage.

Geological Control of Passage Shape

Meso- and microforms are controlled by corrosive and erosive processes (active variables) but have lithologic and tectonic constraints (passive variables). Geology influences passage form through control of:

- The provision of solute pathways.
- Rock strengths.
- Susceptibility to dissolution and erosion (corrasion) (Ford and Williams, 1989).

Isotropy is an important variable. In pure isotropic homogeneous rocks, corrosion and erosion (corrasion) are dominated by hydrodynamic factors. The resulting forms attain their most perfect shapes. These shapes may become distorted or even totally inhibited in anisotropic or impure rock, the irregularity increasing with time. Cave conduits are generally formed along fractures or bedding-plane partings so that anisotropy at this scale is the most important control on passage form.The interaction of two controls on phreatic passage morphology is illustrated in 6.1 Figure 2. In both cases permeability may be regarded as a singularity that allows water access only at the fracture and thus controls passage location. The stratigraphic distribution of solubility, on the other axis, controls the form. In the first case the homogeneity of solubility causes no diversion from the simple hydrodynamically efficient tube form. In the second case variations in solubility are directly apparent in variations of passage cross section.

At the smallest scale, rock purity strongly influences caves. Impure beds have low solubility and form resistant benches in passage walls (as in 6.1 Figure 2). On a larger scale they may confine flow, especially in the vertical plane, and guide passage location and long profile. For example, chert aquicludes in McBride's Cave, Tennessee, U.S.A., cause stepped long profiles with waterfalls at major joint intersections. Impurities within beds in the form of stylolites in places influence cave form. For example, stylolites form waterfalls in Herbert's Pot and Khazad-Dum Caves, Tasmania. They guide phreatic tubes in Peak Cavern, U.K. (Gillieson, 1996).

A special case of geologic control of dissolution rate is apparent in the production of *boxwork* where veins of less soluble material such as extremely pure calcite protrude from the more soluble host rock. Some caves in Norway show such features with calcite veins emergent from marble cave walls. A special type of boxwork, and perhaps the best known, has developed in Wind Cave, South Dakota, U.S.A., where the host rock dolomite is laced through with gypsum veins that have been diagenetically replaced by calcite. The surrounding bedrock was highly altered by sulfuric-acid reaction so it was turned into a friable sandy mass to a depth of several tens of cm. The host rock crumbles away leaving the boxwork protruding up to 100 cm from the cave wall (Palmer and Palmer, 1989; Palmer, Chapter 5.2.2 in this volume).

At the microscopic scale, grain size, surface roughness, and vugginess are important. Typically the finer the grain the more soluble the rock; for example, micrite is more soluble than sparite (Rauch and White, 1977; Dreiss, 1984). However, when the grains become aphanitic (very fine, uniformly packed) then they are not so soluble. Hence, the "Porcellaneous Band" in Gaping Gill Cave, England, causes perched passages (Ford and Williams, 1989). The effect of surface roughness is generally the enhancement of dissolution, by exposing more surface area, but the inhibition of regularity of form. For example, the regularity of scallops (see below) is best developed on fine-grained homogeneous rocks. The vugginess and sugary texture characteristic of dolomitic rocks also inhibits regularity and sharpness of form.

The major controls on passage form are exerted at the macroscopic level. Anisotropy at this scale relates mainly to the secondary porosity, or porosity not related to petrofabric (Choquette and Pray, 1970). Fractures (joints and to a lesser extent faults) and bedding-plane surfaces are the most important controls on the vectors of water flow and on the distribution of drip water and thus on local dissolution and deposition, 6.1 Figure 3. (The guiding fractures are indicated throughout 6.1 Figures 13 to 16 by dotted surfaces). The orientation of planes controls the direction of anisotropy so that dipping strata or fractures, for example, give asymmetric forms (6.1 Figure 3c and f). In places, geologic contacts (6.1 Figure 4), zones of chemical contrast and mechanical weakness can offer the easiest routes for water penetration, so a cave may have a floor in nonkarstic rocks. Waltham (1977) observed that many of the caves of Craven, U.K., are just above shale beds. Figure 7.35g of Ford and Williams (1989) depicts the "contact caves" of Greenbrier Co., West Virginia, U.S.A., developed on a limestone-shale boundary.

Joint planes, bedding planes, and more rarely faults guide most passages. They provide the easiest route for water access and thus guide passage location and to some extent form. Major bedding planes are often continuous over long distances and guide the development of long solutional caves (Ford, 1971). Joints and faults are less continuous and are usually less important for passage guidance.

Bedding planes with depositional disconformities, shale laminae perhaps with pyrite, nodules or sheets of chert, or planes that have sheared during tectonism (causing slickensiding or brecciation) are the most likely to influence passage development (Ford and Williams, 1989). The control may be exerted at the scale of the whole cave. For example, the Mammoth-Flint Ridge-Roppel System, Kentucky, U.S.A. (Palmer, 1981), and Hölloch, Switzerland (Bögli, 1978), are controlled predominantly by bedding planes. Or the control may be apparent at a smaller scale in horizontal sections of systems like parts of the Swildon's Hole-Wookey Hole System, England. Or it may show simply in passage flattening and flat roofs.

The influence of joint planes also may be exerted at the scale of the whole cave. For example, Wind Cave, South Dakota (Ford et al., 1993), and the great gypsum caves in the western Ukraine (Klimchouk, Chapter 5.2.1) are joint-controlled maze caves, and the passage alignment of Ogof Ffynnon Ddu, South Wales, is dominated by a NE and E joint system (Trudgill, 1985). Or the control may simply be at the passage scale in the form of narrow elongate passage cross sections. Many steep passages such as Lancaster Hole, U.K., and Greftkjelen, Norway, are directly related to steeply dipping joints. Most vadose invasion caves exploit the nearest vertical joint and are characterized by steeply dipping passages. Typically, joint guided passages are somewhat angular in section, but it may be more difficult to perceive the original joint control because they become more rounded with enlargement and may contain meandering streams (6.1 Figure 6a).

Faults may guide cave passages, especially tensional normal faults and faults where the rock is brecciated. They are usually made up of a series of closely spaced parallel fractures with lesser ones feathering out from them. These mainly modify rather than control passages, although low-angle thrust faults are often areally extensive and guide flow like bedding planes (Ford and Williams, 1989). Fault control is apparent in the 800 m of slickensided surfaces lining the Great North Road of Dan yr Ogof, Wales (Jennings, 1985), and in the deep pits of Crete and Epirus, Greece (Courbon and Chabert, 1986).

The frequency of bedding planes and faults remains constant, but joints are created anew as karst systems develop. Pressure release in valley walls may produce large tensional joints, especially parallel to steep faces, and compressional joints may develop at the base of cliffs and cave walls (Ford and Williams, 1989). These may also influence passage form. For example, pressure release joints in valley walls cause cave passages to run parallel to the valley wall in Grotte di Castellana, Italy (Renault, 1970), and in Kvithola, Norway (Lauritzen, 1986).

The effect of joints and bedding planes may be magnified where rock has been folded. Typically, the apex of a fold shows a higher frequency of joints, and the bedding planes may be slightly opened. These will then be the preferred sites for cave passages or for large chambers. A classic example is the positioning of the Sarawak Chamber of Lubang Nasib Bagus (Good Luck Cave), Mulu, Sarawak, at the axis of an anticline, 6.1 Figure 4 (Gilli, 1993). Anticlines often direct major trunk passages, such as that of Ogof Ffynnon Ddu, Wales (Lowe, 1989). Preferen-

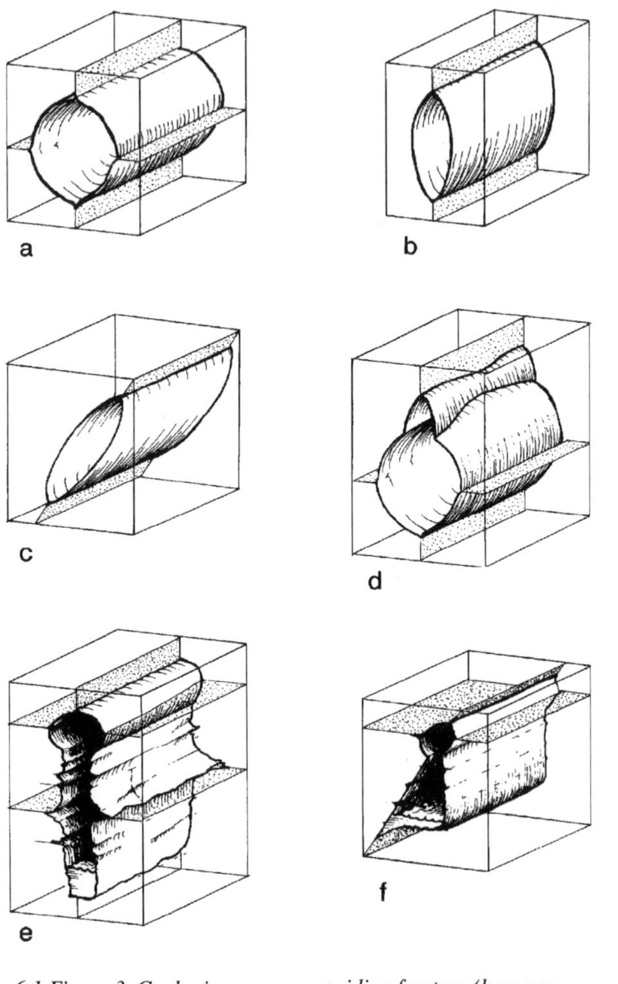

*6.1 Figure 3. Geologic structure and passage morphology. **a** is a symmetrical phreatic tube developed at the intersection between two planes, most commonly a bedding plane (horizontal shaded surface) and a fracture (vertical shaded surface). **b** and **c** are lenticular variants of phreatic tubes, often found in connection with faults or shear fractures. The attitude of the guiding fracture controls the tilt of the profile. **d** are phreatic pockets or arches developed along the guiding fracture (here vertical) where water may seep down and cause mixture corrosion. **e** is a keyhole passage represents a shift from phreatic (upper tube) to vadose (canyon) conditions. The phreatic tube was developed at the intersection of a fracture and the bedding plane (dotted surfaces). The vadose canyon may exploit other fractures and bedding planes as it erodes its way down through the rock mass. **f** is a keyhole passage distorted by inclined shear fracture.*

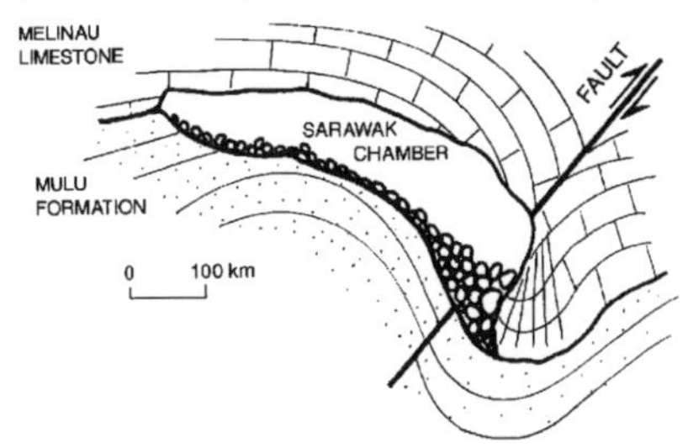

6.1 Figure 4. Large collapse chamber. Sarawak Chamber in Lubang Nasib Bagus (Good Luck Cave), Mulu, Sarawak, is developed at the axis of an anticline and thrust fault, and along the contact between the Mulu Formation and the Melinau Limestone (Gilli, 1993).

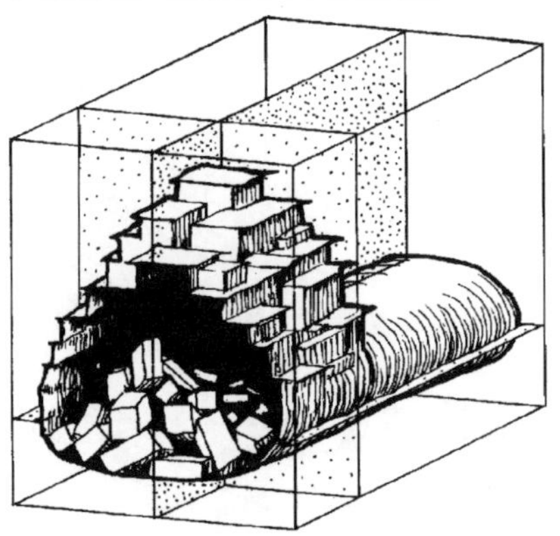

6.1 Figure 5. A collapse dome developed at the junction of two major joints.

tial water flow occurs along synclines, for example, at many of the major vadose passages of the Mammoth Cave System, Kentucky (Palmer, 1981). Passages in the Sistema Purificacion, Mexico, occur downdip and along the trough of a syncline (Hose, 1981), while Réseau Jean-Bernard, Haute-Savoie, France, the world's second deepest known cave, is developed at the flank of a syncline (Lips et al., 1993).

Geologic control is also apparent in the process of breakdown, where the original hydrodynamically created form is modified by mechanical degradation. Where a rock is of low mechanical strength, such as gypsum, breakdown may be the most common process governing passage form. The type of fracture produced as a response to stress will dictate the details of cave-passage shape. For example, blocky fracture will create a much more angular cross section than crumbly fracture. In rocks of great mechanical strength, as in most karst rocks, it is the bedding planes and joints that fail under mechanical stress, so the spacing of joints and bedding planes dictates the occurence and form of breakdown (see White and White, Chapter 6.2).

The stable form for a breakdown passage is a dome. The most stable symmetric domes occur in well bedded and horizontally bedded rocks (6.1 Figure 5). Asymmetric and less stable domes develop where beds dip. Within bedded rocks, breakdown is most likely to lie at passage junctions where prominent joint sets intersect. Where the rock is not well bedded breakdown may not occur, even in large passages. For example, the Big Room of Carlsbad Cavern in nonbedded reefal limestone is more than 50 m wide in places and yet has no breakdown—whereas breakdown occurs readily in adjoining back-reef units (see Hill, 1987, and Chapter 5.2.8). Most of the larger chambers of the world, such as Sarawak Chamber, the largest known in 1999, show some measure of breakdown.

Hydrologic Control of Passage Shape

The hydrologic setting is the other major control on passage meso- and micromorphology. Dissolution can only occur in a liquid medium. On the scale of a cave this usually requires contact with a body of water (although condensation water is sometimes sufficient). The hydrologic behavior, coupled with the physical and chemical properties of the water, govern the form of the passage. Where a rock is completely saturated in the phreatic zone, dissolution is usually omnidirectional, and variations in dissolution rate then depend on the rock properties. However, in the unsaturated aerated vadose zone, water flow is controlled by gravity. Dissolution occurs only where water remains in contact with the rock, and it is predominantly unidirectional and vertical. The hydrologic settings of the meso- and microforms can usually be recognized, and they can be classified into phreatic, vadose, or combination forms. The settings for where, and in what context, vadose and phreatic and combination forms occur are expounded in Ford's Four State Model for Meteoric Caves (Ford and Williams, 1989), which is discussed at length in Chapter 5.3.

Morphology of Meso- and Microscale Features

Meso- and microforms are described below along with a discussion of the effects of geologic and hydrologic controls where these are known.

Mesoforms

They are defined above as forms of the same scale as the cave passage diameter.

Phreatic Passages or "pressure tubes" (6.1 Figure 3a-c and 6.1 Plate 1A) form under totally waterfilled conditions, by corrosion that is symmetric and directed radially out from the passage axis, producing, under-optimum conditions, a nearly perfect tubular cross section. Within the phreatic zone, passage direction is independent of the water table but mainly directed by structural factors, forming phreatic loops. Phreatic tubes, being elements of these loops, are thus rarely horizontal. Diagnostic features are a circular, elliptical, or lenticular cross-sectional shape, and a looping behavior, usually with large scallops (see below) indicating relatively slow water velocity. Impure rock and variable solubility may distort the regular cross section. Phreatic loops within fault zones may in places form strongly lenticular or rift-shaped cross sections.

Epiphreatic Passages form within the epiphreatic or "floodwater zone" (Bögli, 1978), where passages may stay dry for long periods, but when active are completely waterfilled and develop like phreatic tubes. They display the same variation in cross-sectional shape as phreatic passages. The diagnostic difference is due to their development close to the piezometric surface where water movement is relatively rapid, so that epiphreatic passages in suitably homogeneous rock display quite small scallops. Also, due to the same effect, the ceiling often displays smaller scallops than the lower walls and floor. However, if downstream constrictions inhibit velocity in the highest stage flows, then the ceiling may display slower flow scallops than the walls. Epiphreatic forms are also discussed under the section on spongework below.

Vadose Canyon Passages have a fundamentally different form from that of phreatic passages (6.1 Figures 3e and f, and 6.1 Figure 6a, b). Vadose water is driven by gravity. Passage expansion is therefore directed downward or laterally, forming canyon-type passages, the underground equivalent of a surface stream. Vadose canyons may take various shapes, depending on the initial slope and rate of downcutting. They range from tiny extremely sinuous clefts to wide, straight, corridor-shaped galleries that may become tens of meters wide and more than 100 meters deep. A spectacular example of the latter type is the huge streamway canyon of Skocjanska Jame in Slovenia (Kranjc et al., 1997).

As in surface streams, vadose passages are often sinuous, forming *cave meanders* (6.1 Plate 1b). Most cave meanders are formed exclusively by the removal of matter, which makes them different from surface meanders in alluviated streambeds, where material can be *deposited* (point bar) as well as *removed* (outside bend). In this they resemble meanders in ice or in clay (where deposition does

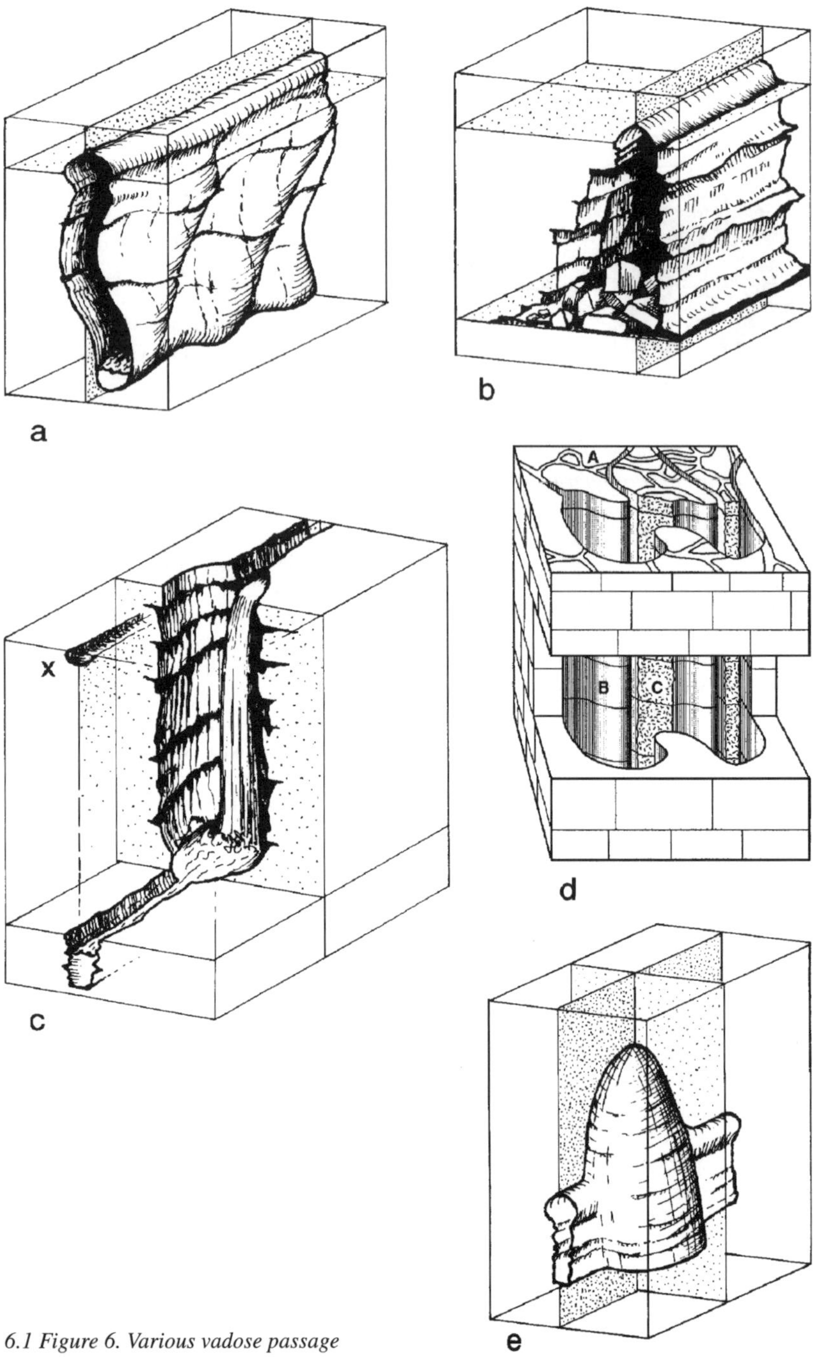

6.1 Figure 6. *Various vadose passage morphologies. **a**—Meandering vadose canyon. The meander loops shift downstream as the canyon deepens (after Ewers, 1972). **b**—When vadose canyons erode down to (local) base level, sediment accumulation initiates lateral erosion (corrasion) (meandering); impervious rocks (dotted plane below the passage) may have the same effect. This may lead to extensive undercutting and collapse. **c**—Vadose waterfall shaft developed by vertical piracy of a high-level passage down to a conduit at lower levels. Note the fluted walls in the shaft. Abandoned high level gallery at X.*

*d—Development of large fluted shafts or domepits, according to Bretz (1942). Small vadose streams from bedding-plane anastomoses–**A**–are captured down to lower levels. As they cut back, they are forced to trace the original anastomotic pattern, which initiates the large flutes–**B**. When the retrogressive streamlets come close enough, blocks may break off–**C**, which then fall down and dissolve away at the bottom of the domepit. e—Vadose domepit developing along the intersection of two fractures.*

not occur because the particle size is too small), which also lack point bars on the inside bends. As in surface streams, vadose passages are cut down under the control of a local base level, but in caves the evidence of former stream positions remains intact. The sinuous bends in cave meanders migrate downstream as the canyon is eroded downward. This results in an extremely complicated morphology, where individual meander bends form semicylinders around axes that plunge in the downstream direction (6.1 Figure 6a). Where a canyon retreats by the headward erosion of a nickpoint, then the meander form does not have this complex morphology. Instead it has the simple vertical wall characteristic of any intrenched meander.

6.1 Plate 1. A—Elliptical phreatic tube in white marble, Fosshølet, Svartisen, north Norway. The oblique guiding fracture with remnants of anastomosing secondary conduits are visible along the upper right and lower left walls. Scallops show flow direction toward the viewer. The coarse clastics along the floor require a much higher flow rate than is indicated by the scallops, hence the boulder fill belongs to a later epiphreatic stage of speleogenesis, which in this case took place during deglaciation. Also, a subset of smaller scallops are detectable on the walls, which might have developed during this stage. The passage is now relict. Plate B—Incised vadose meander in a marble cave, Svartisen, north Norway (photographs by S.-E. Lauritzen).

A vadose cave stream shows a version of many typical fluvial features. For example, when a meander loop is abandoned, it remains as a high level oxbow. Where a flow route divides, a rock pillar is left isolated.

Where the discharge of one stream is captured by another, a higher level dry passage remains in the form of a cave col. One obvious difference is that a meander may undercut a bank but in a cave that bank does not collapse. Instead it forms a meander niche (Bretz, 1942). Vadose passages often show the effects of changing discharge regimes in the form of misfit streams. Because many cave passages formed during the very wet immediately postglacial period, discharge most commonly has decreased, and the cave stream is underfit. The change may be gradual, whereby the meander wavelength gets smaller as the passage is incised, or the floor of a large canyon may simply develop a smaller canyon with a smaller meander wavelength.

When at base level, or ponded by insoluble beds or breakdown, vadose canyons flatten out and accumulate a protective sediment bed that stimulates lateral erosion and dissolution. If extensive, this may initiate breakdown (6.1 Figure 6b).

In addition to and in competition with all these fluvial controls, the vadose canyon may also reflect geologic control, where insoluble beds project as benches into the canyon, the canyon wall is joint or fault guided, or the meander wavelength is constrained by joint spacing.

Vadose shafts develop where water falls steeply, from a surface sinkhole or down a steep joint from one cave level to another (6.1 Figure 6c). Shafts are the passage form associated with vadose invasion caves. They range in size from a narrow solution chimney to a large open vault. They can be several hundred meters deep. Most of the deepest caves of the world are made up of vadose shafts connected by short sections of horizontal passage. Cave waterfalls typically create a semicircular or elliptical pot mainly from the spraying water, with a plunge pool beneath. The form may diverge from a simple clean shaft, because the constant spray carves deep vertical grooves in the walls or encourages the weakening of joints and triggers block collapse from the walls. The flow may later be diverted elsewhere, leaving a dry shaft, or the waterfall may recede in time, producing a typical nickpoint-retreat canyon.

Vadose shafts may also form directly by retrogressive cutting of small vadose streamlets along overlying bedding-plane anastomoses (Bretz, 1942; 6.1 Figure 6d). Vadose shafts can extend up toward the surface by tens of meters (the larger ones are often called *avens*), but generally they are modified into simple vadose shafts as the original source gets widened. (See discussion relevant to vadose shafts in Chapter 5.3).

Domepits develop where water does not fall freely. A trickle of water emerges from a vadose source in the cave roof, but the flow is not strong enough to cause a separation of flow. The trickle of water remains in contact with the cave roof and walls, dissolving surfaces and often carving flutes as it goes. Most of the aggressivity is used up in the first couple of meters of flow so that only the roof around the source is cut back. As in alluvial fans, the locus of flow moves around so that in the long run all parts of the roof around the source tend to get roughly the same amount of dissolution, and a dome of circular cross section is produced (6.1 Figure 6e).

A cave may display a *combination of vadose and phreatic features* as a result of the most common speleogenetic sequence occurring in cave systems, the change from phreatic to vadose conditions. This may occur as a response to external base-level lowering, or simply to lowering of the internal piezometric surface as a result of passage widening and larger flow capacity. This change results in a composite passage cross section, the *keyhole profile*, (6.1 Figure 3f, and 6.1 Figure 6a). The proportion between the phreatic and vadose parts varies over a wide range, depending on the relative time spent in each of the two phases. Trunk conduits that have been drained by the mechanisms mentioned above tend to display regular phreatic tubes above the canyons, while steep-gradient invasion passages (predominantly vadose) tend to display only a minor lenticular phreatic void at roof level. The relative widths of the phreatic and vadose sections of the keyhole passage may also relate to diversion of flow. Where the main water body has been rerouted, leaving behind an underfit stream, only a narrow trench develops, but if the water simply changes from phreatic to vadose, then the canyon may be almost as wide as the original tube.

Sumps or siphons, the points of transition between vadose canyons (keyhole profiles) and exclusively phreatic cross sections are extremely important in speleogenetic analysis, as they indicate the position of local water tables. In looping cave systems (see Chapter 5.3 in this volume), several of these points may link up and describe paleo-piezometric surfaces, which in turn define *external ancient base levels*. Combined with speleothem dating, this makes a powerful tool for calculating external erosion rates (Ford et al., 1981; Atkinson and Rowe, 1992).

Bypass Tubes and Corkscrews may develop where rapid vadose water transports clastic sediment into phreatic loops that act as sediment traps. Sediment accumulation may totally obscure the water flow and reroute it, either in the form of *paragenesis* (see below), or by formation of new shortcuts through the bedrock. These passages, named *bypass tubes* (Ford, 1965a; Ford and Ewers, 1978), are often formed under extremely steep hydraulic gradients due to the sediment plug. The high pressure gradient will force water to exploit and link up every minor fracture that exists inside the rock mass, resulting in an extremely tortuous passage path (6.1 Figure 7). These conditions are perhaps most commonly found in alpine caves where steeply dipping primary speleogenetic guiding voids within tilted or overturned bedrock strata occur together with large surface relief gradients. This effect was also recognized in the Polish Tatras by Jan Rudnicki (1967), who named them "corkscrew passages."

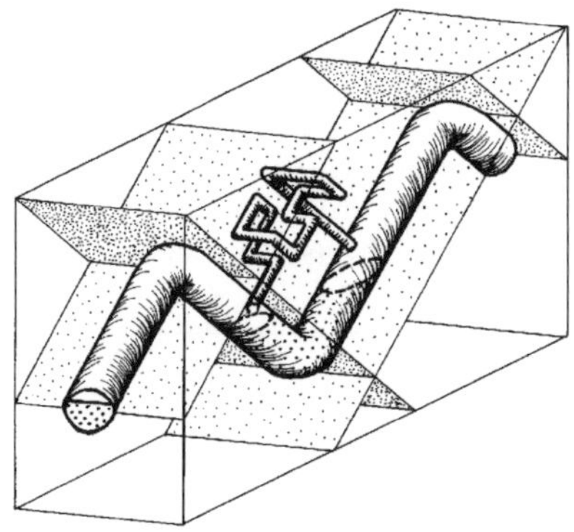

.6.1 Figure 7. Bypass tube with corkscrew morphology. The phreatic loop was choked with impenetrable sediment (indicated by coarse dots) so that a very high hydraulic head was set up across the blockage. This pressure potential forced the water to exploit every available fracture and develop an extremely tortuous bypass tube – a corkscrew passage.

The process of *paragenesis* occurs where sediment fills a phreatic passage (Renault, 1968). Flow velocity increases due to passage constriction, and the response relates to the erodibility of the sediment. As in surface channels, an equilibrium situation develops between flow rate, amount and type of sediment, and shape of the initial ceiling. A single straight or meandering channel, or a system of braided channels may develop (6.1 Figure 8). Further dissolution is then restricted to free rock surfaces above the fill, where fresh water flow can attack. (Yet another mechanism is suggested to form ceiling half-tubes in gypsum caves that is related to density-driven circulation; see Klimchouk, Chapters 5.2.1 and 7.1). The paragenetic process does not require any other guiding void than the primary cave passage. A cycle is initiated whereby the roof is raised, allowing a decrease in fluid velocity, which in turn allows further sediment deposition. Thus the sediment layers build up and the roof recedes. Paragenesis may therefore be seen as a rejuvenation of the speleogenetic process. This is particularly clear when anastomosing ceiling half-tubes and pendants form (see below and 6.1 Figure 8c).

A *paragenetic canyon* develops where flow is restricted to a single channel, and the whole passage ceiling is lifted as an inverted canyon toward the water table (6.1 Figures 8b and 9). Spectacular examples of such passages are the Paläo-traun of Eisriesenwelt, Austria (Bauer, 1961), and the main passage of Storsteinshola, north Norway (Lauritsen and Lauritzen, 1996). Where the flow is braided, smaller-scale features are produced, such as half-tubes and anastomoses (see below).

Paragenetic canyons meander like vadose canyons, and their meanders also migrate as the canyon develops. Once the sediment has been removed, the paragenetic meander axis plunges *upstream* against the direction of flow indicated by scallops, whereas a vadose meander axis plunges downstream (6.1 Figure 10). The two effects are mirror images relative to base level, the water table. However, if the cave is in a fossil stage and heavily fragmented or obscured by breakdown, distinguishing paragenetic from vadose meanders may be difficult. This has important speleogenetic implications, because a paragenetic canyon is formed *below* the water table, whereas a vadose canyon is formed *above* it (6.1 Figure 10).

One diagnostic difference is that a vadose canyon is generally a part of a *keyhole* profile, that originated from a phreatic or epiphreatic ceiling void that may either be a tube or a lenticular void. This primary passage will almost always display the guiding fracture or bedding plane from which it originated. On the other hand, paragenetic passages do not require a guiding fracture at roof level (although they may accidentally intersect one). They originate from a phreatic passage beneath them. The diagnostic differences between scallop and meander morphometry have been studied by Lauritzen and Lauritsen (1995) and developed into a method by which paragenetic and vadose canyons can be uniquely distinguished, 6.1 Figure 10. In looping caves, paragenesis and vadose incision work together from each side

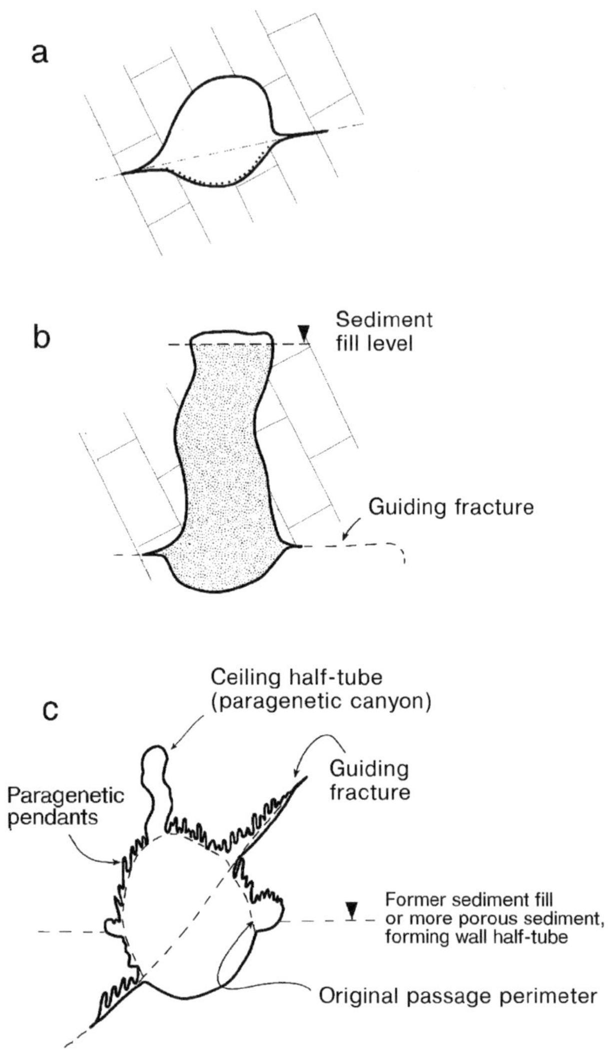

*6.1 Figure 8. Paragenesis occurs when sediment fill directs phreatic dissolution upward. **a** is incipient paragenesis. A thin veneer of insoluble clay isolates the floor of a phreatic passage and forces asymmetric development around the guiding fracture. **b** is a paragenetic canyon. This feature forms when there is a dynamic balance between sediment accumulation rate and roof retreat rate. Often scallop velocity and sediment sizes match in such passages. **c** is a small-scale paragenesis. Several small half-tubes join into an anastomotic network, leaving fields of paragenetic pendants after the sediment fill is removed. Single ceiling half-tubes may develop into "master conduits" (canyons) by competition, just as bedding-plane anastomosis fields do. Wall half-tubes may develop along the top level of a former sediment fill, or along high-permeability layers within the sediment pack.*

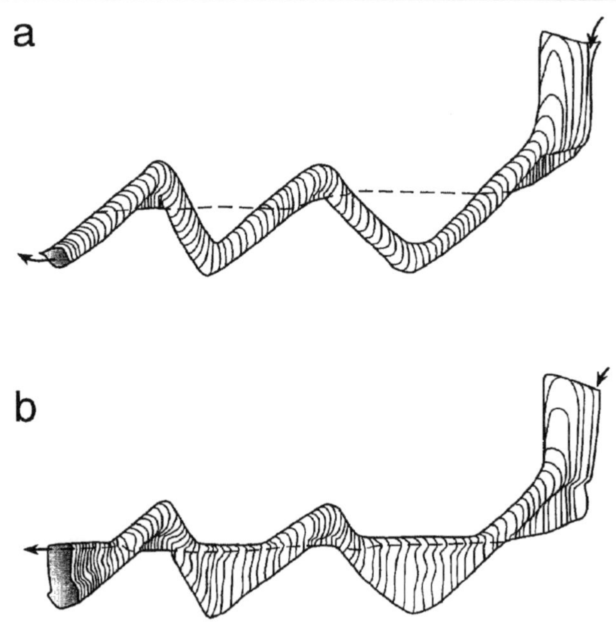

6.1 Figure 9. In a looping cave system, vadose and phreatic canyons (a) erode toward the water table (b), which is often controlled by an external base level. The combination of these forms makes an important tool for paleolandscape analysis and erosion-rate calculation.

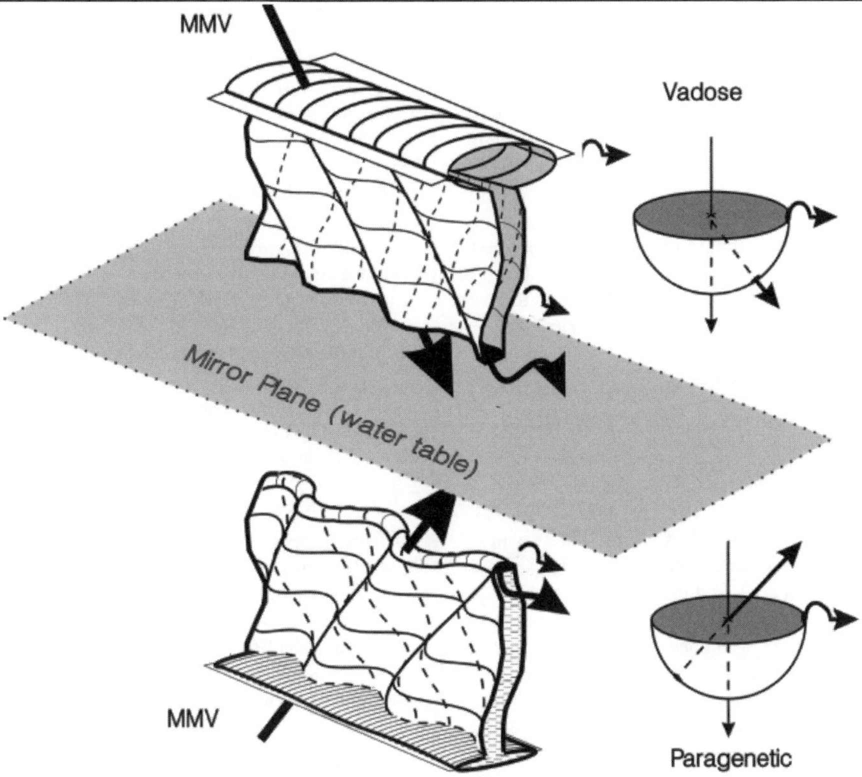

6.1 Figure 10. Differential diagnosis between vadose and paragenetic canyons. Vadose entrenching canyons and paragenetic canyons form a pair of mirror images. They are mirrored through the water table. The meander migration vector (MMV) is shown by the heaviest arrows, and it can be determined from canyon walls and scallop-flow direction. The two vectors plot in the upper and lower hemispheres of a stereonet, respectively. When both vectors are projected into a lower hemisphere, scallop flow direction (hooked arrows) and MMV plot differently for the two cases (right) (from Lauritzen and Lauritsen, 1995).

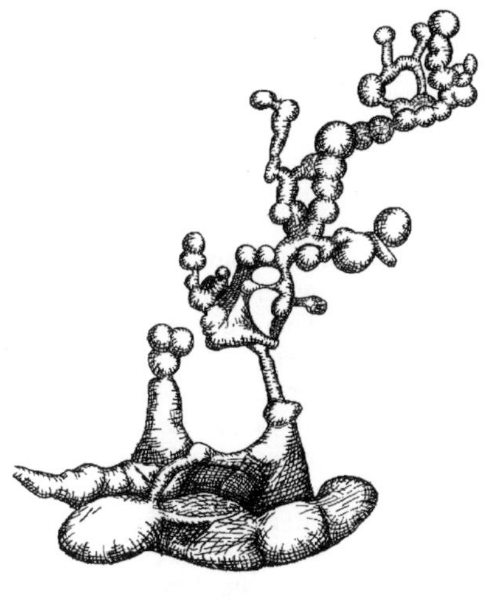

6.1 Figure 11. Cupola morphology in Sátorköpuszta Cave, Hungary (after Müller and Sárváry, 1977).

of the base level to form a water-table cave as the ultimate end product. The combination of these two forms therefore provides a unique definition of a graded base level that can be extended to the surface (6.1 Figure 9).

Convection Cupolas are characteristic of thermal endogenic caves. Thermal water rising through karst may be highly corrosive due to dissolved CO_2 or H_2S. Condensing water vapor and fluids moving in convection cells corrode out deep, rounded, multicuspate solution pockets, above or below the water table (Bac-Moszaszwilli and Rudnicki, 1978; Dublyansky, 1980; Szunyogh, 1984). Spherical or semispherical enlargements that are greater than, or similar to, the diameter of the passage leading into them, are named *cupolas*. Many of these cupolas lack guiding fractures, and their position relates only to the passage from which they developed. Cupolas may comprise merged elements of huge spongeworks (see below), or they may be separate entities organized in a tree-like ascending branchwork (Müller and Sárváry, 1977). Generally they show no evidence of rapid flow. Where water vapor has condensed, the primary cupolas may be modified by flutes. Satorkopuszta Cave, Hungary, has many excellent examples (6.1 Figure 11). The large stomach-like, basal chamber leads upward to branching intestinal passages that end in blind cupolas (Müller and Sárváry, 1977).

Microforms

Microforms are defined above as forms that are much smaller than the passage diameter. For the purposes of this section, the microforms have been placed into rough morphological categories (pockets, notches, tubes, and others), but there may be no genetic connection.

Various Types of Pockets

The normally smooth regular form of phreatic passages may be modified by *phreatic arches, chimneys,* and *pockets*. Where a small discharge of water meets the main phreatic water body, typically as seepage from a fracture, and the two waters have been saturated under different conditions, then enhanced solution through *mixing corrosion* is often possible. If the additional discharge is from a linear source, the whole roof may be modified into a gothic arch, 6.1 Figure 3d. If it is from a point source, it may produce blind shallow pockets, or deeper aven-like chimneys in passage roofs (that may resemble vadose domepits). Pockets are usually up to a few meters wide and deep, but chimneys may extend 30 m upward (Ford and Williams, 1989). Although the form may resemble that of bell holes (see below), simple phreatic chimneys and pockets *differ fundamentally* from bell holes, in that they are always fracture guided, and if they occur in groups then the distribution is linear along the guiding fracture (6.1 Figure 3d).

Bell holes are cigar-shaped vertical cylinders cut into the roofs, almost exclusively, of tropical caves. They are superimposed into the primary, usually phreatic, cave ceiling. There is very little discussion of bell holes in the literature. Bretz (1942) suggested that they were from differential dissolution of rock structures, and Wilford (1964) described them from Sarawak. They have subsequently been noted in most karst textbooks but only recently studied quantitatively (Lauritzen et al., 1997; Tarhule-Lips and Ford, 1997 and in press). They are typically 20-30 cm wide, and 40-80 cm high, but may penetrate several meters into the host rock. They flare out at the base but taper upward either not at all, or only a little, to a paraboloid apex (6.1 Figure 12). They usually occur in groups which seem to have a polygonal distribution (6.1 Plate 2), typical of competitive processes, that is, the units of a bell hole and its associated "personal" space are regularly packed to fill the surface available. They have no water-level control, and the important diagnostic characteristic is that they are *not* fracture guided and have *no* apparent vadose input. In fact they seem to have little geologic or lithologic control. They present a fluted appearance and develop into small vadose dome pits *only* when they become invaded by vadose water and are therefore no longer simple monogenetic forms.

It is hard to assign bell holes firmly either a phreatic or vadose origin, because they seem to have features of both. Ford and Williams (1989), who place them with phreatic features, suggest that they "seem too small to be convectional." Possibly a more pertinent observation is that they seem too deep to be convectional. Phreatic convectional cells could produce a ceiling pocket. But with dissolution proceeding on all surfaces, it would have to have a low width-to-depth ratio.

Modeling of vadose dissolution from a point source produced a form like a "negative" stalagmite, which closely resembled a bell hole (Lauritzen, Lundberg, and Mylroie, unpublished data). Because there is no obvious source of vadose input from above, if this model is correct, then it may be necessary to envisage a source from condensation water, as is postulated by Tarhule-Lips and Ford (in press).

The study of bell holes in Lighthouse Cave, Bahamas (Lauritzen et al., 1997) revealed a feature hitherto unreported:

> Almost every bell hole is associated with a paired *bell pit* in the floor (6.1 Figure 12). They are vertically underneath the apex of the bell hole and morphologically related (pit width is always just a little wider than hole width, although they are never as deep). Whereas the form may superficially resemble a vadose pothole (see below), the distribution indicates no possibility that bell pits were formed by turbulent stream flow, with or without abrasional tools. Some of these bell pits were cut directly into flowstone. For this reason, and because modeling seems to favor it, a vadose origin is suggested. However, a post-vadose, phreatic reexcavation during a subsequent sea level high cannot yet be ruled out.

Vadose potholes are simply the cave equivalent of fluvial abrasional potholes of the ''rock mill'' type. They are characteristic of high-gradient vadose streams. The stream may carry clastic material that gets caught in a random eddy and then provides an abrasional tool to carve a smooth-walled circular hole in the stream bed. Once the hole starts to form there is a positive feedback effect because it will both retain the original tools and attract new ones, thus causing enhanced abrasion. However, the positive feedback continues only until the pothole becomes too deep for turbulent eddies to reach the tools at its base. Vadose potholes in soluble rocks, in caves or in canyons, differ from simple fluvial potholes in that they are usually enhanced by dissolution. So they form faster than and are often bigger than external equivalents in nonkarstic rocks. The dissolutional effect is apparent in scallops along the upper walls above the locus of abrasion, whereas at the base the walls are smooth and lack scalloping.

Potholes remain entrenched into the main passage only if abrasional tools remain in place and if the normal stream downcutting is slower than the abrasion. They are easily destroyed in the cave environment. Where stream downcutting has overtaken potholing, the pothole remnants can be abandoned in the cave wall, as has occured in Swildon's Hole, U.K. (Ford, 1965b). The presence of abandoned "hanging" potholes may be a useful indicator of past flow conditions (6.1 Figure 13).

Various Types of Notches

Waterline notches are the most common type of corrosion notch. They develop where vadose standing water with an open air surface absorbs CO_2 and thereby becomes more aggressive at the top layers than deeper down into the water body. These notches are generally 5 to 10 cm tall, may become quite deep (5 to 100 cm), and may form on each side of the cave passage. In high-latitude caves, they may form by ice contact, that is by the layer of aggressive spring meltwater that collects on the top of winter-frozen bodies of cave water (Lauritzen, unpublished). Analog forms can be formed by warm

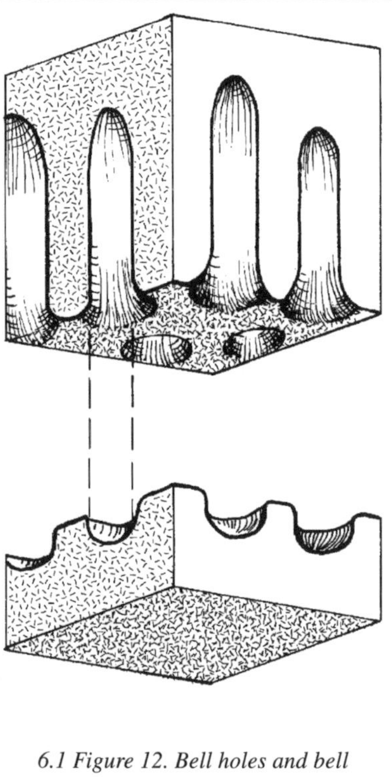

6.1 Figure 12. Bell holes and bell pits (from Lauritzen, Lundberg and Mylroie, in prep).

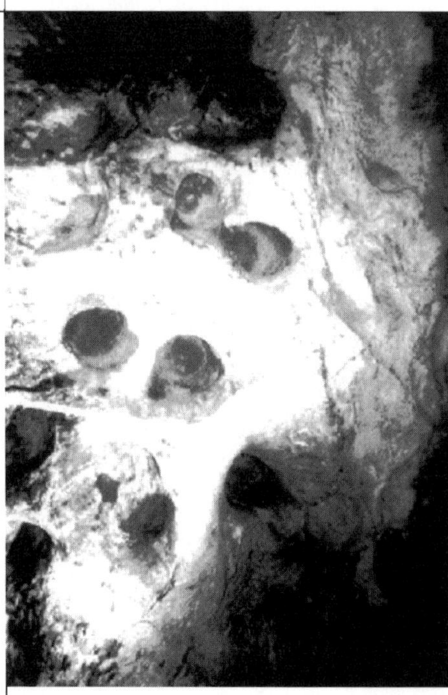

6.1 Plate 2. Polygonal packing of bell holes, Lighthouse Cave, San Salvador, The Bahamas (photo by S.-E. Lauritzen).

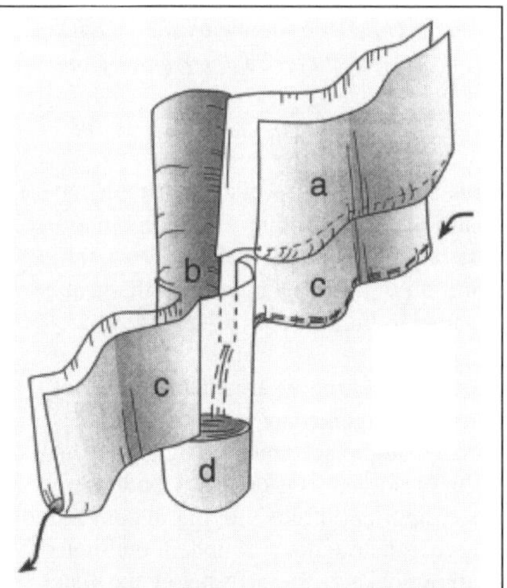

6.1 Figure 13. Vadose canyon entrenchment in a pothole. The original high-discharge canyon (a) drilled out the cylindrical pothole (b). At a later stage, water supply was reduced (either due to external climatic changes, or simply to internal rerouting of the vadose water) causing a smaller canyon (c) to be entrenched. This new discharge has insufficient power to revive the pothole drilling, so that it will become relict and end up as a hanging entrenched pothole in the wall of canyons (a) and (c).

6.1 Plate 3. Water-line notch formed in glacier ice at the level of a small ice-dammed lake. Hardangerjøkulen, south Norway. The melting process (diffusion of heat) is analogous to diffusion of aggressive species, and therefore analog forms are found in karst caves (photo by S.-E. Lauritzen

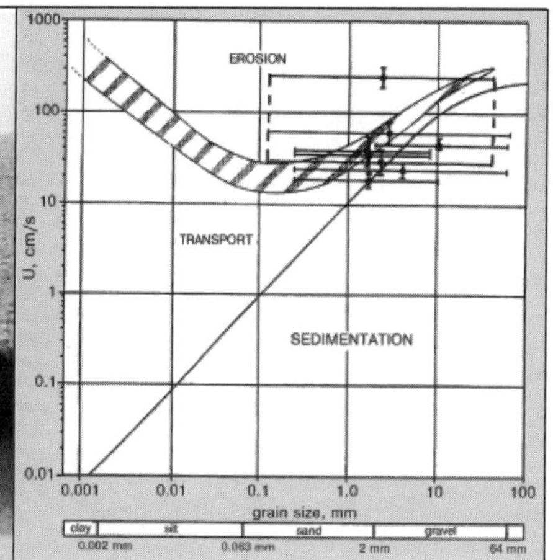

6.1 Figure 14. Hjulstrøm (1935) plot with scallop-dominant velocity (vertical bars) and grain-size range of corresponding sediment (horizontal bars) in paragenetic passages. Matching of water velocities required by scallops and sediments suggests that the two were contemporaneous. The two bars connected by dashed lines represent two scallop subpopulations, of which one velocity (upper) was clearly able to erode the sediment, whereas the other (lower) would probably deposit it (from Lauritsen, 1996).

surface water of ice-dammed lakes on glacier ice, where diffusion of heat replaces the diffusion of CO_2 and its hydrated species (6.1 Plate 3). Waterline notches are important diagnostic tools, as they are formed *at* the water level and thus are perfectly *horizontal.* When uniquely diagnosed, they indicate a *local* relatively stable water surface. If tilted, they may indicate postspeleogenetic tectonic movement or breakdown.

Vadose and paragenetic wall notches have a similar form to that of waterline notches but are *inclined*. They are formed by alluviated streams cutting laterally into passage walls and thus have the same slope as the stream gradient. Such notches are often discontinuous and meander from one wall to the other. Some of them exploit available fracture or bedding-plane partings, but most cut across bedding structures and may cut several meters into the cave wall (White, 1988). They exhibit relatively small scallops (see below), which may match the grain size of fluvial sediment residues in a Hjulstrøm plot (6.1 Figure 14) and thus may be useful for paleoenvironmental reconstruction.

Laugdecke and Laughöhle profiles are standing phreatic water features formed by slowly moving cells of density-driven currents. Dissolution in the phreatic zone is usually omnidirectional. However, where flow is extremely slow, diffusion of solutes from the rock surface into the water body creates a 1 to 3 mm thick sublayer of relatively denser water that in turn sinks or flows downward (at a rate of ~0.5 cm/s) to thereby create flow cells. The upper rock surface is then exposed to more aggressive water, while the dense saturated water impinges on the lower rock surfaces. The least aggressiveness occurs at the base of the cell. These cells create flat ceilings (*Laugdecke*) or inclined walls with a slope of about 45° (*Facetten* or facets) (Kempe et al., 1975). The combination of the two forms a flat-roofed triangular cross section (*Laughöhle*, 6.1 Figure 15). Density differences (relative to aggressive water) can be as high as 30% for halite and 0.1% for gypsum. So Laughöhlen are most common in evaporite karst, although smaller variants are also known from limestone. Kempe et al. (1975) demonstrated supporting evidence for this genetic mechanism in the form of convection currents, or "salt fingers," (Shirtcliffe and Turner, 1969) by means of fluorescein in stagnant pools in gypsum caves. The flat ceiling of Laughöhlen may be taken as a reliable water-table indicator— although Kempe et al. (1975) claim that these caves form within the upper 1-2 meters of the phreatic zone, that is, closely *below* the water level.

Note that it is preferable to retain the German term "Laugdecke" for the flat ceilings in favor of the term "corrosion bevel"

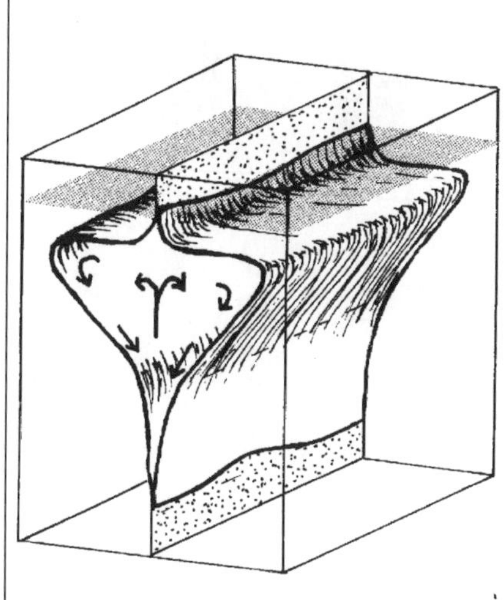

6.1 Figure 15. Laughöhle profile formed by convection cells driven by density currents (after Kempe et al., 1975). The flat ceiling Laugdecke is formed closely below the water level (indicated by the shaded surface), and is therefore a lower boundary indicator of it.

(Ford and Williams, 1989). The term "bevel" usually implies an angle (such as in the beveled edge of a mirror), thus it may be confused with the facets. For discussion of other types of meso- and microfeatures formed by density-driven circulation see Klimchouk, Chapters 5.2.1 and 7.1 in this volume.

Various Types of Tubes

Spongework is a highly complicated three-dimensional cluster of pockets, tubes, and interconnecting cavities of various shapes and sizes (6.1 Plate 4). It resembles the pores of sponge or Swiss cheese. The type locality is in Carlsbad Cavern, New Mexico, U.S.A. (White, 1988). Scallops and other flow indicators are generally absent, suggesting that water flow was slow or stagnant. Floors and ceilings are ungraded and irregular, but the spongework itself may be concentrated at crude levels.

Spongework occurs in a wide variety of lithologies:

- From dolomitic limestone (Tindall Limestone, Northern Territory, Australia) (Lauritzen and Karp, 1993),
- Dense dolomite (Grampians, Scotland), and
- Massive reef limestone (Carlsbad Cavern, New Mexico, U.S.A),
- To poorly consolidated calcarenite (San Salvador Island, The Bahamas).

The wide variety of host rocks suggests that spongework does not share any common geologic setting or pattern. However, Palmer (1975) points out that large-scale spongework morphology (i.e. entire spongework *caves*) is almost invariably connected to caves within isolated uncapped hills of limestone, like reefs, calcarenite dunes, monadnocks or mogotes, or ridges of inclined stripe karst (narrow layers of limestone or marble, overlain and underlain by nonsoluble rocks such as shale or schist, Horn, 1937.)

6.1 Plate 4. Spongework in the passage wall of Cutta Cutta Cave, Northern Territory, Australia (photograph by S.-E. Lauritzen).

Genetic hypotheses on spongework morphology are somewhat divergent, although most workers agree that it is formed within the phreatic zone under hydraulic conditions that in most cases were slow (Bretz, 1942). Ford (1968), based on examples in the Cheddar Caves, Mendips, UK, suggests that phreatic bore passages may later "disintegrate" into a spongework-like morphology if they become stagnant but still reside in the phreatic zone. Stagnant or slow water movement in turn implies sluggish dissolution chemistry close to equilibrium.

Kinetic support for the supposed mechanism of differential corrosion under nearly stagnant flow may be found in saturation-dependent changes *in relative dissolution rate* between rocks of slightly variable composition, 6.1 Figure 16. For instance, rocks that would dissolve at equal rates at low undersaturation, would dissolve at quite different rates (up to a factor of 2, curve 2 and 3 in 6.1 Figure 16) when the system is closer to equilibrium. In other cases inversion of the relative corrosion rate would occur. Such hydraulic and chemical conditions would allow differential corrosion to enhance subtle contrasts in rock solubility and enlarge the primary pores in massive carbonate rocks.

A special case of spongework caves is formed at the fresh/salt-water mixing zone within poorly consolidated calcarenite on carbonate platforms (flank-margin caves; Mylroie and Carew, 1990). Here, waters of different chemical composition are mixed by diffuse phreatic flow where mixing corrosion, in this case of waters of different salinities, and biogenic redox processes certainly play important roles. These spongework shapes are in part a function of density and thermal convection cells and somewhat resemble the cupolas of thermal caves. The typical form is of a larger mother chamber, itself composed of many coalesced cupolas giving a characteristically cuspate wall form, leading horizontally to many small, blind, cupola-like fingers (rather like an over-inflated rubber glove).

The single explanation of slow phreatic flow will not explain all observed cases of spongework. In a very different setting, Palmer (1975, p. 69, Figure 12) recognized scalloped spongework in an epiphreatic floodwater situation that is otherwise known for producing maze caves. It is interesting to

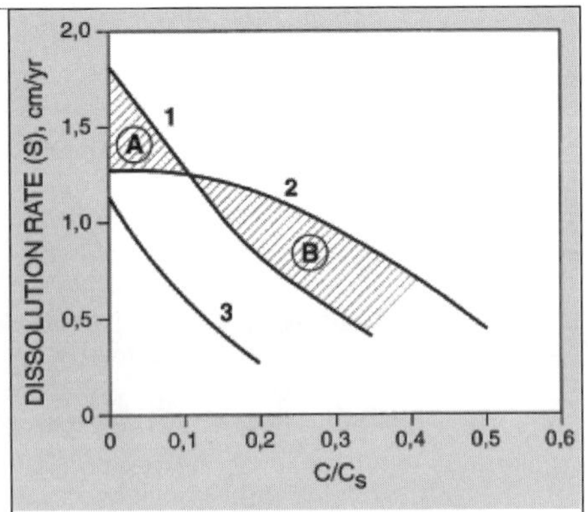

*6.1 Figure 16. Dissolution rate as a function of undersaturation for various rock types; data from Rauch and White (1977) where C/C_s is the solute concentration divided by the concentration of equilibrium. At high undersaturation, rock 1 dissolves faster than rock 2 (region **A**). As the system approaches equilibrium, the relative corrosion rate is inverted (region **B**), where rock 2 dissolves faster than rock 1 (by a factor of 2). At very high undersaturation, rocks 2 and 3 dissolve at nearly the same rate. Closer to equilibrium, rock 2 dissolves almost 5 times faster than rock 3. Behavior like rocks 2 and 3 could explain why condensation water (which is highly undersaturated when it condenses on rock surfaces) tends to cut indiscriminantly and smoothly through speleothems and wallrock. Note that the experiments of Rauch and White (1977) were not run very close to saturation (only to C/C_s = 0.5 for rock 2). Hence experimental data closer to equilibrium are needed to test whether this hypothesis can be extended to "stagnant" conditions and thereby test the hypothesis of spongework development by differential corrosion.*

note that spongework can develop in the waters of meltwater channels in decaying snowpatches (6.1 Plate 5), a situation that is analogous to Palmer's floodwater situation.

Bedding-plane and joint-plane anastomoses (treated at length in Chapter 4.2.1 in this volume) are morphologically the two-dimensional equivalent of three-dimensional spongework. Ewers (1966, 1982) simulated the development of bedding-plane anastomoses in plaster blocks under slow gentle flow conditions. White (1988) suggested that they develop early in the karstification process when flow rate is small and hydraulic head is large. They become inactive once a master passage has emerged and the resistance decreases. They then become active only during flood pulses and thus may continue to form contemporaneously with the main cave. They are best exposed where a bedding-plane block has collapsed. The

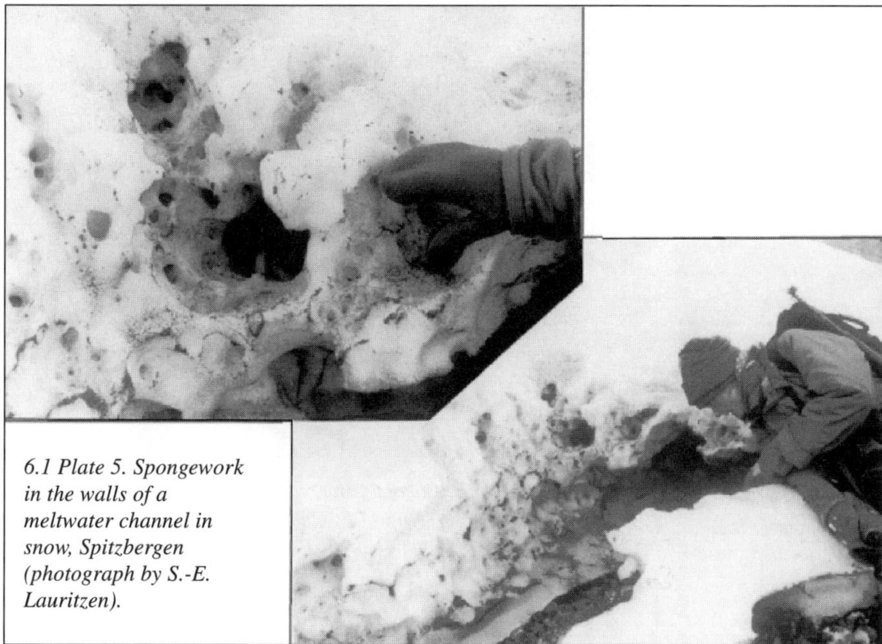

6.1 Plate 5. Spongework in the walls of a meltwater channel in snow, Spitzbergen (photograph by S.-E. Lauritzen).

intervening residual emergent rock *pendants* look like small mesas, the bedding plane remnant forming the flat surface (6.1 Figure 17). Excellent examples of anastomoses are visible in Seawra Cave, Pennsylvania, U.S.A. (White, 1988).

Paragenetic half-tubes, anastomoses, and rock pendants (6.1 Figure 8c and 6.1 Plate 6) are the small-scale features that develop in a cave ceiling where a paragenetic stream becomes braided. Each stream segment produces a *half-tube* in the ceiling, and the network of half-tubes produce an anastamosing pattern.

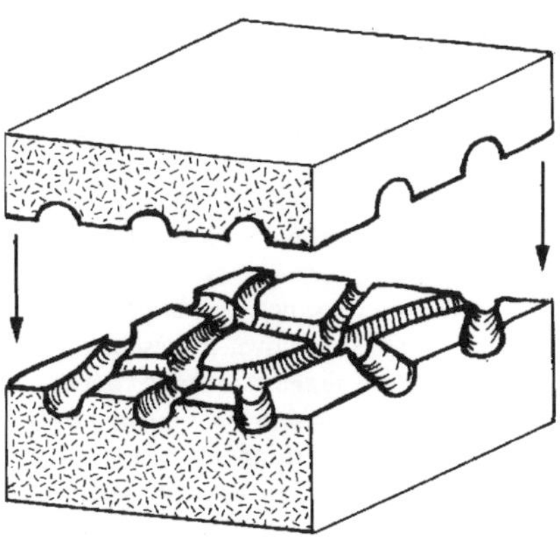

6.1 Figure 17. Bedding-plane anastomoses with remnant upstanding flat-topped sharp-edged pendants exposed by breakdown (the "pendants" here are on the lower fallen block). These pendants, which are almost always exposed by collapse, are different from paragenetic pendants (6.1 Plate 6).

White (1988) likens these to "upside-down stream beds." They are a reasonable analog to a surface anastomosing stream in that the channels, once formed, are not free to change location, as they are in a braided system. The half-tubes are from 2-3 cm up to a meter or so wide. The channels resemble bedding-plane and joint-plane anastomoses but may not be confined by geologic boundaries. Residual bedrock between paragenetic anastomosing half-tubes, when eroded free from the sediment fill, display rounded *paragenetic rock pendants*. In some situations the pendants are many and small (a few centimeters) so that the cave roof looks like the inside of a cow's stomach (6.1 Plate 6). Spectacular half-tubes, anastomoses, and pendants may be viewed in Kentucky Avenue in Mammoth Cave.

It is unfortunate that the term "pendant" is so widespread and so generic. It covers almost any residual form. Thus it is relevant here to discuss yet a third type of pendant, in this case the *vadose pendant*. These are residual emergent rock masses left behind when vadose processes bypass them. Where they emerge from the cave floor or walls they have been called "cer" (Slabe, 1995) or "echinoliths."

Miscellaneous Small-Scale Forms

Condensation corrosion cusps and runnels include all types of small-scale cusps, hollows, and runnels that cut smoothly through contrasting lithologies leaving no physical relief to mark the different materials. They are perhaps the opposite of boxwork, where differential solubility is exploited to an extreme. A typical condensation corrosion cusp cuts through speleothems and wallrock in a single smooth curve such that the junction between the two materials cannot be detected by feel (6.1 Plate 7). Such a situation has often been mistaken for a result of renewed phreatic dissolution. However, many of these forms are in situations that could not have had any kind of phreatic reworking after speleothem deposition.

Condensation corrosion above the water table is well known from thermal caves (Cigna and Forti, 1986) where aggressive vapor condenses on rock faces and produces shallow scallop-like forms and vertical runnels. In Carlsbad Cavern, a hypogenic cave formed by water containing H_2S, Ford and Williams (1989 and references therein) explained the facetted speleothems and bedrock as a result of vapor flow, noting some redeposition at the sheltered downwind edges. However, condensation corrosion forms also occur in nonthermal situations. Perhaps a similar process occurs in any situation where warm moist air is chilled.

There are two problems associated with the production of condensation corrosion forms in nonthermal caves: one is the physics of how a supply of condensing water can be maintained; the other is the extraordinary pattern of dissolution.

There are few studies of condensation corrosion cusps. The finest reported examples are from the tropics, but Dubljansky and Dubljansky (1998, Chapter 3.6 in this volume) have observed condensation from 25° to 70° latitude and from sea level to 2600 m altitude. We have seen excellent examples in the caves of Chillagoe, N Queensland, and Katherine, N Territories, Australia, and in coastal caves of the Caribbean. In such situations warm surface air may be cooled inside the caves and thus condense on walls, although it is difficult to understand how the saturated water can be removed and how heat dissipation through the rock can be rapid enough to maintain the system. In some caves there is evidence that aerosols play a part in the supply of moisture (Pashenko et al., 1996).

The model used to explain spongework may also be applicable here (6.1 Figure 16), but operating in a different part of the graph. The diagram shows that rocks (e.g. 2 and 3) may not display large contrasts in dissolution rate during the early stages of rapid dissolution, in the region far from saturation. Very aggressive condensing water could operate in this part of the system. With the high surface area to water volume ratio, the condensation water would quickly become saturated, but it is continuously replaced by fresh aggressive condensation. This keeps a steady-state reaction rate close to the left side of the graph. Condensation corrosion is discussed in more detail in Chapter 3.6.

Scallops and flutes are perhaps the most well-studied phenomena in cave micromorphology. When turbulent fluid flows in

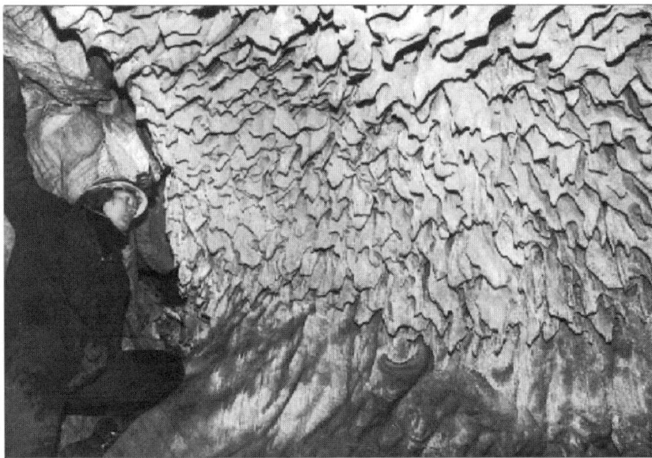

6.1 Plate 6. *Paragenetic pendant field, developed subglacially above a total sediment fill, which was subsequently washed out, Svartisen, north Norway, (photograph by S.-E. Lauritzen).*

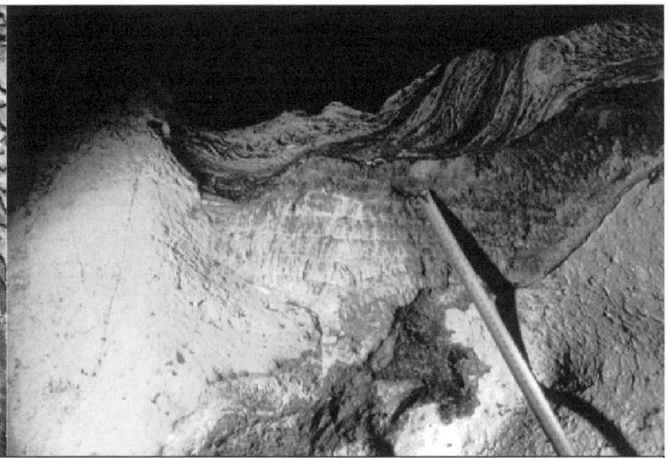

6.1 Plate 7. *Condensation corrosion cusp in a tropical cave, Cutta Cutta Cave, Katherine, Northern Territory, Australia. Speleothems and wallrock have dissolved at closely similar rates, so that the surface is cut smooth (photograph by S.-E. Lauritzen).*

contact with a soluble or volatile solid surface, mass transfer is dependent on the boundary layer shear along stationary eddies, and may result in a surface of regular "facets," either in the form of rounded concavities packed in a polygonal pattern (*scallops*) or runnels transverse to the direction of flow (*flutes*) (Curl, 1974). Scallops and flutes may be regarded as dissolutional analogs to current ripples in unconsolidated sediments (Allen, 1972). Ablational scallops also form on ice by sublimation into turbulent air currents.

"*Solutional facets*" were recognized by many early workers (Lugeon, 1915; Horn, 1935, 1978; Bretz, 1942). But Coleman (1949) was the first to observe that scallops in caves possess asymmetry that is related to the direction of flow. The longitudinal profile of scallops and flutes always has the steepest side facing downstream (6.1 Figure 18). Later, Curl (1966) demonstrated by dimensional analysis that:

- The period, or "wavelength," of scallops is inversely proportional to flow velocity.
- The longitudinal distribution of mass transport causes them to migrate downstream with progressive wall retreat.

The development of scallops has been successfully simulated by flume experiments on plaster blocks by numerous workers (Rudnicki, 1960; Goodchild and Ford, 1971; Allen, 1972; Wigley, 1972; Curl, 1974; Blumberg and Curl, 1974; Thomas, 1979; Gabriel, 1986). These have confirmed the asymmetry principle as well as the size to velocity dependence predicted by Curl's (1966) dimensional analysis. Blumberg and Curl (1974) developed a theoretical equation for scallop length l related to the mean flow velocity in the cave passage m as follows:

$$\pi = \frac{\nu(T)}{\lambda}\mathrm{Re}^{*}\left[\left(2.5\left(\ln\left(\frac{R_h}{\lambda}\right)-\frac{3}{2}\right)+B_L\right)\right] \quad (1)$$

where $n(T)$ is the kinematic viscosity of water at a given temperature, Re* is the scallop Reynolds number, R_h the hydraulic radius of the passage, and B_L Prandtl's bed-roughness constant. Flume experiments made it possible to calibrate Re* = 2,200 and B_L = 9.4. Substituting for these constants (Curl, 1982) and allowing for a compromise of passage cross section (between a circular and parallel-wall conduit) yields

$$Q = \left\{ \frac{e^{\left[\frac{442.8}{T+115}-3.272\right]}}{\lambda_{32}} \left[5.5\ln\left(\frac{4\cdot A}{P\cdot\lambda_{32}}\right)+8.1\right] \right\} \cdot A\cdot 10^{-6} \quad (2)$$

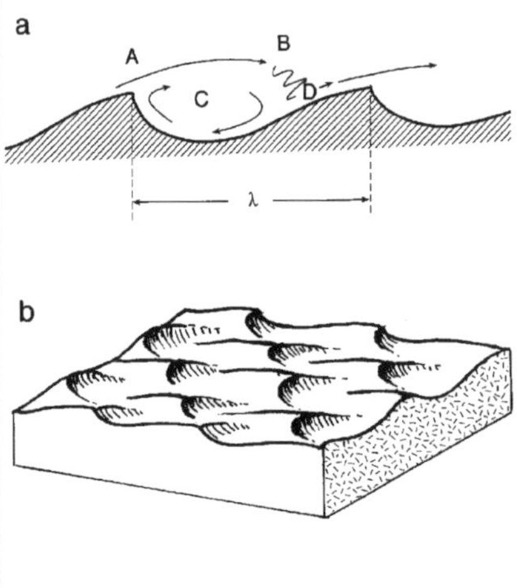

6.1 Figure 18 *(**a**)—Longitudinal profile of a scallop or flute (after Curl, 1974). **A** is the detachment of the laminar sublayer. **B** is the transition to turbulence. **C** is the eddy. **D** is the area of reattachment of flow; the region of highest dissolution rate. The scallop wavelength λ is shown. Figure 18 (**b**)—A population of scallops packed on a rock surface. Flow toward the right.*

where Q is the discharge of the cave passage in m^3/s, T is water temperature in degrees celsius, A is the cross-sectional area of the conduit (cm^2), and P the length (cm) of the wetted perimeter. The mean flow velocity of the passage, μ (cm/s), may be obtained by omitting the last factor Ax10^{-6} from equation (2).

Due to the polygonal packing of a scallop population, small scallops would tend to form at the interface between larger ones, and therefore become over-represented. Curl (1974) therefore suggests that the "Sauter mean," λ_{32} (measured in cm, equation 2) is used instead of the arithmetic mean of a scallop sample to suppress the importance of the smaller features, which are irrelevant to the rate of flow dependence, in which

$$\lambda_{32} = \frac{\sum_{i=1}^{n}\lambda_i^3}{\sum_{i=1}^{n}\lambda_i^2} \quad (3)$$

Curl (pers. comm. 1981, see Lauritzen, 1982)

suggested estimates of the standard deviation of λ_{32}, namely σ_{32}, to be

$$\sigma_{32}^{+} = \lambda_{32} \cdot e^{\sigma_{32}} - \lambda_{32}$$
$$\sigma_{32}^{-} = \lambda_{32} - \frac{\lambda_{32}}{e^{\sigma_{32}}} \tag{4}$$

where

$$\sigma_{32} = \sqrt{\frac{13}{n(n-1)} \sum_{i=1}^{n} \left(\ln \lambda_i - \overline{\ln \lambda_i}\right)^2} \tag{5}$$

and λ_i is the length of the i'th scallop. The length of each individual scallop is measured along the maximum length parallel to the direction of flow (see Lauritzen, 1982, for further discussion). The relationship between λ_{32} and Q was verified by field experiments in an active cave passage to fit within 8% of equations (1 and 2) (Pisarowicz and Maslyn, 1981).

Scallop lengths may range from less than 10 mm to several meters, although recognizable scallops would vanish when λ approaches the diameter of the passage. Then the undulations of the passage itself may be perceived as giant "scallops." Samples of scallop populations usually form unimodal well-sorted log-normal distributions (Lauritzen, 1982; Lauritsen, 1996), 6.1 Figure 19. In rare cases of bimodal distributions, sequential superposition of two flow regimes may be inferred. Because small scallops form faster than large ones (wall retreat rate is proportional to velocity), such patterns should represent a younger population of *small* scallops superimposed onto older *large* ones, rather than the contrary, see for instance 6.1 Plate 1a.

Unimodal, well-sorted, log-normal distributions of scallops indicate a constant rate of flow, or suggest that one discharge is dominant. Realizing that natural water flows are extremely variable, a pertinent question then is which part of a hydrologic regime is represented by a scallop population. That is, what is the *scallop-dominant discharge?* We should expect scallops to represent the discharges that provide the most effective rates of wall retreat. Corrosion rate is generally proportional to flow rate, and hence the highest discharges (small scallops) should intuitively become dominant. Cave divers measured scallops in various allogenic springs in marble stripe karsts and compared the data with independent discharge measurements (Lauritzen et al., 1985; Lauritzen, 1989). Stripe-karst settings provide well-defined boundary conditions of both speleology and hydrology that permit reliable flow calculations. In such allogenic conduits, we found that the scallop-dominant discharges represent flood discharges that occur only 2 to 15% of the time (6.1 Figure 20).

Integration of potential corrosion rate and flow in an allogenic conduit confirm this view, 6.1 Figure 21. Half of all the chemical work (wall-retreat rate) occurs for only 10% of the time, during the highest discharges

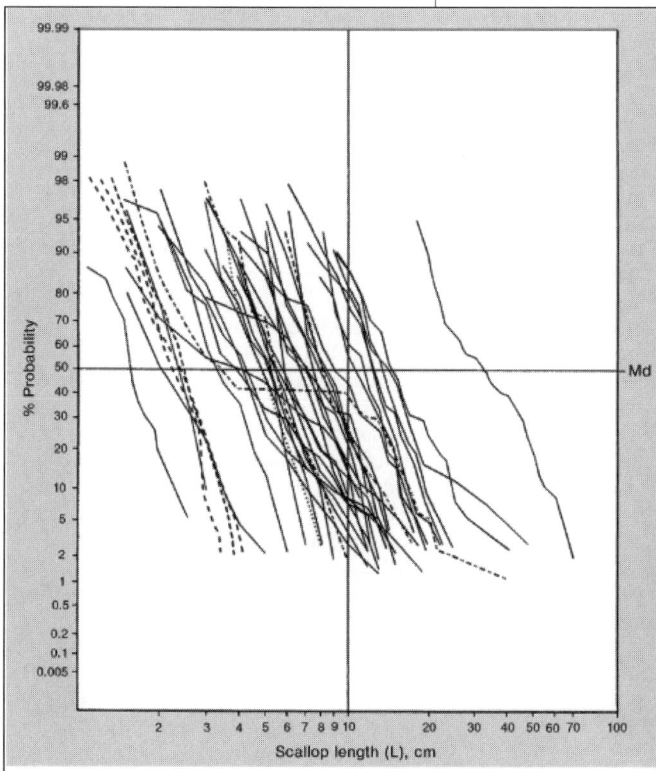

6.1 Figure 19. Log-probability plot of scallop lengths l. Most scallop populations are log-normally distributed (straight lines), but some are bimodal (two, offset straight-line segments). Bimodal scallop populations may be interpreted either as larger slow-flow scallops being overprinted by subsequent rapid flow (the opposite is less likely as small scallops form faster than large ones), or they may be interpreted as indicative of two dominant discharges. (Scallop samples from marble caves in north Norway; Lauritzen, 1996).

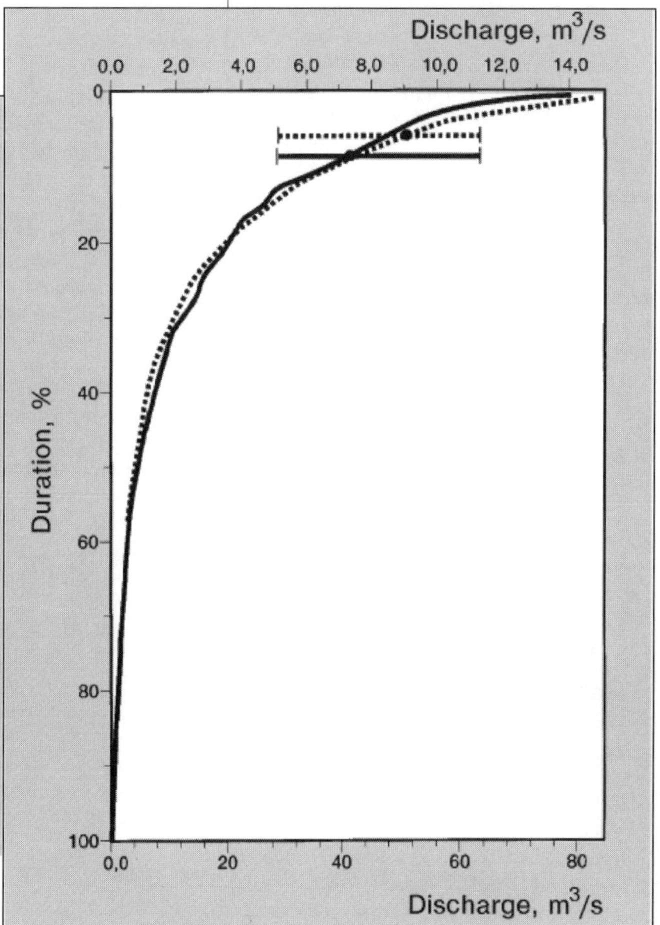

6.1 Figure 20. Long-term flow duration (solid and dashed curves) for two allogenic springs in Norway. Scallop-dominant discharges as determined by scallop morphometry and cave diving are shown with horizontal error bars. The scallops correspond to the 2 to 15% upper flow regimes (from Lauritzen, 1989).

(Lauritzen, 1989). Also, preliminary studies suggest that there is a direct linear relationship between the scallop-dominant discharge and the corresponding drainage area of allogenic springs (6.1 Figure 22).

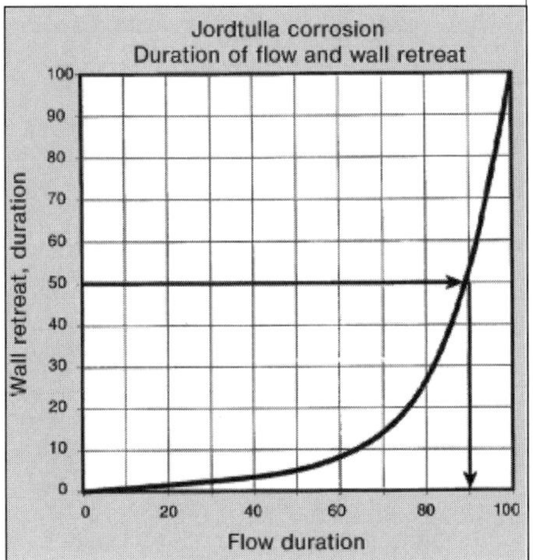

6.1 Figure 21. Relative duration of flow and rate of wall retreat. The arrows indicate that 50% of the total solutional work (upper part of the y-axis) is performed by the very highest discharges (right part of the x-axis) that occur less than 10% of the time. High discharges are therefore much more efficient at eroding the walls, and they also dominate scallop sizes. The diagram is based on integration of flow and water chemistry during two years in the Jordtulla Phreatic System, north Norway (Lauritzen et al., 1985; Lauritzen, 1989).

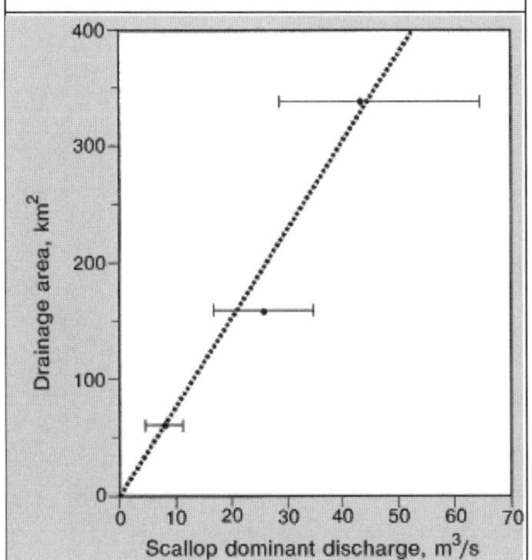

6.1 Figure 22. Scallop-dominant discharge plotted against catchment area for four active (allogenic) phreatic conduits in Norway (Lauritzen, 1989), suggesting a linear relationship between scallop-dominant discharge and catchment area.

This corresponds with the standard fluvial relationship of drainage area and discharge (Strahler 1981, p. 410, Figure 16.8). If this relationship could be confirmed through more data, one would have a powerful tool for deducing *paleowatersheds* from relict cave conduits with scallops.

Scallop morphometry is a powerful tool in speleogenetic analysis. Lauritzen (1982) calculated the flow through a relict phreatic network in marble stripe karst in Norway. First, it was found that closed-conduit discharges deduced from scallops obey flow continuity—the flow rates in confluent and divergent bifurcations add up (6.1 Figure 23). Moreover, it was possible to distinguish between two distinctly different flow regimes—normal phreatic flow when the whole cave system was phreatic, and epiphreatic rapid flow when the system was only partially filled with water.

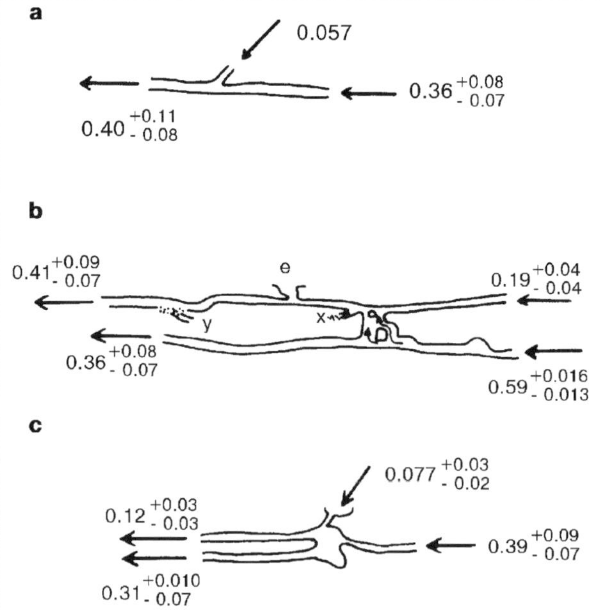

*6.1 Figure 23. Mass balance (continuity) of flow (m³/s) through a branched phreatic network based on scallop discharges, Pikhaugan Caves, Svartisen, north Norway. Within error, flow rates entering and leaving junctions (nodes) add up. The three panels (**a**, **b**, and **c**) are various levels of a much larger (2,000 m in total length) network that as a whole obeyed the continuity principle (from Lauritzen, 1982).*

Various hydraulic parameters, like boundary-shear stress, conduit Reynolds number, boundary friction factor, boundary roughness, and flow power, can be calculated from scallop and sediment sizes in passages (Gale, 1984), which in turn can be used for calculating paleohydraulic gradients (Lauritzen et al., 1985). Recently, these techniques were used on the paleo-current analysis of alpine caves of the Polish Tatras (Kicinska, 1997).

Cave karren, the underground equivalent of the small-scale dissolutional features of karst surfaces that include rills, flutes, pans, and pinnacles, are characteristic only of the vadose environment. Whereas there is a vast range of land-surface karren features, the variety of cave karren is more limited. As in surface karren, a form varies:

- According to the presence or absence of sediment (they grade into paragenetic forms under sediment).
- According to the control on water flow (forms developing under open channel or phreatic flow cannot be called karren).
- According to the predominance of gravitational control versus geologic structural control (forms related to dissolutional etching of lithologic structures cannot be called karren).

In caves there are few situations where water is supplied like rainfall, so many of the standard bare-rock karren are absent from caves. Water in caves is usually stored in some way, so a common cave karren form is the *decantation* form (6.1 Figure 24) (Ford and Lundberg, 1987).

At this stage, a genetic classification of cave karren cannot be presented Many forms are not yet understood—most can only be described. While they exist in most caves—and Slabe (1995) discusses them in some detail—they are rarely noted in the literature. It can be hard to decide which features to include under "cave karren." Some of the forms treated previously in this volume (such as bell holes, pits, and condensation corrosion forms) may be karren. Like other microforms, karren must be superimposed onto a preformed shape. The descriptions below have been roughly categorized according to form.

The simplest circular form is the *splash pit*, controlled by gravity. These form under waterfalls or drips and are equivalent to surface rain pits. Typically they reach diameters of a few centimeters.

Light-oriented photokarren are much more complex circular forms. In the twilight zone of tropical cave entrances, such as in Chillagoe, northern Queensland, Australia, these extremely deeply etched, and yet narrow pits

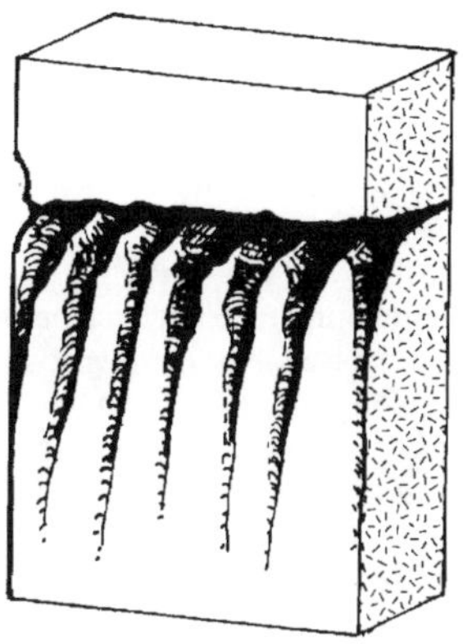

6.1 Figure 24. Decantation runnels developed on a cave wall where water emerges from a bedding plane.

cover the rock surfaces. They may be only a few millimeters wide but up to 10 cm deep and all oriented toward the light. They also have been reported from Sarawak (Bull and Laverty, 1982). They relate to a complex interplay of photosynthetic, and thus also respiratory and biologic activity, and dissolution.

Another quasicircular form is the *etch pit*. These can be produced under a light covering of gravelly sediment where surface tension holds water under each clast and each is then a focus of dissolution (Slabe, 1995). They may reach a centimeter or so in diameter. Etch pits can also be produced on cave walls that are not constantly wetted but which remain damp, as where water can condense. These small circular holes, a few millimeters up to several centimeters in width, usually occur in groups where a whole section of cave wall may seem to have acne, as at Chillagoe.

A circular form clearly related to vadose domepits, gravity controlled, and also developing upward, is the tiny domepits reported from Wales, developing beneath a thin layer of conglomerate (Burke and Bird, 1966). These pits, up to 10 cm in diameter and 17 cm deep, develop from water seeping from the conglomerate.

Most of the linear forms are gravitomorphic. *Rillenkarren*, rows of finger-sized flutes separated by knife-edged ribs, which form from the apex of a watersource and die out with distance, form under spray. Typically, they are apparent only on collapse blocks, because those are the only surfaces available with a suitable slope and under the spray. A special case of what have been termed "rillenkarren" is reported from Lechuguilla Cave, New Mexico, U.S.A. (Ballmann et al., 1991). Here acid-rich water has etched out the limestone bedrock into deep ribs. These should probably be termed *rinnenkarren* or even *spitzkarren*, because they get deeper with distance.

Flutes on a surface that start some distance from a watersource and get deeper with distance are called *rinnenkarren,* and caves often show an equivalent. A version of these occurs where the walls of vertical shafts are gouged by spray into vertical runnels separated by sharp ribs. These may reach sizeable proportions of several meters depth and several decimeters diameter.

Where just a trickle of water runs down an almost vertical face, then clear flutes cannot form. Instead a series of flutes interrupted by numerous horizontal ribs develops. This is called *cockling* (Jennings, 1985) and is essentially a variant on the vertical flute. It is often seen on underhangs or on cave walls close to waterfalls but not in the direct line of spray.

Decantation forms occur where solvent is released from an upslope store. If this is a point source, then single *decantation runnels* are formed. If it is from a diffuse source, then *decantation flutings* are produced, 6.1 Figure 24 (Ford and Lundberg, 1987).

The important diagnostic characteristic is that they peter out with distance from the source as the aggressivity is exhausted. These are commonly observed in caves, often where water runs over ledges or out of bedding-plane fissures. Sometimes these are seen to be active after flood recession, when water ponded above bedding planes is released (White, 1988).

The other main form of karren is the pinnacle. Where these are especially rough, they are termed *spitzkarren.* Surface pinnacles are usually of complex origin: they may be inherited from subsoil forms, or they may develop in situ from complex biologic erosion. Neither of these situations is common in caves, and pinnacles are not common in caves. The closest forms, rock pendants, echinoliths, and cer, have already been treated above.

Summary—Speleogenetic Facies and Agents. In 6.1 Table 1 we have summarized the various morphologic elements discussed above and classified them (where possible) according to the hydrologic zone they belong to. The table is intended as an aid in speleogenetic analysis, although it must be used with care, as not all forms are uniquely associated with only one hydrologic zone.

The active agents in speleogenesis are predominantly water and to some extent air. Each of them may be viewed as a multidimensional vector space, and the various facies or form types as vectors or vector fields within this space. For instance, *water* may accommodate a 5-dimensional space of kinetic energy, chemical aggressiveness, sediment content, temperature, and pressure. In 6.1 Figure 25, several of the most common speleogenetic facies are plotted within a 3-D subspace (velocity, sediment content, and undersaturation), which permits identification of six distinct speleogenetic facies types (Velocity is here nondirectional and only expresses kinetic energy, a scalar. It might also be substituted by the *Reynolds number* (Re) of the moving water),

A: Gross form of passage

Phreatic	Epiphreatic	Vadose	Both	Uncertain
Phreatic tubes	Epiphreatic tubes	Shafts	Keyhole passage	Cupolas
Bypass tubes		Vadose canyon		
Corkscrews		Domepits		
Paragenetic canyon				
Laughöhle-profile				

B: Morphological details inside passages

Phreatic	Water-table	Vadose	Both	Uncertain
Phreatic arches, chimneys, and pockets	Laugdecke(?)	Canyons and potholes	Scallops and flutes	Bell holes and bell pits
Spongework	Water-line notch (i.e. local watertable)	Vadose wall notch		
Bedding plane anastomoses and pendants		Vadose pendants		
Paragenetic half-tubes and rock pendants		Vadose flutings and shafts		
Paragenetic wall notch		Condensation-corrosion forms		
Convection cupolas (?)		Cave karren		
		Convection cupolas (?)		

6.1 Table 1. Cave meso- and micromorphology: Summary table.

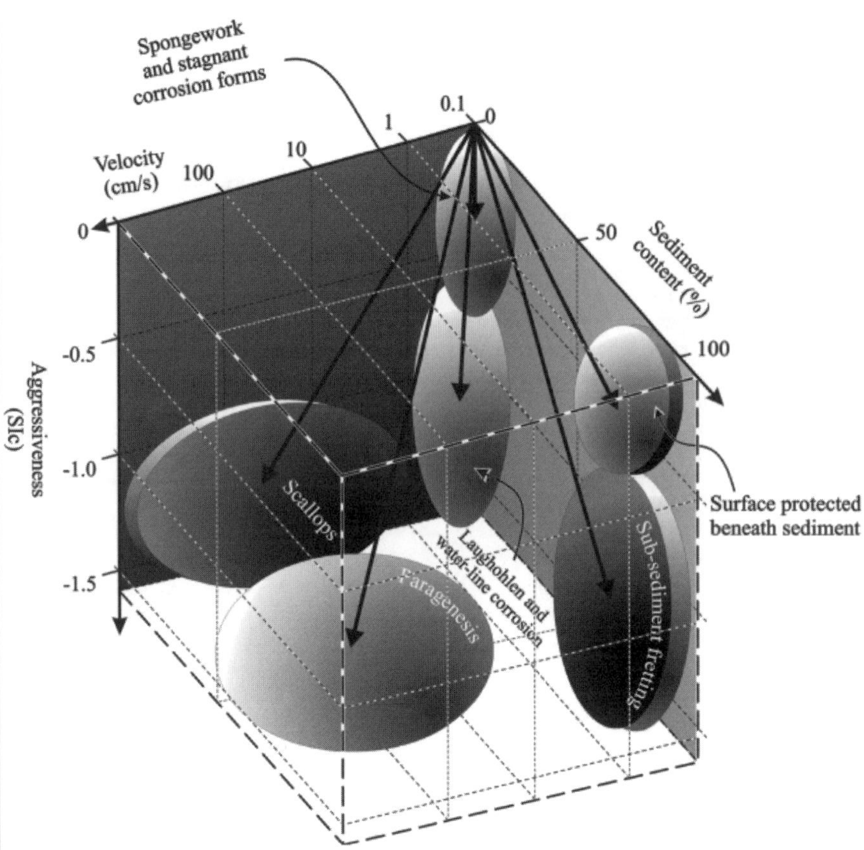

6.1 Figure 25. Various speleogenetic facies in the 3-D vector space of water as a speleogenetic agent. Combinations of water energy (equals velocity, x-axis), sediment content (y-axis), and chemical aggressiveness (z-axis) in various magnitudes, yield the corresponding speleogenetic facies as separate vectors. Note that the boundaries of and between each facies are somewhat arbitrary and must be regarded as fuzzy.

With respect to corrosion (dissolution) rate, an important kinetic boundary is the distinction between aggressive and non-aggressive water. We have here adapted the concept of the kinetic regions of White (1988). We postulate that a reasonable boundary between aggressive and nonaggressive would be between White's region 2 and 3, at $SI_C \geq 0.6$ (corresponding to a $\Delta pH < 0.3$). At this boundary, corrosion rate becomes very sluggish. Spongework and other nonaggressive-type facies may form if the cave walls are exposed to slow-moving water. If covered with sediment, preservation will occur. This is the case for most choked relict cave passages. At higher undersaturation, scallops, paragenesis, water-table features (including Laughöhle profiles), or subsediment fretting will occur, depending on the contributions of velocity and sediment content. In spite of the fact that each axis is quantified (6.1 Figure 25), the diagram must be regarded as semiquantitative, with fuzzy boundaries of and between each facies.

At temperatures less than about 0°C, water forms ice that also creates a set of distinct facies, like ice-contact corrosion forms and frost-related breakdown. At higher temperatures, another set of distinct facies occurs: the hot-water convection facies. A complete but much more complex diagram would result if one added the two dimensions of temperature and pressure (e.g. partial pressure of water vapor in the cave atmosphere). The resulting diagram would then also include ice-contact forms, thermal-water (cupola and convection) facies, aerosol, and condensation-corrosion facies.

References

Allen, J.R.L., 1972, On the origin of cave flutes and scallops by the enlargement of inhomogeneities: *Rassegna Speleologica Italiana,* v. 24, p. 3-23.

Atkinson, T.C., and Rowe, P.J., 1992, Applications of dating to denudation chronology and landscape evolution, *in* Ivanovich, M., and Harmon, R.S., eds., *Uranium-series disequilibrium: Applications to earth, marine, and environmental sciences:* Oxford, Clarendon Press, p. 669-703.

Bac-Moszaszwilli, M., and Rudnicki, J., 1978, On the possible hydrothermal origin of the Dziura Cave (Tatra Mts.): *Kras i Speleologia,* v. 2, no. 11, p. 84-91.

Ballmann, D.S., Davis, D.G., Kambesis, P., Kerbo, R.C., Palmer, A.N., and Palmer, M.V., 1991, *Leuchuguilla: Jewel of the underground:* Basel, Speleo Projects, 144 p.

Bauer, F., 1961, Sedimentation und Konvakuationserweiterung in aufsteigenden Siphonstrecken: *Rassegna Speleologica Italiana,* v. 5, p. 171-175.

Blumberg, P.N., and Curl, R.L., 1974, Experimental and theoretical studies of dissolution roughness: *Journal of Fluid Mechanics,* v. 65, p. 735-751.

Bögli, A., 1978, *Karsthydrographie und physische Speläologie:* Heidelberg, Springer-Verlag, 292 p.

Bretz, J H., 1942, Vadose and phreatic features of limestone caverns: *Journal of Geology,* v. 50, p. 675-811.

Bull, P.A., and Laverty, M., 1982, Observations on phytokarst: *Zeitschrift für Geomorphologie,* v. 26, p. 437-457.

Burke, A.R., and Bird, P.F., 1966, A new mechanism for the formation of vertical shafts in Carboniferous limestone: *Nature,* v. 210, p. 831-832.

Choquette, P.W., and Pray, L.C., 1970, Geological nomenclature and classification of porosity in sedimentary carbonates: *American Association of Petroleum Geologists Bulletin,* v. 54, p. 207-250.

Cigna, A.A., and Forti, P., 1986, The speleogenetic role of airflow caused by convection; 1st contribution: *International Journal of Speleology,* v. 15, p. 41-52.

Coleman, J.F., 1949, An indicator of water-flow in caves: *University of Bristol Spelaeological Society Proceedings,* v. 6, p. 57-67.

Courbon, P., and Chabert, C., 1986, *Atlas des grandes cavites mondiales:* Paris, Union Internationale de Speleologie and Federation Francaise de Speleologie, 255 p.

Curl, R.L., 1966, Scallops and flutes: *Cave Research Group of Great Britain Transactions,* v. 7, p. 121-160.

Curl, R.L., 1974, Deducing flow velocity in cave conduits from scallops: *National*

Speleological Society Bulletin, v. 36, p. 1-5.

Curl, R.L., 1982, *Cave conduit velocity from scallop measurement; HP-65 user program description:* University of Michigan, Department of Civil Engineering, 3 p.

Dreiss, S.J., 1984, Effects of lithology on solution development in carbonate aquifers: *Journal of Hydrology,* v. 70, p. 295-308.

Dublyansky, V.N., 1980, Hydrothermal karst in the alpine folded belt of southern parts of the U.S.S.R.: *Kras i Speleologia,* v. 3, no. 12, p. 18-36.

Dublyansky, V.N., and Dublyansky, Y.V., 1998, The problem of condensation in karst studies: *Journal of Cave and Karst Studies,* v. 60, p. 3-17.

Ewers, R.O., 1966, Bedding-plane anastomoses and their relation to cavern passages: *National Speleological Society Bulletin,* v. 28, p. 133-140.

Ewers, R.O., 1972, *A model for the development of subsurface drainage routes along bedding planes:* University of Cincinnnati, Geology Department, M.Sc., Thesis, 84 p.

Ewers, R.O., 1982, *Cavern development in the dimensions of length and breadth:* McMaster University, Ph.D. dissertation, 398 p.

Ford, D.C., 1965a, The origin of limestone caverns: A model from the central Mendip Hills, England: *National Speleological Society Bulletin,* v. 27, p. 109-132.

Ford, D.C., 1965b, Stream potholes as indicators of erosion phases in limestone caves: *National Speleological Society Bulletin,* v. 27, p. 27-32.

Ford, D.C., 1968, Features of cavern development in central Mendip: *Cave Research Group of Great Britain Transactions,* v. 10, p. 11-25.

Ford, D.C., 1971, Geologic structure and a new explanation of limestone cavern genesis: *Cave Research Group of Great Britain Transactions,* v. 13, p. 81-94.

Ford, D.C., and Ewers, R.O., 1978, The development of limestone cave systems in the dimensions of length and breadth: *Canadian Journal of Earth Sciences,* v. 15, p. 1783-1798.

Ford, D.C., and Lundberg, J., 1987, A review of dissolutional rills in limestone and other soluble rocks: *Catena Supplement,* v. 8, p. 119-140.

Ford, D.C., and Williams, P.W., 1989, *Karst geomorphology and hydrology:* London, Unwin Hyman, 601 p.

Ford, D.C., Schwarcz, H.P., Drake, J.J., Gascoyne, M., Harmon, R.S., and Latham, A.G., 1981, Estimates of the age of the existing relief within the southern Rocky Mountains of Canada: *Arctic and Alpine Research,* v. 13, p. 1-10.

Ford, D.C., Lundberg, J., Palmer, A.N., Palmer, M.V., Dreybrodt, W., and Schwarcz, H.P., 1993, Uranium-series dating of the draining of an aquifer: The example of Wind Cave, Black Hills, South Dakota: *Geological Society of America Bulletin,* v. 105, p. 241-250.

Gabriel, R., 1986, *Contibució a l'estudi de les empremtes de corrent:* Barcelona, Espelo Club de Gracia., 78 p.

Gale, S.J., 1984, The hydraulics of conduit flow in carbonate aquifers: *Journal of Hydrology,* v. 70, p. 309-327.

Gilli, E., 1993, Les grandes volumes souterrains du massif de Mulu (Borneo, Sarawak, Malisie): *Karstologia,* v. 22, p. 58-62.

Gillieson, D., 1996, *Caves; Processes, development, management:* Oxford, Basil Blackwell, 324 p.

Goodchild, M.F., and Ford, D.C., 1971, Analysis of scallop patterns by simulation under controlled conditions: *Journal of Geology,* v. 79, p. 52-62.

Hill, C.A., 1987, Geology of Carlsbad Cavern and other caves in the Guadalupe Mountains, New Mexico and Texas: *New Mexico Bureau of Mines and Mineral Resources Bulletin,* no. 117, 170 p

Hjulstrøm, F., 1935, Studies of the morphological activity of rivers as illustrated by the River Fyris: *Geological Institute of the University of Upsala Bulletin,* v. 25, p. 227-346.

Horn, G., 1935, Über die Bildung von Karsthöhlen unter einem Gletcher: *Norsk Geografisk Tidsskrift,* v. 5, p. 494-498.

Horn, G., 1937, Über einige Karsthöhlen in Norwegen: *Mitteilungen für Höhlen und Karstforschung*, p. 1-15.

Horn, G., 1978, Limestone caves in Nordland: *Cave Geology,* v. 1, p. 124-138.

Hose, L.D., 1981, Fold development in the Anticlinorio Huizachal-Peregrina and its influence on the Sistema Purification, Mexico: *8th International Congress of Speleology,* Bowling Green, v. 1, p. 133-135.

Jennings, J.N., 1985, *Karst Geomorphology:* Oxford, Basil Blackwell, 293 p.

Kempe, S., Brandt, A., Seeger, M., and Vladi, F., 1975, "Facetten" and "Laugdecken," the typical morphological elements of caves developing in standing water: *Annales de Spéléologie,* v. 30, p. 705-708.

Kicinska, D., 1997, Reconstruction of paleocurrents in caves of the Bystra Valley (Tatra Mountains, Poland), on the basis of scallops and deposits analyses: *12th International Congress of Speleology,* La Chaux-de-Fonds, Proceedings, v. 1, p. 168.

Kranjc, A., Slabe, T., and Zupan-Hajna, N., 1997, Speleology, *in* Kranjc, A., ed., *Slovene Classical karst Kras,* Ljubljana, ZRC SAZU, p. 77-101.

Lauritsen, Å., 1996, *Geomorfologisk utvikling av grottesystemene i Kjøpsvik, Nordland (Geomorphological development of cave systems in Kjøpsvik, Nordland):* Bergen University, Department of Geology, M. Sc. Thesis, 220 p.

Lauritsen, Å., and Lauritzen, S.-E., 1996, Quaternary cave and landform development in the Tysfjord region, north Norway: *Karst Waters Institute Special Publication,* no. 2, p. 73-77.

Lauritzen, S.-E., 1982, The paleocurrents and morphology of Pikhåggrottene, Svartisen, north Norway: *Norsk Geografisk Tidsskrift,* v. 36, p. 183-209.

Lauritzen, S.-E., 1986, Kvithola at Fauske; northern Norway: An example of ice-contact speleogenesis: *Norsk Geologisk Tidsskrift,* v. 66, p. 153-161.

Lauritzen, S.-E., 1989, Scallop dominant discharge: *10th International Congress of Speleology*, Budapest, Proceedings, v. 10, p. 123-124.

Lauritzen, S.-E., and Karp, D., 1993, *Speleological assessment of karst aquifers developed within the Tindall Limestone, Katherine, Northern Territory, Australia*: Water Resources Authority, Northern Territory, Australia, 60 p.

Lauritzen, S.-E., and Lauritsen, Å., 1995, Differential diagnosis of paragenetic and vadose canyons: *Cave and Karst Science,* v. 21, p. 55-59.

Lauritzen, S.-E., Abbott, J., Arnesen, R., Crossley, G., Grepperud, D., and Ive, A., 1985, Morphology and hydraulics of an active phreatic conduit: *Cave Science,* v. 12, p. 139-146.

Lauritzen, S.-E., Lundberg, J., Mylroie, J.E., and Dogweiler, T., 1997, Bell hole morphometry of a flank margin cave and possible genetic models: Lighthouse Cave, San Salvador, The Bahamas: *12th International Congress of Speleology,* La Chaux-de-Fonds, Proceedings, v. 1, p. 178.

Lips, B., Gresse, A., Delamette, M., and Maire, R., 1993, Le Gouffre Jean Bernard (1602 m Haute Savoie, Fr.): Ecoulements souterrains et formation du resau: *Karstologia,* v. 21, p. 1-14.

Lowe, D.J., 1989, Limestones and caves of the Forest of Dean, *in* Ford, T.D., ed., *Limestones and caves of Wales:* Cambridge, Cambridge University Press, p. 106-116.

Lugeon, M., 1915, Le striage du lit fluvial: *Annales de Géographie,* v. 24, p. 385-393.

Müller, P., and Sárváry, I., 1977, Some aspects of developments in Hungarian speleology theories during the last ten years: *Karszt es Barlang,* Special issue, p. 53-60.

Mylroie, J.E., and Carew, J.L., 1990, The flank margin model for dissolution cave development in carbonate platforms: *Earth Surface Processes and Landforms,* v. 15, p. 413-424.

Palmer, A.N., 1975, The origin of maze caves. *National Speleological Society Bulletin,* v. 37, p. 57-76.

Palmer, A.N., 1981, *A geological guide to Mammoth Cave National Park*: Teaneck, Zephyrus Press, 196 p.

Palmer, A.N., and Palmer, M.V., 1989, Geology and origin of the Black Hills caves, South Dakota: *National Speleological Society Bulletin,* v. 51, p. 72-99.

Pashenko, S., Dubliansky, Y., Andreichuk, V., and Pashenko, E., 1996, Transformation of fractal atmospheric aerosols moving through a natural cave: *Journal of Aerosol Science,* v. 27, p. S127-S128.

Pisarowicz, J.A., and Maslyn, M., 1981, Empirical confirmation of Curl's flow velocity calculations 1974: *8th International Congress of Speleology,* Bowling Green, Proceedings, v. 2, p. 772-774.

Rauch, H., and White, W.B., 1977, Dissolution kinetics of carbonate rocks: 1, Effect of lithology on dissolution rate: *Water Resources Research,* v. 13, p. 381-394.

Renault, P., 1968, Contribution a l'étude des actions méchaniques et sédimentologiques dans la Spéléogenèse: *Annales de Spéléogie,* v. 23, p. 529-593.

Renault, P., 1970, *La formation des cavernes:* Paris, Presses Universitaires de France, 126 p.

Rudnicki, J., 1960, Experimental work on flutes development: *Speleologica,* v. 2, p. 17-30.

Rudnicki, J., 1967, Origin and age of the western Tatra caverns: *Acta Geologica Polonica,* v. 17, p. 521-591.

Shirtcliffe, T.G.L., and Turner, J.S., 1969, Observations of the cell structure of salt fingers: *Journal of Fluid Mechanics,* v. 41, p. 707-719.

Slabe, T., 1995, *Cave rocky relief and its speleological significance:* Ljubliana, Znanstvenoraziskovalni Center Sazu, 128 p.

Strahler, A.N., 1981, *Physical geology:* New York, Harper and Row, 612 p.

Szunyogh, G., 1984, Theoretical investigation of the origin of spherical caverns of thermal origin (second approach): *Karszt es Barlang,* p. 19-24.

Tarhule-Lips, R., and Ford, D.C., 1997, Morphologic studies of bell hole development on Cayman Brac: *12th International Congress of Speleology,* La Chaux-de-Fonds, Proceedings, v. 1, p. 146.

Tarhule-Lips, R., and Ford, D.C., in press, Bell hole development on Cayman Brac based on morphology: *Cave and Karst Science.*

Thomas, R.M., 1979, Size of scallops and ripples formed by flowing water: *Nature,* v. 277, p. 281-283.

Trudgill, S.T., 1985, *Limestone geomorphology:* London, Longman, 196 p.

Waltham, A.C., 1977, Cave development at the base of the limestone in Yorkshire: *7th International Congress of Speleology* Sheffield, Proceedings, p. 421-423.

White, W.B., 1988, *Geomorphology and hydrology of karst terrains:* New York, Oxford University Press. 464 p.

Wigley, T.M.L., 1972, Analysis of scallop patterns by simulation under controlled conditions: A discussion: *Journal of Geology,* v. 80, p. 121-122.

Wilford, G.E., 1964. The Geology of Sarawak and Sabah caves, *Geological Survey of Borneo Bulletin*, v. 6, p. 1-181.

6.2. Breakdown Morphology

Elizabeth L. White and William B. White

Abstract

The mechanical failure of cave roofs is an important part of the speleogenetic process. For caves to survive, they must be mechanically stable against collapse. Breakdown in caves occurs both early in the speleogenetic process and very late in the speleogenetic process. Early in the process the flow regime shifts from pipe-full conditions to open channel conditions, and the roof first loses buoyant support. Late in the process breakdown becomes part of the overall degradation of the karst system.

Introduction

The term breakdown refers to disarticulated fragments of bedrock that have broken free and fallen into a cave passage. Although usually classified as a form of cave sediment, breakdown also plays a role in the evolution and enlargement of the cave passages themselves. The sections that follow address the mechanism of breakdown formation, the geologic triggers that initiate breakdown, and the role that breakdown plays in the development of caves.

Breakdown Morphology

Breakdown has been traditionally categorized into three main types based on a suggestion made by Davies (1949):

- Block breakdown
- Slab breakdown
- Chip breakdown

The distinctions are that block breakdown consists of masses of fallen rock that span more than one bed. Slab breakdown consists of broken pieces of individual beds. Chip breakdown consists of small pieces of beds that have been fragmented. This classification was taken up by White and White (1969) and found to be generally useful. Block breakdown tends to be associated with massive bedding and also with large scale collapses. Slab breakdown occurs where beds are less thick and where there are well-developed bedding plane partings. Chip breakdown, other than the fragmentation that occurs when larger blocks and slabs fall to the floor, is often indicative of other breakdown processes. Sharp, irregular, blocky fragments are characteristic of ice wedging, and they occur near entrances. Such fragments have also been described in deep alpine caves where they were taken as evidence for very deep freezing temperatures during the Pleistocene glacial maxima (Schmid, 1958). Chip breakdown consisting of thin shards is evidence for breakdown initiated by mineral (usually gypsum) deposition and bedrock replacement.

Accumulations of breakdown are common features within many cave passages. Breakdown pavements litter the floors of many caves. Rock falls from cave walls modify passage cross sections and often disguise the original solution passage shape. More or less continuous stoping of cave roofs can result in a cave passage migrating upward for substantial distances from the original solution passage. Massive collapse produces *breakout domes*. Breakout domes, from which the fallen blocks are removed by subsequent dissolution, form some of the largest underground chambers. If the fallen rock is not entirely removed, *breakdown mountains* may remain. Breakdown collapses truncate passages and dissect long continuous conduits into many short segments. These *terminal breakdowns* are one of the most common ends for cave passages in the Mammoth Cave System (Brucker, 1966).

Fracture Mechanics of Breakdown

Brittle fracture of fixed and cantilever beams provides the simplest model for breakdown, and the concept dates back to the work of Davies (1951) who based his model on the mining literature. The model assumes a rectangular passage formed in well-bedded horizontal limestone. A small amount of elastic sag of the unsupported roof beds causes these beds to separate slightly. The stress distribution takes a domal shape (6.2 Figure 1). The beds act as fixed beams across the cave passage. For the ceiling to be stable, the bending strength of the beams must be greater than the gravitational force acting on the weight of the unsupported span. Thicker beds are stronger than thin beds. There will be a critical thickness for any given passage width at which the strength of the bed is just sufficient to support its weight. When the mechanics are worked out, only the length of the beam, L, and the critical bed thickness, t_{crit}, remain. The width of the beam cancels out. The beam length is set equal to the passage width while the extent of the ceiling bed along the axis of the cave passage does not enter the calculation (6.2 Figure 2).

For a fixed beam, the critical thickness is

$$t_{crit} = \frac{\rho L^2}{2S} \quad (1)$$

where ρ is the density of the bedrock and S is the flexural strength. If the ceiling beds are not supported at both sides of the cave passage, they are treated as cantilever beams for which the critical thickness is

$$t_{crit} = \frac{3\rho L^2}{2S} \quad (2)$$

The fixed beam model implies a completely elastic response of the ceiling beds. It does not allow for plastic deformation and long term creep that could lead to bed failure in the absence of any geologic triggers. It also has no time dependence; a stable ceiling would remain stable until some geologic process destabilized it. Numerous occurrences of recent breakdown have been reported. Some occur near entrances where freeze-thaw cycles may be responsible. Some occur deep in the cave but in areas of active vadose water. However, a fraction of recent breakdown occurrences are in dry passages with no obvious triggering mechanism.

There have been three documented roof failures in Mammoth Cave in the present century. Two were massive rockfalls, but the third involved a plastically deformed ceiling slab that had been mapped in detail (Figure 4 in White and White, 1969). The slab displayed extensive plastic deformation in the 1960s. Sometime in the early 1970s it fell.

The basis for a more comprehensive model was proposed by Tharp (Tharp and Holdrege, 1994; Tharp, 1995). Materials break through a mechanism of crack propagation. Tharp's model is based on the propagation of microcracks which allow deformation and creep. The crack propagation velocity is given by

$$v = c\left(\frac{K_I}{K_{Ic}}\right)^n \quad (3)$$

where K_I is the stress intensity (MPa $m^{0.5}$), K_{Ic} is the fracture toughness of the bedrock, c is a constant related to the activation energy for crack movement. According to the Tharp model, the time scale of crack propagation through beds of nominal thickness can range from thousands of years to as much as 1 million years. The time to failure can be in the same range as the age of the cave passage, implying that breakdown can occur at any time, even in the absence of geologic triggering processes.

The Tharp model introduces the fracture toughness as another parameter in addition to the flexural strength for determining whether particular beds will collapse. The Paleozoic limestones of the eastern United States—where many of the breakdown investigations have been made—are dense fine-grained rocks. Coarse fracturing occurs along joints and bedding-plane partings. But within the rock mass there is little to inhibit crack propagation, and these rocks break mainly by brittle fracture. It is for this reason that the fixed beam model has worked so well.

Porous and vuggy rocks, such as the Tertiary limestones of the Caribbean, may have a lower strength but a higher fracture toughness, because the pores and vugs inhibit crack propagation.

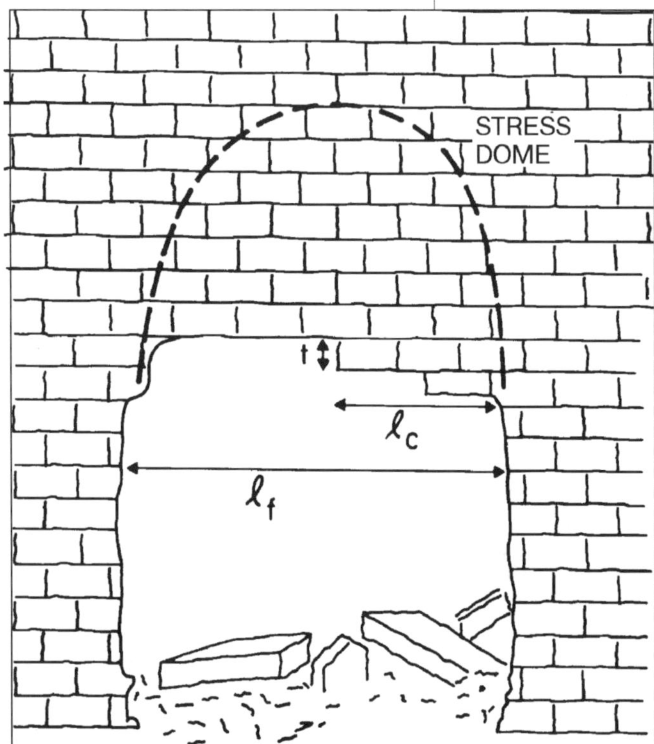

6.2 Figure 1. Sketch showing the parameters of the fixed beam model.

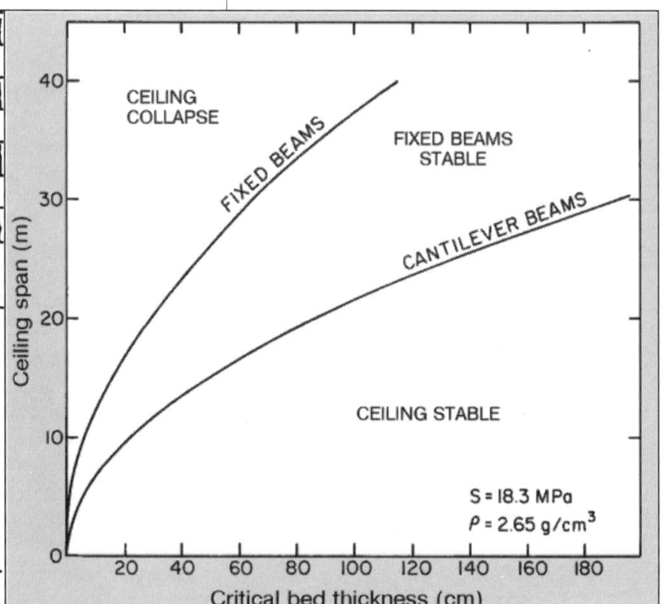

6.2 Figure 2. Cave roof stability according to the fixed beam model.

Geologic Processes That Initiate Breakdown

Within the context of either fixed beam or Tharp models, any geologic process that either lengthens the beams or converts fixed beams into cantilevers can move the ceiling beds from stable to unstable configurations. Beam thickness, flexural strength, and fracture toughness are properties of the bedrock, and they do not change during the evolution of the cave passage. A list of triggering processes, not necessarily complete, is as follows:

- Passage enlargement below the water table. Phreatic passages continue to enlarge as water flows through them. If the hydrologic conditions are such that the passage is not drained, it may continue to enlarge until it becomes mechanically unstable.

- Removal of buoyant support. By Archimedes' principle, the ceiling beds of a water-filled cave passage are buoyed upward by a force proportional to the ratio of the density of water to the density of the bedrock. For limestone with a typical density of 2.4 g/cm^3, 42% of the buoyant support of the ceiling is lost when the cave is drained.

- Effects of base-level back flooding. During the time when the emergent cave passage is in the flood-water zone, rises and falls in base level alternately fill and drain the cave passage. Additional dissolution at this time, particularly dissolution along joints that can turn fixed beams into cantilevers, can destabilize the ceiling.

- Action of vadose water. Formation of vertical shafts, solution chimneys, and solutionally enlarged fractures by the action of unsaturated vadose water often has the effect of cutting ceiling beds, thus changing fixed beams into cantilever beams.

- Ice wedging. Caves that draw in cold winter air can have freezing conditions some distance inside. When water moving through joints and bedding-plane partings freezes, the expansion creates enough force to fracture the bedrock.

- Crystal wedging. Replacement of calcite in the bedrock by other minerals can exert a wedging effect. Gypsum has a greater volume than the calcite it replaces, thus generating enough force to fracture the bedrock.

Role of Breakdown in Speleogenetic Processes

Both geologic triggering processes and slow creep of beds under load assure that breakdown can occur at any time during the evolutionary history of a cave passage. However, breakdown processes are most active during the enlargement phase of cave development and during the decay phase of cave development.

The role of breakdown in the enlargement phase includes the following:

- Breakdown during the enlargement phase exposes more limestone surfaces and thus increases the rate of dissolution.

- Upward stoping by breakdown processes can create large chambers, if actively circulating water removes the breakdown blocks at floor level.

- Upward stoping along fracture zones with removal of fallen blocks results in the formation of stoping shafts. Sotano de las Golondrinas in Mexico is an outstanding example.

- Breakdown in master conduits, particularly during the floodwater stage, can provide a support structure for groundwater dams. Silt and clay that deposit behind the blockage seal the dam, raise hydraulic heads upstream, and thereby generate a hydraulic gradient for the formation of new tap-off passages.

Breakdown continues to play a role during the stagnation and decay phases of cave development as follows:

- Breakdown processes can stope upward to interconnect previously isolated cave levels into an integrated system of passages.

- Truncation of cave passages by the formation of terminal breakdown is a dominant process in the breakup of continuous conduits into the fragments characteristic of the decay stage of cave development.

- The final phase in the decay of caves is the passage collapse that takes place when the eroding land surface intersects the stress dome in the rocks of the cave ceiling.

References

Brucker, R.W., 1966, Truncated cave passages and terminal breakdown in the Central Kentucky Karst: *National Speleological Society Bulletin,* v. 28, p. 171-178.

Davies, W.E., 1949, Features of cavern breakdown: *National Speleological Society Bulletin,* v. 11, p. 34-35.

Davies, W.E., 1951, Mechanics of cavern breakdown: *National Speleological Society Bulletin,* v. 13, p. 36-43.

Schmid, E., 1958, Höhlenforschung und Sedimentanalyse: *Schriften der Institutes für Ur- und Frühgeschichte der Schweiz,* no. 13, 186 p.

Tharp, T.M., and Holdrege, T.J., 1994, Fracture mechanics analysis of limestone cantilevers subject to very long term tensile stress in natural caves, *in* Nelson, P.P., and Laubach, S.E., eds., *1st North American Rock Mechanics Symposium Proceedings:* Rotterdam, A.A. Balkema, p. 817-824.

Tharp, T.M., 1995, Design against collapse of karst caverns, *in* Beck, B.F., ed., *Karst geohazards:* Rotterdam, A.A. Balkema, p. 397-406.

White, E.L., and White, W.B., 1969, Processes of cavern breakdown: *National Speleological Society Bulletin,* v. 31, p. 83-96.

Part 7: Speleogenesis in Noncarbonate Lithologies

7.0 Overview

As the great majority of both theoretical considerations and case studies in this book deal with speleogenesis in carbonate rocks, it is useful to provide a special forum to examine dissolution cave genesis in other rocks. This is the goal of Part 7.

Klimchouk (Chapter 7.1) provides a review of speleogenesis in gypsum. This is a useful arena for testing the validity and limitations of certain general speleogenetic concepts. A difference in solution kinetics between gypsum and calcite impose limitations and peculiar features on the early evolution of conduits in gypsum.

These peculiarities seem to be an extreme and more obvious illustration of some rules of speleogenetic development devised from conceptual and digital modeling of early conduit growth in limestone.

For instance, we have shown (Palmer, 1984, 1991; Dreybrodt, 1996; and also see Chapters 3.4 and 4.2.2) that initiation of early, narrow, and long pathways does not seem feasible under linear dissolution rate laws (n=1) due to exponential decrease of the dissolution rate. Although the dissolution kinetics of gypsum are not well known close to equilibrium, it is generally assumed that they are controlled entirely by diffusion and therefore are linear. If dissolution of gypsum is solely diffusion-controlled, with no change in the kinetic order, conduit initiation could not occur in phreatic settings or by lateral flow through gypsum from distant recharge areas in artesian settings.

Hence, the fact that maze caves are common in gypsum under artesian conditions (see Chapter 5.2.1) gives strong support to a general model of "transverse" artesian speleogenesis where gypsum beds are underlain by or sandwiched between insoluble or low-solubility aquifers (Chapter 5.2). And it suggests that similar processes may be applicable to cave development in carbonates.

In unconfined settings, speleogenesis in gypsum occurs along fissures wide enough to support undersaturated flow throughout their length. Linear or crudely branching caves overwhelmingly predominate, which rapidly adjust to the contemporary geomorphic setting and to the maximum available recharge. Also, if considerable conduit porosity has been created in deep-seated settings, it provides ready paths for more intense groundwater circulation and further cave development when uplift brings the gypsum into the shallow subsurface.

Speleogenesis in salt, reviewed in general and exemplified by the Monte Sedom case in Israel (Frumkin, Chapter 7.2), has been documented only in open, unconfined settings, where it provides a model for simple vadose cave development. Chapter 7.3 deals with speleogenesis in quartzite, illustrated by case studies from southeastern Minas Gerais, Brazil (Corrêa Neto, Chapter 7.3.1), and South Africa (Martini, Chapter 7.3.2). The process involves initial chemical weathering of the quartzite to create zones of friable rocks (sanding or arenisation), which then are removed by piping, with further conduit enlargement due to mechanical erosion by flowing water.

7.0 Plate 1. Selenite gypsum crystal exposed in a ceiling cupola, Optimisticheskaja Cave, western Ukraine (photo by A. Klimchouk).

7.0 Plate 2. Texture of gypsum rock exposed in Dzhurinskaja Cave, western Ukraine (photo by A. Klimchouk).

7.1 Speleogenesis in Gypsum

Alexander Klimchouk

Abstract

The main differences between the solutional properties of gypsum and those of calcite lie in the much higher solubility of gypsum and in the dissolution kinetics of gypsum which is solely diffusion controlled. Unlike calcite, no change of kinetic order occurs with an increase in concentration. Initiation of long lateral flow paths through gypsum is virtually impossible due to the rapid rate of dissolution. No kinetic mechanisms allow slow but uniform dissolutional enlargement throughout the flow paths.

Near the surface, fissures are already wide enough for cave development to occur. This is extremely competitive due to rapid dissolution kinetics and the strong dependence of enlargement rate on flow velocity and discharge. Thus caves in gypsum in exposed settings are mainly linear or crudely branching, rapidly adjusting to the contemporary geomorphic setting and available recharge. Vertical pipes or pits form in the vadose zone. No deep phreatic development and no artesian development by lateral flow from distant recharge areas can occur.

However, cave origin and development does occur in deep-seated confined settings where gypsum beds in stratified sequences are underlain by or sandwiched between poorly soluble aquifers. Two situations support cave origin in gypsum in deep-seated settings:

- Transverse flow across a gypsum bed between overlying and underlying aquifers.
- Lateral flow in an insoluble but permeable aquifer underlying a gypsum bed.

The former situation generates either maze caves, where uniformly distributed fissure networks exist in the gypsum, or discrete voids, where the otherwise low-fissured gypsum is disrupted by prominent tectonic fractures. If considerable conduit porosity has been created in a deep-seated setting, it provides ready paths for more intense groundwater circulation and further cave development when the gypsum is uplifted into the shallow subsurface.

Thick and low-fissured sulfate strata can survive burial with no speleogenesis at all where surrounded by poorly permeable beds. When exposed to the surface, such gypsum deposits undergo speleogenetic development with no inherited features, representing an unmixed pure line of open karst.

Introduction

Certain features of speleogenesis in gypsum, as compared with carbonate rocks, are determined by some geologic peculiarities of gypsiferous formations, and by the chemistry and kinetics of gypsum dissolution. These factors are discussed in detail in Calaforra (1996), Forti (1993), Klimchouk (1996a), and Klimchouk and Andrejchouk (1996a); see also Chapters 3.2 amd 4.1.6.

On the other hand, the type of flow system and the mode of recharge affect speleogenesis equally in gypsum and carbonates.

This chapter briefly summarizes these factors and considers how they influence cave origin and development in gypsum.

Geologic Setting and Physical Characteristics of Gypsum and Anhydrite

The most common sulfate minerals are gypsum ($CaSO_4 \cdot 2H_2O$) and anhydrite ($CaSO_4$). They can occur as thick single beds, but more typically occur interbedded with other sedimentary rocks. Stability of gypsum and anhydrite is considerably affected by changes in physical and chemical parameters occurring within common geological environments. This results in back and forth conversions between these minerals. Theoretical consideration of the processes and mechanisms is given in Chapter 4.1.6, and geologic evidence is reviewed in Klimchouk and Andrejchouk (1996a).

The largest areas of sulfate rocks are located in the Northern Hemisphere, particularly in the United States, where they underlie about 35 to 40 percent of the land area (Johnson, 1996) and the former USSR, where Gorbunova (1977) estimated a figure of 5 million km^2 (about 21 percent). Recent overview of the geographic distribution of gypsum karst is presented in Klimchouk et al. (1996b). Outcrops of sulfate rocks are generally much smaller than those of carbonates. However, gypsum karst develops widely in intrastratal conditions, and this type of karst is similar in extent to carbonate intrastratal karst. Caves in gypsum and anhydrite develop readily and survive through rather prolonged geologic intervals, although most available data deal with caves in gypsum, as anhydrite is less likely to occur in the shallow (speleologically explorable) subsurface.

Evaporite formations containing gypsum and anhydrite occur in various modern tectonic settings, the most important being the following three: cratons, foredeeps, and orogenic regions. Each of these settings imposes specific structural features on a sulfate sequence which thereby determine important properties of the speleogenesis (Klimchouk and Andrejchouk, 1996a).

In Cratonic Regions, sulfate rocks have horizontal to gentle dips (1 to 5°) and extend over large areas ranging up to tens of thousands of km^2. A block-fault structure is common, usually with a system of faults and blocks that have little vertical displacement between them. Gypsum beds are commonly well-fissured, with the fissuring rather uniformly distributed.

Intrastratal karst is by far the most dominant type in this setting. It develops at varying depths beneath a cover bed. Large valleys incised through the cover bed greatly influence the hydrogeologic flow architecture both at a local and a regional scale. Consequently, cave development occurs at considerable depth beneath the valley bottoms. When conduit permeability has evolved, gypsum sequences often behave as good aquifers. Cratonic regions are characterized by a stable tectonic regime and by rather slow groundwater circulation. Cave development may occur over prolonged time spans but is intensified after uplift, forming subjacent and entrenched karst types. Caves are numerous, as shown by cave exploration in entrenched karst and by encounters in mines and boreholes in deep intrastratal settings.

In Foredeeps, the strata are usually gently folded with dips of up to 15°. The rocks are often displaced and broken by faults that disrupt their continuity. Areas of outcrop and near-surface gypsum are linear, elongated along the strike of the foredeep or local fold structures. Sulfates tend to plunge downdip to considerable depth below nonkarstifiable sequences. The karst that develops is limited in area but is often quite intense.

In deep-seated confined aquifers the circulation and karstification are intensified where upward discharge is focused along faults. Large and deep collapse features are common in this structural setting. An outstanding example is the sulfate belt of the Ural Foredeep (Russia).

In Orogenic Regions, sulfate rocks are commonly severely folded with beds becoming vertical or even overturned. Areas where gypsum lies at shallow depth below the surface are commonly rather small but well exposed, with outcrops larger than those in platforms or foredeeps. The rocks are densely fissured, sometimes resulting in breccia. Fissure systems may be of various ages, with their genesis superimposed on each other. Some massifs exhibit an extremely high density of surface karstification expressed as honeycomb or badland landscapes (north Caucasus, central Asia). Others display relatively few point-recharge forms such as dolines and blind valleys because of the development of a crust on the gypsum that prevents dispersed recharge and karstification (Apennines, Sicily, southern Spain).

Recrystallization and other processes that occur in exposed gypsum masses often seal fissures, at least where they are exposed. Caves are usually numerous but rather small. They seem to form in adjustment with the contemporaneous geomorphic system and tend to be linear and to directly connect recharge and discharge points. Data about deep-seated karst in orogenic settings are scarce.

Plasticity and Flow of Sulfate Rocks

Gypsum is often viewed as capable of flow due to plasticity. It is therefore commonly believed that partings and fissures in the gypsum tend to close, thus preventing water circulation and speleogenesis. This view, based largely on laboratory sample tests, is misleading. From published data and my field experience, plasticity in gypsum is important only where the rock mass is fairly homogenous and of considerable thickness.

In contrast, numerous observations in caves, particularly in those occurring in intrastratal settings, prove that open fissures in gypsum beds can survive through lengthy periods of geologic time (Klimchouk and Andrejchouk, 1996a; Klimchouk et al., 1995). Within strata, the structural support by adjacent rocks, by intercalated layers of other lithologies, or by other still unrecognized effects, commonly prevent the plastic flow of gypsum and the sealing of discontinuities.

Recrystallization

Gypsum undergoes recrystallization throughout its early and late diagenetic history. This alters the various properties of the rock, such as its texture, structure, porosity, and strength, and affects speleogenesis in many ways.

In the western Ukraine, recrystallization of gypsum has caused severe textural and structural differentiation, forming three distinct horizons within the bed (Klimchouk et al., 1995). Such differentiation, in turn, has caused the formation of largely independent superimposed networks of late diagenetic fissures confined to each horizon (see 5.2.1 Figure 7). These fissures were exploited by dissolution to generate the structure of multistory maze caves. Thus, textural and structural differentiation of the gypsum by recrystallization was a primary guiding factor in creating the structural prerequisites for speleogenesis.

In some gypsum massifs exposed at the surface (e.g. in Sicily), a distinct crust up to a meter thick is formed in which most fissures are sealed due to recrystallization. The crust prevents diffuse recharge from the surface and the formation of an epikarst. Instead, water is able to penetrate deep into the gypsum only along selected major fissures and faults, thus favoring point recharge and the development of linear caves.

Structural and Hydrostratigraphic Prerequisites for Speleogenesis in Gypsum

Fissures in Gypsum

It is universally accepted that fissures are of primary importance as pathways for initial water circulation in most karst rocks. This is especially true for gypsum and anhydrite, because the effective porosity in these rocks is rather low and bedding partings are often not well preserved. Gypsum and anhydrite can range from little fractured to almost brecciated.

Major tectonic fissures normally cut across

several beds in a succession, providing through paths for cross-formational hydraulic communication. The role in speleogenesis of other types of fissures is commonly overlooked (see more on this topic in Chapter 3.2). In gypsum, perhaps more than in any other soluble rock, lithogenetic (contraction) fissures are very important for karstification. They are commonly confined to certain layers and do not propagate into adjoining beds. They tend to form rather uniform and largely independent networks characterised by a good lateral connectivity but connected vertically only at a relatively small number of discrete points. This provides the structural prerequisites for the development of multistory maze caves (Klimchouk, 1992, 1994; Klimchouk et al., 1995; see also Chapter 5.2.1).

Hydrostratigraphic Arrangement and Initial Permeability

Initial permeabilities of common aquifers (e.g. certain clastic rocks) and of most carbonates in deep settings are normally greater than those of sulfate beds before the onset of speleogenesis. Gypsum beds commonly act initially as low-permeability beds that separate beds of higher permeability.

However, groundwater from adjacent water-bearing formations, or from minor beds of other lithologies such as marl or dolomite, intercalated with the gypsum beds, can be aggressive with respect to gypsum. Moreover, chemical processes in such adjacent or intercalated beds may maintain gypsum's dissolution potential through removal of sulfates from solution (sulfate reduction, dedolomitization).

Late diagenesis or tectonism imposes fissure permeability on gypsum formations, which then begins to play the major role in determining initial flow paths through the gypsiferous sequence. Gradient fields in multistory confined aquifer systems generally promote transverse circulation and speleogenesis (see Chapter 5.2), although the presence of water-conducting intercalations and the discordance of fissure permeability between horizons in a gypsum bed may impose a considerable local lateral component of flow and speleogenetic development.

Gypsum Dissolution Chemistry and Kinetics

The chemistry and kinetics of gypsum dissolution is considered in detail in Chapter 4.1.6. Important properties can be summarized as follows:

- The solubility of gypsum in pure water is roughly an order of magnitude greater than the solubility of calcite in the presence of CO_2. In most relatively shallow environments the influence of temperature variation is minor.

- In deep-seated and subjacent karst settings several chemical and physical factors can considerably increase or renew gypsum solubility:
  - The presence of other salts in ground water (which enhances ionic strength and increases gypsum solubility by up to 3 times).
  - Anaerobic reduction of sulfate in the presence of organic matter.
  - Dedolomitization of intercalated dolomite layers.

- The dissolution kinetics of gypsum, which is solely diffusion-controlled, is described by a first-order equation, whereas anhydrite dissolution obeys a second-order equation. For gypsum, the flow time (distance) at which a solution approaches saturation is very short. Unlike calcite, no change of kinetic order occurs with an increase in concentration. Hence no kinetic mechanism allows slow but uniform dissolutional enlargement throughout long flow paths. Therefore, the initiation of long lateral flow paths through gypsum is virtually impossible.

- Gypsum and anhydrite dissolution proceed even in contact with static water, but dissolution rates increase rapidly with increasing flow velocity. As a turbulent-flow regime sets in, dissolution rates are boosted, probably by an order of magnitude.

- The presence of other salts such as sodium chloride in solution considerably increases the gypsum dissolution rate, and it influences the anhydrite dissolution rate even more drastically.

Origin and Development of Caves in Gypsum

The general conceptual approach to the initiation and evolution of conduits in gypsum is largely the same as that in limestone. However, differences in solution kinetics between gypsum and calcite impose limitations and characteristic features on the early evolution of conduits in gypsum. These characteristics seem to be an extreme and more obvious illustration of speleogenetic development devised from conceptual and digital modeling of early-conduit growth in limestone (e.g. Palmer, 1984, 1991; Dreybrodt, 1996; see also Chapters 3.4 and 4.2). The study of speleogenesis in gypsum is a useful operation for testing the validity and limitations of certain general speleogenetic concepts.

Rapid Dissolution Kinetics and Early Conduit Growth

Palmer (1981, p. 8), stated:

> The origin of a limestone cave hangs by a delicate thread. For it to reach traversable size in a feasible time, dissolution must be rapid; yet too high a rate is detrimental to its early development. In a narrow crack, rapid dissolution would cause near-saturation after only a few centimeters of flow, and enlargement downstream would be negligible. Only slow, uniform dissolution can initiate long passages. This paradox is reconciled by the abrupt slowing of dissolution while the water is still far from saturation. This transition... is expressed by the increase in reaction order (n).

But the problem so clearly outlined by the above quotation—and solved for the initiation of caves in limestone—remains open in the case of speleogenesis in gypsum. Palmer (1984) argued that without this change in dissolution rate, caves, and therefore major karst features, could form only where gradient and initial widths of fissures are exceptionally large (for example, along escarpments).

Dissolution of gypsum is much more rapid than that of limestone. It obeys a first-order equation, with rates controlled solely by diffusion across the boundary layer and kinetics that do not change their order with increasing concentration. However, caves in gypsum are found in diverse settings, including deep-seated confined units. Reconciling this paradox for speleogenesis in gypsum would shed more light on general processes of speleogenesis.

Initiation and Propagation of Early Conduits in Gypsum in a Deep-Seated Setting

Seepage through tight initial fissures is slow and laminar. The rate and configuration of dissolutional widening depends primarily on discharge through the fissure and on the change in solute concentration along its length (Palmer, 1984, 1991; James, 1992, Dreybrodt, 1996). Due to gypsum's fast dissolution kinetics, solute concentration increases rapidly to near saturation, so that the distance through which effective dissolution by slow seepage occurs is very short. Fissures enlarge at their inlets, and the solution rate decreases uniformly downstream, resulting in a tapered geometry. The mode of dissolutional enlargement of the fissure, or through a sequence of interconnected fissures, will change only when breakthrough is achieved, where a flow path enlarges enough to permit an increase in discharge and flow velocity, hence a penetration of significantly undersaturated water beyond the inlet (7.1 Figure 1).

However, because there is no change in the dissolution reaction order, slow enlargement throughout a long lateral path does not seem feasible. The conclusion is that speleogenetic initiation of long primary percolation paths in gypsum is virtually impossible with rapid dissolution kinetics alone. Speleogenesis in gypsum in deep-seated settings can be explained however, for two specific but common hydrogeologic situations.

Transverse Hydraulic Communication Through a Gypsum Bed Sandwiched Between Confined Aquifers—This configuration is depicted in 5.2 Figure 4A and B. It allows for short transverse flow paths to be enlarged to achieve breakthrough within a geologically feasible time even by dissolution at very slow rates under conditions close to saturation. Short distances of flow through the gypsum bed allow for high *Q/L* ratios to be achieved along many alternate flow paths even at sluggish flow conditions, thus favoring maze development. Uniform enlargement of all initiating conduits is also favored by the mode of recharge (from an adjacent nonsoluble formation) and by overall hydrogeologic confinement which prevents accelerated development of successful flow paths (Klimchouk, 1994, 1997a. See also Chapter 5.2 in this volume for a discussion of transverse speleogenesis).

It is likely that in some situations upward propagation of sizeable solution forms, initiated as enlarged sections at the input ends of fissures, occurs due to convection. The contribution of convection effects can be pronounced, however, only where a lower aquifer contains groundwater rather low in total dissolved solids, as compared to a near-saturated sulfate solution. Speleogenesis in this situation is perfectly illustrated by the great gypsum mazes in the western Ukraine (see Chapter 5.2.1 in this volume), and it can serve as a model for the explanation of the origin of many other caves, not necessarily in gypsum.

Where prominent tectonic fissures or faults cross an otherwise low-fissured gypsum bed, single transverse conduits or cavities may initiate easily if the initial widths of these flow paths are large enough to support a through penetration of still undersaturated solution. Even when transverse flow is slow due to sluggish hydrodynamic conditions, cavities can develop due to localized enlargement at the input end and dissolution due to convection. This case is best illustrated by caves of the *schlotten* type in south Harz, Germany (Kempe, 1996; Klimchouk, 1996b) and by vertical collapse structures (*breccia pipes*) initiated in some gypsum beds and propagated by upward stoping through the overlying sequences, examples of which are abundant in many regions of the world.

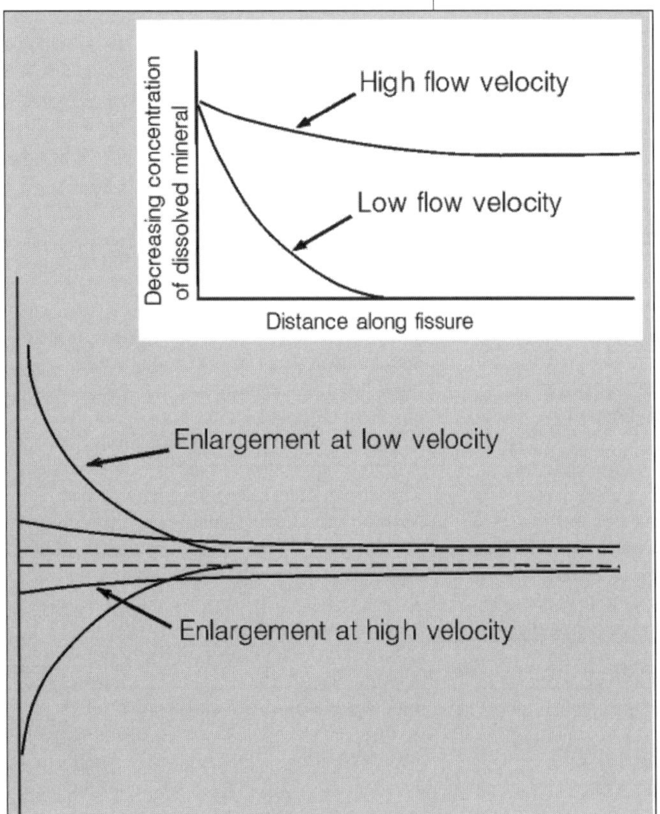

7.1 Figure 1. Effect of flow velocity on configuration of fissure enlargement in gypsum (after James and Lupton, 1978).

Flow in an Insoluble but Permeable Aquifer Underlying a Gypsum Bed—This situation is depicted in 5.2 Figure 4C. If the underlying aquifer is permeable enough, then groundwater can flow along the bottom of the gypsum bed and dissolve it. The dissolved load can be removed from the lithologic interface by diffusion or convection, then flow out with the regional flow.

As the flow path enlarges, the flow velocity increases at the interface between the beds, hence increasing the rate of gypsum dissolution. For a more detailed discussion of this situation see Chapter 5.2. Sizeable conduits or voids likely can develop in this way.

Thus deep-seated gypsum speleogenesis is most likely to occur in stratified sequences where gypsum beds are underlain by or sandwiched between insoluble or comparatively low-solubility aquifers. Initiation of long lateral flow paths in gypsum is prevented by the rapid dissolution kinetics. Thick and low-fissured sulfate strata, especially where surrounded by poorly permeable beds, can survive burial with virtually no speleogenesis. When exposed to the surface, such gypsum deposits undergo speleogenetic development without any inheritance, representing an unmixed pure line of open karst.

Cave Development in Exposed Gypsum—Phreatic, Water Table, and Vadose Conditions

After uplift and erosional entrenchment (subjacent karst setting), gypsum beds are progressively exposed at the surface (exposed karst setting). Conditions for cave development in these new settings depend greatly on the degree of previous speleogenesis. If considerable conduit porosity has already been created in a deep-seated setting, it provides ready paths for groundwater flow. Cave development continues, being greatly intensified because of unconstrained discharge and increased flow velocity.

Enlargement rates locally can be very high, so that flow and preferential cave development readily adjust to the new geomorphic setting and flow configuration. When incising valleys lower a drainage basin to positions within or below the gypsum strata, low-gradient water table or perched (fully drained) conditions are established almost instantly because of the great conduit permeability. Many regions such as the Pre-Urals and Belomorsko-Kulojsky in Russia, and Podol'sky in the western Ukraine, exhibit good examples of cave development under phreatic, water-tab e, and perched conditions that have been superimposed on relict cave systems of artesian origin.

Where cave development occurs in unconfined settings with no major inheritance from previous stages, it responds strongly to patterns of geologic structure and recharge. Because of the kinetic limitations on the initiation of tight flow paths, only those fissures that are originally wide enough to maintain undersaturated flow all the way through can be progressively enlarged. Wide fissures (up to few centimeters) in gypsum readily form, and any inherited structural weaknesses and closed fissures can open up, due to unloading and differential tectonic movements, where relatively thin but areally extensive beds are brought to the surface and are locally entrenched. Thick masses of gypsum, such as those in Sicily or at Emilia-Romagna in Italy, are normally disrupted by only a few localized but prominent fissures and faults, or they display closer-spaced open fissuring due to unloading close to escarpments.

Enlargement rates depend strongly on discharge, flow velocity, and flow regime (laminar vs. turbulent). This dependence is much stronger for gypsum than for limestone, because of the diffusion control of dissolution. Variations in fissure width are even more crucial in determining enlargement rates among alternate flow paths in gypsum than in limestone (7.1 Figure 2). Conduits able to capture more discharge grow in a sharply accelerated manner under phreatic conditions, where the outflow is not constrained. The rapid dissolution kinetics of gypsum causes "runaway" development, so that competition between alternative flow paths is stronger in gypsum that in limestone. This is the main reason for linear cave patterns with no or poorly developed branching being overwhelmingly predominant in unconfined speleogenesis in gypsum (see below).

The development stage for unconfined gypsum speleogenesis is rather short. Passages quickly develop to accommodate the highest possible discharge, and water table or vadose conditions are established. The same reason accounts for the fact that floods due to passage constrictions rarely occur in gypsum caves. Caves in gypsum rapidly adjust to changing base levels.

With the onset of the entrenched karst stage, or in open karst, vadose conditions become increasingly predominant, with the possible continued existence of water-table and phreatic zones in the lower parts of massifs. Rapid enlargement of inherited artesian or newly formed phreatic conduits occurs at the water table, particularly if annual fluctuations of adjacent surface river levels cause periodic backflooding into a cave. Under more stable conditions, such as in the interiors of massifs, extensive horizontal notching may develop in passages at the level of the water table, promoted by density stratification in the water (see below).

In the vadose zone, cave development concentrates along vertical percolation paths and free stream courses, and it is locally very active. As vadose percolation and free-flowing water in gypsum quickly reaches near saturation, most of the dissolution occurs where water first enters the gypsum, or is focused along stream beds during high-flow events. More dispersed dissolutional enlargement may occur through the action of condensation water, but this effect tends to be localized in certain zones (see below).

These characteristics of gypsum speleogenesis cause extreme heterogeneity and anisotropy of karst permeability in unconfined gypsum aquifers where little or no previous karstification has occurred (open karst). Transmissivity is normally high, and storativity is low. No substantial flow occurs at great depth below the water table, as deep-phreatic speleogenesis is normally subdued in gypsum (Forti, 1993; Klimchouk, 1996b). The typical characteristics of unconfined gypsum speleogenesis and hydrology are best exemplified by the gypsum karsts of Sicily and Emilia-Romagna in Italy and Sorbas in Spain.

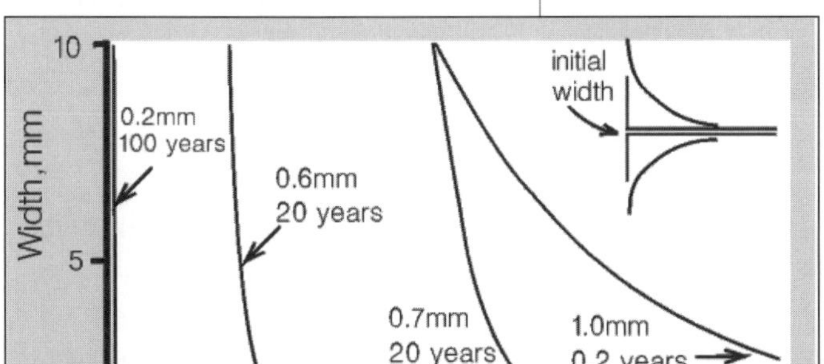

7.1 Figure 2. Penetration distances, or progress of the dissolution front, for massive gypsum, calculated for initial fissure widths ranging from 0.21 to 1.0 mm. Time elapsed since initiation is in years. The hydraulic gradient is 0.2 and water temperature is 10°C. Inset: the form of the dissolution taper into the fissure obtained from theoretical calculations such as these (after James and Lupton, 1978, as adapted by Ford and Williams, 1989).

Speleogenetic Effects of Water-Density Differences

Because dissolution always leads to an increase in the solvent density, gravitational separation of water, and natural convection due to this effect, are inherently involved in cave development. Convection circulation may become noticeable under sluggish forced-flow regimes (Curl, 1966). The relevant dissolutional effects may be significant in limestone (Curl, 1966) although they are much more pronounced in gypsum (Kempe, 1972, 1975; Klimchouk, 1996b, 1997b) and in salt (Frumkin, 1994; Chapter 7.2), due to the higher solubility of these rocks. A necessary condition is continuous or periodic recharge to the soluble rock by fresh water of low dissolved-solids content compared to water near saturation with respect to gypsum.

Dissolution due to convection can be important in cave initiation and development under confined conditions where recharge to gypsum by low-density fresh water occurs from below, such as in the western Ukraine (see Chapter 5.2.1) or in the south Hartz (Kempe, 1996). When abrupt enlargement is created at a fissure input at the base of the gypsum bed, dissolution is further promoted by convection circulation. After dissolving gypsum and increasing in density, the water returns downward into the under-lying aquifer and joins the regional outflow. It is quite possible that, under these conditions, dissolution driven by natural convection contributes even more to the upward propagation of fissure enlargement than does the "penetration distance" mechanism driven by forced seepage.

This view is supported by the common occurrence in many artesian caves of blind cupolas and domepits up to 10 to 15 m high, with very tight fissures visible at their apices. In this way, vertical hydraulic connectivity is promoted between fissures arranged at different levels within a gypsum sequence, so that the effect facilitates the build-up of 3-D cave patterns. Domes of this type, rising from the top of the underlying bed and remaining seemingly isolated from any integrated cave system, were encountered by gypsum quarries in the western Ukraine.

In integrated cave systems, where there is a connection with the upper aquifer, directed (un-looped) buoyant currents operate. This is because less-dense water tends to occupy the uppermost available space. This may account for the formation of at least some keyhole sections and ceiling half-tubes under artesian conditions (Klimchouk, 1997b; see 5.2.1, Figure 9, p. 269).

In the case of large spaces that provide direct hydraulic connections between adjacent aquifers, downward convection is probably the main mechanism of lateral wall retreat (see 5.2 Figure 4B). Also, natural convection due to density differences can likely be important to the initiation and development of cavities along the contact between gypsum and underlying aquifers, even where there are no guiding fissures in the gypsum (see 5.2 Figure 4C).

Under shallow phreatic conditions, characteristic tip-down triangular cross sections develop, with flat ceilings (*Laugdecke* in German), combined with inclined facets. They are quite common in many gypsum caves in Germany, the western Ukraine, the Urals, Siberia, and elsewhere, and have been studied in detail and modeled theoretically by Kempe (1972) and Kempe (1975). They are formed by dissolution in the uppermost aggressive layer of water where there are small up- and downwelling convection cells ("salt-fingers") due to small local density differences (see 5.2.1 Figure 8, p. 268). Inclined facets are formed due to conduit-scale convection circulation, where films of water move downward along the walls with a progressive decrease in solutional potential.

Horizontal notching caused by chemical stratification of water, with the highest dissolution rates in the uppermost layer, may be a common effect in caves within all major soluble rocks (Ford and Williams, 1989; see also Chapter 6.1). It is, however, best displayed in salt (Frumkin, 1994) and in gypsum (Klimchouk and Aksem, 1988; Klimchouk, 1997b).

Speleogenetic Effects of Condensation Dissolution

In the aerated zone of highly karstified entrenched or exposed massifs, condensation can make a significant contribution to groundwater recharge. The role of condensation in karst hydrogeology and speleogenesis has been well studied in the Soviet Union, where Lukin (1962, 1969), Dubljansky (1970), and Dubljansky and Sotzkova (1982) have discussed the theoretical issues and provided assessments and reviews of available field data, including some from gypsum karst areas. Cigna and Forti (1986), Forti (1991, 1993) and Calaforra et al. (1992, 1993) addressed the issue with particular regard to gypsum caves, but they were unaware of the previous Soviet work.

The amount of condensation water that can be formed in caves and fissures depends upon climate, the intensity of air exchange, and temperature differences between the outside and in-cave atmospheres. It is most pronounced during warm seasons, under temperate climatic conditions, and especially in semiarid zones. Water that condenses in transitional microclimatic zones in caves is very aggressive and causes substantial dissolution. The role of condensation corrosion in cave development is more important in gypsum than in carbonates, due to the high solubility and fast dissolution rates of gypsum.

Lukin (1969) estimated that every cubic meter of air leaves 10 g of water condensed on rock surfaces while passing through Kungurskaya Cave (in the Pre-Urals) during the warm season. Calaforra et al. (1993) suggested that in the Cueva del Agua area of the semiarid karst of Sorbas all of the perennial base flow (about 1 L/s, comprising 25 percent of the total discharge of the aquifer) is provided by condensation inside the cave. Forti (1993) estimated that condensation accounts for more than 60 percent of the recharge of the karst aquifer that contains the Cueva del Leon in Argentina.

Estimates by Dubljansky for different regions give warm season rates of condensation-generated flow of 1.4 to 9.7 L/ km^2, comprising from 5.9 to 85 percent of the total recharge (precipitation minus evapotranspiration; see also Chapter 6.3). Klimchouk et al. (1996a) summarized the available data on rates of gypsum dissolution caused by condensation in caves of the western Ukraine and the south of Spain. These range between 0 and 0.13 mm/yr (as measured by the tablet method) and between 0 and 0.45 mm/yr (as measured by microerosion meter). Thus dissolution due to condensation can be a notably active agent of cave development, but only at a local scale, within transitional cave microclimatic zones.

Caves in Gypsum

Occurrence of Caves in Different Types of Gypsum Karst

Currently there are perhaps several thousand gypsum caves known from around the world. A recent review of gypsum karst regions of the world is provided in Klimchouk et al. (1996b). Detailed characteristics of gypsum karst are contained in two recent theme volumes (Klimchouk, Lowe, Cooper, and Sauro, 1996; *Carbonates and Evaporites*, v. 12, no. 1, 1997). In terms of the karst classification adopted here (see Klimchouk and Ford, Chapter 3.1 in this volume), most caves that can be directly explored are in exposed and intrastratal entrenched karst. Caves are rarely accessible in intrastratal subjacent karst and are almost never accessible in deep-seated intrastratal karst, although they are commonly encountered by indirect means (e.g. boreholes). The latter two karst types have by far the greatest areal extent. It can be assumed that known caves represent only a very small portion of all the karst conduits and voids that occur within the upper few hundred meters of the geologic sequence in gypsum karst areas.

The world's largest currently known gypsum caves are listed in 7.1 Table 1. Optimisticheskaja, the longest cave, is the second longest cave of any type known in the world. A striking gap exists between the lengths of the three longest caves and the other caves in the list. The five longest gypsum caves, all located in the western Ukraine, account for well over half of the total known length of gypsum caves. This apparent bias is related partly to the unique artesian setting with a clastic cover that protects the huge mazes. For details see Klimchouk, Chapter 5.2.1 in this volume.

It is far more common, however, for artesian (intrastratal) caves in gypsum to be partly destroyed by later entrenchment and denudation. In addition, all gypsum caves are more easily destroyed by surface exposure than are carbonate caves. This is not only because gypsum is more soluble, but also because of its lower mechanical strength. This accounts for the relative scarcity of 25- to 80-km-long gypsum caves and the generally more modest sizes of the largest chambers and passages in gypsum as compared to those in limestone.

Gypsum caves are commonly found in present-day deep-seated artesian conditions in many intrastratal karst settings. Boreholes and mines have intersected large voids at depths below local base level of 60 to100 m in the "artesian belt" of the gypsum karst in the western Ukraine (see Chapter 5.2.1), of 300 to 400 m or even deeper in the Pre-Urals, the Caspian depression, and the southern Hartz region. Deep dissolution of gypsum is also shown by the presence of collapse forms that have evolved along vertical structures where gypsum lies at depths of several hundred meters (Klimchouk and Andrejchouk, 1996b), and by the widespread occurrence of collapse-breccia horizons, which are commonly the ultimate result of speleogenesis in sulfate beds sandwiched within nonsulfate formations.

Caves formed near the surface are mainly linear or crudely dendritic and commonly up to a few kilometers in length. They may intersect relict gypsum caves. The deepest gypsum cave currently known (Tunel dels Sumidors, Valencia, Spain; Calaforra and Pulido-Bosch, 1996) is only 210 m deep, far shallower than the deepest caves in carbonate karst. The main reasons are geologic. In folded mountainous regions, where the potential drained depth is greatest, gypsum formations are fragmented and do not favor the development of such vertically extensive sequences as do carbonates. Again, a lower mechanical strength and homogeneity of gypsum formations restricts possibilities for deep caves to develop and survive.

Morphology of Caves

Gypsum caves vary greatly in morphology. Several typical patterns can be distinguished:

- Discrete comparatively large voids of irregular often isometric shapes.
- Rectilinear or ramifying mazes in which multiple stories may complicate the structure.
- Caves that are linear or crudely dendritic in plan, with horizontal, inclined, or step-like profiles, and in which multiple storys may complicate the structure,
- Vertical pits and pipes.

Development of these cave types is related to the particular speleogenetic environment and karst type (7.1 Table 2, p. 438). This list does

Longest gypsum caves of the world (1996)			
Name	Length km	Country	Rock Age
Optimisticheskaja	203.00	Ukraine	Neogene
Ozernaja	117.00	Ukraine	Neogene
Zolushka	92.00	Ukraine	Neogene
Mlynki	25.00	Ukraine	Neogene
Kristalnaya	22.00	Ukraine	Neogene
Kulogorskaja-1-2 - Troja	14.10	Russia	Permian
Jester	11.80	USA	Permian
Spipola-Aquafredda	10.50	Italy	Neogene
Slavka	9.10	Ukraine	Neogene
Agua, cueva de	8.35	Spain	Neogene
Verteba	7.82	Ukraine	Neogene
Cater Magara	7.30	Syria	Neogene
Park's Ranch	6.27	USA	Permian
Konstitutzionnaja	5.88	Russia	Permian
Kungurskaya Ledjanaya	5.60	Russia	Permian
Olimpijskaja	5.50	Russia	Permian
Kumichevskaja	5.00	Russia	Permian
Zolotoj Kljutchik	4.38	Russia	Permian
Covadura	4.24	Spain	Neogene
Crystal Caverns	3.81	USA	Permian
Double Barrel	3.72	USA	Permian
Scrooge	3.70	USA	Permian
Umm al Masabih	3.59	Libya	Jurassic
Leningradskaja	3.40	Russia	Permian
Simfonia	3.24	Russia	Permian
Pedro Fernandes	3.21	Spain	Neogene
Pekhorovskaja	3.18	Russia	Permian
Martin	3.15	USA	Permian
Lomonosovskaja	3.13	Russia	Permian
Carcass	2.92	USA	Permian
Vodnaja	2.90	Russia	Permian
Pinezhskaja Terehchenko	2.60	Russia	Permian
Atlantida	2.53	Ukraine	Neogene
Ingh. Ca' Siere	2.50	Italy	Neogene
Wimmelburger Schlotte	2.84	Germany	Permian
Eras'kina 1-2	2.50	Russia	Permian
10-years LSS	2.45	Russia	Permian
Bukovinka	2.41	Ukraine	Neogene
Hyaenlabyrint	2.31	Somalia	Paleogene
Severjanka	2.30	Russia	Permian
Pekhorovskij Proval	2.27	Russia	Permian
Kulogorskaja-5	2.20	Russia	Permian
Geograficheskogo ob-va	2.15	Russia	Permian
Ugryn'	2.12	Ukraine	Neogene
Re Tiberio	2.11	Italy	Neogene
Michele Gortani	2.01	Italy	Neogene
Gostry Govdy	2.00	Ukraine	Neogene

The deepest gypsum caves of the word			
Name	Depth m	Country	Rock age
Tunel dels Sumidors	210	Spain	Triassic
Pozzo A	>200	Italy	Triassic?
Corall, sima del	130	Spain	Neogene
Triple Engle Pit	130	USA	Permian
Covadura	126	Spain	Neogene
Campamento, sima del	122	Spain	Neogene
Aguila, sima del	112	Spain	Triassic
Rio Stella-Rio Basino	100	Italy	Neogene
AB 6	100	Russia	Jurassic

Note: The list has been compiled using data provided by Belski (for USA), Calaforra (for Spain), Forti & Sauro (for Italy), Kempe (for Germany), Klimchouk (for Ukraine), Woigt & Schnadwinkel (for Syria), Malkov & Lavrov (for Russia).

7.1 Table 1. The world's longest (>2000 m) and deepest (>100 m) gypsum caves (as of 1996).

not include cavities formed by differential deformation of layers due to recrystallization (*tumuli*), or gravitational-tectonic caves that are formed, for example, by unloading along escarpments. The major types of gypsum caves are described briefly below.

Discrete Voids—Caves of this type commonly develop under artesian conditions, where the gypsum is underlain by an aquifer, and its own initial permeability is either very inhomogeneous, determined by discrete major tectonic fissures, or is negligible. Their origin and development mechanisms are described above. The best-documented examples are caves in the Sangerhausen and Mansfeld Districts of Germany, encountered through the centuries in the course of mining operations at depths up to 400 m at the base of the Zechstein Gypsum (Kempe, 1996). They are large voids, commonly isometric, or elongated along the major tectonic fissures like the Wimmelburger Schlotten. About 100 cavities of this type are known in the region. Natural convection because of density differences is believed to play an important role in the development of such cavities (Kempe, 1966).

Breakdown of large voids formed in this way is probably the main trigger for the development of vertical through structures (VTS). The latter is the generic term suggested for features that include breccia pipes and collapse columns, which are common in deep-seated intrastratal karsts (Klimchouk and Andrejchouk, 1996b).

Maze Caves—Maze caves constitute 21 percent of the 197 largest recorded gypsum caves in the world, but their proportion increases to 41 percent among gypsum caves over 1000 m long, and to 54 percent over 2000 m long. Considering the length of surveyed passages rather than the number of caves, just the five greatest mazes of the western Ukraine comprise far more then half of all known gypsum passages in the world. Accepting that most if not all known mazes in gypsum are formed according to the mechanism of transverse speleogenesis under a confined setting, the above statistics indirectly reflect limitations imposed by the rapid dissolution kinetics upon the initiation and development of long flow paths in gypsum within the conventional scheme of lateral speleogenesis. In the case of maze caves formed by transverse flow, the total passage length of caves has nothing to do with the actual distances of flow (see Chapter 5.2 for details).

Palmer (1975) distinguished two major situations favorable for maze-cave development:

- Where dispersed aggressive recharge takes place uniformly into all available fissures in a soluble rock unit, entering from an overlying insoluble but permeable

TYPE OF KARST	SPELEOGENETIC SETTINGS			CHARACTERISTICS OF CAVES
	Hydro-geological conditions principal complementary	**Initial permeability (before speleo-genesis)**	**Flow pattern through gypsum and type of recharge**	
Intrastratal deep-seated	**confined (artesian)**	Very inhomo-geneous, generally low to negligible, locally high	1. Ascending flow, localized basal inputs	1. Discrete voids, commonly large and isometric; associated stoping cavities in coverbeds at the top of VTS after breakdowns
		Fairly homogeneous generally low	2. Ascending flow with possible lateral component, dispersed basal inputs	2. Rectilinear 2-D or 3-D (multistory) mazes
Intrastratal subjacent	**confined phreatic water table** vadose	Heterogeneous: low to high	1. Ascending flow with possible considerable lateral component, localized or dispersed basal inputs 2. Descending flow with considerable lateral component, localized inputs from coverbeds and via superficial sink points; possible backflooding	Continuing development of types 1 and 2 3. "Through caves": linear or crudely dendritic in plan, horizontal, inclined, or step-like in profile Enlargement of inherited caves at the water table
Intrastratal entrenched	phreatic **water table vadose**	Heterogeneous: low to high	1. Descending flow with possible considerable lateral component, localized inputs from coverbeds and via superficial sink points; possible backflooding	Continuing or newly started development of type 3 caves 4. Vertical pipes developing downwards from the top of the gypsum
Exposed denuded	phreatic **water table vadose**	Heterogeneous: generally high	1. Descending flow with possible considerable lateral component, localized inputs from coverbeds and via superficial sink points; possible backflooding	Continuing or newly started development of type 3 caves
Exposed open	phreatic **water table vadose**	Heterogeneous: generally low	1. Descending flow with possible considerable lateral component, localized inputs from coverbeds and via superficial sink points; possible backflooding	Development of type 3 caves

7.1 Table 2. Genetic classification of caves in gypsum.

formation, or (less commonly) from an underlying formation.

- Where floodwater recharge to stream caves causes variations in discharge and head in an evolving system to be so great that no fixed passage configuration is allowed to stabilize with respect to flow.

However, I (Klimchouk, 1992, 1994, 1997a; Chapter 5.2) stressed the importance of maze-cave development under confined conditions by upward re-charge into a soluble bed from the basal aquifer in a multistory artesian system. In this way the transverse flow pattern in and through a soluble bed is believed to be responsible for maze development. I know of no unambiguous example of a cave in gypsum that could fit the downward infiltration emphasized by Palmer, although this setting probably could produce mazes in gypsum where transverse flow takes place. The second (floodwater) environment does not produce maze caves in gypsum, because stream caves rapidly adjust their sizes to the highest possible discharge, so that no constrained floods occur. The great majority of maze caves in gypsum are artesian transverse mazes.

For typical rectilinear maze caves to develop in gypsum, not only the artesian flow architecture favoring cross-formational upward hydraulic communication is required, but also a fairly uniform, though not necessarily high, initial fissure permeability within a gypsum bed. Otherwise discrete voids will tend to form rather than maze caves.

Artesian maze caves commonly display a multistory structure. Stories or tiers are not related to cycles of stability and uplift (base-level control) but are controlled stratigraphically. They develop within beds where initial fissures form networks that are well connected laterally but not vertically. It is particularly common for lithogenetic fissures to form largely independent networks confined within certain horizons, with only local connections to adjacent fissured horizons. The most outstanding examples of mazes in gypsum are the great caves of the western Ukraine (see Chapter 5.2.1 for characteristics and maps). They are also known in the Belomorsko-Kulojsky, Povolzhje, Pre-Urals, and some Siberian regions of Russia, in Germany, in France (Paris Basin), and in Spain. Caves of this type are undoubtedly very common in many other regions of intrastratal karst, but favorable conditions are required for them to survive uplift and denudation so as to become accessible and be explored.

Through Caves—Within exposed gypsum karst, or intrastratal entrenched karst, the most common caves developed in adjustment with the contemporary geomorphic configuration (i.e. not inherited from previous stages) are linear or crudely dendritic caves that directly connect sink points to resurgence points. They are collectively termed *through caves* on this account. As shown above, the "runaway" development and competition of alternative flow paths under unconfined conditions is exaggerated in gypsum, so that normally only one passage develops between input and output points (7.1 Figure 3, p. 440). Where there are multiple sink points, a dendritic pattern may develop, as minor flow paths ultimately connect to the nearest major successful conduit that serves as a drain (7.1 Figure 4, p. 441). Speleogenesis in gypsum under unconfined conditions creates extremely anisotropic permeability, with rather simple and strongly hierarchical networks. Forti (1993) claimed the latter to be the principal characteristic of speleogenesis in gypsum, referring to confined speleogenetic environments as rare special cases. The actual situation can be said to be the opposite, considering that exposed or entrenched gypsum karst, with no inherited caves at all, comprise only a minor part of the gypsum formations undergoing karstification (see also Chapter 4.1).

Because of the fast development stage, through caves in gypsum adjust rapidly to the present base level. They also commonly develop along intercalated insoluble and poorly permeable layers (if compared with now-karstified gypsum), or along the top of the basal formation, perched within the vadose zone. Perched streams sink downward into the nearest major tectonic fissure, forming vadose pits that connect different levels (7.1 Figure 5A, p. 441). Intercalations in gypsum sequences play an important role in the early development of conduits under phreatic (semiconfined) conditions. During the vadose stage, erosion of insoluble passage floors may become the predominant mechanism of their further development. This feature is well illustrated in the caves of Sorbas in Spain (see Calaforra and Pulido-Bosch, Chapter 5.3.10 in this volume).

Through caves are common in almost every entrenched and denuded gypsum karst area. The most representative and best-documented examples are in the Emilia-Romagna and Sicily regions of Italy, the Belomorsko-Kulojsky, Pre-Urals, and North Caucasus regions of Russia, at Sorbas in Spain, and in New Mexico and Oklahoma in the United States.

Vertical Pipes and Pits—Vertical dissolution pipes, also known as organ pipes, or *komins* in the Russian literature, are very common in entrenched intrastratal gypsum karst. They develop downward from a suitably protective bed at the top of the gypsum (commonly limestone or dolomite), due to focused dissolution by groundwater that percolates through the overburden, or leaks from perched aquifers above the gypsum (7.1 Figure 6, p. 441). Pipes cut across the whole gypsum stratum, or down to the water table, commonly intersecting relict lateral caves. Their density within a given area depends mainly upon the abundance of percolating trickles in the cover beds, and can be very high in some places, perhaps up to several hundred per km^2. Pipe diameter depends upon the amount of percolating water.

New pipes develop quite rapidly, reaching a diameter of 1 to 3 m before their growth rate slows down. Rapid growth of new pipes is evidenced by an example 1 m in diameter in Zolushka Cave in the western Ukraine, which developed in only 35 years after a borehole drilled from the surface created a new leakage point from the perched aquifer above through an intervening clay bed. Dissolution pipes commonly induce breakdown and vertical through structures (VTS) in cover beds (Klimchouk and Andrejchouk, 1996b). They also provide foci for doline development where cover beds are eroded away.

In areas where substantial local relief and a thick vadose zone are developed in gypsum, vertical pits readily form up to 30 to 40 m deep. They are exaggerated analogies of pipes, formed where there is focused flow at the surface or along an intercalated insoluble layer intercepted by prominent tectonic fissures (7.1 Figure 5A).

Conclusions

Specific features of speleogenesis in gypsum, compared with those in carbonate rocks, are determined by the geologic characteristics of the respective rock formations, and by the chemistry and kinetics of their dissolution. Differences in solution kinetics between gypsum and calcite, namely much higher solubility and solely diffusion-control of gypsum dissolution, with no change in the kinetic order, impose certain limitations and distinctive features in the early evolution of conduits. These characteristics seem to be an extreme and more obvious control of speleogenetic development. Conduit initiation cannot occur in gypsum in deep phreatic (unconfined) settings, and cave initiation by lateral flow through gypsum from distant recharge areas in artesian settings is impossible.

In shallow unconfined settings, speleogenesis occurs along fissures wide enough to support undersaturated flow through them. Cave development in different passages takes place at highly disparate rates because of the rapid dissolution kinetics of gypsum and strong dependence of enlargement rates on flow velocity and discharge. Mainly linear or crudely branching caves form, which rapidly

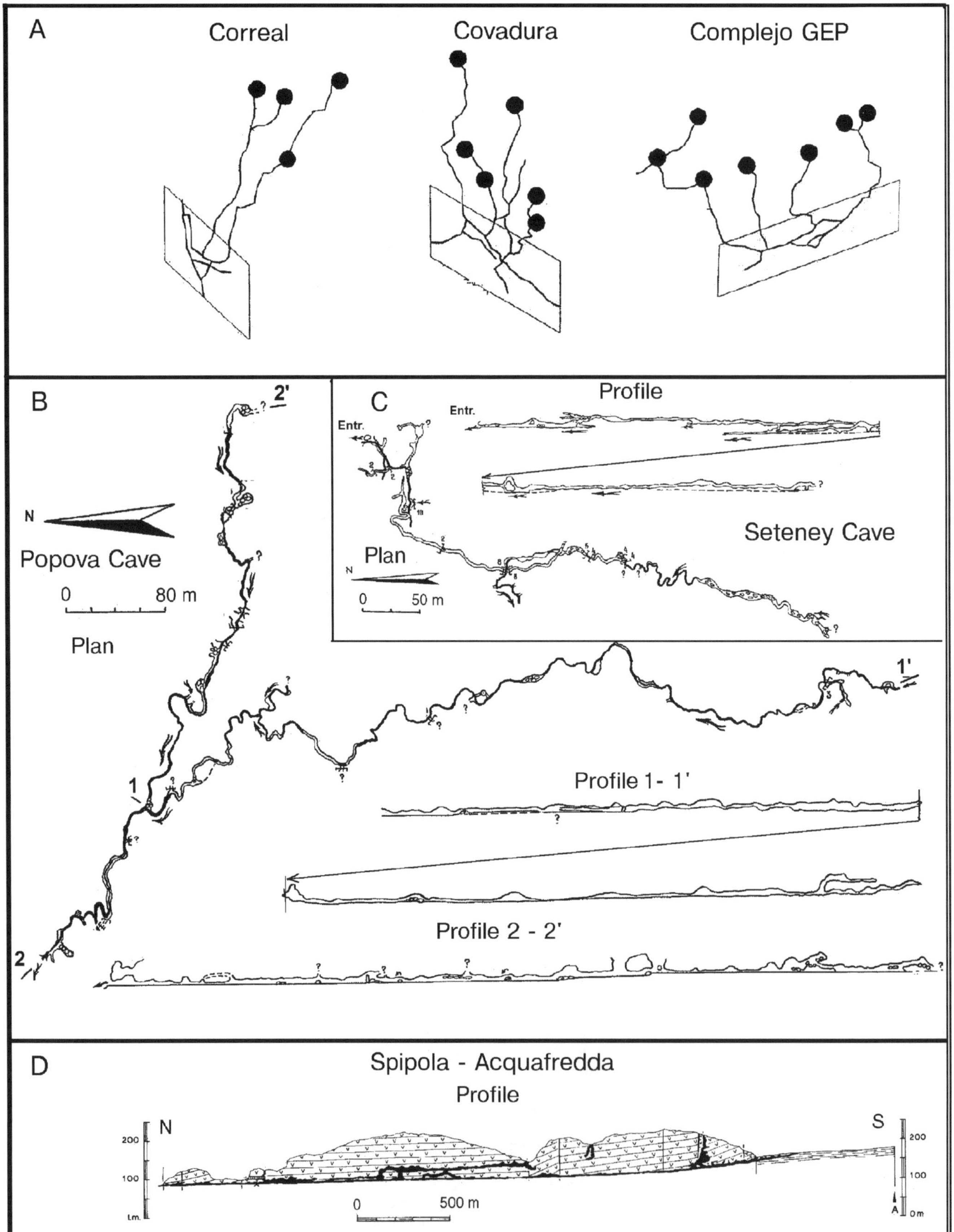

7.1 Figure 3. Generalized sketches, plans, and profiles of some gypsum "through caves:" ***A*** *is Sorbas, Spain, where feeding dolines are indicated by dots (after Calaforra, 1996).* ***B*** *and* ***C*** *are the Ekeptze-Gadyk Massif, North Caucasus, Russia (after Ostapenko, 1994).* ***D*** *is the Spipola-Acquafredda System in the Emilia Romagna, Italy (after Grimandi, 1987).*

adjust to the contemporary geomorphic setting and to the maximum available recharge. Vertical pipes or pits are also common in the vadose zone.

In deep-seated confined settings, cave origin and development occur where gypsum beds in stratified sequences are underlain by, or sandwiched between, insoluble or comparatively low-solubility aquifers. Two hydrogeologic situations supporting cave origin in gypsum in deep-seated settings are:

- Transverse flow through gypsum beds between adjacent aquifers from below to above.
- Lateral flow in an insoluble but permeable aquifer underlying a gypsum bed.

The former situation generates either maze caves where uniformly distributed fissure networks exist in the gypsum, or discrete voids where the otherwise poorly fissured gypsum is disrupted by prominent tectonic fractures. If considerable conduit porosity has been created in the deep-seated setting, it provides ready paths for more intense groundwater circulation and further cave development when uplift brings the gypsum into the shallow subsurface and unconfined setting.

The fact that maze caves readily form in gypsum under artesian conditions gives strong support to the general model of transverse speleogenesis and suggests an even more obvious applicability to artesian cave development in carbonates.

Thick and poorly fissured sulfate strata can survive burial with no speleogenesis at all where surrounded by low-permeability beds. While exposed to the surface, such gypsum deposits undergo speleogenetic development without any inheritance, presenting the "pure line" of barren exposed karst.

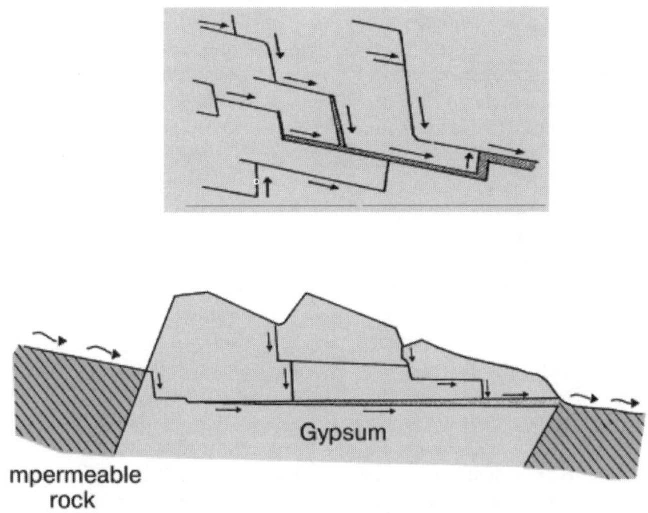

7.1 Figure 4. Sketch of a typical gypsum karst system developed under unconfined conditions, consisting of a principal drainage tube with few and short effluents (after Forti, 1993).

Acknowledgements

My thanks to Derek Ford, Arthur Palmer, and Margaret Palmer for their review of this paper and valuable comments. I am indebted to Arthur and Margaret Palmer for their careful and professional editing of my English, which helped to make the paper more clear and consistent.

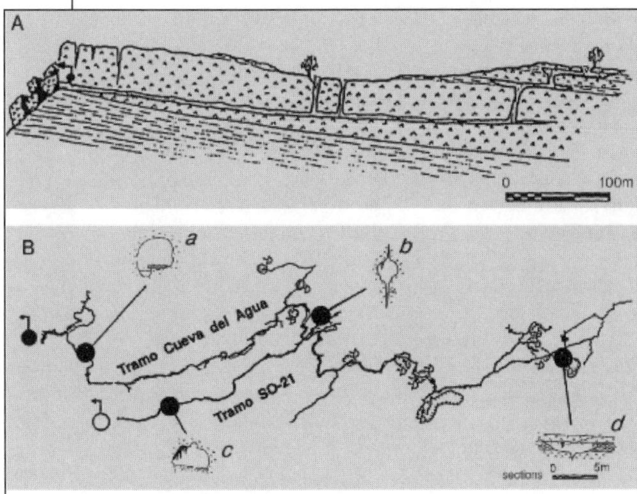

*7.1 Figure 5. Schematic profile **A** and plan **B** of the Cueva del Agua System, a typical "through cave" in the denuded exposed karst of Sorbas, Spain. Cross sections of passages are differentiated throughout the system. **a** is active passages in the downstream section. **b** is relict (abandoned) passages. **c** is phreatic passages with superimposed vadose canyons in the central section. **d** is vadose passages in the upstream section, where they developed by erosion of clastic intercalated layers (adapted from Calaforra, 1996).*

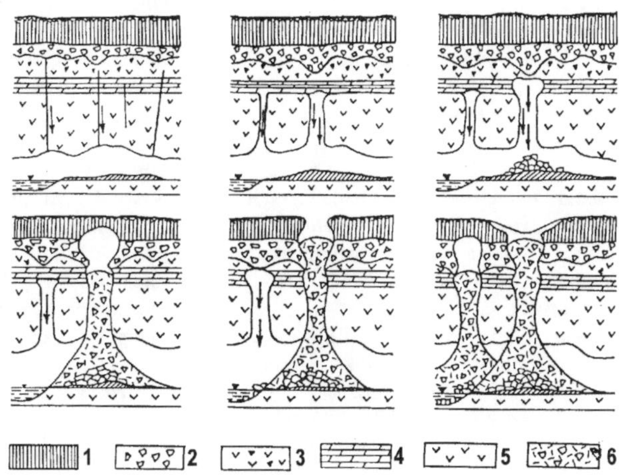

7.1 Figure 6. Successive stages of dissolution-pipe development in gypsum, and VTS (vertical through structure) development in the overburden, with the final appearance of a doline at the surface (based on the example of Kungurskaya Cave in the pre-Urals). 1 is silt, 2 is breakdown slabs, 3 is gypsum, 4 is dolomite, 5 is gypsum/anhydrite, 6 is mixed breakdown clasts forming the bulk of the VTS and talus pipes in the cave (after Klimchouk and Andrejchouk, 1996b).

References

Calaforra, J.M., 1996, *Contribucion al conocimiento de la karstologia de yesos:* Universidad de Granada, Ph.D. Thesis, 350 p.

Calaforra, J.M., Forti, P., and Pulido-Bosch, A., 1992, Nota preliminar sobre la influencia climatica en la evolucion espeleogenetica en yesos con especial referencia a los afloramientos karsticos de Sorbas (Espana) y de Emilia Romagna (Italia): *Espeleotemas*, v. 2, p. 9-18.

Calaforra, J.M., Dell'Aglio, A, and Forti, P., 1993, The role of condensation-corrosion in the development of gypsum karst: The case of the Cueva del Agua (Sorbas, Spain): *11th International Congress of Speleology, Beijing, Proceedings*, p. 63-66.

Calaforra, J.M., and Pulido-Bosch, A., 1996, Some examples of gypsum karsts and the more important gypsum caves in Spain, *in* Klimchouk, A., Lowe, D., Cooper, A., and Sauro, U., eds., Gypsum karst of the world: *International Journal of Speleology*, v. 25, no. 3-4, p. 225-237.

Carbonates and Evaporites, 1997, (Special issue containing papers presented at a Symposium on Evaporite Karst: Processes, Landforms, Examples, and Impacts), v. 12, no. 1.

Cigna, A., and Forti, P., 1986, The speleogenetic role of air flow caused by convection. 1st contribution: *International Journal of Speleology,* v. 15, p. 41-52.

Curl, R.L., 1966, Cave conduit enlargement by natural convection: *Cave Notes*, v. 8, p. 4-8.

Dreybrodt, W., 1996, Principles of early development of karst conduits under natural and man-made conditions revealed by mathematical analysis of numerical models: *Water Resources Research*, v. 32, p. 2923-2935.

Dubljansky, V.N., 1970, Condensation of moisture in the fissure-karst collectors of the Mountainous Crimea, Carpathians and Pridnestrovskaya Podolia. *Doklady Akadamia Nauk Uk. SSR*, ser. B, v. 1, p. 14-17 (in Russian).

Dubljansky, V.N., and Sotzkova, L.M., 1982, Microclimate of karst caves: *Zemlevedenie*, v. 11Y, p. 79-91 (in Russian).

Ford, D.C., and Williams, P.W., 1989, *Karst geomorphology and hydrology:* London, Unwin Hyman. 601 p.

Forti, P., 1991, Il carsismo nei gessi con particolare rigardo a quelli dell'Emilia Romagna: *Speleologia Emiliana*, v. 4, no. 2, p. 11-36.

Forti, P., 1993, Karst evolution and water circulation in gypsum formations: *International Symposium on Water Resources in Karst with Special Emphasis on Arid and Semi-Arid Zones*, Shiraz, Proceedings, p. 791-801.

Frumkin, A., 1994, Morphology and development of salt caves: *National Speleological Society Bulletin*, v. 56, p. 82-95.

Gorbunova, K.A., 1977, *Karst in gypsum of the USSR:* Perm University, 83 p. (in Russian).

Grimandi, P., 1987, Grotta della Spipola: *Ipoantropo*, v. 5, p. 51-64.

James, A.N., 1992, *Soluble materials in civil engineering:* Chichester, Ellis Horwood, 435 p.

James, A.N., and Lupton, A.R.R., 1978, Gypsum and anhydrite in foundations of hydraulic structures: *Geotechnique*, v. 28, p. 249-272.

Johnson, K.S., 1996, Gypsum karst in the United States, *in* Klimchouk, A., Lowe, D., Cooper, A., and Sauro, U., eds., Gypsum karst of the world: *International Journal of Speleology*, v. 25, no. 3-4, p. 183-193.

Kempe, S., 1972, Cave genesis in gypsum with particular reference to underwater conditions: *Cave Science*, v. 49, p. 1-6.

Kempe, S., 1975, "Facetten" and "Laugdecken", the typical morphological elements of caves developing in standing water: *Annales de Speleologie*, v. 4, p. 705-708.

Kempe, S., 1996, Gypsum karst of Germany, *in* Klimchouk, A., Lowe, D., Cooper, A., and Sauro, U., eds., Gypsum karst of the world: *International Journal of Speleology*, v. 25, no. 3-4, p. 209-224.

Klimchouk, A.B., 1992, Large gypsum caves in the western Ukraine and their genesis: *Cave Science,* v. 19, no. 1, p. 3-11.

Klimchouk, A.B., 1994, Speleogenesis under confined conditions, with recharge from adjacent formations: *Publ. Serv. Geol. Luxembourg*, Comptes Rendus du Coll. Int. de Karstol. a Luxembourg, v. 27, p. 85-95.

Klimchouk, A.B., 1996a, Dissolution and conversions of gypsum and anhydrite, *in* Klimchouk, A., Lowe, D., Cooper, A., and Sauro, U., eds., Gypsum karst of the world: *International Journal of Speleology*, v. 25, no. 3-4, p. 21-36.

Klimchouk, A.B., 1996b, Speleogenesis in gypsum, *in* Klimchouk, A., Lowe, D., Cooper, A., and Sauro, U., eds., Gypsum karst of the world, *International Journal of Speleology*, v. 25, no. 3-4, p. 61-82.

Klimchouk, A.B., 1997a, Artesian speleogenetic setting: *12th International Congress of Speleology*, La Chaux-de-Fonds, Proceedings, v. 1, p. 157-160.

Klimchouk, A.B., 1997b, Speleogenetic effects of water density differences: *12th International Congress of Speleology*, La Chaux-de-Fonds, Proceedings, v. 1, p. 161-164.

Klimchouk, A.B., Aksem, S.D., Shestopalov, V.N., and Rud'ko, G.I., 1988, *The study regime of gypsum karst activity in the western Ukraine:* Kiev, Inst. Geol. Nauk, 55 p. (in Russian).

Klimchouk, A.B., and Andrejchouk, V.N., 1996a, Sulphate rocks as an arena for karst development, *in* Klimchouk, A., Lowe, D., Cooper, A., and Sauro, U., eds., Gypsum karst of the world: *International Journal of Speleology*, v. 25, no. 3-4, p. 9-20.

Klimchouk, A.B., and Andrejchouk, V.N., 1996b, Breakdown development in cover beds, and landscape features induced by intrastratal gypsum karst, *in* Klimchouk, A., Lowe, D., Cooper, A., and Sauro, U., eds., Gypsum karst of the world: *International Journal of Speleology*, v. 25, no. 3-4, p. 127-144.

Klimchouk, A.B., et al., 1966a, Dissolution of gypsum from field observations, *in* Klimchouk, A., Lowe, D., Cooper, A., and Sauro, U., eds., Gypsum karst of the world: *International Journal of Speleology*, v. 25, no. 3-4, p. 37-48.

Klimchouk, A.B., et al., 1996b, Gypsum karst of the world, a brief overview. *in* Klimchouk, A., Lowe, D., Cooper, A., and Sauro, U., eds., Gypsum karst of the world: *International Journal of Speleology*, v. 25, no. 3-4, p. 159-181.

Klimchouk, A.B., Lowe, D., Cooper, A., and Sauro, U., eds., 1996, Gypsum karst of the world: *International Journal of Speleology*, v. 25, no. 3-4, 307 p.

Lukin, V.S., 1962, Air regime of karstified massifs, on the example of the Kungursky region, *in* Special'nye voprosy karstovedenija: Moscow, Akadamia Nauk SSSR, p. 58-59 (in Russian).

Lukin, V.S., 1969, Quantification of the processes of evaporation and condensation of vapor in gypsum-anhydrite massifs of the Ufimskoe Plateau: *Zemlevedenie*, v. 9, p. 213-218.

Ostapenko, A.A., 1993, Caves of the Ekeptze-Gadyk Range: *Svet (Light)*, v. 3, no. 9, p. 12-13 (in Russian).

Palmer, A.N., 1975, The origin of maze caves: *National Speleological Society Bulletin*, v. 37, p. 56-76.

Palmer, A.N., 1984, Geomorphic interpretation of karst features, *in* LaFleur, R.G., ed., *Groundwater as a geomorphic agent*: Boston, Allen and Unwin, p. 173-209.

Palmer, A.N., 1991, Origin and morphology of limestone caves: *Geological Society of America Bulletin*, v. 103, p. 1-21.

Woigt, S., and Schnadwinkel, M., 1995, Caving beneath the desert: Cater Magara: *International Caver*, v. 14, p. 15-26.

7.2. Speleogenesis in Salt—The Mount Sedom Area, Israel

Amos Frumkin

Abstract

Salt dissolution often occurs in deeply buried beds, where open caves are unknown. Caves are normally formed by selective dissolution along flow routes, rather than complete dissolution of the bulk salt mass. Most salt caves are found in diapirs, where open fissures drain meteoric water, and then rapidly enlarge to form vadose caves. Salt caves develop faster than other cave types, allowing their use as a natural laboratory for speleogenesis. Salt karst terrains exist mainly in arid climates where rock salt outcrops may escape complete destruction by dissolution. Known salt caves are mostly of Holocene age, whereas older ones are gradually destroyed by dissolution and collapse.

The Mount Sedom salt diapir, with some 20 km of salt caves, is the most studied area of salt karst. Its vadose caves are formed by captured ephemeral streams. Cave profiles are adjusted to base level, allowing reconstruction of the evolutionary history of the region. Some 57 percent of the Mount Sedom surface area is drained by the underground karst system. Water in cave conduits does not reach saturation during flood flow, unless the water is ponded for at least several hours.

Common cave features are vertical shafts close to the cave inlet and subhorizontal passages leading to outlets at base level. Where there is no fissure connection to the edge of the mountain, an inlet cave is formed, capable of absorbing the flood discharge in a terminal pond. Water and solutes escape from the pond by slow seepage through narrow fissures to a regional aquifer.

Speleogenesis in Salt

Rock salt has a few peculiar characteristics apart from its high solubility. The density of compact salt is low (about 2.2 g/cm^3) and it is subject to halokinesis—plastic creeping and deformation of salt under pressure. A thick rock salt sequence deeply buried under denser rocks gives rise to salt diapirs, which comprise much of the known salt outcrops. Although rock salt deposits are common, more than 99 percent of them do not crop out (Kozary et al., 1968). Salt outcrops are not uncommon in deserts (Edgell, 1996; Jackson et al., 1990), but in humid and subhumid regions the bulk of salt layers approaching the surface is commonly dissolved by groundwater down to the depth of active meteoric water circulation. This interstratal dissolution or subrosion process leaves behind only the less soluble constituents of the rock which often form a residual caprock at the top of rising diapirs (Quinlan et al., 1986). The caprock is thus formed below the water table of an aggressive aquifer. Interstratal dissolution is often accompanied by pronounced groundwater salinity, breccia pipes, anomalous dip (draping) of overlying strata, surface subsidence, closed depressions, and bedrock collapse structures (DeMille et al., 1964; Landes, 1961; Miotke, 1971).

Such features may originate at depths of more than 1000 meters below the surface (Ford, 1989). Solute transport at such depth can be driven by geopressure, Bernard-Rayleigh convection, and molecular diffusion (Bennet and Hannor, 1987; Ranganathan and Hanor, 1986). Interstratal karst *caves* are unknown, as they are probably destroyed prior to being observed in boreholes. Features attributed to dynamic dissolution and

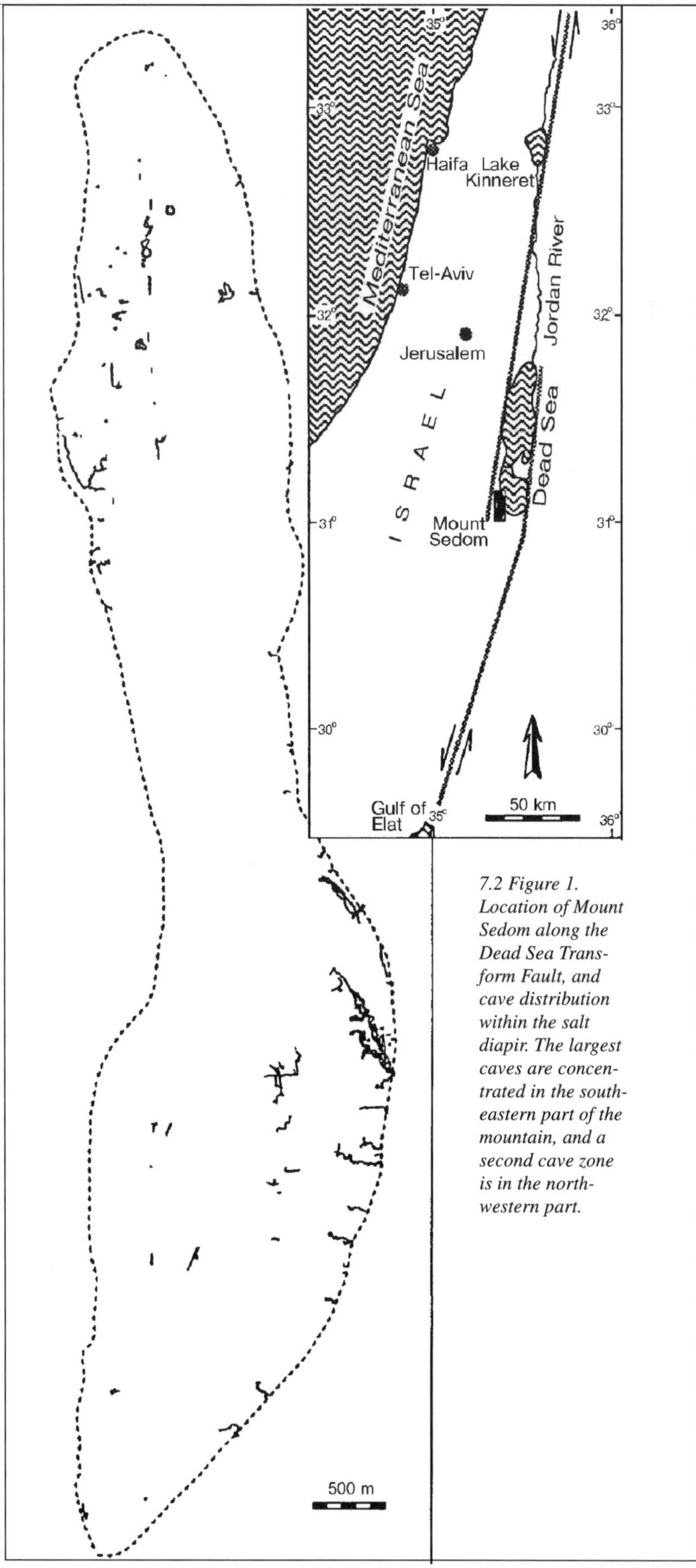

7.2 Figure 1. Location of Mount Sedom along the Dead Sea Transform Fault, and cave distribution within the salt diapir. The largest caves are concentrated in the southeastern part of the mountain, and a second cave zone is in the northwestern part.

recrystallization under heat and pressure at great depth have been observed in the Mount Sedom salt caves. Caves are normally formed due to selective salt dissolution along flow routes, rather than complete dissolution of the bulk salt mass.

Young salt deposits close to the surface may be highly permeable due to lack of cementation between individual halite grains (Yechieli, 1995). Water-filled shafts in recent salt beds are known from the Qinghai Plateau, China (Yuhua and Hua, 1986), and from the Great Plains, Canada (Last, 1993). These caves are apparently formed by ascending aggressive spring water flowing from nearby aquifers into exposed salt beds.

Syngenetic or epigenetic dissolution and precipitation of halite often cements individual halite grains. Following burial, rock salt becomes compact with a negligible porosity below a depth of 40 m (Casas and Lowenstein, 1989). Under such conditions joints and bedding planes may become the only possible initial flow routes. However, fissures in salt tend to anneal at depth by geopressure and halokinesis. When a diapir head rises above its surroundings, lithostatic pressure is released and fissures tend to open in the upper parts of the rock mass, especially near the diapir margins. The fissures may then drain aggressive water, rapidly developing cave passages.

The high solubility of salt allows speleogenesis to occur at a very short time scale, so salt caves are the fastest evolving karst caves (excluding ice cavities). Radiocarbon dates show that an entire cave or a single cave level may develop to accessible dimensions in a mere few hundred years (Frumkin and Ford, 1995; Zoltán, 1994), compared to more than 10,000 years in limestone (Mylroie and Carew, 1987). Human diversion of flow may also form salt caves in a few hundred years (Ford and Williams, 1989, p. 114).

Salt cave profiles evolve in a similar way to vadose limestone caves. The rapid rate of evolution and adjustment renders salt cave profiles to be useful for interpreting regional geomorphic evolution.

Salt brines are very dense, up to about 350 grams/liter. Stagnant cave water from various sources may develop density stratification forming typical dissolution notches, bevels, and facets (Frumkin, 1994b). Similar features are observed in gypsum caves (Pfeiffer and Hahn, 1972; Klimchouk, 1996, 1997) and more rarely in limestone caves (Jennings, 1985, p. 148; Ford and Williams, 1989, p. 307). The density gradient between aggressive and saturated water may reach about 30 percent in salt, about 0.2 percent in gypsum, but only about 0.03 percent in limestone. It favors the development of beveled notches in the more soluble rocks, where water is more likely to reach stable stratification, rather than in limestone.

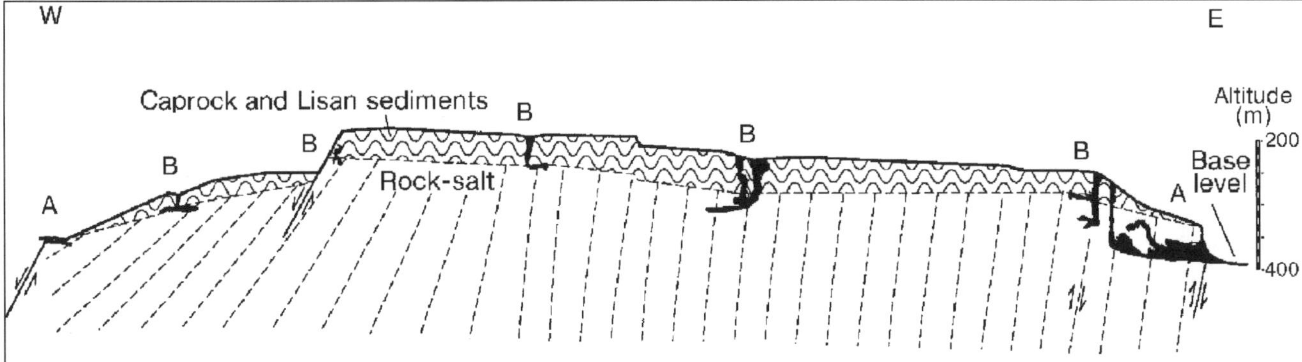

7.2 Figure 2. Geologic cross section of southern Mount Sedom, with Malham Cave–the largest known–in the east. ***A*** *indicates save systems with an open outlet.* ***B*** *indicates inlet caves (see text). Broken lines indicate bedding planes within the salt.*

Few true caves in salt were known until recently. Salt caves have been reported from the Persian Gulf (Shaw, 1979), Romania (Giurgiu, 1985; Ponta, 1986), Algeria (Chabert, 1989), Hungary (Zoltán, 1994), Chile (Salomon, 1995), the former USSR, and Spain. For a salt caving reports bibliography see Choppy (1988). The largest known assemblage of salt caves is found at Mount Sedom, Israel.

Speleogenesis in the Mount Sedom Area

Over 100 caves with more than 20 km of passages have been studied in the Mount Sedom salt diapir, including the largest known salt caves (Donini et al., 1985; Frumkin, 1986). The two longest are Malham Cave, 5.7 km long, and Sedom Cave, 1.8 km (Frumkin, 1994b). The caves are mostly of vadose origin, formed by ephemeral streams.

The high solubility and physical properties of rock salt, the arid climate, and the rapid base-level changes at Mount Sedom seem to be unparalleled in other karst terrains. Mount Sedom salt caves share many morphologic features with vadose allogenic limestone caves, and they share other features with alluvial streams in arid terrains. The range of salt cave types at Mount Sedom seems to include the common morphologic types found in other salt terrains, as well as types unknown from other localities. Each type is represented by enough examples to allow a classification and general discussion.

The salt caves of Mount Sedom formed under seemingly contradictory conditions. The uplift of a large salt mass above base level in an extremely arid climate has favored its preservation without complete dissolution. The high solubility and dissolution rate of salt, together with the allogenic nature of recharge, and the sporadic intense rains focus an immense dissolution energy along the flood routes. The long interval between successive floods allows enough time for some water to seep slowly toward base level. Local rock structure, hydraulic conductivity, and head determine the resulting cave patterns.

A time scale for cave development was obtained by two methods:

- Directly measured downcutting of a young passage from 1985 to 1995 was 23 cm (average 2.3 cm/year.
- Radiocarbon dating of driftwood carried by floods into the caves yielded average downcutting rates of 0.6 to 0.7 cm/year cm during the past 8000 years (Frumkin, 1996).

This long-term rate averages rapid downcutting stages as well as stagnant periods, in which incision practically ceased.

Downcutting variations depend on three groups of variables:

- Passage maturity in terms of reaching an equilibrium profile (see discussion below).
- Diapir uplift rate.
- Climatic variables, such as the Dead Sea level, flood intensity, and rain distribution.

Regional Setting and Geology

Mount Sedom is an exposed salt diapir that forms an elongated ridge 1.5 by 11 km in the Dead Sea Rift Valley (7.2 Figure 1). It rises from -400 m (below sea level) to -160 m, or 250 m above the Dead Sea base level (7.2 Figure 2). The Dead Sea Basin is filled by a thick sequence of detrital sediment and evaporites that propagate salt diapirs (Neev and Hall, 1979). Of these, Mount Sedom has risen to the highest altitude, bringing Neogene beds of marine rock salt high above base level. The average rate of diapir uplift is estimated at several mm/year (Frumkin, 1996).

Borehole evidence indicates that below base level fissures are annealed (closed) under pressure, due to the plastic properties of salt. Bedding planes are vertical or steeply inclined. They often yield under the shear stress induced by the rising diapir and become minor fault planes (Zak and Freund, 1980). Horizontal fissures are rarely found at Mount Sedom.

Before its subaerial extrusion, the top of the rising diapir suffered dissolution by groundwater. Residual relatively insoluble anhydrite, shale, and dolomite accumulated above the salt, forming a caprock that is up to 50 m thick (Zak and Bentor, 1968). The flat near-horizontal contact between the steeply inclined salt layers and the caprock is referred to as the "salt table" (Vroman, 1950). The top of Mount Sedom is approximately flat, with many small catchments up to 0.7 km^2 in area. The caves have formed during the Holocene. The oldest one dated to about 7000 conventional ^{14}C years (Frumkin et al., 1991). Mount Sedom caves thus are some of the youngest known.

Hydrology

The region is extremely arid today, with an average yearly rainfall of 50 mm. The annual precipitation-evaporation deficit exceeds 2 m. Cave streams are fed by allogenic recharge from the caprock catchments during infrequent rainstorms. The eroded caprock contributes large amounts of clastic load to flood water, ranging from 10 percent to 80 percent by weight (Gerson and Inbar, 1974). Flood flow velocities range from 0.5 to 3 m/s, allowing bedload cobbles to be carried along cave passages. A large pile of breakdown material (including some approximately 1000 kg boulders) was observed being swept away by the November, 1994 flash flood, turning it into a debris flow.

Extreme suspended-load concentrations may render the flow dense and highly viscous. Suspended load scarcely changes while the water is flowing through the caves. On the other hand, TDS (total dissolved solids) in flood flow, consisting mainly of halite, increases dramatically within the caves, from about 10 g/L at stream sinks up to 350 g/L. Most flood water remains chemically aggressive while flowing rapidly within cave passages. However, if the flow is stopped or slowed down considerably, saturation is approached within a few hours or days.

Two types of "springs" are known along the margins of Mount Sedom, differing from one another chemically and hydrologically (Frumkin, 1994a).

- Cave outlets where sinking streams reappear during flood events ("resurgence").
- Alluviated springs discharging diffuse continuous flows of brine ("exurgence" of Sweeting, 1972).

The second type of spring emerges along the foot of the eastern escarpment and in the floor of some eastern caves, apparently discharging from a brine aquifer encountered in boreholes (Petroleum Services, 1979). It is saturated in salts, with high concentrations of Mg^{2+}, K^{+}, and Br^{-}, in addition to Cl^{-} and Na^{+} (Frumkin, 1994a). The alluviated exurgences are not further discussed here.

Major Cave Features

Each cave is fed by one or more sinks with a clearly defined catchment developed over the relatively insoluble caprock within Mount Sedom. Often several conduits join underground to form a branchwork cave. Most of the older caves are multiphase, possessing inactive conduits above or beside the modern channels. Relocation of inputs and outputs is common over time.

After a short initiation period, salt caves develop mainly by open-channel flow. The caves retain a downstream slope, developed by gravitational sediment-rich flow.

An active cave typically consists of a vertical shaft and a rock salt passage (7.2 Figure 3), which are discussed below.

Vertical Shafts

Surface channels are captured by subsurface routes through fissures in the caprock or through vertical shafts. Shaft development is not limited by catchment area. The initial requisite for shaft development is a fissure cutting through the caprock to allow meteoric water to reach the salt.

Salt shafts at Mount Sedom are up to 60 m deep with a diameter of up to 15 m. The shaft axis is vertical regardless of bedrock dip or fissure inclination. The shaft morphology is indicative of its water flow regime. Under a thick caprock the vertical section is commonly bell shaped with smooth walls (7.2 Plate 1, left). This form indicates an aggressive film of water flowing downward along the walls during rain events. Any rock protruding from the wall into the shaft attracts more flow which in turn dissolves the protruding rock, favoring a uniform film over the whole rock surface.

When rock salt crops out subaerially or under a thin caprock, shaft walls are directly affected by rainfall. The walls tend to be vertical with vertical grooves cut by rain-drop impact (7.2 Plate 1, right).

Young shafts draining small catchments with low discharge floods may be spindle shaped, that is, they reach their maximum diameter about halfway down, decreasing in (horizontal) cross section toward the shaft floor (7.2 Figure 4, shaft A). The bottom of a spindle-shaped shaft is incised in rock salt, with scarcely any alluvium. Spindle-shaped shafts are believed to develop by descending films of water that approach saturation during their flow. The decreasing diameter toward the bottom of the shaft indicates where the water approaches saturation. During high flood stage, however, a waterfall of aggressive water hits the bottom of the shaft and deepens it.

This model is supported by observations at the bottom of the 70 m-deep Sedom Shaft during the 26 December, 1988 flood. During low discharge, most water descended as a film along the shaft walls. The TDS of the descending film increased by about 300 g/L

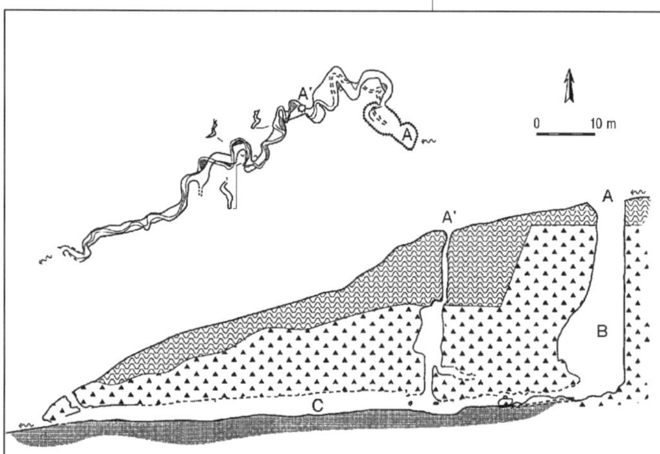

*7.2 Figure 3. Map and profile Beitsim Cave–a typical salt cave–northwestern Mount Sedom. Rocksalt is indicated by black triangles. **A** is an active stream sink. **A'** is an inactive stream sink. **B** is a vertical shaft. **C** is a passage in rock salt.*

7.2 Plate 1. Two shafts converging to become a single shaft. The salt table and caprock occupy the upper quarter. The left shaft developed under a film-flow regime, attaining a half-bell shape. The right shaft has developed vertical grooves indicating direct rain impact.

along the shaft, reaching 317 g/L at the bottom, 96 percent of which was Na^+ and Cl^-. The thin film of water approached saturation with respect to halite at the lower part of the shaft, because of its close contact with the salt walls. During high discharge, most water fell freely without touching the walls of the shaft.

Free-falling high discharge floodwater plays an important role in deepening shafts with a bedrock bottom, due to physical impact combined with aggressive dissolution. In most shafts, however, the floor is covered by alluvial sediment which may shield the bottom from the waterfall impact. The lower parts of these shafts display vertical walls with uniform cross sections (7.2 Figure 4, shafts B, C)

The horizontal cross section of many shafts is elliptical, with a long axis following a caprock fissure that supplies the water.

Some shafts are gradually retreating upstream, in which case only one side of the shaft is active. The mean rate of retreat measured in Dud Shaft from 1982 to 1992 was 5 mm/yr.

In old relict shafts the caprock fissure is often breached or enlarged by erosion and collapse. An example of an active shaft with a narrow aperture in the caprock is shown in 7.2 Figure 5A and a relict shaft widened where the upper part has eroded away in 7.2 Figure 5B. Older shafts become partly or completely destroyed by breakdown and erosion.

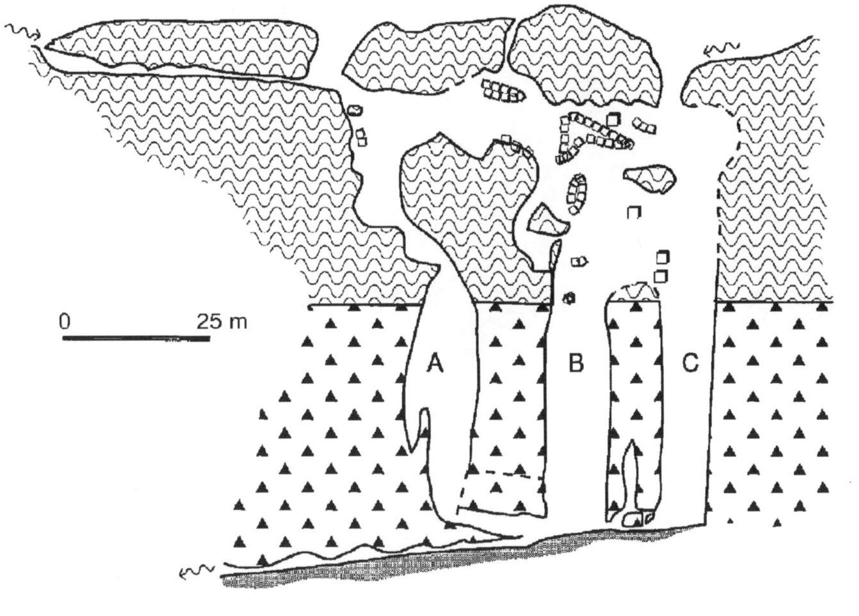

*7.2 Figure 4. Vertical section of a multishaft system, southwestern Mount Sedom. **A** is a spindle-shaped shaft. **B** and **C** are vertically walled shafts.*

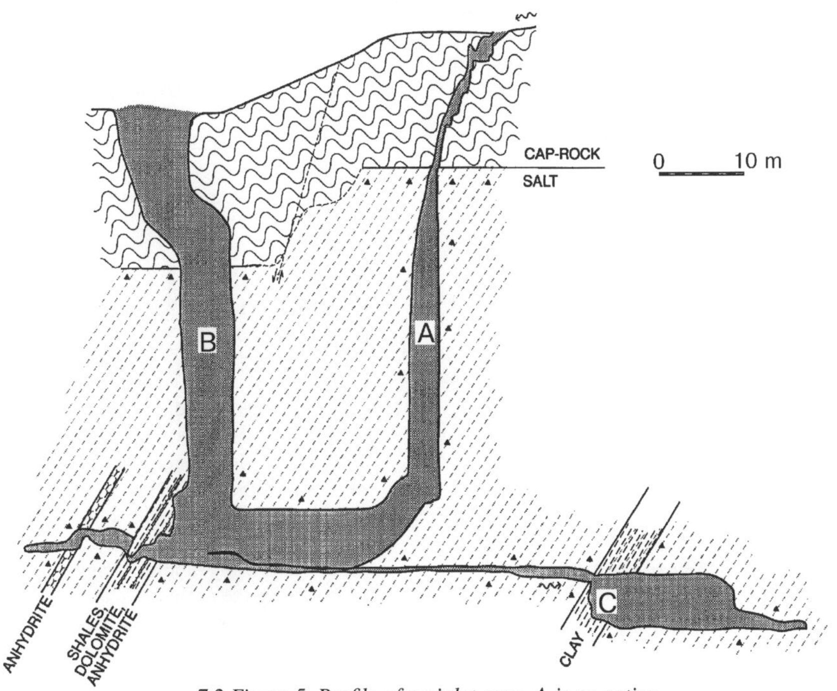

*7.2 Figure 5. Profile of an inlet cave. **A** is an active shaft. **B** is an inactive shaft enlarged by erosion. **C** is a nickpoint developed on insoluble clay beds.*

Passages in Rock Salt

Shafts with a surface catchment area greater than 200 to 300 m^2 are drained by subhorizontal passages, typically vadose canyons with incised meanders. The passages develop along a joint, bedding plane, or fault within the salt. In their early phase, passages follow the initial fissure closely, whereas older subhorizontal passages with alluvial fill are commonly more sinuous (7.2 Plate 2). Some wide passages with highly developed meander notches indicate long periods of flow without vertical downcutting; lateral migration of meanders may cause destruction of earlier canyon walls (Frumkin, 1998).

Downward development of passages is achieved by both incision of a passage floor and its abandonment for a lower-level passage. The older caves in Mount Sedom have several distinct larger levels above the modern channel.

Some 80 percent of the 23 cm of downcutting measured in Tlula Passage (Frumkin and Ford, 1995) from 1985 to 1995 occurred during the major floods of 5 to 6 November, 1994, whereas the rest occurred during two smaller events of 9 February, 1987 and 22 March, 1991. This suggests that the effective downcutting is performed by major flood events with a long recurrence interval. Comparing the measured 10-year downcutting to the vertical extent of the passage (about 10 m) suggests that the entire passage is only a few hundred years old. Salt bedrock is exposed along this passage floor, and its mean gradient is 40 percent. Gradients from 2 to 10 percent are more common in most active passages, however, in which cases the bed is alluviated.

Alluvial clastic deposits are fine-grained to pebbly, including anhydrite, dolomite, and quartz, all locally derived from Mount Sedom. They are usually cemented by salt precipitated by evaporation after each flood. The uppermost inactive story (or "level") of a

7.2 Plate 2. *A narrow meandering cave canyon (0.5 m wide) in salt, looking downward.*

cave is typically steeper than the modern channel. In Sedom Cave, gradients of 27 to 32 percent were measured in the uppermost level above a modern channel of only 5 percent. Downstream, closer to the outlet, upper levels are less steep, becoming either subparallel to modern channels or converging with them.

The development of a cave profile and morphologic features such as meanders and notches is in places constrained by less soluble layers, such as dolomite, shales, and anhydrite, interbedded within the salt. Nickpoints in a passage profile are often associated with insoluble beds (7.2 Figure 5C).

Solid load may settle at the bottom of the cave, shielding the floor and favoring upward dissolution of the roof–a form of paragenesis (Ford and Ewers, 1978). Corrosion bevels—flat roofs not controlled by geologic structure (Ford and Williams, 1989, p. 307)—develop where floodwater touches the ceiling. Any salt protruding from the ceiling into the passage is truncated by the water, and upward solution proceeds with the same rate across the entire bevel. The resulting flat ceiling is an imprint of the water surface. A vadose canyon is often entrenched in the alluvium under a bevel.

Salt caves may be classified into two groups according to the way they terminate:

- Integrated cave systems, having a distinct open outlet.
- Inlet caves, lacking such an outlet.

Integrated Cave Systems

In terms of exploration limits, 38 of the studied caves are "through caves" (as classified by Jennings, 1985, p. 137). They can be physically traversed along the full distance from sink to outlet (7.2 Figures 2A and 3). Of these, 27 cave outlets drain directly to the south end of the Dead Sea, the hydrologic base level of the region. The remainder drain westward into wadis that flow around Mount Sedom toward the Dead Sea. In terms of hydrologic function, the through caves as well as caves that include inaccessible segments but drain freely to base level, may be classified as integrated cave systems as defined by Ford and Ewers (1978). They are large enough to allow turbulent flow from input to output. Turbulent floods flowing through integrated cave systems carry most of their solid and dissolved load through the outlets toward the Dead Sea. The residence time of most floodwater in these caves is short, measured in minutes.

Integrated cave systems are developed along the margins of the rising diapir, where conditions are favorable for inlet-outlet connection. Hydraulic gradients are high, and fissures are long, open, and abundant because of lithostatic stress release (7.2 Figure 1).

Inlet Caves

Caves in the central part of the mountain, accessed through their sinks, seem to have no distinct outlet, and they terminate several meters or tens of meters above the apparent local water table (7.2 Figures 2B and 5). These are referred to as "inlet" caves. The direction of inlet cave passages does not seem to be related to the direction to base level or the water table gradient. Rather, the passages develop in orientations dictated by available open fissures. Some inlet caves seem to flow away from the nearest base level at the Dead Sea. Some inlet caves develop close to one another but do not meet.

The downstream parts of inlet caves often contain steep silt and clay banks with surge marks, similar to those described by Bull (1976) in limestone caves. They indicate ponding of low energy water with variable residence times (see below), in which the silt and clay settled. This suggests that the limit of exploration is also a hydraulic limit between rapid turbulent flow prevailing from input to cave bottom, and diffuse infiltration below the bottom of the explorable passage, down to the outlet. The water infiltrating from inlet caves may recharge the brine aquifer known from boreholes and springs. However, chemical and isotopic differences between the ponded water and the aquifer indicate an additional source and a complex origin of the aquifer (Frumkin, 1994a).

A transition between slow and rapid kinetics, known in limestone to coincide approximately with the laminar to turbulent transition (White, 1977), is not known in salt dissolution. Therefore salt fissures are *not* believed to enlarge over long distances downstream under phreatic conditions. Rather, dissolution seems to be restricted mainly to the vadose zone, forming the tapering passages of inlet caves. Infiltrating water probably approaches saturation rapidly below the bottom of the cave. This may partly explain why true phreatic conduits are not known in Mount Sedom, apart from paragenetic passages developed under full-pipe conditions in the vadose zone. Other arguments discouraging development of salt phreatic caves are:

- Open fissures are hardly available below base level.
- Mixing corrosion (Bögli, 1980, p. 35) is not effective.
- The abundant sediment load settles down in ponded water, blocking fissures and conduits.
- Deep cavities formed in environments sustaining deep groundwater circulation tend to collapse or anneal rapidly under lithostatic pressure and halokinesis.

Silt and clay settling at the bottom of inlet caves impedes downward infiltration, extending the residence time of ponded water. Three of the studied caves in northern Mount Sedom had perennial ponds throughout the period 1984 to 1995. They are perched, without any lithologic control, tens of meters above the nearest potential outlet at the foot of the mountain. The water level in each pond also differs from the others by tens of meters.

Pond waters are highly concentrated with solutes, up to 324 g/L, consisting mainly of NaCl. This is compatible with floodwater chemistry but not with the brine aquifer mentioned above (Frumkin, 1994a). Both dissolution and precipitation features are observed on walls bordering ponds, as well as on cave walls where ponds have dried out (Frumkin and Forti, 1997).

Horizontal notches indicate levels of aggressive water temporarily diluting the pond during floods. A horizontal notch is typically incised 10 to 20 cm into the wall, consisting of a flat corrosion bevel above a sloping sidewall ("facet" of Kempe et al., 1975). These features are formed where ponded water develops a density gradient stratification. The upper aggressive water layer, originating from the latest flood, dissolves the cave wall. A thin film of heavy saturated water descends gravitationally along the wall, initiating convection currents in the water. The saturated water is replaced by aggressive water moving from the center of the pond surface toward the walls (Farkas et al., 1951).

Large secondary halite crystals on some cave walls indicate supersaturation of the ponded water between successive floods. The increasing solute concentration is attributed mainly to direct dissolution of rock salt. However, evaporation also increases the TDS, as indicated by a gradual enrichment of stable isotopes in ponded water following each flood event (Frumkin, 1994a).

Horizontal notches and secondary salt are absent in many inlet caves, indicating that the infiltration rate is high, and the residence time of the ponded water is too short to produce these features.

Cross sections of some inlet cave passages become smaller downstream, tapering to an impenetrable point. There is no lithological constraint. Two alternative explanations may be offered for the tapering:

- The water becomes gradually enriched with solutes along the passage, reducing its dissolution capacity. This is probably the case in long passages with ponded water such as Mifrazim Cave (Frumkin, 1994b).
- The flowing water remains aggressive down to the end of the passage, but flood discharge decreases downstream because of gradual infiltration along the alluviated channel. Ephemeral surface streams in arid environments also display a similar decrease of discharge downstream (Renard and Keppel, 1966). This behavior differs from most caves and fluvial systems in moist environments, where discharge and cross section stay constant or increase downstream (White and Deike, 1989).

Tapering inlet salt caves that terminate with no signs of water ponding, indicate that all flood water infiltrates along the channel. On the other hand, passages in which only a fraction of the floodwater infiltrates along the flow route display sharp changes of cross section where water becomes ponded. Mevokhim Cave is an example, with a large downstream section acting as a temporary reservoir during flood events (7.2 Figure 5C).

Plan views of inlet caves suggest a range that is contained within three end members: elongated conduit, chamber, and maze. The plan views of the three inlet cave types are compared in 7.2 Figure 6, all of which developed along two fissure systems, perpendicular to each other. Some caves display features of two or three end members combined. The most common form is an elongated conduit, consisting of one or more segments developed along fissures (7.2 Figure 6A). Elongation is favored in fissured rock with low hydraulic head, under gravitational flow. The morphologic features of the conduit are similar to those of integrated cave systems (discussed above), down to the the uppermost pond level at the downstream end of the conduit.

A homogeneous massive rock in which fissures are tightly closed gives rise to chamber caves, whose length and width are roughly equal. Chamber caves are more common in northern Mount Sedom, which is less structurally deformed than the southern part of the diapir (7.2 Figure 1). Two radio-carbon dates of wood from Metsah Cave (7.2 Figure 6B) indicate that this chamber cave began to develop in its northern part, gradually expanding southward and westward during the past 4000 years (Frumkin et al., 1991; samples no. Me2, Me3).

Chamber caves are comparable to solution-mined artificial cavities. If aggressive water is artificially introduced to a depth where rock salt joints are tight, then a roughly cylindrical chamber of 100,000 m^3 may be excavated in a few hundred days (Saberian, 1983). In the artificial case the resulting brine is extracted from the cavity by pumping, whereas in natural caves the water carrying the solutes escapes by infiltration.

Close fissure spacing with a considerable hydraulic head supplied by flash floods favors the development of flood network mazes. Karbolot Cave, having the largest catchment among inlet caves and an extensive fissure system, displays a floodwater maze within an elongated cave (7.2 Figure 6C). Infrequent large floods have submerged most of the cave for short periods during recent centuries, as indicated by surge marks, clay deposits, and wood twigs, dated at 480±40 (conventional ^{14}C) years BP (Frumkin et al., 1991; sample no. Ka1).

The salt flood network mazes are comparable to floodwater mazes described in limestone caves (Palmer, 1975). Large artesian maze caves of the type known in gypsum (Klimchouk, 1991) are not known in salt.

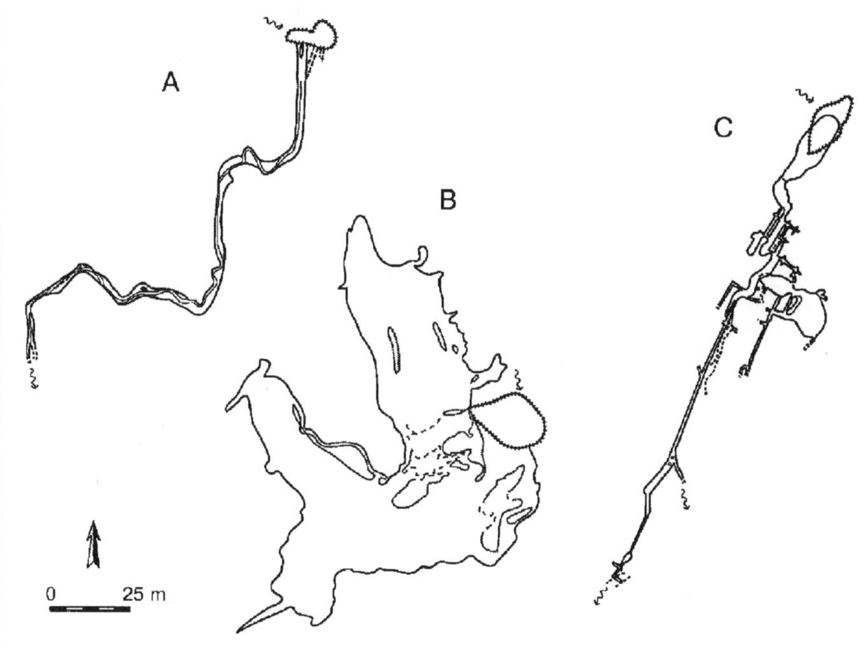

*7.2 Figure 6. Three types of inlet caves in plan view. **A** is an longated stream cave (Notsa Cave). **B** is a chamber cave changing its direction of enlargement through time (Metsah Cave). **C** is an elongated cave with floodwater maze (Karbolot Cave).*

Base Level Control and the Vertical Dimension

Active passages in integrated cave systems are usually adjusted to the present base level of erosion with a subhorizontal passage below which the channel cannot degrade and at which neither net erosion nor deposition occurs (7.2 Figure 3). The downcutting process prior to reaching equilibrium may take some hundred years up to a few thousand years for the measured short term downcutting rate of about 2 cm/year. The equilibrium profile is a time-independent configuration, maintained as long as boundary conditions do not change.

Most integrated cave systems originate as inlet caves, created by capture of surface recharge into a caprock fissure. Diversionary routes are common where earlier conduits are blocked by alluvium, until a conduit connects to an output point. If a connection to the output is not established, the cave continues to evolve as an inlet cave. The earliest integrated flow route can be traced along the highest level of some caves in the form of a fissure widened by dissolution. Early conduits often follow the shortest route available along the vertical fissure from input to output. Young unequilibrated passages cut down in the salt until they reach the minimal gradient allowing transportation of their coarse sediment load. A change of base-level is followed by downcutting, paragenesis, or the development of a completely new cave level. The adjustments of cave profiles to base level changes result in multiphase cave systems. This model is comparable to the vadose theories for the development of limestone caves (Warwick, 1953; Palmer, 1987).

The profile of a subhorizontal passage of an *inlet* cave may be controlled by the width of fissures underlying the cave. The inlet cave can cut down no farther than the lowest level penetrable by turbulent flow carrying suspended load. A later change of the profile is common. Downcutting and diversion of inlet cave passages can be attributed either to uplift of the diapir causing fissures to open and gradients to increase, or to gradual dissolution of the cave floor by infiltrating water, creating steeper local gradients.

References

Bennet, S.S., and Hannor, J.S., 1987, Dynamics of subsurface salt dissolution at the Welch Dome, Louisiana Gulf Coast, *in* Lerch, I., and O'Brien, J.J., eds., *Dynamical geology of salt and related structures:* Orlando, Academic Press, p. 653-677.

Bögli, A., 1980, *Karst hydrology and physical speleology:* Berlin, Springer-Verlag, 284 p.

Bull, P.A., 1976, Dendritic surge marks in caves: *British Cave Research Association Transactions,* v. 3, p. 1-5.

Casas, E., and Lowenstein, T.K., 1989, Diagenesis of saline pan halite: Comparison of petrographic features of modern, Quaternary, and Permian halites: *Journal of Sedimentary Petrology,* v. 59, p. 724-739.

Chabert, C., 1989, Les grandes cavitès mondiales en roches non-calcaires: Union Internationale de Speleologie–Commission des grandes cavitès, 34 p.

Choppy, J., 1988, Roches solubles non carbonatees et karstification: Spèlèo-Club de Paris.

DeMille, G., Shouldice, J.R., and Nelson, H.W., 1964, Collapse structures related to evaporites of the Prairie Formation, Saskatchewan: *Geological Society of America Bulletin,* v. 75, p. 307-316.

Donini, G., Rossi, A., Forti, P., Buzio, A., Calandri, G., and Frumkin, A., 1985, Monte Sedom: *Societa Speleologica Italiana,* 135 p.

Edgell, H.S., 1996, Salt tectonism in the Persian Gulf Basin, *in* Alsop, I., Blundell, D., and Davison, I., eds., Salt Tectonics: *Geological Society of London Special Publication,* no. 100, p. 129-151.

Farkas, L., Litman, H.L., and Bloch, M.R., 1951, The formation of "salt tables" in natural and artificial solar pans: *Research Council of Israel Bulletin,* v. 1, no. 1-2, p. 36-39.

Ford, D.C., 1989, Paleokarst of Canada, *in* Bosák, P., Ford, D., Glazek, J., and Horácek, I., eds., *Paleokarst: A systematic and regional review:* Praha, Academia, p. 313-336.

Ford, D.C., and Ewers, R.O., 1978, The development of limestone cave systems in the dimensions of length and depth: *International Journal of Speleology,* v. 10, p. 213-244.

Ford, D.C., and Williams, P.W., 1989, *Karst geomorphology and hydrology:* London, Unwin Hyman, 601 p.

Frumkin, A., 1986, The cave survey of Moun zionale sul carsismo nelle evaporiti: *Le Grotte d'Italia,* v. 12, p. 305-308.

Frumkin, A., 1994a, Hydrology and denudation rates of halite karst: *Journal of Hydrology,* v. 162, p. 171-189.

Frumkin, A., 1994b, Morphology and development of salt caves: *National Speleological Society Bulletin,* v. 56, p. 82-95.

Frumkin, A., 1996, Uplift rate relative to base level of a salt diapir (Dead Sea, Israel), as indicated by cave levels, *in* Alsop, I., Blundell, D., and Davison, I., eds., Salt Tectonics, *Geological Society of London Special Publication,* no. 100, p. 41-47.

Frumkin, A., 1997, Liquid Crystal Cave, Israel, *in* Cave Minerals of the World: *National Speleological Society*, Huntsville, Alabama, U.S.A., p. 319-322.

Frumkin, A., 1998, Salt cave cross sections and their paleoenvironmental implications: *Geomorphology*, v. 23, p. 183-191.

Frumkin, A., and Ford, D.C., 1995, Rapid entrenchment of stream profiles in the salt caves of Mount Sedom, Israel: *Earth Surface Processes and Landforms*, v. 20, p. 139-152.

Frumkin, A., Magaritz, M., Carmi, I., and Zak, I., 1991, The Holocene climatic record of the salt caves of Mount Sedom, Israel: *The Holocene,* v. 1, p. 191-200.

Gerson, R., and Inbar, M., 1974, The field study program of the Jerusalem-Elat Symposium: *Reviews and summaries of Israeli research projects,* v. 20, p. 7-11.

Giurgiu, I., 1985, Grottes du sel de Roumanie: *Spelunca,* v. 20, p. 34-39.

Jackson, M.P.A., Cornelius, R.R., Craig, C.H., Gansser, A., Stocklin, J., and Talbot, C.J., 1990, Salt diapirs of the Great Kavir, central Iran: *Geological Society of America Memoir,* no. 177, 139 p.

Jennings, J.N., 1985, *Karst geomorphology:* Oxford, Basil Blackwell, 293 p.

Kempe, S., Brandt, A., Seeger, M., and Vladi, F., 1975, "Facetten" and "Laugdecken," the typical morphological elements of caves developing in standing water: *Annales de Spéléologie,* v. 30, p. 705-708.

Klimchouk, A.B., 1991, Large maze caves in gypsum in the western Ukraine: Speleogenesis under artesian conditions: *National Speleological Society Bulletin,* v. 53, p. 71-82.

Klimchouk, A.B., 1996, Speleogenesis in gypsum, *in* Klimchouk, A., Lowe, D., Cooper, A., and Sauro, U., eds., Gypsum karst of the world: *International Journal of Speleology*, v. 25, no. 3-4, p. 61-82.

Klimchouk, A.B., 1997, Speleogenetic effects of water density differences: *12th International Congress of Speleology,* La Chaux-de-Fonds, Proceedings, v. 1, p. 161-164.

Kozary, M.T., Dunlap, J.C., and Humphrey, W.E., 1968, Incidence of saline deposits in geologic time: *Geological Society of America Special Paper,* no. 88, p. 43-57.

Landes, K.K., 1961, Effects of solution of bedrock salt in the Earth's crust, *in* 1st Symposium on Salt, Proceedings, Cleveland, p. 64-73.

Last, W.M., 1993, Salt dissolution features in saline lakes of the northern Great Plains, western Canada: *Geomorphology,* v. 8, p. 321-334.

Miotke, F.-D., 1971, Collapse sinkholes formed by deep-seated solution of salt, Bad Seebruch area, near Hannover, Germany: *Caves and Karst,* v. 13, no. 5, p. 43.

Mylroie, J.E., and Carew, J.L., 1987, Field evidence of the minimum time for speleogenesis: *National Speleological Society Bulletin,* v. 49, p. 67-72.

Neev, D., and Hall, J.K., 1979, Geophysical investigations in the Dead Sea: *Sedimentary Geology,* v. 23, p. 209-238.

Palmer, A.N., 1975, The origin of maze caves: *National Speleological Society Bulletin,* v. 37, p. 56-76.

Palmer, A.N., 1987, Cave levels and their interpretation: *National Speleological Society Bulletin,* v. 49, p. 50-66.

Petroleum Services, 1979, *Solution mined liquid hydrocarbon storage caverns, Mount Sedom:* Tel Aviv, Petroleum Services Ltd., Tel Aviv, Progress Report 2, 12 p.

Pfeiffer, D., and Hahn, J., 1972, Karst of Germany, *in* Herak, M., and Stringfield, V.T., eds., *Karst—Important karst regions of the northern hemisphere:* Amsterdam, Elsevier, p. 189-223.

Ponta, G., 1986, The evaporite karst of Romania, *in* Atti simposio internazionale sul carsismo nelle evaporiti: *Le Grotte d'Italia,* v. 12, p. 407-415.

Quinlan, J.F., Smith, A.R., and Johnson, K.S., 1986, Gypsum karst and salt karst of the United States, *in* Atti simposio internazionale sul carsismo nelle evaporiti: *Le Grotte d'Italia,* v. 12, p. 419-420.

Ranganathan, V., and Hanor, J.S., 1986, Diffusion of dissolved NaCl from evaporites in sedimentary basins: *Eos,* v. 67, p. 274.

Renard, K.G., and Keppel, R.V., 1966, Hydrographs of ephemeral streams in the Southwest: *American Society of Civil Engineers,* Proceedings, v. 92, p. 33-52.

Saberian, A., 1983, Utilization of leaching models in the design of large crude oil storage cavities: *6th International Symposium on Salt,* Alexandria, Proceedings, v. 2, p. 235-348.

Salomon, J.N., 1995, La Chili, Pays de karsts extrémes: *Karstologia,* v. 24, p. 52-56.

Shaw, T.R., 1979, *History of cave science:* Crymych, Oldham, 489 p.

Sweeting, M.M., 1972, *Karst Landforms:* London, Macmillan, 362 p.

Vroman, J., 1950, The movement and solution of salt bodies as applied to Mount Sedom: *Israel Exploration Journal,* v. 1, p. 185-193.

Warwick, G.T., 1953, The origin of limestone caves, *in* Cullingford, C.H.D., ed., *British caving:* London, Routledge and Kegan Paul, p. 41-61.

White, W.B., 1977, Role of solution kinetics in the development of karst aquifers, *in* Tolson, J.S., and Doyle, F.L., eds., *Karst hydrogeology:* 12th Congress of the International Association of Hydrogeologists, p. 503-517.

White, W.B., and Deike, G.H., 1989, Hydraulic geometry of cave passages, *in* White, W.B., and White, E.L., eds., Karst hydrology–Concepts from the Mammoth Cave area: New York, Van Nostrand Reinhold, p. 223-258.

Yechieli, Y., 1995, Are sedimentary salt layers always impermeable?: *Geophysical Research Letters,* v. 22, p. 2761-2764.

Yuhua, G., and Hua, S.L., 1986, Salt karst in Qinghai Plateau, China, *in* Atti simposio internazionale sul carsismo nelle evaporiti, *Le Grotte d'Italia,* v. 12, p. 337-345.

Zak, I., and Bentor, Y.K., 1968, Some new data on the salt deposits of the Dead Sea area, Israel, *in Symposium on the Geology of Saline Deposits:* Earth Sciences, v. 7, p. 137-146.

Zak, I., and Freund, R., 1980, Strain measurements in eastern marginal shear zone of Mount Sedom salt diapir, Israel: *American Association of Petroleum Geologists Bulletin,* v. 64, p. 568-581.

Zoltán, Z., 1994, A berzsenyi D·niel tan·rkÈpz^ f^iskola tudom·nyos k^zlemÈnyei IX. term Èszettudom·nyok: *Szombathely* v. 4, p. 233-248 (in Hungarian).

7.3. Speleogenesis in Quartzite

7.3.1 Speleogenesis in Quartzite in Southeastern Minas Gerais, Brazil

Atlas V. Corrêa Neto

Abstract

Speleogenesis in quartzite from the Andrelândia Group (Proterozoic) began with a long initial period of base level stability when silica solution from quartz and leaching from feldspar and phyllosilicates generated linear zones of friable rock with increased porosity and permeability (sanding, or arenisation). One (or more) uplift episodes followed with lowered base level and increased hydraulic gradients. The faster water flow concentrated in the high-permeability zones, and loose quartz grains were mechanically removed, creating linear conduits (piping). The essential conditions for cave development in southeastern Minas Gerais were:

- A large difference between local and regional base levels.
- The presence of rock layers specifically susceptible to sanding and piping processes (fine-grained micaceous quartzite layers).
- Impermeable layers (schist lenses).
- A sequence of stability–uplift cycles.

Different cave patterns and sizes can be explained by changes in one or more of the above conditions.

Introduction

Quartzite caves in southeastern Brazil are concentrated in Minas Gerais State. Three Proterozoic lithostratigraphic units are their main hosts. The Espinhaço Subgroup (deposited about 1.7 Gyr, Almeida Abreu and Pflug, 1994) is exposed from the north of Belo Horizonte (Minas Gerais State) to the north of Bahia State. Most of the available data about its caves are restricted to topographic surveys. The Minas Subgroup (Lower Proterozoic; Babinski et al., 1991) crops out in the Iron Quadrangle region. This is historically Brazil's most important mining district (iron, maganese, gold, and topaz).

Quartzite caves are known in the surroundings of Ouro Preto (Romero and Lima, 1989) and in the Caraça Range (Brazil's deepest cave, the 360-m-deep Gruta do Centenário, is in the Caraça Range; Rubbioli, 1996). The Andrelândia Group (Ebert, 1955; Trouw et al., 1983), crops out in the south-southeastern part of Minas Gerais. Speleogenesis in the Andrelândia Group is relatively well known (Corrêa Neto et al., 1996). In this contribution, genetic models for quartzite caves from the Andrelândia Group are suggested.

Geology of the Andrelândia Group

The Andrelândia Group comprises highly deformed metasediments of low to medium grade. The quartzite layers which host the Andrelândia Speleological Province (ASP) caves are in the lower parts of the Andrelândia Group (Andreis, 1995, pers. comm.) The quartzite layers form typical cuestas and flat-topped mesas. The isolated quartzitic cuestas stand up well against a background of gentle cliffs composed of metapelitic rocks.

The speleological province (7.3.1 Figure 1)

is composed of five districts, Ibitipoca, Boa Vista, Luminarias, São Tomé, and Carrancas. Climate in the province is humid, tropical, and typically characterized by a rainy summer and dry winter. The Ibitipoca, Carrancas, and Luminarias Districts are the main focus of this work. Data on the geologic controls and speleogenesis of Boa Vista and São Tomé districts are still scarce.

Ibitipoca Speleological District

Initially defined by Perez and Grossi (1986), the Ibitipoca Speleological District (ISD, see 7.3.1 Figure 1 for location) has about 30 presently known caves, all vadose, the largest one being the Gruta das Bromélias (2750 m, 7.3.1 Figure 2a). It is the most important and best known cave of the four ASP districts.

The Ibitipoca Range has an average altitude of 1500-1600 m, reaching a maximum of 1784 m (Ibitipoca or Lombada Peak), and can be described as two cuestas whose flanks dip toward the same valley. The local base level is 100 m to 200 m higher than the regional base level.

Coarse-grained micaceous quartzite is the main rock type, with subordinate fine-grained micaceous quartzite, biotite schist, and muscovite schist lenses. The layers dip SE, rarely exceeding 20°. Metamorphism reached the middle to upper amphibolite facies (Nummer, 1990). The rocks are highly deformed, with a strong foliation (S1) subparallel to the regular sedimentary layering. The regional structure is a large recumbent fold, refolded by open horizontal normal folds. The quartzites are affected by three main fracture systems, NE (predominant), N, and E.

Collapse dolines (for example, Gruta dos Três Arcos) and valleys with steep walls generated by the collapse of cave galleries (Pião Canyon) are common. The largest caves form a hierarchy of underground drainage systems (Corrêa Neto et al., 1993). Three groups of caves have been distinguished according to morphologic and geologic features (Corrêa Neto et al., 1993). Groups I and II (7.3.1 Figure 2, 7.3.1 Plate 1) are described below. Group III is composed of tunnels and arches along surficial rivers and is not relevant to the present discussion.

Group I (7.3.1 Figure 2a, b, 7.3.1 Plate 1a, b) comprises the largest caves (600–2750 m in passage length), forming branchwork patterns with angular passages (cf. Palmer, 1991). The conduits usually have rectangular to oval cross sections (7.3.1 Plate 1a), but some galleries have keyhole cross sections (7.3.1 Plate 1b). The gradient of the passages is about 8°. Semicircular collapse sinkholes and shafts provide access to the caves.

The caves show structural (fracture zones) and stratigraphic controls (a 1.5-m to 2-m-thick layer of fine-grained micaceous quartzite, 7.3.1 Plate 1a). Inactive galleries are found at 1.5 m and 3 m above today's phreatic level. Zones of tubes, conditioned by the rock's foliation, are common.

Thin-section studies show that feldspars from the fine-grained micaceous quartzite are extensively altered to clay minerals. Feldspar is very scarce in the coarse quartzites, but small quantities are found in the fine-grained micaceous quartzite. Fine-grained micaceous quartzite also has a higher porosity than the coarse-grained type.

Group II caves (7.3.1 Figures 2c-e; 7.3.1 Plate 1c, d) have 186–600 m of passage length, with straight or gently meandering galleries (gradient about 5°) forming rudimentary branchwork or network patterns. No clear stratigraphic controls were found. Inactive galleries are 1.5, 7, and 10 m above the present base level. Vertical shafts join the upper and lower conduits, in places reaching the surface. The galleries have simple rectangular or triangular cross sections (7.3.1 Plate 1c, d). A few galleries are narrow and fissure-like. Subvertical pipes are common at the tops of the passages. Old phreatic levels may be indicated by levels of ferruginous concretions. Speleothems of opal-A (formed by the release of capillary water) and allophane (associated with flowing water) are common in both groups of caves.

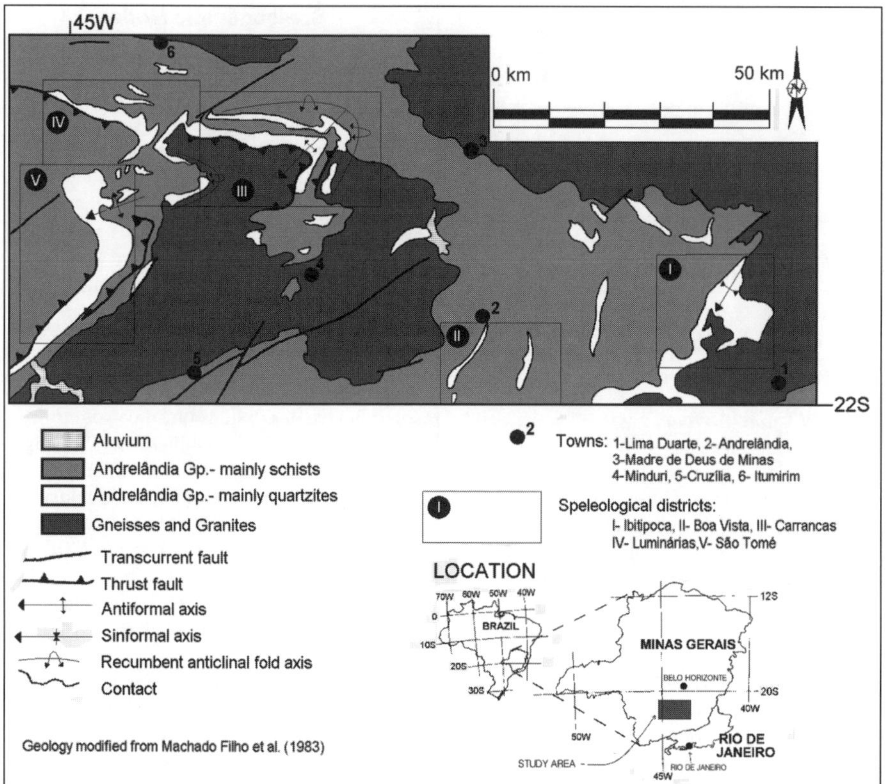

7.3.1 Figure 1. Simplified geologic map of the Andrelândia Speleological Province, with the location of the speleological districts (modified from Machado Filho et al., 1983).

Carrancas Speleological District

The geology of the Carrancas Speleological District (CSD) was summarized by Dayan and Baptista Filho (1984) and Dayan et al. (1987). Over a gneissic basement there is a metasedimentary sequence which comprises, from base to top, a quartzitic unit with lenses and layers of muscovite schist; an intercalated unit with garnet-staurolite-kyanite-muscovite schist (pelitic facies), muscovite-quartzites, graphite-garnet-staurolite-muscovite schist (undivided facies), pure quartzite (psammitic facies), and a unit composed mostly of garnet-kyanite-muscovite-biotite schist. Metamorphic grade has reached the middle amphibolite facies.

The regional structure is explained by at least two deformation phases, the first generating tight isoclinal folds and NE-directed thrust faults, followed by NE-trending folds. The layers dip about 20° toward the NW or S, due to the last refolding phase. The caves were formed in fine- to medium-grained quartzite of the first unit, with centimeter- to decimeter-thick layers of greenish muscovite schist.

The Carrancas Range is a smoothly dipping cuesta. Most of the known caves are on its gently sloping side. The maximum height difference between local and regional base level is approximately 150 m, with a gentle gradient. The study concentrated on

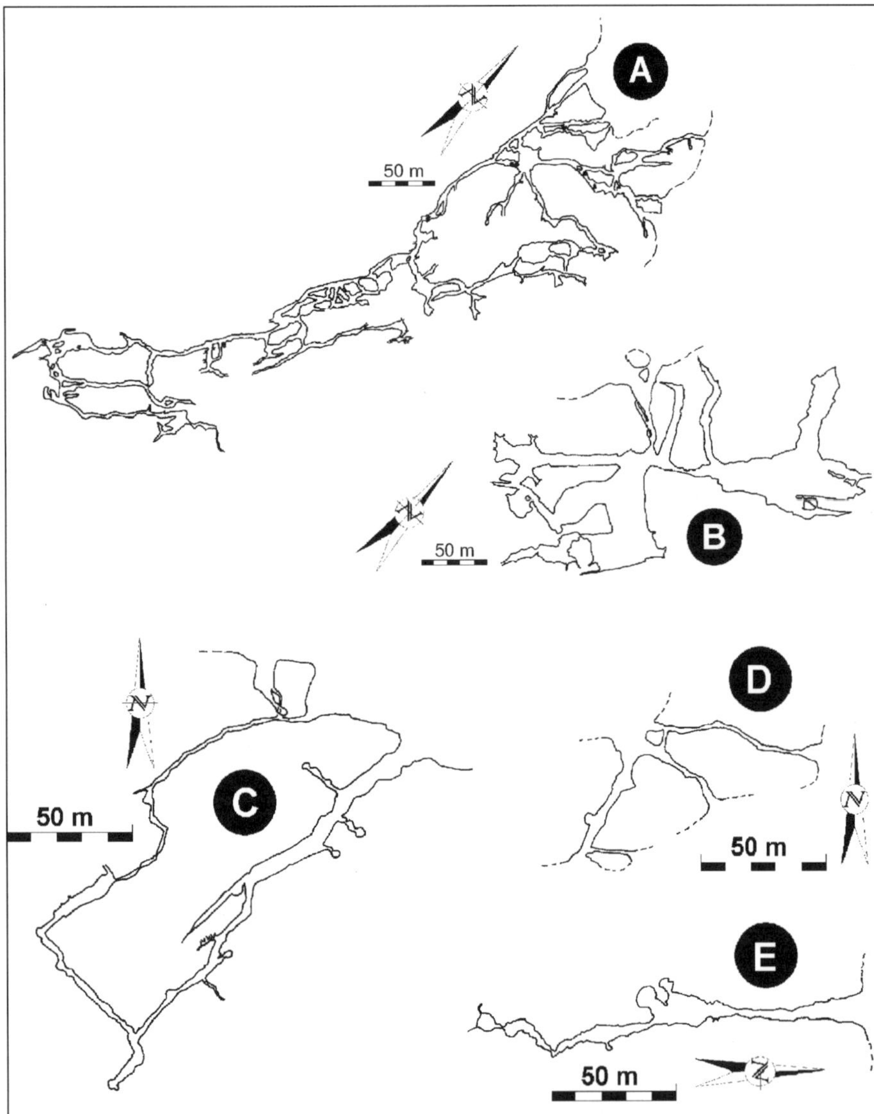

*7.3.1 Figure 2. Simplified cave maps from the Ibitipoca Speleological District: **A** is Gruta das Bromélias, 2750 m. **B** is Gruta das Casas, 600 m. C is Gruta do Bocão, 494 m. **D** is Gruta dos Fugitivos, 186 m. **E** is Gruta do Pião, 126 m. **A** and **B** are Group I caves; **C**, **D** and **E** are Group II caves. See text for details. Maps courtesy of SPEC-Speleogroup.*

Toca de Carrancas (583 m in linear extent, 7.3.1 Figure 3a, 7.3.1 Plate 2a). The other caves seem to be similar. Carrancas is a vadose cave with a rudimentary network pattern, composed of gently meandering galleries, with simple rectangular (7.3.1 Plate 2a), oval, or triangular cross sections. Elliptical collapse sinkholes formed near the surface. Tubes are found above schist lenses and layers. Vertical pipes are common in the ceiling.

Most of the galleries have a centimeter- to decimeter-thick schist layer on the floor. A poorly defined ancient phreatic level is evidenced by erosional forms (small pockets and potholes) 50 cm above today's underground river bed. No abandoned cave levels were found. Coatings of opal are common on the passage walls, as well as allophane flowstone, originating from nearby fractures or pipes.

Luminárias Speleological District
The predominant rock type in the Luminárias Speleological District (LSD) is well-foliated gently-dipping medium- to fine-grained quartzite, with centimeter-thick greenish muscovite schist intercalations. The main regional structures are NE-directed thrusts and isoclinal recumbent folds. The metamorphism of the sediment reached middle amphibolite facies (Almeida, 1992). The overall stratigraphic sequence is similar to that of the Carrancas Range. The quartzite land forms (Mandembe and adjacent ranges) are mesas or gently dipping cuestas. All caves are on the cuestas' gently-dipping slopes.

The caves (7.3.1 Figure 3B, C; 7.3.1 Plate 2B to D) are roughly similar to the ISD (Ibitipoca) group II caves. They also closely resemble some CSD (Carrancas) caves. They are vadose, and their galleries form rudimentary branchwork patterns with rectangular (7.3.1 Plate 2D) or fissure-like (7.3.1 Plate 2B and C) cross sections. Conduit gradients are about 6°. Collapse sinkholes are common along the conduits.

Vertical 1-centimeter-diameter tubes are quite common in the ceilings of the passages. Some foliation-parallel pipe levels were developed above schist layers (7.3.1 Plate 2B). There is no clear stratigraphic control, even though some conduits seem to have been developed directly above the muscovite-schist levels. Passage direction is determined by the intersection between subvertical fractures and foliation planes. An abandoned gallery was found in one of the studied caves about 6 m above today's phreatic level. Small allophane flowstone deposits and opal-A coatings are common in LSD caves.

Speleogenetic Model

Speleogenesis in the ISD (Ibitipoca) is interpreted to be a two-stage process (Corrêa Neto et al., 1993, 1995, 1996), similar to examples from Venezuela (Szczerban and Urbani, 1974; Urbani, 1986) and South Africa (Martini, 1987). The application of these speleogenetic models to quartzite caves from the Andrelândia Group is based on features such as tube levels, silica speleothems, and overall cave patterns.

In the first stage, the phreatic level was stable with a low gradient. Underground water flow was slow, favoring silica dissolution (see discussions on silica solution in White et al., 1966; Jennings, 1983; Young, 1986; Busche and Sponholz, 1992; and Chapter 4.1.8) and alteration of mica and feldspar to argillaceous minerals and hydroxides, which increased porosity and permeability. Concentration of water flow along the intersection of fracture and foliation planes created linear zones of weak altered rock (the *néogrès* of Martini, 1987).

During the second stage, (regional?) uplift increased the phreatic gradient and flow velocity. Flow was concentrated in the linear zones of increased porosity and permeability. Quartz grains were then mechanically removed, and a network of cylindrical conduits (tubes) was created. Some tubes were enlarged and became cave conduits, the

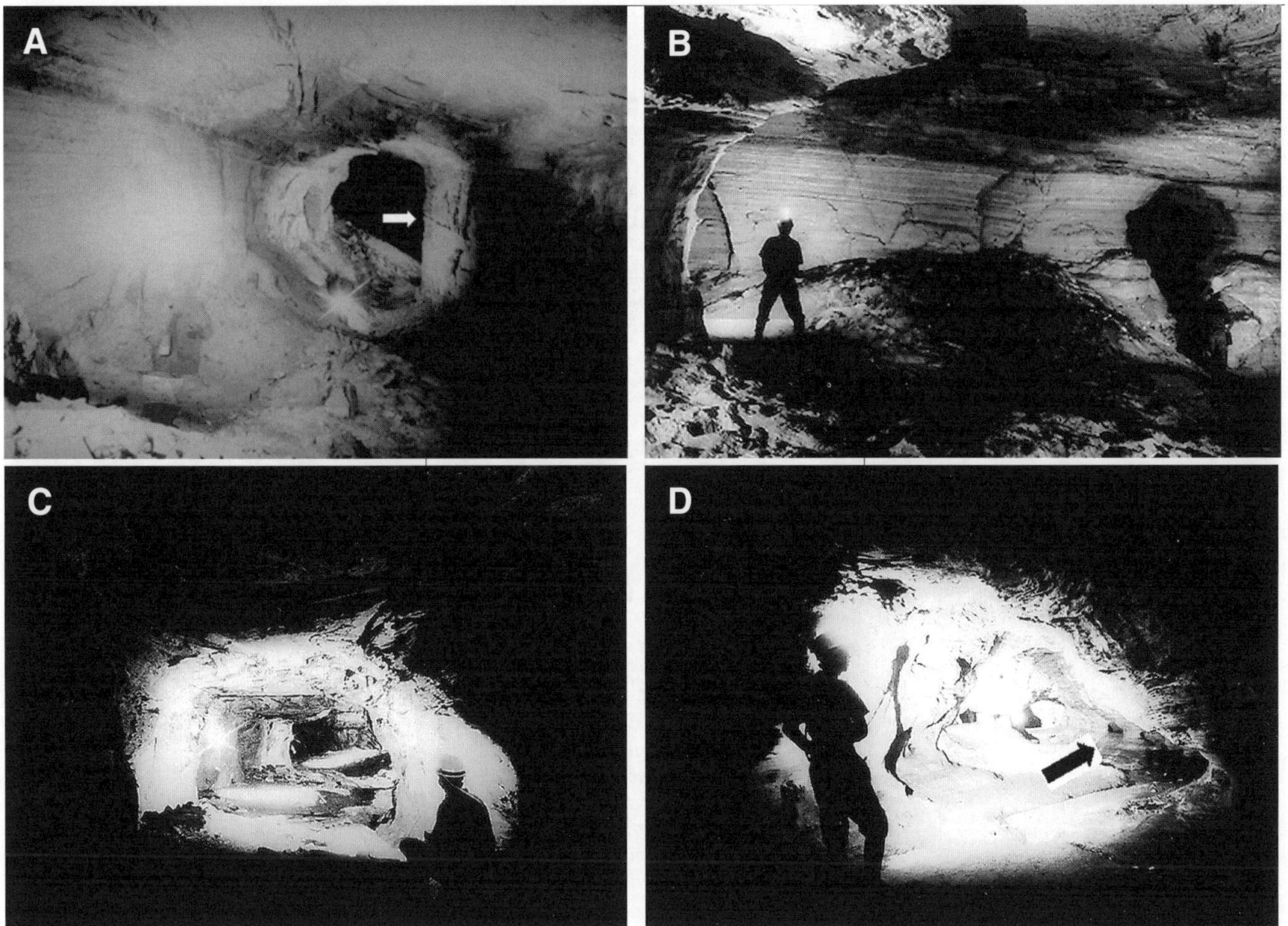

*7.3.1 Plate 1. Caves from the Ibitipoca Speleological District: **A** is an inactive gallery from Gruta das Casas. The top of the fine-grained quartzite layer is indicated by the arrow. Note the overall rectangular cross section. **B** is a typical keyhole-shaped conduit (right side of the photo) from the Gruta das Casas. Note the fine-grained quartzite layer on the wall (light toned). **C** is a conduit from the Gruta do Bocão with a rectangular cross section. **D** is a conduit from the Gruta do Pião with a triangular cross section. Note pipe level on the floor (arrow). **A**, **B** are Group I caves. **C**, **D** are Group II caves.*

main branches of a hierarchy of underground drainage systems. These systems are sequentially composed of interrelated rock pores, 1-centimeter-diameter tubes, 1-decimeter-diameter tubes, and cave galleries.

A sequence of an initial period of phreatic-level stability in a humid climate, with subsequent uplift, seems to be an essential condition for quartzite cave formation, as in Venezuela (Urbani, 1986), South Africa (Martini, 1987), and Libya (Busche and Sponholz, 1992).

The ISD Group I caves are larger than the Group II caves due to the presence of fine-grained micaceous quartzite, because its small grain size increases the surface of contact with water, thus favoring dissolution. Mica alteration has also contributed to the increase in porosity and permeability and to weakening the rock, allowing extensive piping.

Recharge in the Group I caves is both through dolines and diffuse flow. The

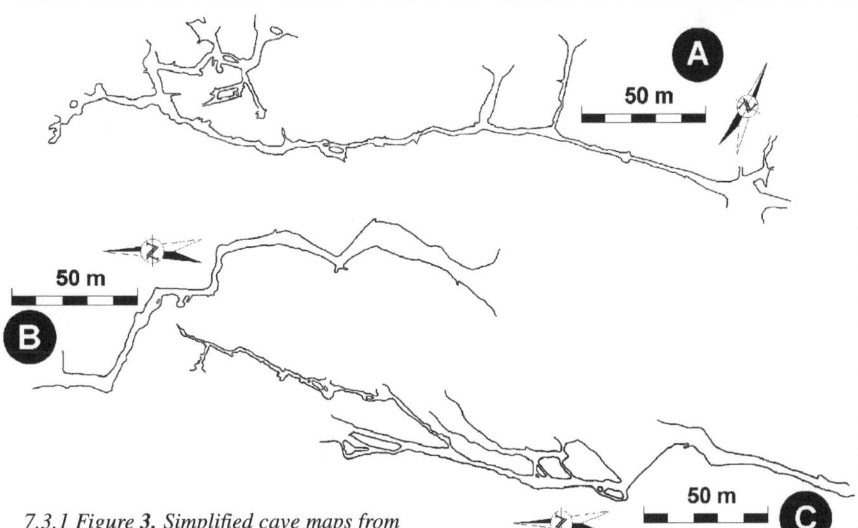

*7.3.1 Figure **3**. Simplified cave maps from the Carrancas (**A**) and the Luminárias (**B** and **C**) Speleological Districts. **A** is Toca de Carrancas, 583 m. **B** is Toca do Chico Lino, 213 m. **C** is Toca Num Sei I and II, 408 and 60 m. Maps courtesy of SPAN (**A**), Bambuí and SPEC Speleogroups (**B** and **C**).*

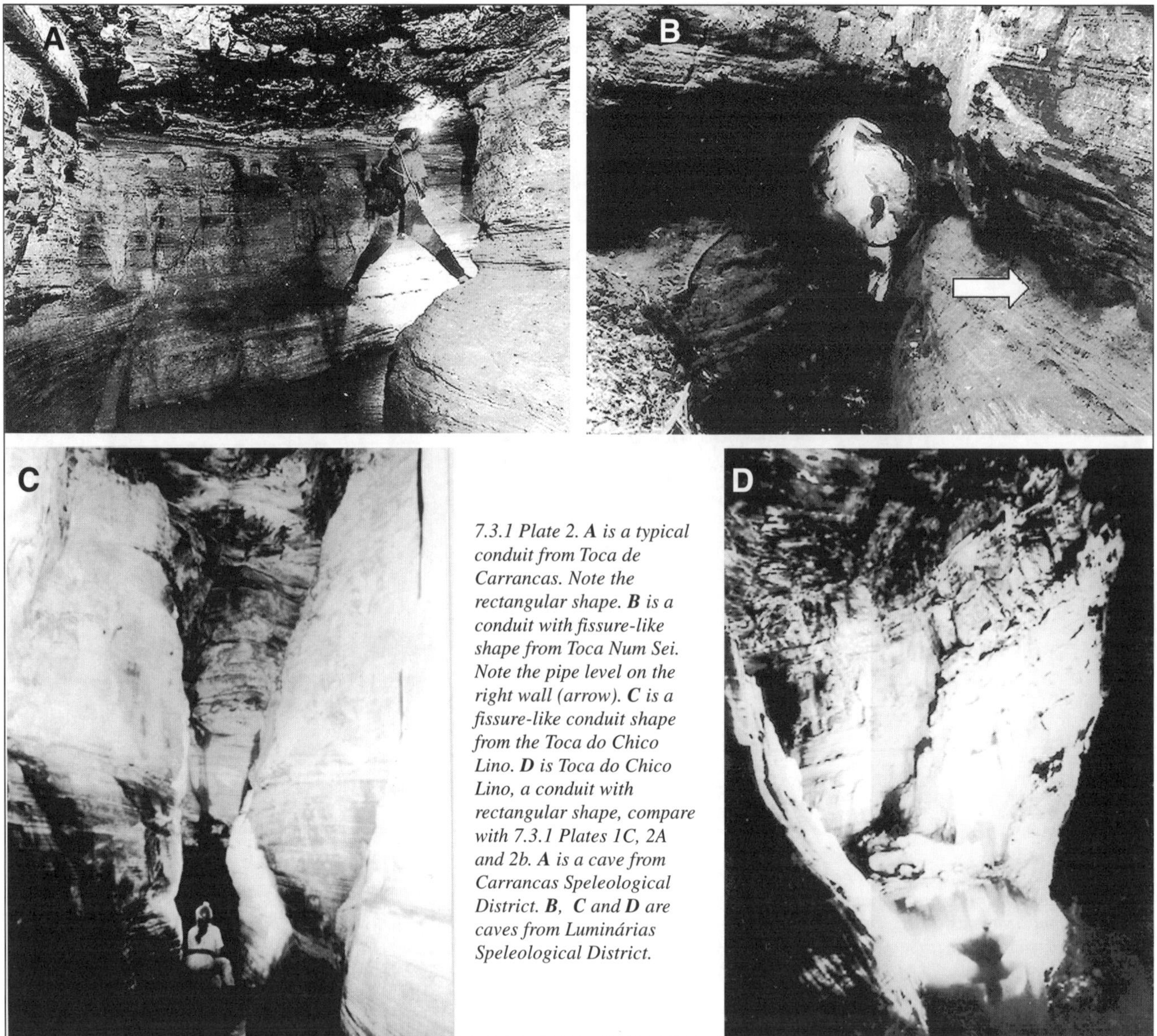

7.3.1 Plate 2. ***A*** *is a typical conduit from Toca de Carrancas. Note the rectangular shape.* ***B*** *is a conduit with fissure-like shape from Toca Num Sei. Note the pipe level on the right wall (arrow).* ***C*** *is a fissure-like conduit shape from the Toca do Chico Lino.* ***D*** *is Toca do Chico Lino, a conduit with rectangular shape, compare with 7.3.1 Plates 1C, 2A and 2b.* ***A*** *is a cave from Carrancas Speleological District.* ***B***, ***C*** *and* ***D*** *are caves from Luminárias Speleological District.*

underground water flow direction is mainly controlled by foliation planes. Group II caves seem to have a greater amount of vertical water flow, fractures being more important as water conduits than in Group I. Diffuse recharge seems to have had a larger contribution than that in Group I.

The presence of correlated cave levels (stories) show that the process is cyclic, with repeated episodes of stability and uplift. If cave levels are used as an evolution and age indicator (Palmer, 1987), then it follows that Group II caves are older than Group I. The formation of the Group I caves started only when the phreatic level coincided with the fine-grained micaceous quartzite layer. The ancient phreatic levels observed inside the caves match those observed outside, marked by alluvial terraces, paleochannels, and fluvial erosional features. Inasmuch as up to the present there has been no absolute age determination of these uplift episodes in the study area, dating can only be relative.

The presence of similar features in caves from the LSD (Luminárias) and CSD (Carrancas) points to comparable speleogenetic processes. Most LSD and CSD caves have cave patterns and cross sections similar to ISD (Ibitipoca) Group II caves, but there are some differences. Cave galleries in the LSD and especially in the CSD have stronger stratigraphic control (the green mica schist layers) than in ISD Group II caves.

Underground water flow in the LSD, STSD (São Tomé), and CSD seems to be quite similar to that of the ISD Group II, but vertical flow through fractures is constrained by impermeable layers, concentrating pipe formation (and subsequent cave conduits) to directly above schist lenses.

LSD and CSD caves do not have well-developed cave levels, compared to ISD caves. In the ISD, repeated cycles of relative base-level stability and uplift do not seem to have affected the LSD and CSD with the same intensity.

Conclusions

The quartzite caves from the Andrelândia Speleological Province share similar characteristics, owing their formation to similar processes involving initial silica dissolution, increase in porosity and permeability, pipe formation, and conduit enlargement due to mechanical erosion by flowing water. Sanding and piping processes seem to be involved in the genesis of most quartzite (and arenitic) caves around the world, as it can be seen in Australia (Jennings, 1983; Young, 1986), Venezuela (Urbani, 1986), South Africa (Martini, 1987), and Libya (Busche and Sponholz, 1992). In the ASP, the essential conditions to quartzite cave formation are as follows:

- One initial long period of phreatic level stability, followed by uplift (probably a sequence of stability–uplift cycles).
- A large difference between local and regional base levels.
- Rock layers specifically susceptible to piping (fine-grained micaceous quartzite layers in ISP), or impermeable layers (schist lenses in LSD and CSD).

Differences in caves of a given district or between districts can be explained by changes in one or more of the above conditions.

Acknowledgments

I thank the administration of the Ibitipoca State Park, the IEF-MG, and the Federal University of Rio de Janeiro for supporting the research since 1990. Field work was made possible thanks to the help of the SPEC and Bambuí speleogroups. I would also like to thank Augusto Auler, Georgette Dutra, Henrique Dayan, Renato R. Andreis, and Allison Sodré-Corrêa for their encouragement and helpful discussions.

References

Almeida Abreu, P.A., and Pflug, R., 1994, The geodynamic evolution of the southern Serra do Espinhaço, Minas Gerais, Brazil: *Zentralblatt für Geologie und Paläontologie,* no. 1-2, p. 21-44

Almeida, J.C.H.,, de, 1992, *Mapeamento geológico da Folha Luminárias-MG (1:50,000), com ênfase na análise estrutural dos metassedimentos do Ciclo Deposicional Andrelândia:* UFRJ, CCMN, Intituto de Geociências, Msc Thesis, 102 p.

Babinski, M., Chemale, J.R., F. and Van Shmus, W.R., 1991, Geocronologia Pb/Pb em rochas carbonáticas do Supergrupo Minas, Quadrlátero Ferrífero, Minas Gerais: 3rd Congresso Brasiliero de Geoquímica, *Abstracts,* p. 628-631.

Busche, D., and Sponholz, B., 1992, Morphological and micromorphological aspects of the sandstone karst of eastern Niger: *Zietschrift für Geomorphogie,* v. 85, p. 1-18.

Corrêa Neto, A.V., Anísio, L.C.C., and Brandão, C.P., 1993, Um endocarste quartzítico na Serra do Ibitipoca, SE de Minas Gerais, *in* 7th Simpósio de Geologia de Minas Gerais: Proceedings, *Boletim SBG núcleo MG,* v. 12, p. 83-86.

Corrêa Neto, A.V., Anísio, L.C.C., and Brandão, C.P., and Cintra, H.P., 1995, Gruta das Bromélias (MG 042), município de Lima Duarte, MG: Uma das maiores cavernas em quartzito do mundo: *Espeleotema,* v. 18. In press.

Corrêa Neto, A.V., Dutra, G., and Baptista Filho, J., 1996, Espeleogênese em quartzitos: Comparações entre as serras de Ibitipoca, Luminárias e Carrancas (MG): *38th Congresso Brasileiro de Geologia,* Salvador, Proceedings, v. 4, p. 555-558.

Dayan, H., and Baptista Filho, J., 1984, Caracterização dos fabrics tridimensionais nos rods de quartzo da base da serra de Carrancas, SE de Minas Gerais: *33rd Congresso Brasileiro de Geologia,* Rio de Janeiro, Proceedings, v. 7, p. 3371-3380.

Dayan, H., Baptista Filho, J., Fonseca, M.S., and Mansur, R.L., 1987, Determinação do estado de deformação finita em quartzitos das serras de Carrancas e Bicas, SE de Minas Gerais: *Anais da Academia Brasileira de Ciências,* v. 79, p. 203-211.

Ebert, H., 1955, Pesquisas na parte suleste do Estado de Minas Gerais: *Relatório Anual da Divisão de Geologia e Mineralogia,* p. 78-89.

Jennings, J.N., 1983, Sandstone pseudokarst or karst? *in* Young, R.W., and Nanson, G.C., eds., *Aspects of Australian sandstone landscapes:* Australia and New Zealand Geomorphology Group, Wollongong, p. 21-30.

Machado Filho, L., Ribeiro, M.W., Gonzales, S.R., Schenini, C.A., dos Santos Neto, A., Palmeira, R.C.B., Pires, J.L., Teixeira, W., and Castro, H.E.F., 1983, Geologia, *in* Projeto Rambrasil, Levantamento de Recursos Minerais, v. 32, Folio 23-24.

Martini, J., 1987, Karst features in quartzite of South Africa: *Karstologia,* v. 9, p. 45-52.

Nummer, A.R., 1990, Estratigrafia e estruturas do Grupo Andrelândia na região de Santa Rita do Ibitipoca-Lima Duarte, Sul de Minas Gerais: *36th Congresso Brasileiro de Geologia,* Natal, Proceedings, v. 6, p. 2786-2797.

Palmer, A.N., 1987, Cave levels and their interpretation: *National Speleological Society Bulletin,* v. 49, p. 50-66.

Palmer, A.N., 1991, Origin and morphology of limestone caves: *Geological Society of America Bulletin,* v. 103, p. 1-21.

Perez, R.C., and Grossi, W.R., 1986, The quartzitic speleological district of the Parque Florestal Estadual do Ibitipoca, Minas Gerais, Brazil: *9th International Congress of Speleology,* Barcelona, Proceedings, p. 12-14.

Romero, A.E., and Lima, M.T., 1989, The Itacolomi Quartzite Karst, Brazil: Application of the underground drainage direction prediction method: *10th International Congress of Speleology,* Budapest, Proceedings, p. 476-480.

Rubbioli, E., 1996, O pico do Inficionado: A exploração da gruta mais profunda do Brasil: *O Carste,* v. 8, no. 3, p. 47-54.

Szczerban, E., and Urbani, F., 1974, Carsos de Venezuela, Parte 4: Formas cársicas en areniscas Precambricas del Território Federal Amazonas y Estado Bolivar: *Boletin Society Venezolana Espeleologica,* v. 5, p. 27-54.

Urbani, F., 1986, Notes on the origin of caves in Precambrian quartzites of Roraima Group, Venezuela: *Interciência,* v. 11, p. 298-300.

Trouw, R.A.J., Ribeiro, A., and Pacciullo, F.V.P., 1983, Geologia estrutural do Grupo São João del Rei, Carrancas e Andrelândia, sul de Minas Gerais: *Anais da Academia Brasileira de Ciências,* v. 55, p. 71-85.

White, W.B., Jefferson, G.L., and Haman, J.F., 1966, Quartzite karst in southeastern Venezuela: *International Journal of Speleology,* v. 2, p. 309-314.

Young, R.W., 1986, Tower karst in sandstone–Bungle Bungle massif, northwestern Australia: *Zeitschrift für Geomorphologie,* v. 30, p. 189-202.

7.3.2 Quartzite Caves in Southern Africa

J.E.J. Martini

Abstract

Several karst systems occur in lower Precambrian and Ordovician quartzite in southern Africa. They are developed only in high rainfall areas. At the surface, pinnacle fields and lapiaz are common. But the drainage remains surficial, except in very localized areas where dolines, swallow-holes, stream caves, and resurgences occur. The underground channels seldom extend more than a kilometer upstream from the resurgences, and generally much less. Caves over 2 km in total passage length have been mapped.

The speleogenesis is by initial weathering of the quartzite. This is transformed into crumbly material and then removed by piping, thus forming caves. Other types of caves have developed by the erosion of deeply weathered diabase sills intrusive into quartzite. Some caves were initiated by gravitational rifting in quartzite and then enlarged by stream action.

Introduction

Karst phenomena in quartzite are well developed in the high-rainfall area of southern Africa, where about 200 caves are known. Some of them have been mentioned in the literature since the beginning of the 20th century (Hall, 1906). But their genesis has been investigated in more detail only since the 1970's (Marker, 1976, Martini, 1979). There are three main cave districts: the Mpumalanga Escarpment, the Cape Peninsula and the Chimanimani Mountains (7.3.2 Figure 1).

7.3.2 Figure 1. Location of main quartzite karst areas.

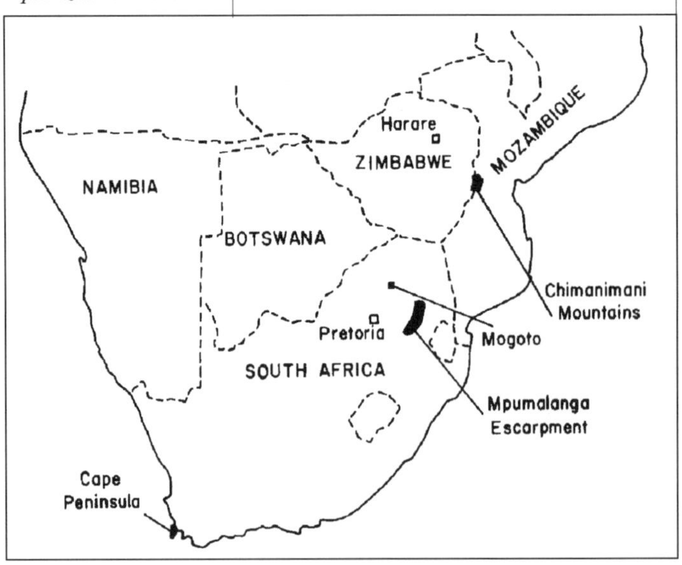

Mpumalanga Escarpment

Morphologically speaking, the Mpumalanga Escarpment represents the mountainous transition from the "High-Veld" (about 1500 m) to the "Low-Veld" (about 500 m). It receives high rainfall, predominantly above 1 m/year. Geologically this is the eastern edge of the Transvaal Basin, filled with Late Archaean to Early Proterozoic sediments deposited in a cratonic environment. About 60 caves have been explored in the quartzite of the Wolkberg Group and Black Reef Formation (Late Archaean), and in the quartzite of the Timeball Hill and Daspoort Formations (Early Proterozoic).

These quartzite beds are affected by low-grade metamorphism, although they are undeformed and monoclinal, forming cuestas and plateaus gently inclined to the west. Their surface is commonly characterized by fields of pinnacles reminiscent of lapiaz. This impression of widely developed karst is misleading, however, as the drainage remains surficial in most areas. The morphology associated with underground drainage—that is dolines, swallow holes, caves, and resurgences—occupies less than 1% of the lapiazed surface. The caves are generally grouped in very localized systems, usually situated at the downdip edge of plateaus, and do not extend more than a few hundred meters from the resurgences.

Berlin Cave Systems

The most spectacular caves and surface karst morphology are developed at Berlin, near Nelspruit (Martini, 1987). The area consists of an inclined plateau of Black Reef Quartzite, 25 m thick, resting on a granite-gneiss basement (7.3.2 Figure 2). It is a fairly pure orthoquartzite. A 0.3-m-thick layer of shale-like tuff lies a few meters above the base of the quartzite. The Black Reef is covered by a thin veneer of deeply weathered impure dolostone. Close to the edges of the plateau, the surface morphology consists of pinnacles, which farther away grade into a typical pavement (7.3.2 Figure 2). The caves are grouped into two systems totalling about 2.5 km of passages. Large dolines collect water which sinks underground, flows downdip, and is forced again to the surface by the damming effect of a basic dike (7.3.2 Figure 2).

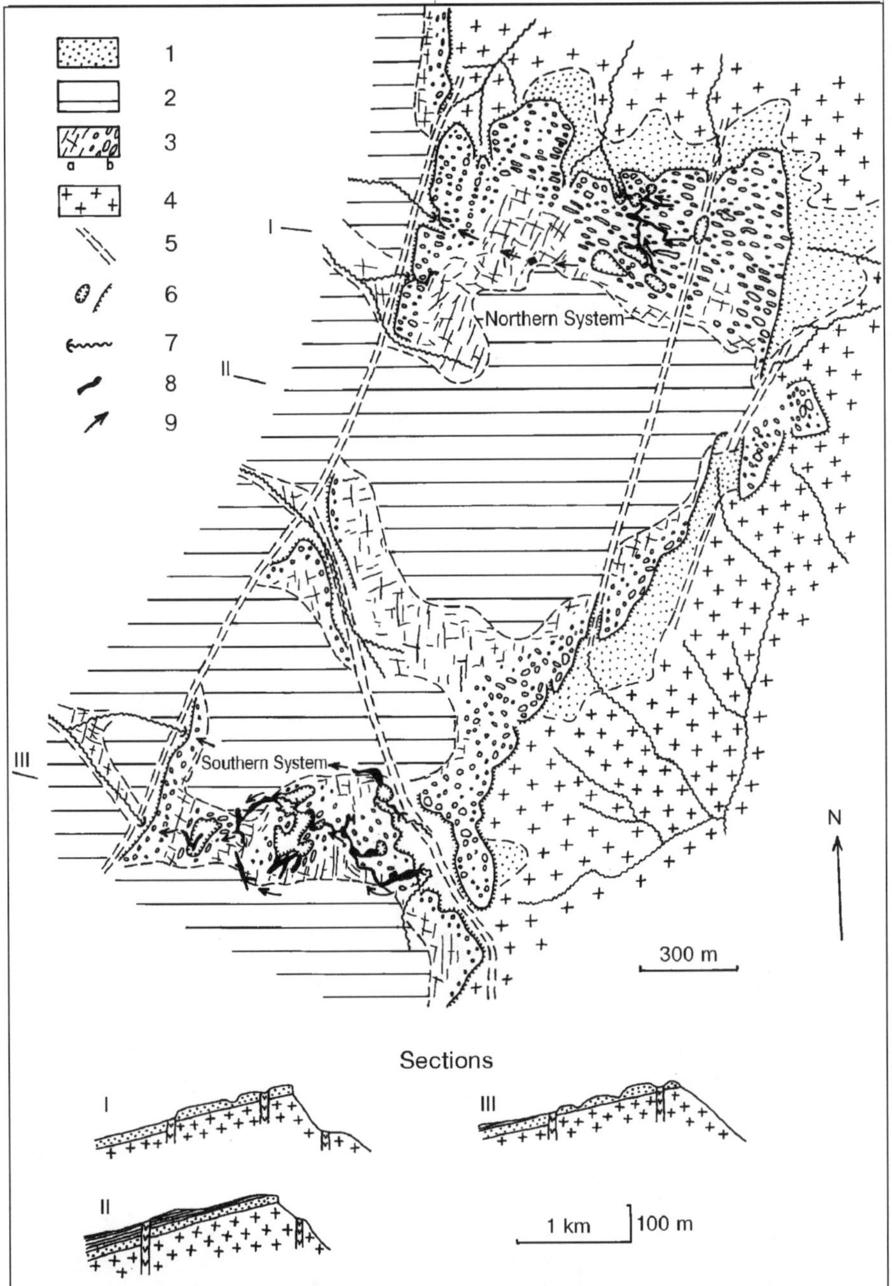

*7.3.2 Figure 2. Map of the Berlin Cave Systems. **1** is scree. **2** is impure dolostone, more or less weathered. **3** is Black Reef Quartzite forming pavement (**a**), and pinnacles (**b**). **4** is granite basement. **5** is diabase dike. **6** is doline and edge of plateau. **7** is stream and swallow hole or resurgence. **8** is cave. **9** is the direction of flow.*

Most of the cave passages are actively vadose and developed on the thin tuff layer, which acts as an impervious horizon. In cross section, the passages predominantly range from wide and low to equidimensional, with arched ceilings and flat floors of tuff constituting the broad beds of small streams. Joint-controlled passages are less common.

Furthermore, the size of a single passage can vary considerably along the strike. For instance, it can range from only a few centimeters of open space above the tuff to chambers as large as 10 by 10 m (7.3.2 Plate 1). Such large voids have been opened in zones of deeply weathered crumbly quartzite, which can be rapidly eroded, whereas the constrictions are controlled by only mildly weathered hard rock. In order to examine the most weathered rock, samples had to be impregnated in situ, otherwise they would have disintegrated into sand before reaching the laboratory. Analyses of water from the caves and resurgences show a low silica content (1 to 9.5 ppm), lower than average for groundwater, suggesting undersaturation to saturation with respect to quartz (Martini, 1979).

Magnet Cave System

The Magnet Cave System is part of another karst area that is somewhat different from the Berlin Karst. It is developed on a 5 to 10° inclined plateau of Daspoort Quartzite. The latter is a 25-m-thick layer of clean well crystallized orthoquartzite, which forms a cliff at the borders of the plateau. A thin layer of shaly tuff is interstratified a few meters below the top, which consists of a l-m-thick magnetite-rich bed. The magnetite quartzite is overlain by 2-3 m of shale and then by a few meters of quartzite (7.3.2 Figure 3).

Several small dolines and sinkholes are observed on the plateau, as well as shallow holes, but pinnacles are poorly developed. The underground channels are vadose and occur at shallow depth, not more than 10 m deep. The most important channel system, Magnet Cave (Martini, 1991), consists of a complex network of low stream passages with a total length of more than 2 km (7.3.2 Figure 3 and 7.3.2 Plate 2). As in the Berlin Caves, the passages are closely controlled by the thin tuff layer. Some of them have perennial streams, whereas others are active only during flood, or are fossil and partly filled with silt. Although the passages are exclusively vadose, distributaries are common with flow being generally downdip. As in most quartzite caves of the region, the degree of weathering of the rock is variable, but the size of the passages is more constant than at Berlin. The roof has collapsed in places, forming the numerous entrances to the cave.

Other cave systems are developed stratigraphically below the tuff, such as the north-westernmost one, or above the shale in the top-most quartzite (7.3.2 Figure 3). One

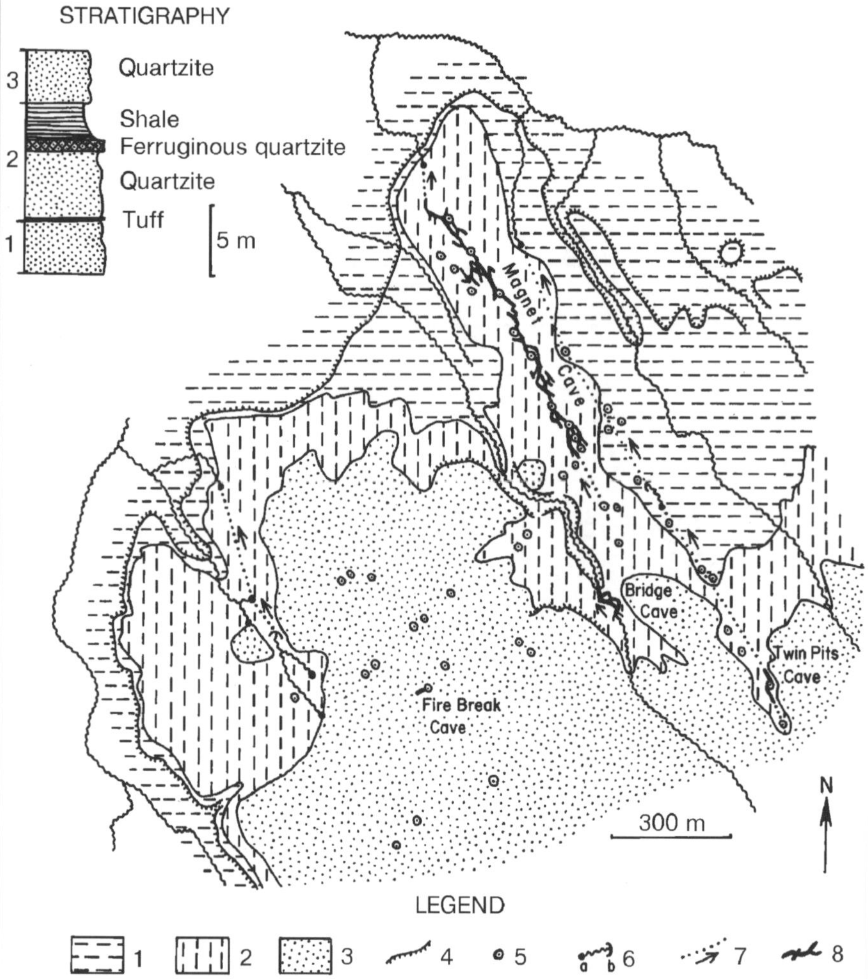

7.3.2 Figure 3. Map of the Magnet Karst System. ***1*** *is the Twin Pit Member,* ***2*** *is the Magnet Member.* ***3*** *is the Fire Break Member.* ***4*** *is an escarpment.* ***5*** *is a doline.* ***6*** *is a surface stream,* ***a*** *is a spring , and* ***b*** *is a swallow hole.* ***7*** *is an inferred underground flow.* ***8*** *is a cave.*

system has reached an advanced maturity stage, because most of the cave roof collapsed, thus forming a gully and leaving only a 60-m-long segment of underground drainage (Bridge Cave, 7.3.2 Figure 3).

Quartzite and Diabase Caves of the Transvaal

In the Transvaal Basin, caves of a different type lie at various stratigraphic levels. They are developed in deeply weathered sills of diabase that intruded into quartzite. Mogoto is the most important cave of this type, totalling 1.6 km of passages (7.3.2 Figure 1). It is controlled by a 3-m-thick sill completely altered into clay, but the Black Reef Quartzite itself, which always forms the ceiling, is practically unweathered (Martini, 1987). Most of the passages are vadose and more or less parallel to the dip of the sill. The size of a passage may vary considerably along the strike: from 15 m wide and 3 m high, with floor and ceiling of quartzite, and side walls of weathered diabase, to impenetrable small channels on top of the sill. These constrictions develop where they are oriented along the strike or obsequent to the dip of the sill. In these places the flow regime was locally phreatic, and the cross sections became very small, due to limited erosive power.

As outlined in Chapter 4.1.8, in the usual quartzite caves, the genesis took place in two phases:

- The weakening of the quartzite cohesion by dissolution of quartz along crystal boundaries, leaving a friable material.
- Speleogenesis proper, by mechanical removal of this friable material.

In the case of quartzite-diabase caves, two phases may also be distinguished:

7.3.2 Plate 1. Large passage in deeply weathered quartzite, Berlin Caves (photo by M. Sefton).

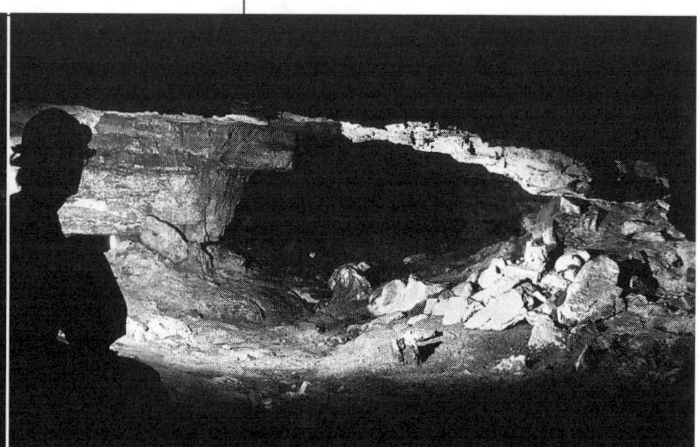

7.3.2 Plate 2. Typical wide passage in Magnet Cave.

- Weathering of diabase, with development of thin planar voids on top of the sill by irregular compaction of the clay.
- Capture of a surface stream flowing through this initial void, which enlarges the void considerably by erosion.

In this process the quartzite plays the role of a protective roof, to which the cave owes its existence.

Cape Peninsula

Most of the Cape Peninsula is underlain by the Table Mountain Group. This consists of fairly pure tabular Ordovician quartzite, several hundred meters thick, affected by very low-grade metamorphism. In this quartzite nearly 100 caves lie in a small area of the northern half of the peninsula (Moore, 1944). Most of them are small, extending no more than a few tens of meters, but the largest of them, Ronan's Well Cave, reaches a total length of 800 m. They are typically vadose, with small streams, and unlike the caves of the Mpumalanga Escarpment the passages are strongly joint controlled. The speleogenesis is also by piping through deeply weathered friable quartzite. The surface karst morphology, however, is not well developed. A few other caves in the same area are of a slightly different nature. They developed along rifts formed at the edge of the tabular plateaus. Their extensions are mostly linear and vertical, and one cave, the Bat-Giant System reaches a total length of 1.65 km. The small streams flowing on the plateau are collected by these cracks and form underground streams. The speleogenesis can be visualized as a combination of initial gravitational tectonics, opening rifts at the edge of the plateau, and subsequent enlargement by weathering and erosion.

Chimanimani Mountains

Deep caves were explored during the early 1990's in the Chimanimani Mountains, which form the border between Zimbabwe and Mozambique (7.3.2 Figure 1). They are on both sides of the border, but are restricted to the Mawenge Plateau, less than 1 km^2 in area on top of steep mountains at about 2200 m. It receives a high rainfall, about 1.3 m/yr. The caves develop in relatively pure white quartzite of the Umkondo Group, which is several hundred meters thick and is Early Proterozoic in age. The plateau is irregular and rugged, crisscrossed by a network of trenches and deep rifts corresponding to fractures (Koliasnikoff, 1994). The latter collect numerous small surface streams into caves, which are all vertical to subvertical crevasse-shaped potholes. A dozen of them have been explored, the deepest reaching 305 m below the surface (7.3.2 Figure 4).

No horizontal passages have been reached at the bottoms, as the caves end in blockages or in fissures too narrow to be negotiated. At a depth of 500 meters below the plateau, however, a resurgence has been explored over a length of 200 m, and it is also strongly rift-controlled. In the caves, the quartzite is compact to friable, indicating a variable degree of weathering. It is possible, as far as genesis is concerned, that these caves are similar to the tectonic caves of the Cape Peninsula mentioned in the previous section, with initial development of open fractures by gravity tectonics fostered by the steep topography, followed by mechanical enlargement by running water.

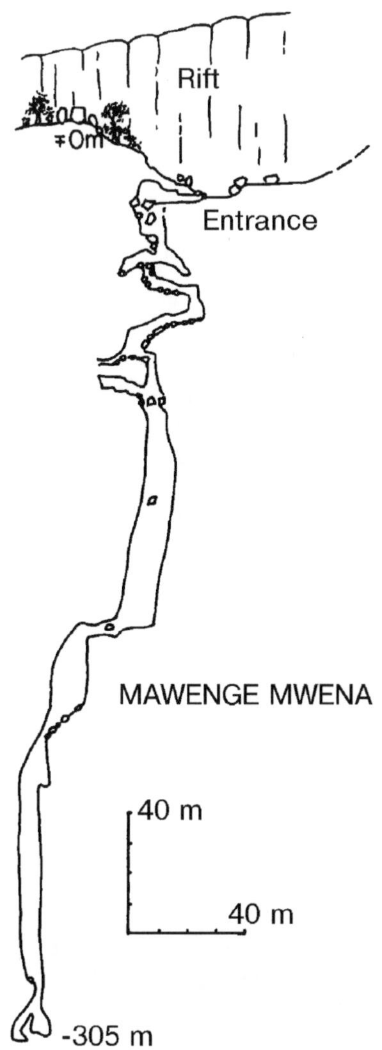

7.3.2 Figure 4. Section of Mawenge Pothole, after Aucamp (1994).

References

Aucamp, J.P., 1994, Mawenge Mwena (Z9): *Bulletin of the South African Spelaeological Association,* v. 35, p. 49-55.

Hall, A.L., 1906, On the geology of the country between Lydenburg and Devil's Kantoor: *Annual Report of the Geological Survey of Transvaal for 1905,* p. 39-57.

Koliasnikoff, A., 1994, Reconnaissance of the Frontier Shaft Area: *Bulletin of the South African Spelaeological Association,* v. 35, p. 8-13.

Marker, M.E., 1976, Note on some South African pseudokarst: *Boletin de la Socidad Venezolana Espeleologia ,* v. 7, p. 5-12.

Martini, J.E.J., 1979, Karst in Black Reef Quartzite near Kaapsehoop, Eastern Transvaal: *Annals of the Geological Survey of South Africa,* v. 13, p. 115-128.

Martini, J.E.J., 1987, Les phénomènes karstiques des quartzites d'Afrique du Sud: *Karstologia,* v. 9, p. 45-52.

Martini, J.E.J., 1991, Magnet: A large quartzite cave system in the eastern Transvaal: *Bulletin of the South African Spelaeological Association,* v. 31, p. 44-59.

Moore, I.W., 1944, Caves of the Muizenberg Range: *Journal of the Mountain Club, South Africa,* v. 47, p. 7-22.

Part 8: Some Implications of Speleogenetic Studies

8.0 Overview

Part 8 combines the theoretical with some applied aspects of speleogenetic studies. Worthington, Ford, and Beddows (Chapter 8.1) show the important implications of what might be termed "speleogenetic wisdom" when studying groundwater behavior in karst. They examine some standard hydrogeologic concepts in the light of knowledge of caves and their patterns, considering a range of case studies to identify the characteristic enhancement of porosity and permeability due to speleogenesis that occurs in carbonate rocks.

The chapter focuses on unconfined carbonate aquifers, as these are the most studied from a speleologic perspective and are most important for water supplies. Four aquifers, differing in rock type, recharge type (allogenic and autogenic), and age (Paleozoic, Mesozoic, and Cenozoic), are described in detail to demonstrate the extent of dissolutional enhancement of porosity and permeability. All four cases are similar in hydraulic function, despite the fact that some of them were previously characterized as different end members of a "karst to non-karst" spectrum. Enhancement of *porosity* by dissolution is relatively minor. Enhancement of *permeability* is considerable. This is because dissolution has created dendritic networks of channels able to convey 94% or more of all flow in the aquifer, with fractures providing a small proportion and the matrix a negligible amount.

These conclusions may be viewed as a warning to hydrogeologists working in carbonate terranes—probably the majority of unconfined aquifers function in a similar manner. Sampling is a major problem in their analysis, because boreholes (the conventional exploration tool in hydrogeology) are unlikely to intersect the major channels that are conveying most of the flow and any contaminants in it. We estimate, using examples of comprehensively mapped caves, that the probability of a borehole intersecting a conduit ranges from 1 in 50 to 1 in 1000 or more. Boreholes simply cannot be relied upon to detect the presence of caves or to "characterize" the hydrologic functioning of cavernous aquifers. Wherever comprehensive evidence has been collected in unconfined carbonate aquifers (cave mapping, plus boreholes, plus lab analysis of core samples), dissolution inexorably results in a similar structure, with channel networks providing most of the permeability of the aquifer, yet occupying a very minor fraction of its volume (Worthington, Ford, and Beddows, Chapter 8.1).

8.0 Plate 1. Danivsky Collapse, formed February 1998, opened an inundated gypsum maze cave, western Ukraine. (photo by A. Klimchouk).

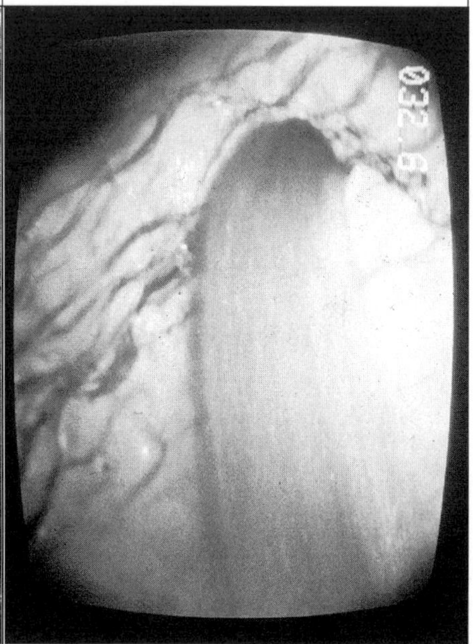

8.0 Plate 2. A borehole camera shows water flowing into the borehole from a small conduit encountered in the wall. (photo supplied by P. Younger).

Lowe (Chapter 8.2) focuses on developments in understanding the vital role played by karstic porosity (broadly viewed as being a product of speleogenesis), in the migration of mineralizing fluids (or hydrocarbons) and in their deposition (or storage). He also comments on the potential role of new speleogenetic concepts in developing greater understanding in the future.

Although some early workers were clearly aware of evidence for a relationship between karstic porosity and economic deposits—and others noted its theoretical likelihood—it has been ignored by many until relatively recent times. This shortfall has gradually been redressed; new understanding of the extent and variety of karst processes is ensuring that new relationships are being recognized and new interpretations and models are being derived. The chapter clearly demonstrates the wide applicability of speleogenetic knowledge to issues in economic geology.

In Chapter 8.3 Aley provides an overview of the water and land-use problems that occur in areas with conduit aquifers. He stresses that sound land management must be based on an understanding that karst is a three-dimensional landscape where the surface and subsurface are intimately and integrally connected.

Failure to recognize that activity at the surface affects the subsurface—and the converse—has long been the root cause of many of the problems of water and land use in karst regions. Karst areas have unique natural-resource problems, whose management can have major economic consequences. Although there is an extensive literature on the nature of particular problems, resource protection, and hazard minimization strategies in karst, it rarely displays an advanced understanding of the processes of conduit formation and their characteristics—yet these are always involved.

8.1. Porosity and Permeability Enhancement in Unconfined Carbonate Aquifers as a Result of Solution

Stephen R.H. Worthington, Derek C. Ford, and Patricia A. Beddows

Abstract

Solution processes in unconfined carbonate aquifers result in a network of channels. To characterize the enhancement of porosity and permeability by dissolution, we examine four contrasting carbonate aquifers:

- Paleozoic dolostone.
- Paleozoic limestone.
- Mesozoic chalk.
- Cenozoic limestone.

In all four cases, the channels add little to the porosity, but they enhance the permeability of fractured rock by one to three orders of magnitude. Similar porosity and permeability changes are predicted for all unconfined carbonate aquifers, in both dolostone and limestone, in both allogenic and autogenic settings, and in carbonate rocks of all ages.

Introduction

Carbonate rocks are chemical precipitates. Chemical processes therefore affect their permeability and porosity at all phases, from deposition through shallow or deep burial to exposure at the surface, as explained in Chapter 3.2.

During burial, porosity reduces from an original 40 to 70 percent (Choquette and Pray, 1970) to only a few percent, or even less in the case of deep burial. Porosity enhancement and the formation of caves does occur to a limited extent in this setting (Chapter 5.2).

But porosity enhancement is much more important on a global scale once the carbonate rock is elevated above sea level and any low-permeability overlying rocks are removed, so that the aquifer becomes unconfined.

This chapter focusses on the setting of unconfined carbonate aquifers. These aquifers contain at least 90 percent of the world's explored caves (Palmer, 1991), provide important water resources, and host almost all groundwater contamination problems in carbonate rocks.

Dissolution by infiltrating meteoric water has long been recognized to result in a significant enhancement of permeability in at least some carbonate aquifers. This is commonly recognized where major caves have been explored or where abundant karst landforms lie at the surface. However, hydrogeologists have no consensus as to the proportion of carbonate aquifers with significant amounts of dissolution-enhanced permeability.

The creation of channels, conduits, and caves results in an increase in porosity, but we find widely contrasting views on the extent of this increase. For example, Freeze and Cherry (1979, p. 27) give a porosity range of 0 to 20 percent for ordinary limestone, and 5 to 50 percent for karst limestone. However, Bonacci (1987, p. 47) compiled data from 20 case studies of different karst aquifers around the world, and found that the effective porosity ranges from 0.17 percent to 10 percent, with a median value of only 1.6 percent.

Karst features such as dolines are often taken to be definitive evidence of a karst aquifer below. But in many areas with sur-

8.1 Plate 1. Infilled dolines in Jurassic limestone revealed by quarrying near Woodstock, England. Away from the quarry, obvious karst features are absent.

ficial sediment the dolines may be infilled and unrecognizable (8.1 Plate 1). Furthermore, caves can exist beneath insoluble caprock, so no evidence for karst features may appear at the surface. Mammoth Cave (Chapter 5.3.8) is a good example, because most of that cave underlies ridges capped by sandstone.

In this chapter we consider a range of case studies in order to identify the characteristic enhancement of porosity and permeability found in carbonate rocks in unconfined aquifers.

Terminology

Confusing and contradictory terminology is a major barrier to communication of ideas regarding flow in carbonate aquifers. Meteoric water circulating through an unconfined carbonate aquifer tends to produce an integrated network of dissolutionally enlarged fractures or fissures (see the chapters of Parts 3 and 4 in this volume). These have been called channels (Choquette and Pray, 1970), macrofissures (Reeves, 1979), or secondary fissures (Price et al., 1993). Larger examples are known as conduits (>1 cm diameter) or caves (~>1 m, enterable by people) (White, 1988; Ford and Williams, 1989). The interconnectivity of the enlarged fractures is of prime importance, resulting in a channel network.

In this account we will use the term channel to refer to all interconnected dissolutional enlargements along joints, faults, and bedding planes, so we include a wide aperture range, from much less than 1 mm up to several meters.

Conceptual Models and Measurement of Porosity and Permeability

In order to measure the porosity and permeability of a carbonate aquifer, we first need a conceptual model of it so that the different porosity and permeability components can be representatively sampled. The different modes of flow in carbonate aquifers have been classified in several of ways. Burdon and Papakis (1963) introduced the term diffuse circulation to describe Darcy flow through pores and fissures in carbonates, differentiating it from a second mode of flow, concentrated circulation. White and Schmidt (1966) suggested the terms diffuse flow and conduit flow. More recently, Atkinson (1985) divided flow into diffuse, fissure, and conduit components, whereas Quinlan and Ewers (1985) suggested granular, diffuse, fracture, and conduit components. Thus a plethora of terms has been used!

The term double porosity is becoming commonly applied in hydrogeology to refer to fractured rock aquifers, where flow in both the fractures and the rock matrix need to be considered. If we add the third element of channels, then we can simply refer to carbonates as triple porosity aquifers (Worthington, 1994; Quinlan et al., 1996).

The rock matrix can be sampled from borehole cores or in fresh quarry faces. This avoids a potential problem of near-surface porosity changes that occur with weathering of natural rock outcrops. Samples can be tested for permeability in a permeameter, and their porosity can also be measured. Borehole packer tests with small separation of the packers (<1 m) can also measure the permeability of the unfractured bedrock.

Fracture permeability is usually assessed by pump, packer, or slug tests in boreholes. These tests measure the aquifer response (usually a change in water level) to a stress such as pumping water or adding or withdrawing a slug of water in a borehole. Fracture porosity can be calculated by using the cubic law with packer test data (Snow, 1968):

$$b = (12\, T \mu / \rho\, g)^{0.33} \qquad (1)$$

where b is fracture aperture, T is transmissivity (determined from packer tests), μ is dynamic viscosity (0.0179 g/cm/s at 0° C, falling to 0.0080 at 30°), ρ is fluid density (1.0 at 0°, 0.996 at 30°), and g is the acceleration due to gravity. Equation 1 assumes that all the flow is in a single smooth-walled fracture. The roughness of fracture walls means that true width will be somewhat larger than the equation indicates. An alternative method of determining fracture porosity is to use slug or pump test results combined with observations of fracture spacing in cores or quarry faces.

Channel porosity (including conduits and caves) can be calculated in several ways. Atkinson (1977) estimated it from the volume of prestorm water flushed out of a spring following heavy rains. Such water is usually clear and has lower suspended-sediment and higher dissolved solid concentrations than the floodwater that succeeds it. There are also accurate survey data for many caves, which can be used to estimate cave volume. A third method is to model a channel network, calculating channel aperture from

$$A = C\, R / v \qquad (2)$$

where A is the channel cross section, C is the catchment area of the channel, R is the runoff (precipitation minus evapotranspiration), and v is channel velocity. Channel velocity may be approximated by an average from tracer tests between sinks and springs in carbonate aquifers. In 8.1 Figure 1 the results of 2877 such tests reported from around the world are shown. The global average channel flow velocity from these tests is 0.022 m/s (78 m/hr).

A critical component of channeling in carbonate rocks is the interconnectivity of the channels. Equation 2 requires the assumption that channels form a tributary pattern. The theoretical studies in Part 4 in this volume and the examples of caves in Part 5 both show that channels are expected to be interconnected. However, patterns of flow often are not clear on cave maps, because the maps show both the passages with active streams and those that have been abandoned.

A clearer pattern emerges if a map displays just the active flowpaths. Three representative channel networks are presented in 8.1 Figure 2, showing the currently active cave streams in particularly well-mapped cave systems (though in each case future cave exploration and mapping may add substantially to each network). In each cave the widths of most passages are exaggerated to make them visible. The smaller headwater passages are exaggerated in width up to 20 times, with minimum true widths being about 0.5 m.

Exaggeration of passage width to aid comprehension is common in small-scale cave maps. But it leads to the false impression that explorable caves occupy a significant fraction of the volume of a limestone aquifer. In each case several tributary cave streams join together, with the combined flow emerging at a spring. In all three caves both discharge and usually passage size increase downstream. Thus channel networks in carbonate aquifers in some respects resemble the flow patterns of surface rivers.

Examples of Porosity and Permeability Measurement in Carbonate Aquifers

The four examples of carbonate aquifers described below illustrate some of the different ways in which the three porosity elements of matrix, fractures, and channels have been characterized. The four aquifers have been chosen to provide a wide range of rock type (limestone and dolostone), recharge (allogenic and autogenic), and age and diagenetic maturity (Paleozoic, Mesozoic, and Cenozoic).

Lower Paleozoic Dolostone: Smithville, Ontario, Canada
The Niagaran dolostone (Silurian) of southern Ontario is in a low-dip cratonic setting like that at Mammoth Cave (Chapter 5.3.8). Surficial karst features are rare, as the area has been glaciated, and the bedrock is mostly covered by several meters of glaciolacustrine silt, clay, and other deposits. Nevertheless, where the dolostone is exposed, it is readily karstified and exhibits good pavement features. Small springs are common and a young cave, 250 m in length, has been mapped close to Smithville.

The hydrogeology was extensively investigated at Smithville (40 km east of Niagara Falls) where PCBs (very toxic polychlorinated biphenyls from electrical transformers) spilled into the bedrock. Tests on 1043 core samples gave an average porosity of 6.6 percent (P. Lapcevic, 1998, personal communication). Packer tests with 2 m spacing in 15 boreholes that penetrate the 38 m thickness of the dolostone gave a geometric mean hydraulic conductivity of 10^{-6} m/s and an arithmetic mean of 10^{-4} m/s (Lapcevic et al., 1997). The geometric mean is used to represent the average permeability in porous media, and it would apply if the fractures were discontinuous and poorly connected (Domenico and Schwartz, 1990, p. 67).

However, the bedding planes may be open for considerable distances here, as they are in most karst regions, so the fracture permeability is likely to be between the arithmetic mean and geometric mean.

The fracture porosity can be calculated using Equation 1, which gives 0.017 percent. This is a minimum value as it assumes smooth fractures and only includes bedding-plane porosity. Porosity due to jointing will add somewhat to this value, but in this low-dip cratonic tectonic setting, jointing is much less important than bedding planes with respect to both permeability and porosity. Some packer-test results were below the lower limit that the test could measure, 10^{-10} m/s, and these reflect the low permeability of the matrix.

Point discharges at bedding planes in quarries and natural rock outcrops around Smithville suggest that channeling is focussed on just four major bedding planes (8.1 Plate 2). With the assumption that 50 percent of runoff infiltrates through the overburden, Equation 2 was used to estimate that the mean channel porosity is 3 x 10^{-4} percent and the hydraulic conductivity is 3 x 10^{-4} m/s. Three pumping tests from a former municipal water well gave average hydraulic conductivities of 3 x 10^{-4} to 10^{-3} m/s. These high values support the theoretical calculations, suggesting that a high-permeability channel network surrounds the spill site.

Upper Paleozoic Limestone: Mammoth Cave Area, Kentucky, U.S.A.
Mammoth Cave (Chapter 5.3.8) is the most extensive known cave in the world. More than 500 km of interconnected passages have been

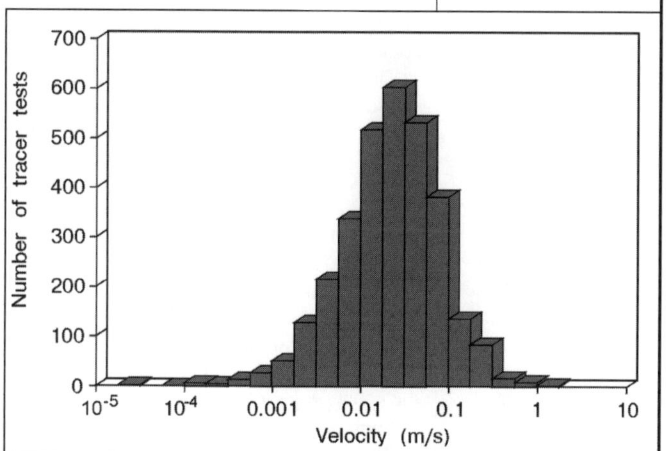

8.1 Figure 1. Velocities of 2877 traces in carbonate bedrock between sinking streams and springs.

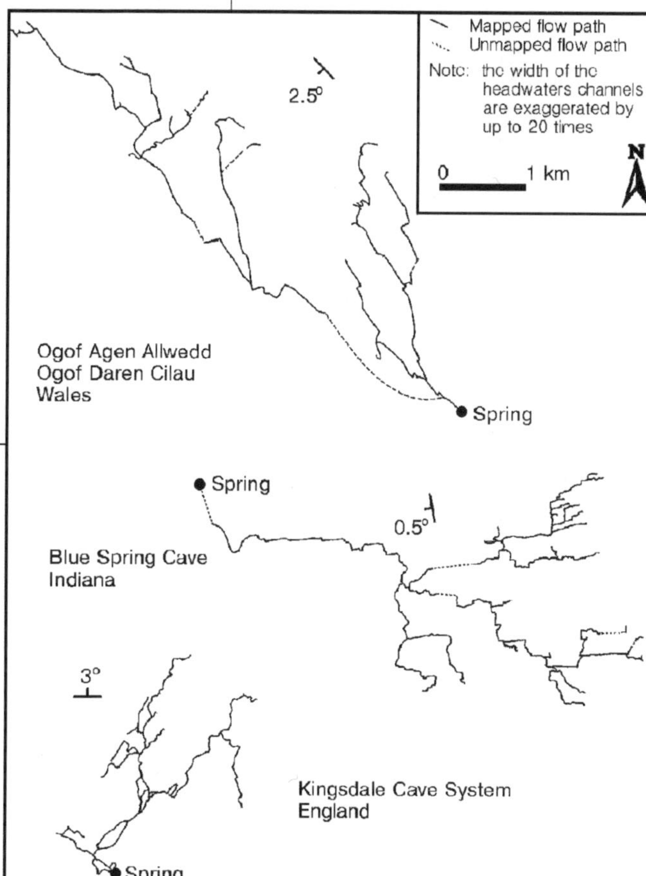

8.1 Figure 2. Dendritic flow networks in three well-mapped groundwater catchments in carbonate bedrock.

8.1 Plate 2. Ice buildups in winter on a dolostone quarry face near Smithville, Ontario, Canada, showing point discharges from three bedding planes. In the foreground, a channel discharge of 0.3 L/s is so great that ice has not formed.

mapped in the cave, which underlies a surface area of about 100 km^2. There is a considerable body of knowledge on the explorable channels and their hydrogeology, both from studies within the cave and from measurements at springs (Palmer, 1981 and Chapter 5.3.8.; White and White, 1989). However, much less is known about matrix and fracture flow in the aquifer.

The limestone in the Mammoth Cave area is Mississippian in age, and dips gently to the northwest. The matrix porosity of the limestone, as measured from rock samples, averages 2.4 percent (Brown and Lambert, 1963; Worthington et al., 1998). Tests on cores from boreholes have given hydraulic conductivity values of 1 x 10^{-11} m/s for the rock matrix. (Brown and Lambert, 1963).

Fracture porosity and permeability have been determined from tests on a series of boreholes around Mill Hole, a karst window into the unexplored trunk channel drainage to Turnhole Spring, one of the largest springs in the region (8.1 Figure 3; Worthington et al., 1998). Downhole videos show that the spacing of fractures (mostly bedding planes) with visible enlargement is 0.7 m, but major open fractures with apertures of 2 to 12 cm have an average spacing of 8.5 m. Slug tests in nine boreholes yielded arithmetic and geometric mean hydraulic conductivities of 3 x 10^{-5} m/s and 6 x 10^{-6} m/s, respectively. Fracture porosity is about 0.03 percent , as calculated with Equation 1, using the slug-test results and fracture spacing from the downhole videos.

Channel storage has been evaluated in two ways for the Turnhole Spring catchment, which has a karst area without surface drainage of 159 km^2 (Worthington et al., 1998). Repeat tracer tests along a 6 km channel from a cave to the karst window gave velocities ranging from 0.0025 m/s to 0.13 m/s (Quinlan and Ewers, 1989), with a geometric mean close to the value of 0.022 m/s used in Equation 2. A total of 34 major flow paths were mapped in the Turnhole Spring catchment by Quinlan and Ray (1981), and the volume of all these major channels was calculated by Equation 2 to be 2.0 x 10^6 m^3.

A second method of calculating channel storage uses the displacement of water of low electrical conductivity following rainstorms. Hess and White (1988) determined an average storm-water travel time of 35 hours for an 11 km distance. The average length of 17 km for the entire groundwater catchment would give a travel time of 54 hours, and this can be combined with discharge data to give an estimated channel volume of 2.6 x 10^6 m^3 for the Turnhole Spring catchment (i.e. this is the approximate volume of active channels during this runoff event). The average of these two estimates for channel storage is 2.3 x 10^6 m^3. Cave passages have been recognized to descend to a depth of at least 23 m below the water table (Palmer, 1989, p. 311). If 23 m is taken to be the effective thickness of the aquifer, then the calculated channel volume represents 0.063 percent of the volume of the limestone aquifer.

The average flow in channels has been calculated by considering a cross section through the aquifer midway between the divide and the springs (8.1 Figure 3). At this point the mean discharge is 3.16 m^3/s. Using Darcy's Law with an average hydraulic gradient of 0.004 (Quinlan and Ray, 1981), matrix flow through the cross section is approximately 2 x 10^{-8} m^3/s, and fracture flow is 0.007 to 0.03 m^3/s, depending on whether the arithmetic or geometric mean of the slug-test data is used. Channel flow constitutes the remaining 3.12 to 3.15 m^3/s. From Darcy's Law, an equivalent hydraulic conductivity for the channel flow is 2.8 x 10^{-3} m/s (Worthington et al., 1998).

Mesozoic Limestone: The English Chalk

Chalk is a distinctive form of limestone that is composed mainly of skeletal calcite from algae. It accumulates in thick homogeneous formations in deeper marine environments and displays relatively little diagenetic alteration. It is rare at the global scale but is prominent in northwestern France, southern England, and a few other regions. The Cretaceous Chalk is the most important aquifer in Britain, and there have been a correspondingly large number of studies of this aquifer. The rock matrix has a high porosity, usually in the range of 20 to 45 percent. But the pore throats are very small (0.1 to 1 μm), which results in very low hydraulic conductivity values (10^{-9} to 10^{-7} m/s). In deep-burial situations (confined flow), such as in the North Sea oil fields, fracturing enhances the hydraulic conductivity to about 10^{-6} m/s (Price et al., 1993).

In the unconfined Chalk aquifer there is further permeability enhancement along fractures. Aquifer tests have shown that the uppermost 30 to 60 m of the saturated Chalk has the highest permeability, which has been attributed to dissolution (Foster and Crease, 1975; Headworth et al., 1982). Furthermore, it has been suggested that permeability is greater along valleys, and that this is due to a combination of stress relief and dissolution.

This results in increasing hydraulic conductivity, from about 4 x 10^{-6} m/s under interfluves to about 6 x 10^{-4} m/s underneath valleys, and an average fracture porosity of 0.01 percent (Price et al., 1993). However, the enhanced hydraulic conductivity values under valleys only represent a small fraction of the total area of the aquifer. If we use the spatial variations in transmissivity for the unconfined chalk in the County of Berkshire (given in Connorton and Reed, 1978), then the areal average for hydraulic conductivity in the upper 50 m of the aquifer is 6 x 10^{-5} m/s.

Some studies suggest that high permeabilities are not restricted to valleys. In Berkshire, for instance, Morel (1980) gives 49 transmissivity values, with the highest occurring under an interfluve. Furthermore, in Dorset, Houston et al. (1986) found that table are discordant to the trends of the major dry valleys there. Gradients along the troughs were found to be up to 2 orders of magnitude lower than elsewhere in the region, implying that hydraulic conductivity in the troughs is correspondingly higher. The troughs terminate at the coast, at major springs. In the Chalk in Yorkshire, recent tracing from wells has demonstrated groundwater velocities of 400 m/day, both along dry valleys and from interfluves to the valleys (Ward et al., 1997).

The geometry of the high-permeability fractures is unclear. Price et al. (1989, p. 5) suggest that these fractures "should not be thought of as simple parallel-plate features." Occasionally, boreholes intersect openings with apertures >1 cm (Price et al., 1982; Price, 1996, p. 86), but more commonly the apertures are less. Dissolution has played the major role in enlarging apertures from their original (tectonic) widths (<0.1 mm) to typically 0.1 to 10 mm. It follows that the positive feedback processes that lead to channeling in other carbonate aquifers (8.1 Figure 2) should also occur here, resulting in interconnected networks of channels that are lenticular in cross section and are elongated along the fractures. This is supported by calculations using packer-test data. These indicate extensive interconnectivity of solutionally enlarged fractures in the Chalk (Price, 1994).

The porosity of these solutionally enlarged fractures, or channels, can be estimated from some studies. Foster and Crease (1975) considered that most of the permeability at a site in Yorkshire was imparted by "a small number (perhaps 4 or 5) of horizontal master conduits" with openings of perhaps 2 mm. Headworth (1978) found two to seven major fissures in each of a series of 19 boreholes in Hampshire.

Using the transmissivity of 2800 m^2/day measured there, one can show with Equation 1 that this transmissivity could be imparted by two fractures with 3 mm apertures or seven fractures with 1.9 mm apertures. If we postulate that the upper 50 m of the saturated zone represents the effective aquifer, then channel porosity from these two studies lies in the range 0.01 to 0.03 percent

Such channeling is supported by observations in quarries (Mortimore, 1993, p. 71) and by the occurrence of springs. Discharge from the Chalk is generally "from small springs and seepages into rivers" (Lloyd, 1993, p. 236). This concentration of discharge at specific points in places can be explained in structural terms (faulting) or lithologic terms (contact with an aquitard). However, the widespread occurrence of springs in the Chalk and in other carbonates most likely represents the discharge points from channel networks. Foster (1978, p. 92) suggested:

> High transmissivity appears to be due to the solution enlargement of fissures, which shows a general tendency to increase down groundwater flow lines to a maximum in the areas of current groundwater discharge.

The widespread intersection of solutionally enlarged fractures in boreholes in the Chalk suggests that these are not single channels such as are found in most caves in most types of limestone. Possibly they occur as series of interconnected anastomosing channels.

A small number of groundwater traces have been conducted from dolines or sinking streams to springs in the Chalk (Atkinson and Smith, 1974; Banks et al., 1995). These have yielded velocities of 0.02 to 0.07 m/s, which are typical of sink-to-spring velocities in limestone (8.1 Figure 2). However, most such dolines and sinking streams occur close to the contact between the Chalk and overlying lower-permeability strata, and thus are associated with the high recharge rates from allogenic runoff.

Cenozoic Limestone: Yucatan, Mexico

The Yucatan Peninsula in Mexico is a low-lying region of Cenozoic limestone, mostly Eocene to Pleistocene in age. Being young, it is diagenetically quite immature and displays high porosities. Harris (1984) measured an average porosity of 17 percent for near-coastal limestone, though the porosity may be somewhat lower for the older limestone farther inland. The pores are large and well connected, with the result that the hydraulic conductivity of rock cores measured in laboratories is in the range 10^{-6} to 5 x 10^{-3} m/s. Pumping tests in wells give much higher hydraulic conductivity values, ranging from 9 x 10^{-4} m/s to 10^{-2} m/s (Gonzalez Herrera, 1992). This increase in hydraulic conductivity of one to two orders of magnitude reflects the high permeability of fractures and possibly of channels. The fracture porosity can be estimated using Equation 1–if fracture apertures are less than 1 mm, then the minimum fracture porosity is 0.1 percent .

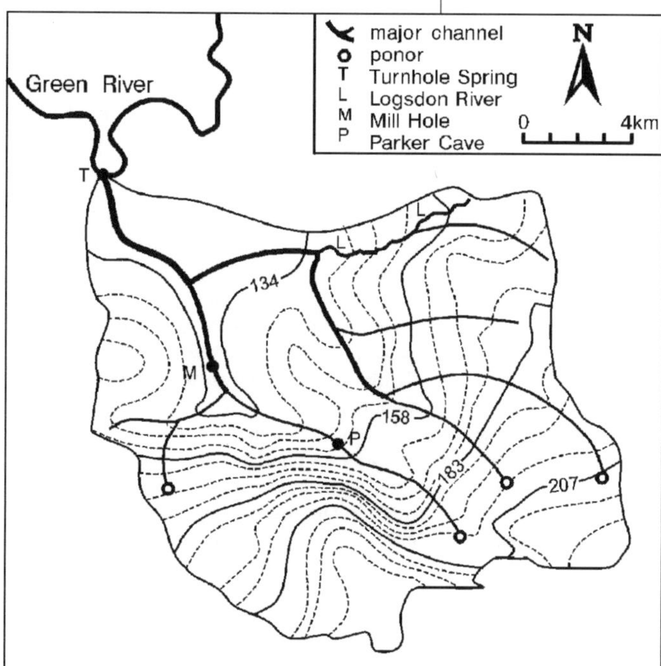

8.1 Figure 3. Groundwater catchment for Turnhole Spring, showing water-level contours and traced flow paths (after Quinlan and Ray, 1981).

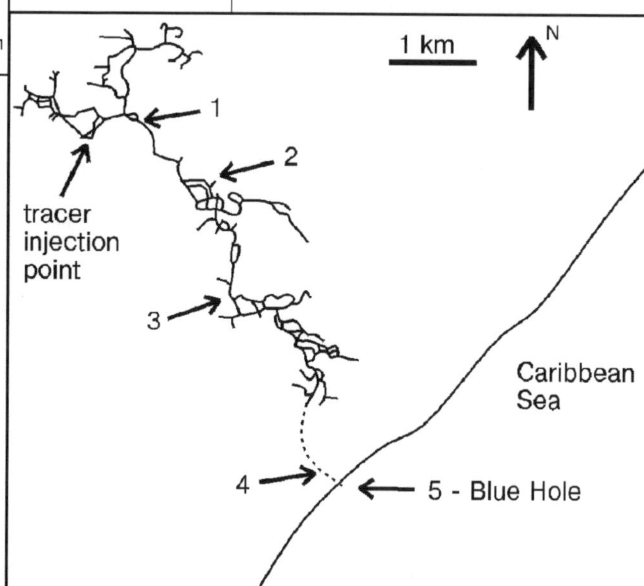

8.1 Figure 4. Map of the major passages in Nohoch Nah Chich (after Hutcheson and Madden, 1994), showing traced flow paths. The whole cave is below the water table.

Two regional computer simulations have been produced for the northern part of the Yucatan Peninsula, with modeled hydraulic conductivity ranging from 0.1 to 1 m/s (Marin, 1990) and from 0.15 to 1.1 m/s (Gonzalez Herrera, 1992). These extremely high values are generated by the exceptionally low, real water-table gradients of 0.000005 to 0.00001 that are fed into the models. The two orders of magnitude increase over pumping-test values suggests that there is a high-permeability channel network that was not sampled by those tests. The difference between the geometric-mean hydraulic conductivities from pumping tests (10^{-3} m/s) and from numerical models (0.36 m/s) is 0.36 m/s, and this represents an approximation of the hydraulic conductivity due to channeling.

It has been known for some time that there are large coastal and submarine springs on both the northern and eastern coasts of the Yucatan Peninsula. Geochemical studies at coastal springs have demonstrated enhanced dissolution in the narrow coastal mixing zone between fresh and salt water (Back et al., 1984). Inland, there are many cenotes, which are roughly circular roof collapses into flooded conduits. Some cenotes are tens of kilometers from the coast, far removed from any simple mixing between fresh water and sea water. However, until re-cently there was no definite proof that large-aperture channels connected these inland cenotes to the coastal springs. In view of the lack of positive proof it has been assumed in a number of papers that no signficant channel networks are present.

Exploration of the flooded caves of the Yucatan Peninsula began in earnest in the mid-1980s. Since then more than 200 km of caves close to the east coast of the peninsula have been explored by scuba divers. At several locations, the coastal springs themselves have been dived inland into major channel networks. The most extensive cave known so far is Nohoch Nah Chich, which has been mapped from the coast to a distance 6 km inland (8.1 Figure 4, p. 427). Measurements have shown that there is a continuous channel over this distance through which there is a discharge >1.3 m^3/s; this channel may extend many kilometers further inland. Tracer testing by Beddows (1998) has measured groundwater velocities of 0.01 to 0.03 m/s along the major channel, which has an average cross section of 120 m^2. The cave exploration and tracing testing demonstrate that dissolution has resulted in a network of channels which impart an extremely high permeability to this aquifer: This explains the hundred-fold increase in permeability over that obtained from pumping tests. A 1994 map of the cave showed that the volume of the 39 km of passages mapped at that time was 4 x 10^6 m^3. The cave underlay a surface area 5.5 km long and 1.9 km wide, and had a vertical range of 80 m. Thus the cave occupies 0.5 percent of that specific volume of limestone.

Discussion

Porosity Enhancement as a Result of Dissolution

Porosity values for the four aquifers described above are presented in 8.1 Table 1. The Paleozoic carbonates in Kentucky and Ontario are diagenetically mature and were buried to great depths beneath cover strata. They have much lower matrix porosities than the younger carbonates of the English Chalk and Yucatan Peninsula, which have not been buried to a significant extent. However, there are no consistent contrasts in fracture or channel porosity between the younger and older carbonates: the values are all much lower than the matrix porosities. Consequently, in all four aquifers almost all the storage is in the matrix (8.1 Table 2).

Permeability Enhancement as a Result of Dissolution

The permeability values for the four aquifers are given in 8.1 Table 3. At least 96 percent of aquifer storage is provided by the matrix of the rock (8.1 Table 1). However, the proportion of the groundwater flow (i.e. mass flux) that takes place through this matrix is minimal in all cases. It is calculated by

$$F_m = K_m / (K_m + K_f + K_c) \qquad (3)$$

where F is the fraction of flow, K is hydraulic conductivity, and the subscripts $m, f,$ and c refer to the matrix, fractures, and channels, respectively. The proportion of flow that takes place through fractures and channels can be similarly calculated.

Results are shown in 8.1 Table 4**.** In each case almost all the flow takes place through the channel network, with fractures providing a small proportion of the flow, and the matrix a negligible amount.

Characterization of Channel Networks

These four sample aquifers display a very wide variety of initial conditions, yet as modern unconfined aquifers they are similar in many important respects. Almost all the storage is in the rock matrix, but almost all the flow takes place through channels. The formation of channel networks is the result of a positive-feedback loop between discharge and dissolution flux in developing channels (see Chapter 4.2.??.). Networks are able to form because of the decrease in dissolution rate as groundwater approachs thermodynamic equilibrium with respect to calcite (Chapters 4.1 and 4.2). The implication of the models discussed in Chapter 4.2 is that channel networks should be found in all carbonates where there is a sufficient flux of infiltrating meteoric water, including all unconfined carbonates.

Major dissolutional enhancement of permeability in carbonate aquifers is dependent on the circulation of groundwater (Chapter 3.4) Thus, the capability must exist for:

- Chemically aggressive groundwater to recharge the system
- The rock unit to transmit water through the fractures.
- Groundwater to drain from the system (Stringfield and LeGrand, 1966).

These three conditions are best satisfied in unconfined carbonate aquifers, which readily allow for recharge to and discharge from the aquifer. The four examples of carbonate aquifers described here represent a great range in rock type, recharge type (allogenic and autogenic), and age (Paleozoic, Mesozoic, and Cenozoic). These three factors will be considered in turn.

Dissolution rates are much slower for dolomite than for calcite. Where dolostone and limestone abut, then karst features are usually found to be better developed on the limestone (Buser et al., 1976). However, long caves are also known in dolostone, including

- The longest cave in South America—Toca del Boa Vista, Brazil, with 78 km of passages mapped.
- Several long caves in Missouri, U.S.A., either partly or wholly formed in dolostone:
    - Mystery Cave, 27 km
    - Berome Moore Cave, 27 km.
    - Rimstone River Cave, 22 km.
    - Carroll Cave, 18 km)

Furthermore, there have been many long-distance dye traces in dolostone. In Missouri, for instance, Aley (1978) described nine traces over distances greater than 40 km. Tracer velocities ranged from 0.009 m/s to 0.059 m/s. Thus the development of efficient dissolutional channel networks in dolostone is not uncommon.

Channeled allogenic water sources are optimum for rapid and efficient cave formation. Surface streams of highly aggressive water flowing from low-permeability rocks onto carbonates usually quickly disappear into sinkholes and caves. Many caves are readily accessible sinking streams. In autogenic settings, the epikarst and other entry channels are small, so caves are less easy to explore, as noted in the cone karst at Caves Branch in Belize, for example (Chapter 5.3.12).

However, more than 50 caves have now

Area	Porosity (%)		
	Matrix	Fracture	Channel
Smithville, Ontario, Canada	6.6	0.02	0.003
Mammoth Cave, Kentucky, U.S.A.	2.4	0.03	0.06
The Chalk, England	30	0.01	0.02
Nohoch Nah Chich, Yucatan, Mexico	17	0.1	0.5

8.1 Table 1. Matrix, fracture, and channel porosity in four carbonate aquifers. The channel porosity in the chalk is equivalent to that of the "secondary fractures" of Price et al., 1993.

Area	Proportion of storage (%)		
	Matrix	Fracture	Channel
Smithville, Ontario, Canada	99.7	0.3	0.05
Mammoth Cave, Kentucky, U.S.A.	96.4	1.2	2.4
The Chalk, England	99.9	0.03	0.07
Nohoch Nah Chich, Yucatan, Mexico	96.6	0.6	2.8

8.1 Table 2. Fractions of storage contributed by matrix, fracture, and channel porosity in four carbonate aquifers.

Area	Hydraulic conductivity ($m\ s^{-1}$)		
	Matrix	Fracture	Channel
Smithville, Ontario, Canada	1×10^{-10}	1×10^{-5}	3×10^{-4}
Mammoth Cave, Kentucky, U.S.A.	2×10^{-11}	1×10^{-5}	3×10^{-3}
The Chalk, England	1×10^{-8}	4×10^{-6}	6×10^{-5}
Nohoch Nah Chich, Yucatan, Mexico	7×10^{-5}	1×10^{-3}	4×10^{-1}

8.1 Table 3. Matrix, fracture, and channel permeability in four carbonate aquifers.

Area	Proportion of flow (%)		
	Matrix	Fracture	Channel
Smithville, Ontario, Canada	0.000003	3.0	97.0
Mammoth Cave, Kentucky, U.S.A.	0.00	0.3	99.7
The Chalk, England	0.02	6.0	94.0
Nohoch Nah Chich, Yucatan, Mexico	0.02	0.2	99.7

8.1 Table 4. Fractions of flow contributed by matrix, fracture, and channel permeability in four carbonate aquifers. The channel flow in the chalk is equivalent to that of the "secondary fractures" of Price et al., 1993.

deep, and almost all of them have no allogenic recharge. In these caves, typically the uppermost 100 to 300 m of passages are small, as they were formed by the small amounts of water that drain from just a few dolines. At greater depths, these surface tributaries join, and passage size and discharge both increase. Thus channel networks and extensive caves are common in autogenic as well as allogenic settings.

As repeatedly noted in this book, the older carbonates have usually been subjected to deep burial, resulting in porosity reduction to a few percent or less. This applies to most Proterozoic and Paleozoic carbonates (e.g. Smithville, Ontario; Mammoth Cave, Kentucky) and to Mesozoic carbonates in alpine fold belts (e.g. Nort of Lake Thun Caves, Chapter 5.3.3; Sistema Cheve, Chapter 5.3.6). Younger carbonates that have escaped folding and deep burial have much higher porosities, such as the English Chalk, the limestone of the Yucatan Peninsula, and young carbonate islands (Chapter 5.1). Channel networks clearly develop in both low-porosity and high-porosity carbonates.

Sampling is a major problem in the hydrologic analysis of carbonate aquifers, because boreholes (the conventional exploration tool in hydrogeology) are unlikely to intersect the major channels that are conveying most of the flow and any contaminants. Some statistics for ten carbonate aquifers containing well-explored caves are provided in 8.1 Table 5, including the cave networks illustrated in 8.1 Figure 2. We will refer to the fraction of the bedrock that is occupied by an explored cave as the cave porosity. In all cases it is less than 0.5 percent. These data thus are comparable to the porosities in 8.1 Table 1, where channel porosities as low as 0.003 percent were calculated.

The final column of 8.1 Table 5 expresses the probability of a standard borehole (which is only 10 to 15 cm in diameter) intercepting the mapped caves. Most values lie between 0.5 percent and 2 percent. The highest value (7.5 percent) is for southern Gunung Api. This is an area that includes many very large cave passages and the largest explored cave room in the world (Sarawak Chamber, volume $1.2 \times 10^7\ m^3$)—thus it represents an extreme value. Even in this extreme case, however, the probability of a borehole intersecting a known cave passage is only 1 in 14. In most cases, the probability ranges from 1 in 50 to 1 in 200. Clearly, boreholes cannot be relied on to detect the presence of caves.

Definition of Channel Networks

Four aquifers have been considered in detail in order to demonstrate the extent of dissolutional enhancement of porosity and permeability in unconfined carbonate strata. In all four cases we have shown that

Cave	Volume of rock: length x width x height m (1)	Volume of cave x 10^6 m^3 (2)	Length of cave km (2)	Cave porosity % (3)	Areal coverage of cave % (4)
Ogof Agen Allwedd - Ogof Daren Cilau, Wales	6200 x 1900 x 50	0.9	75	0.15	1.7
Blue Spring Cave, Indiana, U.S.A.	5100 x 2600 x 45	0.5	32	0.08	1.1
Kingsdale Cave System, England	2600 x 1500 x 100	0.17	20	0.04	1.8
Nohoch Nah Chich, Yucatan, Mexico	5500 x 1900 x 80	4	39	0.48	6.5
Mammoth Cave, Kentucky, U.S.A.	11000 x 9000 x 90	8	550	0.09	1.4
Castleguard Cave, Canada	6500 x 1200 x 400	0.12	20	0.004	0.51
Friars Hole System, West Virginia, U.S.A.	6000 x 2000 x 80	2.7	70	0.28	2.5
McFail's Cave, New York, U.S.A.	3500 x 2300 x 90	0.12	11	0.016	0.37
Skull Cave, New York, U.S.A.	1300 x 940 x 60	0.046	6	0.064	1.2
Caves in Southern Gunung Api, Malaysia	7000 x 2500 x 400	30	110	0.43	7.5

8.1 Table 5. Cave porosity and areal coverage for some well-mapped caves.
Notes:
(1) This represents the minimum rectangular block of rock that can contain the 3-D array of mapped passages in each cave.
(2) These refer to the explored and mapped cave passages. Increases in these values are likely as the caves become more completely explored.
(3) Cave porosity is defined as the volume of mapped cave divided by the minimum rectangular block of rock that can contain the cave, times 100.
(4) The areal coverage is the plan area of the cave divided by the minimum rectangular area which can contain the cave, times 100.

In all four cases we have shown that enhancement of porosity by dissolution is relatively minor. On the other hand, the enhancement of permeability is considerable, because dissolution created dendritic networks of interconnected channels that are able to convey 94 percent or more of the flow in the aquifer.

Opinions range as to whether such high-permeability networks of solutionally enlarged fractures should be termed a karst (or karstic, or karstified) aquifer. Atkinson and Smart (1981, p. 182-183) recognized the problem of defining the term, stating:

> There is no widely accepted definition of karstic groundwater flow, but in our opinion the term should be restricted to solution conduits in which a turbulent flow regime occurs. It is thus possible to envisage a spectrum of aquifer types from the wholly karstic, in which all groundwater flow and storage occurs in solution conduits, to the non-karstic fissured aquifer, in which solutionally-widened fissures are common but turbulent flow in conduits is a rarity.

A much broader definition has been proposed by Huntoon (1995, p. 344):

> A karst aquifer is an aquifer containing soluble rocks with a permeability structure dominated by interconnected conduits dissolved from the host rock which are organized to facilitate the circulation of fluid in a downgradient direction wherein the permeability structure evolved as a consequence of dissolution by the fluid.

Atkinson and Smart (1981) classify the English Chalk as being close to the nonkarstic fissured aquifer end of the spectrum, whereas the Carboniferous Limestone in Britain (in which most of the well-known caves are found) is classified as being closer to the karstic end of the spectrum. However, it is instructive to compare the differences in the responses of these two aquifers where adits have been driven into them below the water table. In the Chalk near Brighton, a total of 13.6 km of adits were driven to provide a water supply (Downing et al., 1993).

In some adits, water-yielding fissures were encountered every 7 to 12 m, but in others the interval was much larger, commonly 150 to 200 m, and in a few, over 500 m. Individual fissures yielded very large flows of the order of 5000 to 15,000 m^3/day (58 to 174 L/s) (Price et al., 1993, p. 48).

In North Wales the Halkyn Deep Level (12 km of adits) and Milwr Sea Level Tunnel (26 km of adits) were driven to dewater lead mines in the Carboniferous limestone. Appleton (1989) describes several inflows to the adits, with the largest inflow being 380 L/s. Many solution cavities were found, most of which were located along mineralized veins. The adits did not intersect any major caves, even though several are known in the limestone outcrop (Appleton, 1989).

In both the Cretaceous chalk and the Carboniferous limestone the inflow into the adits is concentrated at point inputs with discharges up to several hundred liters per second. In both cases dissolution has resulted in a channel network that contributes minimally to enhancing aquifer porosity, but has greatly enhanced aquifer permeability. Therefore these aquifers are very similar in terms of hydraulic functioning.

Wherever comprehensive evidence has been collected in unconfined carbonate aquifers, it suggests that dissolution inexorably results in a similar permeability structure, with channel networks providing most of the permeability of the aquifer, yet occupying a very minor fraction of aquifer volume. We believe that all unconfined carbonate aquifers function in a similar manner.

References

Aley, T., 1978, A predictive hydrologic model for evaluating the effects of land use and management on the quantity and quality of water from Ozark springs: *Missouri Speleology*, v. 18, p. 1-185.

Appleton, P., 1989, Limestones and caves of North Wales, in Ford, T.D., ed., *Limestones and caves of Wales*: Cambridge University Press, p. 217-254.

Atkinson, T.C., 1977, Diffuse flow and conduit flow in limestone terrain in the Mendip Hills, Somerset (Great Britain): *Journal of Hydrology*, v. 35, p. 93-110.

Atkinson, T.C., 1985, Present and future directions in karst hydrogeology: *Annales de la Société géologique de Belgique*, v. 108, p. 293-296.

Atkinson, T.C., and Smart, P.L., 1981, Artificial tracers in hydrogeology, *in A survey of British hydrogeology 1980*: London, Royal Society, p. 173-190.

Atkinson, T.C., and Smith, D.I., 1974, Rapid groundwater flow in fissures in the Chalk–An example from South Hampshire: *Quarterly Journal of Engineering Geology*, v. 7, p. 197-205.

Back, W.B., Hanshaw, B., and van Driel, J.N., 1984, Role of groundwater in shaping the eastern coastline of the Yucatan Peninsula, *in* LaFleur, R.G., ed., *Groundwater as a geomorphic agent*: Boston, Allen and Unwin, p. 280-293.

Banks, D., Davies, C., and Davies, W., 1995, The Chalk as a karstic aquifer: Evidence from a tracer test at Stanford Dingley, Berkshire, UK: *Quarterly Journal of Engineering Geology*, v. 28, p. S31-S38.

Beddows, P.A., 1999, McMaster University, *Conduit hydrology of a Tropical Carbonte Aquifer:Caribbean Coast of the Yucatan Peninsula*, M.Sc. thesis, McMaster University, 162 p.

Bonacci, O., 1987, *Karst hydrology, with special reference to the Dinaric Karst*: Berlin, Springer-Verlag, 184 p.

Brown, R.F., and Lambert, T.W., 1963, Reconnaissance of ground water resources in the Mississippian Plateau region, Kentucky: *U.S. Geological Survey Water Supply Paper* 1603, 58 p.

Burdon, D.J., and Papakis, N., 1963, *Handbook of karst hydrogeology*: Institute for Geology and Subsurface Research, Athens/United Nations Food and Agricultural Organization, 276 p.

Buser, S., Drobne, F., and Gospodarič, R., 1978, Geology and hydrogeology, *in* Gospodarič, R., and Habič, P., eds., *Underground water tracing–Investigations in Slovenia 1972-1975*: Postojna, Institute for Karst Research, p. 27-38.

Choquette, P.W., and Pray, L.C., 1970, Geological nomenclature and classification of porosity in sedimentary carbonates: *American Association of Petroleum Geologists Bulletin*, v. 54, p. 207-250.

Connorton, B.J., and Reed, R.N., 1978, A numerical model for the prediction of long term well yield in an unconfined chalk aquifer: *Quarterly Journal of Engineering Geology*, v. 11, p. 127-138.

Domenico, P.A., and Schwartz, F.W., 1990, *Physical and chemical hydrogeology*: New York, John Wiley, 824 p.

Downing, R.A., Price, M., and Jones, G.P., 1993, *Hydrogeology of the Chalk of North-West Europe*: Oxford, Clarendon Press, 300 p.

Ford, D.C., and Williams, P.W., 1989, *Karst geomorphology and hydrology*: London, Unwin Hyman, 601 p.

Foster, S.S.D., 1978, Discussion, in Thames groundwater scheme: *London, Institution of Civil Engineers*, p. 91-94.

Foster, S.S.D., and Crease, R.I., 1975, Hydraulic behaviour of the Chalk aquifer in the Yorkshire Wolds: *Proceedings of the Institution of Civil Engineers*, v. 59, p. 181-188.

Freeze, R.A., and Cherry, J.A., 1979, *Groundwater*: Englewood Cliffs, Prentice Hall, 604 p.

Gonzalez Herrera, R.A., 1992, *Evolution of groundwater contamination in the Yucatan Karstic Aquifer*: University of Waterloo, M.Sc. thesis, 149 p.

Harris, N.J., 1984, *Diagenesis of upper Pleistocene strand-plain limestones, northeastern Yucatan Peninsula, Mexico*: University of New Orleans, MSc thesis, 130 p.

Headworth, H.G., 1978, Hydrogeological characteristics of artesian boreholes in the Chalk of Hampshire: *Quarterly Journal of Engineering Geology*, v. 11, p. 139-144.

Headworth, H.G., Keating, T., and Packman, M.J., 1982, Evidence for a shallow highly-permeable zone in the Chalk of Hampshire, U.K.: *Journal of Hydrology*, v. 55, p. 93-112.

Hess, J.W., and White, W.B., 1988, Storm response of the karstic carbonate aquifer of southcentral Kentucky: *Journal of Hydrology*, v. 99, p. 235-252.

Houston, J.F.T., Eastwood, J.C., and Cosgrove, T.K.P., 1986, Locating potential borehole sites in a discordant flow regime in the Chalk Aquifer at Lulworth using integrated geophysical surveys: *Quarterly Journal of Engineering Geology*, v. 19, p. 271-282.

Huntoon, P. W., 1995, Is it appropriate to apply porous media groundwater circulation models to karstic aquifers? *in* El-Kadi, A.I., ed., *Groundwater models for resources analysis and management*: Boca Raton, Lewis Publishers, p. 339-358.

Hutcheson, E., and Madden, M., 1994, Unpublished map of Nohoch Nah Chich System.

Lapcevic, P., Novakowski, K., Bickerton, G., and Voralek, J., 1997, Hydraulic characterization of the fracture framework in carbonate rock underlying CWML site, Smithville, Ontario: *Air and Waste Management Association Annual Meeting, Proceedings,* Toronto.

Lloyd, J.W., 1993, The United Kingdom, *in* Downing, R.A., Price, M., and Jones, G.P., eds., *Hydrogeology of the Chalk of North-West Europe*: Oxford, Clarendon Press, p. 220-249.

Marin, L.E., 1990, *Field investigations and numerical simulation of ground-water flow in the karstic aquifer of northwestern Yucatan, Mexico*: Northern Illinois University, Ph.D. dissertation, 183 p.

Morel, E.H., 1980, The use of a numerical model in the management of the Chalk Aquifer in the upper Thames Basin: *Quarterly Journal of Engineering Geology*, v. 13, p. 153-165.

Mortimore, R.N., 1993, Chalk water and engineering geology, *in* Downing, R.A., Price, M., and Jones, G.P., eds., *Hydrogeology of the Chalk of North West Europe*: Oxford, Clarendon Press, p. 67-92.

Palmer, A.N., 1981, *A geological guide to Mammoth Cave National Park*: Teaneck, Zephyrus Press, 210 p.

Palmer, A.N., 1989, Stratigraphic and structural control of cave development and groundwater flow in the Mammoth Cave region, *in* White, W.B., and White, E.L., eds., *Karst hydrology: Concepts from the Mammoth Cave area*, New York, Van Nostrand Reinhold, p. 293-316.

Palmer, A.N., 1991, Origin and morphology of limestone caves: *Geological Society of America Bulletin*, v. 103, p. 1-21.

Price, M., 1994, A method for assessing the extent of fissuring in double-porosity aquifers, using data from packer tests: *International Association of Hydrological Sciences Publication*, no. 222, p. 271-278.

Price, M., 1996, *Introducing groundwater*: London, Chapman and Hall, 278p.

Price, M., Morris, B., and Robertson, A., 1982, A study of intergranular and fissure permeability in Chalk and Permian aquifers, using double packer injection testing: *Journal of Hydrology*, v. 54, p. 401-423.

Price, M., Atkinson, T.C., Wheeler, D., Barker, J.A., and Monkhouse, R.A., 1989, Highway drainage to the Chalk Aquifer: The movement of groundwater in the Chalk near Bricket Wood, Hertfordshire, and its possible pollution by drainage from the M25 motorway: *British Geological Survey Technical Report WD/89/3.*

Price, M., Downing R.A., and Edmunds, W.M., 1993, The Chalk as an aquifer, *in* Downing, R.A., Price, M., and Jones, G.P., eds., *Hydrogeology of the Chalk of North-West Europe*: Oxford, Clarendon Press, p. 35-58.

Quinlan, J.F., and Ewers, R.O., 1989, Subsurface drainage in the Mammoth Cave area, *in* White, W.B., and White, E.L., eds., *Karst hydrology: Concepts from the Mammoth Cave area:* New York, Van Nostrand Reinhold, p. 65-103.

Quinlan, J.F., and Ray, J.A., 1981, Ground-water basins in the Mammoth Cave Region, Kentucky: *Friends of the Karst, Mammoth Cave, Occasional Publication,* no. 1, 1 sheet, scale 1:138,000.

Quinlan, J.F., Davies, G.J., Jones, S.W., and Huntoon, P.W., 1996, The applicability of numerical models to adequately characterize ground-water flow in karstic and other triple-porosity aquifers, *in* Ritchey, J.D., and Rumbaugh, J.O., eds., *Subsurface fluid-flow (ground-water and vadose zone) modeling:* American Society for Testing and Materials, Special Technical Publication, v. 1288, p. 114-133.

Reeves, M.J., 1979, Recharge and pollution of the English Chalk: Some possible mechanisms: *Engineering Geology*, v. 14, p. 231-240.

Snow, D.T., 1968, Rock fracture spacings, openings and porosities: *Journal of the Soil Mechanics and Foundations Division, American Society of Civil Engineers, Proceedings*, v. 94, p. 73-91.

Stringfield, V.T., and Legrand, H.E., 1966, Hydrology of limestone terranes in the Coastal Plain of the southeastern United States: *Geological Society of America Special Paper* 93, 46 p.

Ward, R.S., Williams, A.T., and Chadha, D.S., 1997, The use of groundwater tracers for assessment of protection zones around water supply boreholes–A case study, *in* Kranj, A., ed., *Tracer hydrology 97*: Rotterdam, Balkema, p. 369-376.

White, W.B., 1988, *Geomorphology and hydrology of karst terrains:* Oxford University Press, 464 p.

White, W.B., and Schmidt, V.A., 1966, Hydrology of a karst area in east-central West Virginia: *Water Resources Research,* v. 2, p. 549-560.

White, W.B., and White, E.L., 1989, *Karst hydrology–Concepts from the Mammoth Cave area*: New York, Van Nostrand Reinhold, 346 p.

Worthington, S.R.H., 1994, Flow velocities in unconfined carbonate aquifers, *in* International symposium on changing karst environments: Hydrogeology, geomorphology and conservation, Oxford, England: *Cave and Karst Science*, v. 21, no. 1, p. 21-22.

Worthington, S.R.H., and Ford, D.C., 1997, Strategy for evaluating channeling in the carbonate bedrock at Smithville, Ontario: Air and Waste Management Association Annual Conference, Toronto, Proceedings.

Worthington, S.R.H., Davies, G.J., and Ford, D.C., 1998, Quantification of matrix, fracture, and channel contributions to storage and flow in a Paleozoic carbonate, in Wicks, C., and Sasowsky, I., eds., Approaches to understanding groundwater flow and contaminant transport in carbonate aquifers: *Geological Society of America Special Paper* (in review).

8.2. Role of Speleogenesis in the Development of Hydrocarbon and Mineral Deposits

David J. Lowe

Abstract

Karstic porosity, broadly viewed as being the product of speleogenesis, is now considered to have played a vital role both in the migration of mineralizing fluids (or hydrocarbons) and in their deposition (or storage).

Historically this link was less widely appreciated. Although some early workers were clearly aware of actual evidence for some kind of relationship, and others noted its theoretical likelihood, it has been ignored by many until relatively recent times.

By reference to historical literature from the United Kingdom, and by review of several previously studied and published examples in the U.K. and elsewhere in the world, this chapter provides an overview of developments in the understanding of these fundamental relationships, and comments on the potential role of new speleogenetic concepts in developing a greater understanding in the future.

Introduction

According to Sharp and Kyle (1988, p. 478),

> ...The significance of fluid migration is paramount in many models for the genesis of mineral deposits, but in only a few cases have the actual hydrodynamics of the proposed flow system been intensively investigated.

The same authors suggest (p. 461) that four aspects are critical to ore deposit generation by groundwater:

- Source for the mineral constituents.
- Dissolution of the minerals in water of whatever origin.
- Migration of the fluids.
- Precipitation of the minerals in response to physical and chemical changes.

Arguably, following relatively recent recognition of the potential links between diagenesis, basinal fluid transfer, and early rock-dissolutional processes, some elements of all four aspects can be related in suitable geologic settings to speleogenesis, to the karstic voids that are its inevitable product, or to interaction between mineral-rich fluids and the rock materials within karstic sequences.

Karstic rocks, particularly carbonates such as limestone and dolostone, have long been recognized as providing potentially important aquifers or reservoirs of significant value in supporting the transmission and storage of underground water. More recently, the role of groundwater is acknowledged as fundamental to the migration of hydrocarbons (Tóth, 1988, p. 499) and to a variety of processes leading to mineral deposition (Sharp and Kyle, 1988).

Although a link between these rocks and certain types of mineral deposit has been obvious from the earliest times, only relatively recently have clear views and a wide appreciation of the unique relationships between deep-seated karstification (commonly interpreted as "paleokarstification") and some mineral emplacement mechanisms begun to emerge and be acknowledged. Because of the paramount importance of hydrocarbons in the modern world, an ever-growing effort has been poured into the global search for petroleum in recent times. These explorations and subsequent exploitation of hydrocarbon reserves have cast much light on the nature of the features within potentially karstic rock sequences that are crucial to the movement of some mineral-rich fluids and the emplacement (or storage) of a wide spectrum of solid and fluid natural resources.

Processes and rates of void-system development in karstic rocks vary in detail depending upon initial lithologies (or facies) and subsequent tectonic or metamorphic histories. However, a sweeping generalization is that a potential void development continuum can exist or evolve within potentially karstic rock successions. This ranges from the almost negligible intergranular or matrix porosity exhibited by unfractured evaporites, crystalline limestones, and marbles on the one hand, through enlarged fissure-related voids that can occur in all rock types, to the wide-open dissolutional cavities of fully integrated karst conduit systems.

Thus, the permeability (and storativity) conditions offered by karstic rock successions with different histories can range across orders of magnitude. It is argued elsewhere (see Lowe Chapter 3.3 and Klimchouk Chapter 5.2 in this volume) that the earliest stages of karst development, especially in buried rock sequences, may relate not directly to the soluble lithologies themselves, but to other beds within the succession that initially exhibit higher permeabilities.

Hence, the earliest fluid motion in such sequences may be through impure lithologies or clastic beds that will later take on the role of aquitards or aquicludes. If so, the purer karstic rocks themselves function initially as barriers to fluid transfer, before rock dissolution causes them to undergo a dramatic role reversal. They then become the dominant aquifer rocks within the sequence. Or more accurately, the cavities that develop within them or at their boundaries become the dominant aquifers. The potential significance of nonkarstic (or impure karstic) rocks in this context has long been recognized, though perhaps not fully acknowledged. Such beds were termed, for instance:

- *Carrier beds or aquifers* by Gardner (1935).
- *Favorable beds* (in a variety of contexts) by various authors during the explosion of modern cave development ideas in the 1960's and 1970's.
- *Inception horizons* by Lowe (1992, and Chapter 3.3 in this volume).

In the space of this chapter it is not feasible to revisit the vast accumulation of evidence and many arguments relating to the processes and guidance of karstification in general or underground void (cave) development in particular. Many of the varied points of view, the observations and measurements that underlie them, and the models that hinge upon them are discussed elsewhere in this volume.

Instead, this chapter includes general and specific comments upon historical aspects of the topic. It also refers to several documented examples, each of which could provide a chapter on its own, that describe the importance of karst features to the emplacement and secondary hosting of a variety of natural materials. The choice of what to include in this review of examples is subjective, and the existence of innumerable equally significant alternatives in other parts of the world is acknowledged.

Selective Historical Review

Among potentially karstic rock types, carbonates have long been recognized as valuable aquifer rocks. Locally the link is visible and obvious, where natural springs emerge from within or adjacent to carbonate outcrops. In some areas, where fresh surface water is scarce, caves have been entered and used as water sources for thousands of years. Even now, though modern technology allows for installation of pumps or gravitational pipe systems, this vital resource is still exploited. Elsewhere, carbonate aquifers are tapped by hand-dug wells and by boreholes, to provide artesian or pumped water supplies, while at least some desert oases probably represent "windows," where upward leakage occurs from regional underground karst drainage routes between distant mountain inputs and sea-level (or subsea-level) outputs.

Carbonate (and other karstic) rock terrains have also provided targets for exploitation of minerals, particularly metallic ores, since prehistoric times. Take the outcrop areas of massive Dinantian (Lower Carboniferous) carbonates in Britain as an example. Here ores of lead, zinc, copper, and iron, with lesser amounts of metals such as silver, manganese, nickel, and cobalt, and nonmetallic minerals such as barite and fluorite, have been exploited since the Roman occupation, and perhaps earlier.

Areas such as the Derbyshire Peak District, the Northern Pennines, and the Minera and Halkyn districts of North Wales later supported a thriving lead–zinc mining industry. Carbonate-related copper occurrences in Britain were fewer, but successful mines once operated in North Wales and on the edge of the Peak District. Iron-ore mining was vital to the economy in the Forest of Dean, parts of South Wales, and Cumbria. Smaller quantities of these and other minerals were extracted from Dinantian carbonate host rocks elsewhere, and locally and to a lesser extent, from carbonate rocks of other ages. However, minerals also occur in noncarbonate rock types (including sandstone and volcanic rocks) and are related to igneous intrusions such as the Cornish granite, so links between mineralization and paleo- or buried karstification were not immediately apparent.

Literature describing early geologic investigations in the British Isles abounds with references to karst features, including active cave watercourses and seemingly isolated voids that were intersected underground by miners. Although not referred to as karst, these were recognized as analogous to similar features met in unmineralized caves. Many mines were beset by drainage problems as workings dropped below the level of local natural underground drains, and pumping was needed, or artificial drainage levels were cut, to dewater whole blocks of limestone. Such practices had significant effects upon natural underground drainage, locally causing major water-supply springs and ancient "holy wells" to fail. Elsewhere, formerly flooded cave segments were switched instantaneously from the phreatic zone, becoming abandoned and essentially unmodified relict conduits within new extensions of the vadose zone. Though common in many parts of Britain, particularly striking examples of both effects are documented in the Minera and Halkyn areas of North Wales (Appleton, 1989).

Early geologists recognized that open spaces encountered by miners (sometimes referred to as "accidentals") were natural cavities formed by processes similar to those that had formed caves in unmineralized areas. Examples were common, and some are still well-known today, such as parts of the Speedwell and Blue John Caverns in Derbyshire and Stump Cross Caverns in Yorkshire.

Some geologists realized that orebody form was related to, if not controlled by, rock walls that were indistinguishable from the walls of open karst cavities. For example, describing hematite deposits in North Wales, A. Strahan, later a Director of the British Geological Survey, noted (1885, p. 53):

> It [iron ore] is found in pockets or widened-out spaces....The pockets are irregular in shape, but generally bounded by curving vertical walls like those of a swallow hole.

In the same publication he reported (Strahan, 1885, p. 54) that:

> They [iron, nickel, and cobalt ores] occur in an irregular chamber or cavity in the limestone, originally opened out, as may be inferred from its rounded walls, by the solvent action of carbonated water.

Though describing iron mineralization, which some considered unrelated to processes of base-metal ore emplacement, such observations and conclusions were not unique. Other eminent British Survey geologists later reported similar relationships, in workings for lead and zinc ores in carbonate rocks. It is also significant, although commonly overlooked, that in his *magnum opus* on cave origin W.M. Davis (1930) speculated about the probability of links between deep-seated karstification and mineralization.

Relating Mineralization to Karst. Despite these and presumably similar observations elsewhere in the world, for many years there was a reluctance to relate mineralization to *preexisting* karst features. In Britain, the tendency was toward an almost opposite view, within which observed underground karst features were commonly described as "vein-controlled." This usage suggests that where the two phenomena coexist, the mineral predates and thus helps to define the imprinting of karst features. Such views persisted even after authors such as T.D. Ford and his co-researchers (Ford and King, 1965) described unequivocal examples of base-metal ores within preexisting karstic cavities.

A possible explanation for this misconception in Britain was a failure to recognize the extent of karst development time scales and the potential relationship between early speleogenesis and the slow migration of basinal fluids. Cave development was assumed to be a recent phenomenon, perhaps restricted to the 10,000 years since the last Pleistocene (late Devensian) glaciation.

Paleokarstic features were recognized in their own right, but there was little explicit acknowledgment that karst processes have acted throughout geologic (at least Phanerozoic) time. Nor was it appreciated that much of this activity occurred in confined and buried situations. Tied tacitly to these conceptual shortfalls was a common belief that, even if penecontemporaneous karstification occurred, voids would be obliterated by later tectonism or infilled by younger sediment.

Thus, it was natural to assume that open cavities related to mineral veins must postdate the veins, as some mineralization events could readily be dated as being far older than 10,000 years. In fact, the survival of ancient cavities, with or without subsequent infilling (see for example Osborne, Chapter 3.7), is now well documented.

Equally, it is clear that such voids can continue to provide potential fluid migration routes, whether or not they appear "open." As such they not only provide a key to understanding some aspects of later cave development, but also have a potential role in mineralizing fluid migration and the "storage" of mineral deposits. Rather than open voids in mineralized areas being *a priori* "vein-controlled," there is relative certainty that many mineral veins are "cave-guided." The term "guided" is chosen in preference to "controlled," to acknowledge that some modification of most protocavities during mineralization is inevitable.

Many karst researchers and scientists from other disciplines now accept the potential importance of karst processes through geologic time as a precursor of mineralization. Such processes, active in buried as well as shallow–unconfined conditions, have affected a spectrum of potential host-rock types to various degrees. Dissolutional voids of all sizes, from microfissures, through major cave conduits, to the largest chambers, have contributed to the migration and eventually the emplacement of minerals or storage of hydrocarbons. It appears beyond doubt that examples of such relationships exist, whether at a major or minor scale, in most preserved karst rock sequences. It should not be overlooked, also as a part of modern speleogenetic understanding, that fluids may pass into and out of karstic beds as well as move through them. Thus, mineral deposition may not only occur in the voids of karstic systems, but also in adjacent "aquifers" (such as porous sandstone), through which mineralizing fluids might pass in targeting the favorable migration routes offered by karstic voids in adjacent beds.

Many sequences with mineralized karstic voids have now been studied and described. However, most results have been published in specialized journals and symposium volumes, at worst overlooked entirely by karst scientists, and at best with their full significance to karst research unappreciated. The following subsections record some earlier (although in some cases ongoing) studies, and they are chosen in an attempt to cover a spectrum of mineral types and morphologies met in a selection of geologic situations. Within the first two subsections the European term "Dinantian" describes rocks belonging to the Lower Carboniferous Subsystem, approximating to the Mississippian of North America.

Overview of Selected Examples

Derbyshire White Peak, United Kingdom Many papers have discussed the mineralization and mines in the White Peak of Derbyshire. A well illustrated synthesis was provided by Worley and Ford (1977), from which much of the information below is derived. Since at least the Roman occupation, lead and zinc ores, with fluorite, barite, and calcite, have been mined across most of the Peak District, while copper minerals were obtained in the Ecton District of northern Staffordshire to the west.

The Peak is an eroded inlier of upfolded (overall) Dinantian carbonates, comprising reef, lagoonal, shelf, and basin lithologies, including deep-water mud-mounds. Though commonly termed a "dome," the structure is far more complicated, and folds and faults of several generations are superimposed, giving rise to a complex pattern of fractures and fold-interference effects. It has long been recognized that many fractures have provided migration routes for mineralizing fluids and repositories for a varied suite of minerals.

Most ore bodies fall into four locally overlapping types: veins, scrins, pipes, and flats. Broadly the first two are fissure fills, with veins generally being larger than scrins. Flats are most commonly related to stratigraphic boundaries, while pipes include several more complicated combinations. At least some of the ore bodies, especially the pipes, show signs of emplacement within preexisting karstic cavities.

Worley and Ford (1977) pointed out that the mineralization mechanism was incompletely understood, and this remains the case, although modern theories of karstification and cave development may support a clearer understanding. Broadly, the deposits are hydrothermal and of Mississippi Valley type, having crystallized from moderately high temperature fluids, remote from the influences of contemporary igneous activity (see Sharp and Kyle, 1988, p. 473, for a concise review and references concerning MVT deposits). Crystallization temperatures deduced on the basis of fluid inclusion studies agree well with those of better known MVT ore deposits elsewhere.

As reported by Worley and Ford (1977 and references therein) the most widely accepted view is that ore-bearing saline solutions were driven into the area following its uplift, from younger basins to the east (the present North Sea Basin) and west (now covered by the Cheshire Plain and, perhaps, the Irish Sea Basin). Hydrodynamic flow mechanisms similar to those described in relation to clastic diagenesis (e.g. Schwartz and Longstaffe, 1988) probably drove the fluid transfer. The eastern basin provided the dominant lead–zinc mineralization, whereas copper minerals (now of limited extent) derived from the west.

It is deduced that the main mineralization pulse was late in the Triassic, when Middle to Upper Carboniferous, Permian, and Lower Triassic rocks that once covered the Dinantian succession had been breached by erosion, if not totally removed, to allow for efficient fluid throughput (see Worley and Ford, 1977, Fig. 38).

Mississippi Valley type mineralization has been well studied elsewhere and its mechanisms are not reconsidered here. However, the nature of the preexisting voids that allowed basinal fluids from the east and west to penetrate the White Peak Dinantian rocks is of interest. It is readily acceptable that at least some of the lead–zinc mineralization occupied preexisting (i.e. pre-Late-Triassic) karst voids. It is also clear that many major vein systems are accompanied by voids, sometimes termed "vein cavities," or even by active stream caves.

In view of the preserved evidence, it is surprising, as outlined above, that the common view is that most cavities associated with mineral deposits have formed since mineral emplacement. Yet it is clear that open karst cavities did exist before mineralization, as evidenced by the pipe and flat deposits, and it has been confirmed (Ford, 1952) that karst voids existed in some of the carbonates before the end of the Dinantian. This being so, it is reasonable to assume that early cavities were not isolated but formed parts of widespread probably integrated void systems, which could not have been avoided as migration routes for later mineralizing fluids.

Iron Mineralization in the Forest of Dean, United Kingdom

Deposits of essentially hematitic iron ore, exploited within a Dinantian carbonate sequence in the Forest of Dean, were at least partly emplaced by meteoric circulation within preexisting voids forming parts of an extensive conduit system that originated toward the end of the Dinantian (Lowe, 1974, 1989, 1993). Evidence provided by neptunian infill of latest Dinantian age or slightly younger confirms several phases of emergence and karstification of earlier Dinantian rocks during middle to late Dinantian times. Mid Carboniferous tectonism produced a generally synclinal regional structure, and a long phase of nondeposition or planation of the folded sequence followed.

This phase ended with encroachment of fluviatile and coal-swamp conditions in late Westphalian (Mid to Late Carboniferous) times. Thick sequences of mixed clastic rocks with thin coal seams covered the folded and eroded Dinantian rocks, and the whole sequence was refolded and faulted at the end of the Carboniferous, deepening the synclinal structure and forming an irregular basin with Coal Measures in its core.

Details of the Permian and Triassic geologic history are scanty. No Permian rocks are preserved, and although Triassic rocks are found south of the main basin, none are preserved in stratigraphical contact with the older rocks in the iron ore field. There is, however, a general agreement that during the Late Triassic acidic mineral-rich fluids entered a preexisting conduit system (or perhaps several stratigraphically separated or hydraulically linked systems) within the Dinantian rocks, where iron minerals were deposited as the solutions were neutralized.

Extensive deposits of high-grade iron-ore minerals were precipitated directly into some of the voids. Lesser quantities of lower-grade ore were generated by interaction between the fluids and the carbonate wall rocks. The role played by magnesium in the process is unclear. Some cavity walls pass from being red-stained and heavily laced with iron minerals, into red-stained coarse secondary dolomitic limestone, and then into unstained unaltered calcitic limestone, across a rock thickness of a few millimeters up to about 1 m in heavily jointed situations.

During the Mid to Late Triassic, the Forest of Dean Basin was probably a positive area (reminiscent of an inselberg, but on a grand scale) surrounded by a wide expanse of arid desert or hypersaline lacustrine basins. Analogous conditions also prevailed in the English Midlands, where Dinantian inliers, including that of Breedon Hill, still stand proud of encircling Triassic mudstone, and are pierced by pre-Triassic tunnels, some filled with unconsolidated or poorly consolidated Mid to Late Triassic sediment. In both areas Triassic clastic material accumulated around the Carboniferous "islands," and gentle overspill as well as regionally extensive flash floods of great ferocity carried sediment and dissolved minerals (produced by oxidation on the arid land surface) into open voids in the adjacent carbonate uplands.

Catastrophic flood events during Carnian times were powerful enough to deposit correlated sandstone beds across the full extent of several preserved Triassic basins, and lesser though significant floods occurred periodically throughout the Triassic. Since then the net geologic effect has been erosional. No Triassic sediment is preserved around or above the Dinantian rocks that host the Forest of Dean ores, and no younger rocks are preserved above the Middle Triassic mudstone that surrounds Breedon Hill and other upstanding Midlands inliers, even though the Carboniferous rocks were once covered by Triassic deposits.

In the Forest of Dean area, significant erosion occurred in the wake of renewed uplift in Miocene times. Major rivers such as the Wye and Severn (or their precursors) incised deeply across parts of the area as they accommodated to a new base level. Along parts of its course, the Wye produced steep-sided gorges within which it truncated and exposed relict cave passages, some of which contain (or contained) iron ore of the types described above. Such exposures confirm that infilling by iron minerals was not total, and it seems possible that precipitation was somehow related to the level of a fluctuating regional water table. Elsewhere in the area, major ancient cave passages contain extensive sand and silt that were deposited by large but probably underfit streams. It is tempting to suggest that incising surface streams intersected and reused preexisting trunk passages during post-Miocene incision, and that they were laden with sediment derived from an eroding (and now totally removed) Mesozoic hinterland.

Grand Canyon District of Arizona, Nevada, and Utah, U.S.A.

Uranium and base-metal mineralization within extensive karst cavities in the Mississippian (Lower Carboniferous) Redwall Limestone is described by Wenrich and Sutphin (1994). Most of the details reproduced here are taken from that review, though additional aspects derive from a more recent paper by Huntoon (1996), which provides a more detailed consideration of the mechanism and timescale of karstification that underlies the mineralization. Wenrich and Sutphin emphasize the point that karstic dissolution has commonly been overlooked in relation to ore deposit genesis, in this case with specific respect to ore-bearing breccia-pipe development.

In northwestern Arizona, the 150- to 200-m-thick Redwall Limestone (see also Huntoon, Chapter 5.2.4 in this volume, for consideration of the present groundwater regime) is described as being "...riddled with caves" (Wenrich and Sutphin, 1994). Evidence provided by neptunian infill (Upper Mississippian and Lower Pennsylvanian sediment) confirms that the oldest recognizable dissolutional event occurred before the end of the Mississippian (probably late Dinantian in European terms). This was when a widespread karst landscape developed upon an emergent Redwall Limestone surface, within which valleys up to 120 m deep were incised (Huntoon, 1996). Wenrich and Sutphin (1994) consider that at least some caves have remained open since reaching significant size during this phase. Other caves of this generation were wholly or partially filled by later Carboniferous sediment, or with breakdown material that was subsequently recemented. Such recemented breakdown, including limestone and younger material stoped from above, forms a distinctive breccia, commonly with a matrix of identifiable Late Mississippian sediment.

According to Huntoon (1996) permeability associated with the early karst features was effectively destroyed by pervasive cementation at some stage following mineralization in Triassic times, but it is difficult to reconcile this conclusion with Wenrich and Sutphin's recognition of extant Mississippian cavities.

Another major dissolutional phase affected the Redwall Limestone during the Late Cretaceous–Paleocene (Laramide Uplift). Many cave passages possibly enlarged at this time cut through the breccia deposits that occupy parts of the older karst system. It thus seems that some of the Laramide caves also remain open today. Caves up to 1.6 km long survive from this phase, and many circular breccia bodies are exposed within them, appearing as infilled chimneys and domes in passage ceilings.

In some ways this description is strikingly similar to that of "vertical through structures" (VTS; Klimchouk and Andrejchouck, 1996), and it is tempting to consider whether their origins might be similar, rather than due to the more traditional dissolutional mechanism assumed by Wenrich and Sutphin. Huntoon (1996) provides a convincing description of breccia-pipe development by dissolution and stoping along upward hydraulic gradients as part of a deep basinal circulation system. The mechanisms described have wide implications for considerations of deep-seated/buried karstification in general, and again the model has much in common with cross-formational dissolutional situations described by Klimchouk (1996a, 1997b). See Palmer and Palmer (Chapter 5.2.2) for further discussion of sulfate-carbonate reactions at the same stratigraphic horizons.

A third significant phase of karstification and dissolutional cave development was guided in part by tectonic structures formed during the Miocene, indicating an origin no older than Neogene. Major incision, including that of the Grand Canyon itself, since the Pliocene has been accompanied by modern ongoing karstification (see Huntoon, Chapter 5.2.4 in this volume). Wenrich and Sutphin consider that the three post-Carboniferous dissolutional events have been gradational one into another, rather than being distinct, and they believe that these events had little effect upon mineralization.

An alternative view, not raised by Wenrich and Sutphin but perhaps implicit in Huntoon's (1996) discussion, is that underground karstification may have been continuous since the earliest dissolution of the Redwall Limestone, acting upon the rock under initially confined perhaps deeply buried post Mississippian conditions. Such early on-going dissolutional activity perhaps persisted, running into the later episodes, so that the whole set of events can be viewed as a single though perhaps pulsating, continuum. Whatever the full history of underground karstification, it is clear that relatively recently formed cavities now provide access to extensive karst ramifications that originated during the initial widespread Late Mississippian dissolutional phase.

The Redwall Limestone caves are not distributed uniformly through the full thickness of the formation. Instead they occur grouped at specific stratigraphic levels. The greatest number of caves, described by Wenrich and Sutphin as "...bedding plane controlled," lie within the Mooney Falls Member of the Redwall Limestone, with many passages of lenticular cross section lying along the same horizon. Despite locally significant overprinting of fracture guidance, or guidance by fracture–bedding intersections, the dominant bedding-plane influence may be significant. Clustering of cave passages at specific horizons may reflect that these beds, or their bounding bedding-planes, acted as inception horizons (Lowe, 1992; Chapters 3.3 and 5.2.2 in this volume).

If so, the concept that speleogenesis was ongoing from Late Mississippian times, under buried confined conditions, becomes more readily acceptable, and its possible chemistry and drive are more easily explained. It is interesting to note that, according to Huntoon (Chapter 5.2.4 of this volume), the current groundwater system, active caves, and associated karstification follow joints and normal faults rather than bedding-related features. How the earlier bedding-guided development phases (Wenrich and Sutphin, 1994) relate to the modern and ongoing fracture-guided phase is not described.

Huntoon's (1996) description of the karst history of the region also lays far more stress upon the role of fractures as cave-development guides, as might be expected in a description of a model of vertical or subvertical cross-stratigraphic structures. Theoretical expectation would be that extensive bedding-related ramifications ought to exist in parallel with the pipe structures, as is suggested by Wenrich and Sutphin's (1994) description and as is well shown in the gypsum systems of the Ukraine (Klimchouk, 1996b). The evidence provided by exposed cavities in canyon walls is that at least some such cavities exist, but their full extent can only be guessed at, and it seems likely on the evidence of Huntoon's description that most are infilled by more recent sediment, much of which is pervasively cemented.

In describing the youngest caves in the area, Huntoon (1996) notes that they tend to lie at the base of the Cambrian Muav Limestone sequence (significantly deeper than the Redwall Limestone). He relates this development to the modern level of the floor of the Grand Canyon, stating that the caves must thus be of Pliocene age or younger. Although a clear relationship between the canyon incision level and enlargement of this group of caves is readily acceptable, Huntoon's own deep basinal circulation model implies that inception of karst voids within the Cambrian rocks could be at least as old as (or even older than) the activity that enlarged the dissolutional pipes and related cave systems at higher stratigraphic levels. Such a view is reflected in the Inception Horizon Hypothesis of cave origin (Lowe, 1992), which predicts basin-wide dissolution under deeply buried conditions, imprinting incipient voids that are subsequently enlarged following uplift and the downcutting of surface valleys, which impose local hydraulic gradients.

Wenrich and Sutphin (1994) note that recognition of the role of karst cavities in the genesis of Mississippi Valley type lead–zinc deposits is not a recent development, as a variety of minerals has long been mined from filled karst sinkholes in Missouri. However, the relationship of other mineral occurrences to karstified ground has rarely been discussed.

Breccia Pipes Thousands of breccia pipes, perhaps analogs of Klimchouk and Andrejchouk's (1996) vertical through structures, had developed through the Redwall Limestone and into overlying beds before the end of the Triassic. These pipes now crop out in canyon walls and on the plateau surface in the Grand Canyon region. As parts of a well developed and probably integrated karst conduit system within the Redwall Limestone and adjacent beds, these breccia-filled pipes, together with other open passageways, acted as preferred routes for mineralizing fluids migrating from adjacent basins. Voids within the system provided ideal sites for mineral crystallization.

Some of the pipes in the Grand Canyon area host high-grade uranium ore deposits (a potential emplacement mechanism is described by Huntoon, 1996), as well as sulfides and related minerals containing elements such as cadmium, cobalt, copper, molybdenum, nickel, lead, selenium, silver, vanadium, and zinc. Such pipes were mined for copper from the late 1880s, but at that time their karstic origin was not recognized. Likewise, it is only recently that the full extent and significance of the Redwall Karst has been appreciated. Mineral exploration has been extended to outcrop areas beyond the Colorado Plateau, including areas to the west. Wenrich and Sutphin (1994) describe the specific example of the Apex Mine in the Basin and Range Province, where copper has been mined since the late 1800s. The ore body breccia, with an element suite similar to that found in the Colorado Plateau area, was attributed to fault activity rather than karst processes. Now it has been recognized as a dissolution–collapse breccia. There is clearly potential for finding similar deposits elsewhere in the extensive Redwall Limestone outcrop.

More and more examples of karst-hosted mineralization related to (paleo) karst in Mississippian carbonate sequences are being recognized. Particularly good case studies from elsewhere in North America are described by Sando (1988) and DeVoto (1988). Similar karst-related mineral deposits are found in limestones of other ages, including Triassic rocks in Nevada (where gold is found in several karst-related breccia bodies) and Devonian rocks in the Canadian Northwest Territories (reviewed by Sharp and Kyle, 1988, p. 473-475).

Permian of West Texas, U.S.A.

A review of the karst associated with the Late Permian San Andreas Dolomite, which forms the Yates Oil Field in western Texas, U.S.A., was provided by Craig (1988), and this consideration derives from that source. The Yates Oil Field has demonstrated some unusual though far from unique geologic properties since its discovery in 1926. Bit drops were recorded during well drilling, sudden rushes of oil occurred within the main carbonate bed, dolomitic sand and gravel were brought to the surface by the oil, and sustained high flows were maintained over longer than typical periods.

It was pointed out at an early stage of the field's development that this behavior could be explained by assuming that the flow of oil was through a system of buried caves. However, confirmation was only obtained when cored boreholes were drilled in the 1950s. These boreholes proved the existence of a variety of karst-related features, including collapse breccia and gravity- and water-deposited cave sediment. Subsequently, between 1960-1970, several fully-cored boreholes confirmed the presence of karst-related rock types and more open cavities, and "infill" drilling since that time has enabled a fairly systematic analysis of the caves and the karst lithologies associated with them.

During the Permian Period the Yates Oil Field straddled the edge of a depositional platform, so that its "westside" area is dominated by a platform interior facies of intertidal and lagoonal mudstones and wackestones. In contrast, eastside Yates is dominated by high energy subtidal grainstones and packstones. The area seems to have remained relatively positive throughout much of the Permian, and the marked karstified unconformity at the top of the San Andreas Formation may be related to this, or may reflect more widespread eustatic sea-level changes. It is significant that most of the caves proved to date are in the more open less dense rocks found in eastside Yates, though some have been detected within the denser westside facies.

Modeling of the paleotopography (described in detail by Craig, 1988 and, in still greater detail, by Tinker et al., 1995) has allowed a reconstruction of the Yates Oil Field area at the time of San Andreas Formation karstification. This reconstruction suggests that the area comprised an archipelago of relatively small low-relief islands, on and between which were karst towers and depressions.

Acceptance of this situation allows assumption that freshwater-lens conditions would have been established periodically and that, just as is happening today, the main cave-forming activity was related to mixing-zone activity at the margins of the lenses, and to the effects of meteoric water moving downward through the unsaturated rock above the lenses (see Mylroie and Carew, Chapter 5.1, and Gunn and Lowe, Chapter 5.1.2 in this volume). Subsequently, the Yates Formation (and its contained cavernous porosity) was buried beneath the Grayburg Formation and later Permian deposits, with many of the karst cavities at a spectrum of scales remaining open. At a later stage, these surviving cavities, whether fully open or partially filled by unconsolidated late Permian sediment, acted as migration routes and reservoirs for hydrocarbon deposits. To the time of Craig's 1988 publication, the Yates Oil Field had produced more than 1 billion barrels of oil, with possibly as much as 3 billion barrels still in place.

Sulfur Deposits of the Pre-Carpathian Region of the Ukraine

Aspects of the genesis and distribution of epigenetic sulfur deposits within mixed Miocene evaporite–carbonate sequences in the Pre-Carpathian Basin of the Ukraine were presented in a detailed paper by Klimchouk (1997a), from which this brief review is drawn.

The Pre-Carpathian sulfur-bearing basin lies in the transition zone between the Eastern European and Western European Platforms and the Carpathian Foredeep, where epigenetic limestone and sulfur are associated with Miocene gypsum–anhydrite beds forming part of a composite carbonate–evaporite sequence. All of the economic sulfur deposits under consideration are related to these rocks, both spatially and genetically. Groundwater flow within the Miocene carbonate–evaporite succession is guided by the detailed geologic and geomorphologic relationships of individual beds. Within the mixed succession, extensive cave systems have been formed in sulfate beds by *upward* cross-formational water flow, and these voids govern the water exchange between individual beds in the composite aquifer system (see Chapter 5.2.1 for a more detailed account).

The 15- to 50-m-thick gypsum–anhydrite rock package (GAS) is dominated by gypsum in the platform interior, with increased anhydrite content toward the foredeep. Sulfates extend across about 20,000 km^2 of the platform. The structure and texture of the sulfate beds vary laterally and thin clay and carbonate interbeds occur sporadically throughout.

Near the foredeep the GAS overlies mixed carbonates, argillites, and arenites, which pass into carbonate bioherms and sandy beds toward the platform interior. Near most of the sulfur deposits, lithothamnion-rich limestone dominates, and this functions as a regional "subgypsum" aquifer. A 25-m-thick pelitic and cryptocrystalline limestone layer that generally overlies the GAS seems to be partly evaporitic and partly epigenetic in origin. Details of rock relationships and the epigenetic calcite development mechanism are provided by Klimchouk (1997a). These beds, and still higher sandy carbonates, are viewed as a "supragypsum" aquifer, and the complex relationships between the two aquifers and the evaporites between are significant to the sulfur mineralization situation. Near the foredeep, a thick unbroken clay stratum confines the Miocene aquifer system.

The overall hydrogeology of the basin sequence reflects the individual rock-unit properties and structural relationships. Conditions change broadly between the near-foredeep areas and the platform interior, but lateral rock variation and faulted juxtaposition of permeable and confining beds are important locally. Details are provided by Klimchouk (1997a), who also discusses the role of artesian speleogenesis (see Chapter 5.2.1 in this volume) and the supposed mechanism of biogeochemical epigenetic sulfur formation.

Complementary to the artesian speleogenesis and sulfur genesis model provided by Klimchouk is a demonstration that preexisting theories of sulfur development are untenable. These earlier models were based upon the assumption that the GAS acts as an aquitard or aquiclude. Klimchouk points out that this not only contradicts the available hydrogeologic data (detailed elsewhere in the paper), but also that it ignores the karst features that are so prominent within the GAS and the hydrogeologic function that they perform. The preexisting theories attempted to constrain all geochemical stages of the sulfur formation processes within the supragypsum aquifer, and this led to some rather dubious paleohydrogeologic assumptions. In contrast, the artesian view of speleogenesis explains local characteristics of circulation within the Miocene rocks and supports a karstic model for the origin and distribution of the region's sulfur deposits.

The model hinges upon ascending groundwater from the subgypsum aquifer developing and then further utilizing a system of karstic voids, and then passing on into the supragypsum aquifer above. Development of and flow within the artesian karst systems constrains the spatial and temporal framework of processes of the sulfur cycle by controlling the available geochemical environments, the velocity of reactant migration, and the nature of reaction products at any given time.

Large-scale microbially mediated sulfate-reduction processes and dissolution of sulfates (speleogenesis) are mutually supporting processes, operating against a general background of upward flow through the GAS. Epigenetic calcite precipitates at the top of the GAS, partially replaces sulfate rocks, and is incorporated into the supra-gypsum aquifer. H_2S generated by sulfate reduction escapes with ascending flow into the supragypsum aquifer, where hydrogen sulfide-bearing water mixes with oxygenated water flowing laterally from an adjacent recharge area. Sulfur precipitates under favorable conditions, determined mainly by pH and S^{-2}/O_2 ratios, both of which are controlled by local hydrodynamics governing the interplay of ascending and lateral flows and different geochemical environments.

The ascending pattern of cross-formational hydraulic communication in a multistory artesian system and the cave systems within the GAS are related to potentiometric low zones, commonly associated with valleys incised into the upper confining bed. This underlying explanation explains the well-known association between sulfur orebodies and areas of significant erosional scouring of an upper confining sequence, as recorded associated with epigenetic sulfur deposits in many regions.

Justice cannot be done in this short review to the exhaustive level of detail, both in terms of the underlying observational data and theoretical evaluation, provided in Klimchouk's paper. Important new concepts in several areas are developed, but it is the emphatic demonstration of the role of karstification and its resultant flow patterns and cavities in the sulfur mineralization process that is of prime importance in the present context.

The significance of gypsum–anhydrite as a karstic rock type has commonly been underestimated in the past, particularly in the "western" world, although this narrow view has recently been redressed to some extent (Klimchouk et al., 1996). The synthesis of sulfur mineralization provided by Klimchouk (1997a) for the Pre-Carpathian region contains a wealth of information concerning artesian speleogenesis and its related hydrogeology, as well as details of the complex biogeochemistry involved in sulfur deposition. But above all it provides an up-to-the-minute illustration of the fundamental importance of karst voids and related circulation patterns to processes of mineral migration and emplacement.

It is shown that gypsum karst development is one of the major prerequisites for the formation of epigenetic sulfur deposits. Clearly, many of the ideas woven into the synthesis have a far wider applicability than simply to the sulfur deposits within the evaporitic sequence of the Pre-Carpathians.

Potential Roles of Hydrogen Sulfide and Sulfuric Acid

A relatively brief and preliminary consideration of H_2S-related porosity was provided by Hill (1995), from which this short abstract is derived (see also Chapter 5.2.8 in this volume). Hill emphasised that the H_2S mechanism of porosity creation has not been well studied, and pointed out that the potential importance of hydrogen sulfide to the creation of karstic porosity has hardly even been recognized. Even though porosity creation via H_2S-related reactions has not yet been studied in depth, it seems to offer a means of understanding the origin of some hydrocarbon reservoirs that cannot be explained by traditionally accepted porosity development mechanisms. It is presumably tacit that these same processes offer a means of understanding the origin of some types of void that have not been exploited as hydrocarbon reservoirs, but are identical with those that have.

Porosity may be developed in an H_2S system either by oxidation of hydrogen sulfide or by the mixing of solutions with different H_2S contents (Palmer, 1991). H_2S mixture corrosion operates in much the same way as does CO_2-related mixture corrosion (described in Chapter 4.1.1 in this volume), and Hill considers that this process is probably more important than oxidation as a route toward porosity development in deeply buried sequences. However, in evaporite-rich hydrocarbon environments, porosity is also developed by processes involving the reaction of sulfates with natural gas to generate hydrogen sulfide and the subsequent oxidation of hydrogen sulfide by oxygenated groundwater to produce sulfuric acid.

During the reaction, evaporite rock is replaced by epigenetic carbonate masses such as those referred to as the Castile Buttes (e.g. Hill, 1987). If oxygenated groundwater is accepted as a requirement for sulfuric acid generation by oxidation, this particular reaction chain must occur at shallower depth than does the hydrogen sulfide mixing referred to above.

Having discussed the likelihood of H_2S mechanisms being implicated in carbonate porosity development, Hill goes on to discuss examples of oil field porosity systems that have been evaluated, some of which might be wholly, or at least partially, related to the mechanism. These include three fields in Texas (Ellenburger, Dollarhide, and Yates). Of these, only the Yates Oil Field (discussed previously in this chapter) is considered to be partially related to H_2S processes, perhaps as a secondary effect superimposed on the "island" karst model discussed by Craig (1988). The Lisburne Field in Alaska is also discussed, and here there is a predominance of evidence that the field's porosity is H_2S-generated. Finally, Hill reviews the Middle Eastern Kirkuk Field, again considering it to be a candidate for H_2S-related porosity development.

Summing up, Hill reiterated that this type of porosity is not yet well studied, but that the proposed mechanism can help to answer many questions relating to the nonmeteoric formation of porosity. Though addressed specifically at the question of hydrocarbon emplacement, it should not be overlooked that void genesis by these mechanisms, and on the scale indicated, may have significance both to the karst studies in general and to broader questions of mineral fluid transfer and mineral emplacement. The converse of this is that other, as yet little studied, porosity generation processes may be involved in providing the voids that are so vital to oil field and mineral deposit development.

Conclusions

In the past, the role of karstic features in guiding mineralization and providing host sites for economic mineral deposition or storage has been overlooked, ignored, or underestimated. This shortfall has gradually been redressed, and new understanding of the extent and variety of karst processes is ensuring that new relationships are being recognized and new interpretations and models are continually being derived. Links between karstic voids produced by a spectrum of processes and of mineral deposits or hydrocarbon reservoirs are now readily acceptable to many karst scientists and to specialist workers outside karst science. But things may not end there.

As suggested by Klimchouk (Chapter 5.2 in this volume), "...the concept of ' sulfuric acid oil-field karst' suggested by Hill (1995) should be extended to a broader concept of 'oil-field karst,' with various dissolutional mechanisms involved. It appears that the common views of karst porosity in petroleum geology as paleokarst reservoirs that simply host hydrocarbons, are gradually giving way to the new concept of dynamically linked processes of karstification and the formation of hydrocarbon fields." What is true of the links between hydrocarbons and karstic porosity, perhaps not conceptually different from the links between karstic porosity and underground water resources, may prove to be equally true of the links between other types of natural resources and the voids within which they are found.

References

Appleton, P., 1989, Limestones and caves of North Wales, *in* Ford, T.D., ed., *Limestones and caves of Wales*: Cambridge University Press, p. 217-254.

Craig, D.H., 1988, Caves and other features of Permian karst in San Andreas Dolomite, Yates Field Reservoir, West Texas, *in* James, N.P., and Choquette, P.W., eds., *Paleokarst:* New York, Springer-Verlag, p. 342-363.

Davis, W.M., 1930, The origin of limestone caverns: *Geological Society of America Bulletin,* v. 41, p. 475-628.

DeVoto, R.H., 1988, Late Mississippian paleokarst and related mineral deposits, Leadville Formation, central Colorado, *in* James, N.P., and Choquette, P.W., eds., *Paleokarst:* New York, Springer-Verlag, p. 278-305.

Ford, T.D., 1952, New evidence for the correlation of Lower Carboniferous reefs at Castleton, Derbyshire: *Geological Magazine,* v. 189, p. 346-356.

Ford, T.D., and King, R.J., 1965, Layered epigenetic galena-barite deposits in the Golconda Mine, Brassington, Derbyshire: *Economic Geology,* v. 60, p. 1686-1701.

Gardner, J.H., 1935, Origin and development of limestone caverns: *Geological Society of America Bulletin,* v. 46, p. 1255-1274.

Hill, C.A., 1987, Geology of Carlsbad Cavern and other caves in the Guadalupe Mountains, New Mexico and Texas: *New Mexico Bureau of Mines and Mineral Resources Bulletin,* no. 117, 150 p.

Hill, C.A., 1995, H_2S-related porosity and sulfuric acid oil-field karst, *in* Budd, D.A., Saller, A.H., and Harris, P.M., eds., Unconformities and porosity in carbonate strata: *American Association of Petroleum Geologists Memoir*, no. 63, p. 301-306.

Huntoon, P.W., 1996, Large-basin ground water circulation and paleo-reconstruction of circulation leading to uranium mineralization in Grand Canyon breccia pipes, Arizona: *Mountain Geologist,* v. 33, no. 3, p. 71-84.

Klimchouk, A.B., 1996a, Speleogenesis in gypsum, *in* Klimchouk, A.B., Lowe, D.J., Cooper, A.H., and Sauro, U., eds., Gypsum karst of the world: *International Journal of Speleology,* v. 25, no. 3-4, p. 61-82.

Klimchouk, A.B., 1996b, Gypsum karst in the Western Ukraine, *in* Klimchouk, A.B., Lowe, D.J., Cooper, A.H., and Sauro, U., eds., Gypsum karst of the world: *International Journal of Speleology,* v. 25, no. 3-4, p. 263-278.

Klimchouk, A.B., 1997a, The role of karst in the genesis of sulfur deposits, Pre-Carpathian region, Ukraine: *Environmental Geology,* v. 31, no. 1-2, p. 1-20.

Klimchouk, A.B., 1997b, Artesian speleogenetic setting: *12th International Congress of Speleology,* La Chaux-de-Fonds, Proceedings, v. 1, p. 157-160.

Klimchouk, A.B., Lowe, D., Cooper, A., and Sauro, U., eds., 1996, Gypsum karst of the world: *International Journal of Speleology,* v. 25, no. 3-4, 307 p.

Klimchouk, A.B., and Andrejchouck, V., 1996, Breakdown development in cover beds, and landscape features induced by intrastratal gypsum karst, *in* Klimchouk, A., Lowe, D., Cooper, A., and Sauro, U., eds., Gypsum karst of the world: *International Journal of Speleology,* v. 25, no. 3-4, p. 127-144.

Lowe, D.J., 1974, The geology of the Forest of Dean limestones: *Cave Projects Group Newsletter,* v. 5, p. 73-79.

Lowe, D.J., 1989, Limestones and Caves of the Forest of Dean, *in* Ford, T.D., ed., *Limestones and caves of Wales:* Cambridge University Press, p. 106-116.

Lowe, D.J., 1992, *The origin of limestone caverns: An inception horizon hypothesis:* Manchester Metropolitan University, Ph.D. thesis, 512 p.

Lowe, D.J., 1993, The Forest of Dean caves and karst: Inception horizons and iron-ore deposits: *Cave Science,* v. 20, no. 2, p. 31-43.

Palmer, A.N., 1991, Origin and morphology of limestone caves: *Geological Society of America Bulletin,* v. 103, p. 1-21.

Sando, W.J., 1988, Madison Limestone (Mississippian) paleokarst: A geologic synthesis, *in* James, N.P., and Choquette, P.W., eds., *Paleokarst*: New York, Springer-Verlag, p. 257-277.

Schwartz, F.W., and Longstaffe, F.J., 1988, Ground water and clastic diagenesis, *in* Back, W., Rosenshein, J.S., and Seaber, P.R., eds., *Hydrogeology:* Geological Society of America, Geology of North America, v. O-2, p. 413-434.

Sharp, J.M., Jr., and Kyle, J.R., 1988, The role of ground-water processes in the formation of ore deposits, *in* Back, W., Rosenshein, J.S., and Seaber, P.R., eds., *Hydrogeology:* Geological Society of America, Geology of North America, v. O-2, p. 461-483.

Strahan, A., 1885, The geology of the coasts adjoining Rhyl, Abergele and Colwyn: *Geological Survey of the United Kingdom Memoir*, Old Series, v. 79SW, 77 p.

Tinker, S.W., Ehrets, J.R., and Brondos, M.D., 1995, Multiple karst events related to stratigraphic cyclicity: San Andres Formation, Yates Field, West Texas, *in* Budd, D.A., Saller, A.H., and Harris, P.M., eds., *Unconformities and porosity in carbonate strata:* American Association of Petroleum Geologists Memoir, no. 63, p. 213-237.

Tóth, J., 1988, Ground water and hydrocarbon migration, *in* Back, W., Rosenshein, J.S., and Seaber, P.R., eds., *Hydrogeology:* Geological Society of America, Geology of North America, v. O-2, p. 485-502.

Wenrich, K.J., and Sutphin, H.B., 1994, Grand Canyon caves, breccia pipes, and mineral deposits: *Geology Today,* v. 110, no. 3, p. 97-104.

Worley, N.E., and Ford, T.D., 1977, Minerals and mines, *in* Ford, T.D., ed., *Limestones and caves of the Peak District*: Norwich, Geobooks, p. 143-165.

8.3. Water and Land-Use Problems in Areas of Conduit Aquifers

Thomas Aley

Abstract

Water and land-use problems occur in areas of conduit aquifers because of the intimate interactions that exist between the surface and the subsurface. Water is the agent that most directly links the surface and subsurface.

Karst areas have unique natural resource problems that have major economic consequences. Soils are often of poor agricultural quality. Surface water supplies are limited. Groundwater supplies are often limited, expensive to exploit, and highly subject to contamination because of ineffective natural cleansing processes. The quality of water in a karst spring or well is largely determined by land-use conditions in the recharge area which contributes the water to that well or spring.

The range of land-use activities that can adversely impact groundwater quality in karst aquifers is extensive. Municipal and on-site sewage systems, petroleum storage and distribution facilities, and highways and pipelines have all created widespread problems. Sinkhole collapse has occurred beneath sewage lagoons, highways, and other critical structures.

Recommended resource protection strategies for karst areas are identified. Preventing problems is crucial. Recharge area delineation provides fundamental and essential data, and should be accompanied by hazard-area mapping. Land use in karst areas must be tailored to site conditions.

Introduction

Sound land management in areas of conduit aquifers must be based on an understanding that karst is a three-dimensional landscape where the surface and subsurface are intimately and integrally connected. Failure to recognize that the surface affects the subsurface, and the converse, has long caused many of the problems of water and land use in karst regions.

A characteristic of many karst areas of the world is that they are regions of rural poverty. The poverty results from poor human adaptation to these natural-resource conditions. Key natural-resource conditions are a result of characteristics of the bedrock, the results of its dissolution, and characteristics of the water circulation patterns in karst regions.

Soil Productivity

Karst areas commonly have poor agricultural soils, and rural economies are tightly linked to soil productivity. As the bedrock weathers, only a fraction of it remains on the land to become the mineral fraction of the soil. Much (and often most) of the weathered rock leaves the area dissolved in water. As a result, soil accretion rates in karst areas are commonly much slower than in areas with less soluble rocks. Impacts from past erosion commonly persist longer in karst areas than in areas with less soluble bedrock. This is especially true when the soils are moderately to highly erosive.

Much of the annual water yield from karst areas moves downward into the groundwater system rather than overland via surface runoff. Plant nutrients need only be moved a short distance downward until they are beneath the rooting depth of most agricultural plants and thus are not available to these plants. In contrast, where overland flow is involved, many of the plant nutrients are cycled through the vegetation multiple times before finally being lost from the region in runoff water or in harvested material. The result of this difference in the direction of water yield (downward in karst versus lateral in most other landscapes) is accelerated plant-nutrient loss and resulting nutrient impoverishment of karst soils.

Groundwater Quantity

Water supplies in karst areas are commonly more limited and more expensive to exploit than is the case in nonkarst areas. Even in humid regions, perennial streams in karst areas are relatively rare features, because much of the runoff water moves into and through the groundwater system. This results in greater human dependency on groundwater supplies than is the case in nonkarst regions with similar climates. Surface water is typically less expensive to exploit for livestock and other agricultural uses than is groundwater, and this produces an economic disadvantage for the residents of karst areas and may preclude some otherwise viable land uses.

Very gentle groundwater gradients can readily move groundwater through *conduit* aquifers; steeper gradients are required in most other aquifers. As a result, an upland area underlain by a karst aquifer with preferential groundwater flow routes routinely has a much greater depth to groundwater than in other types of aquifers. The difference is often half an order of magnitude. With deeper well depths the construction and operation costs increase.

Well yields in karst areas commonly vary highly. Adjacent wells constructed to similar depths often have yield differences of an order of magnitude. The yield is a function of the nature and integration of the dissolutional conduits encountered by the individual well. Limited well yields characterise karst areas. A substantial portion of drilled wells fail to provide adequate yields for domestic purposes in some karst areas. Inadequate or marginal water supplies routinely preclude some activities and often have substantial economic impact.

Groundwater Quality

Land-use conditions within the area that contributes water to the spring or well largely determines the groundwater quality at a karst spring or well. The intimate linkage between land use and groundwater quality in karst involves a number of factors (Aley, 1977, 1978).

Karst aquifers that transport water through conduits commonly provide ineffective natural cleansing for the transported water. Important natural cleansing processes include filtration, adsorption, and the passage of time. The common absence of effective natural filtration is demonstrated by the fact that *Lycopodium* spores, an order of magnitude larger than most pathogenic bacteria, have been used as an effective groundwater tracing agent in many karst conduit aquifers (Milanovic, 1981; Gospodarič and Habič, 1976). Straight-line distances through karst aquifers of up to 63.6 km have been traced with *Lycopodium* spores in Missouri, U.S.A. (Aley et al., 1972). The spores travel in suspension. Their passage through a karst groundwater system demonstrates both the absence of effective filtration and the presence of sufficient turbulence within the aquifer to maintain a reasonable number of spores in suspension. The spores settle relatively rapidly, and the author's measurements of stained *Lycopodium* spores indicate that 95% or more of them settle in karst groundwater at rates on the order of 7.0 to 7.5 cm/hour at a water temperature of 21.5 °C.

Adsorption is another important natural cleansing process that is particularly important in the removal of viruses and many hazardous organic compounds from groundwater. Some of the fluorescent groundwater tracer dyes, such as rhodamine WT (Acid Red 388), have moderate sorption tendencies and are not effective groundwater tracing compounds in many aquifers, yet they can be readily transported through many (but not all) karst aquifers because of the lack of effective adsorption. Simultaneous use of several tracer dyes with different sorption characteristics gives qualitative insight into the effectiveness of adsorption within portions of karst aquifers. But such studies are rare.

Time is an important natural cleansing agent, particularly in the case of pathogens. The effectiveness of decomposition and pathogen die off generally increases with the length of residence time in an aquifer. Groundwater flow velocities in karst aquifers are commonly several orders of magnitude more rapid than in other types of aquifers. Consequently, groundwater pollution problems that would be localized in most aquifers tend to affect much larger areas in karst aquifers. At Green Forest, Arkansas (U.S.A.), dye-tracing studies by the author revealed poultry wastewater discharges of about 4 million liters/day routinely reached numerous springs and domestic wells within 2 to 3 weeks of the time of discharge. The affected area encompassed about 156 km^2.

Protection of groundwater quality for a well or spring requires protection of water in the recharge area. In the U.S.A., recharge areas for public water-supply wells are in places delineated as "wellhead protection zones." The wellhead protection zones can then be afforded enhanced water-quality protection. In many nonkarst areas, wellhead protection zones can be credibly delineated on the basis of various measurements and groundwater-modeling approaches. In karst areas, where wellhead protection is particularly important, using conventional wellhead-delineation approaches commonly yield noncredible interpretations. Delineations typically require well-designed and carefully conducted groundwater tracing studies in karst areas. This requires more effort and more hydrologic expertise than more conventional approaches in nonkarst aquifers. Efforts to delineate wellhead protection zones in karst by inappropriate approaches commonly produce misleading information. The resulting errors can be extremely large. As an example, a wellhead protection zone derived through the use of typical nonkarst aquifer calculations was drawn around three municipal wells in a karst aquifer. An inferred 5-year time-of-travel line was mapped. In a subsequent karst-focused study conducted by the author to better delineate the wellhead protection zone, a tracer dye was introduced into the groundwater system at the inferred 5-year time-of-travel line. The tracer dye reached all three municipal wells within 24 hours.

Recharge areas have been delineated for only a tiny fraction of the public wells in karst, and for almost no private wells. Only a few of the most important springs in the U.S.A. have had their recharge areas delineated. Most of these delineations are very generalized and often fail to include large portions of the recharge areas because of logistical or political considerations. Furthermore, groundwater flow routes change under different hydrologic conditions (Aley, 1988). These changes are commonly not assessed in recharge-area delineation. As a result, most karst aquifers lack adequate fundamental information on recharge areas. Instead, groundwater quality protection in important karst aquifers is based on general assumptions and characterizations rather than on relevant site-specific information. It is hardly surprising, then, that the general water-quality strategy in karst areas is to repair problems, if discovered, rather than prevention through a fundamental understanding of the aquifer and appropriate water-quality protection land-use strategies. The discovery of contamination is exacerbated in karst aquifers because many pollutants commonly reach wells and springs as short-duration pulses (Aley, 1982). As a result most contamination of domestic water supplies in karst areas probably remains undiscovered.

Sewage lagoons were popular in karst areas of southern Missouri (U.S.A.) in the 1960's and 1970's, and they were commonly viewed as an inexpensive waste-treatment strategy by engineers and geologists. Some of the lagoons were constructed in areas where surface features clearly demonstrated a history of previous sinkhole collapse. Excessive contaminant leakage into karst groundwater supplies commonly occurred, and catastrophic collapses occurred at several of the lagoon systems. One collapse in a lagoon system serving the city of West Plains, Missouri, resulted in the discharge of 68 million liters of lagoon water into the karst groundwater system on May 5-6, 1978. Intensive surveillance to ascertain the extent of gastroenteritis documented at least 759 cases associated with the ingestion of sewage-contaminated water (Craun, 1984). Three previous collapses had occurred in the same lagoon system, but health-impact data were not developed for the previous failures (Aley et al., 1972; Craun, 1984).

An extensive range of land-use activities can adversely impact groundwater quality in karst aquifers. In the U.S.A., on-site sewage disposal systems are extensively used and probably create the largest number of problems. A study of karst aquifers in Greene County, Missouri (Aley and Thomson, 1984) indicated that about 60% of the on-site sewage systems in the county contributed detectable sewage contaminants to groundwater, and that 15% of the on-site systems were major contributors to groundwater contamination. Municipal sewage treatment facilities and their associated collection lines are commonly implicated in regional groundwater contamination problems (such as those discussed previously at Green Forest, Arkansas, and West Plains, Missouri).

Petroleum storage and distribution facilities commonly experience leaks and spills that have resulted in chemical contamination of hundreds of wells in karst aquifers. It is sometimes incorrectly presumed that such facilities have only minimal impacts on karst groundwater if they are underlain by an appreciable thickness of clay-rich soils or residuum. In many cases an appreciable and rapid macropore flow through such materials and into underlying karst aquifers occurs.

As an illustration, at Lakeview, Arkansas, 375 liters of water tagged with 900 grams of a 75% fluorescein dye (Acid Yellow 73; CI 45350) mixture was added by the author to a tank pit containing a petroleum storage tank with a history of leaks or overfills. A 20-meter-thick clay-rich residuum underlies the pit and the depth to water level in a nearby well with a 21-meter-deep casing was 43 meters. Normal water usage was simulated by pumping the well. Travel time for the first arrival of the tracer dye from the tank pit to the well was 20 hours, and the maximum dye concentration in the well was reached 216 hours after dye introduction into the tank pit.

Leaks and spills from highways and pipelines can seriously impact karst groundwater quality. Some spills (such as gasoline) yield fumes or fluids that may travel through unsaturated karst conduits and reach wells, basements, or caves, where they can explode at certain concentrations. Several such cases in the United States have involved gasoline and propane. It should be recognized that some compounds mixed in air travel more rapidly than liquids underground. They also may follow different flow routes than liquids. A pipeline transporting liquid fertilizer broke at a point about 20.6 km from Maramec Spring, Missouri. Fertilizer rapidly entered the karst groundwater system, where it killed aquatic life in the spring system and subsequently in the stream to which the spring discharged (Vandike, 1984). Travel time through the karst aquifer was 11 to 12 days.

Subsidence and Sinkhole Collapse

Subsidence is a gradual process, whereas sinkholes collapse suddenly. Both commonly occur in karst areas and result in appreciable economic cost. Several fatalities associated with sinkhole collapse induced by mine dewatering have occurred in the Far West Rand of South Africa, and there has been at least one fatal vehicle accident in the U.S.A. resulting from a man driving into a sinkhole that formed catastrophically under a Maryland state highway. Some fatal tractor rollovers occurred when a wheel dropped into a small sinkhole.

Subsidence and sinkhole collapse are natural processes in karst areas, but human actions can induce the subsidence and collapse by changing the water regimen in the area (Aley et al., 1972). In some cases sinkhole collapse is induced by lowering of groundwater levels by pumping of wells or by mining. In other cases, sinkhole collapse is induced by water impoundment or road ditches. The collapses in the West Plains sewage lagoon system mentioned earlier illustrate sinkhole collapse induced by impounding water. Highway drainage has also been correlated with several catastrophic sinkhole collapses. Increased incidence of sinkhole collapse is often associated with urbanization in karst areas. Urbanization dramatically changes the interactions of surface and subsurface waters and focuses the locations of these interactions. Stormwater detention basins, popular for improving surface water quality in many urbanizing areas, have induced many sinkhole collapses in karst areas. And not all of the induced sinkholes are within the detention basins.

Resource Protection Strategies for Karst Areas

The linkage between karst areas and areas of rural poverty was discussed previously. We can do little to change the natural-resource conditions characteristic of karst areas, but we can improve human adaptations to these conditions. Such improved adaptations require an improved recognition, by both the general public and by karst professionals, of the nature and characteristics of interactions between the surface and the subsurface in karst areas. There are important differences between karst lands and other types of landscapes. Failure to recognize and appropriately adjust to these differences has resulted in many of the problems mentioned in this chapter. The previously discussed economic loss associated with the regional pollution around Green Forest, Arkansas, was on the order of $10 million. The loss could have been prevented with a much smaller offsetting cost if the sewage effluents had been conveyed to a more appropriate discharge point. Interactions between the surface and the subsurface in this karst area were not adequately understood and appreciated by those who made the crucial land-use and site-selection decisions.

Preventing land-use and water-quality problems in karst areas is a better strategy than trying to correct the problems once they have occurred. Restoration of groundwater resources is expensive and inefficient, in that it routinely fails to remove most of the contaminants. While restoration of conduit aquifers is often enhanced by large flow volumes through the groundwater system, appreciable and complex contaminant detainment often exists in the epikarstic zone. Furthermore, the area affected by contamination from a single site in a karst region is likely to be much larger than in nonkarst aquifers. Especially in karst areas, emphasis must shift from responding to problems to preventing problems.

Improved human adaptations to karst conditions requires that land-use and land-management decisions simultaneously consider both surface and subsurface resources and conditions. Enhanced assessment and analysis of subsurface resources and their interrelationship with surface conditions is vital. Conditions in karst areas differ from those in nonkarst areas, and karst-specific investigations are needed to appropriately adjust for these differences. Some of these are outlined in the guidelines of ASTM (1995).

Assessment of how a particular land-use activity might affect another parcel of land requires an understanding of water movement between the sites. On the surface in a nonkarst area, this can be ascertained by physically looking at the topography or by using an appropriate map, the fundamental first step in land-use planning. However, in a karst area much of the water moves underground, and accomplishing this fundamental first step in land-use planning is more difficult. This greater difficulty in no way diminishes the necessity of the assessment. Fundamental steps must not be omitted or replaced by unverified assumptions about water flow routes, especially since groundwater flow directions in karst areas are often substantially different from surface flow directions. And multidirectional groundwater flow sometimes occurs (Aley, 1988).

Delineation of recharge areas for important springs and wells requires competently designed and conducted dye-tracing studies. Such data are fundamental to planning and problem-prevention efforts The utility of recharge-area-delineation studies is greatly enhanced if combined with hazard-area (sometimes called land-vulnerability) mapping. The hazard-area mapping is based on the fact that not all lands in a recharge area pose equal hazards for karst water resources.

For example, a sinkhole plain often presents greater hazards than an area with sparse sinkholes. A sinking stream segment downstream from a major highway poses greater potential hazard to groundwater quality than a similar stream segment remote from highways. Many different considerations can be included in hazard-area mapping approaches. Results can be displayed in qualitative categories such as low, moderate, and high hazard conditions. Differences in hazard-area categories can be an important consideration in public land-use decisions.

Many of the water and land-use problems in karst areas can be prevented or mitigated by tailoring of land use to site conditions (Aley and Aley, 1986). For example, many of the on-site sewage systems that contaminate groundwater in karst aquifers are located on poor-quality soils. Often there are suitable places for on-site systems that could be or should have been used. Dramatic differences in the functioning of karst landscapes exist across small distances. Recognition of these differences, and actions to insure that land uses are tailored to site conditions, remain necessary.

References

Aley, T., 1977, A model for relating land use and groundwater quality in southern Missouri, *in* Dilamarter, R.R., and Csallany, S.C., eds., *Hydrologic problems in karst regions:* Bowling Green, Western Kentucky University Press, p. 323-332.

Aley, T., 1978, A predictive hydrologic model for evaluating the effects of land use and management on the quantity and quality of water from Ozark springs: *Missouri Speleology,* v. 18, 185 p.

Aley, T., 1982, Hydrologic impacts of urbanization in the soluble rock lands of Greene County, Missouri: *5th National Cave Management Symposium,* Proceedings, p. 61-69.

Aley, T., 1988, Complex radial flow of ground water in flat-lying residuum-mantled limestone in the Arkansas Ozarks: *2nd Environmental Problems in Karst Terranes and Their Solutions Conference,* National Water Well Association, Proceedings, p. 159-170.

Aley, T., and Aley, C., 1986, Effects of land management on cave and water resources, Dry Medicine Lodge Creek Basin, Bighorn Mountains, Wyoming: *6th National Cave Management Symposium,* Proceedings, p. 79-92.

Aley, T., and Thomson, K.C., 1984, *Septic fields and the protection of groundwater quality in Greene County, Missouri:* Protem, Mo., Ozark Underground Laboratory, Contract report, 162 p.

Aley, T.J., Williams, J.H., and Massello, J.W., 1972, Groundwater contamination and sinkhole collapse induced by leaky impoundments in soluble rock terrain: *Missouri Geological Survey and Water Resources,* Engineering Geology Monographs, ser. 5, 32 p.

ASTM (American Society for Testing and Materials), 1995, *Standard guide for design of ground-water monitoring systems in karst and fractured-rock aquifers:* ASTM Designation D-5717, 17 p.

Craun, G.F., 1984, Health aspects of groundwater pollution, *in* Bitton, G. and Gerba, C.P., eds., *Groundwater pollution microbiology:* New York, John Wiley and Sons, p. 135-180.

Milanovic, P.T., 1981, *Karst hydrogeology:* Water Resources Publications, Littleton, Colorado, 434 p.

Gospodarič, R., and Habič, P., eds., 1976, *Underground water tracing–Investigations in Slovenia 1972-1975:* Postojna, Institute of Karst Research, 309 p.

Vandike, J.E., 1984, Hydrogeologic aspects of the November, 1981, liquid fertilizer pipeline break on groundwater in the Maramec Spring Recharge Area, Phelps County, Missouri: *Missouri Speleology,* v. 25, p. 93-101.

Part 9: Bibliography

A

Akpanika, O.I., Ukpong, E.E., and Olade, M.A., 1987, Mineralogy and geochemical dispersion in tropical soil overlying a talc deposit in south western Nigeria: *Chemical Geology,* v. 63, p. 109-119.

Aksem, S.D., and Klimchouk, A.B., 1991, *Study of equilibria in a rock-solution system and some other problems of sulfate karst hydrochemistry:* Kiev Karst and Speleol. Center, no. 1, 25 p. (in Russian, res. English).

Aley, T., 1977, A model for relating land use and groundwater quality in southern Missouri, *in* Dilamarter, R.R., and Csallany, S.C., eds., *Hydrologic problems in karst regions:* Bowling Green, Western Kentucky University Press, p. 323-332.

Aley, T., 1978, A predictive hydrologic model for evaluating the effects of land use and management on the quantity and quality of water from Ozark springs: *Missouri Speleology,* v. 18, 185 p.

Aley, T., 1982, Hydrologic impacts of urbanization in the soluble rock lands of Greene County, Missouri: *5th National Cave Management Symposium,* Proceedings, p. 61-69.

Aley, T., 1988, Complex radial flow of ground water in flat-lying residuum-mantled limestone in the Arkansas Ozarks: *2nd Environmental Problems in Karst Terranes and Their Solutions Conference,* National Water Well Association, Proceedings, p. 159-170.

Aley, T., and Aley, C., 1986, Effects of land management on cave and water resources, Dry Medicine Lodge Creek Basin, Bighorn Mountains, Wyoming: *6th National Cave Management Symposium,* Proceedings, p. 79-92.

Aley, T., and Thomson, K.C., 1984, *Septic fields and the protection of groundwater quality in Greene County, Missouri:* Protem, Mo., Ozark Underground Laboratory, Contract report, 162 p.

Aley, T.J., Williams, J.H., and Massello, J.W., 1972, Groundwater contamination and sinkhole collapse induced by leaky impoundments in soluble rock terrain: *Missouri Geological Survey and Water Resources,* Engineering Geology Monographs, ser. 5, 32 p.

Alföldi, L., 1979, Budapesti hévizek (Thermal waters of Budapest): *VITUKI Közl.,* no. 20, 102 p.

Allen, J.R.L., 1972, On the origin of cave flutes and scallops by the enlargement of inhomogeneities: *Rassegna Speleologica Italiana,* v. 24, p. 3-23.

Almeida Abreu, P.A., and Pflug, R., 1994, The geodynamic evolution of the southern Serra do Espinhaço, Minas Gerais, Brazil: *Zentralblatt für Geologie und Paläontologie,* no. 1-2, p. 21-44

Almeida, J.C.H., de, 1992, *Mapeamento geológico da Folha Luminárias-MG (1:50,000), com ênfase na análise estrutural dos metassedimentos do Ciclo Deposicional Andrelândia:* UFRJ, CCMN, Intituto de Geociências, M.Sc. Thesis, 102 p.

Anderson, R.Y., and Kirkland, D.W., 1980, Dissolution of salt deposits by brine density flow: *Geology,* v. 8, p. 66-69.

Andrejchouk, V.N., 1988, The tectonic factor and peculiarities of the sulfate karst of Bukovina: Geology, geomorphology, and hydrogeology of karst: Sverdlovsk, 66 p. (in Russian).

Andrieux, C., 1970, Evapo-condensation souterraine: *Annales de Spéléologie*, v. 25, p. 531-559.

Anthony, D.M., Groves, C., and Meiman, J., 1997, Preliminary investigations of seasonal changes in the geochemical evolution of the Logsdon River, Mammoth Cave, Kentucky: *6th Science Conference,* Mammoth Cave National Park, Proceedings, p. 15-23.

Appleton, P., 1989, Limestones and caves of north Wales, *in* Ford, T.D., ed., *Limestones and caves of Wales:* Cambridge University Press, p. 217-254.

Aristotle, 1952, *Meteorologica:* Lee, H.D.P., trans., London, Heinemann, 432 p. [Loeb Classical Library] (Bk. 1, ch. 13).

Ash, D.W., 1984, Evidence for deep-seated groundwater movement in Middle Mississippian carbonate lithologies of south-central Indiana: *National Speleological Society,* Annual Convention, Proceedings, p. 28.

Ashton, M., 1980, The stratigraphy of the Lincolnshire Limestone Formation (Bajocian) in Lincolnshire and Rutland (Leicestershire): *Proceedings of the*

Geological Association, v. 91, p. 203-223.

ASTM (American Society for Testing and Materials), 1995, *Standard guide for design of ground-water monitoring systems in karst and fractured-rock aquifers:* ASTM Designation D-5717, 17 p.

Atkinson, T.C., 1968, The earliest stages of underground drainage in limestones–a speculative discussion: *British Speleological Association Transactions,* v. 6, p. 53-70.

Atkinson, T.C., 1977, Diffuse flow and conduit flow in limestone terrain in the Mendip Hills, Somerset (Great Britain): *Journal of Hydrology,* v. 35, p. 93-110.

Atkinson, T.C., 1985, Present and future directions in karst hydrogeology: *Annales de la Société géologique de Belgique,* v. 108, p. 293-296.

Atkinson, T.C., 1986, Soluble rock terrains, *in* Fookes, P.G., and Vaughan, P.R., eds., A handbook of engineering geomorphology*: Surrey University Press,* p. 241-257.

Atkinson, T.C., and Rowe, P.J., 1992, Applications of dating to denudation chronology and landscape evolution, *in* Ivanovich, M., and Harmon, R.S., eds., *Uranium-series disequilibrium: Applications to earth, marine, and environmental sciences:* Oxford, Clarendon Press, p. 669-703.

Atkinson, T.C., and Smart, P.L., 1981, Artificial tracers in hydrogeology, *in* A survey of British hydrogeology 1980: London, Royal Society, p. 173-190.

Atkinson, T.C., and Smith, D.I., 1974, Rapid groundwater flow in fissures in the Chalk–An example from South Hampshire: *Quarterly Journal of Engineering Geology*, v. 7, p. 197-205.

Atkinson, T.C., and Smith, D.I., 1976, The erosion of limestones. *in* Ford, T.D., and Cullingford, C.H.D., eds., *The Science of Speleology:* London, Academic Press, p. 151-177.

Aubuisson de Voisins, J.F., d'., 1819, Traité de géognosie ...: Strasbourg, Levrault, v. 2, p. 382-383.

Aucamp, J.P., 1994, Mawenge Mwena (Z9): *Bulletin of the South African Spelaeological Association,* v. 35, p. 49-55.

Audra, P., 1990, Une classique en Vercors, le scialet des Sarrasins: *Spelunca,* v. 39, p. 21-24.

Audra, P., 1993, Le creusement dans la zone temporairement noyée: Origine, processus et conséquences sur la morphologie des cavités: *3rd October Meeting,* Montpellier, p. 7-10.

Audra, P., 1994, Karsts alpins. Genèse de grands réseaux souterrains: Exemples le Tennengebirge (Autriche), l'Ile de Crémieu, la Chartreuse et le Vercors (France): *Karstologia Mémoires,* no. 5, 280 p.

Audra, P., 1995a, Karst alpins; Genèse de grands réseaux souterraines; Exemples: le Tennengebirge (Autriche), l'Ile de Crémieu, la Chartreuse et le Vercors (France): *Karstologia Mémoires,* no. 5, 280 p.

Audra, P., 1995b, Looking at alpine karst speleogenesis, case studies from France (Vercors, Chartreuse, Ile de Crémieu) and Austria (Tennengebirge). Changing karst environments: hydrogeology, geomorphology and conservation, International Symposium, Oxford: *Cave and Karst Science,* v. 21, no. 3, p. 75-80.

Audra, P., 1995c, Signification des remplissages des karsts de montagne: *Karstologia,* v. 25, p. 13-20.

Audra, P., 1996, L'apport du paléomagnétisme à la karstogenèse: le cas du chourum du Goutourier. (Massif du Dévoluy, Hautes-Alpes): Mélanges en l'honneur du Professeur Julian, v. 9, p. 109-120.

Audra, P., and Rochette, P., 1993, Premières traces de glaciations du Pléistocène inférieur dans le massif des Alpes. Datation par paléomagnétisme de remplissages à la grotte Vallier (Vercors, Isère, France): *Compte-rendu à l'Académie des sciences,* sec. 2, v. 317, no. 11, p. 1403-1409.

Audra, P., Delannoy, J.-J., and Hobléa, F., 1993, Signification paléogéographique des réseaux perchés des Préalpes françaises du Nord. Exemples en Chartreuse et Vercors: Études de géographie physique de l'URA, v. 903, no. 22, p. 3-18.

Avais, J., 1972, Karst of France, *in* Herak, M., and Springfield, V.T., eds., *Karst–Important karst regions of the northern hemisphere:* Amsterdam, Elsevier, p. 129-188.

B

Babinski, M., Chemale, J.R., F. and Van Shmus, W.R., 1991, Geocronologia Pb/Pb em rochas carbonáticas do Supergrupo Minas, Quadrlátero Ferrífero, Minas Gerais: *Congresso Brasiliero de Geoquímica,* Abstracts v. 3, p. 628-631.

Bac-Moszaszwilli, M., and Rudnicki, J., 1978, On the possible hydrothermal origin of the Dziura Cave (Tatra Mts.): *Kras i Speleologia,* v. 2, no. 11, p. 84-91.

Bachman, G.O., 1984, Regional geology of Ochoan evaporites, northern part of Delaware Basin*: New Mexico Bureau of Mines and Mineral Resources Circular 184,* 22 p.

Back, W.B., Hanshaw, B.B., and van Driel, J.N., 1984, Role of groundwater in shaping the eastern coastline of the Yucatan Peninsula, Mexico, *in* LaFluer, R.G., ed., *Groundwater as a geomorphic agent:* Boston, Allen and Unwin, p. 280-293.

Back, W., Hanshaw, B.B., Herman, J.S., and Van Driel, J.N., 1986, Differential dissolution of a Pleistocene reef in the ground-water mixing zone of coastal Yucatan, Mexico: *Geology*, v. 14, p. 137-140.

Back, W., Hanshaw, B.B., Pyle, T.E., Plummer, L.N., and Weidie, A.E., 1979, Geochemical significance of groundwater discharge and carbonate solution to the formation of Caleta Xel Ha, Quintana Roo, Mexico:*Water Resources Research*, v. 15, p. 1521-1535.

Badino, G., 1995, Fisica del clima sotterraneo: *Memoire dell'Istituto Italiano di Speleologia,* ser. 2, v. 7, 137 p.

Bakalowicz, M., 1975, Géochimie des eaux karstiques et karstification: *Annales de Spéléologie*, v. 30, p. 581-589.

Bakalowicz, M., 1982, La genese de l'aquifer karstique vue par un geochemiste, *in Karst-Larra 82:* Pomplona, Fural Navarra, p. 159-174.

Bakalowicz, M., and Mangin, A., 1980, L'aquifere karstique, sa definition, ses characteristiques et son identification: *Memoires hors Serie Societe Geologique de France*, no. 11, 71 p.

Bakalowicz, M., Blavoux, B., and Mangin, A., 1974, Apports du tracage isotopique naturel a la connaissance du fonctionnement d'un systeme karstique-teneurs en oxygene-18 de trois systemes des Pyrenees, France: *Journal of Hydrology,* v. 23, p. 141-158.

Bakalowicz, M.J., Ford, D.C., Miller, T.E., Palmer, A.N., and Palmer, M.V., 1987, Thermal genesis of dissolution caves in the Black Hills, South Dakota: *Geological Society of America Bulletin*, v. 99, p. 729-738.

Balberg, I., 1986, Connectivity and conductivity in 2d and 3d fracture systems: *Annales of Israel Physical Society,* v. 8, p. 89-101.

Ballmann, D.S., Davis, D.G., Kambesis, P., Kerbo, R.C., Palmer, A.N., and Palmer, M.V., 1991, *Leuchuguilla: Jewel of the underground:* Basel, Speleo Projects, 144 p.

Bandis, S.C., Makurat, A., and Vik, G., 1985, *Predicted and measured hydraulic conductivity of rock joints:* Oslo, Norges Geotekniske Institutt, 16 p.

Banks, D., Davies, C., and Davies, W., 1995, The Chalk as a karstic aquifer: Evidence from a tracer test at Stanford Dingley, Berkshire, UK: *Quarterly Journal of Engineering Geology*, v. 28, p. S31-S38.

Barton, N., and Stephasson, O., eds., 1990, *Rock Joints:* Rotterdam, A.A. Balkema, 994 p.

Baskov, E.A., Petrov, V.V., and Ivanov, T.K., 1996, Hydrogeology of sedimentary basins, *in* Sedimentary basins of Russia, v. 2: St. Petersburg, VSEGEI. 50 p. (in Russian).

Batchelor, G.K., 1967, *Introduction to fluid dynamics:* Cambridge University Press, 186 p.

Bateson, J.H., and Hall, I.H.S., 1971, Revised geologic nomeclature for Pre-Cretaceous rocks of British Honduras: *American Association of Petroleum Geologists Bulletin,* v. 55, p. 529-530.

Bathurst, R.G.C., 1975, *Carbonate Sediments and their Diagenesis, 2nd ed.:* New York, Elsevier, 658 p.

Batsche, H., Bauer, F., Behrens, H., Buchtela, K., Dombrowski, H.J., Geisler, R., Geyh, M.A., Hötzl, H., Hribar, F., Käss, W., Mairhofer, J., Maurin, V., Moser, H., Neumaier, F., Schmitz, J., Schnitzer, W.A., Schreiner, A., Vogg, H. and Zötl, J., 1970, Kombinierte Karstwasseruntersuchungen in Gebiet der Donauversickerung (Baden-Württemberg) in den Jahren 1967-1969: *Steirische Beiträge zur Hydrogeologie*, for 1970, p. 5-165.

Bauer, F., 1961, Sedimentation und Konvakuationserweiterung in aufsteigenden Siphonstrecken: *Rassegna Speleologica Italiana,* v. 5, p. 171-175.

Baumann, J., Buhmann, D., Dreybrodt, W., and Schulz, W., 1985, Calcite dissolution kinetics in porous media: *Chemical Geology*, v. 53, p. 219-228.

Beach, D.K., 1995, Controls and effects of subaerial exposure on cementation and development of secondary porosity in the subsurface of Great Bahama Bank, *in* Budd, A.D., Saller, A.H., and Harris, P.M., eds., Unconformities and porosity in carbonate strata: *American Association of Petroleum Geologists Memoir,* no. 63, p. 1-34.

Bear, J., 1979, *Hydraulics of groundwater:* New York, McGraw-Hill, 567 p.

Beddows, P.A., 1999, McMaster University, *Conduit hydrology of a Tropical Carbonte Aquifer:Caribbean Coast of the Yucatan Peninsula*, M.Sc. thesis, McMaster University, 162 p.

Bedinger, M.S., 1966, Electric-analog study of cave formation: *National Speleological Society Bulletin,* v. 28, p. 127-136.

Beek, W.J., and Mutzall, K.M.K., 1975, *Transport phenomena:* New York, Wiley, 298 p.

Belkin, V.I., and Medvedsky, R.I., 1989, Deep-seated karst and the problem of interconnected ore- and oil-forming processes, *in* Ore-bearing karst of Siberia: Novosibirsk, Siberian Institute of Geology, Geophysics, and Mineral Raw Material, p. 93-103.

Beltiukov, G.V., 1989, On the formation of karst forms due to condensation, *in Problemi kompleksnogo izuchenija karsta gornikh stran:* Tbilisi-Tskhaltubo, p. 91-94 (in Russian).

Beluche, F., Kens, J., and Teyssier, D., 1996, Le reseau Denis Parisis, Bethemont-la-Foret, Val-d'Oise: *Spelunca,* v. 63, p. 31-37.

Bennet, S.S., and Hannor, J.S., 1987, Dynamics of subsurface salt dissolution at the Welch Dome, Louisiana Gulf Coast, *in* Lerch, I., and O'brien, J.J., eds., *Dynamical geology of salt and related structures:* Orlando, Academic Press, p. 653-677.

Bennett, P.C., 1991, Quartz dissolution in organic-rich aqueous systems: *Geochimica et Cosmochimica Acta,* v. 55, p. 1781-1797.

Berkowitz, B., Bear, J., and Braester, C., 1988, Continuum models for contaminant transport in fractured porous formations: *Water Resources Research,* v. 24, p. 1225-1236.

Bernasconi, R., 1966, La condensation interne du karst profond: *Cavernes,* v. 10, no. 2, p. 41-46.

Berner, R.A., 1967, Comparative dissolution characteristics of carbonate minerals in the presence and absence of aqueous magnesium ions: *American Journal of Science,* v. 265, p. 45-70.

Berner, R.A., 1970, Sedimentary pyrite formation: *American Journal of Science,* v. 290, p. 882-911.

Berner, R.A., 1978, Rate control of mineral dissolution under Earth's surface conditions: *American Journal of Science,* v. 278, p. 1235-1252.

Berner, R.A., 1984, Sedimentary pyrite formation: An update: *Geochemica et Cosmochimica Acta,* v. 48, p. 605-615.

Berner, R.A., and Morse, J.W., 1974, Dissolution kinetics of calcium carbonate in seawater, IV: Theory of calcite dissolution: *American Journal of Science,* v. 274, p. 108-134.

Bes, C., 1986. Picos: Sierra de Beza 86: *Lo Bramavenc,* v. 10, p. 114-136.

Bick, K.F., 1962, Geology of the Williamsville Quadrangle, Virginia: *Virginia Division of Mineral Resources Report of Investigations,* no. 2, 40 p.

Bigot, Y.V., 1989, Trave System: Third deepest cave in the world: *Caves and Caving,* v. 46, p. 10-14.

Bischoff, J.L., et al., 1994, Karstification without carbonic acid: Bedrock dissolution by gypsum-driven dedolomitization: *Geology,* v. 22, p. 995-998.

Bischoff, J.L., Juliá, R., Shanks, W.C., and Rosenbauer, R.J., 1994, Karstification without carbonic acid: Bedrock dissolution by gypsum-driven dedolomitization: *Geology,* v. 22, p. 995-998.

Bishop, P.K., and Lloyd, J.W., 1990, Chemical and isotopic evidence for hydrogeochemical processes occurring in the Lincolnshire Limestone: *Journal of Hydrology,* v. 121, p. 293-320.

Bitterli, T., Jeannin, P.-Y., Meyers, K., and Rouiller, P.H., 1990 : Proyecto Cerro Rabon, Mexico, 1985-1989. *Edition Speleo Projects*, Basel:German and English, 58 p.

Bitterli, T., Jeannin, P.-Y., Hapka, R., et al.,1996, Proyecto Cerro Rabón 1990-1994 (Oaxaca, Mexico). *Edition Speleo Projects*, Basel: 192 p.

Bitterli, T. et al., 1990, Réseau des Lausannois: *Höhlenforschung im Gebiet Sieben Hengste-Hohgant,* no. 2, p. 2-65.

Bitterli, T., (in prep.), Le B6.5 (lapiaz des Sieben Hengste, BE, Suisse): Höhlenforschung im Region Hohgant, no. 5.

Bitterli, T., and Jeannin, P.-Y., 1997, Entwicklungs-geschichte der Hohlen im Gebiet Hohgant-Sieben Hengste-Thunersee (Berner Oberland, Schweiz): *12th International Congress of Speleology,* La Chaux-de-Fonds, Proceedings, v. 1, p. 349-354.

Blackburn, G., Bond, R.D., and Clarke, A.R.P., 1965, Soil development associated with stranded beach ridges in south east S.A: *C.S.I.R.O. Soil Publication,* v. 22.

Blount, C.W., and Dickson, F.W., 1973, Gypsum-anhydrite equilibria in systems $CaSO_4$-H_2O and $CaCO_3$-NaCl-H_2O: *American Mineralogy,* v. 58, p. 323-331.

Blumberg, P.N., and Curl, R.L., 1974, Experimental and theoretical studies of dissolution roughness: *Journal of Fluid Mechanics,* v. 65, p. 735-751.

Böcker, T., 1969, Karstic water research in Hungary: *Bull. International Association of Scientific Hydrology,* v. 14, no. 4, p. 7-20.

Bogdanov, A.A., 1947, Dependence of cleavage intensity on the bed thickness: *Soviet Geology*, v. 6, p. 32-41 (in Russian).

Bögli, A., 1964, Mischungskorrosion: Ein Beitrag zur Verkarstungsproblem: *Erdkunde,* v. 18, p. 83-92.

Bögli, A., 1970, Le Hölloch et son karst: Neuchâtel, Editions la Baconnière, 100 p.

Bögli, A., 1978, *Karsthydrographie und physische Speläologie:* Heidelberg, Springer-Verlag, 292 p.

Bögli, A., 1980, *Karst hydrology and physical speleology:* Berlin, Springer-Verlag, 284 p.

Bonacci, O., 1987, *Karst hydrology, with special reference to the Dinaric Karst:* Berlin-Heidelberg, Springer-Verlag, 184 p.

Bond, G.W., 1946, A geochemical survey of the underground water supplies of the Union of South Africa: *South African Geological Survey Memoir,* v. 41, 208 p.

Borevskii, B.V., Khordikainen, M.A., and Yazvin, L.S., 1976, *Exploration and assessment of water reserves in fissure-karst reservoirs:* Moscow, Nedra, 246 p. (in Russian).

Borreguero, M., 1986, *Special Picos:* Puertos de Ondón (unpublished), 15 p.

Bosák, P., 1993, Karst phenomena in the Bandar-e-Khamir area, SE Iran, *in Krasove Sedimenty*, Knihovna Ceske Speleologicke spolechnosti, v. 21, p. 74-78.

Bosák, P., Ford, D.C., and Glazek, J., 1989, Terminology, *in* Bosák, P., Ford, D., Glazek, J., and Horácek, I., eds., *Paleokarst–A systematic and regional review:* Praha, Academia, p. 25-32.

Bosch, X., Custodio, E., and Pascual, M., 1989, Geochemical reactions in carbonate coastal aquifers, Catalonia, Spain, *in*

Simpson, E.S., and Sharp, J.M., eds., *Selected papers on hydrogeology from the 28th International Geological Congress,* Washington, U.S.A.: p. 147-159.

Bottrell, S.H., 1993, Water chemistry in the Xingwen caves, China: *Cave Science,* v. 20, p. 87-92.

Bottrell, S.H., and Raiswell, R., 1998, Sulfur isotopes and sulfur cycling in sediments, *in* Awramik, S.M., and Riding, R., eds., *Microbial Sediments:* Springer-Verlag (in press).

Bottrell, S.H., Carew, J.L., and Mylroie, J.E., 1993, Bacterial sulphate reduction in flank margin environments: Evidence from sulphur isotopes, *in* White, B., ed., *6th Symposium on the Geology of the Bahamas,* Port Charlotte, Florida, Proceedings, Bahamian Field Station, p. 17-21.

Bottrell, S.H., Smart, P.L., Whitaker, F., and Raiswell, R., 1990, Geochemistry and isotope systematics of sulphur in the mixing zone of Bahamian blue holes: *Applied Geochemistry,* v. 6, p. 97-103.

Boutakoff, N., 1963, The geology and geomorphology of the Portland areas: *Geological Survey of Victoria Memoir,* no. 22, 171 p.

Branagan, D.F., 1984, Proterozoic palaeokarst at Torrowangee, western New South Wales?: *Australian Journal of Earth Sciences,* v. 31, p. 155-158.

Branagan, D.F., 1986, Reply to discussion: Proterozoic palaeokarst at Torrowangee, western New South Wales?: *Australian Journal of Earth Sciences,* v. 33, p. 366-367.

Bray, L.G., 1972, Preliminary oxidation studies on some cave waters from south Wales: *Cave Research Group of Great Britain Transactions,* v. 14, p. 59-66.

Brenčič, M., 1992, Koševelc: *Naše Jame,* v. 34, p. 41-51 (in Slovenian with English summary).

Bretz, J H., 1942, Vadose and phreatic features of limestone caverns: *Journal of Geology,* v. 50, p. 675-811.

Bretz, J H., 1949, Carlsbad Caverns and other caves of the Guadalupe block, New Mexico: *Journal of Geology,* v. 57, p. 447-463.

Briceño, H.O., and Schubert, C., 1990, Geomorphology of the Gran Sabana, Guyana Shield, south eastern Venezuela: *Geomorphology,* v. 3, p. 125-141.

Briceño, H.O., Schubert, C., and Paolini, J., 1991, Table-mountain geology and surficial geochemistry: Chimantá Massif, Venezuelan Guyana Shield: *Journal of South American Earth Sciences,* v. 3, p. 179-194.

Brod, L.G., 1964, Artesian origin of fissure caves in Missouri: *National Speleological Society Bulletin,* v. 26, p. 83-114.

Brook, D., 1968, The Three Counties System: *University of Leeds Speleological Association Review,* no. 3, p. 15-19.

Brook, D., 1971, Mudstone beds and cave development in Craven: *University of Leeds Speleological Association Review,* no. 3, p. 15-19.

Brook, D., 1976a, The Porcellanous Band (The Philosopher's Stone): *British Cave Research Association Bulletin,* no. 13, p. 10-11.

Brook, D., 1976b, Hydrological grooves in Kingsdale Master Cave: *British Cave Research Association Bulletin,* no. 14, p. 8-9.

Brook, G.A., Folkoff, M.E., and Box, E.O., 1983, A world model of soil carbon dioxide: *Earth Surface Proc. and Landforms,* v. 8, p. 79-88.

Brown, R.F., and Lambert, T.W., 1963, Reconnaissance of ground water resources in the Mississippian Plateau region, Kentucky: U.S. Geological Survey Water Supply Paper 1603, 58 p.

Brucker, R.W., 1966, Truncated cave passages and terminal breakdown in the Central Kentucky Karst: *National Speleological Society Bulletin,* v. 28, p. 171-178.

Brucker, R.W., Hess, J.W., and White, W.B., 1972, Role of vertical shafts in the movement of ground water in carbonate aquifers: *Ground Water,* v. 10, no. 6, p. 5-13.

Bruent, J., and Vidal, P., 1981, Études climatiques pour la conservation des Grott Ornees deux intervention: *8th International Congress of Speleology,* Bowling Green, Proceedings, p. 659-662.

Buchbinder, B., Magaritz, M., and Buchbinder, L.G., 1983, Turtonian to Neogene palaeokarst in Israel: *Palaeogeography, Palaeoclimatology, Palaeoecology,* v. 43, p. 329-350.

Budd, A.D., Saller, A.H., and Harris, P.M., 1995, Foreword, *in* Budd, A.D., Saller, A.H., and Harris, P.M., eds., Unconformities and porosity in carbonate strata: *American Association of Petroleum Geologists Memoir,* no. 63, p. 7-12.

Budd, A.D., Saller, A.H., and Harris, P.M., eds., 1995, Unconformities and porosity in carbonate strata: *American Association of Petroleum Geologists Memoir,* no. 63, 306 p.

Buffon, G.L., Leclerc de, 1749, Preuves de la théorie de la terre, *in Histoire naturelle, générale et particulière ....:* v. 1, Paris, Imprimerie Royale, p. 125-612.

Buffon, G.L., Leclerc de, 1778, *Histoire naturelle, générale et particulière ....:* Supplement 5, Paris, Imprimerie Royale, p. 463.

Buhmann, D., and Dreybrodt, W., 1985a, The kinetics of calcite dissolution and precipitation in geologically relevant situations of karst areas: 1. Open system: *Chemical Geology,* v. 48, p. 189-211.

Buhmann, D., and Dreybrodt, W., 1985b, The kinetics of calcite dissolution and precipitation in geologically relevant situations of karst areas, 2: Closed systems: *Chemical Geology,* v. 53, p. 109-124.

Buhmann, D., and Dreybrodt, W., 1987, Calcite dissolution kinetics in the system with participation of foreign ions: *Chemical Geology,* v. 64, p. 89-102.

Bull, P.A., 1976, Dendritic surge marks in caves: *British Cave Research Association Transactions,* v. 3, p. 1-5.

Bull, P.A., and Laverty, M., 1982, Observations on phytokarst: *Zeitschrift für Geomorphologie,* v. 26, p. 437-457.

Buneev, A.N., 1932, Aggressive carbon dioxide in mineral waters: *Trudy Tsentral'nogo Instituta Kurortologii,* v. 4.

Burdon, D.J., and Papakis, N., 1963, *Handbook of karst hydrogeology*: Institute for Geology and Subsurface Research, Athens/United Nations Food and Agricultural Organization, 276 p.

Burke, A.R., and Bird, P.F., 1966, A new mechanism for the formation of vertical shafts in Carboniferous limestone: *Nature,* v. 210, p. 831-832.

[Burnet, T.], 1684, *The theory of the earth ....:* London, Kettilby, v. 1, p. 67.

Busche, D., and Sponholz, B., 1992, Morphological and micromorphological aspects of the sandstone karst of eastern Niger: *Zietschrift für Geomorphogie,* v. 85, p. 1-18.

Buser, S., Drobne, F., and Gospodarč, R., 1978, Geology and hydrogeology, *in* Gospodarč, R., and Habič, P., eds., Underground water tracing–Investigations in Slovenia 1972-1975: Postojna, Institute for Karst Research, p. 27-38.

B[rongniart], [A.], 1817, Caverne, *in* Dictionnaire des sciences naturelles, Strasbourg, v. 7, p. 298-309.

C

Calaforra, J.M., 1996, *Contribucion al conocimiento de la karstologia de yesos:* Universidad de Granada, Ph.D. Thesis, 350 p.

Calaforra, J.M., 1998, Karstología de yesos: *Monogr. Univ. (Sc. And Tecnol. Ser.),* Almería, no. 3, 384 p.

Calaforra, J.M., and Forti, P., 1994, Two new types of gypsum speleothems from New Mexico: Gypsum trays and gypsum dust: *National Speleological Society Bulletin,* v. 56, p. 32-37.

Calaforra, J.M., and Pulido-Bosch, A., 1996a, Some examples of gypsum karsts and the more important gypsum caves in Spain, *in* Klimchouk, A., Lowe, D., Cooper, A., and Sauro, U., eds., Gypsum karst of the world: *International Journal of Speleology*, v. 25, no. 3-4, p. 225-237.

Calaforra, J.M., and Pulido-Bosch, A., 1996b,

Un ejemplo de karst de interstratificación: el karst en yeso de Sorbas (Almería, España), *in* Morell, I., and Fagundo, J.R., eds., *Contribuciónes a la Investigación y Gestión del agua subterranea:* Universitat Jaume I Collecció Summa Ciències Experimentals, no. 4. p. 153-160.

Calaforra, J.M., and Pulido-Bosch, A., 1997, Peculiar landforms in the gypsum karst of Sorbas (southeastern Spain): *Carbonates and Evaporites,* v. 12, p. 110-116.

Calaforra, J.M., Dell'Aglio, A., and Forti, P., 1993, The role of condensation-corrosion in the development of gypsum karst: The case of the Cueva del Agua (Sorbas, Spain): *11th International Congress of Speleology*, Beijing, Proceedings, p. 63-66.

Calaforra, J.M., Forti, P., and Pulido-Bosch, A., 1992, Nota preliminar sobre la influencia climatica en la evolucion espeleogenetica en yesos con especial referencia a los afloramientos karsticos de Sorbas (España) y de Emilia Romagna (Italia): *Espeleotemas,* v. 2, p. 9-18.

Cander, H., 1995, Interplay of water-rock interaction efficiency, unconformities, and fluid flow in a carbonate aquifer: Floridan aquifer system, *in* Budd, A.D., Saller, A.H., and Harris, P.M., eds., Unconformities and porosity in carbonate strata: *American Association of Petroleum Geologists Memoir,* no. 63, p. 103-124.

Čar, J., 1982, Geologic setting of the Planina Polje ponor area: *Acta Carsologica,* v. 10, p. 75-105 (in Slovenian with English summary).

Čar, J., and Gospodarič, R., 1984, About geology of karst among Postojna, Planina and Cerknica: *Acta Carsologica,* v. 12, p. 91-106 (in Slovenian with English summary).

Carbonates and Evaporites, 1997, (Special issue containing papers presented at a Symposium on Evaporite Karst: Processes, Landforms, Examples, and Impacts), v. 12, no. 1.

Carew, J., and Mylroie, J., 1987-1988, Rapid karst denudation, San Salvador, Bahamas: *Geo²,* Geology and Geography Section of the National Speleological Society, v. 15, p. 14-15.

Carew, J.L., and Mylroie, J.E., 1995, Quaternary tectonic stability of the Bahamian Archipelago: Evidence from fossil coral reefs and flank margin caves: *Quaternary Science Reviews*, v. 14, p. 144-153.

Carfantan, J.C., 1981, Evolución estructural del sureste de México: Paleogeografía e historía de las zonas internas Mesozoicas: Universidad Nacional Autonómica México, *Instituto Geología Revista,* v. 5, p. 207-216.

Carillo Bravo, J., 1968, Exploracion geologica y posibilidades petroleros de la Plataforma Valles-San Luis Potosí (Sierra Madre Oriental-Altiplano Mexico), *in* Mesa Redonda no. 6: *Problemas de Exploracion en Areas Posiblemente Petroliferas de la Republica Mexicana*, paper no. 5, 20 p.

Carillo Bravo, J., 1971, La plataforma Valles-San Luis Potosí: *Boletin Asociacion Mexicana Geólogos Petroleros,* v. 23, p. 1-112.

Casas, E., and Lowenstein, T.K., 1989, Diagenesis of saline pan halite: Comparison of petrographic features of modern, Quaternary, and Permian halites: *Journal of Sedimentary Petrology,* v. 59, p. 724-739.

Castillo, E., Karadi, G.M., and Krizek, R.J., 1972, Unconfined flow through jointed rock: *Water Resources Bulletin,* v. 8, p. 266-281.

Catcott, A., 1748-1774, Diaries of tours made in England and Wales: Bristol Reference Library MS, B.6495, pt. 3: ff.,16v-17.

Catcott, A., 1761, *A treatise on the deluge; containing ... natural proofs of the deluge ... and ... the cause of caverns or natural grottos ...:* London, Withers, 296 p.

Chabert, C., 1989, Les grandes cavitès mondiales en roches non-calcaires: Union Internationale de Speleologie–Commission des grandes cavitès, 34 p.

Chalcraft, D., and Pye, K., 1984, Humid tropical weathering of quartzite in south eastern Venezuela: *Zietschrift für Geomorphologie*, v. 28, p. 321-332.

Chernyshev, S.N., 1983, *Fissures in rocks:* Moscow, Nauka. 240 p. (in Russian).

Choppy, J., 1988, Roches solubles non carbonatees et karstification: Spèlèo-Club de Paris,

Choquette, P.W., and James, N.P., 1988, Introduction, *in* James, N.P., and Choquette, P.W., eds., *Paleokarst:* New York, Springer, p. 3-21.

Choquette, P.W., and Pray, L.C., 1970, Geologic nomenclature and classification of porosity in sedimentary carbonates: *American Association of Petroleum Geologists Bulletin*, v. 54, p. 207-250.

Cigna, A.A., 1986, Some remarks on phase equilibria of evaporites and other karstifiable rocks, *in Atti simposio internazionale sul carsismo nelle evaporiti: Le Grotte d'Italia,* ser. 4, v. 12, p. 201-208.

Cigna, A.A., and Forti, P., 1986, The speleogenetic role of airflow caused by convection; 1st contribution: *International Journal of Speleology,* v. 15, p. 41-52.

Clemens, T., Hückinghaus, D., Sauter, M., Liedl, R., and Teutsch, G., 1996, A combined continuum and discrete network reactive transport model for the simulation of karst development, *in* Kovar, K. and van der Heijde, P., eds., Calibration and reliability in groundwater modeling: *IAHS Publication*, v. 237, p. 309-318.

Clemens, T., Hückinghaus, D., Sauter, M., Liedl, R., and Teutsch, G., 1997, Simulation of the evolution of maze caves: *12th International Congress of Speleology and 6th Conference on Limestone Hydrology and Fissured Media,* La Chaux-de-Fonds, Proceedings, v. 2, p. 65-68.

Clemens, T., Hückinghaus, D., Sauter, M., Liedl, R., and Teutsch, G., 1998, Simulation of the development of karst aquifers by using a pipe flow model coupled to a continuum model, 2: Model verification and sensitivity analysis: *Water Resources Research,* in preparation.

Coase, A.C., and Judson, D.M., 1977, Dan yr Ogof and its associated caves: *Transactions of the British Cave Research Association,* v. 4, p. 247-344.

Coleman, J.F., 1949, An indicator of water-flow in caves: *University of Bristol Spelaeological Society Proceedings,* v. 6, p. 57-67.

Collignon, B., 1983, Spéléogenése hydrothermale dans les Bibans (Atlas Tellien–Nord de l'Algérie): *Karstologia* (France), v. 2, p. 45-54.

Collignon, B., 1985, Quelques éléments de géologie et D'Hydrogéologie: *Spelunca Supplemént,* no. 19, p. 7-12.

Collignon, B., 1990, Les karst hydrothermaux d'Algérie: *10th International Congress of Speleology,* Budapest, Proceedings, p. 758-760.

Colvée, P., 1973, Cueva en cuarcitas en el cerro Autana, Territorio Federal Amazonas: *Boletin Society Venezolana Espeleologia*, v. 4, p. 5-13.

Combes, P.-J., 1978, Karst precoce et karst secondaire du trioseme horizon de bauxite dans la Zone du Parnasse (Grece*): 4th International Congress I.C.S.O.B.A.*, Athens, Proceedings, p. 101-113.

Connorton, B.J., and Reed, R.N., 1978, A numerical model for the prediction of long term well yield in an unconfined chalk aquifer: *Quarterly Journal of Engineering Geology*, v. 11, p. 127-138.

Cooley, M.E., Harshbarger, J.W., Akers, J.P., and Hardt, W.F., 1969, Regional hydrogeology of the Navajo and Hopi Indian Reservations, Arizona, New Mexico, and Utah: *U.S. Geological Survey, Professional Paper 521A,* 61 p.

Corbel, J., 1959, Vitesse de l'erosion: *Zeitschrift für Geomorphologie*, v. 3, p. 1-28.

Corrêa Neto, A.V., Anísio, L.C.C., and Brandão, C.P., 1993, Um endocarste quartzítico na Serra do Ibitipoca, SE de Minas Gerais, *in* 7th Simpósio de Geologia de Minas Gerais: Proceedings, *Boletim SBG núcleo MG,* v. 12, p. 83-86.

Corrêa Neto, A.V., Anísio, L.C.C., and Brandão, C.P., and Cintra, H.P., 1995, Gruta das Bromélias (MG 042), município de Lima Duarte, MG: Uma das maiores cavernas em quartzito do mundo: *Espeleotema,* v. 18. In press.

Corrêa Neto, A.V., Dutra, G., and Baptista Filho, J., 1996, Espeleogênese em quartzitos: Comparações entre as serras de Ibitipoca, Luminárias e Carrancas (MG): *38th Congresso Brasileiro de Geologia,* Salvador, Proceedings, v. 4, p. 555-558.

Coulson, A.L., 1940, Caves in the Portland District: *Victorian Naturalist,* v. 57, p. 37-9.

Courbon, P., and Chabert, C., 1986, *Atlas des grandes cavites mondiales:* Paris, Union Internationale de Spéléologie and Federation Francaise de Spéléologie, 255 p.

Courbon, P., Chabert, C., Bosted, P., and Lindsley, K., 1989, *Atlas of the great caves of the world:* St. Louis, Cave Books, 369 p.

Crabtree, R.W., and Trudgill, S.T., 1985, Chemical denudation of a magnesian limestone hillslope, field evidence and implications for modelling: *Earth Surface Processes and Landforms,* v. 10, p. 331-341.

Craig, D.H., 1988, Caves and other features of Permian karst in San Andreas Dolomite, Yates Field Reservoir, west Texas, *in* James, N.P., and Choquette, P.W., eds., *Paleokarst:* New York, Springer-Verlag, p. 342-363.

Craun, G.F., 1984, Health aspects of groundwater pollution, *in* Bitton, G. and Gerba, C.P., eds., *Groundwater pollution microbiology:* New York, John Wiley and Sons, p. 135-180.

Crawford, J.E., and Wallace, C.S., 1993, Geology and mineralization of the Culberson sulfur deposit, *in* Love, D.W., et al., eds., Carlsbad region, New Mexico and west Texas: *New Mexico Geological Society Guidebook,* 44th Annual Field Conference, p. 301-316.

Crawford, N.C., 1984, Karst landform development along the Cumberland Plateau escarpment of Tennessee, *in* LaFleur, R.G., ed., *Groundwater as a geomorphic agent:* Boston, Allen and Unwin, p. 294-339.

Cruden, D.M., 1976, Major rock slides in the Rockies: *Canadian Geotechnical Journal,* v. 13, p. 8-20.

Cser, F., and Maucha, L., 1968, Contribution to the origin of "excentric" concretions: *Karszt és Barlangkutatás,* v. 5, p. 83-100.

Cucchi, P., and Forti, P., 1988, Speleogenetic evolution of the "Fiume-Vento" karst system (San Vittore, Genga, Marche, Italy): *International Symposium on Physical-Chemical Hydrogeology of Karst,* Kosice, p. 193-199.

Curl, R.L., 1966a, Cave conduit enlargement by natural convection: *Cave Notes,* v. 8, p. 4-8.

Curl, R.L., 1966b, Scallops and flutes: *Transactions of the Cave Research Group of Great Britain,* v. 7, p. 121-160.

Curl, R.L., 1968, Solution kinetics of calcite: *4th International Congress of Speleology,* Ljubljana, Proceedings, p. 61-66.

Curl, R.L., 1974, Deducing flow velocity in cave conduits from scallops: *National Speleological Society Bulletin,* v. 36, p. 1-5.

Curl, R.L., 1982, *Cave conduit velocity from scallop measurement; HP-65 user program description:* University of Michigan, Department of Civil Engineering, 3 p.

Cvijiç, J., 1893, Das Karstphänomen: *Geographische Abhandlungen herausgegeben von A. Penck,* v. 5, pt. 3, p. 218-329.

Cvijiç, J., 1918, Hydrographie souterraine et évolution morphologique du Karst: *Recueil des Travaux de l'Institute de Géographie Alpine, v.* 4, p. 375-426.

Cvijiç, J., 1960, La géographie des terrains calcaires: *Academie Serbe des Sciences et des Arts Monograph,* no. 26, 212 p.

D

Darton, N.H., 1901, Preliminary description of the geology and water resources of the southern half of the Black Hills and adjoining regions in South Dakota and Wyoming: *U.S. Geological Survey, 21st Annual Report,* Pt. 4, p. 489-599.

Davies, W.E., 1949, Features of cavern breakdown: *National Speleological Society Bulletin,* v. 11, p. 34-35.

Davies, W.E., 1951, Mechanics of cavern breakdown: *National Speleological Society Bulletin,* v. 13, p. 36-43.

Davies, W.E., 1960, Origin of caves in folded limestone: *National Speleological Society Bulletin,* v. 22, p. 5-29.

Davies, W.E., and Chao, E.C.T., 1959, Report on sediments in Mammoth Cave, Kentucky: *U.S. Geological Survey Administrative Report,* 117 p.

Davis, D.G., 1979a, Cave development in the Guadalupe Mountains, New Mexico and Texas: *Caving International,* v. 3, p. 43.

Davis, D.G., 1979b, Geology and speleogenesis of Ogle Cave: Discussion: *National Speleological Society Bulletin,* v. 42, p. 42-48.

Davis, D.G., 1980, Cave development in the Guadalupe Mountains: A critical review of recent hypotheses: *National Speleological Society Bulletin,* v. 42, p. 42-48.

Davis, D.G., 1989, Helictite bushes–a subaqueous speleothem?: *National Speleological Society Bulletin,* v. 51, p. 120-124.

Davis, D.G., and Mosch, C., 1988, Pebble indentations: A new speleogen from a Colorado cave: *National Speleological Society Bulletin,* v. 50, p. 17-20.

Davis, N.W., and Hess, J.W., 1982, Hydrogeology of the drainage system, Burnsville Cove, Virginia: *National Speleological Society Bulletin,* v. 44, p. 78-83.

Davis, S.N., 1966, Initiation of ground-water flow in jointed limestone: *National Speleological Society Bulletin,* v. 28, p. 111-118.

Davis, W. M., 1930, The origin of limestone caverns: *Geological Society of America Bulletin,* v. 41, p. 475-628.

Day, M., 1993, Resource use in the tropical karstlands of central Belize: *Environmental Geology,* v. 21, no. 3, p. 122-128.

Dayan, H., and Baptista Filho, J., 1984, Caracterização dos fabrics tridimensionais nos rods de quartzo da base da serra de Carrancas, SE de Minas Gerais: *33rd Congresso Brasileiro de Geologia,* Rio de Janeiro, Proceedings, v. 7, p. 3371-3380.

Dayan, H., Baptista Filho, J., Fonseca, M.S., and Mansur, R.L., 1987, Determinação do estado de deformação finita em quartzitos das serras de Carrancas e Bicas, SE de Minas Gerais: *Anais da Academia Brasileira de Ciências,* v. 79, p. 203-211.

De Malbos, J., 1854, Mémoire sur les grottes du Vivarais: *Mémoires de l'Académie des Sciences,* Inscriptions et Belles-Lettres de Toulouse, Sér. 4, v. 4, p. 96-114 (p. 100).

De Serres, [P.]M.[T.], 1835, Essai sur les cavernes à ossemens et sur les causes qui les y ont accumulés: *Natuurkundige Verhandelingen van de Hollandsche Maatschappij der Wetenshappen te Haarlem,* v. 22, no. 1, p. 222.

Deák, J., 1979, Environmental isotopes and water chemical studies for groundwater research in Hungary, *in Isotope Hydrology–1978,* Vienna, Interational Atomic Energy Agency, p. 221-249.

Deák, J., 1989, Dating the karst thermal waters in the Eger area: *10th International Congress of Speleology,* Budapest, Proceedings, v. 2, p. 570-572.

Deal, D.E., 1962, *Geology of Jewel Cave National Monument, Custer County, South Dakota, with special reference to cavern formation in the Black Hills:* University of Wyoming, M.S. thesis, 183 p.

[Defoe, D], 1727, *A tour thro' the whole island of Great Britain ...:* London, Strahan, v. 3, p. 59.

Deike, G.H., III, 1960, Origin and geologic relations of Breathing Cave, Virginia: *National Speleological Society Bulletin,* v. 22, p. 30-42

Deike, G.H., 1967, *The development of caverns in the Mammoth Cave region:* Pennsylvania State University, Ph.D. dissertation, 235 p.

DeMille, G., Shouldice, J.R., and Nelson, H.W., 1964, Collapse structures related to evaporites of the Prairie Formation, Saskatchewan: *Geological Society of America Bulletin,* v. 75, p. 307-316.

Dengo, G., and Bohnenberger, O., 1969, Structural development of northern Central America: *American Association of Petroleum and Geologists Memoir,* no. 11, p. 203-220.

Dennis, J.G., 1987, *Structural Geology:* Dubuque, Wm. C. Brown Publishers, 448 p.

DeVoto, R.H., 1988, Late Mississippian paleokarst and related mineral deposits, Leadville Formation, central Colorado, *in* James, N.P., and Choquette, P.W., eds., *Paleokarst:* New York, Springer-Verlag, p. 278-305.

Domenico, P.A., and Schwartz, F.W., 1990, *Physical and chemical hydrogeology:* New York, John Wiley, 824 p.

Donini, G., Rossi, A., Forti, P., Buzio, A., Calandri, G., and Frumkin, A., 1985, Monte Sedom: *Societa Speleologica Italiana,* 135 p.

Douglas, H.H., 1964, *Caves of Virginia:* Falls Church, National Speleological Society Virginia Region, 761 p.

Dove, P.M., and Rimstidt, J.D., 1994, Silica-water interactions, *in* Heaney, P.J., Prewitt, C.T., and Gibbs, G.V., eds., *Silica: Reviews in Minerology,* v. 29, p. 259-308.

Downing, R.A., Price, M., and Jones, G.P., 1993, Hydrogeology of the Chalk of North-West Europe: Oxford, Clarendon Press, 300 p.

Dreiss, S.J., 1984, Effects of lithology on solution development in carbonate aquifers: *Journal of Hydrology,* v. 70, p. 295-308.

Dreiss, S.J., 1989, Regional scale transport in a karst aquifer. 1. Component separation of spring flow hydrographs: *Water Resources Research,* v. 25, p. 117-125.

Drew, D.P., 1975, The caves of Mendip, *in* Smith, D.I., ed., *Limestones and Caves of the Mendip Hills:* Newton Abbot (U.K.), David and Charles.

Dreybrodt, W., 1987, The kinetics of calcite dissolution and its consequences to karst evolution from the initial to the mature state: *National Speleological Society Bulletin,* v. 49, p. 31-49.

Dreybrodt, W., 1988, *Processes in karst systems: Physics, chemistry and geology:* Berlin, Springer-Verlag, 288 p.

Dreybrodt, W., 1990, The role of dissolution kinetics in the development of karst aquifers in limestone: A model simulation of karst evolution: *Journal of Geology,* v. 98, p. 639-655.

Dreybrodt, W., 1992, Dynamics of karstification: A model applied to hydraulic structures in karst terranes: *Applied Hydrogeology,* v. 1, p. 20-32.

Dreybrodt, W., 1995, Principles of karst evolution from initiation to maturity and their relation to physics and chemistry: Contrib. to IGCP 299, Geology, Climate, Hydrology and karst fomation "Global karst."

Dreybrodt, W., 1996, Principles of early development of karst conduits under natural and man-made conditions revealed by mathematical analysis of numerical models: *Water Resources Research,* v. 32, p. 2923-2935.

Dreybrodt, W., and Buhmann, D., 1991, A mass transfer model for dissolution and precipitation of calcite from solutions in turbulent motion: *Chemical Geology,* v. 90, p. 107-122.

Dreybrodt, W., and Siemers, J., 1997, Early evolution of karst aquifers in limestone: Models on two-dimensional percolation clusters: *12th International Congress of Speleology,* La Chaux-de-Fonds, Proceedings, v. 2, p. 75-80.

Dreybrodt, W., Eisenlohr, L., Madry, B., and Ringer, S., 1997, Precipitation kinetics of calcite in the system $CaCO_3 - H_2O - CO_2$: The conversion to CO_2 by the slow process $H^+ + HCO_3^- \rightarrow CO_2 + H_2O$ as a rate limiting step: *Geochimica et Cosmochimica Acta* (in press).

Dreybrodt, W., Lauckner, J., Zaihua, L., Svensson, U., and Buhmann, D., 1996, The kinetics of the reaction $CO_2 + H_2O$ fi $H^+ + HCO_3^-$ as one of the rate-limiting steps for the dissolution of calcite in the system H_2O–CO_2–$CaCO_3$: *Geochimica et Cosmochimica Acta,* v. 60, p. 3375-3381.

Drogue, C., 1980, Essai d'identification d'un type de structure de magasins carbonates fissures: Application a l'interpretation de certains aspects du fonctionnement hydrogeologique: *Mémoires de Société Géologie France,* v. 11, p. 101-108.

Drogue, C., 1992, Hydrodynamics of karstic aquifers: Experimental sites in the Mediterranean karst, southern France, *in* Back, W., ed., *Hydrogeology of Selected Karst Regions:* Hannover, Heise, p. 133-149.

Dronkert, H., 1976, Late Miocene evaporites in the Sorbas basin and adjoining areas: 1st Messinian Seminar, Erice, Proceedings, *Mem. Soc. Geol. Ital.,* v. 16, p. 341-362.

Dronkert, H., 1977a, A preliminary note on the geological setting of the gypsum in the province of Almería (SE Spain), Málaga: *Messinian Seminar,* no. 3, 6 p.

Dronkert, H., 1977b, The evaporites of the Sorbas Basin: *Instituto de Investigaciones Geológicas,* Dip. Prov., Barcelona, no. 33, p. 55-76.

Dronkert, H., and Pagnier, H., 1977, Introduction to the Mio/Pliocene of the Sorbas basin, Málaga: *Messinian Seminar,* no. 3, Field trip no. 3 (Sorbas), p. 21.

Droppa, A., 1957, *Demänovské Jaskyne:* Bratislava, Slovak Academy of Sciences, 287 p.

Droppa, A., 1966, Untersuchungen der Parallelität von Flussterrassen mit horizontalen Höhlen: *3rd International Congress of Speleology,* Vienna, Proceedings, v. 5, p. 79-81.

Dubljansky, V.N., 1969, Methods of calculation of moisture condensation in fissure-karst reservoirs: *Bulletin NTI,* Series Hydrogeol. and Eng.-Geol., no. 6, p. 13-17 (in Russian).

Dubljansky, V.N., 1970, Condensation of moisture in the fissure-karst collectors of the Mountainous Crimea, Carpathians and Pridnestrovskaya Podolia. *Doklady Akadamia Nauk Uk. SSR,* ser. B, v. 1, 14-17 (in Russian).

Dubljansky, V.N., 1974, Giant hydrothermal karst cavern in Rhodopes (Bulgaria): *Peschery (Caves),* v. 14-15, p. 233-237 (in Russian).

Dubljansky, V.N., 1977, *Karst caves and shafts of the Crimea Mountains:* Leningrad, Nauka, 181 p. (in Russian).

Dubljansky, V.N., 1979, Hydrothermal karst in the Alpine folded region of the south of the USSR: *Physical Geography and Geomorphology* (Kiev), v. 21, p. 61-64 (in Russian).

Dubljansky, V.N., 1980, Hydrothermal karst in the Alpine folded region of the southern parts of the U.S.S.R.: *Kras i Speleologia* (Poland), v. 3, no. 12, p. 18-36.

Dubljansky, V.N., and Andrejchouk, V.N., 1991, Superficial, underground, deep-seated karst, *in* Obstanovki karstogenesa: Glubinnyj karst, endokarst, gidrotermokarst: Tesisy dokladov regional'nogo sovehchanija, Kungur, p. 55-64 (in Russian).

Dubljansky, V.N., Dorofeev, E.P., and Borodaeva, L.A., 1984, *Hydrochemistry of Kungur Ice Cave:* Simferopol, Ukr. NIINTI, N603Uk 84, 37 p. (in Russian).

Dubljansky, V.N., Dorofeev, E.P., and Borodaeva, L.A., 1984, *Hydrochemistry of Kungur Ice Cave:* Simferopol, Ukr. NIINTI, N603Uk 84, 37 p. (in Russian).

Dubljansky, V.N., and Dubljansky, Y.V., 1998, The problem of condensation in karst studies: *Journal of Cave and Karst Studies,* v. 60, p. 3-17.

Dubljansky, V.N., and Iliukhin, V.V., 1981, *Travels underground:* Moscow, Physical Culture and Sport, 192 p. (in Russian).

Dubljansky, V.N., and Ivanov, B.N., 1970, Karst of the Podol'sky-Bukovinsky karst region: Kiev University, *Fizicheskaja geografija i geomorfologija,* no. 4 (in Ukrainian).

Dubljansky, V.N., and Kiknadze, T.Z., 1984, *Hydrogeology of karst of the alpine folded region of the south of the USSR:* Moscow, Nauka, 126 p. (in Russian).

Dubljansky, V.N., Kiknadze, T.Z., and Kolomietz, A.L., 1983, Vertical zonation of condensation in the fissure-karst collectors of western Georgia: *Bull. of the Acad. of Sci. of the Georgian SSR,* v. 110, p. 333-336 (in Russian).

Dubljansky, V.N., Klimenko, V.I., Vakhrushev, B.A., and Iliukhin, V.V., 1985, *Karst and groundwaters of mountain massifs of the western Caucasus:* Leningrad, Nauka, 150 p. (in Russian).

Dubljansky, V.N., and Lomaev, A.A., 1980, *Karst caves of the Ukraine:* Kiev, Naukova Dumka, 179 p. (in Russian).

Dubljansky, V.N., and Smol'nikov, B.N., 1969, Karstological and geophysical investigations of karst cavities of the Pridnestrovskaja Podolija i Pokutje: Kiev, Naukova Dumka. 151 p. (in Russian).

Dubljansky, V.N., and Sotskova, L.M., 1977, Microclimate of karst cavities of the Crimea Mountains: *7th International Congress of Speleology,* Sheffield, Proceedings, p. 158-160.

Dubljansky, V.N., and Sotzkova, L.M., 1982, Microclimate of karst caves: *Zemlevedenie,* v. 11Y, p. 79-91 (in Russian).

Dubljansky, V.N., Sotskova, L.M., and Ferbei, G.G., 1989, *Microclimate of karst caves in the Crimea Mountains:* Simferopol, Ukr. NIINTI, N2495Uk 89, 132 p. (in Russian).

Dubljansky, Y.V., 1987, Theoretical modeling of the dynamics of formation of hydrothermal caves, *in* Metodi izuchenija i modelirovanija geologicheskikh yavlenij: Novosibirsk, Institute of Geology and Geophysics, p. 97-111 (in Russian).

Dubljansky, Y.V., 1989, The main principles of development and diagnostic criteria of a carbonate hydrothermal karst: *10th International Congress of Speleology,* Budapest, Proceedings, p. 77-79.

Dubljansky, Y.V., 1990, *Mechanism of formation and modeling of hydrothermal karst:* Novosibirsk, USSR Academy of Sciences, 150 p.

Dubljansky, Y.V., 1990, *Regularities of the formation and modeling of hydrothermal karst:* Novosibirsk, Nauka, 151 p. (in Russian, res. in English).

Dubljansky, Y., 1991a, A Budai-hegység hidrotermális paleokarsztja–A folyadékzárvány-vizsgálatok elsö eredményei (Paleohydrothermal karst of the Buda Hills–The preliminary results of study by fluid inclusion method): *Karszt és Barlang,* no. 1-2, p. 19-24.

Dubljansky, Y.V., 1991b, Zonalnost i mekhanism gidrotermokarsta (Zonation and mechanisms of hydrothermal karst), *in* Environments of karstogenesis: Kungur, p. 32-36.

Dubljansky, Y.V., 1995, Speleogenetic history of the Hungarian hydrothermal karst: *Environmental Geology,* v. 25, p. 24-35.

Dubljansky, Y.V., 1997, Transition between hydrothermal and cold-water karst: *12th International Congress of Speleology,* La Chaux-de-Fonds, Proceedings, v. 2, p. 267-270.

Dubljansky, Y.V., and Dublyansky, V.N., 1984, Obrazovanie gidrotermokarstovikh polostei (Formation of hydrothermal karst caves): *Peshchery,* v. 19, p. 37-42.

Dubljansky, Y.V., and Pashenko, A.E., 1997, Cave popcorn–an aerosol speleothem?: *12th International Congress of Speleology,* Chaux-de-Fonds, Proceedings, v. 1, p. 271-274.

Dublyanskaya, G.N., and Dubljansky, V.N., 1989, The role of condensation in the development of mountain karst, *in Problemi kompleksnogo izuchenija karsta gornikh stran:* Tbilisi-Tskhaltubo, p. 107-108 (in Russian).

Dunham, R.J., 1962, Classification of carbonate rocks: *American Association of Petroleum Geologists Memoir,* no. 1, p. 108-121.

Dunham, R.J., 1971, Meniscus cement, *in* Bricker, O.P., ed., *Carbonate cements:* Baltimore, John Hopkins Press, p. 297-300.

Dupont, E., 1894, Les phénomènes généraux des cavernes en terrains calcareux et la circulation souterraine des eaux dans la région Han-Rochefort: *Bulletin de la Société Belge de Géologie, de Paléontologie et d'Hydrologie,* Mém., v. 7, p. 190-297.

Durov, S.A., 1956, On the question about the origin of the salt composition of karst waters: *Ukrainian Chemical Journal,* v. 22, p. 106-111 (in Russian). (English translation in *Cave Geology,* v. 1, p. 186-190, 1979).

Dyson, N., 1969, The Meal Bank Coal Seam and its influence on cave development: *Bradford Pothole Club Bulletin,* v. 5, no. 6, p. 16-18.

Dyson, N., 1970, Statistical analysis of caves influenced by the Meal Bank Coal Seam: *Bradford Pothole Club Bulletin,* v. 5, no. 7, p. 1.

Dyson, N., 1971, Statistical analysis of potholes influenced by the Meal Bank Coal Seam: *Bradford Pothole Club Bulletin,* v. 5, no. 8, p. 15-16.

Dzuba, A.A., 1984, *Brines discharge in the Siberian Craton:* Novosibirsk, Nauka, 156 p. (in Russian).

Dzuba, A.A., Kustov, Y.I., Pospeev, V.I., and Trofimuk, P.I., 1974, Open and latent discharge of brines in the Angaro-Lensky Artesian Basin, *in Groundwater deposits of the Irkutsk region:* Novosibirsk, Nauka, p. 176-182 (in Russian).

Dzulynsky, S., 1976, Hydrothermal karst and Zn-Pb sulfide ores: *Roczn. Pol. Tow. Geol.,* v. 46, p. 217-230.

Dzulynsky, S., and Sass-Guskiewicz, M., 1985, Hydrothermal karst phenomena as a factor in the formation of Mississippi Valley-type deposits, *in* Handbook of strata-bound and stratiform ore deposits: pt. 4, v. 13, chap. 5, p. 391-439.

E

Ebert, H., 1955, Pesquisas na parte suleste do Estado de Minas Gerais: *Relatório Anual da Divisão de Geologia e Mineralogia,* p. 78-89.

Edgell, H.S., 1996, Salt tectonism in the Persian Gulf Basin, *in* Alsop, I., Blundell, D., and Davison, I., eds., Salt Tectonics: *Geological Society of London Special Publication,* no. 100, p. 129-151.

Egemeier, S.J., 1971, *A comparison of two types of solution caves:* Unpublished report to Carlsbad Caverns National Park. 7 p.

Egemeier, S.J., 1973, *Cavern development by thermal waters with a possible bearing on ore deposition:* Stanford University, Ph.D. dissertation, 88 p.

Egemeier, S.J., 1981, Cavern development by thermal waters: *National Speleological Society Bulletin,* v. 43, p. 31-51.

Eisenlohr, L., Madry, B., and Dreybrodt, W., 1997, Changes in the dissolution kinetics of limestone by intrinsic inhibitors adsorbing to the surface: *12th International Congress of Speleology and 6th Conference on Limestone Hydrology and Fissured Media,* La Chaux-de-Fonds, Proceedings, v. 2, p. 81-84.

Embry, A.F., and Klovan, J.E., 1971, A Late Devonian reef tract in northeastern Banks Island, Northwest Territories: *Canadian Petroleum Geology Bulletin,* v. 19, p. 730-781.

Enjalbert, H., 1964, Phénomènes karstiques Mexique et au Guatemala: *Bulletin Association Geographie Francais* v. 324-325, p. 30-58.

Eraso, A., 1969, La corrosión climática en las cavernas: *Bol. Geol. y Minero,* v. 80, p. 564-581.

Eremenko, Y.P., and Kolpashnikov, G.A., 1974, On the methods of calculations of condensation runoff from spoils of potash mines, *in* Regim, balans i resoursi podzemnikh vod: Minsk, p. 134-141 (in Russian).

Esper, J.F., 1774, *Ausfürliche Nachricht von neuentdeckten Zoolithen ...:* Nürnberg, Knorr, 148 p. (p. 100-107).

Esteban, M., and Klappa, C.F., 1983, Subaerial exposure, *in* Scholle, P.A., Bebout, D.G., and Moore, C.H., eds., Carbonate depositional environments: *American Association of Petroleum Geologists Memoir,* no. 33, p. 1-54.

EUR 16526, 1995, *Hydrogeologic aspects of groundwater protection in karstic areas–Guidelines:* Luxembourg, Office for Official Publications of the EC, 15 p.

Evans, F.G., 1872, The Carboniferous limestone: *Cardiff Naturalists' Society,* Report and Transactions, v. 3, no. 1, for 1870-71, p. 39-47 (p.46-47).

Ewers, R.O., 1966, Bedding-plane anastomoses and their relation to cavern passages: *National Speleological Society Bulletin,* v. 28, p. 133-140.

Ewers, R.O., 1972, *A model for the development of subsurface drainage routes along bedding planes:* University of Cincinnnati, M.Sc. thesis, 84 p.

Ewers, R.O., 1982, *Cavern development in the dimensions of length and breadth:* McMaster University, Ph.D. dissertation, 398 p.

Ewers, R.O., and Quinlan, J.F., 1981, Cavern porosity development in limestone: Low dip model from Mammoth Cave, Ky.: *8th International Congress of Speleology,* Bowling Green, Proceedings, p. 721-731.

Exley, S., 1994, *Caverns measureless to man:* St. Louis, Cave Books, 325 p.

Ezhov, Y.A., Lysenin, G.P., Andreychuk, V.N., and Dublyansky, Y.V., 1992, *Karst in the Earth's crust: Distribution and the main types:* Novosibirsk, United Institute of Geology, Geophysics, and Mineralogy, 76 p. (in Russian).

Ezhov, Y.A., and Lysenin, G.P., 1986, Vertical hydrodynamic zoning of the Earth's crust: *Soviet Geology,* v. 8, p. 111-120 (in Russian).

Ezhov, Y.A., and Lysenin, G.P., 1988, Significance of the zone of transitive pressures in the underground hydrosphere: *Soviet Geology ,* v. 8, p. 107-114 (in Russian).

Ezhov, Y.A., and Vdovin, Y.P., 1970, On the question of vertical hydrodynamic zonality of the Earth's crust: *Soviet Geology,* v. 8, p. 66-76 (in Russian).

Ezhov, Y.A., et al., 1991, Deep-seated karst. Endokarst. Hydrothermal karst (abs.), *in* Karstogenetic settings: Deep-seated karst, endokarst, hydrothermal karst: Regional Symposium, Kungur, p. 4-13 (in Russian).

Ezhov, Y.A., et al., 1992, *Karst in the Earth crust: Distribution and the main types:* Novosibirsk, Institute of Geology, Geophysics, and Mineralogy, 76 p. (in Russian).

Ezhov, Y.A., Lysenin, G.P., and Andreichuk, V.N., 1988, Endokarst, *in* Collection of Papers–1987: *Sverdlovsk Institute of Geology and Geochemistry,* p. 36-40 (in Russian).

Ezhov, Y.A., Lysenin, G.P., Andreychuk, V.N., and Dublyansky, Y.V., 1992, *Karst in the Earth's crust: Distribution and the main types:* Novosibirsk, United Inst. of Geology, Geophysics, and Mineralogy, 76 p. (in Russian).

F

Farias, P., 1982, La estructura del sector central de los Picos de Europa: *Trabajos de Geología Univ. de Oviedo,* no. 12, p. 63-72.

Farkas, L., Litman, H.L., and Bloch, M.R., 1951, The formation of "salt tables" in natural and artificial solar pans: *Research Council of Israel Bulletin,* v. 1, no. 1-2, p. 36-39.

Farr, M., 1991, *The darkness beckons:* St. Louis, Cave Books, 280 p.

Faventies, J., 1561, *De montium origine:* Venice, Academia Veneta, 16 ff.

Fedoseev, N.A., 1975, *The history of study of the major problem of the hydrosphere:* Moscow, Nauka, 206 p. (in Russian).

Fernández-Gibert, E., Palomarès, M., Rossi, C., and Tortosa, A., 1994, Analisis de procedendia en arenas karsticas: Evidencia de una cobertera permotriasica erosionada en el macizo occidental de Picos de Europa: *1st Congreso Internacional sobre Picos de Europa,* Oviedo, Proceedings.

Fernández-Gibert, E., Rossi, C., and Ortiz, I., 1992, Espeleogénesis y procedencia de los sedimentos kársticos en el Pozo del Llastral (Unidad Frontal, Macizo del Cornión, Picos de Europa): *5th Congreso Español de Espeleología,* Actos, p. 416-425.

Filippov, A.G., 1994, Botovskaya cave, *in Problems of physical speleology:* Moscow, p. 142-160 (in Russian).

Finlayson, B., 1982, Granite caves in Girraween Park, south-east Queensland: *Helictite,* v. 20, p. 53-59.

Finnie, J., and Curl, R., 1963, On the functioning of a familiar nonlinear thermodynamic oscillator, *in International Symposium on nonlinear oscillation,* Moscow, v. 3, p. 121-123.

Firsov, L.V., 1990, *Isars:* Novosibirsk, Nauka, 470 p. (in Russian).

Fish, J.E., 1977, *Karst hydrogeology and geomorphology of the Sierra De El Abra and the Valles-San Luis Potosí region, Mexico:* McMaster University, Ph.D. thesis, 469 p.

Flamache, A., 1896, Sur la formation des grottes et des vallées souterraines: *Bulletin de la Société Belge de Géologie, de Paléontologie et d'Hydrologie,* Mém., v. 9, p. 355-367.

Flores, G., 1952, Geology of northern British Honduras *American Association of Petroleum and Geologists Bulletin,* v. 36, p. 404-413.

Ford, D.C., 1965a, The origin of limestone caverns: A model from the central Mendip Hills, England: *National Speleological Society Bulletin,* v. 27, p. 109-132.

Ford, D.C., 1965b, Stream potholes as indicators of erosion phases in limestone caves: *National Speleological Society Bulletin,* v. 27, p. 27-32.

Ford, D.C., 1968, Features of cavern development in central Mendip: *Cave Research Group of Great Britain Transactions,* v. 10, p. 11-25.

Ford, D.C., 1971, Geologic structure and a new explanation of limestone cavern genesis: *Cave Research Group of Great Britain Transactions,* v. 13, p. 81-94.

Ford, D.C., 1986, Genesis of paleokarst and strata-bound zinc-lead sulfide deposits in a Proterozoic dolostone, northern Baffin Island, Canada–A discussion: *Economic Geology,* v. 81, p. 1562-1566.

Ford, D.C., 1987, Effect of glaciations and permafrost upon the development of karst in Canada: *Earth Surface Processes and Landforms,* v. 12, p. 507-521.

Ford, D.C., 1988, Characteristics of dissolution cave systems in carbonate rocks, *in* James, N.P., and Choquette, P.W., eds., *Paleokarst:* New York, Springer Verlag, 601 p.

Ford, D.C., 1989, Paleokarst of Canada, *in* Bosák, P., Ford, D., Glazek, J., and Horácek, I., eds., *Paleokarst: A systematic and regional review:* Praha, Academia, p. 313-336.

Ford, D.C., 1997, Principal features of evaporite karst in Canada: *Carbonates and Evaporites,* v. 12, p. 15-23.

Ford, D.C., and Bakalowicz, M.J., 1983, *Preliminary report upon McMaster University geochemical and isotopic studies in Wind and Jewel Caves:* Hot Springs, S.D., Wind Cave National Park, Open-File Report, 16 p.

Ford, D.C., and Ewers, R.O., 1978, The development of limestone cave systems in the dimensions of length and depth: *Canadian Journal of the Earth Sciences,* v. 15, p. 1783-1798.

Ford, D.C., and Lundberg, J., 1987, A review of dissolutional rills in limestone and other soluble rocks: *Catena Supplement,* v. 8, p. 119-140.

Ford, D.C., and Takácsné Bolner, K., 1991, Abszolút kormeghatározás és stabil izotóp vizsgealatok budai barlangi kalcitmint·kon (U-series dating and stable isotope analyses on calcite precipitates from Buda caves): *Karszt és Barlang,* no. 1-2, p. 11-18.

Ford, D.C., and Williams, P.W., 1989, *Karst geomorphology and hydrology:* London, Unwin Hyman, 601 p.

Ford, D.C., and Zambo, L., 1997, U-series dating of phases of speleothem deposition and erosion in Baradla Cave, Aggtelek National Park, Hungary: *12th International Congress of Speleology,* La Chaux-de-Fonds, Proceedings, v. 1, p. 44.

Ford, D.C., ed., 1983, Castleguard Cave and Karst, Columbia Icefields area, Rocky Mountains of Canada: A symposium: *Arctic and Alpine Research,* v. 15, p. 425-554.

Ford, D.C., Lundberg, J., Palmer, A.N., Palmer, M.V., Dreybrodt, W., and Schwarcz, H.P., 1993, Uranium-series dating of the draining of an aquifer: The example of Wind Cave, Black Hills, South Dakota: *Geological Society of America Bulletin,* v. 105, p. 241-250.

Ford, D.C., Schwarcz, H.P., Drake, J.J., Gascoyne, M., Harmon, R.S., and Latham, A.G., 1981, Estimates of the age of the existing relief within the southern Rocky Mountains of Canada: *Arctic and Alpine*

Research, v. 13, p. 1-10.

Ford, D.C., Smart, P.L., and Ewers, R.O., 1983, The physiography and speleogenesis of Castleguard Cave, Columbia Icefields, Alberta, Canada: *Arctic and Alpine Research,* v. 15, p. 437-450.

Ford, T.D., 1952, New evidence for the correlation of Lower Carboniferous reefs at Castleton, Derbyshire: *Geological Magazine,* v. 189, p. 346-356.

Ford, T.D., 1976, The geology of caves, *in* Ford, T.D., and Cullingford, C.H.D., eds., *The Science of Speleology:* London, Academic Press, p. 11-60.

Ford, T.D., 1989, Paleokarsts of Britain, *in* Bosák, P., Ford, D.C., Glazek, J., and Horácek, I., eds., *Paleokarst–A Systematic and Regional Review:* Amsterdam and Praha, Elsevier and Academia, p. 51-70.

Ford, T.D., 1995, Some thoughts on hydrothermal caves: *Cave and Karst Science,* v. 22, no. 3, p. 107-118.

Ford, T.D., and King, R.J., 1965, Layered epigenetic galena-barite deposits in the Golconda Mine, Brassington, Derbyshire: *Economic Geology,* v. 60, p. 1686-1701.

Forti, P., 1991, Il carsismo nei gessi con particolare rigardo a quelli dell'Emilia Romagna: *Speleologia Emiliana,* v. 4, no. 2, p. 11-36.

Forti, P., 1993, Karst evolution and water circulation in gypsum formations: International Symposium on Water Resources in Karst with Special Emphasis on Arid and Semi-Arid Zones, Shiraz, Proceedings, p. 791-801.

Forti, P., and Rabbi, E., 1981, The role of CO_2 in gypsum speleogenesis: *International Journal of Speleology,* v. 11, p. 207-218.

Forti, P., Benedetto, C., and Costa, G., 1993, Las Brujas Cave (Malargue, Argentina): An example of the oil pools control on speleogenesis: *Theoretical and Applied Karstology,* v. 6, p. 87-93.

Foster, S.S.D., 1978, Discussion, *in* Thames groundwater scheme: London, Institution of Civil Engineers, p. 91-94.

Foster, S.S.D., and Crease, R.I., 1975, Hydraulic behaviour of the Chalk aquifer in the Yorkshire Wolds: *Institution of Civil Engineers,* v. 59, Proceedings, p. 181-188.

Frank, E.F., 1993, *Aspects of karst development and speleogenesis, Isla de Mona, Puerto Rico: An analogue for Pleistocene speleogenesis in the Bahamas:* Mississippi State University, Unpublished M.Sc. thesis, 349 p.

Frank-Kamenetsky, D.A., 1987, *Diffusion and heat transfer in chemical kinetics:* Moscow, Nauka, 502 p. (in Russian).

Freeze, R.A., and Cherry, J.A., 1979, *Groundwater:* Englewood Cliffs, Prentice Hall, 604 p.

Friedman, G.M., 1997, Dissolution-collapse breccias and paleokarst resulting from dissolution of evaporite rocks, especially sulfates: *Carbonates and Evaporites,* v. 12, p. 53-63.

Frolov, N.M., 1976, *Hydrogeothermal studies:* Moscow, Nauka, 280 p. (in Russian).

Frumkin, A., 1986, The cave survey of Mount Sedom, Israel, *in* Atti simposio internazionale sul carsismo nelle evaporiti: *Le Grotte d'Italia,* v. 12, p. 305-308.

Frumkin, A., 1994a, Hydrology and denudation rates of halite karst: *Journal of Hydrology,* v. 162, p. 171-189.

Frumkin, A., 1994b, Morphology and development of salt caves: *National Speleological Society Bulletin,* v. 56, p. 82-95.

Frumkin, A., 1996, Uplift rate relative to base level of a salt diapir (Dead Sea, Israel), as indicated by cave levels, *in* Alsop, I., Blundell, D., and Davison, I., eds., Salt Tectonics, *Geological Society of London Special Publication,* no. 100, p. 41-47.

Frumkin, A., 1997, Liquid Crystal Cave, Israel, *in* Cave Minerals of the World: *National Speleological Society,* Huntsville, p. 319-322.

Frumkin, A., 1998, Salt cave cross sections and their paleoenvironmental implications: *Geomorphology,* v. 23, p. 183-191.

Frumkin, A., and Ford, D.C., 1995, Rapid entrenchment of stream profiles in the salt caves of Mount Sedom, Israel: *Earth Surface Processes and Landforms,* v. 20, p. 139-152.

Frumkin, A., Magaritz, M., Carmi, I., and Zak, I., 1991, The Holocene climatic record of the salt caves of Mount Sedom, Israel: *The Holocene,* v. 1, p. 191-200.

Funcken, L, 1997, Bärenschacht, un exploration unique au monde. 12th International Congress of Speleology, LaChaux-de-Fonds, Switzerland, Proceedings. v. 4, p. 9-12.

G

Gabriel, R., 1986, *Contibució a l'estudi de les empremtes de corrent:* Barcelona, Espelo Club de Gracia., 78 p.

Gádoros, M., and Cser, F., 1986, Aerosols in caves–Theoretical considerations: *9th International Congress of Speleology,* Barcelona, Proceedings, v. 2, p. 90-92.

Gale, S.J., 1984, The hydraulics of conduit flow in carbonate aquifers: *Journal of Hydrology,* v. 70, p. 309-327.

Gams, I., 1959, About the situation and the genesis of the longest slovene caves: *Naše Jame,* v. 1, no. 1, p. 4-10 (in Slovenian with English summary).

Gams, I., 1963, Logarãek: *Acta Carsologica,* v. 3, p. 7-83 (in Slovenian with English summary).

Gams, I., 1965, On the Quarternary geomorphogenesis of the area among the karst poljes of Postojna, Planina and Cerknica: *Geografski vestnik,* v. 37, p. 61-101 (in Slovenian with English summary).

Gams, I., 1966a, Factors and dynamics of corrosion of the carbonatic rocks in the Dinaric and Alpine karst of Slovenia: *Geografski vestnik,* v. 38, p. 11-68.

Gams, I., 1966b, On the hydrology of the territory among the poljes of Postojna, Planina and Cerknica: *Acta Carsologica,* v. 4, p. 5-54 (in Slovenian with English summary).

Gams, I., 1970, Maximisation of the karstic underground water flows in example of the area among the karst poljes of Ceknica and Planina: *Acta Carsologica,* v. 5, p. 171-187 (in Slovenian with English summary).

Gams, I., 1981, Comparative research of limestone solution by means of standard tablets: *8th International Congress of Speleology,* Bowling Green, Proceedings, p. 273-275.

Gams, I., 1994, Types of the poljes in Slovenia, their inundations and land use: *Acta Carsologica,* v. 23, p. 285-302.

Gardner, J.H., 1935, Origin and development of limestone caverns: *Geological Society of America Bulletin,* v. 46, p. 1255-1274.

Gardner, R.A.M., and McLaren, S.J., 1993, Progressive vadose diagenesis in late Quaternary aeolianite deposits?, *in* Pye, K., ed., The dynamics and environmental context of aeolian sedimentary systems: *Geological Society of London Special Publication,* no. 72, p. 219-234.

Garrels, R.M. and Christ, C.L., 1965, *Solutions, minerals, and equilibria:* New York, Harper and Row, 450 p.

Gascoyne, M., Latham, A.G., Harmon, R.S., and Ford, D.C., 1983, The antiquity of Castleguard Cave, Columbia Icefields, Alberta, Canada: *Arctic and Alpine Research,* v. 15, p. 463-470.

Ge, S., 1997, A governing equation for fluid flow in rough fractures: *Water Resources Research,* v. 33, p. 53-61.

Genty, D., 1995, Corrélation entre l'épaisseur des lamines de deux stalagmites modernes et les données météorologiques: *Spéléochronos,* v. 6, p. 35-40.

Genuit, P., and Vidal, B., 1984, La Torca Urriello; Massif Central Picos de Europa; Asturies, Espagne: *Spelunca,* v. 16, p. 14-18.

Gergedava, B.A., 1970, The role of condensation and infiltration waters in cave origin: *Trudy Vses. Geograficheskogo Obshchestva,* v. 102, no. 5, p. 196-198 (in Russian).

Gerson, R., and Inbar, M., 1974, The field study program of the Jerusalem-Elat Symposium: *Reviews and summaries of Israeli research projects,* v. 20, p. 7-11.

Gèze, B., 1958, Sur quelques caractères fondamentaux des circulations karstiques:

Annales de Spéléologie, ser. 3, v. 13, p. 1-22.

Gèze, B., 1965, Les conditions hydrogéologiques des roches calcaires: Bureau de Recherches Geologiques et Minieres, *Chronique d'Hydrogeologie*, v. 7, p. 9-39.

Gilli, E., 1993, Les grandes volumes souterrains du massif de Mulu (Borneo, Sarawak, Malisie): *Karstologia*, v. 22, p. 58-62.

Gillieson, D., 1996, *Caves; Processes, development, management:* Oxford, Basil Blackwell, 324 p.

Ginet, R., 1977, Étude de la condensation atmospherique saisonneire dans la grotte de Hautcourt: *Pr. Acad. Sci.*, v. 18, p. 1615-1618.

Giurgiu, I., 1985, Grottes du sel de Roumanie: *Spelunca*, v. 20, p. 34-39.

Glennie, E.A., 1954a, Artesian flow and cave formation: *Cave Research Group of Great Britain Transactions*, v. 3, p. 55-71.

Glennie, E.A., 1954b, The origin and development of cave systems in limestone: *Cave Research Group of Great Britain Transactions*, v. 3, p. 73-83.

Glover, R.R., 1972, Optical brightners–A new water tracing reagent: *Cave Research Group of Great Britain Transactions*, v. 14, p. 84-88.

Glover, R.R., 1973, Gaping Gill–some underground controls of development: *Craven Pothole Club Journal*, v. 5, p. 8-11 and 58-66.

Glover, R.R., 1974, Cave development in the Gaping Gill System, *in* Waltham, A.C., ed., *Limestones and caves of north-west England:* Newton Abbot, David and Charles, p. 343-384.

Glover, R.R., 1976, The Porcellanous Band in Yorkshire caves: *British Cave Research Association Bulletin*, no. 12, p. 11-12.

Glukhov, I.G., 1963, The role of condensation in the water balance of mountains (on the example of the Crimea Mountains): *Izvestija VUZov*, Geol. Series, v. 3, p. 26-31 (in Russian).

Golod, V.M., 1981, Mathematical model of aerohydrodynamic processes in the aeration zone, *in* Akkumulatsija zimnego kholoda v porodakh i ego ispolzovanije: Perm, p. 6-8 (in Russian).

Golovkinskii, N.A., 1896, *Observations on precipitation in soil:* Attachment to the report of the hydrogeologist for 1896, Simferopol, 6 p. (in Russian).

Gonzalez Herrera, R.A., 1992, Evolution of groundwater contamination in the Yucatan Karstic Aquifer: University of Waterloo, MSc thesis, 149 p.

Goodchild, M.F., and Ford, D.C., 1971, Analysis of scallop patterns by simulation under controlled conditions: *Journal of Geology*, v. 79, p. 52-62.

Gorbunova, K.A., 1977, *Karst in gypsum of the USSR:* Perm University, 83 p. (in Russian).

Gospodarič, R., 1970, Speleological investigations of the Cerknica Cave System: *Acta Carsologica*, v. 5, p. 109-169 (in Slovenian with English summary).

Gospodarič, R., 1976, Quaternary cave development between the Pivka Basin and the Polje of Planina: *Acta Carsologica*, v. 7, p. 5-135 (in Slovenian with English summary).

Gospodarič, R., and Habič, P., 1979, Karst phenomena of Cerkniško Polje: *Acta Carsologica*, v. 8, p. 7-162 (in Slovenian with English summary).

Gospodarič, R., and Habič, P., eds., 1976, *Underground water tracing–investigations in Slovenia 1972-1975:* Postojna, Institute of Karst Research, 309 p.

Granger, D.E., and Palmer, A.N., 1997, A proposed method for determining the age and evolution of the Mammoth Cave System: *6th Science Conference*, Mammoth Cave National Park, Proceedings, p. 145-150.

Grimandi, P., 1987, Grotta della Spipola: *Ipoantropo*, v. 5, p. 51-64.

Groves, C.G., 1993, *Early development of karst systems:* University of Virginia, Ph.D. Thesis, 253 p.

Groves, C.G., and Howard, A.D., 1994a, Early development of karst systems, 1. Preferential flow path enlargement under laminar flow: *Water Resources Research*, v. 30, p. 2837-2846.

Groves, C.G., and Howard, A.D., 1994b, Minimum hydrochemical conditions allowing limestone cave development: *Water Resources Research*, v. 30, p. 607-615.

Grund, A., 1903, Die Karsthydrographie–Studien aus Westbosnien: *Geographische Abhandlungen herausgegeben von A. Penck*, v. 7, p. 103-200.

Grund, A., 1910, Beiträge zur Morphologie des dinarischen Gebirges:*Geographische Abhandlungen herausgegeben von A. Penck*, v. 9.

Grund, A., 1914, Der geographische Zyclus im Karst: *Zeitschrift für Ges. Erdkunde*, v. 52, p. 621-640.

Gunn, J., 1978, *Karst hydrology and solution in the Waitomo District, New Zealand:* University of Auckland, Ph.D. thesis.

Gunn, J., 1981a, Hydrological processes in karst depressions: *Zeitschrift für Geomorphologie*, v. 25, p. 313-331.

Gunn, J., 1981b, Limestone solution rates and processes in the Waitomo District, New Zealand: *Earth Surface Processes and Landforms*, v. 6, p. 427-445.

Gunn, J., 1981c, Prediction of limestone solution rates from rainfall and runoff data: Some comments: *Earth Surface Processes and Landforms*, v. 6, p. 595-597.

Gunn, J., 1983, Point recharge of limestone aquifers–A model from New Zealand karst: *Journal of Hydrology*, v. 61, p. 19-29.

Gurnee, R., and Gurnee, J., 1974, *Discovery at the Rio Camuy:* New York, Crown Publishers, 183 p.

Gutsalo, L.K., 1971, Radiolysis of water as the source of free oxygen in the underground hydrosphere: *Geokhimiya*, v. 12, p. 1473-1481 (in Russian), *Geochemistry International*, v. 8, p. 897-903 (1972).

Gvozdetsky, N.A., 1954, *Karst:* Moscow, Geografgiz, 351 p. (in Russian).

Gvozdetsky, N.A., 1972, *Problems of karst study and a practice:* Moscow, Mysl', 390 p (in Russian).

H

Haas, J.L., 1959, Evaluation of ground water tracing methods used in speleology: *National Speleological Society Bulletin*, v. 21, p. 67-76.

Habe, F., 1976, Die morphologische, hydrographische und speleologische Entwicklung der Flyschbucht von Studeno: *Acta Carsologica*, v. 7, p. 141-213 (in Slovenian with English summary).

Habič, P., 1981, Karst relief and tectonics: *Acta Carsologica*, v. 10, p. 23-44 (in Slovenian with English summary).

Habič, P., 1987, The ponor cave Golobina at Loško Polje: *Acta Carsologica*, v. 16, p. 35-49 (in Slovenian with English summary).

Habič, P., 1989, Pivka karst bifurcation on the Adriatic-Black Sea watershed: *Acta Carsologica*, v. 18, p. 233-264 (in Slovenian with English summary).

Hall, A.L., 1906, On the geology of the country between Lydenburg and Devil's Kantoor: *Annual Report of the Geological Survey of Transvaal for 1905*, p. 39-57.

Halstead, B., and Nicoll, P.G., 1971, Fossilized caves of Mendip: *Studies in Speleology*, v. 2, no. 3-4, p. 93-102.

Hann, J., 1880, Über eine neue Quellentheorie auf meteorologicher Basis: *Ztochr. Osterreichichen Gesellsch. für Meteorol.*, v. 15, p. 482-486.

Harmon, R.S., Atkinson, T.C., and Atkinson, J.L., 1983. The Mineralology of Castleguard Cave, Columbia Ice Fields, Alberta, Canada: *Artic and Alpine Research*, v. 15, p. 503-516.

Harned, H.S., and Hamer, W.J., 1933, The ionization constant of water: *Journal of the American Chemical Society*, v. 55, p. 693-695.

Harris, J.G., Mylroie, J.E., and Carew, J.L., 1995, Banana holes: Unique karst features of the Bahamas: *Carbonates and Evaporites*, v. 10, p. 215-224.

Harris, N.J., 1984, *Diagenesis of upper Pleistocene strand-plain limestones, northeastern Yucatan Peninsula, Mexico*: University of New Orleans, M.Sc. thesis, 130 p.

Häuselmann, P., and Otz, M., 1997, Wasserfärbung im Gebiet Gemmenalp-Sieben Hengste: Ziele und Resultate: *12th International Congress of Speleology,* La Chaux-de-Fonds, Proceedings, v. 2, p. 31-34.

Headworth, H.G., 1978, Hydrogeological characteristics of artesian boreholes in the Chalk of Hampshire: *Quarterly Journal of Engineering Geology*, v. 11, p. 139-144.

Headworth, H.G., Keating, T., and Packman, M.J., 1982, Evidence for a shallow highly-permeable zone in the Chalk of Hampshire, U.K.: *Journal of Hydrology*, v. 55, p. 93-112.

Helgeson, H.C., Murphy, W.M., and Aagaard, P., 1984, Thermodynamic and kinetic constraints on reaction rates among minerals and aqueous solutions. II Rate constants, effective surface area, and the hydrolysis of feldspar: *Geochimica et Cosmochimica Acta,* v. 48, p. 2405-2432.

Helu, P.C., Rodolfo, V.V., and Rodolfo, B.P., 1977, Origin and distribution of Tertiary conglomerates, Veracruz Basin: *American Association of Petroleum Geologists Bulletin,* v. 61, p. 207-226.

Henrich, R., 1984, Facies, dolomitization and karstification of lagoonal carbonates: Triassic of the northern Alps: *Facies,* v. 11, p. 109-156.

Herman, J.S., 1982, *The dissolution kinetics of calcite, dolomite and dolomite rocks in carbon dioxide water system:* Pennsylvania State University, Ph.D. thesis, 218 p.

Hess, J.W., and Harmon, R.S., 1981, Geochronology of speleothems from the Flint Ridge-Mammoth Cave System, Kentucky, USA: *8th International Congress of Speleology,* Bowling Green, Proceedings, v. 2, p. 433-436.

Hess, J.W., and White, W.B., 1988, Storm response of the karstic carbonate aquifer of southcentral Kentucky: *Journal of Hydrology*, v. 99, p. 235-252.

Hess, J.W., and White, W.B., 1989, Chemical hydrology, *in* White, W.B., and White, E.L., eds., *Karst hydrology–Concepts from the Mammoth Cave area:* New York, Van Nostrand Reinhold, 346 p.

Hess, J.W., and White, W.B., 1993, Ground-water geochemistry of the carbonate karst aquifer, south-central Kentucky, USA: *Applied Geochemistry,* v. 8, p. 189-204.

Hevesi, A., 1989, Development and evolution of karst regions in Hungary: *Karszt-es Barlang*, Special Issue, p. 3-16.

High, C., and Hanna, F.K., 1970, A method for the direct measurement of erosion on rock surfaces: *British Geomorphological Research Group Technical Bulletin*, no. 5, 24 p.

High, C.J., 1970, *Aspects of the solutional erosion of limestone, with special consideration of lithological factors:* University of Bristol, Ph.D. dissertation, 228 p.

Hill, A.L., 1957, *The origin of the Kelly Hill Caves, Kangaroo Island, South Australia*: Unpublished paper read to the Cave Exploration Group (South Australia), April 1957.

Hill, C.A., 1981, Speleogenesis of Carlsbad Cavern and other caves of the Guadalupe Mountains*: 8th International Congress of Speleology,* Bowling Green, Proceedings, p. 143-144.

Hill, C.A., 1985, Speleogenesis of Carlsbad Cavern and other caves of the Guadalupe Mountains, New Mexico: *GEO^2, Newsletter of the National Speleological Society Section on Cave Geology and Geography,* v. 12, no. 2, p. 30-31.

Hill, C.A., 1986, Carlsbad Cavern and other caves in the Guadalupe Mountains, New Mexico: A sulfuric acid genesis related to the oil and gas fields of the Delaware basin: *9th International Congress of Speleology*, Barcelona, Proceedings, v. 1, p. 267-269.

Hill, C.A., 1987, Geology of Carlsbad Cavern and other caves in the Guadalupe Mountains, New Mexico and Texas: *New Mexico Bureau of Mines and Mineral Resources Bulletin,* no. 117, 170 p.

Hill, C.A., 1990, Sulfuric acid speleogenesis of Carlsbad Cavern and its relationship to hydrocarbons, Delaware Basin, New Mexico and Texas: *American Association of Petroleum Geologists Bulletin,* v. 74, p. 1685-1694.

Hill, C.A., 1993a, A hydrogen sulfide, basinal-degassing model for the origin of Mississippi Valley-type ore deposits, *in* Shelton, K.L., and Hagni, R.D., eds., *Geology and geochemistry of Mississippi Valley-type ore deposits:* University of Missouri-Rolla Press, p. 121-132.

Hill, C.A., 1993b, Sulfide/barite/fluorite mineral deposits, Guadalupe Mountains, New Mexico and west Texas: *New Mexico Geology,* v. 15, no. 3, p. 56-65.

Hill, C.A., 1995a, H_2S-related porosity and sulfuric acid oil-field karst, *in* Budd, D.A., Saller, A.H., and Harris, P.M., eds., Unconformities in carbonate strata–Their recognition and the significance of associated porosity: *American Association of Petroleum Geologists Memoir,* no. 61, p. 301-306.

Hill, C.A., 1995b, Sulfur redox reactions, native sulfur, Mississippi Valley-type deposits, and sulfuric acid karst, Delaware Basin, New Mexico and Texas: *Environmental Geology,* v. 25, p. 16-23.

Hill, C.A., 1996, Geology of the Delaware Basin, Guadalupe, Apache, and Glass Mountains, New Mexico and West Texas: *Society of Economic Paleontologists and Mineralogists–Permian Basin Section,* Publication 96-39, 480 p.

Hill, C., and Forti, P., 1986, *Cave minerals of the world:* National Speleological Society, Huntsville, Alabama, 238 p.

Hiss, W.L., 1980, Movement of ground water in Permian Guadalupian aquifer systems, southeastern New Mexico and western Texas, *in* Dickerson, P.W., and Hoffer, J.M., eds., *Trans-Pecos region, New Mexico Geological Society Guidebook,* 31st Field Conference, p. 289-294.

Hjulstrøm, F., 1935, Studies of the morphological activity of rivers as illustrated by the River Fyris: *Geological Institute of the University of Upsala Bulletin,* v. 25, p. 227-346.

Holdren, G.R., and Speyer, P.M., 1986, Stoichiometry of alkali feldspar dissolution at room temperature and various pH values, *in* Colman, S.M., and Dethier, P.P., eds., *Rates and chemical weathering of rocks and minerals:* London, Academic Press, 603 p.

Holland, H.D., and Malinin, S.D., 1979, The solubility and occurrence of non-ore minerals, *in* Barnes, H.L., ed., *Geochemistry of hydrothermal ore deposits:* New York, John Wiley and Sons, p. 461-508.

Holland, H.D., Kirsipu, T.-W., Huebner, I.I., and Oxburgh, U.M., 1964, On some aspects of the chemical evolution of cave waters: *Journal of Geology*, v. 72, p. 36-67.

Homza, S., Rajman, L., and Roda, S., 1970, Vznik a vyvoj krasoveho fenomenu Ochtinskej aragonitovej jaskine: *Slovensky kras,* v. 8, p. 21-68 (in Slovak).

Horlacher, H.B., and Lüdecke, H.J., 1992, *Strömungsberechnung für Rohrsysteme:* Expert Verlag, 218 p.

Horn, G., 1935, Über die Bildung von Karsthöhlen unter einem Gletcher: *Norsk Geografisk Tidsskrift,* v. 5, p. 494-498.

Horn, G., 1937, Über einige Karsthöhlen in Norwegen: *Mitteilungen für Höhlen und Karstforschung*, p. 1-15.

Horn, G., 1978, Limestone caves in Nordland: *Cave Geology,* v. 1, p. 124-138.

Hose, L.D., 1981, Fold development in the Anticlinorio Huizachal-Peregrina and its influence on the Sistema Purification, Mexico: *8th International Congress of Speleology,* Bowling Green, Proceedings, v. 1, p. 133-135.

Hose, L.D., 1992, Structural controls on the development of Cueva Cheve: A reconnaissance report: *National Speleological Society Bulletin,* v. 54, p. 89.

Hose, L.D., 1995, Cave in Oaxaca proven to be world's deepest known karst conduit system: *Revista Mexicana de Cíencias Geológias,* v. 12, p. 108-111.

Houston, J.F.T., Eastwood, J.C., and Cosgrove, T.K.P., 1986, Locating potential borehole sites in a discordant flow regime in the Chalk Aquifer at Lulworth using integrated geophysical surveys: *Quarterly Journal of Engineering Geology*, v. 19, p. 271-282.

Howard, A.D., 1960, Geology and origin of the crevice caves of the Iowa, Illinois, and Wisconsin lead-zinc district: *Journal of the Yale Speleological Society*, v. 2, no. 4, pt. 1, p. 61-95.

Howard, A.D., 1964a, Model for cavern development under artesian ground water flow, with special reference to the Black Hills: *National Speleological Society Bulletin*, v. 26, p. 7-16.

Howard, A.D., 1964b, Processes of limestone cave development: *International Journal of Speleology*, v. 1, p. 47-60.

Howard, A.D., and Groves, C.G., 1995, Early development of karst systems, 2. Turbulent flow: *Water Resources Research*, v. 31, p. 19-26.

Hoyos, M., and Herrero, N., 1989, El karst en la Cornisa Cantábrica; El karst en España: *Monografía Soc. Esp. de Geom.*, no. 4, p. 109-120.

Hubbard, D.A., Jr., Herman, J.S., and Bell, P.E., 1986, The role of sulphide oxidation in the genesis of Cesspool Cave, Virginia, U.S.A.: *9th International Congress of Speleology*, Barcelona, Proceedings, v. 1, p. 255-258.

Hubbert, M.K., 1940, The theory of groundwater motion:*Journal of Geology*, v. 48, p. 785-944.

Hückinghaus, D., Clemens, T., Liedl, R., Sauter, M., and Teutsch, G., 1998, Simulation of the development of karst aquifers by using a pipe flow model coupled to a continuum model, 1: Theory: *Water Resources Research*, in preparation.

Huntoon, P.W., 1974, The karstic groundwater basins of the Kaibab Plateau, Arizona: *Water Resources Research*, v. 10, p. 579-590.

Huntoon, P.W., 1981, Fault controlled ground-water circulation under the Colorado River, Marble Canyon, Arizona: *Ground Water*, v. 19, p. 20-27.

Huntoon, P.W., 1990, Phanerozoic structural geology of the Grand Canyon: *in* Beus, S.S., and Morales, M., eds., *Grand Canyon geology:* New York, Oxford University Press, p. 261-309.

Huntoon, P.W., 1993, The influence of Laramide foreland structures on modern ground-water circulation in Wyoming artesian basins, *in* Snoke, A.W., Steidtmann, J.R., and Roberts, S.M., eds., Geology of Wyoming: *Geological Survey of Wyoming Memoir*, no. 5, p. 756-789.

Huntoon, P.W., 1995, Is it appropriate to apply porous media groundwater circulation models to karstic aquifers? *in* El-Kadi, A.I., ed., Groundwater models for resources analysis and management: Boca Raton, Lewis Publishers, p. 339-358.

Huntoon, P.W., 1996, Large-basin ground water circulation and paleo-reconstruction of circulation leading to uranium mineralization in Grand Canyon breccia pipes, Arizona: *Mountain Geologist*, v. 33, no. 3, p. 71-84.

Hutcheson, E., and Madden, M., 1994, Unpublished map of Nohoch Nah Chich System.

[Hutton, J.], 1780, *A tour to the caves, in the environs of Ingleborough and Settle, in the West-Riding of Yorkshire ...:* London, Richardson and Urquhart, 49 p. (p. 38-39).

Hutton, J., 1788, Theory of the earth ...: *Transactions of the Royal Society of Edinburgh*, v. 1, p. 209-304.

I

Iler, R.K., 1977, Hydrogen-bonded complexes of silica with organic compounds, *in* Bentz, G., and Lindquist, I., eds., *Biochemistry of silicon and related problems:* New York, Plenum Press, p. 53-76.

Iliffe, T.M., and Sarbu, S., 1990, Anchialine caves and cave fauna of the South Pacific: *National Speleological Society News*, v. 48, p. 88-96.

Ingri, N., 1977, Aqueous silicic acid, silicates and silicate complexes, *in* Bentz, G., and Lindquist, I., eds., *Biochemistry of silicon and related problems:* New York, Plenum Press, p. 3-52.

ITGE, 1989, Las aguas subterráneas en España: Estudio de síntesis. *Instituto Tecnológico y Geominero de España*, Estudio de síntesis.

Ivanov, B.N., 1956, On the typology of karst landscapes of planes, on the example of the Podol'sko-Bulovinsky karst region, *in* Voprosy izuchenija karsta na juge Evropejskoj chasti SSSR: Yalta, p. 131-156 (in Russian).

Ivanov, B.N., and Dubljansky, V.N., 1966, Superficial and underground karst of the south-west flank of the Russian Platform, *in* Voprosy izuchenija karsta Russlkoj ravniny: Moscow, p. 28-36 (in Russian).

Ivanovich, M., and Harmon, R.S., 1982, *Uranium series disequilibrium: Applications to environmental problems:* Oxford University Press, 571p.

J

Jackson, M.P.A., Cornelius, R.R., Craig, C.H., Gansser, A., Stocklin, J., and Talbot, C.J., 1990, Salt diapirs of the Great Kavir, central Iran: *Geological Society of America Memoir*, no. 177, 139 p.

Jacobs, D., 1977, *The state of the art in numerical analysis:* London, Academic Press, 978 p.

Jagnow, D.H., 1977, *Geologic factors influencing speleogenesis in the Capitan Reef Complex, New Mexico and Texas:* University of New Mexico, M.Sc. thesis, 197 p.

Jagnow, D.H., 1979, *Cavern development in the Guadalupe Mountains:* Columbus, Ohio, Cave Research Foundation, 55 p.

Jagnow, D.H., 1988, The geology of Lechuguilla Cave: *National Speleological Society News*, v. 46, p. 422-425.

Jagnow, D.H., 1996, History of sulfuric acid theory of speleogenesis in the Guadalupe Mountains: *National Speleological Society Convention*, Salida, Program with Abstracts, p. 50-51.

Jakucs, L., 1977a, Morfologicheskie i evolutsionnie typi karstov Vengrii (Morphological and evolutionary types of karst in Hungary): *Acta Geographica* Szeged, v. 17, p. 65-103.

Jakucs, L., 1977b, *Morphogenetics of karst regions: Variants of karst evolution:* Budapest, Akademiai Kiado, 284 p.

James, A.N., 1992, *Soluble materials in civil engineering:* Chichester, Ellis Horwood, 435 p.

James, A.N., and Kirkpatrick, I.M., 1980, Design of foundations of dams containing soluble rock and soils: *Quarterly Journal of Engineering Geology*, v. 13, p. 189-198.

James, A.N., and Lupton, A.R.R., 1978, Gypsum and anhydrite in foundations of hydraulic structures: *Geotechnique*, v. 28, p. 249-272.

James, N.P., and Choquette, P.W., eds., 1988, *Paleokarst:* New York, Springer-Verlag, 601 p.

Jameson, R.A., 1989, Features of condensation corrosion in caves of the Greenbrier Karst, West Virginia (abs.): *National Speleological Society*, Proceedings.

Jameson, R.A., 1995, Condensation, condensation corrosion, and associated features in Snedegar's and Greenville Saltpeter Caves, *in* Underground the Appalachians: *National Speleological Society*, Blacksburg, Virginia, Convention Guidebook, p. 122-125.

Jameson, R.A., and Alexander, E.C., Jr., 1989, Hydrology and chemistry of condensation waters in Snedegar's and Greenville Saltpeter Caves, West Virginia (abs.): *National Speleological Society*, Proceedings

Jeannin, P.-Y., 1990, Néotectonique dans le karst du nord du lac de Thoune (Suisse): *Eclogae Geologicae Helvetiae*, v. 83, p. 323-342.

Jeannin, P.-Y., 1991, Mise en évidence d'importantes glaciations anciennes par l'étude des remplissages karstiques du Réseau des Sieben Hengste (Chaîne bordière helvétique): *Eclogae Geologicae Helvetiae*, v. 84, p. 207-221.

Jenko, F., 1959, *Hidrogeologija i vodno gospodarstvo Krasa:* Ljubljana, 236 p.

Jenko, F., 1959, *The hydrogeology and water*

economy of karst: Ljubljana, Dršavna zalošba Slovenije, 239 p. (in Slovenian with English summary).

Jennings, J.N., 1968, Syngenetic karst in Australia, *in* Williams, P.W. and Jennings, J.N., eds., Contributions to the study of karst: *Australian National University Research School of Pacific Studies Publication,* no. G5, p. 41-110.

Jennings, J.N., 1971, *Karst:* Cambridge, M.I.T. Press, 252 p.

Jennings, J.N., 1972, Observations at the Blue Waterholes, March 1965-April 1969, and limestone solution on Cooleman Plain, N.S.W.: *Helictite*, v. 10, p. 3-46.

Jennings, J.N., 1982, Principles and problems in reconstructing karst history: *Helictite,* v. 20, no. 2, p. 37-52.

Jennings, J.N., 1983, Karst landforms: *American Scientist*, v. 71, p. 578-586.

Jennings, J.N., 1983, Sandstone pseudokarst or karst? *in* Young, R.W., and Nanson, G.C., eds., *Aspects of Australian sandstone landscapes:* Australia and New Zealand Geomorphology Group, Wollongong, p. 21-30.

Jennings, J.N., 1985, *Karst geomorphology:* Oxford, Basil Blackwell, 293 p.

Johns, R.A., Steude, J.S., Castanier, L.M., and Robert, P.V., 1993, Nondestructive measurements of fracture aperture in crystalline rock cores, using X-ray computer tomography: *Journal of Geophysical Research,* v. 98, p. 1889-1900.

Johnson, K.S., 1996, Gypsum karst in the United States, *in* Klimchouk, A., Lowe, D., Cooper, A., and Sauro, U., eds., Gypsum karst of the world: *International Journal of Speleology*, v. 25, no. 3-4, p. 183-193.

Johnson, P.W., and Sanderson, R.B., 1968, Spring flow into the Colorado River, Lees Ferry to Lake Mead, Arizona: Arizona State Land Department, *Water Resources Report,* no. 34, 26 p.

Jones, B., 1992a, Caymanite, a cavity-filling deposit in the Oligocene-Miocene Bluff Formation of the Cayman Islands: *Canadian Journal of Earth Science,* v. 29, p. 720-735.

Jones, B., 1992b, Void-filling deposits in karst terrains of isolated oceanic islands: A case study from Tertiary carbonates of the Cayman Islands: *Sedimentology,* v. 39, p. 857-876.

Jones, B., and Smith, D.S., 1988, Open and filled features on the Cayman Islands: Implications for the recognition of paleokarst: *Canadian Journal of Earth Sciences,* v. 25, p. 1277-1291.

Jones, W.K., 1973, Hydrology of limestone karst: *West Virginia Geological and Economic Survey Bulletin,* no. 36, 49 p.

Julivert, M., Truyols, J., and Garcia Alcalde, J., 1971, Mapa geológico de España, sheet 10 (Mieres): *Instituto Tecnológico y Geominero de España,* scale 1:200,000.

K

Kahle, C.F., 1988, Surface and subsurface paleokarst, Silurian Lockport, and Peebles Dolomites, western Ohio, *in* James, N.P., and Choquette, P.W., eds., *Paleokarst:* New York, Springer-Verlag, p. 229-255.

Kaiser, W., 1976, Behaviour of anhydrite after addition of water: *Society of the International Association of Engineering Geologists Bulletin*, v. 13, p. 68-69.

Kárpát, J., 1982, A Cserszegtomaji-k˙tbarlang (Chersegtomay Pit Cave): *Karszt és Barlang* (Hungary), no. 1, p. 35-40 (in Hungarian).

Kárpát, J., 1983, Az Acheron-Kútbarlang: *Karszt és Barlang* (Hungary), no. 1-2, p. 25-28 (in Hungarian).

Kartsev, A.A., 1972, *Hydrogeology of oil and gas deposits:* Moscow, Nedra, 280 p. (in Russian).

Kartsev, A.A., 1985, Low mineralized waters of oil and gas deposits, their origin and significance, *in* Kolodij, V.V., ed., *Low-mineralized waters of deep-seated horizons of oil- and gas-bearing provinces:* Kiev, Naukova Dumka, p. 4-6 (in Russian).

Kartzev, A.A., Vagin, S.B., and Baskov, E.A., 1969, *Paleohydrogeology:* Moscow, Nedra, 151 p, (in Rusian).

Katzer, F., 1909, Karst und Karsthydrographie: Sarajevo, Zur Kunde der Balkanhalbinsel, Sarajevo, Kajon, no. 8, 94 p.

Kaveev, M.C., 1963, About the influence of carbon dioxide originated during the destruction of oil deposits on development of karst processes: *Doklady Akademii Nauk SSSR,* v. 152, no. 3, p. 254-257. (in Russian).

Kaye, C.A., 1957, The effect of solvent motion on limestone solution: *Journal of Geology*, v. 65, p. 35-46.

Kempe, S., 1972, Cave genesis in gypsum with particular reference to underwater conditions: *Cave Science,* v. 6, no. 49, p. 1-6.

Kempe, S., 1975, “Facetten” and “Laugdecken”, the typical morphological elements of caves developing in standing water: *Annales de Spéléologie*, v. 4, p. 705-708.

Kempe, S., 1996, Gypsum karst of Germany, *in* Klimchouk, A., Lowe, D., Cooper, A., and Sauro, U., eds., Gypsum karst of the world: *International Journal of Speleology*, v. 25, no. 3-4, p. 209-224.

Kempe, S., Brandt, A., Seeger, M., and Vladi, F., 1975, “Facetten”and “Laugdecken”, the typical morphological elements of caves developing in standing water: *Annales de Spéléologie,* v. 30, p. 705-708.

Kenley, P.R., 1976, South western Victoria Quaternary, *in* Douglas, J.G., and Ferguson, J.A., eds., Geology of Victoria: *Geological Society of Australia Special Publication,* no. 5, p. 89-153.

Kent, L.E., 1980, Stratigraphy of South Africa: *Republic of South Africa Geological Survey Handbook 8*, no. 1, p. 603-628.

Khromov, S.P., and Mamontova, L.I., 1963, *Meteorological dictionary:* Leningrad, Gidrometeoizdat, 250 p. (in Russian).

Kicinska, D., 1997, Reconstruction of paleocurrents in caves of the Bystra Valley (Tatra Mountains, Poland), on the basis of scallops and deposits analyses: *12th International Congress of Speleology,* La Chaux-de-Fonds, Proceedings, v. 1, p. 168.

Kieffer, S.W., 1998, Dynamics of the Chicxulub Impact and implications for extinction mechanisms: Geological Association of Canada Annual Meetings Abstract Volume, p. A91.

Kinzikeev, A.R., 1993, Geodynamics and karst process, *in* Ingenernaja geologija karsta (Perm): v. 2, p. 16-20 (in Russian).

Klimchouk, A.B., 1984, The experience of the detailed study of water-mechanical sediments of a large cave system, *in* Peshchery (Caves): Perm University, p. 70-88 (in Russian).

Klimchouk, A.B., 1986, Genesis and development history of the large gypsum caves in the western Ukraine: *Le Grotte d'Italia,* v. 4, no. 13, p. 51-71.

Klimchouk, A.B., 1987, Conditions and peculiarities of karstification in the near-surface zone of carbonaceous massives, *in Caves of Georgia* 11: Tbilisi, Metznierеba, p. 55-65 (in Russian, res. English).

Klimchouk, A.B., 1989, *The significance of the subsurface zone in karst hydrogeology and morphogenesis:* Kiev, Institute of Geologic Science, 43 p. (in Russian, res. English).

Klimchouk, A.B., 1990, Artesian genesis of the large maze caves in the Miocene gypsum of the western Ukraine: *Doklady Akademii Nauk Ukrainskoj SSR,* ser. B, v. 7, p. 28-32 (in Russian, res. in English).

Klimchouk, A.B., 1991, Large maze caves in gypsum in the western Ukraine: Speleogenesis under artesian conditions: *National Speleological Society Bulletin,* v. 53, p. 71-82.

Klimchouk, A.B., 1992, Large gypsum caves in the western Ukraine and their genesis: *Cave Science,* v. 19, p. 3-11.

Klimchouk, A.B., 1994a, Speleogenesis in gypsum and geomicrobiological processes in the Miocene sequence of the Pre-Carpatian region, *in* Sasowsky, I.D., and Palmer, M.V., eds., Breakthroughs in Karst Geomicrobiology and Redox Geochemistry, Abstracts and Field Trip Guide: *Karst Water Institute, Special Publications,* no. 1, p. 40-42.

Klimchouk, A.B., 1994b, Speleogenesis under confined conditions, with recharge from adjacent formations: Publ. Serv.

Geol. Luxembourg, *Comptes Rendus du Coll. Int. de Karstol. a Luxembourg,* v. 27, p. 85-95.

Klimchouk, A.B., 1995, Karst morphogenesis in the epikarstic zone: *Cave and Karst Science,* v. 21, p. 45-50.

Klimchouk, A.B., 1996a, Dissolution and conversions of gypsum and anhydrite, *in* Klimchouk, A., Lowe, D., Cooper, A., and Sauro, U., eds., Gypsum karst of the world: *International Journal of Speleology*, v. 25, no. 3-4, p. 21-36.

Klimchouk, A.B., 1996b, Gypsum karst in the western Ukraine, *in* Klimchouk, A., Lowe, D., Cooper, A., and Sauro, U., eds., Gypsum karst of the world: *International Journal of Speleology*, v. 25, no. 3-4, p. 263-278.

Klimchouk, A.B., 1996c, Hydrogeology of gypsum formations, *in* Klimchouk, A., Lowe, D., Cooper, A., and Sauro, U., eds., Gypsum karst of the world: *International Journal of Speleology*, v. 25, no. 3-4, p. 83-90.

Klimchouk, A.B., 1996d, Speleogenesis in gypsum, *in* Klimchouk, A., Lowe, D., Cooper, A., and Sauro, U., eds., Gypsum karst of the world: *International Journal of Speleology*, v. 25, no. 3-4, p. 61-82.

Klimchouk, A.B., 1996e, The typology of gypsum karst according to its geological and geomorphological evolution, *in* Klimchouk, A., Lowe, D., Cooper, A., and Sauro, U., eds., Gypsum karst of the world: *International Journal of Speleology*, v. 25, no. 3-4, p. 49-60.

Klimchouk, A.B., 1997a, Artesian speleogenetic setting: *12th International Congress of Speleology,* La Chaux-de-Fonds, Proceedings, v. 1, p. 157-160.

Klimchouk, A.B., 1997b, The nature and principal characteristics of epikarst: *12th International Congress of Speleology,* La Chaux-de-Fonds, Proceedings, v. 1, p. 306.

Klimchouk, A.B., 1997c, The role of karst in the genesis of sulfur deposits, Pre-Carpathian region, Ukraine: *Environmental Geology,* v. 31, p. 1-20.

Klimchouk, A.B., 1997d, The role of speleogenesis in the Miocene gypsum in the western Ukraine in groundwater circulation in the multi-story artesian system, *in* Gunay, G., and Johnson, I., eds., *Karst Waters and Environmental Impacts:* Rotterdam, A.A.Balkema., p. 281-292.

Klimchouk, A.B., 1997e, Speleogenetic effects of water density differences: *12th International Congress of Speleology,* La Chaux-de-Fonds, Proceedings, v. 1, p. 161-164.

Klimchouk, A.B., Aksem, S.D., Shestopalov, V.N., and Rud'ko, G.I., 1988, *The study regime of gypsum karst activity in the western Ukraine:* Kiev, Inst. Geol. Nauk, 55 p. (in Russian).

Klimchouk, A.B., and Andrejchouk, V.N., 1988, Geologic and hydrogeologic conditions of development of large gypsum caves in the western Ukraine and their genesis, *in* Peshchery (Caves), Gypsum and anhydrite caves: Perm University, p. 12-25 (in Russian).

Klimchouk, A.B., and Andrejchouk, V.N., 1996a, Breakdown development in cover beds, and landscape features induced by intrastratal gypsum karst, *in* Klimchouk, A., Lowe, D., Cooper, A., and Sauro, U., eds., Gypsum karst of the world, *International Journal of Speleology*, v. 25, no. 3-4, p. 127-144.

Klimchouk, A.B., and Andrejchouk, V.N., 1996b, Sulphate rocks as an arena for karst development, *in* Klimchouk, A., Lowe, D., Cooper, A., and Sauro, U., eds., Gypsum karst of the world: *International Journal of Speleology*, v. 25, no. 3-4, p. 9-20.

Klimchouk, A.B., and Jablokova, N.L., 1989, Evidence of hydrological significance of the epikarstic zone from study of oxygen isotope composition of water, Arabika Massif, western Caucasus: *10th International Congress of Speleology,* Budapest, Proceedings, v. 3, p. 798-799.

Klimchouk, A.B., and Rogozhnikov, V.Y., 1982, *Conjugate analysis of the development history of a large cave system (with the example of Atlantida Cave):* Kiev Institute of Geology Nauk, 58 p. (in Russian).

Klimchouk, A.B., Andrejchouk, V.N., and Turchinov, I.I., 1995, *Structural prerequisites of speleogenesis in gypsum in the western Ukraine:* Kiev, Ukrainian Speleological Association, 104 p.

Klimchouk, A.B., et al., 1985, Geologic and hydrogeologic conditions of karst development of the Pridnestrovsky Podolia, *in* Sokolovskij, I.L., and Klimchouk, A.B., eds., Karst of the Ukraine: *Fizycheskaja Geographija i Geomorphologija,* Kiev, no. 32, p. 47-54 (in Russian).

Klimchouk, A.B., et al., 1996a, Dissolution of gypsum from field observations, *in* Klimchouk, A., Lowe, D., Cooper, A., and Sauro, U., eds., Gypsum karst of the world: *International Journal of Speleology*, v. 25, no. 3-4, p. 37-48.

Klimchouk, A.B., et al., 1996b, Gypsum karst of the world, a brief overview, *in* Klimchouk, A., Lowe, D., Cooper, A., and Sauro, U., eds., Gypsum karst of the world: *International Journal of Speleology*, v. 25, no. 3-4, p. 159-181.

Klimchouk, A.B., Lowe, D., Cooper, A., and Sauro, U., eds., 1996, Gypsum karst of the world: *International Journal of Speleology,* v. 25 no. 3-4, 307 p.

Klimchouk, A.B., Nasedkin, V.M., and Cunningham, K.I., 1995, Speleothems of aerosol origin: *National Speleological Society Bulletin,* v. 57, p. 31-42.

Klimchouk, A.B., Rogozhnikov, V.J., and Lomaev, A.A., 1981, *Karst of the Kyrktau Massif (Zeravshansky Range, Tjan-Shan):* Kiev, Institute of Geologic Science, 54 p. (in Russian).

Klimchouk, A.B., Sauro, U., and Lazzarotto, M., 1996, "Hidden" shafts at the base of the epikarstic zone: A case study from the Sette Communi Plateau, Venetian Pre-Alps, Italy: *Cave and Karst Science,* v. 23, no. 3, p. 101-107.

Klimchouk, A.B., Stotland, A.B., and Lomaev, A.A., 1979, Karst and speleological investigations on the Kyrktau Plateau (Zeravshansky Range, Tjan-Shan): *Izvestija VGO,* v. 3, no. 5, p. 442-448 (in Russian).

Klimochkin, V.V., 1973, The necessity of consideration of condensation processes when determining the balance of groundwaters, *in* Vlagooborot v prirode i ego rol v formirovanii resursov podzemnikh vod: Moscow, Stroiizdat, p. 288-290 (in Russian).

Knez, M., 1996a, *The bedding-plane impact on the development of the karst caves:* Znanstevnoraziskovalni center SAZU, Ljubljana, Ph.D. thesis, 186 p. (in Slovenian with English summary).

Knez, M., 1996b, Vpliv lezik na razvoj kraških jam: primer Velike doline Škocjanska Jame. (The bedding-plane impact on development of karst caves (an example from Velika Doline, Škocjanska Jame): *Zbirka ZRC,* p. 14.

Knuchel, F., 1973, Les cinq principaux essais de coloration de la région karstique situées au Nord des Lacs de Thoune et Brienz: *Cavernes,* v. 17, no. 3, p. 87-93.

Koliasnikoff, A., 1994, Reconnaissance of the Frontier Shaft Area: *Bulletin of the South African Spelaeological Association,* v. 35, p. 8-13.

Kolodij, V.V., 1975, *Underground condensational and solutional waters of oil, gas condensate, and gas deposits:* Kiev, Naukova Dumka, 122 p., (in Russian).

Kolodij, V.V., et al., 1991, *Low-mineralized waters of the deep-seated horizons of oil- and gas-bearing confined basins of Ukraine:* Kiev, Naukova Dumka. 184 p. (in Russian).

Komatina, M., 1975, Development conditions and regionalization of karst: *International Union of Geological Sciences, Series B, Hydrology of Karstic Terrains,* v. 3, p. 21-29.

Korotkevich, G.V., 1970, *Salt karst:* Leningrad, Nedra, 256 p. (in Russian).

Korzhinsky, D.S., 1953, Essay on metasomatic processes, *in* Osnovnye problemy v uchenii o magmaticheskikh rudnykh mestorozhdenij: Moscow, AN SSSR Publ. (in Russian).

Kovács, J., and Müller, P., 1980, A Budai-hegyek hévizes tevékenységének kialakulása és nyomai (Origin and traces of hydrothermal activities in the Buda Range): *Karszt és Barlang,* no. 2, p. 93-98.

Kozary, M.T., Dunlap, J.C., and Humphrey, W.E., 1968, Incidence of saline deposits in geologic time: *Geological Society of America Special Paper,* no. 88, p. 43-57.

Kranjc, A., Slabe, T., and Zupan-Hajna, N., 1997, Speleology, *in* Kranjc, A., ed., *Slovene Classical karst Kras,* Ljubljana, ZRC SAZU, p. 77-101.

Krinsley, D.H., and Doornkamp, J.C., 1973, *Atlas of quartz sand and surface textures:* New York, Cambridge University Press, 91 p.

Krothe, N.C., and Libra, R.D., 1983, Sulfur isotopes and hydrochemical variations in spring waters of southern Indiana, U.S.A.: *Journal of Hydrology,* v. 61, p. 267-283.

Kudrjakov, V.A., 1970, General features of the development of stratally confined groundwater systems, *in* Metodika paleogydrogeologicheskih issledovanij: Ashghabad, p. 88-93 (in Russian).

Kuldzhaev, N.K., 1989, On the theory of a condensation origin for ground waters by A.F. Lebedev: *Izv. of Academy of Sci. of Tadzhik. SSR,* Series Phys-Techn., Chem., and Geol., v. 1, p. 103-106 (in Russian).

Kungur Symposium, 1991, Karstogenetic settings: Deep-seated karst, endokarst, hydrothermal karst: Abstracts of papers of a regional symposium, Kungur, March 1991, 102 p. (in Russian).

Kushnir, S.V., 1988, *Hydrogeochemistry of sulfur deposits of the Pre-Carpathians:* Kiev, Naukova dumka, 179 p. (in Russian).

Kuznetsov, A.M., 1947, On dissolution of gypsum and anhydrite: *Trudy Estestvenno-nauchnogo instituta,* Perm, v. 12, no. 4, p. 127-133 (in Russian).

Kyrle, G., 1923, *Grundriss der Theoretischen Speläologie:* Wein, Östereischischen Staatsdruckerei, 353 p.

L

Landes, K.K., 1961, Effects of solution of bedrock salt in the Earth's crust, *in* 1st Symposium on Salt, Proceedings, Cleveland, p. 64-73.

Lang, K.N., 1708, *Historia lapidum figuratorum Helvetiae:* For the author, Venice, 165 p. (p.7).

Langmuir, D., 1971, The geochemistry of some carbonate ground waters in central Pennsylvania: *Geochimica et Cosmochimica Acta,* v. 35, p. 1023-1045.

Langmuir, D., 1971, The geochemistry of some carbonate ground waters in central Pennsylvania: *Geochimica et Cosmochimica Acta,* v. 35, p. 1023-1045.

Lapcevic, P., Novakowski, K., Bickerton, G., and Voralek, J. 1997, Hydraulic characterization of the fracture framework in carbonate rock underlying CWML site, Smithville, Ontario: *Air and Waste Management Association Annual Meeting Toronto, Proceedings.*

Laptev, F.F., 1939, Aggressive action of water on carbonate rocks, gypsum, and concrete: Moscow-Leningrad, *GONTI,* v. 1, 120 p. (in Russian).

LaRock, E.J., and Cunningham, K.I., 1995, Helictite bush formation and aquifer cooling in Wind Cave, Wind Cave National Park, South Dakota: *National Speleological Society Bulletin,* v. 57, p. 43-51.

Last, W.M., 1993, Salt dissolution features in saline lakes of the northern Great Plains, western Canada: *Geomorphology,* v. 8, p. 321-334.

Latapie, J., 1816, Théorie de la terre déduite de l'organisation des Pyrénées et pays adjacens redigée sur les manuscripts de M. Flamichon: *Tonne,* Pau, 311 p. (p.114-115).

Latham, A.G., 1976, The Porcellanous Band in Yorkshire caves: *British Cave Research Association Bulletin,* no. 13, p. 9-10.

Launay, L., de, and Martel, E.A., 1891, Note sur quelque questions relatives à la géologie des grottes et des eaux souterraines ....: *Bulletin de la Société Géologique de France,* ser. 3, v. 19, p. 142-165 (p. 144-145).

Lauritsen, Å., 1996, *Geomorfologisk utvikling av grottesystemene i Kjøpsvik, Nordland (Geomorphological development of cave systems in Kjøpsvik, Nordland):* Bergen University, Department of Geology, M.Sc. Thesis, 220 p.

Lauritsen, Å., and Lauritzen, S.-E., 1996, Quaternary cave and landform development in the Tysfjord region, north Norway: *Karst Waters Institute Special Publication,* no. 2, p. 73-77.

Lauritzen, S.-E., 1982, The paleocurrents and morphology of Pikhåggrottene, Svartisen, north Norway: *Norsk Geografisk Tidsskrift,* v. 36, p. 183-209.

Lauritzen, S.-E., 1986, Kvithola at Fauske; northern Norway: An example of ice-contact speleogenesis: *Norsk Geologisk Tidsskrift,* v. 66, p. 153-161.

Lauritzen, S.-E., 1989, Scallop dominant discharge: *10th International Congress of Speleology*, Budapest, Proceedings, v. 10, p. 123-124.

Lauritzen, S.-E., Abbott, J., Arnesen, R., Crossley, G., Grepperud, D., and Ive, A., 1985, Morphology and hydraulics of an active phreatic conduit: *Cave Science,* v. 12, p. 139-146.

Lauritzen, S.-E., and Karp, D., 1993, Speleological assessment of karst aquifers developed within the Tindall Limestone, Katherine, Northern Territories, Australia: North Territory, Australia, Water Resources Authority, 60 p.

Lauritzen, S.-E., and Lauritsen, Å., 1995, Differential diagnosis of paragenetic and vadose canyons: *Cave and Karst Science,* v. 21, p. 55-59.

Lauritzen, S.-E., Lundberg, J., Mylroie, J.E., and Dogweiler, T., 1997, Bell hole morphometry of a flank margin cave and possible genetic models: Lighthouse Cave, San Salvador, The Bahamas: *12th International Congress of Speleology,* La Chaux-de-Fonds, Proceedings, v., 1, p. 178.

Lauritzen, S.-.E., Odling, N., and Petersen, J., 1992, Modeling the evolution of channel networks in carbonate rocks, *in* Hudson, J.A., ed., *ISRM Symposium,* Eurock '92: London, Thomas Thelford, p. 57-62.

Lebedev, A.F., 1936, *Soil- and groundwaters:* Moscow-Leningrad, 312 p. (in Russian).

Lehmann, O., 1932, Die Hydrographie des Karstes: *Enzyklopädie der Erdkunde,* Leipzig and Vienna, Frantz Deuticke, 212 p.

Leighton, M.W. and Pendexter, C., 1962, Carbonate rock types. *American Association of Petrololeum Geologists,* Mem. 1, p. 33-61.

Lerman, A., 1970, Chemical equilibria and evolution of chloride brines: *Mineralogical Society of America Special Papers,* no. 3, p. 291-306.

Lesser, F.C., and J.G., 1735, *Anmerckungen über die Baumañs-Höhle wie er sie selbst anno 1734 den 21 May befunden ...:* Hamburg, Stromer, 28 p. (p. 16).

Lewis-Abbott, J.W., 1894, The ossiferous fissures in the valley of the Shode, near Ightham, Kent: *Quarterly Journal of the Geological Society of London,* v. 50, p. 171-187.

Liebe, P., 1989, Thermalkarst-systems in Hungary–Exploration and problems thereof: *10th International Congress of Speleology,* Budapest, Proceedings, v. 2, p. 566-569.

Liebe, P., and Lorberer, Á. 1978, A karsztos héviztárolók áramlási és hömérsékleti viszonyainak vizsg·lata (Study of flow and temperature conditions in a karstic thermal water aquifer): *International Symposium on Karst Hydrology,* Proceedings, v 1, p. 79-110.

Liley, P.E., Touloukian, Y.S., and Gambill, W.R., 1963, Physical and chemical data, *in* Perry, J.H., ed., *Chemical Engineer Handbook:* New York, McGraw-Hill Book Co., 4th edition, p. 30-75.

Lips, B., Gresse, A., Delamette, M., and Maire, R., 1993, Le Gouffre Jean Bernard (1602 m Haute Savoie, Fr.): Ecoulements souterrains et formation du resau: *Karstologia,* v. 21, p. 1-14.

Liu, S.T., and Nancollas, G.H., 1971, The kinetics of dissolution of calcium sulfate dihydrate: *Journal of Inorganic and Nuclear Chemistry,* v. 33, p. 2311-2316.

Liu, Z., and Dreybrodt, W., 1997, Dissolution kinetics of calcium carbonate minerals in

H_2O-CO_2 solutions in turbulent flow: The role of the diffusion boundary layer and the slow reaction $H_2O + CO_2 \leftrightarrow H^+ + HCO_3^-$: *Geochimica et Cosmochimica Acta*, v. 61, p. 2879-2889.

Livingstone, D.A., 1963, Chemical composition of rivers and lakes: *U.S. Geological Survey Professional. Paper 440-G*, 64 p.

Lloyd, J.W., 1993, The United Kingdom, *in* Downing, R.A., Price, M., and Jones, G.P., eds., Hydrogeology of the Chalk of North-West Europe: Oxford, Clarendon Press, p. 220-249.

Loewenthal, R.E., and Marais, G.V.R., 1978, Carbonate chemistry of aquatic systems: *Theory and applications 1:* Ann Arbor, Mich., Ann Arbor Science, 431 p.

Loughlin, W.D., 1983, *The hydrogeologic controls on water quality, ground water circulation, and collapse breccia pipe formation in the western part of the Black Mesa Basin, Coconino County, Arizona:* University of Wyoming, M.S. thesis, 117 p.

Loughlin, W.D., and Huntoon, P.W., 1983, Compilation of available ground water quality data for sources within the Grand Canyon of Arizona: Department of Geology and Geophysics, University of Wyoming, *Report to U.S. National Park Service,* 9 p. plus appendices.

Lovering, T.S., Tweto, O., and Lovering, T.G., 1978, Ore deposits of the Gilman District, Eagle Country, Colorado: *U.S. Geological Survey Professional Paper 1017*, 90 p.

Lowe, D.J., 1974, The geology of the Forest of Dean limestones: *Cave Projects Group Newsletter,* v. 5, p. 73-79.

Lowe, D.J., 1989a, Limestones and Caves of the Forest of Dean, *in* Ford, T.D., ed., *Limestones and caves of Wales:* Cambridge University Press, p. 106-116.

Lowe, D.J., 1989b, *Tonga '87–The report of the 1987 speleological expedition to 'Eua Island, Kingdom of Tonga:* Private publication, 28 p.

Lowe, D.J., 1992a, Chalk caves revisited: *Cave Science,* v. 19, p. 55-58.

Lowe, D.J., 1992b, A historical review of concepts of speleogenesis: *Cave Science*, v. 19, no. 3, p. 63-90.

Lowe, D.J., 1992c, *The origin of limestone caverns: An inception horizon hypothesis:* Manchester Polytechnic University, Ph.D. thesis, 512 p.

Lowe, D.J., 1993, The Forest of Dean caves and karst: Inception Horizons and iron-ore deposits: *Cave Science,* v. 20, p. 31-43.

Lowe, D.J., and Gunn, J., 1986, Caves and limestones of the islands of Tongatapu and 'Eua, Kingdom of Tonga: *British Cave Research Association Transactions,* v, 13, no. 3, p. 105-130.

Lowe, D.J., and Gunn, J., 1995, The role of strong acid in speleo-inception and subsequent cavern development, *in* Barani-Kevel, I., ed., Environmental effects on karst terrains: *Acta Geographica,* (Szeged), v. 34, p. 33-60.

Lugeon, M., 1915, Le striage du lit fluvial: *Annales de Géographie,* v. 24, p. 385-393.

Luiszer, F.G., 1994, Speleogenesis of Cave of the Winds, Manitou Springs, Colorado, *in* Sasowsky, I.D., and Palmer, M.V., eds., Breakthroughs in karst geomicrobiology and redox geochemistry: abs. and field-trip guide, *Karst Waters Institute Special Publication,* no. 1, p. 91-109.

Lukin, V.S., 1962, Air regime of karstified massifs, on the example of the Kungursky region, *in* Special'nye voprosy karstovedenija: Moscow, Akadamia Nauk SSSR, p. 58-59 (in Russian).

Lukin, V.S., 1969, Quantitative expression of evaporation and condensation of water in gypsum-anhydrite massifs of the Ufa Plateau: *Zemlevedenije*, new ser., v. 8, no. 48, p. 213-218 (in Russian).

Lyell, C., 1830, *Principles of geology ...:* London; Murray, v. 1, (p.211).

M

Machado Filho, L., Ribeiro, M.W., Gonzales, S.R., Schenini, C.A., dos Santos Neto, A., Palmeira, R.C.B., Pires, J.L., Teixeira, W., and Castro, H.E.F., 1983, Geologia, *in* Projeto Rambrasil, Levantamento de Recursos Minerais, v. 32, Folio 23-24.

Maire, R., 1990, La haute montagne calcaire; karsts, cavités, remplissages, Quaternaire, paléoclimats: *Karstologia-Mémoires,* no. 3, 73 p.

Mais, K., 1973, Vorläufige Beobachtungen über Kondenswasserkorrosion in Schlenkendurhgangshöle: *6th International Congress of Speleology,* Praha, Proceedings, v. 3, p. 203-207.

Maksimovich, G.A., 1963, *Fundamentals of karst studies, v. 1:* Perm University, 444 p. (in Russian).

Malinin, S.D., 1979, *Fizicheskaja khimija gidrotermalnikh sistem s uglekislotoi* (Physical chemistry of hydrothermal systems with carbon dioxide): Moscow, Nauka, 111 p.

Malkov, V.N., and Frantz, N.A., 1981, On winter condensation, *in* Akkumulatsija zimnego kholoda v porodakh i ego ispolzovanije: Perm, p. 97-99 (in Russian).

Malott, C.A., 1937, The invasion theory of cavern development: *Geological Society of America Proceedings for 1937,* p. 323.

Mangin, A., 1970, Contribution a l'étude des aquifères karstiques a partir de l'analyse des courbes de décrue et tarissement: *Annales de Spéléologie*, v. 25, p. 581-610.

Mangin, A., 1971, Étude des débits classés d'exutoires karstiques portant sur un cycle hydrologique: *Annales de Spéléologie*, v. 26, p. 283-329.

Mangin, A., 1973, Sur la dynamique des transferts en aquifer karstique: *6th International Congress of Speleology,* Olomouc, Proceedings, v. 4, p. 157-162.

Mangin, A., 1975, *Contribution a l'étude hydrodinamique des aquiferes karstiques:* University of Dijon, DES thesis (*Ann. Speleo.* 1974 29(3), p.283-332; 29(4), p.495-601; 1975 30(1), p. 21-124).

Manikhin, V.I., 1966, On the question of solubility of calcium sulfate under high pressures: *Geokhimicheskie Materialy*, v. 13, p. 193-196 (in Russian).

Marin, L.E., 1990, *Field investigations and numerical simulation of ground-water flow in the karstic aquifer of northwestern Yucatan, Mexico:* Northern Illinois University, Ph.D. dissertation, 183 p.

Marker, M.E., 1976, Note on some South African pseudokarst: *Socidad Venezolanade Espeleologia Boletin,* v. 7, p. 5-12.

Markowicz, M., Popov, V., and Pulina, M., 1972, Comments on karst denudation in Bulgaria: *Geographia Polonica*, v. 23, p. 111-139.

Marquínez, J.L., 1997, Mapa geológico de la Región del Cuera y Picos de Europa: *Trabajos de Geología, Univ. de Oviedo, (in press).*

Marsden, A.W., 1991, Aspects of the Peak-Speedwell Cave System 1650-1990: *Cave Science,* v. 18, no. 1, p. 3-1.

Martel, E.A., 1894, *Les Abimes:* Paris, 578 p.

Martel, E.A., 1896, Applications géologiques de la spéléologie ...: *Annales des Mines ...,* Sér. 9, v. 10, Paris, p. 5-100.

Martel, E.A., 1897, *Irlande et cavernes anglaises:* Paris, Delagrave, 403 p. (p.289).

Martel, E.A., 1913, On fluorescein tests at great distances: *Comptes Rendus de l'Academie des Sciences*, v. 157, p. 225-227. (English translation in *Cave Geology*, v. 1, p. 279-281, 1983).

Martel, E.A., 1921, *Nouveau traité des eaux souterraines:* Paris, Delagrave.

Martínez García, E., 1981, El Paleozóico de la Zona Cantábrica oriental (norteoeste de España): *Trabajos de Geología, Univ. de Oviedo,* no. 11, p. 95-127.

Martínez García, E., and Tejerina, L., 1984, Fluorspar deposits associated with Carboniferous and Permian rocks in Asturias and León, northwest Spain: 9th International Congress on Carboniferous Stratigraphy and Geology, Champaign-Urbana, *Compte Rendu,* v. 9, no. 4, p. 467-478.

Martini, J.E.J., 1979, Karst in Black Reef Quartzite near Kaapsehoop, Eastern Transvaal: *Annals of the Geological Survey of South Africa,* v. 13, p. 115-128.

Martini, J.E.J., 1981, The control of karst development with reference to the formation of caves in poorly soluble rocks in the eastern Transvaal, South Africa: *8th International Congress of Speleology,* Bowling Green, Proceedings, p. 4-5.

Martini, J.E.J., 1986, Can caves form in wollastonite?: *Bulletin of South Africa Spelaeological Association,* v. 27, p. 22-30.

Martini, J.E.J., 1987, Les phénomènes karstiques des quartzites d'Afrique du Sud: *Karstologia,* v. 9, p. 45-52.

Martini, J., 1987, Karst features in quartzite of South Africa: *Karstologia,* v. 9, p. 45-52.

Martini, J.E.J., 1991, Magnet: A large quartzite cave system in the eastern Transvaal: *Bulletin of the South African Spelaeological Association,* v. 31, p. 44-59.

Martyn, J., 1729, An account of some observations relating to natural history, made in a journey to the Peak in Derbyshire: *Philosophical Transactions of the Royal Society of London,* v. 36, p. 22-32, 27-28.

Maucci, W., 1975, Il fenomeno dell'erosione inversa come contributo allo studio della speleogenesi: *Le Grotte d'Italia,* ser. 4, v. 4, p. 235-285.

Maucha, L., 1989, Karst water resources research in Hungary and its significance: *Karszt és Barlang,* spec. issue, p. 39-50.

Mavliudov, B., 1989, Snow and ice formations in caves and their regime: *10th International Congress of Speleology,* Budapest, Proceedings, v. 1, p. 295-297.

Maximovich, G.A., 1979, Karstosphere and the types of karst, *in* Karst I gidrogeologia Preduralja: Sverdlovsk, p. 16-21 (in Russian).

Mazzullo, S.J., and Harris, P.M., 1991, Overview of dissolution porosity development in the deep-burial environment, with examples from carbonate reservoirs in the Permian Basin, *in West Texas Geological Society Symposium Publication,* no. 91-89, p. 125-138.

Mazzullo, S.J., and Harris, P.M., 1992, Mesogenetic dissolution: Its role in porosity development in carbonate reservoirs: *American Association Petroleum Geologists Bulletin,* v. 76, p. 607-620.

McDonald, M.G., and Harbaugh, A.W., 1988, A modular three-dimensional finite-difference ground-water flow model: *United States Geological Survey Techniques of Water Resources Investigations,* v. 6, ch. A1.

McGavock, E.H., 1968, Basic ground-water data for southern Coconino County, Arizona: Arizona State Land Department, *Water Resources Report,* no. 33, 49 p.

McLaren, S.J., 1993, Use of cement types in the palaeoenvironmental interpretation of coastal aeolianite sequences, *in* Pye, K., ed, The dynamics and environmental context of aeolian sedimentary systems: *Geological Society of London Special Publication,* no. 72, p. 235-244.

Meiman, J., and Groves, C., 1997, Magnitude/frequency analysis of cave passage development in the Central Kentucky Karst: *6th Science Conference,* Mammoth Cave National Park, Proceedings, p. 11-13.

Mel'nikova, Z.M., and Moshkina, I.A., 1973, Solubility of anhydrite and gypsum in the system Na-Mg-Ca-Cl-SO -H O: *Izvestija Sib. Otdel. AN SSSR,* ser. khim. nauk, v. 4, no. 21, p. 176-182 (in Russian).

Melis, T.S., Phillips, W.M., Webb, R.H., and Bills, D.J., 1996, When the blue-green waters turn red, historical flooding in Havasu Creek, Arizona: *U.S. Geological Survey Water Resources Investigations Report 96-4059,* 136 p.

Merrill, G.K., 1960, Additional notes on vertical shafts in limestone caves: *National Speleological Society Bulletin,* v. 22, p. 101-108.

Meyerhoff, A.A., and Hatten, C.W., 1974, Bahamas salient of North America: Tectonic framework, stratigraphy, and petroleum potential: *American Association of Petroleum Geologists Bulletin,* v. 58, p. 1201-1239.

Mikhailev, V.N., 1989, *Karst of the Kirghizia:* Frunze, Ilim, 146 p. (in Russian).

Milanovič, P.T., 1981, *Karst Hydrogeology:* Fort Littleton, Colorado, Water Resources Publication, 281 p.

Milanovič, P.T., 1981, *Karst hydrogeology:* Water Resources Publications, Littleton, Colorado, 434 p.

Milikhiker, S.H., 1962, Study of karst in the middle part of Chusovaja river in connection with dam constraction, *in* Spetsial'nye Voprosy Karstovednija: Moscow, Akad. Nauk. S.S.S.R. (in Russian).

Millen, T.M., and Dickey, D.N., 1987, A stable isotopic investigation of waters and speleothems in Wind Cave, South Dakota: An application of isotope paleothermometry: *National Speleological Society Bulletin,* v. 49, p. 10-14.

Miller, T.E., 1981, *Hydrochemistry, hydrology and morphology of the Caves Branch Karst, Belize:* McMaster University, Ph.D. thesis, 280 p.

Miller, T.E., 1996, Geologic and hydrologic controls on karst and cave development in Belize: *Journal of Cave and Karst Studies,* v. 58, p. 100-120.

Miotke, F.-D., 1971, Collapse sinkholes formed by deep-seated solution of salt, Bad Seebruch area, near Hannover, Germany: *Caves and Karst,* v. 13, no. 5, p. 43.

Miotke, F.-D., 1975, *Der Karst im zentralen Kentucky bei Mammoth Cave:* Jahrbuch der Geographischen Gesellschaft zu Hannover, 360 p.

Miotke, F.-D., and Palmer, A.N., 1972, *Genetic relationship between caves and landforms in the Mammoth Cave National Park area:* Würtzburg, Böhler Verlag, 69 p.

Mitchell, R.W., Russell, W.H., and Elliott, W.R., 1977, Mexican eyeless Characin fishes, genus *Astyanax:* Environment, distribution, and evolution: *Texas Technical University Museum Special Publications,* no. 12, 89 p.

Mjatiev, A.N., 1947, Confined complex of underground waters and wells: *Izvestija AN SSSR, Otd. Tekhnich. Nauk,* no. 9, p. 33-47 (in Russian).

Molerio, L., 1981, Hidrogeologia y climatologia de la cueva La Mariana: *Volun. hidraul.,* v. 18, no. 57, p. 2-9.

Moncaster, S.J., Bottrell, S.H., Tellam, J.H., and Lloyd, J.W., 1992, Sulfur isotope ratios as tracers of natural and anthropogenic sulfur in the Lincolnshire Limestone aquifer, eastern England, *in* Kharaka, Y., and Maest, A.S., eds., *Water-Rock Interaction:* Rotterdam, Balkema, p. 813-816.

Moneymaker, B.C., 1941, Subriver solution cavities in the Tennessee Valley: *Journal of Geology,* v. 49, p. 74-86.

Moore, C.H., 1989, *Carbonate diagenesis and porosity:* developments in Sedimentology 46. New York: Elsevier. 338 p.

Moore, G.W., and Sullivan, G.N., 1978, *Speleology:* Teaneck, Zephyrus Press, 2nd. ed., 150 p.

Moore, G.W., and Sullivan, N., 1997, *Speleology–Caves and the cave environment:* St. Louis, Cave Books, 3rd ed., 176 p.

Moore, I.W., 1944, Caves of the Muizenberg Range: *Journal of the Mountain Club, South Africa,* v. 47, p. 7-22.

Morehouse, D.F., 1968, Cave development via the sulfuric acid reaction: *National Speleological Society Bulletin,* v. 30, p. 1-10.

Morel, E.H., 1980, The use of a numerical model in the management of the Chalk Aquifer in the upper Thames Basin: *Quarterly Journal of Engineering Geology,* v. 13, p. 153-165.

Morozov, A.T., 1938, On the methods of study of moving vapor moisture in soil-ground: *Transact. of the Inst. of Hydrology and Melioration,* v. 22, p. 18-26 (in Russian).

Morse, J.W., 1978, Dissolution kinetics of calcium carbonate in sea water; VI. The near-equilibrium dissolution kinetics of calcium carbonate rich deep sea sediments: *American Journal of Science,* v. 278, p. 344-353.

Morse, J.W., and Mackenzie, F.T., 1990, *Geochemistry of Sedimentary Carbonates:* Amsterdam and New York, Elsevier, Developments in sedimentology, no. 48, 707 p.

Mortimore, R.N., 1993, Chalk water and engineering geology, *in* Downing, R.A., Price, M., and Jones, G.P., eds., Hydrogeology of the Chalk of North-West Europe: Oxford, Clarendon Press, p. 67-92.

Mossop, G.D., and Shearman, D.J., 1973, Origins of secondary gypsum rocks:

Institute of Mining and Metall., p. B147-B154.

Motyka, I., and Wilk, Z., 1984, Hydraulic structure of karst-fissured Triassic rocks in the vicinity of Olkusz (Poland): *Kras i Speleologia,* v. 14, no. 5, p. 11-24.

Mucci, A., Canuel, R., and Zhong, S., 1989, The solubility of calcite and aragonite in sulfate-free seawater and the seeded growth kinetics and composition of precipitates at 25°C: *Chemical Geology*, v. 74, p. 309-329.

Mucci, A., Morse, J.W., and Kaminsky, M.S., 1985, Auger spectroscopy analysis of magnesian calcite overgrowths precipitated from seawater and solutions of similar composition: *American Journal of Science*, v. 285, p. 289-305.

Mucke, D., and Völker, R., 1978, Kondenswasserkorrosion: *Jahrb. des Höhlenforschers,* p. 7-16.

Mucke, D., Völker, R., and Wadewitz, S., 1983, Cupola formation in occasionally inundated cave roofs, *in Europ. Conf. on Speleology*, Sofia, v. 2, p. 129-132.

Müller, P., 1974, A melegforrás-barlangok és gömbfülkék lebetheziséröl: *Karszt és Barlang* (Budapest), v. 1, p. 7-11 (in Hungarian).

Müller, P., 1989, Hydrothermal paleokarst of Hungary, *in* Bosák, P., Ford, D., Glazek, J., and Horácek, I., eds., Paleokarst–A systematic and regional review: Praha, Academia, p. 155-163.

Müller, P., and Sarvary, I., 1977, Some aspects of developments in Hungarian speleology theories during the last 10 years: *Karszt es Barlang*, special issue, p. 53-60.

Müller, P., and Sárváry, I., 1977, Some aspects of developments in Hungarian speleology theories during the last ten years: *Karszt és Barlang,* Special issue, p. 53-60.

Mylroie, J.E., 1977, *Speleogenesis and karst geomorphology of the Helderberg Plateau, Schoharie County, New York:* Renesselaer Polytechnic Institute, Ph.D. dissertation, 336 p.

Mylroie, J.E., 1984, Speleogenetic contrast between the Bermuda and Bahama Islands, *in* Teeter, J.W., ed., *2nd Symposium on the Geology of the Bahamas,* Fort Lauderdale, Proceedings, CCFL Bahamian Field Station, p. 113-128.

Mylroie, J.E., 1991, Cave development in the glaciated Appalachian karst of New York: Surface-coupled or saline-freshwater mixing hydrology?, *in* Kastning, E.H., and Kastning, K.M., eds., *Appalachian Karst Symposium,* Huntsville, National Speleological Society, Proceedings, p. 85-90.

Mylroie, J.E., and Carew, J.L., 1986, Minimum duration of speleogenesis: *9th International Congress of Speleology,* Barcelona, Proceedings, v. 1, p. 249-251.

Mylroie, J.E., and Carew, J.L., 1987, Field evidence of the minimum time for speleogenesis: *National Speleological Society Bulletin,* v. 49, p. 67-72.

Mylroie, J.E., and Carew, J.L., 1988, Solution conduits as indicators of late Quaternary sea level position: *Quaternary Science Reviews*, v. 7, p. 55-64.

Mylroie, J.E., and Carew, J.L., 1990, The flank margin model for dissolution cave development in carbonate platforms: *Earth Surface Processes and Landforms,* v. 15, p. 413-424.

Mylroie, J.E., and Carew, J.L., 1991, Erosional notches in Bahamian carbonates: Bioerosion or groundwater dissolution?, *in* Bain, R.J., ed., *5th Symposium on the Geology of the Bahamas,* Port Charlotte, Florida, Proceedings, Bahamian Field Station, p. 185-191.

Mylroie, J.E., and Carew, J.L., 1995, Karst development on carbonate islands, *in* Budd, D.A., Harris, P.M., and Saller, A., eds., Unconformities and porosity in carbonate strata: *American Association of Petroleum Geologists Memoir*, no. 63, p. 55-76.

Mylroie, J.E., Carew, J.L., and Moore, A.I., 1995, Blue holes: Definition and genesis: *Carbonates and Evaporites*, v. 10, p. 225-233.

Mylroie, J.E., Carew, J.L., and Vacher, H.L., 1995, Karst development in the Bahamas and Bermuda, *in* Curran, H.A., and White, B., eds., *Terrestrial and shallow marine geology of the Bahamas and Bermuda*: Geological Society of America Special Paper, no. 300, p. 251-267.

Mylroie, J.E., Carew, J.L., Frank, E.F., Panuska, B.C., Taggart, B.E., Troester, J.W., and Carrasquillo, R., 1995b, Comparison of flank margin cave development: San Salvador Island, Bahamas and Isla de Mona, Puerto Rico, *in* Boardman, M.R., ed., *7th Symposium on the Geology of the Bahamas*, San Salvador Island, Proceedings, Bahamian Field Station, p. 49-81.

Mylroie, J.E., Carew, J.L., Sealey, N.E., and Mylroie, J.R., 1991, Cave development on New Providence Island and Long Island, Bahamas: *Cave Science,* v. 18, p. 139-151.

Nádor, A., 1992, *Paleokarsts and long term karst evolution of the Buda Mountains, Hungary:* Eötvös Lóránd University, Ph.D. thesis, 171 p.

N

Ñagy, B., and Pelikán, P., 1973, Metacinnabarit és cinnabarit a Csillaghegyi Rókahegyen (Metacinnabarite and cinnabar in the Roka Hill, Csillaghegy): *MÁFI Évi Jelt.*, p. 51-55.

Neev, D., and Hall, J.K., 1979, Geophysical investigations in the Dead Sea: *Sedimentary Geology,* v. 23, p. 209-238.

Nekrasov, V.V., 1945, Change of systems volume during solidification of hydraulic viscous materials: *Isvestiya AN SSSR*, otd. tekhn. nauk, no. 6 (in Russian).

Nemeriuk, G.E., and Paltsev, V.P., 1969, On the methods of study of the ionic composition of vapor condensed from the air*: 23rd Hydrochem. Symposium.*, Novocherkassk, Proceedings, p. 101-102 (in Russian).

Neuzil, C.E., 1986, Ground water flow in low-permeability environments: *Water Resources Research,* v. 22, p. 1163-1195.

Neuzil, C.E., 1994, How permeable are clays and shales?: *Water Resources Research,* v. 30, p. 145-150.

Nummer, A.R., 1990, Estratigrafia e estruturas do Grupo Andrelândia na região de Santa Rita do Ibitipoca-Lima Duarte, Sul de Minas Gerais: *36th Congresso Brasileiro de Geologia,* Natal, Proceedings, v. 6, p. 2786-2797.

O

Obermaier, H., 1914, Estudio de los glaciares de Los Picos de Europa: *Trab. Mus. Nac. Cienc. Nat., Sr. Geol.*, no. 9, 42 p.

Obolensky, V.N., 1944, *Course of meteorology:* Moscow-Sverdlovsk, 120 p. (in Russian).

Odling, N., and Webman, I., 1991, A "conductance" mesh approach to the permeability of natural and simulated fracture patterns: *Water Resources Research,* v. 27, p. 2633-2643.

Ogden, A.E., 1982, Karst denudation rates for selected spring basins in West Virginia: *National Speleological Society Bulletin*, v. 44, p. 6-10.

Ollier, C.D., 1969, *Weathering:* Edinburgh, Oliver and Boyd, 304 p.

Ollier, C.D., 1975, Coral island geomorphology–The Trobriand Islands: *Zeitschrift für Geomorphologie,* v. 19, p. 164-190.

Olson, R.A., 1984, Genesis of paleokarst and strata-bound zinc-lead sulfide deposits in a Proterozoic dolostone, northern Baffin Island, Canada: *Economic Geology,* v. 70, p. 1056-1103.

Omalius d'Halloy, J.B.J., d', 1831, *Eléments de géologie:* Paris, Levrault.

Osborne, R.A.L., 1983, Cainozoic stratigraphy at Wellington Caves, New South Wales: *Linnaean Society of New South Wales,* Proceedings, v. 107, pt. 2, p. 131-147.

Osborne, R.A.L., 1984, Multiple karstification in the Lachlan Fold Belt in New South Wales: Reconnaissance evidence: *Journal and Proceedings of the Royal Society of New South Wales,* v. 117, p. 15-34.

Osborne, R.A.L., 1986, Cave and landscape

chronology at Timor Caves, New South Wales: *Journal and Proceedings of the Royal Society of New South Wales,* v. 119, p. 55-76.

Osborne, R.A.L., 1991, Palaeokarst deposits at Jenolan Caves, New South Wales: *Journal and Proceedings of the Royal Society of New South Wales,* v. 123, p. 59-73.

Osborne, R.A.L., 1993a, Geological Note: Cave formation by exhumation of Palaeozoic palaeokarst deposits at Jenolan Caves, New South Wales: *Australian Journal of Earth Sciences,* v. 40, p. 591-593.

Osborne, R.A.L., 1993b, The history of karstification at Wombeyan Caves, New South Wales, Australia, as revealed by palaeokarst deposits: *Cave Science,* v. 20, no. 1, p. 1-8.

Osborne, R.A.L., 1993c, A new history of cave development at Bungonia, N.S.W.: *Australian Geographer,* v. 24, p. 62-74.

Osborne, R.A.L., 1994, Caves, dolomite, pyrite, aragonite and gypsum: The karst legacy of the Sydney and Tasmania Basins: *28th Newcastle Symposium on Advances in the Study of the Sydney Basin,* University of Newcastle, Department of Geology Publication, no. 606, p. 322-324.

Osborne, R.A.L., 1995, Evidence for two phases of late Palaeozoic karstification, cave development and sediment filling in south-eastern Australia: *Cave and Karst Science,* v. 22, no. 1, p. 39-44.

Osborne, R.A.L., 1996, Vadose weathering of sulfides and limestone cave development–Evidence from eastern Australia: *Helictite,* v. 34, no. 1, p. 5-15.

Ostapenko, A.A., 1993, Caves of the Ekeptze-Gadyk Range: *Svet (Light),* v. 3, no. 9, p. 12-13 (in Russian).

Ótvós, E.G., 1976, "Pseudokarst" and "pseudokarst terrains": Problems of terminology: *Geological Society of America Bulletin,* v. 87, p. 1021-1027.

Owen, L.A., 1898, *Cave regions of the Ozarks and Black Hills:* New York, Johnson Reprint Corp, (1970 reprint), 228 p.

P

Pace, M.C., Mylroie, J.E., and Carew, J.L., 1993, Petrographic analysis of vertical dissolution features on San Salvador Island, Bahamas, *in* White, B., ed., *6th Symposium on the Geology of the Bahamas,* Port Charlotte, Florida, Proceedings, Bahamian Field Station, p. 109-123.

Pailliet, F.L., Hess, A.E., Cheng, C.H., and Harding, E., 1987, Characterization of fracture permeability with high-resolution vertical flow measurements during borehole pumping: *Ground Water,* v. 25, p. 28-40.

Paine, N., Escobar, E., Hallowes, G.R., Sodha, V.C., and Anagnosei, G., 1982, Surveillance and reevaluation of the Poechos Dam, right wing embankment, Peru: *114th Cong ICOLD,* Rio de Janeiro, Proceedings, Qn 52, R. 19, p. 333-343.

Palmer, A.N., 1974, Geologic influence on cave passage orientation in Ludington Cave, Greenbrier County, West Virginia: *4th Conference on Karst Geology and Hydrology,* West Virginia Geological and Economic Survey, Proceedings, p. 33-40.

Palmer, A.N., 1975, The origin of maze caves: *National Speleological Society Bulletin,* v. 37, p. 57-76.

Palmer, A.N., 1977, Influence of geologic structure on groundwater flow and cave development in Mammoth Cave National Park, U.S.A.: *12th International Association of Hydrogeologists, Memoirs,* p. 405-414.

Palmer, A.N., 1981a, *A geological guide to Mammoth Cave National Park:* Teaneck, Zephyrus Press, 210 p.

Palmer, A.N., 1981b, *Geology of Wind Cave, Wind Cave National Park, South Dakota:* Hot Springs, South Dakota, Wind Cave National History Association, 44 p.

Palmer, A.N., 1981c, Hydrochemical controls in the origin of limestone caves: *8th International Congress of Speleology,* Bowling Green, Proceedings, p. 120-122.

Palmer, A.N., 1984a, Geomorphic interpretation of karst features, *in* LaFleur, R.G., ed., *Groundwater as a geomorphic agent:* Boston, Allen and Unwin, p. 173-209.

Palmer, A.N., 1984b, Recent trends in karst geomorphology: *Journal of Geological Education,* v. 32., p. 247-253.

Palmer, A.N., 1986, Prediction of contaminant paths in karst aquifers: 1st Conference on Environmental Problems in Karst Terranes, *Association of Ground Water Scientists and Engineers,* Proceedings, Dublin, Ohio, p. 32-53.

Palmer, A.N., 1987, Cave levels and their interpretation: *National Speleological Society Bulletin,* v. 49, p. 50-66.

Palmer, A.N., 1988, Solutional enlargement of openings in the vicinity of hydraulic structures in karst regions: 2nd Conference on Environmental Problems in Karst Terranes and Their Solutions, *Association of Groundwater Scientists and Engineers,* Nashville, Tenn., Proceedings, p. 16-18.

Palmer, A.N., 1989a, Stratigraphic and structural control of cave development and groundwater flow in the Mammoth Cave region, *in* White, W.B., and White, E.L., eds., *Karst hydrology–Concepts from the Mammoth Cave area:* New York, Van Nostrand Reinhold, p. 293-316.

Palmer, A.N., 1989b, Geomorphic history of the Mammoth Cave System, *in* White, W.B., and White, E.L., eds., *Karst hydrology–Concepts from the Mammoth Cave area:* New York, Van Nostrand Reinhold, p. 317-337.

Palmer, A.N., 1991, Origin and morphology of limestone caves. *Geological Society of America Bulletin,* no. 103, p. 1-21.

Palmer, A.N., 1995, Geochemical models for the origin of macroscopic solution porosity in carbonate rocks, *in* Budd, A.D., Saller, A.H., and Harris, P.M., eds., Unconformities and porosity in carbonate strata: *American Association of Petroleum Geologists Memoir,* no. 63, p. 77-101.

Palmer, A.N., 1996, Cave patterns in the Guadalupe Mountains: *National Speleological Society,* Salida, Convention Program with Abstracts, p. 51-52.

Palmer, A.N., and Palmer, M.V., 1989, Geologic history of the Black Hills caves, South Dakota: *National Speleological Society Bulletin,* v. 51, p. 72-99.

Palmer, A.N., and Palmer, M.V., 1995a, Geochemistry of capillary seepage in Mammoth Cave: *4th Science Conference,* Mammoth Cave National Park, Proceedings, p. 119-133.

Palmer, A.N., and Palmer, M.V., 1995b, The Kaskaskia paleokarst of the Northern Rocky Mountains and Black Hills, northwestern U.S.A.: *Carbonates and Evaporites,* v. 10, p. 148-160.

Palmer, M.V., 1976, *Groundwater flow paths in limestone solution conduits:* State University of New York, M.A. thesis, Oneonta, NY, 150 p.

Palmer, M.V., and Palmer, A.N., 1975, Landform development in the Mitchell Plain of southern Indiana: *Zeitschrift für Geomorphologie,* v. 19, p. 1-39.

Palmer, M.V., and Palmer, A.N., 1989, Paleokarst of the United States, *in* Bosak, P., Ford, D., Glazek, J., and Horácek, I., eds., *Paleokarst: A systematic and regional review:* Praha, Academia, p. 337-365.

Palmer, R.J., 1989, *Deep into blue holes:* London, Unwin Hyman, Ltd., 164 p.

Parandier, A.N., 1833, Notice sur les causes de l'existence des cavernes ...: *Séance Publique de l'Academie des Sciences Belle-Lettres et Arts de Besançon,* p. 59-85.

Pashchenko, S.E., Dublyansky, Y.V., and Andreichuk, V.N., 1993, Aerosol studies in Kungur Ice Cave: *11th International Congress of Speleology,* Beijing, Proceedings, p. 190-192.

Pashenko, S., Dubliansky, Y., Andreichuk, V., and Pashenko, E., 1996, Transformation of fractal atmospheric aerosol moving through natural cave: *Journal of Aerosol Science,* v. 27, p. S127-S128.

Pasini, G., 1975, Sull'importanza speleogenetica dell' "erosione antigravitativa.": *Le Grotte d'Italia.* v. 4, p. 297-326.

Pasquini, G., 1973, Aggressive condensation: *6th International Congress of Speleology,* Olomouc, Proceedings, v. 8, p. 315-318.

Pávai-Vájna, F., 1931, Über die Rolle heisser Lösungen, Dämpfe und Gase bei der Höhlenbildung: *Hid. Közl.*, no. 10-11, p. 63-69.

Pechorkin, A.I., 1986, *Geodynamics of sulfate karst:* Irkutsk University Publ., 172 p. (in Russian).

Perez, R.C., and Grossi, W.R., 1986, The quartzitic speleological district of the Parque Florestal Estadual do Ibitipoca, Minas Gerais, Brazil: *9th International Congress of Speleology,* Barcelona, Proceedings, p. 12-14.

Petroleum Services, 1979, *Solution mined liquid hydrocarbon storage caverns, Mount Sedom:* Tel Aviv, Petroleum Services Ltd., Tel Aviv, Progress Report 2, 12 p.

Pettijohn, F.J., 1975, *Sedimentary rocks:* New York, Harper and Row, 3rd ed.

Pfeiffer, D., and Hahn, J., 1972, Karst of Germany, *in* Herak, M., and Stringfield, V.T., eds., *Karst—Important karst regions of the northern hemisphere:* Amsterdam, Elsevier, p. 189-223.

Pia, J., 1933, Die Theorien über die Löslichkeit des kohlensauren Kalkes: *Mitteilungen der Geologischen Gesellschaft in Wien*, v. 25, p. 1-93.

Pinder, G.F., and Jones, J.F., 1969, Determination of the ground-water component of peak discharge from the chemistry of total runoff: *Water Resources Research,* v. 5, p. 438-445.

Pinneker, E.V., 1977, *Problems of regional hydrogeology (Regularities of distribution and formation of groundwaters):* Moscow, Nauka, 196 p. (in Russian).

Pinneker, E.V., ed., 1980, *Fundamentals of hydrogeology; General hydrogeology:* Novosibirsk, Nauka, 230 p. (in Russian).

Pinneker, E.V., 1982, Paleohydrogeologic reconstructions, *in* Principles of hydrogeology–geologic activity and history of water in the Earth's subsurface: Novosibirsk, Nauka, p. 153-188 (in Russian).

Pisarowicz, J., 1992, Cueva de Villa Luz–H_2S parameters: National Speleological Society, *Geo²,* v. 20, p. 9.

Pisarowicz, J.A., 1994, Cueva de Villa Luz–An active case of H_2S speleogenesis, *in* Sasowsky, I.D., and Palmer, M.V., eds., Breakthroughs in karst geomicrobiology and redox geochemistry: *Karst Waters Institute Special Publication,* no. 1, p. 60-62.

Pizarowicz, J.A., and Maslyn, M., 1981, Empirical confirmation of Curl's flow velocity calculations 1974: *8th International Speleological Congress,* Bowling Green, Proceedings, v. 2, p. 772-774.

Placer, L., 1981, Geologic structure of southwestern Slovenia: *Geologija,* v. 24, no. 1, p. 27-60 (in Slovenian with English summary).

Plózer, I., 1977, Situation of Hungarian cave diving in 1976: *Karszt és Barlang* (Hungary), Special Issue, p. 67-69.

Pluhar, A., and Ford, D.C., 1970, Dolomite karren of the Niagara Peninsula, Ontario, Canada: *Zeitschrift für Geomorpholgie*, v. 14, p. 392-410.

Plummer, L.N., 1975, Mixing of sea water with calcium carbonate ground water, *in* Whitten, E.H.T., ed., Quantitative studies in geological sciences, *Geological Society of America Memoir*, no. 142, p. 219-236.

Plummer, L.N., and Back, W., 1980, The mass balance approach: Application to interpreting the chemical evolution of hydrogeologic systems: *American Journal of Science,* v. 280, p. 130-142.

Plummer, L.N., and Busenberg, E., 1982, The solubilities of calcite, aragonite, and vaterite in CO_2-H_2O-solutions between 0 and 90°C and an evaluation of the aqueous model for the system $CaCO_3$-CO_2-H_2O: *Geochimica et Cosmochimica Acta,* v. 46, p. 1011-1040.

Plummer, L.N., and Mackenzie, F.T., 1974, Predicting mineral solubility from rate data: Application to the dissolution of magnesian calcites: *American Journal of Science,* v. 274, p. 61-83.

Plummer, L.N., and Wigley, T.M.L., 1976, The dissolution of calcite in CO_2-saturated solutions at 25°C and 1 atmosphere total pressure: *Geochimica et Cosmochimica Acta,* v. 40, p. 191-202.

Plummer, L.N., Wigley, T.M.L., and Parkhurst, D.L., 1978, The kinetics of calcite dissolution in CO_2-water systems at 5° to 60°C and 0.0 to 1.0 atm CO_2: *American Journal of Science,* v. 278, p. 179-216.

Pohl, E.R., 1935, Development of vertical shafts in limestone caves: *Geological Society of America Proceedings for 1935,* p. 96.

Pohl, E.R., 1955, Vertical shafts in limestone caves: *National Speleological Society, Occasional Paper,* no. 2, 24 p.

Polkunov, V.F., 1990, Mineral concentrations of native sulfur, *in* Geotektonika Volyno-Podolii: Kiev, Naukova Dumka, p. 229-232 (in Russian).

Polyak, V.J., and Güven, N., 1996, Mineralization of alunite, natroalunite, and hydrated halloysite in Carlsbad Cavern and Lechuguilla Cave, New Mexico: *Clays and Clay Minerals*, v. 44, p. 843-850.

Polyak, V.J., and Mosch, C.J., 1994, Tyuyamunite and metatyuyamunite?: *National Speleological Society Bulletin,* v. 57, p. 80.

Polyak, V.J., McIntosh, W.C., Güven, N., and Provencio, P., 1998, Age of origin of Carlsbad Cavern and related caves of the Guadalupe Mountains based on $^{40}Ar/^{39}Ar$-dating of alunite: *Science,* v. 279, p. 1919-1922.

Polyak, V.J., McIntosh, W.C., Guven, N., and Provencio, P., 1998, Age and origin of Carlsbad Cavern and related caves from $^{40}AR/^{39}AR$ of alunite: *Science*, v. 279, p. 1919-1922.

Ponta, G., 1986, The evaporite karst of Romania, *in* Atti simposio internazionale sul carsismo nelle evaporiti: *Le Grotte d'Italia,* v. 12, p. 407-415.

Powell, R.L., 1970, Base level, lithologic, and climatic controls of karst groundwater zones in south-central Indiana: *Indiana Academy of Sciences Proceedings*, v. 79, p. 281-291.

Pošpný, F., 1894, The genesis of ore-deposits: *Transactions of the American Institute of Mining Engineers,* v. 23, p. 197-369.

Premru, U., 1982, Geologic structure of southern Slovenia: *Geologija,* v. 25, no. 1, p. 95-126 (in Slovenian with English summary).

Price, M., 1994, A method for assessing the extent of fissuring in double-porosity aquifers, using data from packer tests: *International Association of Hydrological Sciences Publication*, no. 222, p. 271-278.

Price, M., 1996, *Introducing groundwater:* London, Chapman and Hall, 278p.

Price, M., Atkinson, T.C., Wheeler, D., Barker, J.A., and Monkhouse, R.A., 1989, Highway drainage to the Chalk Aquifer: The movement of groundwater in the Chalk near Bricket Wood, Hertfordshire, and its possible pollution by drainage from the M25 motorway: *British Geological Survey Technical Report* WD/89/3.

Price, M., Downing R.A., and Edmunds, W.M., 1993, The Chalk as an aquifer, *in* Downing, R.A., Price, M., and Jones, G.P., eds., *Hydrogeology of the Chalk of North-West Europe*: Oxford, Clarendon Press, p. 35-58.

Price, M., Morris, B., and Robertson, A., 1982, A study of intergranular and fissure permeability in Chalk and Permian aquifers, using double packer injection testing: *Journal of Hydrology*, v. 54, p. 401-423.

Proctor, C.J., 1988, Sea-level related caves on Berry Head, South Devon: *Cave Science,* v. 15, no. 2, p. 39-49.

Prokofiev, S.S., 1964, The role of condensation moisture in creation of karst caves: *Peshchery* (Perm), v. 4, no. 5, p. 35-38 (in Russian).

Pronin, K.K., 1995, Karst caves in the Odessa region: Dep. Paper No. 953 UK95 (Ukraine), 42 p. (in Russian).

Protasov, V.A., 1959, Condensation waters of Crimea Mountains and their role in underground runoff: *3rd Hydrol. Congress*, Leningrad, Proceedings, v. 9, p. 98-103 (in Russian).

Puch, C., Polsson, P., and Chorvot, G., 1985, Présentation géographique et humaine: *Spelunca Supplemént,* no. 19, p. 4-6.

Pulido-Bosch, A., and Calaforra, J.M., 1986,

Formas kársticas en los yesos de Sorbas (Almería): *Estudios sobre Geomorfología del Sur de España,* Univ. Murcia, p. 115-119.

Pulido-Bosch, A., and Calaforra, J.M., 1993, The gypsum karstic aquifer of Sorbas (Almería), *in* Pulido-Bosch, A., ed., *Some Spanish karstic aquifers:* Univ. Granada, p. 225-241.

Q

Queen, J.M., 1981, *A discussion and field guide to the geology of Carlsbad Caverns:* Unpublished report to Carlsbad Caverns National Park, 64 p.

Quinlan, J.F., 1972, Karst-related mineral deposits and possible criteria for the recognition of paleokarsts: A review of preservable characteristics of Holocene and older karst terranes: *24th International Geological Congress,* sec. 6, p. 156-168.

Quinlan, J.F., 1978, *Types of karst, with emphasis on cover beds in their classification and development:* University of Texas at Austin, Ph.D. Thesis. 325 p.

Quinlan, J.F., and Ewers, R.O., 1985, Groundwater flow in limestone terranes: Strategy rationale and procedure for reliable and efficient monitoring of groundwater quality in karst areas: *5th National Symposium and Exp. on Aquifer Restoration and Groundwater Monitoring,* Columbus, p. 197-234.

Quinlan, J.F., and Ewers, R.O., 1989, Subsurface drainage in the Mammoth Cave Area, *in* White, W.B., and White, E.L., eds., *Karst Hydrology: Concepts from the Mammoth Cave Area*, New York, Van Nostrand Reinhold, p. 65-103.

Quinlan, J.F., and Pohl, E.R., 1967, Vertical shafts actively promote slope retreat and dissection of the solution escarpment and the Chester Cuesta in the Central Kentucky Karst: *National Speleological Society Bulletin,* v. 29, p. 109.

Quinlan, J.F., and Ray, J.A., 1981, Groundwater basins in the Mammoth Cave region, Kentucky: Friends of the Karst, Mammoth Cave National Park, Occasional Publication, no. 1, 1 sheet, 1:138,000.

Quinlan, J.F., and Rowe, D.R., 1977, Hydrology and water quality in the central Kentucky karst; Phase I: *University of Kentucky Water Resources Institute Research Report,* no. 101, 93 p.

Quinlan, J.F., and Rowe, D.R., 1978, Hydrology and water quality in the central Kentucky karst: Phase II, Part A: Preliminary summary of the hydrogeology of the Mill Hole sub-basin of the Turnhole Spring groundwater basin*: University of Kentucky Water Resources Institute Research Report,* no. 109, 42 p.

Quinlan, J.F., Davies, G.J., Jones, S.W., and Huntoon, P.W., 1996, The applicability of numerical models to adequately characterize ground-water flow in karstic and other triple-porosity aquifers, *in* Ritchey, J.D., and Rumbaugh, J.O., eds., Subsurface fluid-flow (ground-water and vadose zone) modeling: American Society for Testing and Materials, Special Technical Publication, v. 1288, p. 114-133.

Quinlan, J.F., Smith, A.R., and Johnson, K.S., 1986, Gypsum karst and salt karst of the United States, *in* Atti simposio internazionale sul carsismo nelle evaporiti: *Le Grotte d'Italia,* v. 12, p. 419-420.

Quinlan, J.F., Smith, R.A., and Johnson, K.S., 1987, Gypsum karst and salt karst of the United States of America, *in* Atti symposio international sul carsismo nelle evaporiti: *Le Grotte d'Italia*, ser. 4, v. 13, p. 73–92.

R

Racovitcă, G., and Viemann, J., 1984, Sur le rôle de la condensation souterraine dans la genèse des stalagmites de glace: *Trav. Inst. Spéol. "E. Racovitză,"* v. 23, p. 89-97.

Raeisi, E., and Mylroie, J.E., 1995, Hydrodynamic behavior of caves formed in the fresh-water lens of carbonate islands: *Carbonates and Evaporites,* v. 10, p. 207-214.

Rahn, P.H., and Gries, J.P., 1973, *Large springs in the Black Hills, South Dakota and Wyoming:* South Dakota Geological Survey, Report of Investigations, no. 107, 46 p.

Raines, M.A., and Dewers, T.A., 1997, Dedolomitization as a mechanism for karst generation in Permian Blane Formation, southwestern Oklahoma, USA: *Carbonates and Evaporites,* v. 12, p. 24-31.

Ramsbottom, W.H.C., 1977, Major cycles of regression and transgression (mesothems) in the Namurian: *Yorkshire Geological Society,* Proceedings, v. 41, p. 261-291.

Ranganathan, V., and Hanor, J.S., 1986, Diffusion of dissolved NaCl from evaporites in sedimentary basins: *Eos,* v. 67, p. 274.

Rauch, H., and White, W.B., 1977, Dissolution kinetics of carbonate rocks: 1, Effect of lithology on dissolution rate: *Water Resources Research,* v. 13, p. 381-394.

Ray, J.A., 1996, Fluvial features of the karst-plain erosion surface in the Mammoth Cave region*: 5th Science Conference,* Mammoth Cave, Proceedings, p. 137-156.

Reams, M.W., 1965, Laboratory and field evidence for a vadose origin of foibe (domepits): *International Journal of Speleology,* v, 1, p. 373-389.

Reeves, M.J., 1979, Recharge and pollution of the English Chalk: Some possible mechanisms: *Engineering Geology*, v. 14, p. 231-240.

Renard, K.G., and Keppel, R.V., 1966, Hydrographs of ephemeral streams in the Southwest: *American Society of Civil Engineers,* Proceedings, v. 92, p. 33-52.

Renault, P., 1968, Contribution a l' étude des actions mechaniques et sedimentologiques dans la spéléogènese: *Annales de Speleologie,* no. 22, p. 5-21, 209-267; no. 23, p. 259-307, 529-596; no. 24, p. 313-37.

Renault, P., 1968, Contribution a l'étude des actions méchaniques et sédimentologiques dans la spéléogenèse: *Annales de Spéléologie,* v. 23, p. 529-593.

Renault, P., 1970, *La formation des cavernes:* Paris, Presses Universitaires de France, 126 p.

Rhoades, R., and Sinacori, N.M., 1941, Patterns of groundwater flow and solution: *Journal of Geology,* v. 49, p. 785-794.

Rhodes, D., Lantos, E.A., Lantos, J.A., Webb, R.J., and Owens, D.C., 1984, Pine Point ore bodies and their relationship to the stratigraphy, structure, dolomitisation, and karstification of the Middle Devonian Barrier Complex: *Economic Geology,* v. 79, p. 991-1055.

Riazanova, E.A., 1987, Classification of factors and sources of groundwater level rise, *in* Voprosi ingenerno-geologicheskikh issledovanii na zastraivajemikh territorijakh: Moscow, Stroiizdat, p. 54-59 (in Russian).

Ringer, Á., 1989, The role of cave sites and their chronostratigraphy in the research of the Paleolithic of Hungary: *Karszt-es Barlang*, Special Issue, p. 51-58.

Rogozhnikov, V.Y., 1984, Water-chemical deposits in karst labyrinthic caves of the Podol'sky Pridnestrovje, *in* Peshchery (Caves): Perm University, p. 46-55 (in Russian).

Romero, A.E., and Lima, M.T., 1989, The Itacolomi Quartzite Karst, Brazil: Application of the underground drainage direction prediction method: *10th International Congress of Speleology,* Budapest, Proceedings, p. 476-480.

Rossi, C., Munoz, A., and Cortel, A., 1997, Cave development along the water table in Cobre System (Sierra de Penalabra, Cantabrian Mountain, N. Spain: *12th International Congress of Speleology,* La Chaux-de-Fonds, Proceedings, v. 1, p. 179-182.

Rossi, G., and Sauro, U., 1977, L'Abisso di Lesi: Analisi morfologica e ipotesi genetiche: *Le Grotte d'Italia,* ser. 4, v. 6, p. 73-100.

Rubbioli, E., 1996, O pico do Inficionado: A exploração da gruta mais profunda do Brasil: *O Carste,* v. 8, no. 3, p. 47-54.

Rubin, P.A., 1991, Flow characteristics and scallop-forming hydraulics within the Mill Pond Karst Basin, east-central New York, *in* Kastning, E.H., and Kastning, K.M., eds., Appalachian Karst: *Appalachian*

Karst Symposium, Proceedings, Radford University, p. 75-99.

[Rudder, S.], 1779, *A new history of Gloucestershire ...:* Cirencester, Rudder, 855 p. (p.798).

Rudincky, J., 1978, Role of convection in shaping subterranean karst forms: *Pr. nauk. Usl,* Katowice, Poland, v. 280, p. 92-101.

Rudincky, J., 1989, Relation between natural convection and cave formation in hydrothermal karst: *10th International Congress of Speleology,* Budapest, Proceedings, p. 14-16.

Rudnicki, J., 1960, Experimental work on flutes development: *Speleologica,* v. 2, p. 17-30.

Rudnicki, J., 1967, Origin and age of the western Tatra caverns: *Acta Geologica Polonica,* v. 17, p. 521-591.

Runnels, D.D., 1969, Diagenesis, chemical sediments, and mixing of natural waters: *Journal of Sedimentary Petrology,* v. 39, p. 1188-1201.

Ryder, P.F., 1975, Phreatic network caves in the Swaledale area, Yorkshire*: British Cave Research Association Transactions,* v. 1, p. 101-124.

S

Saberian, A., 1983, Utilization of leaching models in the design of large crude oil storage cavities: *6th International Symposium on Salt,* Alexandria, Proceedings, v. 2, p. 235-348.

Sahimi, M., 1995, *Flow and transport in porous media and fractured rock:* Weinheim, VCH Verlagsges, 482 p.

Saller, A.H., Budd, D.A., and Harris, P.M., 1994, Unconformities and porosity development in carbonate strata: Ideas from a Hedberg Conference: *American Association of Petroleum Geologists Bulletin,* v. 78, p. 857-872.

Salomon, J.N., 1995, La Chili, Pays de karsts extrémes: *Karstologia,* v. 24, p. 52-56.

Sanders, E.M., 1921, The cycle of erosion in a karst region (after Cvijiç): *Geographical Review,* v. 11, p. 593-604 (see Cvijiç, 1918).

Sando, W.J., 1974, Ancient solution phenomena in the Madison Limestone (Mississippian) of north-central Wyoming: *U.S. Geological Survey Journal of Research,* v. 4, p. 133-141.

Sando, W.J., 1988, Madison Limestone (Mississippian) paleokarst: A geologic synthesis, *in* James, N.P., and Choquette, P.W., eds., *Paleokarst:* New York, Springer-Verlag, p. 256-277.

Sanford, W.E., and Konikow, L.F., 1989, Porosity development in coastal carbonate aquifers: *Geology,* v. 17, p. 249-252.

Sangster, D.F., 1988, Breccia-hosted lead-zinc deposits in carbonate rocks, *in* James, N.P., and Choquette, P.W., eds., *Paleokarst:* New York, Springer-Verlag, p. 102-116.

Sasowsky, I.D., and White, W.B., 1994, The role of stress release fracturing in the development of cavernous porosity in carbonate aquifers: *Water Resources Research,* v. 30, p. 3523-3530.

Sauter, M., 1992, Quantification and forecasting of regional groundwater flow and transport in a karst aquifer (Gallusquelle, Malm, SW Germany): Tübinger Geowissenschaftliche Arbeiten, ser. C, v. 13, p. 1-150.

Sauter, M., Liedl, R., Clemens, T., and Hückinghaus, D., 1997, Karst aquifer genesis–Field evidence and modeling approaches: *12th International Congress of Speleology and 6th Conference on Limestone Hydrology and Fissured Media,* La Chaux-de-Fonds, Proceedings, no. 2, p. 107-110.

Schafarzik, F., 1928, Visszapillantás a budai hévforrások fejlödéstörténetére (The overview of thermal springs of Budapest): *Hidr.Közl.,* no. 1, p. 9-14.

Scheurer, G., and Schweitzer, F., 1980, A budai hévizforrások fejlödéstörténete a felsö-pannontól napjainkig (History of development of the Budda thermal springs from the Uupper Pannonian, Miocene-Pliocene, to now): *Hidr. Közl.,* no. 5, p. 391-395.

Schmid, E., 1958, Höhlenforschung und Sedimentanalyse: *Schriften der Institutes für Ur- und Frühgeschichte der Schweiz* no. 13, 186 p.

Schmidt, V.A., 1982, Magnetostratigraphy of sediments in Mammoth Cave, Kentucky: *Science,* v. 217, p. 827-829.

Scholle, P.A., Bebout, D.G., and Moore, C.H., eds., 1983, Carbonate depositional environments: *American Association of Petroleum Geologists Memoir,* no. 33, 365 p.

Scholler, H., 1962, Les eaux souterraines: Paris, Masson and Cie.

Schroeder, J., and Ford, D.C., 1983, Clastic sediments in Castleguard Cave, Columbia Icefields, Alberta, Canada: *Arctic and Alpine Research,* v. 15, p. 451-461.

Schwartz, F.W., and Longstaffe, F.J., 1988, Ground water and clastic diagenesis, *in* Back, W., Rosenshein, J.S., and Seaber, P.R., eds., *Hydrogeology:* Geological Society of America, Geology of North America, v. O-2, p. 413-434.

Šebela, S., 1994, The role of tectonic structures during the formation of cave channels and karst surface phenomena: University of Ljubljana, Ph.D. thesis, 129 p. (in Slovenian with English summary).

Sebev, D.G., 1970, Giant cavern in Rhodopes: *Rodopski Pescherniak,* no. 50 (Chepelare), 34 p. (in Bulgarian).

Secretaría de Recursos Hidráulicos, 1969?, Datos Hidrometricos Cuenca del Río Panuco: *Boletin Hidrológico,* no. 32, sec. 1-2, (data through Dec. 1968).

Seemann, R., 1979, Die sedimentaren Eisenvererzugne der Karstgebiete der Nordlichen Kalkalpen: *Annals Naturhistorisches Museum Wein,* v. 82, p. 209-289.

Sefton, M., 1984, Cave explorations around Tresviso, Picos de Europa, northern Spain, 1975-1983: *Cave Science,* v. 11, p. 199-237.

SEII, 1987, *Memória de Expedición Vega Huerta 87:* Secc. Espel. Ing. Industriales (unpublished), 22 p.

Senior, K.J., 1987, Geology and speleogenesis of the M-2 Cave System, Western Massif, Picos de Europa, northern Spain: *Cave Science,* v. 14, p. 93-103.

Serafimov, S., and Milev, T., 1977, Particularities of geothermal characteristics of the Erma Reka deposit: Fizicheskie Processi Gornogo Proizvodstva (Moscow), v. 3, p. 113-115 (in Russian).

Sharp, J.M., Jr., and Kyle, J.R., 1988, The role of ground-water processes in the formation of ore deposits, *in* Back, W, Rosenshein, J.S., and Seaber, P.R., eds., *Hydrogeology:* Geological Society of America, Geology of North America, v. O-2, p. 461-483.

Shaw, T.R., 1979, *History of cave science:* Crymych, Oldham, 489 p.

Shaw, T.R., 1992, *History of cave science: The exploration and study of limestone caves, to 1900:* Sydney, Sydney Speleological Society, 2nd ed., 338 p.

Shestakov, F.V., 1989, *Condensation of water vapor in soil-ground and in the near-surface layer; Bibliography from 1877 to 1987:* Alma-Ata, Nauka, 77 p. (in Russian).

Shestopalov, V.M., 1981, *Natural resources of underground water of platform artesian basins of the Ukraine:* Kiev, Naukova Dumka, 195 p. (in Russian).

Shestopalov, V.M., 1988, *Methods of study of underground water natural resources:* Moscow, Nedra, 168 p. (in Russian).

Shestopalov, V.M., ed., 1989, *Water exchange in hydrogeological structures of the Ukraine–Water exchange under natural conditions:* Kiev, Naukova Dumka, 288 p. (in Russian).

Shirtcliffe, T.G.L., and Turner, J.S., 1969, Observations of the cell structure of salt fingers: *Journal of Fluid Mechanics,* v. 41, p. 707-719.

Shternina, E.B., 1949, Solubility of gypsum in water solutions of salts: *Izvestija sectora fiz.-him. analiza IONH AN SSSR,* v. 17, p. 203-206 (in Russian).

Shuster, E.T., and White, W.B., 1971, Seasonal fluctuations in the chemistry of limestone springs: A possible means for characterizing carbonate aquifers:*Journal of Hydrology,* v. 14, p. 93-128.

Sibson, R.H., Moore, J.McM., and Rankin, A.H., 1975, Seismic pumping–A hydrothermal fluid transport mechanism: *Journal of the Geological Society of London,* v. 131, p. 653-659.

Siemers, J., and Dreybrodt, W., 1998, Early development of karst aquifers on percolation networks of fractures in limestone: *Water Resources Research,* v. 34, p. 409-419.

Šikiĺ, D., 1979, On the tectonic structure of the watershed of the River Kupa: *Geološki vesnik,* v. 31, p. 151-156 (in Slovenian with English summary).

Simmons, R., 1990, In search of the Cathedral System: *Butler Cave Conservation Society Newsletter,* v. 16, p. 1-8.

Simms, M.J., 1990, Triassic palaeokarst in Britain: *Cave Science,* v. 17, no. 3, p. 93-101.

Singer, P.C., and Stumm, W., 1970, Acid mine drainage: The rate determining step: *Science,* v. 167, p. 1121-1123.

Sjöberg, E.L., 1978, Kinetics and mechanism of calcite dissolution in aqueous solutions at low temperatures: *Acta Univ. Stockholm Contrib. Geol.,* v. 32, no. 1, p. 1-92.

Sjöberg, E.L., and Rickard, D.T., 1984, Temperature dependence of calcite dissolution kinetics between 1 and 62°C at pH 2.7 to 8.4 in aqueous solution: *Geochimica et Cosmochimica Acta,* v. 48, p. 271-288.

Slabe, T., 1995, *Cave rocky relief and its speleological significance:* Ljubliana, Znanstvenoraziskovalni Center Sazu, 128 p.

Slavianov, V.N., 1955, About natural condensors on mountain slopes and about the possibility of their usage for small-scale water supply, *in* Voprosi izuchenija podzemnikh vod i ingenerno-geologicheskikh processov: Moscow, AN SSSR, p. 21-32 (in Russian).

Smart, C.C., 1983a, *The hydrology of a glacierised alpine karst:* McMaster University, Ph.D. thesis, 342 p.

Smart, C.C., 1983b, The hydrology of the Castleguard Karst, Columbia Icefields, Alberta, Canada: *Arctic and Alpine Research,* v. 15, p. 471-486.

Smart, C.C., 1984, Overflow sedimentation in an alpine cave system: *Norsk geografisk Tidsskrift,* v. 38, p. 171-175.

Smart, C.C., 1988, Artificial tracer techniques for the detemination of the structure of conduit aquifers: *Ground Water,* v. 26, p. 445-453.

Smart, C.C., and Ford, D.C., 1986, Structure and function of a conduit aquifer: *Canadian Journal of Earth Sciences,* v. 23, p. 919-929.

Smart, P.L., 1984, The geology, geomorphology, and speleogenesis of the Eastern Massifs, Picos de Europa, Spain: *Cave Science,* v. 11, p. 238-245.

Smart, P.L., 1985, Cave development in the Andara Area, with special reference to 56, *in* La Sima 56 (Picos de Europa, España): *Mem. Lancaster Univ. Spel. Soc. and Secc. Espel. Ing. Industriales,* 86 p.

Smart, P.L., 1986, Origin and development of glacio-karst closed depressions in the Picos de Europa, Spain: *Zeitschreift für Geomorphologie,* v. 30, p. 423-443.

Smart, P.L. and Christopher, N.S.J., 1989, Ogof Ffynnon Ddu, *in* Ford, T.D., ed., *Limestones and caves of Wales:* Cambridge University Press, p. 177-189)

Smart and Christopher, 1989, *in* Ford, T.D., ed., *Limestones and caves of Wales:* Cambridge University Press, p. 177-189)

Smart, P.L., and Friederich, H., 1986, Water movement and storage in the unsaturated zone of a maturely karstified carbonate aquifer, Mendip Hills, England: 1st Conference on Environmental Problems in Karst Terranes, *Association of Ground Water Scientists and Engineers,* Proceedings, Dublin, Ohio, p. 59-87.

Smart, P.L., and Hobbs, S.L., 1986, Characterization of carbonate aquifers: A conceptual base: *Environmental Problems in Karst Terranes and Their Solutions Conference,* Bowling Green, Kentucky, Proceedings, p. 1-14.

Smart, P.L., and Laidlaw, I.M.S., 1977, An evaluation of some fluorescent dyes for water tracing: *Water Resources Research,* v. 13, p. 15-33.

Smart, P.L., and Whitaker, F.F., 1991, Karst processes, hydrology and porosity evolution, *in* Wright, V.P., Esteban, M., and Smart, P.L., eds., Paleokarsts and paleokarstic reservoirs: *P.R.I.S. Occasional Publications Series,* University of Reading, no. 2, p. 1-55.

Smart, P.L., Dawans, J.M., and Whitaker, F., 1988, Carbonate dissolution in a modern mixing zone: *Nature,* v. 335, p. 811-813.

Smith, D.I., 1972, The solution of limestone in an Arctic environment: *Institute of British Geographers Special Publication,* no. 4, p. 187-200.

Smith, D.I., and Atkinson, T.C., 1976, Process, landforms and climate in limestone regions, *in* Derbyshire, E., ed., *Geomorphology and Climate:* London, John Wiley, p. 367-409.

Smith, D.I., and Newson, M.D., 1974, The dynamics of solutional and mechanical erosion in limestone catchments on the Mendip Hills, Somerset, *in* Gregory, K.J., and Walling, D.E., eds., *Fluvial Processes in Instrumented Watersheds:* Institute of British Geographers Special Publication, no. 6, p. 155-167.

Smith, D.I., Greenaway, M.A., Moses, C., and Spate, A.P., 1995, Limestone weathering in eastern Australia, Part 1: Erosion rates: *Earth Surface Processes and Landforms,* v. 20, p. 451-463.

Smith, J., 1994, *Hydrogeology of the Sistema Huautla karst groundwater basin, Sierra Mazateca, Oaxaca, Mexico:* Western Kentucky University, M.S. thesis, 338 p.

Snow, D.T., 1968, Rock fracture spacings, openings and porosities*: Journal of the Soil Mechanics and Foundations Division,* American Society of Civil Engineers, Proceedings, v. 94, p. 73-91.

Sokolov, D.S., 1962, *Principal conditions of karst development:* Moscow, Gosgeolizdat, 321 p. (in Russian).

Sonnenfeld, P., 1984, *Brines and evaporates:* London, Academic Press, 530 p.

Spate, A.P., Jennings, J.N., Smith, D.I., and Greenaway, M.A., 1985, The micro-erosion meter: Use and limitations: *Earth Surface Processes and Landforms,* v. 10, p. 427-440.

Stankevich, E.F., 1970, On the possibility of development of deep-seated karst: *Voprosy Karstovedeniya,* Perm University, no. 2 , p. 43-47 (in Russian).

Stauffer, D., 1985, *Introduction to percolation theory:* London, Taylor and Francis Ltd., 124 p.

Stone, B., and am Ende, B., 1997, The 1994 San Agustin Expedition: *Association for Mexican Cave Studies Activities Newsletter,* v. 21, p. 44-64.

Strahan, A., 1885, The geology of the coasts adjoining Rhyl, Abergele and Colwyn: *Geological Survey of the United Kingdom Memoir, Old Series,* v. 79SW.

Strahler, A.N., 1981, *Physical geology:* New York, Harper and Row, 612 p.

Strakhov, N.M., 1962, *Principles of the theory of lithogenesis–Vol. III Regularities of composition and distribution of aridic sediments:* Moscow, Academy of Sciences Publications, 550 p. (in Russian).

Stringfield, V.T., and LeGrand, H.E., 1966, Hydrology of limestone terranes in the Coastal Plain of the southeastern United States: *Geological Society of America Special Paper 93,* 46 p.

Stringfield, V.T., and LeGrand, H.E., 1969, Hydrology of carbonate rock terranes–A review:*Journal of Hydrology,* v. 8, p. 349-417.

Stumm, W., and Morgan, J.J., 1981, *Aquatic chemistry:* New York, Wiley, 780 p.

Šušteršič, F., 1994b, Classic dolines of classical site: *Acta Carsologica,* v. 23, p. 123-154.

Šušteršič, F., 1978a, Some considerations about stuffed potholes and their filling: *Naše Jame,* v. 17, p. 73-85 (in Slovenian with English summary).

Šušteršič, F., 1978b, A contribution to the knowledge of the inflow part of the Planinsko Polje and the Postojnska Vrata: *Geografski vestnik,* v. 50, p. 51-63 (in Slovenian with English summary).

Šušteršič, F., 1982, Morphology and hydrography of Najdena Jama: *Acta*

Carsologica, v. 10, p. 127-155 (in Slovenian with English summary).

Šušteršič, F., 1987a, The small scale surface karst and solution dolines at the northeastern border of Planinsko Polje: *Acta Carsologica,* v. 16, p. 51-82 (in Slovenian with English summary).

Šušteršič, F., 1987b, An essay on an unconventional geomorphologic representation of Notranjska, *in* Notranjska: Zbornik 14, Postojna, Zborovanja slovenskih geografov, p. 61-72 (in Slovenian with English summary).

Šušteršič, F., 1991, On the objective of speleological science. *Naše Jame,* v. 33, p. 73-85.

Šušteršič, F., 1992a, Delovni seznam jam jugovzhodne Slovenije: *Naše Jame,* v. 34, p. 74 -108.

Šušteršič, F., 1992b, The Dolèek Cave: *Naše Jame,* v. 34, p. 29-39 (in Slovenian with English summary).

Šušteršič, F., 1993, Delovni seznam jam jušnega dela osrednje Slovenije: *Naše Jame,* v. 35, p. 59-99.

Šušteršič, F., 1994a, The Kloka Cave and speleoinception: *Naše Jame,* v. 36, p. 9-30.

Šušteršič, F., 1994c, Delovni seznam jam jušne Slovenije–dopolnila: *Naše Jame,* v. 36, p. 112-140.

Šušteršič, F., 1995, Delovni seznam jam severne Slovenije: *Naše Jame,* v. 37, p. 98-152.

Šušteršič, F., 1996a, The pure karst model: *Cave and Karst Science,* v. 23, no. 1, p. 25-32.

Šušteršič, F., 1996b, Poljes and caves of Notranjska: *Acta Carsologica,* v. 25, p. 251-289.

Šušteršič, F., 1997, Cave patterns north of the Planinsko Polje (Slovenia): *12th International Congress of Speleology,* La Chaux-de-Fonds, Proceedings, v. 1, p. 117-120.

Šušteršič, F., and Puc, M., 1970, The karst underground by the NE nook of the Planinsko Polje: *Acta Carsologica,* v. 5, p. 205-270 (in Slovenian with English summary).

Svensson, U., 1992, *Die Lösungsund Abscheidungskinetik natürlicher Calcitminerale in wässrigen CO_2 -Lösungen nahe dem Gleichgewicht:* Bremen University, Germany, Ph.D. thesis, 123 p.

Svensson, U., and Dreybrodt, W., 1992, Dissolution kinetics of natural calcite minerals in CO_2-water systems approaching calcite equilibrium: *Chemical Geology,* v. 100, p. 129-145.

Sweeting, M.M., 1950, Erosion cycles and limestone caverns in the Ingleborough District: *Geographical Journal,* v. 115, p. 63-78.

Sweeting, M.M., 1966, The weathering of limestone, *in* Dury, G.H., ed., *Essays in Geomorphology:* London, Heinemann, p. 177-210.

Sweeting, M.M., 1972, *Karst Landforms:* London, Macmillan, 362 p.

Sweeting, M.M., 1979, Weathering and solution of the Melinau Limestones in the Gunong Mulu National Park, Sarawak, Malaysia: *Annales de la Société Géologique de Belgique,* v. 102, p. 53-57.

Swinnerton, A.C., 1929, The caves of Bermuda: *Geological Magazine,* v. 66, p. 79-84.

Swinnerton, A.C., 1932, Origin of limestone caverns: *Geological Society of America Bulletin,* v. 43, p. 663-694.

Swinnerton, A.C., 1932, Origin of limestone caverns:*Geological Society of America Bulletin,* v. 43, p. 662-693.

Szczerban, E., and Urbani, F., 1974, Carsos de Venezuela, Parte 4: Formas cársicas en areniscas Precambricas del Território Federal Amazonas y Estado Bolivar: *Boletin Society Venezolana Espeleologica,* v. 5, p. 27-54.

Szczerban, E., Urbani, F., and Colvée, P., 1977, Cuevas y simas en cuarcitas y metalimolitas del Grupo Roraima, meseta de Guaiquinima, estado Bolivar: *Boletin Society Venezolana Espeleologica,* v. 8, p. 127-154.

Szunyogh, G., 1982, A hévizes eredet¸ gömbfölkék kioldódásanak elméleti vizsgálata (Theory of the dissolving of spherical cavities formed by thermal water): *Karszt és Barlang* (Hungary), no. 2, p. 83-88 (in Hungarian).

Szunyogh, G., 1982, Theoretical problems of dissolution related to spherical niches formation: *Karszt és Barlang* (Budapest), v. 2, p. 83-88 (in Hungarian).

Szunyogh, G., 1984, Theoretical investigation of the origin of spherical caverns of thermal origin (second approach): *Karszt és Barlang,* p. 19-24.

Szunyogh, G., 1989, Theoretical investigation of the development of spheroidal niches of thermal water origin– Second approximation: *10th International Congress of Speleology,* Budapest, Proceedings, v. 3, p. 766-768.

T

Takács-Bolner, K., and Kraus, S., 1989, The results of research into caves of thermal water origin: *Karszt és Barlang* (Hungary), Special Issue, p. 31-38.

Takács-Bolner, K., Eszterhás, I., Juhász, M., and Kraus, S., 1989, The Caves: *Karszt és Barlang,* Special Issue, p. 17-30.

Tappin, D.R., and Ballance, P.A., 1991, Contributions to the sedimentary geology of 'Eua Island: Reworking of an oceanic forearc, *in* Ballance, P.A., Herza, R.H., and Stevenson, A.J., eds., *Geology and resources of island arcs–Tonga, Lau, Fiji region*: Circum-Pacific Council for Energy and Mineral Resources Earth Science Studies, p. 1-20.

Tarhule-Lips, R., and Ford, D.C., 1997, Morphologic studies of bell hole development on Cayman Brac: *12th International Congress of Speleology,* La Chaux-de-Fonds, Proceedings, v. 1, p. 146.

Tarhule-Lips, R., and Ford, D.C., in press, Bell hole development on Cayman Brac based on morphology: *Cave and Karst Science.*

Tarhule-Lips, R.F., and Ford, D.C., 1995, Speleothem dissolution on Cayman Brac and Isla de Mona: *Geological Society of America Abstracts with Programs,* v. 27, no. 2, p. 346.

Taylor, B.E., Wheeler, M.C., and Nordstrom, D.K., 1984, Oxygen and sulfur isotope compositions of sulfate in acid mine drainage: Evidence for oxidation mechanisms: *Nature,* v. 308, p. 538-540.

Taylor, B.E., Wheeler, M.C., and Nordstrom, D.K., 1985, Stable isotope geochemistry of acid mine drainage: Experimental oxidation of pyrite: *Geochemica et Cosmochimica Acta,* v. 48, p. 2669-2678.

Taylor, F.W., and Bloom, A.L., 1977, Coral reefs on tectonic blocks–Tonga Island Arc: *3rd International Coral Reef Symposium,* Miami, Proceedings, p. 275-281.

Terjesen, S.G., Erga, O., Thorsen, G., and Ve, A., 1961, Phase boundary processes as rate determining steps in reaction between solids and liquids—The inhibitory effect of metal ions on the formation of calcium bicarbonate by the reaction of calcite with aqueous carbon dioxide: *Chemical Engineers Sci.,* v. 14, p. 267-276.

Tharp, T.M., 1995, Design against collapse of karst caverns, *in* Beck, B.F., ed., *Karst geohazards:* Rotterdam, A.A. Balkema, p. 397-406.

Tharp, T.M., and Holdrege, T.J., 1994, Fracture mechanics analysis of limestone cantilevers subject to very long term tensile stress in natural caves, *in* Nelson, P.P., and Laubach, S.E., eds., *1st North American Rock Mechanics Symposium Proceedings:* Rotterdam, A.A. Balkema, p. 817-824.

Thirria, [C.]E., 1830, Notice sur le terrain Jurassique de département de la Haute-Sâone, et sur quelques-unes des grottes qu'il renferme: *Mémoires de la Société d'Histoire Naturelle de Strasbourg,* v. 1, p. 1-62 (p. 41-42).

Thomas, R.M., 1979, Size of scallops and ripples formed by flowing water: *Nature,* v. 277, p. 281-283.

Thomas, T.M., 1974, The South Wales interstratal karst: *Transactions of the British Cave Research Association,* v. 1, no. 3, p. 131-152.

Thornbury, W.D., 1965, Regional geomorphology of the United States: New York, Wiley, 609 p.

Thrailkill, J., 1968, Chemical and hydrologic factors in the excavation of limestone caves: *Geological Society of America Bulletin*, v. 79, p. 19-46.

Thrailkill, J., and Robl, T.L., 1981, Carbonate geochemistry of vadose water recharging limestone aquifers: *Journal of Hydrology*, v. 54, p. 195-208.

Tinker, S.W., Ehrets, J.R., and Brondos, M.D., 1995, Multiple karst events related to stratigraphic cyclicity: San Andreas Formation, Yates Field, West Texas, *in* Budd, D.A., Saller, A.H., and Harris, P.M., eds., *Unconformities and porosity in carbonate strata:* American Association of Petroleum Geologists Memoir, no. 63, p. 213-237.

Tkachenko, K.D., 1978, Importance of condensation moisture and dew in the water balance of the aeration zone: *Geol. Journ.*, v. 38, no. 3, p. 27-34 (in Russian).

Tkachuk, V.G., ed., 1968, *Hydrogeology of the USSR: Irkutsk region:* Moscow, Nedra, 495 p. (in Russian).

Tóth, J., 1988, Ground water and hydrocarbon migration, *in* Back, W, Rosenshein, J.S., and Seaber, P.R., eds., *Hydrogeology:* Geological Society of America, Geology of North America, v. O-2, p. 485-502.

Töth, J., 1995, Hydraulic continuity in large sedimentary basins: *Hydrogeology Journal*, v. 3, no. 4, p. 4-15.

Tratman, E.K., 1957, A nameless stream: Suggested new term: *Cave Research Group Newsletter*, v. 6, p. 68-69.

Trissino, G.G., 1550, [Letter dated 5 March 1537]. Ff.420-421, *in* Alberti, L., *Descrittione di Tutte Italia ...:* Bologna, Giaccarelli, 469ff. (f.420v).

Troester, J.W., 1994, *The geochemistry, hydrogeology, and geomorphology of the Rio Camuy drainage basin, Puerto Rico: A humid tropical karst:* Pennsylvania State University, Ph.D. dissertation, 345 p.

Trombe, F., 1952, *Traité de Spéléologie:* Paris, Payot, 376 p.

Trouw, R.A.J., Ribeiro, A., and Pacciullo, F.V.P., 1983, Geologia estrutural do Grupo São João del Rei, Carrancas e Andrelândia, sul de Minas Gerais: *Anais da Academia Brasileira de Ciências*, v. 55, p. 71-85.

Trudgill, S.T., 1976, The subaerial and subsoil erosion of limestones on Aldabra Atoll, Indian Ocean: *Zeitschrift für Geomorphologie*, suppl. 26, p. 201-210.

Trudgill, S.T., 1985, *Limestone geomorphology:* London, Longman, 196 p.

Trudgill, S.T., Crabtree, R.W., Ferguson, R.I., Ball, J., and Gent, R., 1994, Ten year remeasurement of chemical denudation on a magnesian limestone hillslope: *Earth Surface Processes and Landforms*, v. 19, p. 109-114.

Tsikarishvili, K.D., 1981, Climatic particularities of karst caves in Georgia: *European Conf. on Speleology*, Sofia, Proceedings, v. 2, p. 386-390 (in Russian).

Tsykin, R.A., 1976, Weathering crust and karst, *in* Weathering crust, Moscow, Nauka, v. 25, p. 186-193.

Tsykin, R.S., 1989, Paleokarst of the Union of Soviet Socialist Republics, *in* Bosák, P., Ford, D., Glazek, J. and Horácek, I., eds., *Paleokarst–A systematic and regional review:* Praha, Academia, p. 337-365.

Tucker, M.E., and Wright, V.P., 1990, *Carbonate sedimentology:* Oxford, Blackwell, 482 p.

Tugarinov, V.V., 1955, Some results of study of water vapor condensation from the air, *in* Voprosi izuchenija podzemnikh vod i ingenerno-geologicheskikh processov: Moscow, AN SSSR, p. 60-78 (in Russian).

Tullis, E.L., and Gries, J.P., 1938, Black Hills caves: *Black Hills Engineer*, v. 24, p. 233-271.

Turchinov, I.I., 1993, Secondary mineral formations of gypsum caves of the western Ukraine: *Svet (Light)*, Kiev, v. 3, no. 9, p. 29-37 (in Russian, English summary).

Turchinov, I.I., 1994, Aerosol mineral formation in gypsum caves of western Ukraine, *in* Voprosi fizicheskoi speleologii: Moscow, MFTI, p. 51-63 (in Russian).

Turyshev, A.V., 1965, About a possible way of the formation of karst cavities at great depths, *in Gidrogeol. Sbornic*, no. 4 (Trudy Instituta Geologii UFAN SSSR, no. 76), Sverdlovsk, p. 65-67 (in Russian).

Twiss, R.J., and Moores, E.M., 1992, *Structural Geology:* New York, W.H. Freeman and Co., 532 p.

U

Uggeri, A., Bini, A., and Quinif, Y., 1990, Datation des sédiments de la grotte Marelli (Italie, Lombardie, Varese): *Spéléochronos*, v. 2, p. 21-28.

Urbani, F., 1986, Notes on the origin of caves in Precambrian quartzites of Roraima Group, Venezuela: *Interciência*, v. 11, p. 298-300.

Usdowski, E., 1982, Reactions and equilibria in the systems $CO_2 - H_2O$ and $CaCO_3 - CO_2 - H_2O$ — A review: *Neues Jahrb. Mineral. Abh.*, v. 144, p. 148-171.

Ustinova, T.I., 1956, Conditions of condensation of atmospheric moisture in caves of the western part of the Crimea Mountains: *Soveshchanije po izucheniju karsta*, Moscow, v. 8, p. 3-4 (in Russian).

V

Vachell, C.T., and Evans, P., 1905, Franklen George Evans, M.R.C.S., F.R.Met. Soc., F.R.A.S. Report and Transactions. *Cardiff Naturalists' Society*, v. 37 for 1904, p. 19-25.

Vacher, H.L., 1988, Dupuit-Ghyben-Herzberg analysis of strip-island lenses: *Geological Society of America Bulletin*, v. 100, p. 580-591.

Vacher, H.L., and Mylroie, J.E., 1991, Geomorphic evolution of topographic lows in Bermudian and Bahamian Islands–Effect of climate: *in* Bain, R.J., ed., *5th Symposium on the Geology of the Bahamas*, Port Charlotte, Florida, Proceedings, Bahamian Field Station, p. 221-234.

Vacher, H.L., and Wallis, T.N., 1992, Comparative hydrology of Bermuda and Great Exuma Island, Bahamas: *Groundwater*, v. 30, p. 15-20.

Vakhrushev, B.A., 1993, Usage of condensation groundwater in ancient and medieval times and at the present time, *in* Dvizhenije v noosphere; Theoreticheskie i regionalnye problemy: Simferopol, p. 92-96 (in Russian).

Van Everingden, R.O., Shakur, M.A., and Krouse, H.R., 1985, Role of corrosion by H_2SO_4 fallout in cave development in a travertine deposit–Evidence from sulfur and oxygen isotopes: *Chemical Geology*, v. 49, p. 205-211.

Vandike, J.E., 1984, Hydrogeologic aspects of the November, 1981, liquid fertilizer pipeline break on groundwater in the Maramec Spring Recharge Area, Phelps County, Missouri: *Missouri Speleology*, v. 25, p. 93-101.

Varnedoe, W.W., 1964, The formation of an extensive maze cave in Alabama: *Alabama Academy of Science Journal*, v. 35, no. 4, p. 143-148.

Vasilievski, P.M., and Zheltov, P.I., 1932, Hydrogeologic studies of Chatirdag Mountain in 1927: *Trudy Vserossiyskogo geologo-razvedochnogo ob'edinenija*, v. 142, p. 1-99 (in Russian).

Vedernikov, V.V., and Larina, L.A., 1985, Calculation of water regime within constructional territories, *in* Raschet vodnogo rezhima zastroennikh territoriy: Moscow, Stroiizdat, p. 52-56 (in Russian).

Vegas, R., and Banda, E., 1982, Tectonic framework and Alpine evolution of the Iberian Peninsula: *Earth Evolution Sciences*, v. 4, p. 320-341.

Vendel, M., and Kisházi, P., 1964, Összefüggések melegforrások és karsztvizek között a Dunántúli Középhegységben (Relationships between thermal springs and karst waters in the Transdanubian Range): *MTA Müszaki–Tudományos Osztály Közl.*

Vidal, B., 1990, Picos de Europa. Sistema del Trave: -1441; T-31, -570: *L'Aven, Spéléo Club de la Seine*, v. 50, p. 63-87.

Vinegra, F., 1966, Paleogeografía y tectónica del Mesozico en la provincia de la Sierra Madre y Macizo de Tezuitlan: *Boletín de la Asociación Mexicana de Geologos*

Petroleros Boletín, v. 18, p. 147-177.

Virlet [d'Aoust], [P.T.], 1834, [Sur les formations des cavernes]: *Bulletin de la Société Géologique de France,* ser. 1, v. 4, p. 345-347.

Vogel, P.N., Mylroie, J.E., and Carew, J.L., 1990, Limestone petrology and cave morphology on San Salvador Island, Bahamas: *Cave Science*, v. 17, no. 1, p. 19-30.

Volger, O., 1877, Über eine neue Quellentheorie auf meteorologicher Basis: *Meteorol. Ztsechz.*

Volkov, S.N., Andrejchouk, V.N., and Janchuk, E.A., 1987, Modern iron-manganese formations of the Zoloushka Cave: *Mineralogicheskij Sbornik Lvov,* v. 4, p. 79-83 (in Russian).

von Bertalanffy, L., 1967, *General Systems Theory; formulations, development, applications:* New York, G. Braziller, 289 p.

Vovk, I.F., 1979a, Natural radiolysis and peculiar features of the composition of groundwaters of oil and gas deposits, *in* Regional'nyje uslovija dinamiki, okhrana I popolnenije podzemnykh vod USSR (collection of papers): Kiev, Naukova Dumka, p. 12-20.

Vovk, I.F., 1979b, *Radiolysis of underground waters and its geochemical role*: Moscow, Nedra, 231 p. (in Russian).

Vroman, J., 1950, The movement and solution of salt bodies as applied to Mount Sedom: *Israel Exploration Journal,* v. 1, p. 185-193.

W

Walkden, G.M., 1974, Paleokarstic surfaces in upper Visean (Carboniferous) limestones of the Derbyshire Block, England: *Journal of Sedimentary Petrology,* v. 44, p. 1232-1247.

Waltham, A.C., 1971a, Controlling factors in the development of caves: *Cave Research Group of Great Britain Transactions*, v. 13, p. 73-80.

Waltham, A.C., 1971b, Shale units in the Great Scar Limestone of the southern Askrigg Block: *Yorkshire Geological Society,* Proceedings, v. 38, p. 285-292.

Waltham, A.C., 1977, Cave development at the base of the limestone in Yorkshire: *7th International Congress of Speleology,* Sheffield, Proceedings, p. 421-423.

Ward, R.S., Williams, A.T., and Chadha, D.S., 1997, The use of groundwater tracers for assessment of protection zones around water supply boreholes–A case study, *in* Kranj, A., ed., *Tracer hydrology 97:* Rotterdam, Balkema, p. 369-376.

Waring, G.A., 1965, Thermal springs of the United States and other countries of the world: *U.S. Geological Survey Professional Paper,* no. 492, 383 p.

Warwick, G.T., 1953, The origin of limestone caves, *in* Cullingford, C.H.D., ed., *British caving:* London, Routledge and Kegan Paul, p. 41-61.

Warwick, G.T., 1962, The origin of limestone caves, *in* Cullingford, C.H.D., ed., *British Caving:* London, Routledge and Kegan Paul, 2nd ed., p. 55-82.

Warwick, G.T., 1968, Some primitive features in British caves: *4th International Congress of Speleology*, Ljubljana, Proceedings, p. 239-252.

Waterman, S.E., 1975, *Simulation of conduit network development on bedding planes:* McMaster University, M.Sc. thesis, 116 p.

Watson, R.A., 1966, Underground solution canyons in the Central Kentucky Karst, USA: *International Journal of Speleology,* v. 2, p. 369-376.

Watson, R.A., and White, W.B., 1985, The history of American theories of cave origin: *Geological Society of America, Centennial Special Volume,* no. 1, p. 109-123.

Webb, J.A., 1986, Discussion: Proterozoic palaeokarst at Torrowangee, western New South Wales?: *Australian Journal of Earth Sciences,* v. 33, p. 365-366.

Wefer, F.L., and Nicholson, I.K., 1982, Exploration and mapping of the Sinking Creek System. *National Speleological Society Bulletin,* v. 44, p. 48-63.

Wein, G., 1977, *A Budai-hegység tektonikája* (Tectonics of the Buda Hills): Budapest, MAFI Alk.Kiadv., 76 p.

Wenrich, K.J., and Sutphin, H.B., 1994, Grand Canyon caves, breccia pipes and mineral deposits: *Geology Today,* v. 110, no. 3, p. 97-104.

Weyl, P.K., 1958, The solution kinetics of calcite: *Journal of Geology,* v. 66, p. 163-176.

Whitaker, F.F., and Smart, P.L., 1990, Circulation of saline groundwaters through carbonate platforms; evidence from the Great Bahama Bank: *Geology*, v. 18, p. 200-204.

White, D.E., 1969, Thermal and mineral waters of the United States–brief review of possible origins: *23rd International Geological Congress,* Prague, Proceedings, v. 19, p. 269-286.

White, E.L., 1989, Flood hydrology, *in* White, W.B., and White, E.L., eds., Karst hydrology–Concepts from the Mammoth Cave area: New York, Van Nostrand Reinhold, p. 127-143.

White, E.L., and White, W.B., 1969, Processes of cavern breakdown: *National Speleological Society Bulletin,* v. 31, p. 83-96.

White, S., 1989, Karst features in Pleistocene dunes, Bats Ridges, western Victoria: *Helictite,* v. 27, no. 2, p. 53-7.

White, S., 1994, Speleogenesis in aeolian calcarenite: A case study in western Victoria: *Environmental Geology,* v. 23, p. 248-255.

White, W.B., 1969, Conceptual models for carbonate aquifers: *Groundwater,* v. 7, p. 15-21.

White, W.B., 1977, Role of solution kinetics in the development of karst aquifers, *in* Tolson, J.S., and Doyle, F.L., eds., *Karst hydrogeology:* 12th Congress of the International Association of Hydrogeologists, p. 503-517.

White, W.B., 1984, Rate processes: Chemical kinetics and karst landform development, *in* La Fleur, R.G., ed., *Groundwater as a geomorphic agent*: London, Allen and Unwin, p. 227-248.

White, W.B., 1988, *Geomorphology and hydrology of karst terrains:* New York and Oxford, Oxford Univ. Press, 464 p.

White, W.B., 1990, Surface and near-surface karst landforms: *Geological Society of America Special Paper,* no. 252, p. 157-175.

White, W.B., 1990, Surface and near-surface landforms, *in* Higgins, C.G., and Coates, D.R., eds., *Groundwater Geomorphology: The Role of Subsurface Water in Earth-Surface Processes and Landforms:* Geological Society of America Special Paper 252, p. 157-175.

White, W.B., 1993, Analysis of karst aquifers, *in* Alley, W.M., ed., *Regional Ground-Water Quality*, New York, Van Nostrand Reinhold, p. 471-489.

White, W.B., and Deike, G.H., 1962, Secondary mineralization in Wind Cave, South Dakota: *National Speleological Society Bulletin,* v. 24, p. 74-87.

White, W.B., and Deike, G.H., 1989, Hydraulic geometry of cave passages, *in* White, W.B., and White, E.L., eds., *Karst hydrology–Concepts from the Mammoth Cave area:* New York, Van Nostrand Reinhold, p. 223-258.

White, W.B., and Hess, J.W., 1982, Geomorphology of Burnsville Cove and the geology of the Butler Cave–Sinking Creek System: *National Speleological Society Bulletin,* v. 44, p. 67-77.

White, W.B., and Longyear, J., 1962, Some limitations on speleo-genetic speculation imposed by the hydraulics of water flow in limestone: *National Speleological Society, Nittany Grotto Newsletter*, v. 10, p. 155-167.

White, W.B., and Schmidt, V.A., 1966, Hydrology of a karst area in east-central West Virginia: *Water Resources Research,* v. 2, p. 549-560.

White, W.B., and White, E.L., 1989, Karst hydrology–Concepts from the Mammoth Cave area: New York, Van Nostrand Reinhold, 346 p.

White, W.B., and White, E.L., 1991, Karst erosion surfaces in the Appalachian Highlands, *in* Kastning, E.H., and

Kastning, K.M., eds., *Appalachian Karst Symposium,* Proceedings: Huntsville, National Speleological Society, p. 1-10.
White, W.B., and White, E.L., eds., 1989, *Karst hydrology–Concepts from the Mammoth Cave area:* New York, Van Nostrand Reinhold, 346 p.
White, W.B., Jefferson, G.L., and Haman, J.F., 1966, Quartzite karst in southeastern Venezuela: *International Journal of Speleology,* v. 2, p. 309-314.
White, W.B., Watson, R.A., Pohl, E.R., and Brucker, R.W., 1970, The Central Kentucky Karst: *Geographical Review,* v. 60, p. 88-115.
Whitehurst, J., 1778, *An inquiry into the original state of formation of the earth* ...:London, Author and Bent, 195 p. (p.49, 51).
Wigley, T.M.L., 1972, Analysis of scallop patterns by simulation under controlled conditions: A discussion: *Journal of Geology,* v. 80, p. 121-122.
Wigley, T.M.L., 1973, Chemical evolution of the system calcite-gypsum-water: *Canadian Journal of Earth Science,* v. 10, p. 306-315.
Wigley, T.M.L., and Plummer, L.N., 1976, Mixing of carbonate waters, *Geochimica et Cosmochimica Acta,* v. 40, p. 989-995.
Wilford, G.E., 1964. The Geology of Sarawak and Sabah caves, *Geological Survey of Borneo Bulletin,* v. 6, p. 1-181.
Wilk, Z., Motyka, J., and Józefko, I., 1984, Investigations of some hydraulic properties of karst solution openings and fractures: *Annales Societatis Geologorum Poloniae Rocznik Polskiegotowarystwa Gelogicznego,* v. 43, no. 1-2, p. 15-43.
Willems, J., 1894, Compte rendu de la session extraordinaire de 1893 du 4 au 9 août, étude de la circulation de l'eau dans les calcaires et visite des grottes de la Lesse: *Bulletin de la Société Belge de Géologie, de Paléontologie et d'Hydrologie Mém.*, Bruxelles, v. 7, p. 298-377.
Williams, D., 1834, [Notebooks, v. 5]: Somerset County Museum MS. TTNCM, 121/1993, Taunton (ff. 35v-36).
Williams, P.W., 1966, Limestone pavements with special reference to western Ireland: *Institute of British Geographers Transactions and Papers*, no. 40, p. 155-172.
Williams, P.W., 1968, An evaluation of the rate and distribution of limestone solution and deposition in the River Fergus Basin, western Ireland, *in* Williams, P.W., and Jennings, J.N., eds., *Contributions to the Study of Karst:* Canberra, Australian National University, p. 1-40.
Williams, P.W., 1983, The role of the subcutaneous zone in karst hydrology: *Journal of Hydrology,* v. 61, p. 45-67.
Williams, P.W., 1983, The role of the subcutaneous zone in karst hydrology: *Journal of Hydrology,* v. 61, p. 45-67.
Williams, P.W., 1985, Subcutaneous hydrology and the development of doline and cockpit karst: *Zeitschrift für Geomorphologie,* v. 29, p. 463-482.
Williams, P.W., and Dowling, R.K., 1979, Solution of marble in the karst of the Pikikiruna Range, Northwest Nelson, New Zealand: *Earth Surface Processes*, v. 4, p. 15-36.
Wilson, J.R., 1977, Lineaments and the origin of caves in the Cumberland Plateau of Alabama: *National Speleological Society Bulletin,* v. 39, p. 9-12.
Wilson, W.L., 1994, Morphology and hydrology of the deepest known cave in the Bahamas: Dean's Blue Hole, Long Island: *7th Symposium on the Geology of the Bahamas*, San Salvador Island, Proceedings, Bahamian Field Station, p. 21.
Wilson, W.L., Mylroie, J.E., and Carew, J.L., 1995, Caves as a geologic hazard: A quantitative analysis from San Salvador Island, Bahamas: *in* Beck, B.F., ed., *Karst geohazards,* Brookfield, Illinois, A.A. Balkema, p. 487-495.
Wissbrun, K.F., French, D.M., and Patterson, A., 1954, The true ionisation constant of carbonic acid in aqueous solution from 5 to 45°C: *Journal Physical Chemistry,* v. 58, p. 693-695.
Woigt, S., and Schnadwinkel, M., 1995, Caving beneath the desert: Cater Magara: *International Caver*, v. 14, p. 15-26.
Wollast, R., 1967, Kinetics of the alteration of K-feldspar in buffered solutions at low temperature: *Geochimica et Cosmochimica Acta,* v. 31, p. 635-648.
Wood, W.W., and Petraitis, M.J., 1984, Origin and distribution of carbon dioxide in the unsaturated zone of the southern High Plains: *Water Resources Research,* v. 20, p. 1193-1208.
Worley, N.E., and Beck, J.S., 1976, Moorfurlong Mine, Bradwell, Derbyshire and its geological evolution: *Transactions of the British Cave Research Association,* v. 3, p. 49-53.
Worley, N.E., and Ford, T.D., 1977, Minerals and mines, *in* Ford, T.D., ed., Limestones and caves of the Peak District: Norwich, Geobooks, p. 143-165.
Worley, N.E., and Nash, D.A., 1977, The geological evolution of the Jugholes Caves, Matlock, Derbyshire: *Transactions of the British Cave Research Association,* v. 4, p. 389-401.
Worthington, S.R.H., 1984, *The palaeodrainage of an Appalachian Fluviokarst: Friar's Hole, West Virginia:* McMaster University, M.Sc. thesis, 218 p.
Worthington, S.R.H., 1991, *Karst hydrogeology of the Canadian Rocky Mountains:* McMaster University, unpublished Ph.D. thesis, 380 p.
Worthington, S.R.H., 1994a, Flow velocities in unconfined carbonate aquifers, *in* International symposium on changing karst environments: Hydrogeology, geomorphology and conservation, Oxford, England: *Cave and Karst Science*, v. 21, no. 1, p. 21-22.
Worthington, S.R.H., 1994b, The possible importance of sulfur minerals in initiating epigenic caves, *in* Sasowsky, I.D., and Palmer, M.V., eds., Breakthroughs in karst geomicrobiology and redox geochemistry: *Karst Waters Institute Special Publication,* no. 1, p. 80-82.
Worthington, S.R.H., and Ford, D.C., 1995, High sulfate concentrations in limestone springs: An important factor in conduit initiation?: *Environmental Geology,* v. 25, p. 9-15.
Worthington, S.R.H., and Ford, D.C., 1997a, Borehole tests for megascale channeling in carbonate aquifers: *12th International Congress of Speleology,* La Chaux-de-Fonds, Proceedings, v. 2, p. 195-198.
Worthington, S.R.H., and Ford, D.C., 1997b, Strategy for evaluating channeling in the carbonate bedrock at Smithville, Ontario: *Air and Waste Management Association Annual Conference,* Toronto, Proceedings.
Worthington, S.R.H., Davies, G.J., and Ford, D.C., 1998, Quantification of matrix, fracture and channel contributions to storage and flow in a Paleozoic carbonate, *in* Wicks, C., and Sasowsky, I., eds., Approaches to understanding groundwater flow and contaminant transport in carbonate aquifers: *Geological Society of America Special Paper* (in review).
Wrenrich, K.J., and Sutphin, H.B., 1994, Grand Canyon caves, breccia pipes and mineral deposits: *Geology Today,* May-June 1994, p. 97-104.
Wright, C.W., 1858, *The Mammoth Cave, Kentucky:* Vincennes, Harvey. 68 p. (p.13-14, 17).
Wright, P.L., 1991, Paleokarst: Types, recognition, controls and associations, *in* Wright, V.P., Esteban, M., and Smart, P.L., eds., Paleokarsts and paleokarstic reservoirs: *P.R.I.S. Occasional Publication Ser.* 2, University of Readings, p. 56-88.
Wright, V.P., 1988, Paleokarsts and paleosols as indicators of paleoclimate and porosity evolution: A case study from the Carboniferous of South Wales, *in* James, N.P., and Choquette, P.W., eds., *Paleokarst:* New York, Springer-Verlag, p. 329-341.
Wright, V.P., 1991, Palaeokarst: Types, recognition, controls and associations, *in* Wright, V.P., ed., *Palaeokarsts and Palaeokarstic Reservoirs:* Postgraduate Research Institute for Sedimentology, University of Reading, Occasional Publication, ser. 2, p. 65-88.

Y

Yakutseni, V.P., 1984, *Intensive gas accumulation in the Earth:* Leningrad, Nauka, 122 p. (in Russian).

Yalin, M., 1971, *Theory of hydraulic models:* London, McMillan.

Yechieli, Y., 1995, Are sedimentary salt layers always impermeable?: *Geophysical Research Letters,* v. 22, p. 2761-2764.

Young, R.W., 1986, Tower karst in sandstone–Bungle Bungle massif, northwestern Australia: *Zeitschrift für Geomorphologie,* v. 30, p. 189-202.

Yuan, D., 1991, *Karst of China:* Beijing, Geological Publishing House, 224 p.

Yuhua, G., and Hua, S.L., 1986, Salt karst in Qinghai Plateau, China, *in* Atti simposio internazionale sul carsismo nelle evaporiti, *Le Grotte d'Italia,* v. 12, p. 337-345.

Z

Zajtzev, I.K., and Tolstikhin, N.I., 1971, Classification of groundwaters and rocks–Basis for hydrogeologic mapping and regionalization, *in* Problemy gidrogeologicheskogo kartirovanija I rajonirovanija: Leningrad, 9. 3-15 (in Russian).

Zak, I., and Bentor, Y.K., 1968, Some new data on the salt deposits of the Dead Sea area, Israel, *in Symposium on the Geology of Saline Deposits:* Earth Sciences, v. 7, p. 137-146.

Zak, I., and Freund, R., 1980, Strain measurements in eastern marginal shear zone of Mount Sedom salt diapir, Israel: *American Association of Petroleum Geologists Bulletin,* v. 64, p. 568-581.

Zamaraev, S.M., et al., 1976, *Structure and development history of the Predbaikalian Foredeep:* Moscow, Nauka, 134 p.

Zambo, L., and Ford, D.C., 1997, Limestone dissolution processes in Beke Doline, Aggtelek National Park, Hungary: *Earth Surface Processes and Landforms,* v. 22, p. 531-543.

Zanbak, C., and Arthur, R.C., 1986, Geochemical and engineering aspects of anhydrite-gypsum phase transitions: *Association of Engineering Geologists Bulletin,* v. 23, p. 419-433.

Zdanovsky, A.B., 1956, *Kinetics of dissolution of natural salts in conditions of forced convection:* Leningrad, Goskhimizdat, 219 p. (in Russian).

Zibold, F., 1904, The role of underground dew in water supply for Feodosia: *Pochvovedenije,* v. 4, p. 21-29 (in Russian).

Žibrik, K., Lewicki, F., and Pičinin, A., 1976, Hydrologic investigations, *in* Gospodarič, R., and Habič, P., eds., *Underground water tracing:* Postojna, Institute for Karst Research, p. 43-55.

Zoltán, Z., 1994, A berzsenyi DŠniel tanŠrkÈpzŲ fŲiskola tudomŠnyos kŲzlemÈnyei IX. termÈszettudomŠnyok: *Szombathely* v. 4, p. 233-248 (in Hungarian).

Zötl, J.G., 1961, Die Hydrographie des nordostalpinen Karstes: *Steierische Beiträg zur Hydrogeologie,* v. 2, p. 54-183.

Zötl, J.G., 1974, *Karsthydrogeologie:* Wien, Springer, 291 p.

Zötl, J., 1989, Paleokarst as an important hydrogeological factor, *in* Bosák, P., Ford, D.C., Glazek, J., and Horácek, I., eds., *Paleokarst–A Systematic and Regional Review:* Amsterdam and Praha, Elsevier and Academia, p. 483-509.

Zverev, V.P., 1967, *Hydrogeochemical investigations of the system gypsum-groundwater:* Moscow, Nauka, 99 p. (in Russian).

Part 10—Index: Authors

Part 10—Index: Subjects

A

B

C

D

E

F

G

H

Q

R

S

T

ID Codes Used in Subject Index

Ala.	Alabama, U.S.A.
Ariz.	Arizona, U.S.A.
Colo.	Colorado, U.S.A.
Ind.	Indiana, U.S.A.
Iowa	Iowa, U.S.A.
Ky.	Kentucky, U.S.A.
Minn.	Minnesota, U.S.A.
Mo.	Missouri, U.S.A.
N.Mex.	New Mexico, U.S.A.
NSW	New South Wales, Australia
N.Y.	New York, U.S.A.
Pa.	Pennsylvania, U.S.A.
P.R.	Puerto Rico, U,S,A,
Tenn.	Tennessee, U.S.A.
S.Dak.	South Dakota, U.S.A.
U.K.	United Kingdom (Great Britain)
Utah	Utah, U.S.A.
Va.	Virginia, U.S.A.
W.Va.	West Virginia, U.S.A.
Wy.	Wyoming, U.S.A.